TABLE 1.3 Conversion Factors from BG and EE Units to SI Units[a]

	To Convert from	to	Multiply by
Acceleration	ft/s²	m/s²	3.048 E − 1
Area	ft²	m²	9.290 E − 2
Density	lbm/ft³	kg/m³	1.602 E + 1
	slugs/ft³	kg/m³	5.154 E + 2
Energy	Btu	J	1.055 E + 3
	ft · lb	J	1.356
Force	lb	N	4.448
Length	ft	m	3.048 E − 1
	in.	m	2.540 E − 2
	mile	m	1.609 E + 3
Mass	lbm	kg	4.536 E − 1
	slug	kg	1.459 E + 1
Power	ft · lb/s	W	1.356
	hp	W	7.457 E + 2
Pressure	in. Hg (60 °F)	N/m²	3.377 E + 3
	lb/ft² (psf)	N/m²	4.788 E + 1
	lb/in.² (psi)	N/m²	6.895 E + 3
Specific weight	lb/ft³	N/m³	1.571 E + 2
Temperature	°F	°C	$T_C = (5/9)(T_F - 32°)$
	°R	K	5.556 E − 1
Velocity	ft/s	m/s	3.048 E − 1
	mi/hr (mph)	m/s	4.470 E − 1
Viscosity (dynamic)	lb · s/ft²	N · s/m²	4.788 E + 1
Viscosity (kinematic)	ft²/s	m²/s	9.290 E − 2
Volume flowrate	ft³/s	m³/s	2.832 E − 2
	gal/min (gpm)	m³/s	6.309 E − 5

[a]For more unit conversions, refer to Appendix E.

TABLE 1.4 Conversion Factors from SI Units to BG and EE Units[a]

	To Convert from	to	Multiply by
Acceleration	m/s²	ft/s²	3.281
Area	m²	ft²	1.076 E + 1
Density	kg/m³	lbm/ft³	6.243 E − 2
	kg/m³	slugs/ft³	1.940 E − 3
Energy	J	Btu	9.478 E − 4
	J	ft · lb	7.376 E − 1
Force	N	lb	2.248 E − 1
Length	m	ft	3.281
	m	in.	3.937 E + 1
	m	mile	6.214 E − 4
Mass	kg	lbm	2.205
	kg	slug	6.852 E − 2
Power	W	ft · lb/s	7.376 E − 1
	W	hp	1.341 E − 3
Pressure	N/m²	in. Hg (60 °F)	2.961 E − 4
	N/m²	lb/ft² (psf)	2.089 E − 2
	N/m²	lb/in.² (psi)	1.450 E − 4
Specific weight	N/m³	lb/ft³	6.366 E − 3
Temperature	°C	°F	$T_F = 1.8 T_C + 32°$
	K	°R	1.800
Velocity	m/s	ft/s	3.281
	m/s	mi/hr (mph)	2.237
Viscosity (dynamic)	N · s/m²	lb · s/ft²	2.089 E − 2
Viscosity (kinematic)	m²/s	ft²/s	1.076 E + 1
Volume flowrate	m³/s	ft³/s	3.531 E + 1
	m³/s	gal/min (gpm)	1.585 E + 4

[a]For more unit conversions, refer to Appendix E.

TABLE 1.5 Approximate Physical Properties of Some Common Liquids (SI Units)

Liquid	Temperature (°C)	Density, ρ (kg/m³)	Specific Weight, γ (kN/m³)	Dynamic Viscosity, μ (N·s/m²)	Kinematic Viscosity, ν (m²/s)	Surface Tension,[a] σ (N/m)	Vapor Pressure, p_v [N/m² (abs)]	Bulk Modulus,[b] E_v (N/m²)
Carbon tetrachloride	20	1,590	15.6	9.58 E − 4	6.03 E − 7	2.69 E − 2	1.3 E + 4	1.31 E + 9
Ethyl alcohol	20	789	7.74	1.19 E − 3	1.51 E − 6	2.28 E − 2	5.9 E + 3	1.06 E + 9
Gasoline[c]	15.6	680	6.67	3.1 E − 4	4.6 E − 7	2.2 E − 2	5.5 E + 4	1.3 E + 9
Glycerin	20	1,260	12.4	1.50 E + 0	1.19 E − 3	6.33 E − 2	1.4 E − 2	4.52 E + 9
Mercury	20	13,600	133	1.57 E − 3	1.15 E − 7	4.66 E − 1	1.6 E − 1	2.85 E + 10
SAE 30 oil[c]	15.6	912	8.95	3.8 E − 1	4.2 E − 4	3.6 E − 2	—	1.5 E + 9
Seawater	15.6	1,030	10.1	1.20 E − 3	1.17 E − 6	7.34 E − 2	1.77 E + 3	2.34 E + 9
Water	15.6	999	9.80	1.12 E − 3	1.12 E − 6	7.34 E − 2	1.77 E + 3	2.15 E + 9

[a]In contact with air.

[b]Isentropic bulk modulus calculated from speed of sound.

[c]Typical values. Properties of petroleum products vary.

TABLE 1.6 Approximate Physical Properties of Some Common Liquids (BG Units)

Liquid	Temperature (°F)	Density, ρ (slugs/ft³)	Specific Weight, γ (lb/ft³)	Dynamic Viscosity, μ (lb·s/ft²)	Kinematic Viscosity, ν (ft²/s)	Surface Tension,[a] σ (lb/ft)	Vapor Pressure, p_v [lb/in.² (abs)]	Bulk Modulus,[b] E_v (lb/in.²)
Carbon tetrachloride	68	3.09	99.5	2.00 E − 5	6.47 E − 6	1.84 E − 3	1.9 E + 0	1.91 E + 5
Ethyl alcohol	68	1.53	49.3	2.49 E − 5	1.63 E − 5	1.56 E − 3	8.5 E − 1	1.54 E + 5
Gasoline[c]	60	1.32	42.5	6.5 E − 6	4.9 E − 6	1.5 E − 3	8.0 E + 0	1.9 E + 5
Glycerin	68	2.44	78.6	3.13 E − 2	1.28 E − 2	4.34 E − 3	2.0 E − 6	6.56 E + 5
Mercury	68	26.3	847	3.28 E − 5	1.25 E − 6	3.19 E − 2	2.3 E − 5	4.14 E + 6
SAE 30 oil[c]	60	1.77	57.0	8.0 E − 3	4.5 E − 3	2.5 E − 3	—	2.2 E + 5
Seawater	60	1.99	64.0	2.51 E − 5	1.26 E − 5	5.03 E − 3	2.56 E − 1	3.39 E + 5
Water	60	1.94	62.4	2.34 E − 5	1.21 E − 5	5.03 E − 3	2.56 E − 1	3.12 E + 5

[a]In contact with air.

[b]Isentropic bulk modulus calculated from speed of sound.

[c]Typical values. Properties of petroleum products vary.

TABLE 1.7 Approximate Physical Properties of Some Common Gases at Standard Atmospheric Pressure (SI Units)

Gas	Temperature (°C)	Density, ρ (kg/m³)	Specific Weight, γ (N/m³)	Dynamic Viscosity, μ (N · s/m²)	Kinematic Viscosity, ν (m²/s)	Gas Constant,[a] R (J/kg · K)	Specific Heat Ratio,[b] k
Air (standard)	15	1.23 E + 0	1.20 E + 1	1.79 E − 5	1.46 E − 5	2.869 E + 2	1.40
Carbon dioxide	20	1.83 E + 0	1.80 E + 1	1.47 E − 5	8.03 E − 6	1.889 E + 2	1.30
Helium	20	1.66 E − 1	1.63 E + 0	1.94 E − 5	1.15 E − 4	2.077 E + 3	1.66
Hydrogen	20	8.38 E − 2	8.22 E − 1	8.84 E − 6	1.05 E − 4	4.124 E + 3	1.41
Methane (natural gas)	20	6.67 E − 1	6.54 E + 0	1.10 E − 5	1.65 E − 5	5.183 E + 2	1.31
Nitrogen	20	1.16 E + 0	1.14 E + 1	1.76 E − 5	1.52 E − 5	2.968 E + 2	1.40
Oxygen	20	1.33 E + 0	1.30 E + 1	2.04 E − 5	1.53 E − 5	2.598 E + 2	1.40

[a]Values of the gas constant are independent of temperature.

[b]Values of the specific heat ratio depend only slightly on temperature.

TABLE 1.8 Approximate Physical Properties of Some Common Gases at Standard Atmospheric Pressure (BG Units)

Gas	Temperature (°F)	Density, ρ (slugs/ft³)	Specific Weight, γ (lb/ft³)	Dynamic Viscosity, μ (lb · s/ft²)	Kinematic Viscosity, ν (ft²/s)	Gas Constant,[a] R (ft · lb/slug · °R)	Specific Heat Ratio,[b] k
Air (standard)	59	2.38 E − 3	7.65 E − 2	3.74 E − 7	1.57 E − 4	1.716 E + 3	1.40
Carbon dioxide	68	3.55 E − 3	1.14 E − 1	3.07 E − 7	8.65 E − 5	1.130 E + 3	1.30
Helium	68	3.23 E − 4	1.04 E − 2	4.09 E − 7	1.27 E − 3	1.242 E + 4	1.66
Hydrogen	68	1.63 E − 4	5.25 E − 3	1.85 E − 7	1.13 E − 3	2.466 E + 4	1.41
Methane (natural gas)	68	1.29 E − 3	4.15 E − 2	2.29 E − 7	1.78 E − 4	3.099 E + 3	1.31
Nitrogen	68	2.26 E − 3	7.28 E − 2	3.68 E − 7	1.63 E − 4	1.775 E + 3	1.40
Oxygen	68	2.58 E − 3	8.31 E − 2	4.25 E − 7	1.65 E − 4	1.554 E + 3	1.40

[a]Values of the gas constant are independent of temperature.

[b]Values of the specific heat ratio depend only slightly on temperature.

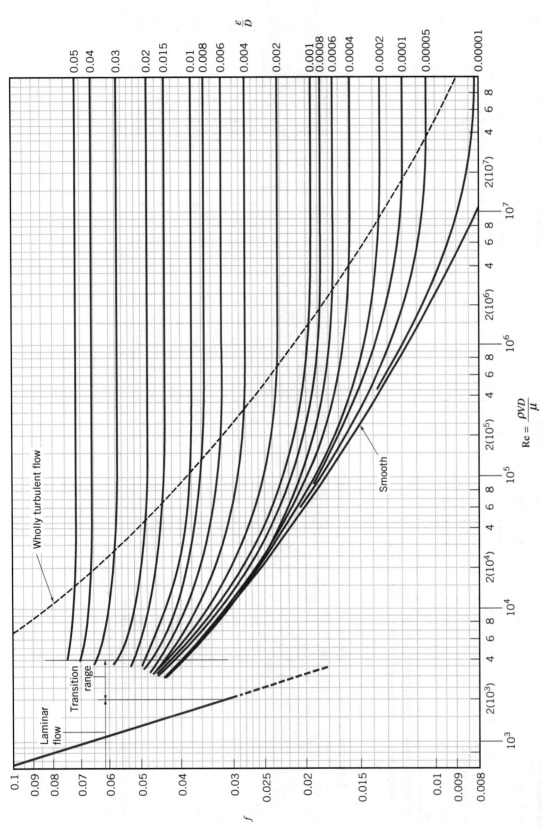

FIGURE 8.20 Friction factor as a function of Reynolds number and relative roughness for round pipes—the Moody chart. (Data from Ref. 7 with permission.)

TABLE 8.1 Equivalent Roughness for New Pipes [Adapted from Moody (Ref. 7) and Colebrook (Ref. 8)]

Pipe	Equivalent Roughness, ε Feet	Equivalent Roughness, ε Millimeters	Pipe	Equivalent Roughness, ε Feet	Equivalent Roughness, ε Millimeters
Riveted steel	0.003–0.03	0.9–9.0	Cast iron	0.00085	0.26
Concrete	0.001–0.01	0.3–3.0	Galvanized iron	0.0005	0.15
Wood stave	0.0006–0.003	0.18–0.9	Commercial steel or wrought iron	0.00015	0.045
			Drawn tubing	0.000005	0.0015
			Plastic, glass	0.0 (smooth)	0.0 (smooth)

Munson, Young, and Okiishi's
Fundamentals of Fluid Mechanics

INTERNATIONAL ADAPTATION

Munson, Young, and Okiishi's

Fundamentals of Fluid Mechanics

SI version

Ninth Edition

INTERNATIONAL ADAPTATION

Andrew L. Gerhart

A. Leon Linton Department of Mechanical, Robotics, and Industrial Engineering
Lawrence Technological University
Southfield, Michigan

John I. Hochstein

Department of Mechanical Engineering
University of Memphis
Memphis, Tennessee

Philip M. Gerhart

WILEY

Munson, Young, and Okiishi's

Fundamentals of Fluid Mechanics

SI version

Ninth Edition

INTERNATIONAL ADAPTATION

Contributing Subject Matter Experts: Dr. Anubhav Rawat, MNIT Allahabad, India; Dr. Abhishek Kundu, MNIT Allahabad, India; Dr Sujit Nath, NIT Silchar, India; Dr. Manish Pandey, NIT Warangal, India

Founded in 1807, John Wiley & Sons, Inc. has been a valued source of knowledge and understanding for more than 200 years, helping people around the world meet their needs and fulfill their aspirations. Our company is built on a foundation of principles that include responsibility to the communities we serve and where we live and work. In 2008, we launched a Corporate Citizenship Initiative, a global effort to address the environmental, social, economic, and ethical challenges we face in our business. Among the issues we are addressing are carbon impact, paper specifications and procurement, ethical conduct within our business and among our vendors, and community and charitable support. For more information, please visit our website: www.wiley.com/go/citizenship.

ISBN: 978-1-119-70326-6

ISBN: 978-1-119-70327-3 (ePub)

ISBN: 978-1-119-76714-5 (ePdf)

Printed and bound by CPI Group (UK) Ltd, Croydon, CR0 4YY

C9781119703266_170724

ANDREW L. GERHART, Professor of Mechanical Engineering at Lawrence Technological University, received his BSME degree from the University of Evansville in 1996, his MSME from the University of Wyoming, and his Ph.D. in Mechanical Engineering from the University of New Mexico.

At Lawrence Tech, Dr. Gerhart has developed both undergraduate and graduate courses in viscous flow, turbulence, creative problem solving, leadership, ethics, professional skills, design, and first-year introductory engineering. He is the supervisor of the Thermal Science and Aerodynamics Laboratories, Coordinator of the Aeronautical Engineering Minor/Certificate, chair of the Interdisciplinary Design and Entrepreneurial Applications four-year engineering curriculum committee, and faculty advisor for the student branch of the American Institute of Aeronautics and Astronautics (AIAA) and the SAE Aero Design team.

Dr. Gerhart facilitates workshops worldwide, having trained hundreds of faculty members in active, collaborative, and problem-based learning, as well as training professional engineers and students in creative problem solving and innovation. He is a member of the American Society for Engineering Education (ASEE) and has received four best paper awards from their Annual Conferences.

Dr. Gerhart was awarded the 2010 Michigan Professor of the Year by the Carnegie Foundation for the Advancement of Teaching and the Council for Advancement and Support of Education, Lawrence Tech's Henry and Barbara Horldt Excellence in Teaching Award, Lawrence Tech's Outstanding Faculty Member Marburger Award, the Engineering Society of Detroit's (ESD) Outstanding Young Engineer, and ESD's Council Leadership Award. He was elected to ESD's College of Fellows, and is actively involved with The American Society of Mechanical Engineers (ASME), serving on the Performance Test Code Committee for Air-Cooled Condensers.

JOHN I. HOCHSTEIN, Professor of Mechanical Engineering at the University of Memphis, received a BE from Stevens Institute of Technology in 1973, an M.S. in Mechanical Engineering from the Pennsylvania State University in 1979, and his Ph.D. in Mechanical Engineering from the University of Akron in 1984. He has been on the faculty of the mechanical engineering department at the University of Memphis since 1991 and served as department chair from 1996 to 2014.

Working as an engineer in nonacademic positions, Dr. Hochstein contributed to the design of the Ohio-Class submarines at the Electric Boat Division of General Dynamics and to the design of the Clinch River Breeder Reactor while an engineer at the Babcock & Wilcox Company. The focus of his doctoral studies was computational modeling of spacecraft cryogenic propellant management systems, and he has remained involved with NASA research on this topic since that time. Dr. Hochstein has twice been a NASA Summer Faculty Fellow for two consecutive summers: once at the NASA Lewis (now Glenn) Research Center, and once at the NASA Marshall Space Flight Center. Dr. Hochstein's current primary research focus is on the capture of hydrokinetic energy to produce electricity.

Dr. Hochstein is an Associate Fellow of the American Institute of Aeronautics and Astronautics. He served in multiple leadership positions during his membership on the Microgravity and Space Processes Technical Committee from 1986 to 2018. Dr. Hochstein has been a member of The American Society of Mechanical Engineers since joining as an undergraduate student. He served from 1996 to 2002 on first the ad hoc K-20 Computational Heat Transfer Committee and then the permanent K-20 Committee. He has been a member of the American Society for Engineering Education since 1985 and he has served the profession as an ABET Program Evaluator since 2002.

PHILIP M. GERHART, former Dean of Engineering and Computer Science and Professor of Mechanical and Civil Engineering at the University of Evansville, received his BSME degree from Rose-Hulman Institute of Technology in 1968 and his M.S. and Ph.D. degrees in Mechanical Engineering from the University of Illinois at Urbana-Champaign in 1969 and 1971. Before becoming Chair of Mechanical and Civil Engineering at the University of Evansville, he was on the Mechanical Engineering faculty at the University of Akron from 1971 to 1984.

Dr. Gerhart taught a variety of courses in fluid mechanics and other thermo-fluid sciences. He consulted widely in the power generation and process industries and authored or coauthored two previous books on fluid mechanics and fluid machinery.

From 1975 to 2017, he was been deeply involved in the development of The American Society of Mechanical Engineers Performance Test Codes. He served as ASME Vice President for Performance Test Codes from 1998 to 2001, and was a member and vice-chair of the Committee on Fans, chair of the Committee on Fired Steam Generators, and a member of the Standing Committee on Performance Test Codes.

Dr. Gerhart served as a member of the American Society for Engineering Education and was a Life Fellow of The American Society of Mechanical Engineers. His honors and awards include the Outstanding Teacher Award from the Faculty Senate of the United Methodist Church, and the Performance Test Codes Medal from ASME.

Dr. Gerhart passed away before the publication of this edition of the textbook. As noted above, his contributions to the profession of engineering were extensive, and his contributions to society were just as numerous (among them, 87 Scouts earned Eagle under his leadership). Beloved by his students, Dr. Gerhart mentored many to successful careers, included authors Andy and John, whose careers were shaped by Dr. Gerhart more than anyone else. He is greatly missed.

BRUCE R. MUNSON, Professor Emeritus of Engineering Mechanics at Iowa State University, received his B.S. and M.S. degrees from Purdue University, and his Ph.D. degree from the Aerospace Engineering and Mechanics Department of the University of Minnesota in 1970.

Prior to joining the Iowa State University faculty in 1974, Dr. Munson was on the mechanical engineering faculty of Duke University from 1970 to 1974. From 1964 to 1966, he worked as an engineer in the jet engine fuel control department of Bendix Aerospace Corporation, South Bend, Indiana.

Dr. Munson's main professional activity has been in the area of fluid mechanics education and research. He has been responsible for the development of many fluid mechanics courses for studies in civil engineering, mechanical engineering, engineering science, and agricultural engineering and is the recipient of an Iowa State University Superior Engineering Teacher Award and the Iowa State University Alumni Association Faculty Citation.

He has authored and coauthored many theoretical and experimental technical papers on hydrodynamic stability, low Reynolds number flow, secondary flow, and the applications of viscous incompressible flow. He is a member of The American Society of Mechanical Engineers.

DONALD F. YOUNG, Anson Marston Distinguished Professor Emeritus in Engineering, received his B.S. degree in mechanical engineering, his M.S. and Ph.D. degrees in theoretical and applied mechanics from Iowa State University, and has taught both undergraduate and graduate courses in fluid mechanics at Iowa State for many years. In addition to being named a Distinguished Professor in the College of Engineering, Dr. Young has also received the Standard Oil Foundation Outstanding Teacher Award and the Iowa State University Alumni Association Faculty Citation. He has been engaged in fluid mechanics research for more than 35 years, with special interests in similitude and modeling and the interdisciplinary field of biomedical fluid mechanics. Dr. Young has contributed to many technical publications and is the author or coauthor of two textbooks on applied mechanics. He is a Fellow of The American Society of Mechanical Engineers.

TED H. OKIISHI, Professor Emeritus of Mechanical Engineering at Iowa State University, joined the faculty there in 1967 after receiving his undergraduate and graduate degrees from that institution.

From 1965 to 1967, Dr. Okiishi served as a U.S. Army officer with duty assignments at the NASA Lewis Research Center, Cleveland, Ohio, where he participated in rocket nozzle heat transfer research, and at the Combined Intelligence Center, Saigon, Republic of South Vietnam, where he studied seasonal river flooding problems.

Professor Okiishi and his students have been active in research on turbomachinery fluid dynamics. Some of these projects have involved significant collaboration with government and industrial laboratory researchers, with two of their papers winning the ASME Melville Medal (in 1989 and 1998).

Dr. Okiishi has received several awards for teaching. He has developed undergraduate and graduate courses in classical fluid dynamics as well as the fluid dynamics of turbomachines.

He is a licensed professional engineer. His professional society activities include having been a vice president of The American Society of Mechanical Engineers and of the American Society for Engineering Education. He is a Life Fellow of The American Society of Mechanical Engineers and past editor of its *Journal of Turbomachinery*. He was recently honored with the ASME R. Tom Sawyer Award.

WADE W. HUEBSCH, Associate Professor in the Department of Mechanical and Aerospace Engineering at West Virginia University, received his B.S. degree in aerospace engineering from San Jose State University where he played college baseball. He received his M.S. degree in mechanical engineering and his Ph.D. in aerospace engineering from Iowa State University in 2000.

Dr. Huebsch specializes in computational fluid dynamics research and has authored multiple journal articles in the areas of aircraft icing, roughness-induced flow phenomena, and boundary layer flow control. He has taught both undergraduate and graduate courses in fluid mechanics and has developed a new undergraduate course in computational fluid dynamics. He has received multiple teaching awards such as Outstanding Teacher and Teacher of the Year from the College of Engineering and Mineral Resources at WVU as well as the Ralph R. Teetor Educational Award from SAE. He was also named as the Young Researcher of the Year from WVU. He is a member of the American Institute of Aeronautics and Astronautics, the Sigma Xi research society, the Society of Automotive Engineers, and the American Society of Engineering Education.

ALRIC P. ROTHMAYER, Professor of Aerospace Engineering at Iowa State University, received his undergraduate and graduate degrees from the Aerospace Engineering Department at the University of Cincinnati, during which time he also worked at NASA Langley Research Center and was a visiting graduate research student at the Imperial College of Science and Technology in London. He joined the faculty at Iowa State University (ISU) in 1985 after a research fellowship sponsored by the Office of Naval Research at University College in London.

Dr. Rothmayer has taught a wide variety of undergraduate fluid mechanics and propulsion courses for over 25 years, ranging from classical low and high speed flows to propulsion cycle analysis.

Dr. Rothmayer was awarded an ISU Engineering Student Council Leadership Award, an ISU Foundation Award for Early Achievement in Research, an ISU Young Engineering Faculty Research Award, and a National Science Foundation Presidential Young Investigator Award. He is an Associate Fellow of the American Institute of Aeronautics and Astronautics (AIAA), and was chair of the 3rd AIAA Theoretical Fluid Mechanics Conference.

Dr. Rothmayer specializes in the integration of computational fluid dynamics with asymptotic methods and low order modeling for viscous flows. His research has been applied to diverse areas ranging from internal flows through compliant tubes to flow control and aircraft icing. In 2001, Dr. Rothmayer won a NASA Turning Goals into Reality (TGIR) Award as a member of the Aircraft Icing Project Team, and also won a NASA Group Achievement Award in 2009 as a member of the LEWICE Ice Accretion Software Development Team. He was also a member of the SAE AC-9C Aircraft Icing Technology Subcommittee of the Aircraft Environmental Systems Committee of SAE and the Fluid Dynamics Technical Committee of AIAA.

Preface

This book is intended to help undergraduate engineering students learn the fundamentals of fluid mechanics. It was developed for use in a first course on fluid mechanics, either one or two semesters/terms. While the principles of this course have been well-established for many years, fluid mechanics education has evolved and improved.

This ninth edition comes with a new look—a standardized format intended to increase accessibility. Concerning the content, the authors strove to continue the distinguished tradition of this text. As it has throughout the past eight editions, the original core prepared by Munson, Young, and Okiishi remains. We have sought to augment this fine text, drawing on our many years of teaching experience. Based on our experience and feedback from colleagues and students, we have made updates to this edition. The changes (listed below, and indicated by the word **New** in descriptions in this preface) are made to clarify, enhance, and expand certain ideas and concepts. In addition, there is revised coverage of some concepts and reorganization at places to foster clarity.

NEW TO THIS EDITION

In addition to the continual effort of updating the scope of the material presented and improving the presentation of all of the material, the following items are new to this edition.

Equilibrium of Moving Fluids (Special Case of Fluid Statics)   A brief **new** section (2.12) has been added to conclude the discussion of static analysis of fluids.

Finite Control Volume Analysis   A **new** section (5.3.5) has been included to enhance student understanding of the energy equation. (The former Section 5.3.5 is now 5.3.6.)

Most Efficient Channel Section   A **new** section (10.5) covering most efficient trapezoidal and triangular channel sections has been included along with **new** solved examples based on them.

Compressible Flow   A **new** section covering compressibility effects on external flow has been added (Section 11.12) and includes a **new** example problem.

Reserve Problems   Approximately 225 **new** Reserve Assessment Problems emphasizing engineering applications have been added to this edition. Reserve Assessment Problems are not available to students unless they are assigned. This allows you, as the instructor, to regulate who sees these questions and the solutions.

KEY FEATURES

Illustrations, Photographs, and Videos

Fluid mechanics has always been a "visual" subject—much can be learned by viewing various characteristics of fluid flow. Fortunately this visual component is becoming easier to incorporate into the learning environment, for both access and delivery, and is an important help in learning fluid mechanics. Thus, many photographs and illustrations have been included in the book. Some of these are within the text material; some are used to enhance the example problems; and some are included as figures adjacent to the text (such as the plot and photograph shown here) to more clearly illustrate various points discussed in the text. Numerous video segments

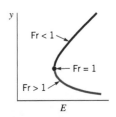

V1.9 Floating razor blade

illustrate many interesting and practical applications of real-world fluid phenomena. Each video segment is identified by a video icon at the appropriate location adjacent to the text material. Each video segment has a separate associated text description of what is shown in the video. There are many homework problems that are directly related to the topics in the videos.

Examples

One of our aims is to represent fluid mechanics as it really is—an exciting and useful discipline. To this end, we include analyses of numerous everyday examples of fluid-flow phenomena to which students and instructors can easily relate. In this edition there are numerous examples that provide detailed solutions to a variety of problems. Many of the examples illustrate engineering applications of fluid mechanics, as is appropriate in an engineering textbook. Several illustrate what happens if one or more of the parameters is changed. This gives the student a better feel for some of the basic principles involved. In addition, many of the examples contain photographs of the actual device or item involved in the example. Also, all of the examples are outlined and carried out with the problem solving methodology of "Given, Find, Solution, and Comment" as discussed in the "Note to User" before Example 1.1.

The Wide World of Fluids

The set of approximately 60 short "The Wide World of Fluids" stories reflect some important and novel ways that fluid mechanics affects our lives. Many of these stories have homework problems associated with them.

Homework Problems

A wide variety of homework problems (more than 50% **new** to this edition) stresses the practical application of principles. The problems are grouped and identified according to topic. The following types of problems are included:

1. "standard" problems,
2. computer problems,
3. discussion problems,
4. supply-your-own-data problems,
5. problems based on "The Wide World of Fluids" topics,
6. problems based on the videos,
7. "Lifelong learning" problems, and
8. problems that require the user to obtain a photograph/image of a given flow situation and write a brief paragraph to describe it.

Computer Problems   Several problems are designated as computer problems. Depending on the preference of the instructor or student, *any* of the problems with numerical data may be solved with the aid of a personal computer, a programmable calculator, or even a smartphone.

Lifelong Learning Problems   Each chapter has lifelong learning problems that involve obtaining additional information about various fluid mechanics topics and writing a brief report about this material.

Well-Paced Concept and Problem-Solving Development

Since this is an introductory text, we have designed the presentation of material to allow for the gradual development of student confidence in fluid mechanics problem solving. Each important concept or notion is considered in terms of simple and easy-to-understand circumstances before more complicated features are introduced. Many pages contain a brief summary (a highlight) sentence adjacent to the text that serves to prepare or remind the reader about an important concept discussed on that page.

Several brief elements have been included in each chapter to help the student see the "big picture" and recognize the central points developed in the chapter. A brief Learning Objectives section is provided at the beginning of each chapter. It is helpful to read through this list prior to reading the chapter to gain a preview of the main concepts presented. Upon completion of the chapter, it is beneficial to look back at the original learning objectives. Additional reinforcement of these learning objectives is provided in the form of a Chapter Summary at the end of each chapter. In this section a brief summary of the key concepts and principles introduced in the chapter is included. All items in the Learning Objectives and the Chapter Summary are "action items" stating something that the student should be able to *do*. A list of the main equations in the chapter is included under Key Equations.

System of Units

The book uses the International System of Units (newtons, kilograms, meters, and seconds), throughout the text, in examples, Questions and Problems. However the British Gravitational System (pounds, slugs, feet, and seconds), and the English Engineering System, sometimes called the U.S. Customary System (pounds (or pounds force), pounds mass, feet, and seconds) are discussed in introductory Chapter 1 to make the reader familiar with them. Comprehensive tables of conversion factor are available in Appendix E.

Prerequisites and Topical Organization

A first course in Fluid Mechanics typically appears in the junior year of a traditional engineering curriculum. Students should have studied statics and dynamics, and mechanics of materials should be at least a corequisite. Prior mathematics should include calculus, with at least the rudiments of vector calculus, and differential equations.

In the first four chapters of this text, the student is made aware of some fundamental aspects of fluid mechanics including important fluid properties, flow regimes, pressure variation in fluids at rest and in motion, fluid kinematics, and methods of flow description and analysis. In Chapter 3 we convey the essential elements of flow kinematics, including Eulerian and Lagrangian descriptions of flow fields, and indicate the vital relationship between the two views. The Bernoulli equation is introduced in Chapter 4 to draw attention, early on, to some of the interesting effects and applications of the relationship between fluid motion and pressure in a flow field. We believe that this early consideration of elementary fluid dynamics increases student enthusiasm for the more complicated material that follows. For instructors who wish to consider kinematics in detail before the material on elementary fluid dynamics, Chapters 3 and 4 can be interchanged without loss of continuity.

Chapters 5, 6, and 7 expand on the basic methods generally used to solve or to begin solving fluid mechanics problems. Emphasis is placed on understanding how flow phenomena are described mathematically and on when and how to use infinitesimal or finite control volumes. The effects of fluid friction on pressure and velocity are also considered in some detail. Although Chapter 5 considers fluid energy and energy dissipation, a formal course in thermodynamics is not a necessary prerequisite. Chapter 7 features the advantages of using dimensional analysis and similitude for organizing data and for planning experiments and the basic techniques involved.

Owing to the growing importance of computational fluid dynamics (CFD) in engineering design and analysis, material on this subject is included in Appendix A. This material may be omitted without any loss of continuity to the rest of the text.

Chapters 8 through 12 offer students opportunities for the further application of the principles learned earlier in the text. Also, where appropriate, additional important notions such as boundary layers, transition from laminar to turbulent flow, turbulence modeling, and flow separation are introduced. Practical concerns such as pipe flow, open-channel flow, flow measurement, drag and lift, the effects of compressibility, and the fundamental fluid mechanics of turbomachinery are included.

Students who study this text and solve a representative set of the problems will have acquired a useful knowledge of the fundamentals of fluid mechanics. Instructors who use this text are provided with numerous topics to select from in order to meet the objectives of their own courses. More material is included than can be reasonably covered in one term. There is sufficient material for a second course, most likely titled "Applied Fluid Mechanics." All are reminded of the fine collection of supplementary material. We have cited throughout the text various articles and books that are available for enrichment.

SUPPORT MATERIAL FOR INSTRUCTORS

In addition to new feature of Reserve Problems, other support material for instructors includes Instructor Solution Manual containing complete and detailed solutions to all the problems in the text. Fluid phenomena videos and figures from the text appropriate for use in lecture slides are media-rich materials available. For support material of Fundamentals of Fluid Mechanics visit www.wiley.com. The materials are available via the Related Resources link found on the product page.

Acknowledgments

First, we wish to express our gratitude to Bruce Munson, Donald Young, and Ted Okiishi for their vision and work in producing the original and subsequent editions of this excellent book. We also thank the people at Wiley, especially Don Fowley and Linda Ratts for their continued trust in us as stewards of this text, along with Kim Eskin and Ashley Patterson for their patience and understanding. Thank you to Steve Gibbs for contributing new homework problems, John Mitchell for his reviews, and the many instructors and students who have offered suggestions. We also thank the following instructors for their work on key tasks for this edition:

Sarah Wilson

Mohd Ali

Joe Schaefer

Jianhu Nie

Orlando Ayala

Michael Boghosian

Narendra Chaganti Subrahmanya Datta

Prahlad Menon

Gianmarco D'Alessandro

Finally, we thank our families for their patience and encouragement during the production of this edition.

Working with students and colleagues over the years has taught us much about fluid mechanics education. We have drawn from this experience for the benefit of users of this book. Obviously we are still learning, and we welcome any suggestions and comments from you.

ANDREW L. GERHART

JOHN I. HOCHSTEIN

PHILIP M. GERHART

Contents

CHAPTER 1

Intoduction

LEARNING OBJECTIVES

After completing this chapter, you should be able to:

- identify the dimensions of physical quantities and correctly use units systems.
- identify the key fluid properties used in the analysis of fluid behavior.

- calculate values for common fluid properties given appropriate information.
- explain the effects of fluid compressibility.
- use the concepts of viscosity, vapor pressure, and surface tension.

Fluid mechanics is the discipline within the broad field of applied mechanics that is concerned with the behavior of liquids and gases at rest or in motion. It covers a vast array of phenomena that occur in nature (with or without human intervention), in biology, and in numerous engineered, invented, or manufactured situations. There are few aspects of our lives that do not involve fluids, either directly or indirectly.

The immense range of different flow conditions is mind-boggling and strongly dependent on the value of the numerous parameters that describe fluid flow. Among the long list of parameters involved are (1) the physical size of the flow, ℓ; (2) the speed of the flow, V; and (3) the pressure, p, as indicated in the adjacent figure for a light aircraft parachute recovery system. These are just three of the important parameters that, along with many others, are discussed in detail in various sections of this book. To get an inkling of the range of some of the parameter values encountered in fluid mechanics, consider the following.

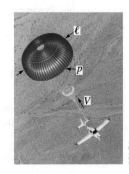

NASA

- Size, ℓ

  Every flow has a characteristic (or typical) length associated with it. For example, for flow of fluid within pipes, the pipe diameter is a characteristic length. Pipe flows include the flow of water in the pipes in our homes, the blood flow in our arteries and veins, and the airflow in our bronchial tree. They also involve pipe sizes that are not within our everyday experiences. Such examples include the flow of oil across Alaska through a 1.22 m-diameter, 1286 km-long pipe and, at the other end of the size scale, the new area of interest involving flow in nanoscale pipes whose diameters are on the order of 10^{-8} m. Each of these pipe flows has important characteristics that are not found in the others.

  Characteristic lengths of some other flows are shown in **Fig. 1.1a**.

- Speed, V

  As we note from The Weather Channel, on a given day the wind speed may cover what we think of as a wide range, from a gentle 8 km/hr breeze to a 100-mph hurricane or a 402.3 km/hr tornado. However, this speed range is small compared to that of the almost

V1.1 Mt. St. Helens eruption

imperceptible flow of the fluid-like magma below the Earth's surface that drives the continental drift motion of the tectonic plates at a speed of about 2×10^{-8} m/s or the hypersonic airflow around a meteor as it streaks through the atmosphere at 3×10^4 m/s.

Characteristic speeds of some other flows are shown in **Fig. 1.1b**.

- Pressure, p

  The pressure within fluids covers an extremely wide range of values. We are accustomed to the 2.4 bar pressure within our car's tires, the "120 over 70" typical blood pressure reading, or the standard 1 bar atmospheric pressure. However, the large 689 bar pressure in the hydraulic ram of an earth mover or the tiny 1.38×10^{-7} bar pressure of a sound wave generated at ordinary talking levels are not easy to comprehend.

  Characteristic pressures of some other flows are shown in **Fig. 1.1c**.

The list of fluid mechanics applications goes on and on. But you get the point. Fluid mechanics is a very important, practical subject that encompasses a wide variety of situations. It is very likely that during your career as an engineer you will be involved in the analysis and design of systems that require a good understanding of fluid mechanics. Although it is not possible to adequately cover all of the important areas of fluid mechanics within one book, it is hoped that this introductory text will provide a sound foundation of the fundamental aspects of fluid mechanics.

V1.2 E. coli swimming

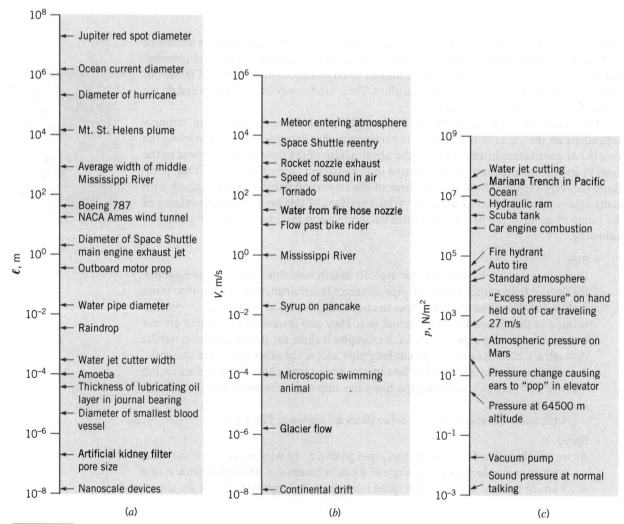

FIGURE 1.1 Characteristic values of some fluid flow parameters for a variety of flows: (a) object size, (b) fluid speed, (c) fluid pressure.

1.1 SOME CHARACTERISTICS OF FLUIDS

One of the first questions we need to explore is—what is a fluid? Or we might ask—what is the difference between a solid and a fluid? We have a general, vague idea of the difference. A solid is "hard" and not easily deformed, whereas a fluid is "soft" and is easily deformed (we can readily move through air). Although quite descriptive, these casual observations of the differences between solids and fluids are not very satisfactory from a scientific or engineering point of view. A closer look at the molecular structure of materials reveals that matter that we commonly think of as a solid (steel, concrete, etc.) has densely spaced molecules with large intermolecular cohesive forces that allow the solid to maintain its shape, and to not be easily deformed. However, for matter that we normally think of as a liquid (water, oil, etc.), the molecules are spaced farther apart, the intermolecular forces are smaller than for solids, and the molecules have more freedom of movement. Thus, liquids can be easily deformed (but not easily compressed) and can be poured into containers or forced through a tube. Gases (air, oxygen, etc.) have even greater molecular spacing and freedom of motion with negligible cohesive intermolecular forces and, as a consequence, are easily deformed (and compressed) and will completely fill the volume of any container in which they are placed. If you drop a block of steel into a container, its shape does not change. If you pour water into a container, it assumes the shape of the container, but its volume does not change. If you introduce air into a container, it expands to fill the container. Liquids and gases are both fluids.

Both liquids and gases are fluids.

The Wide World of Fluids

Will what works in air work in water?

For the past few years, a San Francisco company has been working on small, maneuverable submarines designed to travel through water using wings, controls, and thrusters that are similar to those on jet airplanes. After all, water (for submarines) and air (for airplanes) are both fluids, so it is expected that many of the principles governing the flight of airplanes should carry over to the "flight" of winged submarines. Of course, there are differences. For example, the submarine must be designed to withstand external pressures of nearly 48 bar greater than that inside the vehicle. On the other hand, at high altitude where commercial jets fly, the exterior pressure is 0.24 bar rather than standard sea-level pressure of 1 bar, so the vehicle must be pressurized internally for passenger comfort. In both cases, however, the design of the craft for minimal drag, maximum lift, and efficient thrust is governed by the same fluid dynamic concepts.

Although the differences between solids and fluids can be explained qualitatively on the basis of molecular structure, a more specific distinction is based on how they deform under the action of an external load. Specifically, a **fluid** is defined as a substance that deforms continuously when acted on by a shearing stress of any magnitude. A shearing stress (force per unit area) is created whenever a tangential force acts on a surface, as shown in the adjacent figure. When common solids such as steel or other metals are acted on by a shearing stress, they will initially deform (usually a very small deformation), but they will not continuously deform (flow). However, common fluids such as water, oil, and air satisfy the definition of a fluid—that is, they will flow when acted on by a shearing stress. Some materials, such as slurries, tar, putty, toothpaste, and so on, are not easily classified since they will behave as a solid if the applied shearing stress is small, but if the stress exceeds some critical value, the substance will flow. The study of such materials is called *rheology* and does not fall within the province of classical fluid mechanics. Thus, all the fluids we will be concerned with in this text will conform to the definition of a fluid.

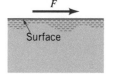

Although the molecular structure of fluids is important in distinguishing one fluid from another, it is not yet practical to study the behavior of individual molecules when trying to describe the behavior of fluids at rest or in motion. Rather, we characterize the behavior by considering the average, or macroscopic, value of the quantity of interest, where the average is evaluated over a small volume containing a large number of molecules. Thus, when we say that the velocity at a certain point in a fluid is so much, we are really indicating the average velocity of the molecules in a small volume surrounding the point. The volume is small compared with the physical dimensions of the system of interest, but large compared with the average distance between molecules. Is this a reasonable way to describe the behavior of a fluid? The answer is generally yes, since the spacing between molecules is typically very small. For gases at normal pressures and temperatures, the spacing is on the order of 10^{-6} mm, and for liquids it is on the order of 10^{-7} mm. The number of molecules per cubic millimeter is on the order of 10^{18} for gases and 10^{21} for liquids. It is clear that the number of molecules in a very tiny volume is huge, and the idea of using average values taken over this volume is certainly reasonable. We assume that all the fluid characteristics we are interested in (pressure, velocity, etc.) vary continuously throughout the fluid—that is, we treat the fluid as a *continuum* and we refer to the very small volume as a point in the flow. This concept will certainly be valid for all the circumstances considered in this text. One area of fluid mechanics for which the continuum concept breaks down is in the study of rarefied gases such as would be encountered at very high altitudes. In this case, the spacing between air molecules can become large and continuum mechanics is not a good model of reality for these flows.

1.2   DIMENSIONS, DIMENSIONAL HOMOGENEITY, AND UNITS

Since in our study of fluid mechanics we will be dealing with a variety of fluid characteristics, it is necessary to develop a system for describing these characteristics both *qualitatively* and *quantitatively*. The qualitative aspect serves to identify the nature, or type, of the characteristics (such as length, time, stress, and velocity), whereas the quantitative aspect provides a numerical measure of the characteristics. The quantitative description requires both a number and a standard by which various quantities can be compared. A standard for length might be a meter or foot, for time an hour or second, and for mass a slug or kilogram. Such standards are called **units,** and several systems of units are in common use as described in the following section. The qualitative description is conveniently given in terms of *primary quantities,* such as length, L, time, T, mass, M, and temperature, Θ. These primary quantities can then be used to provide a qualitative description of any other *secondary quantity*: for example, area $\doteq L^2$, velocity $\doteq LT^{-1}$, density $\doteq ML^{-3}$, and so on, where the symbol $\doteq$ is used to indicate the *dimensions* of the secondary quantity in terms of the primary quantities. Thus, to describe qualitatively a velocity, V, we would write

> Fluid characteristics can be described qualitatively in terms of certain basic quantities such as length, time, and mass.

$$V \doteq LT^{-1}$$

and say that "the dimensions of a velocity equal length divided by time." The primary quantities are also referred to as **basic dimensions.**

For a wide variety of problems involving fluid mechanics, only the three basic dimensions, L, T, and M are required. Alternatively, L, T, and F could be used, where F is the basic dimensions of force. Since Newton's law states that force is equal to mass times acceleration, it follows that $F \doteq MLT^{-2}$ or $M \doteq FL^{-1}T^2$. Thus, secondary quantities expressed in terms of M can be expressed in terms of F through the relationship above. For example, stress, σ, is a force per unit area, so that $\sigma \doteq FL^{-2}$, but an equivalent dimensional equation is $\sigma \doteq ML^{-1}T^{-2}$. **Table 1.1** provides a list of dimensions for a number of common physical quantities.

All theoretically derived equations are **dimensionally homogeneous**—that is, the dimensions of the left side of the equation must be the same as those on the right side, and all additive terms must have the same dimensions. We accept as a fundamental premise that all equations describing physical phenomena must be dimensionally homogeneous. If this were not true, we would be attempting to equate or add unlike physical quantities, which would not make sense. For example, the equation for the velocity, V, of a uniformly accelerated body is

$$V = V_0 + at \tag{1.1}$$

TABLE 1.1 Dimensions Associated with Common Physical Quantities

	FLT System	MLT System		FLT System	MLT System
Acceleration	LT^{-2}	LT^{-2}	Power	FLT^{-1}	ML^2T^{-3}
Angle	$F^0L^0T^0$	$M^0L^0T^0$	Pressure	FL^{-2}	$ML^{-1}T^{-2}$
Angular acceleration	T^{-2}	T^{-2}	Specific heat	$L^2T^{-2}\Theta^{-1}$	$L^2T^{-2}\Theta^{-1}$
Angular velocity	T^{-1}	T^{-1}	Specific weight	FL^{-3}	$ML^{-2}T^{-2}$
Area	L^2	L^2	Strain	$F^0L^0T^0$	$M^0L^0T^0$
Density	$FL^{-4}T^2$	ML^{-3}	Stress	FL^{-2}	$ML^{-1}T^{-2}$
Energy	FL	ML^2T^{-2}	Surface tension	FL^{-1}	MT^{-2}
Force	F	MLT^{-2}	Temperature	Θ	Θ
Frequency	T^{-1}	T^{-1}	Time	T	T
Heat	FL	ML^2T^{-2}	Torque	FL	ML^2T^{-2}
Length	L	L	Velocity	LT^{-1}	LT^{-1}
Mass	$FL^{-1}T^2$	M	Viscosity (dynamic)	$FL^{-2}T$	$ML^{-1}T^{-1}$
Modulus of elasticity	FL^{-2}	$ML^{-1}T^{-2}$	Viscosity (kinematic)	L^2T^{-1}	L^2T^{-1}
Moment of a force	FL	ML^2T^{-2}	Volume	L^3	L^3
Moment of inertia (area)	L^4	L^4	Work	FL	ML^2T^{-2}
Moment of inertia (mass)	FLT^2	ML^2			
Momentum	FT	MLT^{-1}			

where V_0 is the initial velocity, a the acceleration, and t the time interval. In terms of dimensions the equation is

$$LT^{-1} \doteq LT^{-1} + LT^{-2}T$$

and thus Eq. 1.1 is dimensionally homogeneous.

Some equations that are known to be valid contain constants having dimensions. The equation for the distance, d, traveled by a freely falling body can be written as

$$d = 16.1t^2 \tag{1.2}$$

and a check of the dimensions reveals that the constant must have the dimensions of LT^{-2} if the equation is to be dimensionally homogeneous. Actually, Eq. 1.2 is a special form of the well-known equation from physics for freely falling bodies,

$$d = \frac{gt^2}{2} \tag{1.3}$$

in which g is the acceleration of gravity. Equation 1.3 is dimensionally homogeneous and valid in any system of units. For $g = 32.2$ ft/s^2 the equation reduces to Eq. 1.2, and thus Eq. 1.2 is valid only for the system of units using feet and seconds. Equations that are restricted to a particular system of units can be denoted as *restricted homogeneous equations,* as opposed to equations valid in any system of units, which are *general homogeneous equations.* The preceding discussion indicates one rather elementary, but important, use of the concept of dimensions: the determination of one aspect of the generality of a given equation simply based on a consideration of the dimensions of the various terms in the equation. The concept of dimensions also forms the basis for the powerful tool of *dimensional analysis,* which is considered in detail in Chapter 7.

General homogeneous equations are valid in any system of units.

Note to the Users of This Text

All of the examples in the text use a consistent problem-solving methodology, which is similar to that in other engineering courses such as statics. Each example highlights the key elements of analysis: **Given, Find, Solution,** and **Comment.**

The **Given** and **Find** steps clearly identify the information specified to define the problem.

The **Solution** step is where the equations needed to solve the problem are formulated and the problem is actually solved. In this step, there are typically several other tasks that help to set up the solution and are required to solve the problem. The first is a drawing of the problem; where appropriate, it is always helpful to draw a sketch of the problem. Here, the relevant geometry and coordinate system to be used as well as features such as control volumes, forces and pressures, velocities, and mass flowrates are included. This helps in gaining an understanding of the problem. Making appropriate assumptions to solve the problem is the second task. In a realistic engineering problem-solving environment, the appropriate assumptions are developed as an integral part of the solution process. Assumptions can provide simplifications or offer useful constraints, both of which can help in solving the problem. Throughout the examples in this text, the appropriate assumptions are embedded within the **Solution** step, as they are in solving a real-world problem. This provides a realistic problem-solving experience.

The final element in the methodology is the **Comment.** For the examples in the text, this section is used to provide further insight into the problem or the solution. It can also be a point in the analysis at which certain questions are posed. For example: Is the answer reasonable, and does it make physical sense? Are the final units correct? If the value of a parameter was changed, how would the answer change? Adopting this methodology will aid in the development of problem-solving skills for fluid mechanics, as well as other engineering disciplines.

EXAMPLE 1.1 | Restricted and General Homogeneous Equations

Given A liquid flows through an orifice located in the side of a tank, as shown in **Fig. E1.1**. A commonly used equation for determining the volume rate of flow, Q, through the orifice is

$$Q = 0.61 \, A\sqrt{2gh}$$

where A is the area of the orifice, g is the acceleration of gravity, and h is the height of the liquid above the orifice.

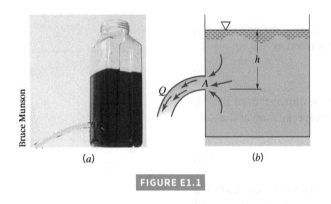

(a)

(b)

Bruce Munson

FIGURE E1.1

Find Investigate the dimensional homogeneity of this formula.

Solution The dimensions of the various terms in the equation are Q = volume/time $\doteq L^3T^{-1}$, A = area $\doteq L^2$, g = acceleration of gravity $\doteq LT^{-2}$, and h = height $\doteq L$.

These terms, when substituted into the equation, yield the dimensional form:

$$(L^3T^{-1}) \doteq (0.61)(L^2)(\sqrt{2})(LT^{-2})^{1/2}(L)^{1/2}$$

or

$$(L^3T^{-1}) \doteq [0.61\sqrt{2}](L^3T^{-1})$$

It is clear from this result that the equation is dimensionally homogeneous (both sides of the formula have the same dimensions of L^3T^{-1}), and the number $0.61\sqrt{2}$ is dimensionless.

If we were going to use this relationship repeatedly, we might be tempted to simplify it by replacing g with its standard value of 9.81 m/s^2 and rewriting the formula as

$$Q = 2.7 \, A\sqrt{h} \tag{1}$$

A quick check of the dimensions reveals that

$$L^3T^{-1} \doteq (2.7)(L^{5/2})$$

and, therefore, the equation expressed as Eq. 1 can only be dimensionally correct if the number 2.7 has the dimensions of $L^{1/2}T^{-1}$. Whenever a number appearing in an equation or formula has dimensions, it means that the specific value of the number will depend on the system of units used. Thus, for the case being considered with feet and seconds used as units, the number 2.7 has units of m$^{1/2}$/s. Equation 1 will only give the correct value for Q (in m^3/s) when A is expressed in square meter and h in meter. Thus, Eq. 1 is a *restricted* homogeneous equation, whereas the original equation is a *general* homogeneous equation that would be valid for any consistent system of units.

Comment A quick check of the dimensions of the various terms in an equation is a useful practice and will often be helpful in detecting errors—that is, as noted previously, all physically meaningful equations must be dimensionally homogeneous. We have briefly alluded to units in this example, and this important topic will be considered in more detail in the next section.

1.2.1 Systems of Units

In addition to the qualitative description of the various quantities of interest, it is generally necessary to have a quantitative measure of a quantity. For example, if we measure the width of this page in the book and say that it is 10 units wide, the statement has no meaning until the unit of length is defined. If we indicate that the unit of length is a meter, and define the meter as some standard length, a unit system for length has been established (and a numerical value can be given to the page width). In addition to length, a unit must be established for each of the remaining basic quantities (force, mass, time, and temperature). There are several systems of units in use, and we shall consider three systems that are commonly used in engineering.

International System (SI, Système Internationale d'Unités)   In 1960, the Eleventh General Conference on Weights and Measures, the international organization responsible for maintaining precise uniform standards of measurements, formally adopted the *International System of Units* as the international standard. This system, commonly termed SI, has been widely adopted worldwide and is widely used (although certainly not exclusively) in the United States. It is expected that the long-term trend will be for all countries to accept SI as the accepted standard, and it is imperative that engineering students become familiar with this system. In SI the unit of length is the meter (m), the time unit is the second (s), the mass unit is the kilogram (kg), and the temperature unit is the kelvin (K). Note that there is no degree symbol used when expressing a temperature in kelvin units. The kelvin temperature scale is an absolute scale and is related to the Celsius (centigrade) scale (°C) through the relationship

$$K = °C + 273.15$$

Although the Celsius scale is not in itself part of SI, it is common practice to specify temperatures in degrees Celsius when using SI units.

The force unit, called the newton (N), is defined from Newton's second law as

$$1 \text{ N} = (1 \text{ kg})(1 \text{ m/s}^2)$$

Thus, a 1-N force acting on a 1-kg mass will give the mass an acceleration of 1 m/s^2. Standard gravity in SI is 9.807 m/s^2 (commonly approximated as 9.81 m/s^2) so that a 1-kg mass weighs 9.81 N under standard gravity. Note that weight and mass are different, both qualitatively and quantitatively! The unit of *work* in SI is the joule (J), which is the work done when the point of application of a 1-N force is displaced through a 1-m distance in the direction of a force. Thus,

> In mechanics it is very important to distinguish between weight and mass.

$$1 \text{ J} = 1 \text{ N} \cdot \text{m}$$

Power is the rate of doing work. The unit of *power* is the watt (W) defined as a joule per second. Thus,

$$1 \text{ W} = 1 \text{ J/s} = 1 \text{ N} \cdot \text{m/s}$$

Prefixes for forming multiples and fractions of SI units are given in **Table 1.2**. For example, the notation kN is read as "kilonewtons" and stands for 10^3 N. Similarly, mm is read as "millimeters" and stands for 10^{-3} m. The centimeter is not an accepted unit of length in the SI system, so for most problems in fluid mechanics in which SI units are used, lengths will be expressed in millimeters or meters.

British Gravitational (BG) System   In the BG system, the unit of length is the foot (ft), the time unit is the second (s), the force unit is the pound (lb), and the temperature unit is the degree Fahrenheit (°F) or the absolute temperature unit is the degree Rankine (°R), where

$$°R = °F + 459.67$$

The mass unit, called the *slug*, is defined from Newton's second law (force = mass × acceleration) as

$$1 \text{ lb} = (1 \text{ slug})(1 \text{ ft/s}^2)$$

TABLE 1.2 Prefixes for SI Units

Factor by Which Unit Is Multiplied	Prefix	Symbol	Factor by Which Unit Is Multiplied	Prefix	Symbol
10^{15}	peta	P	10^{-2}	centi	c
10^{12}	tera	T	10^{-3}	milli	m
10^{9}	giga	G	10^{-6}	micro	μ
10^{6}	mega	M	10^{-9}	nano	n
10^{3}	kilo	k	10^{-12}	pico	p
10^{2}	hecto	h	10^{-15}	femto	f
10	deka	da	10^{-18}	atto	a
10^{-1}	deci	d			

This relationship indicates that a 1-lb force acting on a mass of 1 slug will give the mass an acceleration of 1 ft/s².

The weight, W (which is the force due to gravity, g), of a mass, m, is given by the equation

$$W = mg$$

and in BG units

$$W\ (\text{lb}) = m\ (\text{slugs})\ g\ (\text{ft/s}^2)$$

Since Earth's standard gravity is taken as $g = 32.174$ ft/s² (commonly approximated as 32.2 ft/s²), it follows that a mass of 1 slug weighs 32.2 lb under standard gravity.

> **Three units systems are widely used in engineering:**
> BG – British Gravitational
> EE – English Engineering
> SI – International System

The Wide World of Fluids

How long is a foot?

Today, in the United States, the common length *unit* is the *foot*, but throughout antiquity the unit used to measure length has quite a history. The first length units were based on the lengths of various body parts. One of the earliest units was the Egyptian cubit, first used around 3000 B.C. and defined as the length of the arm from elbow to extended fingertips. Other measures followed, with the foot simply taken as the length of a man's foot. Since this length obviously varies from person to person it was often "standardized" by using the length of the current reigning royalty's foot. In 1791, a special French commission proposed that a new universal length unit called a meter (metre) be defined as the distance of one-quarter of the Earth's meridian (north pole to the equator) divided by 10 million. Although controversial, the meter was accepted in 1799 as the standard. With the development of advanced technology, the length of a meter was redefined in 1983 as the distance traveled by light in a vacuum during the time interval of 1/299,792,458 s. The foot is now defined as 0.3048 meter. Our simple rulers and yardsticks indeed have an intriguing history.

English Engineering (EE) System In the EE system, units for force *and* mass are defined independently; thus, special care must be exercised when using this system in conjunction with Newton's second law. The basic unit of mass is the pound mass (lbm), and the unit of force is the pound (lb).[1] The unit of length is the foot (ft), the unit of time is the second (s), and the absolute temperature scale is the degree Rankine (°R). To make the equation expressing Newton's second law dimensionally homogeneous, we write it as

$$\mathbf{F} = \frac{m\mathbf{a}}{g_c} \tag{1.4}$$

where g_c is a constant of proportionality, which allows us to define units for both force and mass. For the BG system, only the force unit was prescribed and the mass unit defined in a consistent manner such that $g_c = 1$. Similarly, for SI the mass unit was prescribed and the force unit defined in a consistent manner such that $g_c = 1$. For the EE system, a 1-lb force is defined as the force which accelerates a 1 lbm at a rate equal to the standard acceleration due to the earth's gravity, ≈ 32.2 ft/s².

---

[1]It is also common practice to use the notation, lbf, to indicate pound force.

Thus, for Eq. 1.4 to be both numerically and dimensionally correct

$$1 \text{ lb} = \frac{(1 \text{ lbm})(32.174 \text{ ft/s}^2)}{g_c}$$

so that

$$g_c = \frac{(1 \text{ lbm})(32.174 \text{ ft/s}^2)}{(1 \text{ lb})}$$

With the EE system, weight and mass are related through the equation

$$\mathcal{W} = \frac{mg}{g_c}$$

where g is the local acceleration of gravity. Under conditions of standard gravity ($g = g_c$), the weight in pounds and the mass in pound mass are numerically equal. Also, since a 1-lb force gives a mass of 1 lbm an acceleration of 32.174 ft/s^2 and a mass of 1 slug an acceleration of 1 ft/s^2, it follows that

$$1 \text{ slug} = 32.174 \text{ lbm}$$

We cannot overemphasize the importance of paying close attention to units when solving problems. It is very easy to introduce huge errors into problem solutions through the use of incorrect units. Get in the habit of using a *consistent* system of units throughout a given solution. It really makes no difference which system you use as long as you are consistent; for example, don't mix slugs and newtons. If problem data are specified in SI units, then use SI units throughout the solution. If the data are specified in BG units, then use BG units throughout the solution. The relative sizes of the SI, BG, and EE units of length, mass, and force are shown in **Fig. 1.2**.

> When solving problems it is important to use a consistent system of units, e.g., don't mix BG and SI units.

Extensive tables of conversion factors between unit systems, and within unit systems, are provided in Appendix E. For your convenience, abbreviated tables of conversion factors for some quantities commonly encountered in fluid mechanics are presented in **Tables 1.3** and **1.4**, which immediately follow the front cover of this text (using a slightly different format than Appendix E). Note that numbers in these tables are presented in computer exponential notation. For example, the number 5.154 E+2 is the number 5.154×10^2 in scientific notation. You should note that each conversion factor can be thought of as a fraction in which the numerator and denominator are equivalent. For example, an entry for "Length" from Table 1.4 instructs the user "To convert from ... m ... to ... ft ... Multiply by 3.281." Therefore 1 m is the same length as 3.281 ft. Therefore, a fraction formed with a numerator of 1 m and a denominator of 3.281 ft is the very definition of a fraction with a value of one, as is its reciprocal. This may seem obvious when the units of the denominator and numerator are of the same dimension. It is equally true for the more complicated conversion factors that include multiple dimensions and therefore a greater number of units. You already know that you can multiply any quantity by one without changing its value. Likewise, you can multiply (or divide) any quantity by any conversion factor in the tables, provided you use both the number and the units. The result will be correct, even if it does not yield the result you hoped for.

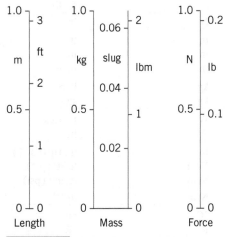

FIGURE 1.2 Comparison of SI, BG, and EE units.

TABLE 1.3 Conversion Factors from BG and EE Units to SI Units[a]

	To Convert from	to	Multiply by
Acceleration	ft/s^2	m/s^2	3.048 E − 1
Area	ft^2	m^2	9.290 E − 2
Density	lbm/ft^3	kg/m^3	1.602 E + 1
	slugs/ft^3	kg/m^3	5.154 E + 2
Energy	Btu	J	1.055 E + 3
	ft · lb	J	1.356
Force	lb	N	4.448
Length	ft	m	3.048 E − 1
	in.	m	2.540 E − 2
	mile	m	1.609 E + 3
Mass	lbm	kg	4.536 E − 1
	slug	kg	1.459 E + 1
Power	ft · lb/s	W	1.356
	hp	W	7.457 E + 2
Pressure	in. Hg (60 °F)	N/m^2	3.377 E + 3
	lb/ft^2 (psf)	N/m^2	4.788 E + 1
	lb/in.^{2} (psi)	N/m^2	6.895 E + 3
Specific weight	lb/ft^3	N/m^3	1.571 E + 2
Temperature	°F	°C	$T_C = (5/9)(T_F − 32°)$
	°R	K	5.556 E − 1
Velocity	ft/s	m/s	3.048 E − 1
	mi/hr (mph)	m/s	4.470 E − 1
Viscosity (dynamic)	lb · s/ft^2	N · s/m^2	4.788 E + 1
Viscosity (kinematic)	ft^2/s	m^2/s	9.290 E − 2
Volume flowrate	ft^3/s	m^3/s	2.832 E − 2
	gal/min (gpm)	m^3/s	6.309 E − 5

[a]For more unit conversions, refer to Appendix E.

TABLE 1.4 Conversion Factors from SI Units to BG and EE Units[a]

	To Convert from	to	Multiply by
Acceleration	m/s^2	ft/s^2	3.281
Area	m^2	ft^2	1.076 E + 1
Density	kg/m^3	lbm/ft^3	6.243 E − 2
	kg/m^3	slugs/ft^3	1.940 E − 3
Energy	J	Btu	9.478 E − 4
	J	ft · lb	7.376 E − 1
Force	N	lb	2.248 E − 1
Length	m	ft	3.281
	m	in.	3.937 E + 1
	m	mile	6.214 E − 4
Mass	kg	lbm	2.205
	kg	slug	6.852 E − 2
Power	W	ft · lb/s	7.376 E − 1
	W	hp	1.341 E − 3
Pressure	N/m^2	in. Hg (60 °F)	2.961 E − 4
	N/m^2	lb/ft^2 (psf)	2.089 E − 2
	N/m^2	lb/in.^{2} (psi)	1.450 E − 4
Specific weight	N/m^3	lb/ft^3	6.366 E − 3
Temperature	°C	°F	$T_F = 1.8 T_C + 32°$
	K	°R	1.800

(Continue)

TABLE 1.4 Conversion Factors from SI Units to BG and EE Units[a] (*Continued*)

	To Convert from	to	Multiply by
Velocity	m/s	ft/s	3.281
	m/s	mi/hr (mph)	2.237
Viscosity (dynamic)	$N \cdot s/m^2$	$lb \cdot s/ft^2$	2.089 E − 2
Viscosity (kinematic)	m^2/s	ft^2/s	1.076 E + 1
Volume flowrate	m^3/s	ft^3/s	3.531 E + 1
	m^3/s	gal/min (gpm)	1.585 E + 4

[a]For more unit conversions, refer to Appendix E.

EXAMPLE 1.2 | BG and SI Units

Given   A tank of liquid having a total mass of 36 kg rests on a support in the equipment bay of the Falcon Heavy.

Find   Determine the force (in newtons) that the tank exerts on the support shortly after lift off when the rocket is accelerating upward as shown in **Fig. E1.2a** at 4.6 m/s².

Kim Shiflett/NASA

FIGURE E1.2a

Solution   A free-body diagram of the tank is shown in **Fig. E1.2b**, where $\mathcal{W}$ is the weight of the tank and liquid, and F_f is the reaction of the floor on the tank. Application of Newton's second law of motion to this body gives

$$\sum F = ma$$

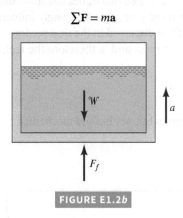

FIGURE E1.2b

or

$$F_f - \mathcal{W} = ma \tag{1}$$

where we have taken upward as the positive direction. Since $\mathcal{W} = mg$, Eq. 1 can be written as

$$F_f = m(g + a) \tag{2}$$

Before substituting any number into Eq. 2, we must decide on a system of units and then be sure all of the data are expressed in these units. Since we want F_f in newtons, we will use SI units so that

$$F_f = 36 \text{ kg}[9.81 \text{ m/s}^2 + 4.6 \text{ m/s}^2]$$
$$= 518 \text{ kg} \cdot \text{m/s}^2$$

Since 1 N = 1 kg · m/s², it follows that

$$F_f = 518 \text{ N} \quad \text{(downward on floor)} \tag{Ans}$$

The direction is downward, since the force shown on the free-body diagram is the force of the support *on the tank* so that the force the tank exerts *on the support* is equal in magnitude but opposite in direction.

Comment   As you work through a large variety of problems in this text, you will find that units play an essential role in arriving at a numerical answer. Be careful! It is easy to mix units and cause large errors. If in the above example the acceleration had been used as 15 ft/s² with m and g expressed in SI units, we would have calculated the force as 893 N and the answer would have been 72% too large!

The Wide World of Fluids

Units and space travel

A NASA spacecraft, the Mars Climate Orbiter, was launched in December 1998 to study the Martian geography and weather patterns. The spacecraft was slated to begin orbiting Mars on September 23, 1999. However, NASA officials lost communication with the spacecraft early that day and it is believed that the spacecraft broke apart or overheated because it came too close to the surface of Mars. Errors in the maneuvering commands sent from Earth caused the Orbiter to sweep within 59.5 km of the surface rather than the intended 149.7 km. The subsequent investigation revealed that the errors were due to a simple mix-up in *units*. One team controlling the Orbiter used SI units, whereas another team used BG units. This costly experience illustrates the importance of using a consistent system of units.

1.3 ANALYSIS OF FLUID BEHAVIOR

The study of fluid mechanics involves the same fundamental laws and principles you have encountered in physics and other mechanics courses. These laws include Newton's laws of motion, conservation of mass, and the first and second laws of thermodynamics. Thus, there are strong similarities between the general approach to fluid mechanics and to rigid-body and deformable-body solid mechanics. This is indeed helpful since many of the concepts and techniques of analysis used in fluid mechanics will be ones you have encountered before in other courses.

The broad subject of fluid mechanics can be generally subdivided into *fluid statics,* in which the fluid is at rest, and *fluid dynamics,* in which the fluid is moving. In the following chapters, we will consider both of these areas in detail. Before we can proceed, however, it will be necessary to define and discuss certain fluid *properties* that are intimately related to fluid behavior. It is obvious that different fluids can have grossly different characteristics. For example, gases are light and compressible, whereas liquids are heavy (by comparison) and relatively incompressible. A syrup flows slowly from a container, but water flows rapidly when poured from the same container. To quantify these differences, certain fluid properties are used. In the following several sections, properties that play an important role in the analysis of fluid behavior are considered.

1.4 MEASURES OF FLUID MASS AND WEIGHT

1.4.1 Density

> The density of a fluid is defined as its mass per unit volume.

The **density** of a fluid, designated by the Greek symbol ρ (rho), is defined as its mass per unit volume. Density is typically used to characterize the mass of a fluid system. In the SI units, ρ has units of kg/m^3, in BG the units are slugs/ft^3, and in EE the units are lbm/ft^3.

The value of density can vary widely between different fluids, but for liquids, variations in pressure and temperature generally have only a small effect on the value of ρ. The small change in the density of water with large variations in temperature is illustrated in **Fig. 1.3. Tables 1.5** and **1.6** list values of density for several common liquids. The density of water at 60 °F is 999 kg/m^3 or 1.94 slugs/ft^3. The large numerical difference between those two values illustrates the importance of paying attention to units! Unlike liquids, the density of a gas is strongly influenced by both pressure and temperature, and this difference will be discussed in the next section.

The *specific volume, v,* is the *volume* per unit mass and is therefore the reciprocal of the density—that is,

$$v = \frac{1}{\rho} \tag{1.5}$$

This property is commonly used in thermodynamics but not in fluid mechanics.

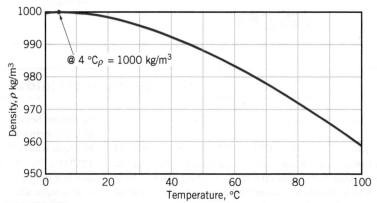

FIGURE 1.3 Density of water as a function of temperature.

TABLE 1.5 Approximate Physical Properties of Some Common Liquids (SI Units)

Liquid	Temperature (°C)	Density, ρ (kg/m³)	Specific Weight, γ (kN/m³)	Dynamic Viscosity, μ (N·s/m²)	Kinematic Viscosity, ν (m²/s)	Surface Tension,[a] σ (N/m)	Vapor Pressure, p_v [N/m² (abs)]	Bulk Modulus,[b] E_v (N/m²)
Carbon tetrachloride	20	1,590	15.6	9.58 E − 4	6.03 E − 7	2.69 E − 2	1.3 E + 4	1.31 E + 9
Ethyl alcohol	20	789	7.74	1.19 E − 3	1.51 E − 6	2.28 E − 2	5.9 E + 3	1.06 E + 9
Gasoline[c]	15.6	680	6.67	3.1 E − 4	4.6 E − 7	2.2 E − 2	5.5 E + 4	1.3 E + 9
Glycerin	20	1,260	12.4	1.50 E + 0	1.19 E − 3	6.33 E − 2	1.4 E − 2	4.52 E + 9
Mercury	20	13,600	133	1.57 E − 3	1.15 E − 7	4.66 E − 1	1.6 E − 1	2.85 E + 10
SAE 30 oil[c]	15.6	912	8.95	3.8 E − 1	4.2 E − 4	3.6 E − 2	—	1.5 E + 9
Seawater	15.6	1,030	10.1	1.20 E − 3	1.17 E − 6	7.34 E − 2	1.77 E + 3	2.34 E + 9
Water	15.6	999	9.80	1.12 E − 3	1.12 E − 6	7.34 E − 2	1.77 E + 3	2.15 E + 9

[a]In contact with air.
[b]Isentropic bulk modulus calculated from speed of sound.
[c]Typical values. Properties of petroleum products vary.

TABLE 1.6 Approximate Physical Properties of Some Common Liquids (BG Units)

Liquid	Temperature (°F)	Density, ρ (slugs/ft³)	Specific Weight, γ (lb/ft³)	Dynamic Viscosity, μ (lb·s/ft²)	Kinematic Viscosity, ν (ft²/s)	Surface Tension,[a] σ (lb/ft)	Vapor Pressure, p_v [lb/in² (abs)]	Bulk Modulus,[b] E_v (lb/in²)
Carbon tetrachloride	68	3.09	99.5	2.00 E − 5	6.47 E − 6	1.84 E − 3	1.9 E + 0	1.91 E + 5
Ethyl alcohol	68	1.53	49.3	2.49 E − 5	1.63 E − 5	1.56 E − 3	8.5 E − 1	1.54 E + 5
Gasoline[c]	60	1.32	42.5	6.5 E − 6	4.9 E − 6	1.5 E − 3	8.0 E + 0	1.9 E + 5
Glycerin	68	2.44	78.6	3.13 E − 2	1.28 E − 2	4.34 E − 3	2.0 E − 6	6.56 E + 5
Mercury	68	26.3	847	3.28 E − 5	1.25 E − 6	3.19 E − 2	2.3 E − 5	4.14 E + 6
SAE 30 oil[c]	60	1.77	57.0	8.0 E − 3	4.5 E − 3	2.5 E − 3	—	2.2 E + 5
Seawater	60	1.99	64.0	2.51 E − 5	1.26 E − 5	5.03 E − 3	2.56 E − 1	3.39 E + 5
Water	60	1.94	62.4	2.34 E − 5	1.21 E − 5	5.03 E − 3	2.56 E − 1	3.12 E + 5

[a]In contact with air.
[b]Isentropic bulk modulus calculated from speed of sound.
[c]Typical values. Properties of petroleum products vary.

1.4.2 Specific Weight

The **specific weight** of a fluid, designated by the Greek symbol γ (gamma), is defined as its *weight* per unit volume. Thus, specific weight is related to density through the equation

$$\gamma = \rho g \tag{1.6}$$

where g is the local acceleration of gravity. Just as density is used to characterize the mass of a fluid system, the specific weight is used to characterize the weight of the system. In the BG and EE systems, γ has units of lb/ft^3 and in SI the units are N/m^3. Under conditions of standard gravity ($g = 32.174$ ft/s^2 = 9.807 m/s^2), water at 15.5 °C (60 °F) has a specific weight of 62.4 lb/ft^3 and 9.80 kN/m^3. Tables 1.5 and 1.6 list values of specific weight for several common liquids (based on standard gravity). More complete tables for water can be found in Appendix B (Tables B.1 and B.2).

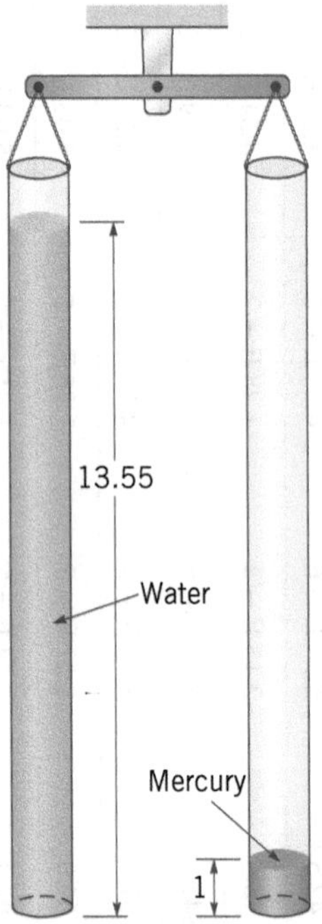

1.4.3 Specific Gravity

The **specific gravity** of a fluid, designated as *SG*, is defined as the ratio of the density of the fluid to the density of water at some specified temperature. Usually the specified temperature is taken as 4 °C (39.2 °F), and at this temperature the density of water is 1000 kg/m^3 or 1.94 slugs/ft^3. In equation form, specific gravity is expressed as

$$SG = \frac{\rho}{\rho_{H_2O@4\,°C}} \tag{1.7}$$

and since it is the *ratio* of densities, the value of *SG* does not depend on the system of units used. For example, the specific gravity of mercury at 20 °C is 13.55. This is illustrated in the adjacent figure. Thus, the density of mercury can be readily calculated in either SI or BG units through the use of Eq. 1.7 as

$$\rho_{Hg} = (13.55)(1000 \text{ kg/m}^3) = 13.6 \times 10^3 \text{ kg/m}^3$$

or

$$\rho_{Hg} = (13.55)(1.94 \text{ slugs/ft}^3) = 26.3 \text{ slugs/ft}^3$$

It is clear that density, specific weight, and specific gravity are all interrelated, and from a knowledge of any one of the three the others can be calculated.

(Figure at left with labels: 13.55, Water, Mercury, 1)

1.5 IDEAL GAS LAW

Compared to liquids, gases are highly compressible. The **ideal gas law** is frequently used as an *equation of state* to define a relationship between pressure, density, and temperature of a gas. In the study of thermodynamics, this relationship is typically expressed by the equation

$$p\upsilon = RT$$

where p is the absolute pressure, υ is the specific volume, R is the gas constant, and T is the absolute temperature.[2] In the study of fluid mechanics, we use density more often than specific volume so an alternative form of this equation more commonly encountered,

$$p = \rho RT \tag{1.8}$$

This relatively simple equation of state provides a good model for gases under typical conditions when they are not approaching liquefaction.

Pressure in a fluid at rest is defined as the normal force per unit area exerted on a plane surface (real or imaginary) immersed in a fluid and is created by the bombardment of the surface with the fluid molecules. From the definition, pressure has the dimension of FL^{-2}. In BG and EE units, pressure is expressed as lb/ft^2 (psf) or lb/in.^{2} (psi) and in SI units as N/m^2. In SI, 1 N/m^2 is defined as a *pascal,* abbreviated as Pa, and pressures are commonly

---

[2]We will use T to represent temperature in thermodynamic relationships, although T is also used to denote the basic dimension of time.

specified in pascals. The pressure in the ideal gas law must be expressed as an **absolute pressure,** denoted (abs), which means that it is measured relative to absolute zero pressure (a pressure that would only occur in a perfect vacuum). Standard sea-level atmospheric pressure (by international agreement) is 101.33 kPa (abs) or 14.696 psi (abs). For most calculations, these pressures can be rounded to 101 kPa, and 14.7 psi respectively. In engineering it is common practice to measure pressure relative to the local atmospheric pressure, and when measured in this fashion, it is called **gage pressure.** Thus, the absolute pressure can be obtained from the gage pressure by adding the value of the atmospheric pressure. For example, as shown in the adjacent figure, a pressure of 206.84 kPa in a tire is equal to 308.17 kPa at standard atmospheric pressure. Pressure is a particularly important fluid characteristic, and it will be discussed more fully in the next chapter.

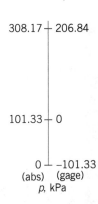

EXAMPLE 1.3 | Ideal Gas Law

Given The compressed air tank shown in **Fig. E1.3a** has a volume of 0.023 m³. The temperature is 21 °C, and the atmospheric pressure is 101 kPa (abs).

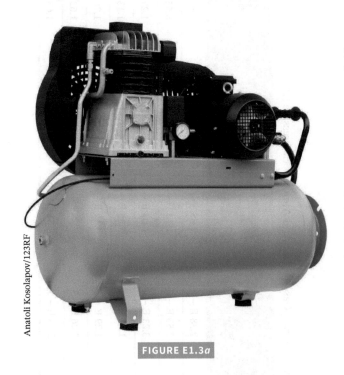

FIGURE E1.3a

Anatoli Kosolapov/123RF

Find When the tank is filled with air at a gage pressure of 3.45 bar, determine the density of the air and the weight of air in the tank.

Solution The air density can be obtained from the ideal gas law (Eq. 1.8)

$$\rho = \frac{p}{RT}$$

so that

$$\rho = \frac{(3.45 \times 15^5\,\text{Pa} + 10^5\,\text{Pa})}{(287\,\text{J/kg·K})[(21 + 273)\text{K}]}$$

$$= 5.27 \text{ kg/m}^3 \qquad\qquad (Ans)$$

Note that both the pressure and temperature were changed to absolute values.

The weight, $\mathcal{W}$, of the air is equal to

$$\mathcal{W} = \rho g \times (\text{volume})$$
$$= (5.27 \text{ kg/m}^3)(9.81 \text{ m/s}^2)$$
$$= 51.7 \text{ N} \qquad\qquad (Ans)$$

Comments By repeating the calculations for various values of the pressure, p, the results shown in **Fig. E1.3b** are obtained. Note that doubling the gage pressure does not double the amount of air in the tank, but doubling the absolute pressure does. Thus, a scuba diving tank at a gage pressure of 689.5 kPa does not contain twice the amount of air as when the gage reads 344.74 kPa.

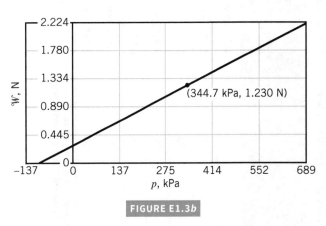

FIGURE E1.3b

The gas constant, R, which appears in Eq. 1.8, depends on the particular gas and is related to the molecular weight of the gas. Values of the gas constant for several common gases are listed in **Tables 1.7** and **1.8**. Also, in these tables, the gas density and specific weight are given for standard atmospheric pressure and gravity and for the temperature listed. More complete tables for air at standard atmospheric pressure can be found in Appendix B (Tables B.3 and B.4).

TABLE 1.7 Approximate Physical Properties of Some Common Gases at Standard Atmospheric Pressure (SI Units)

Gas	Temperature (°C)	Density, ρ (kg/m³)	Specific Weight, γ (N/m³)	Dynamic Viscosity, μ (N·s/m²)	Kinematic Viscosity, ν (m²/s)	Gas Constant,[a] R (J/kg·K)	Specific Heat Ratio,[b] k
Air (standard)	15	1.23 E + 0	1.20 E + 1	1.79 E − 5	1.46 E − 5	2.869 E + 2	1.40
Carbon dioxide	20	1.83 E + 0	1.80 E + 1	1.47 E − 5	8.03 E − 6	1.889 E + 2	1.30
Helium	20	1.66 E − 1	1.63 E + 0	1.94 E − 5	1.15 E − 4	2.077 E + 3	1.66
Hydrogen	20	8.38 E − 2	8.22 E − 1	8.84 E − 6	1.05 E − 4	4.124 E + 3	1.41
Methane (natural gas)	20	6.67 E − 1	6.54 E + 0	1.10 E − 5	1.65 E − 5	5.183 E + 2	1.31
Nitrogen	20	1.16 E + 0	1.14 E + 1	1.76 E − 5	1.52 E − 5	2.968 E + 2	1.40
Oxygen	20	1.33 E + 0	1.30 E + 1	2.04 E − 5	1.53 E − 5	2.598 E + 2	1.40

[a]Values of the gas constant are independent of temperature.
[b]Values of the specific heat ratio depend only slightly on temperature.

TABLE 1.8 Approximate Physical Properties of Some Common Gases at Standard Atmospheric Pressure (BG Units)

Gas	Temperature (°F)	Density, ρ (slugs/ft³)	Specific Weight, γ (lb/ft³)	Dynamic Viscosity, μ (lb·s/ft²)	Kinematic Viscosity, ν (ft²/s)	Gas Constant,[a] R (ft·lb/slug·°R)	Specific Heat Ratio,[b] k
Air (standard)	59	2.38 E − 3	7.65 E − 2	3.74 E − 7	1.57 E − 4	1.716 E + 3	1.40
Carbon dioxide	68	3.55 E − 3	1.14 E − 1	3.07 E − 7	8.65 E − 5	1.130 E + 3	1.30
Helium	68	3.23 E − 4	1.04 E − 2	4.09 E − 7	1.27 E − 3	1.242 E + 4	1.66
Hydrogen	68	1.63 E − 4	5.25 E − 3	1.85 E − 7	1.13 E − 3	2.466 E + 4	1.41
Methane (natural gas)	68	1.29 E − 3	4.15 E − 2	2.29 E − 7	1.78 E − 4	3.099 E + 3	1.31
Nitrogen	68	2.26 E − 3	7.28 E − 2	3.68 E − 7	1.63 E − 4	1.775 E + 3	1.40
Oxygen	68	2.58 E − 3	8.31 E − 2	4.25 E − 7	1.65 E − 4	1.554 E + 3	1.40

[a]Values of the gas constant are independent of temperature.
[b]Values of the specific heat ratio depend only slightly on temperature.

1.6 VISCOSITY

The properties of density and specific weight are measures of the "heaviness" of a fluid. It is clear, however, that these properties are not sufficient to uniquely characterize how fluids behave since two fluids (such as water and oil) can have approximately the same value of density but behave quite differently when flowing. Apparently, some additional property is needed to describe the "fluidity" of the fluid.

To determine this additional property, consider a hypothetical experiment in which a material is placed between two very wide parallel plates, as shown in **Fig. 1.4a**. The bottom plate is rigidly fixed, but the upper plate is free to move. If a solid, such as steel, were placed between the two plates and loaded with the force P as shown, the top plate would be displaced through some small distance, δa (assuming the solid was mechanically attached to the plates). The vertical line AB would be rotated through the small angle, $\delta\beta$, to the new position AB'. We note that to resist the applied force, P, a shearing stress, τ, would be developed at the plate–material interface, and for equilibrium to occur, $P = \tau A$ where A is the effective upper plate area (**Fig. 1.4b**). It is well-known that for elastic solids, such as steel, the small angular displacement, $\delta\beta$ (called the shearing strain), is proportional to the shearing stress, τ, that is developed in the material.

What happens if the solid is replaced with a fluid such as water? We would immediately notice a major difference. When the force P is applied to the upper plate, it will move continuously with a velocity, U (after the initial transient motion has died out), as illustrated in **Fig. 1.5**. This behavior is consistent with the definition of a fluid—that is, if a shearing stress is applied to a fluid, it will deform continuously. A closer inspection of the fluid motion between the two plates would reveal that the fluid in contact with the upper plate moves with the plate velocity, U, and the fluid in contact with the bottom fixed plate has a zero velocity. The fluid between the two plates moves with velocity $u = u(y)$ that would be found to vary linearly, $u = Uy/b$, as illustrated in Fig. 1.5. Thus, a *velocity gradient, du/dy,* is developed in the fluid between the plates. In this particular case, the velocity gradient is a constant because $du/dy = U/b$, but in more complex flow situations, such as that shown in the adjacent photograph, this is not true. The experimental observation that the fluid "sticks" to the solid boundaries is a very important one in fluid mechanics and is usually referred to as the **no-slip condition.** All fluids, both liquids and gases, satisfy this condition for typical flows.

 Video

V1.3 Viscous fluids

 Video

V1.4 No-slip condition

Real fluids, even though they may be moving, always "stick" to the solid boundaries that contain them.

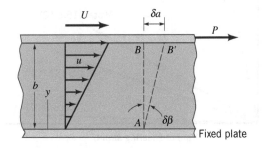

(a)

(b)

FIGURE 1.4 (*a*) Deformation of material placed between two parallel plates. (*b*) Forces acting on upper plate.

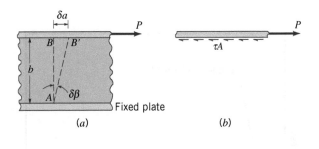

FIGURE 1.5 Behavior of a fluid placed between two parallel plates.

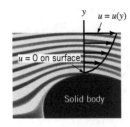

In a small time increment, δt, an imaginary vertical line AB in the fluid would rotate through an angle, $\delta\beta$, so that

$$\delta\beta \approx \tan\delta\beta = \frac{\delta a}{b}$$

Since $\delta a = U\,\delta t$, it follows that

$$\delta\beta = \frac{U\,\delta t}{b}$$

We note that in this case, $\delta\beta$ is a function not only of the force P (which governs U) but also of time. Thus, it is not reasonable to attempt to relate the shearing stress, τ, to $\delta\beta$ as is done for solids. Rather, we consider the *rate* at which $\delta\beta$ is changing and define the **rate of shearing strain**, $\dot{\gamma}$, as

$$\dot{\gamma} = \lim_{\delta t \to 0} \frac{\delta\beta}{\delta t}$$

which in this instance is equal to

$$\dot{\gamma} = \frac{U}{b} = \frac{du}{dy}$$

For many common fluids, continuation of this experiment would reveal that as the shearing stress, τ, is increased by increasing P (recall that $\tau = P/A$), the rate of shearing strain is increased in direct proportion—that is,

$$\tau \propto \dot{\gamma}$$

or

$$\tau \propto \frac{du}{dy}$$

This result indicates that for common fluids such as water, oil, gasoline, and air

$$\tau = \mu \frac{du}{dy} \tag{1.9}$$

where the constant of proportionality between the applied shear stress and the resulting strain rate is designated by the Greek symbol μ (mu) and is called the **absolute viscosity**, *dynamic viscosity,* or simply the *viscosity* of the fluid. In accordance with Eq. 1.9, plots of τ vs. du/dy should be linear with the slope equal to the viscosity, as illustrated in **Fig. 1.6**. The actual value

▶ **Video**

V1.5 Capillary tube viscometer

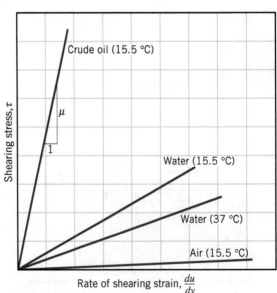

FIGURE 1.6 Linear variation of shearing stress with rate of shearing strain for common fluids.

of the viscosity depends on the particular fluid, and for a particular fluid, the viscosity is also highly dependent on temperature, as illustrated in Fig. 1.6, with the two curves for water. Fluids for which the shearing stress is *linearly* related to the rate of shearing strain (also referred to as the rate of angular deformation) are designated as **Newtonian fluids** after Isaac Newton (1642–1727). Fortunately, most common fluids, both liquids and gases, are Newtonian. A more general formulation of Eq. 1.9, which applies to more complex flows of Newtonian fluids, is given in Section 6.8.1.

> Dynamic viscosity is the fluid property that relates shearing stress and fluid motion.

The Wide World of Fluids

An extremely viscous fluid

Pitch is a derivative of tar once used for waterproofing boats. At elevated temperatures it flows quite readily. At room temperature it feels like a solid—it can even be shattered with a blow from a hammer. However, it is a liquid. In 1927, Professor Parnell heated some pitch and poured it into a funnel. Since that time, it has been allowed to flow freely (or rather, drip slowly) from the funnel. The flowrate is quite small. In fact, to date only seven drops have fallen from the end of the funnel, although the eighth drop is poised ready to fall "soon." While nobody has actually seen a drop fall from the end of the funnel, a beaker below the funnel holds the previous drops that fell over the years. It is estimated that the pitch is about 100 billion times more viscous than water.

Fluids for which the shearing stress is not linearly related to the rate of shearing strain are designated as **non-Newtonian fluids.** Although there is a variety of types of non-Newtonian fluids, the simplest and most common are shown in **Fig. 1.7**. The slope of the shearing stress vs. rate of shearing strain graph is denoted as the *apparent viscosity, μ_{ap}.* For Newtonian fluids, the apparent viscosity is the same as the viscosity and is independent of shear rate.

> For non-Newtonian fluids, the apparent viscosity is a function of the shear rate.

For *shear thinning fluids,* the apparent viscosity decreases with increasing shear rate—the harder the fluid is sheared, the less viscous it becomes. Many colloidal suspensions and polymer solutions are shear thinning. For example, latex paint does not drip from the brush because the shear rate is small and the apparent viscosity is large. However, it flows smoothly onto the wall because the thin layer of paint between the wall and the brush causes a large shear rate and a small apparent viscosity.

> The various types of non-Newtonian fluids are distinguished by how their apparent viscosity changes with shear rate.

For *shear thickening fluids* the apparent viscosity increases with increasing shear rate—the harder the fluid is sheared, the more viscous it becomes. Common examples of this type of fluid include water–corn starch mixture and water–sand mixture ("quicksand"). Thus, the difficulty in removing an object from quicksand increases dramatically as the speed of removal increases.

The other type of behavior indicated in Fig. 1.7 is that of a *Bingham plastic,* which is neither a fluid nor a solid. Such material can withstand a finite, nonzero shear stress, τ_{yield}, the yield stress, without motion (therefore, it is not a fluid), but once the yield stress is exceeded it flows like a fluid (hence, it is not a solid). Toothpaste and mayonnaise are common examples

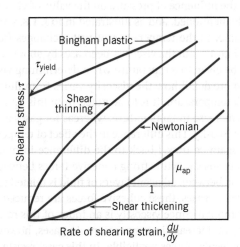

FIGURE 1.7 Variation of shearing stress with rate of shearing strain for several types of fluids, including common non-Newtonian fluids.

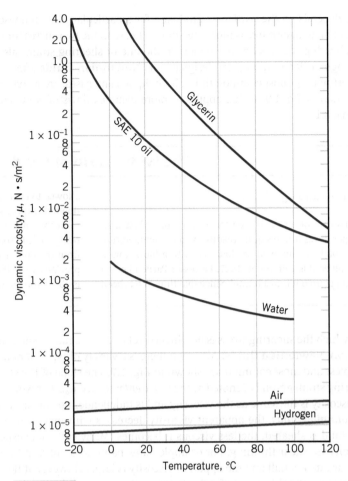

FIGURE 1.8 Dynamic (absolute) viscosity of some common fluids as a function of temperature.

of Bingham plastic materials. As indicated in the adjacent figure, mayonnaise can sit in a pile on a slice of bread (the shear stress less than the yield stress), but it flows smoothly into a thin layer when the knife increases the stress above the yield stress.

From Eq. 1.9 it can be readily deduced that the dimensions of viscosity are FTL^{-2}. Thus, in SI units, viscosity is given as $N \cdot s/m^2$ and in BG and EE units as $lb \cdot s/ft^2$. Values of viscosity for several common liquids and gases are listed in Tables 1.6 through 1.8. A quick glance at these tables reveals the wide variation in viscosity among fluids. Viscosity is only mildly dependent on pressure, and the influence of pressure on the value of viscosity is usually neglected. However, as previously mentioned, and as illustrated in **Fig. 1.8**, viscosity is very sensitive to temperature. For example, as the temperature of water changes from 15.5 to 37 °C, the density decreases by less than 1%, but the viscosity decreases by about 40%. It is thus clear that particular attention must be given to temperature when determining viscosity.

Figure 1.8 shows in more detail how the viscosity varies from fluid to fluid and how for a given fluid it varies with temperature. It is to be noted from this figure that the viscosity of liquids decreases with an increase in temperature, whereas for gases an increase in temperature causes an increase in viscosity. This difference in the effect of temperature on the viscosity of liquids and gases can again be traced back to the difference in molecular structure. The liquid molecules are closely spaced, with strong cohesive forces between molecules, and the resistance to relative motion between adjacent layers of fluid is related to these intermolecular forces. As the temperature increases, these cohesive forces are reduced with a corresponding reduction in resistance to motion. Since viscosity is an index of this resistance, it follows that the viscosity is reduced by an increase in temperature. In gases, however, the molecules are widely spaced and intermolecular forces negligible. In this case, resistance to relative motion arises due to the exchange of momentum of gas molecules between adjacent layers. As molecules are transported by random motion from a region of low bulk velocity to mix with molecules in a region of higher bulk velocity (and vice versa), there is an effective momentum

Video

V1.6 Non-Newtonian behavior

exchange that resists relative motion between the layers. As the temperature of the gas increases, the random molecular activity increases with a corresponding increase in viscosity.

The effect of temperature on viscosity can be closely approximated using two empirical formulas. For gases, the *Sutherland equation* can be expressed as

$$\mu = \frac{CT^{3/2}}{T + S} \qquad (1.10)$$

> **Viscosity is very sensitive to temperature.**

where C and S are empirical constants, and T is absolute temperature. Thus, if the viscosity is known at two temperatures, C and S can be determined. Or, if more than two viscosities are known, the data can be correlated with Eq. 1.10 by using some type of curve-fitting scheme.

For liquids, an empirical equation that has been used is

$$\mu = De^{B/T} \qquad (1.11)$$

where D and B are constants and T is absolute temperature. This equation is often referred to as *Andrade's equation*. As was the case for gases, the viscosity must be known at least for two temperatures so the two constants can be determined. A more detailed discussion of the effect of temperature on fluids can be found in Ref. 1.

EXAMPLE 1.4 | Viscosity and Dimensionless Quantities

Given A dimensionless combination of variables that is important in the study of viscous flow through pipes is called the *Reynolds number*, Re, defined as $\rho VD/\mu$ where, as indicated in **Fig. E1.4**, ρ is the fluid density, V the mean fluid velocity, D the pipe diameter, and μ the fluid viscosity. A Newtonian fluid having a viscosity of $0.38 \ \text{N} \cdot \text{s/m}^2$ and a specific gravity of 0.91 flows through a 25-mm-diameter pipe with a velocity of 2.6 m/s.

nesnelkraM/Getty Images

FIGURE E1.4

Find Determine the value of the Reynolds number using (a) SI units and (b) BG units.

Solution

a. The fluid density is calculated from the specific gravity as

$$\rho = SG\rho_{H_2O@4°C} = 0.91(1000 \ \text{kg/m}^3) = 910 \ \text{kg/m}^3$$

and from the definition of the Reynolds number

$$\text{Re} = \frac{\rho VD}{\mu} = \frac{(910 \ \text{kg/m}^3)(2.6 \ \text{m/s})(25 \ \text{mm})(10^{-3} \ \text{m/mm})}{0.38 \ \text{N} \cdot \text{s/m}^2}$$

$$= 156 \ (\text{kg} \cdot \text{m/s}^2)/\text{N}$$

However, since $1 \ \text{N} = 1 \ \text{kg} \cdot \text{m/s}^2$, it follows that the Reynolds number is unitless—that is,

$$\text{Re} = 156 \qquad (Ans)$$

The value of any dimensionless quantity does not depend on the system of units used if all variables that make up the quantity are expressed in a consistent set of units. To check this, we will calculate the Reynolds number using BG units.

b. We first convert all the SI values of the variables appearing in the Reynolds number to BG values. Thus,

$$\rho = (910 \ \text{kg/m}^3)(1.940 \times 10^{-3}) = 1.77 \ \text{slugs/ft}^3$$
$$V = (2.6 \ \text{m/s})(3.281) = 8.53 \ \text{ft/s}$$
$$D = (0.025 \ \text{m})(3.281) = 8.20 \times 10^{-2} \ \text{ft}$$
$$\mu = (0.38 \ \text{N} \cdot \text{s/m}^2)(2.089 \times 10^{-2}) = 7.94 \times 10^{-3} \ \text{lb} \cdot \text{s/ft}^2$$

and the value of the Reynolds number is

$$\text{Re} = \frac{(1.77 \ \text{slugs/ft}^3)(8.53 \ \text{ft/s})(8.20 \times 10^{-2} \ \text{ft})}{7.94 \times 10^{-3} \ \text{lb} \cdot \text{s/ft}^2}$$

$$= 156 \ (\text{slug} \cdot \text{ft/s}^2)/\text{lb} = 156 \qquad (Ans)$$

since $1 \ \text{lb} = 1 \ \text{slug} \cdot \text{ft/s}^2$.

Comments The values from part (a) and part (b) are the same, as expected. Dimensionless quantities play an important role in fluid mechanics, and the significance of the Reynolds number as well as other important dimensionless combinations will be discussed in detail in Chapter 7. It should be noted that in the Reynolds number it is actually the ratio μ/ρ that is important, and this is the property that is defined as the kinematic viscosity.

EXAMPLE 1.5 | Newtonian Fluid Shear Stress

Given   The velocity distribution for the flow of a Newtonian fluid between two fixed wide, parallel plates (see **Fig. E1.5a**) is given by the equation

$$u = \frac{3V}{2}\left[1 - \left(\frac{y}{h}\right)^2\right]$$

where V is the mean velocity. The fluid has a viscosity of 0.04 kg/m · s. Also, $V = 2$ m/s and $h = 0.5$ cm.

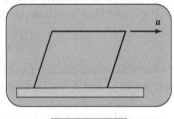

FIGURE E1.5b

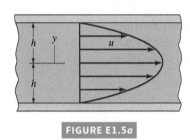

FIGURE E1.5a

Find   Determine: (a) the shearing stress acting on the bottom wall, and (b) the shearing stress acting on a plane parallel to the walls and passing through the centerline (midplane).

Solution   For this type of parallel flow, the shearing stress is obtained from Eq. 1.9,

$$\tau = \mu\frac{du}{dy} \qquad (1)$$

Thus, if the velocity distribution $u = u(y)$ is known, the shearing stress can be determined at all points by evaluating the velocity gradient, du/dy. For the distribution given

$$\frac{du}{dy} = -\frac{3Vy}{h^2} \qquad (2)$$

a. Along the bottom wall $y = -h$ so that (from Eq. 2)

$$\frac{du}{dy} = \frac{3V}{h}$$

and therefore the shearing stress is

$$\tau_{\substack{\text{bottom}\\\text{wall}}} = \mu\left(\frac{3V}{h}\right) = \frac{(0.04\ \text{kg/m · s})(3)(2\ \text{m/s})}{(0.5/100\ \text{m})}$$

$$= 48\ \text{Pa (in direction of flow)} \qquad (Ans)$$

A positive shear stress seems intuitively satisfying. To check your intuition, sketch a sheared fluid particle adjacent to the bottom plate (**Fig. E1.5b**). The fluid motion seems to be trying to drag the plate in the flow direction. Reference to section 1.6 in which the concept of viscosity was introduced, and Fig. 1.4 in particular, makes clear that the shear stress on the fluid at the wall is indeed positive. A review of Eq. 2 and the answers to parts a and b of this example makes it clear that the shear stress varies linearly across the flow field.

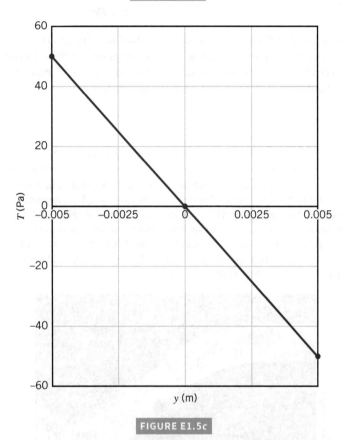

FIGURE E1.5c

b. Along the midplane where $y = 0$ it follows from Eq. 2 that

$$\frac{du}{dy} = 0$$

and thus the shearing stress is

$$\tau_{\text{midplane}} = 0 \qquad (Ans)$$

Comment   Equation 2 has been evaluated at the bottom plate and midplane to provide the answers to parts a and b of this example. If you use it to compute the shear stress at the top plate, you will notice that it produces a shear stress of the same magnitude as at the bottom plate, but of opposite sign. A plot of τ vs. y between the plates, **Fig. E1.5c**, confirms both the difference in sign and the earlier observation of the linear relationship. Can you explain the change of sign? *Hint:* Sketch a sheared fluid particle adjacent to the top plate.

Quite often viscosity appears in fluid flow problems combined with the density in the form

$$\nu = \frac{\mu}{\rho}$$

This ratio is called the **kinematic viscosity** and is denoted with the Greek symbol ν (nu). The dimensions of kinematic viscosity are L^2/T, and the BG units are ft^2/s and SI units are m^2/s. Values of kinematic viscosity for some common liquids and gases are given in Tables 1.5 through 1.8. More extensive tables giving both the dynamic and kinematic viscosities for water and air can be found in Appendix B (Tables B.1 through B.4), and graphs showing the variation in both dynamic and kinematic viscosity with temperature for a variety of fluids are also provided in Appendix B (Figs. B.1 and B.2).

> Kinematic viscosity is defined as the ratio of the absolute viscosity to the fluid density.

Although in this text we are primarily using BG and SI units, dynamic viscosity is often expressed in the metric CGS (centimeter-gram-second) system with units of dyne $\cdot$ s/cm^2. This combination is called a *poise*, abbreviated P. In the CGS system, kinematic viscosity has units of cm^2/s, and this combination is called a *stoke*, abbreviated St.

1.7 COMPRESSIBILITY OF FLUIDS

1.7.1 Bulk Modulus

An important question to answer when considering the behavior of a particular fluid is how easily can the volume (and thus the density) of a given mass of the fluid be changed when there is a change in pressure? That is, how compressible is the fluid? A property that is commonly used to characterize compressibility is the **bulk modulus,** E_v, defined as

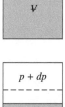

$$E_v = -\frac{dp}{d\forall/\forall} \tag{1.12}$$

where dp is the differential change in pressure needed to create a differential change in volume, $d\forall$, of a volume $\forall$. This is illustrated in the adjacent figure. The negative sign is included since an increase in pressure will cause a decrease in volume. Since a decrease in volume of a given mass, $m = \rho V$, will result in an increase in density, Eq. 1.12 can also be expressed as

$$E_v = \frac{dp}{d\rho/\rho} \tag{1.13}$$

The bulk modulus (also referred to as the *bulk modulus of elasticity*) has dimensions of pressure, FL^{-2}. In SI units, values for E_v are usually given as N/m^2 (Pa) and in BG units as lb/in.^{2} (psi). Large values for the bulk modulus indicate that the fluid is relatively incompressible—that is, it takes a large pressure change to create a small change in volume. As expected, values of E_v for common liquids are large (see Tables 1.5 and 1.6). For example, at atmospheric pressure and a temperature of 15.5 °C, it would require a pressure of 2.5×10^7 Pa to compress a volume of water 1%. This result is representative of the compressibility of liquids. Since such large pressures are required to effect a change in volume, we conclude that liquids can be considered as *incompressible* for most practical engineering applications. As liquids are compressed the bulk modulus increases, but the bulk modulus near atmospheric pressure is usually the one of interest. The use of bulk modulus as a property describing compressibility is most prevalent when dealing with liquids, although the bulk modulus can also be determined for gases.

The Wide World of Fluids

This water jet is a blast

Usually liquids can be treated as incompressible fluids. However, in some applications the *compressibility* of a liquid can play a key role in the operation of a device. For example, a water pulse generator using compressed water has been developed for use in mining operations. It can fracture rock by producing an effect comparable to a conventional explosive such as gunpowder. The device uses the energy stored in a water-filled accumulator to generate an ultrahigh-pressure water pulse ejected through a 10- to 25-mm-diameter discharge valve. At the ultrahigh pressures used (300 to 400 MPa, or 3000 to 4000 atmospheres), the water is compressed (i.e., the volume reduced) by about 10 to 15%. When a fast-opening valve within the pressure vessel is opened, the water expands and produces a jet of water that upon impact with the target material produces an effect similar to the explosive force from conventional explosives. Mining with the water jet can eliminate various hazards that arise with the use of conventional chemical explosives, such as those associated with the storage and use of explosives and the generation of toxic gas by-products that require extensive ventilation.

1.7.2 Compression and Expansion of Gases

When gases are compressed (or expanded), the relationship between pressure and density depends on the nature of the process. If the compression or expansion takes place under constant temperature conditions (*isothermal process*), then from Eq. 1.8

$$\frac{p}{\rho} = \text{constant} \qquad (1.14)$$

If the compression or expansion is frictionless and no heat is exchanged with the surroundings (*isentropic process*), then

$$\frac{p}{\rho^k} = \text{constant} \qquad (1.15)$$

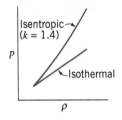

where k is the ratio of the specific heat at constant pressure, c_p, to the specific heat at constant volume, c_v (i.e., $k = c_p/c_v$). The two specific heats are related to the gas constant, R, through the equation $R = c_p - c_v$. As was the case for the ideal gas law, the pressure in both Eqs. 1.14 and 1.15 must be expressed as an absolute pressure. Values of k for some common gases are given in Tables 1.7 and 1.8 and, for air over a range of temperatures, in Appendix B (Tables B.3 and B.4). The pressure–density variations for isothermal and isentropic conditions are illustrated in the adjacent figure.

The value of the bulk modulus depends on the type of process involved.

With explicit equations relating pressure and density, the bulk modulus for gases can be determined by obtaining the derivative $dp/d\rho$ from Eq. 1.14 or 1.15 and substituting the results into Eq. 1.13. It follows that for an isothermal process

$$E_v = p \qquad (1.16)$$

and for an isentropic process,

$$E_v = kp \qquad (1.17)$$

Note that in both cases the bulk modulus varies directly with pressure. For air under standard atmospheric conditions with $p = 1$ bar (abs) and $k = 1.40$, the isentropic bulk modulus is 1.42 bar. A comparison of this figure with that for water under the same conditions ($E_v = 21511$ bar) shows that air is approximately 15,000 times as compressible as water. It is thus clear that in dealing with gases, greater attention will need to be given to the effect of compressibility on fluid behavior. However, as will be discussed further in later sections, gas flows can often be treated as incompressible flows if the changes in pressure are small.

EXAMPLE 1.6 | Isentropic Compression of a Gas

Given A cubic foot of air at an absolute pressure of 10^5 Pa is compressed isentropically to $\frac{1}{2}$ m³ by the tire pump shown in **Fig. E1.6a**.

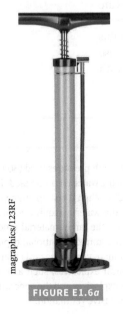

FIGURE E1.6a

Find What is the final pressure?

Solution Assuming the compression occurs so quickly that the effect of heat transfer to the surroundings can be neglected, we use the relationship for isentropic compression

$$\frac{p_i}{\rho_i^k} = \frac{p_f}{\rho_f^k}$$

where the subscripts i and f refer to initial and final states, respectively. Since we are interested in the final pressure, p_f, it follows that

$$p_f = \left(\frac{\rho_f}{\rho_i}\right)^k p_i$$

As the volume, $\mathcal{V}$. is reduced by one-half, the density must double, since the mass, $m = \rho\mathcal{V}$, of the gas remains constant. Thus, with $k = 1.40$ for air

$$p_f = (2)^{1.40}(10^5 \text{ Pa}) = 2.64 \times 10^5 \text{ Pa (abs)} \qquad (Ans)$$

Comment By repeating the calculations for various values of the ratio of the final volume to the initial volume, V_f/V_i, the results shown in **Fig. E1.6b** are obtained. Note that even though

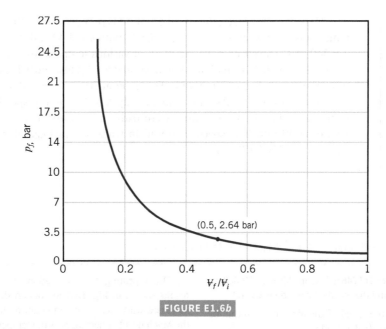

FIGURE E1.6b

air is often considered to be easily compressed (at least compared to liquids), it takes considerable pressure to significantly reduce a given volume of air as is done in an automobile engine where the compression ratio is on the order of $\mathbb{V}_f/\mathbb{V}_i = 1/8 = 0.125$. If the compression is so slow that the isothermal model is appropriate, would the same change in pressure be required to accomplish the specified compression? What would a plot of pressure vs. compression ratio look like for an isothermal compression? Which assumption do you think is a better model for the process in an automobile engine?

1.7.3 Speed of Sound

Another important consequence of the compressibility of fluids is that disturbances introduced at some point in the fluid propagate at a finite velocity. For example, if a fluid is flowing in a pipe and a valve at the outlet is suddenly closed (thereby creating a localized disturbance), the effect of the valve closure is not felt instantaneously upstream. It takes a finite time for the increased pressure created by the valve closure to propagate to an upstream location. Similarly, a loudspeaker diaphragm causes a localized disturbance as it vibrates, and the small change in pressure created by the motion of the diaphragm is propagated through the air with a finite velocity. The velocity at which an infinitesimal disturbance propagates is called the *acoustic velocity* or the **speed of sound**, c. It will be shown in Chapter 11 that the speed of sound is related to changes in pressure and density of the fluid medium through the equation

> The velocity at which small disturbances propagate in a fluid is called the speed of sound.

$$c = \sqrt{\frac{dp}{d\rho}} \tag{1.18}$$

or in terms of the bulk modulus defined by Eq. 1.13

$$c = \sqrt{\frac{E_v}{\rho}} \tag{1.19}$$

Since the disturbance is small, there is negligible heat transfer and the process is assumed to be isentropic. Thus, the pressure–density relationship used in Eq. 1.18 is that for an isentropic process.

For gases undergoing an isentropic process, $E_v = kp$ (Eq. 1.17) so that

$$c = \sqrt{\frac{kp}{\rho}}$$

and making use of the ideal gas law, it follows that

$$c = \sqrt{kRT} \tag{1.20}$$

V1.7 As fast as a speeding bullet

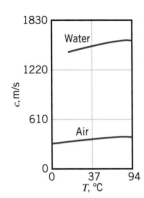

Thus, for ideal gases, the speed of sound is proportional to the square root of the absolute temperature. For example, for air at 15.6 °C with $k = 1.40$ and $R = 287$ J/kg · k, it follows that $c = 340$ m/s. The speed of sound in air at various temperatures can be found in Appendix B (Tables B.3 and B.4). Equation 1.19 is also valid for liquids, and values of E_v can be used to determine the speed of sound in liquids. For water at 20 °C, $E_v = 2.19$ GN/m² and $\rho = 998.2$ kg/m³ so that $c = 1481$ m/s. As shown in the adjacent figure, the speed of sound is much higher in water than in air. If a fluid were truly incompressible ($E_v = \infty$), the speed of sound would be infinite. The speed of sound in water for various temperatures can be found in Appendix B (Tables B.1 and B.2).

EXAMPLE 1.7 | Speed of Sound and Mach Number

Given A jet aircraft flies at a speed of 246 m/s at an altitude of 10668 m, where the temperature is 19 °C and the specific heat ratio is $k = 1.4$.

Find Determine the ratio of the speed of the aircraft, V, to that of the speed of sound, c, at the specified altitude.

Solution From Eq. 1.20, the speed of sound can be calculated as

$$c = \sqrt{kRT}$$

$$= \sqrt{(1.40)(287 \text{J}/\text{kg·K})(-19 + 273)\text{K}}$$

$$= 319 \text{ m/s}$$

Since the air speed is $V = 246$ m/s

the ratio is

$$\frac{V}{c} = \frac{246}{319} = 0.77 \qquad (Ans)$$

Comment This ratio is called the *Mach number*, Ma. If Ma < 1.0, the aircraft is flying at *subsonic* speeds, whereas for Ma > 1.0 it is flying at *supersonic* speeds. The Mach number is an important dimensionless parameter used in the study of the flow of gases at high speeds and will be further discussed in Chapters 7 and 11.

By repeating the calculations for different temperatures, the results shown in **Fig. E1.7** are obtained. Because the speed of sound increases with increasing temperature, for a constant airplane speed, the Mach number decreases as the temperature increases.

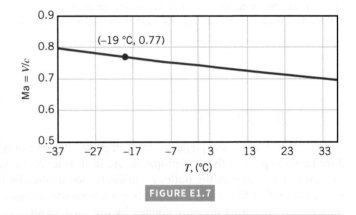

FIGURE E1.7

1.8 VAPOR PRESSURE

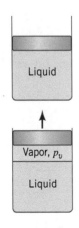

A liquid boils when the pressure is reduced to the vapor pressure.

It is a common observation that liquids such as water and gasoline will evaporate if they are simply placed in a container open to the atmosphere. Evaporation takes place because some liquid molecules at the surface have sufficient momentum to overcome the intermolecular cohesive forces and escape into the atmosphere. If the container is closed with a small air space left above the surface, and this space evacuated to form a vacuum, a pressure will develop in the space as a result of the vapor that is formed by the escaping molecules. When an equilibrium condition is reached so that the number of molecules leaving the surface is equal to the number entering, the vapor is said to be saturated and the pressure that the vapor exerts on the liquid surface is termed the **vapor pressure,** p_v. Similarly, if the end of a completely liquid-filled container is moved, as shown in the adjacent figure, without letting any air into the container, the space between the liquid and the end becomes filled with vapor at a pressure equal to the vapor pressure.

Since the development of a vapor pressure is closely associated with molecular activity, the value of vapor pressure for a particular liquid depends on temperature. Values of vapor pressure for water at various temperatures can be found in Appendix B (Tables B.1 and B.2), and the values of vapor pressure for several common liquids at room temperatures are given in Tables 1.5 and 1.6.

Boiling, which is the formation of vapor bubbles within a fluid mass, is initiated when the absolute pressure in the fluid reaches the vapor pressure. As commonly observed in the kitchen, water at standard atmospheric pressure will boil when the temperature reaches 100 °C—that is, the vapor pressure of water at 100 °C is 1 bar (abs). However, if we attempt to boil water at a higher elevation, say 9144 m above sea level (the approximate elevation of Mt. Everest), where the atmospheric pressure is 1 bar (abs), we find that boiling will start when the temperature is about 69 °C. At this temperature the vapor pressure of water is 0.3 bar. For the U.S. Standard Atmosphere (see Section 2.4), the boiling temperature is a function of altitude as shown in the adjacent figure. Thus, boiling can be induced at a given pressure acting on the fluid by raising the temperature, or at a given fluid temperature by lowering the pressure.

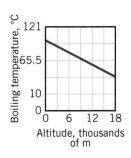

An important reason for our interest in vapor pressure and boiling lies in the common observation that in flowing fluids it is possible to develop very low pressure due to the fluid motion, and if the pressure is lowered to the vapor pressure, boiling will occur. For example, this phenomenon may occur in flow through the irregular, narrowed passages of a valve or pump. When vapor bubbles are formed in a flowing fluid, they are swept along into regions of higher pressure where they suddenly collapse with sufficient intensity to actually cause structural damage. The formation and subsequent collapse of vapor bubbles in a flowing fluid, called *cavitation,* is an important fluid flow phenomenon to be given further attention in Chapters 4, 7, and 12.

In flowing liquids it is possible for the pressure in localized regions to reach vapor pressure, thereby causing cavitation.

1.9 SURFACE TENSION

At the interface between a liquid and a gas, or between two immiscible liquids, forces develop in the liquid surface that cause the surface to behave as if it were a "skin" or "membrane" stretched over the fluid mass. Although such a skin is not actually present, this conceptual analogy allows us to explain several commonly observed phenomena. For example, a steel needle or a razor blade will float on water if placed gently on the surface because the tension developed in the hypothetical skin supports it. Small droplets of mercury will form into spheres when placed on a smooth surface because the cohesive forces in the surface of the mercury tend to hold all the molecules together in a compact shape.

V1.8 Floating razor blade

These various types of surface phenomena are due to the unbalanced cohesive forces acting on the liquid molecules at the fluid surface. Molecules in the interior of the fluid mass are surrounded by molecules that are attracted to each other equally. However, molecules along the surface may experience a net force toward the interior. The apparent physical consequence of this unbalanced force along the surface is to create the hypothetical skin or membrane. A tensile force may be considered to be acting in the plane of the surface along any line in the surface. The intensity of the molecular attraction per unit length along any line in the surface is called the **surface tension** and is designated by the Greek symbol σ (sigma). For a given liquid the surface tension depends on temperature as well as the other fluid it is in contact with at the interface. The dimensions of surface tension are FL^{-1} with SI units of N/m and BG units of lb/ft. Values of surface tension for some common liquids (in contact with air) are given in Table 1.5 (SI) and Table 1.6 (BG). Values for water as a function of temperature are provided in

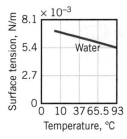

The Wide World of Fluids

Walking on water

Water striders are insects commonly found on ponds, rivers, and lakes that appear to "walk" on water. A typical length of a water strider is about 1 cm., and they can cover 100 body lengths in one second. It has long been recognized that it is *surface tension* that keeps the water strider from sinking below the surface. What has been puzzling is how they propel themselves at such a high speed. They can't pierce the water surface or they would sink. A team of mathematicians and engineers from the Massachusetts Institute of Technology (MIT)

applied conventional flow visualization techniques and high-speed video to examine in detail the movement of the water striders. They found that each stroke of the insect's legs creates dimples on the surface with underwater swirling vortices sufficient to propel it forward. It is the rearward motion of the vortices that propels the water strider forward. To further substantiate their explanation, the MIT team built a working model of a water strider, called Robostrider, which creates surface ripples and underwater vortices as it moves across a water surface. Waterborne creatures, such as the water strider, provide an interesting world dominated by surface tension.

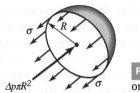

FIGURE 1.9 Forces acting on one-half of a liquid drop.

Appendix B (Tables B.1 and B.2). As indicated in the adjacent figure, the value of the surface tension decreases as the temperature increases.

The pressure inside a drop of fluid can be calculated using the free-body diagram in **Fig. 1.9**. If the spherical drop is cut in half (as shown), the force developed around the edge due to surface tension is $2\pi R\sigma$. This force must be balanced by the pressure difference, Δp, between the internal pressure, p_i, and the external pressure, p_e, acting over the circular area, πR^2. Thus,

$$2\pi R\sigma = \Delta p\, \pi R^2$$

or

$$\Delta p = p_i - p_e = \frac{2\sigma}{R} \tag{1.21}$$

V1.9 Capillary rise

It is apparent from this result that the pressure inside the drop is greater than the pressure surrounding the drop. (Would the pressure on the inside of a bubble of water be the same as that on the inside of a drop of water of the same diameter and at the same temperature?)

Among common phenomena associated with surface tension is the rise (or fall) of a liquid in a capillary tube. If a small open tube is inserted into water, the water level in the tube will rise above the water level outside the tube, as is illustrated in **Fig. 1.10a**. In this situation we have a liquid–gas–solid interface. For the case illustrated there is an attraction (adhesion) between the wall of the tube and liquid molecules, which is strong enough to overcome the mutual attraction (cohesion) of the molecules and pull them up the wall. Hence, the liquid is said to *wet* the solid surface.

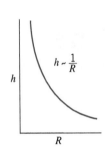

The height, h, is governed by the value of the surface tension, σ, the tube radius, R, the specific weight of the liquid, γ, and the *angle of contact*, θ, between the fluid and tube. From the free-body diagram of **Fig. 1.10b**, we see that the vertical force due to the surface tension is equal to $2\pi R\sigma \cos\theta$ and the weight is $\gamma\pi R^2 h$, and these two forces must balance for equilibrium. Thus,

$$\gamma\pi R^2 h = 2\pi R\sigma \cos\theta$$

V1.10 Contact angle

so that the height is given by the relationship

$$h = \frac{2\sigma \cos\theta}{\gamma R} \tag{1.22}$$

The angle of contact is a function of both the liquid and the surface. For water in contact with clean glass $\theta \approx 0°$. It is clear from Eq. 1.22 that the height is inversely proportional to the tube radius, and therefore, as indicated in the adjacent figure, the rise of a liquid in a tube as a result of capillary action becomes increasingly pronounced as the tube radius is decreased.

FIGURE 1.10 Effect of capillary action in small tubes. (*a*) Rise of column for a liquid that wets the tube. (*b*) Free-body diagram for calculating column height. (*c*) Depression of column for a nonwetting liquid.

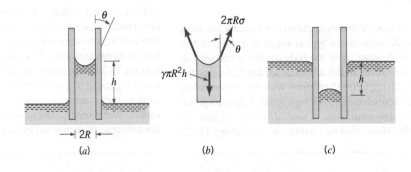

If adhesion of molecules to the solid surface is weak compared to the cohesion between molecules, the liquid will not wet the surface and the level in a tube placed in a nonwetting liquid will actually be depressed, as shown in **Fig. 1.10c**. Mercury is a good example of a nonwetting liquid when it is in contact with a glass tube. For nonwetting liquids, the angle of contact is greater than 90°, and for mercury in contact with clean glass $\theta \approx 130°$.

> **Capillary action in small tubes, which involves a liquid–gas–solid interface, is caused by surface tension.**

EXAMPLE 1.8 | Capillary Rise in a Tube

Given Pressures are sometimes determined by measuring the height of a column of liquid in a vertical tube.

Find What diameter of clean glass tubing is required so that the rise of water at 20 °C in a tube due to capillary action (as opposed to pressure in the tube) is less than $h = 1.5$ mm?

Solution From Eq. 1.22

$$h = \frac{2\sigma \cos \theta}{\gamma R}$$

so that

$$R = \frac{2\sigma \cos \theta}{\gamma h}$$

For water at 20 °C (from Table B.1), $\sigma = 0.0728$ N/m and $\gamma = 9.789$ kN/m^3. Since $\theta \approx 0°$, it follows that for $h = 1.5$ mm,

$$R = \frac{2(0.0728 \text{ N/m})(1)}{(9.789 \times 10^3 \text{ N/m}^3)(1.5 \text{ mm})(10^{-3} \text{ m/mm})}$$

$$= 0.00991 \text{ m}$$

and the minimum required tube diameter, D, is

$$D = 2R = 0.0198 \text{ m} = 19.8 \text{ mm} \qquad (Ans)$$

Comment By repeating the calculations for various values of the capillary rise, h, the results shown in **Fig. E1.8** are obtained. Note that as the allowable capillary rise is decreased, the diameter of the tube must be significantly increased. There is always some capillarity effect, but it can be minimized by using a large enough diameter tube.

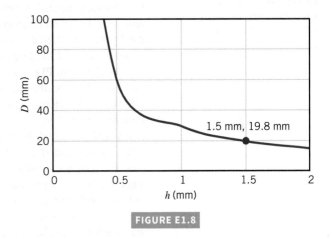

FIGURE E1.8

Surface tension effects play a role in many fluid mechanics problems, including the movement of liquids through soil and other porous media, flow of thin films, formation of drops and bubbles, and the breakup of liquid jets. For example, surface tension is a main factor in the formation of drops from a leaking faucet, as shown in the adjacent photograph. Surface phenomena associated with liquid–gas, liquid–liquid, and liquid–gas–solid interfaces are exceedingly complex, and a more detailed and rigorous discussion of them is beyond the scope of this text. Fortunately, in many fluid mechanics problems, surface phenomena, as characterized by surface tension, are not important because inertial, gravitational, and viscous forces are much more dominant.

photopixel/Shutterstock.com

The Wide World of Fluids

Spreading of oil spills

With the large traffic in oil tankers there is great interest in the prevention of and response to oil spills. As evidenced by the famous *Exxon Valdez* oil spill in Prince William Sound in 1989, oil spills can create disastrous environmental problems. A more recent example of this type of catastrophe is the oil spill that occurred in the Gulf of Mexico in 2010. It is not surprising that much attention is given to the rate at which an oil spill spreads. When spilled, most oils tend to spread horizontally into a smooth and slippery surface, called a slick. There are many factors that influence the ability of an oil slick to spread, including the size of the spill, wind speed and direction, and the physical properties of the oil. These properties include *surface tension, specific gravity,* and *viscosity.* The higher the surface tension the more likely a spill will remain in place. Since the specific gravity of oil is less than one, it floats on top of the water, but the specific gravity of an oil can increase if the lighter substances within the oil evaporate. The higher the viscosity of the oil, the greater the tendency to stay in one place.

1.10 A BRIEF LOOK BACK IN HISTORY

Before proceeding with our study of fluid mechanics, we should pause for a moment to consider the history of this important engineering science. As is true of all basic scientific and engineering disciplines, their actual beginnings are only faintly visible through the haze of early antiquity. But we know that interest in fluid behavior dates back to the ancient civilizations. Through necessity there was a practical concern about the manner in which spears and arrows could be propelled through the air, in the development of water supply and irrigation systems, and in the design of boats and ships. These developments were, of course, based on trial-and-error procedures. However, it was the accumulation of such empirical knowledge that formed the basis for further development during the emergence of the ancient Greek civilization and the subsequent rise of the Roman Empire. Some of the earliest writings that pertain to modern fluid mechanics are those of Archimedes (287–212 B.C.), a Greek mathematician and inventor who first expressed the principles of hydrostatics and flotation. Elaborate water supply systems were built by the Romans during the period from the fourth century B.C. through the early Christian period, and Sextus Julius Frontinus (A.D. 40–103), a Roman engineer, described these systems in detail. However, for the next 1000 years during the Middle Ages (also referred to as the Dark Ages), there appears to have been little added to further understanding of fluid behavior.

> Some of the earliest writings that pertain to modern fluid mechanics can be traced back to the ancient Greek civilization and subsequent Roman Empire.

As shown in **Fig. 1.11**, beginning with the Renaissance period (about the fifteenth century), a rather continuous series of contributions began that forms the basis of what we consider to be the science of fluid mechanics. Leonardo da Vinci (1452–1519) described through sketches and writings many different types of flow phenomena. The work of Galileo Galilei (1564–1642) marked the beginning of experimental mechanics. Following the early Renaissance period and during the seventeenth and eighteenth centuries, numerous significant contributions were made. These include theoretical and mathematical advances associated with the famous names of Newton, Bernoulli, Euler, and d'Alembert. Experimental aspects of fluid mechanics were also advanced during this period, but unfortunately the two different approaches, theoretical and experimental, developed along separate paths. *Hydrodynamics* was the term associated with the theoretical or mathematical study of idealized, frictionless fluid behavior, with the term *hydraulics* being used to describe the applied or experimental aspects of real fluid behavior, particularly the behavior of water. Further contributions and refinements were made to both theoretical hydrodynamics and experimental hydraulics during the nineteenth century, with the general differential equations describing fluid motions that are used in modern fluid mechanics being developed in this period. Experimental hydraulics became more of a science, and many of the results of experiments performed during the nineteenth century are still used today.

At the beginning of the twentieth century, both the fields of theoretical hydrodynamics and experimental hydraulics were highly developed, and attempts were being made to unify the two. In 1904, a classic paper was presented by a German professor, Ludwig Prandtl (1875–1953), who introduced the concept of a "fluid boundary layer," which laid the foundation for the unification

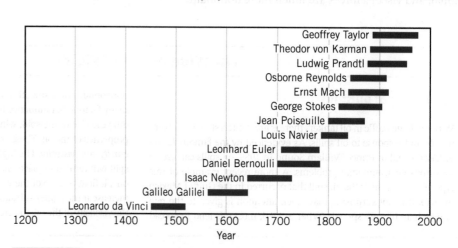

FIGURE 1.11 Time line of some contributors to the science of fluid mechanics.

of the theoretical and experimental aspects of fluid mechanics. Prandtl's idea was that for flow next to a solid boundary a thin fluid layer (boundary layer) develops in which friction (viscous force) is important, but outside this layer the fluid often behaves very much like a frictionless fluid. This relatively simple concept reconciled the seemingly different conclusions between the hydrodynamicists and the hydraulicists. Prandtl is generally accepted as the founder of modern fluid mechanics.

Also, during the first decade of the twentieth century, powered flight was first successfully demonstrated with the subsequent vastly increased interest in *aerodynamics.* Because the design of aircraft required an understanding of fluid flow and an ability to make accurate predictions of the effect of airflow on bodies, the field of aerodynamics provided a great stimulus for the many rapid developments in fluid mechanics that took place during the twentieth century.

As we proceed with our study of the fundamentals of fluid mechanics, we will continue to note the contributions of many of the pioneers in the field. **Table 1.9** provides a chronological

The rich history of fluid mechanics is fascinating, and many of the contributions of the pioneers in the field are noted in the succeeding chapters.

TABLE 1.9 Chronological Listing of Some Contributors to the Science of Fluid Mechanics Noted in the Text

ARCHIMEDES (287– 212 B.C.)
Established elementary principles of buoyancy and flotation.

SEXTUS JULIUS FRONTINUS (A.D. 40–103)
Wrote treatise on Roman methods of water distribution.

LEONARDO da VINCI (1452–1519)
Expressed elementary principle of continuity; observed and sketched many basic flow phenomena; suggested designs for hydraulic machinery.

GALILEO GALILEI (1564–1642)
Indirectly stimulated experimental hydraulics; revised Aristotelian concept of vacuum.

EVANGELISTA TORRICELLI (1608–1647)
Related barometric height to weight of atmosphere, and form of liquid jet to trajectory of free fall.

BLAISE PASCAL (1623–1662)
Finally clarified principles of barometer, hydraulic press, and pressure transmissibility.

ISAAC NEWTON (1642–1727)
Explored various aspects of fluid resistance—inertial, viscous, and wave; discovered jet contraction.

HENRI de PITOT (1695–1771)
Constructed double-tube device to indicate water velocity through differential head.

DANIEL BERNOULLI (1700–1782)
Experimented and wrote on many phases of fluid motion, coining name "hydrodynamics"; devised manometry technique and adapted primitive energy principle to explain velocity-head indication; proposed jet propulsion.

LEONHARD EULER (1707–1783)
First explained role of pressure in fluid flow; formulated basic equations of motion and contributed to the so-called Bernoulli theorem; introduced concept of cavitation and principle of centrifugal machinery.

JEAN le ROND d'ALEMBERT (1717–1783)
Originated notion of velocity and acceleration components, differential expression of continuity, and paradox of zero resistance to steady nonuniform motion.

ANTOINE CHEZY (1718–1798)
Formulated similarity parameter for predicting flow characteristics of one channel from measurements on another.

GIOVANNI BATTISTA VENTURI (1746–1822)
Performed tests on various forms of mouthpieces—in particular, conical contractions and expansions.

LOUIS MARIE HENRI NAVIER (1785–1836)
Extended equations of motion to include "molecular" forces.

AUGUSTIN LOUIS de CAUCHY (1789–1857)
Contributed to the general field of theoretical hydrodynamics and to the study of wave motion.

GOTTHILF HEINRICH LUDWIG HAGEN (1797–1884)
Conducted original studies of resistance in and transition between laminar and turbulent flow.

JEAN LOUIS POISEUILLE (1799–1869)
Performed meticulous tests on resistance of flow through capillary tubes.

HENRI PHILIBERT GASPARD DARCY (1803–1858)
Performed extensive tests on filtration and pipe resistance; initiated open-channel studies carried out by Bazin.

JULIUS WEISBACH (1806–1871)
Incorporated hydraulics in treatise on engineering mechanics, based on original experiments; noteworthy for flow patterns, nondimensional coefficients, weir, and resistance equations.

WILLIAM FROUDE (1810–1879)
Developed many towing-tank techniques, in particular the conversion of wave and boundary layer resistance from model to prototype scale.

(Continue)

TABLE 1.9 Chronological Listing of Some Contributors to the Science of Fluid Mechanics Noted in the Text (*Continued*)

Signal Photos/Alamy Stock Photo

The History Collection/Alamy Stock Photo

The History Collection/Alamy Stock Photo

Ludwig Prandtl

ROBERT MANNING (1816–1897)
Proposed several formulas for open-channel resistance.

GEORGE GABRIEL STOKES (1819–1903)
Derived analytically various flow relationships ranging from wave mechanics to viscous resistance—particularly that for the settling of spheres.

ERNST MACH (1838–1916)
One of the pioneers in the field of supersonic aerodynamics.

OSBORNE REYNOLDS (1842–1912)
Described original experiments in many fields—cavitation, river model similarity, pipe resistance—and devised two parameters for viscous flow; adapted equations of motion of a viscous fluid to mean conditions of turbulent flow.

JOHN WILLIAM STRUTT, LORD RAYLEIGH (1842–1919)
Investigated hydrodynamics of bubble collapse, wave motion, jet instability, laminar flow analogies, and dynamic similarity.

VINCENZ STROUHAL (1850–1922)
Investigated the phenomenon of "singing wires."

EDGAR BUCKINGHAM (1867–1940)
Stimulated interest in the United States in the use of dimensional analysis.

MORITZ WEBER (1871–1951)
Emphasized the use of the principles of similitude in fluid flow studies and formulated a capillarity similarity parameter.

LUDWIG PRANDTL (1875–1953)
Introduced concept of the boundary layer and is generally considered to be the father of present-day fluid mechanics.

LEWIS FERRY MOODY (1880–1953)
Provided many innovations in the field of hydraulic machinery. Proposed a method of correlating pipe resistance data that is widely used.

THEODOR VON KÁRMÁN (1881–1963)
One of the recognized leaders of twentieth century fluid mechanics. Provided major contributions to our understanding of surface resistance, turbulence, and wake phenomena.

PAUL RICHARD HEINRICH BLASIUS (1883–1970)
One of Prandtl's students who provided an analytical solution to the boundary layer equations. Also demonstrated that pipe resistance was related to the Reynolds number.

Source: Used by permission of IIHR Hydroscience & Engineering, The University of Iowa.

listing of some of these contributors and reveals the long journey that makes up the history of fluid mechanics. This list is certainly not comprehensive with regard to all past contributors but includes those who are mentioned in this text. As mention is made in succeeding chapters of the various individuals listed in Table 1.9, a quick glance at this table will reveal where they fit into the historical chain.

It is, of course, impossible to summarize the rich history of fluid mechanics in a few paragraphs. Only a brief glimpse is provided, and we hope it will stir your interest. References 2 to 5 are good starting points for further study, and in particular, Ref. 2 provides an excellent, broad, easily read history. Try it—you might even enjoy it!

CHAPTER SUMMARY

This introductory chapter discussed several fundamental aspects of fluid mechanics. Methods for describing fluid characteristics both quantitatively and qualitatively are considered. For a quantitative description, units are required. The concept of dimensions is introduced in which basic dimensions such as length, L, time, T, and mass, M, are used to provide a description of various quantities of interest. The use of dimensions is helpful in checking the generality of equations, as well as serving as the basis for the powerful tool of dimensional analysis discussed in detail in Chapter 7.

Various important fluid properties are defined, including fluid density, specific weight, specific gravity, viscosity, bulk modulus, speed of sound, vapor pressure, and surface tension. The ideal gas law is introduced to relate pressure, temperature, and density in common

gases, along with a brief discussion of the compression and expansion of gases. The distinction between absolute and gage pressure is introduced. This important idea is explored more fully in Chapter 2.

The following checklist provides a study guide for this chapter. When your study of the entire chapter and end-of-chapter exercises has been completed, you should be able to

- write out meanings of the terms listed and understand each of the related concepts. These terms are particularly important and are set in **bold black** type in the text.

- determine the dimensions of common physical quantities.

- determine whether an equation is a general or restricted homogeneous equation.

- correctly use units and systems of units in your analyses and calculations.
- calculate the density, specific weight, or specific gravity of a fluid from a knowledge of any two of the three.
- calculate the density, pressure, or temperature of an ideal gas (with a given gas constant) from a knowledge of any two of the three.
- relate the pressure and density of a gas as it is compressed or expanded using Eqs. 1.14 and 1.15.

- use the concept of viscosity to calculate the shearing stress in simple fluid flows.
- calculate the speed of sound in fluids using Eq. 1.19 for liquids and Eq. 1.20 for gases.
- determine whether boiling or cavitation will occur in a liquid using the concept of vapor pressure.
- use the concept of surface tension to solve simple problems involving liquid–gas or liquid–solid–gas interfaces.

KEY EQUATIONS

Specific weight	$\gamma = \rho g$	(1.6)
Specific gravity	$SG = \dfrac{\rho}{\rho_{H_2O@4\,°C}}$	(1.7)
Ideal gas law	$p = \rho RT$	(1.8)
Newtonian fluid shear stress	$\tau = \mu \dfrac{du}{dy}$	(1.9)
Bulk modulus	$E_v = -\dfrac{dp}{dV/V}$	(1.12)
Speed of sound in an ideal gas	$c = \sqrt{kRT}$	(1.20)
Capillary rise in a tube	$h = \dfrac{2\sigma \cos\theta}{\gamma R}$	(1.22)

REFERENCES

[1] Reid, R. C., Prausnitz, J. M., and Sherwood, T. K., *The Properties of Gases and Liquids,* 3rd Ed., New York: McGraw-Hill, 1977.

[2] Rouse, H. and Ince, S., *History of Hydraulics,* Iowa Institute of Hydraulic Research, Iowa City, 1957, New York: Dover, 1963.

[3] Tokaty, G. A., *A History and Philosophy of Fluid Mechanics,* Oxfordshire, UK: G. T. Foulis and Co., Ltd., 1971.

[4] Rouse, H., *Hydraulics in the United States 1776–1976,* Iowa Institute of Hydraulic Research, Iowa City, Iowa, 1976.

[5] Garbrecht, G., ed., *Hydraulics and Hydraulic Research—A Historical Review,* Rotterdam, Netherlands: A. A. Balkema, 1987.

[6] Brenner, M. P., Shi, X. D., Eggens, J., and Nagel, S. R., *Physics of Fluids* 7(9): 1995.

[7] Shi, X. D., Brenner, M. P., and Nagel, S. R., *Science* 265: 1994.

QUESTIONS AND PROBLEMS

Ⓒ Problem to be solved with aid of programmable calculator or computer.

Ⓞ Open-ended problem that requires critical thinking. These problems require various assumptions to provide the necessary input data. There are not unique answers to these problems.

Note: Unless specific values of required fluid properties are specified in the problem statement, use the values found in Tables 1.4–1.8 and in the tables in the Appendices.

Section 1.2 Dimensions, Dimensional Homogeneity, and Units

1.2.1 The force, F, of the wind blowing against a building is given by $F = C_D \rho V^2 A/2$, where V is the wind speed, ρ the density of the air, A the cross-sectional area of the building, and C_D is a constant termed the drag coefficient. Determine the dimensions of the drag coefficient.

1.2.2 The Mach number is a dimensionless ratio of the velocity of an object in a fluid to the speed of sound in the fluid. For an airplane flying at velocity V in air at absolute temperature T, the Mach number Ma is

$$\text{Ma} = \frac{V}{\sqrt{kRT}}$$

where k is a dimensionless constant and R is the specific gas constant for air. Show that Ma is dimensionless.

1.2.3 Verify the dimensions, in both the *FLT* and *MLT* systems, of the following quantities that appear in Table 1.1: (a) volume, (b) acceleration, (c) mass, (d) moment of inertia (area), and (e) work.

1.2.4 Verify the dimensions, in both the *FLT* and *MLT* systems, of the following quantities that appear in Table 1.1: (a) angular velocity, (b) energy, (c) moment of inertia (area), (d) power, and (e) pressure.

1.2.5 Verify the dimensions, in both the *FLT* system and the *MLT* system, of the following quantities that appear in Table 1.1: (a) frequency, (b) stress, (c) strain, (d) torque, and (e) work.

1.2.6 If u is a velocity, x a length, and t a time, what are the dimensions (in the *MLT* system) of (a) $\partial u/\partial t$, (b) $\partial^2 u/\partial x\,\partial t$, and (c) $\int(\partial u/\partial t)\,dx$?

1.2.7 Verify the dimensions, in both the *FLT* system and the *MLT* system, of the following quantities that appear in Table 1.1: (a) acceleration, (b) stress, (c) moment of a force, (d) volume, and (e) work.

1.2.8 If p is a pressure, V a velocity, and ρ a fluid density, what are the dimensions (in the *MLT* system) of (a) p/ρ, (b) $pV\rho$, and (c) $p/\rho V^2$?

1.2.9 If P is a force and x a length, what are the dimensions (in the *FLT* system) of (a) dP/dx, (b) d^3P/dx^3, and (c) $\int P\,dx$?

1.2.10 If V is a velocity, ℓ a length, and ν a fluid property (the kinematic viscosity) having dimensions of $L^2 T^{-1}$, which of the following combinations are dimensionless: (a) $V\ell\nu$, (b) $V\ell/\nu$, (c) $V^2\nu$, (d) $V/\ell\nu$?

1.2.11 The momentum flux (discussed in Chapter 5) is given by the product $\dot{m}V$, where $\dot{m}$ is mass flowrate and V is velocity. If mass flowrate is given in units of mass per unit time, show that the momentum flux can be expressed in units of force.

1.2.12 An equation for the frictional pressure loss Δp (mm H_2O) in a circular duct of inside diameter d (mm) and length L (m) for air flowing with velocity V (m/min) is

$$\Delta p = 0.027\left(\frac{L}{d^{1.22}}\right)\left(\frac{V}{V_0}\right)^{1.82}$$

where V_0 is a reference velocity equal to 305 m/min. Find the units of the "constant" 0.027.

1.2.13 The volume rate of flow, Q, through a pipe containing a slowly moving liquid is given by the equation

$$Q = \frac{\pi R^4 \Delta p}{8\mu\ell}$$

where R is the pipe radius, Δp the pressure drop along the pipe, μ a fluid property called viscosity ($FL^{-2}T$), and ℓ the length of pipe. What are the dimensions of the constant $\pi/8$? Would you classify this equation as a general homogeneous equation? Explain.

1.2.14 Show that each term in the following equation has units of N/m³. Consider u a velocity, y a length, x a length, p a pressure, and μ an absolute viscosity.

$$0 = -\frac{\partial p}{\partial x} + \mu\frac{\partial^2 u}{\partial y^2}$$

1.2.15 The pressure difference, Δp, across a partial blockage in an artery (called a *stenosis*) is approximated by the equation

$$\Delta p = K_v \frac{\mu V}{D} + K_u \left(\frac{A_0}{A_1} - 1\right)^2 \rho V^2$$

where V is the blood velocity, μ the blood viscosity ($FL^{-2}T$), ρ the blood density (ML^{-3}), D the artery diameter, A_0 the area of the unobstructed artery, and A_1 the area of the stenosis. Determine the dimensions of the constants K_v and K_u. Would this equation be valid in any system of units?

1.2.16 Assume that the speed of sound, c, in a fluid depends on an elastic modulus, E_v, with dimensions FL^{-2}, and the fluid density, ρ, in the form $c = (E_v)^a(\rho)^b$. If this is to be a dimensionally homogeneous equation, what are the values for a and b? Is your result consistent with the standard formula for the speed of sound? (See Eq. 1.19.)

1.2.17 A formula to estimate the volume rate of flow, Q, flowing over a dam of length, B, is given by the equation

$$Q = 3.09BH^{3/2}$$

where H is the depth of the water above the top of the dam (called the head). This formula gives Q in m³/s when B and H are in meter. Is the constant, 3.09, dimensionless? Would this equation be valid if units other than meter and seconds were used?

1.2.18 A commercial advertisement shows a pearl falling in a bottle of shampoo. If the diameter D of the pearl is quite small and the shampoo sufficiently viscous, the drag $\mathscr{D}$ on the pearl is given by Stokes's law,

$$\mathscr{D} = 3\pi\mu VD$$

where V is the speed of the pearl and μ is the fluid viscosity. Show that the term on the right side of Stokes's law has units of force.

1.2.19 Ⓞ Cite an example of a restricted homogeneous equation contained in a technical article found in an engineering journal in your field of interest. Define all terms in the equation, explain why it is a restricted equation, and provide a complete journal citation (title, date, etc.).

1.2.20 Express the following quantities in SI units: (a) 10.2 in./min, (b) 4.81 slugs, (c) 3.02 lb, (d) 73.1 ft/s², and (e) 0.0234 lb · s/ft².

1.2.21 Express the following quantities in SI units: (a) 160 acres, (b) 15 gallons (U.S.), (c) 240 miles, (d) 79.1 hp, and (e) 60.3 °F.

1.2.22 Water flows from a large drainage pipe at a rate of 1200 gal/min. What is this volume rate of flow in (a) m³/s, (b) liter/min, and (c) ft³/s?

1.2.23 The universal gas constant R_0 is equal to 49,700 ft²/(s² · °R) or 8310 m²/(s² · K). Show that these two magnitudes are equal.

1.2.24 Dimensionless combinations of quantities (commonly called dimensionless parameters) play an important role in fluid mechanics. Make up five possible dimensionless parameters by using combinations of some of the quantities listed in Table 1.1.

1.2.25 An important dimensionless parameter in certain types of fluid flow problems is the *Froude number* defined as $V/\sqrt{g\ell}$, where V is a velocity, g the acceleration of gravity, and ℓ a length. Determine the value of the Froude number for $V = 10$ ft/s, $g = 32.2$ ft/s², and $\ell = 2$ ft. Recalculate the Froude number using SI units for V, g, and ℓ. Explain the significance of the results of these calculations.

Section 1.4 Measures of Fluid Mass and Weight

1.4.1 Obtain a photograph/image of a situation in which the density or specific weight of a fluid is important. Print this photo and write a brief paragraph that describes the situation involved.

1.4.2 A tank contains 500 kg of a liquid whose specific gravity is 2. Determine the volume of the liquid in the tank.

1.4.3 A stick of butter at 25 °C measures 31.75 mm × 31.75 mm × 118.11 mm and weighs 113.4 g. Find its specific weight.

1.4.4 Clouds can weigh thousands of pounds due to their liquid water content. Often this content is measured in grams per cubic meter (g/m^3). Assume that a cumulus cloud occupies a volume of one cubic kilometer, and its liquid water content is 0.2 g/m^3. **(a)** What is the volume of this cloud in cubic miles? **(b)** How much does the water in the cloud weigh in pounds?

1.4.5 A certain object weighs 300 N at the Earth's surface. Determine the mass of the object (in kilograms) and its weight (in newtons) when located on a planet with an acceleration of gravity equal to 1.22 m/s^2.

1.4.6 The density of a certain type of jet fuel is 775 kg/m^3. Determine its specific gravity and specific weight.

1.4.7 At 4 °C a mixture of automobile antifreeze (50% water and 50% ethylene glycol by volume) has a density of 1064 kg/m^3. If the water density is 1000 kg/m^3, find the density of the ethylene glycol.

1.4.8 A *hydrometer* is used to measure the specific gravity of liquids (see **Video V2.8**). For a certain liquid, a hydrometer reading indicates a specific gravity of 1.15. What is the liquid's density and specific weight? Express your answer in SI units.

1.4.9 The information on a can of pop indicates that the can contains 355 mL. The mass of a full can of pop is 0.369 kg, while an empty can weighs 0.153 N. Determine the specific weight, density, and specific gravity of the pop and compare your results with the corresponding values for water at 20 °C. Express your results in SI units.

1.4.10 C The variation in the density of water, ρ, with temperature, T, in the range 20 °C $\leq T \leq$ 50 °C, is given in the following table.

Density (kg/m^3)	998.2	997.1	995.7	994.1	992.2	990.2	988.1
Temperature (°C)	20	25	30	35	40	45	50

Use these data to determine an empirical equation of the form $\rho = c_1 + c_2 T + c_3 T^2$ which can be used to predict the density over the range indicated. Compare the predicted values with the data given. What is the density of water at 42.1 °C?

1.4.11 If 1 cup of cream having a density of 1005 kg/m^3 is turned into 3 cups of whipped cream, determine the specific gravity and specific weight of the whipped cream.

1.4.12 O The presence of raindrops in the air during a heavy rainstorm increases the average density of the air–water mixture. Estimate by what percent the average air–water density is greater than that of just still air. State all assumptions and show calculations.

Section 1.5 Ideal Gas Law

1.5.1 Nitrogen is compressed to a density of 4 kg/m^3 under an absolute pressure of 400 kPa. Determine the temperature in degrees Celsius.

1.5.2 The temperature and pressure at the surface of Mars during a Martian spring day were determined to be −50 °C and 900 Pa, respectively. **(a)** Determine the density of the Martian atmosphere for these conditions if the gas constant for the Martian atmosphere is assumed to be equivalent to that of carbon dioxide. **(b)** Compare the answer from part (a) with the density of the Earth's atmosphere during a spring day when the temperature is 18 °C and the pressure 101.6 kPa (abs).

1.5.3 Assume that the air volume in a small automobile tire is constant and equal to the volume between two concentric cylinders 13 cm high with diameters of 33 cm and 52 cm. The air in the tire is initially at 25 °C and 202 kPa. Immediately after air is pumped into the tire, the temperature is 30 °C and the pressure is 303 kPa. What mass of air was added to the tire? What would be the air pressure after the air has cooled to a temperature of 0 °C?

1.5.4 A compressed air tank contains 5 kg of air at a temperature of 80 °C. A gage on the tank reads 300 kPa. Determine the volume of the tank.

1.5.5 A rigid tank contains air at a pressure of 620.5 kPa and a temperature of 15.5 °C. By how much will the pressure increase as the temperature is increased to 43.3 °C?

1.5.6 The density of oxygen contained in a tank is 2.0 kg/m^3 when the temperature is 25 °C. Determine the gage pressure of the gas if the atmospheric pressure is 97 kPa.

1.5.7 C Develop a computer program for calculating the density of an ideal gas when the gas pressure in pascals (abs), the temperature in degrees Celsius, and the gas constant in J/kg · K are specified. Plot the density of helium as a function of temperature from 0 °C to 200 °C and pressures of 50, 100, 150, and 200 kPa (abs).

Section 1.6 Viscosity

1.6.1 Obtain a photograph/image of a situation in which the viscosity of a fluid is important. Print this photo and write a brief paragraph that describes the situation involved.

1.6.2 For flowing water, what is the magnitude of the velocity gradient needed to produce a shear stress of 1.0 N/m^2?

1.6.3 Make use of the data in Appendix B to determine the dynamic viscosity of glycerin at 85 °F. Express your answer in both SI and BG units.

1.6.4 Video One type of *capillary-tube viscometer* is shown in **Video V1.5** and in **Fig. P1.6.4**. For this device, the liquid to be tested is drawn into the tube to a level above the top etched line. The time is then obtained for the liquid to drain to the bottom etched line. The kinematic viscosity, ν, in m^2/s is then obtained from the equation $\nu = KR^4 t$, where K is a constant, R is the radius of the capillary tube in mm, and t is the drain time in seconds. When glycerin at 20 °C is used as a calibration fluid in a particular viscometer, the drain time is 1430 s. When a liquid having a density of 970 kg/m^3 is tested in the same viscometer, the drain time is 900 s. What is the dynamic viscosity of this liquid?

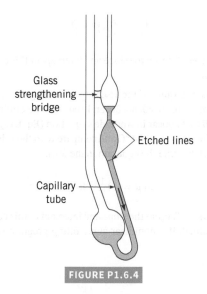

Glass strengthening bridge

Etched lines

Capillary tube

FIGURE P1.6.4

1.6.5 (Video) The viscosity of a soft drink was determined by using a capillary tube viscometer similar to that shown in Fig. P1.6.4 and Video V1.5. For this device, the kinematic viscosity, ν, is directly proportional to the time, t, that it takes for a given amount of liquid to flow through a small capillary tube. That is, $\nu = Kt$. The following data were obtained from regular pop and diet pop. The corresponding measured specific gravities are also given. Based on these data, by what percent is the absolute viscosity, μ, of regular pop greater than that of diet pop?

	Regular pop	Diet pop
t (s)	377.8	300.3
SG	1.044	1.003

1.6.6 The viscosity of a certain fluid is 5×10^{-4} poise. Determine its viscosity in both SI and BG units.

1.6.7 The kinematic viscosity and specific gravity of a liquid are 3.5×10^{-4} m²/s and 0.79, respectively. What is the dynamic viscosity of the liquid in SI units?

1.6.8 A liquid has a specific weight of 9268 N/m³ and a dynamic viscosity of 131.6 N · s/m². Determine its kinematic viscosity.

1.6.9 The kinematic viscosity of oxygen at 20 °C and a pressure of 150 kPa (abs) is 0.104 stokes. Determine the dynamic viscosity of oxygen at this temperature and pressure.

1.6.10 Calculate the Reynolds numbers for the flow of water and for air through a 4-mm-diameter tube, if the mean velocity is 3 m/s and the temperature is 30 °C in both cases (see Example 1.4). Assume the air is at standard atmospheric pressure.

1.6.11 SAE 30 oil at 15.6 °C flows through a 0.05 m-diameter pipe with a mean velocity of 1.5 m/s. Determine the value of the Reynolds number (see Example 1.4).

1.6.12 For air at standard atmospheric pressure the values of the constants that appear in the Sutherland equation (Eq. 1.10) are $C = 1.458 \times 10^{-6}$ kg/(m · s · K$^{1/2}$) and $S = 110.4$ K. Use these values to predict the viscosity of air at 10 °C and 90 °C and compare with values given in Table B.3 in Appendix B.

1.6.13 (C) Use the values of viscosity of air given in Table B.3 at temperatures of 0, 20, 40, 60, 80, and 100 °C to determine the constants C and S that appear in the Sutherland equation (Eq. 1.10). Compare your results with the values given in Problem 1.6.14. *Hint:* Rewrite the equation in the form

$$\frac{T^{3/2}}{\mu} = \left(\frac{1}{C}\right)T + \frac{S}{C}$$

and plot $T^{3/2}/\mu$ vs. T. From the slope and intercept of this curve, C and S can be obtained.

1.6.14 (C) Use the value of the viscosity of water given in Table B.1 at temperatures of 0, 20, 40, 60, 80, and 100 °C to determine the constants D and B that appear in Andrade's equation (Eq. 1.11). Calculate the value of the viscosity at 50 °C and compare with the value given in Table B.1. *Hint:* Rewrite the equation in the form

$$\ln \mu = (B)\frac{1}{T} + \ln D$$

and plot $\ln \mu$ vs. $1/T$. From the slope and intercept of this curve, B and D can be obtained. If a nonlinear curve-fitting program is available,

the constants can be obtained directly from Eq. 1.11 without rewriting the equation.

1.6.15 For a parallel plate arrangement of the type shown in Fig. 1.5, it is found that when the distance between plates is 2 mm, a shearing stress of 150 Pa develops at the upper plate when it is pulled at a velocity of 1 m/s. Determine the viscosity of the fluid between the plates. Express your answer in SI units.

1.6.16 Two flat plates are oriented parallel above a fixed lower plate, as shown in **Fig. P1.6.16**. The top plate, located a distance b above the fixed plate, is pulled along with speed V. The other thin plate is located a distance cb, where $0 < c < 1$, above the fixed plate. This plate moves with speed V_1, which is determined by the viscous shear forces imposed on it by the fluids on its top and bottom. The fluid on the top is twice as viscous as that on the bottom. Plot the ratio V_1/V as a function of c for $0 < c < 1$.

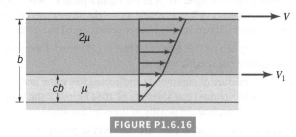

FIGURE P1.6.16

1.6.17 Three large plates are separated by thin layers of ethylene glycol and water, as shown in **Fig. P1.6.17**. The top plate moves to the right at 2 m/s. At what speed and in what direction must the bottom plate be moved to hold the center plate stationary?

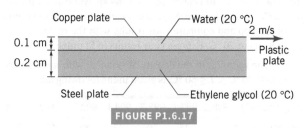

FIGURE P1.6.17

1.6.18 (Video) There are many fluids that exhibit non-Newtonian behavior (see, for example, Video V1.6). For a given fluid, the distinction between Newtonian and non-Newtonian behavior is usually based on measurements of shear stress and rate of shearing strain. Assume that the viscosity of blood is to be determined by measurements of shear stress, τ, and rate of shearing strain, du/dy, obtained from a small blood sample tested in a suitable viscometer. Based on the data provided in the table, determine if the blood is a Newtonian or non-Newtonian fluid. Explain how you arrived at your answer.

τ (N/m²)	0.04	0.06	0.12	0.18	0.30	0.52	1.12	2.10
du/dy (s^{-1})	2.25	4.50	11.25	22.5	45.0	90.0	225	450

1.6.19 The sled shown in **Fig. P1.6.19** slides along on a thin horizontal layer of water between the ice and the runners. The horizontal force that the water puts on the runners is equal to 5.34 N when the sled's speed is 15.24 m/s. The total area of both runners in contact with the water is 0.0074 m², and the viscosity of the water is 1.67×10^{-3} N · s/m². Determine the thickness of the water layer under the runners. Assume a linear velocity distribution in the water layer.

FIGURE P1.6.19

1.6.20 A 25-mm-diameter shaft is pulled through a cylindrical bearing as shown in **Fig. P1.6.20**. The lubricant that fills the 0.3-mm gap between the shaft and bearing is an oil having a kinematic viscosity of 8.0×10^{-4} m^2/s and a specific gravity of 0.91. Determine the force P required to pull the shaft at a velocity of 3 m/s. Assume the velocity distribution in the gap is linear.

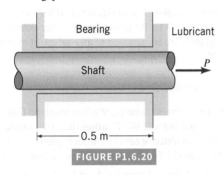

FIGURE P1.6.20

1.6.21 A hydraulic lift in a service station has a 32.50-cm-diameter ram that slides in a 32.52-cm-diameter cylinder. The annular space is filled with SAE 10 oil at 20 °C. The ram is traveling upward at the rate of 0.10 m/s. Find the frictional force when 3.0 m of the ram is engaged in the cylinder.

1.6.22 A piston having a diameter of 0.14 m and a length of 0.24 m slides downward with a velocity V through a vertical pipe. The downward motion is resisted by an oil film between the piston and the pipe wall. The film thickness is 5.08×10^{-5} m and the cylinder weighs 2.225 N. Estimate V if the oil viscosity is 0.766 N $\cdot$ s/m^2. Assume the velocity distribution in the gap is linear.

1.6.23 A 10-kg block slides down a smooth inclined surface as shown in **Fig. P1.6.23**. Determine the terminal velocity of the block if the 0.1-mm gap between the block and the surface contains SAE 30 oil at 15.5 °C. Assume the velocity distribution in the gap is linear, and the area of the block in contact with the oil is 0.1 m^2.

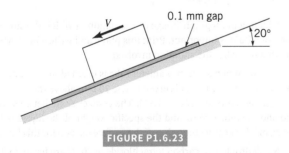

FIGURE P1.6.23

1.6.24 A layer of water flows down an inclined fixed surface with the velocity profile shown in **Fig. P1.6.24**. Determine the magnitude and direction of the shearing stress that the water exerts on the fixed surface for $U = 2$ m/s and $h = 0.1$ m.

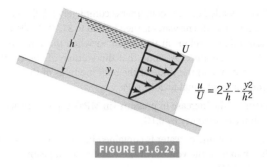

$$\frac{u}{U} = 2\frac{y}{h} - \frac{y^2}{h^2}$$

FIGURE P1.6.24

1.6.25 The concentric cylinder viscometer shown in **Fig. P1.6.25** has a cylinder height of 10.0 cm, a cylinder radius of 3.0 cm, and a uniform gap between the cylinder and the container (bottom and sides) of 0.10 cm. The pulley has a radius of 3.0 cm. Determine the weight required to produce a constant rotational speed of 30 rpm if the gap is filled with: **(a)** water, **(b)** gasoline, and **(c)** glycerin.

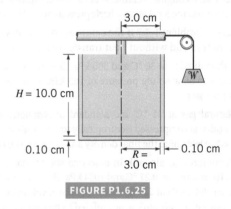

FIGURE P1.6.25

1.6.26 A 0.3 m-diameter circular plate is placed over a fixed bottom plate with a 0.0025 m gap between the two plates filled with glycerin, as shown in **Fig. P1.6.26**. Determine the torque required to rotate the circular plate slowly at 2 rpm. Assume that the velocity distribution in the gap is linear and that the shear stress on the edge of the rotating plate is negligible.

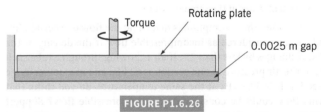

FIGURE P1.6.26

1.6.27 C Vehicle shock absorbers damp out oscillations caused by road roughness. Describe how a temperature change may affect the operation of a shock absorber.

1.6.28 Some measurements on a blood sample at 37 °C indicate a shearing stress of 0.52 N/m^2 for a corresponding rate of shearing strain of 200 s^{-1}. Determine the apparent viscosity of the blood and compare it with the viscosity of water at the same temperature.

Section 1.7 Compressibility of Fluids

1.7.1 Obtain a photograph/image of a situation in which the compressibility of a fluid is important. Print this photo and write a brief paragraph that describes the situation involved.

1.7.2 A sound wave is observed to travel through a liquid with a speed of 1500 m/s. The specific gravity of the liquid is 1.5. Determine the bulk modulus for this fluid.

1.7.3 A rigid-walled cubical container is completely filled with water at 4.45 °C and sealed. The water is then heated to 37.78 °C. Determine the pressure that develops in the container when the water reaches this higher temperature. Assume that the volume of the container remains constant and the value of the bulk modulus of the water remains constant and equal to 2068 MPa.

1.7.4 Estimate the increase in pressure (in MPa) required to decrease a unit volume of mercury by 0.1%.

1.7.5 A 1-m^3 volume of water is contained in a rigid container. Estimate the change in the volume of the water when a piston applies a pressure of 35 MPa.

1.7.6 Determine the speed of sound at 20 °C in (a) air, (b) helium, and (c) natural gas (methane). Express your answer in m/s.

1.7.7 Air is enclosed by a rigid cylinder containing a piston. A pressure gage attached to the cylinder indicates an initial reading of 172.3 kPa. Determine the reading on the gage when the piston has compressed the air to one-third its original volume. Assume the compression process to be isothermal and the local atmospheric pressure to be 101.3 kPa.

1.7.8 Repeat Problem 1.7.7 if the compression process takes place without friction and without heat transfer (isentropic process).

1.7.9 Carbon dioxide at 30 °C and 300 kPa absolute pressure expands isothermally to an absolute pressure of 165 kPa. Determine the final density of the gas.

1.7.10 Natural gas at 21 °C and standard atmospheric pressure of 101.3 kPa (abs) is compressed isentropically to a new absolute pressure of 482.6 kPa. Determine the final density and temperature of the gas.

1.7.11 A compressed air tank in a service station has a volume of 0.283 m^3. It contains air at 21 °C and 1034 kPa. How many tubeless tires can it fill to 308.2 kPa at 21 °C if each tire has a volume of 0.0424 m^3 and the compressed air tank is not refilled? The tank air temperature remains constant at 21 °C because of heat transfer through the tank's large surface area.

1.7.12 Oxygen at 30 °C and 300 kPa absolute pressure expands isothermally to an absolute pressure of 120 kPa. Determine the final density of the gas.

1.7.13 Compare the isentropic bulk modulus of air at 101 kPa (abs) with that of water at the same pressure.

1.7.14 Often the assumption is made that the flow of a certain fluid can be considered as incompressible flow if the density of the fluid changes by less than 2%. If air is flowing through a tube such that the air pressure at one section is 62 kPa and at a downstream section it is 59.3 kPa at the same temperature, do you think that this flow could be considered an incompressible flow? Support your answer with the necessary calculations. Assume standard atmospheric pressure.

1.7.15 An important dimensionless parameter concerned with very high-speed flow is the *Mach number*, defined as V/c, where V is the speed of the object such as an airplane or projectile, and c is the speed of sound in the fluid surrounding the object. For a projectile traveling at 1287.5 kmph through air at 10 °C and standard atmospheric pressure, what is the value of the Mach number?

1.7.16 The "power available in the wind" of velocity V through an area A is

$$\dot{W} = \frac{1}{2}\rho AV^3$$

where ρ is the air density (11.78 N/m^3). For an 29 kmph wind, find the wind area A that will supply a power of 2983 W.

1.7.17 Air enters the converging nozzle shown in **Fig. P1.7.17** at $T_1 = 21$ °C and $V_1 = 15.24$ m/s. At the exit of the nozzle, V_2 is given by

$$V_2 = \sqrt{V_1^2 + 2c_p(T_1 - T_2)}$$

where $c_p = 1.005$ kJ/kg · K and T_2 is the air temperature at the exit of the nozzle. Find the temperature T_2 for which $V_2 = 304.8$ m/s.

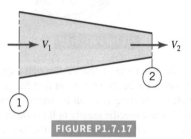

FIGURE P1.7.17

1.7.18 (See The Wide World of Fluids article titled "This water jet is a blast," Section 1.7.1.) By what percent is the volume of water decreased if its pressure is increased to an equivalent to 3000 atmospheres (304 MPa)?

Section 1.8 Vapor Pressure

1.8.1 During a mountain climbing trip it is observed that the water used to cook a meal boils at 90 °C rather than the standard 100 °C at sea level. At what altitude are the climbers preparing their meal? (See Tables B.1 and C.1 for data needed to solve this problem.)

1.8.2 When a fluid flows through a sharp bend, low pressures may develop in localized regions of the bend. Estimate the minimum absolute pressure (in kPa) that can develop without causing cavitation if the fluid is water at 71 °C.

1.8.3 A partially filled closed tank contains ethyl alcohol at 20 °C. If the air above the alcohol is evacuated, what is the minimum absolute pressure that develops in the evacuated space?

1.8.4 Estimate the minimum absolute pressure (in pascals) that can be developed at the inlet of a pump to avoid cavitation if the fluid is carbon tetrachloride at 20 °C.

1.8.5 When water at 70 °C flows through a converging section of pipe, the pressure decreases in the direction of flow. Estimate the minimum absolute pressure that can develop without causing cavitation. Express your answer in both BG and SI units.

1.8.6 At what atmospheric pressure will water boil at 35 °C? Express your answer in both SI and BG units.

Section 1.9 Surface Tension

1.9.1 Obtain a photograph/image of a situation in which the surface tension of a fluid is important. Print this photo and write a brief paragraph that describes the situation involved.

1.9.2 (Video) When a 2-mm-diameter tube is inserted into a liquid in an open tank, the liquid is observed to rise 10 mm above the free surface of the liquid (see Video V1.10). The contact angle between the liquid and the tube is zero, and the specific weight of the liquid is 1.2 × 10^4 N/m^3. Determine the value of the surface tension for this liquid.

1.9.3 Small droplets of carbon tetrachloride at 20 °C are formed with a spray nozzle. If the average diameter of the droplets is 200 μm, what is the difference in pressure between the inside and outside of the droplets?

1.9.4 A 12-mm-diameter jet of water discharges vertically into the atmosphere. Due to surface tension the pressure inside the jet will be slightly higher than the surrounding atmospheric pressure. Determine this difference in pressure.

1.9.5 A method used to determine the surface tension of a liquid is to determine the force necessary to raise a wire ring through the air–liquid interface. What is the value of the surface tension if a force of 0.015 N is required to raise a 4-cm-diameter ring? Consider the ring weightless, as a tensiometer (used to measure the surface tension) "zeroes" out the ring weight.

1.9.6 Calculate the pressure difference between the inside and outside of a spherical water droplet having a diameter of 0.79 mm and a temperature of 10 °C.

1.9.7 (Video) As shown in Video V1.9, surface tension forces can be strong enough to allow a double-edge steel razor blade to "float" on water, but a single-edge blade will sink. Assume that the surface tension forces act at an angle θ relative to the water surface, as shown in **Fig. P1.9.7**. (a) The mass of the double-edge blade is 0.64×10^{-3} kg, and the total length of its sides is 206 mm. Determine the value of θ required to maintain equilibrium between the blade weight and the resultant surface tension force. (b) The mass of the single-edge blade is 2.61×10^{-3} kg, and the total length of its sides is 154 mm. Explain why this blade sinks. Support your answer with the necessary calculations.

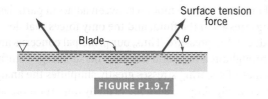

FIGURE P1.9.7

1.9.8 Explain how sweat soldering of copper pipe works from a fluid mechanics viewpoint.

1.9.9 (Video) An open, clean glass tube, having a diameter of 3 mm, is inserted vertically into a dish of mercury at 20 °C (see Video V1.10). How far will the column of mercury in the tube be depressed?

1.9.10 Two vertical, parallel, clean glass plates are spaced a distance of 2 mm apart. If the plates are placed in water, how high will the water rise between the plates due to capillary action?

Lifelong Learning Problems

1.1 LL Although there are numerous non-Newtonian fluids that occur naturally (quicksand and blood among them), with the advent of modern chemistry and chemical processing, many new manufactured non-Newtonian fluids are now available for a variety of novel applications. Obtain information about the discovery and use of newly developed non-Newtonian fluids. Summarize your findings in a brief report.

1.2 LL For years, lubricating oils and greases obtained by refining crude oil have been used to lubricate moving parts in a wide variety of machines, motors, and engines. With the increasing cost of crude oil and the potential for the reduced availability of it, the need for non-petroleum-based lubricants has increased considerably. Obtain information about non-petroleum-based lubricants. Summarize your findings in a brief report.

1.3 LL It is predicted that nanotechnology and the use of nano-sized objects will allow many processes, procedures, and products that, as of now, are difficult for us to comprehend. Among new nanotechnology areas is that of nanoscale fluid mechanics. Fluid behavior at the nanoscale can be entirely different than that for the usual everyday flows with which we are familiar. Obtain information about various aspects of nanofluid mechanics. Summarize your findings in a brief report.

CHAPTER 2

Fluid Statics

LEARNING OBJECTIVES

After completing this chapter, you should be able to:

- determine the pressure at various locations in a fluid at rest.

- explain the concept of manometers and apply appropriate equations to determine pressures.

- calculate the hydrostatic pressure force on a plane or curved submerged surface.

- calculate the buoyant force and determine the stability of floating or submerged objects.

In this chapter we will consider an important class of problems in which the fluid is either at rest or moving in such a manner that there is no relative motion between adjacent particles. In both instances there will be no shearing stresses in the fluid, and the only forces that develop on the surfaces of the particles will be due to the pressure. Thus, our principal concerns are to investigate pressure and its variation throughout a fluid and the force on submerged surfaces due to that pressure variation. The absence of shearing stresses greatly simplifies the analysis and, as we will see, allows us to obtain relatively simple solutions to many important practical problems.

2.1 PRESSURE AT A POINT

As we briefly discussed in Chapter 1, the term *pressure* is used to indicate the normal force per unit area at a given point acting on a given plane within the fluid mass of interest. A question that immediately arises is how the pressure at a point varies with the orientation of the plane passing through the point. To answer this question, consider the free-body diagram, illustrated in **Fig. 2.1**, that represents a small triangular wedge of fluid from some arbitrary location within a fluid mass. Since we are considering the situation in which there are no shearing stresses, the only external forces acting on the wedge are due to the pressure and gravity. For simplicity the forces in the x direction are not shown, and the z axis is taken as the vertical axis so gravity acts in the negative z direction. Although we are primarily interested in fluids at rest, to make the analysis as general as possible, we will allow the fluid element to have accelerated motion. The assumption of zero shearing stresses will still be valid as long as the fluid mass moves as a rigid body; that is, there is no relative motion between adjacent elements.

The equations of motion (Newton's second law, $\mathbf{F} = m\mathbf{a}$) in the y and z directions are, respectively,

$$\sum F_y = p_y\, \delta x\, \delta z - p_s\, \delta x\, \delta s\, \sin\theta = \rho\, \frac{\delta x\, \delta y\, \delta z}{2}\, a_y$$

$$\sum F_z = p_z\, \delta x\, \delta y - p_s\, \delta x\, \delta s\, \cos\theta - \gamma\, \frac{\delta x\, \delta y\, \delta z}{2} = \rho\, \frac{\delta x\, \delta y\, \delta z}{2}\, a_z$$

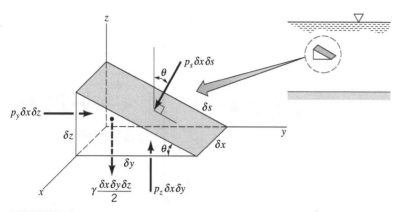

FIGURE 2.1 Forces on an arbitrary wedge-shaped element of fluid.

where p_s, p_y, and p_z are the average pressures on the faces, γ and ρ are the fluid specific weight and density, respectively, and a_y, a_z are the accelerations. Note that a pressure must be multiplied by an appropriate area to obtain the force due to the pressure. It follows from the geometry that

$$\delta y = \delta s \cos\theta \qquad \delta z = \delta s \sin\theta$$

so that the equations of motion can be rewritten as

> The pressure at a point in a fluid at rest is independent of direction.

$$p_y - p_s = \rho a_y \frac{\delta y}{2}$$

$$p_z - p_s = (\rho a_z + \gamma)\frac{\delta z}{2}$$

Since we are really interested in what is happening at a point, we take the limit as δx, δy, and δz approach zero (while maintaining the angle θ), and it follows that

$$p_y = p_s \qquad p_z = p_s$$

or $p_s = p_y = p_z$. The angle θ was arbitrarily chosen so we can conclude that *the pressure at a point in a fluid at rest, or in motion, is independent of direction as long as there are no shearing stresses present*. This important result is known as **Pascal's law**, named in honor of Blaise Pascal (1623–1662), a French mathematician who made important contributions in the field of hydrostatics. Thus, as shown by the adjacent photograph, at the junction of the side and bottom of the beaker, the pressure is the same on the side as it is on the bottom. In Chapter 6, it will be shown that for moving fluids in which there is relative motion between particles (so that shearing stresses develop), the normal stress at a point, which corresponds to pressure in fluids at rest, is not necessarily the same in all directions. In such cases the pressure is defined as the *average* of any three mutually perpendicular normal stresses at the point.

2.2 BASIC EQUATION FOR PRESSURE FIELD

Although we have answered the question of how the pressure at a point varies with direction, we are now faced with an equally important question—how does the pressure in a fluid in which there are no shearing stresses vary from point to point? To answer this question, consider a small rectangular element of fluid removed from some arbitrary position within the mass of fluid of interest as illustrated in **Fig. 2.2**. There are two types of forces acting on this element: **surface forces** due to the pressure and a **body force** equal to the weight of the element. Other possible types of body forces, such as those due to magnetic fields, will not be considered in this text.

If we let the pressure at the center of the element be designated as p, then the average pressure on the various faces can be expressed in terms of p and its derivatives, as shown in Fig. 2.2. We are actually using a Taylor series expansion of the pressure at the element center

> The pressure may vary across a fluid particle.

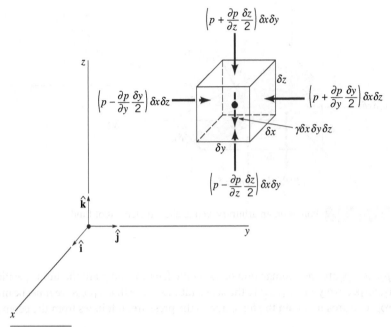

FIGURE 2.2 Surface and body forces acting on small fluid element.

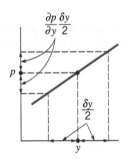

to approximate the pressures a short distance away and neglecting higher order terms that will vanish as we let δx, δy, and δz approach zero. This is illustrated by the adjacent figure. For simplicity the surface forces in the x direction are not shown. The resultant surface force in the y direction is

$$\delta F_y = \left(p - \frac{\partial p}{\partial y}\frac{\delta y}{2}\right)\delta x\,\delta z - \left(p + \frac{\partial p}{\partial y}\frac{\delta y}{2}\right)\delta x\,\delta z$$

or

$$\delta F_y = -\frac{\partial p}{\partial y}\,\delta x\,\delta y\,\delta z$$

Similarly, for the x and z directions the resultant surface forces are

$$\delta F_x = -\frac{\partial p}{\partial x}\,\delta x\,\delta y\,\delta z \qquad \delta F_z = -\frac{\partial p}{\partial z}\,\delta x\,\delta y\,\delta z$$

The resultant surface force acting on the element due to pressure can be expressed in vector form as

$$\delta \mathbf{F}_s = \delta F_x\hat{\mathbf{i}} + \delta F_y\hat{\mathbf{j}} + \delta F_z\hat{\mathbf{k}}$$

or

$$\delta \mathbf{F}_s = -\left(\frac{\partial p}{\partial x}\hat{\mathbf{i}} + \frac{\partial p}{\partial y}\hat{\mathbf{j}} + \frac{\partial p}{\partial z}\hat{\mathbf{k}}\right)\delta x\,\delta y\,\delta z \tag{2.1}$$

where $\hat{\mathbf{i}}$, $\hat{\mathbf{j}}$, and $\hat{\mathbf{k}}$ are the unit vectors along the coordinate axes shown in Fig. 2.2. The group of terms in parentheses in Eq. 2.1 represents in vector form the *pressure gradient* and can be written as

$$\frac{\partial p}{\partial x}\hat{\mathbf{i}} + \frac{\partial p}{\partial y}\hat{\mathbf{j}} + \frac{\partial p}{\partial z}\hat{\mathbf{k}} = \nabla p$$

where

$$\nabla(\ ) = \frac{\partial(\ )}{\partial x}\hat{\mathbf{i}} + \frac{\partial(\ )}{\partial y}\hat{\mathbf{j}} + \frac{\partial(\ )}{\partial z}\hat{\mathbf{k}}$$

and the symbol ∇ is the *gradient* or "del" vector operator. Thus, the resultant surface force per unit volume can be expressed as

$$\frac{\delta \mathbf{F}_s}{\delta x\,\delta y\,\delta z} = -\nabla p$$

> **The resultant surface force acting on a small fluid element depends only on the pressure gradient if there are no shearing stresses present.**

Since the z axis is vertical, the weight of the element is

$$-\delta\mathcal{W}\hat{\mathbf{k}} = -\gamma\,\delta x\,\delta y\,\delta z\,\hat{\mathbf{k}}$$

where the negative sign indicates that the force due to the weight is downward (in the negative z direction). Newton's second law, applied to the fluid element, can be expressed as

$$\sum\delta\mathbf{F} = \delta m\mathbf{a}$$

where $\sum\delta\mathbf{F}$ represents the resultant force acting on the element, $\mathbf{a}$ is the acceleration of the element, and δm is the element mass, which can be written as $\rho\,\delta x\,\delta y\,\delta z$. It follows that

$$\sum\delta\mathbf{F} = \delta\mathbf{F}_s - \delta\mathcal{W}\hat{\mathbf{k}} = \delta m\mathbf{a}$$

or

$$-\nabla p\,\delta x\,\delta y\,\delta z - \gamma\,\delta x\,\delta y\,\delta z\hat{\mathbf{k}} = \rho\,\delta x\,\delta y\,\delta z\mathbf{a}$$

and, therefore,

$$-\nabla p - \gamma\hat{\mathbf{k}} = \rho\mathbf{a} \tag{2.2}$$

Equation 2.2 is the general equation of motion for a fluid in which there are no shearing stresses. We will use this equation in Section 2.11 when we consider the pressure distribution in a moving fluid. For the present, however, we will restrict our attention to the special case of a fluid at rest.

2.3 PRESSURE VARIATION IN A FLUID AT REST

For a fluid at rest, $\mathbf{a} = 0$ and Eq. 2.2 reduces to

$$\nabla p + \gamma\hat{\mathbf{k}} = 0$$

or in component form

$$\frac{\partial p}{\partial x} = 0 \qquad \frac{\partial p}{\partial y} = 0 \qquad \frac{\partial p}{\partial z} = -\gamma \tag{2.3}$$

These equations show that for the coordinates defined, pressure is not a function of x or y. Thus, as we move from point to point in a horizontal plane (any plane parallel to the x–y plane), the pressure does not change. Since p depends only on z, the last of Eqs. 2.3 can be written as the ordinary differential equation

$$\boxed{\frac{dp}{dz} = -\gamma} \tag{2.4}$$

> For liquids or gases at rest, the pressure gradient in the vertical direction at any point in a fluid depends only on the specific weight of the fluid at that point.

Equation 2.4 is the fundamental equation for fluids at rest and can be used to determine how pressure changes with elevation. This equation and the adjacent figure indicate that the pressure gradient in the vertical direction is negative; that is, the pressure decreases as we move upward in a fluid at rest. There is no requirement that γ be a constant. Thus, it is valid for fluids with constant specific weight, such as liquids, as well as fluids whose specific weight may vary with elevation, such as air or other gases. However, to proceed with the integration of Eq. 2.4 it is necessary to stipulate how the specific weight varies with z.

If the fluid is flowing (i.e., not at rest with $\mathbf{a} = 0$), then the pressure variation is usually much more complex than that given by Eq. 2.4. For example, the pressure distribution on your car as it is driven along the road varies in a complex manner with x, y, and z. This idea is addressed in detail in Chapters 4, 6, and 9.

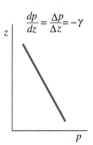

pressure increases with decreasing elevation

2.3.1 Incompressible Fluid

Since the specific weight is equal to the product of fluid density and acceleration due to gravity ($\gamma = \rho g$), changes in γ are caused by a change in either ρ or g. For most engineering applications the variation in g is negligible, so our main concern is with the possible variation in the fluid density. In general, a fluid with constant density is called an **incompressible fluid.** For liquids the variation in density is usually negligible, even for large changes in pressure, so that the assumption of constant specific weight when dealing with liquids is usually a good one. For this instance, Eq. 2.4 can be directly integrated

$$\int_{p_1}^{p_2} dp = -\gamma \int_{z_1}^{z_2} dz$$

to yield

$$p_2 - p_1 = -\gamma(z_2 - z_1)$$

or

$$p_1 - p_2 = \gamma(z_2 - z_1) \tag{2.5}$$

where p_1 and p_2 are pressures at the vertical elevations z_1 and z_2, as is illustrated in **Fig. 2.3**.

Equation 2.5 can be written in the compact form

$$p_1 - p_2 = \gamma h \tag{2.6}$$

or

$$p_1 = \gamma h + p_2 \tag{2.7}$$

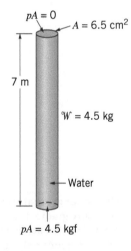

$pA = 0$

$A = 6.5$ cm^2

7 m

$\mathcal{W} = 4.5$ kg

Water

$pA = 4.5$ kgf

where h is the distance, $z_2 - z_1$, which is the depth of fluid measured downward from the location of p_2. This type of pressure distribution is commonly called a **hydrostatic distribution,** and Eq. 2.7 shows that in an incompressible fluid at rest the pressure varies linearly with depth. The pressure must increase with depth to "hold up" the fluid above it.

It can also be observed from Eq. 2.6 that the pressure difference between two points can be specified by the distance h, since

$$h = \frac{p_1 - p_2}{\gamma}$$

In this case h is called the **pressure head** and is interpreted as the height of a column of fluid of specific weight γ required to give a pressure difference $p_1 - p_2$. For example, a pressure difference of 69 kPa can be specified in terms of pressure head as 7 m of water ($\gamma = 9.81$ kN/m^3), or 518 mm of Hg ($\gamma = 133$ kN/m^3). As illustrated by the adjacent figure, a 7 m-tall column of water with a cross-sectional area of 6.5 cm^2 weighs 4.5 kg.

V2.1 Demonstration of atmospheric pressure.

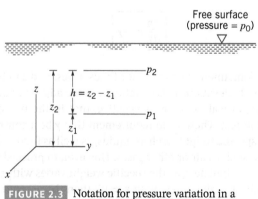

Free surface
(pressure $= p_0$)

p_2

z

$h = z_2 - z_1$

z_2

p_1

z_1

y

x

FIGURE 2.3 Notation for pressure variation in a fluid at rest with a free surface.

The Wide World of Fluids

Giraffe's blood pressure

A giraffe's long neck allows it to graze up to 6 m above the ground. It can also lower its head to drink at ground level. Thus, in the circulatory system, there is a significant *hydrostatic pressure* effect due to this elevation change. To maintain blood flow to its head throughout this change in elevation, the giraffe must maintain a relatively high blood pressure at heart level— approximately two and a half times that in humans. To prevent rupture of blood vessels in the high-pressure lower leg regions, giraffes have a tight sheath of thick skin over their lower limbs that acts like an elastic bandage in exactly the same way as do the g-suits of fighter pilots. In addition, valves in the upper neck prevent backflow into the head when the giraffe lowers its head to ground level. It is also thought that blood vessels in the giraffe's kidney have a special mechanism to prevent large changes in filtration rate when blood pressure increases or decreases with its head movement.

When one works with liquids, there is often a liquid–vapor interface we call a free surface, as is illustrated in Fig. 2.3, and it is convenient to use this surface as a reference plane. The reference pressure p_0 would correspond to the pressure acting on the free surface (which would frequently be atmospheric pressure). If we let $p_2 = p_0$ in Eq. 2.7, it follows that the pressure p at any depth h below the free surface is given by the equation:

$$p = \gamma h + p_0 \tag{2.8}$$

As is demonstrated by Eq. 2.7 or 2.8, the pressure in a homogeneous, incompressible fluid at rest depends only on the depth of the fluid relative to some reference plane. It is *not* influenced by the *size* or *shape* of container in which the fluid is held. Thus, in **Fig. 2.4**, the pressure is the same at all points along the line AB, even though the containers have very irregular shapes. The actual value of the pressure along AB depends only on the depth, h, the surface pressure, p_0, and the specific weight, γ, of the liquid in the container.

FIGURE 2.4 Fluid pressure in containers of arbitrary shape.

EXAMPLE 2.1 | Pressure–Depth Relationship

Given Because of a leak in a buried gasoline storage tank, water has seeped in to the depth shown in **Fig. E2.1**. The specific gravity of the gasoline is $SG = 0.68$.

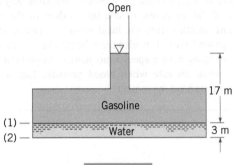

Open

17 m

Gasoline

(1)

(2) — Water — 3 m

FIGURE E2.1

Find Determine the pressure at the gasoline–water interface and at the bottom of the tank. Express the pressure in units of kN/m^2, and as a pressure head in m of water.

Solution Since we are dealing with liquids at rest, the pressure distribution will be hydrostatic, and therefore, the pressure variation can be found from the equation:

$$p = \gamma h + p_0$$

With p_0 corresponding to the pressure at the free surface of the gasoline, the pressure at the interface is

$$p_1 = (SG)(\gamma_{H_2O})h + p_0$$

$$= (0.68)(9.81 \text{ kN/m}^3)(17 \text{ m}) + p_0$$

$$= 113.4 + p_0(\text{kN/m}^2)$$

If we measure the pressure relative to atmospheric pressure (gage pressure), it follows that $p_0 = 0$, and therefore

$$p_1 = 113.4 \text{ kN/m}^2 \qquad (Ans)$$

$$\frac{p_1}{\gamma_{H_2O}} = \frac{113.4 \text{ kN/m}^2}{9.81 \text{ kN/m}^3} = 11.6 \text{ m} \qquad (Ans)$$

We can now apply the same relationship to determine the pressure at the tank bottom; that is,

$$p_2 = \gamma_{H_2O}h_{H_2O} + p_1$$

$$= (9.81 \text{ kN/m}^3)(3 \text{ m}) + 113.4 \text{ kN/m}^2$$

$$= 123.21 \text{ kN/m}^2 \qquad (Ans)$$

$$\frac{p_2}{\gamma_{H_2O}} = \frac{123.21 \text{ kN/m}^2}{9.81 \text{ kN/m}^3} = 12.6 \text{ m} \qquad (Ans)$$

Comment Observe that if we wish to express these pressures in terms of *absolute* pressure, we would have to add the local atmospheric pressure (in appropriate units) to the previous results. A further discussion of gage and absolute pressure is given in Section 2.5.

The transmission of pressure throughout a stationary fluid is the principle upon which many hydraulic devices are based.

The required equality of pressures at a specific elevation throughout a system is important for the operation of hydraulic jacks (see **Fig. 2.5a**), lifts, and presses, as well as hydraulic controls on aircraft and other types of heavy machinery. The fundamental idea behind such devices and systems is demonstrated in **Fig. 2.5b**. A piston located at one end of a closed

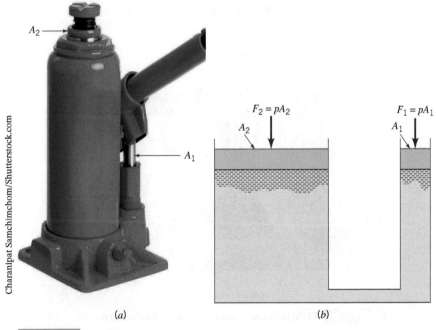

Charanipat Samchimchom/Shutterstock.com

A_2

A_1

$F_2 = pA_2$

A_2

$F_1 = pA_1$

A_1

(a)   (b)

FIGURE 2.5 (a) Hydraulic jack, (b) Transmission of fluid pressure.

system filled with a liquid, such as oil, can be used to change the pressure throughout the system, and thus transmit an applied force F_1 to a second piston where the resulting force is F_2. Since the pressure p acting on the faces of both pistons is the same (the effect of elevation changes is usually negligible for this type of hydraulic device), it follows that $F_2 = (A_2/A_1)F_1$. The piston area A_2 can be made much larger than A_1, and therefore, a large mechanical advantage can be developed; that is, a small force applied at the smaller piston can be used to develop a large force at the larger piston. The applied force could be created manually through some type of mechanical device, such as a hydraulic jack, or through compressed air acting directly on the surface of the liquid, as is done in hydraulic lifts commonly found in service stations.

2.3.2 Compressible Fluid

We normally think of gases such as air, oxygen, and nitrogen as being **compressible fluids** because the density of the gas can change significantly with modest changes in pressure and temperature. Thus, although Eq. 2.4 applies at a point in a gas, it is necessary to consider the possible variation in γ before the equation can be integrated. However, as was discussed in Chapter 1, the specific weights of common gases are small when compared with those of liquids. For example, the specific weight of air at sea level and $156°C$ is 11.99 N/m^3, whereas the specific weight of water under the same conditions is 9.81 kN/m^3. Since the specific weights of gases are comparatively small, it follows from Eq. 2.4 that the pressure gradient in the vertical direction is correspondingly small, and even over distances of several hundred feet the pressure will remain essentially constant for a gas. This means we can neglect the effect of elevation changes on the pressure for stationary gases in tanks, pipes, and so forth in which the distances involved are not large.

For those situations in which the variations in heights are large, on the order of thousands of feet, attention must be given to the variation in the specific weight. As is described in Chapter 1, the equation of state for an ideal gas is

$$p = \rho RT$$

where p is the absolute pressure, R is the gas constant, and T is the absolute temperature. This relationship can be combined with Eq. 2.4 to give

$$\frac{dp}{dz} = -\frac{gp}{RT}$$

and by separating variables

$$\int_{p_1}^{p_2} \frac{dp}{p} = \ln \frac{p_2}{p_1} = -\frac{g}{R} \int_{z_1}^{z_2} \frac{dz}{T} \tag{2.9}$$

where g and R are assumed to be constant over the elevation change from z_1 to z_2. Although the acceleration due to gravity, g, does vary with elevation, the variation is very small (see Tables in Appendix C), and g is usually assumed to be constant at some average value for the range of elevation involved.

Before completing the integration, one must specify the nature of the variation of temperature with elevation. For example, if we assume that the temperature has a constant value T_0 over the range z_1 to z_2 (*isothermal* conditions), it then follows from Eq. 2.9 that

$$p_2 = p_1 \exp\left[-\frac{g(z_2 - z_1)}{RT_0}\right] \tag{2.10}$$

This equation provides the desired pressure–elevation relationship for an isothermal layer. As shown in the adjacent figure, even for a 3048 m altitude change, the difference between the constant temperature (isothermal) and the constant density (incompressible) results is relatively minor. For nonisothermal conditions, a similar procedure can be followed if the temperature–elevation relationship is known, as is discussed in Section 2.4.

> If the specific weight of a fluid varies significantly as we move from point to point, the pressure will no longer vary linearly with depth.

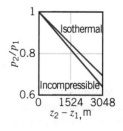

EXAMPLE 2.2 | Incompressible and Isothermal Pressure–Depth Variations

Given In 2010, the world's tallest building, the Burj Khalifa skyscraper, shown in **Fig. E2.2**, was completed and opened in the United Arab Emirates. The final height of the building, which had remained a secret until completion, is 828 m.

Find

a. Estimate the ratio of the pressure at the 828 m top of the building to the pressure at its base, assuming the air to be at a common temperature of 290 °K.

b. Compare the pressure calculated in part (a) with that obtained by assuming the air to be incompressible with $\gamma = 11.99$ N/m^3 at 1 atm (abs) (values for air at standard sea level conditions).

Solution

a. For the assumed isothermal conditions, and treating air as a compressible fluid, Eq. 2.10 can be applied to yield

$$\frac{p_2}{p_1} = \exp\left[-\frac{g(z_2 - z_1)}{RT_0}\right]$$

$$= \exp\left\{-\frac{(9.81 \text{ m/s}^2)(828 \text{ m})}{(8.314 \text{ J/mol} \cdot \text{K})(290 \text{ °K})}\right\}$$

$$= 0.0344 \qquad (Ans)$$

b. If the air is treated as an incompressible fluid, we can apply Eq. 2.5. In this case

$$p_2 = p_1 - \gamma(z_2 - z_1)$$

or

$$\frac{p_2}{p_1} = 1 - \frac{\gamma(z_2 - z_1)}{p_1}$$

$$= 1 - \frac{(11.99 \text{ N/m}^3)(828 \text{ m})}{(101352 \text{ N/m}^2)} = 0.902 \qquad (Ans)$$

Comments Note that there is little difference between the two results. Since the pressure difference between the bottom and top of the building is small, it follows that the variation in fluid density is small and, therefore, the compressible fluid and incompressible fluid analyses yield essentially the same result.

sophiejames/123RF

FIGURE E2.2

We see that for both calculations the pressure decreases by approximately 10% as we go from ground level to the top of this tallest building. It does not require a very large pressure difference to support a 828 m tall column of fluid as light as air. This result supports the earlier statement that the changes in pressures in air and other gases due to elevation changes are very small, even for distances of hundreds of meters. Thus, the pressure differences between the top and bottom of a horizontal pipe carrying a gas, or in a gas storage tank, are negligible because the distances involved are very small.

2.4 STANDARD ATMOSPHERE

An important application of Eq. 2.9 relates to the variation in pressure in the earth's atmosphere. Ideally, we would like to have measurements of pressure versus altitude over the specific range for the specific conditions (temperature, reference pressure) for which the pressure is to be determined. However, this type of information is usually not available. Thus, a "standard atmosphere" has been determined that can be used in the design of aircraft, missiles, and spacecraft and in comparing their performance under standard conditions. The concept of a standard atmosphere was first developed in the 1920s, and since that time, many national and international committees and organizations have pursued the development of such a standard. The currently accepted standard atmosphere is based on a report published in 1962 and updated in 1976 (see Refs. 1 and 2), defining the so-called **U.S. standard atmosphere**, which is an idealized

The standard atmosphere is an idealized representation of mean conditions in the earth's atmosphere.

TABLE 2.1 Properties of U.S. Standard Atmosphere at Sea Level*

Property	SI Units
Temperature, T	288.15 K (15 °C)
Pressure, p	101.33 kPa (abs)
Density, ρ	1.225 kg/m^3
Specific weight, γ	12.014 N/m^3
Viscosity, μ	1.789×10^{-5} N · s/m^2

*Acceleration of gravity at sea level $= 9.807$ m/s²

representation of middle-latitude, year-round mean conditions of the earth's atmosphere. Several important properties for standard atmospheric conditions at *sea level* are listed in **Table 2.1**, and **Fig. 2.6** shows the temperature profile for the U.S. standard atmosphere. As is shown in this figure, the temperature decreases with altitude in the region nearest the earth's surface (*troposphere*), then becomes essentially constant in the next layer (*stratosphere*), and subsequently starts to increase in the next layer. Typical events that occur in the atmosphere are shown in the adjacent figure.

Since the temperature variation is represented by a series of linear segments, it is possible to integrate Eq. 2.9 to obtain the corresponding pressure variation. For example, in the troposphere, which extends to an altitude of about 11 km (~36,000 ft), the temperature variation is of the form

$$T = T_a - \beta z \qquad (2.11)$$

where T_a is the temperature at sea level ($z = 0$) and β is the *lapse rate* (the rate of change of temperature with elevation). For the standard atmosphere in the troposphere, $\beta = 0.00650$ K/m.

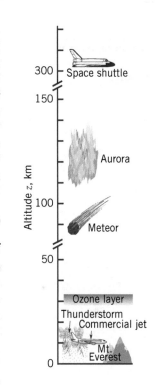

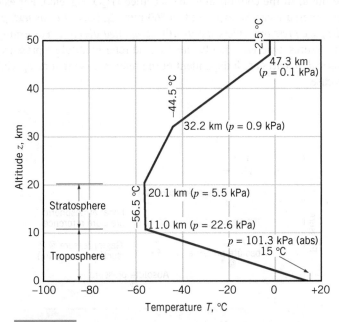

FIGURE 2.6 Variation of temperature with altitude in the U.S. standard atmosphere.

Equation 2.11 used with Eq. 2.9 yields

$$p = p_a \left(1 - \frac{\beta z}{T_a}\right)^{g/R\beta}$$ (2.12)

where p_a is the absolute pressure at $z = 0$. With p_a, T_a, and g obtained from Table 2.1, and with the gas constant $R = 286.9$ J/kg· K the pressure variation throughout the troposphere can be determined from Eq. 2.12. This calculation shows that at the outer edge of the troposphere, where the temperature is -56.5 °C, the absolute pressure is about 23 kPa. It is to be noted that modern jetliners cruise at approximately this altitude. Pressures at other altitudes are shown in Fig. 2.6, and tabulated values for temperature, acceleration of gravity, pressure, density, and viscosity for the U.S. standard atmosphere are given in Tables C.1 and C.2 in Appendix C.

2.5 MEASUREMENT OF PRESSURE

Pressure is designated as either absolute pressure or gage pressure.

Many different devices and techniques have been developed to measure pressure because it is so central to understanding fluid mechanics and to solving problems of practical interest. As is noted briefly in Chapter 1, the pressure at a point within a fluid is reported as either an **absolute pressure** or a **gage pressure.** Absolute pressure is measured relative to a perfect vacuum (absolute zero pressure), whereas gage pressure is measured relative to the local atmospheric pressure. Thus, a gage pressure of zero corresponds to a pressure that is equal to the local atmospheric pressure. Absolute pressures are always positive, but gage pressures can be either positive or negative depending on whether the pressure is above atmospheric pressure (a positive value) or below atmospheric pressure (a negative value). A negative gage pressure is also referred to as a *suction* or **vacuum pressure.** For example, 69 kPa (abs) could be expressed as -31 kPa (gage), if the local atmospheric pressure is 100 kPa, or alternatively 31 kPa suction or 31 kPa vaccum. The concept of gage and absolute pressure is illustrated graphically in **Fig. 2.7** for two typical pressures located at points 1 and 2.

In addition to the reference used for the pressure measurement, the *units* used to express the value are obviously of importance. As is described in Section 1.5, pressure is a force per unit area, and the units in the SI system are N/m^2; this combination is called the pascal and written as Pa (1 N/m$^2 \equiv$ 1 Pa). As noted earlier, pressure can also be expressed as the height of a column of liquid. Then the units will refer to the height of the column (mm, m, etc.), and in addition, the liquid in the column must be specified (H$_2$O, Hg, etc.). For example, standard atmospheric pressure can be expressed as 760 mm Hg (abs). *In this text, pressures will be assumed to be gage pressures unless specifically designated absolute.* For example, 100 kPa would be gage pressures, whereas 100 kPa (abs) would refer to absolute pressures. It is to be noted that *pressure differences* are independent of the reference, so that no special notation is required in this case.

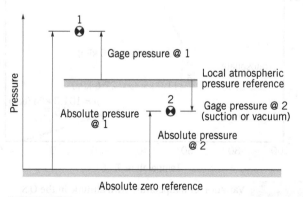

FIGURE 2.7 Graphical representation of gage and absolute pressure.

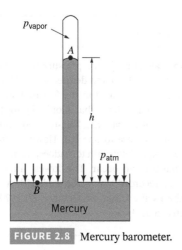

p_{vapor}

A

h

p_{atm}

B

Mercury

FIGURE 2.8 Mercury barometer.

Water

Mercury

The measurement of atmospheric pressure is usually accomplished with a mercury **barometer,** which in its simplest form consists of a glass tube closed at one end with the open end immersed in a container of mercury, as shown in **Fig. 2.8.** The tube is initially filled with mercury (with its open end up) and then turned upside down (open end down), with the open end in the container of mercury. The column of mercury will come to an equilibrium position where its weight plus the force due to the vapor pressure (which develops in the space above the column) balances the force due to the atmospheric pressure. Thus,

$$p_{atm} = \gamma h + p_{vapor} \tag{2.13}$$

where γ is the specific weight of mercury. For most practical purposes, the contribution of the vapor pressure can be neglected because it is very small [for mercury, the fluid most commonly used in barometers, $p_{vapor} = 0.159$ Pa (abs) at a temperature of 20°C], so that $p_{atm} \approx \gamma h$. It is conventional to specify atmospheric pressure in terms of the height, h, in millimeters or inches of mercury. Note that if water were used instead of mercury, the height of the column would have to be approximately 10.36 m rather than 0.759 m of mercury for an atmospheric pressure of 100 kPa (abs)! This is shown to scale in the adjacent figure. The concept of the mercury barometer is an old one, with the invention of this device attributed to Evangelista Torricelli in about 1644.

EXAMPLE 2.3 | Barometric Pressure

Given A mountain lake has an average temperature of 10 °C and a maximum depth of 50 m. The barometric pressure is 600 mm Hg.

Find Determine the absolute pressure (in pascals) at the deepest part of the lake.

Solution The pressure in the lake at any depth, h, is given by the equation

$$p = \gamma h + p_0$$

where p_0 is the pressure at the surface. Since we want the absolute pressure, p_0 will be the local barometric pressure expressed in a consistent system of units; that is

$$\frac{p_{barometric}}{\gamma_{Hg}} = 600 \text{ mm} = 0.6 \text{ m}$$

and for $\gamma_{Hg} = 133$ kN/m^3

$$p_0 = (0.6 \text{ m})(133 \text{ kN/m}^3) = 79.8 \text{ kN/m}^2$$

From Table B.1, $\gamma_{H_2O} = 9.804$ kN/m^3 at 10 °C, and therefore

$$p = (9.804 \text{ kN/m}^3)(50 \text{ m}) + 79.8 \text{ kN/m}^2$$
$$= 490.2 \text{ kN/m}^2 + 79.8 \text{ kN/m}^2$$
$$= 570 \text{ kPa (abs)} \tag{Ans}$$

Comment This simple example illustrates the need for close attention to the units used in the calculation of pressure; that is, be sure to use a *consistent* unit system and be careful not to add a pressure head (m) to a pressure (Pa).

The Wide World of Fluids

Weather, barometers, and bars

One of the most important indicators of weather conditions is *atmospheric pressure*. In general, a falling or low pressure indicates bad weather; rising or high pressure, good weather. During the evening TV weather report in the United States, atmospheric pressure is given as so many inches (commonly around 30 in.). This value is actually the height of the mercury column in a mercury *barometer* adjusted to sea level. To determine the true atmospheric pressure at a particular location, the elevation relative to sea level must be known. Another unit used by meteorologists to

indicate atmospheric pressure is the *bar*, first used in weather reporting in 1914 and defined as $10^5 \, \text{N/m}^2$. The definition of a bar is probably related to the fact that standard sea-level pressure is $1.0133 \times 10^5 \, \text{N/m}^2$, that is, only slightly larger than one bar. For typical weather patterns, "sea-level equivalent" atmospheric pressure remains close to one bar. However, for extreme weather conditions associated with tornadoes, hurricanes, or typhoons, dramatic changes can occur. The lowest atmospheric sea-level pressure ever recorded was associated with a typhoon, Typhoon Tip, in the Pacific Ocean on October 12, 1979. The value was 0.870 bar (0.655 m Hg).

2.6 MANOMETRY

Manometers use vertical or inclined liquid columns to measure pressure.

A standard technique for measuring pressure involves the use of liquid columns in vertical or inclined tubes. Pressure-measuring devices based on this technique are called **manometers.** The mercury barometer is an example of one type of manometer, but there are many other configurations possible depending on the particular application. Three common types of manometers include the piezometer tube, the U-tube manometer, and the inclined-tube manometer.

2.6.1 Piezometer Tube

The simplest type of manometer consists of a vertical tube, open at the top, and attached to the container in which knowledge of the pressure is desired, as illustrated in **Fig. 2.9**. The adjacent figure shows an important device whose operation is based on this principle. It is a sphygmomanometer, the traditional instrument used to measure blood pressure.

Since manometers involve columns of fluids at rest, the fundamental equation describing their function is Eq. 2.8

$$p = \gamma h + p_0$$

which gives the pressure at any elevation within a homogeneous fluid in terms of a reference pressure p_0 and the vertical distance h between p and p_0. Remember that in a fluid at rest pressure will *increase with increasing depth*. Application of this equation to the piezometer tube of Fig. 2.9 indicates that the pressure p_A can be determined by a measurement of h_1 through the relationship

$$p_A = \gamma_1 h_1$$

where γ_1 is the specific weight of the liquid in the container. Note that since the tube is open at the top, the pressure p_0 can be set equal to zero (we are now using gage pressure), with the height h_1 measured from the upper surface to point (1). Since point (1) and point A within the container are at the same elevation, $p_A = p_1$.

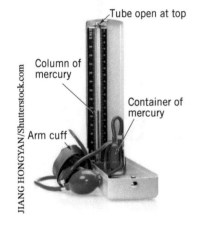

Tube open at top

Column of mercury

Container of mercury

Arm cuff

JIANG HONGYAN/Shutterstock.com

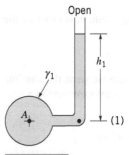

Open

h_1

γ_1

A

(1)

FIGURE 2.9 Piezometer tube.

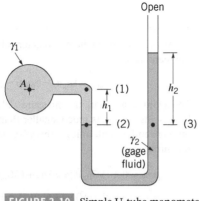

FIGURE 2.10 Simple U-tube manometer.

Although the piezometer tube is a very simple pressure-measuring device, it has several disadvantages. It is suitable only if the pressure in the container is greater than atmospheric pressure (otherwise air would be sucked into the system), and the pressure to be measured must be relatively small so the required height of the column is reasonable. Also the fluid in the container in which the pressure is to be measured must be a liquid rather than a gas.

2.6.2 U-Tube Manometer

Forming the manometer tube in a U-shape, as shown in **Fig. 2.10**, provides the option of introducing a *gauge fluid* with a higher specific weight than the fluid in the container in which the pressure is to be measured. The first, most obvious, capability added by the introduction of the U-shape with a gauge fluid is the ability to measure the pressure in a gas. Again, the pressure to be measured, p_A, is related to the column heights. A numerical value can be computed by repeated application of Eq. 2.8. The following sequence of equations is generated by taking an imaginary "walk" though the manometer from p_A to the open end and repeatedly applying Eq. 2.8.

$$\text{step 1:} \quad p_1 = p_A \quad \text{(same elevation)}$$
$$\text{step 2:} \quad p_2 = p_1 + \gamma_1 h_1 = p_A + \gamma_1 h_1$$
$$\text{step 3:} \quad p_3 = p_2$$
$$\text{step 4:} \quad p_{\text{atm}} = p_3 + \gamma_2(-h_2) = p_A + \gamma_1 h_1 - \gamma_2 h_2$$
$$\text{or}$$
$$p_A = p_{\text{atm}} - \gamma_1 h_1 + \gamma_2 h_2$$

This process can be codified into a simple **manometer rule** by recalling the implications of Eq. 2.8:

1. Write the pressure at either end of the manometer;

2. Proceed through the manometer, adding γh if moving to a greater depth or subtracting γh if moving to a lesser depth;

3. Stop at the far end, or any point in between, and set the expression equal to the local pressure.

If you start to apply the manometer rule at the "far" end of the manometer, the equation you seek is produced in a single line of work. For the manometer of Fig. 2.10:

$$p_{\text{atm}} + \gamma_2 h_2 - \gamma_1 h_1 = p_A \tag{2.14}$$

The contribution of gas columns in manometers is usually negligible because the weight of the gas is so small.

Thinking about this result quickly reveals two useful results. First, if the pressure to be measured is in a gas, $\gamma_2 \gg \gamma_1$, so to a very good approximation: $p_A = p_{\text{atm}} + \gamma_2 h_2$, the same result as was obtained for the piezometer. Second, we can adjust the sensitivity, and therefore the resolution and range, of the pressure measurement device. Increasing the density of the gauge fluid allows practical measurement of much larger pressures but sacrifices resolution. Decreasing the density of the gauge fluid limits the practical range of pressure measurement but increases the resolution. The most common gauge fluids are water, mercury ($SG \approx 13.6$), and oil ($SG \approx 0.8$).

V2.2 Blood pressure measurement

EXAMPLE 2.4 | Simple U-Tube Manometer

Given A closed tank contains compressed air and oil ($SG_{oil} = 0.90$), as is shown in **Fig. E2.4**. A U-tube manometer using mercury ($SG_{Hg} = 13.6$) is connected to the tank as shown. The column heights are $h_1 = 1$ m, $h_2 = 0.15$ m, and $h_3 = 0.25$ m.

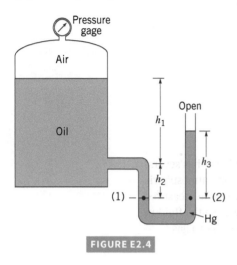

FIGURE E2.4

Find Determine the pressure reading (in Pa) of the gage.

Solution Applying the manometer rule, starting at the open end of the manometer:

$$p_{atm} + \gamma_{Hg}h_3 - \gamma_{oil}h_2 - \gamma_{oil}h_1 = p_{Air}$$

The hydrostatic variation of pressure within the air has been neglected because its density is much smaller than that of either liquid. Using our convention that, unless otherwise stated, pressures will be gauge pressures:

$$p_{Air} = 0 + (SG_{Hg})(\gamma_{H_2O})h_3 + (SG_{oil})(\gamma_{H_2O})(-h_2 - h_1)$$

$$p_{Air} = (9.8\ \text{kN/m}^3)\,[(13.6)\,(0.25\ \text{m}) - (0.9)\,(0.15 + 1\ \text{m})]$$

$$p_{Air} = 23.2\ \text{kN/m}^2 = 23.2 \times 10^3\ \text{Pa} \qquad \text{(Ans)}$$

Comments Note that the air pressure is a function of the height of the mercury in the manometer and the depth of the oil (both in the tank and in the tube). It is not just the mercury in the manometer that is important.

Assume that the gage pressure remains at 23.2 kPa, but the manometer is altered so that it contains only oil. That is, the mercury is replaced by oil. A simple calculation shows that in this case the vertical oil-filled tube would need to be $h_3 = 3.78$ m tall, rather than the original $h_3 = 0.25$ m. There is an obvious advantage of using a heavy fluid such as mercury in manometers.

Manometers are often used to measure the difference in pressure between two points.

The U-tube manometer is also widely used to measure the difference in pressure between two containers or two points in a given system. Consider a manometer connected between containers *A* and *B*, as is shown in **Fig. 2.11**. Application of the manometer rule produces an equation for the pressure difference between the containers. Starting at p_B:

$$p_B + \gamma_3 h_3 + \gamma_2 h_2 - \gamma_1 h_1 = p_A$$

or

$$p_A - p_B = \gamma_3 h_3 + \gamma_2 h_2 - \gamma_1 h_1$$

Carefully include units when replacing symbols with information provided in a problem statement to help ensure a correct solution.

Capillarity due to surface tension at the two-fluid interfaces in the manometer is usually negligible for common gauge fluids, common tube materials, and typical tube bore diameters (see Section 1.9). For highly accurate measurements, it may be necessary to account for the temperature dependence of the fluids' density.

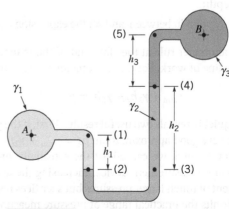

FIGURE 2.11 Differential U-tube manometer.

EXAMPLE 2.5 | U-Tube Manometer

Given As will be discussed in Chapter 4, the volume rate of flow, Q, through a pipe can be determined by means of a flow nozzle located in the pipe, as illustrated in **Fig. E2.5a**. The nozzle creates a pressure drop, $p_A - p_B$, along the pipe that is related to the flow through the equation $Q = K\sqrt{p_A - p_B}$, where K is a constant depending on the pipe and nozzle size. The pressure drop is frequently measured with a differential U-tube manometer of the type illustrated.

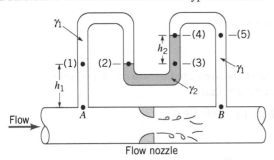

FIGURE E2.5a

Find

a. Determine an equation for $p_A - p_B$ in terms of the specific weight of the flowing fluid, γ_1, the specific weight of the gage fluid, γ_2, and the various heights indicated.

b. For $\gamma_1 = 9.80 \text{ kN/m}^3$, $\gamma_2 = 15.6 \text{ kN/m}^3$, $h_1 = 1.0$ m, and $h_2 = 0.5$ m, what is the value of the pressure drop, $p_A - p_B$?

Solution

a. Although the fluid in the pipe is moving, the fluids in the columns of the manometer are at rest so that the pressure variation in the manometer tubes is hydrostatic. Applying the manometer rule, starting at A:

$$p_A - \gamma_1 h_1 - \gamma_2 h_2 + \gamma_1 (h_1 + h_2) = p_B$$

or

$$p_A - p_B = (\gamma_2 - \gamma_1)h_2 \qquad (Ans)$$

Comment It is to be noted that the only column height of importance is the differential reading, h_2. The differential manometer could be placed 0.5 or 5.0 m above the pipe ($h_1 = 0.5$ m or $h_1 = 5.0$ m), and the value of h_2 would remain the same.

b. The specific value of the pressure drop for the data given is

$$p_A - p_B = (0.5 \text{ m})(15.6 \text{ kN/m}^3 - 9.80 \text{ kN/m}^3)$$
$$= 2.90 \text{ kPa} \qquad (Ans)$$

Comment By repeating the calculations for manometer fluids with different specific weights, γ_2, the results shown in **Fig. E2.5b** are obtained. Note that relatively small pressure differences can be measured if the manometer fluid has nearly the same specific weight as the flowing fluid. It is the difference in the specific weights, $\gamma_2 - \gamma_1$, that is important.

Hence, by rewriting the answer as $h_2 = (p_A - p_B)/(\gamma_2 - \gamma_1)$, it is seen that even if the value of $p_A - p_B$ is small, the value of h_2 can be large enough to provide an accurate reading provided the value of $\gamma_2 - \gamma_1$ is also small.

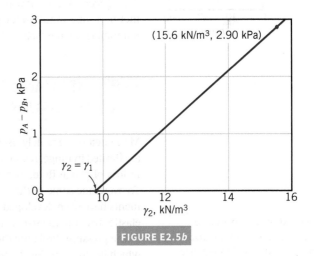

FIGURE E2.5b

2.6.3 Inclined-Tube Manometer

To measure small pressure changes, a manometer of the type shown in **Fig. 2.12** is frequently used. One leg of the manometer is inclined at an angle θ, and the differential reading ℓ_2 is measured along the inclined tube. The difference in pressure $p_A - p_B$ can be expressed as

$$p_A + \gamma_1 h_1 - \gamma_2 \ell_2 \sin\theta - \gamma_3 h_3 = p_B$$

or

$$p_A - p_B = \gamma_2 \ell_2 \sin\theta + \gamma_3 h_3 - \gamma_1 h_1 \qquad (2.15)$$

where it is to be noted the pressure difference between points (1) and (2) is due to the *vertical* distance between the points, which can be expressed as $\ell_2 \sin\theta$. Thus, for relatively small angles, the differential reading along the inclined tube can be made large even for small pressure

Inclined-tube manometers can be used to measure small pressure differences accurately.

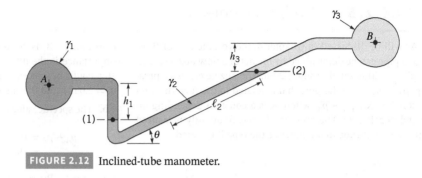

FIGURE 2.12 Inclined-tube manometer.

differences. The inclined-tube manometer is often used to measure small differences in gas pressures so that if pipes A and B contain a gas, then

$$p_A - p_B = \gamma_2 \ell_2 \sin \theta$$

or

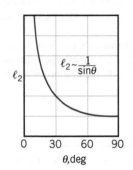

$$\ell_2 = \frac{p_A - p_B}{\gamma_2 \sin \theta} \tag{2.16}$$

where the contributions of the gas columns h_1 and h_3 have been neglected. Equation 2.16 and the adjacent figure show that the differential reading ℓ_2 (for a given pressure difference) of the inclined-tube manometer can be increased over that obtained with a conventional U-tube manometer by the factor $1/\sin \theta$. Recall that $\sin \theta \to 0$ as $\theta \to 0$.

2.7 MECHANICAL AND ELECTRONIC PRESSURE-MEASURING DEVICES

Manometers are widely used because they are simple, inexpensive, and reliable. However, they are limited in range, respond relatively slowly, are unsuitable for environments that might result in loss of gauge fluid, and are not easily interfaced with automated data acquisition systems. To overcome some of these shortcomings, numerous other types of pressure-measuring instruments have been developed. Most of these make use of the idea that when a pressure acts on an elastic structure the structure will deform, and this deformation can be related to the magnitude of the pressure. Probably the most familiar device of this kind is the **Bourdon pressure gage**, which is shown in **Fig. 2.13a**. The essential mechanical element in this gage is the hollow, elastic curved tube (Bourdon tube), which is connected to the pressure source as shown in **Fig. 2.13b**.

A Bourdon tube pressure gage uses a hollow, elastic, and curved tube to measure pressure.

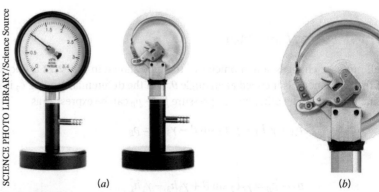

(a)   (b)

FIGURE 2.13 (a) Liquid-filled Bourdon pressure gages for various pressure ranges. (b) Internal elements of Bourdon gages. The "C-shaped" Bourdon tube is shown on the left, and the "coiled spring" Bourdon tube for high pressures of 6.9×10^3 kPa and above is shown on the right. (Photographs courtesy of Weiss Instruments, Inc.)

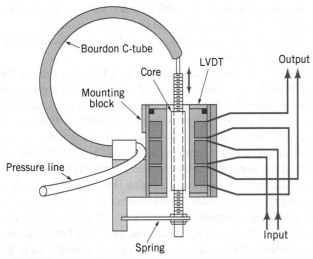

FIGURE 2.14 Pressure transducer that combines a linear variable differential transformer (LVDT) with a Bourdon gage. (From Ref. 4, used by permission.)

As the pressure within the tube increases the tube tends to straighten, and although the deformation is small, it can be translated into the motion of a pointer on a dial as illustrated. Since it is the difference in pressure between the outside of the tube (atmospheric pressure) and the inside of the tube that causes the movement of the tube, the indicated pressure is gage pressure. The Bourdon gage must be calibrated so that the dial reading can directly indicate the pressure in suitable units such as psi, psf, or pascals. A zero reading on the gage indicates that the measured pressure is equal to the local atmospheric pressure. This type of gage can be used to measure a negative gage pressure (vacuum) as well as positive pressures.

The *aneroid* barometer is another type of mechanical gage that is used for measuring atmospheric pressure. Since atmospheric pressure is specified as an absolute pressure, the conventional Bourdon gage is not suitable for this measurement. The common aneroid barometer contains a hollow, closed, elastic element that is evacuated so that the pressure inside the element is near absolute zero. As the external atmospheric pressure changes, the element deflects, and this motion can be translated into the movement of an attached dial. As with the Bourdon gage, the dial can be calibrated to give atmospheric pressure directly, with the usual units being millimeters or inches of mercury.

For many applications in which pressure measurements are required, the pressure must be measured with a device that converts the pressure into an electrical output. For example, it may be desirable to continuously monitor a pressure that is changing with time. This type of pressure-measuring device is called a *pressure transducer,* and many different designs are available. One design simply connects a Bourdon tube to a linear variable differential transformer (LVDT), as illustrated in **Fig. 2.14**, to produce a pressure transducer. The core

V2.3 Bourdon gage

The Wide World of Fluids

Tire pressure warning

Proper tire inflation on vehicles is important for more than ensuring long tread life. It is critical in preventing accidents such as rollover accidents caused by underinflation of tires. The National Highway Traffic Safety Administration is developing a regulation regarding four-tire tire-pressure monitoring systems that can warn a driver when a tire is more than 25% underinflated. In recent years these devices have become a common feature in new vehicles.

A typical tire-pressure monitoring system fits within the tire and contains a *pressure transducer* (usually either a piezo-resistive or a capacitive-type transducer) and a transmitter that sends the information to an electronic control unit within the vehicle. Information about tire pressure and a warning when the tire is underinflated is displayed on the instrument panel. The environment (hot, cold, vibration) in which these devices must operate, their small size, and required low cost provide challenging constraints for the design engineer.

of the LVDT is connected to the free end of the Bourdon tube. As pressure is applied, motion of the tube end moves the core through the coil to produce an output voltage. This voltage is a linear function of the pressure and could be recorded on an oscillograph or digitized for storage or processing on a computer.

One disadvantage of a pressure transducer using a Bourdon tube as the elastic sensing element is that it is limited to the measurement of pressures that are static or only changing slowly (quasistatic). Because of the relatively large mass of the Bourdon tube, it cannot respond to rapid changes in pressure. To overcome this difficulty, a different type of transducer is used in which the sensing element is a thin, elastic diaphragm that is in contact with the fluid. As the pressure changes, the diaphragm deflects, and this deflection can be sensed and converted into an electrical voltage. One way to accomplish this is to locate strain gages either on the surface of the diaphragm not in contact with the fluid or on an element attached to the diaphragm. These gages can accurately sense the small strains induced in the diaphragm and provide an output voltage proportional to pressure. This type of transducer is capable of measuring accurately both small and large pressures, as well as both static and dynamic pressures. For example, strain-gage pressure transducers of the type shown in **Fig. 2.15** are used to measure arterial blood pressure, which is a relatively small pressure that varies periodically with a fundamental frequency of about 1 Hz. The transducer is usually connected to the blood vessel by means of a liquid-filled, small-diameter tube called a pressure catheter. Although the strain-gage type of transducer can be designed to have very good frequency response (up to approximately 10 kHz), they become less sensitive at the higher frequencies, since the diaphragm must be made stiffer to achieve the higher frequency response. As an alternative,

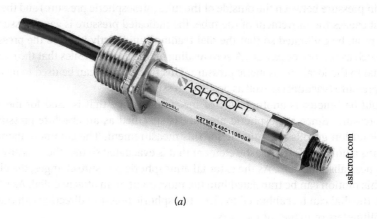

ashcroft.com

(*a*)

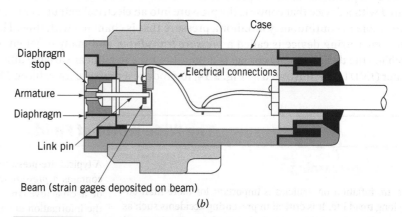

(*b*)

FIGURE 2.15 (*a*) Photograph of a typical pressure transducer with a male thread fitting in front of the diaphragm for system connection and an electrical connector in the rear of the device. (*b*) Schematic diagram of a typical pressure transducer device (male thread connector not shown). Deflection of the diaphragm due to pressure is measured with a silicon beam on which strain gages and an associated bridge circuit have been deposited.

the diaphragm can be constructed of a piezoelectric crystal to be used as both the elastic element and the sensor. When a pressure is applied to the crystal, a voltage develops because of the deformation of the crystal. This voltage is directly related to the applied pressure. Depending on the design, this type of transducer can be used to measure both very low and high pressures (up to approximately 6.9×10^5 kPa) at high frequencies. Additional information on pressure transducers can be found in Refs. 3, 4, and 5.

2.8 HYDROSTATIC FORCE ON A PLANE SURFACE AND PRESSURE DIAGRAM

2.8.1 Hydrostatic Force

When a surface is submerged in a fluid, forces act on the surface due to the fluid. The determination of these forces is important in the design of storage tanks, ships, dams, and other hydraulic structures. For fluids at rest we know that the force must be *perpendicular* to the surface because no shear stresses are present. We also know that the pressure varies linearly with depth as shown in Fig. 2.16 if the fluid is incompressible. For a horizontal surface, such as the bottom of a liquid-filled tank (**Fig. 2.16a**), the magnitude of the resultant force is simply $F_R = pA$, where p is the uniform pressure on the bottom and A is the area of the bottom. For the open tank shown, $p = \gamma h$. Note that if atmospheric pressure acts on both sides of the bottom, as is illustrated, the *resultant* force on the bottom is simply due to the liquid in the tank. Since the pressure is constant and uniformly distributed over the bottom, the resultant force acts through the centroid of the area as shown in Fig. 2.16a. As shown in **Fig. 2.16b**, the pressure on the ends of the tank is not uniformly distributed. Determination of the force due to hydrostatic pressure on the tank ends is more challenging because the pressure is not constant, the ends may not be rectangular plates, and in general they may not be vertical.

Figure 2.17 depicts a submerged inclined plane surface and provides labels for locations and distances that will be required to determine the resultant force and its line of action. For the present we assume that the fluid surface is open to the atmosphere. Let the plane in which the surface lies intersect the free surface at 0 and make an angle θ with this surface as in Fig. 2.17. The x–y coordinate system is defined so that 0 is the origin and $y = 0$ (i.e., the x axis) is directed along the surface as shown. The surface may be of arbitrary shape provided it remains in the x–y plane. We wish to determine the direction, location, and magnitude of the resultant force acting on one side of this area due to the liquid in contact with the area. At any given depth, h, the force acting on dA (the differential area of Fig. 2.17) is $dF = \gamma h\, dA$ and is perpendicular to the surface. Thus, the magnitude of the resultant force can be found by summing these differential forces over the entire surface. In equation form

$$F_R = \int_A \gamma h\, dA = \int_A \gamma y \sin \theta\, dA$$

Video

V2.4 Hoover dam

The resultant force of a static fluid on a plane surface is due to the hydrostatic pressure distribution on the surface.

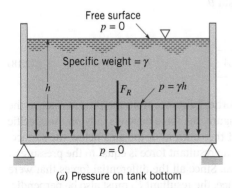

(a) Pressure on tank bottom

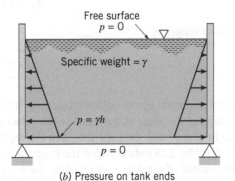

(b) Pressure on tank ends

FIGURE 2.16 (a) Pressure distribution and resultant hydrostatic force on the bottom of an open tank. (b) Pressure distribution on the ends of an open tank.

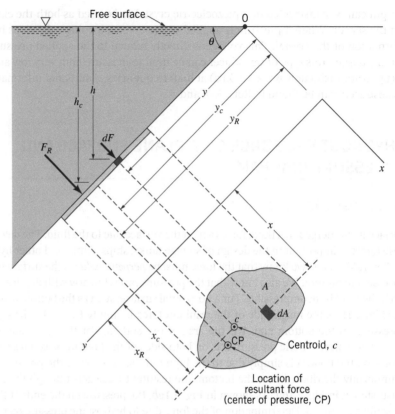

FIGURE 2.17 Notation for hydrostatic force on an inclined plane surface of arbitrary shape.

where $h = y \sin \theta$. For constant γ and θ

$$F_R = \gamma \sin \theta \int_A y \, dA \tag{2.17}$$

The integral appearing in Eq. 2.17 is the *first moment of the area* with respect to the x axis, so we can write

$$\int_A y \, dA = y_c A$$

where y_c is the y coordinate of the centroid of area A measured from the x axis, which passes through 0. Equation 2.17 can thus be written as

$$F_R = \gamma A y_c \sin \theta$$

or more simply as

$$\boxed{F_R = \gamma h_c A} \tag{2.18}$$

> The magnitude of the resultant fluid force depends on only the depth and is equal to the pressure acting at the centroid of the area multiplied by the total area.

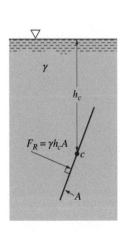

where, as shown by the adjacent figure, h_c is the vertical distance from the fluid surface to the centroid of the area. As indicated in the adjacent figure, the force depends only on the specific weight of the fluid, the total area, and the depth of the centroid of the area below the surface. In effect, Eq. 2.18 indicates that the magnitude of the resultant force is equal to the pressure at the centroid of the area multiplied by the total area. Since all the differential forces that were summed to obtain F_R are perpendicular to the surface, the resultant F_R must also be perpendicular to the surface.

Although our intuition might suggest that the resultant force passes through the centroid of the area, this is not the case. The y coordinate, y_R, of the resultant force can be determined by summation of moments around the x axis. That is, the moment of the resultant force must equal the moment of the distributed pressure force, or

$$F_R y_R = \int_A y \, dF = \int_A \gamma \sin \theta \, y^2 \, dA$$

and, noting that $F_R = \gamma A y_c \sin \theta$,

$$y_R = \frac{\int_A y^2 \, dA}{y_c A}$$

The integral in the numerator is the *second moment of the area (moment of inertia)*, I_x, with respect to an axis formed by the intersection of the plane containing the surface and the free surface (x axis). Thus, we can write

$$y_R = \frac{I_x}{y_c A}$$

Use can now be made of the parallel axis theorem to express I_x as

$$I_x = I_{xc} + A y_c^2$$

where I_{xc} is the second moment of the area with respect to an axis passing through its *centroid* and parallel to the x axis. Thus,

$$\boxed{y_R = \frac{I_{xc}}{y_c A} + y_c} \tag{2.19}$$

As shown by Eq. 2.19 and in the adjacent figure, the resultant force does not pass through the centroid. For nonhorizontal surfaces, the resultant force always acts below the centroid because $I_{xc}/y_c A > 0$.

The x coordinate, x_R, for the resultant force can be determined in a similar manner by summing moments about the y axis. Thus,

$$F_R x_R = \int_A \gamma \sin \theta \, xy \, dA$$

and, therefore,

$$x_R = \frac{\int_A xy \, dA}{y_c A} = \frac{I_{xy}}{y_c A}$$

where I_{xy} is the product of inertia with respect to the x and y axes. Again, using the parallel axis theorem,[1] we can write

$$\boxed{x_R = \frac{I_{xyc}}{y_c A} + x_c} \tag{2.20}$$

where I_{xyc} is the product of inertia with respect to an orthogonal coordinate system passing through the *centroid* of the area and formed by a translation of the x–y coordinate system. If the

The resultant fluid force does not pass through the centroid of the area.

---

[1]Recall that the parallel axis theorem for the product of inertia of an area states that the product of inertia with respect to an orthogonal set of axes (x–y coordinate system) is equal to the product of inertia with respect to an orthogonal set of axes parallel to the original set and passing through the centroid of the area, plus the product of the area and the x and y coordinates of the centroid of the area. Thus, $I_{xy} = I_{xyc} + A x_c y_c$.

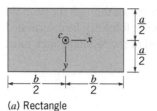

$A = ba$

$I_{xc} = \dfrac{1}{12} ba^3$

$I_{yc} = \dfrac{1}{12} ab^3$

$I_{xyc} = 0$

(a) Rectangle

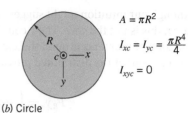

$A = \pi R^2$

$I_{xc} = I_{yc} = \dfrac{\pi R^4}{4}$

$I_{xyc} = 0$

(b) Circle

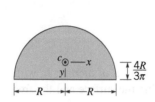

$A = \dfrac{\pi R^2}{2}$

$I_{xc} = 0.1098R^4$

$I_{yc} = 0.3927R^4$

$I_{xyc} = 0$

(c) Semicircle

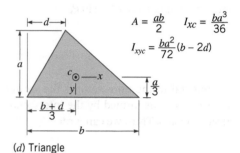

$A = \dfrac{ab}{2}$ $I_{xc} = \dfrac{ba^3}{36}$

$I_{xyc} = \dfrac{ba^2}{72}(b - 2d)$

(d) Triangle

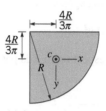

$A = \dfrac{\pi R^2}{4}$

$I_{xc} = I_{yc} = 0.05488R^4$

$I_{xyc} = -0.01647R^4$

(e) Quarter circle

FIGURE 2.18 Geometric properties of some common shapes.

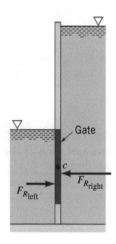

submerged area is symmetrical with respect to an axis passing through the centroid and parallel to either the x or y axis, the resultant force must lie along the line $x = x_c$, since I_{xyc} is identically zero in this case. The point through which the resultant force acts is called the **center of pressure**. It is to be noted from Eqs. 2.19 and 2.20 that as y_c increases the center of pressure moves closer to the centroid of the area. Since $y_c = h_c/\sin \theta$, the distance y_c will increase if the depth of submergence, h_c, increases, or, for a given depth, the area is rotated so that the angle, θ, decreases. Thus, the hydrostatic force on the right-hand side of the gate shown in the adjacent figure acts closer to the centroid of the gate than the force on the left-hand side. Centroidal coordinates and moments of inertia for some common areas are given in **Fig. 2.18**.

The Wide World of Fluids

The Three Gorges Dam

Spanning China's Yangtze River, the Three Gorges Dam became the world's largest power generating facility when the last of its main turbines began operation in July 2012. The dam is of the concrete gravity type, having a length of 2309 m with a height of 185 m. The main elements of the project include the dam, two power plants, and navigation facilities consisting of a ship lock and lift. The power plants contain 26 Francis-type turbines, each with a capacity of 700 megawatts. The spillway section, which is the center section of the dam, is 483 m long with 23 bottom outlets and 22 surface sluice gates. The maximum discharge capacity is 102,500 m³/s. After more than 10 years of construction, the dam gates were finally closed, and on June 10, 2003, the reservoir had been filled to its interim level of 135 m. Due to the large depth of water at the dam and the huge extent of the storage pool, *hydrostatic pressure forces* have been a major factor considered by engineers. When filled to its normal pool level of 175 m, the total reservoir storage capacity is 39.3 billion m³.

EXAMPLE 2.6 | Hydrostatic Force on a Plane Circular Surface

Given The 4-m-diameter circular gate of **Fig. E2.6a** is located in the inclined wall of a large reservoir containing water ($\gamma = 9.80$ kN/m³). The gate is mounted on a shaft along its horizontal diameter, and the water depth is 10 m at the shaft.

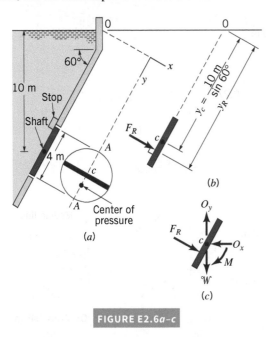

FIGURE E2.6a–c

Find Determine

a. the magnitude and location of the resultant force exerted on the gate by the water and

b. the moment that would have to be applied to the shaft to open the gate.

Solution

a. To find the magnitude of the force of the water, we can apply Eq. 2.18,

$$F_R = \gamma h_c A$$

and since the vertical distance from the fluid surface to the centroid of the area is 10 m, it follows that

$$F_R = (9.80 \times 10^3 \text{ N/m}^3)(10 \text{ m})(4\pi \text{ m}^2)$$

$$= 1230 \times 10^3 \text{ N} = 1.23 \text{ MN} \qquad (Ans)$$

To locate the point (center of pressure) through which F_R acts, we use Eqs. 2.19 and 2.20,

$$x_R = \frac{I_{xyc}}{y_c A} + x_c \qquad y_R = \frac{I_{xc}}{y_c A} + y_c$$

For the coordinate system shown, $x_R = 0$ since the area is symmetric about the y-axis, and the center of pressure

must lie along the diameter A-A. To obtain y_R, we have from Fig. 2.18

$$I_{xc} = \frac{\pi R^4}{4}$$

and y_c is shown in **Fig. E2.6b**. Thus,

$$y_R = \frac{(\pi/4)(2 \text{ m})^4}{(10 \text{ m}/\sin 60°)(4\pi \text{ m}^2)} + \frac{10 \text{ m}}{\sin 60°}$$

$$= 0.0866 \text{ m} + 11.55 \text{ m} = 11.6 \text{ m}$$

and the distance (along the gate) below the shaft to the center of pressure is

$$y_R - y_c = 0.0866 \text{ m} \qquad (Ans)$$

We can conclude from this analysis that the force on the gate due to the water has a magnitude of 1.23 MN and acts through a point along its diameter A-A at a distance of 0.0866 m (along the gate) below the shaft. The force is perpendicular to the gate surface as shown in Fig. E2.6b.

Comment By repeating the calculations for various values of the depth to the centroid, h_c, the results shown in **Fig. E2.6d** are obtained. Note that as the depth increases, the distance between the center of pressure and the centroid decreases.

b. The moment required to open the gate can be obtained with the aid of the free-body diagram of **Fig. E2.6c**. In this diagram, $\mathcal{W}$ is the weight of the gate and O_x and O_y are the horizontal and vertical reactions of the shaft on the gate. We can now sum moments about the shaft

$$\sum M_c = 0$$

and, therefore,

$$M = F_R(y_R - y_c)$$

$$= (1230 \times 10^3 \text{ N})(0.0866 \text{ m})$$

$$= 1.07 \times 10^5 \text{ N} \cdot \text{m} \qquad (Ans)$$

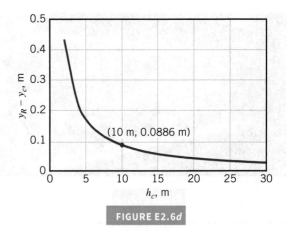

FIGURE E2.6d

EXAMPLE 2.7 | Hydrostatic Pressure Force on a Plane Triangular Surface

Given An aquarium contains seawater ($\gamma = 9.81$ kN/m^3) to a depth of 0.3 m, as shown in **Fig. E2.7a**. To repair some damage to one corner of the tank, a triangular section is replaced with a new section, as illustrated in **Fig. E2.7b**.

Find Determine

a. the magnitude of the force of the seawater on this triangular area, and

b. the location of this force.

Solution

a. The various distances needed to solve this problem are shown in **Fig. E2.7c**. Since the surface of interest lies in a vertical plane, $y_c = h_c = 0.27$ m, and from Eq. 2.18, the magnitude of the force is

$$F_R = \gamma h_c A$$

$$= (9.81 \text{ kN/m}^3)(0.27 \text{ m})[(0.09 \text{ m})^2/2] = 0.011 \text{ kN} \quad (Ans)$$

Comment Note that this force is independent of the tank length. The result is the same if the tank is 0.076 m, 7.6 m, or 76 km long.

b. The y coordinate of the center of pressure (CP) is found from Eq. 2.19,

$$y_R = \frac{I_{xc}}{y_c A} + y_c$$

and from Fig. 2.18

$$I_{xc} = \frac{(0.09 \text{ m})(0.09 \text{ m})^3}{36} = \frac{6.56 \times 10^{-5}}{36} \text{ m}^4$$

$$= 1.82 \times 10^{-6} \text{ m}^4$$

so that

$$y_R = \frac{1.82 \times 10^{-6} \text{ m}^4}{(0.27 \text{ m})(0.09 \text{ m})^2/2} + 0.27 \text{ m}$$

$$= 0.0016 \text{ m} + 0.27 \text{ m} = 0.272 \text{ m} \quad (Ans)$$

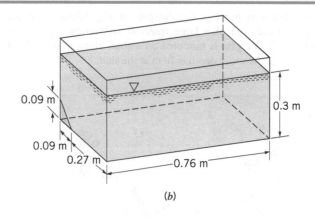

(b)

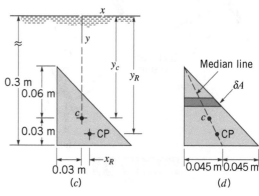

(c)   (d)

FIGURE E2.7b–d

Similarly, from Eq. 2.20

$$x_R = \frac{I_{xyc}}{y_c A} + x_c$$

and from Fig. 2.18

$$I_{xyc} = \frac{(0.09 \text{ m})(0.09 \text{ m})^2}{72}(0.09 \text{ m}) = \frac{0.00073}{72} \text{ m}^4$$

so that

$$x_R = \frac{0.00073/72 \text{ m}^4}{(0.27 \text{ m})(0.09 \text{ m})^2/2} + 0 = 0.00267 \text{ m} \quad (Ans)$$

Comment Thus, we conclude that the center of pressure is 0.00267 m to the right of and 0.0005 m below the centroid of the area. If this point is plotted, we find that it lies on the median line for the area as illustrated in **Fig. E2.7d**. Since we can think of the total area as consisting of a number of small rectangular strips of area δA (and the fluid force on each of these small areas acts through its center), it follows that the resultant of all these parallel forces must lie along the median.

FIGURE E2.7a

2.8.2 Pressure Diagram

In the foregoing section, methods of evaluating hydrostatic forces over different plane surfaces, including the rectangular ones, have been explained. Many a times it is imperative to represent the forces graphically over the rectangular plane volumes. This is done by pressure prisms. To understand this, consider the pressure distribution along a vertical wall of a tank of constant width b, which contains a liquid having a specific weight γ. Since the pressure must vary linearly with depth, we can represent the variation as is shown in **Fig. 2.19a**, where the pressure is equal to zero at the upper surface and equal to γh at the bottom. It is apparent from this diagram that the average pressure occurs at the depth $h/2$ and, therefore, the resultant force acting on the rectangular area $A = bh$ is

$$F_R = p_{av} A = \gamma\left(\frac{h}{2}\right) A$$

which is the same result as obtained from Eq. 2.18. The pressure distribution shown in Fig. 2.19a applies across the vertical surface, so we can draw the three-dimensional representation of the pressure distribution as shown in **Fig. 2.19b**. The base of this "volume" in pressure-area space is the plane surface of interest, and its altitude at each point is the pressure. This volume is called the *pressure prism,* and it is clear that the magnitude of the resultant force acting on the rectangular surface is equal to the volume of the pressure prism. Thus, for the prism of Fig. 2.19b, the fluid force is

$$F_R = \text{volume} = \frac{1}{2}(\gamma h)(bh) = \gamma\left(\frac{h}{2}\right) A$$

where bh is the area of the rectangular surface, A.

The resultant force must pass through the *centroid* of the pressure prism. For the volume under consideration, the centroid is located along the vertical axis of symmetry of the surface and at a distance of $h/3$ above the base (since the centroid of a triangle is located at $h/3$ above its base). This result can readily be shown to be consistent with that obtained from Eqs. 2.19 and 2.20.

This same graphical approach can be used for plane rectangular surfaces that do not extend up to the fluid surface, as illustrated in **Fig. 2.20a**. In this instance, the cross section of the pressure prism is trapezoidal. However, the resultant force is still equal in magnitude to the volume of the pressure prism, and it passes through the centroid of the volume. Specific values can be obtained by decomposing the pressure prism into two parts, $ABDE$ and BCD, as shown in **Fig. 2.20b**. Thus,

$$F_R = F_1 + F_2$$

where the components can readily be determined by inspection for rectangular surfaces. The location of F_R can be determined by summing moments about some convenient axis, such as one passing through A. In this instance

$$F_R y_A = F_1 y_1 + F_2 y_2$$

and y_1 and y_2 can be determined by inspection.

> The magnitude of the resultant fluid force is equal to the volume of the pressure prism and its line of action passes through the centroid of the prism.

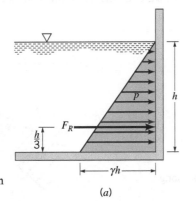
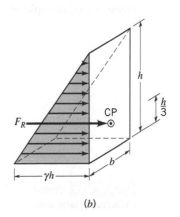

FIGURE 2.19 Pressure prism for vertical rectangular area.

(a)   (b)

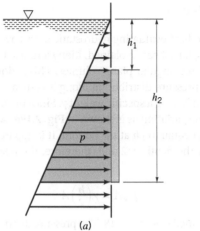
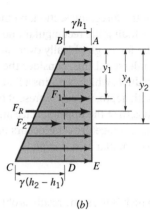

(a)   (b)

> **The use of the pressure prism concept to determine the force on a submerged area is best suited for plane rectangular surfaces.**

For inclined plane rectangular surfaces the pressure prism can still be developed, and the cross section of the prism will generally be trapezoidal, as is shown in **Fig. 2.21**. Although it is usually convenient to measure distances along the inclined surface, the pressures developed depend on the vertical distances as illustrated.

The use of pressure prisms for determining the force on submerged plane areas is convenient if the area is rectangular so the volume and centroid can be easily determined. However, for other nonrectangular shapes, integration would generally be needed to determine the volume and centroid. In these circumstances, it is more convenient to use the equations developed in the previous section in which the necessary integrations have been made and the results presented in a convenient and compact form that is applicable to submerged plane areas of any shape.

The effect of atmospheric pressure on a submerged area has not yet been considered, and we may ask how this pressure will influence the resultant force. If we again consider the pressure distribution on a plane vertical wall, as is shown in **Fig. 2.22a**, the pressure varies from zero at the surface to γh at the bottom. Since we are setting the surface pressure equal to zero, we are using atmospheric pressure as our datum, and thus the pressure used in the determination of the fluid force is gage pressure. If we wish to include atmospheric pressure, the pressure distribution will be as is shown in **Fig. 2.22b**. We note that in this case the force on one side of the wall now consists of F_R as a result of the hydrostatic pressure distribution, plus the contribution of the atmospheric pressure, $p_{\text{atm}}A$, where A is the area of the surface. However, if we are going to include the effect of atmospheric pressure on one side of the wall, we must realize that this same pressure acts on the outside surface (assuming it is exposed to the atmosphere), so that an equal and opposite force will be developed as illustrated in Figure 2.20. Thus, we conclude that the *resultant* fluid force on the surface is only because of the gage pressure contribution of the liquid in contact with the surface—the atmospheric pressure does not contribute to this resultant. Taking this study one step further, if the pressure at the liquid–vapor interface is not atmospheric pressure, both the magnitude and the line of action will differ from that predicted using only gage pressure. An analysis almost identical to that depicted in Fig. 2.20 and described in the associated text is required.

> **The resultant fluid force acting on a submerged area is affected by the pressure at the free surface.**

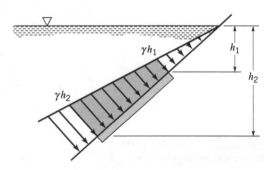

FIGURE 2.21 Pressure variation along an inclined plane area.

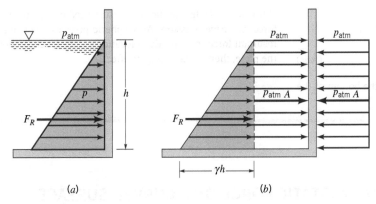

FIGURE 2.22 Effect of atmospheric pressure on the resultant force acting on a plane vertical wall.

EXAMPLE 2.8 | Use of the Pressure Prism Concept

Given A pressurized tank contains oil ($SG = 0.90$) and has a square, 0.6-m by 0.6-m plate bolted to its side, as is illustrated in **Fig. E2.8a**. The pressure gage on the top of the tank reads 50 kPa, and the outside of the tank is at atmospheric pressure.

Find What is the magnitude and location of the resultant force on the attached plate?

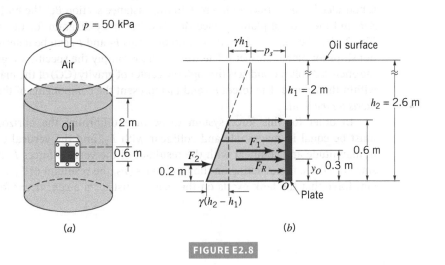

FIGURE E2.8

Solution The pressure distribution acting on the inside surface of the plate is shown in **Fig. E2.8b**. The pressure at a given point on the plate is due to the air pressure, p_s, at the oil surface and the pressure due to the oil, which varies linearly with depth as is shown in the figure. The resultant force on the plate (having an area A) is due to the components, F_1 and F_2, where F_1 and F_2 are due to the rectangular and triangular portions of the pressure distribution, respectively. Thus,

$$F_1 = (p_s + \gamma h_1)A$$

$$= [50 \times 10^3 \text{ N/m}^2$$

$$+ (0.90)(9.81 \times 10^3 \text{ N/m}^3)(2 \text{ m})](0.36 \text{ m}^2)$$

$$= 24.4 \times 10^3 \text{ N}$$

and

$$F_2 = \gamma \left(\frac{h_2 - h_1}{2}\right)A$$

$$= (0.90)(9.81 \times 10^3 \text{ N/m}^3)\left(\frac{0.6 \text{ m}}{2}\right)(0.36 \text{ m}^2)$$

$$= 0.954 \times 10^3 \text{ N}$$

The magnitude of the resultant force, F_R, is therefore

$$F_R = F_1 + F_2 = 25.4 \times 10^3 \text{ N} = 25.4 \text{ kN} \qquad (Ans)$$

The vertical location of F_R can be obtained by summing moments around an axis through point O so that

$$F_R y_O = F_1(0.3 \text{ m}) + F_2(0.2 \text{ m})$$

or

$$y_O = \frac{(24.4 \times 10^3 \text{ N})(0.3 \text{ m}) + (0.954 \times 10^3 \text{ N})(0.2 \text{ m})}{25.4 \times 10^3 \text{ N}}$$

$$= 0.296 \text{ m} \qquad\qquad (Ans)$$

Thus, the force acts at a distance of 0.296 m above the bottom of the plate along the vertical axis of symmetry.

Comment   Note that the air pressure used in the calculation of the force was gage pressure. Atmospheric pressure does not affect the resultant force (magnitude or location), since it acts on both sides of the plate, thereby canceling its effect.

2.9   HYDROSTATIC FORCE ON A CURVED SURFACE

 Video

V2.5 Pop bottle

The equations developed in Section 2.8 for the magnitude and location of the resultant force acting on a submerged surface only apply to plane surfaces. However, many surfaces of interest (such as those associated with dams, pipes, and tanks) are nonplanar. The domed bottom of the beverage bottle shown in the adjacent figure shows a typical curved surface example. Although the resultant fluid force can be determined by integration, as was done for the plane surfaces, this is generally a rather tedious process and no simple, general formulas can be developed. As an alternative approach, we will consider the equilibrium of the fluid volume enclosed by the curved surface of interest and the horizontal and vertical projections of this surface.

For example, consider a curved portion of the swimming pool shown in **Fig. 2.23a**. We wish to find the resultant fluid force acting on section BC (which has a unit length perpendicular to the plane of the paper) shown in **Fig. 2.23b**. We first isolate a volume of fluid that is bounded by the surface of interest, in this instance section BC, the horizontal plane surface AB, and the vertical plane surface AC. The free-body diagram for this volume is shown in **Fig. 2.23c**. The magnitude and location of forces F_1 and F_2 can be determined from the relationships for planar surfaces. The weight, $\mathcal{W}$, is simply the specific weight of the fluid times the enclosed volume and acts through the center of gravity (CG) of the mass of fluid contained within the volume. The forces F_H and F_V represent the components of the force that the tank *exerts on the fluid.*

In order for this force system to be in equilibrium, the horizontal component F_H must be equal in magnitude and collinear with F_2, and the vertical component F_V equal in magnitude and collinear with the resultant of the vertical forces F_1 and $\mathcal{W}$. This follows since the three forces acting on the fluid mass (F_2, the resultant of F_1 and $\mathcal{W}$, and the resultant force that the tank exerts on the mass) must form a *concurrent* force system. That is,

(a)

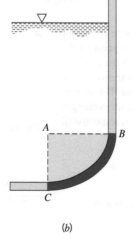

(b)

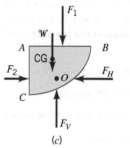

(c)

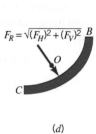

(d)

FIGURE 2.23   Hydrostatic force on a curved surface.

from the principles of statics, it is known that when a body is held in equilibrium by three nonparallel forces, they must be concurrent (their lines of action intersect at a common point) and coplanar. Thus,

$$F_H = F_2$$

$$F_V = F_1 + \mathcal{W}$$

and the magnitude of the resultant is obtained from the equation

$$F_R = \sqrt{(F_H)^2 + (F_V)^2}$$

The resultant F_R passes through the point O, which can be located by summing moments about an appropriate axis. The resultant force of the fluid acting *on the curved surface BC* is equal and opposite in direction to that obtained from the free-body diagram of Fig. 2.23c. The desired fluid force is shown in **Fig. 2.23d**.

EXAMPLE 2.9 | Hydrostatic Pressure Force on a Curved Surface

Given   A 1.8 m diameter drainage conduit of the type shown in **Fig. E2.9a** is half full of water at rest, as shown in **Fig. E2.9b**.

Find   Determine the magnitude and line of action of the resultant force that the water exerts on a 0.3 m length of the curved section BC of the conduit wall.

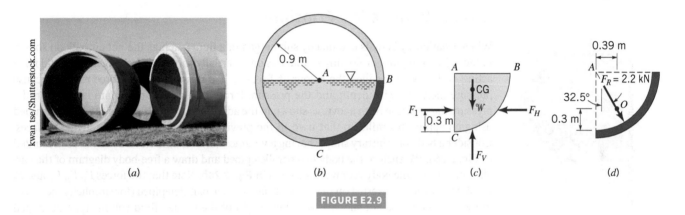

FIGURE E2.9

Solution   We first isolate a volume of fluid bounded by the curved section BC, the horizontal surface AB, and the vertical surface AC, as shown in **Fig. E2.9c**. The volume has a length of 0.3 m. The forces acting on the volume are the horizontal force, F_1, which acts on the vertical surface AC, the weight, $\mathcal{W}$, of the fluid contained within the volume, and the horizontal and vertical components of the force of the conduit wall on the fluid, F_H and F_V, respectively.

The magnitude of F_1 is found from the equation

$$F_1 = \gamma h_c A = (9.81 \text{ kN/m}^3)\left(\frac{0.9}{2}\text{ m}\right)(0.27 \text{ m}^2) = 1.19 \text{ kN}$$

and this force acts 0.3 m above C as shown. The weight $\mathcal{W} = \gamma \mathcal{V}$, where $\mathcal{V}$ is the fluid volume, is

$$\mathcal{W} = \gamma \mathcal{V} = (9.81 \text{ kN/m}^3)(0.81 \pi/4 \text{ m}^2)(0.3 \text{ m}) = 1.87 \text{ kN}$$

and acts through the center of gravity of the mass of fluid, which according to Fig. 2.18 is located 0.39 m to the right of AC as shown. Therefore, to satisfy equilibrium

$$F_H = F_1 = 1.17 \text{ kN} \qquad F_V = \mathcal{W} = 1.87 \text{ kN}$$

and the magnitude of the resultant force is

$$F_R = \sqrt{(F_H)^2 + (F_V)^2}$$
$$= \sqrt{(1.19 \text{ kN})^2 + (1.87 \text{ kN})^2} = 2.22 \text{ kN} \qquad \text{(Ans)}$$

The force the water exerts *on* the conduit wall is equal, but *opposite in direction*, to the forces F_H and F_V shown in Fig. E2.9c. Thus, the resultant force *on the conduit wall* is shown in **Fig. E2.9d**. This force acts through the point O at the angle shown.

Comment   An inspection of this result will show that the line of action of the resultant force passes through the center of the conduit. In retrospect, this is not a surprising result because at each point on the curved surface of the conduit the elemental force due to the pressure is normal to the surface, and each line of action must pass through the center of the conduit. It therefore follows that the resultant of this concurrent force system must also pass through the center of concurrence of the elemental forces that make up the system.

This same general approach can also be used for determining the force on curved surfaces of pressurized, closed tanks. If these tanks contain a gas, the weight of the gas is usually negligible in comparison with the forces developed by the pressure. Thus, the forces (such as F_1 and F_2 in Fig. 2.23*c*) on horizontal and vertical projections of the curved surface of interest can simply be expressed as the internal pressure times the appropriate projected area.

The Wide World of Fluids

Miniature, exploding pressure vessels

Our daily lives are safer because of the effort put forth by engineers to design safe, lightweight pressure vessels such as boilers, propane tanks, and pop bottles. Without proper design, the large *hydrostatic pressure forces on the curved surfaces* of such containers could cause the vessel to explode with disastrous consequences. On the other hand, the world is a friendlier place because of miniature pressure vessels that are designed to explode under the proper conditions—popcorn kernels. Each grain of popcorn

contains a small amount of water within the special, impervious hull (pressure vessel) that, when heated to a proper temperature, turns to steam, causing the kernel to explode and turn itself inside out. Not all popcorn kernels have the proper properties to make them pop well. First, the kernel must be quite close to 13.5% water. With too little moisture, not enough steam will build up to pop the kernel; too much moisture causes the kernel to pop into a dense sphere rather than the light fluffy delicacy expected. Second, to allow the pressure to build up, the kernels must not be cracked or damaged.

2.10 BUOYANCY, FLOTATION, AND STABILITY

2.10.1 Archimedes' Principle

Brandon Bourdages/Shutterstock.com

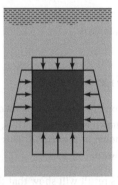

When a stationary body is completely submerged in a fluid (such as the hot air balloon shown in the adjacent figure), or floating so that it is only partially submerged, the resultant fluid force acting on the body is called the **buoyant force.** A net upward vertical force results because pressure increases with depth and the pressure forces acting from below are larger than the pressure forces acting from above, as shown in the adjacent figure. This force can be determined using an approach similar to that used in the previous section for forces on curved surfaces. Consider a body of arbitrary shape, having a volume $\mathcal{V}$, that is immersed in a fluid as illustrated in **Fig. 2.24*a*.** We enclose the body in a parallelepiped and draw a free-body diagram of the parallelepiped with the body removed as shown in **Fig. 2.24*b*.** Note that the forces F_1, F_2, F_3, and F_4 are simply the forces exerted on the plane surfaces of the parallelepiped (for simplicity the forces in the x direction are not shown), $\mathcal{W}$ is the weight of the shaded fluid volume (parallelepiped minus body), and F_B is the force the body is exerting *on the fluid*. The forces on the vertical surfaces, such as F_3 and F_4, are all equal and cancel, so the equilibrium equation of interest is in the z direction and can be expressed as

$$F_B = F_2 - F_1 - \mathcal{W} \tag{2.21}$$

If the specific weight of the fluid is constant, then

$$F_2 - F_1 = \gamma(h_2 - h_1)A$$

where A is the horizontal area of the upper (or lower) surface of the parallelepiped, and Eq. 2.21 can be written as

$$F_B = \gamma(h_2 - h_1)A - \gamma[(h_2 - h_1)A - \mathcal{V}]$$

Simplifying, we arrive at the desired expression for the buoyant force

$$\boxed{F_B = \gamma \mathcal{V}} \tag{2.22}$$

where γ is the specific weight of the fluid and $\mathcal{V}$ is the volume of the body. The effects of the specific weight (or density) of the body as compared to that of the surrounding fluid are illustrated by the adjacent figure. The direction of the buoyant force, which is the force of the

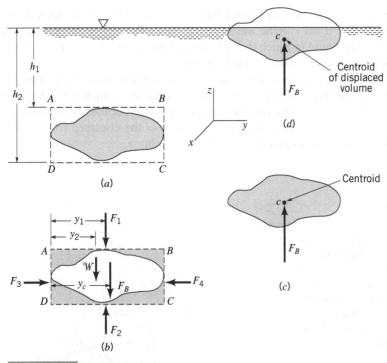

FIGURE 2.24 Buoyant force on submerged and floating bodies.

fluid *on the body,* is opposite to that shown on the free-body diagram. Therefore, the buoyant force has a magnitude equal to the weight of the fluid displaced by the body and is directed vertically upward. This result is commonly referred to as **Archimedes' principle** in honor of Archimedes (287–212 B.C.), a Greek mechanician and mathematician who first enunciated the basic ideas associated with hydrostatics.

The location of the line of action of the buoyant force can be determined by summing moments of the forces shown on the free-body diagram in Fig. 2.24b with respect to some convenient axis. For example, summing moments about an axis perpendicular to the paper through point D we have

$$F_B y_c = F_2 y_1 - F_1 y_1 - \mathcal{W} y_2$$

and on substitution for the various forces

$$\forall y_c = \forall_T y_1 - (\forall_T - \forall) y_2 \tag{2.23}$$

where $\forall_T$ is the total volume $(h_2 - h_1)A$. The right-hand side of Eq. 2.23 is the first moment of the displaced volume $\forall$ with respect to the x–z plane so that y_c is equal to the y coordinate of the centroid of the volume $\forall$. In a similar fashion, it can be shown that the x coordinate of the buoyant force coincides with the x coordinate of the centroid. Thus, we conclude that the *buoyant force passes through the centroid of the displaced volume,* as shown in Fig. 2.24c. The point through which the buoyant force acts is called the **center of buoyancy.**

Archimedes' principle states that the buoyant force has a magnitude equal to the weight of the fluid displaced by the body and is directed vertically upward.

The Wide World of Fluids

Concrete canoes

A solid block of concrete thrown into a pond or lake will obviously sink. But if the concrete is formed into the shape of a canoe it can be made to float. Of course the reason the canoe floats is the development of the *buoyant force* due to the displaced volume of water. With the proper design, this vertical force can be made to balance the weight of the canoe plus passengers—the canoe floats. Each year since 1988 a National Concrete Canoe

Competition for university teams is jointly sponsored by the American Society of Civil Engineers and Master Builders Inc. The canoes must be 90% concrete and are typically designed with the aid of a computer by civil engineering students. Final scoring depends on four components: a design report, an oral presentation, the final product, and racing. In 2011, California Polytechnic State University, San Luis Obispo, won the national championship held at the University of Evansville with its 127 kg, 6 m-long canoe.

V2.7 Hydrometer

These same results apply to floating bodies that are only partially submerged, as illustrated in Fig. 2.24d, if the specific weight of the fluid above the liquid surface is very small compared with the liquid in which the body floats. Since the fluid above the surface is usually air, for practical purposes this condition is satisfied.

In the derivations presented above, the fluid is assumed to have a constant specific weight, γ. If a body is immersed in a fluid in which γ varies with depth, such as in a layered fluid, the magnitude of the buoyant force remains equal to the weight of the displaced fluid. However, the buoyant force does not pass through the centroid of the displaced volume, but rather, it passes through the center of gravity of the displaced volume.

EXAMPLE 2.10 | Buoyant Force on a Submerged Object

Given A Type I offshore life jacket (personal flotation device) of the type worn by commercial fishermen is shown in **Fig. E2.10a**. It is designed for extended survival in rough, open water. According to U.S. Coast Guard regulations, the life jacket must provide a minimum 9.98 kgf net upward force on the user. Consider such a life jacket that uses a foam material with a specific weight of 0.31 kN/m³ for the main flotation material. The remaining material (cloth, straps, fasteners, etc.) weighs 1 kg and is of negligible volume.

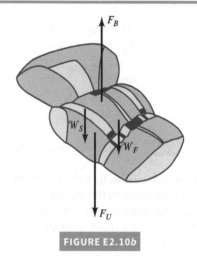

FIGURE E2.10b

$\gamma_{\text{foam}} = 0.31$ kN/m³ is the specific weight of the foam. Thus, from Eq. 1

$$\gamma_{\text{water}}\mathcal{V} = \gamma_{\text{foam}}\mathcal{V} + \mathcal{W}_S + F_U$$

or

$$\mathcal{V} = (\mathcal{W}_S + F_U)/(\gamma_{\text{water}} - \gamma_{\text{foam}})$$
$$= (1 \text{ kgf} + 9.98 \text{ kgf})/(9.81 \text{ kN/m}^3 - 0.31 \text{ kN/m}^3)$$
$$= (166.8 \text{ N})/(9.81 \text{ kN/m}^3 - 0.31 \text{ kN/m}^3)$$
$$= 0.011 \text{ m}^3 \qquad (Ans)$$

FIGURE E2.10a

Find Determine the minimum volume of foam needed for this life jacket.

Solution A free-body diagram of the life jacket is shown in **Fig. E2.10b**, where F_B is the buoyant force acting on the life jacket, $\mathcal{W}_F$ is the weight of the foam, $\mathcal{W}_S = 1$ kgf is the weight of the remaining material, and $F_U = 9.98$ kgf is the required force on the user. For equilibrium it follows that

$$F_B = \mathcal{W}_F + \mathcal{W}_S + F_U \qquad (1)$$

where from Eq. 2.22

$$F_B = \gamma_{\text{water}}\mathcal{V}$$

Here, $\gamma_{\text{water}} = 9.81$ kN/m³ is the specific weight of seawater, and $\mathcal{V}$ is the volume of the foam. Also $\mathcal{W}_{\text{foam}} = \gamma_{\text{foam}}\mathcal{V}$, where

Comments In this example, rather than using difficult-to-calculate hydrostatic pressure force on the irregularly shaped life jacket, we have used the buoyant force. The net effect of the pressure forces on the surface of the life jacket is equal to the upward buoyant force. Do not include both the buoyant force and the hydrostatic pressure effects in your calculations—use one or the other.

There is more to the proper design of a life jacket than just the volume needed for the required buoyancy. According to regulations, a Type I life jacket must also be designed so that it provides proper protection to the user by turning an unconscious person in the water to a face-up position, as shown in Fig. E2.10a. This involves the concept of the stability of a floating object (see Section 2.11.2). The life jacket should also provide minimum interference under ordinary working conditions so as to encourage its use by commercial fishermen.

The effects of buoyancy are not limited to the interaction between a solid body and a fluid. Buoyancy effects can also be seen within fluids alone, as long as a density difference exists. Consider the shaded portion of Fig. 2.24c to be a volume of fluid instead of a solid. This volume of fluid is submerged in the surrounding fluid and therefore has a buoyant force due to the fluid it displaces (like the solid). If this volume contains fluid with a density of ρ_1, then the downward force due to weight is $\mathcal{W} = \rho_1 g V$. In addition, if the surrounding fluid has the same density, then the buoyant force on the volume due to the displaced fluid will be $F_B = \gamma V = \rho_1 g V$. As expected, in this case the weight of the volume is exactly balanced by the buoyant force acting on the volume so there is no net force. However, if the density of fluid in the volume is ρ_2, then $\mathcal{W}$ and F_B will not balance and there will be a net force in the upward or downward direction depending on whether the density in the volume (ρ_2) is less than or greater than, respectively, the density of the surrounding fluid. Note that this difference can develop from two different fluids with different densities or from temperature differences within the same fluid causing density variations in space. For example, smoke from a fire rises because it is lighter (due to its higher temperature) than the surrounding air.

V2.8 Atmospheric buoyancy

V2.9 Density differences in fluids

The Wide World of Fluids

Explosive lake

In 1986, a tremendous explosion of carbon dioxide (CO_2) from Lake Nyos, west of Cameroon, killed more than 1700 people and livestock. The explosion resulted from a buildup of CO_2 that seeped into the high-pressure water at the bottom of the lake from warm springs of CO_2-bearing water. The CO_2-rich water is heavier than pure water and can hold a volume of CO_2 more than five times the water volume. As long as the gas remains dissolved in the water, the stratified lake (i.e., pure water on top, CO_2 water

on the bottom) is stable. But if some mechanism causes the gas bubbles to nucleate, they rise, grow, and cause other bubbles to form, feeding a chain reaction. A related phenomenon often occurs when a pop bottle is shaken and then opened. The pop shoots from the container rather violently. When this set of events occurred in Lake Nyos, the entire lake overturned through a column of rising and expanding *buoyant* bubbles. The heavier-than-air CO_2 then flowed through the long, deep valleys surrounding the lake and asphyxiated human and animal life caught in the gas cloud. One victim was 27 km downstream from the lake.

2.10.2 The Stability of Bodies in Fluids

Another interesting and important problem associated with submerged or floating bodies is concerned with the stability of the bodies. As illustrated in the adjacent figure, a body is said to be in a *stable equilibrium* position if, when displaced, it returns to its equilibrium position. Conversely, it is in an *unstable equilibrium* position if, when displaced (even slightly), it moves to a new equilibrium position. Stability considerations are particularly important for submerged or floating bodies because the centers of buoyancy and gravity do not necessarily coincide. A small rotation can result in either a restoring or overturning couple. For example, for the *completely* submerged body shown in **Fig. 2.25**, which has a center of gravity below the center of buoyancy, a rotation from its equilibrium position will create a restoring couple formed by the weight, $\mathcal{W}$, and the buoyant force, F_B, which causes the body to rotate back to its original position. Thus, for this configuration the body is stable. It is to be noted that as long as the center of gravity is *below* the center of buoyancy, this will always be true; that is, the body is in a *stable equilibrium* position with respect to small rotations. However, as is illustrated in **Fig. 2.26**, if the center of gravity of the completely submerged body is above the center of buoyancy, the resulting couple formed by the weight and the buoyant force will cause the body to overturn and move to a new equilibrium position. Thus, a completely submerged body with its center of gravity *above* its center of buoyancy is in an *unstable equilibrium* position.

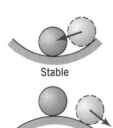

Stable

Unstable

The stability of a body can be determined by considering what happens when it is displaced from its equilibrium position.

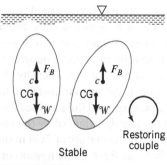

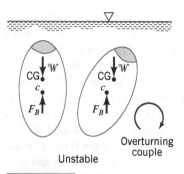

FIGURE 2.25 Stability of a completely immersed body—center of gravity below centroid.

FIGURE 2.26 Stability of a completely immersed body—center of gravity above centroid.

For *floating* bodies, the stability problem is more complicated because as the body rotates the location of the center of buoyancy (which is the centroid of the displaced volume) may change. As is shown in **Fig. 2.27**, a floating body such as a barge that rides low in the water can be stable even though the center of gravity lies above the center of buoyancy. This is true because as the body rotates the buoyant force, F_B, shifts to pass through the centroid of the newly formed displaced volume and, as illustrated, combines with the weight, $\mathcal{W}$, to form a couple that will cause the body to return to its original equilibrium position. However, for the relatively tall, slender body shown in **Fig. 2.28**, a small rotational displacement can cause the buoyant force and the weight to form an overturning couple as illustrated.

We see from even these relatively simple examples that although the concept of floating body stability is not difficult to understand, determining the stability of specific floating bodies can become quite tedious. For even simple shapes, determination of the centroid of irregular displaced volumes can require many calculations as can determination of the center of mass of inhomogeneous bodies. Although both the relatively narrow kayak and the wide houseboat shown in the adjacent photographs are both stable, the kayak will overturn much more easily than the houseboat. The problem can be further complicated by the necessary inclusion of other types of external forces such as those induced by wind gusts or currents. Stability considerations are obviously of great importance in the design of ships, submarines, bathyscaphes, and so forth; such considerations play a significant role in the work of naval architects (see, for example, Ref. 6).

Very stable

▶ **Video**

V2.10 Stability of a floating cube

▶ **Video**

V2.11 Stability of a model barge

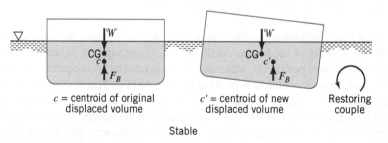

c = centroid of original displaced volume

c' = centroid of new displaced volume

Restoring couple

Stable

FIGURE 2.27 Stability of a floating body—stable configuration.

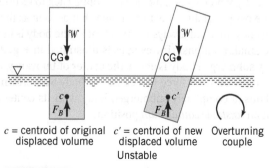

c = centroid of original displaced volume

c' = centroid of new displaced volume

Overturning couple

Unstable

FIGURE 2.28 Stability of a floating body—unstable configuration.

2.11 PRESSURE VARIATION IN A FLUID WITH RIGID-BODY MOTION

Although in this chapter we have been primarily concerned with fluids at rest, the general equation of motion (Eq. 2.2)

$$-\nabla p - \gamma \hat{\mathbf{k}} = \rho \mathbf{a}$$

was developed for both fluids at rest and fluids in motion, with the only stipulation being that there are no shearing stresses present. Equation 2.2 can be expressed in rectangular coordinate components, with the gravity vector in the negative z-direction, as

$$-\frac{\partial p}{\partial x} = \rho a_x \qquad -\frac{\partial p}{\partial y} = \rho a_y \qquad -\frac{\partial p}{\partial z} = \gamma + \rho a_z \qquad (2.24)$$

A general class of problems involving fluid motion in which there are no shearing stresses occurs when a mass of fluid undergoes rigid-body motion. For example, if a container of fluid accelerates along a straight path, the fluid will move as a rigid mass (after the initial sloshing motion has died out) with each particle having the same acceleration. Since there is no deformation, there will be no shearing stresses and, therefore, Eq. 2.2 applies. Similarly, if a fluid is contained in a tank that rotates about a fixed axis, the fluid will simply rotate with the tank as a rigid body, and again Eq. 2.2 can be applied to obtain the pressure distribution throughout the moving fluid. Specific results for these two cases (rigid-body uniform motion and rigid-body rotation) are developed in the following two sections. Although problems relating to fluids having rigid-body motion are not, strictly speaking, "fluid statics" problems, they are included in this chapter because, as we will see, the analysis and resulting pressure relationships are similar to those for fluids at rest.

> There are no shearing stresses in a body of fluid experiencing rigid body translation or rotation.

2.11.1 Linear Motion

We first consider an open container of a liquid that is translating along a straight path with a constant acceleration **a**, as illustrated in **Fig. 2.29**. Since $a_x = 0$, it follows from the first of Eqs. 2.24 that the pressure gradient in the x direction is zero ($\partial p / \partial x = 0$). In the y and z directions

$$\frac{\partial p}{\partial y} = -\rho a_y \qquad (2.25)$$

$$\frac{\partial p}{\partial z} = -\rho(g + a_z) \qquad (2.26)$$

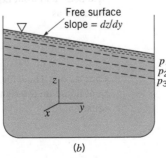

FIGURE 2.29 Linear acceleration of a liquid with a free surface.

The change in pressure between two closely spaced points located at y, z, and $y + dy, z + dz$ can be expressed as

$$dp = \frac{\partial p}{\partial y}\, dy + \frac{\partial p}{\partial z}\, dz$$

or in terms of the results from Eqs. 2.25 and 2.26

$$dp = -\rho a_y\, dy - \rho(g + a_z)\, dz \qquad (2.27)$$

Along a line of *constant* pressure, $dp = 0$, and therefore from Eq. 2.27 it follows that the slope of this line is given by the relationship

$$\frac{dz}{dy} = -\frac{a_y}{g + a_z} \qquad (2.28)$$

This relationship is illustrated by the adjacent figure. Along a free surface the pressure is constant, so that for the accelerating mass shown in Fig. 2.29 the free surface will be inclined if $a_y \neq 0$. In addition, all lines of constant pressure will be parallel to the free surface as illustrated.

For the special circumstance in which $a_y = 0$, $a_z \neq 0$, which corresponds to the mass of fluid accelerating in the vertical direction, Eq. 2.28 indicates that the fluid surface will be horizontal. However, from Eq. 2.26, we see that the pressure distribution is not hydrostatic but is given by the equation

$$\frac{dp}{dz} = -\rho(g + a_z)$$

For fluids of constant density, this equation shows that the pressure will vary linearly with depth, but the variation is due to the combined effects of gravity and the externally induced acceleration, $\rho(g + a_z)$, rather than simply the specific weight ρg. Thus, for example, the pressure along the bottom of a liquid-filled tank that is resting on the floor of an elevator that is accelerating upward will be increased over that which exists when the tank is at rest (or moving with a constant velocity). It is to be noted that for a *freely falling* fluid mass ($a_z = -g$), the pressure gradients in all three coordinate directions are zero, which means that if the pressure surrounding the mass is zero, the pressure throughout will be zero. The pressure throughout a "blob" of orange juice floating in an orbiting space shuttle (a form of free fall) is uniform and is only slightly higher than ambient due to the surface tension holding it together (see Section 1.9).

EXAMPLE 2.11 | Pressure Variation in an Accelerating Tank

Given The cross section for the fuel tank of an experimental vehicle is shown in **Fig. E2.11**. The rectangular tank is vented to the atmosphere, and the specific gravity of the fuel is $SG = 0.65$. A pressure transducer is located in its side as illustrated. During testing of the vehicle, the tank is subjected to a constant linear acceleration, a_y.

Find

a. Determine an expression that relates a_y and the pressure (in kN/m²) at the transducer.

b. What is the maximum acceleration that can occur before the fuel level drops below the transducer?

Solution

a. For a constant horizontal acceleration, the fuel will move as a rigid body, and from Eq. 2.28 the slope of the fuel surface can be expressed as

$$\frac{dz}{dy} = -\frac{a_y}{g}$$

since $a_z = 0$. Thus, for some arbitrary a_y, the change in depth, z_1, of liquid on the right side of the tank can be found from the equation

$$-\frac{z_1}{0.25\ \text{m}} = -\frac{a_y}{g}$$

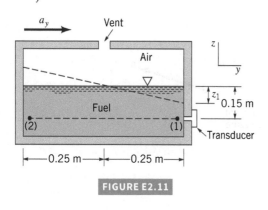

FIGURE E2.11

or

$$z_1 = (0.25 \text{ m})\left(\frac{a_y}{g}\right)$$

Since there is no acceleration in the vertical, z, direction, the pressure along the wall varies hydrostatically as shown by Eq. 2.26. Thus, the pressure at the transducer is given by the relationship

$$p = \gamma h$$

where h is the depth of fuel above the transducer, and therefore

$$p = (0.65)(9.81 \text{ kN/m}^3)[0.15 \text{ m} - (0.25 \text{ m})(a_y/g)]$$

$$= 0.96 - 1.6 \frac{a_y}{g} \qquad (Ans)$$

for $z_1 \leq 0.15$ m. As written, p would be given in kN/m^2.

b. The limiting value for $(a_y)_{max}$ (when the fuel level reaches the transducer) can be found from the equation

$$0.15 \text{ m} = (0.25 \text{ m})\left[\frac{(a_y)_{max}}{g}\right]$$

or

$$(a_y)_{max} = \frac{3g}{5}$$

and for standard acceleration due to gravity

$$(a_y)_{max} = \frac{3}{5}(9.8 \text{ m/s}^2) = 5.9 \text{ m/s}^2 \qquad (Ans)$$

Comment   Note that the pressure in horizontal layers is not constant in this example because $\partial p/\partial y = -\rho a_y \neq 0$. Thus, for example, $p_1 \neq p_2$.

2.11.2 Rigid-Body Rotation

After an initial "start-up" transient, a fluid contained in a tank that rotates with a constant angular velocity ω about an axis, as is shown in **Fig. 2.30**, will rotate with the tank as a rigid body. It is known from elementary particle dynamics that the acceleration of a fluid particle located at a distance r from the axis of rotation is equal in magnitude to $r\omega^2$, and the direction of the acceleration is toward the axis of rotation, as is illustrated in the figure. Since the paths of the fluid particles are circular, it is convenient to use cylindrical polar coordinates r, θ, and z, defined in the insert in Fig. 2.30. It will be shown in Chapter 6 that in terms of cylindrical coordinates the pressure gradient ∇p can be expressed as

> A fluid contained in a tank that is rotating with a constant angular velocity about an axis will rotate as a rigid body.

$$\nabla p = \frac{\partial p}{\partial r}\hat{\mathbf{e}}_r + \frac{1}{r}\frac{\partial p}{\partial \theta}\hat{\mathbf{e}}_\theta + \frac{\partial p}{\partial z}\hat{\mathbf{e}}_z \qquad (2.29)$$

Thus, in terms of this coordinate system

$$\mathbf{a}_r = -r\omega^2\hat{\mathbf{e}}_r \qquad \mathbf{a}_\theta = 0 \qquad \mathbf{a}_z = 0$$

and from Eq. 2.2

$$\frac{\partial p}{\partial r} = \rho r\omega^2 \qquad \frac{\partial p}{\partial \theta} = 0 \qquad \frac{\partial p}{\partial z} = -\gamma \qquad (2.30)$$

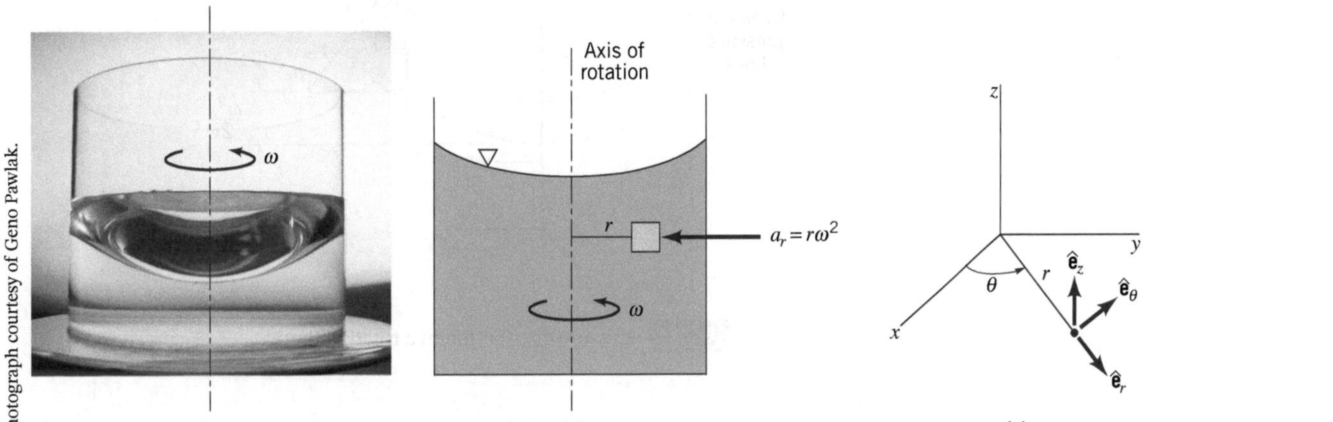

FIGURE 2.30   Rigid-body rotation of a liquid in a tank.

These results show that for this type of rigid-body rotation, the pressure is a function of two variables r and z, and therefore the differential pressure is

$$dp = \frac{\partial p}{\partial r}\, dr + \frac{\partial p}{\partial z}\, dz$$

or

$$dp = \rho r \omega^2\, dr - \gamma\, dz \tag{2.31}$$

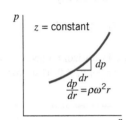

On a horizontal plane ($dz = 0$), it follows from Eq. 2.31 that $dp/dr = \rho \omega^2 r$, which is greater than zero. Hence, as illustrated in the adjacent figure, because of centrifugal acceleration, the pressure increases in the radial direction.

Along a surface of constant pressure, such as the free surface, $dp = 0$, so that from Eq. 2.31 (using $\gamma = \rho g$)

$$\frac{dz}{dr} = \frac{r\omega^2}{g}$$

Integration of this result gives the equation for surfaces of constant pressure as

$$z = \frac{\omega^2 r^2}{2g} + \text{constant} \tag{2.32}$$

The free surface in a rotating liquid is curved rather than flat.

This equation reveals that these surfaces of constant pressure are parabolic, as illustrated in **Fig. 2.31**.

Integration of Eq. 2.31 yields

$$\int dp = \rho \omega^2 \int r\, dr - \gamma \int dz$$

or

$$p = \frac{\rho \omega^2 r^2}{2} - \gamma z + \text{constant} \tag{2.33}$$

where the constant of integration can be expressed in terms of a specified pressure at some arbitrary point r_0, z_0. This result shows that the pressure varies with the distance from the axis of rotation, but at a fixed radius, the pressure varies hydrostatically in the vertical direction, as shown in **Fig. 2.31**.

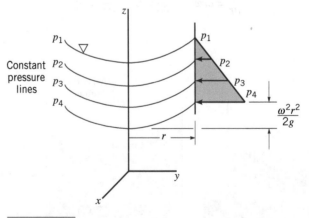

FIGURE 2.31   Pressure distribution in a rotating liquid.

EXAMPLE 2.12 | Free Surface Shape of Liquid in a Rotating Tank

Given It has been suggested that the angular velocity, ω, of a rotating body or shaft can be measured by attaching an open cylinder of liquid, as shown in **Fig. E2.12a**, and measuring with some type of depth gage the change in the fluid level, $H - h_0$, caused by the rotation of the fluid.

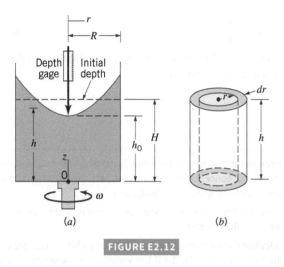

FIGURE E2.12

Find Determine the relationship between this change in fluid level and the angular velocity.

Solution The height, h, of the free surface above the tank bottom can be determined from Eq. 2.32, and it follows that

$$h = \frac{\omega^2 r^2}{2g} + h_0$$

The initial volume of fluid in the tank, V_i, is equal to

$$V_i = \pi R^2 H$$

The volume of the fluid with the rotating tank can be found with the aid of the differential element shown in **Fig. E2.12b**. This cylindrical shell is taken at some arbitrary radius, r, and its volume is

$$dV = 2\pi r h \, dr$$

The total volume is, therefore,

$$V = 2\pi \int_0^R r\left(\frac{\omega^2 r^2}{2g} + h_0\right) dr = \frac{\pi \omega^2 R^4}{4g} + \pi R^2 h_0$$

Since the volume of the fluid in the tank must remain constant (assuming that none spills over the top), it follows that

$$\pi R^2 H = \frac{\pi \omega^2 R^4}{4g} + \pi R^2 h_0$$

or

$$H - h_0 = \frac{\omega^2 R^2}{4g} \qquad (Ans)$$

Comment This is the relationship we were looking for. It shows that the change in depth could indeed be used to determine the rotational speed, although the relationship between the change in depth and speed is not a linear one.

The Wide World of Fluids

Rotating mercury mirror telescope

A telescope mirror has the same shape as the parabolic free surface of a *liquid in a rotating tank*. The liquid mirror telescope (LMT) consists of a pan of liquid (normally mercury because of its excellent reflectivity) rotating to produce the required parabolic shape of the free surface mirror. With recent technological advances, it is possible to obtain the vibration-free rotation and the constant angular velocity necessary to produce a liquid mirror surface precise enough for astronomical use. Construction of the largest LMT, located at the University of British Columbia, was completed in 2003. When it entered service, it was the third largest optical telescope in North America. With a diameter of 1.8 m and a rotation rate of 7 rpm, this mirror uses 30 liters of mercury for its 1-mm thick, parabolic-shaped mirror. One of the major benefits of a LMT (compared to a normal glass mirror telescope) is its low cost. Perhaps the main disadvantage is that a LMT can look only straight up, although there are many galaxies, supernova explosions, and pieces of space junk to view in any part of the sky. The next-generation LMTs may have movable secondary mirrors to allow a larger portion of the sky to be viewed.

2.12 EQUILIBRIUM OF MOVING FLUIDS (SPECIAL CASE OF FLUID STATICS)

Though the equilibrium of moving fluids shall be discussed in detail in Chapter 6 but for the sake of completeness of static analysis of fluids, let us consider the equation of motion obtained from Newton's second law of motion for momentum balance. The equations so obtained will be the equations explained in the form of Eqs. 6.50 in Chapter 6. For inviscid flow, the equations shall be reduced to Euler's equations shown as follows.

$$\rho g_x - \frac{\partial p}{\partial x} = \rho\left(\frac{\partial u}{\partial t} + u\frac{\partial u}{\partial x} + v\frac{\partial u}{\partial y} + w\frac{\partial u}{\partial z}\right) \qquad (2.34a)$$

$$\rho g_y - \frac{\partial p}{\partial y} = \rho\left(\frac{\partial v}{\partial t} + u\frac{\partial v}{\partial x} + v\frac{\partial v}{\partial y} + w\frac{\partial v}{\partial z}\right) \qquad (2.34b)$$

$$\rho g_z - \frac{\partial p}{\partial z} = \rho\left(\frac{\partial w}{\partial t} + u\frac{\partial w}{\partial x} + v\frac{\partial w}{\partial y} + w\frac{\partial w}{\partial z}\right) \qquad (2.34c)$$

Now for a fluid satisfying the condition of static equilibrium, that is, either the fluid should be at rest or moving with constant velocity, the right-hand side of each term of Eq. 2.34 shall become zero and the equation of static equilibrium for a fluid can be written in the following form:

$$\rho g_x - \frac{\partial p}{\partial x} = 0 \tag{2.35a}$$

$$\rho g_y - \frac{\partial p}{\partial y} = 0 \tag{2.35b}$$

$$\rho g_z - \frac{\partial p}{\partial z} = 0 \tag{2.35c}$$

The above equations are the general equations of pressure variation of static fluid under the action of gravitational force only explained earlier in Section 2.3.

CHAPTER SUMMARY

In this chapter, the pressure variation in a fluid at rest is considered, along with some important consequences of this type of pressure variation. It is shown that for incompressible fluids at rest the pressure varies linearly with depth. This type of variation is commonly referred to as a hydrostatic pressure distribution. For compressible fluids at rest, the pressure distribution will not generally be hydrostatic, but Eq. 2.4 remains valid and can be used to determine the pressure distribution if additional information about the variation of the specific weight is specified. The distinction between absolute and gage pressure is discussed along with a consideration of barometers for the measurement of atmospheric pressure.

Pressure-measuring devices called manometers, which utilize static liquid columns, are analyzed in detail. A brief discussion of mechanical and electronic pressure gages is also presented. Equations for determining the magnitude and location of the resultant fluid force acting on a plane surface in contact with a static fluid are developed. A general approach for determining the magnitude and location of the resultant fluid force acting on a curved surface in contact with a static fluid is developed. For submerged or floating bodies, the concept of the buoyant force and the use of Archimedes' principle are reviewed.

The following checklist provides a study guide for this chapter. When your study of the entire chapter has been completed, you should be able to

- write out meanings of the terms listed and understand each of the related concepts. These terms are particularly important and are set in **bold black** type in the text.
- calculate the pressure at various locations within an incompressible fluid at rest.
- calculate the pressure at various locations within a compressible fluid at rest using Eq. 2.4 if the variation in the specific weight is specified.
- use the concept of a hydrostatic pressure distribution to determine pressures from measurements using various types of manometers.
- determine the magnitude, direction, and location of the resultant hydrostatic force acting on a plane surface.
- determine the magnitude, direction, and location of the resultant hydrostatic force acting on a curved surface.
- use Archimedes' principle to calculate the resultant hydrostatic force acting on floating or submerged bodies.
- analyze, based on Eq. 2.2, the motion of fluids moving with simple rigid-body linear motion or simple rigid-body rotation.

KEY EQUATIONS

Pressure gradient in a stationary fluid	$\dfrac{dp}{dz} = -\gamma$	(2.4)
Pressure variation in a stationary incompressible fluid	$p_1 = \gamma h + p_2$	(2.7)
Hydrostatic force on a plane surface	$F_R = \gamma h_c A$	(2.18)
Location of hydrostatic force on a plane surface	$y_R = \dfrac{I_{xc}}{y_c A} + y_c$	(2.19)
	$x_R = \dfrac{I_{xyc}}{y_c A} + x_c$	(2.20)
Buoyant force	$F_B = \gamma \forall$	(2.22)
Pressure gradient in rigid-body motion	$-\dfrac{\partial p}{\partial x} = \rho a_x, \quad -\dfrac{\partial p}{\partial y} = \rho a_y, \quad -\dfrac{\partial p}{\partial z} = \gamma + \rho a_z$	(2.24)
Pressure gradient in rigid-body rotation	$\dfrac{\partial p}{\partial r} = \rho r \omega^2, \quad \dfrac{\partial p}{\partial \theta} = 0, \quad \dfrac{\partial p}{\partial z} = -\gamma$	(2.30)

REFERENCES

[1] *The U.S. Standard Atmosphere, 1962*, U.S. Government Printing Office, Washington, DC, 1962.

[2] *The U.S. Standard Atmosphere, 1976*, U.S. Government Printing Office, Washington, DC, 1976.

[3] Benedict, R. P., *Fundamentals of Temperature, Pressure, and Flow Measurements*, 3rd Ed., Wiley, New York, 1984.

[4] Dally, J. W., Riley, W. F., and McConnell, K. G., *Instrumentation for Engineering Measurements*, 2nd Ed., Wiley, New York, 1993.

[5] Holman, J. P., *Experimental Methods for Engineers*, 4th Ed., McGraw-Hill, New York, 1983.

[6] Comstock, J. P., ed., *Principles of Naval Architecture*, Society of Naval Architects and Marine Engineers, New York, 1967.

QUESTIONS AND PROBLEMS

ⓒ Problem to be solved with aid of programmable calculator or computer.

ⓞ Open-ended problem that requires critical thinking. These problems require various assumptions to provide the necessary input data. There are not unique answers to these problems.

Note: Unless specific values of required fluid properties are specified in the problem statement, use the values found in Tables 1.4–1.8 and in the tables in the Appendices.

Section 2.3 Pressure Variation in a Fluid at Rest

2.3.1 Obtain a photograph/image of a situation in which the fact that in a static fluid the pressure increases with depth is important. Print this photo and write a brief paragraph that describes the situation involved.

2.3.2 The deepest known spot in the oceans is the Challenger Deep in the Mariana Trench of the Pacific Ocean and is approximately 11,000 m below the surface. Assume that the salt water density is constant at 1025 kg/m^3 and determine the pressure at this depth.

2.3.3 A closed tank is partially filled with glycerin. If the air pressure in the tank is 41.5 kPa and the depth of glycerin is 3 m, what is the pressure in kPa at the bottom of the tank?

2.3.4 Ⓥⁱᵈᵉᵒ Blood pressure is usually given as a ratio of the maximum pressure (systolic pressure) to the minimum pressure (diastolic pressure). As shown in Video V2.2, such pressures are commonly measured with a mercury manometer. A typical value for this ratio for a human would be 120/70, where the pressures are in mm Hg. (a) What would these pressures be in pascals? (b) If your car tire was inflated to 120 mm Hg, would it be sufficient for normal driving?

2.3.5 An unknown immiscible liquid seeps into the bottom of an open oil tank. Some measurements indicate that the depth of the unknown liquid is 1.5 m and the depth of the oil (specific weight = 8.5 kN/m^3) floating on top is 5.0 m. A pressure gage connected to the bottom of the tank reads 65 kPa. What is the specific gravity of the unknown liquid?

2.3.6 A 10 m high downspout of a house is clogged at the bottom. Find the pressure at the bottom if the downspout is filled with 15.6 °C rainwater.

2.3.7 How high a column of SAE 30 oil would be required to give the same pressure as 700 mm Hg?

2.3.8 The deepest known spot in the oceans is the Challenger Deep in the Mariana Trench of the Pacific Ocean and is approximately 11,000 m below the surface. For a surface density of 1030 kg/m^3, a constant water temperature, and an isothermal bulk modulus of elasticity of 2.3×10^9 N/m^2, find the pressure at this depth.

2.3.9 A submarine submerges by admitting seawater ($S = 1.03$) into its ballast tanks. The amount of water admitted is controlled by air pressure because seawater will cease to flow into the tank when the internal pressure (at the hull penetration) is equal to the hydrostatic pressure at the depth of the submarine. Consider a ballast tank, which can be modeled as a vertical half-cylinder ($R = 2.5$ m, $L = 6$ m) for which the air pressure control valve has failed shut. The failure occurred at the beginning of a dive from 18 m to 305 m. The tank was initially filled with seawater to a depth of 0.6 m, and the air was at a temperature of 4.44 °C. As the weight of water in the tank is important in maintaining the boat's attitude, determine the weight of water in the tank as a function of depth during the dive. You may assume that tank internal pressure is always in equilibrium with the ocean's hydrostatic pressure and that the inlet pipe to the tank is at the bottom of the tank and penetrates the hull at the "depth" of the submarine.

2.3.10 Determine the pressure at the bottom of an open 5-m-deep tank in which a chemical process is taking place that causes the density of the liquid in the tank to vary as

$$\rho = \rho_{\text{surf}} \sqrt{1 + \sin^2\left(\frac{h}{h_{\text{bot}}} \frac{\pi}{2}\right)}$$

where h is the distance from the free surface and $\rho_{\text{surf}} = 1700$ kg/m^3.

2.3.11 ⓒ In a certain liquid at rest, measurements of the specific weight at various depths show the following variation:

h (m)	γ (kg/m^3)
0	1121
3	1217
6	1346
9	1458
12	1554
15	1634
18	1714
21	1762
24	1794
27	1826
30	1842

The depth $h = 0$ corresponds to a free surface at atmospheric pressure. Determine, through numerical integration of Eq. 2.4, the corresponding variation in pressure and show the results on a plot of pressure (in kPa) versus depth (in meter).

2.3.12 ⓞ Because of elevation differences, the water pressure in the second floor of your house is lower than it is in the first floor. For tall buildings, this pressure difference can become unacceptable. Discuss possible ways to design the water distribution system in very tall buildings so that the hydrostatic pressure difference is within acceptable limits.

2.3.13 ⓒ Under normal conditions, the temperature of the atmosphere decreases with increasing elevation. In some situations, however, a temperature inversion may exist so that the air temperature increases with elevation. A series of temperature probes on a mountain give the elevation–temperature data shown in the table. If the barometric pressure at the base of the mountain is 83.4 kPa, determine by means of numerical integration the pressure at the top of the mountain.

Elevation (m)	Temperature (°C)
1524	10 (base)
1676	12.9
1829	15.7
1951	17
2164	19.4
2255	20.2
2500	21.1
2621	20.8
2804	20
3017	19.5 (top)

2.3.14 Often young children drink milk ($\rho = 1030$ kg/m^3) through a straw. Determine the maximum length of a vertical straw that a child can use to empty a milk container, assuming that the child can develop 75 mm Hg of suction, and use this answer to determine if you think this is a reasonable estimate of the suction that a child can develop.

2.3.15 (See The Wide World of Fluids article titled "Giraffe's blood pressure," Section 2.3.1.) (a) Determine the change in hydrostatic pressure in a giraffe's head as it lowers its head from eating leaves 6 m above the ground to getting a drink of water at ground level, as shown in **Fig. P2.3.15**. Assume the specific gravity of blood is $SG = 1$. (b) Compare the pressure change calculated in part (a) to the normal 120 mm of mercury pressure in a human's heart.

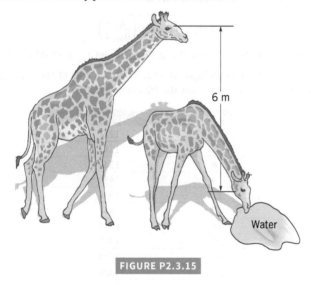

6 m

Water

FIGURE P2.3.15

Section 2.4 Standard Atmosphere

2.4.1 What would be the barometric pressure reading, in mm Hg, at an elevation of 4 km in the U.S. standard atmosphere? (Refer to Table C.1 in Appendix C.)

2.4.2 Denver, Colorado, is called the "mile-high city" because its state capitol stands on land 1609.3 m above sea level. Assuming that the standard atmosphere exists, what is the pressure and temperature of the air in Denver? The temperature follows the lapse rate ($T = T_0 - Bz$).

2.4.3 Assume that a person skiing high in the mountains at an altitude of 4.5 km takes in the same volume of air with each breath as she does while walking at sea level. Determine the ratio of the mass of oxygen inhaled for each breath at this high altitude compared to that at sea level.

2.4.4 Pikes Peak near Denver, Colorado, has an elevation of 4300 m. (a) Determine the pressure at this elevation, based on Eq. 2.12. (b) If the air is assumed to have a constant density of 1.225 kg/m^3, what would the pressure be at this altitude? (c) If the air is assumed

to have a constant temperature of 15 °C, what would the pressure be at this elevation? For all three cases assume standard atmospheric conditions at sea level (see Table 2.1).

2.4.5 Equation 2.12 provides the relationship between pressure and elevation in the atmosphere for those regions in which the temperature varies linearly with elevation. Derive this equation and verify the value of the pressure given in Table C.1 in Appendix C for an elevation of 5 km.

2.4.6 As shown in Fig. 2.6 for the U.S. standard atmosphere, the troposphere extends to an altitude of 11 km where the pressure is 22.6 kPa (abs). In the next layer, called the stratosphere, the temperature remains constant at −56.5 °C. Determine the pressure and density in this layer at an altitude of 15 km. Assume $g = 9.77$ m/s^2 in your calculations. Compare your results with those given in Table C.1 in Appendix C.

2.4.7 The record low sea level barometric pressure ever recorded is 0.655 m of mercury. At what altitude in the standard atmosphere is the pressure equal to this value?

Section 2.5 Measurement of Pressure

2.5.1 On a given day, a barometer at the base of the Washington Monument reads 9.1 m of mercury. What would the barometer reading be when you carry it up to the observation deck 152.4 m above the base of the monument?

2.5.2 Aneroid barometers can be used to measure changes in altitude. If a barometer reads 0.76 m Hg at one elevation, what has been the change in altitude in meters when the barometer reading is 0.72 m Hg? Assume a standard atmosphere and that Eq. 2.12 is applicable over the range of altitudes of interest.

2.5.3 (Video) Bourdon gages (see Video V2.3 and Fig. 2.13) are commonly used to measure pressure. When such a gage is attached to the closed water tank of **Fig. P2.5.3**, the gage reads 34.5 kPa. What is the absolute air pressure in the tank? Assume standard atmospheric pressure of 101.35 kPa.

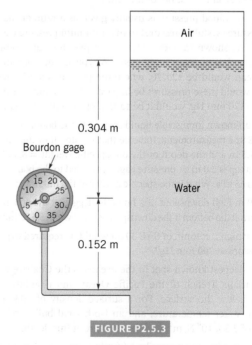

Air

0.304 m

Bourdon gage

Water

0.152 m

FIGURE P2.5.3

2.5.4 A Bourdon pressure gage attached to the outside of a tank containing air reads 530.9 kPa when the local atmospheric pressure is 760 mm Hg. What will be the gage reading if the atmospheric pressure increases to 773 mm Hg?

Section 2.6 Manometry

2.6.1 Obtain a photograph/image of a situation in which the use of a manometer is important. Print this photo and write a brief paragraph that describes the situation involved.

2.6.2 A U-tube manometer is used to check the pressure of natural gas entering a furnace. One side of the manometer is connected to the gas inlet line, and the water level in the other side open to atmospheric pressure rises 762 mm. What is the gage pressure of the natural gas in the inlet line in mm H_2O and in Pa gage?

2.6.3 A barometric pressure of 0.75 m Hg corresponds to what value of atmospheric pressure in psia and in pascals?

2.6.4 For an atmospheric pressure of 101 kPa (abs) determine the heights of the fluid columns in barometers containing one of the following liquids: (a) mercury, (b) water, and (c) ethyl alcohol. Calculate the heights including the effect of vapor pressure and compare the results with those obtained neglecting vapor pressure. Do these results support the widespread use of mercury for barometers? Why?

2.6.5 The closed tank of **Fig. P2.6.5** is filled with water and is 1.5 m long. The pressure gage on the tank reads 48.3 kPa. Determine: (a) the height, h, in the open water column, (b) the gage pressure acting on the bottom tank surface AB, and (c) the absolute pressure of the air in the top of the tank if the local atmospheric pressure is 101.3 kPa.

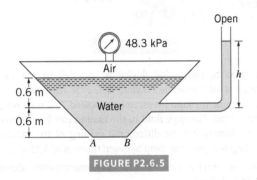

FIGURE P2.6.5

2.6.6 A mercury manometer is connected to a large reservoir of water, as shown in **Fig. P2.6.6**. Determine the ratio, h_w/h_m, of the distances h_w and h_m indicated in the figure.

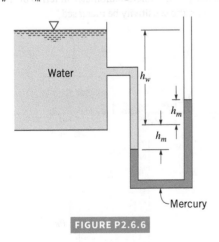

FIGURE P2.6.6

2.6.7 The U-tube manometer shown in **Fig. P2.6.7** has two fluids, water and oil ($S = 0.80$). Find the height difference between the free water surface and the free oil surface with no applied pressure difference.

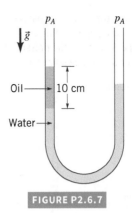

FIGURE P2.6.7

2.6.8 A U-tube manometer is connected to a closed tank containing air and water, as shown in **Fig. P2.6.8**. At the closed end of the manometer, the air pressure is 110.3 kPa. Determine the reading on the pressure gage for a differential reading of 1.2 m on the manometer. Express your answer in kPa (gage). Assume standard atmospheric pressure and neglect the weight of the air columns in the manometer.

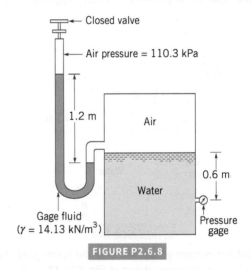

FIGURE P2.6.8

2.6.9 The container shown in **Fig. P2.6.9** holds 15.6 °C water and 15.6 °C air as shown. Find the absolute pressures at locations A, B, and C.

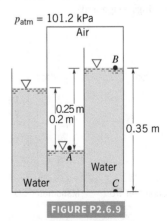

FIGURE P2.6.9

2.6.10 A closed cylindrical tank filled with water has a hemispherical dome and is connected to an inverted piping system, as shown in **Fig. P2.6.10**. The liquid in the top part of the piping system has a specific gravity of 0.8, and the remaining parts of the system are filled with water. If the pressure gage reading at A is 60 kPa, determine (a) the pressure in pipe B, and (b) the pressure head, in millimeters of mercury, at the top of the dome (point C).

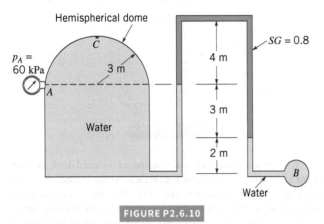

FIGURE P2.6.10

2.6.11 Two pipes are connected by a manometer, as shown in **Fig. P2.6.11**. Determine the pressure difference, $p_A - p_B$, between the pipes.

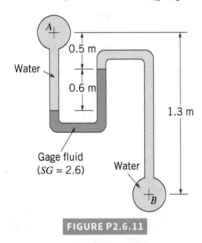

FIGURE P2.6.11

2.6.12 Find the percentage difference in the readings of the two identical U-tube manometers shown in **Fig. P2.6.12**. Manometer 90 uses 90 °C water, and manometer 30 uses 30 °C water. Both have the same applied pressure difference. Does this percentage change with the magnitude of the applied pressure difference? Can the difference between the two readings be ignored?

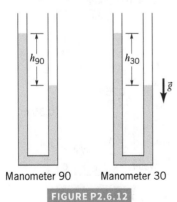

FIGURE P2.6.12

2.6.13 A U-tube manometer is connected to a closed tank, as shown in **Fig. P2.6.13**. The air pressure in the tank is 3.5 kPa, and the liquid in the tank is oil ($\rho_{oil} = 865$ kg/m³). The pressure at point A is 13.8 kPa. Determine: (a) the depth of oil, z, and (b) the differential reading, h, on the manometer.

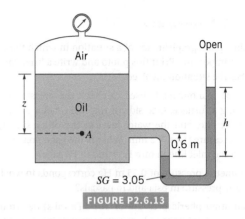

FIGURE P2.6.13

2.6.14 For the inclined-tube manometer of **Fig. P2.6.14**, the pressure in pipe A is 4.1 kPa. The fluid in both pipes A and B is water, and the gage fluid in the manometer has a specific gravity of 2.6. What is the pressure in pipe B corresponding to the differential reading shown?

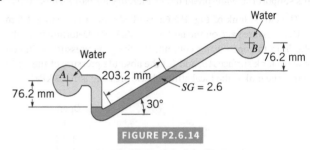

FIGURE P2.6.14

2.6.15 A flowrate measuring device is installed in a horizontal pipe through which water is flowing. A U-tube manometer is connected to the pipe through pressure taps located 76.2 mm. on either side of the device. The gage fluid in the manometer has a density of 1954 kg/m³. Determine the differential reading of the manometer corresponding to a pressure drop between the taps of 3.45 kPa.

2.6.16 The sensitivity *Sen* of the micromanometer shown in **Fig. P2.6.16** is defined as

$$Sen = \frac{H}{p_L - p_R}$$

Find the sensitivity of the micromanometer in terms of the densities ρ_A and ρ_B. How can the sensitivity be increased?

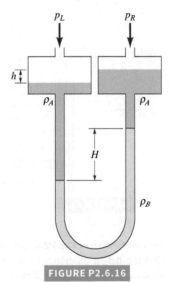

FIGURE P2.6.16

2.6.17 The cylindrical tank with hemispherical ends shown in **Fig. P2.6.17** contains a volatile liquid and its vapor. The liquid density is 800 kg/m³, and its vapor density is negligible. The pressure in the vapor is 120 kPa (abs), and the atmospheric pressure is 101 kPa (abs). Determine: **(a)** the gage pressure reading on the pressure gage, and **(b)** the height, h, of the mercury in the manometer.

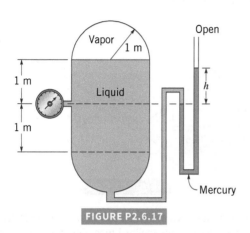

FIGURE P2.6.17

2.6.18 Determine the elevation difference, Δh, between the water levels in the two open tanks shown in **Fig. P2.6.18**.

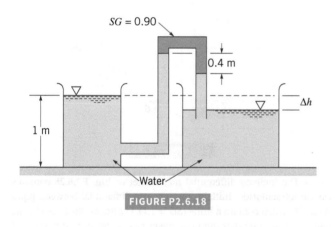

FIGURE P2.6.18

2.6.19 For the configuration shown in **Fig. P2.6.19**, what must be the value of the specific weight of the unknown fluid? Express your answer in kg/m³.

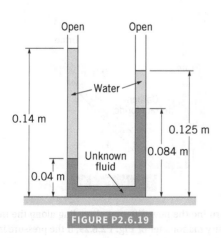

FIGURE P2.6.19

2.6.20 The manometer shown in **Fig. P2.6.20** has an air bubble either in **(a)** the right horizontal line or **(b)** the left vertical leg. Find $h_1 - h_2$ for both cases if $p_A = p_B$.

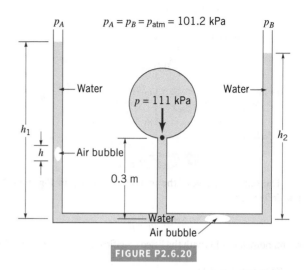

FIGURE P2.6.20

2.6.21 Ⓒ Both ends of the U-tube mercury manometer of **Fig. P2.6.21** are initially open to the atmosphere and under standard atmospheric pressure. When the valve at the top of the right leg is open, the level of mercury below the valve is h_1. After the valve is closed, air pressure is applied to the left leg. Determine the relationship between the differential reading on the manometer and the applied gage pressure, p_g. Show on a plot how the differential reading varies with p_g for $h_1 = 25, 50, 75,$ and 100 mm over the range $0 \le p_g \le 300$ kPa. Assume that the temperature of the trapped air remains constant.

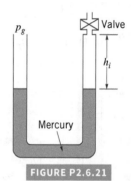

FIGURE P2.6.21

2.6.22 The inverted U-tube manometer of **Fig. P2.6.22** contains oil ($SG = 0.9$) and water as shown. The pressure differential between pipes A and B, $p_A - p_B$, is -5 kPa. Determine the differential reading h.

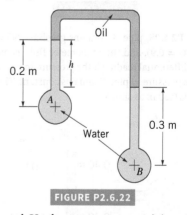

FIGURE P2.6.22

2.6.23 An inverted U-tube manometer containing oil ($SG = 0.8$) is located between two reservoirs, as shown in **Fig. P2.6.23**. The reservoir on the left, which contains carbon tetrachloride, is closed and pressurized to 55.16 kPa. The reservoir on the right contains water and is open to the atmosphere. With the given data, determine the depth of water, h, in the right reservoir.

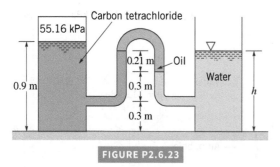

FIGURE P2.6.23

2.6.24 The sensitivity *Sen* of the manometer shown in **Fig. P2.6.24** can be defined as

$$Sen = \frac{h}{p_R - p_L}$$

Three manometer fluids with the listed specific gravities *S* are available:

Kerosene, $S = 0.82$;

SAE 10 oil, $S = 0.87$; and

Normal octane, $S = 0.71$.

Which fluid gives the highest sensitivity? The areas A_R and A_L are much larger than the cross-sectional area of the manometer tube, so $H \ll h$.

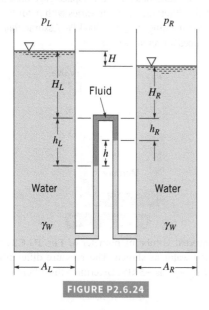

FIGURE P2.6.24

2.6.25 In **Fig. P2.6.25**, pipe *A* contains gasoline ($SG = 0.7$), pipe *B* contains oil ($SG = 0.9$), and the manometer fluid is mercury. Determine the new differential reading if the pressure in pipe *A* is decreased 25 kPa, and the pressure in pipe *B* remains constant. The initial differential reading is 0.30 m as shown.

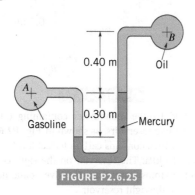

FIGURE P2.6.25

2.6.26 Consider the cistern manometer shown in **Fig. P2.6.26**. The scale is set up on the basis that the cistern area A_1 is infinite. However, A_1 is actually 50 times the internal cross-sectional area A_2 of the inclined tube. Find the percentage error (based on the scale reading) involved in using this scale.

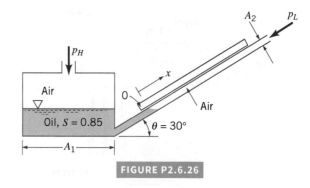

FIGURE P2.6.26

2.6.27 The cistern shown in **Fig. P2.6.27** has a diameter *D* that is four times the diameter *d* of the inclined tube. Find the drop in the fluid level in the cistern and the pressure difference ($p_A - p_B$), if the liquid in the inclined tube rises $\ell = 0.51$ m. The angle θ is 20°. The fluid's specific gravity is 0.85.

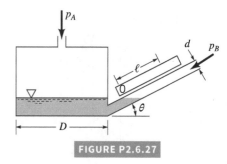

FIGURE P2.6.27

2.6.28 The inclined differential manometer of **Fig. P2.6.28** contains carbon tetrachloride. Initially, the pressure differential between pipes *A* and *B*, which contain a brine ($SG = 1.1$), is zero, as illustrated in the figure. It is desired that the manometer give, a differential reading of 0.3 m (measured along the inclined tube) for a pressure differential of 690 Pa. Determine the required angle of inclination, θ.

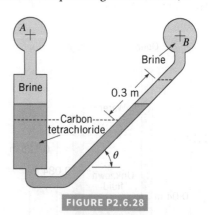

FIGURE P2.6.28

2.6.29 Determine the new differential reading along the inclined leg of the mercury manometer of **Fig. P2.6.29**, if the pressure in pipe *A* is decreased 10 kPa and the pressure in pipe *B* remains unchanged. The fluid in *A* has a specific gravity of 0.9, and the fluid in *B* is water.

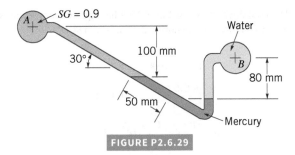

FIGURE P2.6.29

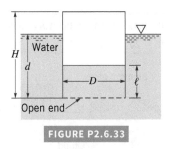

FIGURE P2.6.33

2.6.30 A student needs to measure the air pressure inside a compressed air tank but does not have ready access to a pressure gage. Using materials already in the lab, she builds a U-tube manometer using two clear 0.9 m long plastic tubes, flexible hoses, and a tape measure. The only readily available liquids are water from a tap and a bottle of corn syrup. She selects the corn syrup because it has a larger density ($SG = 1.4$). What is the maximum air pressure, in kPa, that can be measured?

2.6.31 Determine the ratio of areas, A_1/A_2, of the two manometer legs of **Fig. P2.6.31**, if a change in pressure in pipe B of 3.5 kPa gives a corresponding change of 25.4 mm. in the level of the mercury in the right leg. The pressure in pipe A does not change.

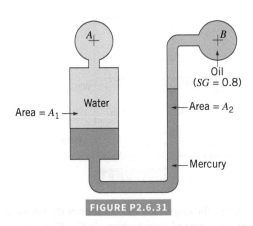

FIGURE P2.6.31

2.6.32 The U-shaped tube shown in **Fig. P2.6.32** initially contains water only. A second liquid with specific weight, γ, less than water is placed on top of the water with no mixing occurring. Can the height, h, of the second liquid be adjusted so that the left and right levels are at the same height? Provide proof of your answer.

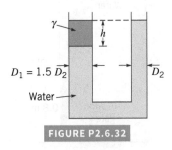

FIGURE P2.6.32

2.6.33 C An inverted hollow cylinder is pushed into the water, as is shown in **Fig. P2.6.33**. Determine the distance, ℓ, that the water rises in the cylinder as a function of the depth, d, of the lower edge of the cylinder. Plot the results for $0 \le d \le H$, when H is equal to 1 m. Assume the temperature of the air within the cylinder remains constant.

Section 2.8 Hydrostatic Force on a Plane Surface

2.8.1 Obtain a photograph/image of a situation in which the hydrostatic force on a plane surface is important. Print this photo and write a brief paragraph that describes the situation involved.

2.8.2 The basic elements of a hydraulic press are shown in **Fig. P2.8.2**. The plunger has an area of 0.00065 m², and a force, F_1, can be applied to the plunger through a lever mechanism having a mechanical advantage of 8 to 1. If the large piston has an area of 0.097 m², what load, F_2, can be raised by a force of 133.5 N applied to the lever? Neglect the hydrostatic pressure variation.

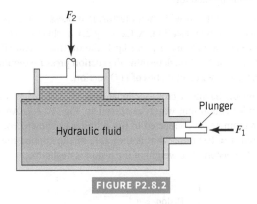

FIGURE P2.8.2

2.8.3 The hydraulic cylinder shown in **Fig. P2.8.3**, with a 0.1 m-diameter piston, is advertised as being capable of providing a force of $F = 196.2$ kN. If the piston has a design pressure (the maximum pressure at which the cylinder should safely operate) of 17.24 MPa gage, can the cylinder safely provide the advertised force?

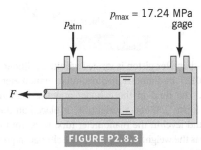

FIGURE P2.8.3

2.8.4 C Video A Bourdon gage (see Fig. 2.13 and Video V2.3) is often used to measure pressure. One way to calibrate this type of gage is to use the arrangement shown in **Fig. P2.8.4a**. The container is filled with a liquid and a weight, $\mathcal{W}$, placed on one side with the gage on the other side. The weight acting on the liquid through a 0.01 m-diameter opening creates a pressure that is transmitted to the gage. This arrangement, with a series of weights, can be used to determine what a change in the dial movement, θ, in **Fig. P2.8.4b**, corresponds to in terms of a change in pressure. Data for a particular gage are provided in the table. Based on a plot of these data, determine the relationship between θ and the pressure, p, where p is measured in kPa.

$\mathcal{W}$ (kg)	0	0.472	0.91	1.46	1.84	2.38	2.86
θ (deg.)	0	20	40	60	80	100	120

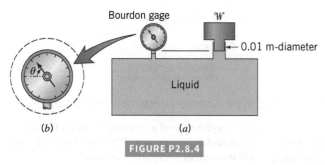

FIGURE P2.8.4

2.8.5 A bottle jack allows an average person to lift one corner of a 1814.4 kg automobile completely off the ground by exerting less than 9 kg of force. Explain how a 9 kg force can be converted into hundreds or thousands of kg of force, and why this does not violate our general perception that you can't get something for nothing (a somewhat loose paraphrase of the first law of thermodynamics). *Hint:* Consider the work done by each force.

2.8.6 Suction is often used in manufacturing processes to lift objects to be moved to a new location. A 1.2 m by 2.4 m sheet of 0.0127 m plywood weighs approximately 16.3 kg. If the machine's end effector has a diameter of 0.127 m, determine the suction pressure required to lift the sheet, expressed in inches of H_2O suction.

2.8.7 A piston having a cross-sectional area of 0.07 m² is located in a cylinder containing water, as shown in **Fig. P2.8.7**. An open U-tube manometer is connected to the cylinder as shown. For $h_1 = 60$ mm and $h = 100$ mm, what is the value of the applied force, P, acting on the piston? The weight of the piston is negligible.

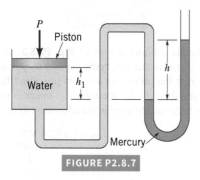

FIGURE P2.8.7

2.8.8 A 0.15 m-diameter piston is located within a cylinder that is connected to a 0.0127 m-diameter inclined-tube manometer, as shown in **Fig. P2.8.8**. The fluid in the cylinder and the manometer is oil (density = 945 kg/m³). When a weight, $\mathcal{W}$, is placed on the top of the cylinder, the fluid level in the manometer tube rises from point (1) to (2). How heavy is the weight? Assume that the change in position of the piston is negligible.

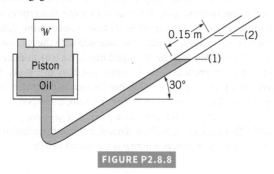

FIGURE P2.8.8

2.8.9 The container shown in **Fig. P2.8.9** has square cross sections. Find the vertical force on the horizontal surface, $ABCD$.

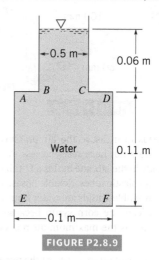

FIGURE P2.8.9

2.8.10 Find the weight $\mathcal{W}$ needed to hold the wall shown in **Fig. P2.8.10** upright. The wall is 10-m wide.

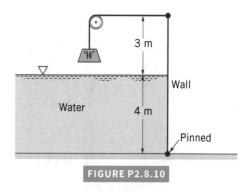

FIGURE P2.8.10

2.8.11 Determine the magnitude and direction of the force that must be applied to the bottom of the gate shown in **Fig. P2.8.11** to keep the gate closed.

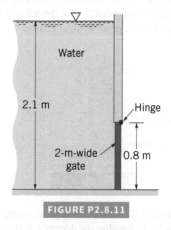

FIGURE P2.8.11

2.8.12 An automobile has just dropped into a river. The car door is approximately a rectangle, measures 0.91 m wide and 1.02 m high, and hinges on a vertical side. The water level inside the car is up to the midheight of the door, and the air inside the car is at atmospheric pressure. Calculate the force required to open the door if the force is applied 0.6 m from the hinge line. See **Fig. P2.8.12**. (The driver did not have the presence of mind to open the window to escape.)

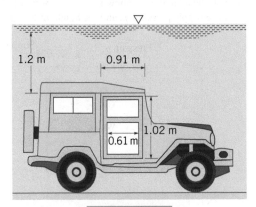

FIGURE P2.8.12

2.8.13 Consider the gate shown in **Fig. P2.8.13**. The gate is massless and has a width b (perpendicular to the paper). The hydrostatic pressure on the vertical side creates a counterclockwise moment about the hinge, and the hydrostatic pressure on the horizontal side (or bottom) creates a clockwise moment about the hinge. Show that the net clockwise moment is

$$\Sigma u = \rho_w g h b \left(\frac{\ell^2}{2} - \frac{h^2}{6} \right)$$

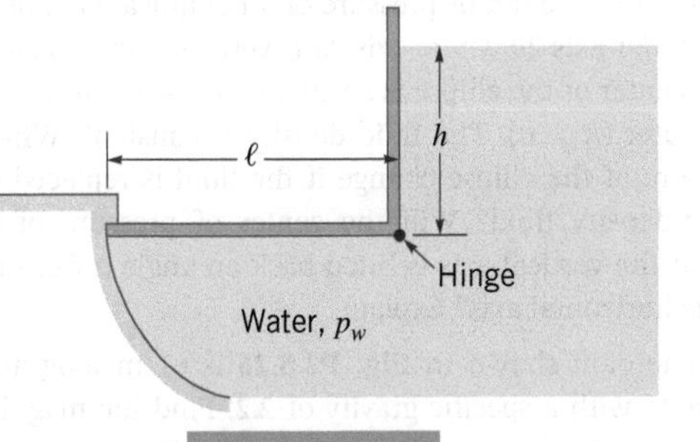

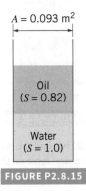

FIGURE P2.8.13

2.8.14 Will the gate in Problem 2.8.13 ever open?

2.8.15 A tank contains 0.15 m of oil ($S = 0.82$) above 0.15 m of water ($S = 1.00$). Find the force on the bottom of the tank. See **Fig. P2.8.15**.

$A = 0.093 \text{ m}^2$

Oil
($S = 0.82$)

Water
($S = 1.0$)

FIGURE P2.8.15

2.8.16 A structure is attached to the ocean floor as shown in **Fig. P2.8.16**. A 2-m-diameter hatch is located in an inclined wall and hinged on one edge. Determine the minimum air pressure, p_1, within the container that will open the hatch. Neglect the weight of the hatch and friction in the hinge.

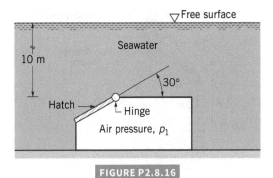

FIGURE P2.8.16

2.8.17 Concrete is poured into the forms, as shown in **Fig. P2.8.17**, to produce a set of steps. Determine the weight of the sandbag needed to keep the bottomless forms from lifting off the ground. The weight of the forms is 38.5 kg, and the density of the concrete is 2402.7 kg/m^3.

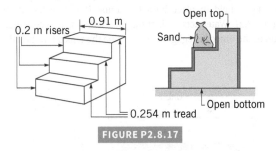

FIGURE P2.8.17

2.8.18 A long, vertical wall separates seawater from fresh water. If the seawater stands at a depth of 7 m, what depth of freshwater is required to give a zero resultant force on the wall? When the resultant force is zero, will the moment due to the fluid forces be zero? Explain.

2.8.19 Forms used to make a concrete basement wall are shown in **Fig. P2.8.19**. Each 1.2 m long form is held together by four ties—two at the top and two at the bottom as indicated. Determine the tension in the upper and lower ties. Assume concrete acts as a fluid with a weight of 2402.7 kg/m^3.

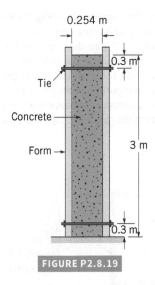

FIGURE P2.8.19

2.8.20 While building a high, tapered concrete wall, builders used the wooden forms shown in **Fig. P2.8.20**. If concrete has a specific gravity of about 2.5, find the total force on each of the three side sections (A, B, and C) of the wooden forms (neglect any restraining force of the two ends of the forms).

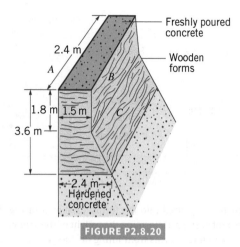

FIGURE P2.8.20

2.8.21 A homogeneous, 1.2 m wide, 2.4 m long rectangular gate weighing 363 kg is held in place by a horizontal flexible cable as shown in **Fig. P2.8.21**. Water acts against the gate, which is hinged at point *A*. Friction in the hinge is negligible. Determine the tension in the cable.

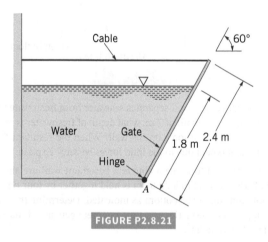

FIGURE P2.8.21

2.8.22 A gate having the shape shown in **Fig. P2.8.22** is located in the vertical side of an open tank containing water. The gate is mounted on a horizontal shaft. **(a)** When the water level is at the top of the gate, determine the magnitude of the fluid force on the rectangular portion of the gate above the shaft and the magnitude of the fluid force on the semicircular portion of the gate below the shaft. **(b)** For this same fluid depth determine the moment of the force acting on the semicircular portion of the gate with respect to an axis that coincides with the shaft.

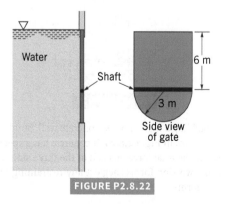

FIGURE P2.8.22

2.8.23 A pump supplies water under pressure to a large tank, as shown in **Fig. P2.8.23**. The circular-plate valve fitted in the short discharge pipe on the tank pivots about its diameter *A–A* and is held shut against the water pressure by a latch at *B*. Show that the force on the latch is independent of the supply pressure, *p*, and the height of the tank, *h*.

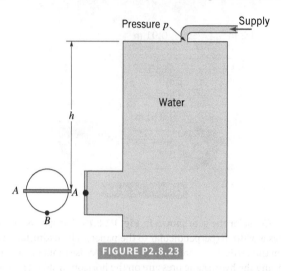

FIGURE P2.8.23

2.8.24 ⓞ Sometimes it is difficult to open an exterior door of a building because the air distribution system maintains a pressure difference between the inside and outside of the building. Estimate how big this pressure difference can be if it is "not too difficult" for an average person to open the door.

2.8.25 Find the center of pressure of an elliptical area of minor axis 2*a* and major axis 2*b* where axis 2*a* is vertical and axis 2*b* is horizontal. The center of the ellipse is a vertical distance *h* below the surface of the water ($h > a$). The fluid density is constant. Will the center of pressure of the ellipse change if the fluid is replaced by another constant-density fluid? Will the center of pressure of the ellipse change if the vertical axis is tilted back an angle α from the vertical about its horizontal axis? Explain.

2.8.26 The dam shown in **Fig. P2.8.26** is 61 m long and is made of concrete with a specific gravity of 2.2. Find the magnitude and *y* coordinate of the line of action of the net horizontal force.

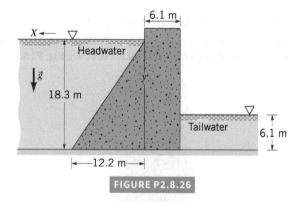

FIGURE P2.8.26

2.8.27 Repeat Problem 2.8.26 but find the magnitude and *x* coordinate of the line of action of the vertical force on the dam resulting from the water.

2.8.28 **Figure P2.8.28** is a representation of the Keswick gravity dam in California. Find the magnitudes and locations of the hydrostatic forces acting on the headwater vertical wall of the dam and on the tailwater inclined wall of the dam. Note that the slope given is the ratio of the run to the rise. Consider a unit length of the dam ($b = 0.3$ m).

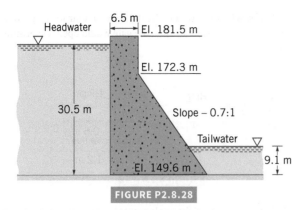

FIGURE P2.8.28

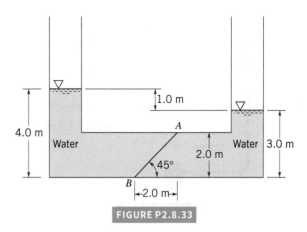

FIGURE P2.8.33

2.8.29 The Keswick dam in Problem 2.8.28 is made of concrete and has a density of 2403 kg/m³. The hydrostatic forces and the weight of the dam produce a total vertical force of the dam on the foundation. Find the magnitude and location of this total vertical force. Consider a unit length of the dam ($b = 0.3$ m).

2.8.30 The Keswick dam in Problem 2.8.28 is made of concrete and has a density of 2403 kg/m³. The coefficient of friction μ between the base of the dam and the foundation is 0.65. Is the dam likely to slide downstream? Consider a unit length of the dam ($b = 0.3$ m).

2.8.31 **Figure P2.8.31** is a representation of the Altus gravity dam in Oklahoma. Find the magnitudes and locations of the horizontal and vertical hydrostatic force components acting on the headwater wall of the dam and on the tailwater wall of the dam. Note that the slope given is the ratio of the run to the rise. Consider a unit length of the dam ($b = 0.3$ m).

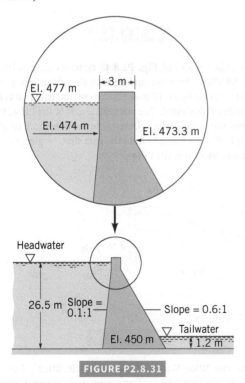

FIGURE P2.8.31

2.8.32 The Altus dam in Problem 2.8.31 is made of concrete with a density of 2403 kg/m³. The coefficient of friction μ between the base of the dam and the foundation is 0.65. Is the dam likely to slide downstream? Consider a unit length of the dam ($b = 0.3$ m).

2.8.33 Find the magnitude and location of the net horizontal force on the gate shown in **Fig. P2.8.33**. The gate width is 5.0 m.

2.8.34 Ⓒ Find the magnitude and location of the net vertical force on the gate in Problem 2.8.33.

2.8.35 A 3-m-wide, 8-m-high rectangular gate is located at the end of a rectangular passage that is connected to a large open tank filled with water, as shown in **Fig. P2.8.35**. The gate is hinged at its bottom and held closed by a horizontal force, F_H, located at the center of the gate. The maximum value for F_H is 3500 kN. (a) Determine the maximum water depth, h, above the center of the gate that can exist without the gate opening. (b) Is the answer the same if the gate is hinged at the top? Explain your answer.

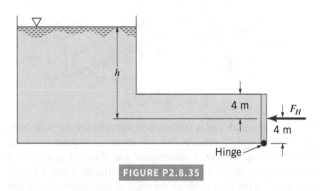

FIGURE P2.8.35

2.8.36 A gate having the cross section shown in **Fig. P2.8.36** is 1.2 m wide and is hinged at C. The gate weighs 8165 kg, and its mass center is 0.5 m to the right of the plane BC. Determine the vertical reaction at A on the gate when the water level is 0.9 m above the base. All contact surfaces are smooth.

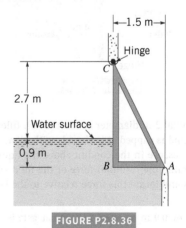

FIGURE P2.8.36

2.8.37 The massless, 1.2 m wide gate shown in **Fig. P2.8.37** pivots about the frictionless hinge O. It is held in place by the 907 kg counterweight, $\mathcal{W}$. Determine the water depth, h.

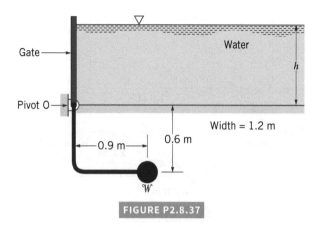

FIGURE P2.8.37

2.8.38 Ⓒ A 90 kg homogeneous gate 3 m wide and 1.5 m long is hinged at point A and held in place by a 3.6 m long brace, as shown in **Fig. P2.8.38**. As the bottom of the brace is moved to the right, the water level remains at the top of the gate. The line of action of the force that the brace exerts on the gate is along the brace. **(a)** Plot the magnitude of the force exerted on the gate by the brace as a function of the angle of the gate, θ, for $0 \leq \theta \leq 90°$. **(b)** Repeat the calculations for the case in which the weight of the gate is negligible. Comment on the result as $\theta \to 0$.

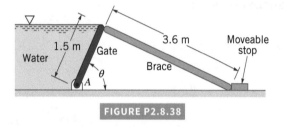

FIGURE P2.8.38

2.8.39 An open tank has a vertical partition and on one side contains gasoline with a density $\rho = 700$ kg/m³ at a depth of 4 m, as shown in **Fig. P2.8.39**. A rectangular gate that is 4 m high and 2 m wide and hinged at one end is located in the partition. Water is slowly added to the empty side of the tank. At what depth, h, will the gate start to open?

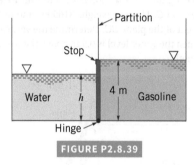

FIGURE P2.8.39

2.8.40 A horizontal 2-m-diameter conduit is half filled with a liquid ($SG = 1.6$) and is capped at both ends with plane vertical surfaces. The air pressure in the conduit above the liquid surface is 200 kPa. Determine the resultant force of the fluid acting on one of the end caps and locate this force relative to the bottom of the conduit.

2.8.41 A 1.2 m by 0.9 m massless rectangular gate is used to close the end of the water tank shown in **Fig. P2.8.41**. A 90 kg weight attached to the arm of the gate at a distance ℓ from the frictionless hinge is just sufficient to keep the gate closed when the water depth

is 0.6 m, that is, when the water fills the semicircular lower portion of the tank. If the water were deeper, the gate would open. Determine the distance ℓ.

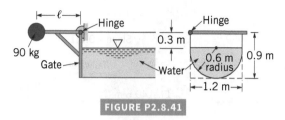

FIGURE P2.8.41

2.8.42 A thin 1.2 m wide, right-angle gate with negligible mass is free to pivot about a frictionless hinge at point O, as shown in **Fig. P2.8.42**. The horizontal portion of the gate covers a 0.3 m-diameter drain pipe that contains air at atmospheric pressure. Determine the minimum water depth, h, at which the gate will pivot to allow water to flow into the pipe.

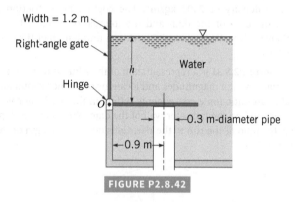

FIGURE P2.8.42

2.8.43 The closed vessel of **Fig. P2.8.43** contains water with an air pressure of 69 kPa at the water surface. One side of the vessel contains a spout that is closed by a 0.15 m-diameter circular gate that is hinged along one side as illustrated. The horizontal axis of the hinge is located 3 m below the water surface. Determine the minimum torque that must be applied at the hinge to hold the gate shut. Neglect the weight of the gate and friction at the hinge.

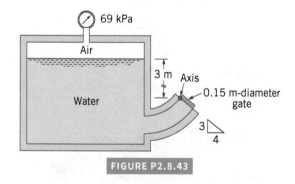

FIGURE P2.8.43

2.8.44 (See The Wide World of Fluids article titled "The Three Gorges Dam," Section 2.8.) **(a)** Determine the horizontal hydrostatic force on the 2309-m-long Three Gorges Dam when the average depth of the water against it is 175 m. **(b)** If all of the 6.4 billion people on Earth were to push horizontally against the Three Gorges Dam, could they generate enough force to hold it in place? Support your answer with appropriate calculations.

Section 2.10 Hydrostatic Force on a Curved Surface

2.10.1 Obtain a photograph/image of a situation in which the hydrostatic force on a curved surface is important. Print this photo and write a brief paragraph that describes the situation involved.

2.10.2 A 0.6 m-diameter hemispherical plexiglass "bubble" is to be used as a special window on the side of an above-ground swimming pool. The window is to be bolted onto the vertical wall of the pool and faces outward, covering a 0.6 m-diameter opening in the wall. The center of the opening is 1.2 m below the surface. Determine the horizontal and vertical components of the force of the water on the hemisphere.

2.10.3 Consider the curved surface shown in **Fig. P2.10.3a** and **b**. The two curved surfaces are identical. How are the vertical forces on the two surfaces alike? How are they different?

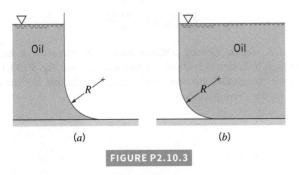

(a)   (b)

FIGURE P2.10.3

2.10.4 **Figure P2.10.4** shows a cross section of a submerged tunnel used by automobiles to travel under a river. Find the magnitude and location of the resultant hydrostatic force on the circular roof of the tunnel. The tunnel is 6440 m long.

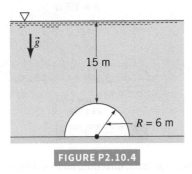

FIGURE P2.10.4

2.10.5 The container shown in **Fig. P2.10.5** has circular cross sections. Find the vertical force on the inclined surface. Also find the net vertical force on the bottom, *EF*. Is the vertical force equal to the weight of the water in the container?

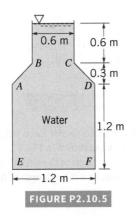

FIGURE P2.10.5

2.10.6 The 5.5 m long lightweight gate of **Fig. P2.10.6** is a quarter circle and is hinged at *H*. Determine the horizontal force, *P*, required to hold the gate in place. Neglect friction at the hinge and the weight of the gate.

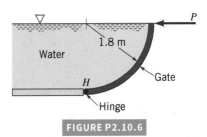

FIGURE P2.10.6

2.10.7 (Video) The air pressure in the top of the 2-liter pop bottle shown in **Video V2.5** and **Fig. P2.10.7** is 275.8 kPa, and the pop depth is 0.25 m. The bottom of the bottle has an irregular shape with a diameter of 0.11 m. (a) If the bottle cap has a diameter of 0.025 m., what is the magnitude of the axial force required to hold the cap in place? (b) Determine the force needed to secure the bottom 0.05 m of the bottle to its cylindrical sides. For this calculation assume the effect of the weight of the pop is negligible. (c) By how much does the weight of the pop increase the pressure 0.05 m above the bottom? Assume the pop has the same specific weight as that of water.

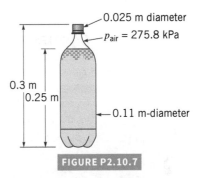

FIGURE P2.10.7

2.10.8 In drilling for oil in the Gulf of Mexico, some divers have to work at a depth of 396 m. (a) Assume that seawater has a constant density of 1025 kg/m^3 and compute the pressure at this depth. The divers breathe a mixture of helium and oxygen stored in cylinders, as shown in **Fig. P2.10.8**, at a pressure of 20.6 MPa. (b) Calculate the force, which trends to blow the end cap off, that the weld must resist while the diver is using the cylinder at 396 m. (c) After emptying a tank, a diver releases it. Will the tank rise or fall, and what is its initial acceleration?

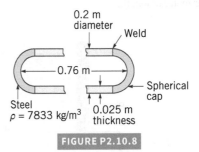

FIGURE P2.10.8

2.10.9 (Video) Hoover Dam (see **Video V2.4**) is the highest arch-gravity type of dam in the United States. A cross section of the dam is shown in **Fig. P2.10.9(a)**. The walls of the canyon in which the dam is located are sloped, and just upstream of the dam the vertical plane shown in **Fig. P2.10.9(b)** approximately represents the cross section of the water acting on the dam. Use this vertical cross section to estimate the resultant horizontal force of the water on the dam and show where this force acts.

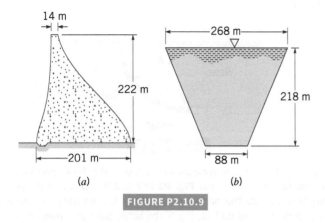

(a)   (b)

FIGURE P2.10.9

2.10.10 A plug in the bottom of a pressurized tank is conical in shape, as shown in **Fig. P2.10.10**. The air pressure is 40 kPa, and the liquid in the tank has a specific weight of 27 kN/m³. Determine the magnitude, direction, and line of action of the force exerted on the curved surface of the cone within the tank due to the 40-kPa pressure and the liquid.

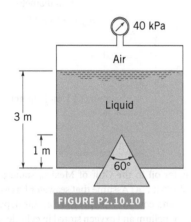

FIGURE P2.10.10

2.10.11 The homogeneous gate shown in **Fig. P2.10.11** consists of one quarter of a circular cylinder and is used to maintain a water depth of 4 m. That is, when the water depth exceeds 4 m, the gate opens slightly and lets the water flow under it. Determine the weight of the gate per meter of length.

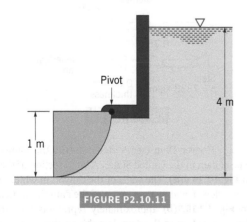

FIGURE P2.10.11

2.10.12 The concrete (density = 2403 kg/m³) seawall of **Fig. P2.10.12** has a curved surface and restrains seawater at a depth of 7.5 m. The trace of the surface is a parabola as illustrated. Determine the moment of the fluid force (per unit length) with respect to an axis through the toe (point A).

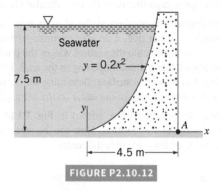

FIGURE P2.10.12

2.10.13 A step-in viewing window having the shape of a half-cylinder is built into the side of a large aquarium. See **Fig. P2.10.13**. Find the magnitude, direction, and location of the net horizontal forces on the viewing window.

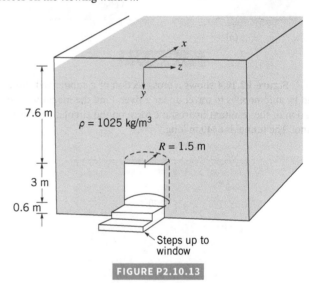

FIGURE P2.10.13

2.10.14 Find the magnitude, direction, and location of the net vertical force acting on the viewing window in Problem 2.10.13.

2.10.15 A 10-m-long log is stuck against a dam, as shown in **Fig. P2.10.15**. Find the magnitudes and locations of both the horizontal force and the vertical force of the water on the log in terms of the diameter D. The center of the log is at the same elevation as the top of the dam.

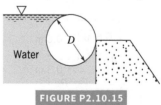

FIGURE P2.10.15

2.10.16 An open tank containing water has a bulge in its vertical side that is semicircular in shape, as shown in **Fig. P2.10.16**. Determine the horizontal and vertical components of the force that the water exerts on the bulge. Base your analysis on a 0.3 m length of the bulge.

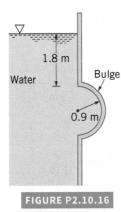

FIGURE P2.10.16

2.10.17 A closed tank is filled with water and has a 1.2 m-diameter hemispherical dome, as shown in **Fig. P2.10.17**. A U-tube manometer is connected to the tank. Determine the vertical force of the water on the dome if the differential manometer reading is 2.1 m and the air pressure at the upper end of the manometer is 86.8 kPa.

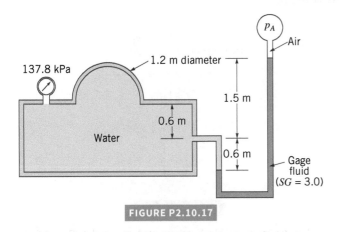

FIGURE P2.10.17

2.10.18 A 3-m-diameter open cylindrical tank contains water and has a hemispherical bottom, as shown in **Fig. P2.10.18**. Determine the magnitude, line of action, and direction of the force of the water on the curved bottom.

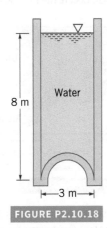

FIGURE P2.10.18

2.10.19 Three gates of negligible weight are used to hold back water in a channel of width b, as shown in **Fig. P2.10.19**. The force of the gate against the block for gate (b) is R. Determine (in terms of R) the force against the blocks for the other two gates.

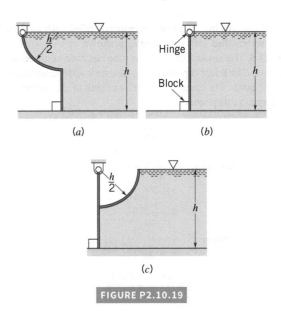

FIGURE P2.10.19

Section 2.11 Buoyancy, Flotation, and Stability

2.11.1 Obtain a photograph/image of a situation in which Archimedes's principle is important. Print this photo and write a brief paragraph that describes the situation involved.

2.11.2 An iceberg (specific gravity 0.917) floats in the ocean (specific gravity 1.025). What percent of the volume of the iceberg is under water?

2.11.3 A floating 1 m thick piece of ice sinks 0.025 m with a 227 kg polar bear in the center of the ice. What is the area of the ice in the plane of the water level? For seawater, $S = 1.03$.

2.11.4 A spherical balloon filled with helium at 4.44 °C and 137.9 kPa has a 7.6 m-diameter. What load can it support in atmospheric air at 4.44 °C and 101.3 kPa? Neglect the balloon's weight.

2.11.5 A river barge, whose cross section is approximately rectangular, carries a load of grain. The barge is 8.5 m wide and 27.5 m long. When unloaded, its draft (depth of submergence) is 1.5 m, and with the load of grain, the draft is 2.1 m. Determine: (a) the unloaded weight of the barge and (b) the weight of the grain.

2.11.6 A barge is 12 m wide by 36.5 m long. The weight of the barge and its cargo is denoted by W. When in salt-free riverwater, it floats 0.076 m deeper than when in seawater ($\gamma = 1025$ kg/m^3). Find the weight W.

2.11.7 When the Tucurui Dam was constructed in northern Brazil, the lake that was created covered a large forest of valuable hardwood trees. It was found that even after 15 years underwater the trees were perfectly preserved and underwater logging was started. During the logging process, a tree is selected, trimmed, and anchored with ropes to prevent it from shooting to the surface like a missile when cut. Assume that a typical large tree can be approximated as a truncated cone with a base diameter of 2.5 m, a top diameter of 0.6 m, and a height of 30 m. Determine the resultant vertical force that the ropes must resist when the completely submerged tree is cut. The specific gravity of the wood is approximately 0.6.

2.11.8 Ⓞ Estimate the minimum water depth needed to float a canoe carrying two people and their camping gear. List all assumptions and show all calculations.

2.11.9 (Video) An inverted test tube partially filled with air floats in a plastic water-filled soft drink bottle, as shown in Video V2.6 and **Fig. P2.11.9**. The amount of air in the tube has been adjusted so that it just floats. The bottle cap is securely fastened. A slight squeezing of the plastic bottle will cause the test tube to sink to the bottom of the bottle. Explain this phenomenon.

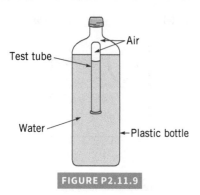

FIGURE P2.11.9

2.11.10 A child's balloon is a sphere 0.3 m in diameter. The balloon is filled with helium ($\rho = 0.224$ kg/m^3). The balloon material weighs 0.04 kg/m^2 of surface area. If the child releases the balloon, how high will it rise in the standard atmosphere? (Neglect expansion of the balloon as it rises.)

2.11.11 A 0.3 m-diameter, 0.6 m long cylinder floats in an open tank containing a liquid having a specific weight γ. A U-tube manometer is connected to the tank, as shown in **Fig. P2.11.11**. When the pressure in pipe A is 690 Pa below atmospheric pressure, the various fluid levels are as shown. Determine the weight of the cylinder. Note that the top of the cylinder is flush with the fluid surface.

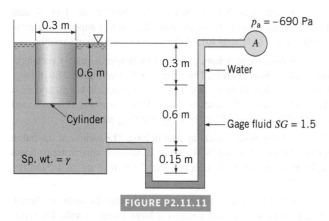

FIGURE P2.11.11

2.11.12 (C) A not-too-honest citizen is thinking of making bogus gold bars by first making a hollow iridium ($S = 22.5$) ingot and plating it with a thin layer of gold ($S = 19.3$) of negligible weight and volume. The bogus bar is to have a mass of 45 kg. What must be the volumes of the bogus bar and of the air space inside the iridium so that an inspector would conclude it was real gold after weighing it in air and water to determine its density? Could lead ($S = 11.35$) or platinum ($S = 21.45$) be used instead of iridium? Would either be a good idea?

2.11.13 A solid cylindrical pine ($S = 0.50$) spar buoy has a cylindrical lead ($S = 11.3$) weight attached, as shown in **Fig. P2.11.13**. Determine the equilibrium position of the spar buoy in seawater (i.e., find d). Is this spar buoy stable or unstable? For seawater, $S = 1.03$.

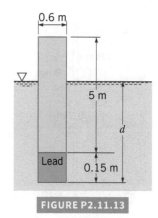

FIGURE P2.11.13

2.11.14 (Video) When a hydrometer (see **Fig. P2.11.14** and Video V2.7) having a stem diameter of 7.6 mm is placed in water, the stem protrudes 80 mm above the water surface. If the water is replaced with a liquid having a specific gravity of 1.10, how much of the stem would protrude above the liquid surface? The hydrometer weighs 0.2 N.

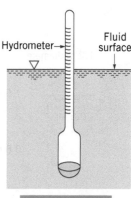

FIGURE P2.11.14

2.11.15 A 0.6 m thick block constructed of wood ($SG = 0.6$) is submerged in oil ($SG = 0.8$) and has a 0.6 m thick aluminum (density = 2691 kg/m^3) plate attached to the bottom as indicated in **Fig. P2.11.15**. Determine completely the force required to hold the block in the position shown. Locate the force with respect to point A.

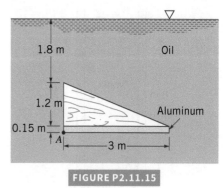

FIGURE P2.11.15

2.11.16 (See The Wide World of Fluids article titled "Concrete canoes," Section 2.11.1.) How much extra water does a 654 N concrete canoe displace compared to an ultralightweight 169 N Kevlar canoe of the same size carrying the same load?

2.11.17 (C) A submarine is modeled as a cylinder with a length of 91.5 m, a diameter of 15.25 m, and a conning tower, as shown in **Fig. P2.11.17**.

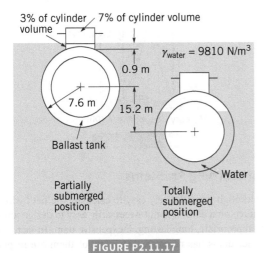

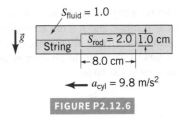

FIGURE P2.12.6

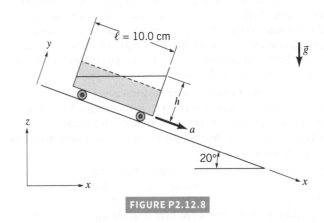

FIGURE P2.11.17

The submarine can dive a distance of 15.2 m from the floating position in about 30 s. Diving is accomplished by taking water into the ballast tank so the submarine will sink. When the submarine reaches the desired depth, some of the water in the ballast tank is discharged leaving the submarine in "neutral buoyancy" (i.e., it will neither rise nor sink). For the conditions illustrated, find (a) the weight of the submarine and (b) the volume (or mass) of the water that must be in the ballast tank when the submarine is in neutral buoyancy. For seawater, S = 1.03.

Section 2.12 Pressure Variation in a Fluid with Rigid-Body Motion

2.12.1 Obtain a photograph/image of a situation in which the pressure variation in a fluid with rigid-body motion is involved. Print this photo and write a brief paragraph that describes the situation involved.

2.12.2 When an automobile brakes, the fuel gage indicates a fuller tank than when the automobile is traveling at a constant speed on a level road. Is the sensor for the fuel gage located near the front or rear of the fuel tank? Assume a constant deceleration.

2.12.3 An open container of oil rests on the flatbed of a truck that is traveling along a horizontal road at 24.58 m/s. As the truck slows uniformly to a complete stop in 5 s, what will be the slope of the oil surface during the period of constant deceleration?

2.12.4 A 19 L, cylindrical open container with a bottom area of 0.0775 m^2 is filled with glycerin and rests on the floor of an elevator. (a) Determine the fluid pressure at the bottom of the container when the elevator has an upward acceleration of 0.9 m/s^2. (b) What resultant force does the container exert on the floor of the elevator during this acceleration? The weight of the container is negligible.

2.12.5 A plastic glass has a square cross section measuring 0.06 m. on a side and is filled to within 0.01 m. of the top with water. The glass is placed in a level spot in a car with two opposite sides parallel to the direction of travel. How fast can the driver of the car accelerate along a level road without spilling any of the water?

2.12.6 The cylinder in **Fig. P2.12.6** accelerates to the left at the rate of 9.80 m/s². Find the tension in the string connecting the rod of circular cross section to the cylinder. The volume between the rod and the cylinder is completely filled with water at 10 °C.

2.12.7 A closed cylindrical tank that is 2.4 m in diameter and 7.5 m long is completely filled with gasoline. The tank, with its long axis horizontal, is pulled by a truck along a horizontal surface. Determine the pressure difference between the ends (along the long axis of the tank) when the truck undergoes an acceleration of 1.5 m/s².

2.12.8 The cart shown in **Fig. P2.12.8** measures 10.0 cm long and 6.0 cm high and has rectangular cross sections. It is half filled with water and accelerates down a 20° incline plane at $a = 1.0 \text{ m/s}^2$. Find the height h.

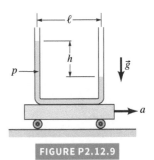

FIGURE P2.12.8

2.12.9 The U-tube manometer in **Fig. P2.12.9** is used to measure the acceleration of the cart on which it sits. Develop an expression for the acceleration of the cart in terms of the liquid height h, the liquid density ρ, the local acceleration of gravity g, and the length ℓ.

FIGURE P2.12.9

2.12.10 Ⓒ A tank has a height of 5.0 cm and a square cross section measuring 5.0 cm on a side. The tank is one-third full of water and is rotated in a horizontal plane with the bottom of the tank 100 cm from the center of rotation and two opposite sides parallel to the ground. What is the maximum rotational speed that the tank of water can be rotated with no water coming out of the tank?

2.12.11 An open 1-m-diameter tank contains water at a depth of 0.7 m when at rest. As the tank is rotated about its vertical axis the center of the fluid surface is depressed. At what angular velocity will the bottom of the tank first be exposed? No water is spilled from the tank.

2.12.12 ⟦C⟧ The U-tube in **Fig. P2.12.12** rotates at 2.0 rev/s. Find the absolute pressures at points C and B if the atmospheric pressure is 101.3 kPa (abs). Recall that 21.1 °C water evaporates at an absolute pressure of 2.5 kPa (abs). Determine the absolute pressures at points C and B if the U-tube rotates at 2.0 rev/s.

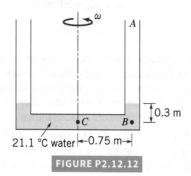

FIGURE P2.12.12

2.12.13 A child riding in a car holds a string attached to a floating, helium-filled balloon. As the car decelerates to a stop, the balloon tilts backward. As the car makes a right-hand turn, the balloon tilts to the right. On the other hand, the child tends to be forced forward as the car decelerates and to the left as the car makes a right-hand turn. Explain these observed effects on the balloon and child.

2.12.14 A closed, 0.4-m-diameter cylindrical tank is completely filled with oil ($SG = 0.9$) and rotates about its vertical longitudinal axis with an angular velocity of 40 rad/s. Determine the difference in pressure just under the vessel cover between a point on the circumference and a point on the axis.

2.12.15 (See The Wide World of Fluids article titled "Rotating mercury mirror telescope," Section 2.12.2.) The largest liquid mirror telescope uses a 1.8 m-diameter tank of mercury rotating at 7 rpm to produce its parabolic-shaped mirror, as shown in **Fig. P2.12.15**. Determine the difference in elevation of the mercury, Δh, between the edge and the center of the mirror.

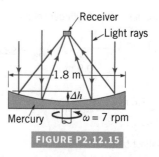

FIGURE P2.12.15

Lifelong Learning Problems

2.1 LL Although it is relatively easy to calculate the net hydrostatic pressure force on a dam, it is not necessarily easy to design and construct an appropriate, long-lasting, inexpensive dam. In fact, inspection of older dams has revealed that many of them are in peril of collapse unless corrective action is soon taken. Obtain information about the severity of the poor conditions of older dams throughout the country. Summarize your findings in a brief report.

2.2 LL Over the years the demand for high-quality, first-growth timber has increased dramatically. Unfortunately, most of the trees that supply such lumber have already been harvested. Recently, however, several companies have started to reclaim the numerous high-quality logs that sank in lakes and oceans during the logging boom times many years ago. Many of these logs are still in excellent condition. Obtain information about the use of fluid mechanics concepts in harvesting sunken logs. Summarize your findings in a brief report.

2.3 LL Liquid-filled manometers and Bourdon tube pressure gages have been the mainstay for measuring pressure for many, many years. However, for many modern applications, these tried-and-true devices are not sufficient. For example, various new uses need small, accurate, inexpensive pressure transducers with digital outputs. Obtain information about some of the new concepts used for pressure measurement. Summarize your findings in a brief report.

CHAPTER 3

Fluid Kinematics

LEARNING OBJECTIVES

After completing this chapter, you should be able to:

- discuss the differences between the Eulerian and Lagrangian descriptions of fluid motion.
- identify various flow characteristics based on the velocity field.
- determine the streamline pattern and acceleration field given a velocity field.
- discuss the differences between a system and a control volume.
- apply the Reynolds transport theorem and the material derivative.

In this chapter we will discuss various aspects of fluid motion without being concerned with the forces necessary to produce the motion. That is, we will consider the *kinematics* of the motion—the velocity and acceleration of the fluid, and the description and visualization of its motion. The analysis of the specific forces necessary to produce the motion (the *dynamics* of the motion) will be discussed in detail in the following chapters. A wide variety of useful information can be gained from a thorough understanding of fluid kinematics. Such an understanding of how to describe and observe fluid motion is an essential step understanding of fluid dynamics.

V3.1 Streaklines

3.1 THE VELOCITY FIELD

Fluids flow. That is, there is a net motion of molecules from one point in space to another as a function of time. As is discussed in Chapter 1, a "particle" of fluid contains so many molecules that it becomes unrealistic (except in special cases) for us to attempt to account for the motion of individual molecules. Rather, we employ the continuum hypothesis and consider fluids to be made up of fluid particles that interact with each other and with their surroundings. Although it is very tiny, each particle contains numerous molecules. We describe the flow of a fluid in terms of the motion of fluid particles rather than individual molecules. This motion can be described in terms of the velocity and acceleration of the fluid particles.

The infinitesimal particles of a fluid are tightly packed together (as is implied by the continuum assumption). Thus, at a given instant, a description of any fluid property (such as density, pressure, velocity, and acceleration) may be specified as a function of the fluid's location. This representation of fluid parameters as functions of the spatial coordinates is termed a **field representation** of the flow. Of course, the specific field representation may be different at different times, so that to describe a fluid flow we must determine the various parameters not only as a function of the spatial coordinates (x, y, z, for example) but also as a function of time, t.

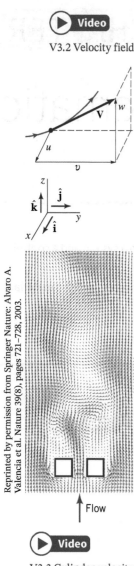

Video

V3.2 Velocity field

Video

V3.3 Cylinder-velocity vectors

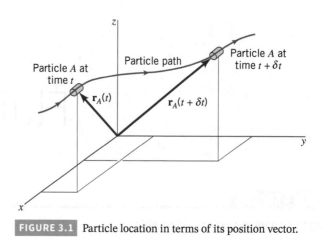

FIGURE 3.1 Particle location in terms of its position vector.

Thus, to completely specify the temperature, T, in a room we must specify the temperature field, $T = T(x, y, z, t)$, throughout the room (from floor to ceiling and wall to wall) at any time of the day or night.

Shown in the adjacent figure is one of the most important fluid variables, the **velocity vector,**

$$\mathbf{V} = u(x, y, z, t)\hat{\mathbf{i}} + v(x, y, z, t)\hat{\mathbf{j}} + w(x, y, z, t)\hat{\mathbf{k}}$$

where u, v, and w are the x, y, and z components of the velocity vector. By definition, the velocity of a particle is the time rate of change of the position vector for that particle. As is illustrated in **Fig. 3.1**, the position of particle A relative to the coordinate system is given by its *position vector*, $\mathbf{r}_A$, which (if the particle is moving) is a function of time. The time derivative of this position gives the *velocity* of the particle, $d\mathbf{r}_A/dt = \mathbf{V}_A$. By writing the velocity for all of the particles, we can obtain the field description of the velocity vector $\mathbf{V} = \mathbf{V}(x, y, z, t)$.

The adjacent figure shows the **velocity field** (i.e., velocity vectors) at a specific instant for flow past two square bars. It is possible to obtain much qualitative and quantitative information for complex flows by using plots such as this.

Because velocity is a vector, it has both a direction and a magnitude. The magnitude of $\mathbf{V}$, denoted $V = |\mathbf{V}| = (u^2 + v^2 + w^2)^{1/2}$, is the speed of the fluid. (It is very common in practical situations to call V velocity rather than speed, i.e., "the velocity of the fluid is 12 m/s.") Because velocity is a vector, there will be a nonzero acceleration if a particle moves at a constant speed but its direction changes as it moves through the flow.

The Wide World of Fluids

Follow those particles

Superimpose two photographs of a bouncing ball taken a short time apart and draw an arrow between the two images of the ball. This arrow represents an approximation of the velocity (displacement/time) of the ball. The particle image velocimeter (PIV) uses this technique to provide the instantaneous *velocity field* for a given cross section of a flow. The flow being studied is seeded with numerous micron-sized particles that are small enough to follow the flow yet big enough to reflect enough light

to be captured by the camera. The flow is illuminated with a light sheet from a double-pulsed laser. A digital camera captures both light pulses on the same image frame, allowing the movement of the particles to be tracked. Using appropriate computer software to carry out a pixel-by-pixel interrogation of the double image, it is possible to track the motion of the particles and determine the two components of velocity in the given cross section of the flow. Using two cameras in a stereoscopic arrangement, it is possible to determine all three components of velocity.

EXAMPLE 3.1 | Velocity Field Representation

Given A velocity field is given by $\mathbf{V} = (V_0/\ell)(-x\hat{\mathbf{i}} + y\hat{\mathbf{j}})$ where V_0 and ℓ are constants.

Find At what location in the flow field is the speed equal to V_0? Make a sketch of the velocity field for $x \geq 0$ by drawing arrows representing the fluid velocity at representative locations.

Solution The x, y, and z components of the velocity are given by $u = -V_0x/\ell, v = V_0y/\ell$, and $w = 0$ so that the fluid speed, V, is

$$V = (u^2 + v^2 + w^2)^{1/2} = \frac{V_0}{\ell}(x^2 + y^2)^{1/2} \qquad (1)$$

The speed is $V = V_0$ at any location on the circle of radius ℓ centered at the origin $[(x^2 + y^2)^{1/2} = \ell]$, as shown in **Fig. E3.1a.** (Ans)

The direction of the fluid velocity relative to the x axis is given in terms of $\theta = \arctan(v/u)$, as shown in **Fig. E3.1b.** For this flow

$$\tan \theta = \frac{v}{u} = \frac{V_0 y/\ell}{-V_0 x/\ell} = \frac{y}{-x}$$

Thus, along the x axis ($y = 0$) we see that $\tan \theta = 0$, so that $\theta = 0°$ or $\theta = 180°$. Similarly, along the y axis ($x = 0$) we obtain $\tan \theta = \pm\infty$ so that $\theta = 90°$ or $\theta = 270°$. Also, for $y = 0$ we find $\mathbf{V} = (-V_0 x/\ell)\hat{\mathbf{i}}$, while for $x = 0$ we have $\mathbf{V} = (V_0 y/\ell)\hat{\mathbf{j}}$, indicating (if $V_0 > 0$) that the flow is directed away from the origin along the y axis and toward the origin along the x axis, as shown in Fig. E3.1a.

By determining $\mathbf{V}$ and θ for other locations in the x–y plane, the velocity field can be sketched as shown in Figure E3.1a. For example, on the line $y = x$ the velocity is at a 45° angle relative to the x axis ($\tan \theta = v/u = -y/x = -1$). At the origin $x = y = 0$ so that $\mathbf{V} = 0$. This point is a stagnation point. The farther from the origin the fluid is, the faster it is flowing (as seen from Eq. 1). By careful consideration of the velocity field it is possible to determine considerable information about the flow.

Comment The velocity field given in this example approximates the flow in the vicinity of the center of the sign shown in **Fig. E3.1c.** When wind blows against the sign, some air flows over the sign, some under it, producing a stagnation point as indicated.

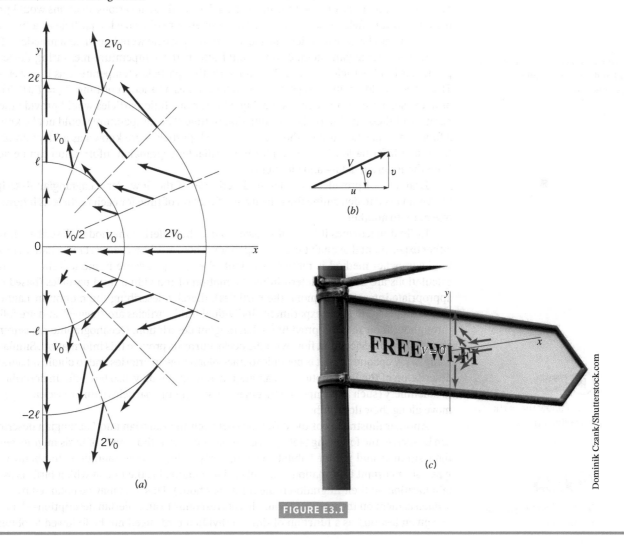

Dominik Czank/Shutterstock.com

FIGURE E3.1

3.1.1 Eulerian and Lagrangian Flow Descriptions

There are two general approaches for analyzing fluid mechanics problems (or problems in other branches of the physical sciences, for that matter). The first method, called the **Eulerian method,** uses the field concept introduced above. In this case, the fluid motion is specified by completely prescribing the necessary properties (pressure, density, velocity, etc.) as functions of space and time. From this method we obtain information about the flow in terms of what happens at fixed points in space as the fluid flows through those points.

A typical Eulerian representation of the flow is shown by the adjacent figure, which involves flow past an airfoil at angle of attack. The pressure field is indicated by using a

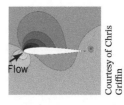

Flow

Courtesy of Chris Griffin

contour plot showing lines of constant pressure, with gray shading indicating the intensity of the pressure.

The second method, called the **Lagrangian method,** involves following individual fluid particles as they move about and determining how the fluid properties associated with these particles change as a function of time. That is, the fluid particles are "tagged" or identified, and their properties determined as they move.

The difference between the two methods of analyzing fluid flow problems can be seen in the example of smoke discharging from a chimney, as is shown in **Fig. 3.2.** In the Eulerian method, one may attach a temperature-measuring device to the top of the chimney (point 0) and record the temperature at that point as a function of time. At different times there are different fluid particles passing by the stationary device. Thus, one would obtain the temperature, T, for that location ($x = x_0$, $y = y_0$, and $z = z_0$) as a function of time. That is, $T = T(x_0, y_0, z_0, t)$. The use of numerous temperature-measuring devices fixed at various locations would provide the temperature field, $T = T(x, y, z, t)$. The temperature of a specific particle as a function of time would not be known unless the location of the particle were known as a function of time.

In the Lagrangian method, one would attach the temperature-measuring device to a particular fluid particle (particle A) and record that particle's temperature as it moves about. Thus, one would obtain that particle's temperature as a function of time, $T_A = T_A(t)$. The use of many such measuring devices moving with various fluid particles would provide the temperature of these fluid particles as a function of time. The temperature would not be known as a function of position unless the location of each particle were known as a function of time. If enough information in Eulerian form is available, Lagrangian information can be derived from the Eulerian data—and vice versa.

Example 3.1 provides an Eulerian description of the flow. For a Lagrangian description, we would need to determine the velocity as a function of time for each particle as it flows from one point to another.

In fluid mechanics it is usually easier to use the Eulerian method to describe a flow—in either experimental or analytical investigations. There are, however, certain instances in which the Lagrangian method is more convenient. For example, some numerical fluid mechanics calculations are based on determining the motion of individual fluid particles (based on the appropriate interactions among the particles), thereby describing the motion in Lagrangian terms. Similarly, in some experiments individual fluid particles are "tagged" and are followed throughout their motion, providing a Lagrangian description. Oceanographic measurements obtained from devices that flow with the ocean currents provide this information. Similarly, by using X-ray opaque dyes, it is possible to trace blood flow in arteries and to obtain a Lagrangian description of the fluid motion. A Lagrangian description may also be useful in describing fluid machinery (such as pumps and turbines) in which fluid particles gain or lose energy as they move along their flow paths.

Another illustration of the difference between the Eulerian and Lagrangian descriptions can be seen in the following biological example. Each year thousands of birds migrate between their summer and winter habitats. Ornithologists study these migrations to obtain various types of important information. One set of data obtained is the rate at which birds pass a certain location on their migration route (birds per hour). This data may be obtained by training a video camera on the fixed location. This corresponds to an Eulerian description—"flowrate" at a given location as a function of time. Individual birds need not be followed to obtain this

---

| Either Eulerian or Lagrangian methods can be used to describe flow fields. |

V3.4 Follow the particles (experiment)

V3.5 Follow the particles (computer)

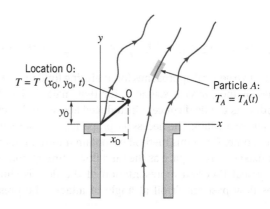

FIGURE 3.2 Eulerian and Lagrangian descriptions of temperature of a flowing fluid.

information. Another type of information is obtained by "tagging" certain birds with radio transmitters and following their motion along the migration route. This corresponds to a Lagrangian description—"position" of a given particle as a function of time.

EXAMPLE 3.2 | The Velocity Field

Given A two-dimensional, unsteady velocity field is given by $u = 4x(1 + t)$ and $v = 4y(-1 + t)$ where u is the x-velocity component and v the y-velocity component.

Find Determine $x(t)$ and $y(t)$ if $x = x_0$ and $y = y_0$ at $t = 0$. Do the velocity components represent an Eulerian description or a Largangian description?

Solution Start with

$$\frac{dx}{dt} = 4x(1+t) \qquad \frac{dy}{dt} = 4y(-1+t) \qquad (1)$$

Separating variables and integrating gives

$$\int_{x_0}^{x} \frac{dx}{x} = \int_{0}^{t} 4(1+t)dt \qquad \int_{y_0}^{y} \frac{dy}{y} = \int_{0}^{t} 4(-1+t)dt$$

$$\ln\left(\frac{x}{x_0}\right) = 4[(1+t)^2]_0^t \qquad \ln\left(\frac{y}{y_0}\right) = 4[(-1+t)^2]_0^t$$

$$= 2[(1+t)^2 - 1] \qquad = 2[(-1+t)^2 - 1]$$

$$x = x_0 e^{2(2t+t^2)} \quad \text{and} \quad y = y_0 e^{2(-2t+t^2)}$$

The velocity components are represented as an Eulerian description.

Comment It is clear from Eq. (1) that x and y-direction velocities are dependent in both space (x or y-coordinate) and time (t). Thus, both the velocity components are represented as an Eulerian description rather than Lagrangian description where velocities might be just function of time.

3.1.2 One-, Two-, and Three-Dimensional Flows

Generally, a fluid flow is a rather complex three-dimensional, time-dependent phenomenon— $\mathbf{V} = \mathbf{V}(x, y, z, t) = u\hat{\mathbf{i}} + v\hat{\mathbf{j}} + w\hat{\mathbf{k}}$. In many situations, however, it is possible to make simplifying assumptions that allow a much easier understanding of the problem without sacrificing too much accuracy. One of these simplifications involves approximating a real flow as a simpler one- or two-dimensional flow.

In almost any flow situation, the velocity field actually contains all three velocity components (u, v, and w, for example). In many situations the **three-dimensional flow** characteristics are important in terms of the physical effects they produce. For these situations it is necessary to analyze the flow in its complete three-dimensional character. Neglect of one or two of the velocity components in these cases would lead to considerable misrepresentation of the effects produced by the actual flow.

The flow of air past an airplane wing provides an example of a complex three-dimensional flow. A feel for the three-dimensional structure of such flows can be obtained by studying **Fig. 3.3**, which is a photograph of the flow past a model wing; the flow has been made visible by using a flow visualization technique.

Most flow fields are actually three-dimensional.

 Video

V3.6 Flow past a wing

National Physical Laboratory/Crown Copyright/Science Source

FIGURE 3.3 Flow visualization of the complex three-dimensional flow past a delta wing.

In many situations one of the velocity components may be small (in some sense) relative to the two other components. In situations of this kind, it may be reasonable to neglect the smaller component and assume **two-dimensional flow**. That is, $\mathbf{V} = u\hat{\mathbf{i}} + v\hat{\mathbf{j}}$, where u and v are functions of x and y (and possibly time, t).

It is sometimes possible to further simplify a flow analysis by assuming that two of the velocity components are negligible, leaving the velocity field to be approximated as a **one-dimensional flow** field. That is, $\mathbf{V} = u\hat{\mathbf{i}}$. As we will learn from examples throughout the remainder of the book, although there are very few, if any, flows that are truly one-dimensional, there are many flow fields for which the one-dimensional flow assumption provides a useful model of the real flow. There are also many flow situations for which use of a one-dimensional flow field assumption would produce give completely erroneous results.

3.1.3 Steady and Unsteady Flows

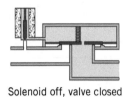

Solenoid off, valve closed

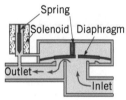

Spring
Solenoid Diaphragm
Outlet
Inlet

Solenoid on, valve open

▶ Video

V3.7 Flow types

▶ Video

V3.8 Jupiter red spot

In the previous discussion we have assumed **steady flow**—the velocity at a given point in space does not vary with time, $\partial \mathbf{V}/\partial t = 0$. In reality, almost all flows are unsteady in some sense. That is, the velocity does vary with time. Of course **unsteady flows** are usually more difficult to analyze (and to investigate experimentally) than are steady flows. Hence, considerable simplification often results if one can make the assumption of steady flow without compromising the usefulness of the results. Among the various types of unsteady flows are nonperiodic flow, periodic flow, and truly random flow. Whether or not unsteadiness of one or more of these types must be included in an analysis is not always immediately obvious.

An example of a nonperiodic, unsteady flow is that produced by turning off a faucet to stop the flow of water. Usually this unsteady flow process is quite mundane, and the forces developed as a result of the unsteady effects need not be considered. However, if the water is turned off suddenly (as with the electrically operated valve in a dishwasher shown in the adjacent figure), the unsteady effects can become important [as in the "water hammer" effects made apparent by the loud banging of the pipes under such conditions (Ref. 1)].

In other flows the unsteady effects may be periodic, occurring time after time in basically the same manner. The periodic injection of the air–gasoline mixture into the cylinder of an automobile engine is such an example. The unsteady effects are quite regular and repeatable and the engine could not operate without them.

In many situations the unsteady character of a flow is seemingly random. That is, there is no repeatable sequence or regular variation to the unsteadiness. This behavior occurs in *turbulent flow* and is absent from *laminar flow*. The "smooth" flow of highly viscous syrup onto a pancake represents a "deterministic" laminar flow. It is quite different from the turbulent flow observed in the "irregular" splashing of water from a faucet onto the sink below it. The "irregular" gustiness of the wind represents another turbulent flow. The differences between these types of flows are examined in considerable detail in Chapters 8 and 9.

It must be understood that the definition of steady or unsteady flow pertains to the behavior of a fluid property as observed at a fixed point in space. For steady flow, the values of all fluid properties (velocity, temperature, density, etc.) at any fixed point are independent of time. However, the value of those properties for a given fluid particle may change with time as the particle flows along, even in steady flow. Thus, the temperature of the exhaust at the exit of a car's exhaust pipe

---

The Wide World of Fluids

New pulsed liquid-jet scalpel

High-speed, liquid-jet cutters are used for cutting a wide variety of materials such as leather goods, jigsaw puzzles, plastic, ceramic, and metal. Typically, compressed air is used to produce a continuous stream of water that is ejected from a tiny nozzle. As this stream impacts the material to be cut, a high pressure (the stagnation pressure) is produced on the surface of the material, thereby cutting the material. Such liquid-jet cutters work well in air but are difficult to control

if the jet must pass through a liquid as often happens in surgery. Researchers have developed a new pulsed jet cutting tool that may allow surgeons to perform microsurgery on tissues that are immersed in water. Rather than using a steady water jet, the system uses *unsteady flow*. A high-energy electrical discharge inside the nozzle momentarily raises the temperature of the microjet to approximately 10,000°C. This creates a rapidly expanding vapor bubble in the nozzle and expels a tiny fluid jet from the nozzle. Each electrical discharge creates a single, brief jet, which makes a small cut in the material.

may be constant for several hours, but the temperature of a fluid particle that left the exhaust pipe five minutes ago is lower now than it was when it left the pipe, even though the flow is steady.

3.1.4 Flow Patterns: Streamlines, Streaklines, and Pathlines

Although fluid motion can be quite complicated, there are various concepts that can be used to help in the visualization and analysis of flow fields. To this end we discuss the use of streamlines, streaklines, and pathlines in flow analysis. The streamline is often used in analytical work, while the streakline and pathline are often used in experimental work.

A **streamline** is a line that is everywhere tangent to the velocity vector. If the flow is steady, nothing at a fixed point (including the velocity direction) changes with time, so the streamlines are fixed lines in space. For unsteady flows the streamlines may change shape with time. Streamlines are obtained analytically by integrating the equations defining lines tangent to the velocity field. As illustrated in the adjacent figure, for two-dimensional flows the slope of the streamline, dy/dx, must be equal to the tangent of the angle that the velocity vector makes with the x axis or

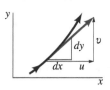

$$\frac{dy}{dx} = \frac{v}{u} \tag{3.1}$$

V3.9 Streamlines

If the velocity field is known as a function of x and y (and t if the flow is unsteady), this equation can be integrated to give the equation of the streamlines.

For unsteady flow there is no easy way to visualize streamlines experimentally in the laboratory. As discussed below, the observation of dye, smoke, or some other tracer injected into a flow can provide useful information, but for unsteady flows it is not necessarily information about the streamlines.

EXAMPLE 3.3 | Streamlines for a Given Velocity Field

Given Consider the two-dimensional steady flow discussed in Example 3.1, $\mathbf{V} = (V_0/\ell)(-x\hat{\mathbf{i}} + y\hat{\mathbf{j}})$.

Find Determine the streamlines for this flow.

Solution

Since

$$u = (-V_0/\ell)x \text{ and } v = (V_0/\ell)y \tag{1}$$

it follows that streamlines are given by solution of the equation

$$\frac{dy}{dx} = \frac{v}{u} = \frac{(V_0/\ell)y}{-(V_0/\ell)x} = -\frac{y}{x}$$

in which variables can be separated and the equation integrated to give

$$\int \frac{dy}{y} = -\int \frac{dx}{x}$$

or

$$\ln y = -\ln x + \text{constant}$$

Thus, along the streamline

$$xy = C, \quad \text{where } C \text{ is a constant} \tag{Ans}$$

By using different values of the constant C, we can plot various lines in the x-y plane—the streamlines. The streamlines for $x \geq 0$ are plotted in **Fig. E3.3**. A comparison of this figure with Fig. E3.1a illustrates the fact that streamlines are lines tangent to the velocity vector.

Comment Note that a flow is not completely specified by the shape of the streamlines alone. For example, the streamlines for the flow with $V_0/\ell = 10$ have the same shape as those for the flow with $V_0/\ell = -10$. However, the direction of the flow is opposite for these two cases. The arrows in Fig. E3.3 representing the flow direction are correct for $V_0/\ell = 10$ because, from Eq. 1, $u = -10x$ and $v = 10y$.

That is, the flow is from right to left. For $V_0/\ell = -10$ the arrows are reversed. The flow is from left to right.

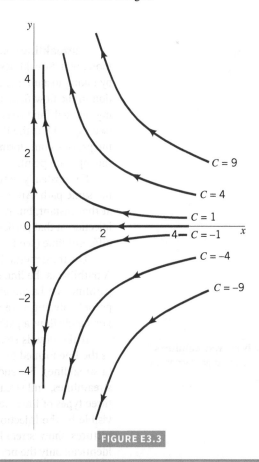

FIGURE E3.3

EXAMPLE 3.4 | Streamlines for Velocity Field at Fixed Coordinate Points

Given   A flow has a velocity field defined by **V**= $(8xy\,\hat{\mathbf{i}} + 4x^2y\,\hat{\mathbf{j}})$ m/s.

Find   Determine an expression for the streamlines passing through $x = 1$ m and $y = 1$ m.

Solution   The equation for a streamline is given by

$$\frac{dy}{dx} = \frac{v}{u}$$

Thus, for $u = 8xy$ and $v = 4x^2y$

$$\frac{dy}{dx} = \frac{4x^2y}{8xy}$$

Separating the variables and integrating yields

$$\int dy = \int \frac{x}{2}\,dx$$

$$y = \frac{x^2}{4} + C$$

where C is a constant

The value of C is determined for the streamline that passes through $(1, 1)$

$$1 = \frac{1^2}{4} + C$$

Thus $C = 3/4$ and the equation for streamlines can be given as

$$y = \frac{x^2}{4} + \frac{3}{4}$$

Comment   If this solution is compared with that of the previous example it is evident that the solution of Example 3.3 is a family of curves (Fig. E3.3) whereas the solution of the current problem is a single particular curve, as shown in Fig E3.4.

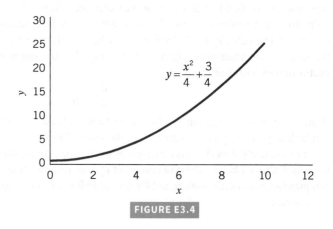

FIGURE E3.4

A **streakline** consists of all particles in a flow that have previously passed through a common point. Streaklines are a laboratory tool rather than an analytical tool. They can be obtained by taking instantaneous photographs of marked particles that all passed through a given location in the flow field at some earlier time. Such a line can be produced by continuously injecting marked fluid (neutrally buoyant smoke in air or dye in water) at a given location (Ref. 2) (see Fig. 9.1). If the flow is steady, each successively injected particle follows precisely behind the previous one, forming a steady streakline that is exactly the same as the streamline through the injection point.

For unsteady flows, particles injected at the same point at different times need not follow the same path. An instantaneous photograph of the marked fluid would show the streakline at that instant, but it would not necessarily coincide with the streamline through the point of injection at that particular time nor with the streamline through the same injection point at a different time (see Example 3.5).

The third method used for visualizing and describing flows involves the use of **pathlines**. A pathline is the line traced out by a given particle as it flows from one point to another. The pathline is a Lagrangian concept that can be produced in the laboratory by marking a fluid particle (dyeing a small fluid element) and taking a time exposure photograph of its motion. For steady flow, a pathline is the same as a streamline; for unsteady flow, they are different.

> For steady flow, streamlines, streaklines, and pathlines are the same.

If the flow is steady, the path taken by a marked particle (a pathline) will be the same as the line formed by all other particles that previously passed through the point of injection (a streakline). For such cases these lines are tangent to the velocity field. Hence, pathlines, streamlines, and streaklines are the same for steady flows. For unsteady flows none of these three types of lines need be the same (Ref. 3). Often one sees pictures of "streamlines" made visible by the injection of smoke or dye into a flow as is shown in **Fig. 3.3**. Actually, such pictures show streaklines rather than streamlines. However, for steady flows the two are identical; only the nomenclature is incorrectly used.

The Wide World of Fluids

Air bridge spanning the oceans

It has long been known that large quantities of material are transported from one location to another by airborne dust particles. It is estimated that 2 billion metric tons of dust are lifted into the atmosphere each year. Most of these particles settle out fairly rapidly, but significant amounts travel large distances. Scientists are beginning to understand the full impact of this phenomenon—it is not only the tonnage transported, but the type of material transported that is significant. In addition to the mundane inert material we all term "dust," it is now known that a wide variety of hazardous materials and organisms are also carried along these literal *particle paths*. Satellite images reveal the amazing rate by which desert soils and other materials are transformed into airborne particles as a result of storms that produce strong winds. Once the tiny particles are aloft, they may travel thousands of miles, crossing the oceans and eventually being deposited on other continents. For the health and safety of all, it is important that we obtain a better understanding of the air bridges that span the oceans and also understand the ramification of such material transport.

EXAMPLE 3.5 | Comparison of Streamlines, Pathlines, and Streaklines

Given Water flowing from the oscillating slit shown in **Fig. E3.5a** produces a velocity field given by $\mathbf{V} = u_0 \sin[\omega(t - y/v_0)]\hat{\mathbf{i}} + v_0\hat{\mathbf{j}}$, where u_0, v_0, and ω are constants. Thus, the y component of velocity remains constant ($v = v_0$), and the x component of velocity at $y = 0$ coincides with the velocity of the oscillating sprinkler head [$u = u_0 \sin(\omega t)$ at $y = 0$].

Find

a. Determine the streamline that passes through the origin at $t = 0$; at $t = \pi/2\omega$.

b. Determine the pathline of the particle that was at the origin at $t = 0$; at $t = \pi/2$.

c. Discuss the shape of the streakline that passes through the origin.

Solution

a. Since $u = u_0 \sin[\omega(t - y/v_0)]$ and $v = v_0$, it follows from Eq. 3.1 that streamlines are given by the solution of

$$\frac{dy}{dx} = \frac{v}{u} = \frac{v_0}{u_0 \sin[\omega(t - y/v_0)]}$$

in which the variables can be separated and the equation integrated (for any given time t) to give

$$u_0 \int \sin\left[\omega\left(t - \frac{y}{v_0}\right)\right] dy = v_0 \int dx$$

or

$$u_0(v_0/\omega) \cos\left[\omega\left(t - \frac{y}{v_0}\right)\right] = v_0 x + C \qquad (1)$$

where C is a constant. For the streamline at $t = 0$ that passes through the origin ($x = y = 0$), the value of C is obtained from Eq. 1 as $C = u_0 v_0/\omega$. Hence, the equation for this streamline is

$$x = \frac{u_0}{\omega}\left[\cos\left(\frac{\omega y}{v_0}\right) - 1\right] \qquad (2) \quad (Ans)$$

Similarly, for the streamline at $t = \pi/2\omega$ that passes through the origin, Eq. 1 gives $C = 0$. Thus, the equation for this streamline is

$$x = \frac{u_0}{\omega} \cos\left[\omega\left(\frac{\pi}{2\omega} - \frac{y}{v_0}\right)\right] = \frac{u_0}{\omega} \cos\left(\frac{\pi}{2} - \frac{\omega y}{v_0}\right)$$

or

$$x = \frac{u_0}{\omega} \sin\left(\frac{\omega y}{v_0}\right) \qquad (3) \quad (Ans)$$

Comment These two streamlines, plotted in **Fig. E3.5b**, are not the same because the flow is unsteady. For example, at the origin ($x = y = 0$) the velocity is $\mathbf{V} = v_0\hat{\mathbf{j}}$ at $t = 0$ and $\mathbf{V} = u_0\hat{\mathbf{i}} + v_0\hat{\mathbf{j}}$ at $t = \pi/2\omega$. Thus, the angle of the streamline passing through the origin changes with time. Similarly, the shape of the entire streamline is a function of time.

b. The pathline of a particle (the location of the particle as a function of time) can be obtained from the velocity field and the definition of the velocity. Since $u = dx/dt$ and $v = dy/dt$ we obtain

$$\frac{dx}{dt} = u_0 \sin\left[\omega\left(t - \frac{y}{v_0}\right)\right] \quad \text{and} \quad \frac{dy}{dt} = v_0$$

The y equation can be integrated (since v_0 = constant) to give the y coordinate of the pathline as

$$y = v_0 t + C_1 \qquad (4)$$

where C_1 is a constant. With this known $y = y(t)$ dependence, the x equation for the pathline becomes

$$\frac{dx}{dt} = u_0 \sin\left[\omega\left(t - \frac{v_0 t + C_1}{v_0}\right)\right] = -u_0 \sin\left(\frac{C_1 \omega}{v_0}\right)$$

This can be integrated to give the x component of the pathline as

$$x = -\left[u_0 \sin\left(\frac{C_1 \omega}{v_0}\right)\right] t + C_2 \qquad (5)$$

where C_2 is a constant. For the particle that was at the origin ($x = y = 0$) at time $t = 0$, Eqs. 4 and 5 give $C_1 = C_2 = 0$. Thus, the pathline is

$$x = 0 \quad \text{and} \quad y = v_0 t \qquad (6) \quad (Ans)$$

Similarly, for the particle that was at the origin at $t = \pi/2\omega$, Eqs. 4 and 5 give $C_1 = -\pi v_0/2\omega$ and $C_2 = -\pi u_0/2\omega$. Thus, the pathline for this particle is

$$x = u_0\left(t - \frac{\pi}{2\omega}\right) \quad \text{and} \quad y = v_0\left(t - \frac{\pi}{2\omega}\right) \qquad (7)$$

The pathline can be drawn by plotting the locus of $x(t)$, $y(t)$ values for $t \geq 0$ or by eliminating the parameter t from Eq. 7 to give

$$y = \frac{v_0}{u_0} x \qquad (8) \quad (Ans)$$

Comment The pathlines given by Eqs. 6 and 8, shown in **Fig. E3.5c**, are straight lines from the origin (rays). The pathlines and streamlines do not coincide because the flow is unsteady.

c. The streakline through the origin at time $t = 0$ is the locus of particles at $t = 0$ that previously ($t < 0$) passed through the origin. The general shape of the streaklines can be seen as follows. Each particle that flows through the origin travels in a straight line (pathlines are rays from the origin), the slope of which lies between $\pm v_0/u_0$, as shown in **Fig. E3.5d**. Particles passing through the origin at different times are located on different rays from the origin and at different distances from the origin. The net result is that a stream of dye continually injected at the origin (a streakline) would have the shape shown in Fig. E3.5d. Because of the unsteadiness, the streakline will vary with time, although it will always have the oscillating, sinuous character shown.

Comment Similar streaklines are given by the stream of water from a garden hose nozzle that oscillates back and forth in a direction normal to the axis of the nozzle.

In this example neither the streamlines, pathlines, nor streaklines coincide. If the flow were steady, all of these lines would be the same.

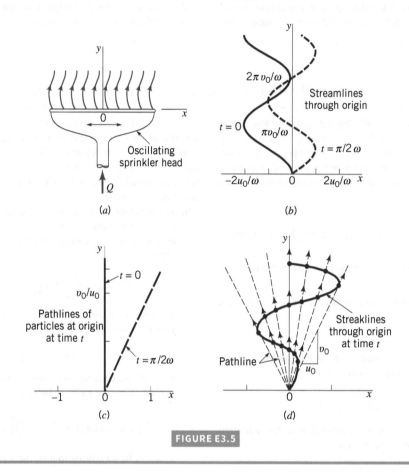

FIGURE E3.5

3.2 THE ACCELERATION FIELD

V3.10 Pathlines

As indicated in the previous section, we can describe fluid motion by either (1) following individual particles (Lagrangian description) or (2) remaining fixed in space and observing different particles as they pass by (Eulerian description). In either case, to apply Newton's second law ($\mathbf{F} = m\mathbf{a}$) we must be able to describe the particle acceleration. For the infrequently used Lagrangian method, we describe the fluid acceleration just as is done in solid body dynamics—$\mathbf{a} = \mathbf{a}(t)$ for each particle. For the Eulerian description, we describe the **acceleration field** as a function of position and time without actually following any particular particle. This is analogous to describing the flow in terms of the velocity field, $\mathbf{V} = \mathbf{V}(x, y, z, t)$, rather than the velocity for particular particles. In this section we will discuss how to obtain the acceleration field if the velocity field is known.

The acceleration of a particle is the time rate of change of its velocity. For unsteady flows the velocity at a given point in space (occupied by different particles at different times) may vary with time, giving rise to a portion of the fluid acceleration. In addition, a fluid particle may

Acceleration is the time rate of change of velocity for a given particle.

experience an acceleration because its velocity changes as it flows from one point to another in space. For example, water flowing through a garden hose nozzle under steady conditions (constant number of gallons per minute from the hose) will experience an acceleration as it changes from its relatively low velocity in the hose to its relatively high velocity at the tip of the nozzle.

3.2.1 Acceleration and the Material Derivative

Consider a fluid particle moving along its pathline as is shown in **Fig. 3.4**. In general, the particle's velocity, denoted $\mathbf{V}_A$ for particle A, is a function of its location and the time. That is,

$$\mathbf{V}_A = \mathbf{V}_A(\mathbf{r}_A, t) = \mathbf{V}_A[x_A(t), y_A(t), z_A(t), t]$$

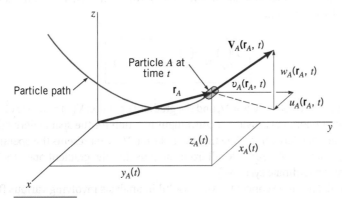

FIGURE 3.4 Velocity and position of particle A at time t.

where $x_A = x_A(t)$, $y_A = y_A(t)$, and $z_A = z_A(t)$ define the location of the moving particle. By definition, the acceleration of a particle is the time rate of change of its velocity. Since the velocity may be a function of both position and time, its value may change because of the change in time as well as a change in the particle's position. Thus, we use the chain rule of differentiation to obtain the acceleration of particle A, denoted $\mathbf{a}_A$, as

$$\mathbf{a}_A(t) = \frac{d\mathbf{V}_A}{dt} = \frac{\partial \mathbf{V}_A}{\partial t} + \frac{\partial \mathbf{V}_A}{\partial x}\frac{dx_A}{dt} + \frac{\partial \mathbf{V}_A}{\partial y}\frac{dy_A}{dt} + \frac{\partial \mathbf{V}_A}{\partial z}\frac{dz_A}{dt} \tag{3.2}$$

Using the fact that the particle velocity components are given by $u_A = dx_A/dt$, $v_A = dy_A/dt$, and $w_A = dz_A/dt$, Eq. 3.2 becomes

$$\mathbf{a}_A = \frac{\partial \mathbf{V}_A}{\partial t} + u_A\frac{\partial \mathbf{V}_A}{\partial x} + v_A\frac{\partial \mathbf{V}_A}{\partial y} + w_A\frac{\partial \mathbf{V}_A}{\partial z}$$

Since the above is valid for any particle, we can drop the reference to particle A and obtain the acceleration field from the velocity field as

$$\boxed{\mathbf{a} = \frac{\partial \mathbf{V}}{\partial t} + u\frac{\partial \mathbf{V}}{\partial x} + v\frac{\partial \mathbf{V}}{\partial y} + w\frac{\partial \mathbf{V}}{\partial z}} \tag{3.3}$$

This is a vector result whose scalar components can be written as

$$a_x = \frac{\partial u}{\partial t} + u\frac{\partial u}{\partial x} + v\frac{\partial u}{\partial y} + w\frac{\partial u}{\partial z}$$

$$a_y = \frac{\partial v}{\partial t} + u\frac{\partial v}{\partial x} + v\frac{\partial v}{\partial y} + w\frac{\partial v}{\partial z} \tag{3.4}$$

and

$$a_z = \frac{\partial w}{\partial t} + u\frac{\partial w}{\partial x} + v\frac{\partial w}{\partial y} + w\frac{\partial w}{\partial z}$$

where a_x, a_y, and a_z are the x, y, and z components of the acceleration.

The material derivative is used to describe time rates of change for a given particle.

The above result is often written in shorthand notation as

$$\mathbf{a} = \frac{D\mathbf{V}}{Dt}$$

where the operator

$$\frac{D(\ )}{Dt} \equiv \frac{\partial(\ )}{\partial t} + u\frac{\partial(\ )}{\partial x} + v\frac{\partial(\ )}{\partial y} + w\frac{\partial(\ )}{\partial z} \tag{3.5}$$

is termed the **material derivative** or *substantial derivative* because it gives the rate of change as we follow a particle of the material (or substance). An often-used shorthand notation for the material derivative operator is

$$\frac{D(\ )}{Dt} = \frac{\partial(\ )}{\partial t} + (\mathbf{V} \cdot \nabla)(\ ) \tag{3.6}$$

The dot product of the velocity vector, $\mathbf{V}$, and the gradient operator, $\nabla(\ ) = \partial(\ )/\partial x\,\hat{\mathbf{i}} + \partial(\ )/\partial y\,\hat{\mathbf{j}} + \partial(\ )/\partial z\,\hat{\mathbf{k}}$ (a vector operator) provides a convenient notation for the spatial derivative terms appearing in the material derivative. Note that the notation $\mathbf{V} \cdot \nabla$ represents the operator $\mathbf{V} \cdot \nabla(\ ) = u\partial(\ )/\partial x + v\partial(\ )/\partial y + w\partial(\ )/\partial z$. (Recall from calculus that the gradient operator has different forms for different coordinate systems.)

The material derivative concept is very useful in analysis involving various fluid parameters, not just the acceleration. The material derivative of any variable is the rate at which that variable changes with time for a given particle as seen by one moving along with the fluid. For example, consider a temperature field $T = T(x, y, z, t)$ associated with a given flow, like the flame shown in the adjacent figure. It may be of interest to determine the time rate of change of temperature of a fluid particle (particle A) as it moves through this temperature field. If the velocity, $\mathbf{V} = \mathbf{V}(x, y, z, t)$, is known, we can apply the chain rule to determine the rate of change of temperature as

$T = T(x, y, z, t)$

Particle A

V

jesadaphorn/Shutterstock.com

$$\frac{dT_A}{dt} = \frac{\partial T_A}{\partial t} + \frac{\partial T_A}{\partial x}\frac{dx_A}{dt} + \frac{\partial T_A}{\partial y}\frac{dy_A}{dt} + \frac{\partial T_A}{\partial z}\frac{dz_A}{dt}$$

This can be written as

$$\frac{DT}{Dt} = \frac{\partial T}{\partial t} + u\frac{\partial T}{\partial x} + v\frac{\partial T}{\partial y} + w\frac{\partial T}{\partial z} = \frac{\partial T}{\partial t} + \mathbf{V} \cdot \nabla T$$

As in the determination of the acceleration, the material derivative operator, $D(\ )/Dt$, appears.

EXAMPLE 3.6 | Acceleration along a Streamline

Given   An incompressible, inviscid fluid flows steadily past a tennis ball of radius R, as shown in **Fig. E3.6a**. According to advanced analysis of the flow, the fluid velocity along streamline A–B is given by

$$\mathbf{V} = u(x)\hat{\mathbf{i}} = V_0\left(1 + \frac{R^3}{x^3}\right)\hat{\mathbf{i}}$$

where V_0 is the upstream velocity far ahead of the sphere.

Find   Determine the acceleration experienced by fluid particles as they flow along this streamline.

Solution   Along streamline A–B there is only one component of velocity ($v = w = 0$) so that from Eq. 3.3

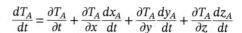

$$\mathbf{a} = \frac{\partial \mathbf{V}}{\partial t} + u\frac{\partial \mathbf{V}}{\partial x} = \left(\frac{\partial u}{\partial t} + u\frac{\partial u}{\partial x}\right)\hat{\mathbf{i}}$$

or

$$a_x = \frac{\partial u}{\partial t} + u\frac{\partial u}{\partial x}, \quad a_y = 0, \quad a_z = 0$$

Since the flow is steady, the velocity at a given point in space does not change with time. Thus, $\partial u/\partial t = 0$. With the given velocity distribution along the streamline, the acceleration becomes

$$a_x = u\frac{\partial u}{\partial x} = V_0\left(1 + \frac{R^3}{x^3}\right)V_0[R^3(-3x^{-4})]$$

or

$$a_x = -3(V_0^2/R)\frac{1+(R/x)^3}{(x/R)^4}$$ (Ans)

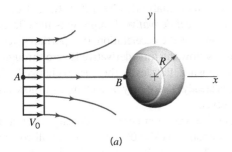

(a)

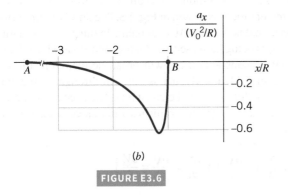

(b)

FIGURE E3.6

Comments Along streamline A–B ($-\infty \leq x \leq -R$ and $y = 0$) the acceleration has only an x component, and it is negative (a deceleration). Thus, the fluid slows down from its upstream velocity of $\mathbf{V} = V_0\hat{\mathbf{i}}$ at $x = -\infty$ to its stagnation point velocity of $\mathbf{V} = 0$ at $x = -R$, the "nose" of the ball. The variation of a_x along streamline A–B is shown in **Fig. E3.6b**. It is the same result as will be obtained in Example 4.1 by using the streamwise component of the acceleration, $a_x = V\,\partial V/\partial s$. The maximum deceleration occurs at $x = -1.205R$ and has a value of $a_{x,max} = -0.610\,V_0^2/R$. Note that this maximum deceleration increases with increasing velocity and decreasing size. As indicated in the following table, typical values of this deceleration can be quite large. For example, the $a_{x,max} = -1.243 \times 10^4$ m/s² value for a pitched baseball is a deceleration approximately 1500 times that of gravity. (Note that this is not the deceleration of the ball; it is the deceleration of the air *relative to the ball*.)

Object	V_0 (m/s)	R (m)	$a_{x,max}$ (m/s²)
Rising weather balloon	0.3	1.22	−0.05
Soccer ball	6.1	0.24	−93
Baseball	27.4	0.037	−1.24 × 10⁴
Tennis ball	30.5	0.032	−1.8 × 10⁴
Golf ball	61	0.021	−1.06 × 10⁵

In general, for fluid particles on streamlines other than A–B, all three components of the acceleration (a_x, a_y, and a_z) will be nonzero.

EXAMPLE 3.7 | Acceleration Along a Streamline at Given Coordinate Points

Given The fluid velocity along the x-axis as shown in **Fig. E3.7** changes from 3 m/s at point A to 9 m/s at point B. It is also known that the velocity is a linear function of distance along the streamline.

Find Determine the acceleration at points A, B, and C. Assume steady flow.

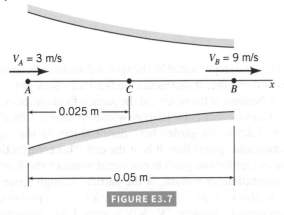

FIGURE E3.7

Solution The generalized material derivative as discussed in Section 3.2.1 can be written as

$$a = \frac{\partial V}{\partial t} + V.\nabla V$$

With $u = u(x)$, $v = 0$, and $w = 0$

this becomes

$$a = \left(\frac{\partial u}{\partial t} + u\frac{\partial u}{\partial x}\right)\hat{i} = u\frac{\partial u}{\partial x}\hat{i}$$ (1)

Since u is a linear function of x, $u = c_1 x + c_2$, where the constants c_1, c_2 are given as

$$u_A = 3 = c_2 \text{ and } u_B = 9 = 0.1C_1 + C_2$$

or $$C_1 = 60 \text{ and } C_2 = 3$$

Thus, $u = (60x + 3)$ m/s with x-m

From Eq. 1

or for $$a = u\frac{\partial u}{\partial x}\hat{i} = (60x+3)\frac{m}{s}\left(60\frac{m}{m\cdot s}\right)\hat{i}$$

$X_A = 0$, $a_A = 180\,\hat{i}$ m/s²

$X_B = 0.05$ m, $a_B = 360\,\hat{i}$ m/s²

$X_C = 0.025$ m, $a_C = 270\,\hat{i}$ m/s²

Comment It is worth mentioning here the solution obtained here is for steady flow and that is why in Eq. (1), the derivative of u with respect to time is neglected. But in real physical world this may not always be true and consideration of the derivative of u with respect to t makes the solution a bit complex. Thus, many a times, for sake of simplicity, the assumption of steady flow is a very good assumption.

3.2.2 Unsteady Effects

As is seen from Eq. 3.5, the material derivative formula contains two types of terms—those involving the time derivative $[\partial(\ )/\partial t]$ and those involving spatial derivatives $[\partial(\ )/\partial x, \partial(\ )/\partial y,$ and $\partial(\ )/\partial z]$. The time derivative portion is termed the *local derivative*. It represents the effects of the unsteadiness of the flow. If the parameter involved is the acceleration, that portion given by $\partial\mathbf{V}/\partial t$ is termed the **local acceleration**. For steady flow the time derivative is zero throughout the flow field $[\partial(\ )/\partial t \equiv 0]$, and the local effect vanishes. Physically, there is no change in flow parameters at a fixed point in space if the flow is steady. There may be a change of those parameters for a fluid particle as it moves about, however.

If a flow is unsteady, its parameter values (velocity, temperature, density, etc.) at any location may change with time. For example, an unstirred ($\mathbf{V} = 0$) cup of coffee will cool over time because of heat transfer to its surroundings. That is, $DT/Dt = \partial T/\partial t + \mathbf{V} \cdot \nabla T = \partial T/\partial t < 0$. Similarly, a fluid particle may have nonzero acceleration as a result of the unsteady effect of the flow. Consider flow in a constant diameter pipe as is shown in **Fig. 3.5**. The flow is assumed to be spatially uniform throughout the pipe. That is, $\mathbf{V} = V_0(t)\hat{\mathbf{i}}$ at all points in the pipe. The value of the acceleration depends on whether V_0 is being increased, $\partial V_0/\partial t > 0$, or decreased, $\partial V_0/\partial t < 0$. Unless V_0 is independent of time ($V_0 \equiv$ constant) there will be an acceleration, the local acceleration term. Thus, the acceleration field, $\mathbf{a} = \partial V_0/\partial t\,\hat{\mathbf{i}}$, is uniform throughout the entire flow, although it may vary with time ($\partial V_0/\partial t$ need not be constant). The acceleration due to the spatial variations of velocity ($u\,\partial u/\partial x, v\,\partial v/\partial y,$ etc.) vanishes automatically for this flow, since $\partial u/\partial x = 0$ and $v = w = 0$. That is,

$$\mathbf{a} = \frac{\partial\mathbf{V}}{\partial t} + u\frac{\partial\mathbf{V}}{\partial x} + v\frac{\partial\mathbf{V}}{\partial y} + w\frac{\partial\mathbf{V}}{\partial z} = \frac{\partial\mathbf{V}}{\partial t} = \frac{\partial V_0}{\partial t}\hat{\mathbf{i}}$$

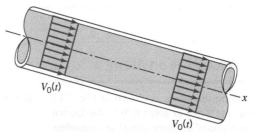

FIGURE 3.5   Uniform, unsteady flow in a constant diameter pipe.

3.2.3 Convective Effects

The portion of the material derivative (Eq. 3.5) represented by the spatial derivatives is termed the *convective derivative* or the *advective derivative*. It represents the fact that a flow property associated with a fluid particle may vary because of the motion of the particle from one point in space where the parameter has one value to another point in space where its value is different. For example, the water velocity at the inlet of the garden hose nozzle shown in the adjacent figure is different (both in direction and speed) than it is at the exit. This contribution to the time rate of change of the parameter for the particle can occur whether the flow is steady or unsteady. It is due to the convection, or motion, of the particle through space in which there is a gradient $[\nabla(\ ) = \partial(\ )/\partial x\,\hat{\mathbf{i}} + \partial(\ )/\partial y\,\hat{\mathbf{j}} + \partial(\ )/\partial z\,\hat{\mathbf{k}}]$ in the parameter value. That portion of the acceleration given by the term $(\mathbf{V} \cdot \nabla)\mathbf{V}$ is termed the **convective acceleration**.

As is illustrated in **Fig. 3.6**, the temperature of a water particle changes as it flows through a water heater. The water entering the heater is always the same cold temperature, and the water leaving the heater is always the same hot temperature. The flow is steady. However, the temperature, T, of each water particle increases as it passes through the heater—$T_{\text{out}} > T_{\text{in}}$. Thus, $DT/Dt \neq 0$ because of the convective term in the total derivative of the temperature. That is, $\partial T/\partial t = 0$, but $u\,\partial T/\partial x \neq 0$ (where x is directed along the streamline), since there is a nonzero temperature gradient along the streamline. A fluid particle

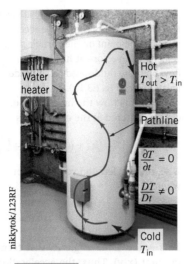

nikkytok/123RF

FIGURE 3.6 Steady-state operation of a water heater.

traveling along this nonconstant temperature path ($\partial T/\partial x \neq 0$) at a specified speed ($u$) will have its temperature change with time at a rate of $DT/Dt = u\,\partial T/\partial x$ even though the flow is steady ($\partial T/\partial t = 0$).

The same types of processes are involved with fluid accelerations. Consider flow in a variable area pipe as shown in **Fig. 3.7**. It is assumed that the flow is steady and one-dimensional with velocity that increases and decreases in the flow direction as indicated. As the fluid flows from section (1) to section (2), its velocity increases from V_1 to V_2. Thus, even though $\partial \mathbf{V}/\partial t = 0$ (steady flow), fluid particles experience an acceleration given by $a_x = u\,\partial u/\partial x$ (convective acceleration). For $x_1 < x < x_2$, it is seen that $\partial u/\partial x > 0$ so that $a_x > 0$—the fluid accelerates. For $x_2 < x < x_3$, it is seen that $\partial u/\partial x < 0$ so that $a_x < 0$—the fluid decelerates. This acceleration and deceleration are shown in the adjacent figure. If $V_1 = V_3$, the amount of acceleration precisely balances the amount of deceleration even though the distances between x_2 and x_1 and x_3 and x_2 are not the same.

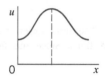

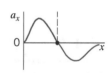

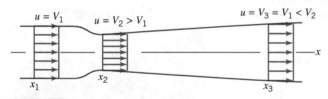

FIGURE 3.7 Uniform, steady flow in a variable area pipe.

The concept of the material derivative can be used to determine the time rate of change of any parameter associated with a particle as it moves about. Its use is not restricted to fluid mechanics alone. The basic ingredients needed to use the material derivative concept are the field description of the parameter, $P = P(x, y, z, t)$, and the rate at which the particle moves through that field, $\mathbf{V} = \mathbf{V}(x, y, z, t)$.

EXAMPLE 3.8 | Acceleration from a Given Velocity Field

Given Consider the steady, two-dimensional flow field discussed in Example 3.3.

Find Determine the acceleration field for this flow.

Solution In general, the acceleration is given by

$$\mathbf{a} = \frac{D\mathbf{V}}{Dt} = \frac{\partial \mathbf{V}}{\partial t} + (\mathbf{V} \cdot \nabla)(\mathbf{V})$$

$$= \frac{\partial \mathbf{V}}{\partial t} + u\frac{\partial \mathbf{V}}{\partial x} + v\frac{\partial \mathbf{V}}{\partial y} + w\frac{\partial \mathbf{V}}{\partial z} \qquad (1)$$

where the velocity is given by $\mathbf{V} = (V_0/\ell)(-x\hat{\mathbf{i}} + y\hat{\mathbf{j}})$ so that $u = -(V_0/\ell)x$ and $v = (V_0/\ell)y$. For steady $[\partial(\ )/\partial t = 0]$, two-dimensional $[w = 0$ and $\partial(\ )/\partial z = 0]$ flow, Eq. 1 becomes

$$\mathbf{a} = u\frac{\partial \mathbf{V}}{\partial x} + v\frac{\partial \mathbf{V}}{\partial y}$$

$$= \left(u\frac{\partial u}{\partial x} + v\frac{\partial u}{\partial y}\right)\hat{\mathbf{i}} + \left(u\frac{\partial v}{\partial x} + v\frac{\partial v}{\partial y}\right)\hat{\mathbf{j}}$$

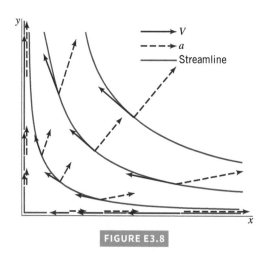

FIGURE E3.8

Hence, for this flow the acceleration is given by

$$\mathbf{a} = \left[\left(-\frac{V_0}{\ell}\right)(x)\left(-\frac{V_0}{\ell}\right) + \left(\frac{V_0}{\ell}\right)(y)(0)\right]\hat{\mathbf{i}}$$
$$+ \left[\left(-\frac{V_0}{\ell}\right)(x)(0) + \left(\frac{V_0}{\ell}\right)(y)\left(\frac{V_0}{\ell}\right)\right]\hat{\mathbf{j}}$$

or

$$a_x = \frac{V_0^2 x}{\ell^2}, \quad a_y = \frac{V_0^2 y}{\ell^2} \qquad (Ans)$$

Comments The fluid experiences an acceleration in both the x and y directions. Since the flow is steady, there is no local

acceleration—the fluid velocity at any given point is constant in time. However, there is a convective acceleration due to the change in velocity from one point on the particle's pathline to another. Recall that the velocity is a vector—it has both a magnitude and a direction. In this flow both the fluid speed (magnitude) and flow direction change with location (see Fig. E3.1a).

For this flow the magnitude of the acceleration is constant on circles centered at the origin, as is seen from the fact that

$$|\mathbf{a}| = (a_x^2 + a_y^2 + a_z^2)^{1/2} = \left(\frac{V_0}{\ell}\right)^2 (x^2 + y^2)^{1/2} \qquad (2)$$

Also, the acceleration vector is oriented at an angle θ from the x axis, where

$$\tan\theta = \frac{a_y}{a_x} = \frac{y}{x}$$

This is the same angle as that formed by a ray from the origin to point (x, y). Thus, the acceleration is directed along rays from the origin and has a magnitude proportional to the distance from the origin. Typical acceleration vectors (from Eq. 2) and velocity vectors (from Example 3.1) are shown in **Fig. E3.8** for the flow in the first quadrant. Note that $\mathbf{a}$ and $\mathbf{V}$ are not parallel except along the x and y axes (a fact that is responsible for the curved pathlines of the flow) and that both the acceleration and velocity are zero at the origin ($x = y = 0$). An infinitesimal fluid particle placed precisely at the origin will remain there, but its neighbors (no matter how close they are to the origin) will drift away.

EXAMPLE 3.9 | The Material Derivative

Given A fluid flows steadily through a two-dimensional nozzle of length ℓ as shown in **Fig. E3.9a**. The nozzle shape is given by

$$y/\ell = \pm 0.5/[1 + (x/\ell)]$$

If viscous and gravitational effects are negligible, the velocity field is approximately

$$u = V_0[1 + x/\ell], v = -V_0 y/\ell \qquad (1)$$

and the pressure field is

$$p - p_0 = -(\rho V_0^2/2)[(x^2 + y^2)/\ell^2 + 2x/\ell]$$

where V_0 and p_0 are the velocity and pressure at the origin, $x = y = 0$. Note that the fluid speed increases as it flows through the nozzle. For example, along the centerline ($y = 0$), $V = V_0$ at $x = 0$ and $V = 2V_0$ at $x = \ell$.

Find Determine, as a function of x and y, the time rate of change of pressure felt by a fluid particle as it flows through the nozzle.

Solution The time rate of change of pressure at any given, fixed point in this steady flow is zero. However, the time rate of change of pressure felt by a particle flowing through the nozzle is given by the material derivative of the pressure and is not zero. Thus,

$$\frac{Dp}{Dt} = \frac{\partial p}{\partial t} + u\frac{\partial p}{\partial x} + v\frac{\partial p}{\partial y} = u\frac{\partial p}{\partial x} + v\frac{\partial p}{\partial y} \qquad (2)$$

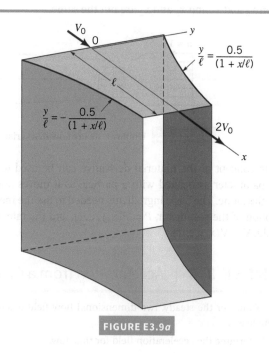

FIGURE E3.9a

where the x and y components of the pressure gradient can be written as

$$\frac{\partial p}{\partial x} = -\frac{\rho V_0^2}{\ell}\left(\frac{x}{\ell}+1\right) \tag{3}$$

and

$$\frac{\partial p}{\partial y} = -\frac{\rho V_0^2}{\ell}\left(\frac{y}{\ell}\right) \tag{4}$$

Therefore, by combining Eqs. (1), (2), (3), and (4) we obtain

$$\frac{Dp}{Dt} = V_0\left(1+\frac{x}{\ell}\right)\left(-\frac{\rho V_0^2}{\ell}\right)\left(\frac{x}{\ell}+1\right) + \left(-V_0\frac{y}{\ell}\right)\left(-\frac{\rho V_0^2}{\ell}\right)\left(\frac{y}{\ell}\right)$$

or

$$\frac{Dp}{Dt} = -\frac{\rho V_0^3}{\ell}\left[\left(\frac{x}{\ell}+1\right)^2 - \left(\frac{y}{\ell}\right)^2\right] \tag{5} \quad (Ans)$$

Comment Lines of constant pressure within the nozzle are indicated in **Fig. E3.9b**, along with some representative streamlines of the flow. Note that as a fluid particle flows along its streamline, it

moves into areas of lower and lower pressure. Hence, even though the flow is steady, the time rate of change of the pressure for any given particle is negative.

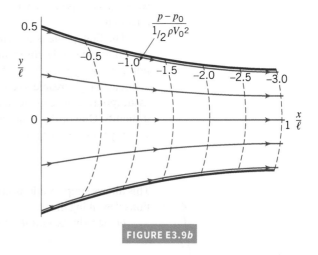

3.2.4 Streamline Coordinates

In many flow situations, it is convenient to use a coordinate system defined in terms of the streamlines of the flow. An example for steady, two-dimensional flows is illustrated in **Fig. 3.8**. Such flows can be described either in terms of the usual x, y Cartesian coordinate system

V3.12 Stream-line coordinates

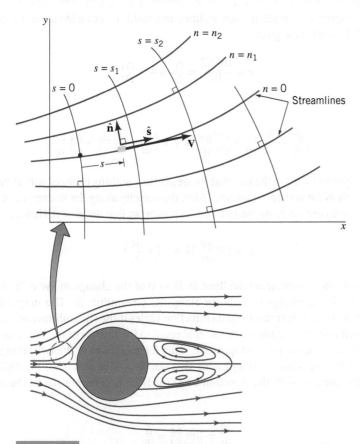

FIGURE 3.8 Streamline coordinate system for two-dimensional flow.

(or some other system such as the r, θ polar coordinate system) or the streamline coordinate system. In the streamline coordinate system the flow is described in terms of one coordinate along the streamlines, denoted s, and the second coordinate normal to the streamlines, denoted n. Unit vectors in these two directions are denoted by $\hat{\mathbf{s}}$ and $\hat{\mathbf{n}}$, as shown in the figure. Care is needed not to confuse the coordinate distance s (a scalar) with the unit vector along the streamline direction, $\hat{\mathbf{s}}$.

The flow plane is therefore covered by an orthogonal curved net of coordinate lines. At any point the s and n directions are perpendicular, but the lines of constant s or constant n are not necessarily straight. Without knowing the actual velocity field (hence, the streamlines) it is not actually possible to construct this flow net. In many situations, appropriate simplifying assumptions can be made so that this lack of information does not present an insurmountable difficulty. One of the major advantages of using the streamline coordinate system is that the velocity is always tangent to the s direction. That is,

$$\mathbf{V} = V\hat{\mathbf{s}}$$

This allows simplifications in describing the fluid particle acceleration and in solving the equations describing the flow.

For steady, two-dimensional flow we can determine the acceleration as

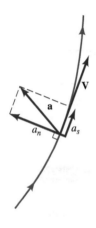

$$\mathbf{a} = \frac{D\mathbf{V}}{Dt} = a_s\hat{\mathbf{s}} + a_n\hat{\mathbf{n}}$$

where a_s and a_n are the streamline and normal components of acceleration, respectively, as indicated in the adjacent figure. We use the material derivative because by definition the acceleration is the time rate of change of the velocity of a given particle as it moves about. If the streamlines are curved, both the speed of the particle and its direction of flow may change from one point to another. In general, for steady flow both the speed and the flow direction are a function of location—$V = V(s, n)$ and $\hat{\mathbf{s}} = \hat{\mathbf{s}}(s, n)$. For a given particle, the value of s changes with time, but the value of n remains fixed because the particle flows along a streamline defined by $n = $ constant. (Recall that streamlines and pathlines coincide in steady flow.) Thus, application of the chain rule gives

$$\mathbf{a} = \frac{D(V\hat{\mathbf{s}})}{Dt} = \frac{DV}{Dt}\hat{\mathbf{s}} + V\frac{D\hat{\mathbf{s}}}{Dt}$$

or

$$\mathbf{a} = \left(\frac{\partial V}{\partial t} + \frac{\partial V}{\partial s}\frac{ds}{dt} + \frac{\partial V}{\partial n}\frac{dn}{dt}\right)\hat{\mathbf{s}} + V\left(\frac{\partial \hat{\mathbf{s}}}{\partial t} + \frac{\partial \hat{\mathbf{s}}}{\partial s}\frac{ds}{dt} + \frac{\partial \hat{\mathbf{s}}}{\partial n}\frac{dn}{dt}\right)$$

This can be simplified by using the fact that for steady flow nothing changes with time at a given point so that both $\partial V/\partial t$ and $\partial \hat{\mathbf{s}}/\partial t$ are zero. Also, the velocity along the streamline is $V = ds/dt$, and the particle remains on its streamline ($n = $ constant) so that $dn/dt = 0$. Hence,

$$\mathbf{a} = \left(V\frac{\partial V}{\partial s}\right)\hat{\mathbf{s}} + V\left(V\frac{\partial \hat{\mathbf{s}}}{\partial s}\right)$$

The orientation of the unit vector along the streamline changes with distance along the streamline.

The quantity $\partial \hat{\mathbf{s}}/\partial s$ represents the limit as $\delta s \to 0$ of the change in the unit vector along the streamline, $\delta \hat{\mathbf{s}}$, per change in distance along the streamline, δs. The magnitude of $\hat{\mathbf{s}}$ is constant ($|\hat{\mathbf{s}}| = 1$; it is a unit vector), but its direction varies if the streamlines are curved. From **Fig. 3.9** it is seen that the magnitude of $\partial \hat{\mathbf{s}}/\partial s$ is equal to the inverse of the radius of curvature of the streamline, $\mathcal{R}$, at the point in question. This follows because the two triangles shown (AOB and $A'O'B'$) are similar triangles so that $\delta s/\mathcal{R} = |\delta \hat{\mathbf{s}}|/|\hat{\mathbf{s}}| = |\delta \hat{\mathbf{s}}|$, or $|\delta \hat{\mathbf{s}}/\delta s| = 1/\mathcal{R}$. Similarly, in the limit $\delta s \to 0$, the direction of $\delta \hat{\mathbf{s}}/\delta s$ is seen to be normal to the streamline. That is,

$$\frac{\partial \hat{\mathbf{s}}}{\partial s} = \lim_{\delta s \to 0} \frac{\delta \hat{\mathbf{s}}}{\delta s} = \frac{\hat{\mathbf{n}}}{\mathcal{R}}$$

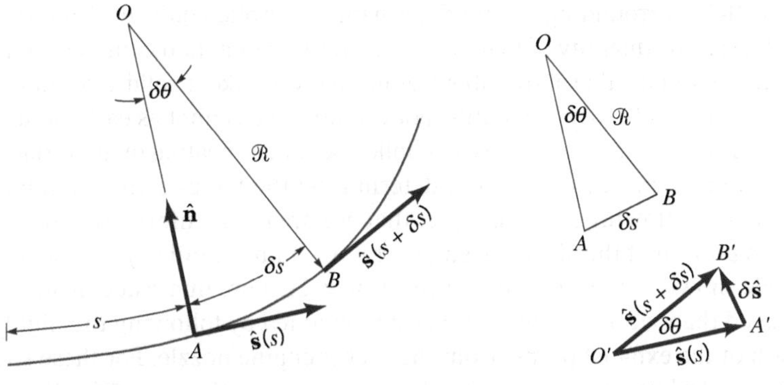

FIGURE 3.9 Relationship between the unit vector along the streamline, $\hat{\mathbf{s}}$, and the radius of curvature of the streamline, $\mathcal{R}$.

Hence, the acceleration for steady, two-dimensional flow can be written in terms of its streamwise and normal components in the form

$$\mathbf{a} = V\frac{\partial V}{\partial s}\hat{\mathbf{s}} + \frac{V^2}{\mathcal{R}}\hat{\mathbf{n}} \quad \text{or} \quad a_s = V\frac{\partial V}{\partial s}, \quad a_n = \frac{V^2}{\mathcal{R}} \tag{3.7}$$

The first term, $a_s = V\,\partial V/\partial s$, represents the convective acceleration along the streamline, and the second term, $a_n = V^2/\mathcal{R}$, represents centripetal acceleration (one type of convective acceleration) normal to the fluid motion. These components can be noted in Fig. E3.5 by resolving the acceleration vector into its components along and normal to the velocity vector. Note that the unit vector $\hat{\mathbf{n}}$ is directed from the streamline toward the center of curvature. These forms of the acceleration will be used in Chapter 4 and are probably familiar from previous dynamics or physics considerations.

3.3   CONTROL VOLUME AND SYSTEM REPRESENTATIONS

As is discussed in Chapter 1, a fluid is a type of matter that is relatively free to move, deform, and interact with its surroundings. As with any matter, a fluid's behavior is described by fundamental physical principles that are modeled by an appropriate set of equations. The application of principles such as the conservation of mass, Newton's laws of motion, and the laws of thermodynamics forms the foundation of fluid mechanics analyses. There are various ways that these governing laws can be applied to a fluid, including the system approach and the control volume approach. By definition, a **system** is a collection of matter of fixed identity (always the same atoms or fluid particles), which may move, flow, and interact with its surroundings. A **control volume,** on the other hand, is a volume in space (a geometric entity, independent of mass) through which fluid may flow.

A system is a specific, identifiable quantity of matter. It may consist of a relatively large amount of mass (such as all of the air in the Earth's atmosphere), or it may be an infinitesimal size (such as a single fluid particle). In any case, the molecules making up the system are "tagged" in some fashion (dyed red, either actually or only in your mind) so that they can be continually identified as they move about. The system may interact with its surroundings by various means (by the transfer of heat or the exertion of a pressure force, for example). It may continually change size and shape, but it always contains the same mass.

A mass of air drawn into an air compressor can be considered as a system. It changes shape and size (it is compressed), its temperature may change, and it is eventually expelled through the outlet of the compressor. The matter associated with the original air drawn into the compressor remains as a system, however. The behavior of this material could be investigated by applying the appropriate equations to this system.

One of the important concepts used in the study of statics and dynamics is that of the free-body diagram. That is, we identify an object, isolate it from its surroundings, replace its surroundings by the equivalent influence on the object, and apply Newton's laws of motion. The body in such cases is our system—an identified portion of matter that we follow during its

Both control volume and system concepts can be used to describe fluid flow.

interactions with its surroundings. In fluid mechanics, it is often quite difficult to identify and keep track of a specific quantity of matter. A finite portion of a fluid contains an uncountable number of fluid particles that move about quite freely, unlike a solid that may deform but usually remains relatively easy to identify. For example, we cannot as easily follow a specific portion of water flowing in a river as we can follow a branch floating on its surface.

We may often be more interested in determining the forces acting on a fan, airplane, or automobile by air flowing past the object than we are in the information obtained by following a given portion of the air (a system) as it flows along. Similarly, for the Space Shuttle launch vehicle shown in the adjacent figure, we may be more interested in determining the thrust produced than we are in the information obtained by following the highly complex, irregular path of the exhaust plume from the rocket engine nozzle. For these situations we often use the control volume approach. We identify a specific volume in space (a volume associated with the fan, airplane, or automobile, for example) and analyze the fluid flow within, through, or around that volume. In general, the control volume can be a moving volume, although for most situations considered in this book we will use only fixed, nondeformable control volumes. The matter within a control volume may change with time as the fluid flows through it. Similarly, the amount of mass within the volume may change with time. The control volume itself is a specific geometric entity, independent of the flowing fluid.

Examples of control volumes and *control surfaces* (the surface of the control volume) are shown in **Fig. 3.10**. For case (*a*), fluid flows through a pipe. The fixed control surface consists of the inside surface of the pipe, the outlet end at section (2), and a section across the pipe at (1). One portion of the control surface is a physical surface (the pipe), while the remainder is simply a surface in space (across the pipe). Fluid flows across part of the control surface but not across all of it.

Another control volume is the rectangular volume surrounding the jet engine shown in Fig. 3.10*b*. If the airplane to which the engine is attached is sitting still on the runway, air flows through this control volume because of the action of the engine within it. The air that was within the engine itself at time $t = t_1$ (a system) has passed through the engine and is outside of the control volume at a later time $t = t_2$ as indicated. At this later time other air (a different system) is within the engine. If the airplane is moving, the control volume is fixed relative to an observer on the airplane, but it is a moving control volume relative to an observer on the ground. In either situation air flows through and around the engine as indicated.

The deflating balloon shown in Fig. 3.10*c* provides an example of a deforming control volume. As time increases, the control volume (whose surface is the inner surface of the balloon) decreases in size. If we do not hold onto the balloon, it becomes a moving, deforming control volume as it darts about the room. The majority of the problems we will analyze can be solved by using a fixed, nondeforming control volume. In some instances, however, it will be advantageous, in fact necessary, to use a moving, deforming control volume.

In many ways the relationship between a system and a control volume is similar to the relationship between the Lagrangian and Eulerian flow description introduced in Section 3.1.1. In the system or Lagrangian description, we follow the fluid and observe its behavior as it moves about. In the control volume or Eulerian description, we remain stationary and observe the fluid's behavior at a fixed location. (If a moving control volume is used, it almost never moves with the system—the system flows through the control volume.) These ideas are discussed in more detail in the next section.

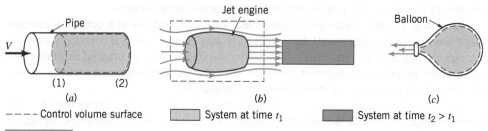

FIGURE 3.10 Typical control volumes: (*a*) fixed control volume, (*b*) fixed or moving control volume, (*c*) deforming control volume.

All of the laws describing the motion of a fluid can be stated in their basic form in terms of a system approach. For example, "the mass of a system remains constant," or "the time rate of change of momentum of a system is equal to the sum of all the forces acting on the system." Note the word system, not control volume, in these statements. To use the governing equations in a control volume approach to problem solving, we must recast the principles in an appropriate manner. To this end we introduce the Reynolds transport theorem in the following section.

The governing laws of fluid motion are stated in terms of fluid systems, not control volumes.

3.4 THE REYNOLDS TRANSPORT THEOREM

We are sometimes interested in what happens to a particular part of the fluid as it moves about. Frequently, we are interested in what effect the fluid has on a particular object or volume in space as fluid interacts with it. Thus, we need to describe the principles describing fluid motion using both system concepts (consider a given mass of the fluid) and control volume concepts (consider a specified volume). To do this we need an analytical tool to shift from one representation to the other. The **Reynolds transport theorem** provides this tool.

All physical principles are stated in terms of physical parameters. Velocity, acceleration, mass, temperature, and momentum are but a few of the more common parameters. Let B represent any of these (or other) fluid parameters and b represent the amount of that parameter per unit mass. That is,

$$B = mb$$

where m is the mass of the portion of fluid of interest. For example, as shown in the adjacent figure, if $B = m$, the mass, it follows that $b = 1$. The mass per unit mass is unity. If $B = mV^2/2$, the kinetic energy of the mass, then $b = V^2/2$, the kinetic energy per unit mass. The parameters B and b may be scalars or vectors. Thus, if $\mathbf{B} = m\mathbf{V}$, the momentum of the mass, then $\mathbf{b} = \mathbf{V}$. (The momentum per unit mass is the velocity.)

The parameter B is termed an *extensive property*, and the parameter b is termed an *intensive property*. The value of B is directly proportional to the amount of the mass being considered, whereas the value of b is independent of the amount of mass. The amount of an extensive property that a system possesses at a given instant, B_{sys}, can be determined by adding up the amount associated with each fluid particle in the system. For infinitesimal fluid particles of volume δV and mass $\rho \, \delta V$, this summation (in the limit of $\delta V \to 0$) takes the form of an integration over all the particles in the system and can be written as

$$B_{sys} = \lim_{\delta V \to 0} \sum_i b_i (\rho_i \, \delta V_i) = \int_{sys} \rho b \, dV$$

The limits of integration cover the entire system—a (usually) moving volume. We have used the fact that the amount of B in a fluid particle of mass $\rho \, \delta V$ is given in terms of b by $\delta B = b\rho \, \delta V$.

Most of the laws governing fluid motion involve the time rate of change of an extensive property of a fluid system—the rate at which the momentum of a system changes with time, the rate at which the mass of a system changes with time, and so on. Thus, we often encounter terms such as

$$\frac{dB_{sys}}{dt} = \frac{d\left(\int_{sys} \rho b \, dV \right)}{dt} \tag{3.8}$$

To formulate the principles in a control volume approach, we must obtain an expression for the time rate of change of an extensive property within a control volume, B_{cv}, not within a system. This can be written as

$$\frac{dB_{cv}}{dt} = \frac{d\left(\int_{cv} \rho b \, dV \right)}{dt} \tag{3.9}$$

where the limits of integration, denoted by cv, cover the control volume of interest. Although Eqs. 3.8 and 3.9 may look very similar, the physical interpretation of each is quite different.

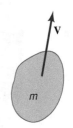

B	$b = B/m$
m	1
$m\mathbf{V}$	$\mathbf{V}$
$\frac{1}{2}mV^2$	$\frac{1}{2}V^2$

Differences between control volume and system concepts are subtle but very important.

Mathematically, the difference is represented by the difference in the limits of integration. Recall that the control volume is a volume in space (in most cases stationary, although if it moves it need not move with the system). On the other hand, the system is an identifiable collection of mass that moves with the fluid (indeed it is a specified portion of the fluid). We will learn that even for those instances when the control volume and the system momentarily occupy the same volume in space, the two quantities dB_{sys}/dt and dB_{cv}/dt need not be the same. The Reynolds transport theorem provides the relationship between the time rate of change of an extensive property for a system and that for a control volume—the relationship between Eqs. 3.8 and 3.9.

EXAMPLE 3.10 | Time Rate of Change for a System and a Control Volume

Given Fluid flows from the fire extinguisher tank shown in **Fig. E3.10a**.

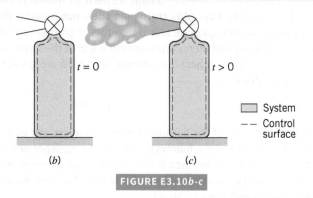

$t = 0$ $t > 0$

☐ System
– – Control surface

(b) (c)

FIGURE E3.10b-c

(a)

FIGURE E3.10a

the tank itself as shown in **Fig. E3.10b**. A short time after the valve is opened, part of the system has moved outside of the control volume as is shown in **Fig. E3.10c**. The control volume remains fixed. The limits of integration are fixed for the control volume; they are a function of time for the system.

Clearly, if mass is to be conserved (one of the basic laws governing fluid motion), the mass of the fluid in the system is constant, so that

$$\frac{d\left(\int_{sys} \rho \, dV\right)}{dt} = 0$$

On the other hand, it is equally clear that some of the fluid has left the control volume through the nozzle on the tank. Hence, the amount of mass within the tank (the control volume) decreases with time, or

$$\frac{d\left(\int_{cv} \rho \, dV\right)}{dt} < 0$$

The actual numerical value of the rate at which the mass in the control volume decreases will depend on the rate at which the fluid flows through the nozzle (i.e., the size of the nozzle and the speed and density of the fluid). Clearly, the meanings of dB_{sys}/dt and dB_{cv}/dt are different. For this example, $dB_{cv}/dt < dB_{sys}/dt$. Other situations may have $dB_{cv}/dt \geq dB_{sys}/dt$.

Find Discuss the differences between dB_{sys}/dt and dB_{cv}/dt if B represents mass.

Solution With $B = m$, the system mass, it follows that $b = 1$ and Eqs. 3.8 and 3.9 can be written as

$$\frac{dB_{sys}}{dt} \equiv \frac{dm_{sys}}{dt} = \frac{d\left(\int_{sys} \rho \, dV\right)}{dt}$$

and

$$\frac{dB_{cv}}{dt} \equiv \frac{dm_{cv}}{dt} = \frac{d\left(\int_{cv} \rho \, dV\right)}{dt}$$

Physically, these represent the time rate of change of mass within the system and the time rate of change of mass within the control volume, respectively. We choose our system to be the fluid within the tank at the time the valve was opened ($t = 0$), and the control volume to be

3.4.1 Derivation of the Reynolds Transport Theorem

A simple version of the Reynolds transport theorem relating system concepts to control volume concepts can be obtained easily for the one-dimensional flow through a fixed control volume such as the variable area duct section shown in **Fig. 3.11a**. We consider the control volume to be that stationary volume within the duct between sections (1) and (2) as indicated in **Fig. 3.11b**. The system that we consider is that fluid occupying the control volume at some initial time t. A short time later, at time $t + \delta t$, the system has moved slightly to the right. The fluid particles that coincided with section (2) of the control surface at time t have moved a distance $\delta \ell_2 = V_2 \delta t$ to the right, where V_2 is the velocity of the fluid as it passes section (2). Similarly, the fluid initially at section (1) has moved a distance $\delta \ell_1 = V_1 \delta t$, where V_1 is the fluid velocity at section (1). We assume the fluid flows across sections (1) and (2) in a direction normal to these surfaces and that V_1 and V_2 are constant across sections (1) and (2).

The moving system flows through the fixed control volume.

As is shown in **Fig. 3.11c**, the outflow from the control volume from time t to $t + \delta t$ is denoted as volume II, the inflow as volume I, and the control volume itself as CV. Thus, the system at time t consists of the fluid in section CV; that is, "SYS = CV" at time t. At time $t + \delta t$ the system consists of the same fluid that now occupies sections (CV − I) + II. That is, "SYS = CV − I + II" at time $t + \delta t$. The control volume remains as section CV for all time.

If B is an extensive parameter of the system, then the value of it for the system at time t is

$$B_{\text{sys}}(t) = B_{\text{cv}}(t)$$

since the system and the fluid within the control volume coincide at this time. Its value at time $t + \delta t$ is

$$B_{\text{sys}}(t + \delta t) = B_{\text{cv}}(t + \delta t) - B_{\text{I}}(t + \delta t) + B_{\text{II}}(t + \delta t)$$

Thus, the change in the amount of B in the system, δB_{sys}, in the time interval δt divided by this time interval is given by

$$\frac{\delta B_{\text{sys}}}{\delta t} = \frac{B_{\text{sys}}(t + \delta t) - B_{\text{sys}}(t)}{\delta t} = \frac{B_{\text{cv}}(t + \delta t) - B_{\text{I}}(t + \delta t) + B_{\text{II}}(t + \delta t) - B_{\text{sys}}(t)}{\delta t}$$

By using the fact that at the initial time t we have $B_{\text{sys}}(t) = B_{\text{cv}}(t)$, this ungainly expression may be rearranged as follows.

The time rate of change of a system property is a Lagrangian concept.

$$\frac{\delta B_{\text{sys}}}{\delta t} = \frac{B_{\text{cv}}(t + \delta t) - B_{\text{cv}}(t)}{\delta t} - \frac{B_{\text{I}}(t + \delta t)}{\delta t} + \frac{B_{\text{II}}(t + \delta t)}{\delta t} \tag{3.10}$$

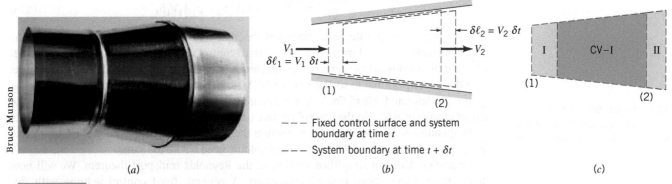

Bruce Munson

(a)

$V_1 \longrightarrow$
$\delta \ell_1 = V_1 \, \delta t \rightarrow$

$\delta \ell_2 = V_2 \, \delta t$

V_2

(1)

(2)

– – – Fixed control surface and system boundary at time t

– – – System boundary at time $t + \delta t$

(b)

I   CV−I   II

(1)

(2)

(c)

FIGURE 3.11   Control volume and system for flow through a variable area pipe.

In the limit $\delta t \to 0$, the left-hand side of Eq. 3.10 is equal to the time rate of change of B for the system and is denoted as DB_{sys}/Dt. We use the material derivative notation, $D(\ )/Dt$, to denote this time rate of change to emphasize the Lagrangian character of this term. (Recall from Section 3.2.1 that the material derivative, DP/Dt, of any quantity P represents the time rate of change of that quantity associated with a given fluid particle as it moves along.) Similarly, the quantity DB_{sys}/Dt represents the time rate of change of property B associated with a system (a given portion of fluid) as it moves along.

In the limit $\delta t \to 0$, the first term on the right-hand side of Eq. 3.10 is seen to be the time rate of change of the amount of B within the control volume

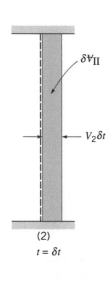

(2)
$t = 0$

$$\lim_{\delta t \to 0} \frac{B_{cv}(t + \delta t) - B_{cv}(t)}{\delta t} = \frac{\partial B_{cv}}{\partial t} = \frac{\partial \left(\int_{cv} \rho b \, d\Psi \right)}{\partial t} \tag{3.11}$$

The third term on the right-hand side of Eq. 3.10 represents the rate at which the extensive parameter B flows from the control volume across the control surface. As indicated in the adjacent figure, during the time interval from $t = 0$ to $t = \delta t$ the volume of fluid that flows across section (2) is given by $\delta \Psi_{II} = A_2 \, \delta \ell_2 = A_2(V_2 \, \delta t)$. Thus, the amount of B within region II, the outflow region, is its amount per unit volume, ρb, times the volume

$$B_{II}(t + \delta t) = (\rho_2 b_2)(\delta \Psi_{II}) = \rho_2 b_2 A_2 V_2 \delta t$$

where b_2 and ρ_2 are the constant values of b and ρ across section (2). Thus, the rate at which this property flows from the control volume, $\dot{B}_{out}$, is given by

$$\dot{B}_{out} = \lim_{\delta t \to 0} \frac{B_{II}(t + \delta t)}{\delta t} = \rho_2 A_2 V_2 b_2 \tag{3.12}$$

Similarly, the inflow of B into the control volume across section (1) during the time interval δt corresponds to that in region I and is given by the amount per unit volume times the volume, $\delta \Psi_I = A_1 \, \delta \ell_1 = A_1(V_1 \, \delta t)$. Hence,

$$B_I(t + \delta t) = (\rho_1 b_1)(\delta \Psi_1) = \rho_1 b_1 A_1 V_1 \, \delta t$$

where b_1 and ρ_1 are the constant values of b and ρ across section (1). Thus, the rate of inflow of the property B into the control volume, $\dot{B}_{in}$, is given by

$$\dot{B}_{in} = \lim_{\delta t \to 0} \frac{B_I(t + \delta t)}{\delta t} = \rho_1 A_1 V_1 b_1 \tag{3.13}$$

If we combine Eqs. 3.10, 3.11, 3.12, and 3.13, we see that the relationship between the time rate of change of B for the system and that for the control volume is given by

$$\frac{DB_{sys}}{Dt} = \frac{\partial B_{cv}}{\partial t} + \dot{B}_{out} - \dot{B}_{in} \tag{3.14}$$

or

$$\frac{DB_{sys}}{Dt} = \frac{\partial B_{cv}}{\partial t} + \rho_2 A_2 V_2 b_2 - \rho_1 A_1 V_1 b_1 \tag{3.15}$$

This is a version of the Reynolds transport theorem valid under the restrictive assumptions associated with the flow shown in Fig. 3.11—fixed control volume with one inlet and one outlet having uniform properties (density, velocity, and the parameter b) across the inlet and outlet with the velocity normal to sections (1) and (2). Note that the time rate of change of B for the system (the left-hand side of Eq. 3.15 or the quantity in Eq. 3.8) is not necessarily the same as the rate of change of B within the control volume (the first term on the right-hand side of Eq. 3.15 or the quantity in Eq. 3.9). This is true because the inflow rate ($b_1 \rho_1 V_1 A_1$) and the outflow rate ($b_2 \rho_2 V_2 A_2$) of the property B for the control volume need not be the same.

Equation 3.15 is a simplified version of the Reynolds transport theorem. We will now derive it for much more general conditions. A general, fixed control volume with fluid flowing through it is shown in **Fig. 3.12**. The flow field may be quite simple (as in the above one-dimensional flow considerations), or it may involve a quite complex, unsteady,

---

The time derivative associated with a system may be different from that for a control volume.

EXAMPLE 3.11 | Use of the Reynolds Transport Theorem

Given Consider again the flow from the fire extinguisher shown in Fig. E3.10. Let the extensive property of interest be the system mass ($B = m$, the system mass, or $b = 1$).

Find Write the appropriate form of the Reynolds transport theorem for this flow.

Solution Again we take the control volume to be the fire extinguisher and the system to be the fluid within it at time $t = 0$. For this case there is no inlet, section (1), across which the fluid flows into the control volume ($A_1 = 0$). There is, however, an outlet, section (2). Thus, the Reynolds transport theorem, Eq. 3.15, along with Eq. 3.9 with $b = 1$ can be written as

$$\frac{Dm_{sys}}{Dt} = \frac{\partial\left(\int_{cv} \rho\,d\Psi\right)}{\partial t} + \rho_2 A_2 V_2 \qquad (1) \qquad (Ans)$$

Comment If we proceed one step further and use the basic principle of conservation of mass, we may set the left-hand side of this equation equal to zero (the amount of mass in a system is constant) and rewrite Eq. 1 in the form

$$\frac{\partial\left(\int_{cv} \rho\,d\Psi\right)}{\partial t} = -\rho_2 A_2 V_2 \qquad (2)$$

The physical interpretation of this result is that the rate at which the mass in the tank decreases in time is equal in magnitude but opposite to the rate of flow of mass from the exit, $\rho_2 A_2 V_2$. Note the units for the two terms of Eq. 2 (kg/s, slugs/s, lb$_m$/s). If there were both an inlet and an outlet to the control volume shown in Fig. E3.10, Eq. 2 would become

$$\frac{\partial\left(\int_{cv} \rho\,d\Psi\right)}{\partial t} = \rho_1 A_1 V_1 - \rho_2 A_2 V_2 \qquad (3)$$

In addition, if the flow were steady, the left-hand side of Eq. 3 would be zero (the amount of mass in the control would be constant in time) and Eq. 3 would become

$$\rho_1 A_1 V_1 = \rho_2 A_2 V_2$$

This is one form of the conservation of mass principle discussed in Section 4.6.2—the mass flowrates into and out of the control volume are equal. Other more general forms are discussed in Chapter 5.

---

three-dimensional situation such as the flow through a human heart as illustrated in the adjacent figure. In any case we again consider the system to be the fluid within the control volume at the initial time t. A short time later a portion of the fluid (region II) has exited from the control volume, and additional fluid (region I, not part of the original system) has entered the control volume.

We consider an extensive fluid property B and seek to determine how the rate of change of B associated with the system is related to the rate of change of B within the control volume at any instant. By repeating the exact steps that we did for the simplified control volume shown in Fig. 3.11, we see that Eq. 3.14 is valid for the general case also, provided that we give the correct interpretation to the terms $\dot{B}_{out}$ and $\dot{B}_{in}$. In general, the control volume may contain more (or less) than one inlet and one outlet. A typical pipe system may contain several inlets and outlets as are shown in **Fig. 3.13**. In such instances we think of all inlets grouped together ($I = I_a + I_b + I_c + \cdots$) and all outlets grouped together ($II = II_a + II_b + II_c + \cdots$), at least conceptually.

The term $\dot{B}_{out}$ represents the net flowrate of the property B from the control volume. Its value can be thought of as arising from the addition (integration) of the contributions through each infinitesimal area element of size δA on the portion of the control surface dividing region II and the control volume. This surface is denoted CS_{out}. As is indicated in **Fig. 3.14**, in time δt the volume of fluid that passes across each area element is given by $\delta\Psi = \delta\ell_n\,\delta A$, where $\delta\ell_n = \delta\ell\cos\theta$ is the height (normal to the base, δA) of the small volume element, and θ is the angle between the velocity vector and the outward pointing normal to the surface, $\hat{n}$. Thus, since $\delta\ell = V\,\delta t$, the amount of the property B carried across the area element δA in the time interval δt is given by

$$\delta B = b\rho\,\delta\Psi = b\rho(V\cos\theta\,\delta t)\,\delta A$$

The rate at which B is carried out of the control volume across the small area element δA, denoted $\delta\dot{B}_{out}$, is

$$\delta\dot{B}_{out} = \lim_{\delta t\to 0}\frac{\rho b\,\delta\Psi}{\delta t} = \lim_{\delta t\to 0}\frac{(\rho b V\cos\theta\,\delta t)\,\delta A}{\delta t} = \rho b V\cos\theta\,\delta A$$

Left Atrium

Right Atrium

Left Ventricle

Right Ventricle

The simplified Reynolds transport theorem can be easily generalized.

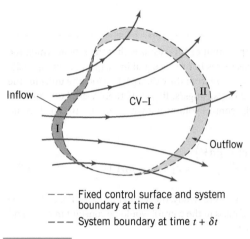

—— Fixed control surface and system
boundary at time t

—— System boundary at time $t + \delta t$

FIGURE 3.12 Control volume and system for flow
through an arbitrary, fixed control volume.

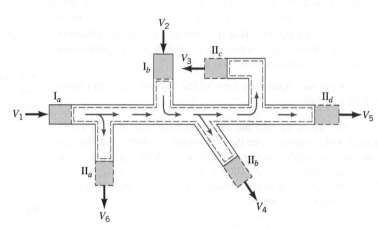

FIGURE 3.13 Typical control volume with more than one inlet and outlet.

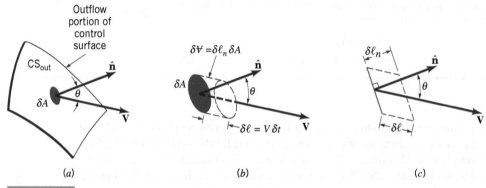

FIGURE 3.14 Outflow across a typical portion of the control surface.

By integrating over the entire outflow portion of the control surface, CS_{out}, we obtain

$$\dot{B}_{\text{out}} = \int_{\text{cs}_{\text{out}}} d\dot{B}_{\text{out}} = \int_{\text{cs}_{\text{out}}} \rho b V \cos \theta \, dA$$

The quantity $V \cos \theta$ is the component of the velocity normal to the area element δA. From the definition of the dot product, this can be written as $V \cos \theta = \mathbf{V} \cdot \hat{\mathbf{n}}$. Hence, an alternate form of the outflow rate is

$$\dot{B}_{\text{out}} = \int_{\text{cs}_{\text{out}}} \rho b \mathbf{V} \cdot \hat{\mathbf{n}} \, dA \tag{3.16}$$

> The flowrate of a parameter across the control surface is written in terms of a surface integral.

In a similar fashion, by considering the inflow portion of the control surface, CS_{in}, as shown in **Fig. 3.15**, we find that the inflow rate of B into the control volume is

$$\dot{B}_{\text{in}} = -\int_{\text{cs}_{\text{in}}} \rho b V \cos \theta \, dA = -\int_{\text{cs}_{\text{in}}} \rho b \mathbf{V} \cdot \hat{\mathbf{n}} \, dA \tag{3.17}$$

We use the standard notation that the unit normal vector to the control surface, $\hat{\mathbf{n}}$, points out from the control volume. Thus, as is shown in **Fig. 3.16**, $-90° < \theta < 90°$ for outflow regions (the normal component of $\mathbf{V}$ is positive; $\mathbf{V} \cdot \hat{\mathbf{n}} > 0$). For inflow regions $90° < \theta < 270°$ (the normal component of $\mathbf{V}$ is negative; $\mathbf{V} \cdot \hat{\mathbf{n}} < 0$). The value of $\cos \theta$ is, therefore, positive on the CV_{out} portions of the control surface and negative on the CV_{in} portions. Over the remainder of the control surface, there is no inflow or outflow, leading to $\mathbf{V} \cdot \hat{\mathbf{n}} = V \cos \theta = 0$ on those portions. On such portions either $V = 0$ (the fluid "sticks" to the surface) or $\cos \theta = 0$ (the fluid "slides" along the surface without

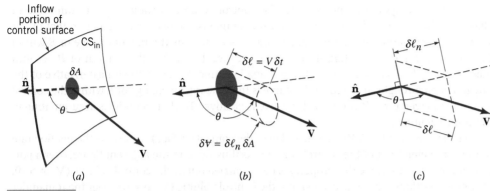

FIGURE 3.15 Inflow across a typical portion of the control surface.

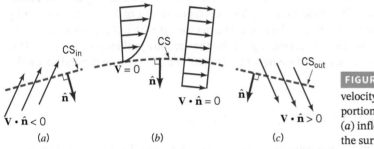

FIGURE 3.16 Possible velocity configurations on portions of the control surface: (a) inflow, (b) no flow across the surface, (c) outflow.

crossing it) (see Fig. 3.16). Therefore, the net flux (flowrate) of parameter B across the entire control surface is

$$\dot{B}_{out} - \dot{B}_{in} = \int_{cs_{out}} \rho b \mathbf{V} \cdot \hat{\mathbf{n}} \, dA - \left(-\int_{cs_{in}} \rho b \mathbf{V} \cdot \hat{\mathbf{n}} \, dA \right)$$

$$= \int_{cs} \rho b \mathbf{V} \cdot \hat{\mathbf{n}} \, dA \tag{3.18}$$

where the integration is over the entire control surface.

By combining Eqs. 3.14 and 3.18, we obtain

$$\frac{DB_{sys}}{Dt} = \frac{\partial B_{cv}}{\partial t} + \int_{cs} \rho b \mathbf{V} \cdot \hat{\mathbf{n}} \, dA$$

This can be written in a slightly different form by using $B_{cv} = \int_{cv} \rho b \, d\forall$ so that

$$\boxed{\frac{DB_{sys}}{Dt} = \frac{\partial}{\partial t} \int_{cv} \rho b \, d\forall + \int_{cs} \rho b \mathbf{V} \cdot \hat{\mathbf{n}} \, dA} \tag{3.19}$$

Equation 3.19 is the general form of the Reynolds transport theorem for a fixed, nondeforming control volume. Its interpretation and use are discussed in the following sections.

3.4.2 Physical Interpretation

The Reynolds transport theorem as given in Eq. 3.19 is widely used in fluid mechanics (and other areas as well). At first it appears to be a rather formidable mathematical expression— perhaps one to avoid if possible. However, a physical understanding of the concepts involved will show that it is a rather straightforward, relatively easy-to-use tool. Its purpose is to provide a link between control volume ideas and system ideas.

The left side of Eq. 3.19 is the time rate of change of an arbitrary extensive parameter of a system. This may represent the rate of change of mass, momentum, energy, or angular momentum of the system, depending on the choice of the parameter B.

Because the system is moving and the control volume is stationary, the time rate of change of the amount of B within the control volume is not necessarily equal to that of the system. The first term on the right side of Eq. 3.19 represents the rate of change of B within the control volume as the fluid flows through it. Recall that b is the amount of B per unit mass, so that $\rho b \, d\mathcal{V}$ is the amount of B in a small volume $d\mathcal{V}$. Thus, the time derivative of the integral of ρb throughout the control volume is the time rate of change of B within the control volume at a given time. If this time derivative is positive, B (whatever it is) is accumulating in the control volume.

The last term in Eq. 3.19 (an integral over the control surface) represents the net flowrate of the parameter B out of the control volume. As illustrated in the adjacent figure, over a portion of the control surface this property is being carried out of the control volume ($\mathbf{V} \cdot \hat{\mathbf{n}} > 0$); over other portions it is being carried into the control volume ($\mathbf{V} \cdot \hat{\mathbf{n}} < 0$). Over the remainder of the control surface there is no transport of B across the surface since $b\mathbf{V} \cdot \hat{\mathbf{n}} = 0$, because either $b = 0$, $\mathbf{V} = 0$, or $\mathbf{V}$ is parallel to the surface at those locations. The mass flowrate through area element δA, given by $\rho \mathbf{V} \cdot \hat{\mathbf{n}} \, \delta A$, is positive for outflow (efflux) and negative for inflow (influx). Each fluid particle or fluid mass carries a certain amount of B with it, as given by the product of B per unit mass, b, and the mass. The net rate at which B is carried out of the control volume across the control surface is given by the area integral term of Eq. 3.19. This net rate across the entire control surface may be negative, zero, or positive depending on the particular situation involved.

3.4.3 Relationship to Material Derivative

In Section 3.2.1 we discussed the concept of the material derivative $D(\ )/Dt = \partial(\ )/\partial t + \mathbf{V} \cdot \nabla(\ ) = \partial(\ )\partial t + u \, \partial(\ )\partial x + v \, \partial(\ )/\partial y + w \, \partial(\ )/\partial z$. The physical interpretation of this derivative is that it provides the time rate of change of a fluid property (temperature, velocity, etc.) associated with a particular fluid particle as it flows. The value of that property for that particle may change because of unsteady effects [the $\partial(\ )/\partial t$ term] or because of effects associated with the particle's motion [the $\mathbf{V} \cdot \nabla(\ )$ term].

Careful consideration of Eq. 3.19 indicates the same type of physical interpretation for the Reynolds transport theorem. The term involving the time derivative of the control volume integral represents unsteady effects associated with the fact that values of the parameter within the control volume may change with time. For steady flow this effect vanishes—fluid flows through the control volume, but the amount of any property, B, within the control volume is constant in time. The term involving the control surface integral represents the convective effects associated with the flow of the system across the fixed control surface. The sum of these two terms gives the rate of change of the parameter B for the system. This corresponds to the interpretation of the material derivative, $D(\ )/Dt = \partial(\ )/\partial t + \mathbf{V} \cdot \nabla(\ )$, in which the sum of the unsteady effect and the convective effect gives the rate of change of a parameter for a fluid particle. As is discussed in Section 3.2, the material derivative operator may be applied to scalars (such as temperature) or vectors (such as velocity). This is also true for the Reynolds transport theorem. The particular parameters of interest, B and b, may be scalars or vectors.

Thus, both the material derivative and the Reynolds transport theorem equations represent ways to transfer from the Lagrangian viewpoint (follow a particle or follow a system) to the Eulerian viewpoint (observe the fluid at a given location in space or observe what happens in the fixed control volume). The material derivative (Eq. 3.5) is essentially the infinitesimal (or derivative) equivalent of the finite size (or integral) Reynolds transport theorem (Eq. 3.19).

3.4.4 Steady and Unsteady Effects

Steady Effects Consider a steady flow [$\partial(\ )/\partial t \equiv 0$] so that Eq. 3.19 reduces to

$$\frac{DB_{\text{sys}}}{Dt} = \int_{\text{cs}} \rho b \mathbf{V} \cdot \hat{\mathbf{n}} \, dA \tag{3.20}$$

In such cases if there is to be a change in the amount of B associated with the system (nonzero left-hand side), there must be a net difference in the rate that B flows into the control volume

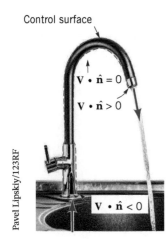

Control surface

$\mathbf{V} \cdot \hat{\mathbf{n}} = 0$

$\mathbf{V} \cdot \hat{\mathbf{n}} > 0$

$\mathbf{V} \cdot \hat{\mathbf{n}} < 0$

Pavel Lipskiy/123RF

The Reynolds transport theorem is the integral counterpart of the material derivative.

compared with the rate that it flows out of the control volume. That is, the integral of $\rho b\mathbf{V} \cdot \hat{\mathbf{n}}$ over the inflow portions of the control surface would not be equal and opposite to that over the outflow portions of the surface.

Consider steady flow through the "black box" control volume that is shown in **Fig. 3.17**. If the parameter B is the mass of the system, the left-hand side of Eq. 3.20 is zero (conservation of mass for the system as discussed in detail in Section 5.1). Hence, the flowrate of mass into the box must be the same as the flowrate of mass out of the box because the right-hand side of Eq. 3.20 represents the net flowrate through the control surface. On the other hand, assume the parameter B is the momentum of the system. The momentum of the system need not be constant. In fact, according to Newton's second law the time rate of change of the system momentum equals the net force, $\mathbf{F}$, acting on the system. In general, the left-hand side of Eq. 3.20 will therefore be nonzero. Thus, the right-hand side, which then represents the net flux of momentum across the control surface, will be nonzero. The flowrate of momentum into the control volume need not be the same as the flowrate of momentum from the control volume. We will investigate these concepts much more fully in Chapter 5. They are the basic principles describing the operation of such devices as jet or rocket engines like the one shown in the adjacent figure.

Control volume

Momentum out

NASA

For steady flows the amount of the property B within the control volume does not change with time. The amount of the property associated with the system may or may not change with time, depending on the particular property considered and the flow situation involved. The difference between that associated with the control volume and that associated with the system is determined by the rate at which B is carried across the control surface—the term $\int_{cs}\rho b\mathbf{V} \cdot \hat{\mathbf{n}}\, dA$.

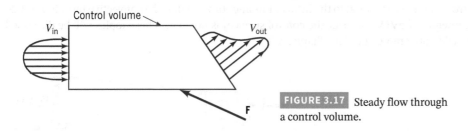

Control volume

V_{in}

V_{out}

F

FIGURE 3.17 Steady flow through a control volume.

Unsteady Effects

Consider unsteady flow $[\partial(\ )/\partial t \neq 0]$ so that all terms in Eq. 3.19 must be retained. When they are viewed from a control volume standpoint, the amount of parameter B within the system may change because the amount of B within the fixed control volume may change with time [the $\partial\left(\int_{cv}\rho b\, d\mathcal{V}\right)/\partial t$ term] and because there may be a net nonzero flow of that parameter across the control surface (the $\int_{cs}\rho b\mathbf{V} \cdot \hat{\mathbf{n}}\, dA$ term).

For the special unsteady situations in which the rate of inflow of parameter B is exactly balanced by its rate of outflow, it follows that $\int_{cs}\rho b\mathbf{V} \cdot \hat{\mathbf{n}}\, dA = 0$, and Eq. 3.19 reduces to

$$\frac{DB_{sys}}{Dt} = \frac{\partial}{\partial t}\int_{cv}\rho b\, d\mathcal{V} \tag{3.21}$$

For such cases, any rate of change in the amount of B associated with the system is equal to the rate of change of B within the control volume. This can be illustrated by considering flow through a constant diameter pipe as is shown in **Fig. 3.18**. The control volume is as shown, and the system is the fluid within this volume at time t_0. We assume the flow is one-dimensional with $\mathbf{V} = V_0\hat{\mathbf{i}}$, where $V_0(t)$ is a function of time, and that the density is constant. At any instant, all particles in the system have the same velocity. We let $\mathbf{B} = $ system momentum $= m\mathbf{V} = mV_0\hat{\mathbf{i}}$, where m is the system mass, so that $\mathbf{b} = \mathbf{B}/m = \mathbf{V} = V_0\hat{\mathbf{i}}$, the fluid velocity. The magnitude of the momentum efflux across the outlet [section (2)] is the same as the magnitude of the momentum influx across the inlet [section (1)]. However, the sign of the efflux is opposite to that of the influx since $\mathbf{V} \cdot \hat{\mathbf{n}} > 0$ for outflow and

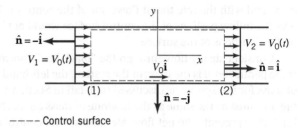

FIGURE 3.18 Unsteady flow through a constant diameter pipe.

$\mathbf{V} \cdot \hat{\mathbf{n}} < 0$ for the inflow. Note that $\mathbf{V} \cdot \hat{\mathbf{n}} = 0$ along the sides of the control volume. Thus, with $\mathbf{V} \cdot \hat{\mathbf{n}} = -V_0$ on section (1), $\mathbf{V} \cdot \hat{\mathbf{n}} = V_0$ on section (2), and $A_1 = A_2$, we obtain

$$\int_{cs} \rho b \mathbf{V} \cdot \hat{\mathbf{n}} \, dA = \int_{cs} \rho(V_0 \hat{\mathbf{i}})(\mathbf{V} \cdot \hat{\mathbf{n}}) \, dA$$

$$= \int_{(1)} \rho(V_0 \hat{\mathbf{i}})(-V_0) \, dA + \int_{(2)} \rho(V_0 \hat{\mathbf{i}})(V_0) \, dA$$

$$= -\rho V_0^2 A_1 \hat{\mathbf{i}} + \rho V_0^2 A_2 \hat{\mathbf{i}} = 0$$

It is seen that for this special case Eq. 3.21 is valid. The rate at which the momentum of the system changes with time is the same as the rate of change of momentum within the control volume. If V_0 is constant in time, there is no rate of change of momentum of the system, and for this special case each of the terms in the Reynolds transport theorem is zero by itself.

Consider the flow through a variable area pipe shown in **Fig. 3.19**. In such cases the fluid velocity is not the same at section (1) as it is at (2). Hence, the efflux of momentum from the control volume is not equal to the influx of momentum, so that the convective term in Eq. 3.20 [the integral of $\rho \mathbf{V}(\mathbf{V} \cdot \hat{\mathbf{n}})$ over the control surface] is not zero. These topics will be discussed in considerably more detail in Chapter 5.

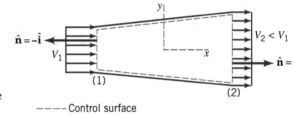

FIGURE 3.19 Flow through a variable area pipe.

3.4.5 Moving Control Volumes

For most problems in fluid mechanics, the control volume may be a fixed volume through which the fluid flows. There are, however, situations for which the analysis is simplified if the control volume is allowed to move or deform. The most general situation would involve a control volume that moves, accelerates, and deforms. As one might expect, the use of these control volumes can become fairly complex.

A number of important problems can be most easily analyzed by using a nondeforming control volume that moves with a constant velocity. An example is shown in **Fig. 3.20** in which a stream of water with velocity $\mathbf{V}_1$ strikes a vane that is moving with constant velocity $\mathbf{V}_0$. It may

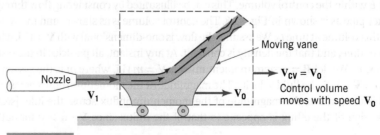

FIGURE 3.20 Example of a moving control volume.

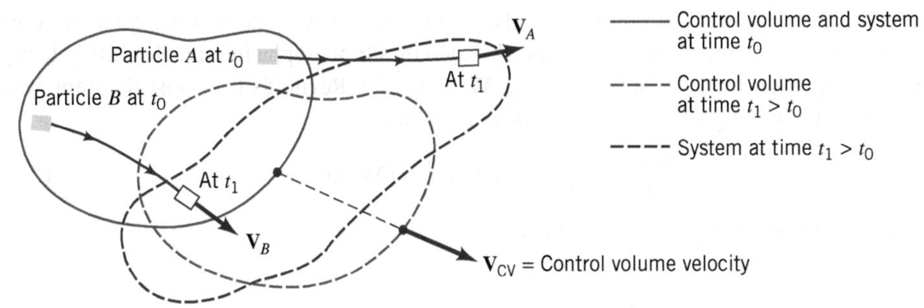

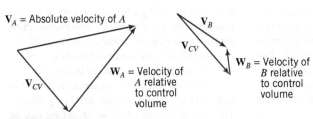

FIGURE 3.21 Typical moving control volume and system.

be of interest to determine the force, **F**, that the water exerts on the vane. This type of problem frequently occurs in turbines where a stream of fluid (water or steam, for example) strikes a series of blades that move past the nozzle. To analyze such problems it is advantageous to use a moving control volume. We will obtain the Reynolds transport theorem for such control volumes.

We consider a control volume that moves with a constant velocity as is shown in **Fig. 3.21**. The shape, size, and orientation of the control volume do not change with time. The control volume merely translates with a constant velocity, **V**$_{cv}$, as shown. In general, the velocity of the control volume and that of the fluid are not the same, so that there is a flow of fluid through the moving control volume just as in the stationary control volume cases discussed in Section 3.4.2. The main difference between the fixed and the moving control volume cases is that it is the *relative velocity*, **W**, that carries fluid across the moving control surface, whereas it is the *absolute velocity*, **V**, that carries the fluid across the fixed control surface. The relative velocity is the fluid velocity relative to the moving control volume—the fluid velocity seen by an observer riding along on the control volume. The absolute velocity is the fluid velocity as seen by a stationary observer in a fixed coordinate system.

The relative velocity is the difference between the absolute velocity and the velocity of the control volume, **W = V − V**$_{cv}$, or

$$\mathbf{V} = \mathbf{W} + \mathbf{V}_{cv} \tag{3.22}$$

Since the velocity is a vector, we must use vector addition as is shown in **Fig. 3.22** to obtain the relative velocity if we know the absolute velocity and the velocity of the control volume. Thus, if the water leaves the nozzle in **Fig. 3.20** with a velocity of **V**$_1$ = 30.48$\hat{\mathbf{i}}$ m/s and the vane has a velocity of **V**$_0$ = 6.09$\hat{\mathbf{i}}$ m/s (the same as the control volume), it appears to an observer riding on the vane that the water approaches the vane with a velocity of **W = V − V**$_{cv}$ = 24.4$\hat{\mathbf{i}}$ m/s. In general, the absolute velocity, **V**, and the control volume velocity, **V**$_{cv}$, will not be in the same direction so that the relative and absolute velocities will have different directions (see **Fig. 3.22**).

The Reynolds transport theorem for a moving, nondeforming control volume can be derived in the same manner that it was obtained for a fixed control volume. As is indicated in **Fig. 3.23**, the only difference that needs be considered is the fact that relative to the moving control volume the fluid velocity observed is the relative velocity, not the absolute velocity. An observer fixed to the moving control volume may or may not even know that he or she is moving relative to some fixed coordinate system. If we follow the derivation that led to Eq. 3.19

The relative velocity is the difference between the absolute velocity and the velocity of the control volume.

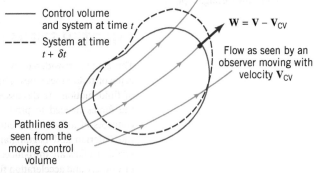

FIGURE 3.22 Relationship between absolute and relative velocities.

FIGURE 3.23 Control volume and system as seen by an observer moving with the control volume.

(the Reynolds transport theorem for a fixed control volume), we note that the corresponding result for a moving control volume can be obtained by simply replacing the absolute velocity, **V**, in that equation by the relative velocity, **W**. Thus, the Reynolds transport theorem for a control volume moving with constant velocity is given by

$$\frac{DB_{sys}}{Dt} = \frac{\partial}{\partial t}\int_{cv}\rho b\, d\mathbf{\Psi} + \int_{cs}\rho b\mathbf{W}\cdot\hat{\mathbf{n}}\, dA \tag{3.23}$$

where the relative velocity is given by Eq. 3.22.

> The Reynolds transport theorem for a moving control volume involves the relative velocity.

3.4.6 Selection of a Control Volume

Any volume in space can be a control volume. It may be of finite size or it may be infinitesimal in size, depending on the type of analysis to be carried out. In most of our cases, the control volume will be a fixed, nondeforming volume. In some situations, we will consider control volumes that move with constant velocity. In either case, it is important that considerable thought go into the selection of the specific control volume to be used.

The selection of an appropriate control volume in fluid mechanics is very similar to the selection of an appropriate free-body diagram in dynamics or statics. In dynamics, we select the body in which we are interested, represent the object in a free-body diagram, and then apply the appropriate principles of mechanics to that body. The ease of solving a given dynamics problem is often very dependent on the specific object that we select for use in our free-body diagram. Similarly, the ease of solving a given fluid mechanics problem is often very dependent on the choice of the control volume used. Only by practice can we develop skill at selecting the "best" control volume. None are "wrong," but some are "much better" than others.

The solution of a typical problem will involve determining parameters such as velocity, pressure, and force at some point in the flow field. It is best to ensure that this point is located on the control surface, not "buried" within the control volume. The unknown will then appear in the convective term (the surface integral) of the Reynolds transport theorem. If possible, the control surface should be normal to the fluid velocity so that the angle θ ($\mathbf{V}\cdot\hat{\mathbf{n}} = V\cos\theta$ as shown by the adjacent figure) in the flux terms of Eq. 3.19 will be 0 or 180°. This will usually simplify the solution process.

Figure 3.24 illustrates three possible control volumes associated with flow through a pipe. If the problem is to determine the pressure at point (1), the selection of the control volume (a) is better than that of (b) because point (1) lies on the control surface. Similarly, control volume (a) is better than (c) because the flow is normal to the inlet and exit portions of the control volume. None of these control volumes are wrong—(a) will be easier to use. Proper control volume selection will become much clearer in Chapter 5 where the Reynolds transport theorem is used to transform the principles we have developed to describe the behavior of systems into equations applying those principles to flow through a control volume.

V3.13 Control volume

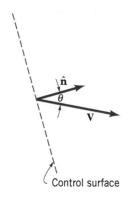

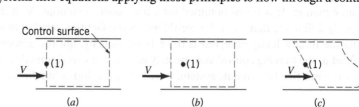

FIGURE 3.24 Various control volumes for flow through a pipe.

CHAPTER SUMMARY

This chapter considered several fundamental concepts of fluid kinematics. That is, various aspects of fluid motion are discussed without regard to the forces needed to produce this motion. The concepts of a field representation of a flow and the Eulerian and Lagrangian approaches to describing a flow are introduced, as are the concepts of velocity and acceleration fields.

The properties of one-, two-, or three-dimensional flows and steady or unsteady flows are introduced along with the concepts of streamlines, streaklines, and pathlines. Streamlines, which are lines tangent to the velocity vectors, are identical to streaklines and pathlines if the flow is steady. For unsteady flows, they need not be identical.

As a fluid particle moves about, its properties (e.g., velocity, density, temperature) may change. The rate of change of these properties can be determined by using the material derivative, which involves both unsteady effects (time rate of change at a fixed location) and convective effects (time rate of change due to the motion of the particle from one location to another).

The concepts of a control volume and a system are introduced, and the Reynolds transport theorem is developed to apply the principles describing the behavior of systems to the flow of fluid through a control volume.

The following checklist provides a study guide for this chapter. When your study of the entire chapter has been completed, you should be able to

- write out meanings of the terms listed and understand each of the related concepts. These terms are particularly important and are set in **bold black** type in the text.
- explain the concept of the field representation of a flow and the difference between Eulerian and Lagrangian methods of describing a flow.
- explain the differences among streamlines, streaklines, and pathlines.
- calculate and plot streamlines for flows with given velocity fields.
- use the concept of the material derivative, with its unsteady and convective effects, to determine time rate of change of a fluid property.
- determine the acceleration field for a flow with a given velocity field.
- discuss the properties of and differences between a system and a control volume.
- interpret, physically and mathematically, the concepts involved in the Reynolds transport theorem.

KEY EQUATIONS

Equation for streamlines	$$\dfrac{dy}{dx} = \dfrac{v}{u}$$	(3.1)
Acceleration	$$\mathbf{a} = \dfrac{\partial \mathbf{V}}{\partial t} + u\dfrac{\partial \mathbf{V}}{\partial x} + v\dfrac{\partial \mathbf{V}}{\partial y} + w\dfrac{\partial \mathbf{V}}{\partial z}$$	(3.3)
Material derivative	$$\dfrac{D(\ )}{Dt} = \dfrac{\partial(\ )}{\partial t} + (\mathbf{V} \cdot \boldsymbol{\nabla})(\ )$$	(3.6)
Streamwise and normal components of acceleration	$$a_s = V\dfrac{\partial V}{\partial s}, \qquad a_n = \dfrac{V^2}{\mathcal{R}}$$	(3.7)
Reynolds transport theorem (restricted form)	$$\dfrac{DB_{\text{sys}}}{Dt} = \dfrac{\partial B_{\text{cv}}}{\partial t} + \rho_2 A_2 V_2 b_2 - \rho_1 A_1 V_1 b_1$$	(3.15)
Reynolds transport theorem (general form)	$$\dfrac{DB_{\text{sys}}}{Dt} = \dfrac{\partial}{\partial t}\int_{\text{cv}} \rho b\, d\mathcal{V} + \int_{\text{cs}} \rho b \mathbf{V} \cdot \hat{\mathbf{n}}\, dA$$	(3.19)
Relative and absolute velocities	$$\mathbf{V} = \mathbf{W} + \mathbf{V}_{\text{cv}}$$	(3.22)

REFERENCES

[1] Streeter, V. L., and Wylie, E. B., *Fluid Mechanics*, 8th Ed., McGraw-Hill, New York, 1985.

[2] Goldstein, R. J., *Fluid Mechanics Measurements*, Hemisphere, New York, 1983.

[3] Homsy, G. M., et al., *Multimedia Fluid Mechanics* CD-ROM, 2nd Ed., Cambridge University Press, New York, 2007.

QUESTIONS AND PROBLEMS

[c] Problem to be solved with aid of programmable calculator or computer.

[o] Open-ended problem that requires critical thinking. These problems require various assumptions to provide the necessary input data. There are not unique answers to these problems.

Note: Unless specific values of required fluid properties are specified in the problem statement, use the values found in Tables 1.4–1.8 and in the tables in the Appendices.

Section 3.1   The Velocity Field

3.1.1 Obtain a photograph/image of a situation in which a fluid is flowing. Print this photo and draw in some lines to represent how you think some streamlines may look. Write a brief paragraph to describe the acceleration of a fluid particle as it flows along one of these streamlines.

3.1.2 The surface velocity of a river is measured at several locations x and can be reasonably represented by

$$V = V_0 + \Delta V(1 - e^{-ax})$$

where V_0, ΔV, and a are constants. Find the Lagrangian description of the velocity of a fluid particle flowing along the surface if $x = 0$ at time $t = 0$.

3.1.3 The velocity field of a flow is given by $\mathbf{V} = 4x^2t\hat{\mathbf{i}} + [6y(t-1) + 4x^2t]\hat{\mathbf{j}}$ m/s, where x and y are in meters and t is in seconds. For fluid particles on the x axis, determine the speed and direction of flow.

3.1.4 A two-dimensional velocity field is given by $u = 2(1 + y)$ and $v = 4$. Determine the equation of the streamline that passes through the origin. On a graph, plot this streamline.

3.1.5 Streamlines are given in Cartesian coordinates by the equation.

$$\psi = U\left(y - \frac{y}{x^2 + y^2}\right) \cdot x^2 + y^2 \geq 1$$

Plot the streamlines for $\psi = 0, \pm 60.0625$ and describe the physical situation represented by this equation. The parameter U is an upstream uniform velocity of 1.0 m/s.

3.1.6 (Video) A flow can be visualized by plotting the velocity field as velocity vectors at representative locations in the flow as shown in Video V3.2 and Fig. E3.1. Consider the velocity field given in polar coordinates by $v_r = -10/r$, and $v_\theta = 10/r$. This flow approximates a fluid swirling into a sink as shown in **Fig. P3.1.6**. Plot the velocity field at locations given by $r = 1, 2$, and 3 with $\theta = 0, 30, 60$, and 90°.

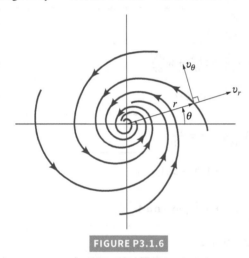

FIGURE P3.1.6

3.1.7 (C) A car accelerates from rest to a final constant velocity V_f and a police officer records the following velocities at various locations x along the highway:

$x = 0$	$V = 0$ kmph
$x = 100$ m	$V = 34.8$ kmph
$x = 200$ m	$V = 47.6$ kmph
$x = 300$ m	$V = 52.3$ kmph
$x = 400$ m	$V = 54.0$ kmph
$x = 1000$ m	$V = 55.0$ kmph $= V_f$

Find a mathematical Eulerian expression for the velocity V traveled by the car as a function of the final velocity V_f of the car and x if $t = 0$ at $V = 0$. *Hint:* Try an exponential fit to the data.

3.1.8 The components of a velocity field are given by $u = x + y$, $v = xy^3 + 16$, and $w = 0$. Determine the location of any stagnation points ($\mathbf{V} = 0$) in the flow field.

3.1.9 The velocity field of a flow is given by $u = -V_0y/(x^2 + y^2)^{1/2}$ and $v = V_0x/(x^2 + y^2)^{1/2}$, where V_0 is a constant. Where in the flow field is the speed equal to V_0? Determine the equation of the streamlines and discuss the various characteristics of this flow.

3.1.10 A velocity field is given by $\mathbf{V} = x\hat{\mathbf{i}} + x(x - 1)(y + 1)\hat{\mathbf{j}}$, where u and v are in m/s and x and y are in meter. Plot the streamline that passes through $x = 0$ and $y = 0$. Compare this streamline with the streakline through the origin.

3.1.11 From time $t = 0$ to $t = 5$ hr radioactive steam is released from a nuclear power plant accident located at $x = -1$ km and $y = 3$ km. The following wind conditions are expected: $\mathbf{V} = 10\hat{\mathbf{i}} - 5\hat{\mathbf{j}}$ kmph for $0 < t < 3$ hr, $\mathbf{V} = 15\hat{\mathbf{i}} + 8\hat{\mathbf{j}}$ kmph for $3 < t < 10$ hr, and $\mathbf{V} = 5\hat{\mathbf{i}}$ kmph for $t > 10$ hr. Draw to scale the expected streakline of the steam for $t = 3, 10$, and 15 hr.

3.1.12 The x and y components of a velocity field are given by $u = 4(x^2y)$ and $v = -4xy^2$. Determine the equation for the streamlines of this flow and compare it with those in Example 3.3. Is the flow in this problem the same as that in Example 3.3? Explain.

3.1.13 In addition to the customary horizontal velocity components of the air in the atmosphere (the "wind"), there often are vertical air currents (thermals) caused by buoyant effects due to uneven heating of the air as indicated in **Fig. P3.1.13**. Assume that the velocity field in a certain region is approximated by $u = u_0, v = v_0(1 - y/h)$ for $0 < y < h$, and $u = u_0, v = 0$ for $y > h$. Plot the shape of the streamline that passes through the origin for values of $u_0/v_0 = 0.5, 1$, and 2.

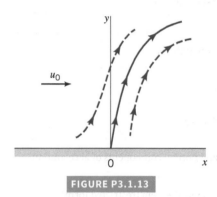

FIGURE P3.1.13

3.1.14 (O) For any steady flow the streamlines and streaklines are the same. For most unsteady flows this is not true. However, there are unsteady flows for which the streamlines and streaklines are the same. Describe a flow field for which this is true.

3.1.15 A tornado has the following velocity components in polar coordinates:

$$V_r = -\frac{C_1}{r} \quad \text{and} \quad V_\theta = -\frac{C_2}{r}$$

Note that the air is spiraling inward. Find an equation for the streamlines. r and θ are polar coordinates.

3.1.16 (See The Wide World of Fluids article titled "Follow those particles," Section 3.1.) Two photographs of four particles in a flow past a sphere are superposed as shown in **Fig. P3.1.16**. The time interval between the photos is $\Delta t = 0.002$ s. The locations of the particles, as determined from the photos, are shown in the table. (a) Determine the fluid velocity for these particles. (b) Plot a graph to compare the results of part (a) with the theoretical velocity, which is given by $V = V_0(1 + a^3/x^3)$, where a is the sphere radius and V_0 is the fluid speed far from the sphere.

Particle	x at $t = 0$ s (m)	x at $t = 0.002$ s (m)
1	−0.500	−0.480
2	−0.250	−0.232
3	−0.140	−0.128
4	−0.120	−0.112

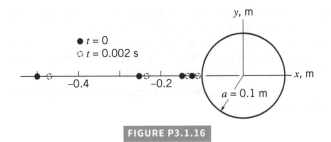

FIGURE P3.1.16

3.1.17 Air flows steadily through a circular, constant-diameter duct. The air is perfectly inviscid, so the velocity profile is flat across each flow area. However, the air density decreases as the air flows down the duct. Is this a one-, two-, or three-dimensional flow?

3.1.18 A constant-density fluid flows in the converging, two-dimensional channel shown in **Fig. P3.1.18**. The width perpendicular to the paper is quite large compared to the channel height. The velocity in the z direction is zero. The channel half-height, Y, and the fluid x velocity, u, are given by

$$Y = \frac{Y_0}{1 + x/\ell} \quad \text{and} \quad u = u_0 \left(1 + \frac{x}{\ell}\right)\left[1 - \left(\frac{y}{Y}\right)^2\right]$$

where x, y, Y, and ℓ are in meters, u is in m/s, $u_0 = 1.0$ m/s, and $Y_0 = 1.0$ m. (a) Is this flow steady or unsteady? Is it one-dimensional, two-dimensional, or three-dimensional? (b) Plot the velocity distribution $u(y)$ at $x/\ell = 0, 0.5$, and 1.0. Use y/Y values of $0, \pm0.2, \pm0.4, \pm0.6, \pm0.8$, and ±1.0.

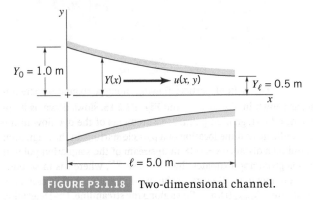

FIGURE P3.1.18 Two-dimensional channel.

3.1.19 ⊙ Pathlines and streaklines provide ways to visualize flows. Another technique would be to instantly inject a line of dye across streamlines and observe how this line moves as time increases. For example, consider the initially straight dye line injected in front of the circular cylinder shown in **Fig. P3.1.19**. Discuss how this dye line would appear at later times. How would you calculate the location of this line as a function of time? This type of line is called a *timeline*.

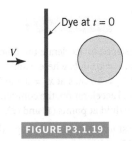

FIGURE P3.1.19

3.1.20 Classify the following flows as one-, two-, or three-dimensional. Sketch a few streamlines for each. (a) Rainwater flow down a wide driveway (b) Flow in a straight horizontal pipe (c) Flow in a straight pipe inclined upward at a 45° angle (d) Flow in a long pipe that

follows the ground in hilly country (e) Flow over an airplane (f) Wind blowing past a tall telephone (g) Flow in the impeller of a centrifugal pump

Section 3.2 The Acceleration Field

3.2.1 The velocity components of u and v of a two-dimensional flow are given by

$$u = ax + \frac{bx}{x^2 y^2} \quad \text{and} \quad v = ay + \frac{by}{x^2 y^2}$$

where a and b are constants. Calculate the acceleration.

3.2.2 Air is delivered through a constant-diameter duct by a fan. The air is inviscid, so the fluid velocity profile is "flat" across each cross section. During the fan start-up, the following velocities were measured at the time t and axial positions x:

	$x = 0$	$x = 10$ m	$x = 20$ m
$t = 0$ s	$V = 0$ m/s	$V = 0$ m/s	$V = 0$ m/s
$t = 1.0$ s	$V = 1.00$ m/s	$V = 1.20$ m/s	$V = 1.40$ m/s
$t = 2.0$ s	$V = 1.70$ m/s	$V = 1.80$ m/s	$V = 1.90$ m/s
$t = 3.0$ s	$V = 2.10$ m/s	$V = 2.15$ m/s	$V = 2.20$ m/s

Calculate the local acceleration, the convective acceleration, and the total acceleration at $t = 1.0$ s and $x = 10$ m. What is the local acceleration after the fan has reached a steady air flowrate?

3.2.3 Water flows through a constant diameter pipe with a uniform velocity given by $\mathbf{V} = (8/t + 5)\hat{\mathbf{j}}$ m/s, where t is in seconds. Determine the acceleration at time $t = 1, 2$, and 10 s.

3.2.4 The velocity of air in the diverging pipe shown in **Fig. P3.2.4** is given by $V_1 = 4t$ m/s and $V_2 = 2t$ m/s, where t is in seconds. (a) Determine the local acceleration at points (1) and (2). (b) Is the average convective acceleration between these two points negative, zero, or positive? Explain.

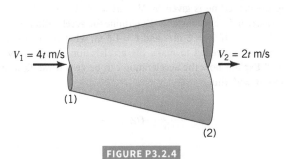

FIGURE P3.2.4

3.2.5 A certain flow field has the velocity vector

$$\mathbf{V} = \frac{-2xyz}{(x^2 + y^2)^2}\hat{\mathbf{i}} + \frac{(x^2 - y^2)z}{(x^2 + y^2)^2}\hat{\mathbf{j}} + \frac{y}{x^2 + y^2}\hat{\mathbf{k}}$$

Find the acceleration vector for this flow.

3.2.6 Determine the x-component of the acceleration, a_x, along the centerline ($y = 0$) for the flow of Problem 3.1.20. Can you determine the acceleration vector at a location not on the centerline? Why or why not?

3.2.7 The velocity of the water in the pipe shown in **Fig. P3.2.7** is given by $V_1 = 0.50t$ m/s and $V_2 = 1.0t$ m/s, where t is in seconds.

Determine the local acceleration at points (1) and (2). Is the average convective acceleration between these two points negative, zero, or positive? Explain.

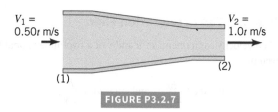

$V_1 = 0.50t$ m/s

$V_2 = 1.0t$ m/s

(1)

(2)

FIGURE P3.2.7

3.2.8 A shock wave is a very thin layer (thickness = ℓ) in a high-speed (supersonic) gas flow across which the flow properties (velocity, density, pressure, etc.) change from state (1) to state (2) as shown in **Fig. P3.2.8**. If $V_1 = 550$ m/s, $V_2 = 215$ m/s, and $\ell = 0.0025$ mm, estimate the average deceleration of the gas as it flows across the shock wave. How many g's deceleration does this represent?

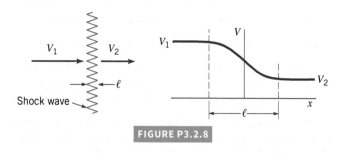

V_1 V_2 ℓ

Shock wave

V_1 V V_2 ℓ x

FIGURE P3.2.8

3.2.9 Ⓞ Estimate the average acceleration of water as it travels through the nozzle on your garden hose. List all assumptions and show all calculations.

3.2.10 Ⓞ A stream of water from the faucet strikes the bottom of the sink. Estimate the maximum acceleration experienced by the water particles. List all assumptions and show calculations.

3.2.11 As a valve is opened, water flows through the diffuser shown in **Fig. P3.2.11** at an increasing flowrate so that the velocity along the centerline is given by $\mathbf{V} = u\hat{\mathbf{i}} = V_0(1 - e^{-ct})(1 - x/\ell)\hat{\mathbf{i}}$, where u_0, c, and ℓ are constants. Determine the acceleration as a function of x and t. If $V_0 = 3$ m/s and $\ell = 1.5$ m, what value of c (other than $c = 0$) is needed to make the acceleration zero for any x at $t = 1$ s? Explain how the acceleration can be zero if the flowrate is increasing with time.

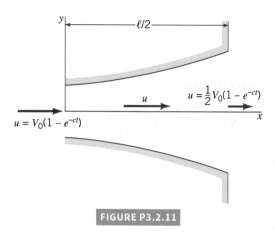

y $\ell/2$

u $u = \frac{1}{2}V_0(1 - e^{-ct})$ x

$u = V_0(1 - e^{-ct})$

FIGURE P3.2.11

3.2.12 A fluid flows along the x axis with a velocity given by $\mathbf{V} = (x/t)\hat{\mathbf{i}}$, where x is in meter and t in seconds. (a) Plot the speed for

$0 \leq x \leq 3$ m and $t = 3$ s. (b) Plot the speed for $x = 2.1$ m and $2 \leq t \leq 4$ s. (c) Determine the local and convective acceleration. (d) Show that the acceleration of any fluid particle in the flow is zero. (e) Explain physically how the velocity of a particle in this unsteady flow remains constant throughout its motion.

3.2.13 A constant-density fluid flows through a converging section having an area A given by

$$A = \frac{A_0}{1 + (x/\ell)}$$

where A_0 is the area at $x = 0$. Determine the velocity and acceleration of the fluid in Eulerian form and then the velocity and acceleration of a fluid particle in Lagrangian form. The velocity is V_0 at $x = 0$ when $t = 0$.

3.2.14 ⟮Video⟯ A hydraulic jump is a rather sudden change in depth of a liquid layer as it flows in an open channel as shown in **Fig. P3.2.14** and Video V10.11. In a relatively short distance (thickness = ℓ) the liquid depth changes from z_1 to z_2, with a corresponding change in velocity from V_1 to V_2. If $V_1 = 0.365$ m/s, $V_2 = 0.09$ m/s, and $\ell = 0.006$ m, estimate the average deceleration of the liquid as it flows across the hydraulic jump. How many g's deceleration does this represent?

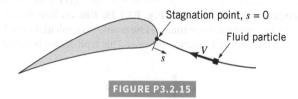

Hydraulic jump

ℓ V_2

V_1

z_1 z_2

FIGURE P3.2.14

3.2.15 ⟮Video⟯ A fluid particle flowing along a stagnation streamline, as shown in Video V3.9 and **Fig. P3.2.15**, slows down as it approaches the stagnation point. Measurements of the dye flow in the video indicate that the location of a particle starting on the stagnation streamline a distance $s = 0.18$ m upstream of the stagnation point at $t = 0$ is given approximately by $s = 0.18e^{-0.5t}$, where t is in seconds and s is in meter. (a) Determine the speed of a fluid particle as a function of time, $V_{particle}(t)$, as it flows along the streamline. (b) Determine the speed of the fluid as a function of position along the streamline, $V = V(s)$. (c) Determine the fluid acceleration along the streamline as a function of position, $a_s = a_s(s)$.

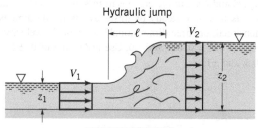

Stagnation point, $s = 0$

Fluid particle

s V

FIGURE P3.2.15

3.2.16 A nozzle is designed to accelerate the fluid from V_1 to V_2 in a linear fashion. That is, $V = ax + b$, where a and b are constants. If the flow is constant with $V_1 = 10$ m/s at $x_1 = 0$ and $V_2 = 25$ m/s at $x_2 = 1$ m, determine the local acceleration, the convective acceleration, and the acceleration of the fluid at points (1) and (2).

3.2.17 An incompressible fluid flows through the converging duct shown in **Fig. P3.2.17a** with velocity V_0 at the entrance. Measurements indicate that the actual velocity of the fluid near the wall of the duct along streamline A–F is as shown in **Fig. P3.2.17b**. Sketch the component of acceleration along this streamline, a, as a function of s. Discuss the important characteristics of your result.

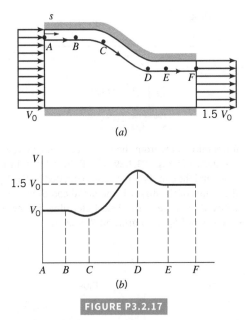

(a)

(b)

FIGURE P3.2.17

3.2.18 C Air flows steadily through a variable area pipe with a velocity of $\mathbf{V} = u(x)\hat{\mathbf{i}}$ m/s, where the approximate measured values of $u(x)$ are given in the table. Plot the acceleration as a function of x for $0 \le x \le 0.3$ m. Plot the acceleration if the flowrate is increased by a factor of N (i.e., the values of u are increased by a factor of N) for $N = 2, 4, 10$.

x (mm)	u (m/s)	x (mm)	u (m/s)
0	3.0	177.8	6.12
25.4	3.1	203.2	5.3
50.8	4.0	228.6	4.1
76.2	6.1	254	3.6
101.6	8.6	279.4	3.1
127	8.65	304.8	3
152.4	7.86	330.2	3

3.2.19 C As is indicated in **Fig. P3.2.19**, the speed of exhaust in a car's exhaust pipe varies in time and distance because of the periodic nature of the engine's operation and the damping effect with distance from the engine. Assume that the speed is given by $V = V_0[1 + ae^{-bx} \sin(\omega t)]$, where $V_0 = 8$ m/s, $a = 0.05$, $b = 0.2$ m^{-1}, and $\omega = 50$ rad/s. Calculate and plot the fluid acceleration at $x = 0, 1, 2, 3, 4$, and 5 m for $0 \le t \le \pi/25$ s.

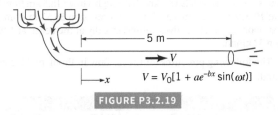

FIGURE P3.2.19

3.2.20 Water flows down the face of the dam shown in **Fig. P3.2.20**. The face of the dam consists of two circular arcs with radii of 3 and 6 m as shown. If the speed of the water along streamline A–B is approximately $V = (2gh)^{1/2}$, where the distance h is as indicated, plot the normal acceleration as a function of distance along the streamline, $a_n = a_n(s)$.

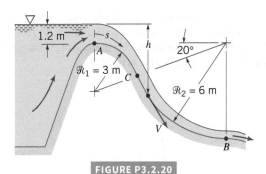

FIGURE P3.2.20

3.2.21 Water flows over the crest of a dam with speed V as shown in **Fig. P3.2.21**. Determine the speed if the magnitude of the normal acceleration at point (1) is to equal the acceleration of gravity, g.

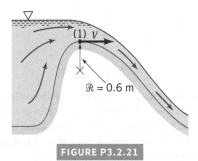

FIGURE P3.2.21

3.2.22 Water flows under the sluice gate shown in **Fig. P3.2.22**. If $V_1 = 3$ m/s, what is the normal acceleration at point (1)?

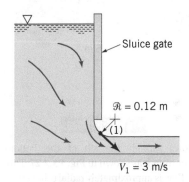

FIGURE P3.2.22

3.2.23 A fluid flows past a sphere with an upstream velocity of $V_0 = 40$ m/s as shown in **Fig. P3.2.23**. From a more advanced theory it is found that the speed of the fluid along the front part of the sphere is $V = \frac{3}{2}V_0 \sin \theta$. Determine the streamwise and normal components of acceleration at point A if the radius of the sphere is $a = 0.20$ m.

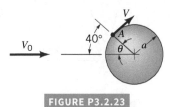

FIGURE P3.2.23

3.2.24 Video Assume that the streamlines for the wingtip vortices from an airplane (see Video V3.6) can be approximated by circles of radius r and that the speed is $V = K/r$, where K is a constant. Determine the streamline acceleration, a_s, and the normal acceleration, a_n, for this flow.

3.2.25 The velocity components for steady flow through the nozzle shown in **Fig. P3.2.25** are $u = -V_0 x/\ell$ and $v = V_0 [1 + (y/\ell)]$, where V_0 and ℓ are constants. Determine the ratio of the magnitude of the acceleration at point (1) to that at point (2).

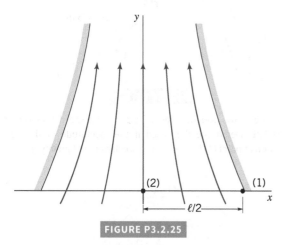

FIGURE P3.2.25

3.2.26 Water flows through the curved hose shown in **Fig. P3.2.26** with an increasing speed of $V = 3t$ m/s, where t is in seconds. For $t = 2$ s determine (a) the component of acceleration along the streamline, (b) the component of acceleration normal to the streamline, and (c) the net acceleration (magnitude and direction).

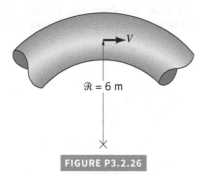

FIGURE P3.2.26

3.2.27 Water flows though the slit at the bottom of a two-dimensional water trough as shown in **Fig. P3.2.27**. Throughout most of the trough the flow is approximately radial (along rays from O) with a velocity of $V = c/r$, where r is the radial coordinate and c is a constant. If the velocity is 0.04 m/s when $r = 0.1$ m, determine the acceleration at points A and B.

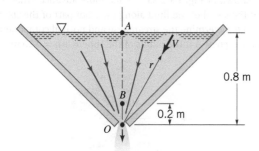

FIGURE P3.2.27

3.2.28 Air flows from a pipe into the region between two parallel circular disks as shown in **Fig. P3.2.28**. The fluid velocity in the gap between the disks is closely approximated by $V = V_0 R/r$, where R is the radius of the disk, r is the radial coordinate, and V_0 is the fluid velocity at the edge of the disk. Determine the acceleration for $r = 0.3$, 0.4 or 0.9 m if $V_0 = 1.5$ m/s and $R = 0.9$ m.

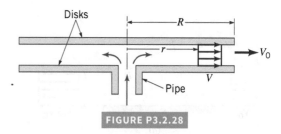

FIGURE P3.2.28

3.2.29 Air flows into a pipe from the region between a circular disk and a cone as shown in **Fig. P3.2.29**. The fluid velocity in the gap between the disk and the cone is closely approximated by $V = V_0 R^2/r^2$, where R is the radius of the disk, r is the radial coordinate, and V_0 is the fluid velocity at the edge of the disk. Determine the acceleration for $r = 0.15$ and 0.6 m if $V_0 = 1.5$ m/s and $R = 0.6$ m.

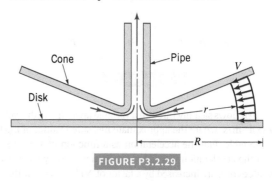

FIGURE P3.2.29

Section 3.2.1 Acceleration and the Material Derivative

3.2.30 A gas flows along the x axis with a speed of $V = 15x$ m/s and a pressure of $p = 20x^2$ N/m^2, where x is in meters. (a) Determine the time rate of change of pressure at the fixed location $x = 1.5$. (b) Determine the time rate of change of pressure for a fluid particle flowing past $x = 1.5$. (c) Explain without using any equations why the answers to parts (a) and (b) are different.

3.2.31 Assume the temperature of the exhaust in an exhaust pipe can be approximated by $T = T_0 (1 + ae^{-bx})[1 + c \cos (\omega t)]$, where $T_0 = 100$ °C, $a = 3, b = 0.03$ m^{-1}, $c = 0.05$, and $\omega = 100$ rad/s. If the exhaust speed is a constant 3 m/s, determine the time rate of change of temperature of the fluid particles at $x = 0$ and $x = 4$ m when $t = 0$.

3.2.32 A bicyclist leaves from her home at 9 A.M. and rides to a beach 74 km away. Because of a breeze off the ocean, the temperature at the beach remains 15.6 °C throughout the day. At the cyclist's home the temperature increases linearly with time, going from 15.6 °C at 9 A.M. to 26.7 °C by 1 P.M. The temperature is assumed to vary linearly as a function of position between the cyclist's home and the beach. Determine the rate of change of temperature observed by the cyclist for the following conditions: (a) as she pedals 18.5 kmph through a town 18.5 km from her home at 10 A.M.; (b) as she eats lunch at a rest stop 55.5 km from her home at noon; (c) as she arrives enthusiastically at the beach at 1 P.M., pedaling 37 kmph.

3.2.33 The following pressures for the air flow in Problem 3.2.2 were measured:

	$x = 0$	$x = 10$ m	$x = 20$ m
$t = 0$ s	$p = 101$ kPa	$p = 101$ kPa	$p = 101$ kPa
$t = 1.0$ s	$p = 121$ kPa	$p = 116$ kPa	$p = 111$ kPa
$t = 2.0$ s	$p = 141$ kPa	$p = 131$ kPa	$p = 121$ kPa
$t = 3.0$ s	$p = 171$ kPa	$p = 151$ kPa	$p = 131$ kPa.

Find the local rate of change of pressure $\partial p/\partial t$ and the convective rate of change of pressure $V \partial p/\partial x$ at $t = 2.0$ s and $x = 10$ m.

Section 3.4 The Reynolds Transport Theorem

3.4.1 Obtain a photograph/image of a situation in which a fluid is flowing. Print this photo and draw a control volume through which the fluid flows. Write a brief paragraph that describes how the fluid flows into and out of this control volume.

3.4.2 [Video] In the region just downstream of a sluice gate, the water may develop a reverse flow region as is indicated in **Fig. P3.4.2** and Video V10.9. The velocity profile is assumed to consist of two uniform regions, one with velocity $V_a = 3$ m/s and the other with $V_b = 0.9$ m/s. Determine the net flowrate of water across the portion of the control surface at section (2) if the channel is 6 m wide.

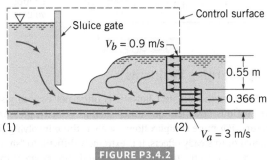

FIGURE P3.4.2

3.4.3 At time $t = 0$ the valve on an initially empty (perfect vacuum, $\rho = 0$) tank is opened and air rushes in. If the tank has a volume of V_0 and the density of air within the tank increases as $\rho = \rho_\infty(1 - e^{-bt})$, where b is a constant, determine the time rate of change of mass within the tank.

3.4.4 [O] From calculus, one obtains the following formula (Leibnitz rule) for the time derivative of an integral that contains time in both the integrand and the limits of the integration:

$$\frac{d}{dt}\int_{x_1(t)}^{x_2(t)} f(x, t)\,dx = \int_{x_1}^{x_2} \frac{\partial f}{\partial t}\,dx + f(x_2, t)\frac{dx_2}{dt} - f(x_1, t)\frac{dx_1}{dt}$$

Discuss how this formula is related to the time derivative of the total amount of a property in a system and to the Reynolds transport theorem.

3.4.5 Air enters an elbow with a uniform speed of 10 m/s as shown in **Fig. P3.4.5**. At the exit of the elbow, the velocity profile is not uniform. In fact, there is a region of separation or reverse flow. The fixed control volume $ABCD$ coincides with the system at time $t = 0$. Make a sketch to indicate (a) the system at time $t = 0.01$ s and (b) the fluid that has entered and exited the control volume in that time period.

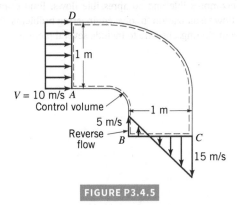

FIGURE P3.4.5

3.4.6 [Video] A layer of oil flows down a vertical plate as shown in **Fig. P3.4.6** with a velocity of $\mathbf{V} = (V_0/h^2)(2hx - x^2)\hat{\mathbf{j}}$, where V_0 and h are constants. (a) Show that the fluid sticks to the plate and that the shear stress at the edge of the layer ($x = h$) is zero. (b) Determine the flowrate across surface AB. Assume the width of the plate is b.

(*Note:* The velocity profile for laminar flow in a pipe has a similar shape. See Video V6.12.)

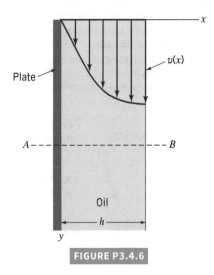

FIGURE P3.4.6

3.4.7 **Figure P3.4.7** shows a fixed control volume. It has a volume $V_0 = 0.028$ m³, a flow area $A = 0.09$ m², and a length $\ell_0 = 0.3$ m. Position x represents the center of the control volume, where the fluid velocity $V_0 = 0.3$ m/s and the density $\rho_0 = 927.7$ kg/m³. Also, at position x the fluid density does not change locally with time but decreases in the axial direction at the linear rate of 450.6 kg/m⁴. Use the system or Lagrangian approach to evaluate $d\rho/dt$. Compare this result with that of the material derivative and flux terms.

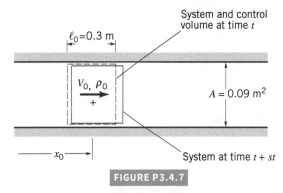

FIGURE P3.4.7

3.4.8 Find DV/Dt for the system in Problem 3.4.7.

3.4.9 Water enters a 1.5 m wide, 0.3 m deep channel as shown in **Fig. P3.4.9**. Across the inlet the water velocity is 1.8 m/s in the center portion of the channel and 0.3 m/s in the remainder of it. Farther downstream the water flows at a uniform 0.6 m/s velocity across the entire channel. The fixed control volume $ABCD$ coincides with the system at time $t = 0$. Make a sketch to indicate (a) the system at time $t = 0.5$ s and (b) the fluid that has entered and exited the control volume in that time period.

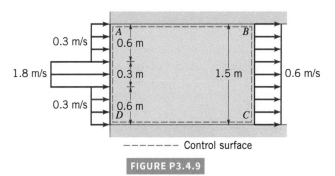

FIGURE P3.4.9

3.4.10 Figure P3.4.10 illustrates a system and fixed control volume at time t and the system at a short time δt later. The system temperature is $T = 310.9$ K at time t and $T = 312.6$ K at time $t + \delta t$, where $\delta t = 0.1$ s. The system mass, m, is 29.2 kg, and 10 percent of it moves out of the control volume in $\delta t = 0.1$ s. The energy per unit mass $\breve{u}$ is $c_v T$, where $c_v = 1.34 \times 10^5$ J/kg · K. The energy U of the system at any time t is $m\breve{u}$. Use the system or Lagrangian approach to evaluate DU/Dt of the system. Compare this result with the DU/Dt evaluated with that of the material time derivative and flux terms.

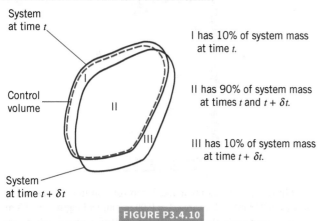

System at time t.

I has 10% of system mass at time t.

Control volume

II has 90% of system mass at times t and $t + \delta t$.

III has 10% of system mass at time $t + \delta t$.

System at time $t + \delta t$

FIGURE P3.4.10

3.4.11 The wind blows across a field with an approximate velocity profile as shown in **Fig. P3.4.11**. Use Eq. 3.16 with the parameter b equal to the velocity to determine the momentum flowrate across the vertical surface A–B, which is of unit depth into the paper.

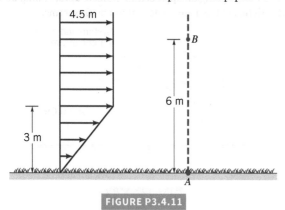

4.5 m

B

6 m

3 m

A

FIGURE P3.4.11

3.4.12 Water flows from a nozzle with a speed of $V = 10$ m/s and is collected in a container that moves toward the nozzle with a speed of $V_{cv} = 2$ m/s as shown in **Fig. P3.4.12**. The moving control surface consists of the inner surface of the container. The system consists of the water in the container at time $t = 0$ and the water between the nozzle and the tank in the constant diameter stream at $t = 0$. At time $t = 0.1$ s what volume of the system remains outside of the control volume? How much water has entered the control volume during this time period? Repeat the problem for $t = 0.3$ s.

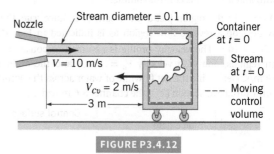

Nozzle　　Stream diameter = 0.1 m　　Container at $t = 0$

$V = 10$ m/s

$V_{cv} = 2$ m/s

3 m

Stream at $t = 0$

Moving control volume

FIGURE P3.4.12

Lifelong Learning Problems

3.1 LL Even for the simplest flows it is often not easy to visually represent various flow field quantities such as velocity, pressure, or temperature. For more complex flows, such as those involving three-dimensional or unsteady effects, it is extremely difficult to "show the data." However, with the use of computers and appropriate software, novel methods are being devised to more effectively illustrate the structure of a given flow. Obtain information about methods used to present complex flow data. Summarize your findings in a brief report.

3.2 LL For centuries people have obtained qualitative and quantitative information about various flow fields by observing the motion of objects or particles in a flow. For example, the speed of the current in a river can be approximated by timing how long it takes a stick to travel a certain distance. The swirling motion of a tornado can be observed by following debris moving within the tornado funnel. Recently, various high-tech methods using lasers and minute particles seeded within the flow have been developed to measure velocity fields. Such techniques include the laser doppler anemometer (LDA), the particle image velocimeter (PIV), and others. Obtain information about new laser-based techniques for measuring velocity fields. Summarize your findings in a brief report.

3.3 LL *Flow visualization* is direct observation of the flow field, usually accomplished in a laboratory. Several photographs in this book (see Figs. 4.1, 9.19, and 11.4) and the majority of the accompanying videos were made with the aid of a flow visualization technique. Do some research on flow visualization techniques, including those for both incompressible and compressible flows. Find examples of the use of flow visualization to solve engineering problems. Prepare a paper on your findings; be sure to include several pictures.

CHAPTER 4

Elementary Fluid Dynamics— The Bernoulli Equation

LEARNING OBJECTIVES

After completing this chapter, you should be able to:

- discuss the application of Newton's second law to fluid flows.
- explain the development, uses, and limitations of the Bernoulli equation.
- use the Bernoulli equation (stand-alone or in combination with the simple continuity equation) to solve flow problems.

- apply the concepts of static, stagnation, dynamic, and total pressures.
- calculate various flow properties using the energy and hydraulic grade lines.

In this chapter we investigate some typical fluid motions (fluid dynamics) in an elementary way. We will discuss in some detail the use of Newton's second law ($\mathbf{F} = m\mathbf{a}$) as it is applied to fluid particle motion that is "ideal" in some sense. We will obtain the celebrated Bernoulli equation and apply it to various flows. Although this equation is one of the oldest in fluid mechanics and the assumptions involved in its derivation are numerous, it can be used effectively to predict, understand, and analyze a variety of flow situations. However, if the equation is applied without proper respect for its restrictions, serious errors can arise. Indeed, the Bernoulli equation is appropriately called the most used and the most abused equation in fluid mechanics.

A thorough understanding of the elementary approach to fluid dynamics involved in this chapter will be useful on its own. It also provides a good foundation for the material in the following chapters where some of the present restrictions are removed and "more nearly exact" results are presented.

> The Bernoulli equation may be the most used and abused equation in fluid mechanics.

4.1 NEWTON'S SECOND LAW

Consider a really tiny volume of fluid, which is still large enough to contain a significant number of molecules. This volume is called a fluid particle (with further description in Section 4.1). As the fluid particle moves from one location to another, it usually experiences an acceleration or deceleration. According to Newton's second law of motion, the net force acting on the fluid particle under consideration must equal its mass times its acceleration,

$$\mathbf{F} = m\mathbf{a}$$

In this chapter we consider fluid motion while neglecting frictional effects. In a flowing fluid, friction is associated with shear stress, and shear stress is associated with viscosity. The simplest way to assure a frictionless flow is to imagine that the fluid has no viscosity. We call this imaginary substance an *inviscid fluid* and the resulting flow an **inviscid flow.**

In practice there are no inviscid fluids, since every fluid supports shear stresses when it is subjected to a rate of strain displacement. For many flow situations the viscous effects are relatively small compared with other effects. As a first approximation for such cases, it is often possible to ignore viscous effects. For example, often the viscous forces developed in flowing water may be several orders of magnitude smaller than forces due to other influences such as gravity or pressure differences. For other water flow situations, however, the viscous effects may be the dominant ones. Similarly, the viscous effects associated with the flow of a gas are often negligible, although in some circumstances they are very important.

With the assumption of inviscid (or frictionless) flow, the fluid motion is governed by pressure and gravity forces only, and Newton's second law as it applies to a fluid particle is

> **Inviscid fluid flow is governed by pressure and gravity forces.**

(Net pressure force on particle) + (net gravity force on particle) =
(particle mass) × (particle acceleration)

The results of the interaction between the pressure, gravity, and acceleration provide numerous useful applications in fluid mechanics.

To apply Newton's second law to a fluid (or any other object), we must define an appropriate coordinate system in which to describe the motion. In general the motion will be three-dimensional and unsteady so that three space coordinates and time are needed to describe it. There are numerous coordinate systems available, including the most often used rectangular (x, y, z) and cylindrical (r, θ, z) systems shown in the adjacent figures. Usually the specific flow geometry dictates which system would be most appropriate.

In this chapter we will be concerned with two-dimensional motion like that confined to the x–z plane as is shown in **Fig. 4.1a**. Clearly, we could choose to describe the flow in terms of the components of acceleration and forces in the x and z coordinate directions. The resulting equations are frequently referred to as a two-dimensional form of the *Euler equations* of motion in rectangular Cartesian coordinates. This approach will be discussed in Chapter 6.

As is done in the study of dynamics (Ref. 1), the motion of each fluid particle is described in terms of its velocity vector, **V**, which is defined as the time rate of change of the position of the particle. The particle's velocity is a vector quantity with a magnitude (the speed, $V = |\mathbf{V}|$) and direction. As the particle moves about, it follows a particular path, the shape of which is governed by the velocity of the particle. The location of the particle along the path is a function of where the particle started at the initial time and its velocity along the path. If it is **steady flow** (i.e., nothing changes with time at a given location in the flow field), each successive particle that passes through a given point [such as point (1) in Fig. 4.1a] will follow the same path. For such cases the path is a fixed line in the x–z plane. Neighboring particles that pass on either side of point (1) follow their own paths, which may be of a different shape than the one passing through (1). The entire x–z plane is filled with such paths.

For steady flows each particle slides along its path, and its velocity vector is everywhere tangent to the path. The lines that are tangent to the velocity vectors throughout the flow field are called **streamlines**. For many situations it is easiest to describe the flow in terms of the "streamline" coordinates based on the streamlines that are illustrated in **Fig. 4.1b**. The particle motion is described in terms of its distance, $s = s(t)$, along the streamline from some convenient origin and the local radius of curvature of the streamline, $\mathcal{R} = \mathcal{R}(s)$. The distance along the

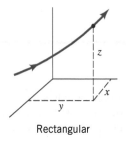

Rectangular

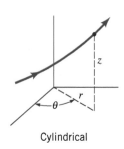

Cylindrical

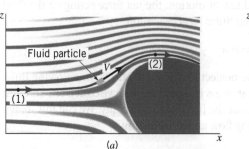

(a)

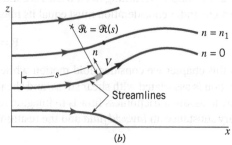

(b)

FIGURE 4.1 (a) Flow in the x–z plane. (b) Flow in terms of streamline and normal coordinates.

streamline is related to the particle's speed by $V = ds/dt$, and the radius of curvature is related to the shape of the streamline. In addition to the coordinate along the streamline, s, the coordinate normal to the streamline, n, as is shown in Fig. 4.1b, will be of use.

Fluid particles accelerate normal to and along streamlines.

To apply Newton's second law to a particle flowing along its streamline, we must write the particle acceleration in terms of the streamline coordinates. By definition, the acceleration is the time rate of change of the velocity of the particle, $\mathbf{a} = d\mathbf{V}/dt$. For two-dimensional flow in the x–z plane, the acceleration has two components—one along the streamline, a_s, the streamwise acceleration, and one normal to the streamline, a_n, the normal acceleration.

Streamwise acceleration results from two facts. First, the speed of the particle may change with time at a point, $V(t)$. Since we are limiting ourselves to steady flow, we will ignore velocity changes with time. (More information is in Section 4.8.2.) Second, the speed of the particle generally varies along the streamline, $V(s)$. For example, in Fig. 4.1a the speed may be 15.24 m/s at point (1) and 30.48 m/s at point (2). Thus, by use of the chain rule of differentiation, the s component of the acceleration is given by $a_s = dV/dt = (\partial V/\partial s)(ds/dt) = (\partial V/\partial s)V$. We have used the fact that speed is the time rate of change of distance, $V = ds/dt$. Note that the streamwise acceleration is the product of the rate of change of speed with distance along the streamline, $\partial V/\partial s$, and the speed, V. Since $\partial V/\partial s$ can be positive, negative, or zero, the streamwise acceleration can, therefore, be positive (acceleration), negative (deceleration), or zero (constant speed).

The normal component of acceleration, the centrifugal acceleration, is given in terms of the particle speed and the radius of curvature of its path. Thus, $a_n = V^2/\mathcal{R}$, where both V and $\mathcal{R}$ may vary along the streamline. These equations for the acceleration should be familiar from the study of particle motion in physics (Ref. 2) or dynamics (Ref. 1). A more complete derivation and discussion of these topics has been covered in Chapter 3.

Thus, the (steady flow) components of acceleration in the s and n directions, a_s and a_n, are given by

$$a_s = V\frac{\partial V}{\partial s}, \qquad a_n = \frac{V^2}{\mathcal{R}} \tag{4.1}$$

where $\mathcal{R}$ is the local radius of curvature of the streamline, and s is the distance measured along the streamline from some arbitrary initial point. In general there can be acceleration along the streamline (because the particle speed changes along its path, $\partial V/\partial s \neq 0$) and acceleration normal to the streamline (because the particle does not flow in a straight line, $\mathcal{R} \neq \infty$). Various flows and the accelerations associated with them are shown in the adjacent figures. As discussed in Section 4.6.2, for incompressible flow the velocity is inversely proportional to the streamline spacing. Hence, converging streamlines produce positive streamwise acceleration. To produce this acceleration there must be a net, nonzero force on the fluid particle.

To determine the forces necessary to produce a given flow (or conversely, what flow results from a given set of forces), we consider the free-body diagram of a small fluid particle as is shown in **Fig. 4.2**. The particle of interest is removed from its surroundings, and the reactions of the surroundings on the particle are indicated by the appropriate forces present, F_1, F_2, and so forth. For the present case, the important forces are assumed to be gravity and pressure. Other forces, such as viscous forces and surface tension effects, are assumed negligible. The acceleration of gravity, g, is assumed to be constant and acts vertically, in the negative z direction, at an angle θ relative to the normal to the streamline.

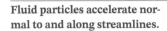

▶ Video

V4.1 Streamlines past an airfoil

$a_s = a_n = 0$

$a_s > 0$

$a_s < 0$

$a_n > 0$

$a_s > 0, \ a_n > 0$

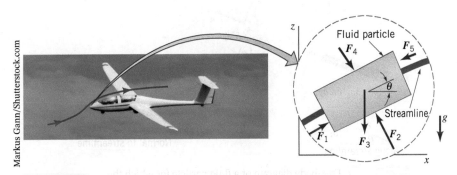

FIGURE 4.2 Isolation of a small fluid particle in a flow field.

4.2 F = ma ALONG A STREAMLINE

Consider the small fluid particle of size δs by δn in the plane of the figure and δy normal to the figure as shown in the free-body diagram of **Fig. 4.3**. Unit vectors along and normal to the streamline are denoted by $\hat{s}$ and $\hat{n}$, respectively. For steady flow, the component of Newton's second law along the streamline direction, s, can be written as

$$\sum \delta F_s = \delta m\, a_s = \delta m\, V \frac{\partial V}{\partial s} = \rho\, \delta\!\!\!\forall\, V \frac{\partial V}{\partial s} \tag{4.2}$$

where $\sum \delta F_s$ represents the sum of the s components of all the forces acting on the particle, which has mass $\delta m = \rho\, \delta\!\!\!\forall$, and $V\, \partial V/\partial s$ is the acceleration in the s direction. Here, $\delta\!\!\!\forall = \delta s\, \delta n\, \delta y$ is the particle volume. Equation 4.2 is valid for both compressible and incompressible fluids. That is, the density need not be constant throughout the flow field.

The gravity force (weight) on the particle can be written as $\delta\!\!\!\;\mathcal{W} = \gamma\, \delta\!\!\!\forall$, where $\gamma = \rho g$ is the specific weight of the fluid (N/m^3). Hence, the component of the weight force in the direction of the streamline is

$$\delta\!\!\!\;\mathcal{W}_s = -\delta\!\!\!\;\mathcal{W} \sin \theta = -\gamma\, \delta\!\!\!\forall \sin \theta$$

If the streamline is horizontal at the point of interest, then $\theta = 0$, and there is no component of particle weight along the streamline to contribute to its acceleration in that direction.

In a flowing fluid the pressure varies from one location to another.

As is indicated in Chapter 2, the pressure is not constant throughout a stationary fluid ($\nabla p \neq 0$) because of the fluid weight. Likewise, in a flowing fluid the pressure is usually not constant. In general, for steady flow, $p = p(s,n)$. If the pressure at the center of the particle shown in Fig. 4.3 is denoted as p, then the pressure at either end may be higher or lower by some amount δp_s. Thus, the pressures on the two end faces that are perpendicular to the streamline are $p + \delta p_s$ and $p - \delta p_s$. Since the particle is "small," we can use a one-term Taylor series expansion for the pressure field (as was done in Chapter 2 for the pressure forces in static fluids) to obtain

$$\delta p_s \approx \frac{\partial p}{\partial s} \frac{\delta s}{2}$$

The net pressure force on a particle is determined by the pressure gradient.

Thus, if δF_{ps} is the net pressure force on the particle in the streamline direction, it follows that

$$\delta F_{ps} = (p - \delta p_s)\, \delta n\, \delta y - (p + \delta p_s)\, \delta n\, \delta y = -2\, \delta p_s\, \delta n\, \delta y$$

$$= -\frac{\partial p}{\partial s}\, \delta s\, \delta n\, \delta y = -\frac{\partial p}{\partial s}\, \delta\!\!\!\forall$$

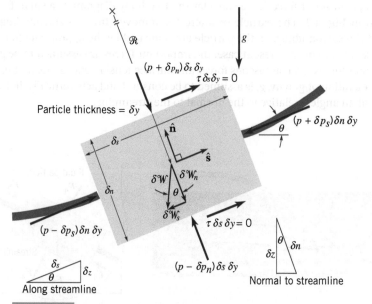

FIGURE 4.3 Free-body diagram of a fluid particle for which the important forces are those due to pressure and gravity.

Note that the actual level of the pressure, p, is not important. What produces a net pressure force is the fact that the pressure is not constant throughout the fluid. The nonzero pressure gradient, $\nabla p = \partial p/\partial s\,\hat{s} + \partial p/\partial n\,\hat{n}$, is what provides a net pressure force on the particle. Viscous forces, represented by $\tau\,\delta s\,\delta y$, are zero, since the fluid is inviscid.

With gravity and pressure forces individually formulated, the net force acting in the streamline direction on the particle shown in Fig. 4.3 is given by

$$\sum \delta F_s = \delta\mathcal{W}_s + \delta F_{ps} = \left(-\gamma \sin\theta - \frac{\partial p}{\partial s}\right)\delta\mathcal{V} \tag{4.3}$$

By combining Eqs. 4.2 and 4.3, we obtain the following equation of motion for steady inviscid flow along the streamline direction:

$$-\gamma \sin\theta - \frac{\partial p}{\partial s} = \rho V\frac{\partial V}{\partial s} = \rho a_s \tag{4.4}$$

We have divided out the common particle volume factor, $\delta\mathcal{V}$, that appears in both the force and the acceleration portions of the equation. This is a representation of the fact that it is the fluid density (mass per unit volume), not the mass, per se, of the fluid particle that is important.

The physical interpretation of Eq. 4.4 is that a change in fluid particle speed is accomplished by the appropriate combination of pressure gradient and particle weight along the streamline. For fluid static situations this balance between pressure and gravity forces is such that no change in particle speed is produced—the right-hand side of Eq. 4.4 is zero, and the particle remains stationary. In a flowing fluid the pressure and weight forces do not necessarily balance—the force unbalance provides the appropriate acceleration and, hence, particle motion.

EXAMPLE 4.1 | Pressure Variation along a Streamline

Given Consider the inviscid, incompressible, steady flow along the horizontal streamline A–B in front of the soccer ball (sphere) of radius a, as shown in **Fig. E4.1a**. From a more advanced theory of flow past a sphere, the fluid velocity along this streamline is

$$V = V_0\left(1 + \frac{a^3}{x^3}\right)$$

as shown in **Fig. E4.1b**.

Find Determine the pressure variation along the streamline from point A far in front of the sphere ($x_A = -\infty$ and $V_A = V_0$) to point B on the sphere ($x_B = -a$ and $V_B = 0$).

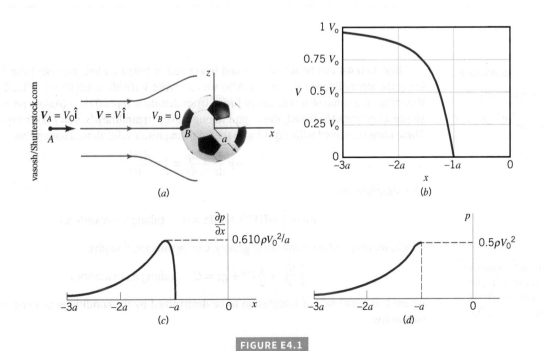

FIGURE E4.1

Solution   Since the flow is steady and inviscid, Eq. 4.4 is valid. In addition, since the streamline is horizontal, $\sin \theta = \sin 0° = 0$ and the equation of motion along the streamline reduces to

$$\frac{\partial p}{\partial s} = -\rho V \frac{\partial V}{\partial s} \tag{1}$$

With the given velocity variation along the streamline, the acceleration term is

$$V \frac{\partial V}{\partial s} = V \frac{\partial V}{\partial x} = V_0 \left(1 + \frac{a^3}{x^3}\right)\left(-\frac{3 V_0 a^3}{x^4}\right)$$
$$= -3 V_0^2 \left(1 + \frac{a^3}{x^3}\right)\frac{a^3}{x^4}$$

where we have replaced s by x, since the two coordinates are identical (within an additive constant) along streamline A–B. It follows that $V\, \partial V/\partial s < 0$ along the streamline. The fluid slows down from V_0 far ahead of the sphere to zero velocity on the "nose" of the sphere ($x = -a$).

Thus, according to Eq. 1, to produce the given motion the pressure gradient along the streamline is

$$\frac{\partial p}{\partial x} = \frac{3\rho a^3 V_0^2 (1 + a^3/x^3)}{x^4} \tag{2}$$

This variation is indicated in **Fig. E4.1c**. It is seen that the pressure increases in the direction of flow ($\partial p/\partial x > 0$) from point A to point B. The maximum pressure gradient ($0.610\ \rho V_0^2/a$) occurs just slightly ahead of the sphere ($x = -1.205a$). It is the pressure gradient that slows the fluid down from $V_A = V_0$ to $V_B = 0$ as shown in Fig. E4.1b.

The pressure distribution along the streamline can be obtained by integrating Eq. 2 from $p = 0$ (gage) at $x = -\infty$ to pressure p at location x. The result, plotted in **Fig. E4.1d**, is

$$p = -\rho V_0^2 \left[\left(\frac{a}{x}\right)^3 + \frac{(a/x)^6}{2}\right] \tag{Ans}$$

Comment   The pressure at B, a stagnation point, since $V_B = 0$, is the highest pressure along the streamline ($p_B = \rho V_0^2/2$). As shown in Chapter 9, this excess pressure on the front of the sphere (i.e., $p_B > 0$) contributes to the net drag force on the sphere. Note that the pressure gradient and pressure are directly proportional to the density of the fluid, a representation of the fact that the fluid inertia is proportional to its mass.

The Wide World of Fluids

Incorrect raindrop shape

The incorrect representation that raindrops are teardrop shaped is found nearly everywhere—from children's books to weather maps on the Weather Channel. About the only time raindrops possess the typical teardrop shape is when they run down a windowpane. The actual shape of a falling raindrop is a function of the size of the drop and results from a balance between surface tension forces and the air pressure exerted on the falling drop. Small drops with a radius less than about 0.5 mm have a spherical shape because the surface tension effect (which is inversely proportional to drop size) wins over the increased pressure, $\rho V_0^2/2$, caused by the motion of the drop and exerted on its bottom. With increasing size, the drops fall faster and the increased pressure causes the drops to flatten. A 2-mm drop, for example, is flattened into a hamburger bun shape. Slightly larger drops are actually concave on the bottom. When the radius is greater than about 4 mm, the depression of the bottom increases and the drop takes on the form of an inverted bag with an annular ring of water around its base. This ring finally breaks up into smaller drops.

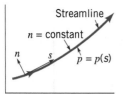

Equation 4.4 can be rearranged and integrated as follows. First, we note from Fig. 4.3 that along the streamline $\sin \theta = dz/ds$. Also we can write $V\, dV/ds = \frac{1}{2}d(V^2)/ds$. Finally, along the streamline the value of n is constant ($dn = 0$) so that $dp = (\partial p/\partial s)\, ds + (\partial p/\partial n)\, dn = (\partial p/\partial s)\, ds$. Hence, as indicated by the adjacent figure, along a given streamline $p(s, n) = p(s)$ and $\partial p/\partial s = dp/ds$. These ideas combined with Eq. 4.4 give the following result valid along a streamline

$$-\gamma \frac{dz}{ds} - \frac{dp}{ds} = \frac{1}{2}\rho \frac{d(V^2)}{ds}$$

This simplifies to

$$dp + \frac{1}{2}\rho d(V^2) + \gamma\, dz = 0 \quad \text{(along a streamline)} \tag{4.5}$$

which, for constant acceleration of gravity, can be integrated to give

$$\int \frac{dp}{\rho} + \frac{1}{2}V^2 + gz = C \quad \text{(along a streamline)} \tag{4.6}$$

For steady, inviscid flow the sum of certain pressure, velocity, and elevation effects is constant along a streamline.

where C is a constant of integration to be determined by the conditions at some point on the streamline.

In general it is not possible to integrate the pressure term because the density may not be constant and, therefore, cannot be removed from under the integral sign. To carry out this integration we must know specifically how the density varies with pressure. This is not always easily determined. For example, for an ideal gas the density, pressure, and temperature are related according to $\rho = p/RT$, where R is the gas constant. To know how the density varies with pressure, we must also know the temperature variation. For now we will assume that the density and specific weight are constant (incompressible flow). The justification for this assumption and the consequences of compressibility will be considered further in Section 4.8.1 and more fully in Chapter 11.

With the additional assumption that the density remains constant (a very good assumption for liquids and also for gases if the speed is "not too high"), Eq. 4.6 assumes the following simple representation for steady, inviscid, incompressible flow.

$$\boxed{p + \frac{1}{2}\rho V^2 + \gamma z = \text{constant along streamline}} \qquad (4.7)$$

This is the celebrated **Bernoulli equation**—a very powerful tool in fluid mechanics. In 1738 Daniel Bernoulli (1700–1782) published his *Hydrodynamics* in which an equivalent of this famous equation first appeared. To use it correctly we must constantly remember the basic assumptions used in its derivation: (1) viscous effects are assumed negligible, (2) the flow is assumed to be steady, (3) the flow is assumed to be incompressible, and (4) the equation is applicable along a streamline. In the derivation of Eq. 4.7, we assume that the flow takes place in a plane (the x–z plane). In general, this equation is valid for both planar and nonplanar (three-dimensional) flows, provided it is applied along the streamline.

We will provide many examples to illustrate the correct use of the Bernoulli equation and will show how a violation of the basic assumptions used in the derivation of this equation can lead to erroneous conclusions. The constant of integration in the Bernoulli equation can be evaluated if sufficient information about the flow is known at one location along the streamline.

V4.2 Balancing ball

V4.3 Flow past a biker

EXAMPLE 4.2 | The Bernoulli Equation

Given Consider the flow of air around a bicyclist moving through still air with velocity V_0, as is shown in **Fig. E4.2**.

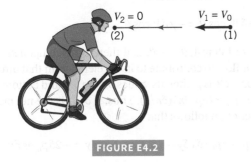

$V_2 = 0$ (2) $V_1 = V_0$ (1)

FIGURE E4.2

Find Determine the difference in the pressure between points (1) and (2).

Solution In a coordinate fixed to the ground, the flow is unsteady as the bicyclist rides by. However, in a coordinate system fixed to the bike, it appears as though the air is flowing steadily toward the bicyclist with speed V_0. Since use of the Bernoulli equation is restricted to steady flows, we select the coordinate system fixed to the bike. If the assumptions of Bernoulli's equation are valid (steady, incompressible, inviscid flow), Eq. 4.7 can be applied as follows along the streamline that passes through (1) and (2)

$$p_1 + \frac{1}{2}\rho V_1^2 + \gamma z_1 = p_2 + \frac{1}{2}\rho V_2^2 + \gamma z_2$$

We consider (1) to be in the free stream so that $V_1 = V_0$ and (2) to be at the tip of the bicyclist's nose and assume that $z_1 = z_2$ and $V_2 = 0$ (both of which, as is discussed in Section 4.5, are reasonable assumptions). It follows that the pressure at (2) is greater than that at (1) by an amount

$$p_2 - p_1 = \frac{1}{2}\rho V_1^2 = \frac{1}{2}\rho V_0^2 \qquad (Ans)$$

Comments A similar result was obtained in Example 4.1 by integrating the pressure gradient, which was known because the velocity distribution along the streamline, $V(s)$, was known. The Bernoulli equation is a general integration of $\mathbf{F} = m\mathbf{a}$. To determine $p_2 - p_1$, knowledge of the detailed velocity distribution is not needed—only the "boundary conditions" at (1) and (2) are required. If, on the other hand, pressure is to be determined at a point between (1) and (2), the value of V must be known at that point along the streamline. Note that if we measure $p_2 - p_1$, we can determine the speed, V_0. As discussed in Section 4.5, this is the principle on which many velocity-measuring devices are based.

If the bicyclist were accelerating or decelerating, the flow would be unsteady (i.e., $V_0 \neq$ constant) and the above analysis would be incorrect because Eq. 4.7 is restricted to steady flow.

The difference in fluid velocity between two points in a flow field, V_1 and V_2, can often be controlled by appropriate geometric constraints of the fluid. For example, a garden hose nozzle is designed to give a much higher velocity at the exit of the nozzle than at its entrance where it is attached to the hose. As is shown by the Bernoulli equation, the pressure within the hose must be larger than that at the exit (for constant elevation, an increase in velocity requires a decrease in pressure if Eq. 4.7 is valid). It is this pressure drop that accelerates the water through the nozzle. Similarly, an airfoil is designed so that the fluid velocity over its upper surface is greater (on the average) than that along its lower surface. From the Bernoulli equation, therefore, the average pressure on the lower surface is greater than that on the upper surface. A net upward force, the lift, results. In essence, the Bernoulli equation represents a balancing act of pressure, velocity, and elevation between points along a streamline.

4.3 F = *ma* NORMAL TO A STREAMLINE

Video

V4.4 Hydrocyclone separator

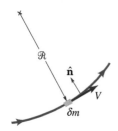

In this section we will consider application of Newton's second law in a direction normal to the streamline. In many flows the streamlines are relatively straight, the flow is essentially one-dimensional, and variations in parameters across streamlines (in the normal direction) can often be neglected when compared to the variations along the streamline. However, in numerous other situations valuable information can be obtained from considering $\mathbf{F} = m\mathbf{a}$ normal to the streamlines. For example, the devastating low-pressure region at the center of a tornado can be explained by applying Newton's second law across the nearly circular streamlines of the tornado.

We again consider the force balance on the fluid particle shown in Fig. 4.3 and the adjacent figure. This time, however, we consider components in the normal direction, $\hat{\mathbf{n}}$. We assume the flow is steady with a normal acceleration $a_n = V^2/\mathcal{R}$, where $\mathcal{R}$ is the local radius of curvature of the streamlines. This acceleration is produced by the change in direction of the particle's velocity as it moves along a curved path. Newton's second law in the normal direction is

$$\sum \delta F_n = \frac{\delta m\, V^2}{\mathcal{R}} = \frac{\rho\, \delta\forall\, V^2}{\mathcal{R}} \tag{4.8}$$

where $\sum \delta F_n$ represents the sum of n components of all the forces acting on the particle and δm is particle mass.

We again assume that the only forces of importance are pressure and gravity. The component of the weight (gravity force) in the normal direction is

$$\delta \mathcal{W}_n = -\delta \mathcal{W} \cos\theta = -\gamma\, \delta\forall \cos\theta$$

To apply F = *ma* normal to streamlines, the normal components of force are needed.

If the streamline is vertical at the point of interest, $\theta = 90°$, and there is no component of the particle weight normal to the direction of flow to contribute to its acceleration in that direction.

If the pressure at the center of the particle is p, then its values on the top and bottom of the particle are $p + \delta p_n$ and $p - \delta p_n$, where $\delta p_n \approx (\partial p/\partial n)(\delta n/2)$. Thus, if δF_{pn} is the net pressure force on the particle in the normal direction, it follows that

$$\delta F_{pn} = (p - \delta p_n)\,\delta s\, \delta y - (p + \delta p_n)\,\delta s\, \delta y = -2\delta p_n\, \delta s\, \delta y$$

$$= -\frac{\partial p}{\partial n}\, \delta s\, \delta n\, \delta y = -\frac{\partial p}{\partial n}\, \delta\forall$$

Hence, the net force acting in the normal direction on the particle shown in Fig 4.3 is given by

Video

V4.5 Aircraft wing tip vortex

$$\sum \delta F_n = \delta \mathcal{W}_n + \delta F_{pn} = \left(-\gamma \cos\theta - \frac{\partial p}{\partial n}\right)\delta\forall \tag{4.9}$$

By combining Eqs. 4.8 and 4.9 and using the fact that along a line normal to the streamline $\cos\theta = dz/dn$ (see Fig. 4.3), we obtain the following equation of motion along the normal direction:

$$-\gamma \frac{dz}{dn} - \frac{\partial p}{\partial n} = \frac{\rho V^2}{\mathcal{R}} \tag{4.10a}$$

The physical interpretation of Eq. 4.10 is that a change in the direction of flow of a fluid particle (i.e., a curved path, $\mathcal{R} < \infty$) is accomplished by the appropriate combination of pressure gradient and particle weight normal to the streamline. A larger speed or density or a smaller radius of curvature of the motion requires a larger force unbalance to produce the motion. For example, if gravity is neglected (as is commonly done for gas flows) or if the flow is in a horizontal ($dz/dn = 0$) plane, Eq. 4.10 becomes

Weight and/or pressure can produce curved streamlines.

$$\frac{\partial p}{\partial n} = -\frac{\rho V^2}{\mathcal{R}} \qquad (4.10b)$$

This indicates that the pressure increases with distance away from the center of curvature ($\partial p/\partial n$ is negative because $\rho V^2/\mathcal{R}$ is positive—the positive n direction points toward the "inside" of the curved streamline). Thus, the pressure outside a tornado (typical atmospheric pressure) is larger than it is near the center of the tornado (where an often dangerously low partial vacuum may occur). This pressure difference is needed to balance the centrifugal acceleration associated with the curved streamlines of the fluid motion (see Fig. E6.6a in Section 6.5.3).

V4.6 Free vortex

EXAMPLE 4.3 | Pressure Variation Normal to a Streamline

Given Shown in **Figs. E4.3a and b** are two flow fields with circular streamlines. The velocity distributions are

$$V(r) = (V_0/r_0)r \qquad \text{for case } (a)$$

and

$$V(r) = \frac{(V_0 r_0)}{r} \qquad \text{for case } (b)$$

where V_0 is the velocity at $r = r_0$.

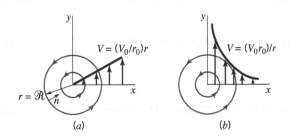

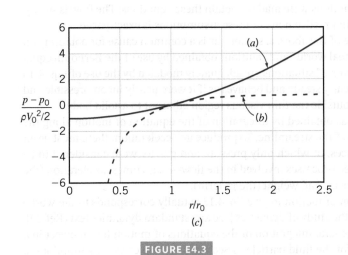

FIGURE E4.3

Find Determine the pressure distributions, $p = p(r)$, for each, given that $p = p_0$ at $r = r_0$.

Solution We assume the flows are steady, inviscid, and incompressible with streamlines in the horizontal plane ($dz/dn = 0$). Because the streamlines are circles, the coordinate n points in a direction opposite that of the radial coordinate, $\partial/\partial n = -\partial/\partial r$, and the radius of curvature is given by $\mathcal{R} = r$. Hence, Eq. 4.10b becomes

$$\frac{\partial p}{\partial r} = \frac{\rho V^2}{r}$$

For case (a) this gives

$$\frac{\partial p}{\partial r} = \rho(V_0/r_0)^2 r$$

whereas for case (b) it gives

$$\frac{\partial p}{\partial r} = \frac{\rho(V_0 r_0)^2}{r^3}$$

For either case the pressure increases as r increases, since $\partial p/\partial r > 0$. Integration of these equations with respect to r, starting with a known pressure $p = p_0$ at $r = r_0$, gives

$$p - p_0 = \left(\rho V_0^2/2\right)\left[(r/r_0)^2 - 1\right] \qquad (Ans)$$

for case (a) and

$$p - p_0 = \left(\rho V_0^2/2\right)\left[1 - (r_0/r)^2\right] \qquad (Ans)$$

for case (b). These pressure distributions are shown in **Fig. E4.3c**.

Comment The pressure distributions needed to balance the centrifugal accelerations in cases (a) and (b) are not the same because the velocity distributions are different. In fact, for case (a) the pressure increases without bound as $r \to \infty$, whereas for case (b) the pressure approaches a finite value as $r \to \infty$. The streamline patterns are the same for each case, however.

Physically, case (a) represents rigid-body rotation (as obtained in a can of water on a turntable after it has been "spun up") and case (b) represents a free vortex (an approximation to a tornado, a hurricane, or the swirl of water in a drain, the "bathtub vortex"). See Fig. E6.6 for an approximation of this type of flow.

The sum of pressure, elevation, and velocity effects is constant across streamlines.

If we multiply Eq. 4.10 by dn, use the fact that $\partial p / \partial n = dp/dn$ if s is constant and integrate across the streamline (in the n direction), we obtain

$$\int \frac{dp}{\rho} + \int \frac{V^2}{\mathcal{R}}\, dn + gz = \text{constant across the streamline} \qquad (4.11)$$

To complete the indicated integrations, we must know how the density varies with pressure and how the fluid speed and radius of curvature vary with n. For incompressible flow the density is constant and the integration involving the pressure term gives simply p/ρ. We are still left, however, with the integration of the second term in Eq. 4.11. Without knowing the n dependence in $V = V(s, n)$ and $\mathcal{R} = \mathcal{R}(s, n)$, this integration cannot be completed.

Thus, the final form of Newton's second law applied across the streamlines for steady, inviscid, incompressible flow is

$$\boxed{\; p + \rho \int \frac{V^2}{\mathcal{R}}\, dn + \gamma z = \text{constant across the streamline} \;} \qquad (4.12)$$

As with the Bernoulli equation, we must be careful that the assumptions involved in the derivation of this equation are not violated when it is used.

4.4   PHYSICAL INTERPRETATIONS AND ALTERNATE FORMS OF THE BERNOULLI EQUATION

In the previous two sections, we developed the basic equations governing fluid motion under a fairly stringent set of restrictions. In spite of the numerous assumptions imposed on these flows, a variety of flows can be readily analyzed with them. A physical interpretation of the equations will be of help in understanding the processes involved. To this end, we rewrite Eqs. 4.7 and 4.12 here and interpret them physically. Application of $\mathbf{F} = m\mathbf{a}$ along and normal to the streamline results in

$$p + \frac{1}{2}\rho V^2 + \gamma z = \text{constant along the streamline} \qquad (4.13)$$

and

$$p + \rho \int \frac{V^2}{\mathcal{R}}\, dn + \gamma z = \text{constant across the streamline} \qquad (4.14)$$

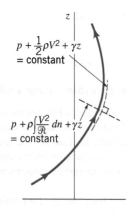

$p + \frac{1}{2}\rho V^2 + \gamma z$ = constant

$p + \rho \int \frac{V^2}{\mathcal{R}}\, dn + \gamma z$ = constant

as indicated by the adjacent figure.

The following basic assumptions were made to obtain these equations: The flow is steady, incompressible, and inviscid. In practice none of these assumptions is exactly true.

A violation of one or more of the above assumptions is a common cause for obtaining an incorrect match between the "real world" and solutions obtained by use of the Bernoulli equation. Fortunately, many "real-world" situations are adequately modeled by the use of Eqs. 4.13 and 4.14, because the flow is nearly steady and behaves as if it were nearly incompressible and inviscid. One example is flow outside the thin viscous region adjacent to a solid body.

The Bernoulli equation was obtained by integration of the equation of motion along the "natural" coordinate direction of the streamline. To produce an acceleration, there must be an unbalance of the resultant forces, of which only pressure and gravity were considered to be important. Thus, there are three processes involved in the flow—mass times acceleration (the $\rho V^2/2$ term), pressure (the p term), and weight (the γz term).

Integration of the equation of motion to give Eq. 4.13 actually corresponds to the work–energy principle often used in the study of dynamics [see any standard dynamics text (Ref. 1)]. This principle results from a general integration of the equations of motion for an object in a way very similar to that done for the fluid particle in Section 4.2. With certain assumptions, a statement of the work–energy principle may be written as follows:

The work done on a particle by all forces acting on the particle is equal to the change of the kinetic energy of the particle.

The Bernoulli equation is a mathematical statement of this principle. Dividing each term by density, we get an alternate but equivalent form of the Bernoulli equation

$$\frac{p}{\rho} + \frac{V^2}{2} + gz = \text{constant along a streamline}$$

Each term in this equation has units of energy per unit mass.

As the fluid particle moves, both gravity and pressure forces do work on the particle. Recall that the work done by a force is equal to the product of the distance the particle travels times the component of force in the direction of travel (i.e., work = $\mathbf{F} \cdot \mathbf{d}$). The terms gz and p/ρ are related to the work done by the weight and pressure forces, respectively. The remaining term, $V^2/2$, is obviously related to the kinetic energy of the particle. In fact, an alternate method of deriving the Bernoulli equation is to use the first and second laws of thermodynamics (the energy and entropy equations) rather than Newton's second law. With the appropriate restrictions, the general energy equation reduces to the Bernoulli equation. This approach is discussed in Section 5.3.3.

An alternate but equivalent form of the Bernoulli equation is obtained by dividing each term of Eq. 4.7 by the specific weight, γ, to obtain

$$\frac{p}{\gamma} + \frac{V^2}{2g} + z = \text{constant on a streamline}$$

> The Bernoulli equation can be written in terms of heights called heads.

Each of the terms in this equation has the units of energy per weight ($LF/F = L$) or length (feet, meters) and represents a certain type of head.

The elevation term, z, is related to the potential energy of the particle and is called the **elevation head.** The pressure term, p/γ, is called the **pressure head** and represents the height of a column of the fluid that is needed to produce the pressure p. The velocity term, $V^2/2g$, is the **velocity head** and represents the vertical distance needed for the fluid to fall freely (neglecting friction) if it is to reach velocity V from rest. The Bernoulli equation states that the sum of the pressure head, the velocity head, and the elevation head is constant along a streamline. The constant is called the **total head.**

EXAMPLE 4.4 | Kinetic, Potential, and Pressure Energy

Given Consider the flow of water from the syringe shown in **Fig. E4.4a**. As indicated in **Fig. E4.4b**, a force, F, applied to the plunger will produce a pressure greater than atmospheric at

Find Discuss the energy of the fluid at points (1), (2), and (3) by using the Bernoulli equation.

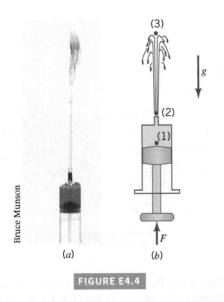

Bruce Munson

(a)   (b)

FIGURE E4.4

	Energy Type		
Point	Kinetic $V^2/2$	Potential gz	Pressure p/ρ
1	Small	Zero	Large
2	Large	Small	Zero
3	Zero	Large	Zero

Solution If the assumptions (steady, inviscid, incompressible flow) of the Bernoulli equation are approximately valid, it then follows that the flow can be explained in terms of the partition of the total energy of the water. According to Eq. 4.13, the sum of the three types of energy (kinetic, potential, and pressure) or heads (velocity, elevation, and pressure) must remain constant. The table above indicates the relative magnitude of each of these energies at the three points shown in the figure.

The motion results in (or is due to) a change in the magnitude of each type of energy as the fluid flows from one location to another. An alternate way to consider this flow is as follows. The pressure

point (1) within the syringe. The water flows from the needle, point (2), with relatively high velocity and coasts up to point (3) at the top of its trajectory.

gradient between (1) and (2) produces an acceleration to eject the water from the needle. Gravity acting on the particle between (2) and (3) produces a deceleration to cause the water to come to a momentary stop at the top of its flight.

Comment If friction (viscous) effects were important, there would be an energy loss between (1) and (3) and for the given p_1 the water would not be able to reach the height indicated in the figure. Such friction may arise in the needle (see Chapter 8 on pipe flow) or between the water stream and the surrounding air (see Chapter 9 on external flow).

Keep in mind that pressure is not energy. It is a stress. Pressure does work *between* points, not *at* a point. It is, however, convenient to think of pressure as a "potential" to do work as we do with gravity.

The Wide World of Fluids

Armed with a water jet for hunting

Archerfish, known for their ability to shoot down insects resting on foliage, are like submarine water pistols. With their snout sticking out of the water, they eject a high-speed water jet at their prey, knocking it onto the water surface where they snare it for their meal. The barrel of their water pistol is formed by placing their tongue against a groove in the roof of their mouth to form a tube. By snapping shut their gills, water is forced through the tube and directed with the tip of their tongue. The archerfish can produce a *pressure head* within their gills large enough so that the jet can reach 2 to 3 m. However, it is accurate to only about 1 m. Recent research has shown that archerfish are very adept at calculating where their prey will fall. Within 100 milliseconds (a reaction time twice as fast as a human's), the fish has extracted all the information needed to predict the point where the prey will hit the water. Without further visual cues it charges directly to that point.

A net force is required to accelerate any mass. For steady flow the acceleration can be interpreted as arising from two distinct occurrences—a change in speed along the streamline and a change in direction if the streamline is not straight. Integration of the equation of motion along the streamline accounts for the change in speed (kinetic energy change) and results in the Bernoulli equation. Integration of the equation of motion normal to the streamline accounts for the centrifugal acceleration ($V^2/\mathscr{R}$) and results in Eq. 4.14.

The pressure variation across straight streamlines is hydrostatic.

When a fluid particle travels along a curved path, a net force directed toward the center of curvature is required. Under the assumptions valid for Eq. 4.14, this force may be either gravity or pressure, or a combination of both. In many instances the streamlines are nearly straight ($\mathscr{R} = \infty$) so that centrifugal effects are negligible and the pressure variation across the streamlines is merely hydrostatic (because of gravity alone), even though the fluid is in motion.

EXAMPLE 4.5 | Pressure Variation in a Flowing Stream

Given Water flows in a curved, undulating waterslide as shown in **Fig. E4.5a**. As an approximation to this flow, consider the inviscid,

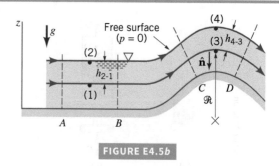

FIGURE E4.5b

kali9/E+/Getty Images

FIGURE E4.5a

incompressible, steady flow shown in **Fig. E4.5b**. From section A to B the streamlines are straight, while from C to D they follow circular paths.

Find Describe the pressure variation between points (1) and (2) and points (3) and (4).

Solution   With the above assumptions and the fact that $\mathcal{R} = \infty$ for the portion from A to B, Eq. 4.14 becomes

$$p + \gamma z = \text{constant}$$

The constant can be determined by evaluating the known variables at the two locations using $p_2 = 0$ (gage), $z_1 = 0$, and $z_2 = h_{2-1}$ to give

$$p_1 = p_2 + \gamma(z_2 - z_1) = p_2 + \gamma h_{2-1} \qquad (Ans)$$

Note that since the radius of curvature of the streamline is infinite, the pressure variation in the vertical direction is the same as if the fluid were stationary.

However, if we apply Eq. 4.14 between points (3) and (4), we obtain (using $dn = -dz$)

$$p_4 + \rho \int_{z_3}^{z_4} \frac{V^2}{\mathcal{R}}(-dz) + \gamma z_4 = p_3 + \gamma z_3$$

With $p_4 = 0$ and $z_4 - z_3 = h_{4-3}$, this becomes

$$p_3 = \gamma h_{4-3} - \rho \int_{z_3}^{z_4} \frac{V^2}{\mathcal{R}} dz \qquad (Ans)$$

To evaluate the integral, we must know the variation of V and $\mathcal{R}$ with z. Even without this detailed information we note that the integral has a positive value. Thus, the pressure at (3) is less than the hydrostatic value, γh_{4-3}, by an amount equal to $\rho \int_{z_3}^{z_4} (V^2/\mathcal{R}) \, dz$. This lower pressure, caused by the curved streamline, is necessary to accelerate the fluid around the curved path.

Comment   Note that we did not apply the Bernoulli equation (Eq. 4.13) across the streamlines from (1) to (2) or (3) to (4). Rather we used Eq. 4.14. As is discussed in Section 4.8, application of the Bernoulli equation across streamlines (rather than along them) may lead to serious errors.

4.5   STATIC, STAGNATION, DYNAMIC, AND TOTAL PRESSURE

A useful concept associated with the Bernoulli equation deals with the stagnation and dynamic pressures. These pressures arise from the conversion of kinetic energy in a flowing fluid into a "pressure rise" as the fluid is brought to rest (as in Example 4.2). In this section we explore various results of this process. Each term of the Bernoulli equation, Eq. 4.13, has the dimensions of force per unit area—N/m². The first term, p, is the actual thermodynamic pressure of the fluid as it flows. To measure its value, one could move along with the fluid, thus being "static" relative to the moving fluid. Hence, it is normally termed the **static pressure**. Another way to measure the static pressure would be to drill a hole in a flat surface and fasten a piezometer tube, as indicated by the location of point (3) in **Fig. 4.4**. As we saw in Example 4.5, the pressure in the flowing fluid of Fig. 4.4 at (1) is $p_1 = \gamma h_{3-1} + p_3$, the same as if the fluid were static. From the manometer considerations of Chapter 2, we know that $p_3 = \gamma h_{4-3}$. Thus, since $h_{3-1} + h_{4-3} = h$, it follows that $p_1 = \gamma h$.

The third term in Eq. 4.13, γz, is termed the *hydrostatic pressure*, in obvious regard to the hydrostatic pressure variation discussed in Chapter 2. It is not actually a pressure but does represent the change in pressure possible due to potential energy variations of the fluid as a result of elevation changes.

The second term in the Bernoulli equation, $\rho V^2/2$, is termed the **dynamic pressure**. Its interpretation can be seen in Fig. 4.4 by considering the pressure at the end of a small tube inserted into the flow and pointing upstream. After the initial transient motion has died out,

> Each term in the Bernoulli equation can be interpreted as a form of pressure.

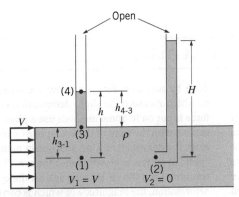

FIGURE 4.4   Measurement of static and stagnation pressures.

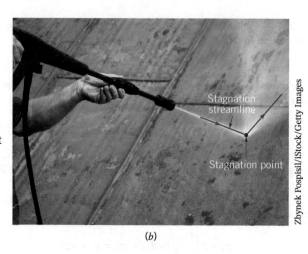

Stagnation point

(a)

Stagnation
streamline

Stagnation point

(b)

FIGURE 4.5　Stagnation points.

the liquid will fill the tube to a height of H, as shown. The fluid in the tube, including that at its tip, (2), will be stationary. That is, $V_2 = 0$, or point (2) is a **stagnation point.**

If we apply the Bernoulli equation between points (1) and (2), using $V_2 = 0$ and assuming that $z_1 = z_2$, we find that

$$p_2 = p_1 + \frac{1}{2}\rho V_1^2$$

Video

V4.7 Stagnation point flow

Hence, the pressure at the stagnation point is greater than the static pressure, p_1, by an amount $\rho V_1^2/2$, the dynamic pressure.

It can be shown that there is a stagnation point on any stationary body that is placed into a flowing fluid. Some of the fluid flows "over" and some "under" the object. The dividing line (or surface for three-dimensional flows) is termed the *stagnation streamline* and terminates at the stagnation point on the body. (See Fig 4.1a indicating dye in water flowing past a blunt object.) For symmetrical nonrotating objects (such as a baseball) the stagnation point is clearly at the tip or front of the object as shown in **Fig. 4.5a.** For other flows such as a water jet against a flat surface as shown in **Fig. 4.5b,** there is also a stagnation point on the surface.

If elevation effects are neglected, the **stagnation pressure,** $p + \rho V^2/2$, is the largest pressure obtainable along a given streamline. It represents the conversion of all of the kinetic energy into a pressure rise. The sum of the static pressure, hydrostatic pressure, and dynamic pressure is termed the **total pressure,** p_T. The Bernoulli equation is a statement that the total pressure remains constant along a streamline. That is,

$$p + \frac{1}{2}\rho V^2 + \gamma z = p_T = \text{constant along a streamline} \tag{4.15}$$

Again, we must be careful that the assumptions used in the derivation of this equation are appropriate for the flow being considered.

The Wide World of Fluids

Pressurized eyes

Our eyes need a certain amount of internal pressure in order to work properly, with the normal range being between 10 and 20 mm of mercury. The pressure is determined by a balance between the fluid entering and leaving the eye. If the pressure is above the normal level, damage may occur to the optic nerve where it leaves the eye, leading to a loss of the visual field termed glaucoma. Measurement of the pressure within the eye can be done by several different noninvasive types of instruments, all of which measure the slight deformation of the eyeball when a force is put on it. Some methods use a physical probe that makes contact with the front of the eye, applies a known force, and measures the deformation. One noncontact method uses a calibrated "puff" of air that is blown against the eye. The *stagnation pressure* resulting from the air blowing against the eyeball causes a slight deformation, the magnitude of which is correlated with the pressure within the eyeball.

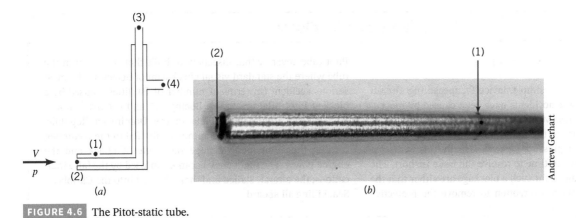

FIGURE 4.6   The Pitot-static tube.

Knowledge of the values of the static and stagnation pressures in a fluid implies that the fluid speed can be calculated. This is the principle on which the **Pitot-static tube** is based [Henri Pitot (1695–1771)]. As shown in **Fig. 4.6**, two concentric tubes are attached to two pressure gages (or a differential gage) so that the values of p_3 and p_4 (or the difference $p_3 - p_4$) can be determined. The center tube measures the stagnation pressure at its open tip. If elevation changes are negligible,

$$p_3 = p + \frac{1}{2}\rho V^2$$

where p and V are the pressure and velocity of the fluid upstream of point (2). The outer tube is made with several small holes at an appropriate distance from the tip so that they measure the static pressure. If the effect of the elevation difference between (1) and (4) is negligible, then

$$p_4 = p_1 = p$$

By combining these two equations, we see that

$$p_3 - p_4 = \frac{1}{2}\rho V^2$$

which can be rearranged to give

$$V = \sqrt{2(p_3 - p_4)/\rho} \qquad (4.16)$$

The actual shape and size of Pitot-static tubes vary considerably. A typical Pitot-static probe used to determine aircraft airspeed is shown in **Fig. 4.7** (see Fig. E4.6a also).

> Pitot-static tubes measure fluid velocity by converting velocity into pressure.

> ▶ **Video**
>
> V4.8 Airspeed indicator

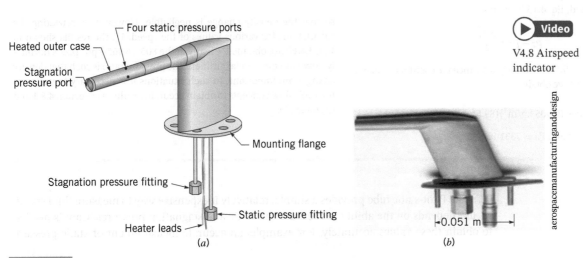

FIGURE 4.7   Airplane Pitot-static probe. (a) Schematic, (b) Photograph.

The Wide World of Fluids

Bugged and plugged Pitot tubes

Although a *Pitot tube* is a simple device for measuring aircraft airspeed, many airplane accidents have been caused by inaccurate Pitot tube readings. Most of these accidents are the result of having one or more of the holes blocked and, therefore, not indicating the correct pressure (speed). Usually this is discovered during takeoff when time to resolve the issue is short. The two most common causes for such a blockage are either that the pilot (or ground crew) has forgotten to remove the protective

Pitot tube cover or that insects have built their nest within the tube where the standard visual check cannot detect it. The most serious accident (in terms of number of fatalities) caused by a blocked Pitot tube involved a Boeing 757 and occurred shortly after takeoff from Puerto Plata in the Dominican Republic. Incorrect airspeed data were automatically fed to the computer, causing the autopilot to change the angle of attack and the engine power. The flight crew became confused by the false indications; the aircraft stalled and then plunged into the Caribbean Sea, killing all aboard.

EXAMPLE 4.6 | Pitot-Static Tube

Given An airplane flies 322 kph at an elevation of 3048 m in a standard atmosphere as shown in **Fig. E4.6a**. The value of standard pressure at the altitude is 69706 Pa.

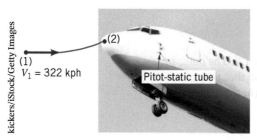

FIGURE E4.6a

Find Determine the pressure at point (1) far ahead of the airplane, the pressure at the stagnation point on the nose of the airplane, point (2), and the pressure difference indicated by a Pitot-static probe attached to the fuselage.

Solution The value of the static pressure at the altitude is given as

$$p_1 = 69706 \text{ Pa} \qquad (Ans)$$

Also the density is $\rho = 0.905 \text{ kg/m}^3$.

If the flow is steady, inviscid, and incompressible and elevation changes are neglected, Eq. 4.13 becomes

$$p_2 = p_1 + \frac{\rho V_1^2}{2}$$

With $V_1 = 322$ kph $= 89.5$ m/s and $V_2 = 0$ (since the coordinate system is fixed to the airplane), we obtain

$$p_2 = 69706 \text{ Pa} + (0.905 \text{ kg/m}^3)(89.5^2 \text{ m}^2/\text{s}^2)/2$$

$$= (69706 + 3265) \text{ Pa} = 7331 \text{ Pa}$$

Thus, the pressure difference indicated by the Pitot-static tube is

$$p_2 - p_1 = \frac{\rho V_1^2}{2} = 3625 \text{ Pa} \qquad (Ans)$$

Comment It was assumed that the flow is incompressible—the density remains constant from (1) to (2). However, since $\rho = p/RT$, a change in pressure (or temperature) will cause a change in density. For this relatively low speed, the ratio of the absolute pressures is nearly unity [i.e., $p_1/p_2 = (69.7 \text{ kPa})/(69.7 \text{ kPa} + 3.62 \text{ kPa}) = 0.951$],

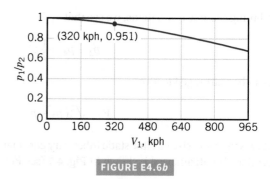

FIGURE E4.6b

so that the density change is negligible. However, by repeating the calculations for various values of the speed, V_1, the results shown in **Fig. E4.6b** are obtained. Clearly at the 805- to 965-kph speeds normally flown by commercial airliners, the pressure ratio is such that density changes are important. In such situations it is necessary to use compressible flow concepts to obtain accurate results (see Section 4.8.1 and Chapter 11).

The Pitot-static tube provides a simple, relatively inexpensive way to measure fluid speed. Its use depends on the ability to measure the static and stagnation pressures. Care is needed to obtain these values accurately. For example, an accurate measurement of static pressure

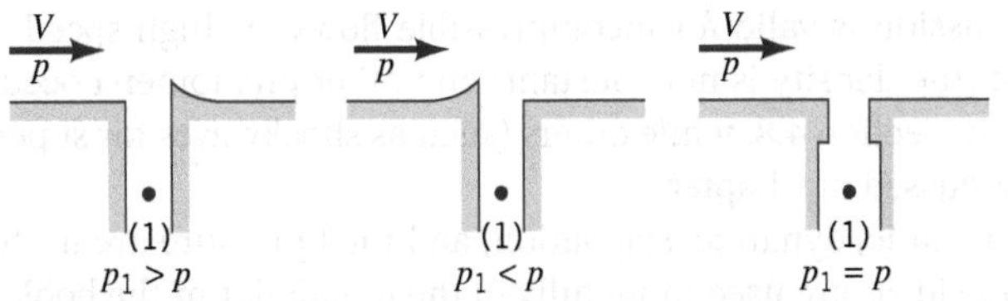

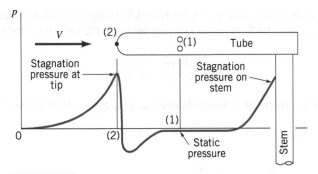

FIGURE 4.8 Incorrect and correct design of static pressure taps.

requires that none of the fluid's kinetic energy be converted into a pressure rise at the point of measurement. This requires a smooth hole with no burrs or imperfections. As indicated in **Fig. 4.8**, such imperfections can cause the measured pressure to be greater or less than the actual static pressure.

Also, the pressure along the surface of an object varies from the stagnation pressure at its stagnation point to values that may be less than the free stream static pressure. A typical pressure variation for a Pitot-static tube is indicated in **Fig. 4.9**. Clearly it is important that the pressure taps be properly located to ensure that (for all practical purposes) the pressure measured is actually the static pressure.

In practice it is often difficult to align the Pitot-static tube directly into the flow direction. Any misalignment will produce a nonsymmetrical flow field that may introduce errors. Typically, yaw angles up to 12° to 20° (depending on the particular probe design) give results that are less than 1% in error from the perfectly aligned results. Generally, it is more difficult to measure static pressure than stagnation pressure.

One method of determining the flow direction and its speed (thus the velocity) is to use a directional-finding Pitot tube, as is illustrated in **Fig. 4.10**. Three pressure taps are drilled into a small circular cylinder, fitted with small tubes, and connected to three pressure transducers. The cylinder is rotated until the pressures in the two side holes are equal, thus indicating that the center hole points directly upstream. The center tap then measures the stagnation pressure. The two side holes are located at an angle (β) specific to the flow conditions, usually between about 33° and about 38°, so that they measure the static pressure. The speed is then obtained from $V = [2(p_2 - p_1)/\rho]^{1/2}$.

> **Accurate measurement of static pressure requires great care.**

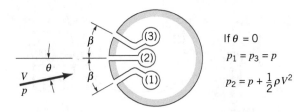

FIGURE 4.9 Typical pressure distribution along a Pitot-static tube.

FIGURE 4.10 Cross section of a directional-finding Pitot-static tube, sometimes called a Fechheimer probe.

The above discussion is valid for incompressible flows. At high speeds, compressibility becomes important (the density is not constant) and other phenomena occur. Some of these ideas are discussed in Section 4.8, while others (such as shockwaves for supersonic Pitot-tube applications) are discussed in Chapter 11.

The concepts of static, dynamic, stagnation, and total pressure are useful in a variety of flow problems. These ideas are used more fully in the remainder of the book.

4.6 APPLICATIONS OF THE BERNOULLI EQUATION

In this section we illustrate various additional applications of the Bernoulli equation. Between any two points, (1) and (2), on a streamline with steady, inviscid, incompressible flow, the Bernoulli equation can be applied in the form

$$p_1 + \frac{1}{2}\rho V_1^2 + \gamma z_1 = p_2 + \frac{1}{2}\rho V_2^2 + \gamma z_2 \qquad (4.17)$$

Obviously, if five of the six variables are known, the remaining one can be determined. In many instances it is necessary to introduce other equations, such as the conservation of mass. Such considerations will be discussed briefly in this section and in more detail in Chapter 5.

4.6.1 Free Jets

One of the oldest equations in fluid mechanics deals with the flow of a liquid from a large reservoir. A modern version of this type of flow involves the flow of coffee from a coffee urn as indicated by the adjacent figure. The basic principles of this type of flow are shown in **Fig. 4.11**, where a jet of liquid of diameter d flows from the nozzle with velocity V. (A nozzle is a device shaped to accelerate a fluid.) Application of Eq. 4.17 between points (1) and (2) on the streamline shown gives

$$\gamma h = \frac{1}{2}\rho V^2$$

We have used the facts that $z_1 = h$, $z_2 = 0$, the reservoir is large ($V_1 \cong 0$) and open to the atmosphere ($p_1 = 0$ gage), and the fluid leaves as a **free jet** ($p_2 = 0$). Thus, we obtain

$$V = \sqrt{2\frac{\gamma h}{\rho}} = \sqrt{2gh} \qquad (4.18)$$

which is the modern version of a result obtained in 1643 by Torricelli (1608–1647), an Italian physicist.

V4.9 Flow from a tank

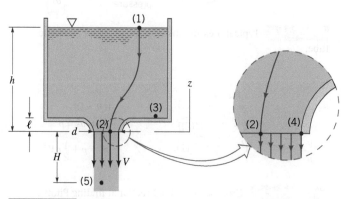

FIGURE 4.11 Vertical flow from a tank.

The fact that the exit pressure equals the surrounding pressure ($p_2 = 0$) can be seen by applying $\mathbf{F} = m\mathbf{a}$, as given by Eq. 4.14, across the streamlines between (2) and (4). If the streamlines at the tip of the nozzle are straight ($\mathcal{R} = \infty$), it follows that $p_2 = p_4$. Since (4) is on the surface of the jet, in contact with the atmosphere, we have $p_4 = 0$. Thus, $p_2 = 0$ also. Since (2) is an arbitrary point in the exit plane of the nozzle, it follows that the pressure is atmospheric across this plane. Physically, since there is no component of the weight force or acceleration in the normal (horizontal) direction, the pressure is constant in that direction.

Once outside the nozzle, the stream continues to fall as a free jet with zero pressure throughout ($p_5 = 0$) and as seen by applying Eq. 4.17 between points (1) and (5), the speed increases according to

$$V = \sqrt{2g(h + H)}$$

where, as shown in Fig. 4.11, H is the distance the fluid has fallen outside the nozzle.

Equation 4.18 could also be obtained by writing the Bernoulli equation between points (3) and (4) using the fact that $z_4 = 0$ and $z_3 = \ell$. Also, $V_3 = 0$, since it is far from the nozzle, and from hydrostatics, $p_3 = \gamma(h - \ell)$.

As learned in physics or dynamics and illustrated in the adjacent figure, any object dropped from rest that falls through a distance h in a vacuum will obtain the speed $V = \sqrt{2gh}$, the same as the water leaving the spout of the watering can shown in the adjacent figure. This is consistent with the fact that all of the particle's potential energy is converted to kinetic energy, provided viscous (friction) effects are negligible. In terms of heads, the elevation head at point (1) is converted into the velocity head at point (2). Recall that for the case shown in Fig. 4.11, the pressure is the same (atmospheric) at points (1) and (2).

For the horizontal nozzle of **Fig. 4.12a**, the velocity of the fluid at the centerline, V_2, will be slightly greater than that at the top, V_1, and slightly less than that at the bottom, V_3, due to the differences in elevation. In general, $d \ll h$, as shown in **Fig. 4.12b**, and we can safely use the centerline velocity as a reasonable "average velocity."

If the exit is not a smooth, well-contoured nozzle, but rather a flat plate as shown in **Fig. 4.13**, the diameter of the jet, d_j, will be less than the diameter of the hole, d_h. This phenomenon, called a *vena contracta* effect, is a result of the inability of the fluid to turn the sharp 90° corner indicated by the dotted lines in the figure.

> The exit pressure for an incompressible fluid jet is equal to the surrounding pressure.

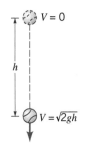

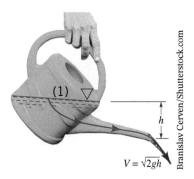

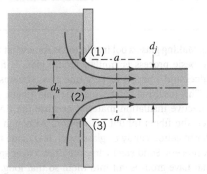

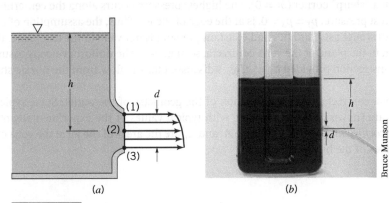

FIGURE 4.12 Horizontal flow from a tank.

FIGURE 4.13 Vena contracta effect for a sharp-edged orifice.

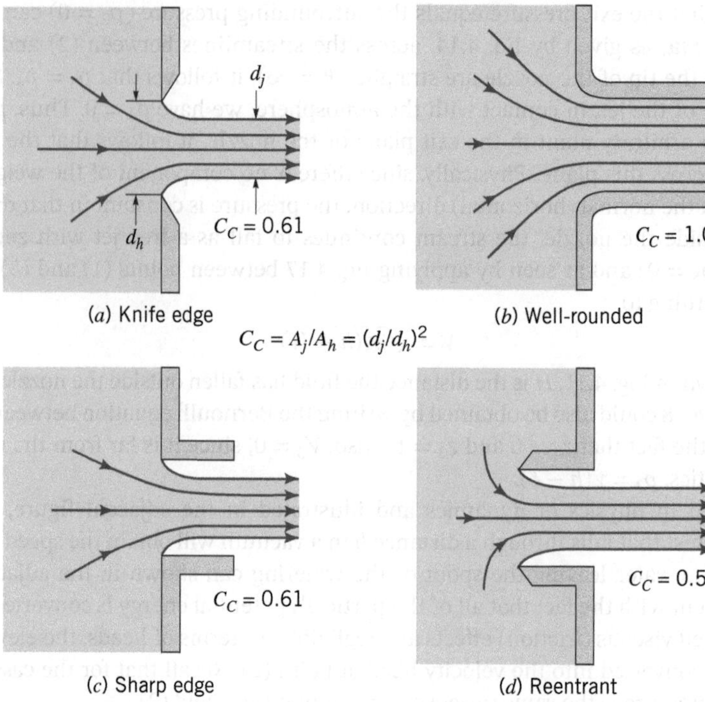

(a) Knife edge $C_C = 0.61$

(b) Well-rounded $C_C = 1.0$

$C_C = A_j/A_h = (d_j/d_h)^2$

(c) Sharp edge $C_C = 0.61$

(d) Reentrant $C_C = 0.50$

FIGURE 4.14 Typical flow patterns and contraction coefficients for various round exit configurations. (a) Knife edge, (b) Well-rounded, (c) Sharp edge, (d) Reentrant.

Since the streamlines in the exit plane are curved ($\mathcal{R} < \infty$), the pressure across them is not constant. It would take an infinite pressure gradient across the streamlines to cause the fluid to turn a "sharp" corner ($\mathcal{R} = 0$). The highest pressure occurs along the centerline at (2) and the lowest pressure, $p_1 = p_3 = 0$, is at the edge of the jet. Thus, the assumption of uniform velocity with straight streamlines and constant pressure is not valid at the exit plane. It is valid, however, in the plane of the vena contracta, section a–a. The uniform velocity assumption is valid at this section provided $d_j \ll h$, as is discussed for the flow from the nozzle shown in Fig. 4.12.

The vena contracta effect is a function of the geometry of the outlet. Some typical configurations are shown in **Fig. 4.14** along with typical values of the experimentally obtained *contraction coefficient*, $C_c = A_j/A_h$, where A_j and A_h are the areas of the jet at the vena contracta and the area of the hole, respectively.

The diameter of a fluid jet is often smaller than that of the hole from which it flows.

The Wide World of Fluids

Cotton candy, glass wool, and steel wool

Although cotton candy and glass wool insulation are made of entirely different materials and have entirely different uses, they are made by similar processes. Cotton candy, invented in 1897, consists of sugar fibers. Glass wool, invented in 1938, consists of glass fibers. In a cotton candy machine, sugar is melted and then forced by centrifugal action to flow through numerous tiny *orifices* in a spinning "bowl." Upon emerging, the thin streams of liquid sugar cool very quickly and become solid threads that are collected on a stick or cone. Making glass wool insulation is somewhat more complex, but the basic process is similar. Liquid glass is forced through tiny orifices and emerges as very fine glass streams that quickly solidify. The resulting intertwined flexible fibers, glass wool, form an effective insulation material because the tiny air "cavities" between the fibers inhibit air motion. Although steel wool looks similar to cotton candy or glass wool, it is made by an entirely different process. Solid steel wires are drawn over special cutting blades that have grooves cut into them so that long, thin threads of steel are peeled off to form the matted steel wool.

4.6.2 Confined Flows

In many cases the fluid is physically constrained within a device so that its pressure cannot be prescribed a priori as was done for the free jet examples above. Such cases include nozzles and pipes of variable diameter for which the fluid velocity changes because the flow area is different from one section to another. For these situations it is necessary to use the concept of conservation of mass (the continuity equation) along with the Bernoulli equation. The complete derivation and use of this equation are discussed in detail in Chapters 3 and 5. For the needs of this chapter we can use a simplified form of the continuity equation obtained from the following intuitive arguments. Consider a fluid flowing through a fixed volume (such as a syringe) that has one inlet and one outlet as shown in **Fig. 4.15a**. If the flow is steady so that there is no additional accumulation of fluid within the volume, the rate at which the fluid flows into the volume must equal the rate at which it flows out of the volume (otherwise, mass would not be conserved).

The continuity equation states that mass cannot be created or destroyed.

The *mass flowrate* from an outlet, $\dot{m}$ (kg/s), is given by $\dot{m} = \rho Q$, where Q (m³/s) is the **volume flowrate**. As depicted in **Fig. 4.15b**, if the outlet area is A and the fluid flows across this area (normal to the area) with an average velocity V, then the volume of the fluid crossing this area in a time interval δt is $VA\,\delta t$, equal to that in a volume of length $V\,\delta t$ and cross-sectional area A. Hence, the volume flowrate (volume per unit time) is $Q = VA$. Thus, $\dot{m} = \rho VA$. To conserve mass, the inflow rate must equal the outflow rate. If the inlet is designated as (1) and the outlet as (2), it follows that $\dot{m}_1 = \dot{m}_2$. Thus, conservation of mass requires

$$\rho_1 A_1 V_1 = \rho_2 A_2 V_2$$

If the density remains constant, then $\rho_1 = \rho_2$, and the above becomes the **continuity equation** for incompressible, *one-dimensional* flow

$$\boxed{A_1 V_1 = A_2 V_2, \text{ or } Q_1 = Q_2} \tag{4.19}$$

For example, if the outlet flow area shown in the adjacent figure is one-half the size of the inlet flow area, it follows that the outlet velocity is twice that of the inlet velocity, since $V_2 = A_1 V_1 / A_2 = 2V_1$. Use of the Bernoulli equation and the flowrate equation (continuity equation) is demonstrated by Example 4.7.

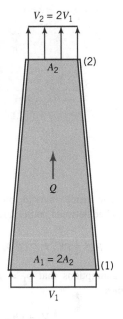

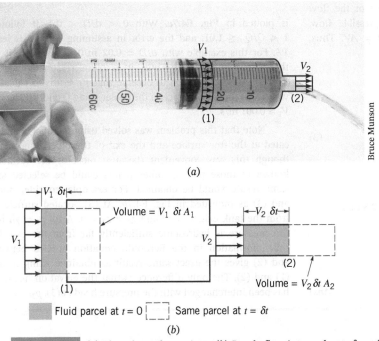

FIGURE 4.15 (a) Flow through a syringe. (b) Steady flow into and out of a volume.

EXAMPLE 4.7 | Flow from a Tank—Gravity Driven

Given A stream of refreshing beverage of diameter $d = 0.02$ m flows steadily from the cooler of diameter $D = 0.4$ m as shown in **Figs. E4.7a** and **b**.

Find Determine the flowrate, Q, from the bottle into the cooler if the depth of beverage in the cooler is to remain constant at $h = 0.4$ m.

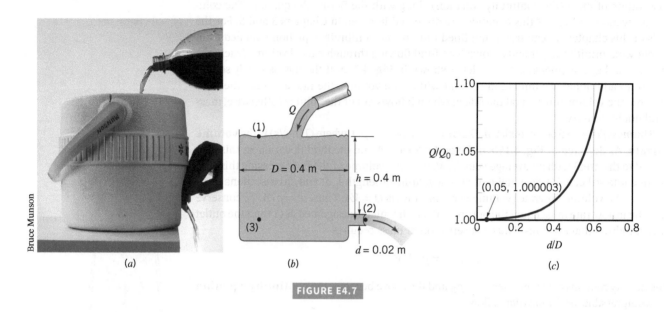

(a)   (b)   (c)

FIGURE E4.7

Solution For steady, inviscid, incompressible flow, the Bernoulli equation applied between points (1) and (2) is

$$p_1 + \frac{1}{2}\rho V_1^2 + \gamma z_1 = p_2 + \frac{1}{2}\rho V_2^2 + \gamma z_2 \tag{1}$$

With the assumptions that $p_1 = p_2 = 0$, $z_1 = h$, and $z_2 = 0$, Eq. 1 becomes

$$\frac{1}{2}V_1^2 + gh = \frac{1}{2}V_2^2 \tag{2}$$

Although the liquid level remains constant ($h = $ constant), there is an average velocity, V_1, across section (1) because of the flow from the tank. From Eq. 4.19 for steady incompressible flow, conservation of mass requires $Q_1 = Q_2$, where $Q = AV$. Thus, $A_1 V_1 = A_2 V_2$, or

$$\frac{\pi}{4}D^2 V_1 = \frac{\pi}{4}d^2 V_2$$

Hence,

$$V_1 = \left(\frac{d}{D}\right)^2 V_2 \tag{3}$$

Equations 1 and 3 can be combined to give

$$V_2 = \sqrt{\frac{2gh}{1 - (d/D)^4}} = \sqrt{\frac{2(9.81 \text{ m/s}^2)(0.4 \text{ m})}{1 - (0.02 \text{ m}/0.4 \text{ m})^4}} = 2.8 \text{ m/s}$$

Thus,

$$Q = A_1 V_1 = A_2 V_2 = \frac{\pi}{4}(0.02 \text{ m})^2 (2.8 \text{ m/s})$$

$$= 8.79 \times 10^{-4} \text{ m}^3/\text{s} \tag{Ans}$$

Comments In this example we have not neglected the kinetic energy of the water in the tank ($V_1 \neq 0$). If the tank diameter is large compared to the jet diameter ($D \gg d$), Eq. 3 indicates that $V_1 \ll V_2$ and the assumption that $V_1 \approx 0$ would be reasonable. The error associated with this assumption can be seen by calculating the ratio of the flowrate assuming $V_1 \neq 0$, denoted Q, to that assuming $V_1 = 0$, denoted Q_0. This ratio, written as

$$\frac{Q}{Q_0} = \frac{V_2}{V_2|_{D=\infty}} = \frac{\sqrt{2gh/[1 - (d/D)^4]}}{\sqrt{2gh}} = \frac{1}{\sqrt{1 - (d/D)^4}}$$

is plotted in **Fig. E4.7c**. With $0 < d/D < 0.4$, it follows that $1 < Q/Q_0 \lesssim 1.01$, and the error in assuming $V_1 = 0$ is less than 1%. For this example with $d/D = 0.02$ m/0.4 m $= 0.05$, it follows that $Q/Q_0 = 1.000003$. Thus, it is often reasonable to assume $V_1 = 0$. Another method to observe that $V_1 \approx 0$ is to calculate the velocity of the fluid surface in the cooler using Q and A_1. The result is $V_1 \approx 0.001$ m/s.

Note that this problem was solved using points (1) and (2) located at the free surface and the exit of the pipe, respectively. Although this was convenient (because most of the variables are known at those points), other points could be selected and the same result would be obtained. For example, consider points (1) and (3) as indicated in Fig. E4.7b. At (3), located sufficiently far from the tank exit, $V_3 = 0$ and $z_3 = z_2 = 0$. Also, $p_3 = \gamma h$ because the pressure is hydrostatic sufficiently far from the exit. Use of this information in the Bernoulli equation applied between (3) and (2) gives the exact same result as obtained using it between (1) and (2). The only difference is that the elevation head, $z_1 = h$, has been interchanged with the pressure head at (3), $p_3/\lambda = h$.

The fact that a kinetic energy change is often accompanied by a change in pressure is shown by Example 4.8.

EXAMPLE 4.8 | Flow from a Tank—Pressure Driven

Given   Air flows steadily from a tank, through a hose of diameter $D = 0.06$ m, and exits to the atmosphere from a nozzle of diameter $d = 0.02$ m as shown in **Fig. E4.8**. The pressure in the tank remains constant at 6.0 kPa (gage), and the atmospheric conditions are standard temperature and pressure.

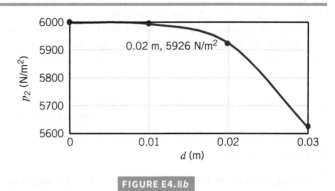

FIGURE E4.8b

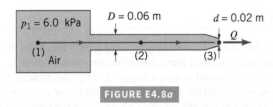

FIGURE E4.8a

Find   Determine

a. the flowrate and

b. the pressure in the hose.

Solution

a. If the flow is assumed steady, inviscid, and incompressible, we can apply the Bernoulli equation along the streamline from (1) to (2) to (3) as

$$p_1 + \tfrac{1}{2}\rho V_1^2 + \gamma z_1 = p_2 + \tfrac{1}{2}\rho V_2^2 + \gamma z_2$$

$$= p_3 + \tfrac{1}{2}\rho V_3^2 + \gamma z_3$$

With the assumption that $z_1 = z_2 = z_3$ (horizontal hose), $V_1 = 0$ (large tank), and $p_3 = 0$ (free jet), this becomes

$$V_3 = \sqrt{\frac{2p_1}{\rho}}$$

and

$$p_2 = p_1 - \tfrac{1}{2}\rho V_2^2 \qquad (1)$$

The density of the air in the tank is obtained from the ideal gas law, using standard absolute pressure and temperature, as

$$\rho = \frac{p_1}{RT_1}$$

$$= \frac{(6.0 + 101)\ \text{kN/m}^2 \times 10^3\ \text{N/kN}}{(286.9\ \text{N} \cdot \text{m/kg} \cdot \text{K})(15 + 273)\ \text{K}}$$

$$= 1.25\ \text{kg/m}^3$$

Thus, we find that

$$V_3 = \sqrt{\frac{2(6.0 \times 10^3\ \text{N/m}^2)}{1.25\ \text{kg/m}^3}} = 98.0\ \text{m/s}$$

or

$$Q = A_3 V_3 = \tfrac{\pi}{4}d^2 V_3 = \tfrac{\pi}{4}(0.02\ \text{m})^2(98.0\ \text{m/s})$$

$$= 0.0308\ \text{m}^3/\text{s} \qquad (Ans)$$

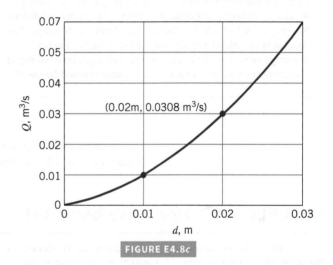

FIGURE E4.8c

Comment   Note that the value of V_3 is determined strictly by the value of p_1 (and the assumptions involved in the Bernoulli equation), independent of the "shape" of the nozzle. The pressure head within the tank, $p_1/\gamma = (6.0\ \text{kPa})/(9.81\ \text{m/s}^2)(1.25\ \text{kg/m}^3) = 489$ m, is converted to the velocity head at the exit, $V_2^2/2g = (98.0\ \text{m/s})^2/(2 \times 9.81\ \text{m/s}^2) = 489$ m. Although we used gage pressure in the Bernoulli equation ($p_3 = 0$), we had to use absolute pressure in the ideal gas law when calculating the density.

b. The pressure within the hose can be obtained from Eq. 1 and the continuity equation (Eq. 4.19)

$$A_2 V_2 = A_3 V_3$$

Hence,

$$V_2 = A_3 V_3/A_2 = \left(\frac{d}{D}\right)^2 V_3$$

$$= \left(\frac{0.02\ \text{m}}{0.06\ \text{m}}\right)^2 (98\ \text{m/s}) = 10.98\ \text{m/s}$$

and from Eq. 1

$$p_2 = 6.0 \times 10^3\ \text{N/m}^2 - \tfrac{1}{2}(1.25\ \text{kg/m}^3)(10.89\ \text{m/s})^2$$

$$= (6000 - 74.1)\,\text{N/m}^2 = 5926\ \text{N/m}^2 \qquad (Ans)$$

Comments In the absence of viscous effects, the pressure throughout the hose is constant and equal to p_2. Physically, the decreases in pressure from p_1 to p_2 to p_3 accelerate the air and increase its kinetic energy from zero in the tank to an intermediate value in the hose and finally to its maximum value at the nozzle exit. Since the air velocity in the nozzle exit is nine times that in the hose, most of the pressure drop occurs across the nozzle ($p_1 = 6000$ N/m^2, $p_2 = 5926$ N/m^2, and $p_3 = 0$).

Since the pressure change from (1) to (3) is not too great [i.e., in terms of absolute pressure $(p_1 - p_3)/p_1 = 6.0/101 = 0.06$],

it follows from the ideal gas law that the density change is also not significant. Hence, the incompressibility assumption is reasonable for this problem. If the tank pressure were considerably larger or if viscous effects were important, application of the Bernoulli equation to this situation would be incorrect.

By repeating the calculations for various nozzle diameters, d, the results shown in **Figs. E4.8b** and **c** are obtained. The flowrate increases as the nozzle is opened (i.e., larger d). Note that if the nozzle diameter is the same as that of the hose ($d = 0.06$ m), the pressure throughout the hose is atmospheric (zero gage).

The Wide World of Fluids

Hi-tech inhaler

The term *inhaler* often brings to mind a treatment for asthma or bronchitis. Work is underway to develop a family of inhalation devices that can do more than treat respiratory ailments. They will be able to deliver medication for diabetes and other conditions by spraying it to reach the bloodstream through the lungs. The concept is to make the spray droplets fine enough to penetrate to the lungs' tiny sacs, the alveoli, where exchanges between blood and the outside world take place. This is accomplished by use of a

laser-machined *nozzle* containing an array of very fine holes that cause the liquid to divide into a mist of micron-scale droplets. The device fits the hand and accepts a disposable strip that contains the medicine solution sealed inside a blister of laminated plastic and the nozzle. An electrically actuated piston drives the liquid from its reservoir through the nozzle array and into the respiratory system. To take the medicine, the patient breathes through the device and a differential pressure transducer in the inhaler senses when the patient's breathing has reached the best condition for receiving the medication. At that point, the piston is automatically triggered.

In many situations the combined effects of kinetic energy, pressure, and gravity are important. Example 4.9 illustrates this.

EXAMPLE 4.9 | Flow in a Variable Area Pipe

Given Water flows through a pipe reducer as is shown in **Fig. E4.9**. The static pressures at (1) and (2) are measured by the inverted U-tube manometer containing oil of specific gravity, SG, less than one.

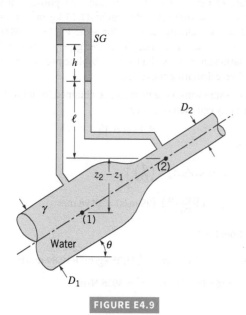

FIGURE E4.9

Find Determine the manometer reading, h.

Solution With the assumptions of steady, inviscid, incompressible flow, the Bernoulli equation can be written as

$$p_1 + \tfrac{1}{2}\rho V_1^2 + \gamma z_1 = p_2 + \tfrac{1}{2}\rho V_2^2 + \gamma z_2$$

The continuity equation (Eq. 4.19) provides a second relationship between V_1 and V_2 if we assume the velocity profiles are uniform at those two locations and the fluid incompressible:

$$Q = A_1 V_1 = A_2 V_2$$

By combining these two equations, we obtain

$$p_1 - p_2 = \gamma(z_2 - z_1) + \tfrac{1}{2}\rho V_2^2[1 - (A_2/A_1)^2] \qquad (1)$$

This pressure difference is measured by the manometer and can be determined by using the pressure–depth ideas (i.e., the manometer rule) developed in Chapter 2. Thus,

$$p_1 - \gamma(z_2 - z_1) - \gamma\ell - \gamma h + SG\gamma h + \gamma\ell = p_2$$

or

$$p_1 - p_2 = \gamma(z_2 - z_1) + (1 - SG)\gamma h \qquad (2)$$

As discussed in Chapter 2, this pressure difference is neither merely γh nor $\gamma(h + z_1 - z_2)$ because the manometer fluid is not water.

Equations 1 and 2 can be combined to give the desired result as follows:

$$(1 - SG)\gamma h = \frac{1}{2}\rho V_2^2\left[1 - \left(\frac{A_2}{A_1}\right)^2\right]$$

or since $V_2 = Q/A_2$

$$h = (Q/A_2)^2\frac{1 - (A_2/A_1)^2}{2g(1 - SG)} \qquad \text{(Ans)}$$

Comment The difference in elevation, $z_1 - z_2$, was not needed because the change in elevation term in the Bernoulli equation exactly cancels the elevation term in the manometer equation. However, the pressure difference, $p_1 - p_2$, depends on the angle θ, because of the elevation, $z_1 - z_2$, in Eq. 1. Thus, for a given flowrate, the pressure difference, $p_1 - p_2$, as measured by a pressure gage would vary with θ, but the manometer reading, h, would be independent of θ.

EXAMPLE 4.10 | Flow through Vertical Variable Area Pipe

Given Water (assumed inviscid and incompressible) flows steadily in the vertical variable-area pipe shown in **Fig. E4.10**.

Find Determine the flowrate if the pressure in each of the gages reads 100 kPa.

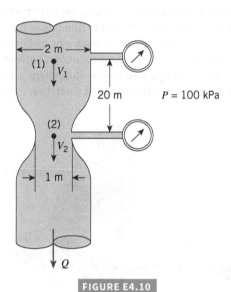

FIGURE E4.10

Solution From the Bernoulli equation,

$$p_1 + \frac{1}{2}\rho V_1^2 + \gamma z_1 = p_2 + \frac{1}{2}\rho V_2^2 + \gamma z_2 \quad \text{where} \quad p_1 = p_2 = 100 \text{ kPa}$$

Thus,

$$\frac{1}{2}\rho(V_2^2 - V_1^2) = \gamma(z_1 - z_2) \qquad (1)$$

Also,

$$A_1 V_1 = A_2 V_2$$

or

$$V_1 = \frac{A_2}{A_1}V_2 = \left[\frac{\frac{\pi}{4}D_2^2}{\frac{\pi}{4}D_1^2}\right]V_2 = \left(\frac{D_2}{D_1}\right)^2 V_2 = \left(\frac{1 \text{ m}}{2 \text{ m}}\right)^2 V_2 = \frac{1}{4}V_2 \qquad (2)$$

Hence, Eq. (1), becomes

$$\frac{1}{2}\rho\left[V_2^2 - \frac{1}{16}V_2^2\right] = \rho g(z_1 - z_2) \qquad (3)$$

or

$$\frac{15}{16}V_2^2 = 2g(z_1 - z_2) = \left(9.81 \frac{\text{m}}{\text{s}}\right)(20 \text{ m})$$

Thus, $V_2 = 20.45$ m/s

$$Q = A_2 V_2 = \frac{\pi}{4}(1 \text{ m})^2(20.45 \text{ m/s}) = 17.38 \text{ m}^3/\text{s}$$

Comment Equation 2 shows that there is no substantial pressure drop between two points (1 and 2) over a vertical pipe, thus the velocity (V_2) shall be controlled by the elevation difference between the points. If there also occurs a pressure drop between points 1 and 2 then V_2 will be mainly controlled by p_1-p_2 and not by z_1-z_2. For even a very small pressure change between the points (which will definitely happen in real physical problems) say 5 kPa, V_2 increases around 3.7 times to 75.84 m/s. Even if we neglect the effect of z_1-z_2, V_2 will be 73.02 m/s. That is why this sensitivity of velocity change due to pressure change across pipe flows is utilized to measure the flow rates across pipes by restriction type flowmeters.

In general, an increase in velocity is accompanied by a decrease in pressure. For example, the velocity of the air flowing over the top surface of an airplane wing is, on the average, faster than that flowing under the bottom surface. Thus, the net pressure force is greater on the bottom than on the top—the wing generates a lift.

If the differences in velocity are considerable, the differences in pressure can also be considerable. For flows of gases, this may introduce compressibility effects as discussed in Section 4.8 and Chapter 11. For flows of liquids, this may result in **cavitation**, a potentially dangerous situation that results when the liquid pressure is reduced to the vapor pressure and the liquid "boils."

Video

V4.10 Venturi channel

Cavitation occurs when the pressure is reduced to the vapor pressure.

As discussed in Chapter 1, the vapor pressure, p_v, is the pressure at which vapor bubbles form in a liquid. It is the pressure at which the liquid starts to boil. Obviously this pressure depends on the type of liquid and its temperature. For example, water, which boils at 100 °C at standard atmospheric pressure, 1 bar, boils at 27 °C if the pressure is 0.035 bar. That is, $p_v = 0.0305$ bar at 27 °C and $p_v = 1$ bar at 100 °C (see Tables B.1 and B.2).

One way to produce cavitation in a flowing liquid is noted from the Bernoulli equation. If the fluid velocity is increased (for example, by a reduction in flow area as shown in **Fig. 4.16**), the pressure will decrease. This pressure decrease (needed to accelerate the fluid through the constriction) can be large enough so that the pressure in the liquid is reduced to its vapor pressure. A simple example of cavitation can be demonstrated with an ordinary garden hose. If the hose is "kinked," a restriction in the flow area in some ways analogous to that shown in Fig. 4.16 will result. The water velocity through this restriction will be relatively large. With a sufficient amount of restriction the sound of the flowing water will change—a definite "hissing" sound is produced. This sound is a result of cavitation.

Cavitation can cause damage to equipment.

In such situations boiling occurs (though the temperature need not be high), vapor bubbles form, and then they collapse as the fluid moves into a region of higher pressure (lower velocity). This process can produce dynamic effects (imploding) that cause very large pressure transients in the vicinity of the bubbles. Pressures as large as 690 MPa are believed to occur. If the bubbles collapse close to a physical boundary they can, over a period of time, cause damage to the surface near the cavitation area. It is common to find pits or holes in the metal surfaces of old propellers or pump impellers. Tip cavitation from a propeller is shown in **Fig. 4.17**. In this case the high-speed rotation of the propeller produced a corresponding low pressure on the propeller. Obviously, proper design and use of equipment are needed to eliminate cavitation damage (see Section 12.4.3 for a discussion of cavitation in pumps).

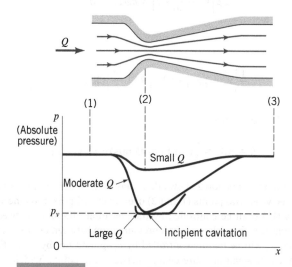

FIGURE 4.16 Pressure variation and cavitation in a variable area pipe.

National Physical Laboratory/Crown Copyright/Science Source

FIGURE 4.17 Tip cavitation from a propeller.

EXAMPLE 4.11 | Siphon and Cavitation

Given A liquid can be siphoned from a container, as shown in **Fig. E4.11a**, provided the end of the tube, point (3), is below the free surface in the container, point (1), and the maximum elevation of the tube, point (2), is "not too great." Consider water at 16 °C being siphoned from a large tank through a constant-diameter hose as shown in **Fig. E4.11b**. The end of the siphon is 1.5 m below the bottom of the tank, and the atmospheric pressure is 1 bar.

Find Determine the maximum height of the hill, H, over which the water can be siphoned without cavitation occurring.

Solution If the flow is steady, inviscid, and incompressible, we can apply the Bernoulli equation along the streamline from (1) to (2) to (3) as follows:

$$p_1 + \frac{1}{2}\rho V_1^2 + \gamma z_1 = p_2 + \frac{1}{2}\rho V_2^2 + \gamma z_2$$
$$= p_3 + \frac{1}{2}\rho V_3^2 + \gamma z_3 \qquad (1)$$

With the tank bottom as the datum, we have $z_1 = 4.6$ m, $z_2 = H$, and $z_3 = -1.5$ m. Also, $V_1 = 0$ (large tank), $p_1 = 0$ (open tank), $p_3 = 0$ (free jet), and from the continuity equation $A_2 V_2 = A_3 V_3$, or because the hose is constant diameter, $V_2 = V_3$. Thus, the speed of the fluid in the hose is determined from Eq. 1 to be

$$V_3 = \sqrt{2g(z_1 - z_3)} = \sqrt{2(9.8 \text{ m/s}^2)[4.6 - (-1.5)]} \text{ m}$$
$$= 10.94 \text{ m/s} = V_2$$

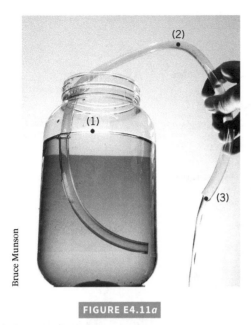

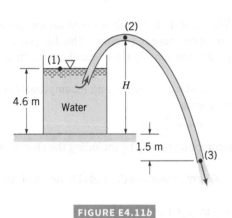

FIGURE E4.11a

FIGURE E4.11b

system will be $p = 1.765$ kPa. Careful consideration of Eq. 2 and Fig. E4.11b will show that this lowest pressure will occur at the top of the hill. Since we have used gage pressure at point (1) ($p_1 = 0$), we must use gage pressure at point (2) also. Thus, $p_2 = 1.765 - 101 = -99.235$ kPa, and Eq. 2 gives

$$(-99.235 \text{ kPa}) = (9810 \text{ N/m}^3)(4.6 - H) \text{ m} - \frac{1}{2}(1000 \text{ kg/m}^3)(10.94 \text{ m})^2$$

or

$$H = 8.6 \text{ m} \qquad (Ans)$$

For larger values of H, vapor bubbles will form at point (2) and the siphon action may stop.

Comments Note that we could have used absolute pressure throughout ($p_2 = 1.765$ kPa and $p_1 = 101$ kPa) and obtained the same result. The lower the elevation of point (3), the larger the flowrate and, therefore, the smaller the value of H allowed.

We could also have used the Bernoulli equation between (2) and (3), with $V_2 = V_3$, to obtain the same value of H. In this case it would not have been necessary to determine V_2 by use of the Bernoulli equation between (1) and (3).

The above results are independent of the diameter and length of the hose (provided viscous effects are not important). Proper design of the hose (or pipe) is needed to ensure that it will not collapse due to the large pressure difference (vacuum) between the inside and outside of the hose.

By using the fluid properties listed in Table 1.5 and repeating the calculations for various fluids, the results shown in **Fig. E4.11c** are obtained. The value of H is a function of both the specific weight of the fluid, γ, and its vapor pressure, p_v.

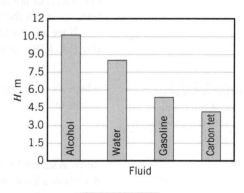

FIGURE E4.11c

Use of Eq. 1 between points (1) and (2) then gives the pressure p_2 at the top of the hill as

$$p_2 = p_1 + \frac{1}{2}\rho V_1^2 + \gamma z_1 - \frac{1}{2}\rho V_2^2 - \gamma z_2$$
$$= \gamma(z_1 - z_2) - \frac{1}{2}\rho V_2^2 \qquad (2)$$

From Table B.1, the vapor pressure of water at 16 °C is 1.765 kPa. Hence, for incipient cavitation the lowest pressure in the

4.6.3 Flowrate Measurement

Many types of devices using principles involved in the Bernoulli equation have been developed to measure fluid velocities and flowrates. The Pitot-static tube discussed in Section 4.5 is an example. Other examples discussed in this section include devices to measure flowrates in pipes and conduits and devices to measure flowrates in open channels. In this chapter we will consider "ideal" **flowmeters**—those devoid of viscous, compressibility, and other "real-world" effects. Corrections for these effects are discussed in Chapters 8 and 10. Our goal here is to understand the basic operating principles of these simple flowmeters.

An effective way to measure the flowrate through a pipe is to place some type of restriction within the pipe as shown in **Fig. 4.18** and to measure the pressure difference between the low-velocity, high-pressure upstream section (1) and the high-velocity, low-pressure downstream section (2). Three commonly used types of flowmeters are illustrated: the *orifice meter*,

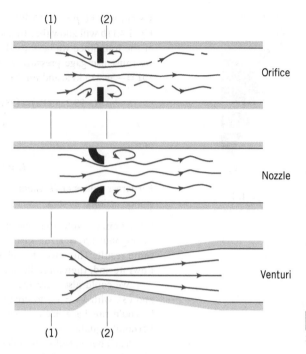

FIGURE 4.18 Typical devices for measuring flowrate in pipes.

the *nozzle meter*, and the *Venturi meter*. The operation of each is based on the same physical principles—an increase in velocity causes a decrease in pressure. The difference between them is a matter of length, cost, accuracy, and how closely their actual operation obeys the idealized flow assumptions.

We assume the flow is horizontal ($z_1 = z_2$), steady, inviscid, and incompressible between points (1) and (2). The Bernoulli equation becomes

$$p_1 + \frac{1}{2}\rho V_1^2 = p_2 + \frac{1}{2}\rho V_2^2$$

(The effect of nonhorizontal flow can be incorporated easily by including the change in elevation, $z_1 - z_2$, in the Bernoulli equation.)

If we assume the velocity profiles are uniform at sections (1) and (2), the continuity equation (Eq. 4.19) can be written as

$$Q = A_1 V_1 = A_2 V_2$$

where A_2 is the small flow area ($A_2 < A_1$) at section (2). Combination of these two equations results in the following theoretical flowrate

$$Q = A_2 \sqrt{\frac{2(p_1 - p_2)}{\rho[1 - (A_2/A_1)^2]}} \tag{4.20}$$

The flowrate varies as the square root of the pressure difference across the flowmeter.

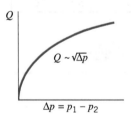

Thus, as shown in the adjacent figure, for a given flow geometry (A_1 and A_2) the flowrate can be determined if the pressure difference, $p_1 - p_2$, is measured. The actual measured flowrate, Q_{actual}, will be smaller than this theoretical result because of various differences between the "real world" and the assumptions used in the derivation of Eq. 4.20. These differences (which are quite consistent and may be as small as 1% to 2% or as large as 40%, depending on the geometry used) can be accounted for by using an empirically obtained discharge coefficient as discussed in Section 8.6.1. A significant portion of the difference between ideal and actual flowrate is due to the vena contracta effect discussed in Sec. 4.6.1.

EXAMPLE 4.12 | Venturi Meter

Given Kerosene (SG = 0.85) flows through the Venturi meter shown in **Fig. E4.12a** with flowrates between 0.001 and 0.01 m^3/s.

Find Determine the range in pressure difference, $p_1 - p_2$, needed to measure these flowrates.

Solution If the flow is assumed to be steady, inviscid, and incompressible, the relationship between flowrate and pressure is given by Eq. 4.20. This can be rearranged to give

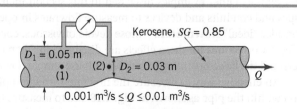

FIGURE E4.12a

$$p_1 - p_2 = \frac{Q^2 \rho [1 - (A_2/A_1)^2]}{2A_2^2}$$

With the density of the flowing fluid

$$\rho = SG\, \rho_{H_2O} = 0.85(1000 \text{ kg/m}^3) = 850 \text{ kg/m}^3$$

and the area ratio

$$A_2/A_1 = (D_2/D_1)^2 = (0.03 \text{ m}/0.05 \text{ m})^2 = 0.6$$

the pressure difference for the smallest flowrate is

$$p_1 - p_2 = (0.001 \text{ m}^3/\text{s})^2(850 \text{ kg/m}^3) \frac{(1 - 0.6^2)}{2[(\pi/4)(0.03 \text{ m})^2]^2}$$

$$= 544 \text{ N/m}^2 = 0.544 \text{ kPa}$$

Likewise, the pressure difference for the largest flowrate is

$$p_1 - p_2 = (0.01)^2(850) \frac{(1 - 0.6^2)}{2[(\pi/4)(0.03)^2]^2}$$

$$= 0.544 \times 10^5 \text{ N/m}^2 = 54.4 \text{ kPa}$$

Thus,

$$0.544 \text{ kPa} \le p_1 - p_2 \le 54.4 \text{ kPa} \qquad (Ans)$$

Comments These values represent the pressure differences for inviscid, steady, incompressible conditions. The ideal results presented here are independent of the particular flowmeter geometry—an orifice, nozzle, or Venturi meter (see Fig. 4.18).

It is seen from Eq. 4.20 that the flowrate varies as the square root of the pressure difference. Hence, as indicated by the numerical results and shown in **Fig. E4.12b**, a 10-fold increase in flowrate requires a 100-fold increase in pressure difference. This nonlinear relationship can cause difficulties when measuring flowrates over a wide range of values. Such measurements would require pressure transducers with a wide range of operation. An alternative is to use two flowmeters in parallel—one for the larger and one for the smaller flowrate ranges.

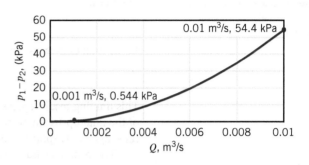

FIGURE E4.12b

Other flowmeters based on the Bernoulli equation are used to measure flowrates in open channels, such as flumes and irrigation ditches. Two of these devices, the *sluice gate* and the *sharp-crested weir*, are discussed next under the assumption of steady, inviscid, incompressible flow. These and other open-channel flow devices are discussed in more detail in Chapter 10.

Sluice gates like those shown in **Fig. 4.19a** are often used to regulate and measure the flowrate in open channels. As indicated in **Fig. 4.19b**, the flowrate, Q, is a function of the water depth upstream, z_1, the width of the gate, b, and the gate opening, a. Application of the Bernoulli equation and continuity equation between points (1) and (2) can provide a good approximation to the actual flowrate obtained. We assume the velocity profiles are uniform sufficiently far upstream and downstream of the gate.

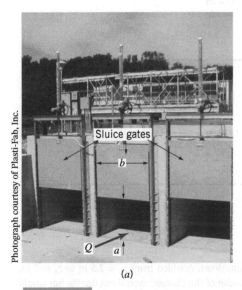

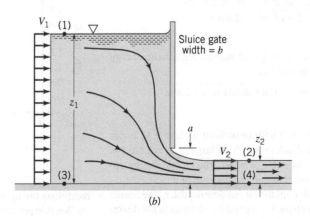

FIGURE 4.19 Sluice gate geometry.

Thus, we apply the Bernoulli equation between points on the free surfaces at (1) and (2) to give

$$p_1 + \frac{1}{2}\rho V_1^2 + \gamma z_1 = p_2 + \frac{1}{2}\rho V_2^2 + \gamma z_2$$

Also, if the gate is the same width as the channel so that $A_1 = bz_1$ and $A_2 = bz_2$, the continuity equation gives

$$Q = A_1 V_1 = bV_1 z_1 = A_2 V_2 = bV_2 z_2$$

The flowrate under a sluice gate depends on the water depths on either side of the gate.

With the fact that $p_1 = p_2 = 0$, these equations can be combined and rearranged to give the flowrate as

$$Q = z_2 b \sqrt{\frac{2g(z_1 - z_2)}{1 - (z_2/z_1)^2}} \qquad (4.21)$$

In the limit of $z_1 \gg z_2$, this result simply becomes

$$Q = z_2 b \sqrt{2gz_1}$$

This limiting result represents the fact that if the depth ratio, z_1/z_2, is large, the kinetic energy of the fluid upstream of the gate is negligible and the fluid velocity after it has fallen a distance $(z_1 - z_2) \approx z_1$ is approximately $V_2 = \sqrt{2gz_1}$.

The results of Eq. 4.21 could also be obtained by using the Bernoulli equation between points (3) and (4) and the fact that $p_3 = \gamma z_1$ and $p_4 = \gamma z_2$ because the streamlines at these sections are straight. In this formulation, rather than the potential energies at (1) and (2), we have the pressure contributions at (3) and (4).

The downstream depth, z_2, not the gate opening, a, was used to obtain the result of Eq. 4.21. As was discussed relative to flow from an orifice (Fig. 4.14), the fluid cannot turn a sharp 90° corner. A vena contracta results with a contraction coefficient, $C_c = z_2/a$, less than 1. Typically C_c is approximately 0.61 over the depth ratio range of $0 < a/z_1 < 0.2$ (see Section 10.7.4 for further discussion). For larger values of a/z_1 the value of C_c increases rapidly.

EXAMPLE 4.13 │ Sluice Gate

Given Water flows under the sluice gate shown in **Fig. E4.13a**.

Find Determine the approximate flowrate per unit width of the channel.

Solution Under the assumptions of steady, inviscid, incompressible flow, we can apply Eq. 4.21 to obtain Q/b, the flowrate per unit width, as

$$\frac{Q}{b} = z_2 \sqrt{\frac{2g(z_1 - z_2)}{1 - (z_2/z_1)^2}}$$

In this instance $z_1 = 2.5$ m and $a = 0.4$ m so the ratio $a/z_1 = 0.16 < 0.20$, and we can assume that the contraction coefficient is approximately $C_c = 0.61$. Thus, $z_2 = C_c a = 0.61(0.4\text{ m}) = 0.244$ m, and we obtain the flowrate

$$\frac{Q}{b} = (0.244\text{ m}) \sqrt{\frac{2(9.81\text{ m/s}^2)(2.5\text{ m} - 0.244\text{ m})}{1 - (0.244\text{ m}/2.5\text{ m})^2}}$$

$$= 1.63\text{ m}^2/\text{s} \qquad \text{(Ans)}$$

Comment If we consider $z_1 \gg z_2$ and neglect the kinetic energy of the upstream fluid, we would have

$$\frac{Q}{b} = z_2 \sqrt{2gz_1} = 0.244\text{ m} \sqrt{2(9.81\text{ m/s}^2)(2.5\text{ m})}$$

$$= 1.71\text{ m}^2/\text{s}$$

In this case the difference in Q with or without including V_1 is not too significant because the depth ratio is fairly large ($z_1/z_2 = 2.5/0.244 = 10.2$). Thus, it is often reasonable to neglect the kinetic energy upstream from the gate compared to that downstream of it.

By repeating the calculations for various flow depths, z_1, the results shown in **Fig. E4.13b** are obtained. Note that the flowrate is not directly proportional to the flow depth. Thus, for example, if during flood

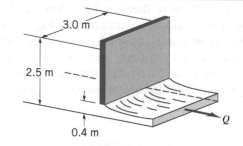

FIGURE E4.13a

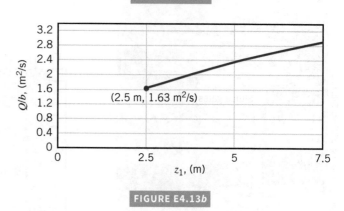

FIGURE E4.13b

conditions the upstream depth doubled from $z_1 = 2.5$ m to $z_1 = 5$ m, the flowrate per unit width of the channel would not double but would increase only from 1.63 m²/s to 2.36 m²/s.

Another device used to measure flow in an open channel is a *weir*. A typical rectangular, sharp-crested weir is shown in **Fig. 4.20**. For such devices the flowrate of liquid over the top of the weir plate is dependent on the weir height, P_w, the width of the channel, b, and the head, H, of the water above the top of the weir. Application of the Bernoulli equation can provide a simple approximation of the flowrate expected for these situations, even though the actual flow is quite complex.

Between points (1) and (2) the pressure and gravitational fields cause the fluid to accelerate from velocity V_1 to velocity V_2. At (1) the pressure is $p_1 = \gamma h$, while at (2) the pressure is essentially atmospheric, $p_2 = 0$. Across the curved streamlines directly above the top of the weir plate (section a–a), the pressure changes from atmospheric on the top surface to some maximum value within the fluid stream and then to atmospheric again at the bottom surface. This distribution is indicated in Fig. 4.20. Such a pressure distribution, combined with the streamline curvature and gravity, produces a rather nonuniform velocity profile across this section. This velocity distribution can be obtained from experiments or a more advanced theory.

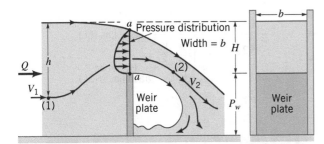

FIGURE 4.20 Rectangular, sharp-crested weir geometry.

For now, we will take a very simple approach and assume that the weir flow is similar in many respects to an orifice-type flow with a free streamline. In this instance we would expect the average velocity across the top of the weir to be proportional to $\sqrt{2gH}$ and the flow area for this rectangular weir to be proportional to Hb. Hence, it follows that

$$Q = C_1 Hb\sqrt{2gH} = C_1 b\sqrt{2g}H^{3/2}$$

where C_1 is a constant to be determined.

Simple use of the Bernoulli equation has provided a method to analyze the relatively complex flow over a weir. The correct functional dependence of Q on H has been obtained ($Q \sim H^{3/2}$, as indicated by the adjacent figure), but the value of the coefficient C_1 is unknown. Even a more advanced analysis cannot predict its value accurately. As is discussed in Chapter 10, experiments are used to determine the value of C_1.

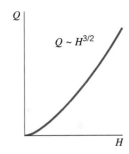

EXAMPLE 4.14 | Weir

Given   Water flows over a triangular weir, as is shown in **Fig. E4.14**.

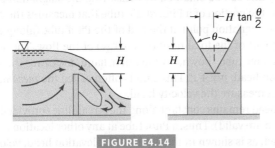

FIGURE E4.14

Find   Based on a simple analysis using the Bernoulli equation, determine the dependence of the flowrate on the depth H. If the flowrate is Q_0 when $H = H_0$, estimate the flowrate when the depth is increased to $H = 3H_0$.

Solution   With the assumption that the flow is steady, inviscid, and incompressible, it is reasonable to assume from Eq. 4.18

that the average speed of the fluid over the triangular notch in the weir plate is proportional to $\sqrt{2gH}$. Also, the flow area for a depth of H is $H[H\tan(\theta/2)]$. The combination of these two ideas gives

$$Q = AV = H^2 \tan\frac{\theta}{2}(C_2\sqrt{2gH}) = C_2 \tan\frac{\theta}{2}\sqrt{2g}\,H^{5/2} \qquad (Ans)$$

where C_2 is an unknown constant to be determined experimentally.

Thus, an increase in the depth by a factor of three (from H_0 to $3H_0$) results in an increase of the flowrate by a factor of

$$\frac{Q_{3H_0}}{Q_{H_0}} = \frac{C_2 \tan(\theta/2)\sqrt{2g}\,(3H_0)^{5/2}}{C_2 \tan(\theta/2)\sqrt{2g}\,(H_0)^{5/2}}$$

$$= 15.6 \qquad (Ans)$$

Comment   Note that for a triangular weir the flowrate is proportional to $H^{5/2}$, whereas for the rectangular weir discussed above, it is proportional to $H^{3/2}$. The triangular weir can be accurately used over a wide range of flowrates.

4.7 THE ENERGY LINE AND THE HYDRAULIC GRADE LINE

As was discussed in Section 4.4, the Bernoulli equation can be considered a form of the energy equation representing the partitioning of energy for an inviscid, incompressible, steady flow. The sum of the various energies of the fluid remains constant as the fluid flows from one section to another. A useful interpretation of the Bernoulli equation can be obtained through use of the concepts of the **hydraulic grade line** (HGL) and the **energy line** (EL). These ideas represent a geometrical interpretation of a flow and can often be effectively used to better grasp the fundamental processes involved.

For steady, inviscid, incompressible flow the total energy remains constant along a streamline. The concept of "head" was introduced by dividing each term in Eq. 4.7 by the specific weight, $\gamma = \rho g$, to give the Bernoulli equation in the following form

$$\frac{p}{\gamma} + \frac{V^2}{2g} + z = \text{ constant on a streamline} = H \qquad (4.22)$$

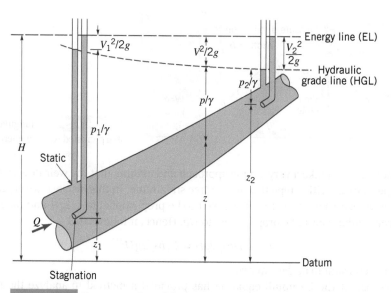

FIGURE 4.21 Representation of the energy line and the hydraulic grade line.

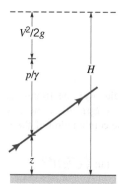

Each of the terms in this equation has the units of length (feet or meters) and represents a certain type of head. The Bernoulli equation states that the sum of the pressure head, the velocity head, and the elevation head is constant along a streamline. This constant is called the *total head*, H, and is shown in the adjacent figure.

The energy line is a line that represents the total head available to the fluid. As shown in **Fig. 4.21**, the elevation of the energy line can be obtained by measuring the stagnation pressure with a Pitot tube. (A Pitot tube is the portion of a Pitot-static tube that measures the stagnation pressure. See Section 4.5.) The stagnation point at the end of the Pitot tube (along with elevation head) provides a measurement of the total head (or energy) of the flow. The static pressure tap connected to the piezometer tube shown, on the other hand, measures the sum of the pressure head and the elevation head, $p/\gamma + z$. This sum is often called the *piezometric head*. The static pressure tap does not measure the velocity head.

According to Eq. 4.22, the total head remains constant along the streamline (provided the assumptions of the Bernoulli equation are valid). Thus, a Pitot tube at any other location in the flow will measure the same total head, as is shown in the figure. The elevation head, velocity head, and pressure head may vary individually along the streamline, however.

The locus of elevations provided by a series of Pitot tubes is termed the energy line, EL. The locus provided by a series of piezometer taps is termed the hydraulic grade line, HGL. Under the assumptions of the Bernoulli equation, the energy line is horizontal. If the fluid velocity changes along the streamline, the hydraulic grade line will not be horizontal. If viscous effects are important (as they often are in pipe flows), the total head does not remain constant due to a loss in useful energy as the fluid flows along its streamline. This

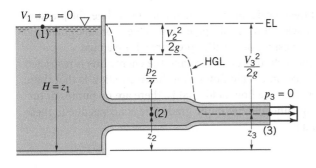

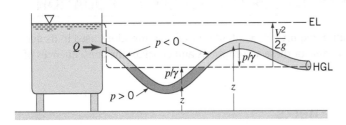

FIGURE 4.22 The energy line and hydraulic grade line for flow from a tank.

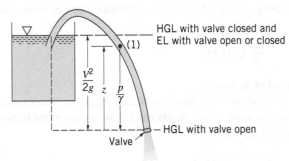

FIGURE 4.23 Use of the energy line and the hydraulic grade line.

means that the energy line is no longer horizontal. Such viscous effects are discussed in Chapters 5 and 8.

The energy line and hydraulic grade line for flow from a large tank are shown in **Fig. 4.22**. If the flow is steady, incompressible, and inviscid, the energy line is horizontal and at the elevation of the liquid in the tank (since the fluid velocity in the tank and the pressure on the surface are zero). The hydraulic grade line lies a distance of one velocity head, $V^2/2g$, below the energy line. Thus, a change in fluid velocity due to a change in the pipe diameter results in a change in the elevation of the hydraulic grade line. At the pipe outlet the pressure head is zero (gage), so the pipe elevation and the hydraulic grade line coincide.

The distance from the pipe to the hydraulic grade line indicates the pressure within the pipe, as is shown in **Fig. 4.23**. If the pipe lies below the hydraulic grade line, the pressure within the

EXAMPLE 4.15 | Energy Line and Hydraulic Grade Line

Given Water is siphoned from the tank shown in **Fig. E4.15** through a hose of constant diameter. A small hole is found in the hose at location (1) as indicated.

FIGURE E4.15

Find When the siphon is used, will water leak out of the hose, or will air leak into the hose, thereby possibly causing the siphon to malfunction?

Solution Whether air will leak into or water will leak out of the hose depends on whether the pressure within the hose at (1) is less than or greater than atmospheric. Which happens can be easily determined by using the energy line and hydraulic grade line concepts.

With the assumption of steady, incompressible, inviscid flow, it follows that the total head is constant—thus, the energy line is horizontal.

Since the hose diameter is constant, it follows from the continuity equation ($AV = $ constant) that the water velocity in the hose is constant throughout. Thus, the hydraulic grade line is a constant distance, $V^2/2g$, below the energy line as shown in Fig. E4.15. Since the pressure at the end of the hose is atmospheric, it follows that the hydraulic grade line is at the same elevation as the end of the hose outlet. The fluid within the hose at any point above the hydraulic grade line will be less than atmospheric pressure.

Thus, air will leak into the hose through the hole at a point (1).   (*Ans*)

Comment In practice, viscous effects may be quite important, making this simple analysis (horizontal energy line) incorrect. However, if the hose is "not too small diameter," "not too long," the fluid "not too viscous," and the flowrate "not too large," the above result may be very accurate. If any of these assumptions are relaxed, a more detailed analysis is required (see Chapter 8). If the end of the hose was closed so that the flowrate were zero, the hydraulic grade line would coincide with the energy line ($V^2/2g = 0$ throughout), the pressure at (1) would be greater than atmospheric, and water would leak through the hole at (1).

For flow below (above) the hydraulic grade line, the pressure is positive (negative).

pipe is positive (above atmospheric). If the pipe lies above the hydraulic grade line, the pressure is negative (below atmospheric). Thus, a scale drawing of a pipeline and the hydraulic grade line can be used to readily indicate regions of positive or negative pressure within a pipe.

In this section, the discussion of the hydraulic grade line and the energy line is restricted to ideal situations involving inviscid, incompressible flows. Another restriction is that there are no "sources" or "sinks" of energy within the flow field; that is, there are no pumps or turbines involved. Alterations in the energy line and hydraulic grade line concepts due to these devices are discussed in Chapters 5 and 8.

4.8 RESTRICTIONS ON USE OF THE BERNOULLI EQUATION

Proper use of the Bernoulli equation requires close attention to the assumptions used in its derivation. In this section we review some of these assumptions and consider the consequences of incorrect use of the equation.

4.8.1 Compressibility Effects

One of the main assumptions is that the fluid is incompressible. Although this is reasonable for most liquid flows, it can, in certain instances, introduce considerable errors for gases.

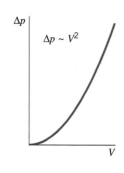

Δp

$\Delta p \sim V^2$

V

In the previous section, we saw that the stagnation pressure, p_{stag}, is greater than the static pressure, p_{static}, by an amount $\Delta p = p_{stag} - p_{static} = \rho V^2/2$ provided that the density remains constant. If this dynamic pressure is not too large compared with the static pressure, the density change between two points is not very large and the flow can be considered incompressible. However, since the dynamic pressure varies as V^2, the error associated with the assumption that a fluid is incompressible increases with the square of the velocity of the fluid, as indicated by the adjacent figure. To account for compressibility effects, we could return to Eq. 4.6 and properly integrate the term $\int dp/\rho$ when ρ is not constant. That method can be more complicated than necessary and will likely introduce new restrictions. Instead we will develop a criterion that allows us to check the validity of the incompressible flow assumption.

From Eq. 1.13,

$$\frac{d\rho}{\rho} = \frac{dp}{E_v}$$

which for a finite change is

$$\frac{\Delta\rho}{\rho} = \frac{\Delta p}{E_v}$$

If we accept a rather arbitrary density change of 4.5% as insignificant, we can model the fluid as incompressible when

$$\frac{\Delta\rho}{\rho} \leq 0.045$$

Of course, this criterion can be decreased or increased.

The Bernoulli equation can help us determine a velocity limit for incompressible flow.

A simple, although specialized, case of compressible flow occurs when the temperature of an ideal gas remains constant along the streamline—isothermal flow. As indicated in Section 1.7.2, Eq. 1.16, for an isothermal flow, $E_v = p$, so

$$\frac{\Delta\rho}{\rho} = \frac{\Delta p}{p} \leq 0.045$$

This criterion should be used with caution because most isothermal flows are accompanied by viscous effects.

A much more common compressible flow condition is that of isentropic (constant entropy) flow of an ideal gas. Such flows are reversible adiabatic processes—"no friction or heat transfer"—and are closely approximated in many physical situations. As indicated in Section 1.7.2, Eq. 1.17, for an isentropic flow, $E_v = kp$, so

$$\frac{\Delta\rho}{\rho} = \frac{\Delta p}{kp}$$

For air, $k = 1.4$, so applying our criterion,

$$\frac{\Delta\rho}{\rho} = \frac{\Delta p}{1.4p} \leq 0.045$$

or

$$\frac{\Delta p}{p} \leq 0.063$$

The Bernoulli equation, as presented in this chapter, applies to an isentropic process. Therefore, we can relate pressure to the fluid velocity with the Bernoulli equation (with negligible elevation change) as $\Delta p = \rho V^2/2$. Thus, we can model a flow as incompressible if

$$\frac{\rho V^2}{p} \leq 0.126$$

For air at standard conditions, this quantity corresponds to a velocity of 102 m/s. As discussed in Chapter 11, this velocity corresponds to about 30% of the speed of sound in standard air (Ma = 0.3). At higher speeds, compressibility may become important (see also Section 11.3, Eq. 11.48).

4.8.2 Unsteady Effects

Another restriction of the Bernoulli equation (Eq. 4.7) is the assumption that the flow is steady. For such flows, on a given streamline, the velocity is a function of only s, the location along the streamline; that is, along a streamline $V = V(s)$. For unsteady flows the velocity is also a function of time so that along a streamline $V = V(s,t)$. Thus, when taking the time derivative of the velocity to obtain the streamwise acceleration, we obtain $a_s = \partial V/\partial t + V\,\partial V/\partial s$ rather than just $a_s = V\,\partial V/\partial s$ as is true for steady flow. For steady flows the acceleration is due to the change in velocity resulting from a change in position of the particle (the $V\,\partial V/\partial s$ term), whereas for unsteady flow there is an additional contribution to the acceleration resulting from a change in velocity with time at a fixed location (the $\partial V/\partial t$ term). These effects have been discussed in detail in Chapter 3. The net effect is that the inclusion of the unsteady term, $\partial V/\partial t$, does not allow the equation of motion to be easily integrated (as was done to obtain the Bernoulli equation) unless additional assumptions are made.

The Bernoulli equation was obtained by integrating the component of Newton's second law (Eq. 4.5) along the streamline. When integrated, the acceleration contribution to this equation, the $\frac{1}{2}\rho\,d(V^2)$ term, gave rise to the kinetic energy term in the Bernoulli equation. If the steps leading to Eq. 4.5 are repeated with the inclusion of the unsteady effect ($\partial V/\partial t \neq 0$), the following is obtained:

The Bernoulli equation can be modified for unsteady flows.

$$\rho\frac{\partial V}{\partial t}\,ds + dp + \frac{1}{2}\rho\,d(V^2) + \gamma\,dz = 0 \quad \text{(along a streamline)}$$

For incompressible flow this can be integrated between points (1) and (2) to give

$$p_1 + \frac{1}{2}\rho V_1^2 + \gamma z_1 = \rho\int_{s_1}^{s_2}\frac{\partial V}{\partial t}\,ds + p_2 + \frac{1}{2}\rho V_2^2 + \gamma z_2 \quad \text{(along a streamline)} \quad (4.23)$$

Equation 4.23 is an unsteady form of the Bernoulli equation valid for unsteady, incompressible, inviscid flow. Except for the integral involving the local acceleration, $\partial V/\partial t$, it is identical to the steady Bernoulli equation. In general, it is not possible to evaluate this integral because the variation of

V4.11 Oscillations in a U-tube

EXAMPLE 4.16 | Unsteady Flow—U-Tube

Given An incompressible, inviscid liquid is placed in a vertical, constant diameter U-tube as indicated in **Fig. E4.16**. When released from the nonequilibrium position shown, the liquid column will oscillate at a specific frequency.

Find Determine this frequency.

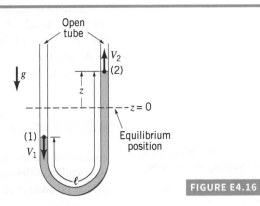

FIGURE E4.16

Solution The frequency of oscillation can be calculated by use of Eq. 4.23 as follows. Let points (1) and (2) be at the air–water interfaces of the two columns of the tube and $z = 0$ correspond to the equilibrium position of these interfaces. Hence, $p_1 = p_2 = 0$, and if $z_2 = z$, then $z_1 = -z$. In general, z is a function of time, $z = z(t)$. For a constant diameter tube, at any instant in time, the fluid speed is constant throughout the tube, $V_1 = V_2 = V$, and the integral representing the unsteady effect in Eq. 4.23 can be written as

$$\int_{s_1}^{s_2} \frac{\partial V}{\partial t}\, ds = \frac{dV}{dt}\int_{s_1}^{s_2} ds = \ell\, \frac{dV}{dt}$$

where ℓ is the total length of the liquid column as shown in the figure. Thus, Eq. 4.23 can be written as

$$\gamma(-z) = \rho\ell\, \frac{dV}{dt} + \gamma z$$

Since $V = dz/dt$ and $\gamma = \rho g$, this can be written as the second-order differential equation describing simple harmonic motion

$$\frac{d^2 z}{dt^2} + \frac{2g}{\ell}\, z = 0$$

which has the solution $z(t) = C_1 \sin(\sqrt{2g/\ell}\, t) + C_2 \cos(\sqrt{2g/\ell}\, t)$. The values of the constants C_1 and C_2 depend on the initial state (velocity and position) of the liquid at $t = 0$. Thus, the liquid oscillates in the tube with a frequency

$$\omega = \sqrt{2g/\ell} \qquad (Ans)$$

Comment This frequency depends on the length of the column and the acceleration of gravity (in a manner very similar to the oscillation of a pendulum). The period of this oscillation (the time required to complete an oscillation) is $t_0 = 2\pi\sqrt{\ell/2g}$.

$\partial V/\partial t$ along the streamline is not known. In some situations the concepts of "irrotational flow" and the "velocity potential" can be used to simplify this integral. These topics are discussed in Chapter 6.

In a few unsteady flow cases, the flow can be made steady by an appropriate selection of the coordinate system. Example 4.17 illustrates this.

EXAMPLE 4.17 | Unsteady or Steady Flow

Given A submarine moves through seawater ($SG = 1.03$) at a depth of 100 m with velocity $V_0 = 2.5$ m/s as shown in **Fig. E4.17**.

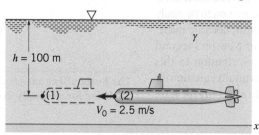

FIGURE E4.17

Find Determine the pressure at the stagnation point (2).

Solution In a coordinate system fixed to the ground, the flow is unsteady. For example, the water velocity at (1) is zero with the submarine in its initial position, but at the instant when the nose, (2), reaches point (1) the velocity there becomes $\mathbf{V}_1 = -V_0\hat{\mathbf{i}}$. Thus, $\partial\mathbf{V}_1/\partial t \neq 0$, and the flow is unsteady. Application of the steady

Bernoulli equation between (1) and (2) would give the incorrect result that $p_1 = p_2 + \rho V_0^2/2$. According to this result, the static pressure is greater than the stagnation pressure—an incorrect use of the Bernoulli equation.

We can either use an unsteady analysis for the flow (which is outside the scope of this text) or redefine the coordinate system so that it is fixed on the submarine, giving steady flow with respect to this system. The correct method would be

$$p_2 = \frac{\rho V_1^2}{2} + \gamma h = [(1.03)(1000)\,\text{kg/m}^3](2.5\ \text{m/s})^2/2$$
$$+ (9.80 \times 10^3\ \text{N/m}^3)(1.03)(100\ \text{m})$$
$$= (3218.75 + 1009400)\ \text{N/m}^2$$
$$= 1013\ \text{kPa} \qquad (Ans)$$

similar to that discussed in Example 4.2.

Comment If the submarine were accelerating, $\partial V_0/\partial t \neq 0$, the flow would be unsteady in either of the coordinate systems mentioned in this example, and we would be forced to use an unsteady form of the Bernoulli equation.

Some unsteady flows may be treated as "quasisteady" and solved approximately by using the steady Bernoulli equation. In these cases the unsteadiness is "not too great" (i.e., $\partial V/\partial t \ll V\partial V/\partial s$), and the steady flow results can be applied at each instant in time as though the flow were steady. The slow draining of a tank filled with liquid provides an example of this type of flow.

4.8.3 Rotational Effects

Care must be used in applying the Bernoulli equation across streamlines.

Another of the restrictions of the Bernoulli equation is that it is applicable along the streamline. Application of the Bernoulli equation across streamlines (i.e., from a point on one streamline to a point on another streamline) can lead to considerable errors, depending on the particular flow conditions involved. In general, the Bernoulli constant varies from streamline to streamline. However, under certain restrictions this constant is the same throughout the entire flow field. Example 4.18 illustrates this fact.

EXAMPLE 4.18 | Use of Bernoulli Equation Across Streamlines

Given Consider the uniform flow in the channel shown in **Fig. E4.18a**. The liquid in the vertical piezometer tube is stationary.

Find Discuss the use of the Bernoulli equation between points (1) and (2), points (3) and (4), and points (4) and (5).

Solution If the flow is steady, inviscid, and incompressible, Eq. 4.7 written between points (1) and (2) gives

$$p_1 + \frac{1}{2}\rho V_1^2 + \gamma z_1 = p_2 + \frac{1}{2}\rho V_2^2 + \gamma z_2$$
$$= \text{constant} = C_{12}$$

Since $V_1 = V_2 = V_0$ and $z_1 = z_2 = 0$, it follows that $p_1 = p_2 = p_0$ and the Bernoulli constant for this streamline, C_{12}, is given by

$$C_{12} = \frac{1}{2}\rho V_0^2 + p_0$$

Along the streamline from (3) to (4) we note that $V_3 = V_4 = V_0$ and $z_3 = z_4 = h$. As was shown in Example 4.6, application of $\mathbf{F} = m\mathbf{a}$ across the streamline (Eq. 4.12) gives $p_3 = p_1 - \gamma h$ because the streamlines are straight and horizontal. The above facts combined with the Bernoulli equation applied between (3) and (4) show that $p_3 = p_4$ and that the Bernoulli constant along this streamline is the same as that along the streamline between (1) and (2). That is, $C_{34} = C_{12}$, or

$$p_3 + \frac{1}{2}\rho V_3^2 + \gamma z_3 = p_4 + \frac{1}{2}\rho V_4^2 + \gamma z_4 = C_{34} = C_{12}$$

Similar reasoning shows that the Bernoulli constant is the same for any streamline in Fig. E4.18. Hence,

$$p + \frac{1}{2}\rho V^2 + \gamma z = \text{constant throughout the flow}$$

Again from Example 4.5 we recall that

$$p_4 = p_5 + \gamma H = \gamma H$$

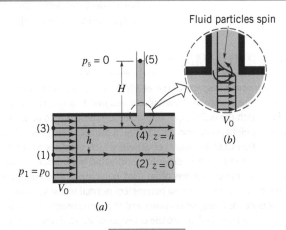

Fluid particles spin

$p_5 = 0$ (5)

H

(3)

h (4) $z = h$ (b)

(1)

$p_1 = p_0$ (2) $z = 0$

V_0

V_0

(a)

FIGURE E4.18

If we apply the Bernoulli equation across streamlines from (4) to (5), we obtain the incorrect result $H = p_4/\gamma + V_4^2/2g$. The correct result is $H = p_4/\gamma$.

In the previous paragraph, we see that we can apply the Bernoulli equation across streamlines (1)–(2) and (3)–(4) (i.e., $C_{12} = C_{34}$) but not across streamlines from (4) to (5). The reason for this is that while the flow in the channel is "irrotational," it is "rotational" between the flowing fluid in the channel and the stationary fluid in the piezometer tube. Because of the uniform velocity profile across the channel, it is seen that the fluid particles do not rotate or "spin" as they move. The flow is "irrotational." However, as seen in **Fig. E4.18b**, there is a very thin shear layer between (4) and (5) in which adjacent fluid particles interact and rotate or "spin." This produces a "rotational" flow. A more complete analysis would show that the Bernoulli equation cannot be applied across streamlines if the flow is "rotational" (see Chapter 6).

As is suggested by Example 4.18, if the flow is "irrotational" (i.e., the fluid particles do not "spin" as they move), it is appropriate to use the Bernoulli equation across streamlines. However, if the flow is "rotational" (fluid particles "spin"), use of the Bernoulli equation is restricted to flow along a streamline. The distinction between irrotational and rotational flow is often a very subtle and confusing one. These topics are discussed in more detail in Chapter 6. A thorough discussion can be found in more advanced texts (Ref. 3).

V4.12 Flow over a cavity

4.8.4 Other Restrictions

Another restriction on the Bernoulli equation is that the flow is inviscid and shear stress is negligible. As is discussed in Section 4.4, the Bernoulli equation is actually a first integral of Newton's second law along a streamline. This general integration was possible because, in the absence of viscous or shearing effects, the fluid system considered was a conservative system. The total energy of the system remains constant. If viscous and/or shearing effects are important the system is nonconservative (dissipative) and useful energy losses occur. A more detailed analysis is needed for these cases. Such material is presented in Chapter 5.

The Bernoulli equation is not valid for flows that involve pumps or turbines.

The final basic restriction on use of the Bernoulli equation is that there are no mechanical devices (pumps or turbines) in the system between the two points for which the equation is applied. These devices necessarily cause unsteady flow or interrupt the streamlines (Ref. 4), if they are present. Analysis of flows with pumps and turbines is covered in Chapters 5 and 12.

In this chapter we have spent considerable time investigating fluid dynamic situations governed by a relatively simple analysis for steady, inviscid, incompressible flows. Many flows can be adequately analyzed by use of these ideas. However, because of the rather severe restrictions imposed, many others cannot. An understanding of these basic ideas will provide a firm foundation for the remainder of the topics in this book.

CHAPTER SUMMARY

In this chapter several aspects of the steady flow of an inviscid, incompressible fluid are discussed. Newton's second law, $\mathbf{F} = m\mathbf{a}$, is applied to flows for which the only important forces are those due to pressure and gravity (weight)—viscous effects are assumed negligible. The result is the often-used Bernoulli equation, which provides a simple relationship among pressure, elevation, and velocity variations along a streamline. A similar but less often used equation is also obtained to describe the variations in these parameters normal to a streamline.

The concept of a stagnation point and the corresponding stagnation pressure is introduced, as are the concepts of static, dynamic, and total pressure and their related heads.

Several applications of the Bernoulli equation are discussed. In some flow situations, such as the use of a Pitot-static tube to measure fluid velocity or the flow of a liquid as a free jet from a tank, a Bernoulli equation alone is sufficient for the analysis. In other instances, such as confined flows in tubes and flowmeters, it is necessary to use both the Bernoulli equation and the continuity equation, which is a statement of the fact that mass is conserved as fluid flows.

The following checklist provides a study guide for this chapter. When your study of the entire chapter has been completed, you should be able to

- write out meanings of the terms listed and understand each of the related concepts. These terms are particularly important and are set in **bold black** type in the text.

- explain the origin of the pressure, elevation, and velocity terms in the Bernoulli equation and how they are related to Newton's second law of motion.

- apply the Bernoulli equation to simple flow situations, including Pitot-static tubes, free jet flows, confined flows, and flowmeters.

- use the concept of conservation of mass (the continuity equation) in conjunction with the Bernoulli equation to solve simple flow problems.

- apply Newton's second law across streamlines for appropriate steady, inviscid, incompressible flows.

- use the concepts of pressure, elevation, velocity, and total heads to solve various flow problems.

- explain and use the concepts of static, stagnation, dynamic, and total pressures.

- use the energy line and the hydraulic grade line concepts to describe various flow problems.

- explain the various restrictions on use of the Bernoulli equation.

KEY EQUATIONS

Streamwise and normal acceleration	$a_s = V\dfrac{\partial V}{\partial s}, \quad a_n = \dfrac{V^2}{\mathcal{R}}$	(4.1)
Force balance along a streamline for steady inviscid flow	$\displaystyle\int \dfrac{dp}{\rho} + \dfrac{1}{2}V^2 + gz = C \quad \text{(along a streamline)}$	(4.6)
Bernoulli equation	$p + \dfrac{1}{2}\rho V^2 + \gamma z = \text{constant along streamline}$	(4.7)
Pressure gradient normal to streamline for inviscid flow in absence of gravity	$\dfrac{\partial p}{\partial n} = -\dfrac{\rho V^2}{\mathcal{R}}$	(4.10b)
Force balance normal to a streamline for steady, inviscid, incompressible flow	$p + \rho \displaystyle\int \dfrac{V^2}{\mathcal{R}}\, dn + \gamma z = \text{constant across the streamline}$	(4.12)
Velocity measurement for a Pitot-static tube	$V = \sqrt{2(p_3 - p_4)/\rho}$	(4.16)
Free jet	$V = \sqrt{2\dfrac{\gamma h}{\rho}} = \sqrt{2gh}$	(4.18)

Continuity equation	$A_1 V_1 = A_2 V_2$, or $Q_1 = Q_2$	(4.19)
Flowmeter equation (ideal)	$Q = A_2 \sqrt{\dfrac{2(p_1 - p_2)}{\rho[1 - (A_2/A_1)^2]}}$	(4.20)
Sluice gate equation (ideal)	$Q = z_2 b \sqrt{\dfrac{2g(z_1 - z_2)}{1 - (z_2/z_1)^2}}$	(4.21)
Total head	$\dfrac{p}{\gamma} + \dfrac{V^2}{2g} + z = \text{constant on a streamline} = H$	(4.22)

REFERENCES

[1] Riley, W. F., and Sturges, L. D., *Engineering Mechanics: Dynamics*, 2nd Ed., Wiley, New York, 1996.

[2] Tipler, P. A., *Physics*, Worth, New York, 1982.

[3] Panton, R. L., *Incompressible Flow*, Wiley, New York, 1984.

[4] Dean, R. C., "On the Necessity of Unsteady Flow in Fluid Machines," *ASME Journal of Basic Engineering* 81D; 24–28, March 1959.

QUESTIONS AND PROBLEMS

[C] Problem to be solved with aid of programmable calculator or computer.

[O] Open-ended problem that requires critical thinking. These problems require various assumptions to provide the necessary input data. There are not unique answers to these problems.

Note: Unless specific values of required fluid properties are specified in the problem statement, use the values found in Tables 1.4–1.8 and in the tables in the Appendices.

Section 4.2 F = ma along a Streamline

4.2.1 Obtain a photograph/image of a situation that can be analyzed by using the Bernoulli equation. Print this photo and write a brief paragraph that describes the situation involved.

4.2.2 Air flows steadily along a streamline from point (1) to point (2) with negligible viscous effects. The following conditions are measured: at point (1) $z_1 = 2$ m and $p_1 = 0$ kPa; at point (2) $z_2 = 10$ m, $p_2 = 20$ N/m², and $V_2 = 0$. Determine the velocity at point (1).

4.2.3 Water flows steadily through the variable area horizontal pipe shown in **Fig. P4.2.3**. The centerline velocity is given by $\mathbf{V} = 3(1 + x)\hat{\mathbf{i}}$ m/s, where x is in feet. Viscous effects are neglected. (a) Determine the pressure gradient, $\partial p/\partial x$ (as a function of x), needed to produce this flow. (b) If the pressure at section (1) is 345 kPa, determine the pressure at (2) by (i) integration of the pressure gradient obtained in (a), (ii) application of the Bernoulli equation.

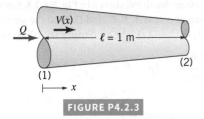

FIGURE P4.2.3

4.2.4 What pressure gradient along the streamline, dp/ds, is required to accelerate water in a horizontal pipe at a rate of 60 m/s²?

4.2.5 At a given location, the airspeed is 40 m/s and the pressure gradient along the streamline is 200 N/m³. Estimate the airspeed at a point 0.5 m farther along the streamline.

4.2.6 What pressure gradient along the streamline, dp/ds, is required to accelerate water upward in a vertical pipe at a rate of 12 m/s²? What is the answer if the flow is downward?

4.2.7 The Bernoulli equation is valid for steady, inviscid, incompressible flows with constant acceleration of gravity. Consider flow on a planet where the acceleration of gravity varies with height so that $g = g_0 - cz$, where g_0 and c are constants. Integrate "F = ma" along a streamline to obtain the equivalent of the Bernoulli equation for this flow.

4.2.8 An incompressible fluid flows steadily past a circular cylinder, as shown in **Fig. P4.2.8**. The fluid velocity along the dividing streamline $(-\infty \leq x \leq -a)$ is found to be $V = V_0(1 - a^2/x^2)$, where a is the radius of the cylinder and V_0 is the upstream velocity. (a) Determine the pressure gradient along this streamline. (b) If the upstream pressure is p_0, integrate the pressure gradient to obtain the pressure $p(x)$ for $-\infty \leq x \leq -a$. (c) Show from the result of part (b) that the pressure at the stagnation point $(x = -a)$ is $p_0 + \rho V_0^2/2$, as expected from the Bernoulli equation.

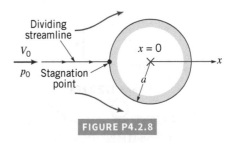

FIGURE P4.2.8

4.2.9 Consider a compressible liquid that has a constant bulk modulus. Integrate "F = ma" along a streamline to obtain the equivalent of the Bernoulli equation for this flow. Assume steady, inviscid flow.

Section 4.3 F = ma Normal to a Streamline

4.3.1 Obtain a photograph/image of a situation in which Newton's second law applied across the streamlines (as given by Eq. 4.12) is important. Print this photo and write a brief paragraph that describes the situation involved.

4.3.2 Air flows along a horizontal, curved streamline with a 6 m radius with a speed of 33.5 m/s. Determine the pressure gradient normal to the streamline.

4.3.3 Water flows around the vertical two-dimensional bend with circular streamlines and constant velocity, as shown in **Fig. P4.3.3**. If the pressure is 50 kPa at point (1), determine the pressures at points (2) and (3). Assume that the velocity profile is uniform as indicated.

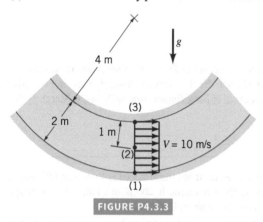

FIGURE P4.3.3

4.3.4 [C] Water flows around the vertical two-dimensional bend with circular streamlines, as is shown in **Fig. P4.3.4**. The pressure at point (1) is measured to be $p_1 = 172.3$ kPa, and the velocity across section a–a is as indicated in the table. Calculate and plot the pressure across section a–a of the channel $[p = p(z)$ for $0 \leq z \leq 0.6$ m].

z (m)	V (m/s)
0	0
0.06	2.4
0.12	4.4
0.18	6
0.24	5.94
0.3	4.75
0.36	2.5
0.42	1.9
0.48	1.1
0.54	0.6
0.6	0

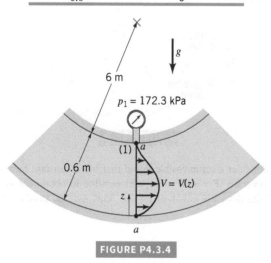

FIGURE P4.3.4

4.3.5 (Video) Water in a container and air in a tornado flow in horizontal circular streamlines of radius r and speed V, as shown in Video V4.6 and **Fig. P4.3.5**. Determine the radial pressure gradient, $\partial p / \partial r$, needed for the following situations: **(a)** The fluid is water with $r = 75$ mm and $V = 0.25$ m/s. **(b)** The fluid is air with $r = 90$ m and $V = 90$ m/s.

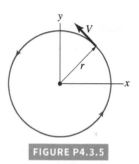

FIGURE P4.3.5

4.3.6 Air flows smoothly over the hood of your car and up past the windshield. However, a bug in the air does not follow the same path; it becomes splattered against the windshield. Explain why this is so.

Section 4.5 Static, Stagnation, Dynamic, and Total Pressure

4.5.1 Obtain a photograph/image of a situation in which the concept of the stagnation pressure is important. Print this photo and write a brief paragraph that describes the situation involved.

4.5.2 At a given point on a horizontal streamline in flowing air, the static pressure is -13.7 kPa (i.e., a vacuum) and the velocity is 45 m/s. Determine the pressure at a stagnation point on that streamline.

4.5.3 [O] A drop of water in a zero-g environment (as in the International Space Station) will assume a spherical shape, as shown in **Fig. P4.5.3a**. A raindrop in the cartoons is typically drawn as in **Fig. P4.5.3b**. The shape of an actual raindrop is more nearly like that shown in **Fig. P4.5.3c**. Discuss why these shapes are as indicated.

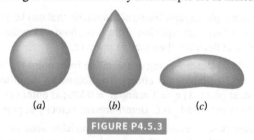

(a) (b) (c)

FIGURE P4.5.3

4.5.4 When an airplane is flying 90 m/s at an altitude of 1524 m in a standard atmosphere, the air velocity at a certain point on the wing is 122 m/s relative to the airplane. **(a)** What suction pressure is developed on the wing at that point? **(b)** What is the pressure at the leading edge (a stagnation point) of the wing?

4.5.5 Air flows over the airfoil shown in **Fig. P4.5.5**. Sensors give the pressures shown at points A, B, C, and D. Find the air velocities just above points A, B, C, and D. The air density is 1.03 kg/m^3.

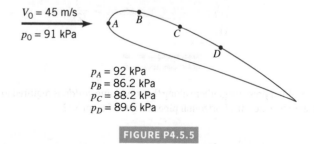

FIGURE P4.5.5

4.5.6 Some animals have learned to take advantage of the Bernoulli effect without having read a fluid mechanics book. For example, a typical prairie dog burrow contains two entrances—a flat front door and a mounded back door, as shown in **Fig. P4.5.6**. When the wind blows with velocity V_0 across the front door, the average velocity across the back door is greater than V_0 because of the mound. Assume the air velocity across the back door is $1.07V_0$. For a wind velocity of 6 m/s, what pressure differences, $p_1 - p_2$, are generated to provide a fresh airflow within the burrow?

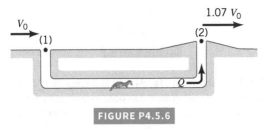

FIGURE P4.5.6

4.5.7 ⃝ Estimate the pressure on your hand when you hold it in the stream of air coming from the air hose at a filling station. List all assumptions and show calculations. Warning: Do not try this experiment; it can be dangerous!

4.5.8 A car racer holds his hand out of his IndyCar while driving through still air with standard atmospheric conditions. (a) For safety, the pit lane speed limit is 27 m/s. At this speed, what is the maximum pressure on his hand? (b) Back on the race track, what is the maximum pressure when he is driving his IndyCar at 100 m/s? (c) On the straightaways, the IndyCar reaches speeds in excess of 105 m/s. For this speed, is your solution method for parts (a) and (b) reasonable? Explain.

4.5.9 What is the minimum height for an oil ($SG = 0.75$) manometer to measure airplane speeds up to 30 m/s at altitudes up to 1500 m? The manometer is connected to a Pitot-static tube, as shown in **Fig. P4.5.9**.

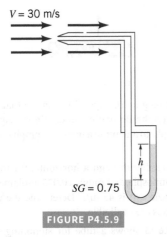

FIGURE P4.5.9

4.5.10 A Pitot-static tube is used to measure the velocity of helium in a pipe. The temperature and pressure are 277.6 K and 170 kPa. A water manometer connected to the Pitot-static tube indicates a reading of 0.5 m. Determine the helium velocity. Is it reasonable to consider the flow as incompressible? Explain.

4.5.11 A Bourdon-type pressure gage is used to measure the pressure from a Pitot tube attached to the leading edge of an airplane wing. The gage is calibrated to read in m/s at standard sea level conditions. If the airspeed meter indicates 67 m/s when flying at an altitude of 3048 m, what is the true airspeed?

4.5.12 ⃝ Estimate the force of a hurricane strength wind against the side of your house. List any assumptions and show all calculations.

4.5.13 A 20 m/s wind blowing past your house speeds up as it flows up and over the roof. If elevation effects are negligible, determine (a) the pressure at the point on the roof where the speed is 30 m/s if the pressure in the free stream blowing toward your house is 101.2 kPa. Would this effect tend to push the roof down against the house, or would it tend to lift the roof? (b) Determine the pressure on a window facing the wind if the window is assumed to be a stagnation point.

4.5.14 (See The Wide World of Fluids article titled "Pressurized eyes," Section 4.5.) Determine the air velocity needed to produce a stagnation pressure equal to 10 mm of mercury.

Section 4.6.1 Free Jets

4.6.1 Water flows through a hole in the bottom of a large, open tank with a speed of 10 m/s. Determine the depth of water in the tank. Viscous effects are negligible.

4.6.2 ⃝ Estimate the pressure needed at the pumper truck in order to shoot water from the street level onto a fire on the roof of a five-story building. List all assumptions and show all calculations.

4.6.3 The tank shown in **Fig. P4.6.3** contains air at atmospheric pressure above the water surface. The velocity of the water flowing from the tank is 7 m/s. Determine the water level (h).

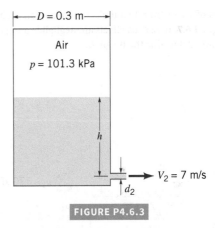

FIGURE P4.6.3

4.6.4 Water flows from the faucet on the first floor of the building shown in **Fig. P4.6.4** with a maximum velocity of 6 m/s. For steady inviscid flow, determine the maximum water velocity from the basement faucet and from the faucet on the second floor (assume each floor is 3.5 m tall).

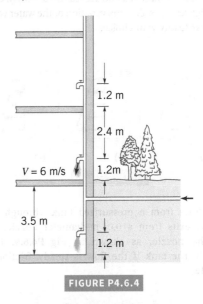

FIGURE P4.6.4

4.6.5 Laboratories containing dangerous materials are often kept at a pressure slightly less than ambient pressure so that contaminants can be filtered through an exhaust system rather than leaked through cracks around doors, etc. If the pressure in such a room is 2.5 mm of water below that of the surrounding rooms, with what velocity will air enter the room through an opening? Assume viscous effects are negligible.

4.6.6 ⓞ The "supersoaker" water gun shown in **Fig. P4.6.6** can shoot more than 9.14 mm in the horizontal direction. Estimate the minimum pressure, p_1, needed in the chamber in order to accomplish this. List all assumptions and show all calculations.

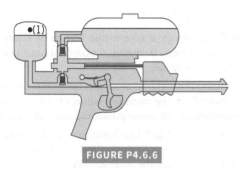

FIGURE P4.6.6

4.6.7 Streams of water from two tanks impinge upon each other, as shown in **Fig. P4.6.7**. If viscous effects are negligible and point A is a stagnation point, determine the height h.

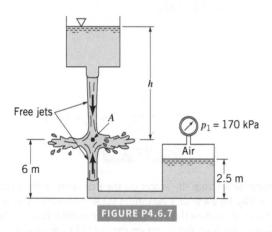

FIGURE P4.6.7

4.6.8 Several holes are punched into a tin can, as shown in **Fig. P4.6.8**. Which of the figures represents the variation of the water velocity as it leaves the holes? Justify your choice.

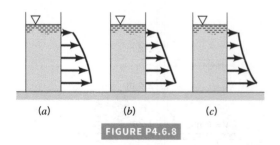

(a)   (b)   (c)

FIGURE P4.6.8

4.6.9 Water flows from a pressurized tank, through a 0.15 m-diameter pipe, exits from a 0.05 m-diameter nozzle, and rises 6 m above the nozzle, as shown in **Fig. P4.6.9**. Determine the pressure in the tank if the flow is steady, frictionless, and incompressible.

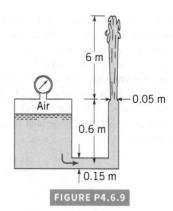

FIGURE P4.6.9

4.6.10 The piston shown in **Fig. P4.6.10** is forcing 21.1 °C water out the exit at 4 m/s. The exit pressure has been measured as $p_e = 275$ kPa. Determine the force on the piston for a piston diameter $D_p = 0.075$ m.

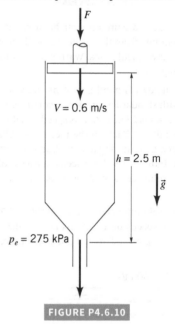

FIGURE P4.6.10

Section 4.6.2 Confined Flows

4.6.11 Obtain a photograph/image of a situation that involves a confined flow for which the Bernoulli and continuity equations are important. Print this photo and write a brief paragraph that describes the situation involved.

4.6.12 Air flows steadily through a horizontal 0.1 m-diameter pipe and exits into the atmosphere through a 0.075 m-diameter nozzle. The velocity at the nozzle exit is 45 m/s. Determine the pressure in the pipe if viscous effects are negligible.

4.6.13 **Figure P4.6.13** shows a tube for siphoning water from an aquarium. Determine the rate at which the water leaves the aquarium for the conditions shown. Is there an advantage to having the large-diameter section? The water flow is inviscid.

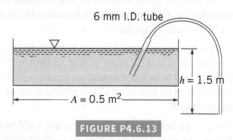

FIGURE P4.6.13

4.6.14 For the pipe enlargement shown in **Fig. P4.6.14**, the pressures at sections (1) and (2) are 388 and 401 kPa, respectively. Determine the weight flowrate of the gasoline in the pipe.

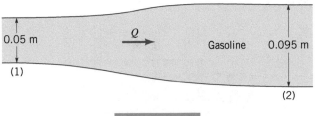

FIGURE P4.6.14

4.6.15 A fire hose nozzle has a diameter of 0.03 m. According to some fire codes, the nozzle must be capable of delivering at least 0.9 m^3/min. If the nozzle is attached to a 0.075 m-diameter hose, what pressure must be maintained just upstream of the nozzle to deliver this flowrate?

4.6.16 (Video) Water flowing from the 0.02 m-diameter outlet shown in Video V8.15 and **Fig. P4.6.16** rises 0.07 m above the outlet. Determine the flowrate.

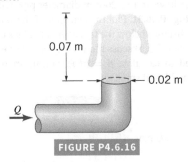

FIGURE P4.6.16

4.6.17 A fire hose has a nozzle outlet velocity of 48 kmph. What is the maximum height the water can reach?

4.6.18 At what rate does oil ($SG = 0.8$) flow from the tank shown in **Fig. P4.6.18**?

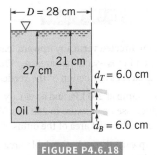

FIGURE P4.6.18

4.6.19 Find the water height h_B in tank B shown in **Fig. P4.6.19** for steady-state conditions.

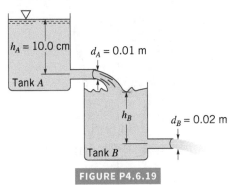

FIGURE P4.6.19

4.6.20 The pressure and average velocity at point A in the pipe shown in **Fig. P4.6.20** are 110 kPa and 1.2 m/s, respectively. Find the height h and the pressure and average velocity at point B. Fluid fills the 0.025 m-diameter discharge pipe.

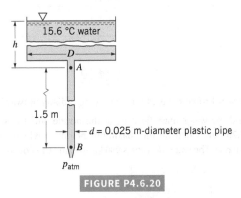

FIGURE P4.6.20

4.6.21 Air is drawn into a wind tunnel used for testing automobiles, as shown in **Fig. P4.6.21**. (a) Determine the manometer reading, h, when the velocity in the test section is 97 kmph. Note that there is a 0.025 m column of oil on the water in the manometer. (b) Determine the difference between the stagnation pressure on the front of the automobile and the pressure in the test section.

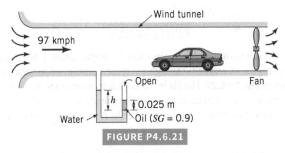

FIGURE P4.6.21

4.6.22 **Figure P4.6.22** shows a duct for testing a centrifugal fan. Air is drawn from the atmosphere ($p_{atm} = 101.3$ kPa, $T_{atm} = 21.1$ °C). The inlet box is 1.2 m × 0.6 m. At section 1, the duct is square in cross-section (0.76 m × 0.76 m). At section 2, the duct is circular and has a diameter of 0.9 m. A water manometer in the inlet box measures a static pressure of -0.05 m of water. Calculate the volume flowrate of air into the fan and the average fluid velocity at both sections 1 and 2. Assume constant density.

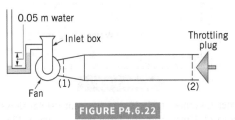

FIGURE P4.6.22

4.6.23 Natural gas (methane) flows from a 0.076 m-diameter gas main, through a 0.025 m-diameter pipe, and into the burner of a furnace at a rate of 2.8 m^3/hr. Determine the pressure in the gas main if the pressure in the 0.025 m pipe is to be 0.15 m of water greater than atmospheric pressure. Neglect viscous effects.

4.6.24 Air flows radially inward between the two parallel circular plates shown in **Fig. P4.6.24**. The pressure at the outer radius $R_o = 5.0$ cm is atmospheric. Find the pressure at the inner radius $R_i = 0.5$ cm if the air density is constant, the air flow is inviscid, the volume flowrate is 4.0 liters/s, and the plate spacing is $h = 0.125$ cm.

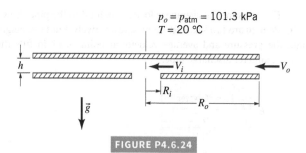

$$p_o = p_{atm} = 101.3 \text{ kPa}$$
$$T = 20 \text{ °C}$$

FIGURE P4.6.24

4.6.25 Repeat Problem 4.6.24 for air flowing radially outward.

4.6.26 Find the water mass flowrate at the nozzle outlet O shown in **Fig. P4.6.26**, and calculate the maximum height to which the water stream will rise. The water density is 980 kg/m³, and the flow is inviscid.

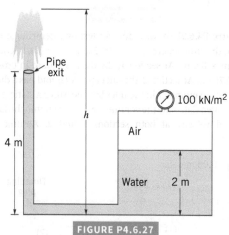

FIGURE P4.6.26

4.6.27 Water (assumed frictionless and incompressible) flows steadily from a large tank and exits through a vertical, constant diameter pipe as shown in **Fig. P4.6.27**. The air in the tank is pressurized to 100 kN/m². Determine **(a)** the height, h, to which the water rises, **(b)** the water velocity in the pipe, and **(c)** the pressure in the horizontal part of the pipe.

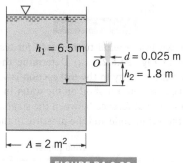

FIGURE P4.6.27

4.6.28 Water (assumed inviscid and incompressible) flows steadily with a speed of 3 m/s from the large tank shown in **Fig. P4.6.28**. Determine the depth, H, of the layer of light liquid (specific weight = 7854 N/m³) that covers the water in the tank.

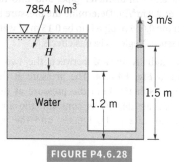

FIGURE P4.6.28

4.6.29 Water flows through the pipe contraction shown in **Fig. P4.6.29**. For the given 0.2-m difference in the manometer level, determine the flowrate as a function of the diameter of the small pipe, D.

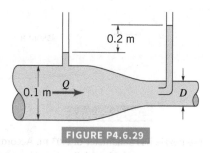

FIGURE P4.6.29

4.6.30 Carbon tetrachloride flows in a pipe of variable diameter with negligible viscous effects. At point A in the pipe, the pressure and velocity are 138 kPa and 9 m/s, respectively. At location B, the pressure and velocity are 158.5 kPa and 4.25 m/s. Which point is at the higher elevation and by how much?

4.6.31 Ⓒ Water flows from a 20-mm-diameter pipe with a flowrate, Q, as shown in **Fig. P4.6.31**. Plot the diameter of the water stream, d, as a function of distance below the faucet, h, for values of $0 \leq h \leq 1$ m and $0 \leq Q \leq 0.004$ m³/s. Discuss the validity of the one-dimensional assumption used to calculate $d = d(h)$, noting, in particular, the conditions of small h and small Q.

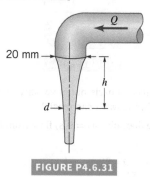

FIGURE P4.6.31

4.6.32 A liquid stream directed vertically upward leaves a nozzle with a steady velocity V_0 and cross-sectional area A_0. Find the velocity V and cross-sectional area A as a function of the vertical position z.

4.6.33 Water leaves a pump at 200 kPa and a velocity of 12 m/s. It then enters a diffuser to increase its pressure to 250 kPa. What must be the ratio of the outlet area to the inlet area of the diffuser?

4.6.34 Water flows upward through a variable area pipe with a constant flowrate, Q, as shown in **Fig. P4.6.34**. If viscous effects are negligible, determine the diameter, $D(z)$, in terms of D_1 if the pressure is to remain constant throughout the pipe. That is, $p(z) = p_1$.

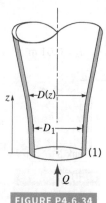

FIGURE P4.6.34

4.6.35 The circular stream of water from a faucet is observed to taper from a diameter of 20 mm to 10 mm in a distance of 50 cm. Determine the flowrate.

4.6.36 Water is siphoned from the tank shown in **Fig. P4.6.36**. The water barometer indicates a reading of 9 m. Determine the maximum value of h allowed without cavitation occurring. Note that the pressure of the vapor in the closed end of the barometer equals the vapor pressure.

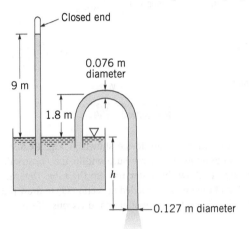

FIGURE P4.6.36

4.6.37 Water is siphoned from a tank, as shown in **Fig. P4.6.37**. Determine the flowrate and the pressure at point A, a stagnation point.

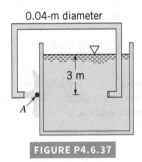

FIGURE P4.6.37

4.6.38 Water is siphoned from the tank shown in **Fig. P4.6.38**. A 0.025 m diameter nozzle is placed at the end of the tube. Determine the flowrate from the tank and the pressure at points (1), (2), and (3) if viscous effects are negligible.

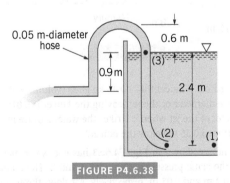

FIGURE P4.6.38

4.6.39 Redo Problem 4.6.38 if there is no nozzle at the end of the tube.

4.6.40 Water exits a pipe as a free jet and flows to a height h above the exit plane, as shown in **Fig. P4.6.40**. The flow is steady, incompressible, and frictionless. (a) Determine the height h. (b) Determine the velocity and pressure at section (1).

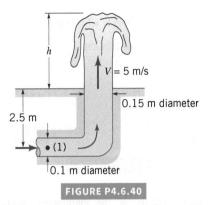

FIGURE P4.6.40

4.6.41 Water flows steadily from a large, closed tank, as shown in **Fig. P4.6.41**. The deflection in the mercury manometer is 0.025 m, and viscous effects are negligible. (a) Determine the volume flowrate. (b) Determine the air pressure in the space above the surface of the water in the tank.

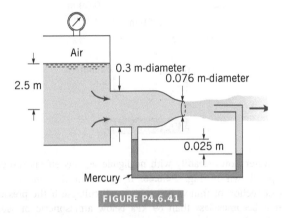

FIGURE P4.6.41

4.6.42 Carbon dioxide flows at a rate of 0.0425 m³/s from a 0.076 m pipe in which the pressure and temperature are 138 kPa (gage) and 58.9 °C into a 0.038 m pipe. If viscous effects are neglected and incompressible conditions are assumed, determine the pressure in the smaller pipe.

4.6.43 Oil of specific gravity 0.83 flows in the pipe shown in **Fig. P4.6.43**. If viscous effects are neglected, what is the flowrate?

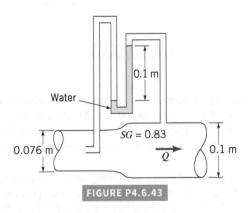

FIGURE P4.6.43

4.6.44 Water flows steadily through the variable area pipe shown in **Fig. P4.6.44** with negligible viscous effects. Determine the manometer reading, H, if the flowrate is 0.4 m³/s and the density of the manometer fluid is 500 kg/m³.

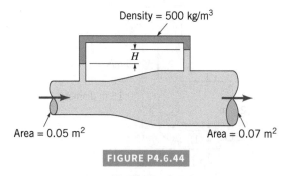

FIGURE P4.6.44

4.6.45 The specific gravity of the manometer fluid shown in **Fig. P4.6.45** is 1.07. Determine the volume flowrate, Q, if the flow is inviscid and incompressible and the flowing fluid is (a) water, (b) gasoline, or (c) air at standard conditions.

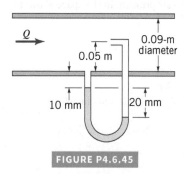

FIGURE P4.6.45

4.6.46 Water flows steadily with negligible viscous effects through the pipe shown in **Fig. P4.6.46**. It is known that the 0.1 m-diameter section of thin-walled tubing will collapse if the pressure within it becomes less than 69 kPa below atmospheric pressure. Determine the maximum value that h can have without causing collapse of the tubing.

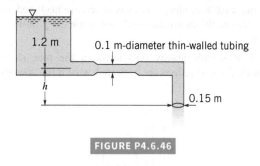

FIGURE P4.6.46

4.6.47 Helium flows through a 0.30-m-diameter horizontal pipe with a temperature of 20 °C and a pressure of 200 kPa (abs) at a rate of 0.30 kg/s. If the pipe reduces to 0.25-m-diameter, determine the pressure difference between these two sections. Assume incompressible, inviscid flow.

4.6.48 Water is pumped from a lake through an 0.2 m pipe at a rate of 0.28 m³/s. If viscous effects are negligible, what is the pressure in the suction pipe (the pipe between the lake and the pump) at an elevation 1.8 m above the lake?

4.6.49 Air is drawn into a small open-circuit wind tunnel, as shown in **Fig. P4.6.49**. Atmospheric pressure is 98.7 kPa (abs), and the

temperature is 27 °C. If viscous effects are negligible, determine the pressure at the stagnation point on the nose of the airplane. Also determine the manometer reading, h, for the manometer attached to the static pressure tap within the test section of the wind tunnel if the air velocity within the test section is 50 m/s.

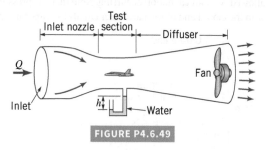

FIGURE P4.6.49

4.6.50 Air flows through the device shown in **Fig. P4.6.50**. If the flowrate is large enough, the pressure within the constriction will be low enough to draw the water up into the tube. Determine the flowrate, Q, and the pressure needed at section (1) to draw the water into section (2). Neglect compressibility and viscous effects.

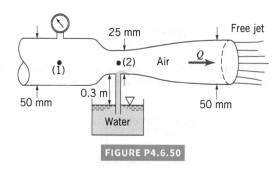

FIGURE P4.6.50

4.6.51 Water flows steadily from the large open tank shown in **Fig. P4.6.51**. If viscous effects are negligible, determine (a) the flowrate, Q, and (b) the manometer reading, h.

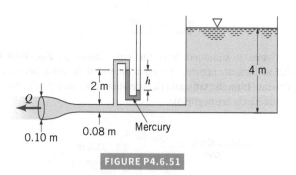

FIGURE P4.6.51

4.6.52 Water from a faucet fills a 0.45 kg glass (volume = 0.0005 m³) in 20 s. If the diameter of the jet leaving the faucet is 0.015 m, what is the diameter of the jet when it strikes the water surface in the glass that is positioned 0.35 m below the faucet?

4.6.53 The nozzle shown in **Fig. P4.6.53** has two water manometers to indicate the static pressures at sections 1 and 2. The diameters D_1 and D_2 are 0.2 m and 0.05 m, respectively. Air flows through the nozzle, and the air and water temperatures are 15.6 °C. Find the air volume flowrate. Assume constant density and inviscid flow. Neglect elevation changes.

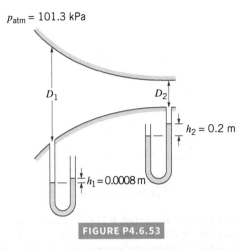

$p_{atm} = 101.3$ kPa

D_1

D_2

$h_2 = 0.2$ m

$h_1 = 0.0008$ m

FIGURE P4.6.53

4.6.54 (Video) Air flows steadily through a converging–diverging rectangular channel of constant width, as shown in **Fig. P4.6.54** and Video V4.10. The height of the channel at the exit and the exit velocity are H_0 and V_0, respectively. The channel is to be shaped so that the distance, d, that water is drawn up into tubes attached to static pressure taps along the channel wall is linear with distance along the channel. That is $d = (d_{max}/L)x$, where L is the channel length and d_{max} is the maximum water depth (at the minimum channel height: $x = L$). Determine the height, $H(x)$, as a function of x and the other important parameters.

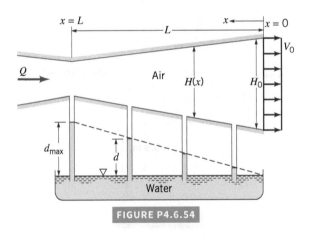

$x = L$

L

x

$x = 0$

V_0

Q

Air

$H(x)$

H_0

d_{max}

d

Water

FIGURE P4.6.54

4.6.55 (C) Water flows from a large tank and through a pipe of variable area, as shown in **Fig. P4.6.55**. The area of the pipe is given by $A = A_0[1 - x(1 - x/\ell)/2\ell]$, where A_0 is the area at the beginning $(x = 0)$ and end $(x = \ell)$ of the pipe. Plot graphs of the pressure within the pipe as a function of distance along the pipe for water depths of $h = 1, 4, 10,$ and 25 m.

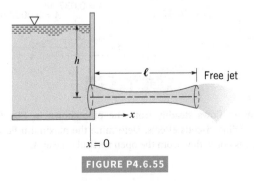

h

ℓ

Free jet

x

$x = 0$

FIGURE P4.6.55

4.6.56 If viscous effects are neglected and the tank is large, determine the flowrate from the tank shown in **Fig. P4.6.56**.

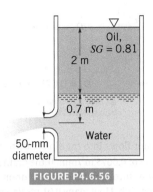

Oil, $SG = 0.81$

2 m

0.7 m

Water

50-mm diameter

FIGURE P4.6.56

4.6.57 Water flows steadily downward in the pipe shown in **Fig. P4.6.57** with negligible losses. Determine the flowrate.

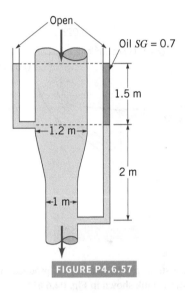

Open

Oil $SG = 0.7$

1.5 m

1.2 m

2 m

1 m

FIGURE P4.6.57

4.6.58 Water flows steadily from a large open tank and discharges into the atmosphere though a 0.076 m-diameter pipe, as shown in **Fig. P4.6.58**. Determine the diameter, d, in the narrowed section of the pipe at A if the pressure gages at A and B indicate the same pressure.

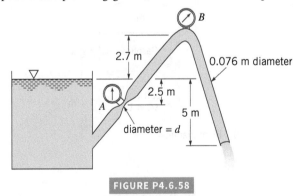

B

2.7 m

0.076 m diameter

2.5 m

A

5 m

diameter $= d$

FIGURE P4.6.58

4.6.59 Water flows from a large tank, as shown in **Fig. P4.6.59**. Atmospheric pressure is 100 kPa, and the vapor pressure is 11 kPa. If viscous effects are neglected, at what height, h, will cavitation begin? To avoid cavitation, should the value of D_1 be increased or decreased? To avoid cavitation, should the value of D_2 be increased or decreased? Explain.

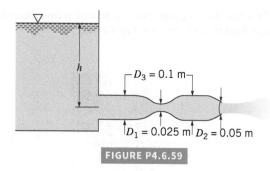

FIGURE P4.6.59

4.6.60 (Video) Water flows into the sink shown in **Fig. P4.6.60** and Video V5.1 at a rate of 0.0075 m³/min. If the drain is closed, the water will eventually flow through the overflow drain holes rather than over the edge of the sink. How many 0.01 m-diameter drain holes are needed to ensure that the water does not overflow the sink? Neglect viscous effects.

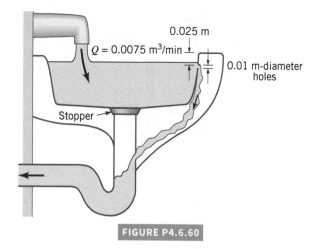

FIGURE P4.6.60

4.6.61 What pressure, p_1, is needed to produce a flowrate of 0.0025 m³/s from the tank shown in **Fig. P4.6.61**?

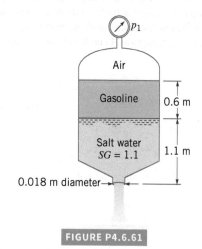

FIGURE P4.6.61

4.6.62 The vent on the tank shown in **Fig. P4.6.62** is closed and the tank pressurized to increase the flowrate. What pressure, p_1, is needed to produce twice the flowrate of that when the vent is open?

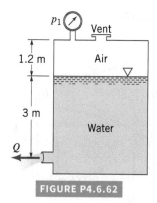

FIGURE P4.6.62

4.6.63 A sump pump is submerged in 15.6 °C ordinary water that vaporizes at a pressure of 1765 Pa. The pump inlet has an inside diameter of 0.0525 m and is 5 m below the water surface. Find the maximum possible flowrate before cavitation (vaporization) occurs.

4.6.64 Water is siphoned from a large tank and discharges into the atmosphere through a 0.05 m-diameter tube, as shown in **Fig. P4.6.64**. The end of the tube is 0.9 m below the tank bottom, and viscous effects are negligible. (a) Determine the volume flowrate from the tank. (b) Determine the maximum height, H, over which the water can be siphoned without cavitation occurring. Atmospheric pressure is 101.3 kPa, and the water vapor pressure is 1.792 kPa.

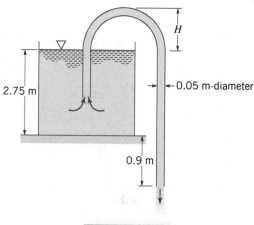

FIGURE P4.6.64

4.6.65 Water flows in the system shown in **Fig. P4.6.65**. Assume frictionless flow. (a) Calculate the rate Q at which water must be added at the inlet to maintain the 4.87 m height. (b) Calculate the height h in feet of water in the static-pressure tube.

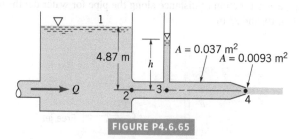

FIGURE P4.6.65

4.6.66 Water flows steadily from the pipe shown in **Fig. P4.6.66** with negligible viscous effects. Determine the maximum flowrate if the water is not to flow from the open vertical tube at A.

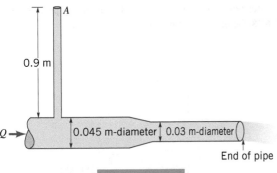

FIGURE P4.6.66

4.6.67 JP-4 fuel ($SG = 0.77$) flows through the Venturi meter shown in **Fig. P4.6.67** with a velocity of 4.57 m/s in the 0.15 m pipe. If viscous effects are negligible, determine the elevation, h, of the fuel in the open tube connected to the throat of the Venturi meter.

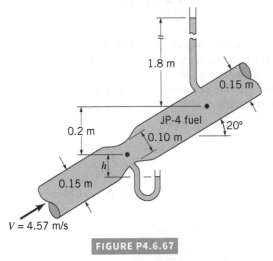

FIGURE P4.6.67

4.6.68 Water, considered an inviscid, incompressible fluid, flows steadily, as shown in **Fig. P4.6.68**. Determine h.

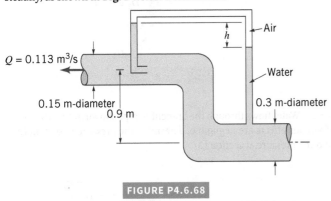

FIGURE P4.6.68

4.6.69 Determine the flowrate through the submerged orifice shown in **Fig. P4.6.69** if the contraction coefficient is $C_c = 0.63$.

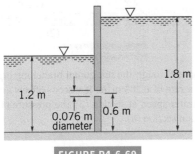

FIGURE P4.6.69

4.6.70 The water clock (clepsydra) shown in **Fig. P4.6.70** is an ancient device for measuring time by the falling water level in a large glass container. The water slowly drains out through a small hole in the bottom. Determine the approximate shape $R(z)$ of a container of circular cross section required for the water level to fall at a constant rate of 5 cm/hr if the drain hole is 1.5 mm in diameter. What size container (h and R_0) permits the clock to run for 24 hr without refilling?

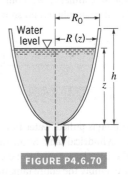

FIGURE P4.6.70

4.6.71 A long water trough of triangular cross section is formed from two planks, as is shown in **Fig. P4.6.71**. A gap of 2.5 mm remains at the junction of the two planks. If the water depth initially was 0.6 m, how long a time does it take for the water depth to reduce to 0.3 m?

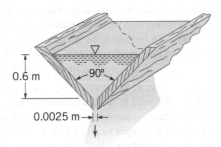

FIGURE P4.6.71

4.6.72 (Video) Pop (with the same properties as water) flows from a 0.1 m-diameter pop container that contains three holes, as shown in **Fig. P4.6.72** (see Video V4.9). The diameter of each fluid stream is 0.004 m and the distance between holes is 0.05 m. If viscous effects are negligible and quasi-steady conditions are assumed, determine the time at which the pop stops draining from the top hole. Assume the pop surface is 0.05 m above the top hole when $t = 0$. Compare your results with the time you measure from the video.

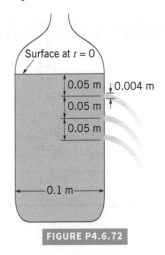

FIGURE P4.6.72

4.6.73 Ⓒ A spherical tank of diameter D has a drain hole of diameter d at its bottom. A vent at the top of the tank maintains atmospheric pressure at the liquid surface within the tank. The flow is quasi-steady and inviscid, and the tank is full of water initially. Determine the water depth as a function of time, $h = h(t)$, and plot graphs of $h(t)$ for tank diameters of 0.3, 1.5, 3, and 6 m if $d = 0.0254$ m.

4.6.74 Ⓞ A small hole develops in the bottom of the stationary rowboat shown in **Fig. P4.6.74**. Estimate the amount of time it will take for the boat to sink. List all assumptions and show all calculations.

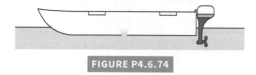

FIGURE P4.6.74

4.6.75 When the drain plug is pulled, water flows from a hole in the bottom of a large, open cylindrical tank. Show that if viscous effects are negligible and if the flow is assumed to be quasi-steady, then it takes 4.41 times longer to empty the entire tank than it does to empty the first half of the tank. Explain why this is so.

4.6.76 Someone siphoned 56.78 liter of gasoline from a gas tank in the middle of the night. The gas tank measures 0.3 m wide × 0.6 m long × 0.45 m high and was full when the thief started. If the siphoning tube has an inside diameter of 0.013 m, find the minimum amount of time needed to siphon the 56.78 liter from the tank. Assume that at any instant of time the steady-state equations are adequate to predict the gasoline velocity in the siphon tube. Also assume that the end of the siphon tube outside the gas tank is at the same level as the bottom of the tank. You may consider the gasoline to be inviscid.

4.6.77 Ⓒ The surface area, A, of the pond shown in **Fig. P4.6.77** varies with the water depth, h, as shown in the table. At time $t = 0$, a valve is opened and the pond is allowed to drain through a pipe of diameter D. If viscous effects are negligible and quasi-steady conditions are assumed, plot the water depth as a function of time from when the valve is opened ($t = 0$) until the pond is drained for pipe diameters of $D = 0.15, 0.3, 0.45, 0.6, 0.75$, and 0.9 m. Assume $h = 5.4$ m at $t = 0$.

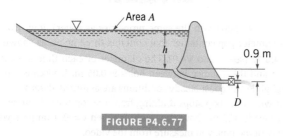

FIGURE P4.6.77

h (m)	A [acres (1 acre = 4047 m^2)]
0	0
0.6	0.3
1.2	0.5
1.8	0.8
2.4	0.9
3.0	1.1
3.6	1.5
4.2	1.8
4.8	2.4
5.4	2.8

4.6.78 Water flows through a horizontal branching pipe, as shown in **Fig. P4.6.78**. Determine the pressure at section (3).

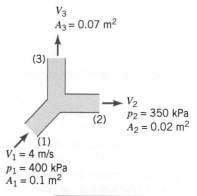

FIGURE P4.6.78

4.6.79 The "wye" fitting shown in **Fig. P4.6.79** lies in a horizontal plane. The fitting splits the inlet flow into two equal parts. At section 1, the water velocity is 3.66 m/s and the pressure is 138 kPa. Calculate the water pressure at sections 2 and 3. Assume inviscid flow and a water temperature of 15.6 °C.

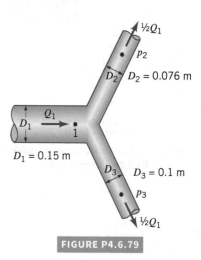

FIGURE P4.6.79

4.6.80 Water flows through the branching pipe shown in **Fig. P4.6.80**. If viscous effects are negligible, determine the pressure at section (2) and the pressure at section (3).

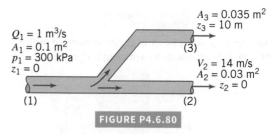

FIGURE P4.6.80

4.6.81 Water flows through the horizontal branching pipe shown in **Fig. P4.6.81** at a rate of 0.283 m^3/s. If viscous effects are negligible, determine the water speed at section (2), the pressure at section (3), and the flowrate at section (4).

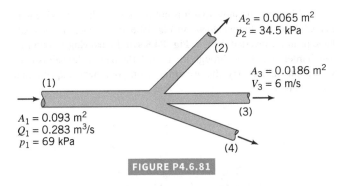

FIGURE P4.6.81

4.6.82 Water flows from a large tank through a large pipe that splits into two smaller pipes, as shown in **Fig. P4.6.82**. If viscous effects are negligible, determine the flowrate from the tank and the pressure at point (1).

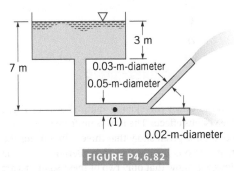

FIGURE P4.6.82

4.6.83 A gutter running along the side of a house is 0.15 m. wide × 12.2 m long. During a hard downpour, water is 0.025 m deep in the gutter. The gutter has only one downspout, and it is 0.076 m in diameter. What is the velocity of the water entering the downspout? The pressure at the downspout entrance is atmospheric pressure.

4.6.84 Air, assumed incompressible and inviscid, flows into the outdoor cooking grill through nine holes of 0.01 m diameter, as shown in **Fig. P4.6.84**. If a flowrate of 39.3 liter/min into the grill is required to maintain the correct cooking conditions, determine the pressure within the grill near the holes.

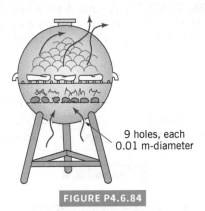

FIGURE P4.6.84

4.6.85 An air cushion vehicle is supported by forcing air into the chamber created by a skirt around the periphery of the vehicle, as shown in **Fig. P4.6.85**. The air escapes through the 0.076 m clearance between the lower end of the skirt and the ground (or water). Assume the vehicle weighs 44.5 kN and is essentially rectangular in shape, 9 m × 20 m. The volume of the chamber is large enough so that the kinetic energy of the air within the chamber is negligible. Determine the flowrate, Q, needed to support the vehicle. If the ground clearance were reduced to 0.05 m, what flowrate would be needed? If the

vehicle weight were reduced to 22.25 kN and the ground clearance maintained at 0.076 m, what flowrate would be needed?

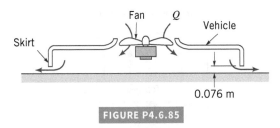

FIGURE P4.6.85

4.6.86 Water flows from the pipe shown in **Fig. P4.6.86** as a free jet and strikes a circular flat plate. The flow geometry shown is axisymmetrical. Determine the flowrate and the manometer reading, H.

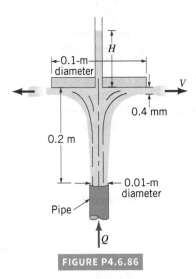

FIGURE P4.6.86

4.6.87 A conical plug is used to regulate the airflow from the pipe shown in **Fig. P4.6.87**. The air leaves the edge of the cone with a uniform thickness of 0.02 m. If viscous effects are negligible and the flowrate is 0.50 m³/s, determine the pressure within the pipe.

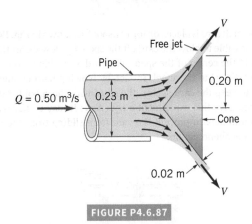

FIGURE P4.6.87

4.6.88 **Figure P4.6.88** shows two tall towers. Air at 10 °C is blowing toward the two towers at $V_0 = 30$ km/hr. If the two towers are 10 m apart and half the air flow approaching the two towers passes between them, find the minimum air pressure between the two towers. Assume constant air density, inviscid flow, and steady-state conditions. The atmospheric pressure is 101 kPa.

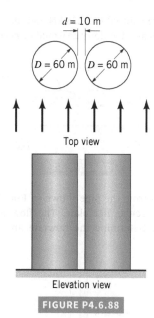

FIGURE P4.6.88

4.6.89 Water flows steadily from a nozzle into a large tank, as shown in **Fig. P4.6.89**. The water then flows from the tank as a jet of diameter d. Determine the value of d if the water level in the tank remains constant. Viscous effects are negligible.

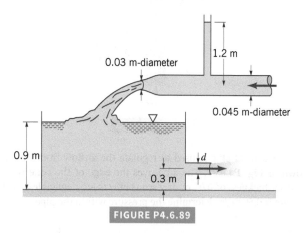

FIGURE P4.6.89

4.6.90 A small card is placed on top of a spool, as shown in **Fig. P4.6.90**. It is not possible to blow the card off the spool by blowing air through the hole in the center of the spool. The harder one blows, the harder the card "sticks" to the spool. In fact, by blowing hard enough, it is possible to keep the card against the spool with the spool turned upside down. Give this experiment a try. (*Note:* It may be necessary to use a thumb tack to prevent the card from sliding from the spool.) Explain this phenomenon.

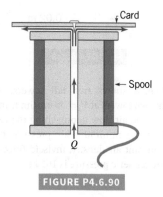

FIGURE P4.6.90

4.6.91 Observations show that it is not possible to blow the table tennis ball from the funnel shown in **Fig. P4.6.91a**. In fact, the ball can be kept in an inverted funnel, **Fig. P4.6.91b**, by blowing through it. The harder one blows through the funnel, the harder the ball is held within the funnel. Try this experiment on your own. Explain this phenomenon.

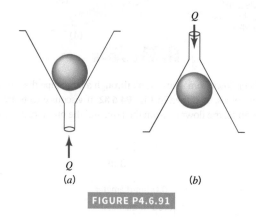

FIGURE P4.6.91

4.6.92 Water flows down the sloping ramp shown in **Fig. P4.6.92** with negligible viscous effects. The flow is uniform at sections (1) and (2). For the conditions given show that three solutions for the downstream depth, h_2, are obtained by use of the Bernoulli and continuity equations. However, show that only two of these solutions are realistic. Determine these values.

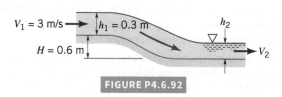

FIGURE P4.6.92

4.6.93 Water flows in a rectangular channel that is 2.0 m wide, as shown in **Fig. P4.6.93**. The upstream depth is 70 mm. The water surface rises 40 mm as it passes over a portion where the channel bottom rises 10 mm. If viscous effects are negligible, what is the flowrate?

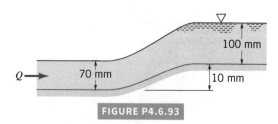

FIGURE P4.6.93

Section 4.6.3 Flowrate Measurement

4.6.94 Obtain a photograph/image of a situation that involves some type of flowmeter. Print this photo and write a brief paragraph that describes the situation involved.

4.6.95 A Venturi meter with a minimum diameter of 0.075 m is to be used to measure the flowrate of water through a 0.1 m-diameter pipe. Determine the pressure difference indicated by the pressure gage attached to the flowmeter if the flowrate is 0.014 m³/s and viscous effects are negligible.

4.6.96 Determine the flowrate through the Venturi meter shown in **Fig. P4.6.96** if ideal conditions exist.

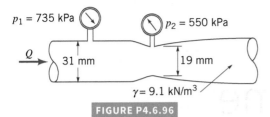

FIGURE P4.6.96

4.6.97 For what flowrate through the Venturi meter of problem 4.6.98 will cavitation begin if $p_1 = 275$ kPa gage, atmospheric pressure is 101 kPa (abs), and the vapor pressure is 3.6 kPa (abs)?

4.6.98 What diameter orifice hole, d, is needed if under ideal conditions the flowrate through the orifice meter of **Fig. P4.6.98** is to be 115 liter/min of seawater with $p_1 - p_2 = 16.34$ kPa? The contraction coefficient is assumed to be 0.63.

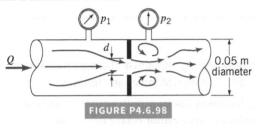

FIGURE P4.6.98

4.6.99 [Video] A weir (see Video V10.13) of trapezoidal cross section is used to measure the flowrate in a channel, as shown in **Fig. P4.6.99**. If the flowrate is Q_0 when $H = \ell/2$, what flowrate is expected when $H = \ell$?

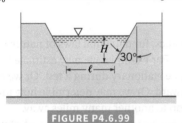

FIGURE P4.6.99

4.6.100 The flowrate in a water channel is sometimes determined by use of a device called a Venturi flume. As shown in **Fig. P4.6.100**, this device consists simply of a bump on the bottom of the channel. If the water surface dips a distance of 0.07 m for the conditions shown, what is the flowrate per width of the channel? Assume the velocity is uniform and viscous effects are negligible.

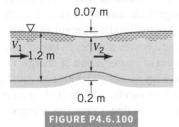

FIGURE P4.6.100

4.6.101 Water flows under the inclined sluice gate shown in **Fig. P4.6.101**. Determine the flowrate if the gate is 2.5 m wide.

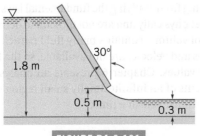

FIGURE P4.6.101

Section 4.7 The Energy Line and the Hydraulic Grade Line

4.7.1 Water flows in a vertical pipe of 0.15-m-diameter at a rate of $0.2 \text{ m}^3/\text{s}$ and a pressure of 200 kPa at an elevation of 25 m. Determine the velocity head and pressure head at elevations of 20 and 55 m.

4.7.2 Draw the energy line and the hydraulic grade line for the flow of Problem 4.6.59.

4.7.3 Draw the energy line and hydraulic grade line for the flow shown in Problem 4.6.46.

Section 4.8 Restrictions on Use of the Bernoulli Equation

4.8.1 Obtain a photograph/image of a flow in which it would not be appropriate to use the Bernoulli equation. Print this photo and write a brief paragraph that describes the situation involved.

4.8.2 The following table lists typical flight speeds for two aircraft. For which of these conditions would it be reasonable to use the incompressible Bernoulli equation to study the aerodynamics associated with their flight? Explain.

Aircraft	Flight speed, km/hr	
	Cruise	Landing approach
Boeing 787	913	214
F-22 fighter	1960	250

4.8.3 A meteorologist uses a Pitot-static tube to measure the wind speed in a tornado. Based on the damage caused by the storm, the tornado is rated as EF5 on the Enhanced Fujita Scale. This means that the wind speed is estimated to be in the range of 420 to 512 kmph. Is it reasonable to use the incompressible Pitot-tube equation (Eq. 4.16) to determine the actual wind speed, or must compressible effects to taken into account? Explain.

Lifelong Learning Problems

4.1 LL The concept of the use of a Pitot-static tube to measure the airspeed of an airplane is rather straightforward. However, the design and manufacture of reliable, accurate, inexpensive Pitot-static tube airspeed indicators is not necessarily simple. Obtain information about the design and construction of modern Pitot-static tubes. Summarize your findings in a brief report.

4.2 LL Orifice, nozzle, or Venturi flowmeters have been used for a long time to accurately measure the flowrate in pipes. However, recently there have been several new concepts suggested or used for such flowrate measurements. Obtain information about new methods to obtain pipe flowrate information. Summarize your findings in a brief report.

4.3 LL Ultra-high-pressure, thin jets of liquids can be used to cut various materials ranging from leather to steel and beyond. Obtain information about new methods and techniques proposed for liquid jet cutting and investigate how they may alter various manufacturing processes. Summarize your findings in a brief report.

CHAPTER 5

Finite Control Volume Analysis

After completing this chapter, you should be able to:

- select an appropriate finite control volume to solve a fluid mechanics problem.

- apply the principles of conservation of mass and energy and Newton's second law of motion to the contents of a finite control volume to get important answers.

- explain how velocity changes and energy transfers in fluid flows are related to forces and torques.

- choose between the Bernoulli equation and the mechanical energy equation for flow along a streamline.

Many fluid mechanics problems can be solved by using finite control volume analysis.

To solve many practical problems in fluid mechanics, fundamental laws of nature are applied to a finite region in space (a finite control volume). For example, we may estimate the maximum anchoring force required to hold a turbojet engine stationary during a test. Or we may design a propeller to move a boat both forward and backward. Or we may determine how much power it would take to move natural gas from one location to another many miles away.

The bases of finite control volume analysis are some fundamental laws of physics, namely, conservation of mass, Newton's second law of motion, and the first and second laws of thermodynamics. While some simplifying approximations are made for practicality, the engineering answers possible with this powerful analysis method have proven valuable in numerous instances.

Conservation of mass is the key to tracking flowing fluid. How much enters and leaves a control volume can be ascertained.

Newton's second law of motion leads to the conclusion that forces can result from or cause changes in a flowing fluid's velocity magnitude and/or direction. Moment of force (torque) can result from or cause changes in a flowing fluid's moment-of-momentum. These forces and torques can be associated with work and power transfer as in pumps and turbines.

The first law of thermodynamics is a statement of conservation of energy. The second law of thermodynamics clarifies the loss of useful energy associated with every real process. The mechanical energy equation based on these two laws can be used to analyze a large variety of steady, incompressible flows in terms of changes in the flow's pressure, elevation, speed, and of shaft work and losses.

Good judgment is required in defining the finite region in space, the control volume, used in solving a problem. What exactly to leave out of and what to leave in the control volume are important considerations. The formulas resulting from applying the fundamental laws to the contents of the control volume are easy to interpret physically and are not difficult to derive and use.

Because a finite region of space, a control volume, contains many fluid particles, detailed knowledge of distributions of fluid properties and velocity are not available, so the fluid properties and characteristics are often average values. Chapter 6 presents an analysis of fluid flow based on what is happening to the contents of an infinitesimally small region of space or control volume through which fluid particles move one at a time.

CONSERVATION OF MASS—THE CONTINUITY EQUATION

5.1.1 Derivation of the Continuity Equation

A system is defined as a collection of unchanging contents, so the **conservation of mass** principle for a system is simply stated as

Time rate of change of the system mass = 0

or

$$\frac{DM_{sys}}{Dt} = 0 \tag{5.1}$$

> The amount of mass in a system does not change with time.

where the system mass, M_{sys}, is more generally expressed as

$$M_{sys} = \int_{sys} \rho \, d\mathcal{V} \tag{5.2}$$

where the integration is over the volume of the system. In words, Eq. 5.2 states that the system mass is equal to the sum of all the density-volume element products for the contents of the system.

For a system and a fixed, nondeforming control volume that are coincident at an instant of time, as illustrated in **Fig. 5.1**, the Reynolds transport theorem (**Eq. 3.19**) with $B = $ mass and $b = 1$ allows us to state that

$$\frac{D}{Dt}\int_{sys} \rho \, d\mathcal{V} = \frac{\partial}{\partial t}\int_{cv} \rho \, d\mathcal{V} + \int_{cs} \rho \mathbf{V} \cdot \hat{\mathbf{n}} \, dA \tag{5.3}$$

or

Time rate of change   time rate of change of the   net rate of flow of
of the mass of the   = mass of the contents of the   + mass through the
coincident system   coincident control volume   control surface

In Eq. 5.3, we express the time rate of change of the system mass as the sum of two control volume quantities, the time rate of change of the mass of the contents of the control volume,

$$\frac{\partial}{\partial t}\int_{cv} \rho \, d\mathcal{V}$$

and the net rate of mass flow through the control surface,

$$\int_{cs} \rho \mathbf{V} \cdot \hat{\mathbf{n}} \, dA$$

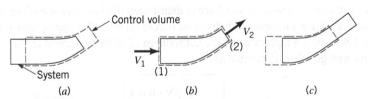

FIGURE 5.1 System and control volume at three different instants of time. (*a*) System and control volume at time $t - \delta t$. (*b*) System and control volume at time t, coincident condition. (*c*) System and control volume at time $t + \delta t$.

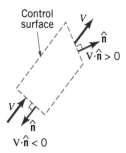

When a flow is steady, all field properties (i.e., properties at any specified point) including density remain constant with time and the time rate of change of the mass of the contents of the control volume is zero. That is,

$$\frac{\partial}{\partial t} \int_{cv} \rho \, d\hspace{-0.35em}\forall = 0$$

The integrand, $\mathbf{V} \cdot \hat{\mathbf{n}} \, dA$, in the mass flowrate integral represents the dot product of the component of velocity, $\mathbf{V}$, perpendicular to the small portion of control surface and the differential area, dA. Thus, $\mathbf{V} \cdot \hat{\mathbf{n}} \, dA$ is the volume flowrate through dA, and $\rho \mathbf{V} \cdot \hat{\mathbf{n}} \, dA$ is the mass flowrate through dA. Furthermore, as shown in the adjacent sketch, the sign of the dot product $\mathbf{V} \cdot \hat{\mathbf{n}}$ is "+" for flow *out* of the control volume and "−" for flow *into* the control volume because $\hat{\mathbf{n}}$ is considered positive when it points out of the control volume. When all of the differential quantities, $\rho \mathbf{V} \cdot \hat{\mathbf{n}} \, dA$, are summed over the entire control surface, as indicated by the integral

$$\int_{cs} \rho \mathbf{V} \cdot \hat{\mathbf{n}} \, dA$$

the result is the net mass flowrate through the control surface, or

$$\int_{cs} \rho \mathbf{V} \cdot \hat{\mathbf{n}} \, dA = \sum \dot{m}_{out} - \sum \dot{m}_{in} \tag{5.4}$$

where $\dot{m}$ is the mass flowrate (kg/s). If the integral in Eq. 5.4 is positive, the net flow is out of the control volume; if the integral is negative, the net flow is into the control volume.

The control volume expression for conservation of mass, which is commonly called the **continuity equation,** for a fixed, nondeforming control volume is obtained by combining Eqs. 5.1, 5.2, and 5.3 to obtain

| The continuity equation is a statement that mass is conserved. |

$$\boxed{\frac{\partial}{\partial t} \int_{cv} \rho \, d\hspace{-0.35em}\forall + \int_{cs} \rho \mathbf{V} \cdot \hat{\mathbf{n}} \, dA = 0} \tag{5.5}$$

In words, Eq. 5.5 states that because mass is conserved, the time rate of change of the mass of the contents of the control volume plus the net rate of mass flow through the control surface must equal zero. The same result could have been obtained more directly by equating the rates of mass flow into and out of the control volume to the rates of accumulation and depletion of mass within the control volume (see Section 4.6.2). It is reassuring, however, to see that the Reynolds transport theorem works for this simple-to-understand case. This confidence will serve us well as we develop control volume expressions for other important principles.

A frequently useful expression for the **mass flowrate** through an area is obtained from the previously derived expression

| Mass flowrate equals the product of density and volume flowrate. |

$$\dot{m} = \int_{A} \rho \mathbf{V} \cdot \hat{\mathbf{n}} \, dA$$

using *representative* or average values for the fluid density, ρ, and the fluid velocity, V, to obtain

$$\boxed{\dot{m} = \rho V A = \rho Q} \tag{5.6}$$

in which the density and velocity are average values and Q is the volume flowrate. For compressible flows, we will normally assume a uniformly distributed fluid density at each section of flow and allow density changes to occur only from section to section. The appropriate fluid velocity to use in Eq. 5.6 is the average value of the component of velocity normal to the section area involved. This average value, $\overline{V}$, defined as

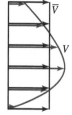

$$\boxed{\overline{V} = \frac{\displaystyle\int_{A} \rho \mathbf{V} \cdot \hat{\mathbf{n}} \, dA}{\rho A}} \tag{5.7}$$

is shown in the adjacent figure.

If the velocity, **V**, is assumed to be uniformly distributed (one-dimensional flow) over the section area, A, then

$$\overline{V} = \frac{\int_A \rho \mathbf{V} \cdot \hat{\mathbf{n}} \, dA}{\rho A} = V \tag{5.8}$$

and the bar notation is not necessary (as in Example 5.1). When the flow is not uniformly distributed over the flow cross-sectional area, the bar notation reminds us that an average velocity is being used (as in Examples 5.2 and 5.4).

V5.1 Sink flow

5.1.2 Fixed, Nondeforming Control Volume

In many applications of fluid mechanics, an appropriate control volume to use is fixed and non-deforming. Several example problems that involve the continuity equation for fixed, nondeforming control volumes (Eq. 5.5) follow in Examples 5.1 through 5.5.

EXAMPLE 5.1 | Conservation of Mass—Steady, Incompressible Flow

Given   A great danger to workers in confined spaces involves the consumption of breathable air (oxygen) and its replacement with other gases such as carbon dioxide (CO_2). To prevent this from happening, confined spaces need to be ventilated. Although there is no standard for air exchange rates, a complete change of the air every 3 minutes has been accepted by the industry as providing effective ventilation.

A worker is performing maintenance in a small rectangular tank with a height of 3.1 m and a square base 1.9 m by 1.9 m. Fresh air enters through an 0.20 m-diameter hose and exits through a 0.10 m-diameter port on the tank wall. The flow is assumed steady and incompressible.

Find   Determine

a. the exchange rate needed (m³/min) for this tank and

b. the velocity of the air entering and exiting the tank at this exchange rate.

Solution

a. The necessary exchange rate, which will provide the flowrate entering and exiting the space, is based on the volume, $\mathcal{V}$, of the tank, where

$$\mathcal{V} = (3.1\ \text{m})(1.9\ \text{m})(1.9\ \text{m}) = 11.2\ \text{m}^3$$

Thus, 11.2 m³ of air is needed to provide one complete air exchange. As described in the problem statement, to provide effective ventilation this must be done every 3 minutes.

Therefore, the required flowrate, Q, is

$$Q = \frac{11.2\ \text{m}^3}{3\ \text{min}} = 3.73\ \text{m}^3/\text{min} \tag{Ans}$$

b. The continuity equation, Eq. 5.5, can be used to calculate the velocities at the inlet and outlet. Thus,

$$\frac{\partial}{\partial t}\int_{cv} \rho \, dV + \sum \rho_{\text{out}} V_{\text{out}} A_{\text{out}} - \sum \rho_{\text{in}} V_{\text{in}} A_{\text{in}} = 0 \tag{1}$$

We consider the volume within the tank to be the control volume, A_{in} the cross-sectional area of the hose as it protrudes through the tank wall, and A_{out} the area of the port in the tank wall. The flow is assumed steady and incompressible, so

$$\frac{\partial}{\partial t}\int_{cv} \rho \, dV = 0$$

and

$$\rho_{\text{out}} = \rho_{\text{in}}$$

Thus, Eq. 1 reduces to

$$V_{\text{out}} A_{\text{out}} - V_{\text{in}} A_{\text{in}} = 0$$

or

$$V_{\text{out}} A_{\text{out}} = V_{\text{in}} A_{\text{in}} = Q \tag{2}$$

which can be rearranged to solve for V_{out} and V_{in}:

$$V_{\text{out}} = \frac{Q}{A_{\text{out}}} = \frac{3.73\ \text{m}^3/\text{min}}{\left(\frac{\pi}{4}\right)(0.10\ \text{m})^2} = 475.15\ \text{m/min} = 7.9\ \text{m/s}$$

$$V_{\text{in}} = \frac{Q}{A_{\text{in}}} = \frac{3.73\ \text{m}^3/\text{min}}{\left(\frac{\pi}{4}\right)(0.20\ \text{m})^2} = 118.8\ \text{m/min} = 1.98\ \text{m/s} \quad (Ans)$$

Comments   In this example, it is quite apparent how much the problem is simplified by assuming the flow is steady and incompressible. It is important for engineers to understand when such assumptions can be made. Also, one can see how the velocity through an outlet/inlet is directly related to the flowrate and geometry of the outlet/inlet. For example, if the air velocity through the outlet causes too much dust to be stirred up, the velocity could be decreased by increasing the diameter of the outlet. The relationship between velocity and diameter is shown in **Fig. E5.1**. As expected, the velocity is inversely proportional to the square of the diameter.

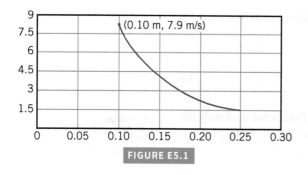

FIGURE E5.1

As mentioned, there is no written standard for air exchange rates, but changing the air every 3 minutes has been accepted by the industry. It is important for engineers to understand that individual industries and companies will have their own established safety precautions, depending on the nature of their work. As an engineer, one must heed these precautions. Safety is always a key component in the design and operation of engineering systems.

EXAMPLE 5.2 | Conservation of Mass—Steady, Compressible Flow

Given Air flows steadily between two sections in a long, straight portion of 0.10 m-inside-diameter pipe, as indicated in **Fig. E5.2**. The uniformly distributed temperature and pressure at each section are given. The average air velocity (nonuniform velocity distribution) at section (2) is 305 m/s.

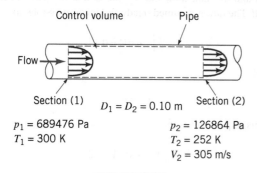

Control volume   Pipe

Flow

Section (1) $D_1 = D_2 = 0.10$ m   Section (2)

$p_1 = 689476$ Pa $p_2 = 126864$ Pa
$T_1 = 300$ K $T_2 = 252$ K
                          $V_2 = 305$ m/s

FIGURE E5.2

Find Calculate the average air velocity at section (1).

Solution The average fluid velocity at any section is the velocity that yields the section mass flowrate when multiplied by the section average fluid density and section area (Eq. 5.7). We relate the flows at sections (1) and (2) with the control volume designated with a dashed line in Fig. E5.2.

Equation 5.5 is applied to the contents of this control volume to obtain

$$\frac{\partial}{\partial t}\int_{cv} \rho\, d\mathcal{V} + \int_{cs} \rho\mathbf{V}\cdot\hat{\mathbf{n}}\, dA = 0 \quad \overset{0 \text{ (flow is steady)}}{}$$

The time rate of change of the mass of the contents of this control volume is zero because the flow is steady. The control surface

integral involves mass flowrates at sections (1) and (2), so that from Eq. 5.4, we get

$$\int_{cs} \rho\mathbf{V}\cdot\hat{\mathbf{n}}\, dA = \dot{m}_2 - \dot{m}_1 = 0$$

or

$$\dot{m}_1 = \dot{m}_2 \qquad (1)$$

and from Eqs. 1, 5.6, and 5.7, we obtain

$$\rho_1 A_1 \bar{V}_1 = \rho_2 A_2 \bar{V}_2 \qquad (2)$$

or since $A_1 = A_2$

$$\bar{V}_1 = \frac{\rho_2}{\rho_1}\bar{V}_2 \qquad (3)$$

Air at the pressures and temperatures involved in this example problem behaves like an ideal gas. The ideal gas equation of state (Eq. 1.8) is

$$\rho = \frac{p}{RT} \qquad (4)$$

Thus, combining Eqs. 3 and 4, we obtain

$$\bar{V}_1 = \frac{p_2 T_1 \bar{V}_2}{p_1 T_2}$$

$$= \frac{(126864 \text{ Pa})(300 \text{ K})(305 \text{ m/s})}{(689476 \text{ Pa})(252 \text{ K})} = 66.81 \text{ m/s} \qquad (Ans)$$

Comment We learn from this example that the continuity equation (Eq. 5.5) is valid for compressible as well as incompressible flows. Also, nonuniform velocity distributions can be handled with the average velocity concept. Significant average velocity changes can occur in pipe flow if the fluid is compressible.

EXAMPLE 5.3 | Conservation of Mass—Two Fluids

Given The inner workings of a dehumidifier are shown in **Fig. E5.3a**. Moist air (a mixture of dry air and water vapor) enters the dehumidifier at the rate of 0.28 m³/hr. Liquid water drains out of the dehumidifier at a rate of 0.0014 m³/hr. A simplified sketch of the process is provided in **Fig. E5.3b**.

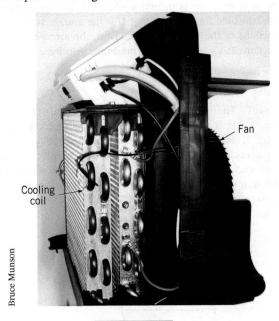

FIGURE E5.3a

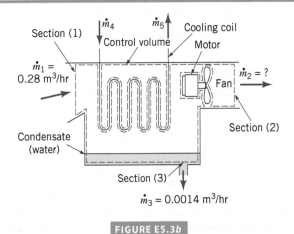

FIGURE E5.3b

way. Thus, the flowrates at sections (1), (2), and (3) appear steady, and the time rate of change of the mass of the contents of the control volume may be considered equal to zero on a time-average basis. The application of Eqs. 5.4 and 5.5 to the control volume contents results in

$$\int_{cs} \rho \mathbf{V} \cdot \hat{\mathbf{n}} \, dA = -\dot{m}_1 + \dot{m}_2 + \dot{m}_3 = 0$$

or

$$\dot{m}_2 = \dot{m}_1 - \dot{m}_3 = 0.28 \text{ m}^3/\text{hr} - 0.0014 \text{ m}^3/\text{hr}$$

$$= 0.278 \text{ m}^3/\text{hr} \qquad\qquad (Ans)$$

Find Determine the mass flowrate of the dry air and the water vapor leaving the dehumidifier.

Solution The unknown mass flowrate at section (2) is linked with the known flowrates at sections (1) and (3) with the control volume designated with a dashed line in Fig. E5.3b. The contents of the control volume are the air and water vapor mixture and the condensate (liquid water) in the dehumidifier at any instant.

Not included in the control volume are the fan and its motor and the condenser coils and refrigerant. Even though the flow in the vicinity of the fan blade is unsteady, it is unsteady in a cyclical

Comment Note that the continuity equation (Eq. 5.5) can be used when there is more than one stream of fluid flowing through the control volume.

The answer is the same with a control volume that includes the cooling coils within the control volume. The continuity equation becomes

$$\dot{m}_2 = \dot{m}_1 - \dot{m}_3 + \dot{m}_4 - \dot{m}_5 \qquad\qquad (1)$$

where $\dot{m}_4$ is the mass flowrate of the cooling fluid flowing into the control volume, and $\dot{m}_5$ is the flowrate out of the control volume through the cooling coil. Since the flow through the coils is steady, it follows that $\dot{m}_4 = \dot{m}_5$. Hence, Eq. 1 gives the same answer as obtained with the original control volume.

EXAMPLE 5.4 | Conservation of Mass—Nonuniform Velocity Profile

Given Incompressible, laminar water flow develops in a straight pipe having a radius, R, as indicated in **Fig. E5.4a**. At section (1), the velocity profile is uniform; the velocity is equal to a constant value U and is parallel to the pipe axis everywhere. At section (2), the velocity profile is axisymmetric and parabolic, with zero velocity at the pipe wall and a maximum value of u_{max} at the centerline.

Find

a. How are U and u_{max} related?

b. How are the average velocity at section (2), $\overline{V}_2$, and u_{max} related?

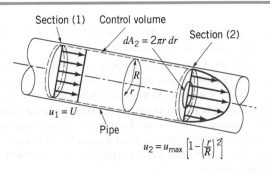

FIGURE E5.4a

Solution

a. An appropriate control volume is sketched (dashed lines) in Fig. E5.4a. The application of Eq. 5.5 to this control volume yields

$$\cancelto{0 \text{ (flow is steady)}}{\frac{\partial}{\partial t}\int_{cv} \rho\, d\Psi} + \int_{cs} \rho\mathbf{V}\cdot\hat{\mathbf{n}}\, dA = 0 \tag{1}$$

At the inlet, section (1), the velocity is uniform with $V_1 = U$, so that

$$\int_{(1)} \rho\mathbf{V}\cdot\hat{\mathbf{n}}\, dA = -\rho_1 A_1 U \tag{2}$$

At the outlet, section (2), the velocity is not uniform. However, the net flowrate through this section is the sum of flows through numerous small washer-shaped areas of size $dA_2 = 2\pi r\, dr$, as shown by the shaded area element in **Fig. E5.4b**. This shape is selected to be the infinitesimal area used in the integrand, because the velocity has the same value, u_2, everywhere in that area. Thus, in the limit of infinitesimal area elements, the summation is replaced by an integration, and the outflow through section (2) is given by

$$\int_{(2)} \rho\mathbf{V}\cdot\hat{\mathbf{n}}\, dA = \rho_2\int_0^R u_2 2\pi r\, dr \tag{3}$$

By combining Eqs. 1, 2, and 3, we get

$$\rho_2\int_0^R u_2 2\pi r\, dr - \rho_1 A_1 U = 0 \tag{4}$$

Since the flow is considered incompressible, $\rho_1 = \rho_2$, the parabolic velocity relationship for flow through section (2) is used in Eq. 4 to yield

$$2\pi u_{max}\int_0^R \left[1 - \left(\frac{r}{R}\right)^2\right] r\, dr - A_1 U = 0 \tag{5}$$

FIGURE E5.4b

Integrating, we get from Eq. 5

$$2\pi u_{max}\left(\frac{r^2}{2} - \frac{r^4}{4R^2}\right)_0^R - \pi R^2 U = 0$$

or

$$u_{max} = 2U \tag{Ans}$$

b. Because this flow is incompressible and the area is constant, we conclude from Eq. 5.7 that U is the average velocity at all sections of the control volume. Thus, the average velocity at section (2), $\bar{V}_2$, is one-half the maximum velocity, u_{max}.

$$\bar{V}_2 = \frac{u_{max}}{2} \tag{Ans}$$

Comment The relationship between u_{max} and the average velocity, $\bar{V}_2$, is a function of the "shape" of the velocity profile, $u(r)$. It does not depend on the magnitude of the velocity. The area enclosed in the parabolic velocity profile is equal to the area enclosed in the uniform velocity profile. As shown in **Fig. E5.4c**, if the velocity profile is a different shape (nonparabolic), the average velocity is not necessarily one-half of the maximum velocity.

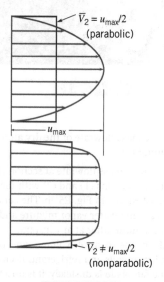

$\bar{V}_2 = u_{max}/2$ (parabolic)

u_{max}

$\bar{V}_2 \neq u_{max}/2$ (nonparabolic)

FIGURE E5.4c

EXAMPLE 5.5 | Conservation of Mass—Unsteady Flow

Given Construction workers in a trench of the type shown in **Fig. E5.5a** are installing a new waterline. The trench is 3.0 m long, 1.5 m wide, and 2.4 m deep. As a result of being near an intersection, carbon dioxide from vehicle exhaust enters the trench at a rate of 2.8 m³/min. Because carbon dioxide has a greater density than air, it will settle to the bottom of the trench and displace the air the workers need to breathe. Assume that there is negligible mixing between the air and carbon dioxide.

Find

a. Estimate the time rate of change of the depth of carbon dioxide in the trench, $\partial h/\partial t$, in meter per minute at any instant.

b. Calculate the time, $t_{h=1.8}$, it would take for the level of carbon dioxide to reach a depth of 1.8 m, the approximate height to fully engulf the utility workers.

FIGURE E5.5a

Solution

a. We use the fixed, nondeforming control volume outlined with a dashed line in **Fig. E5.5b**. This control volume includes, at any instant, the carbon dioxide accumulated in the trench, some of the carbon dioxide flowing from the street into the trench, and some air. Application of Eqs. 5.5 and 5.6 to the contents of the control volume results in

$$\frac{\partial}{\partial t}\int_{\substack{air\\volume}} \rho_{air}\, dV_{air} + \frac{\partial}{\partial t}\int_{\substack{co_2\\volume}} \rho_{co_2}\, dV_{co_2} - \dot{m}_{co_2} + \dot{m}_{air} = 0 \qquad (1)$$

where $\dot{m}_{co_2}$ and $\dot{m}_{air}$ are the mass flowrates of carbon dioxide into and air out of the control volume.

Recall that the mass, dm, of fluid contained in a small volume, dV, is $dm = \rho\, dV$. Hence, the two integrals in Eq. 1 represent the total amount of air and carbon dioxide in the control volume, and the sum of the first two terms is the time rate of change of mass within the control volume.

Note that the time rates of change of air mass and carbon dioxide mass are each not zero. Recognizing, however, that

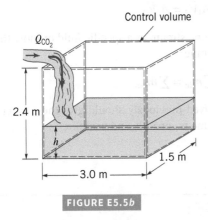

Control volume

Q_{CO_2}

2.4 m

h

1.5 m

3.0 m

FIGURE E5.5b

the air mass must be conserved, we know that the time rate of change of the mass of air in the control volume must be equal to the rate of air mass flow out of the control volume. Thus, applying Eqs. 5.5 and 5.6 to the air only and to the carbon dioxide only, we obtain

$$\frac{\partial}{\partial t}\int_{\substack{air\\volume}} \rho_{air}\, dV_{air} + \dot{m}_{air} = 0$$

for air and

$$\frac{\partial}{\partial t}\int_{\substack{co_2\\volume}} \rho_{co_2}\, dV_{co_2} = \dot{m}_{co_2} \qquad (2)$$

for carbon dioxide. Neglecting the small volume occupied by CO_2 as it flows down to the pool of gas in the trench bottom, the volume occupied by CO_2 inside the control volume is

$$\int_{\substack{co_2\\volume}} \rho_{co_2}\, dV_{co_2} = \rho_{co_2}[h\,(3.0\text{ m})(1.5\text{ m})] \qquad (3)$$

Combining Eqs. 2 and 3, we obtain

$$\rho_{co_2}(4.5\text{ m}^2)\frac{\partial h}{\partial t} = \dot{m}_{co_2}$$

and, thus, since $\dot{m} = \rho Q$,

$$\frac{\partial h}{\partial t} = \frac{Q_{co_2}}{4.5\text{ m}^2}$$

or

$$\frac{\partial h}{\partial t} = \frac{2.8\text{ m}^3/\text{min}}{4.5\text{ m}^2} = 0.622\text{ m/min} \qquad (Ans)$$

b. To find the time it will take to engulf the workers at a depth of $h = 1.8$ m, we need only to divide the height the level needs to reach by the time rate of change of depth. That is,

$$t_{h=1.8} = \frac{1.8\text{ m}}{0.622\,(\text{m/min})} = 2.89\text{ min} \qquad (Ans)$$

Comments As shown in this example, it would not take long for the air within the confined space of the trench to be displaced enough to become a danger to the workers. Furthermore, by the time workers feel the effects of an oxygen-deficient atmosphere, they may be unable to remove themselves from the dangerous space.

Note that the answer to part (b) can be easily obtained by realizing that the accumulated volume, V, of a flow is the flowrate, Q, multiplied by the time it has been flowing, t. That is, $V = Qt$. For this example, the volume of the carbon dioxide in the trench when it is 1.8 m deep (neglecting the volume of the workers and equipment within the trench) is $V = 1.5$ m $\times$ 3.0 m $\times$ 1.8 m = 8.1 m³. Thus, with a flowrate of $Q = 2.8$ m³/min, $t_{h=18} = $ 8.1 m³/2.8 m³/min = 2.89 min is in agreement with the foregoing answer.

The preceding example problems illustrate some important results of applying the conservation of mass principle to the contents of a fixed, nondeforming control volume. The dot product $\mathbf{V} \cdot \hat{\mathbf{n}}$ is "+" for flow out of the control volume and "−" for flow into the control volume. Thus, mass flowrate out of the control volume is "+," and mass flowrate in is "−." When the flow is steady, the time rate of change of the mass of the contents of the control volume

$$\frac{\partial}{\partial t}\int_{cv} \rho\, dV$$

For steady flow, mass flowrate in equals mass flowrate out.

is zero and the net amount of mass flowrate, $\dot{m}$, through the control surface is, therefore, also zero:

$$\sum \dot{m}_{out} - \sum \dot{m}_{in} = 0 \tag{5.9}$$

If the steady flow is also incompressible, the net amount of volume flowrate, Q, through the control surface is also zero:

$$\sum Q_{out} - \sum Q_{in} = 0 \tag{5.10}$$

An unsteady, but cyclical, flow can be considered steady on a time-average basis. When the flow is unsteady, the instantaneous time rate of change of the mass of the contents of the control volume is not necessarily zero and can be an important variable. When the value of

$$\frac{\partial}{\partial t} \int_{cv} \rho \, dV$$

is "+," the mass of the contents of the control volume is increasing. When it is "−," the mass of the contents of the control volume is decreasing.

When the flow is uniformly distributed over the opening in the control surface (one-dimensional flow),

$$\dot{m} = \rho A V$$

where V is the uniform value of the velocity component normal to the section area A. When the velocity is nonuniformly distributed over the opening in the control surface,

$$\dot{m} = \rho A \bar{V} \tag{5.11}$$

where $\bar{V}$ is the average value of the component of velocity normal to the section area A as defined by Eq. 5.7.

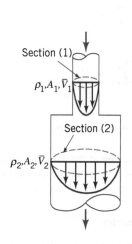

Section (1)

$\rho_1, A_1, \bar{V}_1$

Section (2)

$\rho_2, A_2, \bar{V}_2$

For steady flow involving only one stream of a specific fluid flowing through the control volume at sections (1) and (2), as shown in the adjacent figure

$$\dot{m} = \rho_1 A_1 \bar{V}_1 = \rho_2 A_2 \bar{V}_2 \tag{5.12}$$

and for incompressible flow,

$$Q = A_1 \bar{V}_1 = A_2 \bar{V}_2 \tag{5.13}$$

For steady flow involving more than one stream of a specific fluid or more than one specific fluid flowing through the control volume,

$$\sum \dot{m}_{in} = \sum \dot{m}_{out}$$

The variety of example problems solved previously should give the correct impression that the fixed, nondeforming control volume is versatile and useful.

The Wide World of Fluids

"Green" 6 lpf standards

Toilets account for approximately 40% of all indoor household water use. To conserve water, the new standard is 6 liters of water per flush (lpf). Old toilets use up to 26 lpf; those manufactured after 1980 use 14 lpf. Neither is considered a low-flush toilet. A typical 3.2-person household in which each person flushes a 26 lpf toilet 4 times a day uses 123785 liters of water each year; with a 13 lpf toilet the amount is reduced to 62080 liters. Clearly, the new 6 lpf

toilets will save even more water. However, designing a toilet that flushes properly with such a small amount of water is not simple. Today, there are two basic types involved: those that are gravity powered and those that are pressure powered. Gravity toilets (typical of most currently in use) have rather long cycle times. The water starts flowing under the action of gravity, and the swirling vortex motion initiates the siphon action that builds to a point of discharge. In the newer pressure-assisted models, the *flowrate* is large, but the cycle time is short and the amount of water used is relatively small.

5.1.3 Moving, Nondeforming Control Volume

It is sometimes necessary to use a nondeforming control volume attached to a moving reference frame. Examples include control volumes containing a gas turbine engine on an aircraft in flight, the exhaust stack of a ship at sea, and the gasoline tank of an automobile passing by.

As discussed in Section 3.4.6, when a moving control volume is used, the fluid velocity relative to the moving control volume (relative velocity) is an important flow field variable. The relative velocity, **W**, is the fluid velocity seen by an observer moving with the control volume. The control volume velocity, **V**$_{cv}$, is the velocity of the control volume as seen from a fixed coordinate system. The absolute velocity, **V**, is the fluid velocity seen by a stationary observer in a fixed coordinate system. These velocities are related to each other by the vector equation

$$\mathbf{V} = \mathbf{W} + \mathbf{V}_{cv} \tag{5.14}$$

Some problems are most easily solved by using a moving control volume.

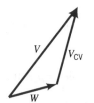

as illustrated in the adjacent figure. This is the same as Eq. 3.22, introduced in Chapter 3.

For a system and a moving, nondeforming control volume that are coincident at an instant of time, the Reynolds transport theorem (Eq. 3.23) for a moving control volume leads to

$$\frac{DM_{sys}}{Dt} = \frac{\partial}{\partial t}\int_{cv} \rho \, d\Psi + \int_{cs} \rho \mathbf{W} \cdot \hat{\mathbf{n}} \, dA \tag{5.15}$$

From Eqs. 5.1 and 5.15, we can get the control volume expression for conservation of mass (the continuity equation) for a moving, nondeforming control volume, namely,

$$\boxed{\frac{\partial}{\partial t}\int_{cv} \rho \, d\Psi + \int_{cs} \rho \mathbf{W} \cdot \hat{\mathbf{n}} \, dA = 0} \tag{5.16}$$

Some examples of the application of Eq. 5.16 follow in Examples 5.6 and 5.7.

EXAMPLE 5.6 | Conservation of Mass—Compressible Flow with a Moving Control Volume

Given An airplane moves forward at a speed of 971 km/hr, as shown in **Fig. E5.6a**. The frontal intake area of the jet engine is 0.80 m², and the entering air density is 0.736 kg/m³. A stationary observer determines that relative to the Earth, the jet engine exhaust gases move away from the engine with a speed of 1050 km/hr.

The engine exhaust area is 0.558 m², and the exhaust gas density is 0.515 kg/m³.

Find Estimate the mass flowrate of fuel into the engine in kg/hr.

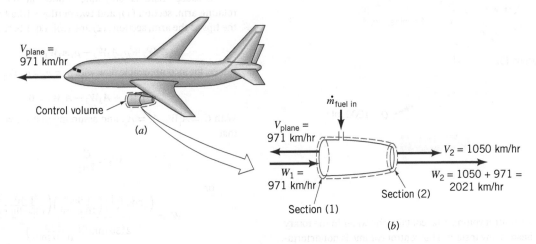

FIGURE E5.6

Solution The control volume, which moves with the airplane (see **Fig. E5.6b**), surrounds the engine and its contents and includes all fluids involved at an instant. The application of Eq. 5.16 to these contents of the control volume yields

$$\underset{\nearrow}{\overset{\text{0 (flow relative to moving control}}{\overset{\text{volume is considered steady on a}}{\overset{\text{time-average basis)}}{}}}$$

$$\frac{\partial}{\partial t}\int_{cv} \rho \, d\forall + \int_{cs} \rho \mathbf{W} \cdot \hat{\mathbf{n}} \, dA = 0 \qquad (1)$$

Assuming one-dimensional flow, we evaluate the surface integral in Eq. 1 and get

$$-\dot{m}_{\text{fuel}} - \rho_1 A_1 W_1 + \rho_2 A_2 W_2 = 0$$
$$\underset{\text{in}}{}$$

or

$$\dot{m}_{\text{fuel}} = \rho_2 A_2 W_2 - \rho_1 A_1 W_1 \qquad (2)$$
$$\underset{\text{in}}{}$$

We consider the intake velocity, W_1, relative to the moving control volume, as being equal in magnitude to the speed of the airplane, 971 km/hr. The exhaust velocity, W_2, also needs to be measured relative

to the moving control volume. Since a fixed observer noted that the exhaust gases were moving away from the engine at a speed of 1050 km/hr, the speed of the exhaust gases relative to the moving control volume, W_2, is determined as follows by using Eq. 5.14

$$V_2 = W_2 + V_{\text{plane}}$$

or

$$W_2 = V_2 - V_{\text{plane}} = 1050 \text{ km/hr} - (-971 \text{ km/hr})$$
$$= 2021 \text{ km/hr}$$

and is shown in Fig. E5.6b.

From Eq. 2,

$$\dot{m}_{\text{fuel}} = \left(0.515 \frac{\text{kg}}{\text{m}^3}\right)(0.558 \text{ m}^2)\left(2021 \frac{\text{km}}{\text{hr}}\right)\left(1000 \frac{\text{m}}{\text{km}}\right)$$
$$\underset{\text{in}}{}$$
$$- \left(0.736 \frac{\text{kg}}{\text{m}^3}\right)(0.80 \text{ m}^2)\left(971 \frac{\text{km}}{\text{hr}}\right)\left(1000 \frac{\text{m}}{\text{km}}\right)$$

$$\dot{m}_{\text{fuel}} = 9100 \frac{\text{kg}}{\text{hr}} \qquad \qquad (Ans)$$
$$\underset{\text{in}}{}$$

Comment Note that the fuel flowrate was obtained as the difference of two large, nearly equal numbers. Very accurate values of W_2 and W_1 are needed to obtain a modestly accurate value of $\dot{m}_{\text{fuel}}$.

EXAMPLE 5.7 | Conservation of Mass—Relative Velocity

Given Water enters a rotating lawn sprinkler through its base at the steady rate of 1000 ml/s, as sketched in **Fig. E5.7**. The exit area of each of the two nozzles is 30 mm².

Find Determine the average speed of the water leaving the nozzle, relative to the nozzle, if

a. the rotary sprinkler head is stationary,

b. the sprinkler head rotates at 600 rpm, and

c. the sprinkler head accelerates from 0 to 600 rpm.

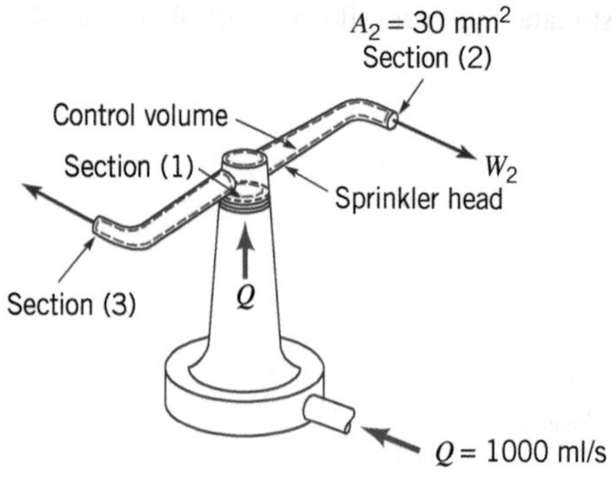

$A_2 = 30 \text{ mm}^2$
Section (2)

Control volume

Section (1)

W_2
Sprinkler head

Section (3)

Q

$Q = 1000 \text{ ml/s}$

FIGURE E5.7

Solution

a. We specify a control volume that contains the water in the rotary sprinkler head at any instant. This control volume is nondeforming, but it moves (rotates) with the sprinkler head.

The application of Eq. 5.16 to the contents of this control volume for situation **(a)**, **(b)**, or **(c)** of the problem results in the same expression, namely,

$$\underset{\nearrow}{\overset{\text{0 flow is steady, or the}}{\overset{\text{control volume is filled with}}{\overset{\text{an incompressible fluid}}{}}}}$$

$$\frac{\partial}{\partial t}\int_{cv} \rho \, d\forall + \int_{cs} \rho \mathbf{W} \cdot \hat{\mathbf{n}} \, dA = 0$$

or

$$\sum \rho_{\text{out}} A_{\text{out}} W_{\text{out}} - \sum \rho_{\text{in}} A_{\text{in}} W_{\text{in}} = 0 \qquad (1)$$

The time rate of change of the mass of water in the control volume is zero because the flow is steady and the control volume is filled with water.

Because there is only one inflow [at the base of the rotating arm, section (1)] and two outflows [the two nozzles at the tips of the arm, sections (2) and (3)], Eq. 1 becomes

$$\rho_2 A_2 W_2 + \rho_3 A_3 W_3 - \rho_1 A_1 W_1 = 0 \qquad (2)$$

Hence, for incompressible flow with $\rho_1 = \rho_2 = \rho_3$, Eq. 2 becomes

$$A_2 W_2 + A_3 W_3 - A_1 W_1 = 0$$

With $Q = A_1 W_1$, $A_2 = A_3$, and assuming that $W_2 = W_3$, it follows that

$$W_2 = \frac{Q}{2A_2}$$

or

$$W_2 = \frac{\left(1000 \frac{\text{ml}}{\text{s}}\right)\left(\frac{1 \text{ liter}}{1000 \text{ ml}}\right)\left(\frac{1 \text{ m}^3}{1000 \text{ liter}}\right)}{2(30 \text{ mm}^2)\left(\frac{1 \text{ m}}{1000 \text{ mm}}\right)^2}$$

$$= 16.7 \text{ m/s} \qquad \qquad (Ans)$$

b. and c. The value of W_2 is independent of the speed of rotation of the sprinkler head and represents the average velocity of the water exiting from each nozzle with respect to the nozzle for cases **(a)**, **(b)**, and **(c)**.

Comment The velocity of water discharging from each nozzle, when viewed from a stationary reference (i.e., V_2), will vary as the rotation speed of the sprinkler head varies, since from Eq. 5.14,

$$V_2 = W_2 - U$$

where $U = \omega R$ is the speed of the nozzle and ω and R are the angular velocity and radius of the sprinkler head, respectively.

When a moving, nondeforming control volume is used, the dot product sign convention used earlier for fixed, nondeforming control volume applications is still valid. Also, if the flow within the moving control volume is steady, or steady on a time-average basis, the time rate of change of the mass of the contents of the control volume is zero. Velocities seen from the control volume reference frame (relative velocities) must be used in the continuity equation. Relative and absolute velocities are related by a vector equation (Eq. 5.14), which also involves the control volume velocity.

5.1.4 Deforming Control Volume

Occasionally, a deforming control volume can simplify the solution of a problem. A deforming control volume involves changing volume size and control surface movement. Thus, the Reynolds transport theorem for a moving control volume can be used for this case, and Eqs. 3.23 and 5.1 lead to

Flowrate is calculated using the velocity relative to the control volume.

$$\boxed{\frac{DM_{sys}}{Dt} = \frac{\partial}{\partial t}\int_{cv} \rho \, d\Psi + \int_{cs} \rho \mathbf{W} \cdot \hat{\mathbf{n}} \, dA = 0} \tag{5.17}$$

The time rate of change term in Eq. 5.17,

$$\frac{\partial}{\partial t}\int_{cv} \rho \, d\Psi$$

is usually nonzero and must be carefully evaluated because the geometry and extent of the control volume vary with time. The mass flowrate term in Eq. 5.17,

$$\int_{cs} \rho \mathbf{W} \cdot \hat{\mathbf{n}} \, dA$$

must be determined using the relative velocity, $\mathbf{W}$, the velocity referenced to the control surface. Since the control volume is deforming, the control surface velocity is not necessarily uniform and identical to the control volume velocity, $\mathbf{V}_{cv}$, as was true for moving, nondeforming control volumes. For the deforming control volume,

The velocity of the surface of a deforming control volume is not the same at all points on the surface.

$$\mathbf{V} = \mathbf{W} + \mathbf{V}_{cs} \tag{5.18}$$

where $\mathbf{V}_{cs}$ is the velocity of the control surface as seen by a fixed observer. The relative velocity, $\mathbf{W}$, must be ascertained with care wherever fluid crosses the control surface. Two example problems that illustrate the use of the continuity equation for a deforming control volume, Eq. 5.17, follow in Examples 5.8 and 5.9.

EXAMPLE 5.8 | Conservation of Mass—Deforming Control Volume

Given A syringe (**Fig. E5.8a**) is used to inoculate a cow. The plunger has a face area of 500 mm². The liquid in the syringe is to be injected steadily at a rate of 300 cm³/min. The leakage rate past the plunger is 0.10 times the volume flowrate out of the needle.

Find With what speed should the plunger be advanced?

Solution The control volume selected for solving this problem is the deforming one illustrated in **Fig. E5.8b**. Section (1) of the control surface moves with the plunger. The surface area of section (1), A_1, is considered equal to the circular area of the face of the plunger, A_p,

although this is not strictly true, since leakage occurs. The difference is small, however. Thus,

$$A_1 = A_p \tag{1}$$

Liquid also leaves the needle through section (2), which involves a fixed area, A_2. The application of Eq. 5.17 to the contents of this control volume gives

$$\frac{\partial}{\partial t}\int_{cv} \rho \, d\Psi + \dot{m}_2 + \rho Q_{leak} = 0 \tag{2}$$

Even though Q_{leak} and the flow through section area A_2 are steady, the time rate of change of the mass of liquid in the shrinking control volume is not zero because the control volume is getting smaller. To evaluate the first term of Eq. 2, we note that

$$\int_{cv} \rho \, d\forall = \rho(\ell A_1 + \forall_{needle}) \tag{3}$$

where ℓ is the changing length of the control volume (see Fig. E5.8) and $\forall_{needle}$ is the volume of the needle. From Eq. 3, we obtain

$$\frac{\partial}{\partial t}\int_{cv} \rho \, d\forall = \rho A_1 \frac{\partial \ell}{\partial t} \tag{4}$$

Note that

$$-\frac{\partial \ell}{\partial t} = V_p \tag{5}$$

where V_p is the speed of the plunger sought in the problem statement. Combining Eqs. 2, 4, and 5, we obtain

$$-\rho A_1 V_p + \dot{m}_2 + \rho Q_{leak} = 0 \tag{6}$$

However, from Eq. 5.6, we see that

$$\dot{m}_2 = \rho Q_2 \tag{7}$$

and Eq. 6 becomes

$$-\rho A_1 V_p + \rho Q_2 + \rho Q_{leak} = 0 \tag{8}$$

Solving Eq. 8 for V_p yields

$$V_p = \frac{Q_2 + Q_{leak}}{A_1} \tag{9}$$

Since $Q_{leak} = 0.1Q_2$, Eq. 9 becomes

$$V_p = \frac{Q_2 + 0.1Q_2}{A_1} = \frac{1.1Q_2}{A_1}$$

and

$$V_p = \frac{(1.1)(300 \text{ cm}^3/\text{min})}{(500 \text{ mm}^2)} \left(\frac{1000 \text{ mm}^3}{\text{cm}^3}\right)$$
$$= 660 \text{ mm/min} \tag{Ans}$$

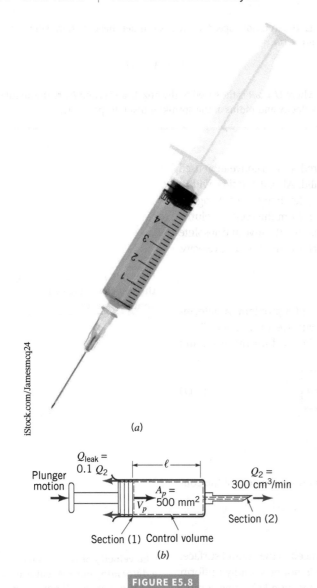

(a)

$Q_{leak} = 0.1 Q_2$

Plunger motion

$A_p = 500 \text{ mm}^2$

V_p

ℓ

$Q_2 = 300 \text{ cm}^3/\text{min}$

Section (2)

Section (1)　Control volume

(b)

FIGURE E5.8

EXAMPLE 5.9 │ Conservation of Mass—Deforming Control Volume

Given　Consider Example 5.5.

Find　Solve the problem of Example 5.5 using a deforming control volume that includes only the carbon dioxide accumulating in the trench.

Solution　For this deforming control volume, Eq. 5.17 leads to

$$\frac{\partial}{\partial t}\int_{\substack{co_2 \\ volume}} \rho \, d\forall + \int_{cs} \rho \mathbf{W} \cdot \hat{\mathbf{n}} \, dA = 0 \tag{1}$$

The first term of Eq. 1 can be evaluated as

$$\frac{\partial}{\partial t}\int_{\substack{co_2 \\ volume}} \rho \, d\forall = \frac{\partial}{\partial t}[\rho h(3.0 \text{ m})(1.5 \text{ m})]$$
$$= \rho(4.5 \text{ m}^2)\frac{\partial h}{\partial t} \tag{2}$$

The second term of Eq. 1 can be evaluated as

$$\int_{cs} \rho \mathbf{W} \cdot \hat{\mathbf{n}} \, dA = -\rho\left(V_{co_2} + \frac{\partial h}{\partial t}\right)A_{co_2} \tag{3}$$

where A_{co_2} and V_{co_2} are the cross-sectional area and velocity of the carbon dioxide flowing from the street down into the trench. Note that the velocity of the carbon dioxide entering the control surface and the velocity of the control surface are summed in Eq. 3, since they are in opposite directions, and therefore, the relative velocity would be the two velocities combined (i.e., like the relative velocity of two cars passing each other going in opposite directions). The overall mass flow term is negative because the carbon dioxide is entering the control volume. Thus, from Eqs. 1, 2, and 3, we obtain

$$\frac{\partial h}{\partial t} = \frac{V_{co_2} A_{co_2}}{(4.5 \text{ m}^2 - A_{co_2})} = \frac{Q_{co_2}}{(4.5 \text{ m}^2 - A_{co_2})}$$

or for $A_{co_2} \ll 4.5 \text{ m}^2$

$$\frac{\partial h}{\partial t} = \frac{2.8 \text{ m}^3/\text{min}}{(4.5 \text{ m}^2)} = 0.622 \text{ m/min} \tag{Ans}$$

Comment　Note that these results using a deforming control volume are the same as those obtained in Example 5.5 with a fixed control volume.

The conservation of mass principle is easily applied to the contents of a control volume. The appropriate selection of a specific kind of control volume (for example, fixed and nondeforming, moving and nondeforming, or deforming) can make the solution of a particular problem less complicated. In general, where fluid flows through the control surface, it is advisable to make the control surface perpendicular to the flow. In the sections ahead, we learn that the conservation of mass principle is primarily used in combination with other important laws to solve problems.

5.2 NEWTON'S SECOND LAW—THE LINEAR MOMENTUM AND MOMENT-OF-MOMENTUM EQUATIONS

5.2.1 Derivation of the Linear Momentum Equation

Newton's second law of motion for a system is

Video

V5.4 Smokestack plume momentum

Time rate of change of the = sum of external forces
linear momentum of the system   acting on the system

The linear momentum of an infinitesimal particle of mass $\rho \, dV$ is the product of its velocity and its mass $\mathbf{V}\rho \, dV$. Thus, the momentum of the entire system is $\int_{\text{sys}} \mathbf{V}\rho \, dV$ and Newton's law becomes

$$\frac{D}{Dt}\int_{\text{sys}} \mathbf{V}\rho \, dV = \sum \mathbf{F}_{\text{sys}} \tag{5.19}$$

Any reference or coordinate system for which this statement is true is called *inertial*. A fixed coordinate system is inertial. A coordinate system that moves in a straight line with constant velocity and is thus without acceleration is also inertial. We proceed to develop the control volume formula for this important law. When a control volume is coincident with a system at an instant of time, the forces acting on the system and the forces acting on the contents of the coincident control volume (see **Fig. 5.2**) are instantaneously identical; that is,

> Forces acting on a flowing fluid can change its velocity magnitude and/or direction.

$$\sum \mathbf{F}_{\text{sys}} = \sum \mathbf{F}_{\substack{\text{contents of the} \\ \text{coincident control volume}}} \tag{5.20}$$

Furthermore, for a system and the contents of a coincident control volume that is fixed and nondeforming, the Reynolds transport theorem [Eq. 3.19 with b set equal to the velocity (i.e., momentum per unit mass), and B_{sys} being the system momentum] allows us to conclude that

$$\frac{D}{Dt}\int_{\text{sys}} \mathbf{V}\rho \, dV = \frac{\partial}{\partial t}\int_{\text{cv}} \mathbf{V}\rho \, dV + \int_{\text{cs}} \mathbf{V}\rho\mathbf{V} \cdot \hat{\mathbf{n}} \, dA \tag{5.21}$$

or

Video

V5.5 Marine propulsion

Time rate of change   time rate of change   net rate of flow
of the linear   of the linear   of linear momentum
momentum of the   =   momentum of the   +   through the
system   contents of the   control surface
                       control volume

Equation 5.21 states that the time rate of change of system linear momentum is expressed as the sum of the two control volume quantities: the time rate of change of the *linear momentum of the contents of the control volume* and the net rate of *linear momentum flow through the control surface*. As particles of mass move into or out of a control volume through the control surface, they carry linear momentum in or out. Thus, linear momentum flow should seem no more unusual than mass flow.

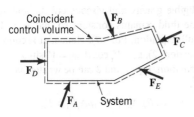

FIGURE 5.2 External forces acting on system and coincident control volume.

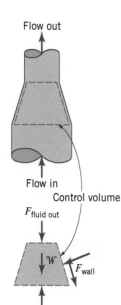

Flow out

Flow in

Control volume

$F_{\text{fluid out}}$

$\mathcal{W}$

F_{wall}

$F_{\text{fluid in}}$

For a control volume that is fixed (and thus inertial) and nondeforming, Eqs. 5.19, 5.20, and 5.21 provide an appropriate mathematical statement of Newton's second law of motion as

$$\frac{\partial}{\partial t}\int_{\text{cv}} \mathbf{V}\rho \, d\mathcal{V} + \int_{\text{cs}} \mathbf{V}\rho\mathbf{V}\cdot\hat{\mathbf{n}} \, dA = \sum\mathbf{F}_{\substack{\text{contents of the}\\\text{control volume}}} \tag{5.22}$$

We call Eq. 5.22 the **linear momentum equation.**

In our application of the linear momentum equation, we initially confine ourselves to fixed, nondeforming control volumes for simplicity. Subsequently, we discuss the use of a moving but inertial, nondeforming control volume. We do not consider here deforming control volumes and accelerating (noninertial) control volumes. If a control volume is noninertial, the acceleration components involved (for example, translation acceleration, Coriolis acceleration, and centrifugal acceleration) require consideration.

The forces involved in Eq. 5.22 are body and surface forces that act on what is contained in the control volume, as shown in the adjacent sketch. The only body force we consider in this chapter is the one associated with the action of gravity. We experience this body force as weight, $\mathcal{W}$. The surface forces are basically exerted on the contents of the control volume by material just outside the control volume in contact with material just inside the control volume. For example, a wall in contact with fluid can exert a reaction surface force on the fluid it bounds. Similarly, fluid just outside the control volume can push on fluid just inside the control volume at a common interface, usually an opening in the control surface through which fluid flow occurs. An immersed object can resist fluid motion with surface forces.

The linear momentum terms in the momentum equation deserve careful explanation. We clarify their physical significance in the following section.

5.2.2 Application of the Linear Momentum Equation

 Video

V5.6 Force due to a water jet

The linear momentum equation for an inertial control volume is a vector equation (Eq. 5.22). In engineering applications, components of this vector equation resolved along orthogonal coordinates, for example, x, y, and z (rectangular coordinate system) or r, θ, and x (cylindrical coordinate system), will normally be used. A simple example involving steady, incompressible flow is considered first.

EXAMPLE 5.10 | Linear Momentum—Change in Flow Direction

Given As shown in **Fig. E5.10a**, a horizontal jet of water exits a nozzle with a uniform speed of $V_1 = 3.1$ m/s, strikes a vane, and is turned through an angle, θ. A similar situation is also shown in Video V5.6.

Find Determine the anchoring force needed to hold the vane stationary if gravity and viscous effects are negligible.

Solution We select a control volume that includes the vane and a portion of the water (see **Figs. E5.10b, c**) and apply the linear momentum equation to this fixed control volume. The only portions of the control surface across which fluid flows are section (1) (the entrance) and section (2) (the exit). Hence, the x and z components of Eq. 5.22 become

$$\frac{\partial}{\partial t}\int_{\text{cv}} u\rho \, d\mathcal{V}^{\,0 \text{ (flow is steady)}} + \int_{\text{cs}} u\rho\mathbf{V}\cdot\hat{\mathbf{n}} \, dA = \sum F_x$$

and

$$\frac{\partial}{\partial t}\int_{\text{cv}} w\rho \, d\mathcal{V}^{\,0 \text{ (flow is steady)}} + \int_{\text{cs}} w\rho\mathbf{V}\cdot\hat{\mathbf{n}} \, dA = \sum F_z$$

or

$$u_2\rho A_2 V_2 - u_1\rho A_1 V_1 = \sum F_x \tag{1}$$

and

$$w_2\rho A_2 V_2 - w_1\rho A_1 V_1 = \sum F_z \tag{2}$$

where $\mathbf{V} = u\hat{\mathbf{i}} + w\hat{\mathbf{k}}$, and $\sum F_x$ and $\sum F_z$ are the net x and z components of force acting on the contents of the control volume. Depending on the particular flow situation being considered and the coordinate system chosen, the x and z components of velocity, u and w, can be positive, negative, or zero. In this example, the flow is in the positive direction at both the inlet and the outlet.

The water enters and leaves the control volume as a free jet at atmospheric pressure. Hence, there is atmospheric pressure surrounding the entire control volume, and the net pressure force on the control volume surface is zero. If we neglect the weight of the water and vane, the only forces applied to the control volume contents are the horizontal and vertical components of the anchoring force, F_{Ax} and F_{Az}, respectively.

With negligible gravity and viscous effects, and since $p_1 = p_2$, the speed of the fluid remains constant so that $V_1 = V_2 = 3.1$ m/s (see the Bernoulli equation, Eq. 4.7). Hence, at section (1), $u_1 = V_1$, $w_1 = 0$, and at section (2), $u_2 = V_1\cos\theta$, $w_2 = V_1\sin\theta$. Using this information, Eqs. 1 and 2 can be written as

$$V_1\cos\theta\,\rho A_2 V_1 - V_1\rho A_1 V_1 = F_{Ax} \tag{3}$$

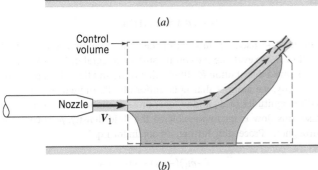

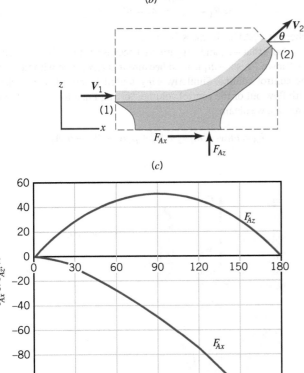

FIGURE E5.10

and

$$V_1 \sin \theta \, \rho A_2 V_1 - 0\rho A_1 V_1 = F_{Az} \tag{4}$$

Equations 3 and 4 can be simplified by using conservation of mass, which states that for this incompressible flow $A_1 V_1 = A_2 V_2$, or $A_1 = A_2$ because $V_1 = V_2$. Thus,

$$F_{Ax} = -\rho A_1 V_1^2 + \rho A_1 V_1^2 \cos \theta = -\rho A_1 V_1^2 (1 - \cos \theta) \tag{5}$$

and

$$F_{Az} = \rho A_1 V_1^2 \sin \theta \tag{6}$$

With the given data, we obtain

$$F_{Ax} = -(1000 \text{ kg/m}^3)(5.57 \times 10^{-3} \text{ m}^2)(3.1 \text{ m/s})^2 (1 - \cos \theta)$$
$$= -53.52(1 - \cos \theta) \text{ N}$$

and

$$F_{Az} = (1000 \text{ kg/m}^3)(5.57 \times 10^{-3} \text{ m}^2)(3.1 \text{ m/s})^2 \sin \theta$$
$$= 53.52 \sin \theta \text{ N} \tag{Ans}$$

Comments The values of F_{Ax} and F_{Az} as a function of θ are shown in **Fig. E5.10d**. Note that if $\theta = 0$ (i.e., the vane does not turn the water), the anchoring force is zero. The inviscid fluid merely slides along the vane without putting any force on it. If $\theta = 90°$, then $F_{Ax} = -53.52$ N and $F_{Az} = 53.52$ N. It is necessary to push on the vane (and, hence, for the vane to push on the water) to the left (F_{Ax} is negative) and up in order to change the direction of flow of the water from horizontal to vertical. This momentum change requires a force. If $\theta = 180°$, the water jet is turned back on itself. This requires no vertical force ($F_{Az} = 0$), but the horizontal force ($F_{Ax} = -107.1$ N) is two times that required if $\theta = 90°$. This horizontal fluid momentum change requires a horizontal force only.

Note that the anchoring force (Eqs. 5 and 6) can be written in terms of the mass flowrate, $\dot{m} = \rho A_1 V_1$, as

$$F_{Ax} = -\dot{m} V_1 (1 - \cos \theta)$$

and

$$F_{Az} = \dot{m} V_1 \sin \theta$$

In this example, exerting a force on a fluid flow resulted in a change in its direction only (i.e., change in its linear momentum).

The Wide World of Fluids

Where the plume goes

Commercial airliners have wheel brakes very similar to those on highway vehicles. In fact, antilock brakes now found on most new cars were first developed for use on airplanes. However, when landing, the major braking force comes from the engine rather than the wheel brakes. Upon touchdown, a piece of engine cowling translates aft and blocker doors drop down, directing the engine airflow into a honeycomb structure called a cascade. The cascade reverses the direction of the high-speed engine exhausts by nearly 180° so that it flows forward. As predicted by the *momentum equation*, the exhaust passing through this system produces a substantial braking force—the reverse thrust. Designers must know the flow pattern of the exhaust plumes to eliminate potential problems. For example, the plumes of hot exhaust must be kept away from parts of the aircraft where repeated heating and cooling could cause premature fatigue. Also, the plumes must not reenter the engine inlet, blow debris from the runway in front of the engine, or envelop the vertical tail.

EXAMPLE 5.11 | Linear Momentum—Weight, Pressure, and Change in Speed

Given As shown in **Fig. E5.11a**, water flows through a nozzle attached to the end of a laboratory sink faucet with a flowrate of 0.6 liters/s. The nozzle inlet and exit diameters are 16 mm and 5 mm, respectively, and the nozzle axis is vertical. The mass of the nozzle is 0.1 kg. The pressure at section (1) is 464 kPa.

Find Determine the anchoring force required to hold the nozzle in place.

Solution The anchoring force sought is the reaction force between the faucet and nozzle threads. To evaluate this force, we select a control volume that includes the entire nozzle and the water contained in the nozzle at an instant, as is indicated in Figs. E5.11a and E5.11b. All of the vertical forces acting on the contents of this control volume are identified in **Fig. E5.11b**. The action of atmospheric pressure cancels out in every direction and is not shown. Gage pressure forces do not cancel out in the vertical direction and are shown. Application of the vertical, or z direction, component of Eq. 5.22 to the contents of this control volume leads to

$$\frac{\partial}{\partial t}\int_{cv} w\rho \, d\mathcal{V} + \int_{cs} w\rho \mathbf{V} \cdot \hat{\mathbf{n}} \, dA = F_A - \mathcal{W}_n - p_1 A_1 \\ - \mathcal{W}_w + p_2 A_2 \qquad (1)$$

where the term 0 (flow is steady) applies over the first integral.

where w is the z direction component of fluid velocity, and the various parameters are identified in the figure.

Note that the positive direction is considered "up" for the forces. We will use this same sign convention for the fluid velocity, w, in Eq. 1. In Eq. 1, the dot product, $\mathbf{V} \cdot \hat{\mathbf{n}}$, is "+" for flow out of the control volume and "−" for flow into the control volume. For this particular example

$$\mathbf{V} \cdot \hat{\mathbf{n}} \, dA = \pm |w| \, dA \qquad (2)$$

with the "+" used for flow out of the control volume and "−" used for flow in. To evaluate the control surface integral in Eq. 1, we need to assume a distribution for fluid velocity, w, and fluid density, ρ. For simplicity, we assume that w is uniformly distributed or constant, with magnitudes of w_1 and w_2 over cross-sectional areas A_1 and A_2. Also, this flow is incompressible so the fluid density, ρ, is constant throughout. Proceeding further, we obtain for Eq. 1

$$(-\dot{m}_1)(-w_1) + \dot{m}_2(-w_2) \\ = F_A - \mathcal{W}_n - p_1 A_1 - \mathcal{W}_w + p_2 A_2 \qquad (3)$$

where $\dot{m} = \rho A V$ is the mass flowrate.

Note that $-w_1$ and $-w_2$ are used because both of these velocities are "down." Also, $-\dot{m}_1$ is used because it is associated with flow into the control volume. Similarly, $+\dot{m}_2$ is used because it is associated with flow out of the control volume. Solving Eq. 3 for the anchoring force, F_A, we obtain

$$F_A = \dot{m}_1 w_1 - \dot{m}_2 w_2 + \mathcal{W}_n + p_1 A_1 + \mathcal{W}_w - p_2 A_2 \qquad (4)$$

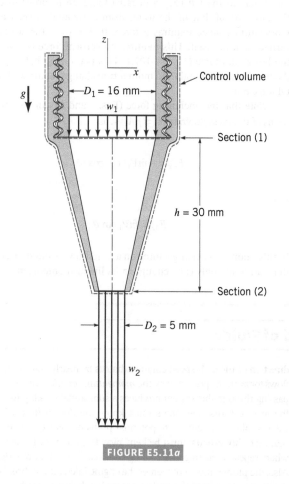

FIGURE E5.11a

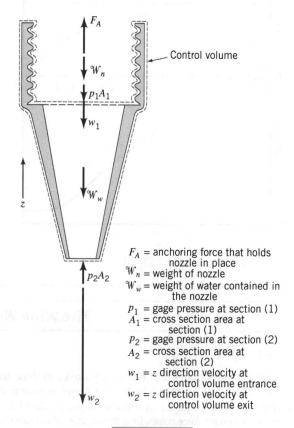

F_A = anchoring force that holds nozzle in place
$\mathcal{W}_n$ = weight of nozzle
$\mathcal{W}_w$ = weight of water contained in the nozzle
p_1 = gage pressure at section (1)
A_1 = cross section area at section (1)
p_2 = gage pressure at section (2)
A_2 = cross section area at section (2)
w_1 = z direction velocity at control volume entrance
w_2 = z direction velocity at control volume exit

FIGURE E5.11b

From the conservation of mass equation, Eq. 5.12, we obtain

$$\dot{m}_1 = \dot{m}_2 = \dot{m} \qquad (5)$$

which when combined with Eq. 4 gives

$$F_A = \dot{m}(w_1 - w_2) + \mathcal{W}_n + p_1 A_1 + \mathcal{W}_w - p_2 A_2 \qquad (6)$$

It is instructive to note how the anchoring force is affected by the different actions involved. As expected, the nozzle weight, $\mathcal{W}_n$, the water weight, $\mathcal{W}_w$, and gage pressure force at section (1), $p_1 A_1$, all increase the anchoring force, while the gage pressure force at section (2), $p_2 A_2$, acts to decrease the anchoring force. The change in the vertical momentum flowrate, $\dot{m}(w_1 - w_2)$, will, in this instance, decrease the anchoring force because this change is negative $(w_2 > w_1)$.

To complete this example, we use quantities given in the problem statement to quantify the terms on the right-hand side of Eq. 6.

From Eq. 5.6,

$$\begin{aligned}
\dot{m} &= \rho w_1 A_1 = \rho Q \\
&= (999 \text{ kg/m}^3)(0.6 \text{ liter/s})(10^{-3} \text{ m}^3/\text{liter}) \\
&= 0.599 \text{ kg/s} \qquad (7)
\end{aligned}$$

and

$$\begin{aligned}
w_1 &= \frac{Q}{A_1} = \frac{Q}{\pi(D_1^2/4)} \\
&= \frac{(0.6 \text{ liter/s})(10^{-3} \text{ m}^3/\text{liter})}{\pi(16 \text{ mm})^2/4(1000^2 \text{ mm}^2/\text{m}^2)} = 2.98 \text{ m/s} \qquad (8)
\end{aligned}$$

Also from Eq. 5.6,

$$\begin{aligned}
w_2 &= \frac{Q}{A_2} = \frac{Q}{\pi(D_2^2/4)} \\
&= \frac{(0.6 \text{ liter/s})(10^{-3} \text{ m}^3/\text{liter})}{\pi(5 \text{ mm})^2/4(1000^2 \text{ mm}^2/\text{m}^2)} = 30.6 \text{ m/s} \qquad (9)
\end{aligned}$$

The weight of the nozzle, $\mathcal{W}_n$, can be obtained from the nozzle mass, m_n, with

$$\mathcal{W}_n = m_n g = (0.1 \text{ kg})(9.81 \text{ m/s}^2) = 0.981 \text{ N} \qquad (10)$$

The weight of the water in the control volume, $\mathcal{W}_w$, can be obtained from the water density, ρ, and the volume of water, $\mathcal{V}_w$, in the truncated cone of height h. That is,

$$\mathcal{W}_w = \rho \mathcal{V}_w g$$

where

$$\begin{aligned}
\mathcal{V}_w &= \frac{1}{12}\pi h (D_1^2 + D_2^2 + D_1 D_2) \\
&= \frac{1}{12}\pi \frac{(30 \text{ mm})}{(1000 \text{ mm/m})} \\
&\quad \times \left[\frac{(16 \text{ mm})^2 + (5 \text{ mm})^2 + (16 \text{ mm})(5 \text{ mm})}{(1000^2 \text{ mm}^2/\text{m}^2)} \right] \\
&= 2.84 \times 10^{-6} \text{ m}^3
\end{aligned}$$

Thus,

$$\begin{aligned}
\mathcal{W}_w &= (999 \text{ kg/m}^3)(2.84 \times 10^{-6} \text{ m}^3)(9.81 \text{ m/s}^2) \\
&= 0.0278 \text{ N} \qquad (11)
\end{aligned}$$

The gage pressure at section (2), p_2, is zero since, as discussed in Section 4.6.1, when a subsonic flow discharges to the atmosphere as in the present situation, the discharge pressure is essentially atmospheric. The anchoring force, F_A, can now be determined from Eqs. 6–11 with

$$\begin{aligned}
F_A &= (0.599 \text{ kg/s})(2.98 \text{ m/s} - 30.6 \text{ m/s}) + 0.981 \text{ N} \\
&\quad + (464 \text{ kPa})(1000 \text{ Pa/kPa}) \frac{\pi(16 \text{ mm})^2}{4(1000^2 \text{ mm}^2/\text{m}^2)} \\
&\quad + 0.0278 \text{ N} - 0
\end{aligned}$$

or

$$\begin{aligned}
F_A &= -16.5 \text{ N} + 0.981 \text{ N} + 93.3 \text{ N} + 0.0278 \text{ N} \\
&= 77.8 \text{ N} \qquad (Ans)
\end{aligned}$$

Since the anchoring force, F_A, is positive, it acts upward in the z direction. The nozzle would be pushed off the pipe if it were not fastened securely.

Comment The control volume selected to solve problems such as these is not unique. The following is an alternate solution that involves two other control volumes—one containing only the nozzle and the other containing only the water in the nozzle. These control volumes are shown in **Figs. E5.11c** and **E5.11d** along with the vertical forces acting on the contents of each control volume. The new force involved, R_z, represents the interaction between the water and the conical inside surface of the nozzle. It includes the net pressure and viscous forces at this interface.

Application of Eq. 5.22 to the contents of the control volume of Fig. E5.11c leads to

$$F_A = \mathcal{W}_n + R_z - p_{\text{atm}}(A_1 - A_2) \qquad (12)$$

The term $p_{\text{atm}}(A_1 - A_2)$ is the resultant force from the atmospheric pressure acting on the exterior surface of the nozzle (i.e., that portion of the surface of the nozzle that is not in contact with the water). Recall that the pressure force on a curved surface (such as the exterior surface of the nozzle) is equal to the pressure times the projection of the surface area on a plane perpendicular to the axis of the nozzle. The projection of this area on a plane perpendicular to the z direction is $A_1 - A_2$. The effect of the atmospheric pressure on the internal area (between the nozzle and the water) is already included in R_z, which represents the net force on this area.

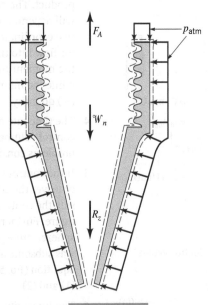

FIGURE E5.11c

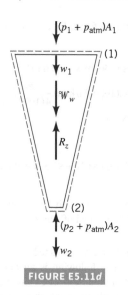

$(p_1 + p_{atm})A_1$

(1)

w_1

$\mathcal{W}_w$

R_z

(2)

$(p_2 + p_{atm})A_2$

w_2

FIGURE E5.11d

Similarly, for the control volume of Fig. E5.11d, we obtain

$$R_z = \dot{m}(w_1 - w_2) + \mathcal{W}_w + (p_1 + p_{atm})A_1 - (p_2 - p_{atm})A_2 \quad (13)$$

where p_1 and p_2 are gage pressures. From Eq. 13, it is clear that the value of R_z depends on the value of the atmospheric pressure, p_{atm}, since $A_1 \neq A_2$. That is, we must use absolute pressure, not gage pressure, to obtain the correct value of R_z. From Eq. 13, we can easily identify which forces acting on the flowing fluid change its velocity magnitude and thus linear momentum.

By combining Eqs. 12 and 13, we obtain the same result for F_A as before (Eq. 6):

$$F_A = \dot{m}(w_1 - w_2) + \mathcal{W}_n + p_1 A_1 + W_w - p_2 A_2$$

Note that although the force between the fluid and the nozzle wall, R_z, is a function of p_{atm}, the anchoring force, F_A, is not. That is, we were correct in using gage pressure when solving for F_A by means of the original control volume shown in Fig. E5.11b.

$\mathbf{V}\rho\mathbf{V} \cdot \hat{\mathbf{n}} > 0$

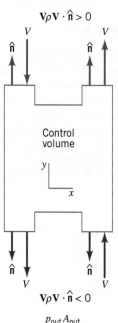

V V

$\hat{\mathbf{n}}$ $\hat{\mathbf{n}}$

Control
volume

$\hat{\mathbf{n}}$ $\hat{\mathbf{n}}$

V V

$\mathbf{V}\rho\mathbf{V} \cdot \hat{\mathbf{n}} < 0$

$p_{out}A_{out}$

V_{out}

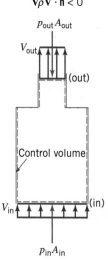

(out)

Control volume

V_{in} (in)

$p_{in}A_{in}$

Several important generalities about the application of the linear momentum equation (Eq. 5.22) are apparent in the example just considered.

1. When the flow is uniformly distributed over a section of the control surface where flow into or out of the control volume occurs, the integral operations are simplified. Thus, one-dimensional flows are easier to work with than flows involving nonuniform velocity distributions.

2. Linear momentum is directional; it can have components in as many as three orthogonal coordinate directions. Furthermore, along any one coordinate, the linear momentum of a fluid particle can be in the positive or negative direction and thus be considered as a positive or a negative quantity. In Example 5.11, only the linear momentum in the z direction was considered (all of it was in the negative z direction and was hence treated as being negative).

3. The flow of positive or negative linear momentum *into* a control volume involves a negative $\mathbf{V} \cdot \hat{\mathbf{n}}$ product. Momentum flow *out* of the control volume involves a positive $\mathbf{V} \cdot \hat{\mathbf{n}}$ product. The correct algebraic sign (+ or −) to assign to momentum flow ($\mathbf{V}\rho\mathbf{V} \cdot \hat{\mathbf{n}} \, dA$) will depend on the sense of the velocity (+ in positive coordinate direction, − in negative coordinate direction) and the $\mathbf{V} \cdot \hat{\mathbf{n}}$ product (+ for flow out of the control volume, − for flow into the control volume). This is shown in the adjacent figure. In Example 5.11, the momentum flow into the control volume past section (1) was a positive (+) quantity, while the momentum flow out of the control volume at section (2) was a negative (−) quantity.

4. The time rate of change of the linear momentum of the contents of a nondeforming control volume (i.e., $\partial/\partial t \int_{cv} \mathbf{V}\rho \, d\forall$) is zero for steady flow. Almost all of the momentum problems considered in this text involve steady flow.

5. If the control surface is selected so that it is perpendicular to the flow where fluid enters or leaves the control volume, the surface force exerted at these locations by fluid outside the control volume on fluid inside will be due to pressure, as shown in the adjacent figure. Furthermore, when subsonic flow exits from a control volume into the atmosphere, atmospheric pressure prevails at the exit cross section. In Example 5.11, the flow was subsonic and so we set the exit flow pressure at the atmospheric level. The continuity equation (Eq. 5.12) allowed us to evaluate the fluid flow velocities w_1 and w_2 at sections (1) and (2).

6. The forces due to atmospheric pressure acting on the control surface may require consideration as indicated by Eq. 13 in Example 5.11 for the reaction force between the nozzle and the fluid. When calculating the anchoring force, F_A, the forces due to atmospheric pressure on the control surface cancel each other (for example, after

combining Eqs. 12 and 13, the atmospheric pressure forces are no longer involved) and gage pressures may be used.

7. The external forces have an algebraic sign—positive if the force is in the assigned positive coordinate direction and negative otherwise.

8. Only external forces acting on the contents of the control volume are considered in the linear momentum equation (Eq. 5.22). If the fluid alone is included in a control volume, reaction forces between the fluid and the surface, or surfaces, in contact with the fluid [wetted surface(s)] will need to be in Eq. 5.22. If the fluid and the wetted surface, or surfaces, are within the control volume, the reaction forces between fluid and the wetted surface(s) do not appear in the linear momentum equation (Eq. 5.22) because they are internal, not external, forces. The anchoring force that holds the wetted surface(s) in place is an external force, however, and must therefore be in Eq. 5.22.

9. The force required to anchor an object will generally exist in response to surface pressure and/or shear forces acting on the control surface, to a change in linear momentum flow through the control volume containing the object, and to the weight of the object and the fluid contained in the control volume. In Example 5.11 the nozzle anchoring force was required mainly because of pressure forces and partly because of a change in linear momentum flow associated with accelerating the fluid in the nozzle. The weight of the water and the nozzle contained in the control volume influenced the size of the anchoring force only slightly.

> A control volume diagram is similar to a free-body diagram.

The Wide World of Fluids

Motorized surfboard

When Bob Montgomery, a former professional surfer, started to design his motorized surfboard (called a jet board), he discovered that there were many engineering challenges. The idea is to provide surfing to anyone, no matter where they live, near or far from the ocean. The rider stands on the device like a surfboard and steers it like a surfboard by shifting his or her body weight. A new, sleek,

compact 33.57 kW engine and pump was designed to fit within the surfboard hull. Thrust is produced in response to the change in *linear momentum* of the water stream as it enters through the inlet passage and exits through an appropriately designed nozzle. Some of the fluid dynamic problems associated with designing the craft included one-way valves so that water does not get into the engine (at both the intake or exhaust ports), buoyancy, hydrodynamic lift, drag, thrust, and hull stability.

To further demonstrate the use of the linear momentum equation (Eq. 5.22), we consider another one-dimensional flow example before moving on to other facets of this important equation.

EXAMPLE 5.12 | Linear Momentum—Pressure and Change in Flow Direction

Given Water flows through a horizontal, 180° pipe bend, as shown in **Fig. E5.12a** and illustrated in **Fig. E5.12b**. The flow cross-sectional area is constant at a value of 9.3×10^{-3} m^2 through the bend. The magnitude of the flow velocity everywhere in the bend is axial and 15.2 m/s. The absolute pressures at the entrance and exit of the bend are 206843 Pa and 165474 Pa, respectively. (Use $p_{atm} = 101353$ Pa)

Find Calculate the horizontal (x and y) components of the anchoring force required to hold the bend in place.

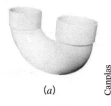

(a)

Camplas

FIGURE E5.12

Solution To determine the components of the anchoring force required to hold the pipe bend in place, a control volume is defined that contains the bend and the water in the bend at an instant (see dashed line in **Fig. E5.12c**). The horizontal forces acting on the contents of this control volume are identified in Fig. E5.12c. Note that the weight of the water is vertical (in the negative z direction) and does not contribute to the x and y components of the anchoring force. All of the horizontal normal and tangential forces exerted on the fluid and the pipe bend are resolved and combined into the two resultant components, F_{Ax} and F_{Ay}. The only portions of the control surface exposed to a pressure other than atmospheric are the inlet and exit planes on which the normal to the surface has no component in the x-direction. Therefore, the x-component of the linear momentum equation can be expressed as

$$\int_{cs} u\rho \mathbf{V} \cdot \hat{\mathbf{n}} \, dA = F_{Ax} \tag{1}$$

To be clear, F_{Ax} is the force exerted on the mass in the control volume that is comprised of both the fluid and the pipe bend. At sections (1) and (2), the flow is in the y direction, and therefore, $u = 0$ at both cross sections. There is no x direction momentum flow into or out of the control volume, and we conclude from Eq. 1 that

$$F_{Ax} = 0 \tag{Ans}$$

For the y direction, we get from Eq. 5.22

$$\int_{cs} v\rho \mathbf{V} \cdot \hat{\mathbf{n}} \, dA = F_{Ay} + p_1 A_1 + p_2 A_2 \tag{2}$$

For one-dimensional flow, the surface integral in Eq. 2 is easy to evaluate, and Eq. 2 becomes

$$(+v_1)(-\dot{m}_1) + (-v_2)(+\dot{m}_2) = F_{Ay} + p_1 A_1 + p_2 A_2 \tag{3}$$

Note that the y component of velocity is positive at section (1) but is negative at section (2). Also, the mass flowrate term is negative at section (1) (flow in) and is positive at section (2) (flow out). From the continuity equation (Eq. 5.12), we get

$$\dot{m} = \dot{m}_1 = \dot{m}_2 \tag{4}$$

and thus Eq. 3 can be written as

$$-\dot{m}(v_1 + v_2) = F_{Ay} + p_1 A_1 + p_2 A_2 \tag{5}$$

Solving Eq. 5 for F_{Ay}, we obtain

$$F_{Ay} = -\dot{m}(v_1 + v_2) - p_1 A_1 - p_2 A_2 \tag{6}$$

From the given data we can calculate the mass flowrate, $\dot{m}$, from Eq. 5.6 as

$$\dot{m} = \rho_1 A_1 v_1 = (1000 \text{ kg/m}^3)(9.3 \times 10^{-3} \text{ m}^2)(15.2 \text{ m/s})$$
$$= 141.36 \text{ kg/s}$$

For determining the anchoring force, F_{Ay}, the effects of atmospheric pressure cancel, and thus gage pressures for p_1 and p_2 are appropriate. By substituting numerical values of variables into Eq. 6, we get

$$F_{Ay} = -(141.36 \text{ kg/s})(15.2 \text{ m/s} + 15.2 \text{ m/s})$$
$$- (206843 \text{ Pa} - 101353 \text{ Pa})(9.3 \times 10^{-3} \text{ m}^2)$$
$$- (165474 \text{ Pa} - 101353 \text{ Pa})(9.3 \times 10^{-3} \text{ m}^2)$$

$$F_{Ay} = -4297 \text{ N} - 981 \text{ N} - 596 \text{ N} = -5874 \text{ N} \tag{Ans}$$

The negative sign for F_{Ay} is interpreted as meaning that the y component of the anchoring force is actually in the negative y direction, not the positive y direction as originally indicated in Fig. E5.12c.

Comment As with Example 5.11, the anchoring force for the pipe bend is independent of the atmospheric pressure. However, the force that the bend exerts on the fluid inside of it, R_y, depends on the atmospheric pressure. We can see this by using a control volume that surrounds only the fluid within the bend, as shown in **Fig. E5.12d**. Application of the momentum equation to this situation gives

$$R_y = -\dot{m}(v_1 + v_2) - p_1 A_1 - p_2 A_2$$

where p_1 and p_2 must be in terms of absolute pressure because the force between the fluid and the pipe wall, R_y, is the complete pressure effect (i.e., absolute pressure). We see that forces exerted on the flowing fluid result in a change in its velocity direction (a change in linear momentum).

Thus, we obtain

$$R_y = -(141.36 \text{ kg/s})(15.2 \text{ m/s} + 15.2 \text{ m/s})$$
$$- (206843 \text{ Pa})(9.3 \times 10^{-3} \text{ m}^2) \tag{7}$$
$$- (165474 \text{ Pa})(9.3 \times 10^{-3} \text{ m}^2)$$
$$= -7760 \text{ N}$$

We can use the control volume that includes just the pipe bend (without the fluid inside it), as shown in **Fig. E5.12e**, to

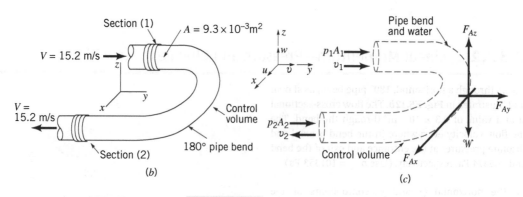

Section (1)
$A = 9.3 \times 10^{-3} \text{m}^2$
$V = 15.2$ m/s
z
$V = 15.2$ m/s
x
y
Section (2)
Control volume
180° pipe bend

(b)

z
w
u v y
x
Pipe bend and water
F_{Az}
$p_1 A_1$
v_1
$p_2 A_2$
v_2
F_{Ay}
W
Control volume
F_{Ax}

(c)

FIGURE E5.12 (Continued)

determine F_{Ay}, the anchoring force component in the y direction necessary to hold the bend stationary. The y component of the momentum equation applied to this control volume gives

$$F_{Ay} = R_y + p_{atm}(A_1 + A_2) \tag{8}$$

where R_y is given by Eq. 7. The $p_{atm}(A_1 + A_2)$ term represents the net pressure force on the outside portion of the control volume. Recall that the pressure force on the inside of the

bend is accounted for by R_y. By combining Eqs. 7 and 8 and we obtain

$$F_{Ay} = -7760 \text{ N} + 101353 \text{ Pa}(9.3 \times 10^{-3} \text{ m}^2 + 9.3 \times 10^{-3} \text{ m}^2)$$

$$= -5874 \text{ N}$$

in agreement with the original answer obtained using the control volume of Fig. E5.12c.

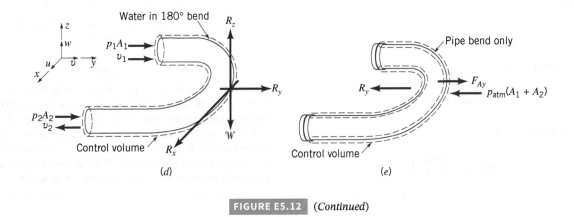

FIGURE E5.12 (Continued)

In Examples 5.10 and 5.12 the force exerted on a flowing fluid resulted in a change in flow direction only. This force was associated with constraining the flow with a vane in Example 5.10 and with a pipe bend in Example 5.12. In Example 5.11 the force exerted on a flowing fluid resulted in a change in velocity magnitude only. This force was associated with a converging nozzle. Anchoring forces are required to hold a vane or conduit stationary. They are most easily estimated with a control volume that contains the vane or conduit and the flowing fluid involved. Alternately, two separate control volumes can be used, one containing the vane or conduit only and one containing the flowing fluid only.

 Video

V5.8 Fire hose

EXAMPLE 5.13 | Linear Momentum—Pressure, Change in Speed, and Friction

Given Air flows steadily between two cross sections in a long, straight portion of 0.10 m.-inside-diameter pipe, as indicated in **Fig. E5.13**, where the uniformly distributed temperature and pressure at each cross section are given. If the average air velocity at section (2) is 305 m/s, we found in Example 5.2 that the average air velocity at section (1) must be 66.81 m/s. Assume uniform velocity distributions at sections (1) and (2).

Find Determine the frictional force exerted by the pipe wall on the airflow between sections (1) and (2).

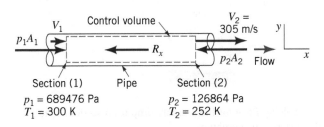

Section (1) **Pipe** **Section (2)**
$p_1 = 689476$ Pa $p_2 = 126864$ Pa
$T_1 = 300$ K $T_2 = 252$ K

FIGURE E5.13

Solution The control volume of Example 5.2 is appropriate for this problem. The forces acting on the air between sections (1) and (2) are identified in Fig. E5.13. The weight of air is considered to be negligibly small. The reaction force between the wetted wall of the pipe and the flowing air, R_x, is the frictional force sought. Application of the axial component of Eq. 5.22 to this control volume yields

$$\int_{cs} u\rho \mathbf{V} \cdot \hat{\mathbf{n}} \, dA = -R_x + p_1 A_1 - p_2 A_2 \tag{1}$$

The positive x direction is set as being to the right. Furthermore, for uniform velocity distributions (one-dimensional flow), Eq. 1 becomes

$$(+u_1)(-\dot{m}_1) + (+u_2)(+\dot{m}_2) = -R_x + p_1 A_1 - p_2 A_2 \tag{2}$$

From conservation of mass (Eq. 5.12), we get

$$\dot{m} = \dot{m}_1 = \dot{m}_2 \tag{3}$$

so that Eq. 2 becomes

$$\dot{m}(u_2 - u_1) = -R_x + A_2(p_1 - p_2) \tag{4}$$

Solving Eq. 4 for R_x, we get

$$R_x = A_2(p_1 - p_2) - \dot{m}(u_2 - u_1) \tag{5}$$

The equation of state gives

$$\rho_2 = \frac{p_2}{RT_2} \tag{6}$$

and the equation for area, A_2, is

$$A_2 = \pi D_2^2 / 4 \tag{7}$$

Thus, from Eqs. 3, 6, and 7

$$\dot{m} = \left(\frac{p_2}{RT_2}\right)\left(\frac{\pi D_2^2}{4}\right)u_2$$

Hence, using $R = 287$ J/kg·K

$$\dot{m} = \frac{126864\,\text{Pa}}{[287\,\text{J/kg·K}](252\text{K})}$$

$$\times \frac{\pi(0.10\,\text{m})^2}{4}\ 305\ \text{m/s} = 4.20\ \text{kg/s} \tag{8}$$

Thus, from Eqs. 5 and 8

$$R_x = \frac{\pi(0.10\,\text{m})^2}{4}\left(689476\ \text{Pa} - 126864\ \text{Pa}\right)$$

$$-4.20\ \text{kg/s}\ (305\ \text{m/s} - 66.81\ \text{m/s})$$
$$= 4419\ \text{N} - 1000\ \text{N}$$

or

$$R_x = 3419\ \text{N} \qquad (Ans)$$

Comment If this were an incompressible flow, the force due to the pressure difference would be equal to the frictional force opposing the flow. The flow would not accelerate. For this compressible flow, the flow must accelerate because the density decreases as the pressure decreases. Therefore, for this compressible flow, the force due to the applied pressure difference must be sufficient to accelerate the flow and overcome the opposing frictional force.

EXAMPLE 5.14 | Linear Momentum—Weight, Pressure, Friction, and Nonuniform Velocity Profile

Given Consider the flow of Example 5.4 to be vertically upward, as shown by **Fig. E5.14**.

Find Develop an expression for the fluid pressure drop that occurs between sections (1) and (2).

Solution A control volume (see dashed lines in Fig. E5.14) that includes only fluid from section (1) to section (2) is selected. The forces acting on the fluid in this control volume are identified in Fig. E5.14. The application of the axial component of Eq. 5.22 to the fluid in this control volume results in

$$\int_{cs} w\rho\mathbf{V}\cdot\hat{\mathbf{n}}\,dA = p_1A_1 - R_z - \mathcal{W} - p_2A_2 \tag{1}$$

where R_z is the resultant force of the wetted pipe wall on the fluid. Furthermore, for uniform flow at section (1), and because the flow at section (2) is out of the control volume, Eq. 1 becomes

$$(+w_1)(-\dot{m}_1) + \int_{A_2} (+w_2)\rho(+w_2\,dA_2) = p_1A_1 - R_z$$

$$- \mathcal{W} - p_2A_2 \tag{2}$$

The positive direction is considered up. The surface integral over the cross-sectional area at section (2), A_2, is evaluated by using the parabolic velocity profile obtained in Example 5.4, $w_2 = 2w_1[1 - (r/R)^2]$, as

$$\int_{A_2} w_2\rho w_2\,dA_2 = \rho\int_0^R w_2^2\,2\pi r\,dr$$

$$= 2\pi\rho\int_0^R (2w_1)^2\left[1 - \left(\frac{r}{R}\right)^2\right]^2 r\,dr$$

or

$$\int_{A_2} w_2\rho w_2\,dA_2 = 4\pi\rho w_1^2 \frac{R^2}{3} \tag{3}$$

Combining Eqs. 2 and 3, we obtain

$$-w_1^2\rho\pi R^2 + \frac{4}{3}w_1^2\rho\pi R^2 = p_1A_1 - R_z - \mathcal{W} - p_2A_2 \tag{4}$$

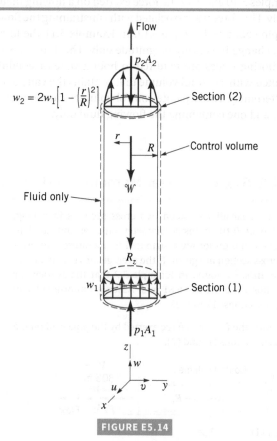

FIGURE E5.14

Solving Eq. 4 for the pressure drop from section (1) to section (2), $p_1 - p_2$, we obtain

$$p_1 - p_2 = \frac{\rho w_1^2}{3} + \frac{R_z}{A_1} + \frac{\mathcal{W}}{A_1} \qquad (Ans)$$

Comment We see that the drop in pressure from section (1) to section (2) occurs because of the following:

1. The change in momentum flow between the two sections associated with going from a uniform velocity profile to a parabolic velocity profile, $\rho w_1^2/3$
2. Pipe wall friction, R_z
3. The weight of the water column, $\mathcal{W}$; a hydrostatic pressure effect

If the velocity profiles had been identically parabolic at sections (1) and (2), the momentum flowrate at each section would have been identical, a condition we call "fully developed" flow. Then the pressure drop, $p_1 - p_2$, would be due only to pipe wall friction and the weight of the water column. If in addition to being fully developed, the flow involved negligible weight effects (for example, horizontal flow of liquids or the flow of gases in any direction), the drop in pressure between any two sections, $p_1 - p_2$, would be a result of pipe wall friction only.

In Example 5.14, the average velocity is the same at section (1) as it is at section (2) ($\overline{V}_1 = \overline{V}_2 = w_1$), but the momentum flux across section (1) is not the same as it is across section (2). If it were, the left-hand side of Eq. 4 would be zero. For a nonuniform flow the momentum flux can be written in terms of the average velocity, $\overline{V}$, and the *momentum coefficient*, β,

> **Momentum flux in nonuniform flow can be evaluated using average velocity and momentum coefficient.**

$$\beta = \frac{\displaystyle\int w\rho \mathbf{V} \cdot \hat{\mathbf{n}}\, dA}{\rho \overline{V}^2 A}$$

Hence, the momentum flux can be written as

$$\int_{cs} w\rho \mathbf{V} \cdot \hat{\mathbf{n}}\, dA = -\beta_1 w_1^2 \rho \pi R^2 + \beta_2 w_1^2 \rho \pi R^2$$

For laminar flow in a pipe, $\beta = 4/3$, and for turbulent flow in a pipe, $\beta \approx 1.02$. For uniform flow, $\beta = 1$.

EXAMPLE 5.15 | Linear Momentum—Thrust

Given A static thrust stand as sketched in **Fig. E5.15** is to be designed for testing a jet engine. The following conditions are known for a typical test: Intake air velocity = 200 m/s; exhaust gas velocity = 500 m/s; intake cross-sectional area = 1 m²; intake static pressure = −22.5 kPa = 78.5 kPa (abs); intake static temperature = 268 K; exhaust static pressure = 0 kPa = 101 kPa (abs).

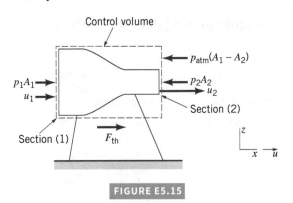

Control volume

$p_{atm}(A_1 - A_2)$

$p_1 A_1$
u_1

$p_2 A_2$
u_2

Section (2)

Section (1) $\quad F_{th}$

FIGURE E5.15

Find Determine the anchoring force required to hold the engine stationary.

Solution The cylindrical control volume outlined with a dashed line in Fig. E5.15 is selected. The external forces acting in the axial direction are also shown. Application of the momentum equation (Eq. 5.22) to the contents of this control volume yields

$$\int_{cs} u\rho \mathbf{V} \cdot \hat{\mathbf{n}}\, dA = p_1 A_1 + F_{th} - p_2 A_2 \\ - p_{atm}(A_1 - A_2) \tag{1}$$

where the pressures are absolute. Thus, for one-dimensional flow, Eq. 1 becomes

$$(+u_1)(-\dot{m}_1) + (+u_2)(+\dot{m}_2) = (p_1 - p_{atm})A_1 \\ - (p_2 - p_{atm})A_2 + F_{th} \tag{2}$$

The positive direction is to the right. The conservation of mass equation (Eq. 5.12) leads to

$$\dot{m} = \dot{m}_1 = \rho_1 A_1 u_1 = \dot{m}_2 = \rho_2 A_2 u_2 \tag{3}$$

Combining Eqs. 2 and 3 and using gage pressure, we obtain

$$\dot{m}(u_2 - u_1) = p_1 A_1 - p_2 A_2 + F_{th} \tag{4}$$

Solving Eq. 4 for the thrust force, F_{th}, we obtain

$$F_{th} = -p_1 A_1 + p_2 A_2 + \dot{m}(u_2 - u_1) \tag{5}$$

We need to determine the mass flowrate, $\dot{m}$, to calculate F_{th}, and to calculate $\dot{m} = \rho_1 A_1 u_1$, we need ρ_1. From the ideal gas equation of state

$$\rho_1 = \frac{p_1}{RT_1} = \frac{(78.5 \text{ kPa})(1000 \text{ Pa/kPa})[1(\text{N/m}^2)/\text{Pa}]}{(286.9 \text{ J/kg} \cdot \text{K})(268 \text{ K})(1 \text{ N} \cdot \text{m/J})}$$

$$= 1.02 \text{ kg/m}^3$$

Thus,

$$\dot{m} = \rho_1 A_1 u_1 = (1.02 \text{ kg/m}^3)(1 \text{ m}^2)(200 \text{ m/s})$$
$$= 204 \text{ kg/s} \tag{6}$$

Finally, combining Eqs. 5 and 6 and substituting given data with $p_2 = 0$, we obtain

$$F_{th} = -(1 \text{ m}^2)(-22.5 \text{ kPa})(1000 \text{ Pa/kPa})[1(\text{N/m}^2)/\text{Pa}] \\ + (204 \text{ kg/s})(500 \text{ m/s} - 200 \text{ m/s})[1 \text{ N}/(\text{kg} \cdot \text{m/s}^2)]$$

or

$$F_{th} = 22{,}500 \text{ N} + 61{,}200 \text{ N} = 83{,}700 \text{ N} \tag{Ans}$$

Comment The force of the thrust stand on the engine is directed toward the right. Conversely, the engine pushes to the left on the thrust stand (or aircraft).

The Wide World of Fluids

Bow thrusters

In the past, large ships required the use of tugboats for precise maneuvering, especially when docking. Nowadays, most large ships (and many moderate to small ones as well) are equipped with bow thrusters to help steer in close quarters. The units consist of a mechanism (usually a ducted propeller mounted at right angles to the fore/aft axis of the ship) that takes water from one side of the bow and ejects it as a water jet on the other side. The *momentum*

flux of this jet produces a thrust to assist in maneuvering the ship toward either port or starboard. Sometimes a second unit is installed in the stern. Initially used in the bows of ferries, these versatile control devices have become popular in offshore oil servicing boats, fishing vessels, and larger oceangoing craft. They permit unassisted maneuvering alongside of oil rigs, vessels, loading platforms, fishing nets, and docks. They also provide precise control at slow speeds through locks, narrow channels, and bridges, where the rudder becomes ineffective.

EXAMPLE 5.16 ┃ Linear Momentum—Nonuniform Pressure

Given　A sluice gate across a channel (**Fig. E5.16a**) of width b is shown in the closed and open positions in **Figs. E5.16b** and **E5.16c**.

Find　Is the anchoring force required to hold the gate in place larger when the gate is closed or when it is open?

(a)

FIGURE E5.16a

Solution　We will answer this question by comparing expressions for the horizontal reaction force, R_x, between the gate and the water when the gate is closed and when the gate is open. The control volume used in each case is indicated with dashed lines in Figs. E5.16b and E5.16c.

　　When the gate is closed, the horizontal forces acting on the contents of the control volume are identified in **Fig. E5.16d**. Application of Eq. 5.22 to the contents of this control volume yields

$$\int_{cs} u\rho \overset{\text{0 (no flow)}}{\cancel{\mathbf{V}}} \cdot \hat{\mathbf{n}}\, dA = \tfrac{1}{2}\gamma H^2 b - R_x \qquad (1)$$

Note that the hydrostatic pressure force, $\gamma H^2 b/2$, is used. From Eq. 1, the force exerted on the water by the gate (which is equal to the force necessary to hold the gate stationary) is

$$R_x = \tfrac{1}{2}\gamma H^2 b \qquad (2)$$

which is equal in magnitude to the hydrostatic force exerted on the gate by the water.

　　When the gate is open, the horizontal forces acting on the contents of the control volume are shown in **Fig. E5.16e**. Application of Eq. 5.22 to the contents of this control volume leads to

$$\int_{cs} u\rho \mathbf{V} \cdot \hat{\mathbf{n}}\, dA = \tfrac{1}{2}\gamma H^2 b - R_x - \tfrac{1}{2}\gamma h^2 b - F_f \qquad (3)$$

Note that because the water at sections (1) and (2) is assumed to be flowing along straight, horizontal streamlines, the pressure distribution at those locations is hydrostatic, varying from zero at the free surface to γ times the water depth at the bottom of the channel (see Chapter 4, Section 4.4). Thus, the pressure forces at sections (1) and (2) (given by the pressure at the centroid times the area) are $\gamma H^2 b/2$ and $\gamma h^2 b/2$, respectively. Also, the frictional force between the channel bottom and the water is specified as F_f. The surface integral in Eq. 3 is nonzero only where there is flow across the control surface. With the assumption of uniform velocity distributions,

$$\int_{cs} u\rho \mathbf{V} \cdot \hat{\mathbf{n}}\, dA = (u_1)\rho(-u_1)Hb + (+u_2)\rho(+u_2)hb \qquad (4)$$

Thus, Eqs. 3 and 4 combine to form

$$-\rho u_1^2 Hb + \rho u_2^2 hb = \tfrac{1}{2}\gamma H^2 b - R_x - \tfrac{1}{2}\gamma h^2 b - F_f \qquad (5)$$

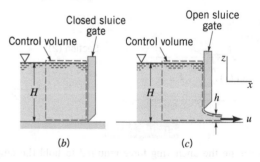

(b)　　　　　　　　　*(c)*

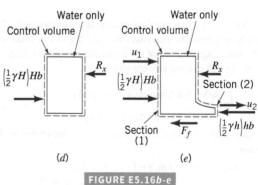

(d)　　　　　　　　　*(e)*

FIGURE E5.16b-e

If $H \gg h$, the upstream velocity, u_1, is much less than u_2, so that the contribution of the incoming momentum flow to the control surface integral can be neglected, and from Eq. 5, we obtain

$$R_x = \frac{1}{2}\gamma H^2 b - \frac{1}{2}\gamma h^2 b - F_f - \rho u_2^2 hb \qquad (6)$$

By using the continuity equation, $\dot{m} = \rho b H u_1 = \rho b h u_2$, Eq. 6 can be rewritten as

$$R_x = \frac{1}{2}\gamma H^2 b - \frac{1}{2}\gamma h^2 b - F_f - \dot{m}(u_2 - u_1) \qquad (7)$$

Hence, since $u_2 > u_1$, by comparing the expressions for R_x (Eqs. 2 and 7), we conclude that the reaction force between the gate and the water (and therefore the anchoring force required to hold the gate in place) is smaller when the gate is open than when it is closed. *(Ans)*

Comment That the anchoring force on the gate is smaller can be deduced from physical reasoning. When the gate is closed, all of the hydrostatic pressure pushes on the gate. When the gate is open, some of the hydrostatic pressure is used to push the water out.

All of the linear momentum examples considered thus far have involved stationary and nondeforming control volumes, which are thus inertial because there is no acceleration. A nondeforming control volume translating in a straight line at constant speed is also inertial because there is no acceleration. For a system and an inertial, moving, nondeforming control volume that are both coincident at an instant of time, the Reynolds transport theorem (Eq. 3.23) is

$$\frac{D}{Dt}\int_{\text{sys}} \mathbf{V}\rho \, d\mathcal{V} = \frac{\partial}{\partial t}\int_{\text{cv}} \mathbf{V}\rho \, d\mathcal{V} + \int_{\text{cs}} \mathbf{V}\rho \mathbf{W} \cdot \hat{\mathbf{n}} \, dA \qquad (5.23)$$

where $\mathbf{W}$ is the relative velocity. When we combine Eq. 5.23 with Eqs. 5.19 and 5.20, we get

$$\frac{\partial}{\partial t}\int_{\text{cv}} \mathbf{V}\rho \, d\mathcal{V} + \int_{\text{cs}} \mathbf{V}\rho \mathbf{W} \cdot \hat{\mathbf{n}} \, dA = \sum \mathbf{F}_{\substack{\text{contents of the} \\ \text{control volume}}} \qquad (5.24)$$

When the equation relating absolute, relative, and control volume velocities (Eq. 5.14) is used with Eq. 5.24, the result is

$$\frac{\partial}{\partial t}\int_{\text{cv}} (\mathbf{W} + \mathbf{V}_{\text{cv}})\rho \, d\mathcal{V} + \int_{\text{cs}} (\mathbf{W} + \mathbf{V}_{\text{cv}})\rho \mathbf{W} \cdot \hat{\mathbf{n}} \, dA = \sum \mathbf{F}_{\substack{\text{contents of the} \\ \text{control volume}}} \qquad (5.25)$$

For a constant control volume velocity, $\mathbf{V}_{\text{cv}}$, and steady flow in the control volume reference frame,

$$\frac{\partial}{\partial t}\int_{\text{cv}} (\mathbf{W} + \mathbf{V}_{\text{cv}})\rho \, dV = 0 \qquad (5.26)$$

Also, for this inertial, nondeforming control volume

$$\int_{\text{cs}} (\mathbf{W} + \mathbf{V}_{\text{cv}})\rho \mathbf{W} \cdot \hat{\mathbf{n}} \, dA = \int_{\text{cs}} \mathbf{W}\rho \mathbf{W} \cdot \hat{\mathbf{n}} \, dA + \mathbf{V}_{\text{cv}}\int_{\text{cs}} \rho \mathbf{W} \cdot \hat{\mathbf{n}} \, dA \qquad (5.27)$$

For steady flow (on an instantaneous or time-average basis), Eq. 5.15 gives

$$\int_{\text{cs}} \rho \mathbf{W} \cdot \hat{\mathbf{n}} \, dA = 0 \qquad (5.28)$$

Combining Eqs. 5.25, 5.26, 5.27, and 5.28, we conclude that the linear momentum equation for an inertial, moving, nondeforming control volume that involves steady (instantaneous or time-average) flow is

$$\boxed{\int_{\text{cs}} \mathbf{W}\rho \mathbf{W} \cdot \hat{\mathbf{n}} \, dA = \sum \mathbf{F}_{\substack{\text{contents of the} \\ \text{control volume}}}} \qquad (5.29)$$

Example 5.17 illustrates the use of Eq. 5.29.

The linear momentum equation can be written for a moving control volume.

▶ Video

V5.9 Jellyfish

▶ Video

V5.10 Running on water

The linear momentum equation for a moving control volume involves the relative velocity.

EXAMPLE 5.17 | Linear Momentum—Moving Control Volume

Given A vane on wheels moves with constant velocity, V_0, when a stream of water having a nozzle exit velocity of V_1 is turned $45°$ by the vane, as indicated in **Fig. E5.17a**. Note that this is the same moving vane considered earlier in Section 3.4.6. The speed of

the water jet leaving the nozzle is 31 m/s, and the vane is moving to the right with a constant speed of 6 m/s.

Find Determine the magnitude and direction of the force, **R**, exerted by the stream of water on the vane surface.

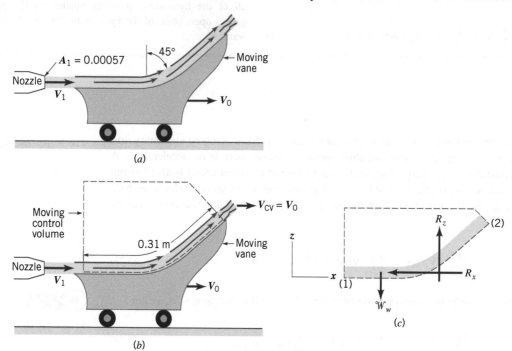

FIGURE E5.17

Solution To determine the magnitude and direction of the force, **R**, exerted by the water on the vane, we apply Eq. 5.29 to the contents of the moving control volume shown in **Fig. E5.17b**. The forces acting on the contents of this control volume are indicated in **Fig. E5.17c**. Note that since the ambient pressure is atmospheric, all pressure forces cancel each other out. Equation 5.29 is applied to the contents of the moving control volume in component directions. For the x direction (positive to the right), we get

$$\int_{cs} W_x \rho \mathbf{W} \cdot \hat{\mathbf{n}} \, dA = -R_x$$

or

$$(+W_1)(-\dot{m}_1) + (+W_2 \cos 45°)(+\dot{m}_2) = -R_x \qquad (1)$$

where

$$\dot{m}_1 = \rho_1 W_1 A_1 \quad \text{and} \quad \dot{m}_2 = \rho_2 W_2 A_2$$

For the vertical or z direction (positive up), we get

$$\int_{cs} W_z \rho \mathbf{W} \cdot \hat{\mathbf{n}} \, dA = R_z - \mathcal{W}_w$$

or

$$(+W_2 \sin 45°)(+\dot{m}_2) = R_z - \mathcal{W}_w \qquad (2)$$

We assume for simplicity that the water flow is frictionless and that the change in water elevation across the vane is negligible. Thus, from the Bernoulli equation (Eq. 4.7), we conclude that the speed of the water relative to the moving control volume, W, is constant or

$$W_1 = W_2$$

The relative speed of the stream of water entering the control volume, W_1, is

$$W_1 = V_1 - V_0 = 31 \text{ m/s} - 6 \text{ m/s} = 25 \text{ m/s} = W_2$$

The water density is constant, so that

$$\rho_1 = \rho_2 = 1000 \text{ kg/m}^3$$

Application of the conservation of mass principle to the contents of the moving control volume (Eq. 5.16) leads to

$$\dot{m}_1 = \rho_1 W_1 A_1 = \rho_2 W_2 A_2 = \dot{m}_2$$

Combining results, we get

$$R_x = \rho W_1^2 A_1 (1 - \cos 45°)$$

or

$$R_x = (1000 \text{ kg/m}^3)(25 \text{ m/s})^2 (0.00057 \text{ m}^2)(1 - \cos 45°)$$
$$= 104.3 \text{ N}$$

Also,

$$R_z = \rho W_1^2 (\sin 45°) A_1 + \mathcal{W}_w$$

where

$$\mathcal{W}_w = \rho g A_1 \ell$$

Thus,

$$R_z = (1000 \text{ kg/m}^3)(25 \text{ m/s})^2 (\sin 45°)(0.00057 \text{ m}^2)$$
$$+ (1000 \times 9.81 \text{ N/m}^3)(0.00057 \text{ m}^2 \times 0.31 \text{ m})$$
$$= 252 \text{ N} + 1.73 \text{ N} = 253.73 \text{ N}$$

Combining the components, we get

$$R = \sqrt{R_x^2 + R_z^2} = \sqrt{(104.3\ \text{N})^2 + (253.73\ \text{N})^2} = 274.3\ \text{N}$$

The angle of **R** from the x direction, α, is

$$\alpha = \tan^{-1}\frac{R_z}{R_x} = \tan^{-1}\frac{253.73}{104.3} = 67.6°$$

The force of the water on the vane is equal in magnitude but opposite in direction from **R**; thus, it points to the right and down at an angle of 67.6° from the x direction and is equal in magnitude to 268 N.

(*Ans*)

Comment The force of the fluid on the vane in the x direction, $R_x = 102$ N, is associated with the x-direction motion of the vane at a constant speed of 6 m/s. Since the vane is not accelerating, this x-direction force is opposed by wheel friction or some other resisting force of the same magnitude. From basic physics, we recall that the power generated by the water jet is the product of the component of the force in the direction of motion and its speed. Thus,

$$\dot{W} = R_x V_0$$
$$= (102\ \text{N})(6\ \text{m/s})$$
$$= 612\ \text{W}$$

It is clear from the preceding examples that forces acting on a flowing fluid may cause it to

V5.11 Thrust from water nozzle

1. change direction
2. speed up or slow down
3. undergo a change in velocity profile
4. do only some or all of the above
5. do none of the above

A net force on the fluid is required for achieving any or all of the first four previously listed results, which all represent a change in fluid momentum. The forces on a flowing fluid balance out with no net force for the fifth.

Typical forces considered in this book include

(a) pressure
(b) friction
(c) weight

and involve some type of constraint such as a vane, channel, or conduit to guide the flowing fluid. A vane, channel, or conduit can exert a net force on the fluid causing it to move or the flowing fluid may exert a net force on it causing the solid boundary to move. When this happens, power is delivered to the fluid (e.g., pump, compressor) or to the vane, channel or conduit (e.g., turbine, sail).

The selection of a control volume is an important matter. For determining anchoring forces, consider including fluid and its constraint in the control volume. For determining force between a fluid and its constraint, consider including only the fluid in the control volume.

5.2.3 Derivation of the Moment-of-Momentum Equation[1]

In many engineering problems, the moment of a force with respect to an axis, namely, *torque*, is important. Newton's second law of motion has already led to a useful relationship between forces and linear momentum flow. The linear momentum equation can also be used to solve problems involving torques. However, by forming the moment of the linear momentum and the moment of the resultant force associated with each particle of fluid with respect to a point in an inertial coordinate system, we will develop a **moment-of-momentum equation** that relates *torques* and *angular momentum flow* for the contents of a control volume. When torques are important, the moment-of-momentum equation is often more convenient to use than the linear momentum equation.

Application of Newton's second law of motion to a particle of fluid yields

$$\frac{D}{Dt}(\mathbf{V}\rho\,\delta\mathcal{V}) = \delta\mathbf{F}_{particle} \tag{5.30}$$

The moment-of-momentum equation is derived from Newton's second law.

where **V** is the particle velocity measured in an inertial reference system, ρ is the particle density, $\delta\mathcal{V}$ is the infinitesimally small particle volume, and $\delta\mathbf{F}_{particle}$ is the infinitesimal resultant

[1]This section may be omitted, along with Sections 5.2.4 and 5.3.6, without loss of continuity in the text material. However, these sections are recommended for those interested in Chapter 12.

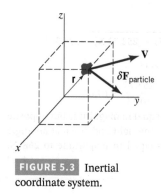

FIGURE 5.3 Inertial coordinate system.

external force acting on the particle. If we form the moment of each side of Eq. 5.30 with respect to the origin of an inertial coordinate system, we obtain

$$\mathbf{r} \times \frac{D}{Dt}(\mathbf{V}\rho\,\delta\forall) = \mathbf{r} \times \delta\mathbf{F}_{\text{particle}} \tag{5.31}$$

where **r** is the position vector from the origin of the inertial coordinate system to the fluid particle (**Fig. 5.3**). We note that

$$\frac{D}{Dt}[(\mathbf{r} \times \mathbf{V})\rho\,\delta\forall] = \frac{D\mathbf{r}}{Dt} \times \mathbf{V}\rho\,\delta\forall + \mathbf{r} \times \frac{D(\mathbf{V}\rho\,\delta\forall)}{Dt} \tag{5.32}$$

and

$$\frac{D\mathbf{r}}{Dt} = \mathbf{V} \tag{5.33}$$

Thus, since

$$\mathbf{V} \times \mathbf{V} = 0 \tag{5.34}$$

by combining Eqs. 5.31, 5.32, 5.33, and 5.34, we obtain the expression

$$\frac{D}{Dt}[(\mathbf{r} \times \mathbf{V})\rho\,\delta\forall] = \mathbf{r} \times \delta\mathbf{F}_{\text{particle}} \tag{5.35}$$

Equation 5.35 is valid for every particle of a system. For a system (collection of fluid particles), we need to use the sum of both sides of Eq. 5.35 to obtain

$$\int_{\text{sys}} \frac{D}{Dt}[(\mathbf{r} \times \mathbf{V})\rho\,d\forall] = \sum(\mathbf{r} \times \mathbf{F})_{\text{sys}} \tag{5.36}$$

where

$$\sum \mathbf{r} \times \delta\mathbf{F}_{\text{particle}} = \sum(\mathbf{r} \times \mathbf{F})_{\text{sys}} \tag{5.37}$$

We note that

$$\frac{D}{Dt}\int_{\text{sys}} (\mathbf{r} \times \mathbf{V})\rho\,d\forall = \int_{\text{sys}} \frac{D}{Dt}[(\mathbf{r} \times \mathbf{V})\rho\,d\forall] \tag{5.38}$$

since the sequential order of differentiation and integration can be reversed without consequence. (Recall that the material derivative, $D(\ )/Dt$, denotes the time derivative following a given system; see Section 3.2.1.) Thus, from Eqs. 5.36 and 5.38, we get

$$\frac{D}{Dt}\int_{\text{sys}} (\mathbf{r} \times \mathbf{V})\rho\,d\forall = \sum(\mathbf{r} \times \mathbf{F})_{\text{sys}} \tag{5.39}$$

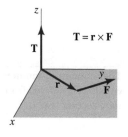

or

Time rate of change of the   = sum of external torques
moment-of-momentum of the system   acting on the system

The adjacent sketch illustrates what torque, $\mathbf{T} = \mathbf{r} \times \mathbf{F}$, is. For a control volume that is instantaneously coincident with the system, the torques acting on the system and on the control volume contents will be identical:

$$\sum(\mathbf{r} \times \mathbf{F})_{\text{sys}} = \sum(\mathbf{r} \times \mathbf{F})_{\text{cv}} \tag{5.40}$$

Furthermore, for the system and the contents of the coincident control volume that is fixed and nondeforming, the Reynolds transport theorem (Eq. 3.19) leads to

$$\frac{D}{Dt}\int_{\text{sys}} (\mathbf{r} \times \mathbf{V})\rho\, dV = \frac{\partial}{\partial t}\int_{\text{cv}} (\mathbf{r} \times \mathbf{V})\rho\, d\Psi + \int_{\text{cs}} (\mathbf{r} \times \mathbf{V})\rho \mathbf{V} \cdot \hat{\mathbf{n}}\, dA \qquad (5.41)$$

or

Time rate of change of the moment-of-momentum of the system	=	time rate of change of the moment-of-momentum of the contents of the control volume	+	net rate of flow of the moment-of-momentum through the control surface

For a control volume that is fixed (and therefore inertial) and nondeforming, we combine Eqs. 5.39, 5.40, and 5.41 to obtain the moment-of-momentum equation:

$$\boxed{\frac{\partial}{\partial t}\int_{\text{cv}} (\mathbf{r} \times \mathbf{V})\rho\, d\Psi + \int_{\text{cs}} (\mathbf{r} \times \mathbf{V})\rho \mathbf{V} \cdot \hat{\mathbf{n}}\, dA = \sum(\mathbf{r} \times \mathbf{F})_{\substack{\text{contents of the}\\\text{control volume}}}} \qquad (5.42)$$

> **For a system, the rate of change of moment-of-momentum equals the net torque.**

An important category of fluid mechanical problems that is readily solved with the help of the moment-of-momentum equation (Eq. 5.42) involves machines that rotate around a single axis. Examples of these machines include rotary lawn sprinklers, ceiling fans, lawn mower blades, wind turbines, turbochargers, and gas turbine engines. As a class, these devices are often called *turbomachines.*

5.2.4 Application of the Moment-of-Momentum Equation[2]

We simplify our use of Eq. 5.42 in several ways:

1. We assume that flows considered are one-dimensional (uniform distributions of average velocity at any section).
2. We confine ourselves to steady or steady-in-the-mean cyclical flows. Thus,

$$\frac{\partial}{\partial t}\int_{\text{cv}} (\mathbf{r} \times \mathbf{V})\rho\, d\Psi = 0$$

   at any instant of time for steady flows or on a time-average basis for cyclical unsteady flows.
3. We work only with the component of Eq. 5.42 resolved along the axis of rotation.

Consider the rotating sprinkler sketched in **Fig. 5.4.** Because the direction and magnitude of the flow through the sprinkler from the inlet [section (1)] to the outlet [section (2)] of the arm changes, the water exerts a torque on the sprinkler head, causing it to tend to rotate or to actually rotate in the direction shown, much like a turbine rotor. In applying the moment-of-momentum equation (Eq. 5.42) to this flow situation, we elect to use the fixed and nondeforming control volume shown in Fig. 5.4. This disk-shaped control volume contains within its boundaries the spinning or stationary sprinkler head and the portion of the water flowing through the sprinkler contained in the control volume at an instant. The control surface cuts through the sprinkler head's solid material so that the shaft torque that resists motion can be clearly identified. When the sprinkler is rotating, the flow field in the stationary control volume is cyclical and unsteady but steady in the mean. We proceed to use the axial component of the moment-of-momentum equation (Eq. 5.42) to analyze this flow.

> **Change in moment of fluid momentum around an axis can result in torque and rotation around that same axis.**

The integrand of the moment-of-momentum flow term in Eq. 5.42,

$$\int_{\text{cs}} (\mathbf{r} \times \mathbf{V})\rho \mathbf{V} \cdot \hat{\mathbf{n}}\, dA$$

can be nonzero only where fluid is crossing the control surface. Everywhere else on the control surface this term will be zero because $\mathbf{V} \cdot \hat{\mathbf{n}} = 0$. Water enters the control volume axially through the hollow stem of the sprinkler at section (1). At this portion of the control surface, the component

---

[2]This section may be omitted, along with Sections 5.2.3 and 5.3.6, without loss of continuity in the text material. However, these sections are recommended for those interested in Chapter 12.

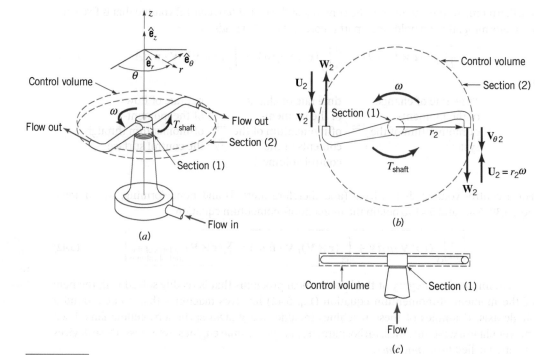

FIGURE 5.4 (*a*) Rotary water sprinkler. (*b*) Rotary water sprinkler, plane view. (*c*) Rotary water sprinkler, side view.

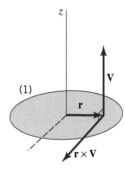

of $\mathbf{r} \times \mathbf{V}$ resolved along the axis of rotation is zero because, as illustrated by the adjacent figure, $\mathbf{r} \times \mathbf{V}$ lies in the plane of section (1), perpendicular to the axis of rotation. Thus, there is no axial moment-of-momentum flow in at section (1). Water leaves the control volume through each of the two nozzle openings at section (2). For the exiting flow, the magnitude of the axial component of $\mathbf{r} \times \mathbf{V}$ is $r_2 V_{\theta 2}$, where r_2 is the radius from the axis of rotation to the nozzle centerline and $V_{\theta 2}$ is the value of the tangential component of the velocity of the flow exiting each nozzle as observed from a frame of reference attached to the fixed and nondeforming control volume. The fluid velocity measured relative to a fixed control surface is an absolute velocity, $\mathbf{V}$. The velocity of the nozzle exit flow as viewed from the nozzle is called the relative velocity, $\mathbf{W}$. The absolute and relative velocities, $\mathbf{V}$ and $\mathbf{W}$, are related by the vector relationship

$$\mathbf{V} = \mathbf{W} + \mathbf{U} \tag{5.43}$$

where $\mathbf{U}$ is the velocity of the moving nozzle as measured relative to the fixed control surface.

The cross product and the dot product involved in the moment-of-momentum flow term of Eq. 5.42,

$$\int_{cs} (\mathbf{r} \times \mathbf{V}) \rho \mathbf{V} \cdot \hat{\mathbf{n}} \, dA$$

can each result in a positive or negative value. For flow into the control volume, $\mathbf{V} \cdot \hat{\mathbf{n}}$ is negative. For flow out, $\mathbf{V} \cdot \hat{\mathbf{n}}$ is positive. The relative directions of $\mathbf{r}$, $\mathbf{V}$, and $\mathbf{r} \times \mathbf{V}$, as well as the correct algebraic sign to assign the axis component of $\mathbf{r} \times \mathbf{V}$ can be ascertained by using the right-hand rule. The positive direction along the axis of rotation is the direction the thumb of the right-hand points when it is extended, and the remaining fingers are curled around the rotation axis in the positive direction of rotation as illustrated in **Fig. 5.5**. The direction of the axial component of $\mathbf{r} \times \mathbf{V}$ is similarly ascertained by noting the direction of the cross product of the radius from the axis of rotation, $r\hat{\mathbf{e}}_r$, and the tangential component of absolute velocity, $V_\theta \hat{\mathbf{e}}_\theta$. Thus, for the sprinkler of Fig. 5.4, we can state that

$$\left[\int_{cs} (\mathbf{r} \times \mathbf{V}) \rho \mathbf{V} \cdot \hat{\mathbf{n}} \, dA \right]_{axial} = (-r_2 V_{\theta 2})(+\dot{m}) \tag{5.44}$$

where, because of mass conservation, $\dot{m}$ is the total mass flowrate through both nozzles. As was demonstrated in Example 5.7, the mass flowrate is the same whether the sprinkler rotates or not. The correct algebraic sign of the axial component of $\mathbf{r} \times \mathbf{V}$ can be easily remembered

The algebraic sign of $\mathbf{r} \times \mathbf{V}$ is obtained by the right-hand rule.

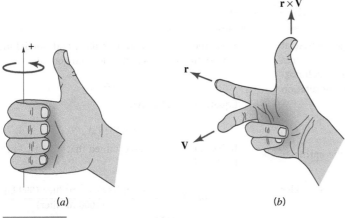

FIGURE 5.5 Right-hand rule convention.

in the following way: if V_θ and U are in the same direction, use $+$; if V_θ and U are in opposite directions, use $-$.

The torque term $[\Sigma(\mathbf{r} \times \mathbf{F})_{\text{contents of the control volume}}]$ of the moment-of-momentum equation (Eq. 5.42) is analyzed next. Confining ourselves to torques acting with respect to the axis of rotation only, we conclude that the shaft torque is important. The net torque with respect to the axis of rotation associated with normal forces exerted on the contents of the control volume will be very small, if not zero. The net axial torque due to fluid tangential forces is also negligibly small for the control volume of Fig. 5.4. Thus, for the sprinkler of Fig. 5.4

$$\sum \left[(\mathbf{r} \times \mathbf{F})_{\substack{\text{contents of the} \\ \text{control volume}}}\right]_{\text{axial}} = \mathbf{T}_{\text{shaft}} \tag{5.45}$$

Note that we have entered T_{shaft} as a positive quantity in Eq. 5.45. This is equivalent to assuming that T_{shaft} is in the same direction as rotation.

For the sprinkler of Fig. 5.4, the axial component of the moment-of-momentum equation (Eq. 5.42) is from Eqs. 5.44 and 5.45

$$-r_2 V_{\theta 2} \dot{m} = T_{\text{shaft}} \tag{5.46}$$

We interpret T_{shaft} being a negative quantity from Eq. 5.46 to mean that the shaft torque actually opposes the rotation of the sprinkler arms, as shown in Fig. 5.4. The shaft torque, T_{shaft}, opposes rotation in all turbine devices.

We evaluate the **shaft power**, $\dot{W}_{\text{shaft}}$, associated with **shaft torque**, T_{shaft}, by forming the product of T_{shaft} and the rotational speed of the shaft, ω. [We use the notation that $W = $ work, $(\dot{}) = d(\ )/dt$, and thus $\dot{W} = $ power.] Thus, from Eq. 5.46, we get

$$\dot{W}_{\text{shaft}} = T_{\text{shaft}}\, \omega = -r_2 V_{\theta 2} \dot{m}\, \omega \tag{5.47}$$

Power is equal to angular velocity times torque.

Since $r_2 \omega$ is the speed of each sprinkler nozzle, U, we can also state Eq. 5.47 in the form

$$\dot{W}_{\text{shaft}} = -U_2 V_{\theta 2} \dot{m} \tag{5.48}$$

Shaft work per unit mass, w_{shaft}, is equal to $\dot{W}_{\text{shaft}}/\dot{m}$. Dividing Eq. 5.48 by the mass flowrate, $\dot{m}$, we obtain

$$w_{\text{shaft}} = -U_2 V_{\theta 2} \tag{5.49}$$

Negative shaft work as in Eqs. 5.47, 5.48, and 5.49 is work out of the control volume; that is, work done by the fluid on the rotor and thus its shaft.

The principles associated with this sprinkler example can be extended to handle most simplified turbomachine flows. The fundamental technique is not difficult. However, the geometry and flow pattern of some turbomachine flows is quite complicated.

Example 5.18 further illustrates how the axial component of the moment-of-momentum equation (Eq. 5.46) can be used.

V5.13 Impulse-type lawn sprinkler

EXAMPLE 5.18 | Moment-of-Momentum—Torque

Given Water enters a rotating lawn sprinkler through its base at the steady rate of 1000 ml/s, as sketched in **Fig. E5.18a**. The exit area of each of the two nozzles is 30 mm², and the flow leaving each nozzle is in the tangential direction. The radius from the axis of rotation to the centerline of each nozzle is 200 mm.

Find

a. Determine the resisting torque required to hold the sprinkler head stationary.

b. Determine the resisting torque associated with the sprinkler rotating with a constant speed of 500 rev/min.

c. Determine the speed of the sprinkler if no resisting torque is applied.

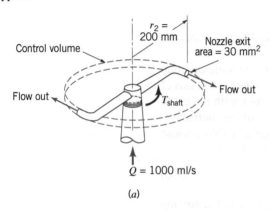

(a)

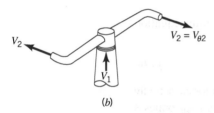

(b)

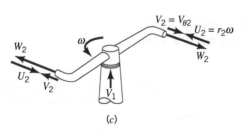

(c)

FIGURE E5.18a-c

Solution To solve parts **(a)**, **(b)**, and **(c)** of this example, we can use the same fixed and nondeforming, disk-shaped control volume illustrated in Fig. 5.4. As indicated in Fig. E5.18a, the only axial torque considered is the one resisting motion, T_{shaft}.

a. When the sprinkler head is held stationary, as specified in part **(a)** of this example problem, the velocities of the fluid entering and leaving the control volume are shown in **Fig. E5.18b**. Equation 5.46 applies to the contents of this control volume. Thus,

$$T_{shaft} = -r_2 V_{\theta 2} \dot{m} \qquad (1)$$

Since the control volume is fixed and nondeforming and the flow exiting from each nozzle is tangential,

$$V_{\theta 2} = V_2 \qquad (2)$$

Equations 1 and 2 give

$$T_{shaft} = -r_2 V_2 \dot{m} \qquad (3)$$

In Example 5.7, we ascertained that $V_2 = 16.7$ m/s. Thus, from Eq. 3 with

$$\dot{m} = Q\rho = \frac{(1000 \text{ ml/s})(10^{-3}\text{m}^3/\text{liter})(999 \text{ kg/m}^3)}{(1000 \text{ ml/liter})}$$

$$= 0.999 \text{ kg/s}$$

we obtain

$$T_{shaft} = -\frac{(200 \text{ mm})(16.7 \text{ m/s})(0.999 \text{ kg/s})[1(\text{N/kg})/(\text{m/s}^2)]}{(1000 \text{ mm/m})}$$

or

$$T_{shaft} = -3.34 \text{ N} \cdot \text{m} \qquad (Ans)$$

b. When the sprinkler is rotating at a constant speed of 500 rpm, the flow field in the control volume is unsteady but cyclical. Thus, the flow field is steady in the mean. The velocities of the flow entering and leaving the control volume are as indicated in **Fig. E5.18c**. The absolute velocity of the fluid leaving each nozzle, V_2, is from Eq. 5.43,

$$V_2 = W_2 - U_2 \qquad (4)$$

where

$$W_2 = 16.7 \text{ m/s}$$

as determined in Example 5.7. The speed of the nozzle, U_2, is obtained from

$$U_2 = r_2\omega \qquad (5)$$

Application of the axial component of the moment-of-momentum equation (Eq. 5.46) leads again to Eq. 3. From Eqs. 4 and 5,

$$V_2 = 16.7 \text{ m/s} - r_2\omega$$

$$= 16.7 \text{ m/s} - \frac{(200 \text{ mm})(500 \text{ rev/min})(2\pi \text{ rad/rev})}{(1000 \text{ mm/m})(60 \text{ s/min})}$$

or

$$V_2 = 16.7 \text{ m/s} - 10.5 \text{ m/s} = 6.2 \text{ m/s}$$

Thus, using Eq. 3 with $\dot{m} = 0.999$ kg/s (as calculated previously), we get

$$T_{shaft} = -\frac{(200 \text{ mm})(6.2 \text{ m/s})0.999 \text{ kg/s}[1(\text{N/kg})/(\text{m/s}^2)]}{(1000 \text{ mm/m})}$$

or

$$T_{shaft} = -1.24 \text{ N} \cdot \text{m} \qquad (Ans)$$

Comment Note that the resisting torque associated with sprinkler head rotation is much less than the resisting torque that is required to hold the sprinkler stationary.

c. When no resisting torque is applied to the rotating sprinkler head, a maximum constant speed of rotation will occur. Application of Eqs. 3, 4, and 5 to the contents of the control volume results in

$$T_{\text{shaft}} = -r_2(W_2 - r_2\omega)\dot{m} \qquad (6)$$

For no resisting torque, Eq. 6 yields

$$0 = -r_2(W_2 - r_2\omega)\dot{m}$$

Thus,

$$\omega = \frac{W_2}{r_2} \qquad (7)$$

In Example 5.4, we learned that the relative velocity of the fluid leaving each nozzle, W_2, is the same regardless of the speed of rotation of the sprinkler head, ω, as long as the mass flowrate of the fluid, $\dot{m}$, remains constant. Thus, by using Eq. 7, we obtain

$$\omega = \frac{W_2}{r_2} = \frac{(16.7 \text{ m/s})(1000 \text{ mm/m})}{(200 \text{ mm})} = 83.5 \text{ rad/s}$$

or

$$\omega = \frac{(83.5 \text{ rad/s})(60 \text{ s/min})}{2\pi \text{ rad/rev}} = 797 \text{ rpm} \qquad (Ans)$$

For this condition ($T_{\text{shaft}} = 0$), the water both enters and leaves the control volume with zero angular momentum.

Comment Note that forcing a change in direction of a flowing fluid, in this case with a sprinkler, resulted in rotary motion and a useful "sprinkling" of water over an area.

By repeating the calculations for various values of the angular velocity, ω, the results shown in **Fig. E5.18d** are obtained. It is seen that the magnitude of the resisting torque associated with rotation is less than the torque required to hold the rotor stationary. Even in the absence of a resisting torque, the rotor maximum speed is finite.

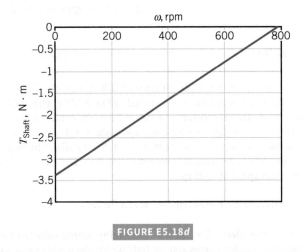

FIGURE E5.18d

When the moment-of-momentum equation (Eq. 5.42) is applied to a more general, one-dimensional flow through a rotating machine, we obtain

$$\boxed{T_{\text{shaft}} = (-\dot{m}_{\text{in}})(\pm r_{\text{in}}V_{\theta\text{in}}) + \dot{m}_{\text{out}}(\pm r_{\text{out}}V_{\theta\text{out}})} \qquad (5.50)$$

by applying the same kind of analysis used with the sprinkler of Fig. 5.4. The "−" is used with mass flowrate into the control volume, $\dot{m}_{\text{in}}$, and the "+" is used with mass flowrate out of the control volume, $\dot{m}_{\text{out}}$, to account for the sign of the dot product, $\mathbf{V} \cdot \hat{\mathbf{n}}$, involved. Whether "+" or "−" is used with the rV_θ product depends on the direction of $(\mathbf{r} \times \mathbf{V})_{\text{axial}}$. A simple way to determine the sign of the rV_θ product is to compare the direction of V_θ and the blade speed, U. As shown in the adjacent figures, if V_θ and U are in the same direction, then the $r V_\theta$ product is positive. If V_θ and U are in opposite directions, the rV_θ product is negative. The sign of the shaft torque is "+" if T_{shaft} is in the same direction along the axis of rotation as ω, and "−" otherwise.

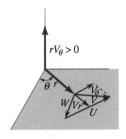

The shaft power, $\dot{W}_{\text{shaft}}$, is related to shaft torque, T_{shaft}, by

$$\dot{W}_{\text{shaft}} = T_{\text{shaft}}\omega \qquad (5.51)$$

Thus, using Eqs. 5.50 and 5.51 with a "+" sign for T_{shaft} in Eq. 5.50, we obtain

$$\dot{W}_{\text{shaft}} = (-\dot{m}_{\text{in}})(\pm r_{\text{in}}\omega V_{\theta\text{in}}) + \dot{m}_{\text{out}}(\pm r_{\text{out}}\omega V_{\theta\text{out}}) \qquad (5.52)$$

or since $r\omega = U$

$$\boxed{\dot{W}_{\text{shaft}} = (-\dot{m}_{\text{in}})(\pm U_{\text{in}}V_{\theta\text{in}}) + \dot{m}_{\text{out}}(\pm U_{\text{out}}V_{\theta\text{out}})} \qquad (5.53)$$

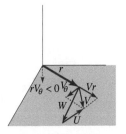

The "+" is used for the UV_θ product when U and V_θ are in the same direction; the "−" is used when U and V_θ are in opposite directions. Also, since $+T_{\text{shaft}}$ was used to obtain Eq. 5.53, when $\dot{W}_{\text{shaft}}$ is positive, power is into the fluid (for example, a pump), and when $\dot{W}_{\text{shaft}}$ is negative, power is out of the fluid (for example, a turbine).

When shaft torque and shaft rotation are in the same (opposite) direction, power is into (out of) the fluid.

The shaft work per unit mass, w_{shaft}, can be obtained from the shaft power, $\dot{W}_{shaft}$, by dividing Eq. 5.53 by the mass flowrate, $\dot{m}$. By conservation of mass,

$$\dot{m} = \dot{m}_{in} = \dot{m}_{out}$$

From Eq. 5.53, we obtain

$$w_{shaft} = -(\pm U_{in} V_{\theta in}) + (\pm U_{out} V_{\theta out}) \tag{5.54}$$

The application of Eqs. 5.50, 5.53, and 5.54 is demonstrated in Example 5.19. More examples of the application of Eqs. 5.50, 5.53, and 5.54 are included in Chapter 12.

EXAMPLE 5.19 | Moment-of-Momentum—Power

Given An air fan has a bladed rotor of 0.30 m-outside-diameter and 0.25 m-inside-diameter, as illustrated in **Fig. E5.19a**. The height of each rotor blade is constant at 0.025 m from blade inlet to outlet. The flowrate is steady, on a time-average basis, at 6.5 m³/min, and the absolute velocity of the air at blade inlet, $\mathbf{V}_1$, is radial. The blade discharge angle is 30° from the tangential direction. The rotor rotates at a constant speed of 1725 rpm.

Find Estimate the power required to run the fan.

Solution We select a fixed and nondeforming control volume that includes the rotating blades and the fluid within the blade row at an instant, as shown with a dashed line in Fig. E5.19a. The flow within this control volume is cyclical but steady in the mean. The only torque we consider is the driving shaft torque, T_{shaft}. This torque is provided by a motor. We assume that the entering and leaving flows are each represented by uniformly distributed velocities and flow properties. To determine the power required to drive the fan, we use Eq. 5.53. Application of this equation to the contents of the control volume in Fig. E5.19 gives

$$\dot{W}_{shaft} = -\dot{m}_1(\pm U_1 \overset{0\,(\mathbf{V}_1\text{ is radial})}{V_{\theta 1}}) + \dot{m}_2(\pm U_2 V_{\theta 2}) \tag{1}$$

From Eq. 1, we see that to calculate fan power, we need a value for the mass flowrate, $\dot{m}$, the rotor exit blade velocity, U_2, and the fluid tangential velocity at blade exit, $V_{\theta 2}$. The mass flowrate, $\dot{m}$, is easily obtained from Eq. 5.6 as

$$\dot{m} = \rho Q = \frac{(1000\,\text{kg/m}^3)(6.5\,\text{m}^3/\text{min})}{(60\,\text{s/min})}$$

$$= 108\,\text{kg/s} \tag{2}$$

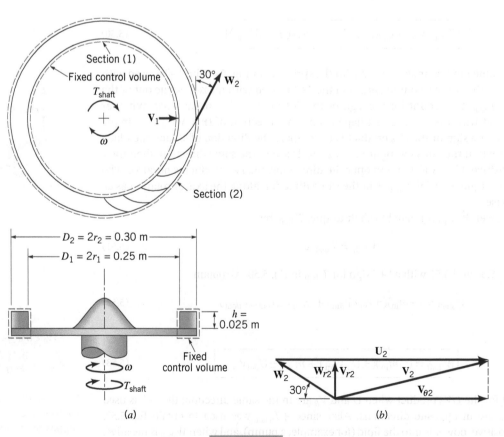

(a)

(b)

FIGURE E5.19

The rotor exit blade speed, U_2, is

$$U_2 = r_2\omega = \frac{(0.15 \text{ m})(1725 \text{ rpm})(2\pi \text{ rad/rev})}{(60 \text{ s/min})}$$

$$= 27 \text{ m/s} \tag{3}$$

To determine the fluid tangential speed at the fan rotor exit, $V_{\theta2}$, we use Eq. 5.43 to get

$$\mathbf{V}_2 = \mathbf{W}_2 + \mathbf{U}_2 \tag{4}$$

The vector addition of Eq. 4 is shown in the form of a "velocity triangle" in **Fig. E5.19b**. From Fig. E5.19b, we can see that

$$V_{\theta2} = U_2 - W_2 \cos 30° \tag{5}$$

To solve Eq. 5 for $V_{\theta2}$, we need a value of W_2, in addition to the value of U_2 already determined (Eq. 3). To compute W_2, we recognize that

$$W_2 \sin 30° = V_{r2} \tag{6}$$

where V_{r2} is the radial component of either $\mathbf{W}_2$ or $\mathbf{V}_2$. Using Eq. 5.6, we obtain

$$\dot{m} = \rho A_2 V_{r2} \tag{7}$$

For the specified geometry, the outflow area is

$$A_2 = 2\pi r_2 h \tag{8}$$

where h is the blade height. Combining Eqs. 7 and 8 yields

$$\dot{m} = \rho 2\pi r_2 h V_{r2} \tag{9}$$

Combining Eqs. 6 and 9 yields

$$W_2 = \frac{\dot{m}}{\rho 2\pi r_2 h \sin 30°} = \frac{\rho Q}{\rho 2\pi r_2 h \sin 30°} \tag{10}$$

$$= \frac{Q}{2\pi r_2 h \sin 30°}$$

Substituting known values into Eq. 10, we obtain

$$W_2 = \frac{6.5 \text{m}^3/\text{min}}{(60 \text{ s/min})2\pi(0.15 \text{ m})(0.025 \text{ m} \sin 30°)}$$

$$= 9.2 \text{ m/s}$$

Using this value of W_2 in Eq. 5, we get

$$V_{\theta2} = U_2 - W_2 \cos 30°$$

$$= 27 \text{ m/s} - (9.2 \text{ m/s})(0.866) = 19 \text{ m/s}$$

Equation 1 can now be used to obtain

$$\dot{W}_{\text{shaft}} = \dot{m}U_2 V_{\theta2} = (108 \text{ kg/s})(27 \text{ m/s})(19 \text{ m/s})$$

$$\dot{W}_{\text{shaft}} = 55.4 \text{ kW} \tag{Ans}$$

Comment Note that the "+" was used with the $U_2 V_{\theta2}$ product because U_2 and $V_{\theta2}$ are in the same direction. This result, 55.4 kW, is the rate of work that must be done on the fluid to produce the specified flow. Ideally, all of this power would go into the flowing air. However, because of "losses," only some of this power will produce useful effects (e.g., movement and pressure rise) on the air. How much useful effect depends on the efficiency of the energy transfer between the fan blades and the fluid. Also, extra power would be needed from the motor to overcome friction in shaft bearings and other mechanical resistance.

5.3 FIRST LAW OF THERMODYNAMICS—THE ENERGY EQUATION

5.3.1 Derivation of the Energy Equation

The **first law of thermodynamics** for a system is, in words,

Time rate of increase of the total stored energy of the system	=	net time rate of energy addition by heat transfer into the system	+	net time rate of energy addition by work transfer into the system

In symbolic form, this statement is

$$\frac{D}{Dt}\int_{\text{sys}} e\rho \, d\mathcal{V} = \left(\sum \dot{Q}_{\text{in}} - \sum \dot{Q}_{\text{out}}\right)_{\text{sys}} + \left(\sum \dot{W}_{\text{in}} - \sum \dot{W}_{\text{out}}\right)_{\text{sys}}$$

or

$$\frac{D}{Dt}\int_{\text{sys}} e\rho \, d\mathcal{V} = \left(\dot{Q}_{\substack{\text{net} \\ \text{in}}} + \dot{W}_{\substack{\text{net} \\ \text{in}}}\right)_{\text{sys}} \tag{5.55}$$

The first law of thermodynamics is a statement of conservation of energy.

Some of these variables deserve a brief explanation before proceeding further. The total stored energy per unit mass for each particle in the system, e, is related to the internal energy per unit mass, $\check{u}$, the kinetic energy per unit mass, $V^2/2$, and the potential energy per unit mass, gz, by the equation

$$e = \check{u} + \frac{V^2}{2} + gz \tag{5.56}$$

In this equation we do not include chemical energy stored in molecular structure and nuclear energy stored in atomic structure because we will not be considering changes of these types of energy.

The net **rate of heat transfer** into the system is denoted with $\dot{Q}_{\text{net in}}$, and the net rate of work done on the system is labeled $\dot{W}_{\text{net in}}$. The sign convention we have used is to assign positive (+) values to heat transfer into the system and work done on the system. Therefore, heat transfer out of the system and work done by the system are assigned negative (−) values. [It should be noted that the opposite sign convention is often used for work (see Ref. 3 for example).] If there is uncertainty regarding the sign convention in use, simply look at the influence of work: work done on the system increases the system's energy.

Equation 5.55 is valid for inertial and noninertial reference systems. We proceed to develop the control volume statement of the first law of thermodynamics. For the control volume that is coincident with the system at an instant of time

$$\left(\dot{Q}_{\substack{\text{net}\\\text{in}}} + \dot{W}_{\substack{\text{net}\\\text{in}}}\right)_{\text{sys}} = \left(\dot{Q}_{\substack{\text{net}\\\text{in}}} + \dot{W}_{\substack{\text{net}\\\text{in}}}\right)_{\substack{\text{coincident}\\\text{control volume}}} \tag{5.57}$$

Furthermore, for the system and the contents of the coincident control volume that is fixed and nondeforming, the Reynolds transport theorem (Eq. 3.19 with the parameter b set equal to e) allows us to conclude that

$$\frac{D}{Dt}\int_{\text{sys}} e\rho \, d\mathcal{V} = \frac{\partial}{\partial t}\int_{\text{cv}} e\rho \, d\mathcal{V} + \int_{\text{cs}} e\rho \mathbf{V} \cdot \hat{\mathbf{n}} \, dA \tag{5.58}$$

or in words,

Time rate of increase of the total stored energy of the system	=	time rate of increase of the total stored energy of the contents of the control volume	+	net rate of flow of the total stored energy out of the control volume through the control surface

Combining Eqs. 5.55, 5.57, and 5.58, we get the control volume formula for the first law of thermodynamics:

$$\frac{\partial}{\partial t}\int_{\text{cv}} e\rho \, d\mathcal{V} + \int_{\text{cs}} e\rho \mathbf{V} \cdot \hat{\mathbf{n}} \, dA = \left(\dot{Q}_{\substack{\text{net}\\\text{in}}} + \dot{W}_{\substack{\text{net}\\\text{in}}}\right)_{\text{cv}} \tag{5.59}$$

The total stored energy per unit mass, e, in Eq. 5.59 is for fluid particles entering, leaving, and within the control volume. Further explanation of the heat transfer and work transfer involved in this equation follows in the next paragraphs.

The heat transfer rate, $\dot{Q}$, represents all of the ways in which energy is exchanged between the control volume contents and surroundings because of a temperature difference. Thus, radiation, conduction, and/or convection are possible. As shown by the adjacent figure, heat transfer into the control volume is considered positive; heat transfer out is negative. In many engineering applications, the process is *adiabatic*; the heat transfer rate, $\dot{Q}$, is zero. The net heat transfer rate, $\dot{Q}_{\text{net in}}$, can also be zero when $\Sigma\dot{Q}_{\text{in}} - \Sigma\dot{Q}_{\text{out}} = 0$.

The rate at which work is done, $\dot{W}$, also called *power*, is positive when work is done on the contents of the control volume by the surroundings. In the following paragraphs, we consider several different ways in which work is done on/by the control volume contents and develop an expression for each interaction.

It is common to have a work interaction between the control volume contents and a rotating shaft. In rotary devices such as turbines, fans, and propellers, a rotating shaft transfers work across that portion of the control surface that slices through the shaft. Even in reciprocating machines like positive displacement internal combustion engines and compressors that utilize piston-in-cylinder arrangements, a rotating crankshaft is used. Since work is the dot product of

---

The energy equation relates stored energy to the rate of heat transfer and the rate of work done.

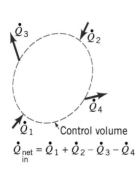

$\dot{Q}_{\substack{\text{net}\\\text{in}}} = \dot{Q}_1 + \dot{Q}_2 - \dot{Q}_3 - \dot{Q}_4$

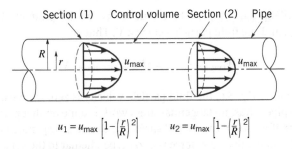

Section (1)   Control volume Section (2)   Pipe

$$u_1 = u_{max}\left[1-\left(\frac{r}{R}\right)^2\right] \qquad u_2 = u_{max}\left[1-\left(\frac{r}{R}\right)^2\right]$$

FIGURE 5.6 Simple, fully developed pipe flow.

force and related displacement, rate of work (or power) is the dot product of force and related displacement per unit time. For a rotating shaft, the power transfer, $\dot{W}_{shaft}$, is related to the shaft torque that causes the rotation, T_{shaft}, and the angular velocity of the shaft, ω, by the relationship

$$\dot{W}_{shaft} = T_{shaft}\omega$$

When the control surface cuts through the shaft material, the shaft torque is exerted by shaft material at the control surface. To allow for consideration of problems involving more than one shaft, we use the notation

$$\dot{W}_{\substack{shaft \\ net\ in}} = \sum_{in}\dot{W}_{shaft} - \sum_{out}\dot{W}_{shaft} \tag{5.60}$$

Work transfer can also occur at the control surface when a force associated with fluid normal stress acts over a distance. Consider the simple pipe flow illustrated in **Fig. 5.6** and the control volume shown. For this situation, the fluid normal stress, σ, is simply equal to the negative of fluid pressure, p, in all directions; that is,

$$\sigma = -p \tag{5.61}$$

This relationship can be used with varying amounts of approximation for many engineering problems (see Chapter 6).

As shown in the adjacent figure, the power, $\dot{W}$, associated with a force, $\mathbf{F}$, acting on an object moving with velocity, $\mathbf{V}$, is given by the dot product $\mathbf{F} \cdot \mathbf{V}$. Hence, the power associated with normal stresses acting on a single fluid particle, $\delta\dot{W}_{normal\ stress}$, can be evaluated as the dot product of the normal stress force, $\delta\mathbf{F}_{normal\ stress}$, and the fluid particle velocity, $\mathbf{V}$, as

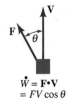

$$\dot{W} = \mathbf{F}\cdot\mathbf{V}$$
$$= FV\cos\theta$$

$$\delta\dot{W}_{normal\ stress} = \delta\mathbf{F}_{normal\ stress} \cdot \mathbf{V}$$

If the normal stress force is expressed as the product of local normal stress, $\sigma = -p$, and fluid particle surface area, $\hat{\mathbf{n}}\,\delta A$, the result is

$$\delta\dot{W}_{normal\ stress} = \sigma\hat{\mathbf{n}}\,\delta A \cdot \mathbf{V} = -p\hat{\mathbf{n}}\,\delta A \cdot \mathbf{V} = -p\mathbf{V}\cdot\hat{\mathbf{n}}\,\delta A$$

For all fluid particles on the control surface of Fig. 5.6 at the instant considered, the rate of work done on the control volume contents due to fluid normal stress, $\dot{W}_{normal\ stress}$, is

$$\dot{W}_{\substack{normal \\ stress}} = \int_{cs}\sigma\mathbf{V}\cdot\hat{\mathbf{n}}\,dA = \int_{cs}-p\mathbf{V}\cdot\hat{\mathbf{n}}\,dA \tag{5.62}$$

The work represented by the pressure integral in Eq. 5.62 is often called **flow work**. Physically, it represents the sum of work done on the system by fluid particles pushing into the control volume (+) and work done on the surrounding fluid by particles pushing out from the control volume (−).

Note that the value of $\dot{W}_{normal\ stress}$ for particles on the wetted inside surface of the pipe is zero because $\mathbf{V} \cdot \hat{\mathbf{n}}$ is zero there. Thus, $\dot{W}_{normal\ stress}$ can be nonzero only where fluid enters and leaves the control volume. Although only a simple pipe flow was considered, Eq. 5.62 is quite general, and the control volume used in this example can serve as a general model for other cases.

Work transfer can also occur at the control surface because of tangential stress. Rotating shaft work is transferred by tangential stresses in the shaft material. For a fluid particle, shear

Work interactions frequently involve rotating shafts, normal stresses, and tangential stresses.

stress force power, $\delta\dot{W}_{\text{tangential stress}}$, can be evaluated as the dot product of tangential stress force, $\delta\mathbf{F}_{\text{tangential stress}}$, and the fluid particle velocity, $\mathbf{V}$. That is,

$$\delta\dot{W}_{\text{tangential stress}} = \delta\mathbf{F}_{\text{tangential stress}} \cdot \mathbf{V}$$

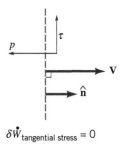

$\delta\dot{W}_{\text{tangential stress}} = 0$

For the control volume of Fig. 5.6, the fluid particle velocity is zero everywhere on the wetted inside surface of the pipe. Thus, no tangential stress work is done on that portion of the control surface. Furthermore, if we select the control surface so that it is perpendicular to the fluid particle velocity, then the tangential stress force is also perpendicular to the velocity. Therefore, the tangential stress work transfer is zero on that part of the control surface. This is illustrated in the adjacent figure. Thus, in general, we select control volumes like the one of Fig. 5.6 and consider fluid tangential stress power transfer to be negligibly small.

Using the information we have developed about power, we can express the first law of thermodynamics for the contents of a control volume by combining Eqs. 5.59, 5.60, and 5.62 to obtain

$$\frac{\partial}{\partial t}\int_{cv} e\rho \, d\mathcal{V} + \int_{cs} e\rho\mathbf{V}\cdot\hat{\mathbf{n}} \, dA = \dot{Q}_{\substack{\text{net}\\\text{in}}} + \dot{W}_{\substack{\text{shaft}\\\text{net in}}} - \int_{cs} p\mathbf{V}\cdot\hat{\mathbf{n}} \, dA \qquad (5.63)$$

Finally, we multiply and divide by ρ inside the pressure integral, substitute the stored energy from Eq. 5.56, and group terms to obtain the **energy equation**:

$$\frac{\partial}{\partial t}\int_{cv} e\rho \, d\mathcal{V} + \int_{cs}\left(\breve{u} + \frac{p}{\rho} + \frac{V^2}{2} + gz\right)\rho\mathbf{V}\cdot\hat{\mathbf{n}} \, dA = \dot{Q}_{\substack{\text{net}\\\text{in}}} + \dot{W}_{\substack{\text{shaft}\\\text{net in}}} \qquad (5.64)$$

5.3.2 Application of the Energy Equation

In Eq. 5.64, the term $\partial/\partial t \int_{cv} e\rho \, d\mathcal{V}$ represents the time rate of change of the total stored energy, e, of the contents of the control volume. This term is zero when the flow is steady. This term is also zero in the mean when the flow is steady in the mean (cyclical).

In Eq. 5.64, the integrand of

$$\int_{cs}\left(\breve{u} + \frac{p}{\rho} + \frac{V^2}{2} + gz\right)\rho\mathbf{V}\cdot\hat{\mathbf{n}} \, dA$$

> The term p/ρ in the energy equation represents flow work, not stored energy.

can be nonzero only where fluid crosses the control surface ($\mathbf{V}\cdot\hat{\mathbf{n}} \neq 0$). Otherwise, $\mathbf{V}\cdot\hat{\mathbf{n}}$ is zero and the integrand is zero for that portion of the control surface. If the properties within parentheses, $\breve{u}$, p/ρ, $V^2/2$, and gz, are all assumed to be uniformly distributed over the flow cross-sectional areas involved, the integration becomes simple and gives

$$\int_{cs}\left(\breve{u} + \frac{p}{\rho} + \frac{V^2}{2} + gz\right)\rho\mathbf{V}\cdot\hat{\mathbf{n}} \, dA = \sum_{\substack{\text{flow}\\\text{out}}}\left(\breve{u} + \frac{p}{\rho} + \frac{V^2}{2} + gz\right)\dot{m}$$

$$- \sum_{\substack{\text{flow}\\\text{in}}}\left(\breve{u} + \frac{p}{\rho} + \frac{V^2}{2} + gz\right)\dot{m} \qquad (5.65)$$

Furthermore, if there is only one stream entering and leaving the control volume, then

$$\int_{cs}\left(\breve{u} + \frac{p}{\rho} + \frac{V^2}{2} + gz\right)\rho\mathbf{V}\cdot\hat{\mathbf{n}} \, dA =$$

$$\left(\breve{u} + \frac{p}{\rho} + \frac{V^2}{2} + gz\right)_{\text{out}}\dot{m}_{\text{out}} - \left(\breve{u} + \frac{p}{\rho} + \frac{V^2}{2} + gz\right)_{\text{in}}\dot{m}_{\text{in}} \qquad (5.66)$$

Uniform flow, as described previously, will occur in an infinitesimally small diameter streamtube as illustrated in **Fig. 5.7**. This kind of streamtube flow is representative of the steady flow of a particle of fluid along a pathline. We can also idealize actual conditions by disregarding nonuniformities in a finite cross section of flow. We call this one-dimensional flow, and although such uniform flow rarely occurs in reality, the simplicity achieved with the one-dimensional approximation often justifies its use. More details about the effects of nonuniform distributions of velocities and other fluid flow variables are considered in Section 5.3.4 and in Chapters 8, 9, and 10.

If shaft work is involved, the flow must be unsteady, at least locally (see Refs. 1 and 2). The flow in any fluid machine that involves shaft work is unsteady within that machine. For example, the velocity and pressure at a fixed location near the rotating blades of a fan are unsteady

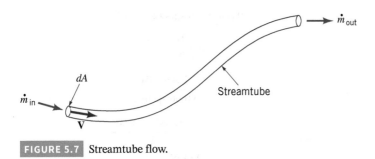

FIGURE 5.7 Streamtube flow.

owing to blade passing. However, upstream and downstream of the machine, the flow may be steady. Most often shaft work is associated with flow that is unsteady in a recurring or cyclical way. On a time-average basis for flow that is one-dimensional, cyclical, and involves only one stream of fluid entering and leaving the control volume, such as the hair dryer shown in the adjacent figure, Eq. 5.64 can be simplified with the help of Eqs. 5.9 and 5.66 to form

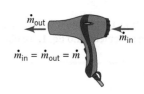

$$\dot{m}\left[\check{u}_{out} - \check{u}_{in} + \left(\frac{p}{\rho}\right)_{out} - \left(\frac{p}{\rho}\right)_{in} + \frac{V_{out}^2 - V_{in}^2}{2} + g(z_{out} - z_{in})\right] = \dot{Q}_{net \atop in} + \dot{W}_{shaft \atop net\,in} \quad (5.67)$$

$$\dot{m}_{in} = \dot{m}_{out} = \dot{m}$$

We call Eq. 5.67 the *one-dimensional energy equation for steady-in-the-mean flow*. Note that Eq. 5.67 is valid for incompressible and compressible flows. Often, the fluid property called *enthalpy*, $\check{h}$, where

$$\check{h} = \check{u} + \frac{p}{\rho} \quad (5.68)$$

is used in Eq. 5.67. With enthalpy, the one-dimensional energy equation for steady-in-the-mean flow (Eq. 5.67) is

$$\dot{m}\left[\check{h}_{out} - \check{h}_{in} + \frac{V_{out}^2 - V_{in}^2}{2} + g(z_{out} - z_{in})\right] = \dot{Q}_{net \atop in} + \dot{W}_{shaft \atop net\,in} \quad (5.69)$$

> The energy equation is sometimes written in terms of enthalpy.

Equation 5.69 is often used for solving compressible flow problems. Examples 5.20 and 5.21 illustrate how Eqs. 5.67 and 5.69 can be used.

EXAMPLE 5.20 | Energy—Pump Power

Given A pump delivers water at a steady rate of 1.14 m³/min, as shown in **Fig. E5.20**. Just upstream of the pump [section (1)] where the pipe diameter is 0.09 m, the pressure is 124106 Pa. Just downstream of the pump [section (2)] where the pipe diameter is 0.03 m, the pressure is 413686 Pa. The change in water elevation across the pump is zero. The rise in internal energy of water, $\check{u}_2 - \check{u}_1$, associated with a temperature rise across the pump is 28.34 m/s. The pumping process is considered to be adiabatic.

Find Determine the power (hp) required by the pump.

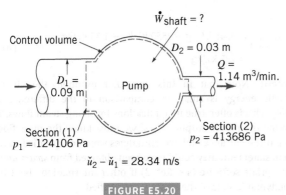

$\dot{W}_{shaft} = ?$

Control volume

$D_2 = 0.03$ m

$D_1 = 0.09$ m

Pump

$Q = 1.14$ m³/min.

Section (1)
$p_1 = 124106$ Pa

Section (2)
$p_2 = 413686$ Pa

$\check{u}_2 - \check{u}_1 = 28.34$ m/s

FIGURE E5.20

Solution We include in our control volume the water contained in the pump between its entrance and exit sections. Application of Eq. 5.67 to the contents of this control volume on a time-average basis yields

$$\dot{m}\left[\check{u}_2 - \check{u}_1 + \left(\frac{p}{\rho}\right)_2 - \left(\frac{p}{\rho}\right)_1 + \frac{V_2^2 - V_1^2}{2} + \overbrace{g(z_2 - z_1)}^{\text{0 (no elevation change)}}\right]$$

$$= \dot{Q}_{net \atop in} + \dot{W}_{shaft \atop net\,in} \quad (1)$$

(0 (adiabatic flow))

We can solve directly for the power required by the pump, $\dot{W}_{shaft\,net\,in}$, from Eq. 1, after we first determine the mass flowrate, $\dot{m}$, the speed of flow into the pump, V_1, and the speed of the flow out of the pump, V_2. All other quantities in Eq. 1 are given in the problem statement. From Eq. 5.6, we get

$$\dot{m} = \rho Q = \frac{(1000\,\text{kg/m}^3)(1.14\,\text{m}^3/\text{min})}{(60\,\text{s/min})}$$

$$= 19\,\text{kg/s}$$

Also from Eq. 5.6,

$$V = \frac{Q}{A} = \frac{Q}{\pi D^2/4}$$

so

$$V_1 = \frac{Q}{A_1} = \frac{(1.14 \text{ m}^3/\text{min})4}{(60 \text{ s/min})\pi(0.09 \text{ m})^2}$$

$$= 3 \text{ m/s} \qquad (3)$$

and

$$V_2 = \frac{Q}{A_2} = \frac{(1.14 \text{ m}^3/\text{min})4}{(60 \text{ s/min})\pi(0.03 \text{ m})^2}$$

$$= 27 \text{ m/s} \qquad (4)$$

Substituting the values of Eqs. 2, 3, and 4 and values from the problem statement into Eq. 1, we obtain

$$\dot{W}_{\substack{\text{shaft} \\ \text{net in}}} = (19 \text{ kg/s})(28.34 \text{ m/s})$$

$$+ \frac{413686 \text{ Pa}}{(1000 \text{ kg/m}^3)}$$

$$- \frac{124106 \text{ Pa}}{(1000 \text{ kg/m}^3)}$$

$$+ \frac{(27\text{m/s})-(3\text{m/s})^2}{2}$$

$$= 13 \text{ kW} \qquad (Ans)$$

Comment Of the total 13 kW, internal energy change accounts for 8.54 kW, the pressure rise accounts for 5.5 kW, and the kinetic energy increase accounts for 6.84 kW.

An actual pump would require slightly more than 13 kW owing to mechanical friction loss in bearings and shaft seals. The sum of the pressure rise, the kinetic energy increase, and any increase in elevation is sometimes called *water horsepower*.

EXAMPLE 5.21 | Energy—Turbine Power per Unit Mass of Flow

Given A steam turbine generator unit used to produce electricity is shown in **Fig. E5.21a**. The steam enters the turbine with a velocity of 30 m/s and enthalpy, $\check{h}_1$, of 3348 kJ/kg (see **Fig. E5.21b**). The steam leaves the turbine as a mixture of vapor and liquid having a velocity of 60 m/s and an enthalpy of 2550 kJ/kg. Assume that the flow through the turbine is adiabatic and that changes in elevation are negligible.

Find Determine the work output per unit mass of steam through-flow.

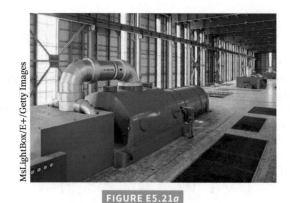

MsLightBox/E+/Getty Images

FIGURE E5.21a

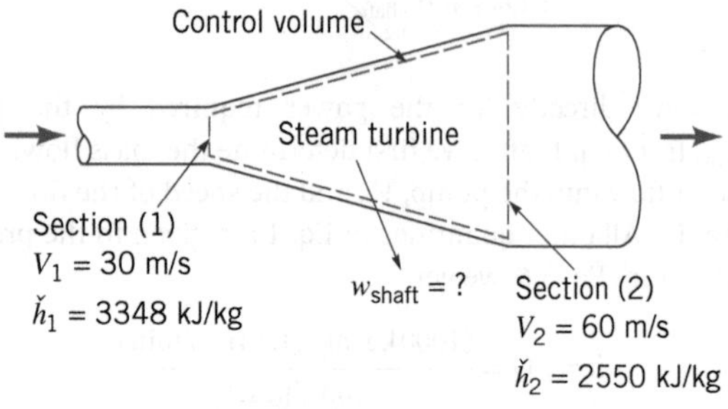

Control volume

Steam turbine

Section (1)
$V_1 = 30$ m/s
$\check{h}_1 = 3348$ kJ/kg

$w_{\text{shaft}} = ?$

Section (2)
$V_2 = 60$ m/s
$\check{h}_2 = 2550$ kJ/kg

FIGURE E5.21b

Solution We use a control volume that includes the steam in the turbine from the entrance to the exit, as shown in Fig. E5.21b. Applying Eq. 5.69 to the steam in this control volume, we get

0 (elevation change is negligible)

0 (adiabatic flow)

$$\dot{m}\left[\check{h}_2 - \check{h}_1 + \frac{V_2^2 - V_1^2}{2} + g(z_2 - z_1)\right] = \dot{Q}_{\substack{\text{net} \\ \text{in}}} + \dot{W}_{\substack{\text{shaft} \\ \text{net in}}} \qquad (1)$$

The work output per unit mass of steam through-flow, $w_{\text{shaft net in}}$, can be obtained by dividing Eq. 1 by the mass flowrate, $\dot{m}$, to obtain

$$w_{\substack{\text{shaft} \\ \text{net in}}} = \frac{\dot{W}_{\substack{\text{shaft} \\ \text{net in}}}}{\dot{m}} = \check{h}_2 - \check{h}_1 + \frac{V_2^2 - V_1^2}{2} \qquad (2)$$

Since $w_{\text{shaft net out}} = -w_{\text{shaft net in}}$, we obtain

$$w_{\substack{\text{shaft} \\ \text{net out}}} = \check{h}_1 - \check{h}_2 + \frac{V_1^2 - V_2^2}{2}$$

or

$$w_{\substack{\text{shaft} \\ \text{net out}}} = 3348 \text{ kJ/kg} - 2550 \text{ kJ/kg}$$

$$+ \frac{[(30 \text{ m/s})^2 - (60 \text{ m/s})^2][1 \text{ J/(N} \cdot \text{m})]}{2[1(\text{kg} \cdot \text{m})/(\text{N} \cdot \text{s}^2)](1000 \text{ J/kJ})}$$

Thus,

$$w_{\substack{\text{shaft} \\ \text{net out}}} = 3348 \text{ kJ/kg} - 2550 \text{ kJ/kg} - 1.35 \text{ kJ/kg}$$

$$= 797 \text{ kJ/kg} \qquad (Ans)$$

Comment Note that in this particular example, the change in kinetic energy is small in comparison to the difference in enthalpy. This is often true in applications involving steam flows. To determine the power output, $\dot{W}_{\text{shaft}}$, we must know the mass flowrate, $\dot{m}$. If the entrance and exit enthalpies were not specified in the problem statement, they could have been obtained from *steam tables* or appropriate software (see Ref. 3) if other information about the thermodynamic state of the steam was specified.

If the flow is truly steady throughout, so that no work is done, one-dimensional, and only one fluid stream is involved, then setting the shaft work to zero, the energy equation becomes

$$\dot{m}\left[\breve{u}_{\text{out}} - \breve{u}_{\text{in}} + \left(\frac{p}{\rho}\right)_{\text{out}} - \left(\frac{p}{\rho}\right)_{\text{in}} + \frac{V_{\text{out}}^2 - V_{\text{in}}^2}{2} + g(z_{\text{out}} - z_{\text{in}})\right] = \dot{Q}_{\substack{\text{net} \\ \text{in}}} \quad (5.70)$$

We call Eq. 5.70 the *one-dimensional, steady-flow energy equation*. This equation is valid for incompressible and compressible flows. For compressible flows, enthalpy is most often used in the one-dimensional, steady-flow energy equation; thus, we have

$$\dot{m}\left[\breve{h}_{\text{out}} - \breve{h}_{\text{in}} + \frac{V_{\text{out}}^2 - V_{\text{in}}^2}{2} + g(z_{\text{out}} - z_{\text{in}})\right] = \dot{Q}_{\substack{\text{net} \\ \text{in}}} \quad (5.71)$$

An application of Eq. 5.70 is illustrated in Example 5.22.

V5.14 Pelton wheel turbine

EXAMPLE 5.22 | Energy—Temperature Change

Given The 128 m waterfall shown in **Fig. E5.22a** involves steady flow from one large body of water to another.

Find Determine the temperature change associated with this flow.

David Hayes/123RF

FIGURE E5.22a

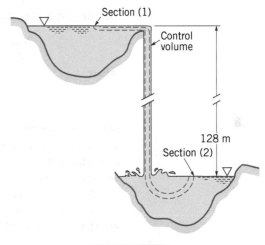

FIGURE E5.22b

We assume that the flow is adiabatic. Thus, $\dot{Q}_{\text{net in}} = 0$. Also, because the flow is incompressible ($\rho_1 = \rho_2$) and atmospheric pressure prevails at sections (1) and (2) ($p_1 = p_2$),

$$\left(\frac{p}{\rho}\right)_1 = \left(\frac{p}{\rho}\right)_2 \quad (3)$$

Furthermore,

$$V_1 = V_2 = 0 \quad (4)$$

because the surface of each large body of water is considered motionless. Thus, Eqs. 1–4 combine to yield

$$T_2 - T_1 = \frac{g(z_1 - z_2)}{\breve{c}}$$

so that with

$$\breve{c} = 4.2 \text{ J/kg} \cdot \text{K}$$

$$T_1 - T_2 = \frac{(9.81 \text{ m/s}^2)(128 \text{ m})}{(4.2 \text{ J/kg·K})}$$

$$= 300 \text{ K}$$

Comment Note that it takes a considerable change of potential energy to produce even a small increase in temperature.

Solution To solve this problem, we consider a control volume consisting of a small cross-sectional streamtube from the nearly motionless surface of the upper body of water to the nearly motionless surface of the lower body of water, as is sketched in **Fig. E5.22b**. We need to determine $T_2 - T_1$. This temperature change is related to the change of internal energy of the water, $\breve{u}_2 - \breve{u}_1$, by the relationship

$$T_2 - T_1 = \frac{\breve{u}_2 - \breve{u}_1}{\breve{c}} \quad (1)$$

Where $\breve{c}$ is the specipic heat of water. The application of Eq. 5.70 to the contents of this control volume leads to

$$\dot{m}\left[\breve{u}_2 - \breve{u}_1 + \left(\frac{p}{\rho}\right)_2 - \left(\frac{p}{\rho}\right)_1 + \frac{V_2^2 - V_1^2}{2} + g(z_2 - z_1)\right]$$

$$= \dot{Q}_{\substack{\text{net} \\ \text{in}}} \quad (2)$$

A form of the energy equation that is most often used to solve incompressible flow problems is developed in Section 5.3.3.

5.3.3 The Mechanical Energy Equation and the Bernoulli Equation

When the one-dimensional energy equation for steady-in-the-mean flow, Eq. 5.67, is applied to a flow that is truly steady, Eq. 5.67 becomes the one-dimensional, steady-flow energy equation, Eq. 5.70. The only difference between Eq. 5.67 and Eq. 5.70 is that shaft power, $\dot{W}_{\text{shaft net in}}$, is zero if the flow is steady throughout the control volume. (Fluid machines involve locally unsteady flow.) If in addition to being steady, the flow is incompressible ($\rho_{\text{out}} = \rho_{\text{in}} = \rho$), we get from Eq. 5.70

$$\dot{m}\left[\check{u}_{\text{out}} - \check{u}_{\text{in}} + \frac{p_{\text{out}}}{\rho} - \frac{p_{\text{in}}}{\rho} + \frac{V_{\text{out}}^2 - V_{\text{in}}^2}{2} + g(z_{\text{out}} - z_{\text{in}})\right] = \dot{Q}_{\substack{\text{net} \\ \text{in}}} \tag{5.72}$$

Dividing Eq. 5.72 by the mass flowrate, $\dot{m}$, and rearranging terms, we obtain

$$\frac{p_{\text{out}}}{\rho} + \frac{V_{\text{out}}^2}{2} + gz_{\text{out}} = \frac{p_{\text{in}}}{\rho} + \frac{V_{\text{in}}^2}{2} + gz_{\text{in}} - \left(\check{u}_{\text{out}} - \check{u}_{\text{in}} - q_{\substack{\text{net} \\ \text{in}}}\right) \tag{5.73}$$

where

$$q_{\substack{\text{net} \\ \text{in}}} = \frac{\dot{Q}_{\text{net in}}}{\dot{m}}$$

is the heat transfer rate per mass flowrate, or heat transfer per unit mass. Note that Eq. 5.73 involves energy per unit mass and is applicable to one-dimensional, steady, incompressible flow of a single stream of fluid between two sections, or flow through a stream tube between the two sections, or flow along a streamline between the two sections.

Now assume that the steady, incompressible flow that we are considering is also frictionless. In this case, the Bernoulli equation, Eq. 4.17, can also be used to describe what happens along a streamline between two sections in the flow

$$p_{\text{out}} + \frac{\rho V_{\text{out}}^2}{2} + \gamma z_{\text{out}} = p_{\text{in}} + \frac{\rho V_{\text{in}}^2}{2} + \gamma z_{\text{in}} \tag{5.74}$$

We divide Eq. 5.74 by density, ρ, to obtain

$$\frac{p_{\text{out}}}{\rho} + \frac{V_{\text{out}}^2}{2} + gz_{\text{out}} = \frac{p_{\text{in}}}{\rho} + \frac{V_{\text{in}}^2}{2} + gz_{\text{in}} \tag{5.75}$$

Equation 5.75 is the Bernoulli equation in energy form (see Section 4.4). It is derived from Newton's second law of motion, which is a different physical principle from conservation of energy, so Eqs. 5.73 and 5.75 are independent of each other (i.e., independently true).

A comparison of Eqs. 5.73 and 5.75 prompts us to conclude that

$$\check{u}_{\text{out}} - \check{u}_{\text{in}} - q_{\substack{\text{net} \\ \text{in}}} = 0 \tag{5.76}$$

when the steady incompressible flow is frictionless. For steady incompressible flow with friction, we learn from experience (the second law of thermodynamics; see Section 5.4 for details) that

$$\check{u}_{\text{out}} - \check{u}_{\text{in}} - q_{\substack{\text{net} \\ \text{in}}} > 0 \tag{5.77}$$

In Eqs. 5.73 and 5.75, we can consider the combination of variables

$$\frac{p}{\rho} + \frac{V^2}{2} + gz$$

as equal to *useful energy*. This energy is useful because it can be completely used to produce work. Useful energy is sometimes called *available energy*. Thus, from inspection of Eqs. 5.73 and 5.75, we can conclude that $\check{u}_{\text{out}} - \check{u}_{\text{in}} - q_{\text{net in}}$ represents the **loss** of useful or available energy that occurs in an incompressible fluid flow because of friction. In equation form we have

$$\check{u}_{\text{out}} - \check{u}_{\text{in}} - q_{\substack{\text{net} \\ \text{in}}} = \text{loss} \tag{5.78}$$

For a frictionless flow, Eqs. 5.73 and 5.75 tell us that loss equals zero.

It is often convenient to express Eq. 5.73 in terms of loss as

$$\frac{p_{out}}{\rho} + \frac{V_{out}^2}{2} + gz_{out} = \frac{p_{in}}{\rho} + \frac{V_{in}^2}{2} + gz_{in} - \text{loss} \tag{5.79}$$

An application of Eq. 5.79 is illustrated in Example 5.23.

EXAMPLE 5.23 | Energy—Effect of Loss of Available Energy

Given As shown in **Fig. E5.23a**, air flows from a room through two different vent configurations: a cylindrical hole in the wall having a diameter of 120 mm and the same diameter cylindrical hole in the wall but with a well-rounded entrance. The room pressure is held constant at 1.0 kPa above atmospheric pressure. Both vents exhaust into the atmosphere. As discussed in Section 8.4.2, the loss in available energy associated with flow through the cylindrical vent from the room to the vent exit is $0.5 V_2^2/2$, where V_2 is the uniformly distributed exit velocity of air. The loss in available energy associated with flow through the rounded entrance vent from the room to the vent exit is $0.05 V_2^2/2$, where V_2 is the uniformly distributed exit velocity of air.

Find Compare the volume flowrates associated with the two different vent configurations.

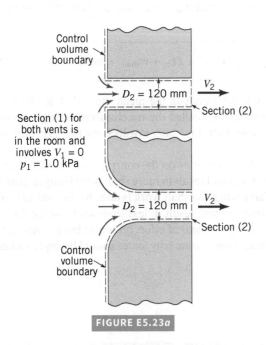

Control volume boundary

V_2

$D_2 = 120$ mm

Section (2)

Section (1) for both vents is in the room and involves $V_1 = 0$
$p_1 = 1.0$ kPa

$D_2 = 120$ mm V_2

Section (2)

Control volume boundary

FIGURE E5.23a

Solution We use the control volume for each vent sketched in Fig. E5.23a. What is sought is the flowrate, $Q = A_2 V_2$, where A_2 is the vent exit cross-sectional area, and V_2 is the uniformly distributed exit velocity. For both vents, application of Eq. 5.79 leads to

$$\frac{p_2}{\rho} + \frac{V_2^2}{2} + \cancel{gz_2}^{\ 0\ (\text{no elevation change})} = \frac{p_1}{\rho} + \cancel{\frac{V_1^2}{2}}^{\ 0\,(V_1 \approx 0)} + \cancel{gz_1} - {}_1\text{loss}_2 \tag{1}$$

where ${}_1\text{loss}_2$ is the loss between sections (1) and (2). Solving Eq. 1 for V_2, we get

$$V_2 = \sqrt{2\left[\left(\frac{p_1 - p_2}{\rho}\right) - {}_1\text{loss}_2\right]} \tag{2}$$

Since

$${}_1\text{loss}_2 = K_L \frac{V_2^2}{2} \tag{3}$$

where K_L is the loss coefficient ($K_L = 0.5$ and 0.05 for the two vent configurations involved), we can combine Eqs. 2 and 3 to get

$$V_2 = \sqrt{2\left[\left(\frac{p_1 - p_2}{\rho}\right) - K_L\frac{V_2^2}{2}\right]} \tag{4}$$

Solving Eq. 4 for V_2, we obtain

$$V_2 = \sqrt{\frac{p_1 - p_2}{\rho[(1 + K_L)/2]}} \tag{5}$$

Therefore, for flowrate, Q, we obtain

$$Q = A_2 V_2 = \frac{\pi D_2^2}{4}\sqrt{\frac{p_1 - p_2}{\rho[(1 + K_L)/2]}} \tag{6}$$

For the rounded entrance cylindrical vent, Eq. 6 gives

$$Q = \frac{\pi (120 \text{ mm})^2}{4(1000 \text{ mm/m})^2}$$

$$\times \sqrt{\frac{(1.0 \text{ kPa})(1000 \text{ Pa/kPa})[1(\text{N/m}^2)/(\text{Pa})]}{(1.23 \text{ kg/m}^3)[(1 + 0.05)/2][1(\text{N} \cdot \text{s}^2)/(\text{kg} \cdot \text{m})]}}$$

or

$$Q = 0.445 \text{ m}^3/\text{s} \tag{Ans}$$

For the cylindrical vent, Eq. 6 gives us

$$Q = \frac{\pi (120 \text{ mm})^2}{4(1000 \text{ mm/m})^2}$$

$$\times \sqrt{\frac{(1.0 \text{ kPa})(1000 \text{ Pa/kPa})[1(\text{N/m}^2)/(\text{Pa})]}{(1.23 \text{ kg/m}^3)[(1 + 0.5)/2][1(\text{N} \cdot \text{s}^2)/(\text{kg} \cdot \text{m})]}}$$

or

$$Q = 0.372 \ \text{m}^3/\text{s} \qquad (Ans)$$

Comment   By repeating the calculations for various values of the loss coefficient, K_L, the results shown in **Fig. E5.23b** are obtained. Note that the rounded entrance vent allows the passage of more air than does the cylindrical vent because the loss associated with the rounded entrance vent is less than that for the cylindrical one. For this flow, the pressure drop, $p_1 - p_2$, has two purposes: (1) overcome the loss associated with the flow and (2) produce the kinetic energy at the exit. Even if there were no loss (i.e., $K_L = 0$), a pressure drop would be needed to accelerate the fluid through the vent.

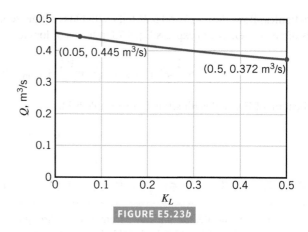

FIGURE E5.23b

An important group of fluid mechanics problems involves one-dimensional, incompressible, steady-in-the-mean flow with friction and shaft work. Included in this category are flows through pumps, blowers, fans, and turbines. For this kind of flow, Eq. 5.67 becomes

$$\dot{m}\left[\check{u}_{\text{out}} - \check{u}_{\text{in}} + \frac{p_{\text{out}}}{\rho} - \frac{p_{\text{in}}}{\rho} + \frac{V_{\text{out}}^2 - V_{\text{in}}^2}{2} + g(z_{\text{out}} - z_{\text{in}})\right] = \dot{Q}_{\substack{\text{net} \\ \text{in}}} + \dot{W}_{\substack{\text{shaft} \\ \text{net in}}} \qquad (5.80)$$

Dividing Eq. 5.80 by mass flowrate and using the work per unit mass, $w_{\substack{\text{shaft} \\ \text{net in}}} = \dot{W}_{\substack{\text{shaft} \\ \text{net in}}} / \dot{m}$, we obtain

$$\frac{p_{\text{out}}}{\rho} + \frac{V_{\text{out}}^2}{2} + gz_{\text{out}} = \frac{p_{\text{in}}}{\rho} + \frac{V_{\text{in}}^2}{2} + gz_{\text{in}} + w_{\substack{\text{shaft} \\ \text{net in}}} - \left(\check{u}_{\text{out}} - \check{u}_{\text{in}} - q_{\substack{\text{net} \\ \text{in}}}\right) \qquad (5.81)$$

Since the flow is incompressible, Eq. 5.78 shows that $\check{u}_{\text{out}} - \check{u}_{\text{in}} - q_{\text{net in}}$ equals the loss of useful energy and Eq. 5.81 can be expressed as

$$\boxed{\frac{p_{\text{out}}}{\rho} + \frac{V_{\text{out}}^2}{2} + gz_{\text{out}} = \frac{p_{\text{in}}}{\rho} + \frac{V_{\text{in}}^2}{2} + gz_{\text{in}} + w_{\substack{\text{shaft} \\ \text{net in}}} - \text{loss}} \qquad (5.82)$$

V5.15 Energy transfer

This is a form of the energy equation for steady-in-the-mean flow that is often used for incompressible flow problems. It is sometimes called the **mechanical energy equation,** or the *extended Bernoulli equation*. Note that Eq. 5.82 involves energy per unit mass $(\text{N} \cdot \text{m/kg} = \text{m}^2/\text{s}^2)$.

According to Eq. 5.82, when the shaft work is done on the control volume contents, as for example with a pump, a larger amount of loss will result in more shaft work being required for the same rise in available energy. Similarly, when the shaft work is done by the control volume contents (for example, a turbine), a larger loss will result in less shaft work out for the same drop in available energy. Designers spend a great deal of effort on minimizing losses in fluid flow components. The following examples demonstrate why losses should be kept as small as possible in fluid systems.

Minimizing loss is a central goal of fluid mechanical design.

EXAMPLE 5.24 | Energy—Fan Work and Efficiency

Given   An axial-flow ventilating fan driven by a motor that delivers 0.4 kW of power to the fan blades produces a 0.6-m-diameter axial stream of air having a speed of 12 m/s. The flow far upstream of the fan has negligible speed.

Find   Determine how much of the work to the air actually produces useful effects, that is, fluid motion and a rise in available energy. Estimate the aerodynamic efficiency of this fan.

Solution   We select a fixed and nondeforming control volume, as is illustrated in **Fig. E5.24**. The application of Eq. 5.82 to the contents of this control volume leads to

$$w_{\substack{\text{shaft} \\ \text{net in}}} - \text{loss} = \left(\frac{\cancel{p_2}}{\rho} + \frac{V_2^2}{2} + \cancel{gz_2}\right) - \left(\cancel{\frac{p_1}{\rho}} + \cancel{\frac{V_1^2}{2}} + \cancel{gz_1}\right) \qquad (1)$$

0 (atmospheric pressures cancel)   0 $(V_1 \approx 0)$

0 (no elevation change)

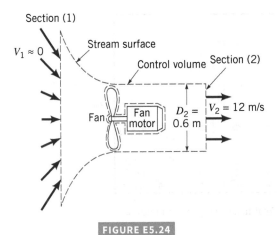

Section (1)

$V_1 \approx 0$

Stream surface

Control volume   Section (2)

Fan   Fan motor $D_2 = 0.6$ m $V_2 = 12$ m/s

FIGURE E5.24

where $w_{\text{shaft net in}}$ − loss is the amount of work added to the air that produces a useful effect. Equation 1 leads to

$$w_{\substack{\text{shaft}\\ \text{net in}}} - \text{loss} = \frac{V_2^2}{2} = \frac{(12 \text{ m/s})^2}{2[1(\text{kg} \cdot \text{m})/(\text{N} \cdot \text{s}^2)]}$$

$$= 72.0 \text{ N} \cdot \text{m/kg} \qquad (2) \quad (Ans)$$

A reasonable estimate of efficiency, η, would be the ratio of amount of work that produces a useful effect, Eq. 2, to the amount of work delivered to the fan blades. That is,

$$\eta = \frac{w_{\substack{\text{shaft}\\ \text{net in}}} - \text{loss}}{w_{\substack{\text{shaft}\\ \text{net in}}}} \qquad (3)$$

To calculate the efficiency, we need a value of $w_{\text{shaft net in}}$, which is related to the power delivered to the blades, $\dot{W}_{\text{shaft net in}}$. We note that

$$w_{\substack{\text{shaft}\\ \text{net in}}} = \frac{\dot{W}_{\substack{\text{shaft}\\ \text{net in}}}}{\dot{m}} \qquad (4)$$

where the mass flowrate, $\dot{m}$, is (from Eq. 5.6)

$$\dot{m} = \rho AV = \rho \frac{\pi D_2^2}{4} V_2 \qquad (5)$$

For fluid density, ρ, we use 1.23 kg/m^3 (standard air) and, thus, from Eqs. 4 and 5, we obtain

$$w_{\substack{\text{shaft}\\ \text{net in}}} = \frac{\dot{W}_{\substack{\text{shaft}\\ \text{net in}}}}{(\rho \pi D_2^2/4) V_2}$$

$$= \frac{(0.4 \text{ kW})[1000 \text{ (Nm)/(skW)}]}{(1.23 \text{ kg/m}^3)[(\pi)(0.6 \text{ m})^2/4](12 \text{ m/s})}$$

or

$$w_{\substack{\text{shaft}\\ \text{net in}}} = 95.8 \text{ N} \cdot \text{m/kg} \qquad (6)$$

From Eqs. 2, 3, and 6, we obtain

$$\eta = \frac{72.0 \text{ N} \cdot \text{m/kg}}{95.8 \text{ N} \cdot \text{m/kg}} = 0.752 \qquad (Ans)$$

Comment Note that only 75% of the power that was delivered to the air resulted in useful effects and, thus, 25% of the shaft power is lost to air friction. The power input to the motor would be more than 0.4 kW because of electrical losses in the motor and mechanical friction in the bearings.

The Wide World of Fluids

Curtain of air

An air curtain is produced by blowing air through a long rectangular nozzle to produce a high-velocity sheet of air, or a "curtain of air." This air curtain is typically directed over a doorway or opening as a replacement for a conventional door. The air curtain can be used for such things as keeping warm air from infiltrating dedicated cold spaces, preventing dust and other contaminants from entering a clean environment, and even just keeping insects out of the

workplace, still allowing people to enter or exit. A disadvantage over conventional doors is the added *power requirements* to operate the air curtain, although the advantages can outweigh the disadvantage for various industrial applications. New applications for current air curtain designs continue to be developed. For example, the use of air curtains as a means of road tunnel fire security is currently being investigated. In such an application, the air curtain would act to isolate a portion of the tunnel where fire has broken out and not allow smoke and fumes to infiltrate the entire tunnel system.

If Eq. 5.82, which involves energy per unit mass, is multiplied by fluid density, ρ, we obtain

$$p_{\text{out}} + \frac{\rho V_{\text{out}}^2}{2} + \gamma z_{\text{out}} = p_{\text{in}} + \frac{\rho V_{\text{in}}^2}{2} + \gamma z_{\text{in}} + \rho w_{\substack{\text{shaft}\\ \text{net in}}} - \rho(\text{loss}) \qquad (5.83)$$

where $\gamma = \rho g$ is the specific weight of the fluid. Equation 5.83 involves *energy per unit volume*, and the units involved are identical with those used for pressure (N · m/m^3 = N/m^2).

If Eq. 5.82 is divided by the acceleration of gravity, g, we get

$$\boxed{\frac{p_{\text{out}}}{\gamma} + \frac{V_{\text{out}}^2}{2g} + z_{\text{out}} = \frac{p_{\text{in}}}{\gamma} + \frac{V_{\text{in}}^2}{2g} + z_{\text{in}} + h_s - h_L} \qquad (5.84)$$

where

$$h_s = w_{\text{shaft net in}}/g = \frac{\dot{W}_{\substack{\text{shaft}\\ \text{net in}}}}{\dot{m}g} = \frac{\dot{W}_{\substack{\text{shaft}\\ \text{net in}}}}{\gamma Q} \qquad (5.85)$$

▶ **Video**

V5.16 Water plant aerator

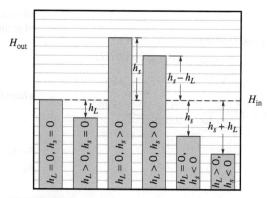

FIGURE 5.8   Total-head change in fluid flows.

Head is energy per unit weight.

is the **shaft work head** and $h_L = loss/g$ is called the **head loss**. Equation 5.84 involves *energy per unit weight* (N · m/N = m). In Section 4.7, we introduced the notion of "head," which is energy per unit weight. Units of length (for example, m) are used to quantify the amount of head involved[3]. If a turbine is in the control volume, h_s is negative because it is associated with shaft work done by the fluid. For a pump in the control volume, h_s is positive because it is associated with shaft work done on the fluid.

We can define a total head, H, as follows:

$$H = \frac{p}{\gamma} + \frac{V^2}{2g} + z$$

Then Eq. 5.84 can be expressed as

$$H_{out} = H_{in} + h_s - h_L$$

Recall that h_s is positive for a pump and negative for a turbine. Some important possible values of H_{out} in comparison to H_{in} are shown in **Fig. 5.8**. Note that h_L (head loss) always reduces the value of H_{out}, except in the ideal case when it is zero. Note also that h_L lessens the amount of shaft work that can be extracted from a fluid. When $h_L = 0$ (ideal condition), the shaft work head, h_s, and the change in total head are the same. This head change is sometimes called ideal head change. The corresponding ideal shaft work head is minimum required to achieve a desired effect. For work out, it is the maximum possible. Designers usually strive to minimize loss.

EXAMPLE 5.25 | Energy—Head Loss and Power Loss

Given   The pump shown in **Fig. E5.25a** adds 7.5 kW to the water as it pumps water from the lower lake to the upper lake. The elevation difference between the lake surfaces is 9 m, and the head loss is 4.6 m.

Find   Determine

  a. the flowrate and

  b. the power loss associated with this flow.

Solution

  a. The energy equation (Eq. 5.84) for this flow is

$$\frac{p_2}{\gamma} + \frac{V_2^2}{2g} + z_2 = \frac{p_1}{\gamma} + \frac{V_1^2}{2g} + z_1 + h_s - h_L \qquad (1)$$

where points 2 and 1 (corresponding to "out" and "in" in Eq. 5.84) are located on the lake surfaces. Thus, $p_2 = p_1 = 0$ and $V_2 = V_1 = 0$ so that Eq. 1 becomes

$$h_s = h_L + z_2 - z_1 \qquad (2)$$

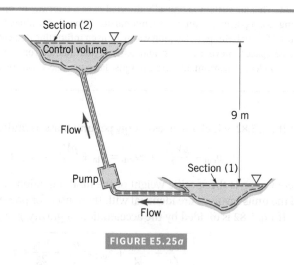

FIGURE E5.25a

---

[3]When working with energy per unit mass, as in Eq. 5.82, the term "loss" is sometimes replaced by the term "gh_L," which represents the energy loss per unit mass as a loss of potential energy.

where $z_2 = 9$ m, $z_1 = 0$, and $h_L = 4.6$ m. The pump head is obtained from Eq. 5.85 as

$$h_s = \dot{W}_{\text{shaft net in}}/\gamma Q$$
$$= (7.5 \times 10^3 \text{ W})/(1000 \times 9.81 \text{ N/m}^3)\,Q$$
$$= (0.76 \text{ m}^4/\text{s})/Q$$

where h_s is in m when Q is in m^3/s.

Hence, from Eq. 2,

$$0.76/Q = 4.6 \text{ m} + 9 \text{ m}$$

or

$$Q = 0.05 \text{ m}^3/\text{s} \qquad \text{(Ans)}$$

b. The power lost due to friction can be obtained from Eq. 5.85 as

$$\dot{W}_{\text{loss}} = \gamma Q h_L = (9810 \text{ N/m}^3)(0.05 \text{ m}^3/\text{s})(4.6 \text{ m})$$
$$= 2.26 \text{ kW} \qquad \text{(Ans)}$$

Comments The remaining 7.5 kW − 2.26 kW = 5.24 kW that the pump adds to the water is used to lift the water from the lower to the upper lake. This energy is not "lost," but it is stored as potential energy.

By repeating the calculations for various head losses, h_L, the results shown in **Fig. E5.25b** are obtained. Note that as the head loss increases, the flowrate decreases because an increasing portion of the 7.5 kW supplied by the pump is lost and, therefore, not available to lift the fluid to the higher elevation.

Note that in this example, the purpose of the pump is to lift the water (a 9 m head) and overcome the head loss (a 4.6 m head); it does not, overall, alter the water's pressure or velocity.

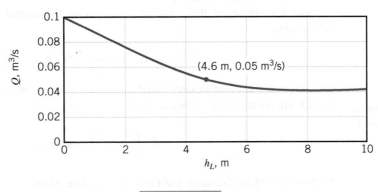

FIGURE E5.25b

A comparison of the energy equation and the Bernoulli equation has led to the concept of loss of available energy in incompressible fluid flows with friction. In Chapter 8, we discuss in detail some methods for estimating loss in incompressible flows with friction. In Section 5.4, we demonstrate that loss of available energy is also an important factor to consider in compressible flows with friction.

The Wide World of Fluids

Smart shocks

Vehicle shock absorbers are dampers used to provide a smooth, controllable ride. When going over a bump, the relative motion between the tires and the vehicle body displaces a piston in the shock and forces a viscous fluid through a small orifice or channel. The viscosity of the fluid produces a *head loss* that dissipates energy to dampen the vertical motion. Current shocks use a fluid with fixed viscosity. However, recent technology has been developed that uses a synthetic oil with millions of tiny iron balls suspended in it. These tiny balls react to a magnetic field generated by an electric coil on the shock piston in a manner that changes the fluid viscosity, going anywhere from essentially no damping to a solid almost instantly. A computer adjusts the current to the coil to select the proper viscosity for the given conditions (i.e., wheel speed, vehicle speed, steering-wheel angle, lateral acceleration, brake application, and temperature). The goal of these adjustments is an optimally tuned shock that keeps the vehicle on a smooth, even keel while maximizing the contact of the tires with the pavement for any road conditions.

5.3.4 Application of the Energy Equation to Nonuniform Flows

The forms of the energy equation discussed in Sections 5.3.2 and 5.3.3 are applicable to one-dimensional flows, flows that are approximated with uniform velocity distributions where fluid crosses the control surface.

If the velocity profile at any section where flow crosses the control surface is not uniform, inspection of the energy equation for a control volume, Eq. 5.64, suggests that the integral

$$\int_{cs} \frac{V^2}{2} \rho \mathbf{V} \cdot \hat{\mathbf{n}} \, dA$$

will require special attention. The other terms of Eq. 5.64 can be accounted for as already discussed in Sections 5.3.2 and 5.3.3.

For one stream of fluid entering and leaving the control volume, we can define the relationship

$$\int_{cs} \frac{V^2}{2} \rho \mathbf{V} \cdot \hat{\mathbf{n}} \, dA = \dot{m}\left(\frac{\alpha_{out}\overline{V}^2_{out}}{2} - \frac{\alpha_{in}\overline{V}^2_{in}}{2} \right)$$

where α is the **kinetic energy coefficient** and $\overline{V}$ is the average velocity defined earlier in Eq. 5.7. Therefore, we can conclude that

$$\frac{\dot{m}\alpha\overline{V}^2}{2} = \int_A \frac{V^2}{2} \rho \mathbf{V} \cdot \hat{\mathbf{n}} \, dA$$

for flow through surface area A of the control surface. Thus,

$$\alpha = \frac{\int_A (V^2/2)\rho \mathbf{V} \cdot \hat{\mathbf{n}} \, dA}{\dot{m}\overline{V}^2/2} \tag{5.86}$$

It can be shown that for any velocity profile, $\alpha \geq 1$, with $\alpha = 1$ only for uniform flow. Some typical velocity profile examples for flow in a conventional pipe are shown in the adjacent sketch. Therefore, for nonuniform velocity profiles, the energy equation on an energy per unit mass basis for the incompressible flow of one stream of fluid through a control volume that is steady in the mean is

$$\boxed{\frac{p_{out}}{\rho} + \frac{\alpha_{out}\overline{V}^2_{out}}{2} + gz_{out} = \frac{p_{in}}{\rho} + \frac{\alpha_{in}\overline{V}^2_{in}}{2} + gz_{in} + w_{\substack{shaft \\ net\,in}} - loss} \tag{5.87}$$

On an energy per unit volume basis, we have

$$p_{out} + \frac{\rho\alpha_{out}\overline{V}^2_{out}}{2} + \gamma z_{out} = p_{in} + \frac{\rho\alpha_{in}\overline{V}^2_{in}}{2} + \gamma z_{in} + \rho w_{\substack{shaft \\ net\,in}} - \rho(loss) \tag{5.88}$$

and on an energy per unit weight or head basis, we have

$$\boxed{\frac{p_{out}}{\gamma} + \frac{\alpha_{out}\overline{V}^2_{out}}{2g} + z_{out} = \frac{p_{in}}{\gamma} + \frac{\alpha_{in}\overline{V}^2_{in}}{2g} + z_{in} + \frac{w_{\substack{shaft \\ net\,in}}}{g} - h_L} \tag{5.89}$$

When considering highly turbulent flow, α is nearly 1 and can therefore be neglected from the energy equation without a loss of accuracy. Also, when nonvelocity terms dominate the energy equation, the use of α can again be neglected without a loss of accuracy. Examples 5.26 and 5.27 illustrate the use of the kinetic energy coefficient.

The kinetic energy coefficient is used to account for nonuniform flows.

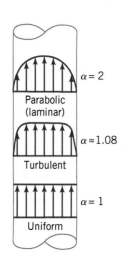

$\alpha = 2$
Parabolic (laminar)

$\alpha \approx 1.08$
Turbulent

$\alpha = 1$
Uniform

EXAMPLE 5.26 | Energy—Effect of Nonuniform Velocity Profile

Given The small fan shown in **Fig. E5.26** moves air at a mass flow-rate of 0.1 kg/min. Upstream of the fan, the pipe diameter is 60 mm, the flow is laminar, the velocity distribution is parabolic, and the kinetic energy coefficient, α_1, is equal to 2.0. Downstream of the fan, the pipe diameter is 30 mm, the flow is turbulent, the velocity profile is quite uniform, and the kinetic energy coefficient, α_2, is equal to 1.08. The rise in static pressure across the fan is 0.1 kPa, and the fan motor draws 0.14 W.

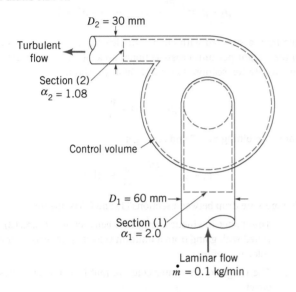

FIGURE E5.26

Find Compare the value of loss calculated: **(a)** assuming uniform velocity distributions and **(b)** considering actual velocity distributions.

Solution Application of Eq. 5.87 to the contents of the control volume shown in Fig. E5.26 leads to

$$\frac{p_2}{\rho} + \frac{\alpha_2 \overline{V}_2^2}{2} + \cancel{gz_2}^{\text{0 (change in } gz \text{ is negligible)}} = \frac{p_1}{\rho} + \frac{\alpha_1 \overline{V}_1^2}{2} + \cancel{gz_1}$$
$$- \text{loss} + w_{\substack{\text{shaft} \\ \text{net in}}} \tag{1}$$

or solving Eq. 1 for loss, we get

$$\text{loss} = w_{\substack{\text{shaft} \\ \text{net in}}} - \left(\frac{p_2 - p_1}{\rho}\right) + \frac{\alpha_1 \overline{V}_1^2}{2} - \frac{\alpha_2 \overline{V}_2^2}{2} \tag{2}$$

To proceed further, we need values of $w_{\text{shaft net in}}$, $\overline{V}_1$, and $\overline{V}_2$. These quantities can be obtained as follows. For shaft work

$$w_{\substack{\text{shaft} \\ \text{net in}}} = \frac{\text{power to fan motor}}{\dot{m}}$$

or

$$w_{\substack{\text{shaft} \\ \text{net in}}} = \frac{(0.14 \text{ W})[(1 \text{ N} \cdot \text{m/s})/\text{W}]}{0.1 \text{ kg/min}}(60 \text{ s/min})$$
$$= 84.0 \text{ N} \cdot \text{m/kg} \tag{3}$$

For the average velocity at section (1), $\overline{V}_1$, from Eq. 5.11, we obtain

$$\overline{V}_1 = \frac{\dot{m}}{\rho A_1}$$

$$= \frac{\dot{m}}{\rho (\pi D_1^2/4)} \tag{4}$$

$$= \frac{(0.1 \text{ kg/min})(1 \text{ min/60 s})(1000 \text{ mm/m})^2}{(1.23 \text{ kg/m}^3)[\pi (60 \text{ mm})^2/4]}$$

$$= 0.479 \text{ m/s}$$

For the average velocity at section (2), $\overline{V}_2$

$$\overline{V}_2 = \frac{(0.1 \text{ kg/min})(1 \text{ min/60 s})(1000 \text{ mm/m})^2}{(1.23 \text{ kg/m}^3)[\pi (30 \text{ mm})^2/4]}$$
$$= 1.92 \text{ m/s} \tag{5}$$

a. For the assumed uniform velocity profiles ($\alpha_1 = \alpha_2 = 1.0$), Eq. 2 yields

$$\text{loss} = w_{\substack{\text{shaft} \\ \text{net in}}} - \left(\frac{p_2 - p_1}{\rho}\right) + \frac{\overline{V}_1^2}{2} - \frac{\overline{V}_2^2}{2} \tag{6}$$

Using Eqs. 3, 4, and 5 and the pressure rise given in the problem statement, Eq. 6 gives

$$\text{loss} = 84.0 \frac{\text{N} \cdot \text{m}}{\text{kg}} - \frac{(0.1 \text{ kPa})(1000 \text{ Pa/kPa})(1 \text{ N/m}^2/\text{Pa})}{1.23 \text{ kg/m}^3}$$

$$+ \frac{(0.479 \text{ m/s})^2}{2[1(\text{kg} \cdot \text{m})/(\text{N} \cdot \text{s}^2)]} - \frac{(1.92 \text{ m/s})^2}{2[1(\text{kg} \cdot \text{m})/(\text{N} \cdot \text{s}^2)]}$$

or

$$\text{loss} = 84.0 \text{ N} \cdot \text{m/kg} - 81.3 \text{ N} \cdot \text{m/kg}$$
$$+ 0.115 \text{ N} \cdot \text{m/kg} - 1.84 \text{ N} \cdot \text{m/kg}$$
$$= 0.975 \text{ N} \cdot \text{m/kg} \qquad (Ans)$$

b. For the actual velocity profiles ($\alpha_1 = 2, \alpha_2 = 1.08$), Eq. 1 gives

$$\text{loss} = w_{\substack{\text{shaft} \\ \text{net in}}} - \left(\frac{p_2 - p_1}{\rho}\right) + \alpha_1 \frac{\overline{V}_1^2}{2} - \alpha_2 \frac{\overline{V}_2^2}{2} \tag{7}$$

If we use Eqs. 3, 4, and 5 and the given pressure rise, Eq. 7 yields

$$\text{loss} = 84 \text{ N} \cdot \text{m/kg} - \frac{(0.1 \text{ kPa})(1000 \text{ Pa/kPa})(1 \text{ N/m}^2/\text{Pa})}{1.23 \text{ kg/m}^3}$$

$$+ \frac{2(0.479 \text{ m/s})^2}{2[1(\text{kg} \cdot \text{m})/(\text{N} \cdot \text{s}^2)]} - \frac{1.08(1.92 \text{ m/s})^2}{2[1(\text{kg} \cdot \text{m})/(\text{N} \cdot \text{s}^2)]}$$

or

$$\text{loss} = 84.0 \text{ N} \cdot \text{m/kg} - 81.3 \text{ N} \cdot \text{m/kg}$$
$$+ 0.230 \text{ N} \cdot \text{m/kg} - 1.99 \text{ N} \cdot \text{m/kg}$$
$$= 0.940 \text{ N} \cdot \text{m/kg} \qquad (Ans)$$

Comment The difference in loss calculated assuming uniform velocity profiles and actual velocity profiles is not large compared to $w_{\text{shaft net in}}$ for this fluid flow situation.

EXAMPLE 5.27 | Energy—Effect of Nonuniform Velocity Profile

Given Consider the flow situation of Example 5.14.

Find Apply Eq. 5.87 to develop an expression for the fluid pressure drop that occurs between sections (1) and (2). By comparing the equation for pressure drop obtained presently with the result of Example 5.14, obtain an expression for loss between sections (1) and (2).

Solution Application of Eq. 5.87 to the flow of Example 5.14 (see Fig. E5.14) leads to

$$\frac{p_2}{\rho} + \frac{\alpha_2 \overline{w}_2^2}{2} + gz_2 = \frac{p_1}{\rho} + \frac{\alpha_1 \overline{w}_1^2}{2} + gz_1 - \text{loss} + \overset{0 \ (\text{no shaft work})}{w_{\text{shaft}}} \quad (1)$$

Solving Eq. 1 for the pressure drop, $p_1 - p_2$, we obtain

$$p_1 - p_2 = \rho \left[\frac{\alpha_2 \overline{w}_2^2}{2} - \frac{\alpha_1 \overline{w}_1^2}{2} + g(z_2 - z_1) + \text{loss} \right] \quad (2)$$

Since the fluid velocity at section (1), w_1, is uniformly distributed over cross-sectional area A_1, the corresponding kinetic energy coefficient, α_1, is equal to 1.0. The kinetic energy coefficient at section (2), α_2, needs to be determined from the velocity profile distribution given in Example 5.14. Using Eq. 5.86, we get

$$\alpha_2 = \frac{\displaystyle\int_{A_2} \rho w_2^3 \, dA_2}{\dot{m} \overline{w}_2^2} \quad (3)$$

Substituting the parabolic velocity profile equation into Eq. 3, we obtain

$$\alpha_2 = \frac{\rho \displaystyle\int_0^R (2w_1)^3 [1 - (r/R)^2]^3 2\pi r \, dr}{(\rho A_2 \overline{w}_2) \overline{w}_2^2}$$

From conservation of mass, since $A_1 = A_2$

$$w_1 = \overline{w}_2 \quad (4)$$

Then, substituting Eq. 4 into Eq. 3, we obtain

$$\alpha_2 = \frac{\rho 8 \overline{w}_2^3 \, 2\pi \displaystyle\int_0^R [1 - (r/R)^2]^3 r \, dr}{\rho \pi R^2 \overline{w}_2^3}$$

or

$$\alpha_2 = \frac{16}{R^2} \int_0^R [1 - 3(r/R)^2 + 3(r/R)^4 - (r/R)^6] r \, dr$$
$$= 2 \quad (5)$$

Now, we combine Eqs. 2 and 5 to get

$$p_1 - p_2 = \rho \left[\frac{2.0 \overline{w}_2^2}{2} - \frac{1.0 \overline{w}_1^2}{2} + g(z_2 - z_1) + \text{loss} \right] \quad (6)$$

However, from conservation of mass, $\overline{w}_2 = \overline{w}_1 = \overline{w}$, so that Eq. 6 becomes

$$p_1 - p_2 = \frac{\rho \overline{w}^2}{2} + \rho g(z_2 - z_1) + \rho(\text{loss}) \quad (7)$$

The term associated with change in elevation, $\rho g(z_2 - z_1)$, is equal to the weight per unit cross-sectional area, $\mathcal{W}/A$, of the water contained between sections (1) and (2) at any instant,

$$\rho g(z_2 - z_1) = \frac{\mathcal{W}}{A} \quad (8)$$

Thus, combining Eqs. 7 and 8, we get

$$p_1 - p_2 = \frac{\rho \overline{w}^2}{2} + \frac{\mathcal{W}}{A} + \rho(\text{loss}) \quad (9)$$

The pressure drop between sections (1) and (2) is due to:

1. The change in kinetic energy between sections (1) and (2) associated with going from a uniform velocity profile to a parabolic velocity profile.

2. The weight of the water column, that is, hydrostatic pressure effect.

3. Viscous loss.

Comparing Eq. 9 for pressure drop with the one obtained in Example 5.14 (i.e., the answer of Example 5.14), we obtain

$$\frac{\rho \overline{w}^2}{2} + \frac{\mathcal{W}}{A} + \rho(\text{loss}) = \frac{\rho \overline{w}^2}{3} + \frac{R_z}{A} + \frac{\mathcal{W}}{A} \quad (10)$$

or

$$\text{loss} = \frac{R_z}{\rho A} - \frac{\overline{w}^2}{6} \quad (Ans)$$

Comment We conclude that while some of the pipe wall friction force, R_z, resulted in loss of available energy, a portion of this friction, $\rho A \overline{w}^2/6$, led to the velocity profile change.

5.3.5 Comparison of Various Forms of the Energy Equation

While the Bernoulli equation was derived from Newton's second law in Chapter 3, its assumptions (steady, incompressible, inviscid flow along a streamline) allow it to be expressed in terms of energy per unit mass (Eq. 5.75). It could in fact be derived from the general energy equation (Eq. 5.67) or the mechanical energy equation (Eq. 5.82). Therefore, when applying the energy equation for steady flow along a streamline, you essentially have three forms of the energy equation from which to choose:

General energy equation (Eq. 5.67):

$$\dot{m} \left[\check{u}_{\text{out}} - \check{u}_{\text{in}} + \left(\frac{p}{\rho} \right)_{\text{out}} - \left(\frac{p}{\rho} \right)_{\text{in}} + \frac{V_{\text{out}}^2 - V_{\text{in}}^2}{2} + g(z_{\text{out}} - z_{\text{in}}) \right] = \dot{Q}_{\text{net}\atop\text{in}} + \dot{W}_{\text{shaft}\atop\text{net in}}$$

Mechanical energy equation (Eq. 5.82):

$$\left(\frac{p_{out}}{\rho}\right) + \left(\frac{V_{out}^2}{2}\right) + gz_{out} = \left(\frac{p_{in}}{\rho}\right) + \left(\frac{V_{in}^2}{2}\right) + gz_{in} + w_{shaft \atop net \; in} - loss$$

Bernoulli equation (Eq. 5.75):

$$\left(\frac{p_{out}}{\rho}\right) + \left(\frac{V_{out}^2}{2}\right) + gz_{out} = \left(\frac{p_{in}}{\rho}\right) + \left(\frac{V_{in}^2}{2}\right) + gz_{in}$$

You should choose the simplest equation that accurately represents the actual flow. You may use only *one* of the three equations because either the assumptions and restrictions on a particular equation make it inconsistent with the others or the equations are actually identical.

Table 5.1 summarizes the assumptions restrictions and conditions on these equations. Suppose first that the flow in a pipe is compressible. You cannot use the mechanical energy equation or the Bernoulli equation; the general energy equation is the only possible choice. Next, assume that the flow is incompressible but has shear stresses (or friction). The Bernoulli equation does not apply, but the general energy equation and mechanical energy equation do; however, if the flow is incompressible, $\rho_{out} = \rho_{in}$ and from Eq. 5.78,

$$\check{u}_{out} - \check{u}_{in} - q_{net \atop in} = loss$$

so the general energy and the mechanical energy equations are actually the same. The mechanical energy equation is generally simpler because one quantity (loss) replaces three $\left(\check{u}_{out}, \check{u}_{in}, q_{net \atop in}\right)$. Finally, suppose that the flow is incompressible, with no friction and no work. You will recognize that the Bernoulli equation applies. A check of Table 5.1 shows that the other two equations are also valid; however, if there is no friction then

$$loss = 0 \quad \text{and} \quad \check{u}_{out} - \check{u}_{in} = q_{net \atop in}$$

Because $W_{shaft \atop net \; in} = 0$, the general energy and mechanical energy equations reduce to the Bernoulli equation automatically.

A crucial point in choosing the proper form of the energy equation is the fluid's compressibility. Obviously, liquids usually behave as if they were incompressible. Sometimes you can treat a flowing gas as incompressible. The choice of an incompressible fluid model depends as much on the specific flow situation as it does on the fluid involved. Two criteria should be considered when deciding on the compressibility (or lack of compressibility) of a flow.

The first and most obvious criterion for incompressible flow is that the largest possible density change in the flow field be negligibly small. Section 4.8.1 considers the coupling between density and pressure and suggests that you may neglect density variation if the dynamic pressure of the flow is considerably smaller than the bulk modulus of elasticity of the fluid.

Another criterion for the incompressible flow model involves the magnitude of mechanical versus thermal energies. If compressibility is significant, mechanical and thermal energies may be of comparable magnitude, and you must use the general energy equation. On the other hand, one form of energy could dominate. You can assume incompressible flow if the ratio

$$\frac{\text{Typical mechanical energy}}{\text{Typical thermal energy}}$$

TABLE 5.1   Assumptions and conditions for the general, mechanical, and Bernoulli equations

	General Energy	Mechanical Energy	Bernoulli
Steady/unsteady	Steady only	Steady only	Steady only
Compressibility	Compressible or incompressible	Incompressible	Incompressible
Friction	Allowed	Allowed	None
Heat transfer	Allowed	Allowed	Usually none, but allowed
Shaft work	Allowed	Allowed	None

is very small. The typical mechanical energy is $V^2/2$, and the typical thermal energy is $\check{u}_{\text{in}}$. For air flowing at 50 m/s, $V^2/2$ is 1250 m²/s². If the air temperature is 20 °C, its internal energy, u, is approximately 210,000 m²/s², so the ratio $\dfrac{V^2/2}{\check{u}} \approx 0.006$ is quite small. Thus, you may treat the air as if it were incompressible. At 150 m/s, however, compressibility may be significant.

The information in Table 5.1 regarding compressibility, friction, heat transfer, and work applies to any of the control volume, mass flow averaged, or streamline forms of the energy equation (e.g., Eq. 5.70 or Eq. 5.73). It may also be applied to unsteady flow (Eq. 5.64). Each equation has certain restrictions and conditions and should be used only if they are valid. For example, if the fluid velocity and pressure vary across streamlines at a given cross-sectional plane, the mass averaged control volume forms of an equation are more appropriate than the streamline form.

You still must never mix energy equations. You cannot use both the Bernoulli equation and the finite control volume general energy equation for the same region of flow; either the equations will not give independent information or one of them is invalid. The streamline and mass averaged forms of an equation are either identical (if the properties and velocity are truly uniform) or inconsistent. The finite control volume and mass flow averaged forms are redundant because one may be obtained from the other simply by multiplying or dividing by the mass flowrate. In summary:

> **One and only one form of the energy equation may be used for a particular region of the flow.**

5.3.6 Combination of the Energy Equation and the Moment-of-Momentum Equation[4]

If Eq. 5.82 is used for one-dimensional incompressible flow through a turbomachine, we can use Eq. 5.54, developed in Section 5.2.4 from the moment-of-momentum equation (Eq. 5.42), to evaluate shaft work. This application of both Eqs. 5.54 and 5.82 allows us to ascertain the amount of loss that occurs in incompressible turbomachine flows as is demonstrated in Example 5.28.

EXAMPLE 5.28 | Energy—Fan Performance

Given Consider the fan of Example 5.19.

Find Show that only some of the shaft power delivered to the air is converted into useful effects. Develop a meaningful efficiency equation and a practical means for estimating lost shaft energy.

Solution We use the same control volume used in Example 5.19. Application of Eq. 5.82 to the contents of this control volume yields

$$\frac{p_2}{\rho} + \frac{V_2^2}{2} + gz_2 = \frac{p_1}{\rho} + \frac{V_1^2}{2} + gz_1 + w_{\substack{\text{shaft} \\ \text{net in}}} - \text{loss} \tag{1}$$

As in Example 5.26, we can see with Eq. 1 that a "useful effect" in this fan can be defined as

$$\text{useful effect} = w_{\substack{\text{shaft} \\ \text{net in}}} - \text{loss}$$

$$= \left(\frac{p_2}{\rho} + \frac{V_2^2}{2} + gz_2\right) - \left(\frac{p_1}{\rho} + \frac{V_1^2}{2} + gz_1\right) \tag{2} \quad (Ans)$$

In other words, only a portion of the shaft work delivered to the air by the fan blades is used to increase the available energy of the air; the rest is lost because of fluid friction.

A meaningful efficiency equation involves the ratio of shaft work converted into a useful effect (Eq. 2) to shaft work delivered to the air, $w_{\text{shaft net in}}$. Thus, we can express efficiency, η, as

$$\eta = \frac{w_{\substack{\text{shaft} \\ \text{net in}}} - \text{loss}}{w_{\substack{\text{shaft} \\ \text{net in}}}} \tag{3}$$

When Eq. 5.54, which was developed from the moment-of-momentum equation (Eq. 5.42), is applied to the contents of the control volume of Fig. E5.19, we obtain

$$w_{\substack{\text{shaft} \\ \text{net in}}} = +U_2 V_{\theta 2} \tag{4}$$

Combining Eqs. 2, 3, and 4, we obtain

$$\eta = \big\{ \left[(p_2/\rho) + (V_2^2/2) + gz_2\right] - \left[(p_1/\rho) + (V_1^2/2) + gz_1\right] \big\} / U_2 V_{\theta 2} \tag{5} \quad (Ans)$$

---

[4]This section may be omitted without loss of continuity in the text material. This section should not be considered without prior study of Sections 5.2.3 and 5.2.4. All of these sections are recommended for those interested in Chapter 12.

Equation 5 provides us with a practical means to evaluate the efficiency of the fan of Example 5.19.

Combining Eqs. 2 and 4, we obtain

$$\text{loss} = U_2 V_{\theta 2} - \left[\left(\frac{p_2}{\rho} + \frac{V_2^2}{2} + gz_2 \right) \right.$$
$$\left. - \left(\frac{p_1}{\rho} + \frac{V_1^2}{2} + gz_1 \right) \right] \qquad (6) \quad (Ans)$$

Comment   Equation 6 provides us with a useful method of evaluating the loss due to fluid friction in the fan of Example 5.19 in terms of fluid mechanical variables that can be measured. The efficiency calculated by Eq. 3 is called *aerodynamic efficiency* if the machine is a fan and *hydraulic efficiency* if the machine is a pump or water turbine.

CHAPTER SUMMARY

In this chapter the flow of a fluid is analyzed by using important principles including conservation of mass, Newton's second law of motion, and the first law of thermodynamics as applied to control volumes. The Reynolds transport theorem is used to convert basic system-oriented laws into corresponding control volume formulations.

The continuity equation, a statement of the fact that mass is conserved, is obtained in a form that can be applied to any flow—steady or unsteady, incompressible or compressible. Simplified forms of the continuity equation help us to keep track of fluid in a control volume, where it enters, where it leaves, and within. Mass or volume flowrates of fluid entering or leaving a control volume and rate of accumulation or depletion of fluid within a control volume can be evaluated.

The linear momentum equation, a form of Newton's second law of motion applicable to flow of fluid through a control volume, is obtained and used to solve flow problems. Net force results from or causes changes in linear momentum (velocity magnitude and/or direction) of fluid flowing through a control volume.

The moment-of-momentum equation, which involves the relationship between torque and changes in angular momentum, is obtained and used to solve flow problems dealing with turbines (energy extracted from a fluid) and pumps (energy supplied to a fluid).

The steady-state energy equation, obtained from the first law of thermodynamics (conservation of energy), is written in several forms. The first (Eq. 5.67 or Eq. 5.69) involves power terms. The second form (Eq. 5.82 or Eq. 5.84) is termed the mechanical energy equation or the extended Bernoulli equation. This equation is used for different applications than the Bernoulli equation because it contains an extra term that accounts for energy losses due to friction in the flow, as well as terms accounting for the work of pumps or turbines.

The following checklist provides a study guide for this chapter. When your study of the entire chapter has been completed, you should be able to

- write out meanings of the terms listed and understand each of the related concepts. These terms are particularly important and are set in **bold black** type in the text.

- select an appropriate control volume for a given problem and draw an accurately labeled control volume diagram.

- use the continuity equation and a control volume to solve problems involving mass or volume flowrate.

- use the linear momentum equation and a control volume, in conjunction with the continuity equation as necessary, to solve problems involving forces related to linear momentum change.

- use the moment-of-momentum equation to solve problems involving torque and related work and power due to angular momentum change.

- use the energy equation, in one of its appropriate forms, to solve problems involving losses due to friction (head loss) and energy input by pumps or extraction by turbines.

- use the momentum coefficient in the linear momentum equation and the kinetic energy coefficient in the energy equation to account for nonuniform flows.

KEY EQUATIONS

Conservation of mass	$\dfrac{\partial}{\partial t}\displaystyle\int_{cv} \rho \, d\mathcal{V} + \int_{cs} \rho \mathbf{V} \cdot \hat{\mathbf{n}} \, dA = 0$	(5.5)
Mass flowrate	$\dot{m} = \rho Q = \rho A V$	(5.6)
Average velocity	$\overline{V} = \dfrac{\displaystyle\int_A \rho \mathbf{V} \cdot \hat{\mathbf{n}} \, dA}{\rho A}$	(5.7)
Steady-flow mass conservation	$\sum \dot{m}_{\text{out}} - \sum \dot{m}_{\text{in}} = 0$	(5.9)
Moving control volume mass conservation	$\dfrac{\partial}{\partial t}\displaystyle\int_{cv} \rho \, d\mathcal{V} + \int_{cs} \rho \mathbf{W} \cdot \hat{\mathbf{n}} \, dA = 0$	(5.16)
Deforming control volume mass conservation	$\dfrac{DM_{\text{sys}}}{Dt} = \dfrac{\partial}{\partial t}\displaystyle\int_{cv} \rho \, d\mathcal{V} + \int_{cs} \rho \mathbf{W} \cdot \hat{\mathbf{n}} \, dA = 0$	(5.17)
Force related to change in linear momentum	$\dfrac{\partial}{\partial t}\displaystyle\int_{cv} \mathbf{V} \rho \, d\mathcal{V} + \int_{cs} \mathbf{V} \rho \mathbf{V} \cdot \hat{\mathbf{n}} \, dA = \sum \mathbf{F}_{\substack{\text{contents of the} \\ \text{control volume}}}$	(5.22)

Moving control volume force related to change in linear momentum	$\int_{cs} \mathbf{W}\rho\mathbf{W} \cdot \hat{\mathbf{n}}\, dA = \sum \mathbf{F}_{\substack{\text{contents of the}\\ \text{control volume}}}$	(5.29)
Vector addition of absolute and relative velocities	$\mathbf{V} = \mathbf{W} + \mathbf{U}$	(5.43)
Shaft torque from force	$\sum\left[\left(\mathbf{r} \times \mathbf{F}\right)_{\substack{\text{contents of the}\\ \text{control volume}}}\right]_{\text{axial}} = \mathbf{T}_{\text{shaft}}$	(5.45)
Shaft torque related to change in moment-of-momentum (angular momentum)	$T_{\text{shaft}} = (-\dot{m}_{\text{in}})(\pm r_{\text{in}}V_{\theta\text{in}}) + \dot{m}_{\text{out}}(\pm r_{\text{out}}V_{\theta\text{out}})$	(5.50)
Shaft power related to change in moment-of-momentum (angular momentum)	$\dot{W}_{\text{shaft}} = (-\dot{m}_{\text{in}})(\pm U_{\text{in}}V_{\theta\text{in}}) + \dot{m}_{\text{out}}(\pm U_{\text{out}}V_{\theta\text{out}})$	(5.53)
First law of thermodynamics (conservation of energy)	$\dfrac{\partial}{\partial t}\int_{cv} e\rho\, d\mathcal{V} + \int_{cs}\left(\breve{u} + \dfrac{p}{\rho} + \dfrac{V^2}{2} + gz\right)\rho\mathbf{V} \cdot \hat{\mathbf{n}}\, dA = \dot{Q}_{\substack{\text{net}\\ \text{in}}} + \dot{W}_{\substack{\text{shaft}\\ \text{net in}}}$	(5.64)
First law of thermodynamics steady flow (simplified form)	$\dot{m}\left[\breve{h}_{\text{out}} - \breve{h}_{\text{in}} + \dfrac{V_{\text{out}}^2 - V_{\text{in}}^2}{2} + g(z_{\text{out}} - z_{\text{in}})\right] = \dot{Q}_{\substack{\text{net}\\ \text{in}}} + \dot{W}_{\substack{\text{shaft}\\ \text{net in}}}$	(5.69)
Mechanical energy equation (incompressible flow)	$\dfrac{p_{\text{out}}}{\rho} + \dfrac{V_{\text{out}}^2}{2} + gz_{\text{out}} = \dfrac{p_{\text{in}}}{\rho} + \dfrac{V_{\text{in}}^2}{2} + gz_{\text{in}} + w_{\substack{\text{shaft}\\ \text{net in}}} - \text{loss}$	(5.82)

REFERENCES

[1] Eck, B., *Technische Stromungslehre*, Springer-Verlag, Berlin, Germany, 1957.

[2] Dean, R. C., "On the Necessity of Unsteady Flow in Fluid Machines," *ASME Journal of Basic Engineering* 81D; 24–28, March 1959.

[3] Moran, M. J., Shapiro, H. N., Boettner, D. D., and Bailey, M. B., *Fundamentals of Engineering Thermodynamics*, 9th Ed., Wiley, New York, 2018.

QUESTIONS AND PROBLEMS

Ⓒ Problem to be solved with aid of programmable calculator or computer.

Ⓞ Open-ended problem that requires critical thinking. These problems require various assumptions to provide the necessary input data. There are not unique answers to these problems.

Note: Unless specific values of required fluid properties are specified in the problem statement, use the values found in Tables 1.4–1.8 and in the tables in the Appendices.

Section 5.1.1 Derivation of the Continuity Equation

5.1.1 Use the Reynolds transport theorem (Eq. 3.19) with B = volume and, therefore, b = volume/mass = 1/density to obtain the continuity equation for steady or unsteady incompressible flow through a fixed control volume: $\int_{cv} \mathbf{V} \cdot \hat{\mathbf{n}}\, dA = 0$.

5.1.2 An incompressible fluid flows horizontally in the x–y plane with a velocity given by

$$u = 30\left(1 - e^{-4\frac{y}{h}}\right) \text{ m/s}, v = 0$$

where y and h are in meters and h is a constant. Determine the average velocity for the portion of the flow between $y = 0$ and $y = h$.

Section 5.1.2 Fixed, Nondeforming Control Volume—Uniform Velocity Profile or Average Velocity

5.1.3 Water flows steadily through the horizontal piping system shown in **Fig. P5.1.3**. The velocity is uniform at section (1), the mass flowrate is 146 kg/s at section (2), and the velocity is nonuniform at section (3). (a) Determine the value of the quantity $\dfrac{D}{Dt}\int_{\text{sys}} \rho\, d\mathcal{V}$, where the system is the water contained in the pipe bounded by sections (1), (2), and (3). (b) Determine the mean velocity at section (2). (c) Determine, if possible, the value of the integral $\int_{(3)} \rho\mathbf{V} \cdot \hat{\mathbf{n}}\, dA$ over section (3). If it is not possible, explain what additional information is needed to do so.

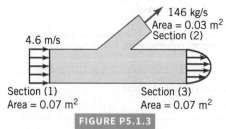

146 kg/s
Area = 0.03 m²
Section (2)

4.6 m/s

Section (1)
Area = 0.07 m²

Section (3)
Area = 0.07 m²

FIGURE P5.1.3

5.1.4 Water flows out through a set of thin, closely spaced blades, as shown in **Fig. P5.1.4**, with a speed of $V = 3$ m/s around the entire circumference of the outlet. Determine the mass flowrate through the inlet pipe.

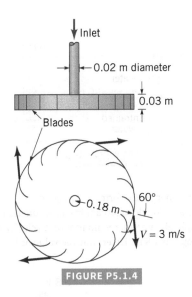

FIGURE P5.1.4

5.1.5 Ⓞ Estimate the rate (in liter/hr) that your car uses gasoline when it is being driven on an interstate highway. Determine how long it would take to empty a 340 gm soft-drink container at this flowrate. List all assumptions and show calculations.

5.1.6 The pump shown in **Fig. P5.1.6** produces a steady flow of 0.04 m³/s through the nozzle. Determine the nozzle exit diameter, D_2, if the exit velocity is to be $V_2 = 30$ m/s.

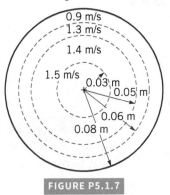

FIGURE P5.1.6

5.1.7 The fluid axial velocities shown in **Fig. P5.1.7** are the average velocities measured in m/s in each annular area of a duct. Find the volume flowrate for the flowing fluid.

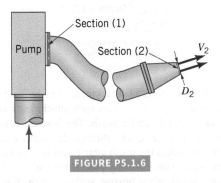

FIGURE P5.1.7

5.1.8 The human circulatory system consists of a complex branching pipe network ranging in diameter from the aorta (largest) to the capillaries (smallest). The average radii and the number of these vessels are shown in the table. Does the average blood velocity increase, decrease, or remain constant as it travels from the aorta to the capillaries?

Vessel	Average Radius (mm)	Number
Aorta	12.5	1
Arteries	2.0	159
Arterioles	0.03	1.4×10^7
Capillaries	0.006	3.9×10^9

5.1.9 Air flows steadily between two cross sections in a long, straight section of 0.1-m inside diameter pipe. The static temperature and pressure at each section are indicated in **Fig. P5.1.9**. If the average air velocity at section (1) is 205 m/s, determine the average air velocity at section (2).

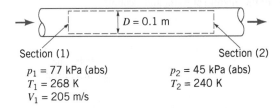

Section (1)
$p_1 = 77$ kPa (abs)
$T_1 = 268$ K
$V_1 = 205$ m/s

Section (2)
$p_2 = 45$ kPa (abs)
$T_2 = 240$ K

FIGURE P5.1.9

5.1.10 [Video] A hydraulic jump (see Video V10.11) is in place downstream from a spillway, as indicated in **Fig. P5.1.10**. Upstream of the jump, the depth of the stream is 0.18 m and the average stream velocity is 5.5 m/s. Just downstream of the jump, the average stream velocity is 1.04 m/s. Calculate the depth of the stream, h, just downstream of the jump.

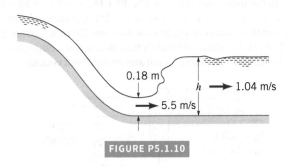

FIGURE P5.1.10

5.1.11 A woman is emptying her aquarium at a steady rate with a small pump. The water is pumped into a 0.30 m-diameter cylindrical bucket, and its depth is increasing at a rate of 0.1 m per minute. Find the rate at which the aquarium water level is dropping if the aquarium measures 0.6 m (wide) × 0.9 m (long) × 0.45 m (high).

5.1.12 An evaporative cooling tower (see **Fig. P5.1.12**) is used to cool water from 43 °C to 27 °C. Water enters the tower at a rate of 32 kg/hr. Dry air (no water vapor) flows into the tower at a rate of 19 kg/hr. If the rate of wet airflow out of the tower is 20 kg/hr, determine the rate of water evaporation and the rate of cooled water flow in kg/hr.

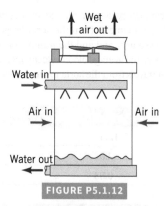

FIGURE P5.1.12

5.1.13 At cruise conditions, air flows into a jet engine at a steady rate of 29 kg/s. Fuel enters the engine at a steady rate of 0.27 kg/s. The average velocity of the exhaust gases is 457 m/s relative to the engine. If the effective cross-sectional area of the engine exhaust is 0.33 m^2, estimate the density of the exhaust gases in kg/m^3.

5.1.14 Water at 0.1 m^3/s and alcohol ($SG = 0.8$) at 0.3 m^3/s are mixed in a y-duct, as shown in **Fig. P5.1.14**. What is the average density of the mixture of water and alcohol?

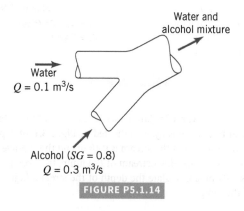

FIGURE P5.1.14

5.1.15 In the vortex tube shown in **Fig. P5.1.15**, air enters at 200 kPa absolute and 280 K. Hot air leaves at 120 kPa absolute and 320 K, whereas cold air leaves at 100 kPa absolute and 220 m/s. The hot air mass flowrate, $\dot{m}_H$, equals the cold air mass flowrate, $\dot{m}_C$. Find the ratio of the hot air exit area to cold air exit area for equal exit velocities.

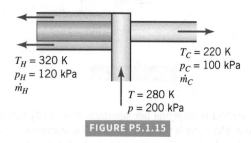

FIGURE P5.1.15

Section 5.1.2 Fixed, Nondeforming Control Volume—Nonuniform Velocity Profile

5.1.16 A water jet pump (see **Fig. P5.1.16**) involves a jet cross-sectional area of 0.01 m^2 and a jet velocity of 30 m/s. The jet is surrounded by entrained water. The total cross-sectional area associated

with the jet and entrained streams is 0.075 m^2. These two fluid streams leave the pump thoroughly mixed with an average velocity of 6 m/s through a cross-sectional area of 0.075 m^2. Determine the pumping rate (i.e., the entrained fluid flowrate) involved in liters/s.

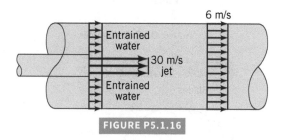

FIGURE P5.1.16

5.1.17 Two rivers merge to form a larger river, as shown in **Fig. P5.1.17**. At a location downstream from the junction (before the two streams completely merge), the nonuniform velocity profile is as shown and the depth is 1.8 m. Determine the value of V.

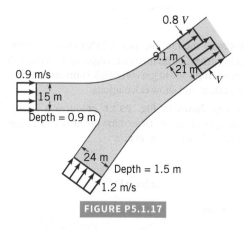

FIGURE P5.1.17

5.1.18 (Video) Various types of attachments can be used with the shop vac shown in **Video V5.2**. Two such attachments are shown in **Fig. P5.1.18**—a nozzle and a brush. The flowrate is 0.03 m^3/s. (a) Determine the average velocity through the nozzle entrance, V_n. (b) Assume the air enters the brush attachment in a radial direction all around the brush with a velocity profile that varies linearly from 0 to V_b along the length of the bristles, as shown in the figure. Determine the value of V_b.

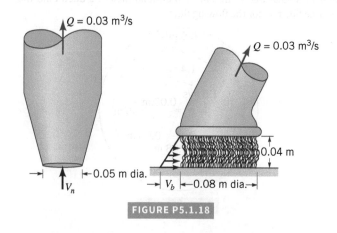

FIGURE P5.1.18

5.1.19 An appropriate turbulent pipe flow velocity profile is

$$\mathbf{V} = u_c \left(\frac{R - r}{R} \right)^{1/n} \hat{\mathbf{i}}$$

where u_c = centerline velocity, r = local radius, R = pipe radius, and $\hat{\mathbf{i}}$ = unit vector along pipe centerline. Determine the ratio of average velocity, $\bar{u}$, to centerline velocity, u_c, for (a) $n = 4$, (b) $n = 6$, (c) $n = 8$, (d) $n = 10$. Compare the different velocity profiles.

5.1.20 As shown in **Fig. P5.1.20**, at the entrance to a 0.9 m-wide channel the velocity distribution is uniform with a velocity V. Further downstream, the velocity profile is given by $u = 4y - 2y^2$, where u is in m/s and y is in m. Determine the value of V.

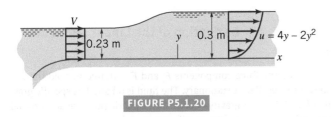

FIGURE P5.1.20

5.1.21 The cross-sectional area of a rectangular duct is divided into 16 equal rectangular areas, as shown in **Fig. P5.1.21**. The axial fluid velocity measured in m/s in each smaller area is given in the figure. Estimate the volume flowrate and average axial velocity.

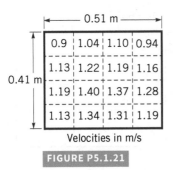

FIGURE P5.1.21

5.1.22 Oil for lubricating the thrust bearing shown in **Fig. P5.1.22** flows into the space between the bearing surfaces through a circular inlet pipe with velocity

$$u = U_0 \left[1 - \left(\frac{r}{R} \right)^2 \right]$$

where $R = 1.5$ mm. The oil has a specific gravity $S = 0.86$ and flows in the inlet pipe at a rate of 0.006 kg/s. Compute the *average* velocity, V_1, of the oil in the inlet pipe and the average velocity, V_2, at the outlet (plane 2) and the maximum velocity in the oil inlet pipe (U_0). Assume radial flow.

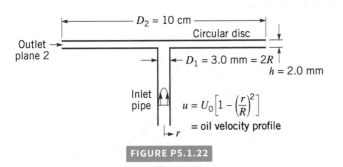

FIGURE P5.1.22

Section 5.1.2 Fixed, Nondeforming Control Volume—Unsteady Flow

5.1.23 Air at standard conditions enters the compressor shown in **Fig. P5.1.23** at a rate of 0.28 m³/s. It leaves the tank through a 0.03 m-diameter pipe with a density of 1.80 kg/m³ and a uniform speed of 213 m/s. (a) Determine the rate (kg/s) at which the mass of air in the tank is increasing or decreasing. (b) Determine the average time rate of change of air density within the tank.

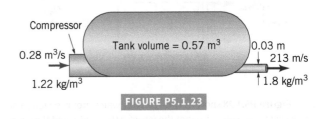

FIGURE P5.1.23

5.1.24 Estimate the time required to fill with water a cone-shaped container (see **Fig. P5.1.24**) 1.5 m high and 1.5 m across at the top if the filling rate is 0.08 m³/min.

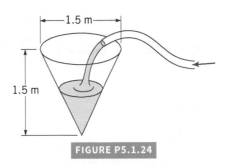

FIGURE P5.1.24

Section 5.1.3 Moving, Nondeforming Control Volume

5.1.25 ⊙ For an automobile moving along a highway, describe the control volume you would use to estimate the flowrate of air across the radiator. Explain how you would estimate the velocity of that air.

5.1.26 A water jet leaves a fixed nozzle with a velocity of 10 m/s as shown in **Fig. P5.1.26**. The jet diameter is 10 cm. A 30° cone is pushed into the water jet at a speed of 5 m/s. The water impinges on the cone with the jet axis and the cone axis in perfect alignment so that the water is divided evenly by the cone. Bernoulli's equation suggests that the water velocity *relative to the cone surface* is constant because the pressure on the jet boundary is constant. Determine the thickness of the water stream when it reaches the base of the cone.

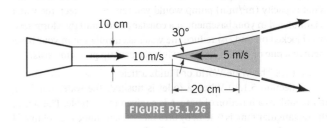

FIGURE P5.1.26

Section 5.1.4 Deforming Control Volume

5.1.27 A hypodermic syringe (see **Fig. P5.1.27**) is used to apply a vaccine. If the plunger is moved forward at the steady rate of 20 mm/s and if vaccine leaks past the plunger at 0.1 of the volume flowrate out the needle opening, calculate the average velocity of the needle exit flow. The inside diameters of the syringe and the needle are 20 mm and 0.7 mm.

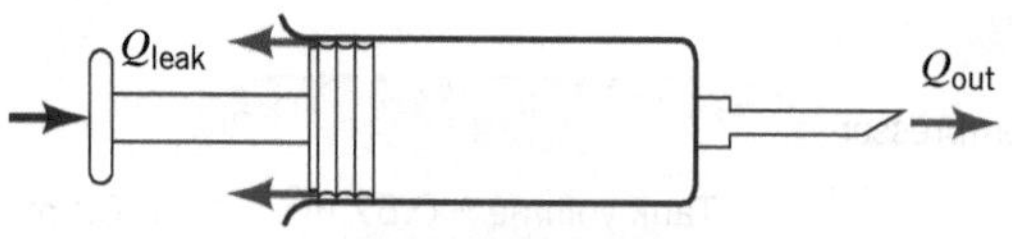

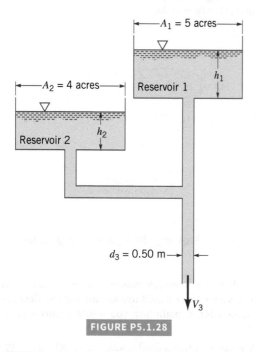

FIGURE P5.1.27

5.1.28 **Figure P5.1.28** shows a two-reservoir water supply system. The water level in reservoir 1 drops at the rate of 0.01 m/min, and the water level in reservoir 2 drops at the rate of 0.015 m/min. Calculate the average velocity, V_3, in the 0.50-m-diameter pipe.

FIGURE P5.1.28

5.1.29 (Video) The Hoover Dam (see Video V2.5) backs up the Colorado River and creates Lake Mead, which is approximately 185 km long and has a surface area of approximately 362 km². If during flood conditions the Colorado River flows into the lake at a rate of 1274 m³/s and the outflow from the dam is 226 m³/s, how many meter per 24-hour day will the lake level rise?

5.1.30 Storm sewer backup causes your basement to flood at the steady rate of 0.025 m of depth per hour. The basement floor area is 139 m². What capacity (m³/min) pump would you rent to (a) keep the water accumulated in your basement at a constant level until the storm sewer is blocked off and (b) reduce the water accumulation in your basement at a rate of 0.08 m/hr even while the backup problem exists?

5.1.31 (See The Wide World of Fluids article "'Green' 6-lpf standards," Section 5.1.2.) When a toilet is flushed, the water depth, h, in the tank as a function of time, t, is as given in the table. The size of the rectangular tank is 0.48 m by 0.19 m. (a) Determine the volume of water used per flush (m³). (b) Plot the flowrate for $0 \leq t \leq 6$ s.

t (s)	h (m)
0	0.14
0.5	0.13
1.0	0.12
2.0	0.09
3.0	0.06
4.0	0.04
5.0	0.02
6.0	0

Section 5.2.1 Derivation of the Linear Momentum Equation

5.2.1 Find the force components F_x and F_y required to hold the box shown in **Fig. P5.2.1** stationary. The fluid is oil and has specific gravity of 0.85. Neglect gravity effects. Atmospheric pressure acts around the entire box. Steady flow conditions prevail.

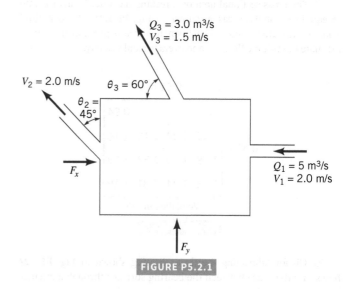

FIGURE P5.2.1

Section 5.2.2 Application of the Linear Momentum Equation

5.2.2 ⊙ When a baseball player catches a ball, the force of the ball on her glove is as shown as a function of time in **Fig. P5.2.2**. Describe how this situation is similar to the force generated by the deflection of a jet of water by a vane. *Note:* Consider many baseballs being caught in quick succession.

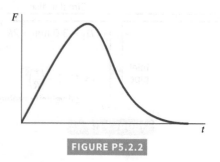

FIGURE P5.2.2

5.2.3 Find the horizontal and vertical forces to hold stationary the nozzle shown in **Fig. P5.2.3**. The fluid flowing through it is 10 °C liquid water; $A_1 = 1.0$ m², $A_2 = 0.25$ m², $V_1 = 20$ m/s, $p_2 = p_{atm}$, and $p_1 = p_{atm} + 30$ kPa. Neglect gravity.

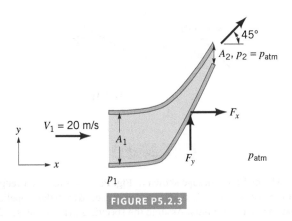

FIGURE P5.2.3

5.2.4 Water flows through a horizontal bend and discharges into the atmosphere, as shown in **Fig. P5.2.4**. When the pressure gage reads 68948 Pa, the resultant x-direction anchoring force, F_{Ax}, in the horizontal plane required to hold the bend in place is shown on the figure. Determine the flowrate through the bend and the y-direction anchoring force, F_{Ay}, required to hold the bend in place. The flow is not frictionless.

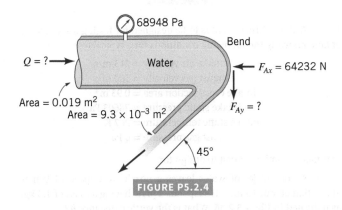

FIGURE P5.2.4

5.2.5 Find the magnitude of the force, F, required to hold the plate in **Fig. P5.2.5** stationary.

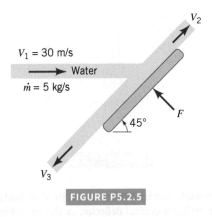

FIGURE P5.2.5

5.2.6 Water enters the horizontal, circular cross-sectional, sudden contraction nozzle sketched in **Fig. P5.2.6** at section (1) with a uniformly distributed velocity of 7.6 m/s and a pressure of 517125 Pa. The water exits from the nozzle into the atmosphere at section (2), where the uniformly distributed velocity is 30 m/s. Determine the axial component of the anchoring force required to hold the contraction in place.

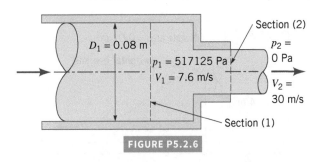

FIGURE P5.2.6

5.2.7 Ⓞ A truck carrying chickens is too heavy for a bridge that it needs to cross. The empty truck is within the weight limits but with the chickens it is overweight. It is suggested that if one could get the chickens to fly around the truck (i.e., by banging on the truck side) it would be safe to cross the bridge. Do you agree? Explain.

5.2.8 ⟨Video⟩ Exhaust (assumed to have the properties of standard air) leaves the 1.22 m-diameter chimney shown in **Video V5.4** and **Fig. P5.2.8** with a speed of 1.82 m/s. Because of the wind, after a few diameters downstream the exhaust flows in a horizontal direction with the speed of the wind 4.6 m/s. Determine the horizontal component of the force that the blowing wind exerts on the exhaust gases.

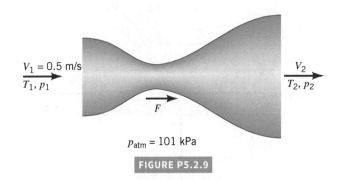

FIGURE P5.2.8

5.2.9 Air at $T_1 = 300$ K, $p_1 = 303$ kPa, and $V_1 = 0.5$ m/s enters the Venturi shown in **Fig. P5.2.9**. The air leaves at $T_2 = 220$ K and $p_2 = 101$ kPa; $A_1 = 0.6$ m² and $A_2 = 1.0$ m². Calculate the horizontal force required to hold the Venturi stationary.

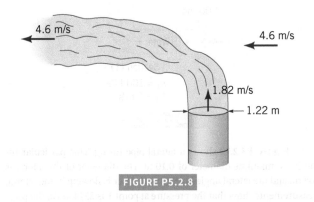

FIGURE P5.2.9

5.2.10 Water flows steadily from a tank mounted on a cart, as shown in **Fig. P5.2.10**. After the water jet leaves the nozzle of the tank, it falls and strikes a vane attached to another cart. The cart's wheels are frictionless, and the fluid is inviscid. (a) Determine the speed of the water leaving the tank, V_1, and the water speed leaving the cart, V_2. (b) Determine the tension in rope A. (c) Determine the tension in rope B.

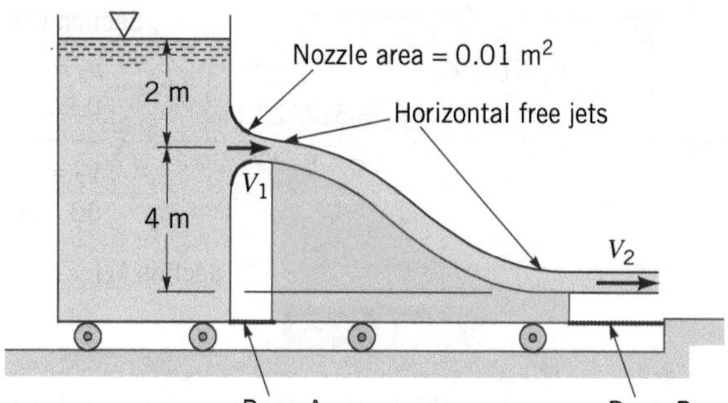

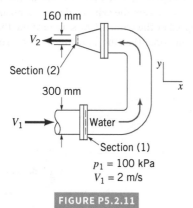

Nozzle area = 0.01 m²

Horizontal free jets

2 m

4 m

V_1

V_2

Rope A   Rope B

FIGURE P5.2.10

5.2.11 Determine the magnitude and direction of the anchoring force needed to hold the horizontal elbow and nozzle combination shown in **Fig. P5.2.11** in place. Atmospheric pressure is 100 kPa(abs). The gage pressure at section (1) is 100 kPa. At section (2), the water exits to the atmosphere.

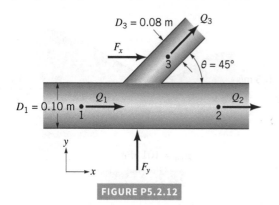

160 mm

V_2

Section (2)

300 mm

V_1

Water

Section (1)

$p_1 = 100$ kPa
$V_1 = 2$ m/s

FIGURE P5.2.11

5.2.12 **Figure P5.2.12** shows a lateral pipe fitting. This particular fitting has a mainline diameter of 0.10 m. The diameter of the lateral is 0.08 m, and the lateral angle is 45°; 16 °C water is flowing in the lateral. Measurements show that the pressure at point 1 is 234430 Pa, the pressure at point 2 is 241325 Pa, the pressure at point 3 is 230983 Pa, and the flowrate at point 2 is 0.03 m³/s. Determine the horizontal and vertical force components (F_x and F_y) required to hold the lateral fitting stationary. Neglect gravity. $Q_1 = 0.05$ m³/s.

$D_3 = 0.08$ m Q_3

F_x

3

$\theta = 45°$

$D_1 = 0.10$ m Q_1 Q_2

1   2

y

x F_y

FIGURE P5.2.12

5.2.13 Water flows steadily between fixed vanes, as shown in **Fig. P5.2.13**. Find the x and y components of the water's force on the vanes. The total volume flowrate is 100 m³/s, pressure $p_1 = 150$ kPa, and pressure $p_2 = 101$ kPa.

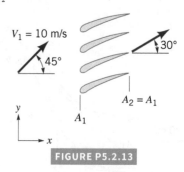

$V_1 = 10$ m/s

30°

45°

y

$A_2 = A_1$

A_1

x

FIGURE P5.2.13

5.2.14 The hydraulic dredge shown in **Fig. P5.2.14** is used to dredge sand from a river bottom. Estimate the thrust needed from the propeller to hold the boat stationary. Assume the specific gravity of the sand/water mixture is $SG = 1.4$.

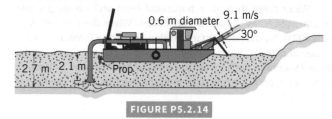

0.6 m diameter   9.1 m/s

30°

2.7 m   2.1 m   Prop

FIGURE P5.2.14

5.2.15 A static thrust stand is to be designed for testing a specific jet engine knowing the following conditions for a typical test.

$$\text{intake air velocity} = 213 \text{ m/s}$$
$$\text{exhaust gas velocity} = 500 \text{ m/s}$$
$$\text{intake cross section area} = 0.93 \text{ m}^2$$
$$\text{intake static pressure} = 78603 \text{ Pa}$$
$$\text{intake static temperature} = 267 \text{ K}$$
$$\text{exhaust gas pressure} = 0 \text{ Pa}$$

Estimate a nominal thrust to design for.

5.2.16 A vertical jet of water leaves a nozzle at a speed of 9 m/s and a diameter of 15 mm. It suspends a plate having a mass of 1.3 kg, as indicated in **Fig. P5.2.16**. What is the vertical distance h?

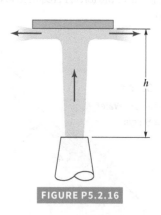

h

FIGURE P5.2.16

5.2.17 A horizontal, circular cross-sectional jet of air having a diameter of 0.15 m strikes a conical deflector, as shown in **Fig. P5.2.17**. A horizontal anchoring force of 22 N is required to hold the cone in place. Estimate the nozzle flowrate in m³/s. The magnitude of the velocity of the air remains constant.

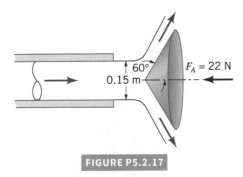

FIGURE P5.2.17

5.2.18 A jet pump works by injecting a higher speed flow into a slower flow as depicted in **Fig. P5.2.18**. Assuming a uniform velocity profile at station 2, and neglecting losses, determine the pressure increase, $p_2 - p_1$. Assume the fluid is 5 °C water.

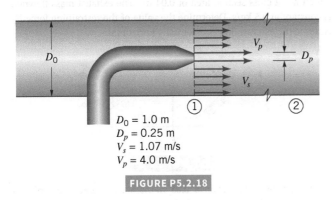

$D_0 = 1.0$ m
$D_p = 0.25$ m
$V_s = 1.07$ m/s
$V_p = 4.0$ m/s

FIGURE P5.2.18

5.2.19 Air flows into the atmosphere from a nozzle and strikes a vertical plate, as shown in **Fig. P5.2.19**. A horizontal force of 12 N is required to hold the plate in place. Determine the reading on the pressure gage. Assume the flow to be incompressible and frictionless.

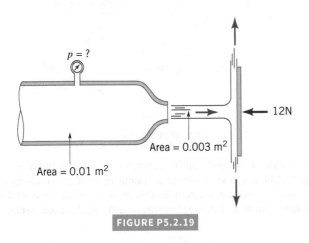

$p = ?$

12N

Area = 0.003 m²

Area = 0.01 m²

FIGURE P5.2.19

5.2.20 Water flows from a large tank into a dish, as shown in **Fig. P5.2.20**. (a) If at the instant shown the tank and the water in it weigh W_1 N, what is the tension, T_1, in the cable supporting the tank? (b) If at the instant shown the dish and the water in it weigh W_2 N, what is the force, F_2, needed to support the dish?

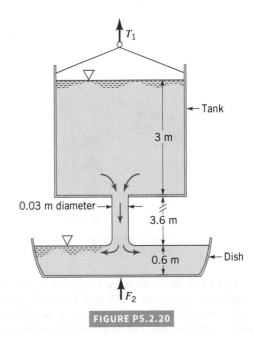

T_1

Tank

3 m

0.03 m diameter

3.6 m

0.6 m — Dish

F_2

FIGURE P5.2.20

5.2.21 **Figure P5.2.21** shows the configuration of the center (tail-mounted) jet engine on an airliner. The airliner is cruising at altitude, and the velocities shown are relative to an observer on board. Calculate the thrust force that the engine exerts on the airplane.

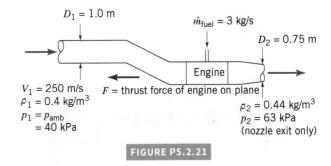

$D_1 = 1.0$ m

$\dot{m}_{fuel} = 3$ kg/s

$D_2 = 0.75$ m

Engine

$V_1 = 250$ m/s F = thrust force of engine on plane
$\rho_1 = 0.4$ kg/m³
$p_1 = p_{amb}$
     $= 40$ kPa

$\rho_2 = 0.44$ kg/m³
$p_2 = 63$ kPa
(nozzle exit only)

FIGURE P5.2.21

5.2.22 The plate shown in **Fig. P5.2.22** is 0.5 m wide perpendicular to the paper. Calculate the velocity of the water jet required to hold the plate upright.

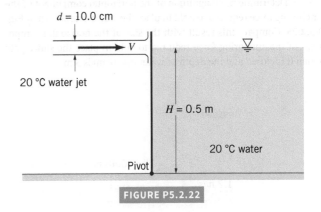

$d = 10.0$ cm

V

20 °C water jet

$H = 0.5$ m

20 °C water

Pivot

FIGURE P5.2.22

5.2.23 Two water jets of equal size and speed strike each other, as shown in **Fig. P5.2.23**. Determine the speed, V, and direction, θ, of the resulting combined jet. Gravity is negligible.

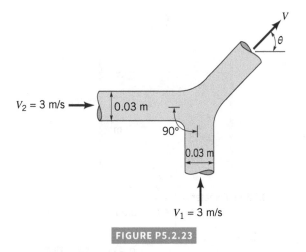

FIGURE P5.2.23

5.2.24 **Figure P5.2.24** shows coal being dropped from a hopper onto a conveyor belt at a constant rate of 0.70 m³/s. The coal has a specific gravity ranging from 1.12 to 1.50. The belt has a speed of 4.6 m/s and a loaded length of 1.5 m/s. Estimate the torque required to turn the drive pulley of the conveyor belt. The drive pulley diameter is 0.60 m. Assume that there is no friction between the belt and the other rollers.

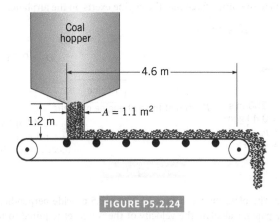

FIGURE P5.2.24

5.2.25 Determine the magnitude of the horizontal component of the anchoring force required to hold in place the sluice gate shown in **Fig. P5.2.25**. Compare this result with the size of the horizontal component of the anchoring force required to hold in place the sluice gate when it is closed and the depth of water upstream is 3 m.

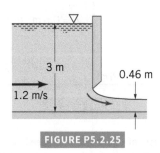

FIGURE P5.2.25

5.2.26 Water flows steadily into and out of a tank that sits on frictionless wheels, as shown in **Fig. P5.2.26**. Determine the diameter D so that the tank remains motionless if $F = 0$.

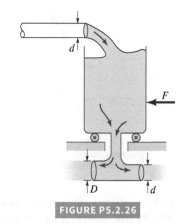

FIGURE P5.2.26

5.2.27 The rocket shown in **Fig. P5.2.27** is held stationary by the horizontal force, F_x, and the vertical force, F_z. The velocity and pressure of the exhaust gas are 1540 m/s and 137900 Pa at the nozzle exit, which has a cross section area of 0.04 m². The exhaust mass flowrate is constant at 9.5 kg/s. Determine the value of the restraining force F_x. Assume the exhaust flow is essentially horizontal.

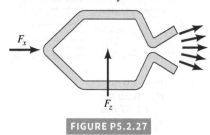

FIGURE P5.2.27

5.2.28 Air discharges from a 0.05 m-diameter nozzle and strikes a curved vane, which is in a vertical plane, as shown in **Fig. P5.2.28**. A stagnation tube connected to a water U-tube manometer is located in the free air jet. Determine the horizontal component of the force that the air jet exerts on the vane. Neglect the weight of the air and all friction.

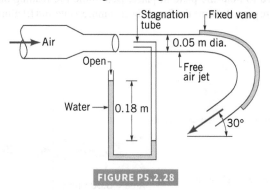

FIGURE P5.2.28

5.2.29 Water is sprayed radially outward over 180°, as indicated in **Fig. P5.2.29**. The jet sheet is in the horizontal plane. If the jet velocity at the nozzle exit is 6.1 m/s, determine the direction and magnitude of the resultant horizontal anchoring force required to hold the nozzle in place.

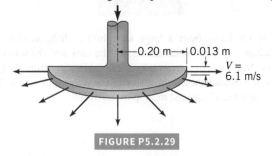

FIGURE P5.2.29

5.2.30 A sheet of water of uniform thickness ($h = 0.01$ m) flows from the device shown in **Fig. P5.2.30**. The water enters vertically through the inlet pipe and exits horizontally with a speed that varies linearly from 0 to 10 m/s along the 0.2-m length of the slit. Determine the y component of anchoring force necessary to hold this device stationary.

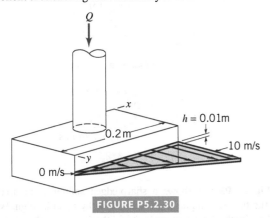

FIGURE P5.2.30

5.2.31 The results of a wind tunnel test to determine the drag on a body (see **Fig. P5.2.31**) are summarized below. The upstream [section (1)] velocity is uniform at 30 m/s. The static pressures are given by $p_1 = p_2 = 101357$ Pa. The downstream velocity distribution, which is symmetrical about the centerline, is given by

$$u = 100 - 30\left(1 - \frac{|y|}{3}\right) \quad |y| \leq 0.92 \text{ m}$$

$$u = 100 \quad |y| > 0.92 \text{ m}$$

where u is the velocity in m/s and y is the distance on either side of the centerline in feet (see Fig. P5.2.31). Assume that the body shape does not change in the direction normal to the paper. Calculate the drag force (reaction force in x direction) exerted on the air by the body per unit length normal to the plane of the sketch.

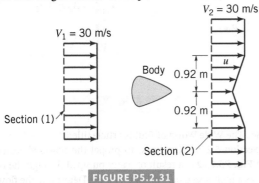

FIGURE P5.2.31

5.2.32 A variable mesh screen produces a linear and axisymmetric velocity profile, as indicated in **Fig. P5.2.32** in the airflow through a 0.61 m-diameter circular cross-sectional duct. The static pressures upstream and downstream of the screen are 1379 Pa and 1034 Pa and are uniformly distributed over the flow cross-sectional area. Neglecting the force exerted by the duct wall on the flowing air, calculate the screen drag force.

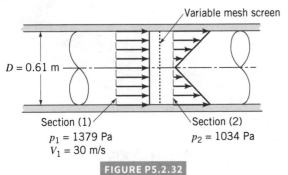

FIGURE P5.2.32

5.2.33 Consider unsteady flow in the constant diameter, horizontal pipe shown in **Fig. P5.2.33**. The velocity is uniform throughout the entire pipe, but it is a function of time: $\mathbf{V} = u(t)\hat{\mathbf{i}}$. Use the x component of the unsteady momentum equation to determine the pressure difference $p_1 - p_2$. Discuss how this result is related to $F_x = ma_x$.

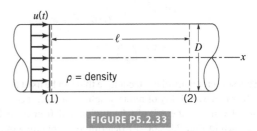

FIGURE P5.2.33

5.2.34 In a turbulent pipe flow that is fully developed, the axial velocity profile is,

$$u = u_c\left[1 - \left(\frac{r}{R}\right)\right]^{1/7}$$

as is illustrated in **Fig. P5.2.34**. Compare the axial direction momentum flowrate calculated with the average velocity, $\bar{u}$, with the axial direction momentum flowrate calculated with the nonuniform velocity distribution taken into account.

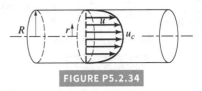

FIGURE P5.2.34

5.2.35 Water from a garden hose is sprayed against your car to rinse dirt from it. Estimate the force that the water exerts on the car. List all assumptions and show calculations.

5.2.36 [Video] A Pelton wheel vane directs a horizontal, circular cross-sectional jet of water symmetrically, as indicated in **Fig. P5.2.36** and Video V5.6. The jet leaves the nozzle with a velocity of 30 m/s. Determine the x-direction component of anchoring force required to (a) hold the vane stationary and (b) confine the speed of the vane to a value of 3 m/s to the right. The fluid speed magnitude remains constant along the vane surface.

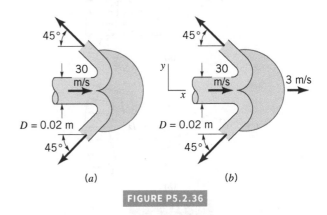

FIGURE P5.2.36

5.2.37 [Video] The thrust developed to propel the jet ski shown in Video V9.18 and **Fig. P5.2.37** is a result of water pumped through the vehicle and exiting as a high-speed water jet. For the conditions shown in the figure, what flowrate is needed to produce a 1335 N thrust? Assume the inlet and outlet jets of water are free jets.

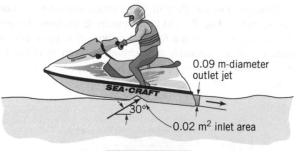

FIGURE P5.2.37

5.2.38 Thrust vector control is a technique that can be used to greatly improve the maneuverability of military fighter aircraft. It consists of using a set of vanes in the exit of a jet engine to deflect the exhaust gases, as shown in **Fig. P5.2.38**. **(a)** Determine the pitching moment (the moment tending to rotate the nose of the aircraft up) about the aircraft's mass center (cg) for the conditions indicated in the figure. **(b)** By how much is the thrust (force along the centerline of the aircraft) reduced for the case indicated compared to normal flight when the exhaust is parallel to the centerline?

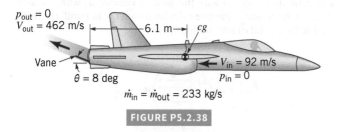

FIGURE P5.2.38

5.2.39 The exhaust gas from the rocket shown in **Fig. P5.2.39a** leaves the nozzle with a uniform velocity parallel to the x axis. The gas is assumed to be discharged from the nozzle as a free jet. **(a)** Show that the thrust that is developed is equal to $\rho A V^2$, where $A = \pi D^2/4$. **(b)** The exhaust gas from the rocket nozzle shown in **Fig. P5.2.39b** is also uniform, but rather than being directed along the x axis, it is directed along rays from point 0 as indicated. Determine the thrust for this rocket.

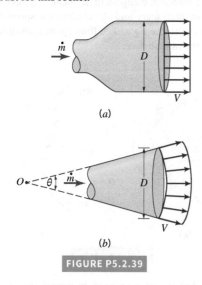

FIGURE P5.2.39

5.2.40 (See The Wide World of Fluids article titled "Where the plume goes," Section 5.2.2.) Air flows into the jet engine shown in **Fig. P5.2.40** at a rate of 131 kg/s and a speed of 91 m/s. Upon landing, the engine exhaust exits through the reverse thrust mechanism with a speed of 274 m/s in the direction indicated. Determine the reverse thrust applied by the engine to the airplane. Assume the inlet and exit pressures are atmospheric and that the mass flowrate of fuel is negligible compared to the air flowrate through the engine.

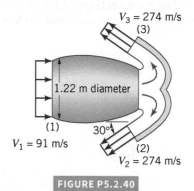

FIGURE P5.2.40

5.2.41 **Figure P5.2.41** shows a sharp-edged splitter plate used to control the flow of a liquid jet W units wide by H_1 units high. Write expressions for the deflection angle, θ, and the force, F, of the jet on the splitter plate as a function of the fluid density ρ, H_1, W, V, and plate insertion h. This force, F, has no components parallel to the plate. Assume that the jet flow is inviscid, that the jet width W remains constant, $H_2/H_3 = H_2'/H_3' = h/(H_1 - h)$, and that the fluid density is constant.

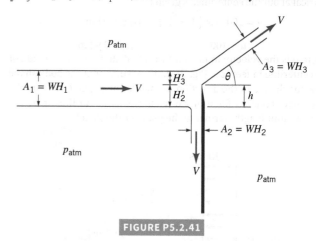

FIGURE P5.2.41

5.2.42 (See The Wide World of Fluids article titled "Motorized surfboard," Section 5.2.2.) The thrust to propel the powered surfboard shown in **Fig. P5.2.42** is a result of water pumped through the board that exits as a high-speed 0.07 m-diameter jet. Determine the flowrate and the velocity of the exiting jet if the thrust is to be 1335 N. Neglect the momentum of the water entering the pump.

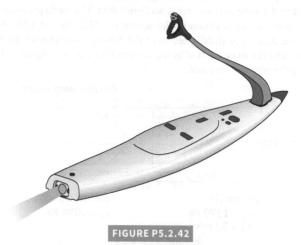

FIGURE P5.2.42

5.2.43 (See The Wide World of Fluids article titled "Bow thrusters," Section 5.2.2.) The bow thruster on the boat shown in **Fig. P5.2.43** is used to turn the boat. The thruster produces a 1-m-diameter jet of water with a velocity of 10 m/s. Determine the force produced by the thruster. Assume that the inlet and outlet pressures are zero and that the momentum of the water entering the thruster is negligible.

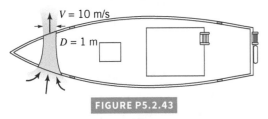

FIGURE P5.2.43

5.2.44 Water flows from a two-dimensional open channel and is diverted by an inclined plate, as illustrated in **Fig. P5.2.44**. When the velocity at section (1) is 3 m/s, what horizontal force (per unit width) is required to hold the plate in position? At section (1) the pressure distribution is hydrostatic, and the fluid acts as a free jet at section (2). Neglect friction.

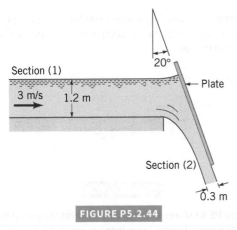

FIGURE P5.2.44

5.2.45 If a valve in a pipe is suddenly closed, a large pressure surge may develop. For example, when the electrically operated shut-off valve in a dishwasher closes quickly, the pipes supplying the dishwasher may rattle or "bang" because of this large pressure pulse. Explain the physical mechanism for this "water hammer" phenomenon. How could this phenomenon be analyzed?

5.2.46 A snowplow mounted on a truck clears a path 3.6 m through heavy wet snow, as shown in **Fig. P5.2.46**. The snow is 2.4 m deep, and its density is 160 kg/m³. The truck travels at 48 km/hr. The snow is discharged from the plow at an angle of 45° from the direction of travel and 45° above the horizontal, as shown in Fig. P5.2.46. Estimate the force required to push the plow.

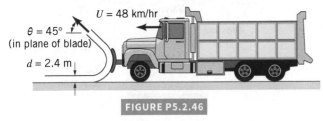

FIGURE P5.2.46

Section 5.2.3 Derivation of the Moment-of-Momentum Equation

5.2.47 An incompressible fluid flows outward through a blower, as indicated in **Fig. P5.2.47**. The shaft torque involved, T_{shaft}, is estimated with the following relationship:

$$T_{\text{shaft}} = \dot{m} r_2 V_{\theta 2}$$

where $\dot{m}$ = mass flowrate through the blower, r_2 = outer radius of blower, and $V_{\theta 2}$ = tangential component of absolute fluid velocity leaving the blower. State the flow conditions that make this formula valid.

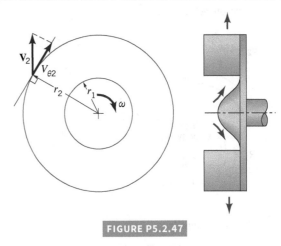

FIGURE P5.2.47

Section 5.2.4 Application of the Moment-of-Momentum Equation

5.2.48 (Video) Five liters/s of water enter the rotor shown in Video V5.12 and **Fig. P5.2.48** along the axis of rotation. The cross-sectional area of each of the three nozzle exits normal to the relative velocity is 18 mm². How large is the resisting torque required to hold the rotor stationary? How fast will the rotor spin steadily if the resisting torque is reduced to zero and (a) $\theta = 0°$, (b) $\theta = 30°$, (c) $\theta = 60°$?

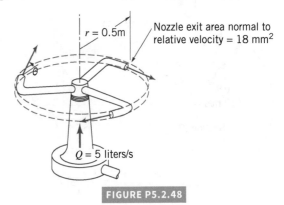

FIGURE P5.2.48

5.2.49 The hydraulic turbine shown in **Fig. P5.2.49** has a 10 °C water flowrate of 36.4 m³/s, an inlet radius $R_1 = 1.0$ m, an outlet radius $R_2 = 0.50$ m, a blade depth (perpendicular to paper) $h = 0.50$ m, and a rotational speed $N = 360$ rpm; $V_1 = 50$ m/s, $V_2 = 30$ m/s, $\alpha_1 = 13.4°$, and $\alpha_2 = 40°$. Calculate the power transferred by the fluid to the rotor, the inlet relative velocity W_1, and the direction W_1 makes with the radius at the inlet.

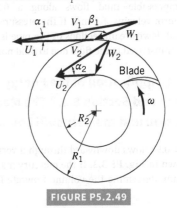

FIGURE P5.2.49

5.2.50 Calculate the torque required to drive the pump shown in **Fig. P5.2.50** at 30 Hz and to deliver 20 °C water at 3.0 m³/min.

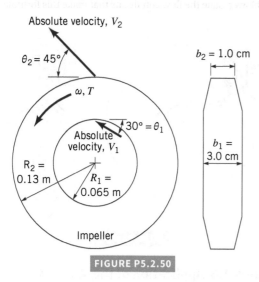

FIGURE P5.2.50

5.2.51 An inward flow radial turbine (see **Fig. P5.2.51**) involves a nozzle angle, α_1, of 60° and an inlet rotor tip speed, U_1, of 6 m/s. The ratio of rotor inlet to outlet diameters is 1.8. The absolute velocity leaving the rotor at section (2) is radial with a magnitude of 12 m/s. Determine the energy transfer per unit mass of fluid flowing through this turbine if the fluid is **(a)** air, **(b)** water.

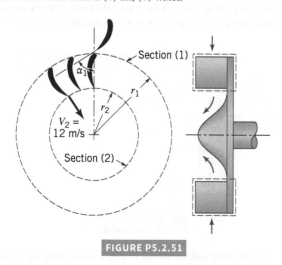

FIGURE P5.2.51

Section 5.3.1 Derivation of the Energy Equation

5.3.1 Distinguish between shaft work and other kinds of work associated with a flowing fluid.

5.3.2 An incompressible fluid flows along a 0.20-m-diameter pipe with a uniform velocity of 3 m/s. If the pressure drop between the upstream and downstream sections of the pipe is 20 kPa, determine the power transferred to the fluid due to fluid normal stresses.

Section 5.3.2 Application of the Energy Equation— No Shaft Work and Section 5.3.3 The Mechanical Energy Equation and the Bernoulli Equation

5.3.3 Oil ($SG = 0.9$) flows downward through a vertical pipe contraction, as shown in **Fig. P5.3.3**. If the mercury manometer reading, h, is 100 mm, determine the volume flowrate for frictionless

flow. Is the actual flowrate more or less than the frictionless value? Explain.

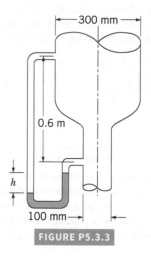

FIGURE P5.3.3

5.3.4 An incompressible liquid flows steadily along the pipe shown in **Fig. P5.3.4**. Determine the direction of flow and the head loss over the 6-m length of pipe.

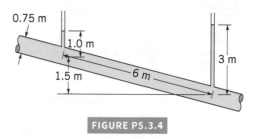

FIGURE P5.3.4

5.3.5 **Figure P5.3.5** shows a test rig for evaluating the loss coefficient, K, for a valve. Mechanical energy losses in valves are modeled by the equation:

$$gh_L = K\left(\frac{V^2}{2}\right)$$

where gh_L is the mechanical energy loss and V is the flow velocity entering the valve. In a particular test, the pressure gage reads 40 kPa, gage, and the 1.5-m³ catch tank fills in 2 min 55 s. Calculate the loss coefficient for a water temperature of 20 °C.

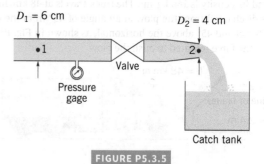

FIGURE P5.3.5

5.3.6 An automobile engine will work best when the back pressure at the interface of the exhaust manifold and the engine block is minimized. Show how reduction of losses in the exhaust manifold, piping, and muffler will also reduce the back pressure. How could losses in the exhaust system be reduced? What primarily limits the minimization of exhaust system losses?

Section 5.3.2 Application of the Energy Equation— With Shaft Work

5.3.7 [O] Based on flowrate and pressure rise information, estimate the power output of a human heart.

5.3.8 Water is pumped from the tank shown in **Fig. P5.3.8**. The head loss is known to be $1.2V^2/2g$, where V is the average velocity in the pipe. According to the pump manufacturer, the relationship between the pump head and the flowrate is as shown in **Fig. P5.3.8**: $h_p = 20 - 2000Q^2$, where h_p is in meters and Q is in m³/s. Determine the flowrate, Q.

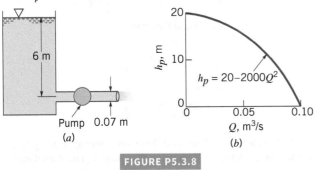

FIGURE P5.3.8

5.3.9 A pump moves water horizontally at a rate of 0.02 m³/s. Upstream of the pump where the pipe diameter is 90 mm, the pressure is 120 kPa. Downstream of the pump where the pipe diameter is 30 mm, the pressure is 400 kPa. If the loss in energy across the pump due to fluid friction effects is 170 N · m/kg, determine the hydraulic efficiency of the pump.

5.3.10 Water is to be pumped from the large tank shown in **Fig. P5.3.10** with an exit velocity of 6 m/s. It was determined that the original pump (pump 1) that supplies 1 kW of power to the water did not produce the desired velocity. Hence, it is proposed that an additional pump (pump 2) be installed as indicated to increase the flowrate to the desired value. How much power must pump 2 add to the water? The head loss for this flow is $h_L = 250Q^2$, where h_L is in m when Q is in m³/s.

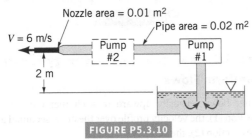

FIGURE P5.3.10

5.3.11 (See The Wide World of Fluids article titled "Curtain of air," Section 5.3.3.) The fan shown in **Fig. P5.3.11** produces an air curtain to separate a loading dock from a cold storage room. The air curtain is a jet of air 3 m wide, 0.15 m thick moving with speed $V = 9.1$ m/s. The loss associated with this flow is loss $= K_L V^2/2$, where $K_L = 5$. How much power must the fan supply to the air to produce this flow?

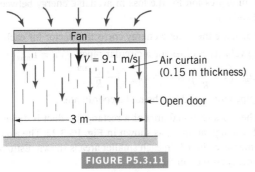

FIGURE P5.3.11

Section 5.3.3 The Mechanical Energy Equation and the Bernoulli Equation—Combined with Linear Momentum

5.3.12 When the pump shown in **Fig. P5.3.12** is stopped, there is no flow through the system and the spring force is zero. With the pump running, a 0.15 m-diameter jet leaves the pipe and the spring force is 1869 N. The water surface elevation in the tank is constant. Determine the water flowrate and the power consumed by the pump. Assume quasi-steady flow.

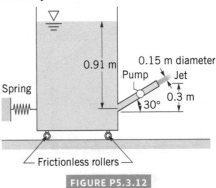

FIGURE P5.3.12

5.3.13 Air flows past an object in a pipe of 2-m diameter and exits as a free jet, as shown in **Fig. P5.3.13**. The velocity and pressure upstream are uniform at 10 m/s and 50 N/m², respectively. At the pipe exit the velocity is nonuniform as indicated. The shear stress along the pipe wall is negligible. **(a)** Determine the head loss associated with a particle as it flows from the uniform velocity upstream of the object to a location in the wake at the exit plane of the pipe. **(b)** Determine the force that the air exerts on the object.

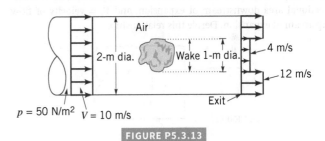

FIGURE P5.3.13

5.3.14 [Video] Near the downstream end of a river spillway, a hydraulic jump often forms, as illustrated in **Fig. P5.3.14** and **Video V10.11**. The velocity of the channel flow is reduced abruptly across the jump. Using the conservation of mass and linear momentum principles, derive the following expression for h_2:

$$h_2 = -\frac{h_1}{2} + \sqrt{\left(\frac{h_1}{2}\right)^2 + \frac{2V_1^2 h_1}{g}}$$

The loss of available energy across the jump can also be determined if energy conservation is considered. Derive the loss expression

$$\text{jump loss} = \frac{g(h_2 - h_1)^3}{4 h_1 h_2}$$

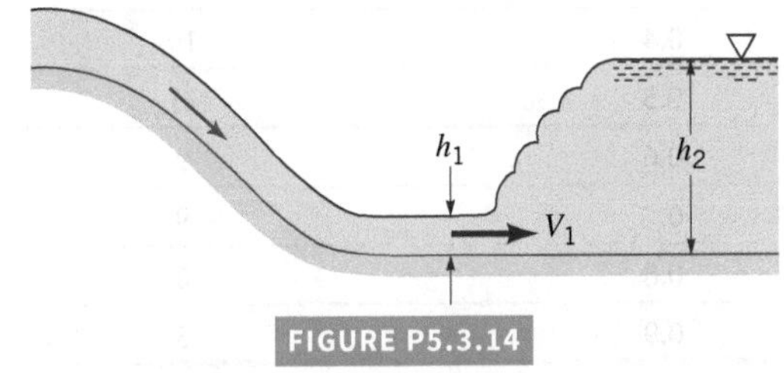

FIGURE P5.3.14

5.3.15 Water flows steadily down the inclined pipe, as indicated in **Fig. P5.3.15**. Determine the following: (a) the difference in pressure $p_1 - p_2$, (b) the loss between sections (1) and (2), and (c) the net axial force exerted by the pipe wall on the flowing water between sections (1) and (2).

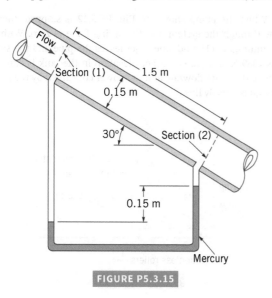

FIGURE P5.3.15

5.3.16 When fluid flows through an abrupt expansion, as indicated in **Fig. P5.3.16**, the loss in available energy across the expansion, loss_{ex}, is often expressed as

$$\text{loss}_{ex} = \left(1 - \frac{A_1}{A_2}\right)^2 \frac{V_1^2}{2}$$

where A_1 = cross-sectional area upstream of expansion, A_2 = cross-sectional area downstream of expansion, and V_1 = velocity of flow upstream of expansion. Derive this relationship.

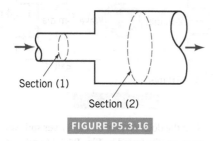

FIGURE P5.3.16

5.3.17 ⓒ Water (15.5 °C) flows through an annular space formed by inserting a 0.02 m-radius solid cylinder into a 0.04 m-radius tube. The following axial velocities were measured in the annular space.

$(r - r_i)/(r_o - r_i)$	u (m/s)
0.1	4
0.2	7
0.3	9
0.4	10
0.5	11
0.6	9
0.7	9
0.8	6
0.9	3

Assume that the no-slip condition ($u = 0$) exists at the solid boundaries. What are the rates of mass, momentum, and kinetic energy flow through the annular space?

5.3.18 Find the acceleration of the cart shown in **Fig. P5.3.18** as a function of the water height in the cart, which varies with time. The initial total mass is m_0, and the fluid density is ρ_0. Assume frictionless bearings, a frictionless surface, constant fluid density, uniform velocity over area A_N, all the fluid in the cart has the cart velocity, no air drag, and $A \gg A_N$.

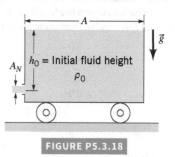

FIGURE P5.3.18

5.3.19 Two water jets collide and form one homogeneous jet, as shown in **Fig. P5.3.19**. (a) Determine the speed, V, and direction, θ, of the combined jet. (b) Determine the loss for a fluid particle flowing from (1) to (3), from (2) to (3). Gravity is negligible.

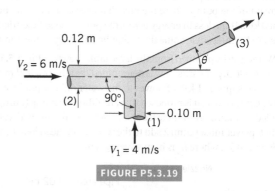

FIGURE P5.3.19

Section 5.3.4 Application of the Energy Equation to Nonuniform Flows

5.3.20 Water flows vertically upward in a circular cross-sectional pipe. At section (1), the velocity profile over the cross-sectional area is uniform. At section (2), the velocity profile is

$$\mathbf{V} = w_c \left(\frac{R - r}{R}\right)^{1/7} \hat{\mathbf{k}}$$

where $\mathbf{V}$ = local velocity vector, w_c = centerline velocity in the axial direction, R = pipe inside radius, and r = radius from pipe axis. Develop an expression for the loss in available energy between sections (1) and (2).

5.3.21 Calculate the kinetic energy correction factor for each of the following velocity profiles for a circular pipe: (a) $u = u_{max}\left(1 - \frac{r}{R}\right)$; (b) $u = u_{max}\left(1 - \frac{r^2}{R^2}\right)$; (c) $u = u_{max}\left(1 - \frac{r}{R}\right)^{1/9}$.

R is the pipe radius, and r is the radial coordinate.

5.3.22 The cross-sectional area of a rectangular duct is divided into 16 equal rectangular areas, as shown in **Fig. P5.3.22**. The axial fluid velocity measured in m/s in each smaller area is shown. Estimate the kinetic energy correction factor.

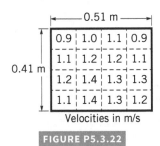

FIGURE P5.3.22

5.3.23 A small fan moves air at a mass flowrate of 1.82×10^{-3} kg/s. Upstream of the fan, the pipe diameter is 0.06 m, the flow is laminar, the velocity distribution is parabolic, and the kinetic energy coefficient, α_1, is equal to 2.0. Downstream of the fan, the pipe diameter is 0.02 m, the flow is turbulent, the velocity profile is quite flat, and the kinetic energy coefficient, α_2, is equal to 1.08. If the rise in static pressure across the fan is 103 Pa and the fan shaft draws 0.18 Watt, compare the value of loss calculated: (a) assuming uniform velocity distributions and (b) considering actual velocity distributions.

Section 5.3.5 Combination of the Energy Equation and the Moment-of-Momentum Equation

5.3.24 Air enters a radial blower with zero angular momentum. It leaves with an absolute tangential velocity, V_θ, of 61 m/s. The rotor blade speed at rotor exit is 52 m/s. If the stagnation pressure rise across the rotor is 2758 Pa, calculate the loss of available energy across the rotor and the rotor efficiency.

5.3.25 Water enters a pump impeller radially. It leaves the impeller with a tangential component of absolute velocity of 10 m/s. The impeller exit diameter is 60 mm, and the impeller speed is 1800 rpm. If the stagnation pressure that rises across the impeller is 45 kPa, determine the loss of available energy across the impeller and the hydraulic efficiency of the pump.

5.3.26 Water enters an axial-flow turbine rotor with an absolute velocity tangential component, V_θ, of 4.6 m/s. The corresponding blade velocity, U, is 15.2 m/s. The water leaves the rotor blade row with no angular momentum. If the stagnation pressure drop across the turbine is 82740 Pa, determine the hydraulic efficiency of the turbine.

5.3.27 An inward flow radial turbine (see **Fig. P5.3.27**) involves a nozzle angle, α_1, of 60° and an inlet rotor tip speed, U_1, of 9.1 m/s. The ratio of rotor inlet to outlet diameters is 2.0. The radial component of velocity remains constant at 6.1 m/s through the rotor, and the flow leaving the rotor at section (2) is without angular momentum. If the flowing fluid is water and the stagnation pressure drop across the rotor is 110320 Pa, determine the loss of available energy across the rotor and the hydraulic efficiency involved.

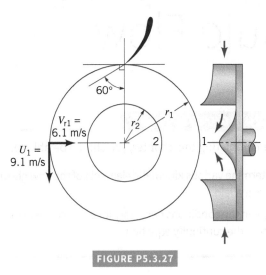

FIGURE P5.3.27

Lifelong Learning Problems

5.1 LL What are typical efficiencies associated with swimming, and how can they be improved?

5.2 LL *Magnetohydrodynamics* (MHD) is the study of the dynamics of electrically conducting fluids. Discuss the special features of MHD flows and some engineering applications for these flows.

5.3 LL Discuss the main causes of loss of available energy in a turbo pump and how they can be minimized. What are typical turbo-pump efficiencies?

5.4 LL Most of the loss of useful energy in fluid flows is due to turbulence. Explain the mechanisms by which turbulence and fluid viscosity work together to dissipate kinetic energy into heat/internal energy. Find a short "poem" that describes this process.

CHAPTER 6

Differential Analysis of Fluid Flow

LEARNING OBJECTIVES

After completing this chapter, you should be able to:

- determine various kinematic elements of the flow given the velocity field.
- explain the conditions necessary for a velocity field to satisfy the continuity equation.
- apply the concepts of stream function and velocity potential.
- characterize simple potential flow fields.
- analyze certain types of flows using the Navier–Stokes equations.

The previous chapter focused on the use of finite control volumes for solving a variety of fluid mechanics problems. This approach is very practical and useful because it does not require detailed knowledge of the pressure and velocity variations within the control volume. Typically, we found that only conditions on the surface of the control volume were needed, and thus problems could be solved without detailed knowledge of the flow field. However, many situations arise in which the details of the flow are important and the finite control volume approach will not yield the desired information. For example, we may need to know how the velocity varies over the cross section of a pipe, or how the pressure and shear stress vary along the surface of an airplane wing. In these circumstances we need to develop relationships that apply at a point, or at least in a very small region within a given flow field. This approach, which involves an *infinitesimal control volume*, as distinguished from a finite control volume, is commonly referred to as *differential analysis* because (as we will soon discover) the governing equations are differential equations.

In this chapter we will provide an introduction to the differential equations that describe (in detail) the motion of fluids. Unfortunately, a general solution for this set of nonlinear partial differential equations has not been obtained. Although exact solutions for special cases have been obtained, each is for relatively simple flow. Thus, although differential analysis has the potential for supplying very detailed information about flow fields, this information is not easily extracted. Nevertheless, this approach provides a fundamental basis for the study of fluid mechanics. You should not be too discouraged. The solutions that have been obtained provide significant insight into the physical processes occurring in a flow. A few are presented in this chapter. Assuming that viscous forces are negligible in comparison with the others results in equations that are much easier solve. Although this may seem to be a drastic assumption, the solutions obtained have been very useful is solving problems of practical interest. Some examples of these so-called *inviscid flow* solutions are also presented in this chapter.

It is known that for certain types of flows the flow field can be conceptually divided into two regions—a very thin region near the boundaries of the system in which viscous effects are important, and a region away from the boundaries in which the flow is essentially inviscid. By making certain assumptions about the behavior of the fluid in the thin layer near the boundaries, and using the assumption of inviscid flow outside this layer, a large class of problems can be solved using differential analysis. These *boundary layer problems* are discussed in Chapter 9.

Finally, with the availability of powerful computers, it is feasible to obtain approximate solutions to the differential equations using the techniques of numerical analysis. Although it is beyond the scope of this book to delve extensively into this approach, which is generally referred to as *computational fluid dynamics* (CFD), the reader should be aware of this approach to complex flow problems. CFD has become a common engineering tool, and a brief introduction can be found in Appendix A. The two adjacent video links present animations of CFD simulations of a complicated flow to show how CFD can help us to understand that flow.

We begin our introduction to differential analysis by reviewing and extending some of the ideas associated with fluid kinematics that were introduced in Chapter 3. With this background the remainder of the chapter will be devoted to the derivation of the basic differential equations (which will be based on the principle of conservation of mass and Newton's second law of motion) and to some applications.

V6.1 Spinning football-pressure contours

V6.2 Spinning football-velocity vectors

6.1 FLUID ELEMENT KINEMATICS

In this section we will be concerned with the mathematical description of the motion of fluid elements moving in a flow field. A small fluid element in the shape of a cube that is initially in one position will move to another position during a short time interval δt, as illustrated in **Fig. 6.1a**. Because of the generally complex velocity variation within the field, we expect the element to undergo deformation and rotation during its translation from one position to another. Even though they occur simultaneously, we can break the element's complex motion into four components: *translation, rotation, linear deformation,* and *angular deformation*, as shown in **Fig. 6.1b**. Since element motion and deformation are intimately related to the velocity and variation of velocity throughout the flow field, we will briefly review the manner in which velocity and acceleration fields can be described.

Fluid element motion consists of translation, linear deformation, rotation, and angular deformation.

6.1.1 Velocity and Acceleration Fields Revisited

As discussed in detail in Section 3.1, the velocity field can be described by specifying the velocity **V** at all points, and at all times, within the flow field of interest. Thus, in terms of rectangular coordinates, the notation **V** (x, y, z, t) means that the velocity of a fluid particle depends on where it is located within the flow field (as determined by its coordinates, $x, y,$ and z) and when it occupies the particular point (as determined by the time, t). As is pointed out in Section 3.1.1,

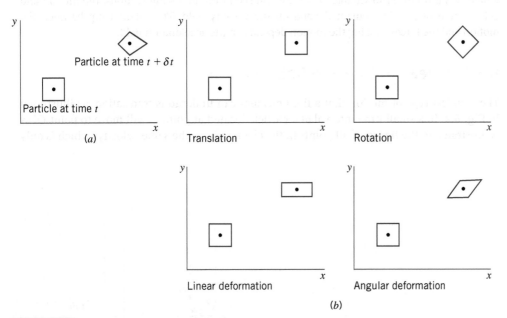

FIGURE 6.1 General fluid element motion and its components: (*a*) total element motion; (*b*) components of element motion.

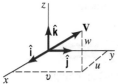

this method of describing the fluid motion is called the Eulerian method. It is also convenient to express the velocity in terms of three rectangular components so that

$$\mathbf{V} = u\hat{\mathbf{i}} + v\hat{\mathbf{j}} + w\hat{\mathbf{k}} \tag{6.1}$$

where u, v, and w are the velocity components in the x, y, and z directions, respectively, and $\hat{\mathbf{i}}$, $\hat{\mathbf{j}}$, and $\hat{\mathbf{k}}$ are the corresponding unit vectors, as shown in the adjacent figure. Of course, each of these components will, in general, be a function of x, y, z, and t. One of the goals of differential analysis is to determine how these velocity components depend on x, y, z, and t for a particular problem.

With this description of the velocity field, it was also shown in Section 3.2.1 that the acceleration of a fluid particle can be expressed as

$$\mathbf{a} = \frac{\partial \mathbf{V}}{\partial t} + u\frac{\partial \mathbf{V}}{\partial x} + v\frac{\partial \mathbf{V}}{\partial y} + w\frac{\partial \mathbf{V}}{\partial z} \tag{6.2}$$

and in component form:

$$a_x = \frac{\partial u}{\partial t} + u\frac{\partial u}{\partial x} + v\frac{\partial u}{\partial y} + w\frac{\partial u}{\partial z} \tag{6.3a}$$

$$a_y = \frac{\partial v}{\partial t} + u\frac{\partial v}{\partial x} + v\frac{\partial v}{\partial y} + w\frac{\partial v}{\partial z} \tag{6.3b}$$

$$a_z = \frac{\partial w}{\partial t} + u\frac{\partial w}{\partial x} + v\frac{\partial w}{\partial y} + w\frac{\partial w}{\partial z} \tag{6.3c}$$

The acceleration is also concisely expressed as

$$\mathbf{a} = \frac{D\mathbf{V}}{Dt} \tag{6.4}$$

The acceleration of a fluid particle is described using the material derivative.

where the operator

$$\frac{D(\ )}{Dt} = \frac{\partial(\ )}{\partial t} + u\frac{\partial(\ )}{\partial x} + v\frac{\partial(\ )}{\partial y} + w\frac{\partial(\ )}{\partial z} \tag{6.5}$$

is termed the *material derivative*, or *substantial derivative*. In vector notation

$$\frac{D(\ )}{Dt} = \frac{\partial(\ )}{\partial t} + (\mathbf{V} \cdot \nabla)(\ ) \tag{6.6}$$

where the gradient operator, $\nabla(\ )$, is

$$\nabla(\ ) = \frac{\partial(\ )}{\partial x}\hat{\mathbf{i}} + \frac{\partial(\ )}{\partial y}\hat{\mathbf{j}} + \frac{\partial(\ )}{\partial z}\hat{\mathbf{k}} \tag{6.7}$$

which was introduced in Chapter 2. As we will see in the following sections, the motion and deformation of a fluid element depend on the velocity field. The relationship between the motion and the forces causing the motion depends on the acceleration field.

6.1.2 Linear Motion and Deformation

The simplest type of motion that a fluid element can undergo is translation, as illustrated in **Fig. 6.2**. In a small time interval δt a particle located at point O will move to point O' as is illustrated in the figure. If all points in the element have the same velocity (which is only

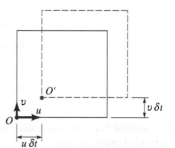

FIGURE 6.2 Translation of a fluid element.

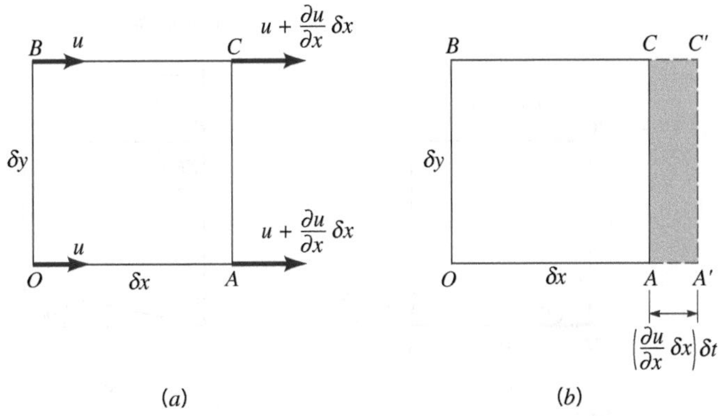

$$u + \frac{\partial u}{\partial x}\, \delta x$$

$$u + \frac{\partial u}{\partial x}\, \delta x$$

$$\left(\frac{\partial u}{\partial x}\, \delta x\right) \delta t$$

FIGURE 6.3 Linear deformation of a fluid element.

(a)   (b)

true if there are no velocity gradients), then the element will simply translate from one position to another. However, because of the presence of velocity gradients, the element will generally be deformed and rotated as it moves. For example, consider the effect of a single velocity gradient, $\partial u/\partial x$, on a small cube having sides δx, δy, and δz. As is shown in **Fig. 6.3a**, if the x component of velocity of O and B is u, then at nearby points A and C the x component of the velocity can be expressed as $u + (\partial u/\partial x)\, \delta x$. This difference in velocity causes a "stretching" of the volume element by an amount $(\partial u/\partial x)(\delta x)(\delta t)$ during the short time interval δt in which line OA stretches to OA' and BC to BC' (**Fig. 6.3b**). The corresponding change in the original volume, $\delta V = \delta x\, \delta y\, \delta z$, would be

$$\text{Change is } \delta V = \left(\frac{\partial u}{\partial x}\, \delta x\right)(\delta y\, \delta z)(\delta t)$$

and the *rate* at which the volume δV is changing *per unit volume* due to the gradient $\partial u/\partial x$ is

$$\frac{1}{\delta V}\frac{d(\delta V)}{dt} = \lim_{\delta t \to 0}\left[\frac{(\partial u/\partial x)\, \delta t}{\delta t}\right] = \frac{\partial u}{\partial x} \tag{6.8}$$

If velocity gradients $\partial v/\partial y$ and $\partial w/\partial z$ are also present, then using a similar analysis it follows that, in the general case,

$$\frac{1}{\delta V}\frac{d(\delta V)}{dt} = \frac{\partial u}{\partial x} + \frac{\partial v}{\partial y} + \frac{\partial w}{\partial z} = \nabla \cdot \mathbf{V} \tag{6.9}$$

The rate of volume change per unit volume is related to the velocity gradients.

This rate of change of the volume per unit volume is called the **volumetric dilatation rate**. Thus, we see that the volume of a fluid element may change as the element moves from one location to another in the flow field. However, for an *incompressible fluid* the volumetric dilatation rate must be zero, since the element volume cannot change without a change in fluid density (the element mass must be conserved). Variations in the velocity in the direction of the velocity, as represented by the derivatives $\partial u/\partial x$, $\partial v/\partial y$, and $\partial w/\partial z$, simply cause a *linear deformation* of the element in the sense that the shape of the element does not change. Cross derivatives, such as $\partial u/\partial y$ and $\partial v/\partial x$, will cause the element to rotate and generally to undergo an *angular deformation*, which changes the shape of the element.

V6.3 Shear deformation

6.1.3 Angular Motion and Deformation

For simplicity we will consider motion in the x–y plane, but the results can be readily extended to the more general three-dimensional case. The velocity variation that causes rotation and angular deformation is illustrated in **Fig. 6.4a**. In a short time interval δt the line segments OA and OB will rotate through the angles $\delta\alpha$ and $\delta\beta$ to the new positions OA' and OB', as is shown in **Fig. 6.4b**. The angular velocity of line OA, ω_{OA}, is

$$\omega_{OA} = \lim_{\delta t \to 0}\frac{\delta\alpha}{\delta t}$$

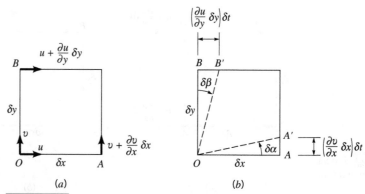

FIGURE 6.4   Angular motion and deformation of a fluid element.

For small angles

$$\tan \delta\alpha \approx \delta\alpha = \frac{(\partial v/\partial x)\,\delta x\,\delta t}{\delta x} = \frac{\partial v}{\partial x}\,\delta t \tag{6.10}$$

so that

$$\omega_{OA} = \lim_{\delta t \to 0}\left[\frac{(\partial v/\partial x)\,\delta t}{\delta t}\right] = \frac{\partial v}{\partial x}$$

Note that if $\partial v/\partial x$ is positive, ω_{OA} will be counterclockwise. Similarly, the angular velocity of the line OB is

$$\omega_{OB} = \lim_{\delta t \to 0}\frac{\delta\beta}{\delta t}$$

and

$$\tan \delta\beta \approx \delta\beta = \frac{(\partial u/\partial y)\,\delta y\,\delta t}{\delta y} = \frac{\partial u}{\partial y}\,\delta t \tag{6.11}$$

so that

$$\omega_{OB} = \lim_{\delta t \to 0}\left[\frac{(\partial u/\partial y)\,\delta t}{\delta t}\right] = \frac{\partial u}{\partial y}$$

Rotation of fluid particles is related to certain velocity gradients in the flow field.

In this instance if $\partial u/\partial y$ is positive, ω_{OB} will be clockwise. The angular velocity of the element is defined as the average angular velocity of two mutually perpendicular lines (ω_{OA} and ω_{OB}).[1] Following the typical convention for a Cartesian coordinate system in which the z-coordinate is perpendicular to both the x- and y-coordinates and the right-hand rule is used to determine the positive direction of rotation about a coordinate, the rotation of the fluid element of Fig. 6.4 can be expressed as

$$\omega_z = \frac{1}{2}\left(\frac{\partial v}{\partial x} - \frac{\partial u}{\partial y}\right) \tag{6.12}$$

Rotation of the fluid element about the other two coordinate axes can be obtained in a similar manner with the result that for rotation about the x axis

$$\omega_x = \frac{1}{2}\left(\frac{\partial w}{\partial y} - \frac{\partial v}{\partial z}\right) \tag{6.13}$$

and for rotation about the y axis

$$\omega_y = \frac{1}{2}\left(\frac{\partial u}{\partial z} - \frac{\partial w}{\partial x}\right) \tag{6.14}$$

[1]With this definition ω_z can also be interpreted to be the angular velocity of the bisector of the angle between the lines OA and OB.

The three components, ω_x, ω_y, and ω_z can be combined to give the rotation vector, $\boldsymbol{\omega}$, in the form

$$\boldsymbol{\omega} = \omega_x \hat{\mathbf{i}} + \omega_y \hat{\mathbf{j}} + \omega_z \hat{\mathbf{k}} \qquad (6.15)$$

An examination of this result reveals that $\boldsymbol{\omega}$ is equal to one-half the curl of the velocity vector. That is,

$$\boldsymbol{\omega} = \tfrac{1}{2} \operatorname{curl} \mathbf{V} = \tfrac{1}{2} \boldsymbol{\nabla} \times \mathbf{V} \qquad (6.16)$$

because in Cartesian coordinates the vector operation $\boldsymbol{\nabla} \times \mathbf{V}$ is

$$\tfrac{1}{2} \boldsymbol{\nabla} \times \mathbf{V} = \tfrac{1}{2} \begin{vmatrix} \hat{\mathbf{i}} & \hat{\mathbf{j}} & \hat{\mathbf{k}} \\ \dfrac{\partial}{\partial x} & \dfrac{\partial}{\partial y} & \dfrac{\partial}{\partial z} \\ u & v & w \end{vmatrix} = \tfrac{1}{2}\left(\dfrac{\partial w}{\partial y} - \dfrac{\partial v}{\partial z}\right)\hat{\mathbf{i}} + \tfrac{1}{2}\left(\dfrac{\partial u}{\partial z} - \dfrac{\partial w}{\partial x}\right)\hat{\mathbf{j}} + \tfrac{1}{2}\left(\dfrac{\partial v}{\partial x} - \dfrac{\partial u}{\partial y}\right)\hat{\mathbf{k}}$$

Vorticity in a flow field is twice the fluid particle rotation rate.

The **vorticity**, $\boldsymbol{\zeta}$, is defined as a vector that is twice the rotation vector; that is,

$$\boxed{\boldsymbol{\zeta} = 2\,\boldsymbol{\omega} = \boldsymbol{\nabla} \times \mathbf{V}} \qquad (6.17)$$

The use of the vorticity to describe the rotational characteristics of the fluid simply eliminates the $\left(\tfrac{1}{2}\right)$ factor associated with the rotation vector. The adjacent figure shows vorticity contours of the wing tip vortex flow shortly after an aircraft has passed. The lighter colors indicate stronger vorticity (see also Fig. 3.3).

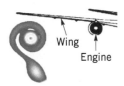

Wing

Engine

We observe from Eq. 6.12 that the fluid element will rotate about the z axis as an *undeformed* block (i.e., $\omega_{OA} = -\omega_{OB}$) only when $\partial u/\partial y = -\partial v/\partial x$. Otherwise the rotation will be associated with an angular deformation. We also note from Eq. 6.12 that when $\partial u/\partial y = \partial v/\partial x$ the rotation around the z axis is zero. More generally if $\boldsymbol{\nabla} \times \mathbf{V} = 0$, then the rotation (and the vorticity) are zero, and flow fields for which this condition applies are termed **irrotational flow**. We will find in Section 6.4 that the condition of irrotationality often greatly simplifies the analysis of complex flow fields. However, it is probably not immediately obvious why some flow fields would be irrotational, and we will need to examine this concept more fully in Section 6.4.

EXAMPLE 6.1 | Vorticity

Given For a certain two-dimensional flow field the velocity is given by the equation

$$\mathbf{V} = (x^2 - y^2)\hat{\mathbf{i}} - 2xy\hat{\mathbf{j}}$$

Find Is this flow irrotational?

Solution For an irrotational flow the rotation vector, $\boldsymbol{\omega}$, having the components given by Eqs. 6.12, 6.13, and 6.14 must be zero. For the prescribed velocity field

$$u = x^2 - y^2 \qquad v = -2xy \qquad w = 0$$

and therefore

$$\omega_x = \tfrac{1}{2}\left(\dfrac{\partial w}{\partial y} - \dfrac{\partial v}{\partial z}\right) = 0$$

$$\omega_y = \tfrac{1}{2}\left(\dfrac{\partial u}{\partial z} - \dfrac{\partial w}{\partial x}\right) = 0$$

$$\omega_z = \tfrac{1}{2}\left(\dfrac{\partial v}{\partial x} - \dfrac{\partial u}{\partial y}\right) = \tfrac{1}{2}[(-2y) - (-2y)] = 0$$

Thus, the flow is irrotational. (Ans)

Comments It is to be noted that for a two-dimensional flow field (where the flow is in the x–y plane) ω_x and ω_y will always be zero because, by definition of two-dimensional flow, u and v are not functions of z, and w is zero. In this instance the condition for irrotationality simply becomes $\omega_z = 0$ or $\partial v/\partial x = \partial u/\partial y$.

The streamlines for the steady, two-dimensional flow of this example are shown in **Fig. E6.1**. (Information about how to calculate streamlines for a given velocity field is given in Sections 4.1.4 and 6.2.3.) It is noted that all of the streamlines (except for the one through the origin) are curved. However, because the flow is irrotational, there is no rotation of the fluid elements. That is, lines OA and OB of Fig. 6.4 rotate with the same speed but in opposite directions.

As shown by Eq. 6.17, the condition of irrotationality is equivalent to the fact that the vorticity, $\boldsymbol{\zeta}$, is zero or the curl of the velocity is zero.

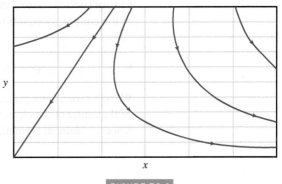

y

x

FIGURE E6.1

In addition to the rotation associated with the derivatives $\partial u/\partial y$ and $\partial v/\partial x$, it is observed from Fig. 6.4b that these derivatives can cause the fluid element to undergo an *angular deformation*, which results in a change in shape of the element. The change in the original right angle formed by the lines OA and OB is termed the shearing strain, $\delta\gamma$, and from Fig. 6.4b

$$\delta\gamma = \delta\alpha + \delta\beta$$

where $\delta\gamma$ is considered to be positive if the original right angle is decreasing. The rate of change of $\delta\gamma$ is called the *rate of shearing strain*, or the *rate of angular deformation*, and is commonly denoted with the symbol $\dot\gamma$. The angles $\delta\alpha$ and $\delta\beta$ are related to the velocity gradients through Eqs. 6.10 and 6.11 so that

$$\dot\gamma = \lim_{\delta t \to 0} \frac{\delta\gamma}{\delta t} = \lim_{\delta t \to 0}\left[\frac{(\partial v/\partial x)\,\delta t + (\partial u/\partial y)\,\delta t}{\delta t}\right]$$

and, therefore,

$$\dot\gamma = \frac{\partial v}{\partial x} + \frac{\partial u}{\partial y} \tag{6.18}$$

As we will learn in Section 6.8, the rate of angular deformation is related to a corresponding shearing stress that causes the fluid element to change in shape. From Eq. 6.18 we note that if $\partial u/\partial y = -\partial v/\partial x$, the rate of angular deformation is zero, and this condition corresponds to the case in which the element is simply rotating as an undeformed block (Eq. 6.12). In the remainder of this chapter we will see how the various kinematic relationships developed in this section play an important role in the development and subsequent analysis of the differential equations that model fluid motion.

6.2   CONSERVATION OF MASS

Conservation of mass requires that the mass of a system remain constant.

As is discussed in Section 5.1, conservation of mass requires that the mass, M, of a system remain constant as the system moves through the flow field. In equation form this principle is expressed as

$$\frac{DM_{\text{sys}}}{Dt} = 0$$

We found it convenient to use the control volume approach for fluid flow problems, with the control volume representation of the conservation of mass written as

$$\frac{\partial}{\partial t}\int_{\text{cv}} \rho\, dV + \int_{\text{cs}} \rho\mathbf{V}\cdot\hat{\mathbf{n}}\, dA = 0 \tag{6.19}$$

where the equation (commonly called the **continuity equation**) can be applied to a finite control volume (CV), which is bounded by a control surface (CS). The first integral on the left side of Eq. 6.19 represents the rate at which the mass within the control volume is changing, and the second integral represents the net rate at which mass is flowing out through the control surface (rate of mass outflow − rate of mass inflow). To obtain the differential form of the continuity equation, Eq. 6.19 is applied to an infinitesimal control volume.

6.2.1 Differential Form of Continuity Equation

We will take as our control volume the small, stationary cubical element shown in **Fig. 6.5a**. At the center of the element the fluid density is ρ, and the velocity has components u, v, and w. Since the element is small, the volume integral in Eq. 6.19 can be expressed as

$$\frac{\partial}{\partial t}\int_{\text{cv}} \rho\, dV \approx \frac{\partial\rho}{\partial t}\,\delta x\,\delta y\,\delta z \tag{6.20}$$

The rate of mass flow through the surfaces of the element can be obtained by considering the flow in each of the coordinate directions separately. For example, in **Fig. 6.5b** flow in the x

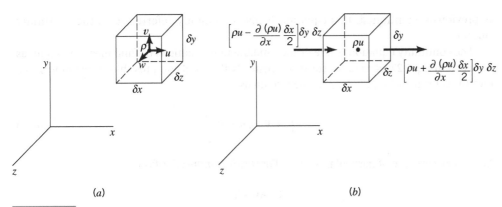

FIGURE 6.5 A differential element for the development of conservation of mass equation.

direction is depicted. If we let ρu represent the x component of the mass rate of flow per unit area at the center of the element, then on the right face

$$\rho u\big|_{x+(\delta x/2)} = \rho u + \frac{\partial(\rho u)}{\partial x}\frac{\delta x}{2} \tag{6.21}$$

and on the left face

$$\rho u\big|_{x-(\delta x/2)} = \rho u - \frac{\partial(\rho u)}{\partial x}\frac{\delta x}{2} \tag{6.22}$$

Note that we are really using a Taylor series expansion of ρu and neglecting higher order terms such as $(\delta x)^2$, $(\delta x)^3$, and so on. The adjacent figure is a depiction of the use of the truncated Taylor series to represent the velocity at the face of the element. When the right sides of Eqs. 6.21 and 6.22 are multiplied by the area $\delta y\,\delta z$, the rate at which mass is crossing the right and left sides of the element is obtained as is illustrated in Fig. 6.5b. When these two expressions are combined, the net rate of mass flowing from the element through the two surfaces can be expressed as

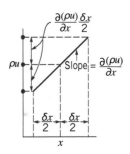

$$\begin{aligned}\text{Net rate of mass}\atop\text{outflow in } x \text{ direction} &= \left[\rho u + \frac{\partial(\rho u)}{\partial x}\frac{\delta x}{2}\right]\delta y\,\delta z \\ &- \left[\rho u - \frac{\partial(\rho u)}{\partial x}\frac{\delta x}{2}\right]\delta y\,\delta z = \frac{\partial(\rho u)}{\partial x}\,\delta x\,\delta y\,\delta z\end{aligned} \tag{6.23}$$

For simplicity, only flow in the x direction has been considered in Fig. 6.5b. In general, there will also be flow in the y and z directions. An analysis similar to the one used for flow in the x direction shows that

$$\text{Net rate of mass}\atop\text{outflow in } y \text{ direction} = \frac{\partial(\rho v)}{\partial y}\,\delta x\,\delta y\,\delta z \tag{6.24}$$

and

$$\text{Net rate of mass}\atop\text{outflow in } z \text{ direction} = \frac{\partial(\rho w)}{\partial z}\,\delta x\,\delta y\,\delta z \tag{6.25}$$

Thus,

$$\text{Net rate of}\atop\text{mass outflow} = \left[\frac{\partial(\rho u)}{\partial x} + \frac{\partial(\rho v)}{\partial y} + \frac{\partial(\rho w)}{\partial z}\right]\delta x\,\delta y\,\delta z \tag{6.26}$$

We now combine Eqs. 6.19, 6.20, and 6.26, cancel the common term $\delta x\,\delta y\,\delta z$, and take the limit as δz, δy, and δz, approach zero, shrinking the tiny control volume to a point. The result is the differential equation for conservation of mass

> The continuity equation is one of the fundamental equations of fluid mechanics.

$$\boxed{\dfrac{\partial\rho}{\partial t} + \dfrac{\partial(\rho u)}{\partial x} + \dfrac{\partial(\rho v)}{\partial y} + \dfrac{\partial(\rho w)}{\partial z} = 0} \tag{6.27}$$

As previously mentioned, this equation is also commonly referred to as the continuity equation.

The continuity equation is one of the fundamental equations of fluid mechanics and, as expressed in Eq. 6.27, is valid for steady or unsteady flow and compressible or incompressible flow. In vector notation, Eq. 6.27 can be written as

$$\frac{\partial \rho}{\partial t} + \nabla \cdot \rho \mathbf{V} = 0 \tag{6.28}$$

Two special cases are of particular interest. For *steady compressible* flow

$$\nabla \cdot \rho \mathbf{V} = 0$$

or

$$\frac{\partial(\rho u)}{\partial x} + \frac{\partial(\rho v)}{\partial y} + \frac{\partial(\rho w)}{\partial z} = 0 \tag{6.29}$$

For incompressible flow the continuity equation reduces to a simple relationship involving certain velocity gradients.

This follows since by definition ρ is not a function of time for steady flow but could be a function of position. For *incompressible* flow the fluid density, ρ, is a constant throughout the flow field so that Eq. 6.28 becomes

$$\nabla \cdot \mathbf{V} = 0 \tag{6.30}$$

or

$$\frac{\partial u}{\partial x} + \frac{\partial v}{\partial y} + \frac{\partial w}{\partial z} = 0 \tag{6.31}$$

Equation 6.31 applies to both steady and unsteady incompressible flow. Note that Eq. 6.31 is the same as that obtained by setting the volumetric dilatation rate (Eq. 6.9) equal to zero. This result should not be surprising because both relationships are based on conservation of mass for incompressible fluids. However, the expression for the volumetric dilatation rate was developed from a system approach, whereas Eq. 6.31 was developed from a control volume approach. In the former case the deformation of a particular differential mass of fluid was studied, and in the latter case mass flow through a fixed differential volume was studied.

EXAMPLE 6.2 | Continuity Equation

Given   The velocity components for a certain incompressible, steady-flow field are

$$u = x^2 + y^2 + z^2$$
$$v = xy + yz + z$$
$$w = ?$$

Find   Determine the form of the z component of the velocity, w, required to satisfy the continuity equation.

Solution

Any physically possible velocity distribution must for an incompressible flow satisfy conservation of mass as expressed by the continuity equation

$$\frac{\partial u}{\partial x} + \frac{\partial v}{\partial y} + \frac{\partial w}{\partial z} = 0$$

For the given velocity distribution

$$\frac{\partial u}{\partial x} = 2x \quad \text{and} \quad \frac{\partial v}{\partial y} = x + z$$

so that the required expression for $\partial w / \partial z$ is

$$\frac{\partial w}{\partial z} = -2x - (x + z) = -3x - z$$

Integration with respect to z yields

$$w = -3xz - \frac{z^2}{2} + f(x, y) \tag{Ans}$$

Comment   The third velocity component cannot be explicitly determined because the function $f(x, y)$ can have any form and conservation of mass will still be satisfied. The specific form of this function will be governed by the flow field described by these velocity components—that is, some additional information is needed to completely determine w.

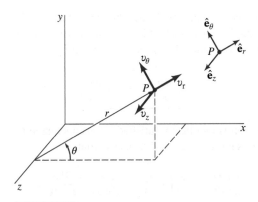

FIGURE 6.6 The representation of velocity components in cylindrical polar coordinates.

6.2.2 Cylindrical Polar Coordinates

For some problems it is more convenient to express the various differential relationships in cylindrical polar coordinates rather than Cartesian coordinates. As is shown in **Fig. 6.6**, with cylindrical coordinates a point is located by specifying the coordinates r, θ, and z. The coordinate r is the radial distance from the z axis, θ is the angle measured from a line parallel to the x axis (with counterclockwise taken as positive), and z is the coordinate along the z axis. The velocity components, as sketched in Fig. 6.6, are the radial velocity, v_r, the tangential velocity, v_θ, and the axial velocity, v_z. Thus, the velocity at some arbitrary point P can be expressed as

For some problems, velocity components expressed in cylindrical polar coordinates will be convenient.

$$\mathbf{V} = v_r\hat{\mathbf{e}}_r + v_\theta\hat{\mathbf{e}}_\theta + v_z\hat{\mathbf{e}}_z \tag{6.32}$$

where $\hat{\mathbf{e}}_r$, $\hat{\mathbf{e}}_\theta$, and $\hat{\mathbf{e}}_z$ are the unit vectors in the r, θ, and z directions, respectively, as are illustrated in Fig. 6.6. The use of cylindrical coordinates is particularly convenient when the boundaries of the flow system are cylindrical. Several examples illustrating the use of cylindrical coordinates will be provided in succeeding sections of this chapter.

The differential form of the continuity equation in cylindrical coordinates is

$$\boxed{\frac{\partial\rho}{\partial t} + \frac{1}{r}\frac{\partial(r\rho v_r)}{\partial r} + \frac{1}{r}\frac{\partial(\rho v_\theta)}{\partial\theta} + \frac{\partial(\rho v_z)}{\partial z} = 0} \tag{6.33}$$

This equation can be derived by following the same procedure used in the preceding section. For steady, compressible flow

$$\frac{1}{r}\frac{\partial(r\rho v_r)}{\partial r} + \frac{1}{r}\frac{\partial(\rho v_\theta)}{\partial\theta} + \frac{\partial(\rho v_z)}{\partial z} = 0 \tag{6.34}$$

For incompressible flow (steady or unsteady)

$$\frac{1}{r}\frac{\partial(r v_r)}{\partial r} + \frac{1}{r}\frac{\partial v_\theta}{\partial\theta} + \frac{\partial v_z}{\partial z} = 0 \tag{6.35}$$

6.2.3 The Stream Function

Steady, incompressible, plane, two-dimensional flow represents one of the simplest types of flow of practical importance. By plane, two-dimensional flow we mean that there are only two velocity components, such as u and v, when the flow is considered to be in the x–y plane. For this flow the continuity equation, Eq. 6.31, reduces to

$$\frac{\partial u}{\partial x} + \frac{\partial v}{\partial y} = 0 \tag{6.36}$$

Velocity components in a two-dimensional flow field can be expressed in terms of a stream function.

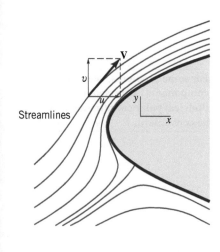

We have two variables, u and v, to deal with, but they must be related in a special way as indicated by Eq. 6.36. This equation suggests that if we define a function $\psi(x, y)$, called the **stream function**, which relates the velocities shown in the adjacent figure as

$$u = \frac{\partial \psi}{\partial y} \qquad v = -\frac{\partial \psi}{\partial x} \qquad\qquad (6.37)$$

then the continuity equation is identically satisfied. This conclusion can be verified by simply substituting the expressions for u and v into Eq. 6.36 so that

$$\frac{\partial}{\partial x}\left(\frac{\partial \psi}{\partial y}\right) + \frac{\partial}{\partial y}\left(-\frac{\partial \psi}{\partial x}\right) = \frac{\partial^2 \psi}{\partial x\, \partial y} - \frac{\partial^2 \psi}{\partial y\, \partial x} = 0$$

Thus, whenever the velocity components are defined in terms of the stream function, we know that conservation of mass will be satisfied. Of course, we still do not know what $\psi(x, y)$ is for in a particular problem, but at least we have simplified the analysis by having to determine only one unknown function, $\psi(x, y)$, rather than the two functions, $u(x, y)$ and $v(x, y)$.

Another particular advantage of using the stream function is related to the fact that *lines along which ψ is constant are streamlines*. Recall from Section 3.1.4 that streamlines are lines in the flow field that are everywhere tangent to the velocities, as is illustrated in **Fig. 6.7** and in the adjacent figure. It follows from the definition of the streamline that the slope at any point along a streamline is given by

$$\frac{dy}{dx} = \frac{v}{u}$$

The change in the value of ψ as we move from one point (x, y) to a nearby point $(x + dx, y + dy)$ is given by the relationship:

$$d\psi = \frac{\partial \psi}{\partial x}\, dx + \frac{\partial \psi}{\partial y}\, dy = -v\, dx + u\, dy$$

Along a line of constant ψ we have $d\psi = 0$ so that

$$-v\, dx + u\, dy = 0$$

and, therefore, along a line of constant ψ

$$\frac{dy}{dx} = \frac{v}{u}$$

which is the defining equation for a streamline. Thus, if we know the function $\psi(x, y)$, we can plot lines of constant ψ to provide the family of streamlines that are helpful in visualizing the

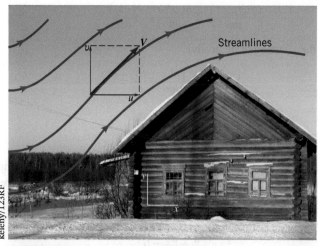

FIGURE 6.7   Velocity and velocity components along a streamline.

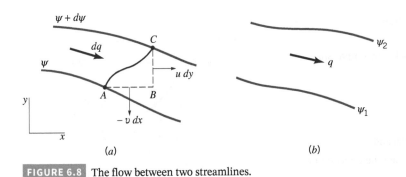

FIGURE 6.8 The flow between two streamlines.

flow pattern. An infinite number of streamlines make up a particular flow field, since for each constant value assigned to ψ a streamline can be drawn.

The actual numerical value associated with a particular streamline is not of particular significance, but the change in the value of ψ is related to the volume rate of flow. Consider two closely spaced streamlines, shown in **Fig. 6.8a**. The lower streamline is designated ψ and the upper one $\psi + d\psi$. Let dq represent the volume rate of flow (per unit width perpendicular to the x–y plane) passing between the two streamlines. Note that flow never crosses streamlines, since by definition the velocity is tangent to the streamline. From conservation of mass we know that the inflow, dq, crossing the arbitrary surface AC of Fig. 6.8a must equal the net outflow through surfaces AB and BC. Thus,

$$dq = u\,dy - v\,dx$$

or in terms of the stream function

$$dq = \frac{\partial \psi}{\partial y}\,dy + \frac{\partial \psi}{\partial x}\,dx \tag{6.38}$$

The right-hand side of Eq. 6.38 is equal to $d\psi$ so that

$$dq = d\psi \tag{6.39}$$

Thus, the volume rate of flow, q, between two streamlines such as ψ_1 and ψ_2 of **Fig. 6.8b** can be determined by integrating Eq. 6.39 to yield

$$q = \int_{\psi_1}^{\psi_2} d\psi = \psi_2 - \psi_1 \tag{6.40}$$

The relative value of ψ_2 with respect to ψ_1 determines the direction of flow, as shown in the adjacent figure.

In cylindrical coordinates the continuity equation (Eq. 6.35) for incompressible, plane, two-dimensional flow reduces to

$$\frac{1}{r}\frac{\partial(r v_r)}{\partial r} + \frac{1}{r}\frac{\partial v_\theta}{\partial \theta} = 0 \tag{6.41}$$

and the velocity components, v_r and v_θ, can be related to the stream function, $\psi(r, \theta)$, through the equations

$$v_r = \frac{1}{r}\frac{\partial \psi}{\partial \theta} \qquad v_\theta = -\frac{\partial \psi}{\partial r} \tag{6.42}$$

as shown in the adjacent figure.

Substitution of these expressions for the velocity components into Eq. 6.41 shows that the continuity equation is identically satisfied. The stream function concept can be extended to axisymmetric flows, such as flow in pipes or flow around bodies of revolution, and to two-dimensional compressible flows. However, the concept is not applicable to general three-dimensional flows using a single stream function.

The change in the value of the stream function is related to the volume rate of flow.

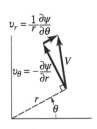

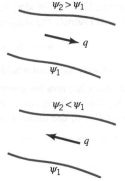

EXAMPLE 6.3 | Stream Function

Given The velocity components in a steady, incompressible, two-dimensional flow field are

$$u = 2y$$
$$v = 4x$$

Find

a. Determine the corresponding stream function and

b. Show on a sketch several streamlines. Indicate the direction of flow along the streamlines.

Solution

a. From the definition of the stream function (Eqs. 6.37)

$$u = \frac{\partial \psi}{\partial y} = 2y$$

and

$$v = -\frac{\partial \psi}{\partial x} = 4x$$

The first of these equations can be integrated to give

$$\psi = y^2 + f_1(x)$$

where $f_1(x)$ is an arbitrary function of x. Similarly from the second equation

$$\psi = -2x^2 + f_2(y)$$

where $f_2(y)$ is an arbitrary function of y. It now follows that in order to satisfy both expressions for the stream function

$$\psi = -2x^2 + y^2 + C \qquad (Ans)$$

where C is an arbitrary constant.

Since the velocities are related to the derivatives of the stream function, an arbitrary constant can always be added to the function, and the value of the constant is actually of no consequence. Usually, for simplicity, we set $C = 0$ so that for this particular example the simplest form for the stream function is

$$\psi = -2x^2 + y^2 \qquad (1) \quad (Ans)$$

Either answer indicated would be acceptable.

b. Streamlines can now be determined by setting ψ = constant and plotting the resulting curve. With the preceding expression for ψ (with $C = 0$) the value of ψ at the origin is zero, so that the

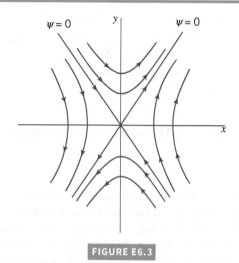

FIGURE E6.3

equation of the streamline passing through the origin (the $\psi = 0$ streamline) is

$$0 = -2x^2 + y^2$$

or

$$y = \pm \sqrt{2}x$$

Other streamlines can be obtained by setting ψ equal to various constants. It follows from Eq. 1 that the equations of these streamlines (for $\psi \neq 0$) can be expressed in the form

$$\frac{y^2}{\psi} - \frac{x^2}{\psi/2} = 1$$

which we recognize as the equation of a hyperbola. Thus, the streamlines are a family of hyperbolas with the $\psi = 0$ streamlines as asymptotes. Several of the streamlines are plotted in **Fig. E6.3**. Because the velocities can be calculated at any point, the direction of flow along a given streamline can be easily deduced. For example, $v = -\partial \psi / \partial x = 4x$ so that $v > 0$ if $x > 0$ and $v < 0$ if $x < 0$. The direction of flow is indicated on the figure.

Comment One could sketch the streamlines by drawing several velocity vectors and then sketching curves tangent to them.

6.3 THE LINEAR MOMENTUM EQUATION

To develop the differential momentum equations we can start with the linear momentum equation (Newton's Second Law)

$$\mathbf{F} = \frac{D\mathbf{P}}{Dt}\bigg|_{sys} \qquad (6.43)$$

where **F** is the resultant force acting on a fluid mass, **P** is the linear momentum defined as

$$\mathbf{P} = \int_{sys} \mathbf{V}\, dm$$

and the operator $D(\ )/Dt$ is the material derivative (see Section 3.2.1 and Eqs. 6.5–6.7). In the last chapter it was demonstrated how Eq. 6.43 in the form

$$\sum \mathbf{F}_{\substack{\text{contents of the} \\ \text{control volume}}} = \frac{\partial}{\partial t} \int_{\text{cv}} \mathbf{V}\rho \, d\Psi + \int_{\text{cs}} \mathbf{V}\rho\mathbf{V} \cdot \hat{\mathbf{n}} \, dA \tag{6.44}$$

could be applied to a finite control volume to solve a variety of flow problems. To obtain the differential form of the linear momentum equation, we can either apply Eq. 6.43 to a differential system, consisting of a mass, δm, or apply Eq. 6.44 to an infinitesimal control volume, δV, which initially bounds the mass δm. It is probably simpler to use the system approach since application of Eq. 6.43 to the differential mass, δm, yields

$$\delta \mathbf{F} = \frac{D(\mathbf{V}\,\delta m)}{Dt}$$

where $\delta \mathbf{F}$ is the resultant force acting on δm. Using this system approach δm can be treated as a constant so that

$$\delta \mathbf{F} = \delta m \frac{D\mathbf{V}}{Dt}$$

But $D\mathbf{V}/Dt$ is the acceleration, $\mathbf{a}$, of the element. Thus,

$$\delta \mathbf{F} = \delta m \, \mathbf{a} \tag{6.45}$$

which is simply Newton's second law applied to the mass δm. This is the same result that would be obtained by applying Eq. 6.44 to an infinitesimal control volume (see Ref. 1). Before we can proceed, it is necessary to examine how the force $\delta \mathbf{F}$ can be most conveniently expressed.

6.3.1 Description of Forces Acting on the Differential Element

In general, two types of forces need to be considered: *surface forces*, which act on the surface of the differential element, and *body forces*, which are distributed throughout the element. For our purpose, the only body force, $\delta \mathbf{F}_b$, of interest is the weight of the element, which can be expressed as

Both surface forces and body forces generally act on fluid particles.

$$\delta \mathbf{F}_b = \delta m \, \mathbf{g} \tag{6.46}$$

where $\mathbf{g}$ is the vector representation of the acceleration of gravity. In component form

$$\delta F_{bx} = \delta m \, g_x \tag{6.47a}$$
$$\delta F_{by} = \delta m \, g_y \tag{6.47b}$$
$$\delta F_{bz} = \delta m \, g_z \tag{6.47c}$$

where g_x, g_y, and g_z are the components of the acceleration of gravity vector in the x, y, and z directions, respectively.

Surface forces act on the element as a result of its interaction with its surroundings. At any arbitrary location within a fluid mass, the force acting on a small area, δA, which lies in an arbitrary surface, can be represented by $\delta \mathbf{F}_s$, as is shown in **Fig. 6.9**. In general, $\delta \mathbf{F}_s$ will be inclined with respect to the surface. The force $\delta \mathbf{F}_s$ can be resolved into three components, δF_n, δF_1, and δF_2, where δF_n is normal to the area, δA, and δF_1 and δF_2 are parallel to the area and orthogonal to each other. The *normal stress*, σ_n, is defined as

$$\sigma_n = \lim_{\delta A \to 0} \frac{\delta F_n}{\delta A}$$

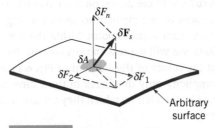

FIGURE 6.9 Components of force acting on an arbitrary differential area.

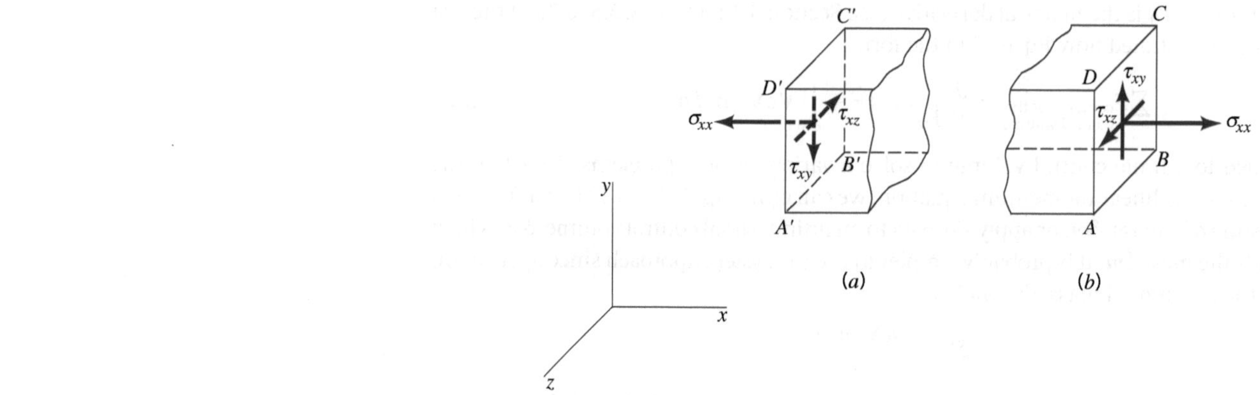

FIGURE 6.10 Double subscript notation for stresses.

and the *shearing* stresses are defined as

$$\tau_1 = \lim_{\delta A \to 0} \frac{\delta F_1}{\delta A}$$

and

$$\tau_2 = \lim_{\delta A \to 0} \frac{\delta F_2}{\delta A}$$

We will use σ for normal stresses and τ for shearing stresses. The intensity of the force per unit area at a point in a body can thus be characterized by a normal stress and two shearing stresses, if the orientation of the area is specified. For purposes of analysis it is usually convenient to reference the area to the coordinate system. For example, for the rectangular coordinate system shown in **Fig. 6.10,** we choose to consider the stresses acting on planes parallel to the coordinate planes. On the plane *ABCD* of Fig. 6.10a, which is parallel to the *y–z* plane, the normal stress is denoted σ_{xx} and the shearing stresses are denoted as τ_{xy} and τ_{xz}. To easily identify the particular stress component we use a double subscript notation. The first subscript indicates the direction of the *normal* to the plane on which the stress acts, and the second subscript indicates the direction of the stress. Thus, normal stresses have repeated subscripts, whereas the subscripts for the shearing stresses are always different.

It is also necessary to establish a sign convention for the stresses. We define the positive direction for the stress as the positive coordinate direction on the surfaces for which the outward normal is in the positive coordinate direction. This is the case illustrated in Fig. 6.10a where the outward normal to the area *ABCD* is in the positive *x* direction. The positive directions for σ_{xx}, τ_{xy}, and τ_{xz} are as shown in Fig. 6.10a. If the outward normal points in the negative coordinate direction, as in Fig. 6.10b for the area *A'B'C'D'*, then the stresses are considered positive if directed in the negative coordinate directions. Thus, the stresses shown in Fig. 6.10b are considered to be positive when directed as shown. Note that positive normal stresses are tensile stresses; that is, they tend to "stretch" the material.

It should be emphasized that the state of stress at a point in a material is not completely defined by simply three components of a "stress vector." This follows, since any particular stress vector depends on the orientation of the plane passing through the point. However, it can be shown that the normal and shearing stresses acting on *any* plane passing through a point can be expressed in terms of the stresses acting on three orthogonal planes passing through the point (Ref. 2).

Surface forces can be expressed in terms of the shear and normal stresses.

We now can express the surface forces acting on a small cubical element of fluid in terms of the stresses acting on the faces of the element as shown in **Fig. 6.11**. It is expected that in general the stresses will vary from point to point within the flow field. Thus, through the use of Taylor series expansions, we will express the stresses on the various faces in terms of the corresponding stresses at the center of the element of Fig. 6.11 and their gradients in the coordinate directions. For simplicity only the forces in the *x* direction are shown. Note that the stresses must be multiplied by the area on which they act to obtain the force. Summing all these forces in the *x* direction yields

$$\delta F_{sx} = \left(\frac{\partial \sigma_{xx}}{\partial x} + \frac{\partial \tau_{yx}}{\partial y} + \frac{\partial \tau_{zx}}{\partial z} \right) \delta x \, \delta y \, \delta z \qquad (6.48a)$$

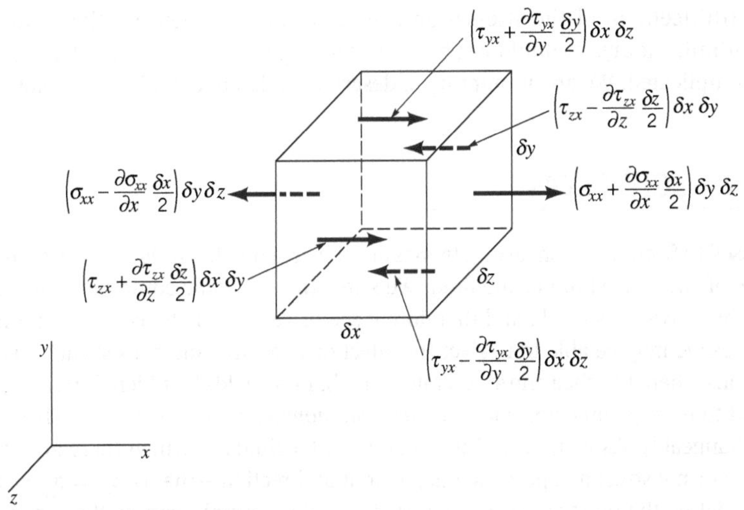

FIGURE 6.11 Surface forces in the x direction acting on a fluid element.

for the resultant surface force in the x direction. In a similar manner the resultant surface forces in the y and z directions can be expressed as

$$\delta F_{sy} = \left(\frac{\partial \tau_{xy}}{\partial x} + \frac{\partial \sigma_{yy}}{\partial y} + \frac{\partial \tau_{zy}}{\partial z}\right) \delta x \, \delta y \, \delta z \tag{6.48b}$$

$$\delta F_{sz} = \left(\frac{\partial \tau_{xz}}{\partial x} + \frac{\partial \tau_{yz}}{\partial y} + \frac{\partial \sigma_{zz}}{\partial z}\right) \delta x \, \delta y \, \delta z \tag{6.48c}$$

The resultant surface force can now be expressed as

$$\delta \mathbf{F}_s = \delta F_{sx} \hat{\mathbf{i}} + \delta F_{sy} \hat{\mathbf{j}} + \delta F_{sz} \hat{\mathbf{k}} \tag{6.49}$$

and this force combined with the body force, $\delta \mathbf{F}_b$, yields the resultant force, $\delta \mathbf{F}$, acting on the differential mass, δm. That is, $\delta \mathbf{F} = \delta \mathbf{F}_s + \delta \mathbf{F}_b$.

6.3.2 Equations of Motion

The expressions for the body and surface forces can now be used in conjunction with Eq. 6.45 to develop the equations of motion. In component form Eq. 6.45 can be written as

$$\delta F_x = \delta m \, a_x$$
$$\delta F_y = \delta m \, a_y$$
$$\delta F_z = \delta m \, a_z$$

where $\delta m = \rho \, \delta x \, \delta y \, \delta z$ and the acceleration components are given by Eq. 6.3. It now follows (using Eqs. 6.47 and 6.48 for the forces on the element) that

$$\rho g_x + \frac{\partial \sigma_{xx}}{\partial x} + \frac{\partial \tau_{yx}}{\partial y} + \frac{\partial \tau_{zx}}{\partial z} = \rho\left(\frac{\partial u}{\partial t} + u\frac{\partial u}{\partial x} + v\frac{\partial u}{\partial y} + w\frac{\partial u}{\partial z}\right) \tag{6.50a}$$

$$\rho g_y + \frac{\partial \tau_{xy}}{\partial x} + \frac{\partial \sigma_{yy}}{\partial y} + \frac{\partial \tau_{zy}}{\partial z} = \rho\left(\frac{\partial v}{\partial t} + u\frac{\partial v}{\partial x} + v\frac{\partial v}{\partial y} + w\frac{\partial v}{\partial z}\right) \tag{6.50b}$$

$$\rho g_z + \frac{\partial \tau_{xz}}{\partial x} + \frac{\partial \tau_{yz}}{\partial y} + \frac{\partial \sigma_{zz}}{\partial z} = \rho\left(\frac{\partial w}{\partial t} + u\frac{\partial w}{\partial x} + v\frac{\partial w}{\partial y} + w\frac{\partial w}{\partial z}\right) \tag{6.50c}$$

The motion of a fluid is governed by a set of nonlinear differential equations.

where the element volume $\delta x \, \delta y \, \delta z$ cancels out before taking the limit, shrinking the tiny particle to a point.

Equations 6.50, sometimes called the *Cauchy equations*, are the general differential equations of motion for a fluid. In fact, they are applicable to any continuum (solid or fluid) in motion or at rest. However, before we can use the equations to solve specific problems, some additional information about the stresses must be obtained. Otherwise, we will have more unknowns (thirteen, all of the stresses and velocities and the density) than equations (including continuity, four). It should not be too surprising that the differential analysis of fluid motion is complicated. We are attempting to describe, in detail, complex fluid motion.

6.4   INVISCID FLOW

As is discussed in Section 1.6, shearing stresses develop in a moving fluid because of the viscosity and the rate of shearing strain of the fluid. We know that for some common fluids, such as air and water, the viscosity is small, and therefore it seems reasonable to assume that under some circumstances we may be able to neglect the effect of viscosity (and thus shearing stresses), at least in regions where the shearing strain rate is small. Flow fields in which the shearing stresses are assumed to be negligible are said to be *inviscid, nonviscous,* or *frictionless.* These terms are used interchangeably. As is discussed in Section 2.1, for fluids in which there are no shearing stresses, the normal stress at a point is independent of direction—that is, $\sigma_{xx} = \sigma_{yy} = \sigma_{zz}$. In this instance we define the pressure, p, as the negative of the normal stress so that, as indicated by the adjacent figure,

$$-p = \sigma_{xx} = \sigma_{yy} = \sigma_{zz}$$

The negative sign is used so that a *compressive* normal stress (which is what we expect in a fluid) will give a *positive* value for p.

In Chapter 4 the inviscid flow concept was used in the development of the Bernoulli equation, and numerous applications of this important equation were considered. In this section we will again consider the Bernoulli equation and will show how it can be derived from the general equations of motion for inviscid flow.

6.4.1   Euler's Equations of Motion

For an inviscid flow in which all the shearing stresses are zero, and the normal stresses are replaced by $-p$, the general equations of motion (Eqs. 6.50) reduce to

$$\rho g_x - \frac{\partial p}{\partial x} = \rho\left(\frac{\partial u}{\partial t} + u\frac{\partial u}{\partial x} + v\frac{\partial u}{\partial y} + w\frac{\partial u}{\partial z}\right) \tag{6.51a}$$

$$\rho g_y - \frac{\partial p}{\partial y} = \rho\left(\frac{\partial v}{\partial t} + u\frac{\partial v}{\partial x} + v\frac{\partial v}{\partial y} + w\frac{\partial v}{\partial z}\right) \tag{6.51b}$$

$$\rho g_z - \frac{\partial p}{\partial z} = \rho\left(\frac{\partial w}{\partial t} + u\frac{\partial w}{\partial x} + v\frac{\partial w}{\partial y} + w\frac{\partial w}{\partial z}\right) \tag{6.51c}$$

These equations are commonly referred to as **Euler's equations of motion**, named in honor of Leonhard Euler (1707–1783), a famous Swiss mathematician who pioneered work on the relationship between pressure and flow. In vector notation Euler's equations can be expressed as

Euler's equations of motion apply to an inviscid flow field.

$$\rho \mathbf{g} - \nabla p = \rho\left[\frac{\partial \mathbf{V}}{\partial t} + (\mathbf{V} \cdot \nabla)\mathbf{V}\right] \tag{6.52}$$

Although Eqs. 6.51 are considerably simpler than the general equations of motion, Eqs. 6.50, they are still not amenable to a general analytical solution that would allow us to determine the pressure and velocity at all points within an inviscid flow field. (At least, when the continuity equation is included, the number of equations equals the number of unknowns.) The main difficulty in solving the equations arises from the nonlinear velocity terms ($u\,\partial u/\partial x$, $v\,\partial u/\partial y$, etc.), which appear in the convective acceleration, or the $(\mathbf{V} \cdot \nabla)\mathbf{V}$ term. Because of these terms, Euler's equations are nonlinear partial differential equations for which we do not have a general solution. However, under some circumstances we can use them to obtain useful information about inviscid flow fields. For example, as shown in the following section we can integrate Eq. 6.52 to obtain a relationship (the Bernoulli equation) between elevation, pressure, and velocity along a streamline.

6.4.2 The Bernoulli Equation

In Section 3.2 the Bernoulli equation was derived by a direct application of Newton's second law to a fluid particle moving along a streamline. In this section we will again derive this important equation, starting from Euler's equations. Of course, we should obtain the same result since Euler's equations simply represent a statement of Newton's second law expressed in a general form that maintains the restriction of zero viscosity. We will restrict our attention to steady flow so Euler's equation in vector form becomes

$$\rho \mathbf{g} - \nabla p = \rho (\mathbf{V} \cdot \nabla) \mathbf{V} \tag{6.53}$$

We wish to integrate this differential equation along some arbitrary streamline (**Fig. 6.12**) and select the coordinate system with the z axis vertical (with "up" being positive) so that, as shown in the adjacent figure, the acceleration due to gravity vector can be expressed as

$$\mathbf{g} = -g\nabla z$$

where g is the magnitude of that vector. Also, it will be convenient to use the vector identity

$$(\mathbf{V} \cdot \nabla)\mathbf{V} = \frac{1}{2}\nabla(\mathbf{V} \cdot \mathbf{V}) - \mathbf{V} \times (\nabla \times \mathbf{V})$$

Equation 6.53 can now be written in the form

$$-\rho g \nabla z - \nabla p = \frac{\rho}{2}\nabla(\mathbf{V} \cdot \mathbf{V}) - \rho \mathbf{V} \times (\nabla \times \mathbf{V})$$

and this equation can be rearranged to yield

$$\frac{\nabla p}{\rho} + \frac{1}{2}\nabla(V^2) + g\nabla z = \mathbf{V} \times (\nabla \times \mathbf{V})$$

We next take the dot product of each term with a differential length $d\mathbf{s}$ along a streamline (Fig. 6.12). Thus,

$$\frac{\nabla p}{\rho} \cdot d\mathbf{s} + \frac{1}{2}\nabla(V^2) \cdot d\mathbf{s} + g\nabla z \cdot d\mathbf{s} = [\mathbf{V} \times (\nabla \times \mathbf{V})] \cdot d\mathbf{s} \tag{6.54}$$

Since $d\mathbf{s}$ has a direction along the streamline, the vectors $d\mathbf{s}$ and $\mathbf{V}$ are parallel. However, as shown in the adjacent figure, the vector $\mathbf{V} \times (\nabla \times \mathbf{V})$ is perpendicular to $\mathbf{V}$ (why?), so it follows that

$$[\mathbf{V} \times (\nabla \times \mathbf{V})] \cdot d\mathbf{s} = 0$$

Recall also that the dot product of the gradient of a scalar and a differential length gives the differential change in the scalar in the direction of the differential length. That is, with $d\mathbf{s} = dx\,\hat{\mathbf{i}} + dy\,\hat{\mathbf{j}} + dz\,\hat{\mathbf{k}}$, we can write $\nabla p \cdot d\mathbf{s} = (\partial p/\partial x)\,dx + (\partial p/\partial y)\,dy + (\partial p/\partial z)\,dz = dp$. Thus, Eq. 6.54 becomes

$$\frac{dp}{\rho} + \frac{1}{2}d(V^2) + g\,dz = 0 \tag{6.55}$$

where the change in p, V, and z is along the streamline. Equation 6.55 can now be integrated to give

$$\boxed{\int \frac{dp}{\rho} + \frac{V^2}{2} + gz = \text{constant}} \tag{6.56}$$

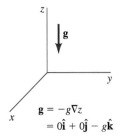

$$\mathbf{g} = -g\nabla z$$
$$= 0\hat{\mathbf{i}} + 0\hat{\mathbf{j}} - g\hat{\mathbf{k}}$$

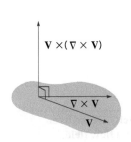

$$\mathbf{V} \times (\nabla \times \mathbf{V})$$

Euler's equations can be integrated to give the relationship among pressure, velocity, and elevation for inviscid flow.

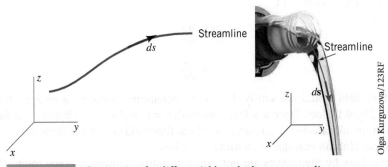

Streamline

Streamline

Olga Kurguzova/123RF

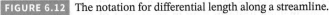

FIGURE 6.12 The notation for differential length along a streamline.

which indicates that the sum of the three terms on the left side of the equation must remain a constant along a given streamline. Equation 6.56 is valid for both compressible and incompressible inviscid flows, but for compressible fluids the variation in ρ with p must be specified before the first term in Eq. 6.56 can be evaluated.

For inviscid, incompressible fluids (commonly called **ideal fluids**) Eq. 6.56 can be written as

$$\frac{p}{\rho} + \frac{V^2}{2} + gz = \text{constant along a streamline} \tag{6.57}$$

and this equation is the **Bernoulli equation** used extensively in Chapter 4. It is often convenient to write Eq. 6.57 between two points (1) and (2) along a streamline and to express the equation in the "head" form by dividing each term by g so that

$$\frac{p_1}{\gamma} + \frac{V_1^2}{2g} + z_1 = \frac{p_2}{\gamma} + \frac{V_2^2}{2g} + z_2 \tag{6.58}$$

It should be again emphasized that the Bernoulli equation, as expressed by Eqs. 6.57 and 6.58, is restricted to the following:

- inviscid flow
- steady flow
- incompressible flow
- flow along a streamline

Note that these are precisely the same restrictions that appeared in Chapter 4. You may want to go back and review some of the examples in Chapter 4 that illustrate the use of the Bernoulli equation.

6.4.3 Irrotational Flow

If we make one additional assumption—that the flow is *irrotational*—the analysis of inviscid flow problems is further simplified. Recall from Section 6.1.3 that the rotation of a fluid element is equal to $\frac{1}{2}(\boldsymbol{\nabla} \times \mathbf{V})$, and an irrotational flow field is one for which $\boldsymbol{\nabla} \times \mathbf{V} = 0$ (i.e., the curl of velocity is zero). Since the vorticity, ζ, is defined as $\boldsymbol{\nabla} \times \mathbf{V}$, it also follows that in an irrotational flow field the vorticity is zero. The concept of irrotationality may seem to be a rather strange condition for a flow field. Why would a flow field be irrotational? To answer this question we note that if $\frac{1}{2}(\boldsymbol{\nabla} \times \mathbf{V}) = 0$, then each of the components of this vector, as are given by Eqs. 6.12, 6.13, and 6.14, must be equal to zero. Since these components include the various velocity gradients in the flow field, the condition of irrotationality imposes specific relationships among these velocity gradients. For example, for rotation about the z axis to be zero, it follows from Eq. 6.12 that

> The vorticity is zero in an irrotational flow field.

$$\omega_z = \frac{1}{2}\left(\frac{\partial v}{\partial x} - \frac{\partial u}{\partial y}\right) = 0$$

and, therefore,

$$\frac{\partial v}{\partial x} = \frac{\partial u}{\partial y} \tag{6.59}$$

Similarly from Eqs. 6.13 and 6.14

$$\frac{\partial w}{\partial y} = \frac{\partial v}{\partial z} \tag{6.60}$$

$$\frac{\partial u}{\partial z} = \frac{\partial w}{\partial x} \tag{6.61}$$

A general flow field would not satisfy these three equations. However, a uniform flow as is illustrated in **Fig. 6.13** does. Since $u = U$ (a constant), $v = 0$, and $w = 0$, it follows that Eqs. 6.59, 6.60, and 6.61 are all satisfied. Therefore, a uniform flow field (in which there are no velocity gradients) is certainly an example of an irrotational flow.

Uniform flows by themselves are not very interesting. However, many interesting and important flow problems include uniform flow in some part of the flow field. Two examples are shown in **Fig. 6.14**. In Fig. 6.14a a solid body is placed in a uniform stream of fluid.

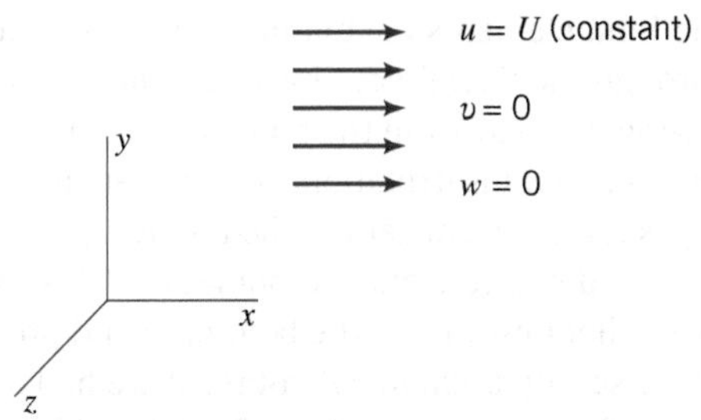

$u = U$ (constant)

$v = 0$

$w = 0$

FIGURE 6.13 Uniform flow in the *x* direction.

Far away from the body the flow remains uniform, and in this far region the flow is irrotational. In Fig. 6.14c, flow from a large reservoir enters a pipe through a well-rounded entrance where the velocity distribution is essentially uniform. Thus, at the entrance the flow is irrotational.

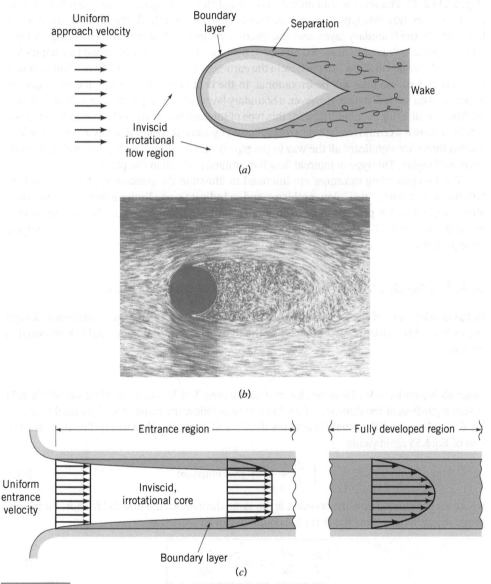

Uniform approach velocity

Boundary layer

Separation

Inviscid irrotational flow region

Wake

(a)

(b)

Entrance region

Fully developed region

Uniform entrance velocity

Inviscid, irrotational core

Boundary layer

(c)

FIGURE 6.14 Various regions of flow: (*a*) around bodies; (*b*) actual flow over a circular cylinder showing wake behind the cylinder; (*c*) through channels.

For an inviscid fluid there are no shearing stresses—the only forces acting on a fluid element are its weight and pressure forces. Since the weight acts through the element center of gravity, and the pressure acts in a direction normal to the element surface, neither of these forces can cause the element to rotate. Therefore, for an inviscid fluid, if some part of the

Flow fields involving real fluids often include both regions of negligible shearing stresses and regions of significant shearing stresses.

flow field is irrotational, the fluid elements emanating from this region will not take on any rotation as they progress through the flow field. This phenomenon is illustrated in Fig. 6.14*a* in which fluid elements flowing far away from the body have irrotational motion, and as they flow around the body the motion remains irrotational except very near the boundary. Near the boundary the velocity changes rapidly from zero at the boundary (no-slip condition) to some relatively large value in a short distance from the boundary. This rapid change in velocity gives rise to a large velocity gradient normal to the boundary and produces significant shearing stresses, even though the viscosity is small. Of course if we had a truly inviscid fluid, the fluid would simply "slide" past the boundary and the flow would be irrotational everywhere. But this is not the case for real fluids, so we will typically have a layer (usually relatively thin) near any fixed surface in a moving stream in which shearing stresses are not negligible. This layer is called the *boundary layer*. Outside the boundary layer the flow is well represented by the irrotational flow model. Another possible consequence of the boundary layer is that the main stream may "separate" from the surface and form a *wake* downstream from the body (Fig. 6.14*a* & *b*). The wake would include a region of slow, perhaps randomly moving fluid. To completely analyze this type of flow it is necessary to consider both the inviscid, irrotational flow outside the boundary layer, and the viscous, rotational flow within the boundary layer and to somehow "match" these two regions. This type of analysis is developed in Chapter 9.

As is illustrated in Fig. 6.14*c*, the flow in the entrance to a pipe may be uniform (if the entrance is well-rounded) and thus will be irrotational. In the central core of the pipe the flow remains irrotational for some distance. However, a boundary layer will develop along the wall and grow in thickness until it fills the pipe. Thus, for this type of internal flow there will be an *entrance region* in which there is a central irrotational core, followed by a so-called *fully developed region* in which viscous forces are significant all the way to the pipe centerline. The flow is rotational in the fully developed region. This type of internal flow is examined in detail in Chapter 8.

The two preceding examples are intended to illustrate the possible applicability of irrotational flow to some "real" flow problems and to indicate some limitations of the irrotationality assumption. We proceed to develop some useful equations based on the assumptions of inviscid, incompressible, irrotational flow, with the admonition to use caution when applying the equations.

6.4.4 The Bernoulli Equation for Irrotational Flow

In the development of the Bernoulli equation in Section 6.4.2, Eq. 6.54 was integrated along a streamline. This restriction was imposed so the right side of the equation could be set equal to zero; that is,

$$[\mathbf{V} \times (\nabla \times \mathbf{V})] \cdot d\mathbf{s} = 0$$

(since $d\mathbf{s}$ is parallel to $\mathbf{V}$). However, for irrotational flow, $\nabla \times \mathbf{V} = 0$, so the right side of Eq. 6.54 is zero regardless of the direction of $d\mathbf{s}$. We can now follow the same procedure used to obtain Eq. 6.55, where the differential changes dp, $d(V^2)$, and dz can be taken in any direction. Integration of Eq. 6.55 again yields

$$\int \frac{dp}{\rho} + \frac{V^2}{2} + gz = \text{constant} \tag{6.62}$$

where for irrotational flow the constant is the same throughout the flow field. Thus, for incompressible, irrotational flow the Bernoulli equation can be written as

$$\boxed{\frac{p_1}{\gamma} + \frac{V_1^2}{2g} + z_1 = \frac{p_2}{\gamma} + \frac{V_2^2}{2g} + z_2} \tag{6.63}$$

The Bernoulli equation can be applied between any two points in an irrotational flow field.

between *any two points in the flow field*. Equation 6.63 is exactly the same form as Eq. 6.58 but is not limited to application along a streamline. However, Eq. 6.63 is restricted to

- inviscid flow
- steady flow
- incompressible flow
- irrotational flow

It may be worthwhile to review the use and misuse of the Bernoulli equation for rotational flow as is illustrated in Example 4.17.

6.4.5 The Velocity Potential

For an irrotational flow the velocity gradients are related through Eqs. 6.59, 6.60, and 6.61. In this case the velocity components can be expressed in terms of a scalar function $\phi(x, y, z, t)$ as

$$u = \frac{\partial \phi}{\partial x} \qquad v = \frac{\partial \phi}{\partial y} \qquad w = \frac{\partial \phi}{\partial z} \tag{6.64}$$

where ϕ is called the **velocity potential**. Direct substitution of these expressions for the velocity components into Eqs. 6.59, 6.60, and 6.61 will verify that a velocity field defined by Eqs. 6.64 is indeed irrotational. In vector form, Eqs. 6.64 can be written as

$$\boxed{\mathbf{V} = \nabla \phi} \tag{6.65}$$

so that for an irrotational flow the velocity is expressible as the gradient of a scalar function ϕ.

The velocity potential is a consequence of the irrotationality of the flow field, whereas the stream function is a consequence of conservation of mass (see Section 6.2.3). It is to be noted, however, that the velocity potential can be defined for a general three-dimensional flow, whereas a single stream function is restricted to two-dimensional flows.

For an incompressible fluid we know from conservation of mass that

$$\boxed{\nabla \cdot \mathbf{V} = 0}$$

and therefore for incompressible, irrotational flow (with $\mathbf{V} = \nabla \phi$) it follows that

$$\nabla^2 \phi = 0 \tag{6.66}$$

where $\nabla^2(\ ) = \nabla \cdot \nabla(\ )$ is the *Laplacian operator*. In Cartesian coordinates

$$\frac{\partial^2 \phi}{\partial x^2} + \frac{\partial^2 \phi}{\partial y^2} + \frac{\partial^2 \phi}{\partial z^2} = 0$$

This differential equation arises in many different areas of engineering and physics and is called *Laplace's equation*. Thus, inviscid, incompressible, irrotational flow fields are governed by Laplace's equation. This type of flow is commonly called a **potential flow**. To complete the mathematical formulation of a given problem, boundary conditions must be specified. These are usually velocities specified on the boundaries of the flow field of interest. It follows that if the potential function can be determined, then the velocity at all points in the flow field can be determined from Eq. 6.64, and the pressure at all points can be determined from the Bernoulli equation (Eq. 6.63). Although the concept of the velocity potential is applicable to both steady and unsteady flow, we will confine our attention to steady flow.

Potential flows, governed by Eqs. 6.64 and 6.66, are irrotational flows. That is, the vorticity is zero throughout. If vorticity is present (e.g., boundary layer, wake), then the flow cannot be described by Laplace's equation. The adjacent figure illustrates a flow in which the vorticity is not zero in two regions—the separated region behind the bump and the boundary layer next to the solid surface. The details of this kind of flow field are examined in detail in Chapter 9.

For some problems it will be convenient to use cylindrical coordinates, r, θ, and z. In this coordinate system the gradient operator is

$$\nabla(\ ) = \frac{\partial(\ )}{\partial r} \hat{\mathbf{e}}_r + \frac{1}{r} \frac{\partial(\ )}{\partial \theta} \hat{\mathbf{e}}_\theta + \frac{\partial(\ )}{\partial z} \hat{\mathbf{e}}_z \tag{6.67}$$

so that

$$\nabla \phi = \frac{\partial \phi}{\partial r} \hat{\mathbf{e}}_r + \frac{1}{r} \frac{\partial \phi}{\partial \theta} \hat{\mathbf{e}}_\theta + \frac{\partial \phi}{\partial z} \hat{\mathbf{e}}_z \tag{6.68}$$

where $\phi = \phi(r, \theta, z)$. Since

$$\mathbf{V} = v_r \hat{\mathbf{e}}_r + v_\theta \hat{\mathbf{e}}_\theta + v_z \hat{\mathbf{e}}_z \tag{6.69}$$

it follows for an irrotational flow (with $\mathbf{V} = \nabla \phi$)

$$v_r = \frac{\partial \phi}{\partial r} \qquad v_\theta = \frac{1}{r} \frac{\partial \phi}{\partial \theta} \qquad v_z = \frac{\partial \phi}{\partial z} \tag{6.70}$$

Inviscid, incompressible, irrotational flow fields are governed by Laplace's equation and are called potential flows.

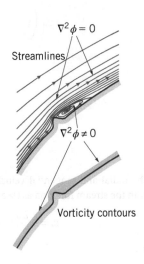

Streamlines

$\nabla^2 \phi = 0$

$\nabla^2 \phi \neq 0$

Vorticity contours

Also, Laplace's equation in cylindrical coordinates is

$$\frac{1}{r}\frac{\partial}{\partial r}\left(r\frac{\partial \phi}{\partial r}\right) + \frac{1}{r^2}\frac{\partial^2 \phi}{\partial \theta^2} + \frac{\partial^2 \phi}{\partial z^2} = 0 \qquad (6.71)$$

EXAMPLE 6.4 | Velocity Potential and Inviscid Flow Pressure

Given The two-dimensional flow of a nonviscous, incompressible fluid in the vicinity of the 90° corner of **Fig. E6.4a** is described by the stream function

$$\psi = 2r^2 \sin 2\theta$$

where ψ has units of m^2/s when r is in meters. Assume the fluid density is 10^3 kg/m^3 and the x–y plane is horizontal—

that is, there is no difference in elevation between points (1) and (2).

Find

  a. Determine, if possible, the corresponding velocity potential.

  b. If the pressure at point (1) on the wall is 30 kPa, what is the pressure at point (2)?

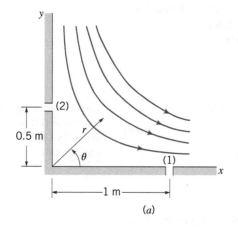

(a)

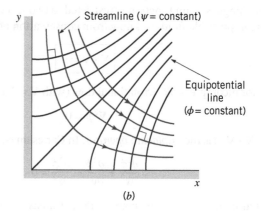

(b)

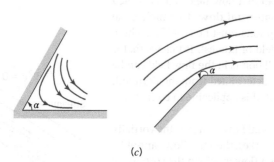

(c)

FIGURE E6.4

Solution

a. The radial and tangential velocity components can be obtained from the stream function as (see Eq. 6.42)

$$v_r = \frac{1}{r}\frac{\partial \psi}{\partial \theta} = 4r\cos 2\theta$$

and

$$v_\theta = -\frac{\partial \psi}{\partial r} = -4r\sin 2\theta$$

Because

$$v_r = \frac{\partial \phi}{\partial r}$$

it follows that

$$\frac{\partial \phi}{\partial r} = 4r\cos 2\theta$$

and therefore by integration

$$\phi = 2r^2\cos 2\theta + f_1(\theta) \qquad (1)$$

where $f_1(\theta)$ is an arbitrary function of θ. Similarly

$$v_\theta = \frac{1}{r}\frac{\partial \phi}{\partial \theta} = -4r\sin 2\theta$$

and integration yields

$$\phi = 2r^2\cos 2\theta + f_2(r) \qquad (2)$$

where $f_2(r)$ is an arbitrary function of r. To satisfy both Eqs. 1 and 2, the velocity potential must have the form

$$\phi = 2r^2\cos 2\theta + C \qquad (Ans)$$

where C is an arbitrary constant. As is the case for stream functions, the specific value of C is not important, and it is customary to let $C = 0$ so that the velocity potential for this corner flow is

$$\phi = 2r^2 \cos 2\theta \qquad (Ans)$$

Comment In the statement of this problem, it was implied by the wording "if possible" that we might not be able to find a corresponding velocity potential. The reason for this concern is that we can always define a stream function for two-dimensional flow, but the flow must be *irrotational* if there is a corresponding velocity potential. Thus, the fact that we were able to determine a velocity potential means that the flow is irrotational. Several streamlines and lines of constant ϕ are plotted in **Fig. E6.4b**. These two sets of lines are *orthogonal*. The reason why streamlines and lines of constant ϕ are always orthogonal is explained in Section 6.5.

b. Since we have an irrotational flow of a nonviscous, incompressible fluid, the Bernoulli equation can be applied between any two points. Thus, between points (1) and (2) with no elevation change

$$\frac{p_1}{\gamma} + \frac{V_1^2}{2g} = \frac{p_2}{\gamma} + \frac{V_2^2}{2g}$$

or

$$p_2 = p_1 + \frac{\rho}{2}(V_1^2 - V_2^2) \qquad (3)$$

Because

$$V^2 = v_r^2 + v_\theta^2$$

it follows that for any point within the flow field

$$V^2 = (4r \cos 2\theta)^2 + (-4r \sin 2\theta)^2$$
$$= 16r^2(\cos^2 2\theta + \sin^2 2\theta)$$
$$= 16r^2$$

This result indicates that the square of the velocity at any point depends only on the radial distance, r, to the point. Note that the constant, 16, has units of s^{-2}. Thus,

$$V_1^2 = (16\ s^{-2})(1\ m)^2 = 16\ m^2/s^2$$

and

$$V_2^2 = (16\ s^{-2})(0.5\ m)^2 = 4\ m^2/s^2$$

Substitution of these velocities into Eq. 3 gives

$$p_2 = 30 \times 10^3\ N/m^2 + \frac{10^3\ kg/m^3}{2}(16\ m^2/s^2 - 4\ m^2/s^2)$$
$$= 36\ kPa \qquad (Ans)$$

Comment The stream function used in this example could also be expressed in Cartesian coordinates as

$$\psi = 2r^2 \sin 2\theta = 4r^2 \sin \theta \cos \theta$$

or

$$\psi = 4xy$$

since $x = r \cos \theta$ and $y = r \sin \theta$. However, in the cylindrical polar form the results can be generalized to describe flow in the vicinity of a corner of angle α (see **Fig. E6.4c**) with the equations

$$\psi = Ar^{\pi/\alpha} \sin \frac{\pi\theta}{\alpha}$$

and

$$\phi = Ar^{\pi/\alpha} \cos \frac{\pi\theta}{\alpha}$$

where A is a constant.

6.5 SOME BASIC, PLANE POTENTIAL FLOWS

A major advantage of Laplace's equation is that it is a linear partial differential equation. Since it is linear, various solutions can be added to obtain other solutions—that is, if $\phi_1(x, y, z)$ and $\phi_2(x, y, z)$ are two solutions to Laplace's equation, then $\phi_3 = \phi_1 + \phi_2$ is also a solution. The practical implication of this result is that if we have certain basic solutions we can combine them to obtain more complicated and interesting solutions. In this section several basic velocity potentials, which describe some relatively simple flows, will be determined. In the next section these basic potentials will be combined to represent more complicated flows.

> For potential flow, basic solutions can be simply added to obtain more complicated solutions.

For simplicity, only plane (two-dimensional) flows will be considered. In this case, by using Cartesian coordinates

$$u = \frac{\partial\phi}{\partial x} \qquad v = \frac{\partial\phi}{\partial y} \qquad (6.72)$$

or by using cylindrical coordinates

$$v_r = \frac{\partial\phi}{\partial r} \qquad v_\theta = \frac{1}{r}\frac{\partial\phi}{\partial\theta} \qquad (6.73)$$

as shown in the adjacent figure. Because we can define a stream function for plane flow, we can also let

$$u = \frac{\partial\psi}{\partial y} \qquad v = -\frac{\partial\psi}{\partial x} \qquad (6.74)$$

or

$$v_r = \frac{1}{r}\frac{\partial\psi}{\partial\theta} \qquad v_\theta = -\frac{\partial\psi}{\partial r} \qquad (6.75)$$

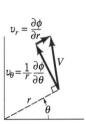

where the stream function was previously defined in Eqs. 6.37 and 6.42. We know that by defining the velocities in terms of the stream function conservation of mass is identically satisfied. If we now impose the condition of irrotationality, it follows from Eq. 6.59 that

$$\frac{\partial u}{\partial y} = \frac{\partial v}{\partial x}$$

and in terms of the stream function

$$\frac{\partial}{\partial y}\left(\frac{\partial \psi}{\partial y}\right) = \frac{\partial}{\partial x}\left(-\frac{\partial \psi}{\partial x}\right)$$

or

$$\frac{\partial^2 \psi}{\partial x^2} + \frac{\partial^2 \psi}{\partial y^2} = 0$$

Thus, for a plane irrotational flow we can use either the velocity potential or the stream function—both must satisfy Laplace's equation in two dimensions. It is apparent from these results that the velocity potential and the stream function are somehow related. We have previously shown that lines of constant ψ are streamlines; that is,

$$\left.\frac{dy}{dx}\right|_{\text{along } \psi=\text{constant}} = \frac{v}{u} \tag{6.76}$$

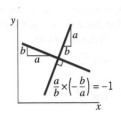

The change in ϕ as we move from one point (x, y) to a nearby point $(x + dx, y + dy)$ is given by the relationship

$$d\phi = \frac{\partial \phi}{\partial x}\, dx + \frac{\partial \phi}{\partial y}\, dy = u\, dx + v\, dy$$

Along a line of constant ϕ we have $d\phi = 0$ so that

$$\left.\frac{dy}{dx}\right|_{\text{along } \phi=\text{constant}} = -\frac{u}{v} \tag{6.77}$$

A comparison of Eqs. 6.76 and 6.77 shows that lines of constant ϕ (called **equipotential lines**) are orthogonal to lines of constant ψ (streamlines) at all points where they intersect. (Recall that two lines are orthogonal if the product of their slopes is -1, as illustrated in the adjacent figure.) For any potential flow field, a **flow net** can be drawn that consists of a family of streamlines and equipotential lines. The flow net is useful in visualizing flow patterns and can be used to obtain graphical solutions by sketching streamlines and equipotential lines and adjusting the lines until the lines are approximately orthogonal at all points where they intersect. An example of a flow net is shown in **Fig. 6.15**. Velocities can be estimated from the flow net because the velocity is inversely proportional to the streamline spacing, as illustrated in the adjacent figure. Thus, for example, from Fig. 6.15 we can see that the velocity near the inside corner will be higher than the velocity along the outer part of the bend.

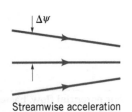

Streamwise acceleration

Streamwise deceleration

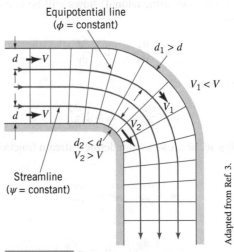

FIGURE 6.15 Flow net for a 90° bend.

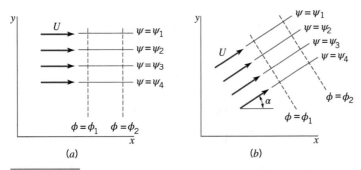

FIGURE 6.16 Uniform flow: (*a*) in the *x* direction; (*b*) in an arbitrary direction, α.

6.5.1 Uniform Flow

The simplest plane flow is one for which the streamlines are all straight and parallel, and the magnitude of the velocity is constant. This type of flow is called a **uniform flow**. For example, consider a uniform flow in the positive *x* direction as is illustrated in **Fig. 6.16a**. In this instance, $u = U$ and $v = 0$, and in terms of the velocity potential

$$\frac{\partial \phi}{\partial x} = U \qquad \frac{\partial \phi}{\partial y} = 0$$

These two equations can be integrated to yield

$$\phi = Ux + C$$

where C is an arbitrary constant, which can be set equal to zero. Thus, for a uniform flow in the positive *x* direction

$$\phi = Ux \tag{6.78}$$

The corresponding stream function can be obtained in a similar manner, since

$$\frac{\partial \psi}{\partial y} = U \qquad \frac{\partial \psi}{\partial x} = 0$$

and, therefore,

$$\psi = Uy \tag{6.79}$$

These results can be generalized to provide the velocity potential and stream function for a uniform flow at an angle α with the *x* axis, as in **Fig. 6.16b**. For this case

$$\phi = U(x \cos \alpha + y \sin \alpha) \tag{6.80}$$

and

$$\psi = U(y \cos \alpha - x \sin \alpha) \tag{6.81}$$

6.5.2 Source and Sink

Consider a fluid flowing radially outward from a line through the origin perpendicular to the *x*–*y* plane as is shown in **Fig. 6.17**. Let *m* be the volume rate of flow emanating from the line (per unit length), and therefore to satisfy conservation of mass

$$(2\pi r)v_r = m$$

or

$$v_r = \frac{m}{2\pi r}$$

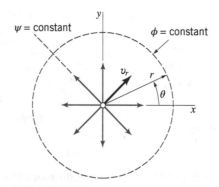

FIGURE 6.17 The streamline pattern for a source.

A source or sink represents a purely radial flow.

Also, because the flow is a purely radial flow, $v_\theta = 0$, the corresponding velocity potential can be obtained by integrating the equations

$$\frac{\partial \phi}{\partial r} = \frac{m}{2\pi r} \qquad \frac{1}{r}\frac{\partial \phi}{\partial \theta} = 0$$

It follows that

$$\phi = \frac{m}{2\pi} \ln r \tag{6.82}$$

If m is positive, the flow is radially outward, and the flow is considered to be a **source** flow. If m is negative, the flow is toward the origin, and the flow is considered to be a **sink** flow. The flowrate, m, is the *strength* of the source or sink.

As shown in the adjacent figure, at the origin where $r = 0$ the velocity becomes infinite, which is of course physically impossible. Thus, sources and sinks do not exist in real flow fields, and the line representing the source or sink is a mathematical *singularity* in the flow field. However, some real flows can be approximated at points away from the origin by using sources or sinks. Also, the velocity potential representing this hypothetical flow can be combined with other basic velocity potentials to approximately describe some real flow fields. This idea is further discussed in Section 6.6.

The stream function for the source can be obtained by integrating the relationships

$$v_r = \frac{1}{r}\frac{\partial \psi}{\partial \theta} = \frac{m}{2\pi r} \qquad v_\theta = -\frac{\partial \psi}{\partial r} = 0$$

to yield

$$\psi = \frac{m}{2\pi} \theta \tag{6.83}$$

It is apparent from Eq. 6.83 that the streamlines (lines of ψ = constant) are radial lines, and from Eq. 6.82 that the equipotential lines (lines of ϕ = constant) are concentric circles centered at the origin.

EXAMPLE 6.5 | Potential Flow—Sink

Given A nonviscous, incompressible fluid flows between wedge-shaped walls into a small opening as shown in **Fig. E6.5**. The velocity potential (in m²/s) that approximately describes this flow is

$$\phi = -2 \ln r$$

Find Determine the volume rate of flow (per unit length) into the opening.

Solution The components of velocity are

$$v_r = \frac{\partial \phi}{\partial r} = -\frac{2}{r} \qquad v_\theta = \frac{1}{r}\frac{\partial \phi}{\partial \theta} = 0$$

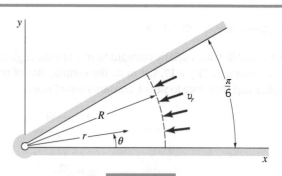

FIGURE E6.5

which indicates we have a purely radial flow. The flowrate per unit width, q, crossing the arc of length $R\pi/6$ can thus be obtained by integrating the expression

$$q = \int_0^{\pi/6} v_r R\, d\theta = -\int_0^{\pi/6} \left(\frac{2}{R}\right) R\, d\theta$$

$$= -\frac{\pi}{3} = -1.05 \text{ m}^2/\text{s} \qquad (Ans)$$

Comment Note that the radius R is arbitrary since the flowrate crossing any curve between the two walls must be the same. The negative sign indicates that the flow is toward the opening, that is, in the negative radial direction. The units are m²/s because q is the flowrate per unit width.

6.5.3 Vortex

We next consider a flow field in which the streamlines are concentric circles—that is, we interchange the velocity potential and stream function for the source. Thus, let

$$\phi = K\theta \qquad (6.84)$$

and

$$\psi = -K \ln r \qquad (6.85)$$

> A vortex represents a flow in which the streamlines are concentric circles.

where K is a constant. In this case the streamlines are concentric circles, as are illustrated in **Fig. 6.18**, with $v_r = 0$ and

$$v_\theta = \frac{1}{r}\frac{\partial \phi}{\partial \theta} = -\frac{\partial \psi}{\partial r} = \frac{K}{r} \qquad (6.86)$$

This result indicates that the tangential velocity varies inversely with the distance from the origin, as shown in the adjacent figure, with a singularity occurring at $r = 0$ (where the velocity becomes infinite).

It may seem strange that this **vortex** motion is irrotational (and it is since the flow field is described by a velocity potential). However, it must be recalled that rotation refers to the orientation of a fluid element and not the path followed by the element. Thus, for an irrotational vortex, if a pair of small sticks were placed in the flow field at location A, as indicated in **Fig. 6.19a**, the sticks would rotate as they move to location B. One of the sticks, the one that is aligned along the streamline, would follow a circular path and rotate in a counterclockwise direction. The other stick would rotate in a clockwise direction due to the nature of the flow field—that is, the part of the stick nearest the origin moves faster than the opposite end. Although both sticks are rotating, the average angular velocity of the two sticks is zero and the flow is irrotational.

If the fluid were rotating as a rigid body, such that $v_\theta = K_1 r$, where K_1 is a constant, then sticks similarly placed in the flow field would rotate as is illustrated in **Fig. 6.19b**. This type of vortex motion is *rotational* and cannot be described with a velocity potential. The rotational

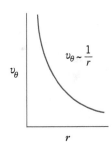

> Vortex motion can be either rotational or irrotational.

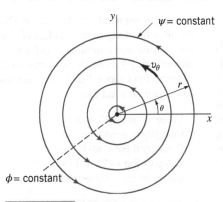

FIGURE 6.18 The streamline pattern for a vortex.

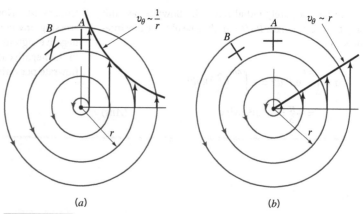

FIGURE 6.19 Motion of fluid element from A to B: (a) for irrotational (free) vortex; (b) for rotational (forced) vortex.

vortex is commonly called a *forced vortex*, whereas the irrotational vortex is usually called a *free vortex*. The swirling motion of the water as it drains from a bathtub is similar to that of a free vortex, whereas the motion of a liquid contained in a tank that is rotated about its axis with angular velocity ω corresponds to a forced vortex.

A *combined vortex* is one with a forced vortex as a central core and a velocity distribution corresponding to that of a free vortex outside the core. Thus, for a combined vortex

$$v_\theta = \omega r \qquad r \leq r_0 \tag{6.87}$$

and

$$v_\theta = \frac{K}{r} \qquad r > r_0 \tag{6.88}$$

where K and ω are constants and r_0 corresponds to the radius of the central core. The pressure distribution in both the free and forced vortex was previously considered in Example 4.3 (see Fig. E6.6a for an approximation of this type of flow).

A mathematical concept commonly associated with vortex motion is that of **circulation.** The circulation, Γ, is defined as the line integral of the tangential component of the velocity taken around a closed curve in the flow field. In equation form, Γ can be expressed as

$$\Gamma = \oint_C \mathbf{V} \cdot d\mathbf{s} \tag{6.89}$$

The Wide World of Fluids

Some hurricane facts

One of the most interesting, yet potentially devastating, naturally occurring fluid flow phenomena is a hurricane. Broadly speaking, a hurricane is a rotating mass of air circulating around a low-pressure central core. In some respects the motion is similar to that of a *free vortex*. The Caribbean and Gulf of Mexico experience the most hurricanes, with the official hurricane season being from June 1 to November 30. Hurricanes are usually 480 to 640 km wide and are structured around a central eye in which the air is relatively calm. The eye is surrounded by an eye wall, which is the region of strongest winds and precipitation. As one goes from the eye wall

to the eye, the wind speeds decrease sharply and within the eye the air is relatively calm and clear of clouds. However, in the eye the pressure is at a minimum and may be 10% less than standard atmospheric pressure. This low pressure creates strong downdrafts of dry air from above. Hurricanes are classified into five categories based on their wind speeds:

Category one—33–42 m/s
Category two—43–49 m/s
Category three—50–58 m/s
Category four—59–69 m/s
Category five—greater than 69 m/s

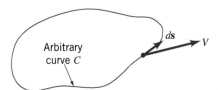

FIGURE 6.20 The notation for determining circulation around a closed curve *C*.

where the integral sign means that the integration is taken around a closed curve, *C*, in the counterclockwise direction, and **ds** is a differential length along the curve as is illustrated in **Fig. 6.20**. For an irrotational flow, **V** = **∇**φ so that **V** · **ds** = **∇**φ · **ds** = *d*φ and, therefore,

$$\Gamma = \oint_C d\phi = 0$$

This result indicates that for an irrotational flow the circulation will generally be zero. (Chapter 9 has further discussion of circulation in real flows.) However, if there are singularities enclosed within the curve the circulation may not be zero. For example, for the free vortex with $v_\theta = K/r$ the circulation around the circular path of radius *r* shown in **Fig. 6.21** is

$$\Gamma = \int_0^{2\pi} \frac{K}{r}(r\, d\theta) = 2\pi K$$

which shows that the circulation is nonzero and the constant $K = \Gamma/2\pi$. However, for irrotational flows the circulation around any path that does not include a singular point will be zero. This can be easily confirmed for the closed path *ABCD* of Fig. 6.21 by evaluating the circulation around that path.

The velocity potential and stream function for the free vortex are commonly expressed in terms of the circulation as

$$\phi = \frac{\Gamma}{2\pi}\theta \qquad (6.90)$$

and

$$\psi = -\frac{\Gamma}{2\pi}\ln r \qquad (6.91)$$

The concept of circulation is often useful when evaluating the forces developed on bodies immersed in moving fluids. This application will be considered in Section 6.6.3.

The numerical value of the circulation may depend on the particular closed path considered.

V6.4 Vortex in a beaker

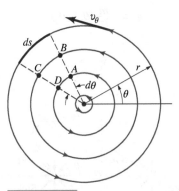

FIGURE 6.21 Circulation around various paths in a free vortex.

EXAMPLE 6.6 | Potential Flow—Free Vortex

Given A liquid drains from a large tank through a small opening as illustrated in **Fig. E6.6a**. A vortex forms whose velocity distribution away from the tank opening can be approximated as that of a free vortex having a velocity potential

$$\phi = \frac{\Gamma}{2\pi}\theta$$

Find Determine an expression relating the surface shape to the strength of the vortex as specified by the circulation Γ.

Solution Since the free vortex represents an irrotational flow field, the Bernoulli equation

$$\frac{p_1}{\gamma} + \frac{V_1^2}{2g} + z_1 = \frac{p_2}{\gamma} + \frac{V_2^2}{2g} + z_2$$

can be written between any two points. If the points are selected at the free surface, $p_1 = p_2 = 0$, so that

$$\frac{V_1^2}{2g} = z_s + \frac{V_2^2}{2g} \tag{1}$$

where the free surface elevation, z_s, is measured relative to a datum passing through point (1) as shown in **Fig. E6.6b**.

The velocity is given by the equation

$$v_\theta = \frac{1}{r}\frac{\partial \phi}{\partial \theta} = \frac{\Gamma}{2\pi r}$$

We note that far from the origin at point (1), $V_1 = v_\theta \approx 0$ so that Eq. 1 becomes

$$z_s = -\frac{\Gamma^2}{8\pi^2 r^2 g} \tag{Ans}$$

which is the desired equation for the surface profile.

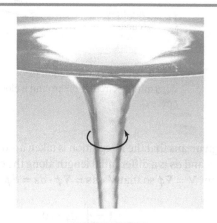

FIGURE E6.6a

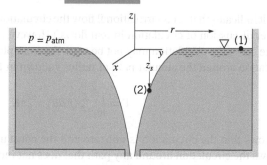

FIGURE E6.6b

Comment The negative sign indicates that the surface falls as the origin is approached as shown in Fig. E6.6. This solution is not valid very near the origin since the predicted velocity becomes excessively large as the origin is approached.

6.5.4 Doublet

A doublet is formed by an appropriate source–sink pair.

The final, basic potential flow to be considered is one that is formed by combining a source and sink in a special way. Consider the equal strength, source–sink pair of **Fig. 6.22**. The combined stream function for the pair is

$$\psi = -\frac{m}{2\pi}(\theta_1 - \theta_2)$$

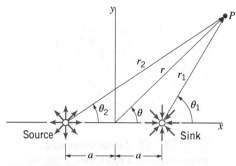

FIGURE 6.22 The combination of a source and sink of equal strength located along the x axis.

which can be rewritten as

$$\tan\left(-\frac{2\pi\psi}{m}\right) = \tan(\theta_1 - \theta_2) = \frac{\tan\theta_1 - \tan\theta_2}{1 + \tan\theta_1 \tan\theta_2} \tag{6.92}$$

From Fig. 6.22 it follows that

$$\tan\theta_1 = \frac{r\sin\theta}{r\cos\theta - a}$$

and

$$\tan\theta_2 = \frac{r\sin\theta}{r\cos\theta + a}$$

These results substituted into Eq. 6.92 give

$$\tan\left(-\frac{2\pi\psi}{m}\right) = \frac{2ar\sin\theta}{r^2 - a^2}$$

so that

$$\psi = -\frac{m}{2\pi}\tan^{-1}\left(\frac{2ar\sin\theta}{r^2 - a^2}\right) \tag{6.93}$$

The adjacent figure shows typical streamlines for this flow. For small values of the distance a

$$\psi = -\frac{m}{2\pi}\frac{2ar\sin\theta}{r^2 - a^2} = -\frac{mar\sin\theta}{\pi(r^2 - a^2)} \tag{6.94}$$

since the tangent of an angle approaches the value of the angle for small angles.

The so-called **doublet** is formed by letting the source and sink approach one another ($a \to 0$) while increasing the strength m ($m \to \infty$) so that the product ma/π remains constant. In this case, because $r/(r^2 - a^2) \to 1/r$, Eq. 6.94 reduces to

$$\psi = -\frac{K\sin\theta}{r} \tag{6.95}$$

where K, a constant equal to ma/π, is called the *strength* of the doublet. The corresponding velocity potential for the doublet is

$$\phi = \frac{K\cos\theta}{r} \tag{6.96}$$

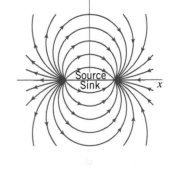

A doublet is formed by letting a source and sink approach one another.

Plots of lines of constant ψ reveal that the streamlines for a doublet are circles through the origin tangent to the x axis as shown in **Fig. 6.23**. Just as sources and sinks are not physically realistic entities, neither are doublets. However, the doublet when combined with other basic potential flows provides a useful representation of some flow fields of practical interest. For example, we will show in Section 6.6.3 that the combination of a uniform flow and a doublet can be used to represent the flow around a circular cylinder. **Table 6.1** provides a summary of the pertinent equations for the basic, plane potential flows considered in the preceding sections.

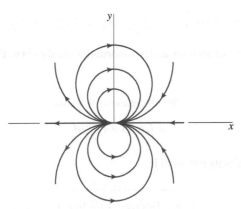

FIGURE 6.23 Streamlines for a doublet.

TABLE 6.1 Summary of Basic, Plane Potential Flows

Description of Flow Field	Velocity Potential	Stream Function	Velocity Components[a]
Uniform flow at angle α with the x axis (see Fig. 6.16b)	$\phi = U(x\cos\alpha + y\sin\alpha)$	$\psi = U(y\cos\alpha - x\sin\alpha)$	$u = U\cos\alpha$ $v = U\sin\alpha$
Source or sink (see Fig. 6.17) $m > 0$ source $m < 0$ sink	$\phi = \dfrac{m}{2\pi}\ln r$	$\psi = \dfrac{m}{2\pi}\theta$	$v_r = \dfrac{m}{2\pi r}$ $v_\theta = 0$
Free vortex (see Fig. 6.18) $\Gamma > 0$ counterclockwise motion $\Gamma < 0$ clockwise motion	$\phi = \dfrac{\Gamma}{2\pi}\theta$	$\psi = -\dfrac{\Gamma}{2\pi}\ln r$	$v_r = 0$ $v_\theta = \dfrac{\Gamma}{2\pi r}$
Doublet (see Fig. 6.23)	$\phi = \dfrac{K\cos\theta}{r}$	$\psi = -\dfrac{K\sin\theta}{r}$	$v_r = -\dfrac{K\cos\theta}{r^2}$ $v_\theta = -\dfrac{K\sin\theta}{r^2}$

[a]Velocity components are related to the velocity potential and stream function through the relationships:

$$u = \frac{\partial\phi}{\partial x} = \frac{\partial\psi}{\partial y} \qquad v = \frac{\partial\phi}{\partial y} = -\frac{\partial\psi}{\partial x} \qquad v_r = \frac{\partial\phi}{\partial r} = \frac{1}{r}\frac{\partial\psi}{\partial\theta} \qquad v_\theta = \frac{1}{r}\frac{\partial\phi}{\partial\theta} = -\frac{\partial\psi}{\partial r}.$$

6.6 SUPERPOSITION OF BASIC, PLANE POTENTIAL FLOWS

As was discussed in the previous section, potential flows satisfy Laplace's equation, which is a linear partial differential equation. It therefore follows that the various basic velocity potentials and stream functions can be combined to form new potentials and stream functions. (Why is this true?) Whether such combinations yield useful results remains to be seen. It is noteworthy that for inviscid flow the velocity vector is tangent to both a solid boundary and a streamline. Therefore, no flow can cross either, and any streamline can be thought of as a solid boundary in an inviscid flow. Thus, if we can combine some of the basic velocity potentials or stream functions to yield a streamline that corresponds to a particular body shape of interest, that combination can be used to describe in detail the inviscid flow around that body. This method of solving some interesting flow problems, commonly called the **method of superposition**, is illustrated in the following three sections.

6.6.1 Source in a Uniform Stream—Half-Body

Consider the superposition of a source and a uniform flow as shown in **Fig. 6.24a**. The resulting stream function is

Flow around a half-body is obtained by the addition of a source to a uniform flow.

$$\psi = \psi_{\text{uniform flow}} + \psi_{\text{source}}$$

$$= Ur\sin\theta + \frac{m}{2\pi}\theta \tag{6.97}$$

and the corresponding velocity potential is

$$\phi = Ur\cos\theta + \frac{m}{2\pi}\ln r \tag{6.98}$$

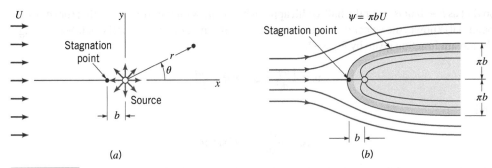

FIGURE 6.24 The flow around a half-body: (*a*) superposition of a source and a uniform flow; (*b*) replacement of streamline $\psi = \pi bU$ with solid boundary to form a half-body.

It is clear that at some point along the negative x axis the velocity due to the source will be of the same magnitude as the velocity due to the uniform stream, but each will be in the direction opposite to the other. Superposition of the uniform flow and the source produces a stagnation point at that location. For the source alone

Video

V6.5 Half-body

$$v_r = \frac{m}{2\pi r}$$

so that the stagnation point will occur at $x = -b$ where

$$U = \frac{m}{2\pi b}$$

or

$$b = \frac{m}{2\pi U} \tag{6.99}$$

The value of the stream function at the stagnation point can be obtained by evaluating ψ at $r = b$ and $\theta = \pi$, which yields from Eq. 6.97

$$\psi_{stagnation} = \frac{m}{2}$$

Because $m/2 = \pi bU$ (from Eq. 6.99), it follows that the equation of the streamline passing through the stagnation point is

$$\pi bU = Ur \sin \theta + bU\theta$$

or

$$r = \frac{b(\pi - \theta)}{\sin \theta} \tag{6.100}$$

> For inviscid flow, a streamline can be replaced by a solid boundary.

where θ can vary between 0 and 2π. A plot of this streamline is shown in **Fig. 6.24b**. If we replace this streamline with a solid boundary, as indicated in the figure, then it is clear that this combination of a uniform flow and a source can be used to describe the flow around the nose of a streamlined body placed in a uniform stream. The body is open at the downstream end and thus is called a **half-body**. Other streamlines in the flow field can be obtained by setting $\psi =$ constant in Eq. 6.97 and plotting the resulting equation. A number of these streamlines are shown in Fig. 6.24b. For completeness, Fig. 6.24b includes a few streamlines emanating from the source but remaining within the body. These are not of interest in the present study. It should be noted that the singularity in the flow field (the source) occurs inside the body, and there are no singularities in the flow field of interest (outside the body).

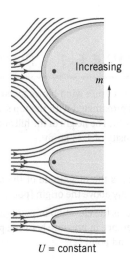

Increasing
m

$U =$ constant

The width of the half-body asymptotically approaches $2\pi b = m/U$. Thus, as shown in the adjacent figure, for a given free stream velocity, U, the width of the half-body increases with increasing source strength. This follows from Eq. 6.100, which can be written as

$$y = b(\pi - \theta)$$

so that as $\theta \to 0$ or $\theta \to 2\pi$ the half-width approaches $\pm b\pi$. With the stream function (or velocity potential) known, the velocity components at any point can be obtained. For the half-body, using the stream function given by Eq. 6.97,

$$v_r = \frac{1}{r}\frac{\partial \psi}{\partial \theta} = U \cos \theta + \frac{m}{2\pi r}$$

and

$$v_\theta = -\frac{\partial \psi}{\partial r} = -U \sin \theta$$

Thus, the square of the magnitude of the velocity, V, at any point is

$$V^2 = v_r^2 + v_\theta^2 = U^2 + \frac{Um \cos \theta}{\pi r} + \left(\frac{m}{2\pi r}\right)^2$$

Because $b = m/2\pi U$

$$V^2 = U^2\left(1 + 2\frac{b}{r}\cos\theta + \frac{b^2}{r^2}\right) \tag{6.101}$$

With the velocity known, the pressure at any point can be determined from the Bernoulli equation, which can be written between any two points in the flow field because the flow is irrotational. Thus, applying the Bernoulli equation between a point far from the body, where the pressure is p_0 and the velocity is U, and some arbitrary point with pressure p and velocity V, it follows that

$$p_0 + \frac{1}{2}\rho U^2 = p + \frac{1}{2}\rho V^2 \tag{6.102}$$

where elevation changes have been neglected. Equation 6.101 can now be substituted into Eq. 6.102 to obtain the pressure at any point in terms of the reference pressure, p_0, and the upstream velocity, U.

This relatively simple potential flow provides some useful information about the flow around the front part of a streamlined body, such as a bridge pier or strut placed in a uniform stream. An important point to be noted is that the velocity tangent to the surface of the body is not zero; that is, the fluid "slips" by the boundary. This result is a consequence of neglecting viscosity, the fluid property that causes real fluids to stick to the boundary, thus creating the "no-slip" condition. All potential flows differ from the flow of real fluids in this respect and do not accurately represent the velocity very near the boundary. However, outside this very thin boundary layer, the flow field predicted by potential flow theory will provide a good approximation of the real flow if flow separation does not occur (see Section 9.2.6). Also, the pressure distribution along the surface will closely approximate that predicted from the potential flow theory because the boundary layer is thin and there is little opportunity for the pressure to vary through the thin layer. In fact, as discussed in more detail in Chapter 9, the pressure distribution obtained from potential flow theory is used in conjunction with viscous flow theory to determine the nature of flow within the boundary layer.

For a potential flow the fluid is allowed to slip past a fixed solid boundary.

EXAMPLE 6.7 | Potential Flow—Half-body

Given A 18 m/s wind blows toward a hill arising from a plain that can be approximated with the top section of a half-body as illustrated in **Fig. E6.7a**. The height of the hill approaches 61 m as shown. Assume an air density of 1.23 kg/m³.

Find

a. What is the magnitude of the air velocity at a point on the hill directly above the origin [point (2)]?

b. What is the elevation of point (2) above the plain and what is the difference in pressure between point (1) on the plain far from the hill and point (2)?

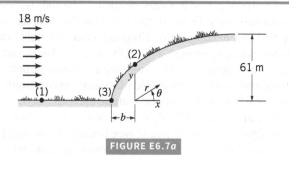

FIGURE E6.7a

Solution

a. The velocity is given by Eq. 6.101 as

$$V^2 = U^2\left(1 + 2\frac{b}{r}\cos\theta + \frac{b^2}{r^2}\right)$$

At point (2), $\theta = \pi/2$, and because this point is on the surface (Eq. 6.100),

$$r = \frac{b(\pi - \theta)}{\sin\theta} = \frac{\pi b}{2} \tag{1}$$

Thus,

$$V_2^2 = U^2\left[1 + \frac{b^2}{(\pi b/2)^2}\right] = U^2\left(1 + \frac{4}{\pi^2}\right)$$

and the magnitude of the velocity at (2) for a 18 m/s approaching wind is

$$V_2 = \left(1 + \frac{4}{\pi^2}\right)^{1/2}(18 \text{ m/s}) = 21.3 \text{ m/s} \qquad (Ans)$$

b. The elevation at (2) above the plain is given by Eq. 1 as

$$y_2 = \frac{\pi b}{2}$$

The height of the hill approaches 61 m and is equal to πb. Therefore

$$y_2 = \frac{61 \text{ m}}{2} = 30.5 \text{ m} \qquad (Ans)$$

From the Bernoulli equation (with the y axis the vertical axis)

$$\frac{p_1}{\gamma} + \frac{V_1^2}{2g} + y_1 = \frac{p_2}{\gamma} + \frac{V_2^2}{2g} + y_2$$

so that

$$p_1 - p_2 = \frac{\rho}{2}\left(V_2^2 - V_1^2\right) + \gamma(y_2 - y_1)$$

and with

$$V_1 = 18 \text{ m/s}$$

and

$$V_2 = 21.3 \text{ m/s}$$

it follows that

$$p_1 - p_2 = \frac{1.23 \text{ kg/m}^3}{2}[(21.3 \text{ m/s})^2 - (18 \text{ m/s})^2]$$
$$+ (1.23 \text{ kg/m}^3)(9.81 \text{ m/s}^2)(30.5 \text{ m} - 0 \text{ m})$$
$$= 448 \text{ N/m}^2 \qquad (Ans)$$

Comments This result indicates that the pressure on the hill at point (2) is slightly lower than the pressure on the plain at some distance from the base of the hill with a 368 N/m² difference due to the elevation increase and a 80 N/m² difference due to the velocity increase.

By repeating the calculations for various values of the upstream wind speed, U, the results shown in **Fig. E6.7b** are obtained. Note that as the wind speed increases, the pressure difference increases from the calm conditions of $p_1 - p_2 = 368$ N/m².

The maximum velocity along the hill surface does not occur at point (2) but farther up the hill at $\theta = 63°$. At this point $V_{\text{surface}} = 1.26U$. The minimum velocity ($V = 0$) and maximum pressure occur at point (3), the stagnation point.

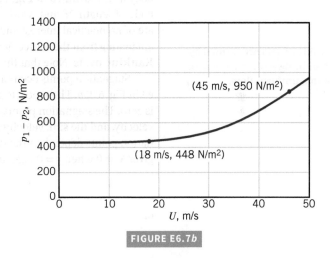

FIGURE E6.7b

6.6.2 Rankine Ovals

The half-body described in the previous section is a body that is "open" at one end. To simulate the flow around a closed body, a source and a sink of equal strength can be combined with a uniform flow as shown in **Fig. 6.25a**. The stream function for this combination is

$$\psi = Ur\sin\theta - \frac{m}{2\pi}(\theta_1 - \theta_2) \tag{6.103}$$

and the velocity potential is

$$\phi = Ur\cos\theta - \frac{m}{2\pi}(\ln r_1 - \ln r_2) \tag{6.104}$$

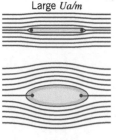

FIGURE 6.25 The flow around a Rankine oval: (*a*) superposition of source–sink pair and a uniform flow; (*b*) replacement of streamline $\psi = 0$ with solid boundary to form Rankine oval.

As discussed in Section 6.5.4, the stream function for the source–sink pair can be expressed as in Eq. 6.93, and, therefore, Eq. 6.103 can also be written as

$$\psi = Ur \sin\theta - \frac{m}{2\pi} \tan^{-1}\left(\frac{2ar \sin\theta}{r^2 - a^2}\right)$$

or

$$\psi = Uy - \frac{m}{2\pi} \tan^{-1}\left(\frac{2ay}{x^2 + y^2 - a^2}\right) \qquad (6.105)$$

The corresponding streamlines for this flow field are obtained by setting $\psi = $ constant. If several of these streamlines are plotted, it will be discovered that the streamline $\psi = 0$ forms a closed body as is illustrated in **Fig. 6.25***b*. We can think of this streamline as forming the surface of a body of length 2ℓ and width $2h$ placed in a uniform stream. The streamlines inside the body are of no practical interest and are not shown. Note that since the body is closed, all of the flow emanating from the source flows into the sink. These bodies have an oval shape and are termed **Rankine ovals**. Note that the Rankine oval is not an ellipse; it is more blunt.

Stagnation points occur at the upstream and downstream ends of the body, as are indicated in Fig. 6.25*b*. These points can be located by determining where along the *x* axis the velocity is zero. The stagnation points correspond to the points where the uniform velocity, the source velocity, and the sink velocity all combine to produce a zero velocity. The locations of the stagnation points depend on the value of *a*, *m*, and *U*. The body half-length, ℓ (the value of $|x|$ that gives $\mathbf{V} = 0$ when $y = 0$), can be expressed as

> Rankine ovals are formed by combining a source and sink with a uniform flow.

$$\ell = \left(\frac{ma}{\pi U} + a^2\right)^{1/2} \qquad (6.106)$$

or

$$\frac{\ell}{a} = \left(\frac{m}{\pi U a} + 1\right)^{1/2} \qquad (6.107)$$

> Large *Ua/m*

The body half-width, *h*, can be obtained by determining the value of *y* where the *y* axis intersects the $\psi = 0$ streamline. Thus, from Eq. 6.105 with $\psi = 0$, $x = 0$, and $y = h$, it follows that

$$h = \frac{h^2 - a^2}{2a} \tan \frac{2\pi U h}{m} \qquad (6.108)$$

or

$$\frac{h}{a} = \frac{1}{2}\left[\left(\frac{h}{a}\right)^2 - 1\right] \tan\left[2\left(\frac{\pi U a}{m}\right)\frac{h}{a}\right] \qquad (6.109)$$

> Small *Ua/m*

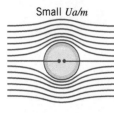

Equations 6.107 and 6.109 show that both ℓ/a and h/a are functions of the dimensionless parameter, $\pi U a/m$. Although for a given value of Ua/m the corresponding value of ℓ/a can be determined directly from Eq. 6.107, h/a must be determined by a trial-and-error solution of Eq. 6.109.

A large variety of body shapes with different length to width ratios can be obtained by using different values of Ua/m, as shown in the adjacent figure. As this parameter becomes large, flow around a long slender body is described, whereas for small values of the parameter, flow around a more blunt shape is obtained. Downstream from the point of maximum body width, the surface pressure increases with distance along the surface. In a real flow, this condition (called an adverse pressure gradient) typically leads to separation of the flow from the surface, resulting in a large low-pressure wake on the downstream side of the body. Separation is not predicted by potential theory (which simply indicates a symmetrical flow). This is illustrated in the adjacent figure for an extreme blunt shape, showing the differences between potential and viscous flow. Therefore, outside the very thin viscous boundary layer, the velocity (and pressure) field of a Rankine oval provides a good approximation of the real flow over the front part of the body only.

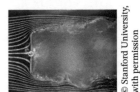

Potential Flow

© S. T. Thoroddsen & Stanford University

6.6.3 Flow Around a Circular Cylinder

As was noted in the previous section, when the distance between the source–sink pair approaches zero, the shape of the Rankine oval becomes more blunt and in fact approaches a circular shape. At the same time, the size of the body gets smaller and smaller. Since the doublet described in Section 6.5.4 was developed by letting a source–sink pair approach one another, it might be expected that a uniform flow in the positive x direction combined with a doublet could be used to represent flow around a circular cylinder. This combination gives for the stream function

$$\psi = Ur\sin\theta - \frac{K\sin\theta}{r} \tag{6.110}$$

and for the velocity potential

$$\phi = Ur\cos\theta + \frac{K\cos\theta}{r} \tag{6.111}$$

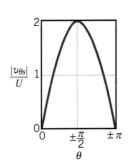

Viscous Flow

© Stanford University, with permission

In order for the stream function to represent flow around a circular cylinder, it is necessary that $\psi = $ constant for $r = a$, where a is the radius of the cylinder. Equation 6.110 can be written as

$$\psi = \left(U - \frac{K}{r^2}\right) r\sin\theta$$

Therefore, $\psi = 0$ for $r = a$ if

$$U - \frac{K}{a^2} = 0$$

which indicates that the doublet strength, K, must be equal to Ua^2 for the flow field to represent uniform flow over a cylinder. Thus, the stream function for flow around a circular cylinder can be expressed as

$$\boxed{\psi = Ur\left(1 - \frac{a^2}{r^2}\right)\sin\theta} \tag{6.112}$$

and the corresponding velocity potential is

$$\boxed{\phi = Ur\left(1 + \frac{a^2}{r^2}\right)\cos\theta} \tag{6.113}$$

A sketch of the streamlines for this flow field is shown in **Fig. 6.26**.

The velocity components can be obtained from either Eq. 6.112 or 6.113 as

$$v_r = \frac{\partial\phi}{\partial r} = \frac{1}{r}\frac{\partial\psi}{\partial\theta} = U\left(1 - \frac{a^2}{r^2}\right)\cos\theta \tag{6.114}$$

and

$$v_\theta = \frac{1}{r}\frac{\partial\phi}{\partial\theta} = -\frac{\partial\psi}{\partial r} = -U\left(1 + \frac{a^2}{r^2}\right)\sin\theta \tag{6.115}$$

It is clear from either Eq. 6.114 or 6.115 that on the surface of the cylinder where $r = a$:

$$v_r = 0, \, v_{\theta s} = -2U\sin\theta$$

As shown in the adjacent figure, the maximum velocity occurs at the top and bottom of the cylinder $(\theta = \pm\pi/2)$ and has a magnitude of twice the upstream velocity, U. As we move away from the cylinder along the ray $\theta = \pi/2$ the velocity varies, as is illustrated in Fig. 6.26.

A doublet combined with a uniform flow can be used to represent flow around a circular cylinder.

 Video

V6.6 Circular cylinder

 Video

V6.7 Ellipse

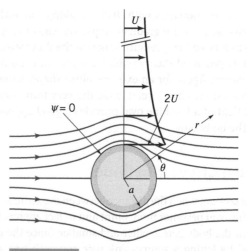

FIGURE 6.26 The flow around a circular cylinder.

The pressure distribution on the cylinder surface is obtained from the Bernoulli equation.

The pressure distribution on the cylinder surface is obtained from the Bernoulli equation written from a point far from the cylinder where the pressure is p_0 and the velocity is U so that

$$p_0 + \frac{1}{2}\rho U^2 = p_s + \frac{1}{2}\rho v_{\theta s}^2$$

where p_s is the surface pressure. If the flow is in a horizontal plane or the influence of elevation changes are neglected, noting that $v_{\theta s} = -2U \sin \theta$, the surface pressure can be expressed as

$$p_s = p_0 + \frac{1}{2}\rho U^2(1 - 4\sin^2 \theta) \tag{6.116}$$

A comparison of this theoretical, symmetrical pressure distribution expressed in dimensionless form with a typical measured distribution is shown in **Fig. 6.27**. This figure clearly reveals that only on the upstream part of the cylinder is there approximate agreement between the potential flow and the experimental results. Because of the viscous boundary layer that develops on the cylinder, the main flow separates from the surface of the cylinder, leading to the large difference between the theoretical, frictionless fluid solution and the experimental results on the downstream side of the cylinder (see Chapter 9).

V6.8 Forces on sus-
pended ball

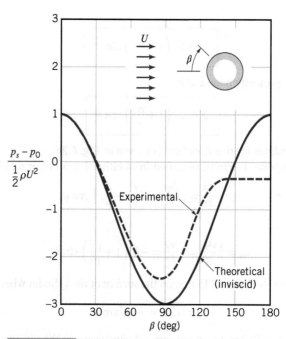

FIGURE 6.27 A comparison of theoretical (inviscid) pressure distribution on the surface of a circular cylinder with typical experimental distribution.

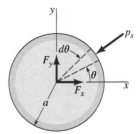

FIGURE 6.28 The notation for determining lift and drag on a circular cylinder.

The resultant force (per unit length) developed on the cylinder can be determined by integrating the pressure over the surface. From **Fig. 6.28** it can be seen that

$$F_x = -\int_0^{2\pi} p_s \cos \theta \, a \, d\theta \tag{6.117}$$

and

$$F_y = -\int_0^{2\pi} p_s \sin \theta \, a \, d\theta \tag{6.118}$$

where F_x is the *drag* (force parallel to direction of the uniform flow) and F_y is the *lift* (force perpendicular to the direction of the uniform flow). Substitution for p_s from Eq. 6.116 into these two equations, and subsequent integration, reveals that $F_x = 0$ and $F_y = 0$. These results indicate that both the drag and lift as predicted by potential theory for a fixed cylinder in a uniform stream are zero. Since the pressure distribution is symmetrical around the cylinder, this is not really a surprising result. However, we know from experience that there is a significant drag developed on a cylinder when it is placed in a moving fluid. This discrepancy is known as d'Alembert's paradox. The paradox is named after Jean le Rond d'Alembert (1717–1783), a French mathematician and philosopher, who first showed that the drag on bodies immersed in inviscid fluids is zero. It was not until the latter part of the nineteenth century and the early part of the twentieth century that the role viscosity plays in the steady fluid motion was understood and d'Alembert's paradox explained (see Section 9.1).

V6.9 Potential and viscous flow

Potential theory incorrectly predicts that the drag on a cylinder is zero.

EXAMPLE 6.8 | Potential Flow—Cylinder

Given When a circular cylinder is placed in a uniform stream, a stagnation point is created on the cylinder as is shown in **Fig. E6.8a**. If a small hole is located at this point, the stagnation pressure, p_{stag}, can be measured and used to determine the approach velocity, U.

Find

a. Show how p_{stag} and U are related.

b. If the cylinder is misaligned by an angle α (**Fig. E6.8b**), but the measured pressure is still interpreted as the stagnation pressure, determine an expression for the ratio of the true velocity, U, to the predicted velocity, U'. Plot this ratio as a function of α for the range $-20° \leq \alpha \leq 20°$.

Solution

a. The velocity at the stagnation point is zero, so the Bernoulli equation written between a point on the stagnation

streamline upstream from the cylinder and the stagnation point gives

$$\frac{p_0}{\gamma} + \frac{U^2}{2g} = \frac{p_{stag}}{\gamma}$$

Thus,

$$U = \left[\frac{2}{\rho} (p_{stag} - p_0) \right]^{1/2} \tag{Ans}$$

Comment A measurement of the difference between the pressure at the stagnation point and the upstream pressure can be used to measure the approach velocity. This is, of course, the same result that was obtained in Section 4.5 for Pitot-static tubes.

b. If the direction of the fluid approaching the cylinder is not known precisely, it is possible that the cylinder is misaligned by some angle, α. In this instance the pressure actually measured, p_α, will be different from the stagnation pressure, but if the misalignment is

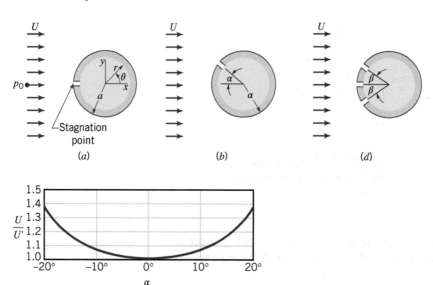

FIGURE E6.8

not recognized the predicted approach velocity, U', would still be calculated as

$$U' = \left[\frac{2}{\rho}(p_\alpha - p_0)\right]^{1/2}$$

Thus,

$$\frac{U(\text{true})}{U'(\text{predicted})} = \left(\frac{p_{\text{stag}} - p_0}{p_\alpha - p_0}\right)^{1/2} \tag{1}$$

The velocity on the surface of the cylinder, v_θ, where $r = a$, is obtained from Eq. 6.115 as

$$v_\theta = -2U \sin \theta$$

If we now write the Bernoulli equation between a point upstream of the cylinder and the point on the cylinder where $r = a, \theta = \alpha$, it follows that

$$p_0 + \frac{1}{2}\rho U^2 = p_\alpha + \frac{1}{2}\rho(-2U \sin \alpha)^2$$

and, therefore,

$$p_\alpha - p_0 = \frac{1}{2}\rho U^2(1 - 4 \sin^2 \alpha) \tag{2}$$

Since $p_{\text{stag}} - p_0 = \frac{1}{2}\rho U^2$ it follows from Eqs. 1 and 2 that

$$\frac{U(\text{true})}{U'(\text{predicted})} = (1 - 4 \sin^2 \alpha)^{-1/2} \tag{Ans}$$

This velocity ratio is plotted as a function of the misalignment angle α in **Fig. E6.8c**.

Comments   It is clear from these results that significant errors can arise if the stagnation pressure tap is not aligned with the stagnation streamline. As is discussed in Section 4.5, if two additional, symmetrically located holes are drilled into the cylinder, as are illustrated in **Fig. E6.8d**, the correct orientation of the cylinder can be determined. The cylinder is rotated until the pressures in the two symmetrically placed holes are equal, thus indicating that the center hole coincides with the stagnation streamline. For $\beta = 30°$ the pressure at the two holes theoretically corresponds to the upstream pressure, p_0. With this orientation a measurement of the difference in pressure between the center hole and the side holes can be used to determine U.

An additional, interesting potential flow can be developed by adding a free vortex to the stream function or velocity potential for the flow around a cylinder. In this case

$$\psi = Ur\left(1 - \frac{a^2}{r^2}\right) \sin \theta - \frac{\Gamma}{2\pi} \ln r \tag{6.119}$$

and

$$\phi = Ur\left(1 + \frac{a^2}{r^2}\right) \cos \theta + \frac{\Gamma}{2\pi}\theta \tag{6.120}$$

where Γ is the circulation. We note that the circle $r = a$ will still be a streamline (and thus can be replaced with a solid cylinder) because the streamlines for the added free vortex are all circular. However, the tangential velocity, v_θ, on the surface of the cylinder ($r = a$) now becomes

$$v_{\theta s} = -\left.\frac{\partial \psi}{\partial r}\right|_{r=a} = -2U \sin \theta + \frac{\Gamma}{2\pi a} \tag{6.121}$$

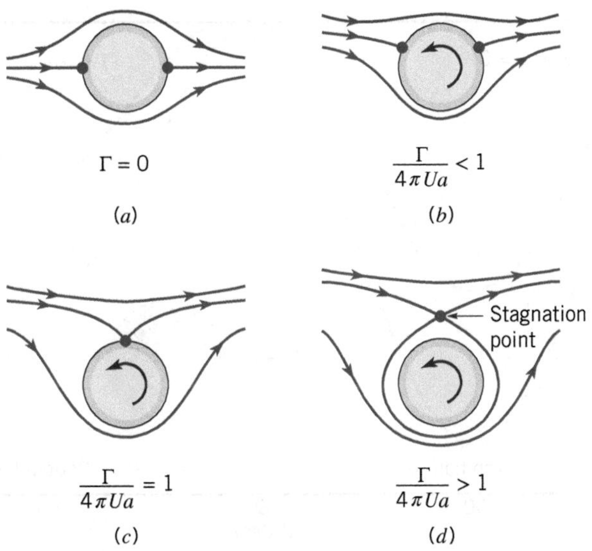

FIGURE 6.29 The location of stagnation points on a circular cylinder: (*a*) without circulation; (*b, c, d*) with circulation.

Therefore, it appears that the potential flow produced by superposition of a uniform flow, a doublet, and a free vortex might be a good approximation of the uniform flow over a rotating cylinder of radius a. Because of the presence of viscosity in any real fluid, the fluid in contact with the rotating cylinder would rotate with the same velocity as the cylinder, and the resulting flow field would resemble that of the potential flow solution.

Flow around a rotating cylinder is approximated by the addition of a free vortex.

A variety of streamline patterns can be developed, depending on the vortex strength, Γ. For example, from Eq. 6.121 we can determine the location of stagnation points on the surface of the cylinder. These points will occur at $\theta = \theta_{\text{stag}}$, where $v_\theta = 0$ and therefore from Eq. 6.121

$$\sin \theta_{\text{stag}} = \frac{\Gamma}{4\pi Ua} \tag{6.122}$$

If $\Gamma = 0$, then $\theta_{\text{stag}} = 0$ or π—that is, the stagnation points occur at the front and rear of the cylinder as are shown in **Fig. 6.29a**. However, for $-1 \leq \Gamma/4\pi Ua \leq 1$, the stagnation points will occur at some other location on the surface as illustrated in **Figs. 6.29b** and **c**. If the absolute value of the parameter $\Gamma/4\pi Ua$ exceeds 1, Eq. 6.122 cannot be satisfied, and the stagnation point is located away from the cylinder as shown in **Fig. 6.29d**.

The force per unit length developed on the cylinder can again be obtained by integrating the pressure force around the circumference as in Eqs. 6.117 and 6.118. For the cylinder with circulation, the surface pressure, p_s, is obtained from the Bernoulli equation (with the surface velocity given by Eq. 6.121)

$$p_0 + \frac{1}{2}\rho U^2 = p_s + \frac{1}{2}\rho\left(-2U \sin\theta + \frac{\Gamma}{2\pi a}\right)^2$$

or

$$p_s = p_0 + \frac{1}{2}\rho U^2\left(1 - 4\sin^2\theta + \frac{2\Gamma \sin\theta}{\pi aU} - \frac{\Gamma^2}{4\pi^2 a^2 U^2}\right) \tag{6.123}$$

Equation 6.123 substituted into Eq. 6.117 for the drag, and integrated, again yields

$$F_x = 0$$

That is, even for the rotating cylinder no force in the direction of the uniform flow is developed. This result could have been inferred from the fore-aft symmetry of the streamlines of the flow. However, the streamlines are not symmetric in the cross-flow direction. Substitution of Eq. 123 into Eq. 6.118 yields

Potential flow past a cylinder with circulation gives zero drag but nonzero lift.

$$F_y = -\rho U\Gamma \tag{6.124}$$

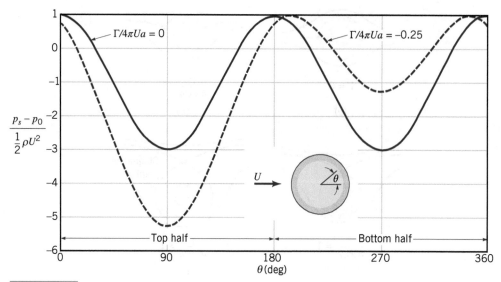

FIGURE 6.30 Pressure distribution on a circular cylinder with and without rotation.

Thus, for the cylinder with circulation, lift is developed equal to the product of the fluid density, the upstream velocity, and the circulation. The negative sign means that if U is positive (in the positive x direction) and Γ is positive (a free vortex with counterclockwise rotation), the direction of the F_y is downward.

Of course, if the cylinder is rotated in the clockwise direction ($\Gamma < 0$), the direction of F_y would be upward. This can be seen by studying the surface pressure distribution (Eq. 6.123), which is plotted in **Fig. 6.30** for two situations. One has $\Gamma/4pUa = 0$, which corresponds to no rotation of the cylinder. The other has $\Gamma/4pUa = -0.25$, which corresponds to clockwise rotation of the cylinder. With no rotation, the fore-aft and cross-flow symmetries of the pressure distributions are obvious. With rotation, the fore-aft symmetry of the pressure distribution is still clear but the cross-flow symmetry of the pressure distribution is no longer present. In this case the two stagnation points [i.e., $(p_s - p_0)/(rU^2/2) = 1$] are located on the bottom of the cylinder, and the average pressure on the top half of the cylinder is less than that on the bottom half. The result is an upward lift force. It is this force acting in a direction perpendicular to the direction of the approach velocity that causes baseballs and golf balls to curve when they spin as they are propelled through the air. The development of this lift on rotating bodies is called the *Magnus effect* (see Section 9.4 for further comments).

Although Eq. 6.124 was developed for a cylinder with circulation, it gives the lift per unit length for any two-dimensional object of any cross-sectional shape placed in a uniform, inviscid stream. The circulation is determined around any closed curve containing the body. The generalized equation relating lift to fluid density, velocity, and circulation is called the *Kutta–Joukowski law* and is commonly used to determine the lift on airfoils (see Section 9.4.2 and Refs. 2–6).

The Wide World of Fluids

A sailing ship without sails

A sphere or cylinder spinning about its axis when placed in an airstream develops a force at right angles to the direction of the airstream. This phenomenon is commonly referred to as the *Magnus effect* and is responsible for the curved paths of baseballs and golf balls. Another lesser-known application of the Magnus effect was proposed by a German physicist and engineer, Anton Flettner, in the 1920s. Flettner's idea was to use the Magnus effect to make a ship move. To demonstrate the practicality of the "rotor-ship" he

purchased a sailing schooner and replaced the ship's masts and rigging with two vertical cylinders that were 15 m high and 3 m in diameter. The cylinders looked like smokestacks on the ship. Their spinning motion was developed by 34 kW motors. The combination of a wind and the rotating cylinders created a force (Magnus effect) to push the ship forward. The ship, named the *Baden Baden*, made a successful voyage across the Atlantic, arriving in New York Harbor on May 9, 1926. Although the feasibility of the rotor-ship was clearly demonstrated, it proved to be less efficient and practical than more conventional vessels and the idea was not pursued.

6.7 OTHER ASPECTS OF POTENTIAL FLOW

In the preceding section the method of superposition of basic potential functions and stream functions has been used to obtain detailed descriptions of irrotational flow around certain body shapes immersed in a uniform stream. For the cases considered, two or more of the basic potentials were combined, and the question was asked: What kind of flow does this combination represent? This approach is relatively simple and does not require the use of advanced mathematical techniques. It is, however, restrictive in its general applicability. It does not allow us to specify a priori the body shape and then determine the velocity potential or stream function that describes the flow around that body. Determining the velocity potential or stream function for a given body shape is a much more complicated problem.

It is possible to extend the idea of superposition by considering a *distribution* of sources, sinks, vortexes, or doublets, which when combined with a uniform flow can describe the flow around bodies of arbitrary shape. Singularities may be distributed along an axis or over a surface. Techniques are available to determine the required distribution to produce a prescribed body shape. Also, for plane potential flow problems it can be shown that complex variable theory (the use of real and imaginary numbers) can be effectively used to obtain solutions to a great variety of important flow problems. Numerical techniques can be used to solve not only plane two-dimensional problems but also the more general three-dimensional problems. Because potential flow must satisfy Laplace's equation, any procedure available for solving this equation can be applied to the analysis of the irrotational flow of frictionless fluids. Potential flow theory is an old and well-established discipline within the general field of fluid mechanics. The interested reader can find many detailed references on this subject, including Refs. 2–7 given at the end of this chapter.

An important point to remember is that regardless of the particular technique used to obtain a solution to a potential flow problem, the solution remains approximate because of the fundamental assumption of a frictionless fluid. Thus, "exact" solutions based on potential flow theory represent, at best, only approximate solutions to real fluid problems. The applicability of potential flow theory to real fluid problems has been alluded to in a number of examples in the previous section. As a rule of thumb, potential flow theory will usually provide a reasonable approximation in those circumstances when we are dealing with a low viscosity fluid moving at a relatively high velocity, in regions of the flow field in which the flow is accelerating. Under these circumstances we generally find that the effect of viscosity is confined to the thin boundary layer that develops at a solid boundary. Outside the boundary layer, the velocity distribution and the pressure distribution are closely approximated by the potential flow solution. However, in those regions of the flow field in which the flow is decelerating (for example, in the rearward portion of a bluff body or in the expanding region of a conduit), the pressure near a solid boundary will increase in the direction of flow. The adverse pressure gradient associated with the decreasing flow velocity can lead to flow separation, a phenomenon that causes dramatic changes in the flow field which are generally not accounted for by potential theory. However, as discussed in Chapter 9, in which boundary layer theory is developed, it is found that potential flow theory is frequently used to obtain the appropriate pressure distribution near the body's surface in combination with the boundary layer approximation to the viscous flow equations to obtain very useful solutions near the boundary (and also to predict separation). The general differential equations that describe viscous fluid behavior and some simple solutions to these equations are considered in the remaining sections of this chapter.

Potential flow solutions are always approximate because the fluid is assumed to be frictionless.

V6.10 Potential flow

6.8 VISCOUS FLOW

To include viscous effects into the differential analysis of fluid motion we must return to the previously derived general equations of motion, Eqs. 6.50. These equations include both stresses and velocities. There are more unknowns than equations. Before proceeding, a relationship between the stresses and velocities must be specified.

6.8.1 Stress–Deformation Relationships

For incompressible Newtonian fluids the stresses are linearly related to the rates of deformation and can be expressed in Cartesian coordinates as (for normal stresses)

$$\sigma_{xx} = -p + 2\mu\frac{\partial u}{\partial x} \tag{6.125a}$$

$$\sigma_{yy} = -p + 2\mu\frac{\partial v}{\partial y} \tag{6.125b}$$

$$\sigma_{zz} = -p + 2\mu\frac{\partial w}{\partial z} \tag{6.125c}$$

(for shearing stresses)

$$\tau_{xy} = \tau_{yx} = \mu\left(\frac{\partial u}{\partial y} + \frac{\partial v}{\partial x}\right) \tag{6.125d}$$

$$\tau_{yz} = \tau_{zy} = \mu\left(\frac{\partial v}{\partial z} + \frac{\partial w}{\partial y}\right) \tag{6.125e}$$

$$\tau_{zx} = \tau_{xz} = \mu\left(\frac{\partial w}{\partial x} + \frac{\partial u}{\partial z}\right) \tag{6.125f}$$

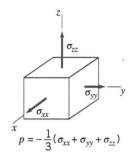

$$p = -\frac{1}{3}(\sigma_{xx} + \sigma_{yy} + \sigma_{zz})$$

where p is the pressure, the negative of the average of the three normal stresses; that is, as indicated in the adjacent figure, $-p = (\frac{1}{3})(\sigma_{xx} + \sigma_{yy} + \sigma_{zz})$. For viscous fluids in motion the normal stresses are not necessarily the same in different directions, thus, the need to define the pressure as the average of the three normal stresses. For fluids at rest, or frictionless fluids, the normal stresses are equal in all directions. Detailed discussions of the development of these stress–velocity gradient relationships can be found in Refs. 3, 8, and 9. An important point to note is that whereas for elastic solids the stresses are linearly related to the deformation (or strain), for Newtonian fluids the stresses are linearly related to the *rate* of deformation (or rate of strain).

In cylindrical polar coordinates the stresses for incompressible Newtonian fluids are expressed as (for normal stresses)

> **For Newtonian fluids, stresses are linearly related to the rate of strain.**

$$\sigma_{rr} = -p + 2\mu\frac{\partial v_r}{\partial r} \tag{6.126a}$$

$$\sigma_{\theta\theta} = -p + 2\mu\left(\frac{1}{r}\frac{\partial v_\theta}{\partial\theta} + \frac{v_r}{r}\right) \tag{6.126b}$$

$$\sigma_{zz} = -p + 2\mu\frac{\partial v_z}{\partial z} \tag{6.126c}$$

(for shearing stresses)

$$\tau_{r\theta} = \tau_{\theta r} = \mu\left[r\frac{\partial}{\partial r}\left(\frac{v_\theta}{r}\right) + \frac{1}{r}\frac{\partial v_r}{\partial\theta}\right] \tag{6.126d}$$

$$\tau_{\theta z} = \tau_{z\theta} = \mu\left(\frac{\partial v_\theta}{\partial z} + \frac{1}{r}\frac{\partial v_z}{\partial\theta}\right) \tag{6.126e}$$

$$\tau_{zr} = \tau_{rz} = \mu\left(\frac{\partial v_r}{\partial z} + \frac{\partial v_z}{\partial r}\right) \tag{6.126f}$$

The double subscript has a meaning similar to that of stresses expressed in Cartesian coordinates—that is, the first subscript indicates the plane on which the stress acts and the second subscript the direction. Thus, for example, σ_{rr} refers to a stress acting on a plane perpendicular to the radial direction and in the radial direction (thus a normal stress). Similarly, $\tau_{r\theta}$ refers to a stress acting on a plane perpendicular to the radial direction but in the tangential (θ direction) and is therefore a shearing stress.

6.8.2 The Navier–Stokes Equations

The stresses as defined in the preceding section can be substituted into the differential equations of motion (Eqs. 6.50) and simplified by using the continuity equation for incompressible flow (Eq. 6.31). For rectangular coordinates (see the adjacent figure) the results are

(*x* direction)

$$\rho\left(\frac{\partial u}{\partial t} + u\frac{\partial u}{\partial x} + v\frac{\partial u}{\partial y} + w\frac{\partial u}{\partial z}\right) = -\frac{\partial p}{\partial x} + \rho g_x + \mu\left(\frac{\partial^2 u}{\partial x^2} + \frac{\partial^2 u}{\partial y^2} + \frac{\partial^2 u}{\partial z^2}\right) \qquad (6.127a)$$

(*y* direction)

$$\rho\left(\frac{\partial v}{\partial t} + u\frac{\partial v}{\partial x} + v\frac{\partial v}{\partial y} + w\frac{\partial v}{\partial z}\right) = -\frac{\partial p}{\partial y} + \rho g_y + \mu\left(\frac{\partial^2 v}{\partial x^2} + \frac{\partial^2 v}{\partial y^2} + \frac{\partial^2 v}{\partial z^2}\right) \qquad (6.127b)$$

(*z* direction)

$$\rho\left(\frac{\partial w}{\partial t} + u\frac{\partial w}{\partial x} + v\frac{\partial w}{\partial y} + w\frac{\partial w}{\partial z}\right) = -\frac{\partial p}{\partial x} + \rho g_z + \mu\left(\frac{\partial^2 w}{\partial x^2} + \frac{\partial^2 w}{\partial y^2} + \frac{\partial^2 w}{\partial z^2}\right) \qquad (6.127c)$$

where u, v, and w are the x, y, and z components of velocity as shown in the adjacent figure. These equations appear to be formidable, and indeed, they are; however, it is important to recall that they are just Newton's second law, $F = ma$. We have rearranged the equations so that the acceleration terms are on the left side and the force terms are on the right. These equations are commonly called the **Navier–Stokes equations**, named in honor of the French mathematician L. M. H. Navier (1785–1836) and the English mechanician Sir G. G. Stokes (1819–1903), who were responsible for their formulation. These three equations of motion, when combined with the conservation of mass equation (Eq. 6.31), provide a complete mathematical description of the flow of incompressible Newtonian fluids. We have four equations (three momentum equations plus continuity) and four unknowns (u, v, w, and p). Therefore, when a complete set of boundary conditions and initial conditions are provided, the problem is "well-posed" in mathematical terms. Unfortunately, because of the complexity of the Navier–Stokes equations (they are nonlinear, second-order, partial differential equations), they are not amenable to an exact mathematical solution except in a few instances. However, in those few instances in which solutions have been obtained and compared with experimental results, the results have been in close agreement.

The Navier–Stokes equations are the basic differential equations describing the flow of Newtonian fluids.

In terms of cylindrical polar coordinates (see the adjacent figure), the Navier–Stokes equations can be written as

(*r* direction)

$$\rho\left(\partial v_r/\partial t + v_r\frac{\partial v_r}{\partial r} + \frac{v_\theta}{r}\frac{\partial v_r}{\partial \theta} - \frac{v_\theta^2}{r} + v_z\frac{\partial v_r}{\partial z}\right)$$
$$= -\frac{\partial p}{\partial r} + \rho g_r + \mu\left[\frac{1}{r}\frac{\partial}{\partial r}\left(r\frac{\partial v_r}{\partial r}\right) - \frac{v_r}{r^2} + \frac{1}{r^2}\frac{\partial^2 v_r}{\partial \theta^2} - \frac{2}{r^2}\frac{\partial v_\theta}{\partial \theta} + \frac{\partial^2 v_r}{\partial z^2}\right] \qquad (6.128a)$$

(*θ* direction)

$$\rho\left(\frac{\partial v_\theta}{\partial t} + v_r\frac{\partial v_\theta}{\partial r} + \frac{v_\theta}{r}\frac{\partial v_\theta}{\partial \theta} + \frac{v_r v_\theta}{r} + v_z\frac{\partial v_\theta}{\partial z}\right)$$
$$= -\frac{1}{r}\frac{\partial p}{\partial \theta} + \rho g_\theta + \mu\left[\frac{1}{r}\frac{\partial}{\partial r}\left(r\frac{\partial v_\theta}{\partial r}\right) - \frac{v_\theta}{r^2} + \frac{1}{r^2}\frac{\partial^2 v_\theta}{\partial \theta^2} + \frac{2}{r^2}\frac{\partial v_r}{\partial \theta} + \frac{\partial^2 v_\theta}{\partial z^2}\right] \qquad (6.128b)$$

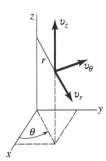

(*z* direction)

$$\rho\left(\frac{\partial v_z}{\partial t} + v_r\frac{\partial v_z}{\partial r} + \frac{v_\theta}{r}\frac{\partial v_z}{\partial \theta} + v_z\frac{\partial v_z}{\partial z}\right)$$
$$= -\frac{\partial p}{\partial z} + \rho g_z + \mu\left[\frac{1}{r}\frac{\partial}{\partial r}\left(r\frac{\partial v_z}{\partial r}\right) - \frac{1}{r^2}\frac{\partial^2 v_z}{\partial \theta^2} + \frac{\partial^2 v_z}{\partial z^2}\right] \qquad (6.128c)$$

To provide a brief introduction to the use of the Navier–Stokes equations, a few of the simplest exact solutions are developed in the next section. Although these solutions will prove to be relatively straightforward, this is not the case in general. In fact, since these equations were first formulated, relatively few exact solutions have been obtained.

6.9 SOME SIMPLE SOLUTIONS FOR LAMINAR, VISCOUS, INCOMPRESSIBLE FLOWS

A principal difficulty in solving the Navier–Stokes equations is their nonlinearity arising from the convective acceleration terms (i.e., $u\,\partial u/\partial x$, $w\,\partial v/\partial z$, etc.). There are no general analytical schemes for solving nonlinear partial differential equations (e.g., superposition of solutions cannot be used), and each problem must be considered individually. For most practical flow problems, fluid particles do have accelerated motion as they move from one location to another in the flow field. Thus, the convective acceleration terms are usually important. However, there are a few special cases for which the convective acceleration vanishes because of the geometry of the flow system. In these cases exact solutions are often possible. Although the Navier–Stokes equations apply to both laminar and turbulent flow, for turbulent flow each velocity component fluctuates in an apparently random fashion, with a very short time scale, and this added complication makes an analytical solution intractable. Thus, the exact solutions are for laminar flows in which the velocity is either independent of time (steady flow) or dependent on time (unsteady flow) in a well-defined manner.

6.9.1 Steady, Laminar Flow Between Fixed Parallel Plates

An exact solution can be obtained for steady laminar flow between fixed parallel plates.

We first consider flow between the two horizontal, infinite parallel plates of **Fig. 6.31a**. For this geometry the fluid particles move in the x direction parallel to the plates, and there is no velocity in the y or z direction—that is, $v = 0$ and $w = 0$. In this case it follows from the continuity equation (Eq. 6.31) that $\partial u/\partial x = 0$. Furthermore, there would be no variation of u in the z direction for infinite plates, and for steady flow $\partial u/\partial t = 0$ so that $u = u(y)$. If these conditions are applied to the Navier–Stokes equations (Eqs. 6.127), they reduce to

$$0 = -\frac{\partial p}{\partial x} + \mu\left(\frac{\partial^2 u}{\partial y^2}\right) \tag{6.129}$$

$$0 = -\frac{\partial p}{\partial y} - \rho g \tag{6.130}$$

$$0 = -\frac{\partial p}{\partial z} \tag{6.131}$$

where we have set $g_x = 0$, $g_y = -g$, and $g_z = 0$. That is, the y axis points up. We see that for this particular problem the Navier–Stokes equations reduce to some rather simple equations.

Equations 6.130 and 6.131 can be integrated to yield

$$p = -\rho g y + f_1(x) \tag{6.132}$$

which shows that the pressure varies hydrostatically in the y direction. Equation 6.129, rewritten as

$$\frac{d^2 u}{dy^2} = \frac{1}{\mu}\frac{\partial p}{\partial x}$$

can be integrated to obtain

$$\frac{du}{dy} = \frac{1}{\mu}\left(\frac{\partial p}{\partial x}\right)y + c_1$$

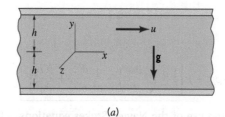
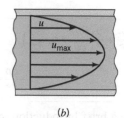

(a)   (b)

FIGURE 6.31 The viscous flow between parallel plates: (a) coordinate system and notation used in analysis; (b) parabolic velocity distribution for flow between parallel fixed plates.

and integrated again to yield

$$u = \frac{1}{2\mu}\left(\frac{\partial p}{\partial x}\right)y^2 + c_1 y + c_2 \qquad (6.133)$$

Note that for this simple flow the pressure gradient, $\partial p/\partial x$, is treated as constant, as far as the integration is concerned, because (as shown in Eq. 6.132) it is not a function of y. The two constants c_1 and c_2 must be determined from the boundary conditions. For example, if the two plates are stationary, $u = 0$ for $y = \pm h$ (because of the no-slip condition for viscous fluids). To satisfy this condition $c_1 = 0$ and

$$c_2 = -\frac{1}{2\mu}\left(\frac{\partial p}{\partial x}\right)h^2$$

Thus, the velocity distribution becomes

$$u = \frac{1}{2\mu}\left(\frac{\partial p}{\partial x}\right)(y^2 - h^2) = \frac{h^2}{2\mu}\left(-\frac{\partial p}{\partial x}\right)\left(1-\left(\frac{y}{h}\right)^2\right) \qquad (6.134)$$

where the second form of the equation emphasizes that the shape of the velocity distribution is independent of the size of the gap between the plates and that flow in the positive direction is associated with a negative pressure gradient in that direction. Equation 6.134 shows that the velocity profile between the two fixed plates is parabolic as illustrated in **Fig. 6.31b**.

The volume rate of flow, q, passing between the plates (for a unit width in the z direction) is obtained from the relationship

$$q = \int_{-h}^{h} u \, dy = \int_{-h}^{h} \frac{1}{2\mu}\left(\frac{\partial p}{\partial x}\right)(y^2 - h^2) \, dy$$

or

$$q = -\frac{2h^3}{3\mu}\left(\frac{\partial p}{\partial x}\right) \qquad (6.135)$$

Using the convention that Δp represents the pressure drop between an upstream point and a point at distance ℓ downstream, and recognizing that the pressure gradient is constant,

$$\frac{\Delta p}{\ell} = -\frac{\partial p}{\partial x}$$

and Eq. 6.135 can be expressed as

$$q = \frac{2h^3 \Delta p}{3\mu\ell} \qquad (6.136)$$

The flowrate is proportional to the pressure gradient, inversely proportional to the viscosity, and strongly dependent ($\sim h^3$) on the gap width. In terms of the mean velocity, V, where $V = q/2h$, Eq. 6.136 becomes

$$\boxed{V = \frac{h^2 \Delta p}{3\mu\ell}} \qquad (6.137)$$

Equations 6.136 and 6.137 provide convenient relationships for relating the pressure drop along a parallel-plate channel and the rate of flow or average velocity. The maximum velocity, u_{max}, occurs midway ($y = 0$) between the two plates, as shown in Fig. 6.31b, so that from Eq. 6.134

$$u_{max} = -\frac{h^2}{2\mu}\left(\frac{\partial p}{\partial x}\right)$$

or

$$u_{max} = \frac{3}{2}V \qquad (6.138)$$

The details of the steady laminar flow between infinite parallel plates are completely predicted by this solution to the Navier–Stokes equations. For example, if the pressure gradient, viscosity, and plate spacing are specified, then from Eq. 6.134 the velocity profile can be determined, and from Eqs. 6.136 and 6.137 the corresponding flowrate and mean velocity determined. In addition, because the pressure gradient in the x direction is constant, from Eq. 6.132 it follows that

The Navier–Stokes equations provide detailed flow characteristics for laminar flow between fixed parallel plates.

$$f_1(x) = \left(\frac{\partial p}{\partial x}\right)x + p_0$$

where p_0 is a reference pressure at $x = y = 0$, and the pressure variation throughout the fluid can be obtained from

$$p = -\rho g y + \left(\frac{\partial p}{\partial x}\right) x + p_0 \qquad (6.139)$$

For a specific fluid and reference pressure, p_0, the pressure at any point can be predicted. This relatively simple example of an exact solution illustrates the detailed information about the flow field that can be obtained from a solution to the Navier–Stokes equations. The flow will be laminar if the Reynolds number, $\text{Re} = \rho V(2h)/\mu$, remains below about 1400. For flow with larger Reynolds numbers, the flow becomes turbulent and the preceding analysis is not valid because the flow field is complex, three-dimensional, and unsteady.

The Wide World of Fluids

9072 kg on 55160 N/m²

Place a golf ball on the end of a garden hose and then slowly turn the water on a small amount until the ball just barely lifts off the end of the hose, leaving a small gap between the ball and the hose. The ball is free to rotate. This is the idea behind the new "floating ball water fountains" developed in Finland. Massive, 9072 kg, 1.83 m-diameter stone spheres are supported by the pressure force of the water on the curved surface within a pedestal and rotate so easily that even a small child can change their direction of rotation. The key to the fountain design is the ability to grind and polish stone to an accuracy of a few thousandths of an inch. This allows the gap between the ball and its pedestal to be very small (on the order of 1.27×10^{-4} m) and the water flowrate correspondingly small (on the order of 19 liter/min). Due to the small gap, the flow in the gap is essentially that of *flow between parallel plates*. Although the sphere is very heavy, the pressure under the sphere within the pedestal needs to be only about 55160 N/m². A similar fountain can be seen in the Inner Harbor area in Baltimore, Maryland.

6.9.2 Couette Flow

For a given flow geometry, the character and details of the flow are strongly dependent on the boundary conditions.

Another simple parallel-plate flow can be developed by fixing one plate and letting the other plate move with a constant velocity, U, as is illustrated in **Fig. 6.32a**. The Navier–Stokes equations reduce to the same form as those in the preceding section, and the general solution for the pressure and velocity distribution are still given by Eqs. 6.132 and 6.133, respectively. However, for the moving plate problem the boundary conditions for the velocity are different. For this case we locate the origin of the coordinate system at the bottom plate and designate the distance between the two plates as b (see Fig. 6.32a). The two constants c_1 and c_2 in Eq. 6.133 can be determined from the no-slip boundary conditions, $u = 0$ at $y = 0$ and $u = U$ at $y = b$. It follows that

$$u = U\frac{y}{b} + \frac{1}{2\mu}\left(\frac{\partial p}{\partial x}\right)(y^2 - by) \qquad (6.140)$$

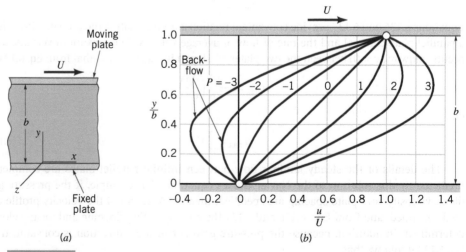

(a)   (b)

FIGURE 6.32   The viscous flow between parallel plates with bottom plate fixed and upper plate moving (Couette flow): (a) coordinate system and notation used in analysis; (b) velocity distribution as a function of parameter, P, where $P = -(b^2/2\mu U)\,\partial p/\partial x$.

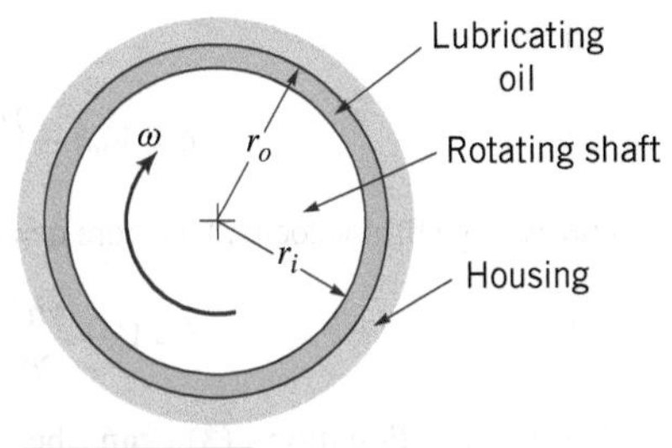

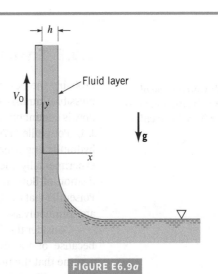

FIGURE 6.33 Flow in the narrow gap of a journal bearing.

or, in dimensionless form,

$$\frac{u}{U} = \frac{y}{b} - \frac{b^2}{2\mu U}\left(\frac{\partial p}{\partial x}\right)\left(\frac{y}{b}\right)\left(1 - \frac{y}{b}\right) \qquad (6.141)$$

The velocity profile will depend on the dimensionless parameter

$$P = -\frac{b^2}{2\mu U}\left(\frac{\partial p}{\partial x}\right)$$

Several profiles are shown in **Fig. 6.32b**. This type of flow is called **Couette flow.**

The simplest type of Couette flow is one for which the pressure gradient is zero; that is, the fluid motion is caused by the fluid being dragged along by the moving boundary. In this case, with $\partial p/\partial x = 0$, Eq. 6.140 simply reduces to

> Flow between parallel plates with one plate fixed and the other moving is called Couette flow.

$$u = U\frac{y}{b} \qquad (6.142)$$

which indicates that the velocity varies linearly between the two plates as shown in Fig. 6.31b for $P = 0$. This velocity profile was developed back in Chapter 1 when the concept of viscosity was first introduced. This velocity profile is a good approximation to the flow between closely spaced concentric cylinders in which one cylinder is fixed and the other cylinder rotates with a constant angular velocity, ω. As illustrated in **Fig. 6.33**, the flow in an unloaded journal bearing might be approximated by this simple Couette flow if the gap width is very small (i.e., $r_o - r_i \ll r_i$). In this case $U = r_i\omega, b = r_o - r_i$, and the shearing stress resisting the rotation of the shaft can be simply calculated as $\tau = \mu r_i\omega/(r_o - r_i)$. When the bearing is loaded (i.e., a force applied normal to the axis of rotation), the shaft will no longer remain concentric with the housing and the flow cannot be treated as flow between parallel boundaries. Such problems are addressed in lubrication theory (see, for example, Ref. 10).

EXAMPLE 6.9 | Plane Couette Flow

Given A wide moving belt passes through a container of a viscous liquid. The belt moves vertically upward with a constant velocity, V_0, as illustrated in **Fig. E6.9a**. Because of viscous forces the belt picks up a film of fluid of thickness h. Gravity tends to make the fluid drain down the belt. Assume that the flow is laminar, steady, and fully developed.

Find Use the Navier–Stokes equations to determine an expression for the average velocity of the fluid film as it is dragged up the belt.

Solution Since the flow is assumed to be fully developed, the only velocity component is in the y direction (the v component) so that $u = w = 0$. It follows from the continuity equation that $\partial v/\partial y = 0$, and for steady flow $\partial v/\partial t = 0$, so that $v = v(x)$. Under these conditions the Navier–Stokes equations for the x direction (Eq. 6.127a) and the z direction (perpendicular to the paper) (Eq.6.127c) simply reduce to

$$\frac{\partial p}{\partial x} = 0 \qquad \frac{\partial p}{\partial z} = 0$$

FIGURE E6.9a

This result indicates that the pressure does not vary over a horizontal plane, and because the pressure on the surface of the film ($x = h$) is atmospheric, the pressure throughout the film must be atmospheric (or zero gage pressure). The equation of motion in the y direction (Eq. 6.127b) thus reduces to

$$0 = -\rho g + \mu \frac{d^2 v}{dx^2}$$

or

$$\frac{d^2 v}{dx^2} = \frac{\gamma}{\mu} \qquad (1)$$

Integration of Eq. 1 yields

$$\frac{dv}{dx} = \frac{\gamma}{\mu} x + c_1 \qquad (2)$$

On the film surface ($x = h$) we assume the shearing stress is zero—that is, the drag of the air on the film is negligible. The shearing stress at the free surface (or any interior parallel surface) is designated as τ_{xy}, where from Eq. 6.125d

$$\tau_{xy} = \mu \left(\frac{dv}{dx} \right)$$

Thus, if $\tau_{xy} = 0$ at $x = h$, it follows from Eq. 2 that

$$c_1 = -\frac{\gamma h}{\mu}$$

A second integration of Eq. 2 provides the velocity distribution in the film as

$$v = \frac{\gamma}{2\mu} x^2 - \frac{\gamma h}{\mu} x + c_2$$

At the belt ($x = 0$) the fluid velocity must match the belt velocity, V_0, so that

$$c_2 = V_0$$

and the velocity distribution is therefore

$$v = \frac{\gamma}{2\mu} x^2 - \frac{\gamma h}{\mu} x + V_0 \qquad (3)$$

With the velocity distribution known we can determine the flowrate per unit width, q, from the relationship

$$q = \int_0^h v \, dx = \int_0^h \left(\frac{\gamma}{2\mu} x^2 - \frac{\gamma h}{\mu} x + V_0 \right) dx$$

and thus

$$q = V_0 h - \frac{\gamma h^3}{3\mu}$$

The average film velocity, V (where $q = Vh$), is therefore

$$V = V_0 - \frac{\gamma h^2}{3\mu} \qquad (Ans)$$

Comment Equation (3) can be written in dimensionless form as

$$\frac{v}{V_0} = c \left(\frac{x}{h} \right)^2 - 2c \left(\frac{x}{h} \right) + 1$$

where $c = \gamma h^2 / 2\mu V_0$. This velocity profile is shown in **Fig. E6.9b**. Note that even though the belt is moving upward, for $c > 1$ (e.g., for fluids with small enough viscosity or with a small enough belt speed) there are portions of the fluid that flow downward (as indicated by $v/V_0 < 0$).

It is interesting to note from this result that there will be a net upward flow of liquid (positive V) only if $V_0 > \gamma h^2 / 3\mu$. It takes a relatively large belt speed to lift a small viscosity fluid.

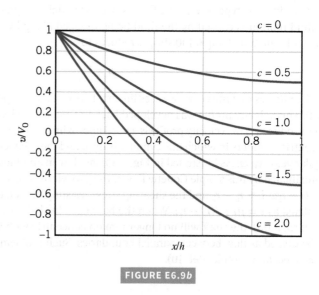

FIGURE E6.9b

6.9.3 Steady, Laminar Flow in Circular Tubes

An exact solution can be obtained for steady, incompressible, laminar flow in circular tubes.

Probably the best known exact solution to the Navier–Stokes equations is for steady, incompressible, laminar flow through a straight circular tube of constant cross section. This type of flow is commonly called *Hagen–Poiseuille flow*, or simply *Poiseuille flow*. It is named in honor of J. L. Poiseuille (1799–1869), a French physician, and G. H. L. Hagen (1797–1884), a German hydraulic engineer. Poiseuille was interested in blood flow through capillaries and deduced experimentally the resistance laws for laminar flow through circular tubes. Hagen's investigation of flow in tubes was also experimental. It was actually after the work of Hagen and Poiseuille that the theoretical results presented in this section were determined, but their names are commonly associated with the solution of this problem.

Consider the flow through a horizontal circular tube of radius R as is shown in **Fig. 6.34a**. Because of the cylindrical geometry it is convenient to use cylindrical coordinates. We assume that the flow is parallel to the walls so that $v_r = 0$ and $v_\theta = 0$, and from the continuity equation (Eq. 6.34) $\partial v_z / \partial z = 0$. Also, for steady, axisymmetric flow, v_z is not a function of t or θ,

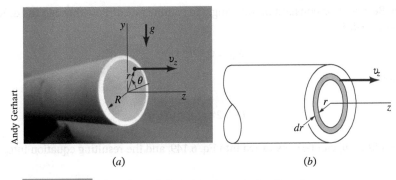

(a)   (b)

FIGURE 6.34 The viscous flow in a horizontal, circular tube:
(a) coordinate system and notation used in analysis; (b) flow through
differential annular ring.

so the velocity, v_z, is only a function of the radial position within the tube—that is, $v_z = v_z(r)$.
Under these conditions the Navier–Stokes equations (Eqs. 6.128) reduce to

$$0 = -\rho g \sin \theta - \frac{\partial p}{\partial r} \tag{6.143}$$

$$0 = -\rho g \cos \theta - \frac{1}{r}\frac{\partial p}{\partial \theta} \tag{6.144}$$

$$0 = -\frac{\partial p}{\partial z} + \mu \left[\frac{1}{r}\frac{\partial}{\partial r}\left(r\frac{\partial v_z}{\partial r} \right) \right] \tag{6.145}$$

where we have used the relationships $g_r = -g \sin \theta$ and $g_\theta = -g \cos \theta$ (with θ measured from
the horizontal plane).

Equations 6.143 and 6.144 can be integrated to produce

$$p = -\rho g (r \sin \theta) + f_1(z)$$

or

$$p = -\rho g y + f_1(z) \tag{6.146}$$

Equation 6.146 indicates that the pressure is hydrostatically distributed at any particular cross
section, and the z component of the pressure gradient, $\partial p / \partial z$, is not a function of r or θ.

The equation of motion in the z direction (Eq. 6.145) can be written in the form

$$\frac{1}{r}\frac{\partial}{\partial r}\left(r\frac{\partial v_z}{\partial r} \right) = \frac{1}{\mu}\frac{\partial p}{\partial z}$$

and integrated (using the fact that $\partial p / \partial z = \text{constant}$) to produce

$$r\frac{\partial v_z}{\partial r} = \frac{1}{2\mu}\left(\frac{\partial p}{\partial z} \right)r^2 + c_1$$

Integrating again we obtain

$$v_z = \frac{1}{4\mu}\left(\frac{\partial p}{\partial z} \right)r^2 + c_1 \ln r + c_2 \tag{6.147}$$

Requiring this solution to produce a finite velocity at the tube's centerline, $(r = 0)$, dictates
that $c_1 = 0$. At the wall $(r = R)$ the no-slip condition requires that the velocity must be zero so that

$$c_2 = -\frac{1}{4\mu}\left(\frac{\partial p}{\partial z} \right)R^2$$

and the velocity distribution becomes

$$v_z = \frac{1}{4\mu}\left(\frac{\partial p}{\partial x} \right)(r^2 - R^2) = \frac{R^2}{4\mu}\left(-\frac{\partial p}{\partial x} \right)\left(1 - \left(\frac{r}{R} \right)^2 \right) \tag{6.148}$$

Thus, at any cross section the velocity distribution is parabolic.

To obtain a relationship between the volume rate of flow, Q, passing through the tube and
the pressure gradient, we consider the flow through the differential, washer-shaped ring of

V6.13 Laminar flow

The velocity distribution is
parabolic for steady, laminar
flow in circular tubes.

Fig. 6.34b. Because v_z is constant on this ring, the volume rate of flow through the differential area $dA = (2\pi r)\,dr$ is

$$dQ = v_z(2\pi r)\,dr$$

and therefore

$$Q = 2\pi \int_0^R v_z r\,dr \qquad (6.149)$$

V6.14 Complex pipe flow

Equation 6.148 for v_z can be substituted into Eq. 6.149, and the resulting equation integrated to yield

$$Q = -\frac{\pi R^4}{8\mu}\left(\frac{\partial p}{\partial z}\right) \qquad (6.150)$$

This relationship can be expressed in terms of the pressure drop, Δp, which occurs over a length, ℓ, along the tube,

$$\frac{\Delta p}{\ell} = -\frac{\partial p}{\partial z}$$

and therefore

$$\boxed{Q = \frac{\pi R^4 \Delta p}{8\mu\ell}} \qquad (6.151)$$

Poiseuille's law relates pressure drop and flowrate for steady, laminar flow in circular tubes.

For a given pressure drop per unit length, the volume rate of flow is inversely proportional to the viscosity and proportional to the tube radius to the fourth power. A doubling of the tube radius produces a 16-fold increase in flow! Equation 6.151 is commonly called **Poiseuille's law**.

In terms of the mean velocity, V, where $V = Q/\pi R^2$, Eq. 6.151 becomes

$$V = \frac{R^2 \Delta p}{8\mu\ell} \qquad (6.152)$$

The maximum velocity v_{max} occurs at the center of the tube, where from Eq. 6.148

$$v_{max} = -\frac{R^2}{4\mu}\left(\frac{\partial p}{\partial z}\right) = \frac{R^2 \Delta p}{4\mu\ell} \qquad (6.153)$$

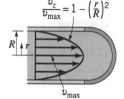

$$\frac{v_z}{v_{max}} = 1 - \left(\frac{r}{R}\right)^2$$

so that

$$v_{max} = 2V$$

The velocity distribution, as shown in the adjacent figure, can be written in terms of v_{max} as

$$\frac{v_z}{v_{max}} = 1 - \left(\frac{r}{R}\right)^2 \qquad (6.154)$$

As was true for the similar case of flow between parallel plates (sometimes referred to as *plane Poiseuille flow*), a very detailed description of the pressure and velocity distribution in tube flow results from this solution of the Navier–Stokes equations. Numerous experiments performed to substantiate the theoretical results show that the theory and experiment are in agreement for the laminar flow of Newtonian fluids in circular tubes or pipes. In general, the flow remains laminar for Reynolds numbers, Re $= \rho V(2R)/\mu$, below about 2100. Turbulent flow in tubes is considered in Chapter 8.

The Wide World of Fluids

Poiseuille's law revisited

Poiseuille's law for the *laminar flow* of fluids in tubes has an unusual history. It was developed in 1842 by a French physician, J. L. M. Poiseuille, who was interested in the flow of blood in capillaries. Poiseuille, through a series of carefully conducted experiments using water flowing through very small tubes, arrived at the formula, $Q = K\Delta p D^4/\ell$. In this formula Q is the flowrate, K an empirical constant, Δp the pressure drop over the length ℓ, and D the tube diameter. Another formula was given for the value of K as a function of the water temperature. It was not until the concept of viscosity was introduced at a later date that Poiseuille's law was derived mathematically and the constant K found to be equal to $\pi/8\mu$, where μ is the fluid viscosity. The experiments by Poiseuille have long been admired for their accuracy and completeness considering the laboratory instrumentation available in the mid-nineteenth century.

6.9.4 Steady, Axial, Laminar Flow in an Annulus

The differential equations (Eqs. 6.143, 6.144, 6.145) used in the preceding section for flow in a tube also apply to the axial flow in the annular space between two fixed, concentric cylinders (**Fig. 6.35**). Equation 6.147 for the velocity distribution still applies, but for the stationary annulus the boundary conditions become $v_z = 0$ at $r = r_o$ and $v_z = 0$ for $r = r_i$. With these two conditions the constants c_1 and c_2 in Eq. 6.147 can be determined and the velocity distribution becomes

An exact solution can be obtained for axial flow in the annular space between two fixed, concentric cylinders.

$$v_z = \frac{1}{4\mu}\left(\frac{\partial p}{\partial z}\right)\left[r^2 - r_o^2 + \frac{r_i^2 - r_o^2}{\ln(r_o/r_i)}\ln\frac{r}{r_o}\right] \tag{6.155}$$

The corresponding volume rate of flow is

$$Q = \int_{r_i}^{r_o} v_z(2\pi r)\,dr = -\frac{\pi}{8\mu}\left(\frac{\partial p}{\partial z}\right)\left[r_o^4 - r_i^4 - \frac{(r_o^2 - r_i^2)^2}{\ln(r_o/r_i)}\right]$$

or in terms of the pressure drop, Δp, in length ℓ of the annulus

$$\boxed{Q = \frac{\pi\Delta p}{8\mu\ell}\left[r_o^4 - r_i^4 - \frac{(r_o^2 - r_i^2)^2}{\ln(r_o/r_i)}\right]} \tag{6.156}$$

The velocity at any radial location within the annular space can be obtained from Eq. 6.155. The maximum velocity occurs at the radius $r = r_m$ where $\partial v_z/\partial r = 0$. Thus,

$$r_m = \left[\frac{r_o^2 - r_i^2}{2\ln(r_o/r_i)}\right]^{1/2} \tag{6.157}$$

An inspection of this result shows that the maximum velocity does not occur at the midpoint of the annular space, but rather it occurs nearer the inner cylinder. The specific location depends on r_o and r_i.

These results for flow through an annulus are valid only if the flow is laminar. A criterion based on the conventional Reynolds number (which is defined in terms of the tube diameter) cannot be directly applied to the annulus, since there are really "two" diameters involved. For tube cross sections other than simple circular tubes it is common practice to use an "effective" diameter, termed the *hydraulic diameter*, D_h, which is defined as

$$D_h = \frac{4 \times \text{cross-sectional area}}{\text{wetted perimeter}}$$

The wetted perimeter is the perimeter in contact with the fluid. For an annulus

$$D_h = \frac{4\pi(r_o^2 - r_i^2)}{2\pi(r_o + r_i)} = 2(r_o - r_i)$$

Using the average velocity computed from $V = Q/(\text{cross-sectional area})$ and the hydraulic diameter, the Reynolds number is $\text{Re} = \rho D_h V/\mu$. It is commonly assumed that if this Reynolds number remains below about 2100 the flow will be laminar. A more comprehensive examination of the use of hydraulic diameter for modeling flows in noncircular cross sections is conducted in Section 8.4.3.

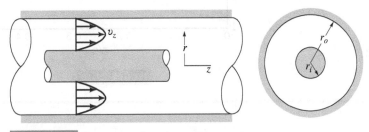

FIGURE 6.35 The viscous flow through an annulus.

EXAMPLE 6.10 | Laminar Flow in an Annulus

Given   A viscous liquid ($\rho = 1.18 \times 10^3$ kg/m³; $\mu = 0.0045$ N · s/m²) flows at a rate of 12 ml/s through a horizontal, 4-mm-diameter tube.

Find

a. Determine the pressure drop along a l-m length of the tube, which is far from the tube entrance, so that the only component of velocity is parallel to the tube axis.

b. If a 2-mm-diameter rod is placed in the 4-mm-diameter tube to form a symmetric annulus, what is the pressure drop along a l-m length if the flowrate remains the same as in part (a)?

Solution

a. We first calculate the Reynolds number, Re, to determine whether or not the flow is laminar. With the diameter $D = 4$ mm $= 0.004$ m, the mean velocity is

$$V = \frac{Q}{(\pi/4)D^2} = \frac{(12 \text{ ml/s})(10^{-6} \text{ m}^3/\text{ml})}{(\pi/4)(0.004 \text{ m})^2}$$

$$= 0.955 \text{ m/s}$$

and, therefore,

$$\text{Re} = \frac{\rho V D}{\mu} = \frac{(1.18 \times 10^3 \text{ kg/m}^3)(0.955 \text{ m/s})(0.004 \text{ m})}{0.0045 \text{ N} \cdot \text{s/m}^2}$$

$$= 1000$$

The Reynolds number is well below the critical value of 2100, so we can safely assume that the flow is laminar. Thus, we can apply Eq. 6.151 to obtain

$$\Delta p = \frac{8 \mu \ell Q}{\pi R^4}$$

$$= \frac{8(0.0045 \text{ N} \cdot \text{s/m}^2)(1 \text{ m})(12 \times 10^{-6} \text{ m}^3/\text{s})}{\pi (0.002 \text{ m})^4}$$

$$= 8.59 \text{ kPa} \qquad\qquad (Ans)$$

b. For flow in the annulus with an outer radius $r_o = 0.002$ m and an inner radius $r_i = 0.001$ m, the mean velocity is

$$V = \frac{Q}{\pi (r_o^2 - r_i^2)} = \frac{12 \times 10^{-6} \text{ m}^3/\text{s}}{(\pi)[(0.002 \text{ m})^2 - (0.001 \text{ m})^2]}$$

$$= 1.27 \text{ m/s}$$

and the Reynolds number [based on the hydraulic diameter, $D_h = 2(r_o - r_i) = 2(0.002 \text{ m} - 0.001 \text{ m}) = 0.002$ m] is

$$\text{Re} = \frac{\rho D_h V}{\mu}$$

$$= \frac{(1.18 \times 10^3 \text{ kg/m}^3)(0.002 \text{ m})(1.27 \text{ m/s})}{0.0045 \text{ N} \cdot \text{s/m}^2}$$

$$= 666$$

This value is also well below 2100, so the flow in the annulus should also be laminar. From Eq. 6.156,

$$\Delta p = \frac{8 \mu \ell Q}{\pi} \left[r_o^4 - r_i^4 - \frac{(r_o^2 - r_i^2)^2}{\ln(r_o/r_i)} \right]^{-1}$$

so that

$$\Delta p = \frac{8(0.0045 \text{ N} \cdot \text{s/m}^2)(1 \text{ m})(12 \times 10^{-6} \text{ m}^3/\text{s})}{\pi}$$

$$\times \left\{ (0.002 \text{ m})^4 - (0.001 \text{ m})^4 \right.$$

$$\left. - \frac{[(0.002 \text{ m})^2 - (0.001 \text{ m})^2]^2}{\ln(0.002 \text{ m}/0.001 \text{ m})} \right\}^{-1}$$

$$= 68.2 \text{ kPa} \qquad\qquad (Ans)$$

Comments   The pressure drop in the annulus is much larger than that in the tube. This is not a surprising result because, to maintain the same flow in the annulus as that in the open tube, the average velocity must be larger (the cross-sectional area is smaller) and the pressure difference along the annulus must overcome the shearing stresses that develop along both an inner and an outer wall.

By repeating the calculations for various radius ratios, r_i/r_o, the results shown in **Fig. E6.10** are obtained. It is seen that the pressure drop ratio, $\Delta p_{\text{annulus}} / \Delta p_{\text{tube}}$ (i.e., the pressure drop in the annulus compared to that in a tube with a radius equal to the outer radius of the annulus, r_o), is a strong function of the radius ratio. Even an annulus with a very small inner radius will have a pressure drop significantly larger than that of a tube. For example, if the inner radius is only 1/100 of the outer radius, $\Delta p_{\text{annulus}} / \Delta p_{\text{tube}} = 1.28$. As shown in the figure, for larger inner radii, the pressure drop ratio is much larger [i.e., $\Delta p_{\text{annulus}} / \Delta p_{\text{tube}} = 7.94$ for $r_i/r_o = 0.50$ as in part (b) of this example].

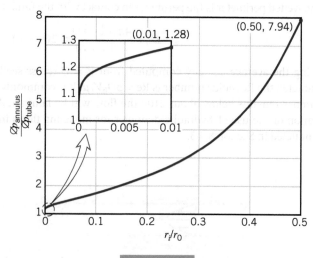

FIGURE E6.10

6.10 OTHER ASPECTS OF DIFFERENTIAL ANALYSIS

In this chapter the basic differential equations that govern the flow of incompressible fluids have been developed. The Navier–Stokes equations, which can be compactly expressed in vector notation as

$$\rho\left(\frac{\partial \mathbf{V}}{\partial t} + \mathbf{V} \cdot \nabla \mathbf{V}\right) = -\nabla p + \rho \mathbf{g} + \mu \nabla^2 \mathbf{V} \tag{6.158}$$

along with the continuity equation

$$\nabla \cdot \mathbf{V} = 0 \tag{6.159}$$

are the general equations of motion for incompressible Newtonian fluids. Although we have restricted our attention to incompressible fluids, these equations can be readily extended to include compressible fluids. It is well beyond the scope of this introductory text to consider in depth the variety of analytical and numerical techniques that can be used to obtain both exact and approximate solutions to the Navier–Stokes equations. Students, however, should be aware of the existence of these very general equations, which are frequently used as the basis for many advanced analyses of fluid motion. A few relatively simple solutions have been obtained and discussed in this chapter to indicate the type of detailed flow information that can be obtained by using differential analysis. However, it is hoped that the relative ease with which these solutions were obtained does not give the false impression that solutions to the Navier–Stokes equations are readily available. This is certainly not true, and as previously mentioned there are actually very few practical fluid flow problems that can be solved by using an exact analytical approach. In fact, there are no known analytical solutions to Eq. 6.158 for flow past any object such as a sphere, cube, or airplane.

> Very few practical fluid flow problems can be solved using an exact analytical approach.

Because of the difficulty in solving the Navier–Stokes equations, much attention has been given to various types of approximate solutions. For example, if the viscosity is set equal to zero, the Navier–Stokes equations reduce to Euler's equations. Thus, the frictionless fluid solutions discussed previously are actually approximate solutions to the Navier–Stokes equations. At the other extreme, for problems involving slowly moving fluids, viscous effects may be dominant and the nonlinear (convective) acceleration terms can be neglected. This assumption greatly simplifies the analysis because the equations now become linear. There are numerous analytical solutions to these "*slow flow*" or "*creeping flow*" problems. Another broad class of approximate solutions is concerned with flow in the very thin boundary layer. L. Prandtl showed in 1904 how the Navier–Stokes equations could be simplified to study flow in boundary layers. Such "boundary layer solutions" play a very important role in the study of fluid mechanics. By "patching together" an inviscid flow solution and a boundary layer solution, a relatively complete solution can be found for many important flows. A further discussion of boundary layers is conducted in Chapter 9.

6.10.1 Numerical Methods

Numerical methods using digital computers are commonly utilized to solve a wide variety of flow problems. As discussed previously, although the differential equations that model the flow of Newtonian fluids [the Navier–Stokes equations (6.158)] were derived many years ago, there are few known analytical solutions to them. With the advent of high-speed digital computers it has become possible to obtain approximate numerical solutions to these (and other fluid mechanics) equations for many different types of problems, including both inviscid flows and boundary layer flows. A brief introduction to computational fluid dynamics (CFD) is provided in Appendix A.

V6.15 CFD example

The Wide World of Fluids

Fluids in the Academy Awards

A computer science professor at Stanford University and his colleagues were awarded a Scientific and Technical Academy Award for applying the Navier–Stokes equations for use in Hollywood movies. These researchers make use of computational algorithms to numerically solve the Navier–Stokes equations (also termed computational fluid dynamics, or CFD) and simulate complex liquid flows. The realism of the simulations has found application in the entertainment industry. Movie producers have used the power of these numerical tools to simulate flows from ocean waves in *Pirates of the Caribbean* to lava flows in the final duel in *Star Wars: Revenge of the Sith*. Therefore, even Hollywood has recognized the usefulness of CFD.

CHAPTER SUMMARY

Differential analysis of fluid flow is concerned with the development of concepts and techniques that can be used to provide a detailed, point-by-point description of a flow field. Concepts related to the motion and deformation of a fluid element are introduced, including the Eulerian method for describing the velocity and acceleration of fluid particles. Linear deformation and angular deformation of a fluid element are described through the use of flow characteristics such as the volumetric dilatation rate, the rate of angular deformation, and vorticity. The differential form of the conservation of mass equation (continuity equation) is derived in both rectangular and cylindrical polar coordinates.

The stream function for steady, incompressible, plane, two-dimensional flow is introduced. The general equations of motion are developed, and for inviscid flow these equations are reduced to the simpler Euler equations. The Euler equations are integrated to produce the Bernoulli equation, and the concept of irrotational flow is introduced. Use of the velocity potential for describing irrotational flow is considered in detail, and several basic velocity potentials are described, including those for a uniform flow, source or sink, vortex, and doublet. The technique of using various combinations of these basic velocity potentials, by superposition, to form new potentials is described. Flows around a half-body, a Rankine oval, and a circular cylinder are obtained using this superposition technique.

Basic differential equations describing incompressible, viscous flow (the Navier–Stokes equations) are introduced. Several relatively simple solutions for steady, viscous, laminar flow between parallel plates and through circular tubes are developed.

The following checklist provides a study guide for this chapter. When your study of the entire chapter has been completed, you should be able to

- write out meanings of the terms listed and understand each of the related concepts. These terms are particularly important and are set in **bold black** type in the text.
- determine the acceleration of a fluid particle, given the equation for the velocity field.
- determine the volumetric dilatation rate, vorticity, and rate of angular deformation for a fluid element, given the equation for the velocity field.
- show that a velocity field satisfies the continuity equation.
- use the stream function to describe a flow field.
- use the velocity potential to describe a flow field.
- use superposition of basic velocity potentials to describe simple potential flow fields.
- use the Navier–Stokes equations to determine the detailed flow characteristics of incompressible, steady, laminar, viscous flow between parallel plates and through circular tubes.

KEY EQUATIONS

Acceleration of fluid particle	$\mathbf{a} = \dfrac{\partial \mathbf{V}}{\partial t} + u\dfrac{\partial \mathbf{V}}{\partial x} + v\dfrac{\partial \mathbf{V}}{\partial y} + w\dfrac{\partial \mathbf{V}}{\partial z}$	(6.2)
Vorticity	$\zeta = 2\boldsymbol{\omega} = \boldsymbol{\nabla} \times \mathbf{V}$	(6.17)
Conservation of mass	$\dfrac{\partial \rho}{\partial t} + \dfrac{\partial(\rho u)}{\partial x} + \dfrac{\partial(\rho v)}{\partial y} + \dfrac{\partial(\rho w)}{\partial z} = 0$	(6.27)
Stream function	$u = \dfrac{\partial \psi}{\partial y} \qquad v = -\dfrac{\partial \psi}{\partial x}$	(6.37)
Euler's equations of motion	$\rho g_x - \dfrac{\partial p}{\partial x} = \rho\left(\dfrac{\partial u}{\partial t} + u\dfrac{\partial u}{\partial x} + v\dfrac{\partial u}{\partial y} + w\dfrac{\partial u}{\partial z}\right)$	(6.51a)
	$\rho g_y - \dfrac{\partial p}{\partial y} = \rho\left(\dfrac{\partial v}{\partial t} + u\dfrac{\partial v}{\partial x} + v\dfrac{\partial v}{\partial y} + w\dfrac{\partial v}{\partial z}\right)$	(6.51b)
	$\rho g_z - \dfrac{\partial p}{\partial z} = \rho\left(\dfrac{\partial w}{\partial t} + u\dfrac{\partial w}{\partial x} + v\dfrac{\partial w}{\partial y} + w\dfrac{\partial w}{\partial z}\right)$	(6.51c)
Velocity potential	$\mathbf{V} = \boldsymbol{\nabla}\phi$	(6.65)
Laplace's equation	$\nabla^2 \phi = 0$	(6.66)
Uniform potential flow $\quad \phi = U(x\cos\alpha + y\sin\alpha)$	$\psi = U(y\cos\alpha - x\sin\alpha)$ $\quad u = U\cos\alpha \quad v = U\sin\alpha$	
Source and sink $\quad \phi = \dfrac{m}{2\pi}\ln r$	$\psi = \dfrac{m}{2\pi}\theta \qquad\qquad \begin{aligned} v_r &= \dfrac{m}{2\pi r} \\ v_\theta &= 0 \end{aligned}$	
Vortex $\quad \phi = \dfrac{\Gamma}{2\pi}\theta$	$\psi = -\dfrac{\Gamma}{2\pi}\ln r \qquad \begin{aligned} v_r &= 0 \\ v_\theta &= \dfrac{\Gamma}{2\pi r} \end{aligned}$	

Doublet	$\phi = \dfrac{K\cos\theta}{r}$	$\psi = -\dfrac{K\sin\theta}{r}$	$v_r = -\dfrac{K\cos\theta}{r^2}$
			$v_\theta = -\dfrac{K\sin\theta}{r^2}$

The Navier–Stokes equations

(x direction)

$$\rho\left(\frac{\partial u}{\partial t} + u\frac{\partial u}{\partial x} + v\frac{\partial u}{\partial y} + w\frac{\partial u}{\partial z}\right) = -\frac{\partial p}{\partial x} + \rho g_x + \mu\left(\frac{\partial^2 u}{\partial x^2} + \frac{\partial^2 u}{\partial y^2} + \frac{\partial^2 u}{\partial z^2}\right) \tag{6.127a}$$

(y direction)

$$\rho\left(\frac{\partial v}{\partial t} + u\frac{\partial v}{\partial x} + v\frac{\partial v}{\partial y} + w\frac{\partial v}{\partial z}\right) = -\frac{\partial p}{\partial y} + \rho g_y + \mu\left(\frac{\partial^2 v}{\partial x^2} + \frac{\partial^2 v}{\partial y^2} + \frac{\partial^2 v}{\partial z^2}\right) \tag{6.127b}$$

(z direction)

$$\rho\left(\frac{\partial w}{\partial t} + u\frac{\partial w}{\partial x} + v\frac{\partial w}{\partial y} + w\frac{\partial w}{\partial z}\right) = -\frac{\partial p}{\partial z} + \rho g_z + \mu\left(\frac{\partial^2 w}{\partial x^2} + \frac{\partial^2 w}{\partial y^2} + \frac{\partial^2 w}{\partial z^2}\right) \tag{6.127c}$$

REFERENCES

1 White, F. M., *Fluid Mechanics*, 5th Ed., McGraw-Hill, New York, 2003.

2 Streeter, V. L., *Fluid Dynamics*, McGraw-Hill, New York, 1948.

3 Rouse, H., *Advanced Mechanics of Fluids*, Wiley, New York, 1959.

4 Milne-Thomson, L. M., *Theoretical Hydrodynamics*, 4th Ed., Macmillan, New York, 1960.

5 Robertson, J. M., *Hydrodynamics in Theory and Application*, Prentice-Hall, Englewood Cliffs, N.J., 1965.

6 Panton, R. L., *Incompressible Flow*, 3rd Ed., Wiley, New York, 2005.

7 Hess, J. L., "Review of Integral Equation Techniques for Solving Potential-Flow Problems with Emphasis on the Surface-Source Method," *Computer Methods in Applied Mechanics and Engineering*, 5, 1975, pp. 145–196.

8 Li, W. H., and Lam, S. H., *Principles of Fluid Mechanics*, Addison-Wesley, Reading, Mass., 1964.

9 Schlichting, H., *Boundary-Layer Theory*, 8th Ed., McGraw-Hill, New York, 2000.

10 Fuller, D. D., *Theory and Practice of Lubrication for Engineers*, Wiley, New York, 1984.

QUESTIONS AND PROBLEMS

© Problem to be solved with aid of programmable calculator or computer.

Ⓞ Open-ended problem that requires critical thinking. These problems require various assumptions to provide the necessary input data. There are not unique answers to these problems.

Note: Unless specific values of required fluid properties are specified in the problem statement, use the values found in Tables 1.4–1.8 and in the tables in the Appendices.

Section 6.1   Fluid Element Kinematics

6.1.1 The velocity in a certain two-dimensional flow field is given by the equation

$$\mathbf{V} = 2xt\hat{\mathbf{i}} - 2yt\hat{\mathbf{j}}$$

where the velocity is in m/s when x, y, and t are in meter and seconds, respectively. Determine expressions for the local and convective components of acceleration in the x and y directions. What is the magnitude and direction of the velocity and the acceleration at the point $x = y = 0.61$ m at the time $t = 0$?

6.1.2 The velocity in a certain flow field is given by the equation

$$\mathbf{V} = yz\hat{\mathbf{i}} + x^2z\hat{\mathbf{j}} + x\hat{\mathbf{k}}$$

Determine the expressions for the three rectangular components of acceleration.

6.1.3 A two-dimensional flow field described by

$$\mathbf{V} = (2x^2y + x)\hat{\mathbf{i}} + (2xy^2 + y + 1)\hat{\mathbf{j}}$$

where the velocity is in m/s when x and y are in meters. Determine the angular rotation of a fluid element located at $x = 0.5$ m, $y = 1.0$ m.

6.1.4 Determine the vorticity field for the following velocity vector:

$$\mathbf{V} = (x^2 - y^2)\hat{\mathbf{i}} - 2xy\hat{\mathbf{j}}$$

6.1.5 Determine an expression for the vorticity of the flow field described by

$$\mathbf{V} = -xy^3\hat{\mathbf{i}} + y^4\hat{\mathbf{j}}$$

Is the flow irrotational?

6.1.6 According to Eq. 6.134, the x-velocity in fully developed laminar flow between parallel plates is given by

$$u = \frac{1}{2\mu}\left(\frac{\partial p}{\partial x}\right)(y^2 - h^2) = \frac{h^2}{2\mu}\left(-\frac{\partial p}{\partial x}\right)\left(1 - \left(\frac{y}{h}\right)^2\right)$$

The y-velocity is $v = 0$. Determine the volumetric strain rate, the vorticity, and the rate of angular deformation. What is the shear stress at the plate surface?

6.1.7 For a certain incompressible, two-dimensional flow field the velocity component in the y direction is given by the equation

$$v = 3xy + x^2y$$

Determine the velocity component in the x direction so that the volumetric dilatation rate is zero.

6.1.8 An incompressible viscous fluid is placed between two large parallel plates as shown in **Fig. P6.1.8**. The bottom plate is fixed and the upper plate moves with a constant velocity, U. For these conditions the velocity distribution between the plates is linear and can be expressed as

$$u = U\frac{y}{b}$$

Determine: (a) the volumetric dilatation rate, (b) the rotation vector, (c) the vorticity, and (d) the rate of angular deformation.

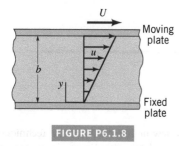

FIGURE P6.1.8

6.1.9 ⟮Video⟯ A viscous fluid is contained in the space between concentric cylinders. The inner wall is fixed, and the outer wall rotates with an angular velocity ω (see **Fig. P6.1.9a** and **Video V6.3**). Assume that the velocity distribution in the gap is linear as illustrated in **Fig. P6.1.9b**. For the small rectangular element shown in Fig. P6.1.9b, determine the rate of change of the right angle γ due to the fluid motion. Express your answer in terms of r_o, r_i, and ω.

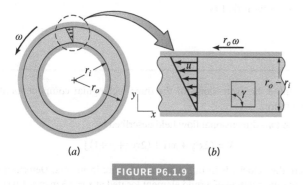

(a)   (b)

FIGURE P6.1.9

6.1.10 Air is delivered through a constant-diameter duct by a fan. The air is inviscid, so the fluid velocity profile is "flat" across each cross section. During the fan start-up, the following velocities were measured at the time t and axial positions x:

	$x = 0$	$x = 10$ m	$x = 20$ m
$t = 0$ s	$V = 0$ m/s	$V = 0$ m/s	$V = 0$ m/s
$t = 1.0$ s	$V = 1.00$ m/s	$V = 1.20$ m/s	$V = 1.40$ m/s
$t = 2.0$ s	$V = 1.70$ m/s	$V = 1.80$ m/s	$V = 1.90$ m/s
$t = 3.0$ s	$V = 2.10$ m/s	$V = 2.15$ m/s	$V = 2.20$ m/s

Estimate the local acceleration, the convective acceleration, and the total acceleration at $t = 1.0$ s and $x = 10$ m. What is the local acceleration after the fan has reached a steady air flowrate?

Section 6.2 Conservation of Mass

6.2.1 The velocity components of an incompressible, two-dimensional velocity field are given by the equations

$$u = 2xy$$
$$v = x^2 - y^2$$

Show that the flow is irrotational and satisfies conservation of mass.

6.2.2 For incompressible fluids the volumetric dilatation rate must be zero; that is, $\nabla \cdot V = 0$. For what combination of constants, a, b, c, and e can the velocity components

$$u = ax + by$$
$$u = cx + ey$$
$$w = 0$$

be used to describe an incompressible flow field?

6.2.3 Consider the two-dimensional channel flow of Problem 6.1.3. Show that the given velocities satisfy conservation of mass in both differential and control volume forms.

6.2.4 The x-velocity profile in a certain laminar boundary layer is approximated as follows

$$u = U_0 \sin\left(\frac{\pi}{2}\frac{y}{0.1\sqrt{x}}\right)$$

Determine the y-velocity, $v(x, y)$.

6.2.5 For each of the following stream functions, with units of m²/s, determine the magnitude and the angle the velocity vector makes with the x axis at $x = 1$ m, $y = 2$ m. Locate any stagnation points in the flow field. (a) $\psi = xy$ (b) $\psi = -2x^2 + y$

6.2.6 The stream function for an incompressible, two-dimensional flow field is

$$\psi = ay - by^3$$

where a and b are constants. Is this an irrotational flow? Explain.

6.2.7 For a certain two-dimensional flow field

$$u = 0$$
$$v = V$$

(a) What are the corresponding radial and tangential velocity components? (b) Determine the corresponding stream function expressed in Cartesian coordinates and in cylindrical polar coordinates.

6.2.8 In a certain steady, two-dimensional flow field the fluid density varies linearly with respect to the coordinate x: that is, $\rho = Ax$ where A is a constant. If the x component of velocity u is given by the equation $u = y$, determine an expression for v.

6.2.9 In a two-dimensional, incompressible flow field, the x component of velocity is given by the equation $u = 2x$. (a) Determine the corresponding equation for the y component of velocity if $v = 0$ along the x axis. (b) For this flow field, what is the magnitude of the average velocity of the fluid crossing the surface OA of **Fig. P6.2.9**? Assume that the velocities are in feet per second when x and y are in feet.

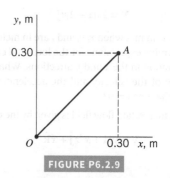

FIGURE P6.2.9

6.2.10 The stream function for an incompressible flow field is given by the equation

$$\psi = 3x^2 y - y^3$$

where the stream function has the units of m²/s with x and y in meters. **(a)** Sketch the streamline(s) passing through the origin. **(b)** Determine the rate of flow across the straight path AB shown in **Fig. P6.2.10**.

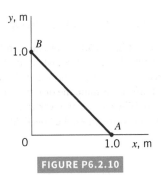

FIGURE P6.2.10

6.2.11 The stream function for an incompressible, two-dimensional flow field is

$$\psi = 3x^2 y + y$$

For this flow field, plot several streamlines.

6.2.12 Consider the incompressible, two-dimensional flow of a nonviscous fluid between the boundaries shown in **Fig. P6.2.12**. The velocity potential for this flow field is

$$\phi = x^2 - y^2$$

(a) Determine the corresponding stream function. **(b)** What is the relationship between the discharge, q (per unit width normal to plane of paper) passing between the walls and the coordinates x_i, y_i of any point on the curved wall? Neglect body forces.

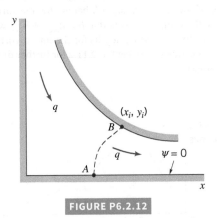

FIGURE P6.2.12

Section 6.3 The Linear Momentum Equation

6.3.1 A fluid with a density of 2000 kg/m³ flows steadily between two flat plates as shown in **Fig. P6.3.1**. The bottom plate is fixed and the top one moves at a constant speed in the x direction. The velocity is $V = 0.20y\hat{i}$ m/s where y is in meters. The acceleration of gravity is $g = -9.8\hat{j}$ m/s². The only nonzero shear stresses, $\tau_{yx} = \tau_{xy}$, are constant throughout the flow with a value of 5 N/m². The normal stress at the origin ($x = y = 0$) is $\sigma_{xx} = -100$ kPa. Use the x and y components of the

equations of motion (Eqs. 6.50a and b) to determine the normal stress throughout the fluid. Assume that $\sigma_{xx} = \sigma_{yy}$.

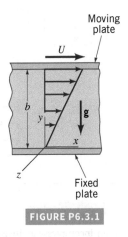

FIGURE P6.3.1

Section 6.4 Inviscid Flow

6.4.1 Given the stream function for a flow as $\psi = 4x^2 - 4y^2$, show that the Bernoulli equation can be applied between any two points in the flow field.

6.4.2 A steady, uniform, incompressible, inviscid, two-dimensional flow makes an angle of 30° with the horizontal x axis. **(a)** Determine the velocity potential and the stream function for this flow. **(b)** Determine an expression for the pressure gradient in the vertical y direction. What is the physical interpretation of this result?

6.4.3 The velocity potential for a certain inviscid flow field is

$$\phi = -(3x^2 y - y^3)$$

where ϕ has the units of m²/s when x and y are in meter. Determine the pressure difference (in Pa) between the points (1, 2) and (4, 4), where the coordinates are in meter, if the fluid is water and elevation changes are negligible.

6.4.4 The stream function for a two-dimensional, nonviscous, incompressible flow field is given by the expression

$$\psi = -2(x - y)$$

where the stream function has the units of m²/s with x and y in feet. **(a)** Is the continuity equation satisfied? **(b)** Is the flow field irrotational? If, so, determine the corresponding velocity potential. **(c)** Determine the pressure gradient in the horizontal x direction at the point $x = 0.61$ m, $y = 0.61$ m.

6.4.5 The velocity potential for a certain inviscid, incompressible flow field is given by the equation

$$\phi = 2x^2 y - \left(\frac{2}{3}\right)y^3$$

where ϕ has the units of m²/s when x and y are in meters. Determine the pressure at the point $x = 2$ m, $y = 2$ m if the pressure at $x = 1$ m, $y = 1$ m is 200 kPa. Elevation changes can be neglected, and the fluid is water.

6.4.6 Consider the two-dimensional air flow around the corner, as shown in **Fig. P6.4.6**. The x- and y-direction velocities are

$$u = \frac{v_0}{L} \sin h\left(\frac{x}{L}\right) \cos h\left(\frac{y}{L}\right)$$

and

$$v = -\frac{v_0}{L} \cos h\left(\frac{x}{L}\right) \sin h\left(\frac{y}{L}\right)$$

respectively. Assume constant density, steady flow, negligible gravity, and inviscid flow. Find $p(x, y)$.

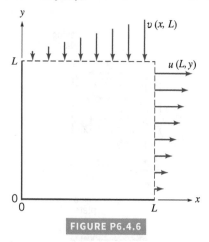

FIGURE P6.4.6

6.4.7 The streamlines for an incompressible, inviscid, two-dimensional flow field are all concentric circles, and the velocity varies directly with the distance from the common center of the streamlines; that is

$$v_\theta = Kr$$

where K is a constant. **(a)** For this *rotational* flow, determine, if possible, the stream function. **(b)** Can the pressure difference between the origin and any other point be determined from the Bernoulli equation? Explain.

6.4.8 The velocity potential

$$\phi = -k(x^2 - y^2) \quad (k = \text{constant})$$

may be used to represent the flow against an infinite plane boundary, as illustrated in **Fig. P6.4.8**. For flow in the vicinity of a stagnation point, it is frequently assumed that the pressure gradient along the surface is of the form

$$\frac{\partial p}{\partial x} = Ax$$

where A is a constant. Use the given velocity potential to show that this is true.

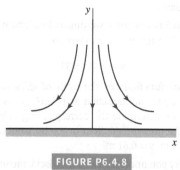

FIGURE P6.4.8

6.4.9 Water is flowing between wedge-shaped walls into a small opening as shown in **Fig. P6.4.9**. The velocity potential with units m²/s for this flow is $\phi = -2 \ln r$ with r in meters. Determine the pressure differential between points A and B.

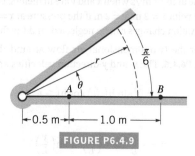

FIGURE P6.4.9

6.4.10 The velocity potential for a given two-dimensional flow field is

$$\phi = \left(\frac{5}{3}\right)x^3 - 5xy^2$$

Show that the continuity equation is satisfied and determine the corresponding stream function.

Section 6.5 Some Basic, Plane Potential Flows

6.5.1 As illustrated in **Fig. P6.5.1**, a tornado can be approximated by a free vortex of strength Γ for $r > R_c$, where R_c is the radius of the core. Velocity measurements at points A and B indicate that $V_A = 38$ m/s and $V_B = 18$ m/s. Determine the distance from point A to the center of the tornado. Why can the free vortex model not be used to approximate the tornado throughout the flow field ($r \geq 0$)?

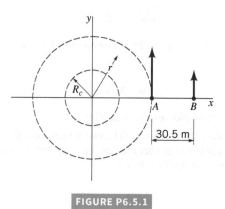

FIGURE P6.5.1

6.5.2 The motion of a liquid in an open tank is that of a combined vortex consisting of a forced vortex for $0 \leq r \leq 0.61$ m and a free vortex for $r > 0.61$ m. The velocity profile and the corresponding shape of the free surface are shown in **Fig. P6.5.2**. The free surface at the center of the tank is a depth h below the free surface at $r = \infty$. Determine the value of h. Note that $h = h_{\text{forced}} + h_{\text{free}}$, where h_{forced} and h_{free} are the corresponding depths for the forced vortex and the free vortex, respectively (see Section 2.11.2 for further discussion of the forced vortex).

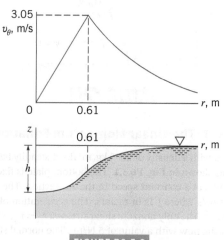

FIGURE P6.5.2

6.5.3 (Video) When water discharges from a tank through an opening in its bottom, a vortex may form with a curved surface profile, as shown in **Fig. P6.5.3** and Video V6.4. Assume that the velocity distribution in the vortex is the same as that for a free vortex. At the same time the water is being discharged from the tank at point A, it is desired to discharge a small quantity of water through the pipe B. As the discharge through A is increased, the strength of the vortex, as indicated by its circulation, is increased. Determine the maximum strength that the vortex can have in order that no air is sucked in at B. Express your answer in terms of the circulation. Assume that the fluid level in the tank at a large distance from the opening at A remains constant and viscous effects are negligible.

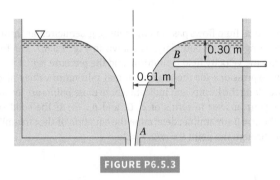

FIGURE P6.5.3

6.5.4 Water flows over a flat surface at 1.22 m/s, as shown in **Fig. P6.5.4**. A pump draws off water through a narrow slit at a volume rate of 2.8×10^{-3} m³/s per meter length of the slit. Assume that the fluid is incompressible and inviscid and can be represented by the combination of a uniform flow and a sink. Locate the stagnation point on the wall (point A) and determine the equation for the stagnation streamline. How far above the surface, H, must the fluid be so that it does not get sucked into the slit?

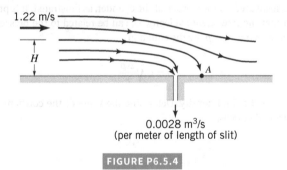

1.22 m/s

H

A

0.0028 m³/s
(per meter of length of slit)

FIGURE P6.5.4

6.5.5 A source of strength m is located a distance ℓ from a vertical solid wall as shown in **Fig. P6.5.5**. The velocity potential for this incompressible, irrotational flow is given by

$$\phi = \frac{m}{4\pi}\{\ln[(x-\ell)^2 + y^2] + \ln[(x+\ell)^2 + y^2]\}$$

(a) Show that there is no flow through the wall. **(b)** Determine the velocity distribution along the wall. **(c)** Determine the pressure distribution along the wall, assuming $p = p_0$ far from the source. Neglect the effect of the fluid weight on the pressure.

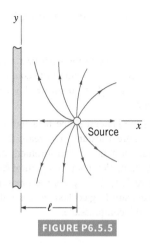

Source

x

y

ℓ

FIGURE P6.5.5

6.5.6 If the velocity field is given by $\mathbf{V} = ax\hat{\mathbf{i}} - ay\hat{\mathbf{j}}$, and a is a constant, find the circulation around the closed curve shown in **Fig. P6.5.6**

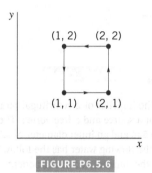

y

(1, 2)   (2, 2)

(1, 1)   (2, 1)

x

FIGURE P6.5.6

6.5.7 (See The Wide World of Fluids article titled "Some Hurricane Facts," Section 6.5.3.) Consider a category five hurricane that has a maximum wind speed of 72 m/s at the eye wall, 16.1 km from the center of the hurricane. If the flow in the hurricane outside of the hurricane's eye is approximated as a free vortex, determine the wind speeds at locations 32.2 km, 48.3 km, and 64.4 km from the center of the storm.

6.5.8 Consider the flow of a liquid of viscosity μ and density ρ down an inclined plate making an angle θ with the horizontal. The film thickness is t and is constant. The fluid velocity parallel to the plate is given by

$$V_x = \frac{\rho t^2 g \cos\theta}{2\mu}\left[1 - \left(\frac{y}{t}\right)^2\right]$$

where y is the coordinate normal to the plate. Calculate Φ and Ψ for this flow and show that neither satisfies Laplace's equation. Why not?

6.5.9 A horizontal oil-hearing stratum is 3 m high and, after hydraulic fracturing ("fracking"), provides a volume flowrate of $Q = 1000$ m³/hr. The flow moves radially inward and is collected by a vertical porous pipe having an outer radius of 1.0 m. The Laplace equation for the potential flow model of the oil is

$$\frac{1}{r}\left(\frac{\partial\Phi}{\partial r}\right) + \frac{\partial^2\Phi}{\partial r^2} + \frac{1}{r^2}\left(\frac{\partial^2\Phi}{\partial\theta^2}\right) = 0$$

Find Φ.

6.5.10 Show that the circulation of a free vortex for any closed path that does not enclose the origin is zero.

6.5.11 Show that the circulation of a free vortex for any closed path that encloses the origin is Γ.

Section 6.6 Superposition of Basic, Plane Potential Flows

6.6.1 Consider a uniform flow in the positive x direction combined with a free vortex located at the origin of the coordinate system. The streamline $\psi = 0$ passes through the point $x = 4, y = 0$. Determine the equation of this streamline.

6.6.2 A body having the general shape of a half-body is placed in a stream of fluid. At a great distance upstream the velocity is U as shown in **Fig. P6.6.2**. Show how a measurement of the differential pressure between the stagnation point and point A can be used to predict the free-stream velocity, U. Express the pressure differential in terms of U and fluid density. Neglect body forces and assume that the fluid is nonviscous and incompressible.

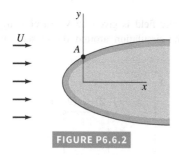

FIGURE P6.6.2

6.6.3 The flow in the impeller of a centrifugal pump is modeled by the superposition of a source and a free vortex. The impeller has an outer diameter of 0.5 m and an inner diameter of 0.3 m. At the outlet from the impeller, the flowing water has the following velocity components, relative to the impeller: radial component 2 m/s and tangential component 7 m/s. (a) Find the strength of the source and the vortex required to model this flow. (b) Assume that the impeller blades are shaped like the streamlines and plot an impeller blade shape. (c) Find the radial and tangential components of velocity at the inlet to the impeller.

6.6.4 [Video] One end of a pond has a shoreline that resembles a half-body as shown in **Fig. P6.6.4**. A vertical porous pipe is located near the end of the pond so that water can be pumped out. When water is pumped at the rate of 0.08 m³/s through a 3-m-long pipe, what will be the velocity at point A? *Hint*: Consider the flow *inside* a half-body (see **Video V6.5**).

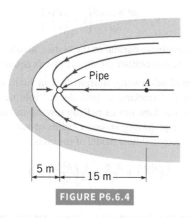

FIGURE P6.6.4

6.6.5 A Rankine oval is formed by combining a source–sink pair, each having a strength of 3.3 m²/s and separated by a distance of 4 m along the x axis, with a uniform velocity of 3.05 m/s (in the positive x direction). Determine the length and thickness of the oval.

6.6.6 A 6.7 m/s wind flows over a Quonset hut having a radius of 4 m and a length of 18 m, as shown in **Fig. P6.6.6**. The upstream pressure and temperature are equal to those inside the Quonset hut: 101329 Pa and 21 °C. Estimate the upward force on the Quonset hut. Find the location θ on the roof of the Quonset hut where the pressure is p_∞.

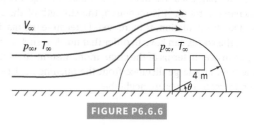

FIGURE P6.6.6

6.6.7 An ideal fluid flows past an infinitely long, semicircular "hump" located along a plane boundary, as shown in **Fig. P6.6.7**. Far from the hump the velocity field is uniform, and the pressure is p_0. (a) Determine expressions for the maximum and minimum values of the pressure along the hump, and indicate where these points are located. Express your answer in terms of ρ, U, and p_0. (b) If the solid surface is the $\psi = 0$ streamline, determine the equation of the streamline passing through the point $\theta = \pi/2, r = 2a$.

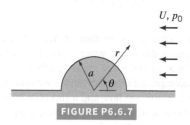

FIGURE P6.6.7

6.6.8 Assume that the flow around the long circular cylinder of **Fig. P6.6.8** is nonviscous and incompressible. Two pressures, p_1 and p_2, are measured on the surface of the cylinder, as illustrated. It is proposed that the free-stream velocity, U, can be related to the pressure difference $\Delta p = p_1 - p_2$ by the equation

$$U = C \sqrt{\frac{\Delta p}{\rho}}$$

where ρ is the fluid density. Determine the value of the constant C. Neglect body forces.

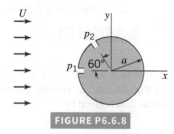

FIGURE P6.6.8

6.6.9 [C] Consider the steady potential flow around the circular cylinder shown in **Fig. 6.26**. On a plot show the variation of the magnitude of the dimensionless fluid velocity, V/U, along the positive y axis. At what distance, y/a (along the y axis), is the velocity within 2% of the free-stream velocity?

6.6.10 A highway sign has the shape of a half-cylinder with radius 2.0 m and length 30.0 m. A rearward 20-km/hr wind is blowing over the sign as shown in **Fig. P6.6.10**.

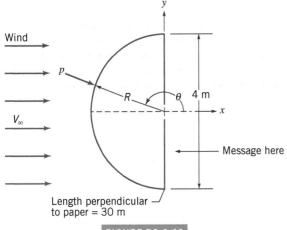

FIGURE P6.6.10

The pressure on the flat face is constant and equal to the pressure at the corner. Find the force on the sign. Neglect end effects.

6.6.11 Air at 25 °C flows normal to the axis of an infinitely long cylinder of 1.0-m radius. The cylinder is rotating at 10 rad/s, and the approach velocity is 100 km/hr. Find the maximum and minimum pressures on the cylinder surface. Assume potential flow.

6.6.12 The velocity potential for a cylinder (**Fig. P6.6.12**) rotating in a uniform stream of fluid is

$$\phi = Ur\left(1 + \frac{a^2}{r^2}\right)\cos\theta + \frac{\Gamma}{2\pi}\theta$$

where Γ is the circulation. For what value of the circulation will the stagnation point be located at: (a) point A; (b) point B?

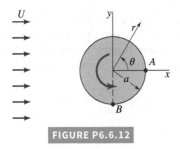

FIGURE P6.6.12

6.6.13 (See The Wide World of Fluids article titled "A Sailing Ship without Sails," Section 6.6.3.) Determine the magnitude of the total force developed by the two rotating cylinders on the Flettner "rotor-ship" due to the Magnus effect. Assume a wind-speed relative to the ship of (a) 4.4 m/s and (b) 13.4 m/s. Each cylinder has a diameter of 2.7 m, a length of 15.2 m, and rotates at 750 rev/min. Use Eq. 6.124 and calculate the circulation by assuming the air sticks to the rotating cylinders. *Note:* This calculated force is at right angles to the direction of the wind and it is the component of this force in the direction of motion of the ship that gives the propulsive thrust. Also, due to viscous effects, the actual propulsive thrust will be smaller than that calculated from Eq. 6.124, which is based on inviscid flow theory.

6.6.14 Consider the possibility of using two rotating cylinders to replace the conventional wings on an airplane for lift. Consider an airplane flying at 600 km/hr through the Standard Atmosphere at 5,000 m. Each "wing cylinder" has a 4.0-m radius. The surface velocity of each cylinder is 30 km/hr. Find the length ℓ of each wing to develop a lift of 1500 kN. Assume potential flow and neglect end effects.

6.6.15 Air at 25 °C flows normal to the axis of an infinitely long cylinder of 1.0-m radius. The approach velocity is 100 km/hr. Find the

rotational velocity of the cylinder so that a single stagnation point occurs on the lower-most part of the cylinder. Assume potential flow.

Section 6.8 Viscous Flow

6.8.1 The stream function for a certain incompressible, two-dimensional flow field is

$$\psi = 3r^2\sin 2\theta + 2\theta$$

where ψ is in m²/s when r is in meter and θ in radians. Determine the shearing stress, $\tau_{r\theta}$, at the point $r = 0.61$ m, $\theta = \pi/3$ radians if the fluid is water.

6.8.2 Determine the shearing stress for an incompressible Newtonian fluid with a velocity distribution of

$$\mathbf{V} = (3xy^2 - 4x^3)\hat{\mathbf{i}} + (12x^2y - y^3)\hat{\mathbf{j}}$$

where V in m/s, $x = 5$ m and $y = 4$ m. Assume that the viscosity

$$\mu = 1.3 \times 10^{-2}\,\text{N}\cdot\text{s/m}^2.$$

6.8.3 The velocity of a fluid particle moving along a horizontal streamline that coincides with the x axis in a plane, two-dimensional, incompressible flow field was experimentally found to be described by the equation $u = x^2$. Along this streamline determine an expression for (a) the rate of change of the v component of velocity with respect to y, (b) the acceleration of the particle, and (c) the pressure gradient in the x direction. The fluid is Newtonian.

6.8.4 "Stokes's first problem" involves the instantaneous acceleration at time $t = 0$ of a flat plate to a constant velocity U_0 while in contact with a "semi-infinite," static fluid as shown in **Fig. P6.8.4**. For a constant fluid density and viscosity, the simplified momentum equation is

$$\frac{\partial u}{\partial t} = v\frac{\partial^2 u}{\partial y^2}$$

where u is the fluid velocity in the x or velocity U_0 direction and y is a coordinate normal to the plate. Determine the appropriate boundary conditions and initial conditions for this problem and then solve the differential equation to determine the velocity distribution $u/U_0 = f(y, t)$. *Hint:* Assume that f is a function of a single variable η where $\eta = y/2\sqrt{vt}$.

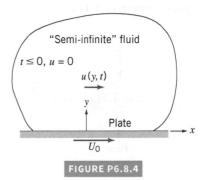

FIGURE P6.8.4

Section 6.9.1 Steady, Laminar Flow between Fixed Parallel Plates

6.9.1 The velocity distribution for steady, laminar flow in circular tubes is parabolic. Ethyl alcohol at 30 °C flows with a steady mean velocity 0.070 m/s through a 7.4-mm-diameter horizontal tube. (a) Would you expect the velocity distribution to be parabolic in this case? (b) Assuming the velocity profile is parabolic, what is the pressure drop per unit length along the tube?

6.9.2 Two fixed, horizontal, parallel plates are spaced 0.01 m apart. A viscous liquid ($\mu = 5.4 \times 10^{-3}$ N · s/m², $SG = 0.9$) flows between the plates with a mean velocity of 0.15 m/s. The flow is laminar. Determine the pressure drop per unit length in the direction of flow. What is the maximum velocity in the channel?

6.9.3 A liquid of constant density ρ and constant viscosity μ flows down a wide, long inclined flat plate. The plate makes an angle θ with the horizontal. The velocity components do not change in the direction of the plate, and the fluid depth, h, normal to the plate is constant. There is negligible shear stress by the air on the fluid. Find the velocity profile $u(y)$, where u is the velocity parallel to the plate and y is measured perpendicular to the plate. Write an expression for the volume flowrate per unit width of the plate.

6.9.4 We will see in Chapter 8 that the pressure drop in fully developed pipe flow is sometimes computed with the aid of a friction factor, defined by

$$f = \frac{\Delta p}{\frac{1}{2}\rho V^2} \frac{D}{\ell}$$

where V is the average velocity and ℓ is the length of pipe over which Δp occurs. For laminar fully developed flow, f can be evaluated from

$$f = \frac{C}{\text{Re}}$$

where C is a constant and Re is the Reynolds number, given by Re $= \rho V D/\mu$. Determine the constant C for a parallel plate channel. For the diameter D, use twice the plate spacing, h.

6.9.5 The bearing shown in **Fig. P6.9.5** consists of two parallel discs of radius R separated from each other by a small distance h ($h \ll R$). The lower disc is made of a porous material, and an incompressible, viscous fluid is pumped through it and into the gap, filling the space between the discs completely. The pores of the lower disc are closely spaced, so that the velocity of the fluid as it leaves the surface of the porous disc may be regarded as a uniform value w_0, that is, small compared to the mean radial velocity. Find the radial velocity u and the load L that the bearing can support as a function of w_0, μ, R, and h. The inertia terms are negligible, the circumferential velocity is zero ($V_\theta = 0$), and angular symmetry exists ($\partial/\partial\theta = 0$).

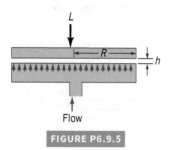

FIGURE P6.9.5

6.9.6 A Bingham plastic is a fluid in which the stress τ is related to the rate of strain du/dy by

$$\tau = \tau_0 + \mu\left(\frac{du}{dy}\right)$$

where τ_0 and μ are constants. Consider the flow of a Bingham plastic between two fixed, horizontal, infinitely wide, flat plates. For fully developed flow with $du/dx = 0$ and dp/dx constant, find $u(y)$.

Section 6.9.2 Couette Flow

6.9.7 A viscous, incompressible fluid flows between the two infinite, vertical, parallel plates of **Fig. P6.9.7**. Determine, by use of the Navier-Stokes equations, an expression for the pressure gradient in the direction of flow. Express your answer in terms of the mean velocity. Assume that the flow is laminar, steady, and uniform.

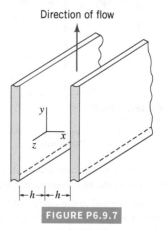

FIGURE P6.9.7

6.9.8 A layer of viscous liquid of constant thickness (no velocity perpendicular to plate) flows steadily down an infinite, inclined plane. Determine, by means of the Navier–Stokes equations, the relationship between the thickness of the layer and the discharge per unit width. The flow is laminar, and assume air resistance is negligible so that the shearing stress at the free surface is zero.

6.9.9 The viscous, incompressible flow between the parallel plates shown in **Fig. P6.9.9** is caused by both the motion of the bottom plate and pressure gradient, $\partial p/\partial x$. As noted in Section 6.9.2, an important dimensionless parameter for this type of problem is $P = -(b^2/2\,\mu U)(\partial p/\partial x)$ where μ is the fluid viscosity. Make a plot of the dimensionless velocity distribution (similar to that shown in Fig. 6.32b) for $P = 3$. For this case where does the maximum velocity occur?

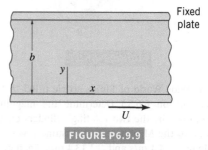

FIGURE P6.9.9

6.9.10 The disc shown in **Fig. P6.9.10** has a diameter D of 1 m and a rotational speed of 1800 rpm. It is positioned 4 mm from a solid boundary. The gap is filled with 15 °C SAE 10 oil. Find the torque required to overcome the frictional resistance of the oil on the disc. Assume that the circumferential velocity (V_θ) is linear across the gap at the each radius r.

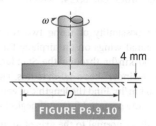

FIGURE P6.9.10

6.9.11 A viscous fluid (specific weight = 1280 kg/m³; viscosity = 0.02 N · s/m²) is contained between two infinite, horizontal parallel plates as shown in **Fig. P6.9.11**. The fluid moves between the plates under the action of a pressure gradient, and the upper plate moves with a velocity U while the bottom plate is fixed. A U-tube manometer connected between two points along the bottom indicates a differential reading of 0.003 m. If the upper plate moves with a velocity of 0.006 m/s, at what distance from the bottom plate does the maximum velocity in the gap between the two plates occur? Assume laminar flow.

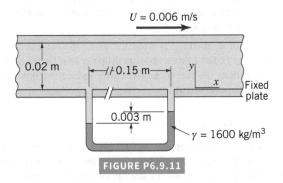

FIGURE P6.9.11

6.9.12 A flat block is pulled along a horizontal flat surface by a horizontal rope perpendicular to one of the sides. The block measures 1.0 m × 1.0 m, has a mass of 100 kg and a constant velocity of 1.0 m/s, and is separated from the flat surface by a 0.10-cm-thick oil layer of 15 °C SAE 20 crankcase oil. Find the coefficient of sliding friction for the block. Does this coefficient of friction change with the velocity?

6.9.13 A vertical shaft passes through a bearing and is lubricated with an oil having a viscosity of 0.2 N · s/m² as shown in **Fig. P6.9.13**. Assume that the flow characteristics in the gap between the shaft and bearing are the same as those for laminar flow between infinite parallel plates with zero pressure gradient in the direction of flow. Estimate the torque required to overcome viscous resistance when the shaft is turning at 80 rev/min.

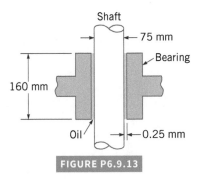

FIGURE P6.9.13

6.9.14 (Video) A viscous fluid is contained between two long concentric cylinders. The geometry of the system is such that the flow between the cylinders is approximately the same as the laminar flow between two infinite parallel plates. (a) Determine an expression for the torque required to rotate the outer cylinder with an angular velocity ω. The inner cylinder is fixed. Express your answer in terms of the geometry of the system, the viscosity of the fluid, and the angular velocity. (b) For a small, rectangular element located at the fixed wall, determine an expression for the rate of angular deformation of this element (see Video V6.3).

Section 6.9.3 Steady, Laminar Flow in Circular Tubes

6.9.15 Oil (SAE 30) flows between parallel plates spaced 5 mm apart. The bottom plate is fixed, but the upper plate moves with a velocity of 0.2 m/s in the positive x direction. The pressure gradient is 60 kPa/m, and it is negative. Compute the velocity at various points across the channel and show the results on a plot. Assume laminar flow.

6.9.16 Ethyl alcohol flows through a horizontal tube having a diameter of 10 mm. If the mean velocity is 0.15 m/s, what is the pressure drop per unit length along the tube? What is the velocity at a distance of 2 mm from the tube axis?

6.9.17 An infinitely long, solid, vertical cylinder of radius R is located in an infinite mass of an incompressible fluid. Start with the Navier–Stokes equation in the θ direction and derive an expression for the velocity distribution for the steady-flow case in which the cylinder is rotating about a fixed axis with a constant angular velocity ω. You need not consider body forces. Assume that the flow is axisymmetric and the fluid is at rest at infinity.

6.9.18 We will see in Chapter 8 that the pressure drop in fully developed pipe flow is sometimes computed with the aid of a friction factor, defined by

$$f = \frac{\Delta p}{\frac{1}{2}\rho V^2} \frac{D}{\ell}$$

where V is the average velocity, D is the pipe diameter, and ℓ is the length of pipe over which Δp occurs. For laminar fully developed flow, f can be evaluated from

$$f = \frac{C}{\text{Re}}$$

where C is a constant and Re is the Reynolds number, given by Re = $\rho VD/\mu$. Find the constant C for flow in a circular tube.

6.9.19 A liquid (viscosity = 0.002 N · s/m²; density = 1000 kg/m³) is forced through the circular tube shown in **Fig. P6.9.19**. A differential manometer is connected to the tube as shown to measure the pressure drop along the tube. When the differential reading, Δh, is 9 mm, what is the mean velocity in the tube?

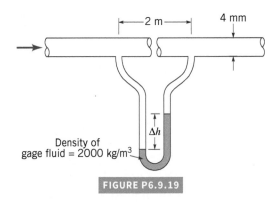

FIGURE P6.9.19

6.9.20 Fluid with kinematic viscosity υ flows down an inclined circular pipe of length ℓ and diameter D with flowrate Q. Find the vertical drop per unit length of the pipe so that the pressure drop $(p_1 - p_2)$ is zero for laminar flow.

6.9.21 Blood flows at volume rate Q in a circular tube of radius R. The blood cells concentrate and flow near the center of the tube, while the cell-free fluid (plasma) flows in the outer region. The center core of radius R_c has a viscosity μ_c, and the plasma has a viscosity μ_p. Assume laminar, fully developed flow for both the core and plasma flows and show that an "apparent" viscosity is defined by

$$\mu_{app} = \frac{\pi R^4 \Delta p}{8LQ}$$

is given by

$$\mu_{app} = \frac{\mu_p}{1 - (R_c/R)^4 (1 - \mu_p/\mu_c)}$$

Section 6.9.4   Steady, Axial, Laminar Flow in an Annulus

6.9.22 Consider a steady, laminar flow through a straight horizontal tube having the constant elliptical cross section given by the equation

$$\frac{x^2}{a^2} + \frac{y^2}{b^2} = 1$$

The streamlines are all straight and parallel. Investigate the possibility of using an equation for the z component of velocity of the form

$$w = A\left(1 - \frac{x^2}{a^2} - \frac{y^2}{b^2}\right)$$

as an exact solution to this problem. With this velocity distribution, what is the relationship between the pressure gradient along the tube and the volume flowrate through the tube?

6.9.23 A viscous fluid is contained between two infinitely long, vertical, concentric cylinders. The outer cylinder has a radius r_o and rotates with an angular velocity ω. The inner cylinder is fixed and has a radius r_i. Make use of the Navier–Stokes equations to obtain an exact solution for the velocity distribution in the gap. Assume that the flow in the gap is axisymmetric (neither velocity nor pressure are functions of angular position θ within the gap) and that there are no velocity components other than the tangential component. The only body force is the weight.

6.9.24 A double-pipe heat exchanger consists of two concentric tubes with one fluid flowing in the central tube and the other flowing in the annulus between the tubes. In a particular exchanger, cold (15 °C) water flows at 300 liters/min through a 10 cm diameter inner tube and hot (50 °C) SAE 30 oil flows at 400 liters/min in the annulus between the inner tube and a 12 cm diameter outer tube. The heat exchanger is 2 m in length. (a) Calculate the pressure drop in the oil stream (b) Can you calculate the pressure drop in the water using the same equations? Why or why not?

6.9.25 A wire of diameter d is stretched along the centerline of a pipe of diameter D. For a given pressure drop per unit length of pipe, by how much does the presence of the wire reduce the flowrate if (a) $d/D = 0.1$; (b) $d/D = 0.01$?

Lifelong Learning Problems

6.1 LL What sometimes appear at first glance to be simple fluid flows can contain subtle, complex fluid mechanics. One such example is the stirring of tea leaves in a teacup. Obtain information about "Einstein's tea leaves" and investigate some of the complex fluid motions interacting with the leaves. Summarize your findings in a brief report.

6.2 LL Find information about the (proposed) Madaras Rotor Powerplant. What is it? How would it work? When was it first proposed? Has there been interest more recently? Do you think it is feasible? Summarize your findings in a short report.

CHAPTER 7

Dimensional Analysis, Similitude, and Modeling

After completing this chapter, you should be able to:

- apply the Buckingham pi theorem.
- develop a set of dimensionless variables for a given flow situation.
- recognize common dimensionless groups as an aid to interpret and correlate data.

- discuss the use of dimensionless variables in data analysis.
- apply the concepts of modeling and similitude to develop prediction equations.

Although many practical engineering problems involving fluid mechanics can be solved by using the equations and analytical procedures described in the preceding chapters, there remain a large number of problems that rely on experimental data for their solution. In fact, it is probably fair to say that very few problems involving real flows can be solved by analysis alone. The solution to many problems is achieved through the use of a combination of theoretical and numerical (computer) analysis and experimental data. Thus, engineers working on fluid mechanics problems should be familiar with the experimental approach to these problems so that they can interpret and make use of data obtained by others, such as might appear in handbooks, or be able to plan and execute the necessary experiments in their own laboratories. In this chapter we consider some techniques and ideas that are important in the planning and execution of experiments, as well as in understanding and correlating data that may have been obtained by other experimenters.

An obvious goal of any experiment is to make the results as widely applicable as possible. To achieve this end, the concept of **similitude** is often used so that measurements made on one system (for example, in the laboratory) can be used to describe the behavior of other similar systems (outside the laboratory). The laboratory systems are usually thought of as *models* and are used to study the phenomenon of interest under carefully controlled conditions. From these model studies, empirical formulations can be developed, or specific predictions of one or more characteristics of some other similar system can be made. To do this, it is necessary to establish the relationship between the laboratory model and the "other" system. In the following sections, we find out how this can be accomplished in a systematic manner.

The specific topic of this chapter is *dimensional analysis.* Dimensional analysis is a packaging or compacting technique used to reduce the complexity of experimental programs and at the same time increase the generality of experimental information. The information packaging produced by dimensional analysis also has other uses. It is equally applicable to theoretical (analytic) and computational investigations and results. Also, dimensional analysis gives us the tools to decide a priori whether a flow is likely to be laminar or turbulent or whether simplifying assumptions, such as neglecting compressibility, are appropriate. These decisions in turn guide the engineer to the appropriate design or problem solution technique.

Experimentation and modeling are widely used techniques in fluid mechanics.

 Video

V7.1 Real and model flies

Before we discuss the subject in detail, we need to point out the following about dimensional analysis:

- Dimensional analysis is sometimes very confusing to beginners. The basic principle involved seems obvious, almost trivial. The execution of the technique, however, often seems almost "magical."

- Confusion arises because we usually expect theories to be complete, that is, to begin at the beginning and carry through to the end. Dimensional analysis stops far short of providing a final answer to a particular engineering problem; it simply leads to a more efficient organization of the variables. The ultimate answer comes only in the laboratory or from analytic or computational methods.

- Dimensional analysis tells us that we *can* organize variables efficiently and suggests possible ways that the organization can be carried out; however, it does not allow us to find unique results.

Two analogies help illustrate the utility of dimensional analysis. Dimensional analysis is similar to a trash compactor. Trash thrown into the compactor is crushed into a smaller space so it is easier to handle. The compactor does not solve the trash problem nor does it ultimately dispose of any of the trash put into it; it only reduces it to a more convenient package. In the same way, dimensional analysis reduces the number of variables that we must consider in an analytic or experimental investigation by "compacting" several of the raw variables into convenient packages.

Dimensional analysis also is similar to packing a suitcase for a trip. The traveler must select the items to take. Any item that is not selected will not be in the suitcase. The traveler organizes the items in the suitcase in one of several equally convenient ways. Different travelers prefer different arrangements, but that does not make any arrangement superior. After the suitcase is packed, it is up to the traveler to carry it to the ultimate destination. The suitcase does not go on its own, and just packing it does not accomplish the objective of the trip.

The Wide World of Fluids

Model study of New Orleans levee breach caused by Hurricane Katrina

Much of the devastation to New Orleans from Hurricane Katrina in 2005 was a result of flood waters that surged through a breach of the 17th Street Outfall Canal. To better understand why this occurred and to determine what can be done to prevent future occurrences, the U.S. Army Engineer Research and Development Center Coastal and Hydraulics Laboratory conducted tests on a large (1:50 length scale) 1394 square meter hydraulic *model* that replicates 0.8 km of the canal surrounding the breach and more than a km of the adjacent Lake Pontchartrain front. The objective of the study was to obtain information regarding the effect that waves had on the breaching of the canal and to investigate the surging water currents within the canals. The waves are generated by computer-controlled wave generators that can produce waves of varying heights, periods, and directions similar to the storm conditions that occurred during the hurricane. Data from the study was used to calibrate and validate information that was fed into various numerical model studies of the disaster.

7.1 THE NEED FOR DIMENSIONAL ANALYSIS

To illustrate a typical fluid mechanics problem in which experimentation is required, consider the steady flow of an incompressible Newtonian fluid through a long, smooth-walled, horizontal, circular pipe. An important characteristic of this system, which would be of interest to an engineer designing a pipeline, is the pressure drop per unit length that develops along the pipe as a result of friction. Although this would appear to be a relatively simple flow problem, it cannot generally be solved analytically (even with the aid of large computers) without the use of experimental data.

The first step in planning an experiment to study this problem would be to decide on the variables that will have an effect on the pressure drop per unit length, $\Delta p_\ell\,[(\text{N/m}^2)/\text{m} = \text{N/m}^3]$. We expect the list to include the pipe diameter, D, the fluid density, ρ, fluid viscosity, μ, and the mean velocity, V, at which the fluid is flowing through the pipe. Thus, we can express this relationship as

$$\Delta p_\ell = f(D, \rho, \mu, V) \tag{7.1}$$

which simply indicates mathematically that we expect the pressure drop per unit length to be some function of the factors contained within the parentheses. At this point the nature of the function is unknown, and the objective of the experiments to be performed is to determine the nature of this function.

To perform the experiments in a meaningful and systematic manner, it would be necessary to change one of the variables, such as the velocity, while holding all others constant, and measure the corresponding pressure drop. This series of tests would yield data that could be represented graphically as is illustrated in **Fig. 7.1a**. It is to be noted that this plot would only be valid for the specific pipe and for the specific fluid used in the tests; this certainly does not give us the general formulation we are looking for. We could repeat the process by varying each of the other variables in turn, as is illustrated in **Figs. 7.1b**, **7.1c**, and **7.1d**. This approach to determining the functional relationship between the pressure drop and the various factors that influence it, although logical in concept, is fraught with difficulties. Some of the experiments would be hard to carry out; for example, to obtain the data illustrated in Fig. 7.1c it would be necessary to vary fluid density while holding viscosity constant. How would you do this? Finally, once we obtained the various curves shown in Figs. 7.1a, 7.1b, 7.1c, and 7.1d, how could we combine these data to obtain the desired general functional relationship between Δp_ℓ, D, ρ, μ, and V that would be valid for any similar pipe system?

Fortunately, there is a much simpler approach to this problem that will minimize such difficulties. In the following sections we will show that rather than working with the original list of variables, as described in Eq. 7.1, we can collect these into two nondimensional combinations of variables (called **dimensionless products**, **dimensionless variables**, or **dimensionless groups**) so that

$$\frac{D\,\Delta p_\ell}{\rho V^2} = \phi\left(\frac{\rho V D}{\mu}\right) \tag{7.2}$$

Thus, instead of having to work with five variables, we now have only two. The necessary experiment would simply consist of varying the dimensionless product $\rho V D / \mu$ and determining the corresponding value of $D\,\Delta p_\ell / \rho V^2$. The results of the experiment could then be represented by a single, universal curve as is illustrated in **Fig. 7.2**. This curve would be valid for any combination

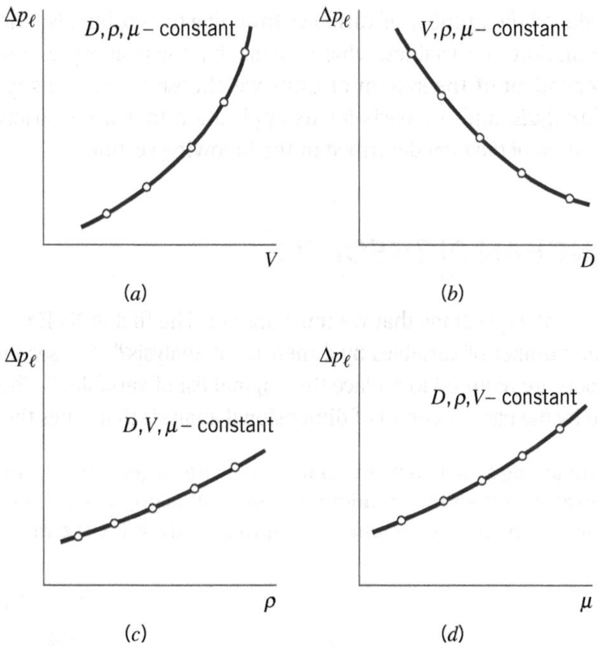

> It is important to develop a meaningful and systematic way to perform an experiment.

FIGURE 7.1 Illustrative plots showing how the pressure drop in a pipe may be affected by several different factors.

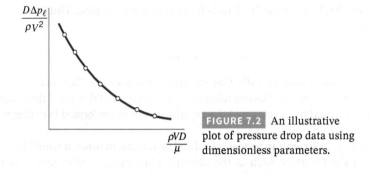

FIGURE 7.2 An illustrative plot of pressure drop data using dimensionless parameters.

Dimensionless products are important and useful in the planning, execution, and interpretation of experiments.

of smooth-walled pipe and incompressible Newtonian fluid. To obtain this curve we could choose a pipe of convenient size and a fluid that is easy to work with. Note that we wouldn't have to use different pipe sizes or even different fluids. It is clear that the experiment would be much simpler, easier to do, and less expensive (which would certainly make an impression on your boss). And think how much less paper (or disk space) would be required to present the results.

The basis for this simplification lies in a consideration of the dimensions of the variables involved. As was discussed in Chapter 1, a qualitative description of physical quantities can be given in terms of **basic dimensions** such as mass, M, length, L, and time, T.[1] Alternatively, we could use force, F, L, and T as basic dimensions, since from Newton's second law

$$F \doteq MLT^{-2}$$

(Recall from Chapter 1 that the notation $\doteq$ is used to indicate dimensional equality.) The dimensions of the variables in the pipe flow example are $\Delta p_\ell \doteq FL^{-3}$, $D \doteq L$, $\rho \doteq FL^{-4}T^2$, $\mu \doteq FL^{-2}T$, and $V \doteq LT^{-1}$. [Note that the pressure drop per unit length has the dimensions of $(F/L^2)/L = FL^{-3}$. Also note that you can find common physical quantities expressed in their basic dimensions in Table 1.1 of Chapter 1.] A quick check of the dimensions of the two groups that appear in Eq. 7.2 shows that they are in fact *dimensionless* products; that is,

$$\frac{D \, \Delta p_\ell}{\rho V^2} \doteq \frac{L(F/L^3)}{(FL^{-4}T^2)(LT^{-1})^2} \doteq F^0 L^0 T^0$$

and

$$\frac{\rho V D}{\mu} \doteq \frac{(FL^{-4}T^2)(LT^{-1})(L)}{(FL^{-2}T)} \doteq F^0 L^0 T^0$$

Not only have we reduced the number of variables from five to two, but also the new groups are dimensionless combinations of variables, which means that the results presented in the form of Fig. 7.2 will be independent of the system of units we choose to use. This type of analysis is called *dimensional analysis*, and the basis for its application to a wide variety of problems is found in the *Buckingham pi theorem* described in the following section.

7.2   BUCKINGHAM PI THEOREM

There are two fundamental questions that we must answer. The first is "Is Eq. 7.2 a fluke, or can we always reduce the number of variables by dimensional analysis?" The second is "How many dimensionless products are required to replace the original list of variables?" The answer to these questions is supplied by the basic theorem of dimensional analysis that states the following:

> If an equation involving k variables is dimensionally homogeneous, it can be reduced to a relationship among $k - r$ independent dimensionless products, where r is the minimum number of reference dimensions required to describe the variables.

[1]As noted in Chapter 1, we will use T to represent the basic dimension of time, although T is also used for temperature in thermodynamic relationships (such as the ideal gas law). The dimension of temperature is often represented by θ.

The dimensionless products are frequently referred to as **pi terms**, and the theorem is called the **Buckingham pi theorem**.[2] Edgar Buckingham used the symbol Π to represent a dimensionless product, and this notation is commonly used. Although the pi theorem is a simple one, its proof is not so simple, and we will not include it here. Many entire books have been devoted to the subject of similitude and dimensional analysis, and a number of these are listed at the end of this chapter (Refs. 1–15). Students interested in pursuing the subject in more depth (including the proof of the pi theorem) can refer to one of these books.

> Dimensional analysis is based on the Buckingham pi theorem.

The pi theorem is based on the idea of dimensional homogeneity, which was introduced in Chapter 1. Essentially for any physically meaningful equation involving k variables, such as

$$u_1 = f(u_2, u_3, \ldots, u_k)$$

the dimensions of the variable on the left side of the equal sign must be equal to the dimensions of any term that stands by itself on the right side of the equal sign. The pi theorem then guarantees that we can rearrange the equation into a set of dimensionless products (pi terms) so that

$$\Pi_1 = \phi(\Pi_2, \Pi_3, \ldots, \Pi_{k-r})$$

where $\phi(\Pi_2, \Pi_3, \ldots, \Pi_{k-r})$ is a function of Π_2 through Π_{k-r}.

The required number of pi terms is fewer than the number of original variables by r, where r is determined by the minimum number of reference dimensions required to describe the original list of variables. Usually the reference dimensions required to describe the variables in fluid mechanics will be the basic dimensions M, L, and T or F, L, and T. However, in some instances perhaps only two dimensions, such as L and T, are required, or maybe just one, such as L. Also, in a few rare cases the variables may be described by some combination of basic dimensions, such as M/T^2 and L, and in this case r would be equal to two rather than three. Although the use of the pi theorem may appear to be a little mysterious and complicated, we will develop a simple, systematic procedure for developing the pi terms for a given problem.

7.3 DETERMINATION OF PI TERMS

Several methods can be used to form the dimensionless products, or pi terms, that arise in a dimensional analysis. To begin we will consider a method that will allow us to *systematically* form the pi terms so that we are sure that they are dimensionless and independent and that we have the right number. The method we will describe in detail in this section is called the **method of repeating variables.**

It will be helpful to break the repeating variable method down into a series of distinct steps that can be followed for any given problem. With a little practice you will be able to readily complete a dimensional analysis for your problem.

> A dimensional analysis can be performed using a series of distinct steps.

   STEP 1   List all the variables that are involved in the problem. This step is the most difficult one, and it is, of course, vitally important that all pertinent variables be included. Otherwise the dimensional analysis will not be correct! We are using the term *variable* to include any quantity, including dimensional and nondimensional constants, which play a role in the phenomenon under investigation. (The term variable could be replaced with parameter or property considering that constants by definition are not variable.) All such quantities should be included in the list of variables to be considered for the dimensional analysis. The determination of the variables must be accomplished by knowledge of the problem and the physical laws that govern the phenomenon. Typically the variables will include those that are necessary to describe the *geometry* of the system (such as a pipe diameter), to define any *fluid properties* (such as a fluid viscosity), and to indicate *external*

---

[2]Although several early investigators, including Lord Rayleigh (1842–1919) in the nineteenth century, contributed to the development of dimensional analysis, Edgar Buckingham's (1867–1940) name is usually associated with the basic theorem. He stimulated interest in the subject in the United States through his publications during the early part of the twentieth century. See, for example, E. Buckingham, "On Physically Similar Systems: Illustrations of the Use of Dimensional Equations", *Phys. Rev.*, 4 (1914), 345–376.

effects that influence the system (such as a driving pressure drop per unit length). These general classes of variables are intended as broad categories that should be helpful in identifying variables. It is likely, however, that there will be variables that do not fit easily into one of these categories, and each problem needs to be carefully analyzed.

Any dimension must appear in at least two parameters, or obtaining dimensionless groups would not be possible. Whenever a dimension is represented in only one parameter, either that parameter is irrelevant or you must include another parameter that also includes the unique dimension.

Since we wish to keep the number of variables to a minimum, so that we can minimize the amount of work, it is important that all variables be independent. For example, if in a certain problem the cross-sectional area of a pipe is an important variable, either the area or the pipe diameter could be used, but not both, since they are obviously not independent. Similarly, if both fluid density, ρ, and specific weight, γ, are important variables, we could list ρ and γ, or ρ and g (acceleration of gravity), or γ and g. However, it would be incorrect to use all three since $\gamma = \rho g$; that is, ρ, γ, and g are not independent. Note that although g would normally be constant in a given experiment, that fact is irrelevant as far as a dimensional analysis is concerned.

STEP 2   Express each of the variables in terms of basic dimensions. For the typical fluid mechanics problem the basic dimensions will be either M, L, and T or F, L, and T. Dimensionally these two sets are related through Newton's second law ($\mathbf{F} = m\mathbf{a}$) so that $F \doteq MLT^{-2}$. For example, $\rho \doteq ML^{-3}$ or $\rho \doteq FL^{-4}T^2$. Thus, either set can be used. The basic dimensions for typical variables found in fluid mechanics problems are listed in Table 1.1 in Chapter 1.

STEP 3   Determine the required number of pi terms. This can be accomplished by means of the Buckingham pi theorem, which indicates that the number of pi terms is equal to $k - r$, where k is the number of variables in the problem (which is determined from Step 1) and r is the number of reference dimensions required to describe these variables (which is determined from Step 2). The reference dimensions usually correspond to the basic dimensions and can be determined by an inspection of the dimensions of the variables obtained in Step 2. As previously noted, there may be occasions (usually rare) in which the basic dimensions appear in combinations so that the number of reference dimensions is less than the number of basic dimensions. In this case, the number of dimensions in the MLT and FLT systems will be different. Then r will be the smaller of the two. This possibility is illustrated in Example 7.2.

STEP 4   Select a number of repeating variables, where the number required is equal to the number of reference dimensions. Essentially what we are doing here is selecting a set of variables from the original list of variables which can be combined with each of the remaining variables to form a pi term. After completing subsequent steps, the repeating variables will appear in multiple pi terms to produce dimensionless groups from the non-repeating variables. All of the required reference dimensions must be included within the group of repeating variables, and each repeating variable must be dimensionally independent of the others (i.e., the dimensions of one repeating variable cannot be reproduced by some combination of products of powers of the remaining repeating variables). This means that the repeating variables cannot themselves be combined to form a dimensionless product.

For any given problem we usually are interested in determining how one particular variable is influenced by the other variables. We would consider this variable to be the dependent variable, and we would want this to appear in only one pi term. Thus, do *not* choose the dependent variable as one of the repeating variables, since the repeating variables will generally appear in more than one pi term. Other tips for choosing repeating variables are do not choose a variable whose effect on the problem is questionable or important only part of the time and select variables as close to dimensionally pure as possible (e.g., pipe diameter is pure length, fluid density is almost pure mass, etc.). Also do not select parameters that are already dimensionless (e.g., angle), and whenever possible choose dimensional constants over dimensional variables so that only one pi term contains the dimensional variable.

STEP 5   Form a pi term by multiplying one of the nonrepeating variables by the product of the repeating variables, each raised to an exponent that will make the combination dimensionless. Essentially each pi term will be of the form $u_i u_1^{a_i} u_2^{b_i} u_3^{c_i}$ where u_i is one of the nonrepeating variables; $u_1, u_2,$ and u_3 are the repeating variables; and the exponents $a_i, b_i,$ and c_i are determined so that the combination is dimensionless. Although this appears difficult, the process is straightforward as will be illustrated in an example involving pipe flow that follows this list of steps.

STEP 6   Repeat Step 5 for each of the remaining nonrepeating variables. The resulting set of pi terms will correspond to the required number obtained from Step 3. If not, check your work—you have made a mistake!

STEP 7   Check all the resulting pi terms to make sure they are dimensionless. It is easy to make a mistake in forming the pi terms. However, this can be checked by simply substituting the dimensions of the variables into the pi terms to confirm that they are all dimensionless. One good way to do this is to express the variables in terms of $M, L,$ and T if the basic dimensions $F, L,$ and T were used initially, or vice versa, and then verify that all the dimensions cancel for each pi term.

STEP 8   Rearrange the pi terms to simplify the problem or to correspond to customary usage. Later, we will see that certain dimensionless groups are commonly used in fluid mechanics and you can communicate better with other engineers if you use these groups. Any pi term of the problem can be replaced using the following operations without sacrificing its dimensionless nature:

- Multiply by a constant.
- Raise the pi term to any positive or negative power (which of course includes taking the reciprocal).
- Multiply any power of the pi term with any power of any of the other pi terms.
- Form a product or quotient of any pi term with any other pi term in the problem to replace one of the pi terms.
- Any combination of the above.
- Substitute a dimensional parameter in a pi term with another parameter of the same dimensions (e.g., substitute diameter for length).

STEP 9   Express the final form as a relationship among the pi terms, and think about what it means. Typically the final form can be written as

$$\Pi_1 = \phi(\Pi_2, \Pi_3, \ldots, \Pi_{k-r})$$

where Π_1 would contain the dependent variable in the numerator. It should be emphasized that if you started out with the correct list of variables (and the other steps were completed correctly), then the relationship in terms of the pi terms can be used to describe the problem. You need only work with the pi terms—not with the individual variables. However, it should be clearly noted that this is as far as we can go with the dimensional analysis; that is, the actual functional relationship among the pi terms must be determined by experiment or computation.

By using dimensional analysis, the original problem is simplified and defined with pi terms.

To illustrate these steps we will again consider the problem discussed earlier in this chapter: the steady flow of an incompressible Newtonian fluid through a long, smooth-walled, horizontal circular pipe. We are interested in the pressure drop per unit length, Δp_ℓ, along the pipe as illustrated by the adjacent figure. First (Step 1) we must list all of the pertinent variables that are involved based on the experimenter's knowledge of the problem. In this problem we assume that

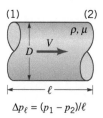

$$\Delta p_\ell = (p_1 - p_2)/\ell$$

$$\Delta p_\ell = f(D, \rho, \mu, V)$$

where D is the pipe diameter, ρ and μ are the fluid density and viscosity, respectively, and V is the mean velocity.

Next (Step 2) we express all the variables in terms of basic dimensions. Using F, L, and T as basic dimensions it follows that

$$\Delta p_\ell \doteq FL^{-3}$$
$$D \doteq L$$
$$\rho \doteq FL^{-4}T^2$$
$$\mu \doteq FL^{-2}T$$
$$V \doteq LT^{-1}$$

We could also use M, L, and T as basic dimensions if desired—the final result will be the same. Note that for density, which is a mass per unit volume (ML^{-3}), we have used the relationship $F \doteq MLT^{-2}$ to express the density in terms of F, L, and T. Do not mix the basic dimensions; that is, use either F, L, and T or M, L, and T.

We can now apply the pi theorem to determine the required number of pi terms (Step 3). An inspection of the dimensions of the variables from Step 2 reveals that all three basic dimensions (F, L, and T) are required to describe the variables. Since there are five ($k = 5$) variables (do not forget to count the dependent variable, Δp_ℓ) and three required reference dimensions ($r = 3$), then according to the pi theorem there will be ($5 - 3$), or two pi terms required.

The repeating variables to be used to form the pi terms (Step 4) need to be selected from the list D, ρ, μ, and V. Remember, we do not want to use the dependent variable as one of the repeating variables. Since three reference dimensions are required, we will need to select three repeating variables. Generally, we would try to select as repeating variables those that are the simplest, dimensionally. For example, if one of the variables has the dimension of a length, choose it as one of the repeating variables. In this example we will use D, V, and ρ as repeating variables. Note that these are dimensionally independent, since D is a length, V involves both length and time, and ρ involves force, length, and time. This means that we cannot form a dimensionless product from this set.

We are now ready to form the two pi terms (Step 5). Typically, we would start with the dependent variable and combine it with the repeating variables to form the first pi term; that is,

$$\Pi_1 = \Delta p_\ell D^a V^b \rho^c$$

Since this combination is to be dimensionless, it follows that

$$(FL^{-3})(L)^a(LT^{-1})^b(FL^{-4}T^2)^c \doteq F^0L^0T^0$$

The exponents, a, b, and c must be determined such that the resulting exponent for each of the basic dimensions—F, L, and T—must be zero (so that the resulting combination is dimensionless). Taking inventory of each of the dimension's exponents (while watching closely for positive and negative signs), we can write

$$1 + c = 0 \qquad \text{(for } F\text{)}$$
$$-3 + a + b - 4c = 0 \qquad \text{(for } L\text{)}$$
$$-b + 2c = 0 \qquad \text{(for } T\text{)}$$

The solution of this system of algebraic equations gives the desired values for a, b, and c. It follows that $a = 1$, $b = -2$, $c = -1$, and, therefore,

$$\Pi_1 = \frac{\Delta p_\ell D}{\rho V^2}$$

The process is now repeated for the remaining nonrepeating variables (Step 6). In this example there is only one additional variable (μ) so that

$$\Pi_2 = \mu D^a V^b \rho^c$$

or

$$(FL^{-2}T)(L)^a(LT^{-1})^b(FL^{-4}T^2)^c \doteq F^0L^0T^0$$

and, therefore,

$$1 + c = 0 \qquad \text{(for } F\text{)}$$
$$-2 + a + b - 4c = 0 \qquad \text{(for } L\text{)}$$
$$1 - b + 2c = 0 \qquad \text{(for } T\text{)}$$

Solving these equations simultaneously it follows that $a = -1$, $b = -1$, $c = -1$ so that

$$\Pi_2 = \frac{\mu}{DV\rho}$$

Note that we end up with the correct number of pi terms as determined from Step 3.

At this point stop and check to make sure the pi terms are actually dimensionless (Step 7). We will check using both *FLT* and *MLT* dimensions. Thus,

$$\Pi_1 = \frac{\Delta p_\ell D}{\rho V^2} \doteq \frac{(FL^{-3})(L)}{(FL^{-4}T^2)(LT^{-1})^2} \doteq F^0 L^0 T^0$$

$$\Pi_2 = \frac{\mu}{DV\rho} \doteq \frac{(FL^{-2}T)}{(L)(LT^{-1})(FL^{-4}T^2)} \doteq F^0 L^0 T^0$$

or alternatively,

$$\Pi_1 = \frac{\Delta p_\ell D}{\rho V^2} \doteq \frac{(ML^{-2}T^{-2})(L)}{(ML^{-3})(LT^{-1})^2} \doteq M^0 L^0 T^0$$

$$\Pi_2 = \frac{\mu}{DV\rho} \doteq \frac{(ML^{-1}T^{-1})}{(L)(LT^{-1})(ML^{-3})} \doteq M^0 L^0 T^0$$

Soon we will discover that a standard dimensionless parameter, called the Reynolds number, is defined by $\text{Re} = \rho V D/\mu$. To conform to customary usage (Step 8), we replace Π_2 by its reciprocal. Finally (Step 9), we can express the result of the dimensional analysis as

$$\frac{\Delta p_\ell D}{\rho V^2} = \phi\left(\frac{\rho V D}{\mu}\right)$$

This result indicates that this problem can be studied in terms of these two pi terms, rather than the original five variables we started with.

Dimensional analysis will *not* provide the form of the function ϕ. All that we know about the relationship between Π_1 and Π_2 is

$$\frac{\Delta p_\ell D}{\rho V^2} = \phi\left(\frac{\rho V D}{\mu}\right)$$

The form of the function can only be obtained from a suitable set of experiments or computations that may yield a relationship such as the example shown in the adjacent figure.

To summarize, the steps to be followed in performing a dimensional analysis using the method of repeating variables are as follows:

STEP 1   List all the variables that are involved in the problem.

STEP 2   Express each of the variables in terms of basic dimensions.

STEP 3   Determine the required number of pi terms.

STEP 4   Select a number of repeating variables, where the number required is equal to the number of reference dimensions (usually the same as the number of basic dimensions).

STEP 5   Form a pi term by multiplying one of the nonrepeating variables by the product of repeating variables each raised to an exponent that will make the combination dimensionless.

STEP 6   Repeat Step 5 for each of the remaining nonrepeating variables.

STEP 7   Check all the resulting pi terms to make sure they are dimensionless and independent.

STEP 8   Rearrange the pi terms to simplify the problem or to correspond to customary usage.

STEP 9   Express the final form as a relationship among the pi terms and think about what it means.

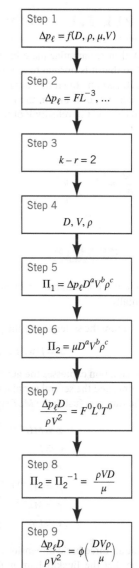

Step 1
$\Delta p_\ell = f(D, \rho, \mu, V)$

Step 2
$\Delta p_\ell = FL^{-3}, \dots$

Step 3
$k - r = 2$

Step 4
D, V, ρ

Step 5
$\Pi_1 = \Delta p_\ell D^a V^b \rho^c$

Step 6
$\Pi_2 = \mu D^a V^b \rho^c$

Step 7
$\dfrac{\Delta p_\ell D}{\rho V^2} = F^0 L^0 T^0$

Step 8
$\Pi_2 = \Pi_2^{-1} = \dfrac{\rho V D}{\mu}$

Step 9
$\dfrac{\Delta p_\ell D}{\rho V^2} = \phi\left(\dfrac{D V \rho}{\mu}\right)$

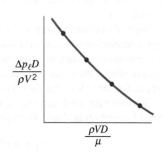

The method of repeating variables can be most easily carried out by following a step-by-step procedure.

EXAMPLE 7.1 | Method of Repeating Variables

Given   A thin rectangular plate having a width w and a height h is located so that it is normal to a moving stream of fluid as shown in **Fig. E7.1**. Assume the force (drag), $\mathcal{D}$, that the fluid exerts on the plate is a function of w and h, the fluid viscosity and density, μ and ρ, respectively, and the velocity V of the fluid approaching the plate.

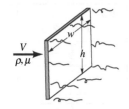

V7.2 Flow past a flat plate

FIGURE E7.1

Find   Determine a suitable set of pi terms to study this problem experimentally.

Solution   From the statement of the problem we can write

$$\mathcal{D} = f(w, h, \mu, \rho, V)$$

where this equation expresses the general functional relationship between the drag and the several variables that will affect it. The dimensions of the variables (using the MLT system) are

$$\mathcal{D} \doteq MLT^{-2}$$
$$w \doteq L$$
$$h \doteq L$$
$$\mu \doteq ML^{-1}T^{-1}$$
$$\rho \doteq ML^{-3}$$
$$V \doteq LT^{-1}$$

We see that all three basic dimensions are required to define the six variables so that the Buckingham pi theorem tells us that three pi terms will be needed (six variables minus three reference dimensions, $k - r = 6 - 3$).

We will next select three repeating variables such as w, V, and ρ. A quick inspection of these three reveals that they are dimensionally independent, since each one contains a basic dimension not included in the others. Note that it would be incorrect to use both w and h as repeating variables since they have the same dimensions.

Starting with the dependent variable, $\mathcal{D}$, the first pi term can be formed by combining $\mathcal{D}$ with the repeating variables such that

$$\Pi_1 = \mathcal{D} w^a V^b \rho^c$$

and in terms of dimensions

$$(MLT^{-2})(L)^a (LT^{-1})^b (ML^{-3})^c \doteq M^0 L^0 T^0$$

Thus, for Π_1 to be dimensionless it follows that

$$\begin{aligned} 1 + c &= 0 & \text{(for } M) \\ 1 + a + b - 3c &= 0 & \text{(for } L) \\ -2 - b &= 0 & \text{(for } T) \end{aligned}$$

and, therefore, $a = -2$, $b = -2$, and $c = -1$. The pi term then becomes

$$\Pi_1 = \frac{\mathcal{D}}{w^2 V^2 \rho}$$

Next the procedure is repeated with the second nonrepeating variable, h, so that

$$\Pi_2 = h w^a V^b \rho^c$$

It follows that

$$(L)(L)^a (LT^{-1})^b (ML^{-3})^c \doteq M^0 L^0 T^0$$

and

$$\begin{aligned} c &= 0 & \text{(for } M) \\ 1 + a + b - 3c &= 0 & \text{(for } L) \\ b &= 0 & \text{(for } T) \end{aligned}$$

so that $a = -1, b = 0, c = 0$, and therefore

$$\Pi_2 = \frac{h}{w}$$

The remaining nonrepeating variable is μ so that

$$\Pi_3 = \mu w^a V^b \rho^c$$

with

$$(ML^{-1}T^{-1})(L)^a (LT^{-1})^b (ML^{-3})^c \doteq M^0 L^0 T^0$$

and, therefore,

$$\begin{aligned} 1 + c &= 0 & \text{(for } M) \\ -1 + a + b - 3c &= 0 & \text{(for } L) \\ -1 - b &= 0 & \text{(for } T) \end{aligned}$$

Solving for the exponents, we obtain $a = -1, b = -1, c = -1$ so that

$$\Pi_3 = \frac{\mu}{wV\rho}$$

Now that we have the three required pi terms we should check to make sure they are dimensionless. To make this check we use F, L, and T, which will also verify the correctness of the original dimensions used for the variables. Thus,

$$\Pi_1 = \frac{\mathcal{D}}{w^2 V^2 \rho} \doteq \frac{(F)}{(L)^2 (LT^{-1})^2 (FL^{-4}T^2)} \doteq F^0 L^0 T^0$$

$$\Pi_2 = \frac{h}{w} \doteq \frac{(L)}{(L)} \doteq F^0 L^0 T^0$$

$$\Pi_3 = \frac{\mu}{wV\rho} \doteq \frac{(FL^{-2}T)}{(L)(LT^{-1})(FL^{-4}T^2)} \doteq F^0 L^0 T^0$$

If these do not check, go back to the original list of variables and make sure you have the correct dimensions for each of the variables and then check the algebra you used to obtain the exponents a, b, and c.

Next, we see that Π_3 is the reciprocal of the Reynolds number, so we invert the fraction. We also change h/w to w/h.

Finally, we can express the results of the dimensional analysis in the form

$$\frac{\mathcal{D}}{w^2 \rho V^2} = \phi \left(\frac{w}{h}, \frac{\rho V w}{\mu} \right) \tag{Ans}$$

The ratio of the plate width to height, w/h, is called the *aspect ratio*, and $\rho V w/\mu$ is the Reynolds number.

Comment Π_2 is simply a ratio of two lengths and might have been intuitively obvious. To proceed, it would be necessary to perform a set of experiments to determine the nature of the function ϕ, as discussed in Section 7.7.

7.4 SOME DIRECTIONS ABOUT DIMENSIONAL ANALYSIS

The preceding section provides a systematic approach for determining pi terms. Other methods, such as formation by inspection, as is discussed in Sec. 7.5, could be used, although we think the method of repeating variables is the easiest for the beginning student to use because, once the significant dimensional variables are known, the procedure becomes algorithmic. In this section we will elaborate on some of the more subtle points that often prove to be puzzling to students.

7.4.1 Selection of Variables

One of the most important and difficult steps in applying dimensional analysis is the selection of the variables. As noted previously, for convenience we will use the term variable to indicate any quantity involved, including dimensional and nondimensional constants. There is no simple procedure whereby the variables can be easily identified. Generally, one must rely on a good understanding of the phenomenon involved and the governing physical laws. If extraneous variables are included, then too many pi terms appear in the final solution, and it may be difficult, time consuming, and expensive to eliminate these experimentally. If important variables are omitted, then an incomplete result will be obtained; and again, this may prove to be costly and difficult to ascertain. It is, therefore, imperative that sufficient time and attention be given to this first step in which the variables are determined.

Most engineering problems involve certain simplifying assumptions that have an influence on the variables to be considered. Usually we wish to keep the problem as simple as possible, perhaps even if some accuracy is sacrificed. A suitable balance between simplicity and accuracy is a desirable goal. How "accurate" the solution must be depends on the objective of the study; that is, we may be only concerned with general trends and, therefore, some variables that are thought to have only a minor influence in the problem may be neglected for simplicity.

For most engineering problems (including areas outside of fluid mechanics), pertinent variables can be classified into three general groups—geometry, material properties, and external effects.

> It is often helpful to classify variables into three groups—geometry, material properties, and external effects.

Geometry The geometric characteristics can usually be described by a series of lengths and angles. In most problems the geometry of the system plays an important role, and a sufficient number of geometric variables must be included to describe the system. These variables can usually be readily identified.

Material Properties Since the response of a system to applied external effects such as forces, pressures, and changes in temperature is dependent on the nature of the materials involved in the system, the material properties that relate the external effects and the responses must be included as variables. For example, for Newtonian fluids the viscosity of the fluid is the property that relates the applied forces to the rates of deformation of the fluid. As the material behavior becomes more complex, such as would be true for non-Newtonian fluids, the determination of material properties becomes difficult, and this class of variables can be troublesome to identify.

External Effects This terminology is used to denote any variable that produces, or tends to produce, a change in the system. For example, in structural mechanics, forces (either concentrated or distributed) applied to a system tend to change its geometry, and such forces would need to be considered as pertinent variables. For fluid mechanics, variables in this class would be related to pressures, velocities, or gravity.

These three general classes of variables are intended as broad categories that should be helpful in identifying variables. It is likely, however, that there will be important variables that do not fit easily into one of the three categories so each problem needs to be carefully analyzed.

Since we wish to keep the number of variables to a minimum, it is important that all variables are independent. For example, if in a given problem we know that the moment of inertia of the area of a circular plate is an important variable, we could list either the moment of inertia or the plate diameter as the pertinent variable. However, it would be unnecessary to include both moment of inertia and diameter, assuming that the diameter enters the problem only through the moment of inertia. In more general terms, if we have a problem in which the variables are

$$f(p, q, r, \ldots, u, v, w, \ldots) = 0 \tag{7.3}$$

and it is known that there is an additional relationship among some of the variables, for example,

$$q = f_1(u, v, w, \ldots) \tag{7.4}$$

then q is not required and can be omitted. Conversely, if it is known that the only way the variables $u, v, w, \ldots$ enter the problem is through the relationship expressed by Eq. 7.4, then the variables $u, v, w, \ldots$ can be replaced by the single variable q, therefore reducing the number of variables.

In summary, the following points should be considered in the selection of variables:

1. Clearly define the problem. What is the main variable of interest (the dependent variable)?

2. Consider the basic laws that govern the phenomenon. Even a crude theory that describes the essential aspects of the system may be helpful.

3. Start the variable selection process by grouping the variables into three broad classes: geometry, material properties, and external effects.

4. Consider other variables that may not fall into one of the above categories. For example, time will be an important variable if any of the variables are time dependent.

5. Be sure to include all quantities that enter the problem even though some of them may be held constant (e.g., the acceleration of gravity, g). For a dimensional analysis it is the dimensions of the quantities that are important—not specific values!

6. Make sure that all variables are independent. Look for relationships among subsets of the variables.

7.4.2 Determination of Reference Dimensions

For any given problem it is obviously desirable to reduce the number of pi terms to a minimum; therefore, we wish to reduce the number of variables to a minimum—that is, we certainly do not want to include extraneous variables. It is also important to know how many reference dimensions are required to describe the variables. As we have seen in the preceding examples, F, L, and T appear to be a convenient set of basic dimensions for characterizing fluid-mechanical quantities. There is, however, really nothing "fundamental" about this set and, as previously noted, M, L, and T would also be suitable. Actually any set of measurable quantities could be used as basic dimensions provided that the selected combination can be used to describe all secondary quantities. However, the use of *FLT* or *MLT* as basic dimensions is the simplest, and these dimensions can be used to describe fluid-mechanical phenomena. Of course, in some problems only one or two of these are required. In addition, we occasionally find that the number of reference dimensions needed to describe all variables is smaller than the number of basic dimensions. This point is illustrated in Example 7.2. Interesting discussions, both practical and philosophical, relative to the concept of basic dimensions can be found in the books by Huntley (Ref. 4) and by Isaacson and Isaacson (Ref. 12).

Typically, in fluid mechanics, the required number of reference dimensions is three, but in some problems only one or two are required.

7.4.3 Uniqueness of Pi Terms

A little reflection on the process used to determine pi terms by the method of repeating variables reveals that the specific pi terms obtained depend on the somewhat arbitrary selection of repeating variables. For example, in the problem of studying the pressure drop in a pipe,

EXAMPLE 7.2 | Determination of Pi Terms

Given An open, cylindrical paint can having a diameter D, shown in **Fig. E7.2**, is filled to a depth h with paint having a specific weight γ. The vertical deflection, δ, of the center of the bottom is a function of D, h, d, γ, and E, where d is the thickness of the bottom and E is the modulus of elasticity of the bottom material.

FIGURE E7.2

Bruce Munson

Find Determine the functional relationship between the vertical deflection, δ, and the independent variables using dimensional analysis.

Solution From the statement of the problem

$$\delta = f(D, h, d, \gamma, E)$$

and the dimensions of the variables are

$$\delta \doteq L$$
$$D \doteq L$$
$$h \doteq L$$
$$d \doteq L$$
$$\gamma \doteq FL^{-3} \doteq ML^{-2}T^{-2}$$
$$E \doteq FL^{-2} \doteq ML^{-1}T^{-2}$$

where the dimensions have been expressed in terms of both the *FLT* and *MLT* systems.

We now apply the pi theorem to determine the required number of pi terms. First, let us use F, L, and T as our system of basic dimensions. There are six variables and two reference dimensions (F and L) required so that four pi terms are needed. For repeating variables, we can select D and γ so that

$$\Pi_1 = \delta D^a \gamma^b$$

$$(L)(L)^a (FL^{-3})^b \doteq F^0 L^0$$

and

$$1 + a - 3b = 0 \quad \text{(for } L\text{)}$$
$$b = 0 \quad \text{(for } F\text{)}$$

Therefore, $a = -1, b = 0$, and

$$\Pi_1 = \frac{\delta}{D}$$

Similarly,

$$\Pi_2 = hD^a \gamma^b$$

and following the same procedure as above, $a = -1, b = 0$ so that

$$\Pi_2 = \frac{h}{D}$$

The remaining two pi terms can be found using the same procedure, with the result

$$\Pi_3 = \frac{d}{D} \quad \Pi_4 = \frac{E}{D\gamma}$$

Thus, this problem can be studied by using the relationship

$$\frac{\delta}{D} = \phi\left(\frac{h}{D}, \frac{d}{D}, \frac{E}{D\gamma}\right) \qquad \text{(Ans)}$$

Comments Let us now solve the same problem using the *MLT* system. Although the number of variables is obviously the same, it would seem that there are three reference dimensions required, rather than two. If this were indeed true, it would certainly be fortuitous, since we would reduce the number of required pi terms from four to three. Does this seem right? How can we reduce the number of required pi terms by simply using the *MLT* system of basic dimensions? The answer is that we cannot, and a closer look at the dimensions of the variables listed above reveals that actually only two reference dimensions, MT^{-2} and L, are required.

This is an example of the situation in which the number of reference dimensions differs from the number of basic dimensions. It does not happen very often and can be detected by looking at the dimensions of the variables (regardless of the systems used) and making sure how many reference dimensions are actually required to describe the variables. Once the number of reference dimensions has been determined, we can proceed as before. Since the number of repeating variables must equal the number of reference dimensions, it follows that two reference dimensions are still required and we could again use D and γ as repeating variables. The pi terms would be determined in the same manner. For example, the pi term containing E would be developed as

$$\Pi_4 = ED^a \gamma^b$$
$$(ML^{-1}T^{-2})(L)^a (ML^{-2}T^{-2})^b \doteq (MT^{-2})^0 L^0$$
$$1 + b = 0 \quad \text{(for } MT^{-2}\text{)}$$
$$-1 + a - 2b = 0 \quad \text{(for } L\text{)}$$

and, therefore, $a = -1, b = -1$ so that

$$\Pi_4 = \frac{E}{D\gamma}$$

which is the same as Π_4 obtained using the *FLT* system. The other pi terms would be the same, and the final result is the same; that is,

$$\frac{\delta}{D} = \phi\left(\frac{h}{D}, \frac{d}{D}, \frac{E}{D\gamma}\right) \qquad \text{(Ans)}$$

This will always be true—you cannot affect the required number of pi terms by using M, L, and T instead of F, L, and T, or vice versa.

A handy "rule of thumb" is to check the number of reference dimensions in both the FLT system and the MLT system. The *smaller* number of reference dimensions is subtracted from the number of variables to determine the number of pi terms.

we selected D, V, and ρ as repeating variables. This led to the formulation of the problem in terms of pi terms as

$$\frac{\Delta p_\ell D}{\rho V^2} = \phi\left(\frac{\rho V D}{\mu}\right) \tag{7.5}$$

What if we had selected D, V, and μ as repeating variables? A quick check will reveal that the pi term involving Δp_ℓ becomes

$$\frac{\Delta p_\ell D^2}{V\mu}$$

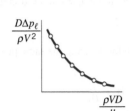

and the second pi term remains the same. Thus, we can express the final result as

$$\frac{\Delta p_\ell D^2}{V\mu} = \phi_1\left(\frac{\rho V D}{\mu}\right) \tag{7.6}$$

Both results are correct, and both would lead to the same final relationship for Δp_ℓ. Note, however, that the functions ϕ and ϕ_1 in Eqs. 7.5 and 7.6 will be different because the dependent pi terms are different for the two relationships. As shown by the adjacent figures, the resulting graph of dimensionless data will be different for the two formulations. However, when extracting the physical variable, Δp_ℓ, from the two results, the values will be the same.

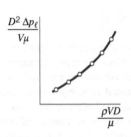

We can conclude from this illustration that there is *not* a unique set of pi terms which arises from a dimensional analysis. However, the required *number* of pi terms is fixed, and once a correct set is determined, all other possible sets can be developed from this set by combinations of products of powers of the original set. Thus, if we have a problem involving, say, three pi terms,

$$\Pi_1 = \phi(\Pi_2, \Pi_3)$$

we could always form a new set from this one by combining the pi terms. For example, we could form a new pi term, Π'_2, by letting

$$\Pi'_2 = \Pi_2^a\, \Pi_3^b$$

Once a correct set of pi terms is obtained, any other set can be obtained by manipulation of the original set.

where a and b are arbitrary exponents. Then the relationship could be expressed as

$$\Pi_1 = \phi_1(\Pi'_2, \Pi_3)$$

or

$$\Pi_1 = \phi_2(\Pi_2, \Pi'_2)$$

All of these would be correct. It should be emphasized, however, that the required number of pi terms cannot be reduced by this manipulation; only the form of the pi terms is altered. By using this technique, we see that the pi terms in Eq. 7.6 could be obtained from those in Eq. 7.5; that is, we multiply Π_1 in Eq. 7.5 by Π_2 so that

$$\left(\frac{\Delta p_\ell D}{\rho V^2}\right)\left(\frac{\rho V D}{\mu}\right) = \frac{\Delta p_\ell D^2}{V\mu}$$

which is the Π_1 of Eq. 7.6.

7.5 DETERMINATION OF PI TERMS BY INSPECTION

The method of repeating variables for forming pi terms has been presented in Section 7.3. This method provides a step-by-step procedure that if executed properly will provide a correct and complete set of pi terms. Although this method is simple and straightforward, it is rather tedious, particularly for problems in which large numbers of variables are involved. Since the only restrictions placed on the pi terms are that they be (1) correct in number, (2) dimensionless, and (3) independent, it is possible to simply form the pi terms by inspection, without resorting to the more formal procedure.

To illustrate this approach, we again consider the pressure drop per unit length along a smooth pipe. Regardless of the technique to be used, the starting point remains the same—determine the variables, which in this case are

$$\Delta p_\ell = f(D, \rho, \mu, V)$$

Next, the dimensions of the variables are listed:

$$\Delta p_\ell \doteq FL^{-3}$$
$$D \doteq L$$
$$\rho \doteq FL^{-4}T^2$$
$$\mu \doteq FL^{-2}T$$
$$V \doteq LT^{-1}$$

and subsequently the number of reference dimensions determined. The application of the pi theorem then tells us how many pi terms are required. In this problem, since there are five variables and three reference dimensions, two pi terms are needed. Thus, the required number of pi terms can be easily obtained. The determination of this number should always be done at the beginning of the analysis.

Once the number of pi terms is known, we can form each pi term by inspection, simply making use of the fact that each pi term must be dimensionless. We will always let Π_1 contain the dependent variable, which in this example is Δp_ℓ. Since this variable has the dimensions FL^{-3}, we need to combine it with other variables so that a nondimensional product will result. One possibility is to first divide Δp_ℓ by ρ so that

> Pi terms can be formed by inspection by simply making use of the fact that each pi term must be dimensionless.

$$\frac{\Delta p_\ell}{\rho} \doteq \frac{(FL^{-3})}{(FL^{-4}T^2)} \doteq \frac{L}{T^2} \qquad \text{(cancels } F\text{)}$$

The dependence on F has been eliminated, but $\Delta p_\ell/\rho$ is obviously not dimensionless. To eliminate the dependence on T, we can divide by V^2 so that

$$\left(\frac{\Delta p_\ell}{\rho}\right)\frac{1}{V^2} \doteq \left(\frac{L}{T^2}\right)\frac{1}{(LT^{-1})^2} \doteq \frac{1}{L} \qquad \text{(cancels } T\text{)}$$

Finally, to make the combination dimensionless we multiply by D so that

$$\left(\frac{\Delta p_\ell}{\rho V^2}\right)D \doteq \left(\frac{1}{L}\right)(L) \doteq L^0 \qquad \text{(cancels } L\text{)}$$

Thus,

$$\Pi_1 = \frac{\Delta p_\ell D}{\rho V^2}$$

Next, we will form the second pi term by selecting the variable that was not used in Π_1, which in this case is μ. We simply combine μ with the other variables to make the combination dimensionless (but do not use Δp_ℓ in Π_2, since we want the dependent variable to appear only in Π_1). For example, divide μ by ρ (to eliminate F), then by V (to eliminate T), and finally by D (to eliminate L). Thus,

$$\Pi_2 = \frac{\mu}{\rho V D} \doteq \frac{(FL^{-2}T)}{(FL^{-4}T^2)(LT^{-1})(L)} \doteq F^0 L^0 T^0$$

and, therefore,

$$\frac{\Delta p_\ell D}{\rho V^2} = \phi\left(\frac{\mu}{\rho V D}\right)$$

which is, of course, the same result we obtained by using the method of repeating variables. Sometimes some of the pi terms are obvious, such as when two variables have the same dimensions. For flow in a circular pipe, both the pipe length L and the pipe diameter D appear and their ratio, L/D, is an obvious pi term.

An additional concern, when one is forming pi terms by inspection, is to make certain that they are all independent. In the pipe flow example, Π_2 contains μ, which does not appear in Π_1, and therefore these two pi terms are obviously independent. In a more general case a pi term would not be independent of the others in a given problem if it can be formed by some combination of the others. For example, if Π_2 can be formed by a combination of say Π_3, Π_4, and Π_5 such as

$$\Pi_2 = \frac{\Pi_3^2\,\Pi_4}{\Pi_5}$$

then Π_2 is not an independent pi term. We can ensure that each pi term is independent of those preceding it by incorporating a new variable in each pi term.

Although forming pi terms by inspection is essentially equivalent to the repeating variable method, it is less structured. With a little practice the pi terms can be readily formed by inspection, and this method offers an alternative to more formal procedures.

7.6　COMMON DIMENSIONLESS GROUPS IN FLUID MECHANICS

Above, in Section 7.4, we learned two important facts: (1) The number of dimensional variables necessary to describe any problem in fluid mechanics is limited, and (2) the number of different dimensionless groups (pi terms) that could be formed from the dimensional variables is unlimited. If we consider only the field of fluid mechanics, including both compressible and incompressible flows, a rather complete list of geometric and dimensional variables emerges: lengths ($\ell_1, \ell_2, \ell_3 \ldots$), angles ($\alpha_1, \alpha_2, \alpha_3 \ldots$), surface roughness ($\varepsilon$), density ($\rho$), viscosity ($\mu$), surface tension ($\sigma$), modulus of elasticity (E_v), speed of sound (c), gravity (g), fluid velocity (V), boundary velocity (U), frequency of oscillation or boundary angular velocity (ω), pressure (p or Δp), wall shear stress (τ_w), head loss (h_L), lift ($\mathscr{L}$), and drag ($\mathscr{D}$). Of course not all of these variables appear in every problem.

Rather than perform a dimensional analysis for every new problem, we could do the analysis once and for all using this list of variables to create a set of dimensionless groups that could be used for (almost) any fluid mechanics problem. If we were to do so, we would end up with a complete set of pi terms, but, for the most part, we would have different pi terms from those that are commonly used in fluid mechanics. **Table 7.1** shows **common dimensionless groups** that are used in fluid mechanics. With a few exceptions, each dimensionless group is either referred to as a "number" or a "coefficient." The numbers are named in honor of people from the history of fluid mechanics (see Section 1.10) and usually represent independent variables such as velocity or viscosity. The "coefficients" generally represent dependent variables such as pressure or drag.

It is usually possible to provide a physical interpretation to the dimensionless groups, which can be helpful in assessing their influence in a particular application. For example, the Froude number is an index of the ratio of the mass times acceleration of a fluid particle (sometimes called the "inertia force") to the force of gravity (weight). This can be demonstrated by considering a fluid particle moving along a streamline (**Fig. 7.3**). The magnitude of the component of inertia force F_1 along the streamline can be expressed as $F_1 = a_s m$, where a_s is the magnitude of the acceleration along the streamline for a particle having a mass m. From our study of particle motion along a curved path (see Section 3.1) we know that

$$a_s = \frac{dV_s}{dt} = V_s \frac{dV_s}{ds}$$

> A useful physical interpretation can often be given to dimensionless groups.

where s is measured along the streamline. If we write the velocity, V_s and length, s, in dimensionless form, that is,

$$V_s^* = \frac{V_s}{V} \qquad s^* = \frac{s}{\ell}$$

where V and ℓ represent some characteristic velocity and length, respectively, then

$$a_s = \frac{V^2}{\ell} V_s^* \frac{dV_s^*}{ds^*}$$

TABLE 7.1 Some Common Variables and Dimensionless Groups in Fluid Mechanics

Variables: Acceleration of gravity, g; Bulk modulus, E_v; Characteristic length, ℓ; Density, ρ; Frequency of oscillating flow or rotation of boundary, ω; Pressure, p (or Δp); Speed of sound, c; Surface tension, σ; Velocity, V; Viscosity, μ; Boundary velocity, U; Surface roughness, ε; Wall shear stress, τ_w; head loss, h_L; Drag, $\mathcal{D}$; Lift, $\mathcal{L}$.

Dimensionless Group	Name	Interpretation (Index of Force Ratio Indicated)	Types of Applications
$\dfrac{\rho V \ell}{\mu}$	Reynolds number, Re	$\dfrac{\text{inertia force}}{\text{viscous force}}$	Generally of importance in all types of fluid dynamics problems
$\dfrac{V}{\sqrt{g\ell}}$	Froude number, Fr	$\dfrac{\text{inertia force}}{\text{gravitational force}}$	Flow with a free surface
$\dfrac{p}{\rho V^2}$	Euler number, Eu	$\dfrac{\text{pressure force}}{\text{inertia force}}$	Problems in which pressure, or pressure differences, are of interest
$\dfrac{\rho V^2}{E_v}$	Cauchy number,[a] Ca	$\dfrac{\text{inertia force}}{\text{compressibility force}}$	Flows in which the compressibility of the fluid is important
$\dfrac{V}{c}$	Mach number,[a] Ma	$\dfrac{\text{inertia force}}{\text{compressibility force}}$	Flows in which the compressibility of the fluid is important
$\dfrac{\omega \ell}{V}$	Strouhal number, St	$\dfrac{\text{inertia (local) force}}{\text{inertia (convective) force}}$	Unsteady flow with a characteristic frequency of oscillation or a rotating boundary
$\dfrac{\rho V^2 \ell}{\sigma}$	Weber number, We	$\dfrac{\text{inertia force}}{\text{surface tension force}}$	Problems in which surface tension is important
$\dfrac{V}{U}$	Velocity ratio, Φ	$\dfrac{\text{fluid velocity}}{\text{boundary velocity}}$	Flows with a moving boundary
$\dfrac{\varepsilon}{\ell}$	Relative roughness	$\dfrac{\text{surface roughness}}{\text{characteristic length}}$	Turbulent flow over a rough surface
$\dfrac{\ell_1}{\ell}$	Length ratio	$\dfrac{\text{length}}{\text{reference length}}$	Flow with two (or more) relevant dimensions
α	Angle of attack	Angle between approach velocity and reference axis	Flow over immersed bodies
$\dfrac{h_L}{V^2/2g}$	Loss coefficient, K	$\dfrac{\text{energy loss}}{\text{kinetic energy}}$	Internal flow, especially through valves and fittings
$\dfrac{\Delta p}{\frac{1}{2}\rho V^2}$	Pressure coefficient, C_p	$\dfrac{\text{pressure}}{\text{dynamic pressure}}$	Euler number with ½ in honor of Bernoulli
$\dfrac{\tau_w}{\frac{1}{2}\rho V^2}$	Friction coefficient, C_f	$\dfrac{\text{shear force}}{\text{inertia force}}$	Laminar or turbulent internal and external flow with viscous stress
$\dfrac{\mathcal{D}}{\frac{1}{2}\rho V^2 \ell^2}$	Drag coefficient, C_D	$\dfrac{\text{drag force}}{\text{inertia force}}$	Flow over immersed bodies
$\dfrac{\mathcal{L}}{\frac{1}{2}\rho V^2 \ell^2}$	Lift coefficient, C_L	$\dfrac{\text{lift force}}{\text{inertia force}}$	Flow over airfoils, wings, and turbomachine blades

[a]The Cauchy number and the Mach number are related, and either can be used as an index of the relative effects of inertia and compressibility. See accompanying discussion.

Special names along with physical interpretations are given to the most common dimensionless groups.

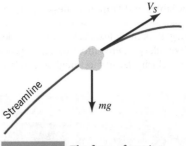

FIGURE 7.3 The force of gravity acting on a fluid particle moving along a streamline.

and

$$F_I = \frac{V^2}{\ell} V_s^* \frac{dV_s^*}{ds^*} m$$

The magnitude of the weight of the particle, F_G, is $F_G = gm$, so the ratio of the inertia to the gravitational force is

$$\frac{F_I}{F_G} = \frac{V^2}{g\ell} V_s^* \frac{dV_s^*}{ds^*}$$

Thus, the force ratio F_I/F_G is proportional to $V^2/g\ell$, and the square root of this ratio, $V/\sqrt{g\ell}$, is called the *Froude number*. We see that a physical interpretation of the Froude number is that it is a measure of, or an index of, the relative importance of inertial forces acting on fluid particles to the weight of the particle. Note that the Froude number is not really *equal* to this force ratio, but is simply some type of average measure of the relative influence of these two forces. In a problem in which gravity (or weight) is not important, the Froude number would not appear as an important pi term. A similar interpretation in terms of indices of force ratios can be given to the other dimensionless groups, as indicated in Table 7.1, and a further discussion of the basis for this type of interpretation is given in the last section in this chapter. These force ratios can aid in determining the types of application or problem in which a dimensionless group may arise and is briefly noted in the last column of Table 7.1. Thus, for many instances, Table 7.1 can serve as a guide for choosing relevant dimensionless groups for a problem. Some additional details about these important dimensionless groups are provided next.

V7.3 Reynolds number

Reynolds Number

The Reynolds number is undoubtedly the most famous dimensionless parameter in fluid mechanics. It is named in honor of Osborne Reynolds (1842–1912), a British engineer who first demonstrated that this combination of variables could be used as a criterion to distinguish between laminar and turbulent flow. In most fluid flow problems there will be a characteristic length, ℓ, and a velocity, V, as well as the fluid properties density, ρ, and viscosity, μ, which are relevant variables in the problem. Thus, with these variables the Reynolds number

$$\mathrm{Re} = \frac{\rho V \ell}{\mu}$$

arises naturally from the dimensional analysis. The Reynolds number is a measure of the ratio of the inertia force on an element of fluid to the viscous force on an element. When these two types of forces are important in a given problem, the Reynolds number will play an important role. However, if the Reynolds number is very small ($\mathrm{Re} \ll 1$), this is an indication that the viscous forces are dominant in the problem, and it may be possible to neglect the inertial effects; that is, the density of the fluid will not be an important variable. Flows at very small Reynolds numbers are commonly referred to as "creeping flows" as discussed in Section 6.10. For large Reynolds number flows, viscous effects are small relative to inertial effects and for these cases one might expect that it may be possible to neglect the effect of viscosity and consider the problem as one involving a "nonviscous" fluid, as considered in detail in Sections 6.4 through 6.7. This is not always the case, however. At high Reynolds numbers, viscous effects are confined to a thin layer near a bounding surface. For accelerating flow, the boundary layer does have a small effect on the flow, and most of the flow can be assumed inviscid; but for decelerating flows the boundary layer can separate from the surface with profound effects, as shown by the adjacent figures.

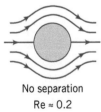

No separation
Re ≈ 0.2

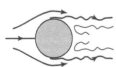

Laminar boundary layer, wide turbulent wake
Re ≈ 20,000

A high Reynolds number also warns us to expect turbulence, which is dominated by the inertial effects of the disorderly motion.

Froude Number   The Froude number

$$\text{Fr} = \frac{V}{\sqrt{g\ell}}$$

is distinguished from the other dimensionless groups in Table 7.1 in that it contains the acceleration of gravity, g. The acceleration of gravity becomes an important variable in a fluid dynamics problem in which the fluid weight is an important force. As discussed, the Froude number is a measure of the ratio of the inertia force on an element of fluid to the weight of the element. It will generally be important in problems involving flows with free surfaces since gravity principally affects this type of flow. Typical problems would include the study of the flow of water around ships (with the resulting wave action) or flow through rivers or open conduits. The Froude number is named in honor of William Froude (1810–1879), a British civil engineer, mathematician, and naval architect who pioneered the use of towing tanks for the study of ship design. It is to be noted that the Froude number is sometimes defined as the square of the Froude number listed in Table 7.1.

V7.4 Froude number

Euler Number   The Euler number

$$\text{Eu} = \frac{p}{\rho V^2}$$

can be interpreted as a measure of the ratio of pressure forces to inertial forces, where p is some characteristic pressure in the flow field. Very often the Euler number is written in terms of a pressure difference, Δp, so that $\text{Eu} = \Delta p/\rho V^2$. Also, this combination expressed as $\Delta p/\frac{1}{2}\rho V^2$ is called the *pressure coefficient*. Some form of the Euler number would normally be used in problems in which pressure or the pressure difference between two points is an important variable. The Euler number is named in honor of Leonhard Euler (1707–1783), a famous Swiss mathematician who pioneered work on many phenomena in science, engineering, and mathematics, including the relationship between pressure and flow. For problems in which cavitation is of concern, the dimensionless group $(p_r - p_v)/\frac{1}{2}\rho V^2$ is commonly used, where p_v is the vapor pressure and p_r is some reference pressure. Although this dimensionless group has the same form as the pressure coefficient, it is generally referred to as the *cavitation number*.

Cauchy Number and Mach Number   The Cauchy number

$$\text{Ca} = \frac{\rho V^2}{E_v}$$

and the Mach number

$$\text{Ma} = \frac{V}{c}$$

are important dimensionless groups in problems in which fluid compressibility is a significant factor. Since the speed of sound, c, in a fluid is equal to $c = \sqrt{E_v/\rho}$ (see Section 1.7.3), it follows that

$$\text{Ma} = V\sqrt{\frac{\rho}{E_v}}$$

and the square of the Mach number

$$\text{Ma}^2 = \frac{\rho V^2}{E_v} = \text{Ca}$$

> The Mach number is a commonly used dimensionless parameter in compressible flow problems.

is equal to the Cauchy number. Thus, either number (but not both) may be used in problems in which fluid compressibility is important. Both numbers can be interpreted as representing an index of the ratio of inertial forces to forces resisting compression. When the Mach number is relatively small (say, less than 0.3), the inertial forces induced by the fluid motion are not sufficiently large to cause a significant change in the fluid density, and in this case the compressibility of the fluid can be neglected. The Mach number is the more commonly used parameter in compressible flow problems, particularly in the fields of gas dynamics and aerodynamics. The Cauchy number is named in honor of Augustin Louis de Cauchy (1789–1857), a French engineer, mathematician, and hydrodynamicist. The Mach number is named in honor of Ernst Mach (1838–1916), an Austrian physicist and philosopher.

Strouhal Number   The Strouhal number

$$\text{St} = \frac{\omega \ell}{V}$$

V7.5 Strouhal number

is a dimensionless parameter that is likely to be important in unsteady, oscillating flow problems in which the frequency of the oscillation is ω. It represents a measure of the ratio of inertial forces due to the unsteadiness of the flow (local acceleration) to the inertial forces due to changes in velocity from point to point in the flow field (convective acceleration). This type of unsteady flow may develop when a fluid flows past a solid body (such as a wire or cable) placed in the moving stream. For example, in a certain Reynolds number range, a periodic flow will develop downstream from a cylinder placed in a moving fluid due to a regular pattern of vortices that are shed from the body (see Fig. 9.23). This system of vortices, called a *Kármán vortex street* [named after Theodor von Kármán (1881–1963), a famous Hungarian-American fluid mechanician], creates an oscillating flow at a discrete frequency, ω, such that the Strouhal number can be closely correlated with the Reynolds number. When the frequency is in the audible range, a sound can be heard and the bodies appear to "sing." In fact, the Strouhal number is named in honor of Vincenz Strouhal (1850–1922), who used this parameter in his study of "singing wires." The most dramatic evidence of this phenomenon occurred in 1940 with the collapse of the Tacoma Narrows Bridge. The shedding frequency of the vortices coincided with the natural frequency of the bridge, thereby setting up a resonant condition that eventually led to the collapse of the bridge.

There are, of course, other types of oscillating flows. For example, blood flow in arteries is periodic and can be analyzed by breaking up the periodic motion into a series of harmonic components (Fourier series analysis), with each component having a frequency that is a multiple of the fundamental frequency, ω (the pulse rate). Rather than use the Strouhal number in this type of problem, a dimensionless group formed by the product of St and Re is used; that is

$$\text{St} \times \text{Re} = \frac{\rho \omega \ell^2}{\mu}$$

The square root of this dimensionless group is often referred to as the *frequency parameter*.

Something like the Strouhal number (actually, its reciprocal) appears in the field of fluid machinery, where ω represents the rotational speed of a rotor.

Weber Number   The Weber number

$$\text{We} = \frac{\rho V^2 \ell}{\sigma}$$

may be important in problems in which there is an interface between two fluids. In this situation the surface tension may play an important role in the phenomenon of interest. The Weber number can be thought of as an index of the inertial force to the surface tension force acting on a fluid element. Common examples of problems in which this parameter may be important include the flow of thin films of liquid, or the formation of droplets or bubbles. Clearly, not all problems involving flows with an interface will require the inclusion of surface tension. The flow of water in a river is not affected significantly by surface tension, since inertial and gravitational effects are dominant ($\text{We} \ll 1$). However, as discussed in a later section, for river models (which may have small depths) caution is required so that surface tension does not become important in the model, whereas it is not important in the actual river. The Weber number is named after Moritz Weber (1871–1951), a German professor of naval mechanics who was instrumental in formalizing the general use of common dimensionless groups as a basis for similitude studies.

The Coefficients   The remainder of the dimensionless groups in Table 7.1, most of which are "coefficients" rather than "numbers," generally represent dependent variables. Some additional information about these groups will be discussed in Section 7.9 and their significance will be apparent in Chapter 8 and Chapter 9. Several of the groups in Table 7.1 will reappear (often in disguise) in Chapter 12 on turbomachinery.

We close this discussion of common dimensionless groups with two interesting observations. First, it is usually possible to interpret the groups as ratios of *energies* as well as ratios of *forces*.

The Reynolds number, for example, can be shown to represent the ratio of kinetic energy to the dissipation of energy associated with viscosity. Second, the list of dimensionless groups in Table 7.1 represents only a small fraction of the common dimensionless groups (if one is willing to stretch the meaning of "common" just a bit). *The Land Chart of Dimensionless Numbers* (Ref. 16) lists 154 dimensionless groups, most of which are named for a prominent engineer or scientist.

The Wide World of Fluids

Slip at the micro scale

A goal in chemical and biological analyses is to miniaturize the experiment, which has many advantages including reduction in sample size. In recent years, there has been significant work on integrating these tests on a single microchip to form the "lab-on-a-chip" system. These devices are on the millimeter scale with complex passages for fluid flow on the micron scale (or smaller). While there are advantages to miniaturization, care must be taken in moving to smaller and smaller flow regimes,

as you will eventually bump into the continuum assumption. To characterize this situation, a dimensionless number termed the Knudsen number, $Kn = \lambda/\ell$, is commonly employed. Here λ is the mean free path, and ℓ is the characteristic length of the system. If Kn is smaller than 0.01, then the flow can be described by the Navier–Stokes equations with no-slip at the walls. For $0.01 < Kn < 0.3$, the same equations can be used, but there can be slip between the fluid and the wall so the boundary conditions need to be adjusted. For $Kn > 10$, the continuum assumption breaks down and the Navier–Stokes equations are no longer valid.

7.7 CORRELATION OF EXPERIMENTAL DATA

One of the most important uses of dimensional analysis is as an aid in the efficient handling, interpretation, and correlation of experimental data. Since the field of fluid mechanics relies heavily on empirical data, it is not surprising that dimensional analysis is such an important tool in this field. As noted previously, a dimensional analysis cannot provide a complete answer to any given problem, since the analysis only provides the dimensionless groups describing the phenomenon, and not the specific relationship among the groups. To determine this relationship, suitable experimental data must be obtained. The degree of difficulty involved in this process depends on the number of pi terms, and the nature of the experiments (How hard is it to obtain the measurements?). The simplest problems are obviously those involving the fewest pi terms, and the following sections indicate how the complexity of the analysis increases with the increasing number of pi terms.

7.7.1 Problems with One Pi Term

Application of the pi theorem indicates that if the number of variables minus the number of reference dimensions is equal to unity, then only *one* pi term is required to describe the phenomenon. The functional relationship that must exist for one pi term is

$$\Pi_1 = C$$

where C is a constant. This is one situation in which a dimensional analysis reveals the specific form of the relationship and, as is illustrated by the following example, shows how the individual variables are related. The value of the constant, however, must still be determined by experiment, analysis, or computation.

> If only one pi term is involved in a problem, it must be equal to a constant.

EXAMPLE 7.3 | Flow with Only One Pi Term

Given As shown in **Fig. E7.3**, assume that the drag, $\mathcal{D}$, acting on a spherical particle that falls very slowly through a viscous fluid, is a function of the particle diameter, D, the particle velocity, V, and the fluid viscosity, μ.

Find Determine, with the aid of dimensional analysis, how the drag depends on the particle velocity.

Solution From the information given, it follows that

$$\mathcal{D} = f(D, V, \mu)$$

and the dimensions of the variables are

$$\mathcal{D} \doteq F$$
$$D \doteq L$$
$$V \doteq LT^{-1}$$
$$\mu \doteq FL^{-2}T$$

We see that there are four variables and three reference dimensions (F, L, and T) required to describe the variables. Thus, according to the pi

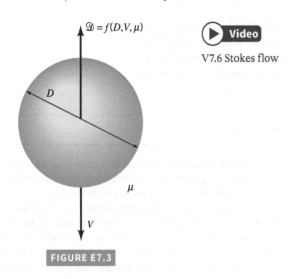

$$\mathscr{D} = f(D, V, \mu)$$

▶ **Video**

V7.6 Stokes flow

FIGURE E7.3

theorem, one pi term is required. This pi term can be easily formed by inspection and can be expressed as

$$\Pi_1 = \frac{\mathscr{D}}{\mu V D}$$

Because there is only one pi term, it follows that

$$\frac{\mathscr{D}}{\mu V D} = C$$

where C is a constant. Thus,

$$\mathscr{D} = C \mu V D$$

Thus, for a given particle and fluid, the drag varies directly with the velocity so that

$$\mathscr{D} \propto V \qquad\qquad (Ans)$$

Comments Actually, the dimensional analysis reveals that the drag not only varies directly with the velocity, but it also varies directly with the particle diameter and the fluid viscosity. We could not, however, predict the value of the drag, since the

constant, C, is unknown. An experiment would have to be performed in which the drag and the corresponding velocity are measured for a given particle and fluid. Although in principle we would only have to run a single test, we would certainly want to repeat it several times to obtain a reliable value for C. It should be emphasized that once the value of C is determined it is not necessary to run similar tests by using different spherical particles and fluids; that is, C is a universal constant as long as the drag is a function only of particle diameter, velocity, and fluid viscosity.

An approximate solution to this problem can also be obtained theoretically, from which it is found that $C = 3\pi$ so that

$$\mathscr{D} = 3\pi\mu V D$$

This equation is commonly called *Stokes law* and is used in the study of the settling of particles. Our experiments would reveal that this result is only valid for small Reynolds numbers ($\rho V D/\mu \ll 1$). This follows, since in the original list of variables, we have neglected inertial effects (fluid density is not included as a variable). The inclusion of an additional variable would lead to another pi term so that there would be two pi terms rather than one.

Consider a free-body diagram of a sphere in Stokes flow; there would be a buoyant force in the same direction as the drag in Fig. E7.3, as well as a weight force in the opposite direction. As shown above, the drag force is proportional to the product of the diameter and fall velocity, $\mathscr{D} \propto V D$. The weight and buoyant force are proportional to the diameter cubed, W and $F_B \propto D^3$. Given equilibrium conditions, the force balance can be written as

$$\mathscr{D} = W - F_B$$

Based on the scaling laws for these terms, it follows that $V D \propto D^3$. Hence, the fall velocity will be proportional to the square of the diameter, $V \propto D^2$. Therefore, for two spheres, one having twice the diameter of the other, and falling through the same fluid, the sphere with the larger diameter will fall four times faster (see Video V7.7).

This problem demonstrates that dimensional analysis is helpful in theoretical studies as well as experimental studies.

7.7.2 Problems with Two or More Pi Terms

If a given phenomenon can be described with two pi terms such that

$$\Pi_1 = \phi(\Pi_2)$$

For problems involving only two pi terms, results of an experiment can be conveniently presented in a simple graph.

the functional relationship among the variables can then be determined by varying Π_2 and measuring the corresponding values of Π_1. For this case the results can be conveniently presented in graphical form by plotting Π_1 versus Π_2 as is illustrated in **Fig. 7.4**. It should be emphasized that the curve shown in Fig. 7.4 would be a "universal" one for the particular phenomenon studied. This means that if the variables and the resulting dimensional analysis are correct, then there is

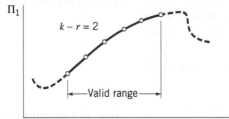

FIGURE 7.4 The graphical presentation of data for problems involving two pi terms, with an illustration of the potential danger of extrapolation of data.

only a single relationship between Π_1 and Π_2, as illustrated in Fig. 7.4. However, since this is an empirical relationship, we can only say that it is valid over the range of Π_2 covered by the experiments. It would be unwise to extrapolate beyond this range, since as illustrated with the dashed lines in the figure, the nature of the phenomenon could dramatically change as the range of Π_2 is extended. In addition to presenting the data graphically, it may be possible (and desirable) to obtain an empirical equation relating Π_1 and Π_2 by using a standard curve-fitting technique. This equation is the function, ϕ, covering the valid range of Π_2.

EXAMPLE 7.4 | Dimensionless Correlation of Experimental Data

Given The relationship between the pressure drop per unit length along a smooth-walled, horizontal pipe and the variables that affect the pressure drop is to be determined experimentally. In the laboratory the pressure drop was measured over a 1.5 m length of smooth-walled pipe having an inside diameter of 1.25 cm The fluid used was water at 15 °C ($\mu = 112 \times 10^{-5}$ Pa · s, $\rho = 1000$ kg/m^3). Tests were run in which the velocity was varied and the corresponding pressure drop was measured. The results of these tests are shown in the given table:

Velocity (m/s)	Pressure drop for 1.5 m length (kPa)
0.36	0.29
0.59	0.75
0.89	1.48
1.78	5.07
3.39	15.75
5.15	32.60
7.1	57.45
8.75	82.83

Find Make use of these data to obtain a general relationship between the pressure drop per unit length and the other variables.

Solution The first step is to perform a dimensional analysis during the planning stage *before* the experiments are actually run. As was discussed in Section 7.3, we will assume that the pressure drop per unit length, Δp_ℓ, is a function of the pipe diameter, D, fluid density, ρ, fluid viscosity, μ, and the velocity, V. Thus,

$$\Delta p_\ell = f(D, \rho, \mu, V)$$

and application of the pi theorem yields two pi terms

$$\Pi_1 = \frac{D \,\Delta p_\ell}{\rho V^2} \quad \text{and} \quad \Pi_2 = \frac{\rho V D}{\mu}$$

Hence,

$$\frac{D \,\Delta p_\ell}{\rho V^2} = \phi\left(\frac{\rho V D}{\mu}\right)$$

To determine the form of the relationship, we need to vary the Reynolds number, Re = $\rho V D/\mu$, and to measure the corresponding values of $D \,\Delta p_\ell/\rho V^2$. The Reynolds number could be varied by changing any one of the variables, ρ, V, D, or μ, or any combination of them. However, the simplest way to do this is to vary the velocity, since this will allow us to use the same fluid and pipe. Based on the data given, values for the two pi terms can be computed, with the result:

$D \,\Delta p_\ell/\rho V^2$	$\rho V D/\mu$
0.0195	4.01×10^3
0.0175	6.68×10^3
0.0155	9.97×10^3
0.0132	2.00×10^4
0.0113	3.81×10^4
0.0101	5.80×10^4
0.00939	8.00×10^4
0.00893	9.85×10^4

These are dimensionless groups so that their values are independent of the system of units used so long as a consistent system is used. For example, if the velocity is in m/s, then the diameter should be in feet, not inches or meters. Note that since the Reynolds numbers are all greater than 2100, the flow in the pipe is turbulent (see Section 8.1.1).

A plot of these two pi terms can now be made with the results shown in **Fig. E7.4a**. The correlation appears to be quite good, and if it was not, this would suggest that either we had large experimental measurement errors or that we had perhaps omitted

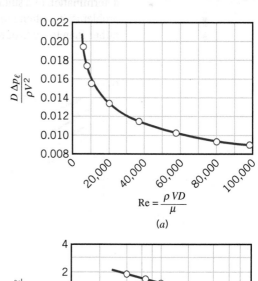

(a)

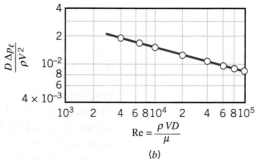

(b)

FIGURE E7.4

an important variable. The curve shown in Fig. E7.4a represents the general relationship between the pressure drop and the other factors in the range of Reynolds numbers between 4.01×10^3 and 9.85×10^4. Thus, for this range of Reynolds numbers it is *not* necessary to repeat the tests for other pipe sizes or other fluids provided the assumed independent variables (D, ρ, μ, V) are the only important ones.

Since the relationship between Π_1 and Π_2 is nonlinear, it is not immediately obvious what form of empirical equation might be used to describe the relationship. If, however, the same data are plotted on a logarithmic scale, as is shown in **Fig. E7.4b**, the data form a straight line, suggesting that a suitable equation is of the form $\Pi_1 = A \Pi_2^n$ where A and n are empirical constants to be determined from the data by using a suitable curve-fitting technique, such as a nonlinear regression program. For the data given in this example, a good fit of the data is obtained with the equation

$$\Pi_1 = 0.150\ \Pi_2^{-0.25} \qquad\qquad (Ans)$$

Comment In 1911, H. Blasius (1883–1970), a German fluid mechanician, established a similar empirical equation that is used widely for predicting the pressure drop in smooth pipes in the range $4 \times 10^3 < \mathrm{Re} < 10^5$ (Ref. 17). This equation can be expressed in the form

$$\frac{D\,\Delta p_\ell}{\rho V^2} = 0.1582 \left(\frac{\rho V D}{\mu}\right)^{-1/4}$$

The so-called Blasius formula is based on numerous experimental results of the type used in this example. Flow in pipes is discussed in more detail in the next chapter, where it is shown how pipe roughness (which introduces another variable) may affect the results given in this example (which is for smooth-walled pipes).

It is not really necessary to do a full dimensional analysis for this problem; we can use common dimensionless groups from Table 7.1. We would expect a Reynolds number and an Euler number to describe the problem. Since we are dealing with pressure drop per unit length, the pipe length ℓ is implicit in Δp_ℓ. Then $\Pi_1 = (D/\ell) \times \mathrm{Eu}$.

As the number of required pi terms increases, it becomes more difficult to display the results in a convenient graphical form and to determine a specific empirical equation that describes the phenomenon. For problems involving three pi terms

$$\Pi_1 = \phi(\Pi_2, \Pi_3)$$

it is still possible to show data correlations on simple graphs by plotting families of curves as illustrated in **Fig. 7.5**. This is an informative and useful way of representing the data in a general way. It may also be possible to determine a suitable empirical equation relating the three pi terms. However, as the number of pi terms continues to increase, corresponding to an increase in the general complexity of the problem of interest, both the graphical presentation and the determination of a suitable empirical equation become intractable. For these more complicated problems, it is often more feasible to use models to predict specific characteristics of the system rather than to try to develop general correlations.

> For problems involving more than two or three pi terms, it is often necessary to use a model to predict specific characteristics.

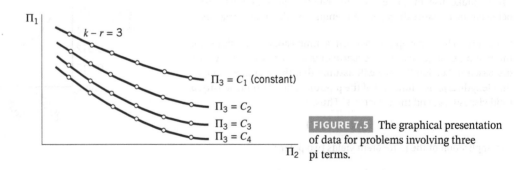

FIGURE 7.5 The graphical presentation of data for problems involving three pi terms.

7.8 MODELING AND SIMILITUDE

Physical models are widely used in fluid mechanics. Major engineering projects involving structures, aircraft, ships, rivers, harbors, dams, air and water pollution, and so on, frequently involve the use of models. Although the term *model* is used in many different contexts, the "engineering model" generally conforms to the following definition. *A* **model** *is a representation of a physical system that may be used to predict the behavior of the system in some desired respect.* The full-sized physical system for which the predictions are to be made is called the

prototype. Although *mathematical* or *computer* models may also conform to this definition, our interest will be in physical models; that is, models that resemble the prototype but are generally of a different size, may involve different fluids, and often operate under different conditions (pressures, velocities, etc.). As shown in the adjacent figure, usually a model is smaller than the prototype. Therefore, it is more easily handled in the laboratory and less expensive to construct and operate than a large prototype. (It should be noted that variables or pi terms without a subscript will refer to the prototype, whereas the subscript m will be used to designate the model variables or pi terms.) Occasionally, if the prototype is very small, it may be advantageous to have a model that is larger than the prototype so that it can be more easily studied. For example, large models have been used to study the motion of red blood cells, which are approximately 8 μm in diameter. With the successful development of a valid model, it is possible to predict the behavior of the prototype under a certain set of conditions. We may also wish to examine a priori the effect of possible design changes that are proposed for a hydraulic structure or fluid flow system. There is, of course, an inherent danger in the use of models in that predictions can be made that are in error and the error not detected until the prototype is found not to perform as predicted. It is, therefore, imperative that the model be properly designed and tested and that the results be interpreted correctly. In the following sections we will develop the procedures for designing models so that the model and prototype will behave in a similar fashion.

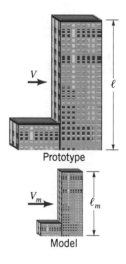

Prototype

Model

Video

V7.7 Model airplane

7.8.1 Theory of Models

The theory of models can be readily developed by using the principles of dimensional analysis. It has been shown that any given problem can be described in terms of a set of pi terms as

$$\Pi_1 = \phi(\Pi_2, \Pi_3, \ldots, \Pi_n) \tag{7.7}$$

In formulating this relationship, only a knowledge of the general nature of the physical phenomenon, and the variables involved, is required. Specific values for variables (size of components, fluid properties, and so on) are not needed to perform the dimensional analysis. Thus, Eq. 7.7 applies to any system that is governed by the same variables. If Eq. 7.7 describes the behavior of a particular prototype, a similar relationship can be written for a model of this prototype; that is,

$$\Pi_{1m} = \phi(\Pi_{2m}, \Pi_{3m}, \ldots, \Pi_{nm}) \tag{7.8}$$

where the form of the function will be the same as long as the same phenomenon is involved in both the prototype and the model. Variables, or pi terms, without a subscript will refer to the prototype, whereas the subscript m will be used to designate the model variables or pi terms.

The pi terms can be developed so that Π_1 contains the variable that is to be predicted from observations made on the model. Therefore, if the model is designed and operated under the following conditions,

$$\Pi_{2m} = \Pi_2$$
$$\Pi_{3m} = \Pi_3$$
$$\vdots \tag{7.9}$$
$$\Pi_{nm} = \Pi_n$$

then with the presumption that the form of ϕ is the same for model and prototype, it follows that

$$\Pi_1 = \Pi_{1m} \tag{7.10}$$

Equation 7.10 is the desired **prediction equation** and indicates that the measured value of Π_{1m} obtained with the model will be equal to the corresponding Π_1 for the prototype as long as the other pi terms are equal. The conditions specified by Eqs. 7.9 provide the **model design conditions,** also called **similarity requirements** or **modeling laws.**

The similarity requirements for a model can be readily obtained with the aid of dimensional analysis.

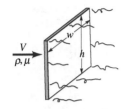

As an example of the procedure, consider the problem of determining the drag, $\mathscr{D}$, on a thin rectangular plate ($w \times h$ in size) placed normal to a fluid with velocity, V, as shown in the adjacent figure. The dimensional analysis of this problem was performed in Example 7.1, where it was assumed that

$$\mathscr{D} = f(w, h, \mu, \rho, V)$$

Application of the pi theorem yielded

$$\frac{\mathscr{D}}{w^2 \rho V^2} = \phi\left(\frac{w}{h}, \frac{\rho V w}{\mu}\right) \tag{7.11}$$

(Do you recognize a drag coefficient, a length ratio, and a Reynolds number?)

We are now concerned with designing a model that could be used to predict the drag on a certain prototype (which presumably has a different size than the model). Since the relationship expressed by Eq. 7.11 applies to both prototype and model, Eq. 7.11 is assumed to govern the prototype, with a similar relationship

$$\frac{\mathscr{D}_m}{w_m^2 \rho_m V_m^2} = \phi\left(\frac{w_m}{h_m}, \frac{\rho_m V_m w_m}{\mu_m}\right) \tag{7.12}$$

for the model. The model design conditions, or similarity requirements, are therefore

$$\frac{w_m}{h_m} = \frac{w}{h} \quad \text{and} \quad \frac{\rho_m V_m w_m}{\mu_m} = \frac{\rho V w}{\mu}$$

The size of the model is obtained from the first requirement, which indicates that

$$w_m = \frac{h_m}{h} w \tag{7.13}$$

We are free to establish the height ratio between the model and prototype, h_m/h, but then the model plate width, w_m, is fixed in accordance with Eq. 7.13.

The second similarity requirement indicates that the model and prototype must be operated at the same Reynolds number. Thus, the required velocity for the model is obtained from the relationship

$$V_m = \frac{\mu_m}{\mu} \frac{\rho}{\rho_m} \frac{w}{w_m} V \tag{7.14}$$

Note that this model design requires not only geometric scaling, as specified by Eq. 7.13, but also the correct scaling of the velocity in accordance with Eq. 7.14. This result is typical of most model designs—there is more to the design than simply scaling the geometry!

With the foregoing similarity requirements satisfied, the prediction equation for the drag is

$$\frac{\mathscr{D}}{w^2 \rho V^2} = \frac{\mathscr{D}_m}{w_m^2 \rho_m V_m^2}$$

or

$$\mathscr{D} = \left(\frac{w}{w_m}\right)^2 \left(\frac{\rho}{\rho_m}\right) \left(\frac{V}{V_m}\right)^2 \mathscr{D}_m$$

Similarity between a model and a prototype is achieved by equating pi terms.

Thus, a measured drag on the model, $\mathscr{D}_m$, must be multiplied by the ratio of the square of the plate widths, the ratio of the fluid densities, and the ratio of the square of the velocities to obtain the predicted value of the prototype drag, $\mathscr{D}$.

Generally, as is illustrated in this example, to achieve similarity between model and prototype behavior, *all the corresponding pi terms must be equated between model and prototype*. Usually, one or more of these pi terms will involve ratios of important lengths (such as w/h in the foregoing example); that is, they are purely geometrical. Thus, when we equate the pi terms involving length ratios, we are requiring that there be complete *geometric similarity* between the model and prototype. This means that the model must be a scaled version of the prototype. Geometric scaling should extend to the finest features of the system, such as surface roughness, or small protuberances on a structure, since these kinds of geometric features may significantly influence the flow. Any deviation from complete geometric similarity for a model must be carefully considered. Sometimes complete geometric scaling may be difficult to achieve, particularly when dealing with surface roughness, since roughness is difficult to characterize and control.

Another group of typical pi terms (such as the Reynolds number in the foregoing example) involves force ratios as noted in Table 7.1. The equality of these pi terms requires the ratio of like forces in model and prototype to be the same. Thus, for flows in which the Reynolds numbers are equal, the ratio of viscous forces in model and prototype is equal to the ratio of inertia forces. This is further illustrated by the adjacent photographs. The top photograph is of flow past a cylinder with diameter, D, and velocity, V_o, while the velocity is doubled and the diameter halved in the bottom photograph using the same fluid (image shown in 2x enlargement). Since the Reynolds numbers match, the flows look and behave the same. If other pi terms are involved, such as the Froude number or Weber number, a similar conclusion can be drawn; that is, the equality of these pi terms requires the ratio of like forces in model and prototype to be the same. Thus, when these types of pi terms are equal in model and prototype, we have *dynamic similarity* between model and prototype. It follows that with both geometric and dynamic similarity the streamline patterns will be the same and corresponding velocity ratios (V_m/V) and acceleration ratios (a_m/a) are constant throughout the flow field. Thus, *kinematic similarity* exists between model and prototype. To have complete similarity between model and prototype, we must maintain geometric, kinematic, and dynamic similarity between the two systems. This will automatically follow if all the important variables are included in the dimensional analysis, and if all the similarity requirements based on the resulting pi terms are satisfied.

V7.8 Environmental models

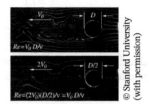

© Stanford University (with permission)

The Wide World of Fluids

Modeling parachutes in a water tunnel

The first use of a parachute with a free-fall jump from an aircraft occurred in 1914, although parachute jumps from hot-air balloons had occurred since the late 1700s. In more modern times parachutes are commonly used by the military and for safety and sport. It is not surprising that there remains interest in the design and characteristics of parachutes, and recently researchers at the Worcester Polytechnic Institute studied various aspects of the aerodynamics associated with parachutes. An unusual part of their study was that they used small-scale parachutes tested in a *water tunnel*. The *model parachutes* were reduced in size by a factor of 30 to 60 times. Various types of tests could be performed, ranging from the study of the velocity fields in the wake of the canopy with a steady free-stream velocity to the study of conditions during rapid deployment of the canopy. According to the researchers, the advantage of using water as the working fluid, rather than air, was that the velocities and deployment dynamics were slower than in the atmosphere, thus providing more time to collect detailed experimental data.

EXAMPLE 7.5 | Prediction of Prototype Performance from Model Data

Given A long structural component of a bridge has an elliptical cross section shown in **Fig. E7.5**. It is known that when a steady wind blows past this type of bluff body, vortices may develop on the downwind side that are shed in a regular fashion at some definite frequency. Since these vortices can create harmful periodic forces acting on the structure, it is important to determine the shedding frequency. For the specific structure of interest, $D = 0.1$ m, $H = 0.3$ m, and a representative wind velocity is 50 km/hr. Standard air can be assumed. The shedding frequency is to be determined through the use of a small-scale model that is to be tested in a water tunnel. For the model $D_m = 20$ mm and the water temperature is 20 °C.

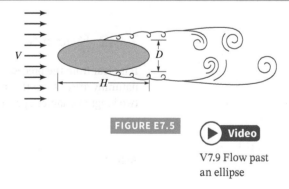

FIGURE E7.5

V7.9 Flow past an ellipse

Find Determine the model dimension, H_m, and the velocity at which the test should be performed. If the shedding frequency for the model is found to be 49.9 Hz, what is the corresponding frequency for the prototype?

Solution   We expect the shedding frequency, ω, to depend on the lengths D and H, the approach velocity, V, and the fluid density, ρ, and viscosity, μ. Thus,

$$\omega = f(D, H, V, \rho, \mu)$$

where

$$\omega \doteq T^{-1}$$
$$D \doteq L$$
$$H \doteq L$$
$$V \doteq LT^{-1}$$
$$\rho \doteq ML^{-3}$$
$$\mu \doteq ML^{-1}T^{-1}$$

Since there are six variables and three reference dimensions (MLT), three pi terms are required. Application of the pi theorem yields

$$\frac{\omega D}{V} = \phi\left(\frac{D}{H}, \frac{\rho V D}{\mu}\right)$$

We recognize the pi term on the left as the Strouhal number, and the dimensional analysis indicates that the Strouhal number is a function of the length ratio, D/H, and the Reynolds number. Thus, to maintain similarity between model and prototype

$$\frac{D_m}{H_m} = \frac{D}{H}$$

and

$$\frac{\rho_m V_m D_m}{\mu_m} = \frac{\rho V D}{\mu}$$

From the first similarity requirement

$$H_m = \frac{D_m}{D} H$$
$$= \frac{(20 \times 10^{-3}\ \text{m})}{(0.1\ \text{m})}(0.3\ \text{m})$$
$$H_m = 60 \times 10^{-3}\ \text{m} = 60\ \text{mm} \qquad (Ans)$$

The second similarity requirement indicates that the Reynolds number must be the same for model and prototype so that the model velocity must satisfy the condition

$$V_m = \frac{\mu_m}{\mu}\frac{\rho}{\rho_m}\frac{D}{D_m} V \qquad (1)$$

For air at standard conditions, $\mu = 1.79 \times 10^{-5}\,\text{kg/m} \cdot \text{s}$, $\rho = 1.23\ \text{kg/m}^3$, and for water at 20 °C, $\mu = 1.00 \times 10^{-3}\,\text{kg/m} \cdot \text{s}$, $\rho = 998\ \text{kg/m}^3$. The fluid velocity for the prototype is

$$V = \frac{(50 \times 10^3\ \text{m/hr})}{(3600\ \text{s/hr})} = 13.9\ \text{m/s}$$

The required velocity can now be calculated from Eq. 1 as

$$V_m = \frac{[1.00 \times 10^{-3}\,\text{kg/(m} \cdot \text{s)}]}{[1.79 \times 10^{-5}\,\text{kg/(m} \cdot \text{s)}]}\frac{(1.23\ \text{kg/m}^3)}{(998\ \text{kg/m}^3)}$$
$$\times \frac{(0.1\ \text{m})}{(20 \times 10^{-3}\,\text{m})}(13.9\ \text{m/s})$$
$$V_m = 4.79\ \text{m/s} \qquad (Ans)$$

This is a reasonable velocity that could be readily achieved in a water tunnel.

With the two similarity requirements satisfied, it follows that the Strouhal numbers for prototype and model will be the same so that

$$\frac{\omega D}{V} = \frac{\omega_m D_m}{V_m}$$

and the predicted prototype vortex shedding frequency is

$$\omega = \frac{V}{V_m}\frac{D_m}{D}\omega_m$$
$$= \frac{(13.9\ \text{m/s})}{(4.79\ \text{m/s})}\frac{(20 \times 10^{-3}\,\text{m})}{(0.1\ \text{m})}(49.9\ \text{Hz})$$
$$\omega = 29.0\ \text{Hz} \qquad (Ans)$$

Comment   This same model could also be used to predict the drag per unit length, $\mathcal{D}_\ell$ (N/m), on the prototype, since the drag would depend on the same variables as those used for the frequency. Thus, the similarity requirements would be the same and with these requirements satisfied it follows that the drag coefficient would be equal in model and prototype. The measured drag per unit length on the model could then be related to the corresponding drag per unit length on the prototype through the relationship

$$\mathcal{D}_\ell = \left(\frac{D}{D_m}\right)\left(\frac{\rho}{\rho_m}\right)\left(\frac{V}{V_m}\right)^2 \mathcal{D}_{\ell m}$$

Of course it is not necessary to do the formal dimensional analysis; one could simply choose Strouhal number, Reynolds number, and length ratio from Table 7.1.

7.8.2 Model Scales

It is clear from the preceding section that the ratio of like quantities for the model and prototype naturally arises from the similarity requirements. For example, if in a given problem there are two length variables ℓ_1 and ℓ_2, the resulting similarity requirement based on a length ratio is

$$\frac{\ell_1}{\ell_2} = \frac{\ell_{1m}}{\ell_{2m}}$$

so that

$$\frac{\ell_{1m}}{\ell_1} = \frac{\ell_{2m}}{\ell_2}$$

We define the ratio ℓ_{1m}/ℓ_1 or ℓ_{2m}/ℓ_2 as the **length scale**. For true models there will be only one length scale, and all lengths are fixed in accordance with this scale. There are, however, other scales such as the velocity scale, V_m/V, density scale, ρ_m/ρ, viscosity scale, μ_m/μ, and so on. In fact, we can define a scale for each of the variables in the problem. Thus, it is actually meaningless to talk about a "scale" of a model without specifying which scale.

We will designate the length scale as λ_ℓ, and other scales as $\lambda_V, \lambda_\rho, \lambda_\mu$, and so on, where the subscript indicates the particular scale. Also, we will take the ratio of the model value to the prototype value as the scale (rather than the inverse). Length scales are often specified, for example, as 1 : 10 or as a $\frac{1}{10}$ scale model. The meaning of this specification is that the model is one-tenth the size of the prototype, and the tacit assumption is that all relevant lengths are scaled accordingly so the model is geometrically similar to the prototype.

> The ratio of a model variable to the corresponding prototype variable is called the scale for that variable.

The Wide World of Fluids

Galloping Gertie

One of the most dramatic bridge collapses occurred in 1940 when the Tacoma Narrows Bridge, located near Tacoma, Washington, failed due to aerodynamic instability. The bridge had been nicknamed "Galloping Gertie" due to its tendency to sway and move in high winds. On the fateful day of the collapse, the wind speed was 65 km/hr. This particular combination of a high wind and the aeroelastic properties of the bridge created large oscillations leading to its failure. The bridge was replaced in 1950, and a second

bridge parallel to the existing structure was opened in 2007. To determine possible wind interference effects due to two bridges in close proximity, wind tunnel tests were run in a 9 m × 9 m wind tunnel operated by the National Research Council of Canada. *Models* of the two side-by-side bridges, each having a length scale of 1:211, were tested under various wind conditions. Since the failure of the original Tacoma Narrows Bridge, it is now common practice to use wind tunnel model studies during the design process to evaluate any bridge that is to be subjected to wind-induced vibrations.

7.8.3 Practical Aspects of Using Models

Validation of Model Design   Most model studies involve simplifying assumptions with regard to the variables to be considered. Although the assumptions are frequently less stringent than required for mathematical models, they nevertheless introduce some uncertainty in the model design. It is, therefore, desirable to check the design experimentally whenever possible. In some situations the purpose of the model is to predict the effects of certain proposed changes in a given prototype, and in this instance some actual prototype data may be available. The model can be designed, constructed, and tested, and the model prediction can be compared with these data. If the agreement is satisfactory, then the model can be changed in the desired manner, and the corresponding effect on the prototype can be predicted with increased confidence.

Another useful and informative procedure is to run tests with a series of models of different sizes, where one of the models can be thought of as the prototype and the others as "models" of this prototype. With the models designed and operated on the basis of the proposed design, a necessary condition for the validity of the model design is that an accurate prediction be made between any pair of models, since one can always be considered as a model of the other. Although suitable agreement in validation tests of this type does not unequivocally indicate a correct model design (e.g., the length scales between laboratory models may be significantly different than required for actual prototype prediction), it is certainly true that if agreement between models cannot be achieved in these tests, there is no reason to expect that the same model design can be used to predict prototype behavior correctly.

V7.10 Model of fish hatchery pond

Distorted Models   Although the general idea behind establishing similarity requirements for models is straightforward (we simply equate pi terms), it is not always possible to satisfy all the known requirements. If one or more of the similarity requirements are not met, for example, if $\Pi_{2m} \neq \Pi_2$, then it follows that the prediction equation $\Pi_1 = \Pi_{1m}$ is not true; that is, $\Pi_1 \neq \Pi_{1m}$. Models for which one or more of the similarity requirements are not satisfied are called **distorted models.**

Models for which one or more similarity requirements are not satisfied are called distorted models.

Distorted models are rather commonplace, and they are used for a variety of reasons. For example, perhaps a suitable fluid cannot be found for the model. The classic example of a distorted model occurs in the study of open channel or free-surface flows. Typically, in these problems both the Reynolds number, $\rho V \ell / \mu$, and the Froude number, $V / \sqrt{g\ell}$, are involved.

Froude number similarity requires

$$\frac{V_m}{\sqrt{g_m \ell_m}} = \frac{V}{\sqrt{g\ell}}$$

If the model and prototype are operated in the same gravitational field, then the required velocity scale is

$$\frac{V_m}{V} = \sqrt{\frac{\ell_m}{\ell}} = \sqrt{\lambda_\ell}$$

Reynolds number similarity requires

$$\frac{\rho_m V_m \ell_m}{\mu_m} = \frac{\rho V \ell}{\mu}$$

and the velocity scale is

$$\frac{V_m}{V} = \frac{\mu_m}{\mu} \frac{\rho}{\rho_m} \frac{\ell}{\ell_m}$$

V7.11 Distorted river model

Since the velocity scale must be equal to the square root of the length scale, it follows that

$$\frac{\mu_m / \rho_m}{\mu / \rho} = \frac{\nu_m}{\nu} = (\lambda_\ell)^{3/2} \tag{7.15}$$

where the ratio μ/ρ is the kinematic viscosity, ν. Although in principle it may be possible to satisfy this design condition, it may be quite difficult, if not impossible, to find a suitable model fluid, particularly for small length scales. For problems involving rivers, spillways, and harbors, for which the prototype fluid is water, the models are also relatively large so that the only practical model fluid is water. However, in this case (with the kinematic viscosity scale equal to unity) Eq. 7.15 will not be satisfied, and a distorted model will result. Generally, hydraulic models of this type are distorted and are designed on the basis of the Froude number, with the Reynolds number different between model and prototype.

Distorted models can be successfully used, but the interpretation of results obtained with this type of model is obviously more difficult than the interpretation of results obtained with **true models** for which all similarity requirements are met. There are no general rules for handling distorted models, and essentially each problem must be considered on its own merits. The success of using distorted models depends to a large extent on the skill and experience of the investigator responsible for the design of the model and in the interpretation of experimental data obtained from the model. Distorted models are widely used, and additional information can be found in the references at the end of the chapter. References 14 and 15 contain detailed discussions of several practical examples of distorted fluid flow and hydraulic models.

The Wide World of Fluids

Old Man River in (large) miniature

One of the world's largest scale models, a Mississippi River model, resides near Jackson, Mississippi. It is a detailed, complex model that covers many acres and replicates the 1,250,000-acre Mississippi River basin. Built by the Army Corps of Engineers and used from 1943 to 1973, today it has mostly gone to ruin. As with many hydraulic models, this is a *distorted model*, with a horizontal scale of 1:2000 and a vertical scale of 1:100. One step along the model river corresponds to one mile along the river.

All essential river basin elements such as geological features, levees, and railroad embankments were sculpted by hand to match the actual contours. The main purpose of the model was to predict floods. This was done by supplying specific amounts of water at prescribed locations along the model and then measuring the water depths up and down the model river. Because of the length scale, there is a difference in the time taken by the corresponding model and prototype events. Although it takes days for the actual floodwaters to travel from Sioux City, Iowa, to Omaha, Nebraska, it would take only minutes for the simulated flow in the model.

7.9 TYPICAL MODEL STUDIES

Models are used to investigate many different types of fluid mechanics problems, and it is difficult to characterize in a general way all necessary similarity requirements, since each problem is unique. We can, however, broadly classify many of the problems on the basis of the general nature of the flow and subsequently develop some general characteristics of model designs in each of these classifications. In the following sections we will consider models for the study of (1) flow through closed conduits, (2) flow around immersed bodies, and (3) flow with a free surface. Turbomachine models are considered in Chapter 12.

7.9.1 Flow Through Closed Conduits

Common examples of this type of flow include pipe flow and flow through valves, fittings, and metering devices. Although the conduit cross sections are often circular, they could have other shapes as well and may contain expansions or contractions. Since there are no fluid interfaces or free surfaces, the dominant forces are inertial and viscous so that the Reynolds number is an important similarity parameter. For low Mach numbers ($Ma < 0.3$), compressibility effects are usually negligible for both the flow of liquids and gases. For this class of problems, geometric similarity between model and prototype must be maintained. Generally the geometric characteristics can be described by a series of length terms, $\ell_1, \ell_2, \ell_3, \ldots, \ell_i$, and ℓ, where ℓ is some particular length dimension for the system. Such a series of length terms leads to a set of length ratios

$$\frac{\ell_i}{\ell}$$

where $i = 1, 2, \ldots$, and so on. In addition to the basic geometry of the system, the roughness of the internal surface in contact with the fluid may be important. If the average height of surface roughness elements is defined as ε, then the pi term representing roughness will be the relative roughness, ε/ℓ. This parameter indicates that for complete geometric similarity, surface roughness would also have to be scaled. Note that this implies that for length scales less than 1, the model surfaces should be smoother than those in the prototype since $\varepsilon_m = \lambda_\ell \varepsilon$. To further complicate matters, the pattern of roughness elements in model and prototype would have to be similar. These are conditions that are virtually impossible to satisfy exactly. Fortunately, in some problems the surface roughness plays a minor role and can be neglected. However, in other problems (such as turbulent flow through pipes) roughness can be very important.

It follows from this discussion that for flow in closed conduits at low Mach numbers, any dependent pi term (the one that contains the particular variable of interest, such as pressure drop) can be expressed as

$$\text{Dependent pi term} = \phi\left(\frac{\ell_i}{\ell}, \frac{\varepsilon}{\ell}, \frac{\rho V \ell}{\mu}\right) \tag{7.16}$$

This is a general formulation for this type of problem. The first two pi terms of the right side of Eq. 7.16 lead to the requirement of geometric similarity so that

$$\frac{\ell_{im}}{\ell_m} = \frac{\ell_i}{\ell} \qquad \frac{\varepsilon_m}{\ell_m} = \frac{\varepsilon}{\ell}$$

or

$$\frac{\ell_{im}}{\ell_i} = \frac{\varepsilon_m}{\varepsilon} = \frac{\ell_m}{\ell} = \lambda_\ell$$

This result indicates that the investigator is free to choose a length scale, λ_ℓ, but once this scale is selected, all other pertinent lengths must be scaled in the same ratio.

The additional similarity requirement arises from the equality of Reynolds numbers

$$\frac{\rho_m V_m \ell_m}{\mu_m} = \frac{\rho V \ell}{\mu}$$

> Geometric and Reynolds number similarity is usually required for models involving flow through closed conduits.

Accurate predictions of flow behavior require the correct scaling of velocities.

From this condition the velocity scale is established so that

$$\frac{V_m}{V} = \frac{\mu_m}{\mu} \frac{\rho}{\rho_m} \frac{\ell}{\ell_m} \qquad (7.17)$$

and the actual value of the velocity scale depends on the viscosity and density scales, as well as the length scale. Different fluids can be used in model and prototype. However, if the same fluid is used (with $\mu_m = \mu$ and $\rho_m = \rho$), then

$$\frac{V_m}{V} = \frac{\ell}{\ell_m}$$

Thus, $V_m = V/\lambda_\ell$, which indicates that the fluid velocity in the model will be larger than that in the prototype for any length scale less than 1. Since length scales are typically much less than unity, Reynolds number similarity may be difficult to achieve because of the large model velocities required.

With these similarity requirements satisfied, it follows that the dependent pi term will be equal in model and prototype. For example, if the dependent variable of interest is the pressure differential,[3] Δp, between two points along a closed conduit, then the dependent dimensionless group could be the Euler number

$$\text{Eu} = \frac{\Delta p}{\rho V^2}$$

The prototype pressure drop would then be obtained from the relationship

$$\Delta p = \frac{\rho}{\rho_m} \left(\frac{V}{V_m}\right)^2 \Delta p_m$$

so that from a measured pressure differential in the model, Δp_m, the corresponding pressure differential for the prototype could be predicted. Note that in general $\Delta p \ne \Delta p_m$.

EXAMPLE 7.6 | Reynolds Number Similarity

Given Model tests are to be performed to study the flow through a large check valve having a 0.6 m-diameter inlet and carrying water at a flowrate of 0.85 m³/s as shown in **Fig. E7.6a**. The working fluid in the model is water at the same temperature as that in the prototype. Complete geometric similarity exists between model and prototype, and the model inlet diameter is 7.5 cm.

Find Determine the required flowrate in the model.

Solution To ensure dynamic similarity, the model tests should be run so that

$$\text{Re}_m = \text{Re}$$

or

$$\frac{V_m D_m}{\nu_m} = \frac{VD}{\nu}$$

where V and D correspond to the inlet velocity and diameter, respectively. Since the same fluid is to be used in model and prototype, $\nu = \nu_m$, and therefore

$$\frac{V_m}{V} = \frac{D}{D_m}$$

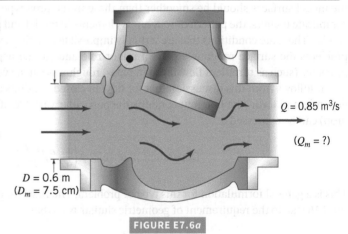

$Q = 0.85$ m³/s

$(Q_m = ?)$

$D = 0.6$ m
$(D_m = 7.5$ cm$)$

FIGURE E7.6a

The discharge, Q, is equal to VA, where A is the inlet area, so

$$\frac{Q_m}{Q} = \frac{V_m A_m}{VA} = \left(\frac{D}{D_m}\right) \frac{[(\pi/4)D_m^2]}{[(\pi/4)D^2]}$$

$$= \frac{D_m}{D}$$

---

[3]In some previous examples the pressure differential *per unit length*, Δp_ℓ, was used. This is appropriate for flow in long pipes or conduits in which the pressure would vary linearly with distance. However, in the more general situation the pressure may not vary linearly with position so that it is necessary to consider the pressure differential, Δp, as the dependent variable. In this case the distance between pressure taps is an additional variable (as well as the distance of one of the taps measured from some reference point within the flow system).

and for the data given

$$Q_m = \frac{(7.5/100 \text{ m})}{(0.6 \text{ m})}(0.85 \text{ m}^3/\text{s})$$

$$Q_m = 0.106 \text{ m}^3/\text{s} \qquad (Ans)$$

Comment As indicated by the foregoing analysis, to maintain Reynolds number similarity using the same fluid in model and prototype, the required velocity scale is inversely proportional to the length scale, that is, $V_m/V = (D_m/D)^{-1}$. This strong influence of the length scale on the velocity scale is shown in **Fig. E7.6b**. For this particular example, $D_m/D = 0.125$, and the corresponding velocity scale is 8 (see Fig. E7.6b). Thus, with the prototype velocity equal to $V = (0.85 \text{ m}^3/\text{s})/(\pi/4)(0.6 \text{ m})^2 = 3 \text{ m/s}$, the required model velocity is $V_m = 24 \text{ m/s}$. Although this is a relatively large velocity, it could be attained in a laboratory facility. It is to be noted that if we tried to use a smaller model, say one with $D_m = 2.5 \text{ cm}$, the required model velocity is 72 m/s, a very high velocity that would be difficult to achieve. These results are indicative of one of the difficulties encountered in maintaining Reynolds

number similarity—the required model velocities may be impractical to obtain.

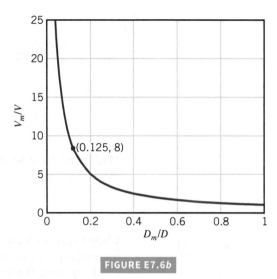

FIGURE E7.6b

Two additional points should be made with regard to modeling flows in closed conduits. First, for large Reynolds numbers, inertial forces are much larger than viscous forces, and in this case it may be possible to neglect viscous effects. (This would imply turbulent flow, not inviscid flow.) The important practical consequence of this is that it would not be necessary to maintain Reynolds number similarity between model and prototype. However, *both* model and prototype would have to operate at large Reynolds numbers. Since we do not know, a priori, what a "large Reynolds number" is, the effect of Reynolds numbers would have to be determined from the model. This could be accomplished by varying the model Reynolds number to determine the range (if any) over which the dependent pi term ceases to be affected by changes in Reynolds number.

The second point relates to the possibility of cavitation in flow through closed conduits. For example, flow through the complex passages that may exist in valves may lead to local regions of high velocity (and thus low pressure), which can cause the fluid to cavitate. If the model is to be used to study cavitation phenomena, then the vapor pressure, p_v, becomes an important variable and an additional similarity requirement such as equality of the cavitation number $(p_r - p_v)/\frac{1}{2}\rho V^2$ is required, where p_r is some reference pressure. The use of models to study cavitation is complicated, since it is not fully understood how vapor bubbles form and grow. The initiation of bubbles seems to be influenced by the microscopic particles that exist in most liquids, and how this aspect of the problem influences model studies is not clear. Additional details can be found in Ref. 18.

In some problems Reynolds number similarity may be relaxed.

7.9.2 Flow Around Immersed Bodies

Models have been widely used to study the flow characteristics associated with bodies that are completely immersed in a moving fluid. Examples include flow around aircraft, submarines, automobiles, golf balls, and buildings. (These types of models are usually tested in wind tunnels as is illustrated in **Fig. 7.6**.) Modeling laws for these problems are similar to those described in the preceding section; that is, geometric and Reynolds number similarity is required. Since there are no fluid interfaces, surface tension (and therefore the Weber number) is not important. Also, gravity will not affect the flow patterns, so the Froude number need not be considered. The Mach number will be important for high-speed flows in which compressibility becomes an important factor, but for incompressible flows (such as those in liquids or gases at relatively low speeds) the

Geometric and Reynolds number similarity is usually required for models involving flow around bodies.

FIGURE 7.6   Model of the National Bank of Commerce, San Antonio, Texas, for measurement of peak, rms, and mean pressure distributions. The model is located in a long-test-section, meteorological wind tunnel.

Mach number can be omitted as a similarity requirement. In this case, a general formulation for these problems is

$$\text{Dependent pi term} = \phi\left(\frac{\ell_i}{\ell}, \frac{\varepsilon}{\ell}, \frac{\rho V \ell}{\mu}\right) \tag{7.18}$$

V7.12 Wind engineering models

where ℓ is some characteristic length of the system and ℓ_i represents other pertinent lengths, ε/ℓ is the relative roughness of the surface (or surfaces), and $\rho V \ell/\mu$ is the Reynolds number.

Frequently, the dependent variable of interest for this type of problem is the drag, $\mathcal{D}$, developed on the body, and in this situation the dependent dimensionless group would usually be expressed in the form of a drag coefficient,

$$C_D = \frac{\mathcal{D}}{\frac{1}{2}\rho V^2 \ell^2}$$

ℓ^2 is usually taken as some representative area of the object. Thus, drag studies can be undertaken with the formulation

$$\frac{\mathcal{D}}{\frac{1}{2}\rho V^2 \ell^2} = C_D = \phi\left(\frac{\ell_i}{\ell}, \frac{\varepsilon}{\ell}, \frac{\rho V \ell}{\mu}\right) \tag{7.19}$$

It is clear from Eq. 7.19 that geometric similarity

$$\frac{\ell_{im}}{\ell_m} = \frac{\ell_i}{\ell} \qquad \frac{\varepsilon_m}{\ell_m} = \frac{\varepsilon}{\ell}$$

as well as Reynolds number similarity

$$\frac{\rho_m V_m \ell_m}{\mu_m} = \frac{\rho V \ell}{\mu}$$

For flow around bodies, drag is often the dependent variable of interest.

must be maintained. If these conditions are met, then

$$\frac{\mathcal{D}}{\frac{1}{2}\rho V^2 \ell^2} = \frac{\mathcal{D}_m}{\frac{1}{2}\rho_m V_m^2 \ell_m^2}$$

or

$$\mathcal{D} = \frac{\rho}{\rho_m}\left(\frac{V}{V_m}\right)^2 \left(\frac{\ell}{\ell_m}\right)^2 \mathcal{D}_m$$

Measurements of model drag, $\mathcal{D}_m$, can then be used to predict the corresponding drag, $\mathcal{D}$, on the prototype from this relationship.

As was discussed in the previous section, one of the common difficulties with models is related to the Reynolds number similarity requirement, which establishes the model velocity as

$$V_m = \frac{\mu_m}{\mu}\frac{\rho}{\rho_m}\frac{\ell}{\ell_m}V \tag{7.20}$$

or

$$V_m = \frac{\nu_m}{\nu}\frac{\ell}{\ell_m}V \tag{7.21}$$

where ν_m/ν is the ratio of kinematic viscosities. If the same fluid is used for model and prototype so that $\nu_m = \nu$, then

$$V_m = \frac{\ell}{\ell_m} V$$

and, therefore, the required model velocity will be higher than the prototype velocity for ℓ/ℓ_m greater than 1. Since this ratio is often relatively large, the required value of V_m may be large. For example, for a $\frac{1}{10}$ length scale, and a prototype velocity of 80 km/hr, the required model velocity is 800 km/hr. This is a value that is unreasonably high to achieve with liquids, and for gas flows this would be in the range where compressibility would be important in the model (but not in the prototype).

As an alternative, we see from Eq. 7.21 that V_m could be reduced by using a different fluid in the model such that $\nu_m/\nu < 1$. For example, the ratio of the kinematic viscosity of water to that of air is approximately $\frac{1}{10}$, so that if the prototype fluid were air, tests might be run on the model using water. This would reduce the required model velocity, but it still may be difficult to achieve the necessary velocity in a suitable test facility, such as a water tunnel.

Another possibility for wind tunnel tests would be to increase the air pressure in the tunnel so that $\rho_m > \rho$, thus reducing the required model velocity as specified by Eq. 7.20. Fluid viscosity is not strongly influenced by pressure. Although pressurized tunnels have been built, they are obviously complicated and expensive.

The required model velocity can also be reduced if the length scale is modest; that is, the model is relatively large. For wind tunnel testing, this requires a large test section, which greatly increases the cost of the facility. However, large wind tunnels suitable for testing very large models (or prototypes) are in use. One such tunnel, located at the NASA Ames Research Center, Moffett Field, California, has a test section that is 12 m by 24 m and can accommodate test speeds to 552 km/hr. Such a large and expensive test facility is obviously not feasible for university or industrial laboratories, so most model testing has to be accomplished with relatively small models.

Video

V7.13 Model airplane test in water

Video

V7.14 Large-scale wind tunnel

EXAMPLE 7.7 | Model Design Conditions and Predicted Prototype Performance

---

Given The drag on the airplane shown in **Fig. E7.7** cruising at 386 km/hr in standard air is to be determined from tests on a 1:10 scale model placed in a pressurized wind tunnel. To minimize compressibility effects, the airspeed in the wind tunnel is also to be 386 km/hr.

$V = 386$ km/hr

NASA

FIGURE E7.7

Find Determine

a. the required air pressure in the tunnel (assuming the same air temperature for model and prototype) and

b. the drag on the prototype corresponding to a measured force of 4.4 N on the model.

Solution

a. From Eq. 7.19 it follows that drag can be predicted from a geometrically similar model if the Reynolds numbers in model and prototype are the same. Thus,

$$\frac{\rho_m V_m \ell_m}{\mu_m} = \frac{\rho V \ell}{\mu}$$

For this example, $V_m = V$ and $\ell_m/\ell = \frac{1}{10}$ so that

$$\frac{\rho_m}{\rho} = \frac{\mu_m}{\mu} \frac{V}{V_m} \frac{\ell}{\ell_m}$$

$$= \frac{\mu_m}{\mu}(1)(10)$$

and therefore

$$\frac{\rho_m}{\rho} = 10 \frac{\mu_m}{\mu}$$

This result shows that the same fluid with $\rho_m = \rho$ and $\mu_m = \mu$ cannot be used if Reynolds number similarity is to be maintained. One possibility is to pressurize the wind tunnel to increase the density of the air. We assume that an increase in pressure does not significantly change the viscosity so that the required increase in density is given by the relationship

$$\frac{\rho_m}{\rho} = 10$$

For an ideal gas, $p = \rho RT$ so that

$$\frac{p_m}{p} = \frac{\rho_m}{\rho}$$

for constant temperature $(T = T_m)$. Therefore, the wind tunnel would need to be pressurized so that

$$\frac{p_m}{p} = 10$$

Since the prototype operates at standard atmospheric pressure, the required pressure in the wind tunnel is 10 atmospheres or

$$p_m = 10 \, (101.3 \text{ kPa})$$
$$= 1013 \text{ kPa} \qquad (Ans)$$

Comment Thus, we see that a high pressure would be required and this could not be achieved easily or inexpensively. However, under these conditions, Reynolds similarity would be attained.

b. The drag could be obtained from Eq. 7.19 so that

$$\frac{\mathscr{D}}{\frac{1}{2}\rho V^2 \ell^2} = \frac{\mathscr{D}_m}{\frac{1}{2}\rho_m V_m^2 \ell_m^2}$$

or

$$\mathscr{D} = \frac{\rho}{\rho_m}\left(\frac{V}{V_m}\right)^2\left(\frac{\ell}{\ell_m}\right)^2 \mathscr{D}_m$$

$$= \left(\frac{1}{10}\right)(1)^2(10)^2 \mathscr{D}_m$$

$$= 10\mathscr{D}_m$$

Thus, for a drag of 4.4 N on the model the corresponding drag on the prototype is

$$\mathscr{D} = 44 \text{ N} \qquad (Ans)$$

V7.15 Full-scale race car model

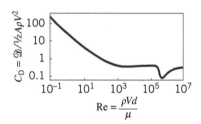

At high Reynolds numbers the drag is often essentially independent of the Reynolds number.

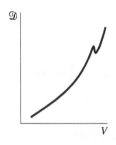

Fortunately, in many situations the flow characteristics are not strongly influenced by the Reynolds number over the operating range of interest. In these cases we can avoid the rather stringent similarity requirement of matching Reynolds numbers. To illustrate this point, consider the variation in the drag coefficient with the Reynolds number for a smooth sphere of diameter d placed in a uniform stream with approach velocity, V. Some typical data are shown in the adjacent figure. (A larger version is shown in Fig. 9.23a.) For Reynolds numbers between approximately 10^3 and 2×10^5 the drag coefficient is relatively constant and does not strongly depend on the specific value of the Reynolds number. Thus, exact Reynolds number similarity is not required in this range. For other geometric shapes we would typically find that for high Reynolds numbers, inertial forces are dominant (rather than viscous forces), and the drag is essentially independent of the Reynolds number.

Finally, consider the interpretation of trends in the *dimensional* data when plotting pi terms. For example, if ρ, μ, and d remain constant, then an increase in Re comes from an increase in V. Intuitively, it would seem that if V increases, the drag would increase. However, as shown in the figure, the drag *coefficient* generally decreases with increasing Re. When interpreting data, one needs to be aware if the variables are nondimensional. In this case, the physical drag force is proportional to the drag coefficient times the velocity squared. Thus, as shown by the adjacent figure, the drag force does, as expected, increase with increasing velocity. The exception occurs in the Reynolds number range $2 \times 10^5 < \text{Re} < 4 \times 10^5$ where the drag coefficient decreases dramatically with increasing Reynolds number. This phenomenon is discussed in Section 9.3.

For problems involving high velocities in which the Mach number is greater than about 0.3, the influence of compressibility, and therefore the Mach number (or Cauchy number), becomes significant. In this case complete similarity requires not only geometric and Reynolds number similarity but also Mach number similarity so that

$$\frac{V_m}{c_m} = \frac{V}{c} \qquad (7.22)$$

This similarity requirement, when combined with that for Reynolds number similarity (Eq. 7.21), yields

$$\frac{c}{c_m} = \frac{\nu}{\nu_m}\frac{\ell_m}{\ell} \qquad (7.23)$$

Clearly the same fluid with $c = c_m$ and $\nu = \nu_m$ cannot be used in model and prototype unless the length scale is unity (which means that we are running tests on the prototype). In high-speed aerodynamics the prototype fluid is usually air, and it is difficult to satisfy Eq. 7.23 for reasonable length scales. Thus, models involving high-speed flows are often distorted with respect to Reynolds number similarity, but Mach number similarity is maintained.

7.9.3 Flow with a Free Surface

Flows in canals, rivers, spillways, and stilling basins, as well as flow around ships, are all examples of flow phenomena involving a free surface. For this class of problems, both gravitational and inertial forces are important and, therefore, the Froude number becomes an important similarity parameter. Also, since there is a free surface with a liquid–air interface, forces due to surface tension may be significant, and the Weber number becomes another similarity parameter that needs to be considered along with the Reynolds number. Geometric variables will obviously still be important. Thus, a general formulation for problems involving flow with a free surface can be expressed as

$$\text{Dependent pi term} = \phi\left(\frac{\ell_i}{\ell}, \frac{\varepsilon}{\ell}, \frac{\rho V \ell}{\mu}, \frac{V}{\sqrt{g\ell}}, \frac{\rho V^2 \ell}{\sigma}\right) \qquad (7.24)$$

As discussed previously, ℓ is some characteristic length of the system, ℓ_i represents other pertinent lengths, and ε/ℓ is the relative roughness of the various surfaces. Since gravity is the driving force in these problems, Froude number similarity is definitely required so that

$$\frac{V_m}{\sqrt{g_m \ell_m}} = \frac{V}{\sqrt{g\ell}}$$

The model and prototype are expected to operate in the same gravitational field ($g_m = g$), and therefore it follows that

$$\frac{V_m}{V} = \sqrt{\frac{\ell_m}{\ell}} = \sqrt{\lambda_\ell} \qquad (7.25)$$

Thus, when models are designed on the basis of Froude number similarity, the velocity scale is determined by the square root of the length scale. As is discussed in Section 7.8.3, to simultaneously have Reynolds and Froude number similarity it is necessary that the kinematic viscosity scale be related to the length scale as

$$\frac{\nu_m}{\nu} = (\lambda_\ell)^{3/2} \qquad (7.26)$$

The working fluid for the prototype is normally either freshwater or seawater and the length scale is small. Under these circumstances it is virtually impossible to satisfy Eq. 7.26, so models involving free-surface flows are usually distorted. The problem is further complicated if an attempt is made to model surface tension effects, since this requires the equality of Weber numbers, which leads to the condition

$$\frac{\sigma_m/\rho_m}{\sigma/\rho} = (\lambda_\ell)^2 \qquad (7.27)$$

for the kinematic surface tension (σ/ρ). It is again evident that the same fluid cannot be used in model and prototype if we are to have similitude with respect to surface tension effects for $\lambda_\ell \neq 1$.

Fortunately, in many problems involving free-surface flows, both surface tension and viscous effects are small and consequently strict adherence to Weber and Reynolds number similarity is not required. Certainly, surface tension is not important in large hydraulic structures and rivers. Our only concern would be if in a model the depths were reduced to the point where surface tension becomes an important factor, whereas it is not in the prototype. This is of particular importance in the design of river models, since the length scales are typically small (so that the width of the model is reasonable), but with a small length scale the required model depth may be very small. To overcome this problem, different horizontal and vertical length scales are often used for river models. Although this approach eliminates surface tension effects in the model, it introduces geometric distortion that must be accounted for empirically, usually by increasing the model surface roughness. It is important in these circumstances that verification tests with the model be performed (if possible) in which model data are compared with available prototype river flow data. Model roughness can be adjusted to give satisfactory agreement between model and prototype, and then the model subsequently used to predict the effect of proposed changes on river characteristics (such as velocity patterns or surface elevations).

Froude number similarity is usually required for models involving free-surface flows.

V7.16 River flow model

V7.17 Boat model

FIGURE 7.7 A scale hydraulic model (1:197) of the Guri Dam in Venezuela, which is used to simulate the characteristics of the flow over and below the spillway and the erosion below the spillway.

V7.18 Dam model

For large hydraulic structures, such as dam spillways, the Reynolds numbers are large so that viscous forces are small in comparison to the forces due to gravity and inertia. In this case, Reynolds number similarity is not maintained and models are designed on the basis of Froude number similarity. Care must be taken to ensure that the model Reynolds numbers are also large, but they are not required to be equal to those of the prototype. This type of hydraulic model is usually made as large as possible so that the Reynolds number will be large. A spillway model is shown in **Fig. 7.7**. Also, for relatively large models the geometric features of the prototype can be accurately scaled, as well as surface roughness. Note that $\varepsilon_m = \lambda_\ell \varepsilon$, which indicates that the model surfaces must be smoother than the corresponding prototype surfaces for $\lambda_\ell < 1$.

The Wide World of Fluids

Ice engineering

Various types of models have been studied in wind tunnels, water tunnels, and towing tanks for many years. But another type of facility is needed to study ice and ice-related problems. The U.S. Army Cold Regions Research and Engineering Laboratory has developed a unique complex that houses research facilities for studies related to the mechanical behavior of ice and ice–structure interactions. The laboratory contains three separate cold-rooms— a test basin, a flume, and a general research area. In the test basin,

large-scale model studies of ice forces on structures such as dams, piers, ships, and offshore platforms can be performed. Ambient temperatures can be controlled as low as −28.8 °C, and at this temperature a 2-mm-per-hour ice growth rate can be achieved. It is also possible to control the mechanical properties of the ice to properly match the physical scale of the model. Tests run in the recirculating flume can simulate river processes during ice formation. And in the large research area, scale models of lakes and rivers can be built and operated to model ice interactions with various types of engineering projects.

EXAMPLE 7.8 | Froude Number Similarity

Given The spillway for the dam shown in **Fig. E7.8a** is 20 m wide and is designed to carry 125 m³/s at flood stage. A 1:15 model is constructed to study the flow characteristics through the spillway. The effects of surface tension and viscosity are to be neglected.

Find

a. Determine the required model width and flowrate.

b. What operating time for the model corresponds to a 24-hr period in the prototype?

Solution The width, w_m, of the model spillway is obtained from the length scale, λ_ℓ, so that

$$\frac{w_m}{w} = \lambda_\ell = \frac{1}{15}$$

Marty Melchior

FIGURE E7.8a

Therefore,

$$w_m = \frac{20 \text{ m}}{15} = 1.33 \text{ m} \qquad (Ans)$$

Of course, all other geometric features (including surface roughness) of the spillway must be scaled in accordance with the same length scale.

With the neglect of surface tension and viscosity, Eq. 7.24 indicates that dynamic similarity will be achieved if the Froude numbers are equal between model and prototype. Thus,

$$\frac{V_m}{\sqrt{g_m \ell_m}} = \frac{V}{\sqrt{g\ell}}$$

and for $g_m = g$

$$\frac{V_m}{V} = \sqrt{\frac{\ell_m}{\ell}}$$

Since the flowrate is given by $Q = VA$, where A is an appropriate cross-sectional area, it follows that

$$\frac{Q_m}{Q} = \frac{V_m A_m}{VA} = \sqrt{\frac{\ell_m}{\ell}}\left(\frac{\ell_m}{\ell}\right)^2$$

$$= (\lambda_\ell)^{5/2}$$

where we have made use of the relationship $A_m/A = (\ell_m/\ell)^2$. For $\lambda_\ell = \frac{1}{15}$ and $Q = 125 \text{ m}^3/\text{s}$

$$Q_m = \left(\frac{1}{15}\right)^{5/2}(125 \text{ m}^2/\text{s}) = 0.143 \text{ m}^3/\text{s} \qquad (Ans)$$

The time scale can be obtained from the velocity scale, since the velocity is distance divided by time ($V = \ell/t$), and therefore

$$\frac{V}{V_m} = \frac{\ell}{t}\frac{t_m}{\ell_m}$$

or

$$\frac{t_m}{t} = \frac{V}{V_m}\frac{\ell_m}{\ell} = \sqrt{\frac{\ell_m}{\ell}} = \sqrt{\lambda_\ell}$$

This result indicates that time intervals in the model will be smaller than the corresponding intervals in the prototype if $\lambda_\ell < 1$. For $\lambda_\ell = \frac{1}{15}$ and a prototype time interval of 24 hr

$$t_m = \sqrt{\frac{1}{15}}(24 \text{ hr}) = 6.20 \text{ hr} \qquad (Ans)$$

Comment As indicated by the foregoing analysis, the time scale varies directly as the square root of the length scale. Thus, as shown in **Fig. E7.8b**, the model time interval, t_m, corresponding to a 24-hr prototype time interval can be varied by changing the length scale, λ_ℓ. The ability to scale times may be very useful, since it is possible to "speed up" events in the model that may occur over a relatively long time in the prototype. There is, of course, a practical limit to how small the length scale (and the corresponding time scale) can become. For example, if the length scale is too small then surface tension effects may become important in the model whereas they are not in the prototype. In such a case the present model design, based simply on Froude number similarity, would not be adequate.

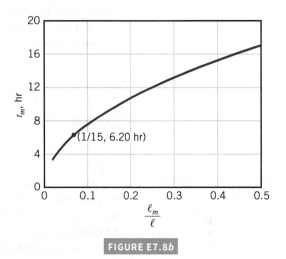

FIGURE E7.8b

There are, unfortunately, problems involving flow with a free surface in which viscous, inertial, and gravitational forces are all important. The drag on a ship as it moves through water is due to the viscous shearing stresses that develop along its hull, as well as a pressure-induced component of drag caused by both the shape of the hull and wave action. The shear drag is a function of the Reynolds number, whereas the pressure drag is a function of the Froude number. Since both Reynolds number and Froude number similarity cannot be simultaneously achieved by using water as the model fluid (which is the only practical fluid for ship models), some technique other than a straightforward model test must be employed. One common approach is to measure the total drag on a small, geometrically similar model as it is towed through a model basin at Froude numbers matching those of the prototype. The shear drag on the model is calculated using analytical techniques of the type described in Chapter 9. This calculated value is then subtracted from the total drag to obtain pressure drag, and using Froude number scaling the pressure drag on the prototype can then be predicted. The experimentally determined value can then be combined with a calculated value of the shear drag (again using analytical techniques) to provide the desired total drag on the ship. Ship models are widely used to study new designs, but the tests require extensive facilities (see **Fig. 7.8**).

It is clear from this brief discussion of various types of models involving free-surface flows that the design and use of such models requires considerable ingenuity, as well as a good

Video

V7.19 Testing of large yacht model

Photograph courtesy of the U.S. Navy's David W. Taylor Research Center.

FIGURE 7.8 Instrumented, small-waterplane-area, twin hull (SWATH) model suspended from a towing carriage.

understanding of the physical phenomena involved. This is generally true for most model studies. Modeling is both an art and a science. Motion picture producers make extensive use of model ships, fires, explosions, and the like. It is interesting to attempt to observe the flow differences between these distorted model flows and the real thing.

7.10   SIMILITUDE BASED ON GOVERNING DIFFERENTIAL EQUATIONS

Similarity laws can be directly developed from the equations governing the phenomenon of interest.

In the preceding sections of this chapter, dimensional analysis has been used to obtain similarity laws. This is a simple, straightforward approach to modeling, which is widely used. The use of dimensional analysis requires only a knowledge of the variables that influence the phenomenon of interest. Although the simplicity of this approach is attractive, it must be recognized that omission of one or more important variables may lead to serious errors in the model design. An alternative approach is available if the equations (usually differential equations) governing the phenomenon are known. In this situation similarity laws can be developed from the governing equations, even though it may not be possible to obtain analytic solutions to the equations.

To illustrate the procedure, consider the flow of an incompressible Newtonian fluid. For simplicity we will restrict our attention to two-dimensional flow, although the results are applicable to the general three-dimensional case. From Chapter 6 we know that the governing equations are the continuity equation

$$\frac{\partial u}{\partial x} + \frac{\partial v}{\partial y} = 0 \tag{7.28}$$

and the Navier–Stokes equations

$$\rho\left(\frac{\partial u}{\partial t} + u\frac{\partial u}{\partial x} + v\frac{\partial u}{\partial y}\right) = -\frac{\partial p}{\partial x} + \mu\left(\frac{\partial^2 u}{\partial x^2} + \frac{\partial^2 u}{\partial^2 y}\right) \tag{7.29}$$

$$\rho\left(\frac{\partial v}{\partial t} + u\frac{\partial v}{\partial x} + v\frac{\partial v}{\partial y}\right) = -\frac{\partial p}{\partial y} - \rho g + \mu\left(\frac{\partial^2 v}{\partial x^2} + \frac{\partial^2 v}{\partial y^2}\right) \tag{7.30}$$

where the y axis is vertical, so that the gravitational body force, ρg, only appears in the "y equation." To continue the mathematical description of the problem, boundary conditions are required. For example, velocities on all boundaries may be specified; that is, $u = u_B$ and $v = v_B$ at all boundary points $x = x_B$ and $y = y_B$. In some types of problems it may be necessary to specify the pressure over some part of the boundary. For time-dependent problems, initial conditions would also have to be provided, which means that the values of all dependent variables would be given at some time (usually taken at $t = 0$).

Once the governing equations, including boundary and initial conditions, are known, we are ready to proceed to develop similarity requirements. The next step is to define a new set of variables that are dimensionless. To do this we select a reference quantity for each type of variable. In this problem the variables are u, v, p, x, y, and t so we will need a reference velocity, V, a reference pressure, p_0, a reference length, ℓ, and a reference time, τ. These reference quantities should be parameters that appear in the problem. For example, ℓ may be a characteristic length of a body immersed in a fluid or the width of a channel through which a fluid is flowing. The velocity, V, may be the free-stream velocity or the inlet velocity. The new dimensionless (starred) variables can be expressed as

$$u^* = \frac{u}{V} \qquad v^* = \frac{v}{V} \qquad p^* = \frac{p}{p_0}$$

$$x^* = \frac{x}{\ell} \qquad y^* = \frac{y}{\ell} \qquad t^* = \frac{t}{\tau}$$

as shown in the adjacent figure.

The governing equations can now be rewritten in terms of these new variables. For example,

$$\frac{\partial u}{\partial x} = \frac{\partial V u^*}{\partial x^*}\frac{\partial x^*}{\partial x} = \frac{V}{\ell}\frac{\partial u^*}{\partial x^*}$$

and

$$\frac{\partial^2 u}{\partial x^2} = \frac{V}{\ell}\frac{\partial}{\partial x^*}\left(\frac{\partial u^*}{\partial x^*}\right)\frac{\partial x^*}{\partial x} = \frac{V}{\ell^2}\frac{\partial^2 u^*}{\partial x^{*2}}$$

The other terms that appear in the equations can be expressed in a similar fashion. Thus, in terms of the new variables the governing equations become

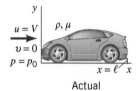

Actual

Dimensionless

$$\frac{\partial u^*}{\partial x^*} + \frac{\partial v^*}{\partial y^*} = 0 \tag{7.31}$$

and

$$\left[\frac{\rho V}{\tau}\right]\frac{\partial u^*}{\partial t^*} + \left[\frac{\rho V^2}{\ell}\right]\left(u^*\frac{\partial u^*}{\partial x^*} + v^*\frac{\partial u^*}{\partial y^*}\right) = -\left[\frac{p_0}{\ell}\right]\frac{\partial p^*}{\partial x^*} + \left[\frac{\mu V}{\ell^2}\right]\left(\frac{\partial^2 u^*}{\partial x^{*2}} + \frac{\partial^2 u^*}{\partial y^{*2}}\right) \tag{7.32}$$

$$\underbrace{\left[\frac{\rho V}{\tau}\right]\frac{\partial v^*}{\partial t^*}}_{F_{I\ell}} + \underbrace{\left[\frac{\rho V^2}{\ell}\right]\left(u^*\frac{\partial v^*}{\partial x^*} + v^*\frac{\partial v^*}{\partial y^*}\right)}_{F_{Ic}}$$

$$= -\underbrace{\left[\frac{p_0}{\ell}\right]\frac{\partial p^*}{\partial y^*}}_{F_P} - \underbrace{[\rho g]}_{F_G} + \underbrace{\left[\frac{\mu V}{\ell^2}\right]\left(\frac{\partial^2 v^*}{\partial x^{*2}} + \frac{\partial^2 v^*}{\partial y^{*2}}\right)}_{F_V} \tag{7.33}$$

The terms appearing in brackets contain the reference quantities and can be interpreted as indices of the various forces (per unit volume) that are involved. Thus, as is indicated in Eq. 7.33, $F_{I\ell}$ = inertia (local) force, F_{Ic} = inertia (convective) force, F_p = pressure force, F_G = gravitational force, and F_V = viscous force. As the final step in the nondimensionalization process, we will divide each term in Eqs. 7.32 and 7.33 by one of the bracketed quantities. Although any one of these quantities could be used, it is conventional to divide by the bracketed quantity $\rho V^2/\ell$, which is the index of the convective inertia force. The final nondimensional form then becomes

$$\left[\frac{\ell}{\tau V}\right]\frac{\partial u^*}{\partial t^*} + u^*\frac{\partial u^*}{\partial x^*} + v^*\frac{\partial u^*}{\partial y^*} = -\left[\frac{p_0}{\rho V^2}\right]\frac{\partial p^*}{\partial x^*} + \left[\frac{\mu}{\rho V\ell}\right]\left(\frac{\partial^2 u^*}{\partial x^{*2}} + \frac{\partial^2 u^*}{\partial y^{*2}}\right) \tag{7.34}$$

$$\left[\frac{\ell}{\tau V}\right]\frac{\partial v^*}{\partial t^*} + u^*\frac{\partial v^*}{\partial x^*} + v^*\frac{\partial v^*}{\partial y^*} = -\left[\frac{p_0}{\rho V^2}\right]\frac{\partial p^*}{\partial y^*} - \left[\frac{g\ell}{V^2}\right] + \left[\frac{u}{\rho V\ell}\right]\left(\frac{\partial^2 v^*}{\partial x^{*2}} + \frac{\partial^2 v^*}{\partial y^{*2}}\right) \tag{7.35}$$

We see that bracketed terms are standard dimensionless groups (or their reciprocals) which were developed from dimensional analysis; that is, $\ell/\tau V$ is a form of the Strouhal number, $p_0/\rho V^2$ the Euler number, $g\ell/V^2$ the reciprocal of the square of the Froude number, and $\mu/\rho V\ell$ the reciprocal of the Reynolds number. From this analysis it is now clear how the dimensionless groups can be interpreted as the ratio of two forces, and how these groups arise naturally from the governing equations.

Governing equations expressed in terms of dimensionless variables lead to the appropriate dimensionless groups.

Although we really have not helped ourselves with regard to obtaining an analytical solution to these equations (they are still complicated and not amenable to an analytical solution), the dimensionless forms of the equations, Eqs. 7.31, 7.34, and 7.35, can be used to establish similarity requirements. From these equations it follows that if two systems are governed by these equations, then the solutions (in terms of u^*, v^*, p^*, x^*, y^*, and t^*) will be the same if the four parameters $\ell/\tau V$, $p_0/\rho V^2$, $V^2/g\ell$, and $\rho V\ell/\mu$ are equal for the two systems. The two systems will be dynamically similar. Of course, boundary and initial conditions expressed in dimensionless form must also be equal for the two systems, and this will require complete geometric similarity. These are the same similarity requirements that would be determined by a dimensional analysis if the same variables were considered. However, the advantage of working with the governing equations is that the variables appear naturally in the equations, and we do not have to worry about omitting an important one, provided the governing equations are correctly specified. We can thus use this method to deduce the conditions under which two solutions will be similar even though one of the solutions will most likely be obtained experimentally.

In the foregoing analysis we have considered a general case in which the flow may be unsteady, and both the actual pressure level, p_0, and the effect of gravity are important. A reduction in the number of similarity requirements can be achieved if one or more of these conditions is removed. For example, if the flow is steady, the dimensionless group, $\ell/\tau V$, can be eliminated.

The actual pressure level will only be of importance if we are concerned with cavitation. If not, the flow patterns and the pressure differences will not depend on the pressure level. In this case, p_0 can be taken as ρV^2 (or $\frac{1}{2}\rho V^2$), and the Euler number can be eliminated as a similarity requirement. However, if we are concerned about cavitation (which will occur in the flow field if the pressure at certain points reaches the vapor pressure, p_v), then the actual pressure level is important. Usually, in this case, the characteristic pressure, p_0, is defined relative to the vapor pressure such that $p_0 = p_r - p_v$ where p_r is some reference pressure within the flow field. With p_0 defined in this manner, the similarity parameter $p_0/\rho V^2$ becomes $(p_r - p_v)/\rho V^2$. This parameter is frequently written as $(p_r - p_v)/\frac{1}{2}\rho V^2$, and in this form, as was noted previously in Section 7.6, is called the cavitation number. Thus we can conclude that if cavitation is not of concern we do not need a similarity parameter involving p_0, but if cavitation is to be modeled, then the cavitation number becomes an important similarity parameter.

The Froude number, which arises because of the inclusion of gravity, is important for problems in which there is a free surface. Examples of these types of problems include the study of rivers, flow through hydraulic structures such as spillways, and the drag on ships. In these situations the shape of the free surface is influenced by gravity, and therefore the Froude number becomes an important similarity parameter. However, if there are no free surfaces, the only effect of gravity is to superimpose a hydrostatic pressure distribution on the pressure distribution created by the fluid motion. The hydrostatic distribution can be eliminated from the governing equation (Eq. 7.30) by defining a new pressure, $p' = p - \rho g y$, and with this change the Froude number does not appear in the nondimensional governing equations.

We conclude from this discussion that for the steady flow of an incompressible fluid without free surfaces, dynamic and kinematic similarity will be achieved if (for geometrically similar systems) Reynolds number similarity exists. If free surfaces are involved, Froude number similarity must also be maintained. For free-surface flows we have tacitly assumed that surface tension is not important. We would find, however, that if surface tension is included, its effect would appear in the free-surface boundary condition, and the Weber number, $\rho V^2\ell/\sigma$, would become an additional similarity parameter. In addition, if the governing equations for compressible fluids are considered, the Mach number, V/c, would appear as an additional similarity parameter.

It is clear that all the common dimensionless groups that we previously developed by using dimensional analysis appear in the governing equations that describe fluid motion when these equations are expressed in terms of dimensionless variables. Thus, use of the governing equations to obtain similarity laws provides an alternative to dimensional analysis. This approach has the advantage that the variables are known and the assumptions involved are clearly identified. In addition, a physical interpretation of the various dimensionless groups can often be obtained.

CHAPTER SUMMARY

Many practical engineering problems involving fluid mechanics require experimental data for their solution. Thus, laboratory studies and experimentation play a significant role in this field. It is important to develop good procedures for the design of experiments so they can be efficiently completed with as broad applicability as possible. To achieve this end the concept of similitude is often used in which measurements made in the laboratory can be utilized for predicting the behavior of other similar systems. In this chapter, dimensional analysis is used for designing such experiments, as an aid for correlating experimental data, and as the basis for the design of physical models. As the name implies, dimensional analysis is based on a consideration of the dimensions required to describe the variables in a given problem. A discussion of the use of dimensions and the concept of dimensional homogeneity (which forms the basis for dimensional analysis) was included in Chapter 1.

Essentially, dimensional analysis simplifies a given problem described by a certain set of variables by reducing the number of variables that need to be considered. In addition to being fewer in number, the new variables are dimensionless products of the original variables. Typically these new dimensionless variables are much simpler to work with in performing the desired experiments. The Buckingham pi theorem, which forms the theoretical basis for dimensional analysis, is introduced. This theorem establishes the framework for reducing a given problem described in terms of a set of variables to a new set of fewer dimensionless variables. A simple method, called the repeating variable method, is described for actually forming the dimensionless variables (often called pi terms). Forming dimensionless variables by inspection is also considered. It is shown how the use of dimensionless variables can be of assistance in planning experiments and as an aid in correlating experimental data.

The repeating variable method (or any other method for forming dimensionless groups) can be applied to a generic fluid mechanics problem in which nearly all of the pertinent dimensional variables appear. The result is a set of common dimensionless groups (called "numbers" and "coefficients") that can be used to describe almost any problem. These common dimensionless groups represent the ratios of relevant forces or energies in a flow.

For problems in which there are a large number of variables, the use of physical models is described. Models are used to make specific predictions from laboratory tests rather than formulating a general relationship for the phenomenon of interest. The correct design of a model is obviously imperative for the accurate predictions of other similar, but usually larger, systems. It is shown how dimensional analysis can be used to establish a valid model design. An alternative approach for establishing similarity requirements using governing equations (usually differential equations) is presented.

The following checklist provides a study guide for this chapter. When your study of the entire chapter has been completed, you should be able to

- write out meanings of the terms listed and understand each of the related concepts. These terms are particularly important and are set in **bold black** type in the text.

- use the Buckingham pi theorem to determine the number of independent dimensionless variables needed for a given flow problem.

- form a set of dimensionless variables using the method of repeating variables.

- form a set of dimensionless variables by inspection.

- recognize common dimensionless groups used in fluid mechanics and list the physical phenomena they represent.

- use dimensionless variables as an aid in interpreting and correlating experimental data.

- use dimensional analysis to establish a set of similarity requirements (and prediction equation) for a model and use them to predict the behavior of another similar system (the prototype).

- rewrite a given governing equation in a suitable nondimensional form and deduce similarity requirements from the nondimensional form of the equation.

KEY EQUATIONS

Reynolds number	$\mathrm{Re} = \dfrac{\rho V \ell}{\mu}$
Froude number	$\mathrm{Fr} = \dfrac{V}{\sqrt{g\ell}}$
Euler number	$\mathrm{Eu} = \dfrac{p}{\rho V^2}$
Cauchy number	$\mathrm{Ca} = \dfrac{\rho V^2}{E_\nu}$
Mach number	$\mathrm{Ma} = \dfrac{V}{c}$
Strouhal number	$\mathrm{St} = \dfrac{\omega \ell}{V}$
Weber number	$\mathrm{We} = \dfrac{\rho V^2 \ell}{\sigma}$
Relative roughness	$\dfrac{\varepsilon}{\ell}$
Loss coefficient	$K = \dfrac{h_L}{V^2/2g}$

Pressure coefficient	$C_p = \dfrac{\Delta p}{1/2\rho V^2}$
Friction coefficient	$C_f = \dfrac{\tau_w}{1/2\rho V^2}$
Drag coefficient	$C_D = \dfrac{\mathscr{D}}{1/2\rho V^2 \ell^2}$
Lift coefficient	$C_L = \dfrac{\mathscr{L}}{1/2\rho V^2 \ell^2}$

REFERENCES

[1] Bridgman, P. W., *Dimensional Analysis*, Yale University Press, New Haven, Connecticut, 1922.

[2] Murphy, G., *Similitude in Engineering*, Ronald Press, New York, 1950.

[3] Langhaar, H. L., *Dimensional Analysis and Theory of Models*, Wiley, New York, 1951.

[4] Huntley, H. E., *Dimensional Analysis*, Macdonald, London, 1952.

[5] Duncan, W. J., *Physical Similarity and Dimensional Analysis: An Elementary Treatise*, Edward Arnold, London, 1953.

[6] Sedov, K. I., *Similarity and Dimensional Methods in Mechanics*, Academic Press, New York, 1959.

[7] Ipsen, D. C., *Units, Dimensions, and Dimensionless Numbers*, McGraw-Hill, New York, 1960.

[8] Kline, S. J., *Similitude and Approximation Theory*, McGraw-Hill, New York, 1965.

[9] Skoglund, V. J., *Similitude—Theory and Applications*, International Textbook, Scranton, Pennsylvania, 1967.

[10] Baker, W. E., Westline, P. S., and Dodge, F. T., *Similarity Methods in Engineering Dynamics—Theory and Practice of Scale Modeling*, Hayden (Spartan Books), Rochelle Park, New Jersey, 1973.

[11] Taylor, E. S., *Dimensional Analysis for Engineers*, Clarendon Press, Oxford, UK, 1974.

[12] Isaacson, E. de St. Q., and Isaacson, M. de St. Q., *Dimensional Methods in Engineering and Physics*, Wiley, New York, 1975.

[13] Schuring, D. J., *Scale Models in Engineering*, Pergamon Press, New York, 1977.

[14] Yalin, M. S., *Theory of Hydraulic Models*, Macmillan, London, 1971.

[15] Sharp, J. J., *Hydraulic Modeling*, Butterworth, London, 1981.

[16] Omega Engineering Inc., *The Land Chart of Dimensionless Numbers*, Model DLN-1, 1991.

[17] Schlichting, H., *Boundary-Layer Theory*, 7th Ed., McGraw-Hill, New York, 1979.

[18] Knapp, R. T., Daily, J. W., and Hammitt, F. G., *Cavitation*, McGraw-Hill, New York, 1970.

QUESTIONS AND PROBLEMS

Ⓒ Problem to be solved with aid of programmable calculator or computer.

Ⓞ Open-ended problem that requires critical thinking. These problems require various assumptions to provide the necessary input data. There are not unique answers to these problems.

Note: Unless specific values of required fluid properties are specified in the problem statement, use the values found in Tables 1.4–1.8 and in the tables in the Appendices.

Section 7.1 The Need for Dimensional Analysis

7.1.1 What are the dimensions of density, pressure, specific weight, surface tension, and dynamic viscosity in (a) the *FLT* system, and (b) the *MLT* system? Compare your results with those given in Table 1.1 in Chapter 1.

7.1.2 An equation used to evaluate vacuum filtration is

$$Q = \frac{\Delta p A^2}{\alpha(\Psi R w + A R_f)}$$

where $Q \doteq L^3/T$ is the filtrate volume flowrate, $\Delta p \doteq F/L^2$ the vacuum pressure differential, $A \doteq L^2$ the filter area, α the filtrate "viscosity," $\Psi \doteq L^3$ the filtrate volume, $R \doteq L/F$ the sludge specific resistance, $w \doteq F/L^3$ the weight of dry sludge per unit volume of filtrate, and R_f the specific resistance of the filter medium. What are the dimensions of R_f and α?

7.1.3 Verify that the left-hand side of Eq. 7.2 is dimensionless using the *MLT* system.

7.1.4 The Reynolds number, $\rho VD/\mu$, is a very important parameter in fluid mechanics. Verify that the Reynolds number is dimensionless, using both the *FLT* system and the *MLT* system for basic dimensions, and determine its value for ethyl alcohol flowing at a velocity of 3 m/s through a 0.0505 m-diameter pipe.

7.1.5 For the flow of a thin film of a liquid with a depth h and a free surface, two important dimensionless parameters are the Froude number, $V/\sqrt{gh}$, and the Weber number, $\rho V^2 h/\sigma$. Determine the value of these two parameters for glycerin (at 20 °C) flowing with a velocity of 0.7 m/s at a depth of 3 mm.

7.1.6 The Mach number for a body moving through a fluid with velocity V is defined as V/c, where c is the speed of sound in the fluid. This dimensionless parameter is usually considered to be important in fluid dynamics problems when its value exceeds 0.3. What would be the velocity of a body at a Mach number of 0.3 if the fluid is (a) air at standard atmospheric pressure and 20 °C, and (b) water at the same temperature and pressure?

Section 7.3 Determination of Pi Terms

7.3.1 A mixing basin in a sewage filtration plant is stirred by a mechanical agitator with a power input $\dot{W} \doteq F \cdot L/T$. Other parameters describing the performance of the mixing process are the fluid absolute viscosity $\mu \doteq F \cdot T/L^2$, the basin volume $V \doteq L^3$, and the velocity gradient $G \doteq 1/T$. Determine the form of the dimensionless relationship.

7.3.2 The excess pressure inside a bubble (discussed in Chapter 1) is known to be dependent on bubble radius and surface tension. After finding the pi terms, determine the variation in excess pressure if we (a) double the radius and (b) double the surface tension.

7.3.3 At a sudden contraction in a pipe the diameter changes from D_1 to D_2. The pressure drop, Δp, which develops across the contraction is a function of D_1 and D_2, as well as the velocity, V, in the larger pipe, and the fluid density, ρ, and viscosity, μ. Use D_1, V, and μ as repeating variables to determine a suitable set of dimensionless parameters. Why would it be incorrect to include the velocity in the smaller pipe as an additional variable?

7.3.4 Water sloshes back and forth in a tank as shown in **Fig. P7.3.4**. The frequency of sloshing, ω, is assumed to be a function of the acceleration of gravity, g, the average depth of the water, h, and the length of the tank, ℓ. Develop a suitable set of dimensionless parameters for this problem using g and ℓ as repeating variables.

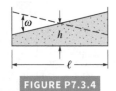

FIGURE P7.3.4

7.3.5 It is desired to determine the wave height when wind blows across a lake. The wave height, H, is assumed to be a function of the wind speed, V, the water density, ρ, the air density, ρ_a, the water depth, d, the distance from the shore, ℓ, and the acceleration of gravity, g, as shown in **Fig. P7.3.5**. Use d, V, and ρ as repeating variables to determine a suitable set of pi terms that could be used to describe this problem.

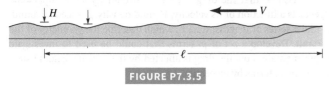

FIGURE P7.3.5

7.3.6 The pressure rise, Δp, across a pump can be expressed as

$$\Delta p = f(D, \rho, \omega, Q)$$

where D is the impeller diameter, ρ the fluid density, ω the rotational speed, and Q the flowrate. Determine a suitable set of dimensionless parameters.

7.3.7 A thin elastic wire is placed between rigid supports. A fluid flows past the wire, and it is desired to study the static deflection, δ, at the center of the wire due to the fluid drag. Assume that

$$\delta = f(\ell, d, \rho, \mu, V, E)$$

where ℓ is the wire length, d the wire diameter, ρ the fluid density, μ the fluid viscosity, V the fluid velocity, and E the modulus of elasticity of the wire material. Develop a suitable set of pi terms for this problem.

7.3.8 (Video) Because of surface tension, it is possible, with care, to support an object heavier than water on the water surface as shown in **Fig. P7.3.8** (see **Video V1.9**). The maximum thickness, h, of a square of material that can be supported is assumed to be a function of the length of the side of the square, ℓ, the density of the material, ρ, the acceleration of gravity, g, and the surface tension of the liquid, σ. Develop a suitable set of dimensionless parameters for this problem.

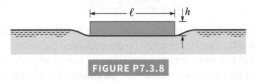

FIGURE P7.3.8

7.3.9 (Video) Under certain conditions, wind blowing past a rectangular speed limit sign can cause the sign to oscillate with a frequency ω (see **Fig. P7.3.9** and **Video V9.9**). Assume that ω is a function of the sign width, b, sign height, h, wind velocity, V, air density, ρ, and an elastic constant, k, for the supporting pole. The constant, k, has dimensions of FL. Develop a suitable set of pi terms for this problem.

FIGURE P7.3.9

7.3.10 A cone and plate viscometer consists of a cone with a very small angle α that rotates above a flat surface as shown in **Fig. P7.3.10**. The torque, $\mathcal{T}$, required to rotate the cone at an angular velocity ω is a function of the radius, R, the cone angle, α, and the fluid viscosity, μ, in addition to ω. With the aid of dimensional analysis, determine how the torque will change if both the viscosity and angular velocity are doubled.

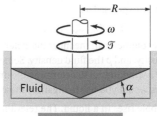

FIGURE P7.3.10

7.3.11 The speed of sound in a gas, c, is a function of the gas pressure, p, and density, ρ. Determine, with the aid of dimensional analysis, how the velocity is related to the pressure and density. Be careful when you decide on how many reference dimensions are required.

7.3.12 A cylinder with a diameter D floats upright in a liquid as shown in **Fig. P7.3.12**. When the cylinder is displaced slightly along its vertical axis it will oscillate about its equilibrium position with a frequency, ω. Assume that this frequency is a function of the diameter, D, the mass of the cylinder, m, and the specific weight, γ, of the liquid. Determine, with the aid of dimensional analysis, how the frequency is related to these variables. If the mass of the cylinder were increased, would the frequency increase or decrease?

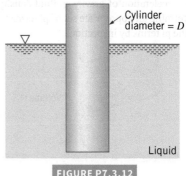

FIGURE P7.3.12

7.3.13 Ⓞ Consider a typical situation involving the flow of a fluid that you encounter almost every day. List what you think are the important physical variables involved in this flow and determine an appropriate set of pi terms for this situation.

7.3.14 The weir shown in **Fig. P7.3.14** is used to measure the volume flowrate Q. The height H is a measure of this flowrate. The weir has length L (perpendicular to the paper). Select and include relevant fluid properties and find the appropriate dimensionless parameters.

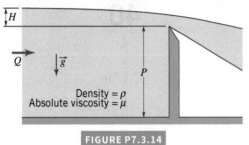

Density $= \rho$
Absolute viscosity $= \mu$

FIGURE P7.3.14

7.3.15 Experiments are conducted on a washing machine agitator. The relevant dimensional parameters are the driving torque, $\mathcal{T}$, the oscillation frequency, f, the angular velocity, ω, the number of paddles, N, the paddle height, H, and the paddle width, w. Specify the relevant fluid properties and find the appropriate dimensionless parameters.

7.3.16 The input power, $\dot{W}$, to a large industrial fan depends on the fan impeller diameter D, fluid viscosity μ, fluid density ρ, volumetric flow Q, and blade rotational speed ω. What are the appropriate dimensionless parameters?

7.3.17 The drag, $\mathcal{D}$, on a washer-shaped plate placed normal to a stream of fluid can be expressed as

$$\mathcal{D} = f(d_1, d_2, V, \mu, \rho)$$

where d_1 is the outer diameter, d_2 the inner diameter, V the fluid velocity, μ the fluid viscosity, and ρ the fluid density. Some experiments are to be performed in a wind tunnel to determine the drag. What dimensionless parameters would you use to organize these data?

7.3.18 A vapor bubble rises in a liquid. The relevant dimensional parameters are the liquid specific weight, γ_ℓ, the vapor specific weight, γ_v, bubble velocity, V, bubble diameter, d, surface tension, σ, and liquid viscosity, μ. Find appropriate dimensionless parameters.

7.3.19 The flowrate, Q, of water in an open channel is assumed to be a function of the cross-sectional area of the channel, A, the height of the roughness of the channel surface, ε, the acceleration of gravity, g, and the slope, S_o, of the hill on which the channel sits. Put this relationship into dimensionless form.

Section 7.5 Determination of Pi Terms by Inspection

7.5.1 A viscous fluid is poured onto a horizontal plate as shown in **Fig. P7.5.1**. Assume that the time, t, required for the fluid to flow a certain distance, d, along the plate is a function of the volume of fluid poured, $\mathcal{V}$, acceleration of gravity, g, fluid density, ρ, and fluid viscosity, μ. Determine an appropriate set of pi terms to describe this process. Form the pi terms by inspection.

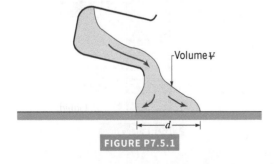

Volume $\mathcal{V}$

d

FIGURE P7.5.1

7.5.2 The velocity, c, at which pressure pulses travel through arteries (pulse-wave velocity) is a function of the artery diameter, D, and wall thickness, h, the density of blood, ρ, and the modulus of elasticity, E, of the arterial wall. Determine a set of nondimensional parameters that can be used to study experimentally the relationship between the pulse-wave velocity and the variables listed. Form the nondimensional parameters by inspection.

7.5.3 Video As shown in **Fig. P7.5.3** and Video V5.6, a jet of liquid directed against a block can tip over the block. Assume that the velocity, V, needed to tip over the block is a function of the fluid density, ρ, the diameter of the jet, D, the weight of the block, $\mathcal{W}$, the width of the block, b, and the distance, d, between the jet and the bottom of the block. (a) Determine a set of dimensionless parameters for this problem. Form the dimensionless parameters by inspection. (b) Use the momentum equation to determine an equation for V in terms of the other variables. (c) Compare the results of parts (a) and (b).

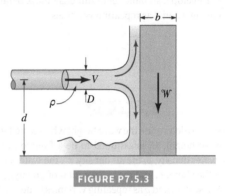

FIGURE P7.5.3

7.5.4 Assume that the drag, $\mathcal{D}$, on an aircraft flying at supersonic speeds is a function of its velocity, V, fluid density, ρ, speed of sound, c, and a series of lengths, $\ell_1, ..., \ell_i$, which describe the geometry of the aircraft. Develop a set of pi terms that could be used to investigate experimentally how the drag is affected by the various factors listed. Form the pi terms by inspection.

7.5.5 A screw propeller has the following relevant dimensional parameters: axial thrust, F, propeller diameter, D, fluid kinematic viscosity, ν, fluid density, ρ, gravitational acceleration, g, advance velocity, V, and rotational speed, N. Find appropriate dimensionless parameters to present the test data.

Section 7.6 Common Dimensionless Groups in Fluid Mechanics

7.6.1 A large, hot plate hangs vertically in a room. Heat transfer from the plate to the air near it causes the air density to decrease. This lighter air rises upward with a velocity proportional to the group

$$[\alpha_T g L |T_p - T_f|]^{1/2}$$

where α_T is the coefficient of thermal expansion [units $=$ K^{-1}], g is the acceleration of gravity, L is the plate length, and T_p and T_f are the temperatures of the plate and fluid, respectively, in K. The resulting heat transfer process depends on the Grashof number, defined by

$$G = \frac{\rho^2 \alpha_T g |T_p - T_f| L^3}{\mu^2}$$

where ρ is density and μ is dynamic viscosity. Show that G is dimensionless. How is G related to the common dimensionless parameters of fluid mechanics? Is it equivalent to a combination, power, or product of one or more of the common dimensionless parameters listed in Table 7.1?

7.6.2 Develop the Weber number by starting with estimates for the inertia and surface tension forces.

7.6.3 Develop the Froude number by starting with estimates of the fluid kinetic energy and fluid potential energy.

7.6.4 The following dimensionless groups are often used to present data on centrifugal pumps: flow coefficient $\varphi = \dfrac{Q}{\omega D^3}$, head coefficient $\psi = \dfrac{gH}{\omega^2 D^2}$, power coefficient $\xi = \dfrac{\dot{W}}{\omega^3 D^5}$, efficiency $\eta = \dfrac{\rho g Q H}{\dot{W}}$, specific speed $N_s = \dfrac{\omega \sqrt{Q}}{(gh)^{3/4}}$, specific diameter $D_s = \dfrac{D(gH)^{1/4}}{\sqrt{Q}}$. Show that the last three groups can be formed from combinations of the first three groups.

7.6.5 A liquid spray nozzle is designed to produce a specific size droplet with diameter, d. The droplet size depends on the nozzle diameter, D, nozzle velocity, V, and the liquid properties ρ, μ, σ. Using the common dimensionless terms found in Table 7.1, determine the functional relationship for the dependent diameter ratio of d/D.

Section 7.7 Correlation of Experimental Data

7.7.1 C The pressure rise, $\Delta p = p_2 - p_1$, across the abrupt expansion of **Fig. P7.7.1** through which a liquid is flowing can be expressed as

$$\Delta p = f(A_1, A_2, \rho, V_1)$$

where A_1 and A_2 are the upstream and downstream cross-sectional areas, respectively, ρ is the fluid density, and V_1 is the upstream velocity. Some experimental data obtained with $A_2 = 0.12$ m^2, $V_1 = 1.5$ m/s, and using water with $\rho = 1000$ kg/m^3 are given in the following table:

A_1 (m^2)	9.3×10^{-3}	2.3×10^{-2}	3.4×10^{-2}	4.8×10^{-2}	5.6×10^{-2}
Δp (N/m^2)	156	376	493	555	589

Plot the results of these tests using suitable dimensionless parameters. With the aid of a standard curve fitting program determine a general equation for Δp and use this equation to predict Δp for water flowing through an abrupt expansion with an area ratio $A_1/A_2 = 0.35$ at a velocity $V_1 = 1.14$ m/s.

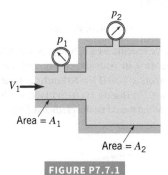

FIGURE P7.7.1

7.7.2 C The pressure drop, Δp, over a certain length of horizontal pipe is assumed to be a function of the velocity, V, of the fluid in the pipe, the pipe diameter, D, and the fluid density and viscosity, ρ and μ. (a) Show that this flow can be described in dimensionless form as a "pressure coefficient," $C_p = \Delta p/(0.5\rho V^2)$ that depends on the Reynolds number, $Re = \rho V D/\mu$. (b) The following data were obtained in an experiment involving a fluid with $\rho = 1030$ kg/m^3, $\mu = 0.1$ Pa $\cdot$ s, and $D = 0.03$ m. Plot a dimensionless graph and use a power law equation to determine the functional relationship between the pressure coefficient and the Reynolds number. (c) What are the limitations on the applicability of your equation obtained in part (b)?

V, m/s	Δp, kPa
1	9
3	34
5	52
6	61

7.7.3 C Video As shown in Fig. 2.26, **Fig. P7.7.3**, and Video V2.12, a rectangular barge floats in a stable configuration provided the distance between the center of gravity, CG, of the object (boat and load) and the center of buoyancy, C, is less than a certain amount, H. If this distance is greater than H, the boat will tip over. Assume H is a function of the boat's width, b, length, ℓ, and draft, h. (a) Put this relationship into dimensionless form. (b) The results of a set of experiments with a model barge with a width of 1.0 m are shown in the table. Plot these data in dimensionless form and determine a power-law equation relating the dimensionless parameters.

ℓ, m	h, m	H, m
2.0	0.10	0.833
4.0	0.10	0.833
2.0	0.20	0.417
4.0	0.20	0.417
2.0	0.35	0.238
4.0	0.35	0.238

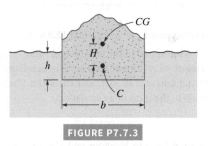

FIGURE P7.7.3

7.7.4 Video The time, t, it takes to pour a certain volume of liquid from a cylindrical container depends on several factors, including the viscosity of the liquid (see Video V1.3). Assume that for very viscous liquids the time it takes to pour out two-thirds of the initial volume depends on the initial liquid depth, ℓ, the cylinder diameter, D, the liquid viscosity, μ, and the liquid specific weight, γ. The data shown in the following table were obtained in the laboratory. For these tests $\ell = 45$ mm, $D = 67$ mm, and $\gamma = 9.60$ kN/m^3. (a) Perform a dimensional analysis, and based on the data given, determine if variables used for this problem appear to be correct. Explain how you arrived at your answer. (b) If possible, determine an equation relating the pouring time and viscosity for the cylinder and liquids used in these tests. If it is not possible, indicate what additional information is needed.

μ (N $\cdot$ s/m^2)	11	17	39	61	107
t (s)	15	23	53	83	145

7.7.5 In order to maintain uniform flight, smaller birds must beat their wings faster than larger birds. It is suggested that the relationship between the wingbeat frequency, ω, beats per second, and the bird's wingspan, ℓ, is given by a power law relationship, $\omega \sim \ell^n$. **(a)** Use dimensional analysis with the assumption that the wingbeat frequency is a function of the wingspan, the specific weight of the bird, γ, the acceleration of gravity, g, and the density of the air, ρ_a, to determine the value of the exponent n. **(b)** Some typical data for various birds are given in the following table. Do these data support your result obtained in part (a)? Provide appropriate analysis to show how you arrived at your conclusion.

Bird	Wingspan, m	Wingbeat frequency, beats/s
Purple martin	0.28	5.3
Robin	0.36	4.3
Mourning dove	0.46	3.2
Crow	1.00	2.2
Canada goose	1.50	2.6
Great blue heron	1.80	2.0

7.7.6 A 250-m-long ship has a wetted area of 8000 m^2. A $\frac{1}{100}$ scale model is tested in a towing tank with the prototype fluid, and the results are:

Model velocity (m/s)	0.57	1.02	1.40
Model drag (N)	0.50	1.02	1.65

Calculate the prototype drag at 7.5 m/s and 12.0 m/s.

7.7.7 A student is interested in the aerodynamic drag on spheres. She conducts a series of wind tunnel tests on a 10-cm-diameter sphere. The air in the wind tunnel is at 50 °C and 101.3 kPa. She presents the results of her tests in the form of a correlation,

$$\mathscr{D} = 0.0015V^{1.92}$$

where $\mathscr{D}$ is the drag force in Newtons and V is the velocity in m/s. Another student realizes that the results would have been more effectively presented in terms of dimensionless parameters, say,

$$\Pi_1 = f(\Pi_2) = C\Pi_2^a$$

where Π_1 and Π_2 are dimensionless parameters and C and a are constants. What are the most appropriate dimensionless parameters (Π_1 and Π_2) and the corresponding values of C and a? What would be the drag on a 2-cm-diameter sphere placed in a 1 m/s, 20 °C water stream?

Section 7.8 Modeling and Similitude

7.8.1 SAE 30 oil at 15.5 °C is pumped through a 0.91 m-diameter pipeline at a rate of 1690 liter/min. A model of this pipeline is to be designed using a 0.076 m-diameter pipe and water at 15.5 °C as the working fluid. To maintain Reynolds number similarity between these two systems, what fluid velocity will be required in the model?

7.8.2 (Video) You are to conduct wind tunnel testing of a new football design that has a smaller lace height than previous designs (see Videos V6.1 and V6.2). It is known that you will need to maintain Re and St similarity for the testing. Based on standard college quarterbacks, the prototype parameters are set at $V = 65$ km/hr and $\omega = 200$ rpm, where V and ω are the speed and angular velocity of the football. The prototype football has a 20 cm diameter. Due to instrumentation required to measure pressure and shear stress on the surface of the football, the model will require a length scale of 2:1 (the model will be larger than the prototype). Determine the required model freestream velocity and model angular velocity.

7.8.3 A model of a submarine, 1:15 scale, is to be tested at 55 m/s in a wind tunnel with standard sea-level air, while the prototype will be operated in seawater. Determine the speed of the prototype to ensure Reynolds number similarity.

7.8.4 The drag characteristics of a torpedo are to be studied in a water tunnel using a 1:5 scale model. The tunnel operates with fresh water at 20 °C, whereas the prototype torpedo is to be used in seawater at 15.6 °C. To correctly simulate the behavior of the prototype moving with a velocity of 30 m/s, what velocity is required in the water tunnel?

7.8.5 The water velocity at a certain point along a 1:10 scale model of a dam spillway is 3 m/s. What is the corresponding prototype velocity if the model and prototype operate in accordance with Froude number similarity?

7.8.6 The fluid dynamic characteristics of an airplane flying 390 km/hr at 3000 m are to be investigated with the aid of a 1:20 scale model. If the model tests are to be performed in a wind tunnel using standard air, what is the required air velocity in the wind tunnel? Is this a realistic velocity?

7.8.7 If an airplane travels at a speed of 1120 km/hr at an altitude of 15 km, what is the required speed at an altitude of 8 km to satisfy Mach number similarity? Assume the air properties correspond to those for the U.S. standard atmosphere.

7.8.8 When small particles of diameter d are transported by a moving fluid having a velocity V, they settle to the ground at some distance ℓ after starting from a height h as shown in **Fig. P7.8.8**. The variation in ℓ with various factors is to be studied with a model having a length scale of $\frac{1}{10}$. Assume that

$$\ell = f(h, d, V, \gamma, \mu)$$

where γ is the particle specific weight and μ is the fluid viscosity. The same fluid is to be used in both the model and the prototype, but γ(model) $= 9 \times \gamma$(prototype). **(a)** If $V = 80$ km/hr, at what velocity should the model tests be run? **(b)** During a certain model test it was found that ℓ (model) $= 0.24$ m. What would be the predicted ℓ for this test?

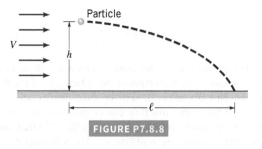

FIGURE P7.8.8

7.8.9 A thin layer of an incompressible fluid flows steadily over a horizontal smooth plate as shown in **Fig. P7.8.9**. The fluid surface is open to the atmosphere, and an obstruction having a square cross section is placed on the plate as shown. A model with a length scale of $\frac{1}{4}$ and a fluid density scale of 1.0 is to be designed to predict the depth of fluid, y, along the plate. Assume that inertial, gravitational, surface tension, and viscous effects are all important. What are the required viscosity and surface tension scales?

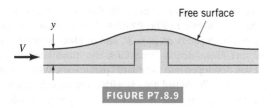

Free surface

FIGURE P7.8.9

7.8.10 (Video) During a storm, a snow drift is formed behind a snow fence as shown in **Fig. P7.8.10** (see Video V9.9). Assume that the height of the drift, h, is a function of the number of inches of snow deposited by the storm, d, height of the fence, H, width of slats in the fence, b, wind speed, V, acceleration of gravity, g, air density, ρ, and specific weight of snow, γ_s. (a) If this problem is to be studied with a model, determine the similarity requirements for the model and the relationship between the drift depth for model and prototype (prediction equation). (b) A storm with winds of 48 km/hr deposits 40 cm of snow having a specific weight of 785 N/m³. A $\frac{1}{2}$-sized scale model is to be used to investigate the effectiveness of a proposed snow fence. If the air density is the same for the model and the storm, determine the required specific weight for the model snow and required wind speed for the model.

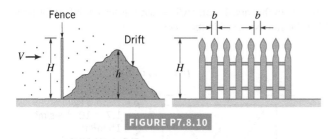

FIGURE P7.8.10

7.8.11 The pressure drop between the entrance and exit of a 150-mm-diameter 90° elbow, through which ethyl alcohol at 20 °C is flowing, is to be determined with a geometrically similar model. The velocity of the alcohol is 5 m/s. The model fluid is to be water at 20 °C, and the model velocity is limited to 10 m/s. (a) What is the required diameter of the model elbow to maintain dynamic similarity? (b) A measured pressure drop of 20 kPa in the model will correspond to what prototype value?

7.8.12 For a certain model study involving a 1:5 scale model, it is known that Froude number similarity must be maintained. The possibility of cavitation is also to be investigated, and it is assumed that the cavitation number must be the same for model and prototype. The prototype fluid is water at 30 °C, and the model fluid is water at 70 °C. If the prototype operates at an ambient pressure of 101 kPa (abs), what is the required ambient pressure for the model system?

7.8.13 A solid sphere having a diameter d and specific weight γ_s is immersed in a liquid having a specific weight $\gamma_f (\gamma_f > \gamma_s)$ and then released. It is desired to use a model system to determine the maximum height, h, above the liquid surface that the sphere will rise upon release from a depth H. It can be assumed that the important liquid properties are the density, γ_f/g, specific weight, γ_f, and viscosity, μ_f. Establish the model design conditions and the prediction equation, and determine whether the same liquid can be used in both the model and prototype systems.

7.8.14 A thin layer of particles rests on the bottom of a horizontal tube as shown in **Fig. P7.8.14**. When an incompressible fluid flows through the tube, it is observed that at some critical velocity the particles will rise and be transported along the tube. A model is to be used to determine this critical velocity. Assume the critical velocity, V_c, to be a function of the pipe diameter, D, particle diameter, d, the fluid density, ρ, and viscosity, μ, the density of the particles, ρ_p, and the acceleration of gravity, g. (a) Determine the similarity requirements for the model, and the relationship between the critical velocity for model and prototype (the prediction equation). (b) For a length scale of $\frac{1}{2}$ and a fluid density scale of 1.0, what will be the critical velocity scale (assuming all similarity requirements are satisfied)?

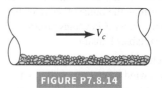

V_c

FIGURE P7.8.14

7.8.15 (Video) The pressure rise, Δp, across a blast wave, as shown in **Fig. P7.8.15**, and Video V11.8, is assumed to be a function of the amount of energy released in the explosion, E, the air density, ρ, the speed of sound, c, and the distance from the blast, d. (a) Put this relationship in dimensionless form. (b) Consider two blasts: the prototype blast with energy release E and a model blast with 1/1000th the energy release ($E_m = 0.001E$). At what distance from the model blast will the pressure rise be the same as that at a distance of 1.6 km from the prototype blast?

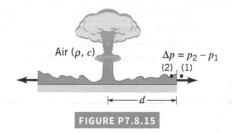

Air (ρ, c) $\Delta p = p_2 - p_1$
                         (2) (1)
$\leftarrow$ d $\rightarrow$

FIGURE P7.8.15

7.8.16 An incompressible fluid oscillates harmonically ($V = V_0 \sin \omega t$, where V is the velocity) with a frequency of 10 rad/s in a 10 cm-diameter pipe. A $\frac{1}{4}$ scale model is to be used to determine the pressure difference per unit length, Δp_ℓ (at any instant) along the pipe. Assume that

$$\Delta p_\ell = f(D, V_0, \omega, t, \mu, \rho)$$

where D is the pipe diameter, ω the frequency, t the time, μ the fluid viscosity, and ρ the fluid density. (a) Determine the similarity requirements for the model and the prediction equation for Δp_ℓ. (b) If the same fluid is used in the model and the prototype, at what frequency should the model operate?

7.8.17 The drag, D, on a sphere located in a pipe through which a fluid is flowing is to be determined experimentally (see Fig. P7.8.17). Assume that the drag is a function of the sphere diameter, d, the pipe diameter, D, the fluid velocity, V, and the fluid density, ρ. (a) What dimensionless parameters would you use for this problem? (b) Some experiments using water indicate that for $d = 0.51$ cm, $D = 1.27$ cm, and $V = 0.61$ m/s, the drag is 6.67×10^{-3} N. If possible, estimate the drag on a sphere located in a 0.609 m/s-diameter pipe through which water is flowing with a velocity of 1.82 m/s. The sphere diameter is such that geometric similarity is maintained. If it is not possible, explain why not.

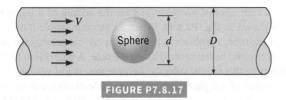

FIGURE P7.8.17

7.8.18 The drag characteristics for a newly designed automobile having a maximum characteristic length of 8 m are to be determined through a model study. The characteristics at both low speed (approximately 30 km/hr) and high speed (140 km/hr) are of interest. For a series of projected model tests, an unpressurized wind tunnel that will accommodate a model with a maximum characteristic length of 1.5 m is to be used. Determine the range of air velocities that would be required for the wind tunnel if Reynolds number similarity is desired. Are the velocities suitable? Explain.

7.8.19 The drag characteristics of an airplane are to be determined by model tests in a wind tunnel operated at an absolute pressure of 1300 kPa. If the prototype is to cruise in standard air at 385 km/hr, and the corresponding speed of the model is not to differ by more than 20% from this (so that compressibility effects may be ignored), what range of length scales may be used if Reynolds number similarity is to be maintained? Assume the viscosity of air is unaffected by pressure, and the temperature of air in the tunnel is equal to the temperature of the air in which the airplane will fly.

7.8.20 The drag on a sphere moving in a fluid is known to be a function of the sphere diameter, the velocity, and the fluid viscosity and density. Laboratory tests on a 10 cm-diameter sphere were performed in a water tunnel and some model data are plotted in **Fig. P7.8.20**. For these tests the viscosity of the water was 1×10^{-3} Pa · s and the water density was 1000 kg/m³. Estimate the drag on an 2.4 m-diameter balloon moving in air at a velocity of 1 m/s. Assume the air to have a viscosity of 1.8×10^{-5} Pa · s and a density of 1.2 kg/m³.

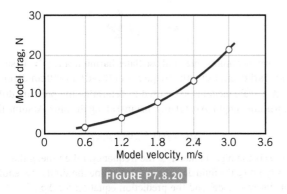

FIGURE P7.8.20

7.8.21 The design of a river model is to be based on Froude number similarity, and a river depth of 3 m is to correspond to a model depth of 100 mm. Under these conditions what is the prototype velocity corresponding to a model velocity of 2 m/s?

7.8.22 Glycerin at 20 °C flows with a velocity of 4 m/s through a 30 mm-diameter tube. A model of this system is to be developed using standard air as the model fluid. The air velocity is to be 2 m/s. What tube diameter is required for the model if dynamic similarity is to be maintained between model and prototype?

7.8.23 To test the aerodynamics of a new prototype automobile, a scale model will be tested in a wind tunnel. For dynamic similarity, it will be required to match Reynolds number between model and prototype.

Assuming that you will be testing a one-tenth-scale model and both model and prototype will be exposed to standard air pressure, will it be better for the wind tunnel air to be colder or hotter than standard sea-level air temperature of 15°C? Why?

7.8.24 In low-speed external flow over a bluff object, vortices are shed from the object (as shown in **Fig. P7.8.24**). The frequency of vortex shedding, $f \doteq 1/T$, depends on ρ, μ, V, and d. Make a dimensional analysis of this phenomenon. Transform the dimensionless parameters that you developed into "standard" parameters. What are their names? Two identically shaped objects with a size ratio 2:1 are tested in air flow. What velocity ratio is necessary for dynamic similarity? What is the ratio of vortex-shedding frequencies?

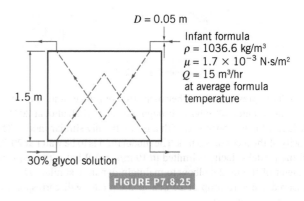

FIGURE P7.8.24

7.8.25 In a nutritional products plant, a plate heat exchanger cools infant formula by approximately 22 °C from an inlet temperature of 68 °C as it flows through the plates shown in **Fig. P7.8.25**. The process engineer would like to be able to cool the infant formula by as much as 30 °C as it passes through the plates. The plate heat exchanger is currently in use, so he would like to build a $\frac{1}{2}$-scale model for testing purposes. In the model, water will be used instead of infant formula. What flowrate is needed in the model?

D = 0.05 m

Infant formula
ρ = 1036.6 kg/m³
μ = 1.7 × 10⁻³ N·s/m²
Q = 15 m³/hr
at average formula temperature

1.5 m

30% glycol solution

FIGURE P7.8.25

Section 7.9 Typical Model Studies

7.9.1 The pressure rise, Δp, across a centrifugal pump of a given shape can be expressed as

$$\Delta p = f(D, \omega, \rho, Q)$$

where D is the impeller diameter, ω the angular velocity of the impeller, ρ the fluid density, and Q the volume rate of flow through the pump. A model pump having a diameter of 20 cm is tested in the laboratory using water. When operated at an angular velocity of 40π rad/s the model pressure rise as a function of Q is shown in **Fig. P7.9.1**. Use this curve to predict the pressure rise across a geometrically similar pump (prototype) for a prototype flowrate of 0.2 m³/s. The prototype has a diameter of 30 cm and operates at an angular velocity of 60π rad/s. The prototype fluid is also water.

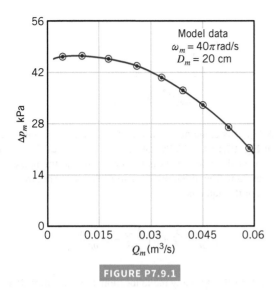

FIGURE P7.9.1

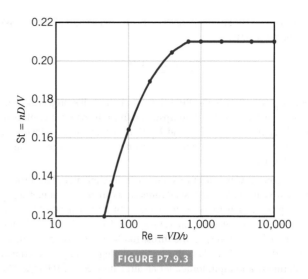

FIGURE P7.9.3

7.9.2 (Video) At a large fish hatchery the fish are reared in open, water-filled tanks. Each tank is approximately square in shape with curved corners, and the walls are smooth. To create motion in the tanks, water is supplied through a pipe at the edge of the tank. The water is drained from the tank through an opening at the center (see Video V7.8). A model with a length scale of 1:13 is to be used to determine the velocity, V, at various locations within the tank. Assume that $V = f(\ell, \ell_i, \rho, \mu, g, Q)$ where ℓ is some characteristic length such as the tank width, ℓ_i represents a series of other pertinent lengths, such as inlet pipe diameter, fluid depth, etc., ρ is the fluid density, μ is the fluid viscosity, g is the acceleration of gravity, and Q is the discharge through the tank. (a) Determine a suitable set of dimensionless parameters for this problem and the prediction equation for the velocity. If water is to be used for the model, can all of the similarity requirements be satisfied? Explain and support your answer with the necessary calculations. (b) If the flowrate into the full-sized tank is 950 liter/min, determine the required value for the model discharge assuming Froude number similarity. What model depth will correspond to a depth of 80 cm in the full-sized tank?

7.9.3 (Video) (See The Wide World of Fluids article titled "Galloping Gertie," Section 7.8.2.) The Tacoma Narrows Bridge failure is a dramatic example of the possible serious effects of wind-induced vibrations. As a fluid flows around a body, vortices may be created that are shed periodically, creating an oscillating force on the body. If the frequency of the shedding vortices coincides with the natural frequency of the body, large displacements of the body can be induced as was the case with the Tacoma Narrows Bridge. To illustrate this type of phenomenon, consider fluid flow past a circular cylinder (see Video V7.5). Assume the frequency, n, of the shedding vortices behind the cylinder is a function of the cylinder diameter, D, the fluid velocity, V, and the fluid kinematic viscosity, v. (a) Determine a suitable set of dimensionless variables for this problem. One of the dimensionless variables should be the Strouhal number, nD/V. (b) Some results of experiments in which the shedding frequency of the vortices (in Hz) was measured, using a particular cylinder and Newtonian, incompressible fluid, are shown in **Fig. P7.9.3**. Is this a "universal curve" that can be used to predict the shedding frequency for any cylinder placed in any fluid? Explain. (c) A certain structural component in the form of a 2.5 cm-diameter, 4 m-long rod acts as a cantilever beam with a natural frequency of 19 Hz. Based on the data in Fig. P7.9.1, estimate the wind speed that may cause the rod to oscillate at its natural frequency. *Hint*: Use a trial-and-error solution.

7.9.4 (See The Wide World of Fluids article titled "Ice Engineering," Section 7.9.3.) A model study is to be developed to determine the force exerted on bridge piers due to floating chunks of ice in a river. The piers of interest have square cross sections. Assume that the force, R, is a function of the pier width, b, the thickness of the ice, d, the velocity of the ice, V, the acceleration of gravity, g, the density of the ice, ρ_i, and a measure of the strength of the ice, E_i, where E_i has the dimensions FL^{-2}. (a) Based on these variables determine a suitable set of dimensionless variables for this problem. (b) The prototype conditions of interest include an ice thickness of 30 cm and an ice velocity of 1.8 m/s. What model ice thickness and velocity would be required if the length scale is to be 1/10? (c) If the model and prototype ice have the same density, can the model ice have the same strength properties as that of the prototype ice? Explain.

7.9.5 (Video) As illustrated in Video V7.8, models are commonly used to study the dispersion of a gaseous pollutant from an exhaust stack located near a building complex. Similarity requirements for the pollutant source involve the following independent variables: the stack gas speed, V, the wind speed, U, the density of the atmospheric air, ρ, the difference in densities between the air and the stack gas, $\rho - \rho_s$, the acceleration of gravity, g, the kinematic viscosity of the stack gas, v_s, and the stack diameter, D. (a) Based on these variables, determine a suitable set of similarity requirements for modeling the pollutant source. (b) For this type of model a typical length scale might be 1:200. If the same fluids were used in model and prototype, would the similarity requirements be satisfied? Explain and support your answer with the necessary calculations.

7.9.6 Drag measurements were taken for a sphere with a diameter of 5 cm, moving at 4 m/s in water at 20 °C. The resulting drag on the sphere was 10 N. For a balloon with 1-m diameter rising in air with standard temperature and pressure, determine (a) the velocity if Reynolds number similarity is enforced and (b) the drag force if the drag coefficient (Eq. 7.19) is the dependent pi term.

7.9.7 A prototype automobile is designed to travel at 65 km/hr. A model of this design is tested in a wind tunnel with identical standard sea-level air properties at a 1:5 scale. The measured model drag is 400 N, enforcing dynamic similarity. Determine (a) the drag force on the prototype and (b) the power required to overcome this drag. See Eq. 7.19.

7.9.8 A stream of atmospheric air is used to keep a ping-pong ball aloft by blowing the air upward over the ball. The ping-pong ball has a mass of 3 g and a diameter $D_1 = 4$ cm, and the air stream has an upward velocity of $V_1 = 1$ m/s. This system is to be modeled by pumping water upward with a velocity V_2 over a solid ball of diameter D_2 and density $\rho_{b_2} = 2710 \text{ kg/m}^3$. In both cases, the net weight of the ball W_b is equal to the drag,

$$W_b = \frac{C_D \rho A V^2}{2}$$

where $C_D = 0.60$, ρ is the fluid density, A the ball's projected area, and V the velocity of the fluid upstream from the ball. Determine all possible combinations of V_2 and D_2. *Hint:* A force balance involving the drag on the ball, the buoyant force on the ball, and the weight of the ball is needed.

7.9.9 A company manufactures geometrically similar airplane propellers up to 5.0 m in diameter. Wind tunnel tests are run on geometrically similar propellers up to 0.25 m in diameter. In the test of a 0.33-m model of a 5.0-m propeller, the air velocity was 50 m/s. The model rotational speed was 2500 rpm, the thrust 100 N, and the propeller input torque 30 N · m. Calculate the corresponding prototype rotational speed, thrust, and input torque for an airplane speed of 100 m/s. The Reynolds numbers and Mach numbers are of minor importance.

7.9.10 A new blimp will move at 6 m/s in 20 °C air, and we want to predict the drag force. Using a 1:13-scale model in water at 20 °C and measuring a 2500-N drag force on the model, determine (a) the required water velocity, (b) the drag on the prototype blimp and, (c) the power that will be required to propel it through the air.

Section 7.10 Similitude Based on Governing Differential Equations

7.10.1 An incompressible fluid is contained between two infinite parallel plates as illustrated in **Fig. P7.10.1**. Under the influence of a harmonically varying pressure gradient in the x direction, the fluid oscillates harmonically with a frequency ω. The differential equation describing the fluid motion is

$$\rho \frac{\partial u}{\partial t} = X \cos \omega t + \mu \frac{\partial^2 u}{\partial y^2}$$

where X is the amplitude of the pressure gradient. Express this equation in nondimensional form using h and ω as reference parameters.

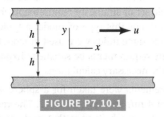

FIGURE P7.10.1

7.10.2 The deflection of the cantilever beam of **Fig. P7.10.2** is governed by the differential equation.

$$EI \frac{d^2 y}{dx^2} = P(x - \ell)$$

where E is the modulus of elasticity and I is the moment of inertia of the beam cross section. The boundary conditions are $y = 0$ at $x = 0$ and $dy/dx = 0$ at $x = 0$. **(a)** Rewrite the equation and boundary conditions in dimensionless form using the beam length, ℓ, as the reference length. **(b)** Based on the results of part (a), what are the similarity requirements and the prediction equation for a model to predict deflections?

FIGURE P7.10.2

7.10.3 A liquid is contained in a pipe that is closed at one end as shown in **Fig. P7.10.3**. Initially the liquid is at rest, but if the end is suddenly opened the liquid starts to move. Assume the pressure p_1 remains constant. The differential equation that describes the resulting motion of the liquid is

$$\rho \frac{\partial v_z}{\partial t} = \frac{p_1}{\ell} + \mu \left(\frac{\partial^2 v_z}{\partial r^2} + \frac{1}{r} \frac{\partial v_z}{\partial r} \right)$$

where v_z is the velocity at any radial location, r, and t is time. Rewrite this equation in dimensionless form using the liquid density, ρ, the viscosity, μ, and the pipe radius, R, as reference parameters.

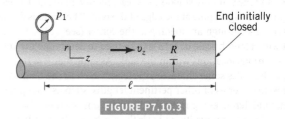

FIGURE P7.10.3

7.10.4 A student drops two spherical balls of different diameters and different densities. She has a stroboscopic photograph showing the positions of each ball as a function of time. However, she wants to express the velocity of each as a function of time in dimensionless form. Develop the dimensionless group. The equation of motion for each ball is

$$mg - \frac{C_D}{2} \rho A V^2 = m \frac{dV}{dt}$$

where m is ball mass, g is acceleration of gravity, C_D is a dimensionless and constant drag coefficient, ρ is air mass density, A is ball cross-sectional area ($= \pi D^2 / 4$ with D ball diameter), V is ball velocity, and t is time.

7.10.5 The basic equation that describes the motion of the fluid above a large oscillating flat plate is

$$\frac{\partial u}{\partial t} = v \frac{\partial^2 u}{\partial y^2}$$

where u is the fluid velocity component parallel to the plate, t is time, y is the spatial coordinate perpendicular to the plate, and v is the fluid kinematic viscosity. The plate oscillating velocity is given by $U = U_0 \sin \omega t$. Find appropriate dimensionless parameters and the dimensionless differential equation.

7.10.6 The dimensionless parameters for a ball released and falling from rest in a fluid are

$$C_D, \quad \frac{gt^2}{D}, \quad \frac{\rho}{\rho_b}, \quad \text{and} \quad \frac{Vt}{D}$$

where C_D is a drag coefficient (assumed to be constant at 0.4), g is the acceleration of gravity, D is the ball diameter, t is the time after it is released, ρ is the density of the fluid in which it is dropped, and ρ_b is the density of the ball. Ball 1, an aluminum ball ($\rho_{b_1} = 2710$ kg/m^3) having a diameter $D_1 = 1.5$ cm, is dropped in water ($\rho = 1000$ kg/m^3). The ball velocity V_1 is 0.75 m/s at $t_1 = 0.10$ s. Find the corresponding velocity V_2 and time t_2 for ball 2 having $D_2 = 2.5$ cm and $\rho_{b_2} = \rho_{b_1}$. Next, use the computed value of V_2 and the equation of motion,

$$\rho_b V g - \frac{C_D}{2} \rho A V^2 = \rho_b V \frac{dV}{dt}$$

where V is the volume of the ball and A is its cross-sectional area, to verify the value of t_2. Should the two values of t_2 agree?

Lifelong Learning Problems

7.1 LL Microfluidics is the study of fluid flow in fabricated devices at the micro scale. Advances in microfluidics have enhanced the ability of scientists and engineers to perform laboratory experiments using miniaturized devices known as a "lab-on-a-chip." Obtain information about a lab-on-a-chip device that is available commercially and investigate its capabilities. Summarize your findings in a brief report.

7.2 LL For some types of aerodynamic wind tunnel testing, it is difficult to simultaneously match both the Reynolds number and Mach number between model and prototype. Engineers have developed several potential solutions to the problem including pressurized wind tunnels and lowering the temperature of the flow. Obtain information about cryogenic wind tunnels and explain the advantages and disadvantages. Summarize your findings in a brief report.

7.3 LL As mentioned in the text, there is a large number of common, named, dimensionless groups. Do some research and find at least 20 common dimensionless groups that have not been considered in this chapter. Report the group name, group definition, physical significance of the group, and whether the group is similar to another group or can be formed by combining other groups.

CHAPTER 8

Viscous Flow in Pipes

V8.1 Turbulent jet

Pipe flow is very important in our daily operations.

In the previous chapters we have considered a variety of topics concerning the motion of fluids. The basic governing principles concerning mass, momentum, and energy were developed and applied, in conjunction with rather severe assumptions, to numerous flow situations. In this chapter we will apply the basic principles to a specific, important topic—the incompressible flow of viscous fluids in pipes and ducts.

The transport of a fluid (liquid or gas) in a closed conduit (commonly called a *pipe* if it is of round cross section or a *duct* if it is not round) is extremely important in our daily operations. A brief consideration of the world around us will indicate that there is a wide variety of applications of pipe flow. Such applications range from the large, man-made Alaskan pipeline that carries crude oil almost 1287.5 km across Alaska, to the more complex (and certainly not less useful) natural systems of "pipes" that carry blood throughout our body and air into and out of our lungs. Other examples include the water pipes in our homes and the distribution system that delivers the water from the city well to the house. Numerous hoses and pipes carry hydraulic fluid or other fluids to various components of vehicles and machines. The air quality within our buildings is maintained at comfortable levels by the distribution of conditioned (heated, cooled, humidified/dehumidified) air through a maze of pipes and ducts. Although all of these systems are different, the fluid mechanics principles governing the fluid motions are common. The purpose of this chapter is to understand the basic processes involved in such flows.

Some of the basic components of a typical *pipe system* are shown in **Fig. 8.1**. They include the pipes themselves (perhaps of more than one diameter), the various fittings used to connect the individual pipes to form the desired system, the flowrate control devices (valves), and the pumps or turbines that add energy to or remove energy from the fluid. Even the most simple pipe systems are actually quite complex when they are viewed in terms of rigorous analytical considerations. We will use an "exact" analysis of the simplest pipe flow topics (such as laminar flow in long, straight, constant diameter pipes) and dimensional analysis considerations combined with experimental results for the other pipe flow topics. Such an approach is not unusual in fluid mechanics investigations. When "real-world" effects are important (such as viscous effects in pipe flows), it is often difficult or "impossible" to use only theoretical methods to obtain the desired results. A judicious combination of experimental data with theoretical considerations and dimensional analysis often provides the desired results. The flow in pipes discussed in this chapter is an example of such an analysis.

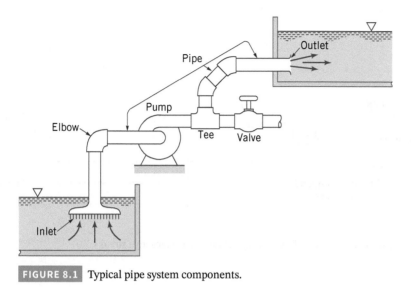

FIGURE 8.1 Typical pipe system components.

8.1 GENERAL CHARACTERISTICS OF PIPE FLOW

Before we apply the various governing equations to pipe flow examples, we will discuss some of the basic concepts of pipe flow. With these ground rules established, we can then proceed to formulate and solve various important flow problems.

Although not all conduits used to transport fluid from one location to another are round in cross section, most of the common ones are. These include typical water pipes, hydraulic hoses, and other conduits that are designed to withstand a considerable pressure difference across their walls without undue distortion of their shape. Typical conduits of noncircular cross section include heating and air conditioning ducts that are often of rectangular cross section. Normally the pressure difference between the inside and outside of these ducts is relatively small. Most of the basic principles involved are independent of the cross-sectional shape, although the details of the flow may be dependent on it. Unless otherwise specified, we will assume that the conduit is round, although we will show how to account for other shapes.

For all flows involved in this chapter, we assume that the pipe is completely filled with the fluid being transported as is shown in **Fig. 8.2a**. Thus, we will not consider a concrete pipe through which rainwater flows without completely filling the pipe, as is shown in **Fig. 8.2b**. Such flows, called open-channel flow, are treated in Chapter 10. The difference between open-channel flow and the pipe flow of this chapter is in the fundamental mechanism that drives the flow. For open-channel flow, gravity alone is the driving force—the water flows down a hill. For pipe flow, gravity may be important (the pipe need not be horizontal), but the main driving force is likely to be a pressure gradient along the pipe. If the pipe is not full, it is not possible to maintain this pressure difference, $p_1 - p_2$.

> The pipe is assumed to be completely full of the flowing fluid.

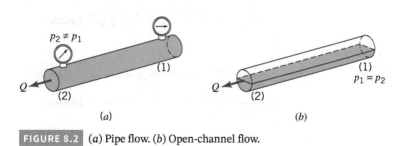

FIGURE 8.2 (a) Pipe flow. (b) Open-channel flow.

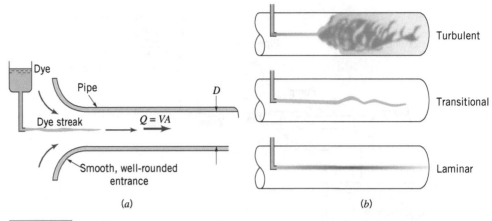

FIGURE 8.3 (*a*) Experiment to illustrate type of flow. (*b*) Typical dye streaks.

8.1.1 Laminar or Turbulent Flow

▶ **Video**

V8.2 Laminar/turbulent pipe flow

A flow may be laminar, transitional, or turbulent.

The flow of a fluid in a pipe may be laminar flow or it may be turbulent flow. Osborne Reynolds (1842–1912), a British scientist and mathematician, was the first to distinguish the difference between these two classifications of flow by using a simple apparatus as shown by the adjacent figure, which is a sketch of Reynolds' dye experiment. Reynolds injected dye into a pipe in which water flowed due to gravity. The entrance region of the pipe is depicted in **Fig. 8.3a**. If water runs through a pipe of diameter D with an average velocity V, the following characteristics are observed by injecting neutrally buoyant dye as shown. For "small enough flowrates" the dye streak (a streakline) will remain as a well-defined line as it flows along, with only slight blurring due to molecular diffusion of the dye into the surrounding water. For a somewhat larger "intermediate flowrate" the dye streak fluctuates in time and space, and intermittent bursts of irregular behavior appear along the streak. On the other hand, for "large enough flowrates" the dye streak almost immediately becomes blurred and spreads across the entire pipe in a random fashion. These three characteristics, denoted as **laminar, transitional,** and **turbulent** flow, respectively, are illustrated in **Fig. 8.3b**.

The curves shown in **Fig. 8.4** represent the x component of the velocity as a function of time at a point A in the flow. The random fluctuations of the turbulent flow (with the associated particle mixing) are what disperse the dye throughout the pipe and cause the blurred appearance illustrated in Fig. 8.3b. For laminar flow in a pipe there is only one component of velocity, $\mathbf{V} = u\hat{\mathbf{i}}$. For turbulent flow the predominant component of velocity is also along the pipe, but it is unsteady (fluctuating) and accompanied by seemingly random components normal to the pipe axis, $\mathbf{V} = u\hat{\mathbf{i}} + v\hat{\mathbf{j}} + w\hat{\mathbf{k}}$. Such motion in a typical flow occurs too fast for our eyes to follow. Slow-motion pictures of the flow can more clearly reveal the irregular, seemingly random, turbulent nature of the flow.

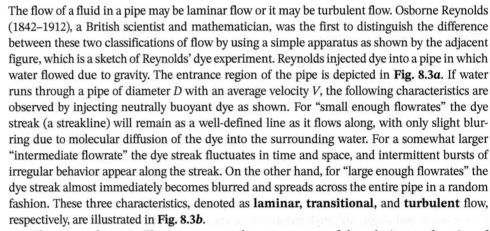

FIGURE 8.4 Time dependence of fluid velocity at a point.

As was discussed in Chapter 7, we should not label dimensional quantities as being "large" or "small," such as "small enough flowrates" in the preceding paragraphs. Rather, the appropriate dimensionless quantity should be identified and the "small" or "large" character attached to it. A quantity is "large" or "small" only relative to a reference quantity. The ratio of those quantities results in a dimensionless quantity. For pipe flow the most important dimensionless parameter is the Reynolds number, Re—the ratio of the inertia to viscous effects in the flow. Hence, in the previous paragraph the term flowrate should be replaced by Reynolds number, $Re = \rho VD/\mu$, where V is the average velocity in the pipe. That is, the flow in a pipe is laminar, transitional, or turbulent provided the Reynolds number is "small enough," "intermediate," or "large enough." It is not only the fluid velocity that determines the character of the flow—its density, viscosity, and the pipe size are of equal importance. These parameters combine to produce the Reynolds number. The distinction between laminar and turbulent pipe flow and its dependence on an appropriate dimensionless quantity was first pointed out by Osborne Reynolds in 1883.

The Reynolds number ranges for which laminar, transitional, or turbulent pipe flows are obtained cannot be precisely given. The actual transition from laminar to turbulent flow may take place at various Reynolds numbers, depending on how much the flow is disturbed by vibrations of the pipe, roughness of the entrance region, and the like. For general engineering purposes (i.e., without undue precautions to eliminate such disturbances), the following values are appropriate: The flow in a round pipe is laminar if the Reynolds number is less than approximately 2100. The flow in a round pipe is turbulent if the Reynolds number is greater than approximately 4000. For Reynolds numbers between these two limits, the flow may switch between laminar and turbulent conditions in an apparently random fashion (transitional flow).

Pipe flow characteristics are dependent on the value of the Reynolds number.

V8.3 Intermittent turbulent burst in pipe flow

The Wide World of Fluids

Nanoscale flows

The term *nanoscale* generally refers to objects with characteristic lengths from atomic dimensions up to a few hundred nanometers (nm). (Recall that 1 nm = 10^{-9} m.) Nanoscale fluid mechanics research has recently uncovered many surprising and useful phenomena. No doubt many more remain to be discovered. For example, in the future researchers envision using nanoscale tubes to push tiny amounts of water-soluble drugs to exactly where they are needed in the human body. Because of the tiny diameters involved, the *Reynolds numbers* for such flows are extremely small and the flow is definitely laminar. In addition, some standard properties of everyday flows (for example, the fact that a fluid sticks to a solid boundary) may not be valid for nanoscale flows. Also, ultratiny mechanical pumps and valves are difficult to manufacture and may become clogged by tiny particles such as biological molecules. As a possible solution to such problems, researchers have investigated the possibility of using a system that does not rely on mechanical parts. It involves using light-sensitive molecules attached to the surface of the tubes. By shining light onto the molecules, the light-responsive molecules attract water and cause motion of water through the tube.

EXAMPLE 8.1 | Laminar or Turbulent Flow

Given Water at a temperature of 10 °C flows through a pipe of diameter $D = 0.0185$ m and into a glass as shown in **Fig. E8.1a**.

Find Determine

a. the minimum time taken to fill a 12-oz glass (volume = 0.00035 m³) with water if the flow in the pipe is to be laminar. Repeat the calculations if the water temperature is 60 °C.

b. the maximum time taken to fill the glass if the flow is to be turbulent. Repeat the calculations if the water temperature is 60 °C.

Solution

a. If the flow in the pipe is to remain laminar, the minimum time to fill the glass will occur if the Reynolds number is the maximum allowed for laminar flow, typically Re = $\rho VD/\mu$ = 2100. Thus, $V = 2100\ \mu/\rho D$, where from Table B.1,

FIGURE E8.1a

Somchai Jongmeesuk/123RF

$\rho = 999.57$ kg/m^3 and $\mu = 1.307 \times 10^{-3}$ Pa·s at 10 °C, while $\rho = 983.2$ kg/m^3 and $\mu = 4.665 \times 10^{-4}$ Pa·s at 60 °C. Thus, the maximum average velocity for laminar flow in the pipe is

$$V = \frac{2100\mu}{\rho D} = \frac{2100(1.307 \times 10^{-3} \text{ Pa}\cdot\text{s})}{(999.57 \text{ kg/m}^3)(0.0185 \text{ m})}$$

$$= 0.148 \text{ m/s}$$

Similarly, $V = 0.0538$ m/s at 60 °C. With Ψ = volume of glass and $\Psi = Qt$ we obtain

$$t = \frac{\Psi}{Q} = \frac{\Psi}{(\pi/4)D^2V} = \frac{4(0.00035 \text{ m}^3)}{(\pi(0.0185)^2(0.148) \text{ m/s}}$$

$$= 8.805 \text{ s at } T = 10 \text{ °C}$$

Similarly, $t = 24.4$ s at 60 °C. To maintain laminar flow, the less viscous hot water requires a lower flowrate than the cold water.

b. If the flow in the pipe is to be turbulent, the maximum time to fill the glass will occur if the Reynolds number is the minimum allowed for turbulent flow, Re = 4000. Thus, $V = 4000\mu/\rho D = 0.283$ m/s and

$$t = 4.65 \text{ s at } 10 \text{ °C} \qquad\qquad (Ans)$$

Similarly, $V = 0.103$ m/s and $t = 12.8$ s at 60 °C.

Comments Note that because water is "not very viscous," the velocity must be "fairly small" to maintain laminar flow. In general, turbulent flows are encountered more often than laminar flows because of the relatively small viscosity of most common fluids (water, gasoline,

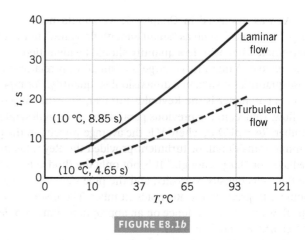

air). By repeating the calculations at various water temperatures, T (i.e., with different densities and viscosities), the results shown in **Fig. E8.1***b* are obtained. As the water temperature increases, the kinematic viscosity, $\upsilon = \mu/\rho$, decreases and the corresponding times to fill the glass increase as indicated. (Temperature effects on the viscosity of gases are the opposite; increase in temperature causes an increase in viscosity.)

If the flowing fluid had been honey with a kinematic viscosity ($\upsilon = \mu/\rho$) 3000 times greater than that of water, the velocities given earlier would be increased by a factor of 3000 and the times reduced by the same factor. As shown later in this chapter, the pressure needed to force a very viscous fluid through a pipe at such a high velocity may be unreasonably large.

8.1.2 Entrance Region and Fully Developed Flow

Any fluid flowing in a pipe had to enter the pipe at some location. The region of flow near where the fluid enters the pipe is termed the *entrance region* and is illustrated in **Fig. 8.5**. It may be the first few feet of a pipe connected to a tank or the initial portion of a long run of a hot air duct coming from a furnace.

As is shown in Fig. 8.5, the fluid typically enters the pipe with a nearly uniform velocity profile at section (1). As the fluid moves through the pipe, viscous effects cause it to stick to the pipe wall (the no-slip boundary condition). This is true whether the fluid is relatively inviscid air or a very viscous oil. Thus, a *boundary layer* in which viscous effects are important is produced along

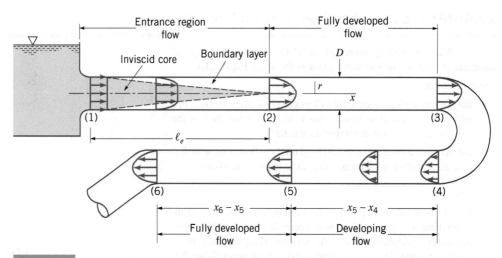

FIGURE 8.5 Entrance region, developing flow, and fully developed flow in a pipe system.

the pipe wall such that the initial velocity profile changes with distance along the pipe, x (see also Fig. 6.14b). The boundary layer grows in thickness as the fluid proceeds downstream. For fluid outside the boundary layer [within the *inviscid core* surrounding the centerline from (1) to 2)], viscous effects are negligible. Since the fluid velocity within the boundary layer is reduced, the inviscid core fluid must accelerate as it moves downstream to maintain constant mass flow at all cross sections. Once the fluid reaches the end of the entrance length, section (2), the velocity profile does not vary with x. The boundary layer has grown in thickness to completely fill the pipe.

The shape of the velocity profile in the pipe depends on whether the flow is laminar or turbulent, as does the length of the entrance region, ℓ_e. As with many other properties of pipe flow, the dimensionless **entrance length**, ℓ_e/D, correlates quite well with the Reynolds number. Analytic and experimental investigations have demonstrated that typical entrance lengths are given by

$$\frac{\ell_e}{D} = 0.06 \, \text{Re for laminar flow} \tag{8.1}$$

and

$$\frac{\ell_e}{D} = 4.4 \, (\text{Re})^{1/6} \text{ for turbulent flow} \tag{8.2}$$

The entrance length is a function of the Reynolds number.

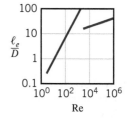

For very low Reynolds number flows the entrance length can be quite short ($\ell_e = 0.6D$ if Re = 10), whereas for laminar flow with large Reynolds number it may take a length equal to many pipe diameters before the end of the entrance region is reached ($\ell_e = 120D$ for Re = 2000). For many practical engineering problems, $10^4 < \text{Re} < 10^5$ so that as shown by the adjacent figure, $20D < \ell_e < 30D$.

Calculation of the velocity profile and pressure distribution within the entrance region is quite complex. However, once the fluid reaches the end of the entrance region, section (2) of Fig. 8.5, the flow is simpler to describe because the velocity is a function of only the distance from the pipe centerline, r, and independent of x. This is true until the character of the pipe changes in some way, such as a change in diameter, or the fluid flows through a bend, valve, or some other component at section (3). The flow between (2) and (3) is termed **fully developed flow**. Beyond the interruption of the fully developed flow [at section (4)], the flow gradually begins its return to its fully developed character [section (5)] and continues with this profile until the next pipe system component (bend, tee, valve, etc.) is reached [section (6)]. In general, developing flow length caused by a component is shorter than an entrance length. In many cases the pipe is long enough so that there is a considerable length of fully developed flow compared with the developing flow length [$(x_3 - x_2) \gg \ell_e$ and $(x_6 - x_5) \gg (x_5 - x_4)$]. In other cases the distance between one component of the pipe system and the next component is so short that fully developed flow is never achieved.

8.1.3 Pressure and Shear Stress

Fully developed steady flow in a constant diameter pipe may be driven by gravity and/or pressure forces. For horizontal pipe flow, gravity has no effect except for a hydrostatic pressure variation across the pipe, γD, that is usually negligible. It is the pressure difference, $\Delta p = p_1 - p_2$, between one section of the horizontal pipe and another which forces the fluid through the pipe. Viscous effects provide the restraining force that exactly balances the pressure force, thereby allowing the fluid to flow through the pipe with no acceleration. If viscous effects were absent in such flows, the pressure would be constant throughout the pipe, except for the hydrostatic variation.

In non-fully developed flow regions, such as the entrance region of a pipe, the fluid accelerates or decelerates as it flows. (The velocity profile changes from a uniform profile at the entrance of the pipe to its fully developed profile at the end of the entrance region.) Thus, in the entrance region there is a balance between pressure, viscous, and inertia (acceleration) forces. The result is a pressure distribution along the horizontal pipe as shown in **Fig. 8.6**. The magnitude of the pressure gradient, $\partial p/\partial x$, is larger in the entrance region than in the fully developed region, where it is a constant, $\partial p/\partial x = -\Delta p/\ell < 0$.

The fact that there is a nonzero pressure gradient along the horizontal pipe is a result of viscous effects. As is discussed in Chapter 3, if the viscosity were zero, the pressure would not

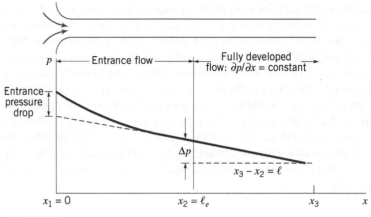

FIGURE 8.6 Pressure distribution along a horizontal pipe.

vary with x. The need for the pressure drop can be viewed from two different standpoints. In terms of a force balance, the pressure force is needed to overcome the viscous forces generated. In terms of an energy balance, the work done by the pressure force is needed to overcome the viscous dissipation of energy throughout the fluid. If the pipe is not horizontal, the pressure gradient along it is due in part to the component of weight in that direction. As is discussed in Section 8.2.1, this contribution due to the weight either enhances or retards the flow, depending on whether the flow is downhill or uphill.

The nature of the pipe flow is strongly dependent on whether the flow is laminar or turbulent. This is a direct consequence of the differences in the nature of the shear stress in laminar and turbulent flows. As is discussed in some detail in Section 8.3.3, the shear stress in laminar flow is a direct result of momentum transfer among the randomly moving molecules (a microscopic phenomenon). The shear stress in turbulent flow is largely a result of momentum transfer among the randomly moving, finite-sized fluid particles (a macroscopic phenomenon). The net result is that the physical properties of the shear stress are quite different for laminar flow than for turbulent flow.

> **Laminar flow characteristics are different than those for turbulent flow.**

8.2 FULLY DEVELOPED LAMINAR FLOW

As is indicated in the previous section, the flow in long, straight, constant diameter sections of a pipe becomes fully developed. That is, the velocity profile is the same at any cross section of the pipe. Although this is true whether the flow is laminar or turbulent, the details of the velocity profile (and other flow properties) are quite different for these two types of flow. As will be seen in the remainder of this chapter, knowledge of the velocity profile can lead directly to other useful information such as pressure drop, head loss, flowrate, and the like. Thus, we begin by developing the equation for the velocity profile in fully developed laminar flow. If the flow is not fully developed, a theoretical analysis becomes much more complex and is outside the scope of this text. If the flow is turbulent, a rigorous theoretical analysis is as yet not possible.

Although most flows are turbulent rather than laminar, and many pipes are not long enough to allow the attainment of fully developed flow, a theoretical treatment and full understanding of fully developed laminar flow is of considerable importance. First, it represents one of the few theoretical viscous analyses that can be carried out "exactly" (within the framework of quite general assumptions) without using other ad hoc assumptions or approximations. An understanding of the method of analysis and the results obtained provides a foundation from which to carry out more complicated analyses. Second, there are many practical situations involving the use of fully developed laminar pipe flow.

There are numerous ways to derive important results pertaining to fully developed laminar flow. Three alternatives include: (1) from $\mathbf{F} = m\mathbf{a}$ applied directly to a fluid element, (2) from the Navier–Stokes equations of motion, and (3) from dimensional analysis methods.

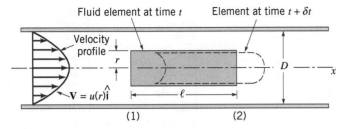

FIGURE 8.7 Motion of a cylindrical fluid element within a pipe.

8.2.1 From F = ma Applied Directly to a Fluid Element

We consider the fluid element at time t as is shown in **Fig. 8.7**. It is a circular cylinder of fluid of length ℓ and radius r centered on the axis of a horizontal pipe of diameter D. Because the velocity is not uniform across the pipe, the initially flat ends of the cylinder of fluid at time t become distorted at time $t + \delta t$ when the fluid element has moved to its new location along the pipe as shown in the figure. If the flow is fully developed and steady, the distortion on each end of the fluid element is the same, and no part of the fluid experiences any acceleration as it flows, as shown in the adjacent figure. The local acceleration is zero ($\partial \mathbf{V}/\partial t = 0$) because the flow is steady, and the convective acceleration is zero ($\mathbf{V} \cdot \nabla \mathbf{V} = u\, \partial u/\partial x\, \hat{\mathbf{i}} = 0$) because the flow is fully developed. Thus, every part of the fluid merely flows along its streamline parallel to the pipe walls with constant velocity, although neighboring particles have slightly different velocities. The velocity varies from one pathline to the next. This velocity variation, combined with the fluid viscosity, produces the shear stress.

If gravitational effects are neglected, the pressure is constant across any vertical cross section of the pipe, although it varies along the pipe from one section to the next. Thus, if the pressure is $p = p_1$ at section (1), it is $p_2 = p_1 - \Delta p$ at section (2) where Δp is the pressure drop between sections (1) and (2). We anticipate the fact that the pressure decreases in the direction of flow so that $\Delta p > 0$. A shear stress, τ, acts on the surface of the cylinder of fluid. This viscous stress is a function of the radius of the cylinder, $\tau = \tau(r)$.

As was done in fluid statics analysis (Chapter 2), we isolate the cylinder of fluid as is shown in **Fig. 8.8** and apply Newton's second law, $F_x = ma_x$. In this case, even though the fluid is moving, it is not accelerating, so that $a_x = 0$. Thus, fully developed horizontal pipe flow is merely a balance between pressure and viscous forces—the pressure difference acting on the end of the cylinder of area πr^2, and the shear stress acting on the lateral surface of the cylinder of area $2\pi r \ell$. This force balance can be written as

$$(p_1)\pi r^2 - (p_1 - \Delta p)\pi r^2 - (\tau)2\pi r\ell = 0$$

which can be simplified to give

$$\frac{\Delta p}{\ell} = \frac{2\tau}{r} \tag{8.3}$$

Equation 8.3 represents the basic balance in forces needed to drive each fluid particle along the pipe with constant velocity. Since neither Δp nor ℓ are functions of the radial coordinate, r, it follows that $2\tau/r$ must also be independent of r. That is, $\tau = Cr$, where C is a constant. At $r = 0$ (the centerline of the pipe) there is no shear stress ($\tau = 0$). At $r = D/2$ (the pipe wall) the shear stress is a maximum, denoted τ_w, the **wall shear stress**. Hence, $C = 2\tau_w/D$ and the shear stress distribution throughout the pipe is a linear function of the radial coordinate

$$\tau = \frac{2\tau_w r}{D} \tag{8.4}$$

Steady, fully developed pipe flow experiences no acceleration.

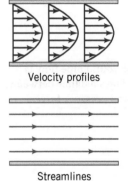

Velocity profiles

Streamlines

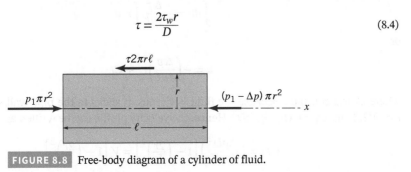

FIGURE 8.8 Free-body diagram of a cylinder of fluid.

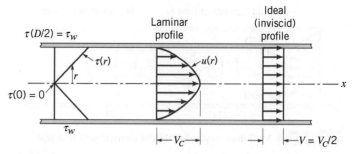

FIGURE 8.9   Shear stress distribution within the fluid in a pipe·(laminar or turbulent flow) and typical velocity profiles.

as is indicated in **Fig. 8.9**. The linear dependence of τ on r is a result of the pressure force being proportional to r^2 (i.e., the pressure acts on the end of the fluid cylinder; area $= \pi r^2$) and the shear force being proportional to r (i.e., the shear stress acts on the lateral sides of the cylinder; area $= 2\pi r\ell$). If the viscosity were zero there would be no shear stress, and the pressure would be constant throughout the horizontal pipe ($\Delta p = 0$). As is seen from Eqs. 8.3 and 8.4, the pressure drop and wall shear stress are related by

Basic horizontal pipe flow is governed by a balance between viscous and pressure forces.

$$\Delta p = \frac{4\ell\tau_w}{D} \tag{8.5}$$

A small shear stress can produce a large pressure difference if the pipe is relatively long ($\ell/D \gg 1$).

Although we are discussing laminar flow, a closer consideration of the assumptions involved in the derivation of Eqs. 8.3, 8.4, and 8.5 reveals that these equations are valid for both laminar and turbulent flow. To carry the analysis further we must prescribe how the shear stress is related to the velocity. This is the critical step that separates the analysis of laminar from that of turbulent flow—from being able to solve for the laminar flow properties and not being able to solve for the turbulent flow properties without additional ad hoc assumptions. As is discussed in Section 8.3, the shear stress dependence for turbulent flow is very complex. However, for laminar flow of a Newtonian fluid, the shear stress is simply proportional to the velocity gradient, "$\tau = \mu\, du/dy$" (see Section 1.6). In the notation associated with our pipe flow, this becomes

$$\tau = -\mu\frac{du}{dr} \tag{8.6}$$

The negative sign is included to give $\tau > 0$ with $du/dr < 0$. (The velocity decreases from the pipe centerline to the pipe wall.)

Equations 8.3 and 8.6 represent the two governing laws for fully developed laminar flow of a Newtonian fluid within a horizontal pipe. The one is Newton's second law of motion, and the other is the definition of a Newtonian fluid. By combining these two equations, we obtain

$$\frac{du}{dr} = -\left(\frac{\Delta p}{2\mu\ell}\right)r$$

which can be integrated to give the velocity profile as follows:

$$\int du = -\frac{\Delta p}{2\mu\ell}\int r\,dr$$

or

$$u = -\left(\frac{\Delta p}{4\mu\ell}\right)r^2 + C_1$$

where C_1 is a constant. Because the fluid is viscous, it sticks to the pipe wall so that $u = 0$ at $r = D/2$. Thus, $C_1 = (\Delta p/16\mu\ell)D^2$. Hence, the velocity profile can be written as

$$u(r) = \left(\frac{\Delta pD^2}{16\mu\ell}\right)\left[1 - \left(\frac{2r}{D}\right)^2\right] = V_c\left[1 - \left(\frac{2r}{D}\right)^2\right] \tag{8.7}$$

where $V_c = \Delta p D^2/(16\mu\ell)$ is the centerline velocity. (Simply set $r = 0$ to obtain V_c.) An alternative expression can be written by using the relationship between the wall shear stress and the pressure gradient (Eqs. 8.5 and 8.7) to give

$$u(r) = \frac{\tau_w D}{4\mu}\left[1 - \left(\frac{r}{R}\right)^2\right]$$

where $R = D/2$ is the pipe radius.

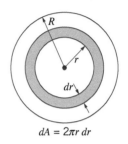

Under certain restrictions the velocity profile in a pipe is parabolic.

This velocity profile, plotted in Fig. 8.9, is parabolic in the radial coordinate, r, has a maximum velocity, V_c, at the pipe centerline, and a minimum velocity (zero) at the pipe wall. The volume flowrate through the pipe can be obtained by integrating the velocity profile across the pipe. Since the flow is axisymmetric about the centerline, the velocity is constant on small area elements consisting of rings of radius r and thickness dr, as shown in the adjacent figure. Thus,

$$Q = \int u \, dA = \int_{r=0}^{r=R} u(r) 2\pi r \, dr = 2\pi V_c \int_0^R \left[1 - \left(\frac{r}{R}\right)^2\right] r \, dr$$

$dA = 2\pi r \, dr$

or

$$Q = \frac{\pi R^2 V_c}{2}$$

By definition, the average velocity is the flowrate divided by the cross-sectional area, $V = Q/A = Q/\pi R^2$, so that for this flow

$$V = \frac{\pi R^2 V_c}{2\pi R^2} = \frac{V_c}{2} = \frac{\Delta p D^2}{32\mu\ell} \tag{8.8}$$

and

$$\boxed{Q = \frac{\pi D^4 \Delta p}{128\mu\ell}} \tag{8.9}$$

As is indicated in Eq. 8.8, the average velocity is one-half of the maximum velocity. In general, for velocity profiles of other shapes (such as for turbulent pipe flow), the average velocity is not merely the average of the maximum (V_c) and minimum (0) velocities as it is for the laminar parabolic profile. The two velocity profiles indicated in Fig. 8.9 provide the same flowrate—one is the fictitious ideal ($\mu = 0$) profile; the other is the actual laminar flow profile.

The above results confirm the following properties of laminar pipe flow. For a horizontal pipe the flowrate is (a) directly proportional to the pressure drop, (b) inversely proportional to the viscosity, (c) inversely proportional to the pipe length, and (d) proportional to the pipe diameter to the fourth power. With all other parameters fixed, an increase in diameter by a factor of 2 will increase the flowrate by a factor of $2^4 = 16$— the flowrate is very strongly dependent on pipe size. This dependence is shown in the adjacent figure. Likewise, a small error in pipe diameter can cause a relatively large error in flowrate. For example, a 2% error in diameter gives an 8% error in flowrate ($Q \sim D^4$ or $\delta Q \sim 4D^3\delta D$, so that $\delta Q/Q = 4\,\delta D/D$). This flow, the properties of which were first established experimentally by two independent workers, G. Hagen (1797–1884) in 1839 and J. Poiseuille (1799–1869) in 1840, is termed *Hagen–Poiseuille flow*. Equation 8.9 is commonly referred to as **Poiseuille's law.** Recall that all of these results are restricted to laminar flow (those with Reynolds numbers less than approximately 2100) in a horizontal pipe.

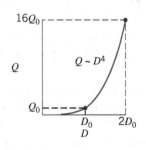

Poiseuille's law is valid for laminar flow only.

The adjustment necessary to account for nonhorizontal pipes, as shown in **Fig. 8.10**, can be easily included by replacing the pressure drop, Δp, by the combined effect of pressure and gravity, $\Delta p - \gamma\ell \sin\theta$, where θ is the angle between the pipe and the horizontal. (Note that $\theta > 0$ if the flow is uphill, while $\theta < 0$ if the flow is downhill.) This can be seen from the force balance in the x direction (along the pipe axis) on the cylinder of fluid shown in Fig. 8.10b. The method is exactly analogous to that used to obtain the Bernoulli equation (Eq. 3.6) when the streamline is not horizontal. The net force in the x direction is a combination of the pressure force in that direction, $\Delta p \pi r^2$, and the component of weight in that direction, $-\gamma \pi r^2 \ell \sin\theta$. The result is a slightly modified form of Eq. 8.3 given by

$$\frac{\Delta p - \gamma\ell \sin\theta}{\ell} = \frac{2\tau}{r} \tag{8.10}$$

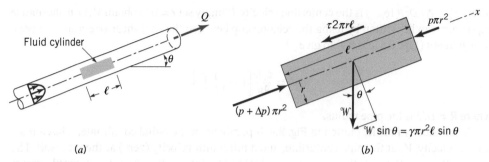

FIGURE 8.10   Free-body diagram of a fluid cylinder for flow in a nonhorizontal pipe.

Thus, all of the results for the horizontal pipe are valid provided the pressure gradient is adjusted for the elevation term, that is, Δp is replaced by $\Delta p - \gamma\ell \sin\theta$ so that

$$V = \frac{(\Delta p - \gamma\ell \sin\theta)D^2}{32\mu\ell} \tag{8.11}$$

and

$$Q = \frac{\pi(\Delta p - \gamma\ell \sin\theta)D^4}{128\mu\ell} \tag{8.12}$$

It is seen that the driving force for pipe flow can be either a pressure drop in the flow direction, Δp, or the component of weight in the flow direction, $-\gamma\ell \sin\theta$. If the flow is downhill (i.e., $\sin\theta < 0$), gravity helps the flow, so a smaller pressure drop is required. If the flow is uphill (i.e., $\sin\theta > 0$), gravity works against the flow, so a larger pressure drop is required. Note that $\gamma\ell \sin\theta = \gamma\Delta z$ (where Δz is the change in elevation) is a hydrostatic type pressure term. If there is no flow, $V = 0$ and $\Delta p = \gamma\ell \sin\theta = \gamma\Delta z$, as expected for fluid statics.

EXAMPLE 8.2 | Laminar Pipe Flow

Given   An oil with a viscosity of $\mu = 0.40$ N · s/m^2 and density $\rho = 900$ kg/m^3 flows in a pipe of diameter $D = 0.020$ m.

Find

a. What pressure drop, $p_1 - p_2$, is needed to produce a flowrate of $Q = 2.0 \times 10^{-5}$ m^3/s if the pipe is horizontal with $x_1 = 0$ and $x_2 = 10$ m?

b. How steep of a hill, θ, must the pipe be on if the oil is to flow through the pipe at the same rate as in part (a), but with $p_1 = p_2$?

c. For the conditions of part (b), if $p_1 = 200$ kPa, what is the pressure at section $x_3 = 5$ m, where x is measured along the pipe?

Solution

a. If the Reynolds number is less than 2100, the flow is laminar and the equations derived in this section are valid. Since the average velocity is $V = Q/A = (2.0 \times 10^{-5}$ m^3/s)/ $[\pi(0.020)^2$m$^2/4] = 0.0637$ m/s, the Reynolds number is $Re = \rho V D/\mu = 2.87 < 2100$. Hence, the flow is laminar, and from Eq. 8.9 with $\ell = x_2 - x_1 = 10$ m, the pressure drop is

$$\Delta p = p_1 - p_2 = \frac{128\mu\ell Q}{\pi D^4}$$

$$= \frac{128(0.40 \text{ N} \cdot \text{s/m}^2)(10.0 \text{ m})(2.0 \times 10^{-5} \text{ m}^3/\text{s})}{\pi(0.020 \text{ m})^4}$$

or

$$\Delta p = 20{,}400 \text{ N/m}^2 = 20.4 \text{ kPa} \qquad (Ans)$$

b. If the pipe is on a hill of angle θ such that $\Delta p = p_1 - p_2 = 0$, Eq. 8.12 gives

$$\sin\theta = -\frac{128\mu Q}{\pi\rho g D^4} \tag{1}$$

or

$$\sin\theta = \frac{-128(0.40 \text{ N} \cdot \text{s/m}^2)(2.0 \times 10^{-5} \text{ m}^3/\text{s})}{\pi(900 \text{ kg/m}^3)(9.81 \text{ m/s}^2)(0.020 \text{ m})^4} \qquad (Ans)$$

Thus, $\theta = -13.34°$.

Comment   This checks with the previous horizontal result as is seen from the fact that a change in elevation of $\Delta z = \ell \sin\theta = (10 \text{ m}) \sin(-13.34°) = -2.31$ m is equivalent to a pressure change of $\Delta p = \rho g \Delta z = (900 \text{ kg/m}^3)(9.81 \text{ m/s}^2)(2.31 \text{ m}) = 20{,}400$ N/m^2, which is equivalent to that needed for the horizontal pipe. For the horizontal pipe it is the work done by the pressure forces that overcomes the viscous dissipation. For the zero-pressure-drop pipe on the hill, it is the change in potential energy of the fluid "falling" down the hill that is converted to the energy lost by viscous dissipation.

Note that if it is desired to increase the flowrate to $Q = 1.0 \times 10^{-4} \, m^3/s$ with $p_1 = p_2$, the value of θ given by Eq. 1 is $\sin \theta = -1.15$. Since the sine of an angle cannot be greater than 1, this flow would not be possible. The weight of the fluid would not be large enough to offset the viscous force generated for the flowrate desired. A larger-diameter pipe would be needed.

   c. With $p_1 = p_2$ the length of the pipe, ℓ, does not appear in the flowrate equation (Eq. 1). This is a statement of the fact that for such cases the pressure is constant all along the pipe (provided the pipe lies on a hill of constant slope). This can be seen by substituting the values of Q and θ from case (b) into Eq. 8.12 and noting that $\Delta p = 0$ for any ℓ. For example, $\Delta p = p_1 - p_3 = 0$

if $\ell = x_3 - x_1 = 5$ m. Thus, $p_1 = p_2 = p_3$ so that

$$p_3 = 200 \, kPa \qquad (Ans)$$

Comment Note that if the fluid were gasoline ($\mu = 3.1 \times 10^{-4}$ N · s/m^2 and $\rho = 680$ kg/m^3), the Reynolds number would be Re = 2790, the flow would probably not be laminar, and use of Eqs. 8.9 and 8.12 would give incorrect results. Also note from Eq. 1 that the kinematic viscosity, $v = \mu/\rho$, is the important viscous parameter. This is a statement of the fact that with constant pressure along the pipe, it is the ratio of the viscous force ($\sim\mu$) to the weight force ($\sim\gamma = \rho g$) that determines the value of θ.

8.2.2 From the Navier–Stokes Equations

In the previous section we obtained results for fully developed laminar pipe flow by applying Newton's second law and the assumption of a Newtonian fluid to a specific portion of the fluid—a cylinder of fluid centered on the axis of a long, round pipe. When this governing law and assumptions are applied to a general fluid flow (not restricted to pipe flow), the result is the Navier–Stokes equations as discussed in Chapter 6. In Section 6.9.3 these equations were solved for the specific geometry of fully developed laminar flow in a round pipe. The results are the same as those given in Eq. 8.7.

> *Poiseuille's law can be obtained from the Navier–Stokes equations.*

   We will not repeat the detailed steps used to obtain the laminar pipe flow from the Navier–Stokes equations (see Section 6.9.3) but will indicate how the various assumptions used and steps applied in the derivation correlate with the analysis used in the previous section.

   General motion of an incompressible Newtonian fluid is governed by the continuity equation (conservation of mass, Eq. 6.31) and the momentum equation (Eq. 6.127), which are rewritten here for convenience:

$$\nabla \cdot \mathbf{V} = 0 \qquad (8.13)$$

$$\frac{\partial \mathbf{V}}{\partial t} + (\mathbf{V} \cdot \nabla)\mathbf{V} = -\frac{\nabla p}{\rho} + \mathbf{g} + v\nabla^2\mathbf{V} \qquad (8.14)$$

For steady, fully developed flow in a pipe, the velocity contains only an axial component, which is a function of only the radial coordinate [$\mathbf{V} = u(r)\hat{\mathbf{i}}$]. For such conditions, the left side of Eq. 8.14 is zero. This is equivalent to saying that the fluid experiences no acceleration as it flows along. The same constraint was used in the previous section when considering $\mathbf{F} = m\mathbf{a}$ for the fluid cylinder. Thus, with $\mathbf{g} = -g\hat{\mathbf{k}}$ the Navier–Stokes equations become

$$\nabla \cdot \mathbf{V} = 0$$
$$\nabla p + \rho g\hat{\mathbf{k}} = \mu\nabla^2\mathbf{V} \qquad (8.15)$$

The flow is governed by a balance of pressure, weight, and viscous forces in the flow direction, similar to that shown in Fig. 8.10 and Eq. 8.10. If the flow were not fully developed (as in an entrance region, for example), it would not be possible to simplify the Navier–Stokes equations to that form given in Eq. 8.15 (the nonlinear term $(\mathbf{V} \cdot \nabla)\mathbf{V}$ would not be zero), and the solution would be very difficult to obtain.

   Because of the assumption that $\mathbf{V} = u(r)\hat{\mathbf{i}}$, the continuity equation, Eq. 8.13, is automatically satisfied. This conservation of mass condition was also automatically satisfied by the incompressible flow assumption in the derivation in the previous section. The fluid flows across one section of the pipe at the same rate that it flows across any other section.

   When it is written in terms of polar coordinates (as was done in Section 6.9.3), the component of Eq. 8.15 along the pipe becomes

> *The governing differential equations can be simplified by appropriate assumptions.*

$$\frac{\partial p}{\partial x} + \rho g \sin \theta = \mu \frac{1}{r}\frac{\partial}{\partial r}\left(r\frac{\partial u}{\partial r}\right) \qquad (8.16)$$

Since the flow is fully developed, $u = u(r)$ and the right side is a function of, at most, only r. The left side is a function of, at most, only x. It was shown that this leads to the condition that the pressure gradient in the x direction is a constant—$\partial p/\partial x = -\Delta p/\ell$. The same condition was used in the derivation of the previous section (Eq. 8.3).

It is seen from Eq. 8.16 that the effect of a nonhorizontal pipe enters into the Navier–Stokes equations in the same manner as was discussed in the previous section. The pressure gradient in the flow direction is coupled with the effect of the weight in that direction to produce an effective pressure gradient of $-\Delta p/\ell + \rho g \sin \theta$.

The velocity profile is obtained by integration of Eq. 8.16. Since it is a second-order equation, two boundary conditions are needed—(1) the fluid sticks to the pipe wall (as was also done in Eq. 8.7) and (2) either of the equivalent forms that the velocity remains finite throughout the flow (in particular $u < \infty$ at $r = 0$) or, because of symmetry, that $du/dr = 0$ at $r = 0$. In the derivation of the previous section, only one boundary condition (the no-slip condition at the wall) was needed because the equation integrated was a first-order equation. The other condition ($du/dr = 0$ at $r = 0$) was automatically built into the analysis because of the fact that $\tau = -\mu \, du/dr$ and $\tau = 2\tau_w r/D = 0$ at $r = 0$.

The results obtained by either applying $\mathbf{F} = m\mathbf{a}$ to a fluid cylinder (Section 8.2.1) or solving the Navier–Stokes equations (Section 6.9.3) are exactly the same. Similarly, the basic assumptions regarding the flow structure are the same. This should not be surprising because the two methods are based on the same principle—Newton's second law. One is restricted to fully developed laminar pipe flow from the beginning (the drawing of the free-body diagram), and the other starts with the general governing equations (the Navier–Stokes equations) with the appropriate restrictions concerning fully developed laminar flow applied as the solution process progresses.

8.2.3 From Dimensional Analysis

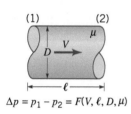

(1)   (2)

$\Delta p = p_1 - p_2 = F(V, \ell, D, \mu)$

Although fully developed laminar pipe flow is simple enough to allow the rather straightforward solutions discussed in the previous two sections, it is worthwhile to consider this flow from a dimensional analysis standpoint as will be necessary for turbulent flow in Section 8.4. Thus, we assume that the pressure drop in the horizontal pipe, Δp, is a function of the average velocity of the fluid in the pipe, V, the length of the pipe, ℓ, the pipe diameter, D, and the viscosity of the fluid, μ, as shown in the adjacent figure. We have not included the density or the specific weight of the fluid as parameters because for such flows they are not important parameters. There is neither mass (density) times acceleration nor a component of weight (specific weight times volume) in the flow direction involved. (Note that density or specific weight should be included for non-horizontal pipe flow.) Thus,

$$\Delta p = F(V, \ell, D, \mu)$$

Five variables can be described in terms of three reference dimensions (M, L, T). According to the results of dimensional analysis (Chapter 7), this flow can be described in terms of $k - r = 5 - 3 = 2$ dimensionless groups. One such representation is

$$\frac{D \, \Delta p}{\mu V} = \phi\left(\frac{\ell}{D}\right) \tag{8.17}$$

where $\phi(\ell/D)$ is an unknown function of the length to diameter ratio of the pipe.

Although this is as far as dimensional analysis can take us, it seems reasonable to impose a further assumption that the pressure drop is directly proportional to the pipe length. That is, it takes twice the pressure drop to force fluid through a pipe if its length is doubled. The only way that this can be true is if $\phi(\ell/D) = C\ell/D$, where C is a constant. Thus, Eq. 8.17 becomes

$$\frac{D \, \Delta p}{\mu V} = \frac{C\ell}{D}$$

which can be rewritten as

$$\frac{\Delta p}{\ell} = \frac{C\mu V}{D^2}$$

or

$$Q = AV = \frac{(\pi/4C) \, \Delta p D^4}{\mu \ell} \tag{8.18}$$

The basic functional dependence for laminar pipe flow given by Eq. 8.18 is the same as that obtained by the analysis of the two previous sections. The value of C must be determined by theory (as done in the previous two sections) or experiment. For a round pipe, $C = 32$. For ducts of other cross-sectional shapes, the value of C is different (see Section 8.4.3).

It is usually advantageous to describe a process in terms of dimensionless quantities. To this end we rewrite the pressure drop equation for laminar horizontal pipe flow, Eq. 8.8, as $\Delta p = 32\mu\ell V/D^2$ and divide both sides by the dynamic pressure, $\rho V^2/2$, to obtain the dimensionless form as

Dimensional analysis can be used to put pipe flow parameters into dimensionless form.

$$\frac{\Delta p}{\frac{1}{2}\rho V^2} = \frac{(32\mu\ell V/D^2)}{\frac{1}{2}\rho V^2} = 64\left(\frac{\mu}{\rho VD}\right)\left(\frac{\ell}{D}\right) = \frac{64}{\text{Re}}\left(\frac{\ell}{D}\right)$$

This is often written as

$$\Delta p = f\frac{\ell}{D}\frac{\rho V^2}{2}$$

where the dimensionless quantity

$$f = \Delta p(D/\ell)/(\rho V^2/2)$$

is termed the **friction factor,** or sometimes the *Darcy friction factor* [H. P. G. Darcy (1803–1858)]. (This parameter should not be confused with the less-used Fanning friction factor, which is defined to be $f/4$. In this text we will use only the Darcy friction factor.) Thus, the friction factor for laminar fully developed round pipe flow is simply

$$f = \frac{64}{\text{Re}} \tag{8.19}$$

as shown in the adjacent figure.

By substituting the pressure drop in terms of the wall shear stress (Eq. 8.5), we obtain an alternate expression for the friction factor as a dimensionless wall shear stress

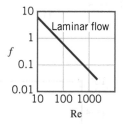

$$f = \frac{8\tau_w}{\rho V^2} \tag{8.20}$$

Knowledge of the friction factor will allow us to obtain a variety of information regarding pipe flow. For turbulent flow the dependence of the friction factor on the Reynolds number is much more complex than that given by Eq. 8.19 for laminar flow. This is discussed in detail in Section 8.4.

8.2.4 Energy Considerations

In the previous three sections we derived the basic laminar flow results from application of $\mathbf{F} = m\mathbf{a}$ or dimensional analysis considerations. It is equally important to understand the implications of energy considerations of such flows. To this end we consider the energy equation for incompressible, steady flow between two locations as is given in Eq. 5.89 and presented here without shaft work head

$$\frac{p_1}{\gamma} + \alpha_1\frac{V_1^2}{2g} + z_1 = \frac{p_2}{\gamma} + \alpha_2\frac{V_2^2}{2g} + z_2 + h_L \tag{8.21}$$

Recall that the kinetic energy coefficients, α_1 and α_2, compensate for the fact that the velocity profile across the pipe is not uniform. For uniform velocity profiles, $\alpha = 1$, whereas for any non-uniform profile, $\alpha > 1$. The head loss term, h_L, accounts for any energy loss associated with the flow. This loss is a direct consequence of the viscous dissipation that occurs throughout the fluid in the pipe. For the ideal (inviscid) cases discussed in previous chapters, $\alpha_1 = \alpha_2 = 1$, $h_L = 0$, and the energy equation reduces to the familiar Bernoulli equation discussed in Chapter 3 (Eq. 3.7).

Even though the velocity profile in viscous pipe flow is not uniform, for fully developed flow it does not change from section (1) to section (2) so that $\alpha_1 = \alpha_2$. Thus, the kinetic energy is the same at any section ($\alpha_1 V_1^2/2 = \alpha_2 V_2^2/2$) and the energy equation becomes

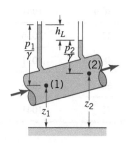

$$\left(\frac{p_1}{\gamma} + z_1\right) - \left(\frac{p_2}{\gamma} + z_2\right) = h_L \tag{8.22}$$

The energy dissipated by the viscous forces within the fluid is supplied by the excess work done by the pressure and gravity forces as shown in the adjacent figure.

A comparison of Eqs. 8.22 and 8.10 shows that the head loss is given by

$$h_L = \frac{2\tau\ell}{\gamma r}$$

(recall $p_1 = p_2 + \Delta p$ and $z_2 - z_1 = \ell \sin\theta$), which, by use of Eq. 8.4, can be rewritten in the form

$$h_L = \frac{4\ell\tau_w}{\gamma D} \tag{8.23}$$

The head loss in a pipe is a result of the viscous shear stress on the wall.

It is the shear stress at the wall (which is directly related to the viscosity and the shear stress throughout the fluid) that is responsible for the head loss. A closer consideration of the assumptions involved in the derivation of Eq. 8.23 will show that it is valid for both laminar and turbulent flow.

EXAMPLE 8.3 | Laminar Pipe Flow Properties

Given   The flowrate, Q, of corn syrup through the horizontal pipe shown in **Fig. E8.3a** is to be monitored by measuring the pressure difference between sections (1) and (2). It is proposed that $Q = K\Delta p$, where the calibration constant, K, is a function of temperature, T, because of the variation of the syrup's viscosity and density with temperature. These variations are given in **Table E8.3**.

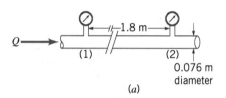

(a)

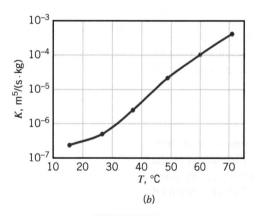

(b)

FIGURE E8.3

TABLE E8.3

T (°C)	ρ (kg/m^3)	μ (kg/m · s)
15.5	1067	1.91
26.6	1061	9.1×10^{-1}
37	1056	18.2×10^{-2}
48.8	1051	21.0×10^{-3}
60	1046	44.0×10^{-4}
71	1041	11.0×10^{-4}

Find

a. Plot $K(T)$ versus T for 15.5 °C $\le T \le$ 71 °C.

b. Determine the wall shear stress and the pressure drop, $\Delta p = p_1 - p_2$, for $Q = 0.014$ m^3/s and $T = 37$ °C.

c. For the conditions of part (b), determine the net pressure force, $(\pi D^2/4)\,\Delta p$, and the net shear force, $\pi D\ell\tau_w$, on the fluid within the pipe between the sections (1) and (2).

Solution

a. If the flow is laminar, it follows from Eq. 8.9 that

$$Q = \frac{\pi D^4 \Delta p}{128\mu\ell} = \frac{\pi(0.076 \text{ m})^4 \Delta p}{128\mu (1.8 \text{ m})}$$

or

$$Q = K\Delta p = \frac{4.55 \times 10^{-7}}{\mu}\Delta p \tag{1}$$

where the units on Q, Δp, and μ are m^3/s, N/m^2, and kg/m · s, respectively. Thus

$$K = \frac{4.55 \times 10^{-7}}{\mu} \tag{Ans}$$

where the units of K are m^3/s · kg. By using values of the viscosity from Table E8.3, the calibration curve shown in **Fig. E8.3b** is obtained. This result is valid only if the flow is laminar.

Comment   As shown in Section 8.5, for turbulent flow the flowrate is not linearly related to the pressure drop so it would not be possible to have $Q = K\Delta p$. Note also that the value of K is independent of the syrup density (ρ was not used in the calculations) because laminar pipe flow is governed by pressure and viscous effects; inertia is not important.

b. For $T = 37$ °C, the viscosity is $\mu = 0.1819$ kg/m · s so that with a flowrate of $Q = 0.014$ m^3/s the pressure drop (according to Eq. 8.9) is

$$\Delta p = \frac{128\mu\ell Q}{\pi D^4}$$

$$= \frac{128(0.1819 \text{ kg/m} \cdot \text{s})(1.8 \text{ m})(0.014 \text{ m}^3/\text{s})}{\pi(0.76 \text{ m})^4}$$

$$= 43.76 \text{ N/m}^2 \tag{Ans}$$

provided the flow is laminar. For this case

$$V = \frac{Q}{A} = \frac{0.014 \text{ m}^3/\text{s}}{0.076} = 3.087 \text{ m/s}$$

so that

$$Re = \frac{\rho VD}{\mu} = \frac{(1056 \text{ kg/m}^3)(3.087 \text{ m/s})(0.076 \text{ m})}{(0.1819 \text{ kg/m} \cdot \text{s})}$$

$$= 1362 < 2100$$

Hence, the flow is laminar. From Eq. 8.5 the wall shear stress is

$$\tau_w = \frac{\Delta p D}{4\ell} = \frac{(43.76 \text{ N/m}^2)(0.076 \text{ m})}{4(1.8 \text{ m})} = 1.84 \text{ N/m}^2 \qquad (Ans)$$

c. For the conditions of part (b), the net pressure force, F_p, on the fluid within the pipe between sections (1) and (2) is

$$F_p = \frac{\pi}{4} D^2 \Delta p = \frac{\pi}{4}(0.076 \text{ m})^2 (43.76 \text{ N/m}^2) = 0.198 \text{ N} \qquad (Ans)$$

Similarly, the net viscous force, F_v, on that portion of the fluid is

$$F_v = 2\pi \left(\frac{D}{2}\right) \ell \tau_w$$

$$= 2\pi \left[\frac{0.076}{2}\right] (1.8 \text{ m})(1.84 \text{ N/m}^2) = 0.79 \text{ N} \qquad (Ans)$$

Comment   Note that the values of these two forces are the same. The net force is zero; there is no acceleration.

8.3 FULLY DEVELOPED TURBULENT FLOW

In the previous section various properties of fully developed laminar pipe flow were discussed. Since turbulent pipe flow is actually more likely to occur than laminar flow in practical situations, it is necessary to obtain similar information for turbulent pipe flow. However, turbulent flow is a very complex process. Numerous persons have devoted considerable effort in attempting to understand the variety of baffling aspects of turbulence. Although a considerable amount of knowledge about the topic has been developed, the field of turbulent flow still remains the least understood area of fluid mechanics. In this book we can provide only some of the very basic ideas concerning turbulence. The interested reader should consult some of the many books available for further reading (Refs. 1–3).

8.3.1 Transition from Laminar to Turbulent Flow

When viscous effects are present in a flow, the flow may be laminar, turbulent, or in transition between the two. Whether a flow will be laminar or turbulent can be inferred from the value of the Reynolds number. The value that distinguishes between the two, called the critical value, depends on the specific flow situation involved. For example, flow in a pipe and flow along a flat plate (boundary layer flow, as is discussed in Section 9.2.4) can be laminar or turbulent, depending on the value of the Reynolds number. As a general rule for pipe flow, the value of the Reynolds number must be less than approximately 2100 for laminar flow and greater than approximately 4000 for turbulent flow. For flow along a flat plate the transition between laminar and turbulent flow occurs at a Reynolds number of approximately 500,000 (see Section 9.2.4). The value is much larger for the flat plate because the length term in the Reynolds number is the distance measured from the leading edge of the plate.

Consider a long section of pipe that is initially filled with a fluid at rest. As the valve is opened to start the flow, the flow velocity and, hence, the Reynolds number increase from zero (no flow) to their maximum steady-state flow values, as is shown in **Fig. 8.11**. Assume this transient process is slow enough so that unsteady effects are negligible (quasi-steady flow). For an initial time period the Reynolds number is small enough for laminar flow to occur. At some time the Reynolds number reaches 2100, and the flow begins its transition to turbulent conditions. Intermittent spots or bursts of turbulence appear. As the Reynolds number is increased, the entire flow field becomes turbulent. The flow remains turbulent as long as the Reynolds number exceeds approximately 4000.

A typical trace of the axial component of velocity measured at a given location in the flow, $u = u(t)$, is shown in **Fig. 8.12**. Its irregular, seemingly random nature is the distinguishing feature of turbulent flow. (It is currently unknown whether turbulent flow properties are truly random. Since they appear random, we will use that descriptor in this text.) The character of many of the important properties of the flow depends strongly on the existence and nature of

Turbulent flows involve randomly fluctuating parameters.

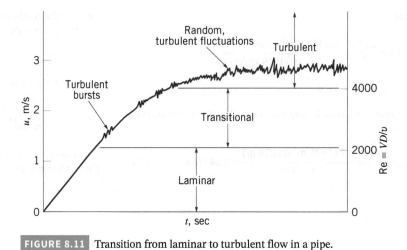

FIGURE 8.11   Transition from laminar to turbulent flow in a pipe.

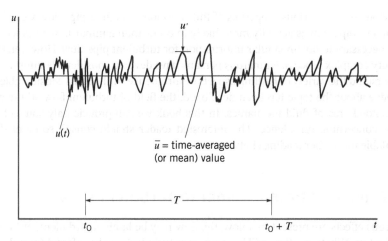

FIGURE 8.12   The time-averaged, $\bar{u}$, and fluctuating, u', description of a parameter for turbulent flow.

the turbulent fluctuations or randomness indicated. Calculation of the heat transfer, pressure drop, and many other parameters would not be possible without inclusion of the seemingly small, but very important, effects associated with the irregular motion of the flow. Also shown in Fig. 8.12 is the time-averaged velocity, $\bar{u}$. This is the velocity most commonly needed for pipe flows and many other engineering applications.

Consider flow in a pan of water placed on a stove. With the stove turned off, the fluid is stationary. The initial sloshing has died out because of viscous dissipation within the water. With the stove turned on, a temperature gradient in the vertical direction, $\partial T/\partial z$, is produced. The water temperature is greatest near the pan bottom and decreases toward the top of the fluid layer. If the temperature difference is very small, the water will remain stationary, even though the water density is smallest near the bottom of the pan because of the decrease in density with an increase in temperature. A further increase in the temperature gradient will cause a buoyancy-driven instability that results in fluid motion—the light, warm water rises to the top, and the heavy, cold water sinks to the bottom. This slow, regular "turning over" increases the heat transfer from the pan to the water and promotes mixing within the pan. As the temperature gradient increases still further, the fluid motion becomes more vigorous and eventually turns into a chaotic, random, turbulent flow with considerable mixing, vaporization (boiling), and greatly increased heat transfer rate. The flow has progressed from a stationary fluid, to laminar flow, and finally to turbulent, multi-phase (liquid and vapor) flow.

Mixing processes and heat and mass transfer processes are considerably enhanced in turbulent flow compared to laminar flow. This is due to the macroscopic scale of the randomness in turbulent flow. We are all familiar with the "rolling," vigorous eddy-type motion of the water in a pan being heated on the stove (even if it is not heated to boiling). Such finite-sized random mixing is very effective in transporting energy and mass throughout the flow field, thereby increasing the various rate processes involved. Laminar flow, on the other hand, can be thought of as very small but finite-sized fluid particles flowing smoothly in layers, one over another. The only randomness and mixing take place on the molecular scale and result in relatively small heat, mass, and momentum transfer rates.

Without turbulence it would be virtually impossible to carry out life as we now know it. Mixing is one positive application of turbulence, as discussed above, but there are other situations where turbulent flow is desirable. To transfer the required heat between a solid and an adjacent fluid (such as in the cooling coils of an air conditioner or a boiler of a power plant) would require an enormously large heat exchanger if the flow were laminar. Similarly, the required mass transfer of a liquid state to a vapor state (such as is needed in the evaporated cooling system associated with sweating) would require very large surfaces if the fluid flowing past the surface were laminar rather than turbulent. As shown in Chapter 9, turbulence can also aid in delaying flow separation to reduce drag forces.

V8.4 Laminar and turbulent mixing

The Wide World of Fluids

Smaller heat exchangers

Automobile radiators, air conditioners, and refrigerators contain heat exchangers that transfer energy from (to) the hot (cold) fluid within the heat exchanger tubes to (from) the colder (hotter) surrounding fluid. These units can be made smaller and more efficient by increasing the heat transfer rate across the tubes' surfaces. If the flow through the tubes is laminar, the heat transfer rate is relatively small. Significantly larger heat transfer rates are obtained if the flow within the tubes is turbulent. Even greater heat transfer rates

can be obtained by the use of turbulence promoters, sometimes termed "turbulators," which provide additional *turbulent mixing* motion than would normally occur. Such enhancement mechanisms include internal fins, spiral wire or ribbon inserts, and ribs or grooves on the inner surface of the tube. While these mechanisms can increase the heat transfer rate by 1.5 to 3 times over that for a bare tube at the same flowrate, they also increase the pressure drop (and therefore the power) needed to produce the flow within the tube. Thus, a compromise involving increased heat transfer rate and increased power consumption is often needed.

Turbulence is also of importance in the mixing of fluids. Smoke from a stack would continue for miles as a ribbon of pollutant without rapid dispersion within the surrounding air if the flow were laminar rather than turbulent. Under certain atmospheric conditions this is observed to occur. Although there is mixing on a molecular scale (laminar flow), it is several orders of magnitude slower and less effective than the mixing on a macroscopic scale (turbulent flow). It is considerably easier to mix cream into a cup of coffee (turbulent flow) than to thoroughly mix two colors of a viscous paint (laminar flow).

In other situations laminar (rather than turbulent) flow is desirable. The pressure drop in pipes (hence, the power requirements for pumping) can be considerably lower if the flow is laminar rather than turbulent. Fortunately, the blood flow through a person's arteries is normally laminar, except in the largest arteries with high blood flowrates. The aerodynamic drag on an airplane wing can be considerably smaller with laminar flow past it than with turbulent flow.

V8.5 Stirring color into paint

8.3.2 Turbulent Shear Stress

The fundamental difference between laminar and turbulent flow lies in the chaotic, random behavior of the various fluid parameters. Such variations occur in the three components of velocity, the pressure, the shear stress, the temperature, and any other variable that has a field description. Turbulent flow is characterized by random, three-dimensional vorticity (i.e., fluid particle rotation or spin; see Section 6.1.3). As is indicated in Fig. 8.12, such flows can be described in terms of their mean values (denoted with an overbar) on which are superimposed the fluctuations (denoted with a prime). Thus, if $u = u(x, y, z, t)$ is the x component of instantaneous velocity, then its time mean (or *time-average*) value, $\bar{u}$, is

Turbulent flow parameters can be described in terms of mean and fluctuating portions.

$$\bar{u} = \frac{1}{T} \int_{t_0}^{t_0+T} u(x, y, z, t)\, dt \tag{8.24}$$

where the time interval, T, is considerably longer than the period of the longest fluctuations, but considerably shorter than any unsteadiness of the average velocity. This is illustrated in Fig. 8.12.

The *fluctuating part* of the velocity, u', is that time-varying portion that differs from the average value

$$u = \bar{u} + u' \quad \text{or} \quad u' = u - \bar{u} \tag{8.25}$$

The structure and characteristics of turbulence may vary from one flow situation to another. One turbulent flow may have very large velocity fluctuations and another may have quite small fluctuations. Unfortunately, simply calculating the time average of the fluctuations will not tell us the general magnitude of the fluctuations as it will always be zero, since

$$\bar{u'} = \frac{1}{T} \int_{t_0}^{t_0+T} (u - \bar{u})\, dt = \frac{1}{T} \left(\int_{t_0}^{t_0+T} u\, dt - \bar{u} \int_{t_0}^{t_0+T} dt \right)$$
$$= \frac{1}{T} (T\bar{u} - T\bar{u}) = 0$$

The fluctuations are equally distributed on either side of the average. Since the square of a fluctuation quantity cannot be negative $[(u')^2 \geq 0]$, as indicated in **Fig. 8.13**, its average value is positive. Thus,

$$\overline{(u')^2} = \frac{1}{T} \int_{t_0}^{t_0+T} (u')^2\, dt > 0$$

This concept can be applied to the other fluctuating velocity components, v' or w', as well. On the other hand, it may be that the average of products of the fluctuations, such as $\overline{u'v'}$, are zero or nonzero (either positive or negative). The time average of square of velocity fluctuation is used to measure the *turbulence intensity* (or level of the turbulence), which may be larger in a very gusty wind than it is in a relatively steady (although turbulent) wind. The turbulence intensity, $\mathcal{I}$, is often defined as the square root of the mean square of the fluctuating velocity divided by the time-averaged velocity, or

$$\mathcal{I} = \frac{\sqrt{\overline{(u')^2}}}{\bar{u}} = \frac{\left[\frac{1}{T} \int_{t_0}^{t_0+T} (u')^2\, dt \right]^{1/2}}{\bar{u}}$$

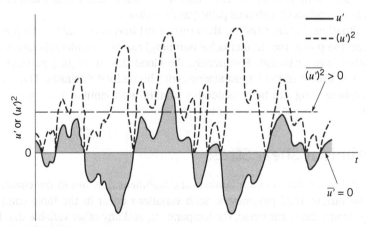

FIGURE 8.13   Average of the fluctuations and average of the square of the fluctuations.

The larger the turbulence intensity, the larger the fluctuations of the velocity (and other flow parameters). Well-designed wind tunnels have typical values of $\mathscr{I} \approx 0.01$, although with extreme care, values as low as $\mathscr{I} = 0.0002$ have been obtained. On the other hand, values of $\mathscr{I} \gtrsim 0.1$ are found for the flow in the atmosphere and rivers. A typical atmospheric wind speed graph is shown in the adjacent figure.

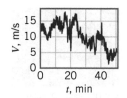

Another turbulence parameter that is different from one flow situation to another is the period of the fluctuations—the *time scale* of the fluctuations shown in Fig. 8.12. In many flows, such as the flow of water from a faucet, typical frequencies are on the order of 10, 100, or 1000 cycles per second (cps). For other flows, such as the Gulf Stream current in the Atlantic Ocean or flow of the atmosphere of Jupiter, characteristic random oscillations may have a period on the order of hours, days, or more.

It is tempting to extend the concept of viscous shear stress for laminar flow ($\tau = \mu\, du/dy$) to that of turbulent flow by replacing u, the instantaneous velocity, by $\bar{u}$, the time-averaged velocity. However, numerous experimental and theoretical studies have shown that such an approach leads to completely incorrect results. That is, $\tau \neq \mu\, d\bar{u}/dy$. A physical explanation for this behavior can be found in the concept of what produces a shear stress.

> The relationship between fluid motion and shear stress is very complex for turbulent flow.

Laminar flow is modeled as fluid particles that flow smoothly along in layers, gliding past the slightly slower or faster ones on either side. As is discussed in Chapter 1, the fluid actually consists of numerous molecules darting about in an almost random fashion as is indicated in **Fig. 8.14a**. The motion is not entirely random—a slight bias in one direction produces the flowrate we associate with the motion of fluid particles, $\bar{u}$. As the molecules dart across a given plane (plane A–A, for example), the ones moving upward have come from an area of smaller average x component of velocity than the ones moving downward, which have come from an area of larger velocity.

The momentum flux in the x direction across plane A–A of Fig. 8.14a gives rise to a drag (to the left) of the lower fluid on the upper fluid and an equal but opposite effect of the upper fluid on the lower fluid. The sluggish molecules moving upward across plane A–A must be accelerated by the fluid above this plane. The rate of change of momentum in this process produces (on the macroscopic scale) a shear force. Similarly, the more energetic molecules moving down across plane A–A must be slowed down by the fluid below that plane. This shear force is present only if there is a gradient in $u = u(y)$, otherwise the average x component of velocity (and momentum) of the upward and downward molecules is exactly the same. In addition, there are attractive forces between molecules. By combining these effects we obtain the well-known Newton viscosity law: $\tau = \mu\, du/dy$, where on a molecular basis μ is related to the mass and speed (temperature) of the random motion of the molecules.

> Turbulent flow shear stress is larger than laminar flow shear stress because of the irregular, random motion.

Although the above random motion of the molecules is also present in turbulent flow, there is another factor that is generally more pronounced and therefore more important. A simplistic way of thinking about turbulent flow is to consider it as consisting of a series of random,

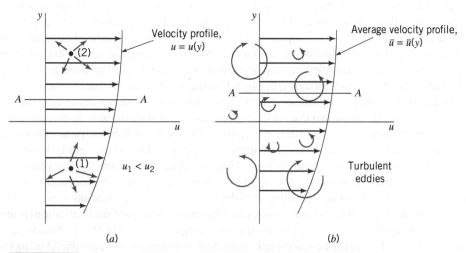

FIGURE 8.14 (a) Laminar flow shear stress caused by random motion of molecules. (b) Turbulent flow as a series of random, three-dimensional eddies.

© P. Dimotakis. Dimotakis, P.E., R.C. Miake-Lye, and D.A. Papantoniou 1983 Structure and Dynamics of Round Turbulent Jets. Phys. Fluids 26(11):3185–3192.

three-dimensional eddy type motions as is depicted (in one dimension only) in **Fig. 8.14b** and apparent in the adjacent photograph. These eddies range in size from very small diameter (on the order of the size of a fluid particle) to fairly large diameter (on the order of the size of the object or flow geometry considered). They move about randomly, conveying mass with an average velocity $\bar{u} = \bar{u}(y)$. This eddy structure greatly promotes mixing within the fluid. It also greatly increases the transport of x momentum across plane A–A. That is, finite "chunks" of fluid (not merely individual molecules as in laminar flow) are randomly transported across this plane, resulting in a relatively large (when compared with laminar flow) shear force. These "chunks" vary in size but are much larger than molecules.

The Wide World of Fluids

Listen to the flowrate

Sonar systems are designed to listen to transmitted and reflected sound waves in order to locate submerged objects. They have been used successfully for many years to detect and track underwater objects such as submarines and aquatic animals. Recently, sonar techniques have been refined so that they can be used to determine the flowrate in pipes. These new flowmeters work for turbulent, not laminar, pipe flows because their operation depends strictly on the

existence of *turbulent eddies* within the flow. (Ultrasonic flowmeters, on the other hand, work for both laminar and turbulent flow. See Section 8.6.1.) The flowmeters contain a sonar-based array that listens to and interprets pressure fields generated by the turbulent motion in pipes. By listening to the pressure fields associated with the movement of the turbulent eddies, the device can determine the speed at which the eddies travel past an array of sensors. The flowrate is determined by using a calibration procedure that links the speed of the turbulent structures to the volumetric flowrate.

V8.6 Turbulence in a bowl

The random velocity components that account for this momentum transfer (hence, the shear force) are u' (for the x component of velocity) and v' (for the rate of mass transfer crossing the plane). A more detailed consideration of the processes involved will show that the apparent shear stress on plane A–A is given by the following (Ref. 2):

$$\tau = \mu \frac{d\bar{u}}{dy} - \rho \overline{u'v'} = \tau_{\text{lam}} + \tau_{\text{turb}} \tag{8.26}$$

The shear stress is the sum of a laminar portion and a turbulent portion.

Note that if the flow is laminar, $u' = v' = 0$, so that $\overline{u'v'} = 0$ and Eq. 8.26 reduces to the customary random molecule-motion-induced *laminar shear stress*, $\tau_{\text{lam}} = \mu \, d\bar{u}/dy$. For turbulent flow it is found that the **turbulent shear stress**, $\tau_{\text{turb}} = -\rho \overline{u'v'}$, is positive (a positive v' is usually correlated with a negative u'). Hence, the shear stress is greater in turbulent flow than in laminar flow. Note the unit on τ_{turb} are (density)(velocity)^2 = N/m^2, as expected. Terms of the form $-\rho \overline{u'v'}$ (or $-\rho \overline{v'w'}$, etc.) are called *Reynolds stresses* in honor of Osborne Reynolds who first discussed them in 1895.

It is seen from Eq. 8.26 that the shear stress in turbulent flow is not merely proportional to the gradient of the time-averaged velocity, $\bar{u}(y)$. It also contains a contribution due to the random fluctuations of the x and y components of velocity. The density is involved because of the momentum transfer of the fluid within the random eddies. Although the relative magnitude of τ_{lam} compared to τ_{turb} is a complex function dependent on the specific flow involved, typical measurements indicate the structure shown in **Fig. 8.15a**. (Recall from Eq. 8.4 that the shear stress is proportional to the distance from the centerline of the pipe.) In a very narrow region near the wall (the *viscous sublayer*), the laminar shear stress is dominant. Away from the wall (in the *outer layer*) the turbulent portion of the shear stress is dominant. The transition between these two regions occurs in the *overlap layer*. The corresponding typical velocity profile is shown in **Fig. 8.15b**.

The scale of the sketches shown in Fig. 8.15 is not necessarily correct. Typically the value of τ_{turb} is 100 to 1000 times greater than τ_{lam} in the outer region, while the converse is true in the viscous sublayer. A correct modeling of turbulent flow is strongly dependent on an accurate knowledge of τ_{turb}. This, in turn, requires an accurate knowledge of the fluctuations u' and v', or $\rho \overline{u'v'}$. As yet it is not possible to solve the governing equations (the Navier–Stokes equations) for these details of the flow, although numerical techniques (see Appendix A) using the largest and fastest computers available have produced important information about some of the characteristics of turbulence. Considerable effort has gone into the study of turbulence.

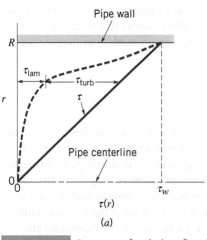
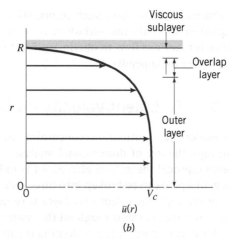

FIGURE 8.15 Structure of turbulent flow in a pipe. (*a*) Shear stress. (*b*) Average velocity.

Much remains to be learned. Perhaps studies in the new areas of chaos and fractal geometry will provide the tools for a better understanding of turbulence (see Section 8.3.5).

The vertical scale of Fig. 8.15 is also distorted. The viscous sublayer is usually a very thin layer adjacent to the wall. For example, for water flow in a 76 mm-diameter pipe with an average velocity of 3.04 m/s, the viscous sublayer is approximately 0.51 mm thick. Since the fluid motion within this thin layer is critical in terms of the overall flow (the no-slip condition and the wall shear stress occur in this layer), it is not surprising to find that turbulent pipe flow properties can be quite dependent on the roughness of the pipe wall, unlike laminar pipe flow, which is independent of roughness. Small roughness elements (scratches, rust, sand or dirt particles, etc.) can easily disturb this viscous sublayer (see Section 8.4), thereby affecting the entire flow.

One model for the shear stress for turbulent flow is given in terms of the *eddy viscosity, η*, where

$$\tau_{\text{turb}} = \eta \frac{d\bar{u}}{dy} \tag{8.27}$$

This extension of laminar flow terminology was introduced by J. Boussinesq, a French scientist, in 1877. Although the concept of an eddy viscosity is intriguing, in practice it is not an easy parameter to use. Unlike the absolute viscosity, μ, which is a known value for a given fluid, the eddy viscosity is a function of both the fluid and the flow conditions. That is, the eddy viscosity of water cannot be looked up in handbooks—its value changes from one turbulent flow condition to another and from one point in a turbulent flow to another.

The inability to accurately determine the Reynolds stress, $-\rho\overline{u'v'}$, is equivalent to not knowing the eddy viscosity. Several semiempirical theories have been proposed (Ref. 3) to determine approximate values of η. L. Prandtl (1875–1953), a German physicist and aerodynamicist, proposed that the turbulent process could be viewed as the random transport of bundles of fluid particles over a certain distance, ℓ_m, the *mixing length*, from a region of one velocity to another region of a different velocity. By the use of some ad hoc assumptions and physical reasoning, it was concluded that the eddy viscosity was given by

$$\eta = \rho \ell_m^2 \left| \frac{d\bar{u}}{dy} \right|$$

Thus, the turbulent shear stress is

$$\tau_{\text{turb}} = \rho \ell_m^2 \left(\frac{d\bar{u}}{dy} \right)^2 \tag{8.28}$$

> Various ad hoc assumptions have been used to approximate turbulent shear stresses.

The problem is thus shifted to that of determining the mixing length, ℓ_m. Further considerations indicate that ℓ_m is not a constant throughout the flow field. Near a solid surface the turbulence is dependent on the distance from the surface. Thus, additional assumptions are made regarding how the mixing length varies throughout the flow.

The net result is that as yet there is no general, all-encompassing, useful model that can accurately predict the shear stress throughout a general incompressible, viscous turbulent flow. Without such information it is impossible to integrate the force balance equation to obtain the turbulent velocity profile and other useful information, as was done for laminar flow to obtain Eq. 8.7. Further information on models of turbulence is presented in Appendix A.

8.3.3 Turbulent Velocity Profile

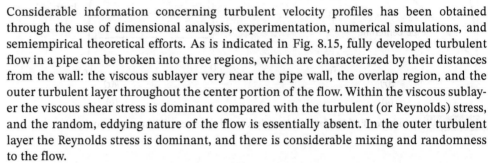

V8.7 Turbulent channel flow

Considerable information concerning turbulent velocity profiles has been obtained through the use of dimensional analysis, experimentation, numerical simulations, and semiempirical theoretical efforts. As is indicated in Fig. 8.15, fully developed turbulent flow in a pipe can be broken into three regions, which are characterized by their distances from the wall: the viscous sublayer very near the pipe wall, the overlap region, and the outer turbulent layer throughout the center portion of the flow. Within the viscous sublayer the viscous shear stress is dominant compared with the turbulent (or Reynolds) stress, and the random, eddying nature of the flow is essentially absent. In the outer turbulent layer the Reynolds stress is dominant, and there is considerable mixing and randomness to the flow.

The character of the flow within these two regions is entirely different. For example, within the viscous sublayer the fluid viscosity is an important parameter; the density is unimportant. In the outer layer the opposite is true. By a careful use of dimensional analysis arguments for the flow in each layer and by a matching of the results in the common overlap layer, it has been possible to obtain the following conclusions about the turbulent velocity profile in a smooth pipe (Ref. 5).

In the viscous sublayer the velocity profile can be written in dimensionless form as

$$\frac{\bar{u}}{u^*} = \frac{yu^*}{v} \tag{8.29}$$

where $y = R - r$ is the distance measured from the wall, $\bar{u}$ is the time-averaged x component of velocity, and $u^* = (\tau_w/\rho)^{1/2}$ is termed the *friction velocity*. Note that u^* is not an actual velocity of the fluid—it is merely a quantity that has dimensions of velocity. As is indicated in **Fig. 8.16**, Eq. 8.29 (commonly called the *law of the wall*) is valid very near the smooth wall, for $0 \leq yu^*/v \lesssim 5$.

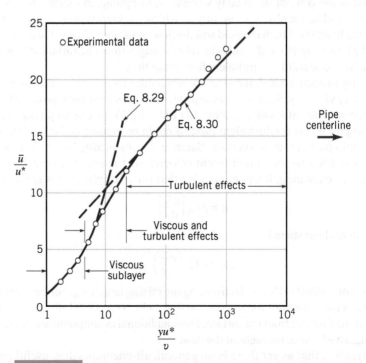

FIGURE 8.16 Typical structure of the turbulent velocity profile in a pipe.

Dimensional analysis arguments indicate that in the overlap region the velocity should vary as the logarithm of y. Thus, the following expression has been proposed:

$$\frac{\bar{u}}{u^*} = 2.44 \ln\left(\frac{yu^*}{v}\right) + 5.0 \qquad (8.30)$$

A turbulent flow velocity profile can be divided into various regions.

where the constants 2.44 and 5.0 have been determined experimentally. As is indicated in Fig. 8.16, for regions not too close to the smooth wall, but not all the way out to the pipe center, Eq. 8.30 gives a reasonable correlation with the experimental data. Note that the horizontal scale is a logarithmic scale. This tends to exaggerate the size of the viscous sublayer relative to the remainder of the flow. As is shown in Example 8.4, the viscous sublayer is usually quite thin. Similar results can be obtained for turbulent flow past rough walls (Ref. 17).

A number of other correlations exist for the velocity profile in turbulent pipe flow. In the central region (the outer turbulent layer) the expression $(V_c - \bar{u})/u^* = 2.5 \ln(R/y)$, where V_c is the centerline velocity, is often suggested as a good correlation with experimental data. Another often-used (and relatively easy to use) correlation is the empirical *power-law velocity profile*

V8.8 Laminar to turbulent flow from a pipe

$$\frac{\bar{u}}{V_c} = \left(1 - \frac{r}{R}\right)^{1/n} \qquad (8.31)$$

In this representation, the value of n is a function of the Reynolds number, as is indicated in **Fig. 8.17**. The one-seventh power-law velocity profile ($n = 7$) is often used as a reasonable approximation for many practical flows. Typical turbulent velocity profiles based on this power-law representation are shown in **Fig. 8.18**.

A closer examination of Eq. 8.31 shows that the power-law profile cannot be valid near the wall, since according to this equation the velocity gradient is infinite there. In addition, Eq. 8.31 cannot be precisely valid near the centerline because it does not give $d\bar{u}/dr = 0$ at $r = 0$. However, it does provide a reasonable approximation to the measured velocity profiles across most of the pipe.

Note from Fig. 8.18 that the turbulent profiles are much "flatter" than the laminar profile and that this flatness increases with Reynolds number (i.e., with n). The reason for this flatness can be inferred from the discussion concerning Fig. 8.14 in Section 8.3.2. Increased momentum transport due to turbulence, in essence, homogenizes the velocity profile; high velocities interchange places with low velocities resulting in a flattened time average profile. Recall from Chapter 3 that reasonable approximate results are often obtained by using the inviscid Bernoulli equation and by assuming a fictitious uniform velocity profile. Since most flows are turbulent and turbulent flows tend to have nearly uniform velocity profiles, the usefulness of the Bernoulli equation and the uniform profile assumption is not unexpected. Of course, many properties of the flow cannot be accounted for without including viscous effects.

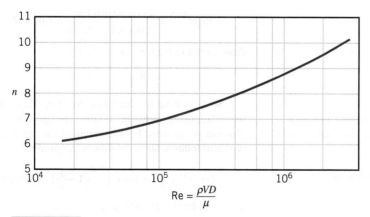

FIGURE 8.17 Exponent, n, for power-law velocity profiles. (Adapted from Ref. 1.)

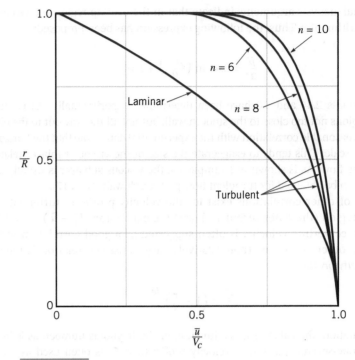

FIGURE 8.18 Typical laminar flow and turbulent flow velocity profiles.

EXAMPLE 8.4 | Turbulent Pipe Flow Properties

Given Water at 20 °C ($\rho = 998$ kg/m^3 and $v = 1.004 \times 10^{-6}$ m^2/s) flows through a horizontal pipe of 0.1-m diameter with a flowrate of $Q = 4 \times 10^{-2}$ m^3/s and a pressure gradient of 2.59 kPa/m.

Find

a. Determine the approximate thickness of the viscous sublayer.

b. Determine the approximate centerline velocity, V_c.

c. Determine the ratio of the turbulent to laminar shear stress, $\tau_{\text{turb}}/\tau_{\text{lam}}$, at a point midway between the centerline and the pipe wall (i.e., at $r = 0.025$ m).

Solution

a. According to Fig. 8.16, the thickness of the viscous sublayer, δ_s, is approximately

$$\frac{\delta_s u^*}{v} = 5$$

Therefore,

$$\delta_s = 5\frac{v}{u^*}$$

where

$$u^* = \left(\frac{\tau_w}{\rho}\right)^{1/2} \tag{1}$$

The wall shear stress can be obtained from the pressure drop data and Eq. 8.5, which is valid for either laminar or turbulent flow. Thus,

$$\tau_w = \frac{D\,\Delta p}{4\ell} = \frac{(0.1\text{ m})(2.59 \times 10^3\text{ N/m}^2)}{4(1\text{ m})} = 64.8\text{ N/m}^2$$

Hence, from Eq. 1 we obtain

$$u^* = \left(\frac{64.8\text{ N/m}^2}{998\text{ kg/m}^3}\right)^{1/2} = 0.255\text{ m/s}$$

so that

$$\delta_s = \frac{5(1.004 \times 10^{-6}\text{ m}^2/\text{s})}{0.255\text{ m/s}}$$

$$= 1.97 \times 10^{-5}\text{ m} \simeq 0.02\text{ mm} \qquad (Ans)$$

Comment As stated previously, the viscous sublayer is very thin. Minute imperfections on the pipe wall will protrude into this sublayer and affect some of the characteristics of the flow (i.e., wall shear stress and pressure drop).

b. The centerline velocity can be obtained from the average velocity and the assumption of a power-law velocity profile as follows. For this flow with

$$V = \frac{Q}{A} = \frac{0.04\text{ m}^3/\text{s}}{\pi(0.1\text{ m})^2/4} = 5.09\text{ m/s}$$

the Reynolds number is

$$\text{Re} = \frac{VD}{v} = \frac{(5.09\text{ m/s})(0.1\text{ m})}{(1.004 \times 10^{-6}\text{ m}^2/\text{s})} = 5.07 \times 10^5$$

Thus, from Fig. 8.17, $n = 8.4$ so that

$$\frac{\bar{u}}{V_c} \approx \left(1 - \frac{r}{R}\right)^{1/8.4}$$

To determine the centerline velocity, V_c, we must know the relationship between V (the average velocity) and V_c. This can be obtained by integration of the power-law velocity profile as follows. Since the flow is axisymmetric,

$$Q = AV = \int \bar{u}\,dA = V_c \int_{r=0}^{r=R} \left(1 - \frac{r}{R}\right)^{1/n}(2\pi r)\,dr$$

which can be integrated to give

$$Q = 2\pi R^2 V_c \frac{n^2}{(n+1)(2n+1)}$$

Thus, since $Q = \pi R^2 V$, we obtain

$$\frac{V}{V_c} = \frac{2n^2}{(n+1)(2n+1)}$$

With $n = 8.4$ in the present case, this gives

$$V_c = \frac{(n+1)(2n+1)}{2n^2} V = 1.186V = 1.186\,(5.09 \text{ m/s})$$

$$= 6.04 \text{ m/s} \qquad\qquad\qquad (Ans)$$

Recall that $V_c = 2V$ for laminar pipe flow.

c. From Eq. 8.4, which is valid for laminar or turbulent flow, the shear stress at $r = 0.025$ m is

$$\tau = \frac{2\tau_w r}{D} = \frac{2(64.8 \text{ N/m}^2)(0.025 \text{ m})}{(0.1 \text{ m})}$$

or

$$\tau = \tau_{\text{lam}} + \tau_{\text{turb}} = 32.4 \text{ N/m}^2$$

where $\tau_{\text{lam}} = -\mu\, d\bar{u}/dr$. From the power-law velocity profile (Eq. 8.31) we obtain the gradient of the average velocity as

$$\frac{d\bar{u}}{dr} = -\frac{V_c}{nR}\left(1 - \frac{r}{R}\right)^{(1-n)/n}$$

which gives

$$\frac{d\bar{u}}{dr} = -\frac{(6.04 \text{ m/s})}{8.4(0.05 \text{ m})}\left(1 - \frac{0.025 \text{ m}}{0.05 \text{ m}}\right)^{(1-8.4)/8.4}$$

$$= -26.5/\text{s}$$

Thus,

$$\tau_{\text{lam}} = -\mu\frac{d\bar{u}}{dr} = -(\nu\rho)\frac{d\bar{u}}{dr}$$

$$= -(1.004 \times 10^{-6} \text{ m}^2/\text{s})(998 \text{ kg/m}^3)(-26.5/\text{s})$$

$$= 0.0266 \text{ N/m}^2$$

Thus, the ratio of turbulent to laminar shear stress is given by

$$\frac{\tau_{\text{turb}}}{\tau_{\text{lam}}} = \frac{\tau - \tau_{\text{lam}}}{\tau_{\text{lam}}} = \frac{32.4 - 0.0266}{0.0266} = 1220 \qquad (Ans)$$

Comment   As expected, most of the shear stress at this location in the turbulent flow is due to the turbulent shear stress.

The turbulent flow characteristics discussed in this section are not unique to turbulent flow in round pipes. Many of the characteristics introduced (e.g., the Reynolds stress, the viscous sublayer, the overlap layer, the outer layer, the general characteristics of the velocity profile, etc.) are found in other turbulent flows. In particular, turbulent pipe flow and turbulent flow past a solid wall (boundary layer flow) share many of these common traits. Such ideas are discussed more fully in Chapter 9.

8.3.4 Turbulence Modeling

Although it is not yet possible to theoretically predict the random, irregular details of turbulent flows, it would be useful to be able to predict the time-averaged flow fields (pressure, velocity, etc.) directly from the basic governing equations. To this end one can time average the governing Navier–Stokes equations (Eqs. 6.31 and 6.127) to obtain equations for the average velocity and pressure. However, because the Navier–Stokes equations are nonlinear, the resulting time-averaged differential equations contain not only the desired average pressure and velocity as variables, but also averages of products of the fluctuations—terms of the type that one tried to eliminate by averaging the equations! For example, the Reynolds stress $-\rho\overline{u'v'}$ (see Eq. 8.26) occurs in the time-averaged momentum equation.

Thus, it is not possible to merely average the basic differential equations and obtain governing equations involving only the desired averaged quantities. This is the reason for the variety of ad hoc assumptions that have been proposed to provide "closure" to the equations governing the average flow. That is, the set of governing equations must be a complete or closed set of equations—the same number of equations as unknowns.

Various attempts have been made to solve this closure problem (Refs. 1, 32). Such schemes involving the introduction of an eddy viscosity or the mixing length (as introduced in Section 8.3.2) are termed algebraic or zero-equation models. Other methods include the one-equation and two-equation models, where zero-, one- and two- are the number of differential equations that describe the model. Because of the complexity of these turbulence models, as well as the Navier-Stokes equations themselves, the only viable path to solving the equations is to apply numerical methods on a computer. This field, called *Computational Fluid Dynamics*, is discussed briefly in Appendix A.

Turbulence modeling is an important topic. Although considerable progress has been made, much remains to be done in this area.

8.3.5 Chaos and Turbulence

Chaos theory is a relatively new branch of mathematical physics that may provide insight into the complex nature of turbulence. This method combines mathematics and numerical (computer) techniques to provide a new way to analyze certain problems. Chaos theory, which is quite complex and is currently under development, involves the behavior of nonlinear dynamical systems and their response to initial and boundary conditions. The flow of a viscous fluid, which is governed by the nonlinear Navier–Stokes equations (Eq. 6.127), may be such a system.

To solve the Navier–Stokes equations for the velocity and pressure fields in a viscous flow, one must specify the particular flow geometry being considered (the boundary conditions) and the condition of the flow at some particular time (the initial conditions). If, as some researchers predict, the Navier–Stokes equations allow chaotic behavior, then the state of the flow at times after the initial time may be very sensitive to the initial conditions. A slight variation to the initial flow conditions may cause the flow at later times to be quite different than it would have been with the original, only slightly different initial conditions. When carried to the extreme, the flow may be "chaotic," "random," or perhaps (in current terminology), "turbulent."

The occurrence of such behavior would depend on the value of the Reynolds number. For example, it may be found that for sufficiently small Reynolds numbers the flow is not chaotic (i.e., it is laminar), while for large Reynolds numbers it is chaotic with turbulent characteristics.

Thus, with the advancement of chaos theory it may be found that the numerous ad hoc turbulence ideas mentioned in previous sections (e.g., eddy viscosity, mixing length, law of the wall, etc.) may not be needed. It may be that chaos theory can provide the turbulence properties and structure directly from the governing equations. As of now we must wait until this exciting topic is developed further. The interested reader is encouraged to consult Ref. 4 for a general introduction to chaos or Ref. 33 for additional material.

8.4   PIPE FLOW LOSSES VIA DIMENSIONAL ANALYSIS

As noted previously, turbulent flow can be a very complex, difficult topic—one that as yet has defied a rigorous theoretical treatment. Thus, most turbulent pipe flow analyses are based on experimental data and semi-empirical formulas. These data are expressed conveniently in dimensionless form.

It is often necessary to determine the head loss, h_L, that occurs in a pipe flow so that the mechanical energy equation, Eq. 5.84 or 5.89, can be used in the analysis of pipe flow problems. As shown in Fig. 8.1, a typical pipe system usually consists of various lengths of straight pipe interspersed with various types of components (valves, elbows, etc.). The overall head loss for the pipe system consists of the head loss due to viscous effects in the straight pipes, termed the **major loss** and denoted $h_{L\,\text{major}}$, and the head loss in the various pipe components, termed the **minor loss** and denoted $h_{L\,\text{minor}}$. That is,

$$h_L = h_{L\,\text{major}} + h_{L\,\text{minor}}$$

The head loss designations of "major" and "minor" do not necessarily reflect the relative importance of each type of loss. For a pipe system that contains many components and a relatively short length of pipe, the minor loss may actually be larger than the major loss.

8.4.1 Major Losses

A dimensional analysis treatment of pipe flow provides the most convenient base from which to consider turbulent, fully developed pipe flow. An introduction to this topic was given in Section 8.3. As is discussed in Sections 8.2.1 and 8.2.4, the pressure drop and head loss in a pipe are dependent on the wall shear stress, τ_w, between the fluid and pipe surface. A fundamental difference between laminar and turbulent flow is that the shear stress for turbulent flow is a function of the density of the fluid, ρ, as shown in Eq. 8.26. For laminar flow, the shear stress is independent of the density, leaving the viscosity, μ, as the only important fluid property.

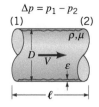

Thus, as indicated in the adjacent figure, the pressure drop, Δp for steady, incompressible turbulent flow in a horizontal round pipe of diameter D can be written in functional form as

$$\Delta p = F(V, D, \ell, \varepsilon, \mu, \rho) \qquad (8.32)$$

where V is the average velocity, ℓ is the pipe length, and ε is a measure of the roughness of the pipe wall. It is clear that Δp should be a function of V, D, and ℓ. The dependence of Δp on the fluid properties μ and ρ is expected because of the dependence of τ on these parameters.

Although the pressure drop for laminar pipe flow is found to be independent of the roughness of the pipe, it is necessary to include this parameter when considering turbulent flow. As is discussed in Section 8.3.3 and illustrated in **Fig. 8.19**, for turbulent flow there is a relatively thin viscous sublayer formed in the fluid near the pipe wall. In many instances this layer is very thin; $\delta_s/D \ll 1$, where δ_s is the sublayer thickness. If a typical wall roughness element protrudes sufficiently far into (or even through) this layer, the structure and properties of the viscous sublayer (along with Δp and τ_w) will be different than if the wall were smooth. Thus, for turbulent flow the pressure drop is expected to be a function of the wall roughness. For laminar flow there is no thin viscous layer—viscous effects are important across the entire pipe. Thus, relatively small roughness elements have completely negligible effects on laminar pipe flow. Of course, for pipes with very large wall "roughness" ($\varepsilon/D \gtrsim 0.1$), such as that in corrugated pipes, the flowrate may be a function of the "roughness." We will consider only typical constant diameter pipes with relative roughnesses in the range $0 \leq \varepsilon/D \lesssim 0.05$. Analysis of flow in corrugated pipes does not fit into the standard constant diameter pipe category, although experimental results for such pipes are available (Ref. 30).

> **Unlike laminar pipe flow characteristics, turbulent pipe flow characteristics depend on the fluid density and the pipe roughness.**

The list of parameters given in Eq. 8.32 is apparently a complete one. That is, experiments have shown that other parameters (such as surface tension, vapor pressure, etc.) do not affect the pressure drop for the conditions stated (steady, incompressible flow; round, horizontal pipe). Since there are seven variables ($k = 7$) that can be written in terms of the three reference dimensions MLT ($r = 3$), Eq. 8.32 can be written in dimensionless form in terms of $k - r = 4$ dimensionless groups. As was discussed in Section 7.9.1, one such representation is

$$\frac{\Delta p}{\frac{1}{2}\rho V^2} = \tilde{\phi}\left(\frac{\rho V D}{\mu}, \frac{\ell}{D}, \frac{\varepsilon}{D}\right)$$

This result differs from that used for laminar flow (see Eq. 8.17) in two ways. First, we have chosen to make the pressure dimensionless by dividing by the dynamic pressure, $\rho V^2/2$, rather than a characteristic viscous shear stress, $\mu V/D$. This convention was chosen in recognition of the fact

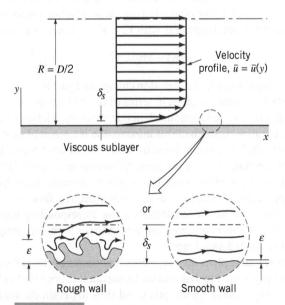

FIGURE 8.19 Flow in the viscous sublayer near rough and smooth walls.

that the shear stress for turbulent flow is normally dominated by τ_{turb}, which is a stronger function of the density than it is of viscosity. Second, we have introduced two additional dimensionless parameters, the Reynolds number, $Re = \rho VD/\mu$, and the **relative roughness**, ε/D, which are not present in the laminar formulation because the two parameters ρ and ε are not important in fully developed laminar pipe flow.

As was done for laminar flow, the functional representation can be simplified by imposing the reasonable assumption that the pressure drop should be proportional to the pipe length. (Such a step is not within the realm of dimensional analysis. It is suggested by theory and experiments.) The only way that this can be true is if the ℓ/D dependence is factored out as

$$\frac{\Delta p}{\frac{1}{2}\rho V^2} = \frac{\ell}{D}\phi\left(Re, \frac{\varepsilon}{D}\right)$$

As was discussed in Section 8.2.3, the quantity $\Delta pD/(\ell\rho V^2/2)$ is termed the friction factor, f. Thus, for a horizontal pipe

$$\Delta p = f\frac{\ell}{D}\frac{\rho V^2}{2} \tag{8.33}$$

where

$$f = \phi\left(Re, \frac{\varepsilon}{D}\right)$$

For laminar fully developed flow, the value of f is simply $f = 64/Re$, independent of ε/D. For turbulent flow, the functional dependence of the friction factor on the Reynolds number and the relative roughness, $f = \phi(Re, \varepsilon/D)$, is a rather complex one that cannot, as yet, be obtained from a theoretical analysis. The results are obtained from an exhaustive set of experiments and usually presented in terms of a curve-fitting formula or the equivalent graphical form.

From Eq. 5.89, the head form of the mechanical energy equation for steady incompressible flow with no shaft work head is

$$\frac{p_1}{\gamma} + \alpha_1\frac{V_1^2}{2g} + z_1 = \frac{p_2}{\gamma} + \alpha_2\frac{V_2^2}{2g} + z_2 + h_L$$

where h_L is the head loss between sections (1) and (2). With the assumption of a constant diameter ($D_1 = D_2$ so that $V_1 = V_2$), horizontal ($z_1 = z_2$) pipe with fully developed flow ($\alpha_1 = \alpha_2$), this becomes $\Delta p = p_1 - p_2 = \gamma h_L$, which can be combined with Eq. 8.33 to give

| The major head loss in pipe flow is given in terms of the friction factor. |

$$h_{L\,major} = f\frac{\ell}{D}\frac{V^2}{2g} \tag{8.34}$$

Equation 8.34, called the *Darcy–Weisbach equation*, is valid for any fully developed, steady, incompressible pipe flow—whether the pipe is horizontal or on a hill. On the other hand, Eq. 8.33 is valid only for horizontal pipes. In general, with $V_1 = V_2$ the energy equation gives

$$p_1 - p_2 = \gamma(z_2 - z_1) + \gamma h_L = \gamma(z_2 - z_1) + f\frac{\ell}{D}\frac{\rho V^2}{2}$$

Part of the pressure change is due to the elevation change and part is due to the head loss associated with frictional effects, which are given in terms of the friction factor, f.

It is not easy to determine the functional dependence of the friction factor on the Reynolds number and relative roughness. Much of this information is a result of experiments conducted by J. Nikuradse in 1933 (Ref. 6) and amplified by many others since then. One difficulty lies in the determination of the roughness of the pipe. Nikuradse used artificially roughened pipes produced by gluing sand grains of known size onto pipe walls to produce pipes with sandpaper-type surfaces. The pressure drop needed to produce a desired flowrate was measured, and the data were converted into the friction factor for the corresponding Reynolds number and relative roughness. The tests were repeated numerous times for a wide range of Re and ε/D to determine the $f = \phi(Re, \varepsilon/D)$ dependence.

In commercially available pipes the roughness is not as uniform and well defined as in the artificially roughened pipes used by Nikuradse. However, it is possible to obtain a measure of the effective relative roughness of typical pipes and thus to obtain the friction factor. Typical roughness values for various pipe surfaces are given in **Table 8.1**. **Figure 8.20** shows the functional dependence of f on Re and ε/D and is called the **Moody chart** in honor of L. F. Moody,

TABLE 8.1 Equivalent Roughness for New Pipes

Pipe	Equivalent Roughness, ε	
	Millimeters	Feet
Riveted steel	0.9–9.0	0.003–0.03
Concrete	0.3–3.0	0.001–0.01
Wood stave	0.18–0.9	0.0006–0.003
Cast iron	0.26	0.00085
Galvanized iron	0.15	0.0005
Commercial steel or wrought iron	0.045	0.00015
Drawn tubing	0.0015	0.000005
Plastic, glass	0.0 (smooth)	0.0 (smooth)

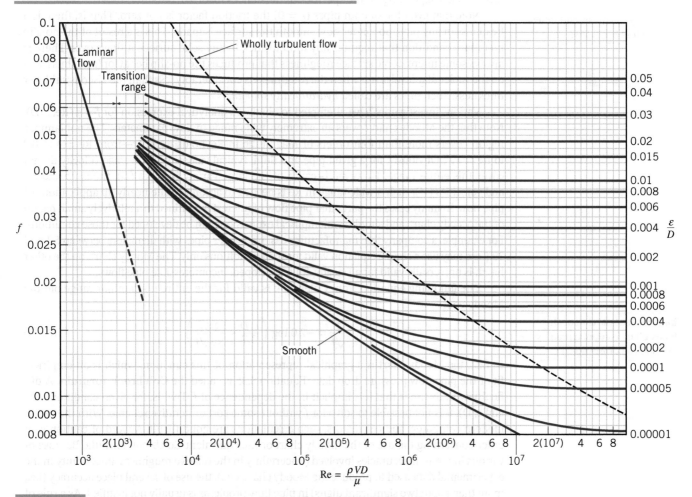

FIGURE 8.20 Friction factor as a function of Reynolds number and relative roughness for round pipes—the Moody chart. (Source: From Moody, L.F. (1944) "Friction Factors for Pipe Flow," *Transactions of the ASME* 66, 671–684. Reprinted with permission of The American Society of Mechanical Engineers.)

who, along with C. F. Colebrook, correlated the original data of Nikuradse in terms of the relative roughness of commercially available pipe materials. It should be noted that the values of ε/D do not necessarily correspond to the actual values obtained by a microscopic determination of the average height of the roughness of the surface. They do, however, provide the correct correlation for $f = \phi(\text{Re}, \varepsilon/D)$.

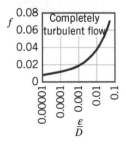

For any pipe, even smooth ones, the head loss is not zero.

It is important to observe that the values of relative roughness given pertain to new, clean pipes. After considerable use, most pipes (because of a buildup of corrosion or scale) may have a relative roughness that is considerably larger (perhaps by an order of magnitude) than that given. As shown by the adjacent photo, very old pipes may have enough scale buildup to not only alter the value of ε but also to change their effective diameter by a considerable amount.

The following characteristics are observed from the data of Fig. 8.20. For laminar flow, $f = 64/\text{Re}$, which is independent of relative roughness. For turbulent flows with very large Reynolds numbers, $f = \phi(\varepsilon/D)$, which, as shown in the adjacent figure, is independent of the Reynolds number. For such flows, commonly termed *completely turbulent flow* (or *wholly turbulent flow*), the laminar sublayer is so thin (its thickness decreases with increasing Re) that the surface roughness completely dominates the character of the flow near the wall. Hence, the pressure drop required is a result of an inertia-dominated turbulent shear stress rather than the viscosity-dominated laminar shear stress normally found in the viscous sublayer. For flows with moderate values of Re, the friction factor is indeed dependent on both the Reynolds number and relative roughness—$f = \phi(\text{Re}, \varepsilon/D)$. The gap in the figure for which no values of f are given (the $2100 < \text{Re} < 4000$ range) is a result of the fact that the flow in this transition range may be laminar or turbulent (or an unsteady mix of both) depending on the specific circumstances involved.

Note that even for smooth pipes ($\varepsilon = 0$) the friction factor is not zero. That is, there is a head loss in any pipe, no matter how smooth the surface is made. This is a result of the no-slip boundary condition that requires any fluid to stick to any solid surface it flows over. There is always some microscopic surface roughness that produces the no-slip behavior (and thus $f \neq 0$) on the molecular level, even when the roughness is considerably less than the viscous sublayer thickness. Such pipes are called *hydraulically smooth*.

Various investigators have attempted to obtain an analytical expression for $f = \phi(\text{Re}, \varepsilon/D)$. Note that the Moody chart covers an extremely wide range in flow parameters. The nonlaminar region covers more than four orders of magnitude in Reynolds number—from $\text{Re} = 4 \times 10^3$ to $\text{Re} = 10^8$. Obviously, for a given pipe and fluid, typical values of the average velocity do not cover this entire range. However, because of the large variety in pipes (D), fluids (ρ and μ), and velocities (V), such a wide range in Re is needed to accommodate nearly all applications of pipe flow. In many cases the particular pipe flow of interest is confined to a relatively small region of the Moody chart, and simple semiempirical expressions can be developed for those conditions. For example, a company that manufactures cast iron water pipes with diameters between 5.1 and 30.5 cm may use a simple equation valid for their conditions only. The Moody chart, on the other hand, is universally valid for all steady, fully developed, incompressible pipe flows.

The turbulent portion of the Moody chart is represented by the Colebrook formula.

The following equation from Colebrook is valid for the entire nonlaminar range of$a$ the Moody chart

$$\frac{1}{\sqrt{f}} = -2.0 \log \left(\frac{\varepsilon/D}{3.7} + \frac{2.51}{\text{Re}\sqrt{f}} \right) \tag{8.35a}$$

In fact, the Moody chart is a graphical representation of this equation, which is an empirical fit of the pipe flow pressure drop data. Equation 8.35 is called the **Colebrook formula**. A difficulty with its use is that it is implicit in the dependence of f. That is, for given conditions (Re and ε/D), it is not possible to solve for f without some sort of iterative scheme. With the use of modern computers and calculators, such calculations are not difficult. A word of caution is in order concerning the use of the Moody chart or the equivalent Colebrook formula. Because of various inherent inaccuracies involved (uncertainty in the relative roughness, uncertainty in the experimental data used to produce the Moody chart, etc.), the use of several place accuracy (i.e., more than about two significant digits) in pipe flow problems is usually not justified. As a rule of thumb, a 10% accuracy is the best expected. It is possible to obtain an equation that adequately approximates the Colebrook/Moody chart relationship but does not require an iterative scheme. For example, the Haaland equation (Ref. 34), which is easier to use, is given by

$$\frac{1}{\sqrt{f}} = -1.8 \log \left[\left(\frac{\varepsilon/D}{3.7} \right)^{1.11} + \frac{6.9}{\text{Re}} \right] \tag{8.35b}$$

where one can solve for f explicitly. The Swanee–Jain and Churchill formulas are also explicit in f and are given in the problem set for Section 8.4.1. Von Kármán provided a formula that is exclusively for wholly turbulent flow and is also given in the problems.

EXAMPLE 8.5 | Comparison of Laminar or Turbulent Pressure Drop

Given Air under standard conditions flow through a 4.0 mm-diameter drawn tubing with an average velocity $V = 50$ m/s. For such conditions the flow would normally be turbulent. However, if prcautions are taken to eliminate disturbances to the flow (the entrance to the tube is very smooth, the air is dust free, the tube does not vibrate, etc.), it may be possible to maintain laminar flow.

Find

a. Determine the pressure drop in a 0.1 m section of the tube if the flow is laminar.

b. Repeat the calculations if the flow is turbulent.

Solution Under standard temperature and pressure conditions the density and viscosity a $\rho = 1.23$ kg/m³ and $\mu = 1.79 \times 10^{-5}$ N · s/m². Thus, the Reynolds number is

$$\text{Re} = \frac{\rho V D}{\mu} = \frac{(1.23 \text{ kg/m}^3)(50 \text{ m/s})(0.004 \text{ m})}{1.79 \times 10^{-5} \text{ N} \cdot \text{s/m}^2} = 13,700$$

which would normally indicate turbulent flow.

a. If the flow were laminar, then $f = 64/\text{Re} = 64/13,700 = 0.00467$, and the pressure drop in a 0.1 m-long horizontal section of the pipe would be

$$\Delta p = f \frac{\ell}{D} \frac{1}{2} \rho V^2$$

$$= (0.00467) \frac{(0.1 \text{ m})}{(0.004 \text{ m})} \frac{1}{2} (1.23 \text{ kg/m}^3)(50 \text{ m/s})^2$$

or

$$\Delta p = 0.179 \text{ kPa} \qquad (Ans)$$

Comment Note that the same result is obtained from Eq. 8.8:

$$\Delta p = \frac{32 \mu \ell}{D^2} V$$

$$= \frac{32(1.79 \times 10^{-5} \text{ N} \cdot \text{s/m}^2)(0.1 \text{ m})(50 \text{ m/s})}{(0.004 \text{ m})^2}$$

$$= 179 \text{ N/m}^2$$

b. If the flow were turbulent, then $f = \phi(\text{Re}, \varepsilon/D)$, where from Table 8.1, $\varepsilon = 0.0015$ mm so that $\varepsilon/D = 0.0015$ mm/4.0 mm = 0.000375. From the Moody chart with Re = 1.37×10^4 and $\varepsilon/D = 0.000375$ we obtain $f = 0.028$. Thus, the pressure drop in this case would be approximately

$$\Delta p = f \frac{\ell}{D} \frac{1}{2} \rho V^2 = (0.028) \frac{(0.1 \text{ m})}{(0.004 \text{ m})} \frac{1}{2} (1.23 \text{ kg/m}^3)(50 \text{ m/s})^2$$

or

$$\Delta p = 1.076 \text{ kPa} \qquad (Ans)$$

Comment A considerable savings in effort to force the fluid through the pipe could be realized (0.179 kPa rather than 1.076 kPa) if the flow could be maintained as laminar flow at this Reynolds number. In general this is very difficult to do, although laminar flow in pipes has been maintained up to Re ≈ 100,000 in rare instances.

An alternate method to determine the friction factor for the turbulent flow would be to use the Colebrook formula, Eq. 8.35a. Thus,

$$\frac{1}{\sqrt{f}} = -2.0 \log \left(\frac{\varepsilon/D}{3.7} + \frac{2.51}{\text{Re} \sqrt{f}} \right)$$

$$= -2.0 \log \left(\frac{0.000375}{3.7} + \frac{2.51}{1.37 \times 10^4 \sqrt{f}} \right)$$

or

$$\frac{1}{\sqrt{f}} = -2.0 \log \left(1.01 \times 10^{-4} + \frac{1.83 \times 10^{-4}}{\sqrt{f}} \right) \qquad (1)$$

By using a root-finding technique on a computer or calculator, the solution to Eq. 1 is determined to be $f = 0.0291$, in agreement (within the accuracy of reading the graph) with the Moody chart method of $f = 0.028$.

Equation 8.35b provides an alternate form to the Colebrook formula that can be used to solve for the friction factor directly.

$$\frac{1}{\sqrt{f}} = -1.8 \log \left[\left(\frac{\varepsilon/D}{3.7} \right)^{1.11} + \frac{6.9}{\text{Re}} \right]$$

$$= -1.8 \log \left[\left(\frac{0.000375}{3.7} \right)^{1.11} + \frac{6.9}{1.37 \times 10^4} \right]$$

$$= 0.0289$$

This agrees with the Colebrook formula and Moody chart values obtained above.

Numerous other empirical formulas can be found in the literature (Ref. 5) for portions of the Moody chart. For example, an often-used equation, commonly referred to as the Blasius formula, for turbulent flow in smooth pipes ($\varepsilon/D = 0$) with Re < 10^5 is

$$f = \frac{0.316}{\text{Re}^{1/4}}$$

For our case this gives

$$f = 0.316 (13,700)^{-0.25} = 0.0292$$

which is in agreement with the previous results. Note that the value of f is relatively insensitive to ε/D for this particular situation. Whether the tube was smooth glass ($\varepsilon/D = 0$) or the drawn tubing ($\varepsilon/D = 0.000375$) would not make much difference in the pressure drop. For this flow, an increase in relative roughness by a factor of 30 to $\varepsilon/D = 0.0113$ (equivalent to a commercial steel surface; see Table 8.1) would give $f = 0.043$. This would represent an increase in pressure drop and head loss by a factor of $0.043/0.0291 = 1.48$ compared with that for the original drawn tubing.

The pressure drop of 1.076 kPa in a length of 0.1 m of pipe corresponds to a change in absolute pressure [assuming $p = 101$ kPa (abs) at $x = 0$] of approximately $1.076/101 = 0.0107$, or about 1%. Thus, the incompressible flow assumption on which the above calculations (and all of the formulas in this chapter) are based is reasonable. However, if the pipe were 2 m long the pressure drop would be 21.5 kPa, approximately 20% of the original pressure. In this case the density would not be approximately constant along the pipe, and a compressible flow analysis would be needed. Such considerations are discussed in Chapter 11.

8.4.2 Minor Losses

As discussed in the previous section, the head loss in long, straight sections of pipe, the major losses, can be calculated by use of the friction factor obtained from either the Moody chart or the Colebrook equation. Most pipe systems, however, consist of considerably more than straight pipes. These additional components (valves, bends, tees, and the like) add to the overall head loss of the system. Such losses are generally termed *minor losses*, with the corresponding head loss denoted $h_{L\,minor}$. In this section we indicate how to determine the various minor losses that commonly occur in pipe systems.

The head loss associated with flow through a valve is a common minor loss. The purpose of a valve is to provide a means to regulate the flowrate. This is accomplished by changing the geometry of the system (i.e., closing or opening the valve alters the flow pattern through the valve), which in turn alters the losses associated with the flow through the valve. The flow resistance or head loss through the valve may be a significant portion of the resistance in the system. In fact, with the valve closed, the resistance to the flow is infinite—the fluid cannot flow. Such minor losses may be very important indeed. With the valve wide open the extra resistance due to the presence of the valve may or may not be negligible.

The flow pattern through a typical component such as a valve is shown in **Fig. 8.21**. It is not difficult to realize that a theoretical analysis to predict the details of such flows to obtain the head loss for these components is not, as yet, possible. Thus, the head loss information for essentially all components is given in dimensionless form and based on experimental data. The most common method used to determine these head losses or pressure drops is to specify the **loss coefficient**, K_L, which is defined as

Losses due to pipe system components are given in terms of loss coefficients.

$$K_L = \frac{h_{L\,minor}}{(V^2/2g)} = \frac{\Delta p}{\frac{1}{2}\rho V^2}$$

so that

$$\Delta p = K_L \frac{1}{2}\rho V^2$$

or

$$\boxed{h_{L\,minor} = K_L \frac{V^2}{2g}}$$

(8.36)

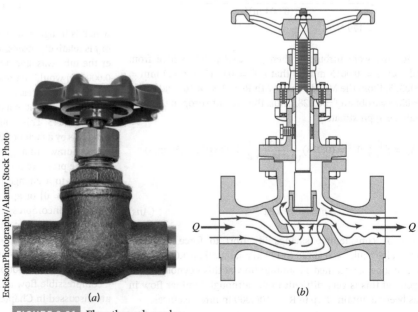

EricksonPhotography/Alamy Stock Photo

(a)

$Q \longrightarrow$ $\longrightarrow Q$

(b)

FIGURE 8.21 Flow through a valve.

The pressure drop across a component that has a loss coefficient of $K_L = 1$ is equal to the dynamic pressure, $\rho V^2/2$. As shown by Eq. 8.36 and the adjacent figure, for a given value of K_L the head loss is proportional to the square of the velocity.

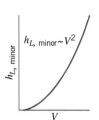

The actual value of K_L is strongly dependent on the geometry of the component considered. It may also be dependent on the fluid properties. That is,

$$K_L = \phi \,(\text{geometry, Re})$$

where $\text{Re} = \rho VD/\mu$ is the pipe Reynolds number. For many practical applications the Reynolds number is large enough so that the flow through the component is dominated by inertia effects, with viscous effects being of secondary importance. This is true because of the relatively large accelerations and decelerations experienced by the fluid as it flows along a rather curved, variable area (perhaps even torturous) path through the component (see Fig. 8.21). In a flow that is dominated by inertia effects rather than viscous effects, it is usually found that pressure drops and head losses correlate directly with the dynamic pressure. This is why the friction factor for very large Reynolds number, fully developed pipe flow is independent of the Reynolds number. The same condition is found to be true for flow through pipe components. Thus, in most cases of practical interest the loss coefficients for components are a function of geometry only, $K_L = \phi\,(\text{geometry})$.

> For most flows the loss coefficient is independent of the Reynolds number.

Minor losses are sometimes given in terms of an *equivalent length*, ℓ_{eq}. In this terminology, the head loss through a component is given in terms of the equivalent length of pipe that would produce the same head loss as the component. That is,

$$h_{L\,\text{minor}} = K_L \frac{V^2}{2g} = f \frac{\ell_{eq}}{D}\frac{V^2}{2g}$$

or

$$\ell_{eq} = \frac{K_L D}{f}$$

where D and f are based on the pipe containing the component. The head loss of the pipe system is the same as that produced in a straight pipe whose length is equal to the pipes of the original system plus the sum of the additional equivalent lengths of all of the components of the system. Most pipe flow analyses, including those in this book, use the loss coefficient method rather than the equivalent length method to determine the minor losses.

Many pipe systems contain various transition sections in which the pipe diameter changes from one size to another. Such changes may occur abruptly or rather smoothly through some type of area change section. Any change in flow area contributes losses that are not accounted for by friction alone. By continuity, the area change causes a velocity change in incompressible flow. Thus, the $V^2/2g$ terms do not cancel from the energy equation given in Eq. 5.84. The extreme cases involve flow into a pipe from a reservoir (an entrance) or out of a pipe into a reservoir (an exit).

A fluid may flow from a reservoir into a pipe through any number of differently shaped entrance regions as are sketched in **Fig. 8.22**. Each geometry has an associated loss coefficient. A typical flow pattern for flow entering a pipe through a square-edged entrance is sketched in **Fig. 8.23a**. As was discussed in Chapter 3, a vena contracta region may result because the fluid cannot turn a sharp right-angle corner. The flow is said to separate from the sharp corner. The maximum velocity at section (2) is greater than that in the pipe at section (3), and the pressure there is lower. If this high-speed fluid could slow down efficiently, the kinetic energy could be converted into pressure (the Bernoulli effect), and the ideal pressure distribution indicated in **Fig. 8.23b** would result. The head loss for the entrance would be essentially zero.

Such is not the case. Although a fluid may be accelerated very efficiently, it is very difficult to slow down (decelerate) a fluid efficiently. Thus, the extra kinetic energy of the fluid at section (2) is partially lost because of viscous dissipation, so that the pressure does not return to the ideal value. An entrance head loss (pressure drop) is produced as is indicated in Fig. 8.23b. The majority of this loss is due to inertia effects that are eventually dissipated by the shear stresses within the fluid. Only a small portion of the loss is due to the wall shear stress within the entrance region. The net effect is that the loss coefficient for a square-edged entrance is approximately $K_L = 0.50$. One-half of a velocity head is lost as the fluid enters the pipe. If the pipe protrudes into the tank (a reentrant entrance), as is shown in Fig. 8.22a, the losses are even greater.

> Minor head losses are usually a result of the dissipation of kinetic energy.

V8.11 Entrance/exit flows

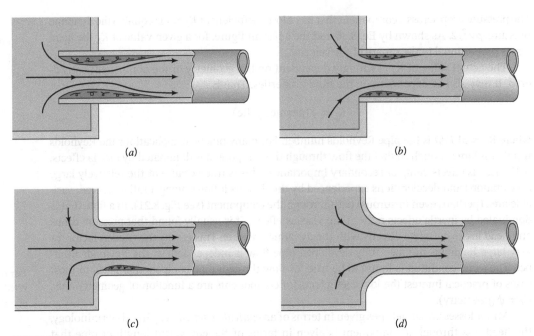

FIGURE 8.22 Entrance flow conditions and loss coefficient (Data from Refs. 28, 29). (a) Reentrant, $K_L = 0.8$, (b) sharp-edged, $K_L = 0.5$, (c) slightly rounded, $K_L = 0.2$, (see Fig. 8.24), (d) well-rounded, $K_L = 0.04$, (see Fig. 8.24).

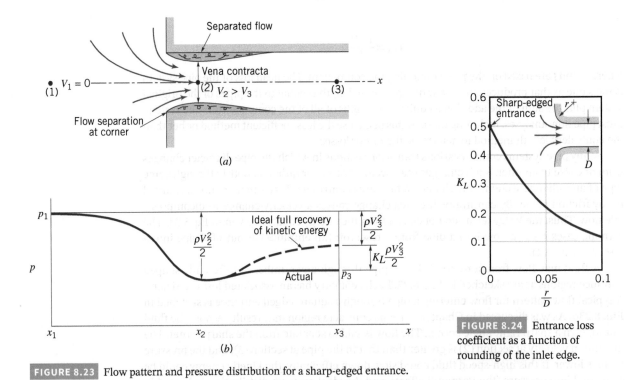

FIGURE 8.24 Entrance loss coefficient as a function of rounding of the inlet edge.

FIGURE 8.23 Flow pattern and pressure distribution for a sharp-edged entrance.

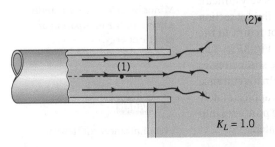

$K_L = 1.0$

An obvious way to reduce the entrance loss is to round the entrance region as is shown in Figs. 8.22c and 8.22d, thereby reducing or eliminating the vena contracta effect. Typical values for the loss coefficient for entrances with various amounts of rounding of the lip are shown in **Fig. 8.24**. A significant reduction in K_L can be obtained with only slight rounding.

A head loss (the exit loss) is also produced when a fluid flows from a pipe into a tank as is shown in the adjacent figure. In these cases the entire kinetic energy of the exiting fluid (velocity V_1) is dissipated through viscous effects as the stream of fluid mixes with the fluid in the tank and eventually comes to rest ($V_2 = 0$). The exit loss between points (1) and (2) is therefore equivalent to one velocity head, or $K_L = 1$. No matter what its shape, the loss coefficient for an exit is 1.0. A word of caution: Exit loss should be considered only when the downstream point in the analysis lies a sufficient distance beyond the exit to ensure that the exit kinetic energy has been dissipated. Consider calculation of flow through a fire hose and nozzle as shown in the adjacent figure. In the energy equation, the exit kinetic energy will be included as $V_2^2/2$, so it must not be included in the losses.

Losses also occur because of a change in pipe diameter as is shown in **Figs. 8.25** and **8.26**. The sharp-edged entrance and exit flows discussed in the previous paragraphs are limiting cases of this type of flow with either $A_1/A_2 = \infty$, or $A_1/A_2 = 0$, respectively. The loss coefficient for a sudden contraction, $K_L = h_L/(V_2^2/2g)$, is a function of the area ratio, A_2/A_1, as is shown in Fig. 8.25. The value of K_L changes gradually from one extreme of a sharp-edged entrance ($A_2/A_1 = 0$ with $K_L = 0.50$) to the other extreme of no area change ($A_2/A_1 = 1$ with $K_L = 0$).

In many ways, the flow in a sudden expansion is similar to exit flow. As is indicated in **Fig. 8.27**, the fluid leaves the smaller pipe and initially forms a jet-type structure as it enters the larger pipe. Within a few diameters downstream of the expansion, the jet becomes dispersed across the pipe, and fully developed flow becomes established again. In this process [between sections (2) and (3)] a portion of the kinetic energy of the fluid is dissipated as a result of viscous effects. A square-edged exit is the limiting case with $A_1/A_2 = 0$.

A sudden expansion is one of the few components (perhaps the only one) for which the loss coefficient can be obtained by means of a simple analysis. To do this we consider the continuity and momentum equations for the control volume shown in Fig. 8.27 and the energy equation applied between (2) and (3). We assume that the flow is uniform at sections (1), (2), and (3) and the pressure is constant across the left side of the control volume ($p_a = p_b = p_c = p_1$). The resulting three governing equations (mass, momentum, and energy) are

$$A_1 V_1 = A_3 V_3$$
$$p_1 A_3 - p_3 A_3 = \rho A_3 V_3 (V_3 - V_1)$$

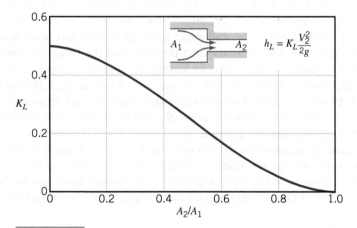

FIGURE 8.25 Loss coefficient for a sudden contraction (Ref. 10).

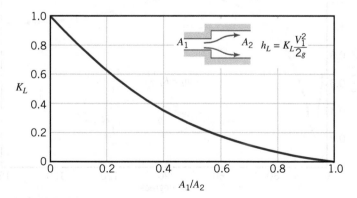

FIGURE 8.26 Loss coefficient for a sudden expansion (Ref. 10).

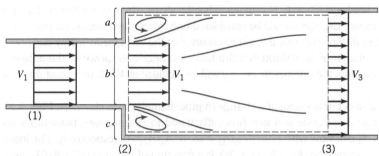

FIGURE 8.27 Control volume used to calculate the loss coefficient for a sudden expansion.

and

$$\frac{p_1}{\gamma} + \frac{V_1^2}{2g} = \frac{p_3}{\gamma} + \frac{V_3^2}{2g} + h_L$$

These can be combined to give the loss coefficient, $K_L = h_L/(V_1^2/2g)$, as

$$K_L = \left(1 - \frac{A_1}{A_2}\right)^2$$

where we have used the fact that $A_2 = A_3$. This result, plotted in Fig. 8.26, is in good agreement with experimental data. As with so many minor loss situations, it is not the viscous effects directly (i.e., the wall shear stress) that cause the loss. Rather, it is the dissipation of kinetic energy (another type of viscous effect) as the fluid decelerates inefficiently.

The losses may be quite different if the contraction or expansion is gradual. Typical results for a conical *diffuser* with a given area ratio, A_2/A_1, are shown in **Fig. 8.28**. (A diffuser is a device shaped to decelerate a fluid.) Clearly the included angle of the diffuser, θ, is a very important parameter. For very small angles, the diffuser is excessively long and most of the head loss is due to the wall shear stress as in fully developed flow. For moderate or large angles, the flow separates from the walls and the losses are due mainly to a dissipation of the kinetic energy of the jet leaving the smaller diameter pipe. In fact, for moderate or large values of θ (i.e., $\theta > 35°$ for the case shown in Fig. 8.28), the conical diffuser is, perhaps unexpectedly, less efficient than a sharp-edged expansion which has $K_L = (1 - A_1/A_2)^2$. There is an optimum angle ($\theta \approx 8°$ for the case illustrated) for which the loss coefficient is a minimum. The relatively small value of θ for the minimum K_L results in a long diffuser and is an indication of the fact that it is difficult to efficiently decelerate a fluid.

It must be noted that the conditions indicated in Fig. 8.28 represent typical results only. Flow through a diffuser is very complicated and may be strongly dependent on the area ratio A_2/A_1, specific details of the geometry, and the Reynolds number. The data are often presented

V8.12 Separated flow in a diffuser

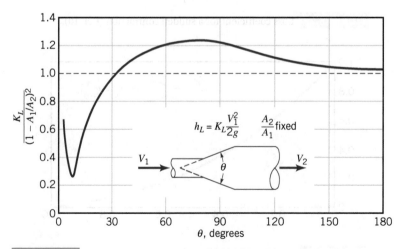

FIGURE 8.28 Loss coefficient for a typical conical diffuser (Ref. 5).

in terms of a *pressure recovery coefficient*, $C_p = (p_2 - p_1)/(\rho V_1^2/2)$, which is the ratio of the static pressure rise across the diffuser to the inlet dynamic pressure. Considerable effort has gone into understanding this important topic (Refs. 11, 12).

Flow in a conical contraction (a nozzle; reverse the flow direction shown in Fig. 8.28) is less complex than that in a conical expansion. Typical loss coefficients based on the downstream (high-speed) velocity can be quite small, ranging from $K_L = 0.02$ for $\theta = 30°$, to $K_L = 0.07$ for $\theta = 60°$, for example. It is relatively easy to accelerate a fluid efficiently.

Bends in pipes produce a greater head loss than if the pipe were straight. The losses are due to the separated region of flow near the inside of the bend (especially if the bend is sharp) and the swirling secondary flow that occurs because of the imbalance of centripetal forces as a result of the curvature of the pipe centerline. These effects and the associated values of K_L for large Reynolds number flows through a 90° bend are shown in **Fig. 8.29**. The friction loss due to the axial length of the pipe bend must be calculated and added to that given by the loss coefficient of Fig. 8.29.

V8.13 Car exhaust system

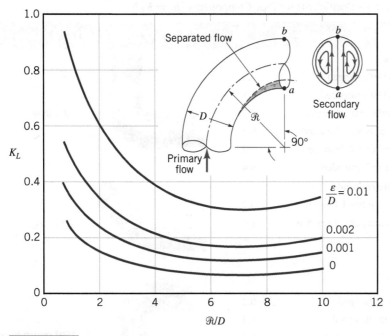

FIGURE 8.29 Character of the flow in a 90° bend and the associated loss coefficient (Ref. 5).

For situations in which space is limited, a flow direction change is often accomplished by use of miter bends, as is shown in **Fig. 8.30**, rather than smooth bends. The considerable losses in such bends can be reduced by the use of carefully designed guide vanes that help direct the flow with less unwanted swirl and disturbances.

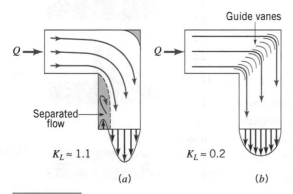

FIGURE 8.30 Character of the flow in a 90° mitered bend and the associated loss coefficient: (a) without guide vanes, (b) with guide vanes.

Another important category of pipe system components is that of commercially available pipe fittings such as elbows, tees, reducers, valves, and filters. The values of K_L for such components depend strongly on the shape of the component and only very weakly on the Reynolds number for typical large Re flows. Thus, the loss coefficient for a 90° elbow depends on whether the pipe joints are threaded or flanged but is, within the accuracy of the data, fairly independent of the pipe diameter, flowrate, or fluid properties (i.e., the Reynolds number effect). Typical values of K_L for such components are given in **Table 8.2**. These typical components are designed more for ease of manufacturing and costs than for reduction of the head losses that they produce. The flowrate from a faucet in a typical house is sufficient whether the value of K_L for an elbow is the typical $K_L = 1.5$, or it is reduced to $K_L = 0.2$ by use of a more expensive long-radius, gradual bend (Fig. 8.29).

Valves control the flowrate by providing a means to adjust the overall system loss coefficient to the desired value. When the valve is closed, the value of K_L is infinite and no fluid

Extensive tables are available for loss coefficients of standard pipe components.

TABLE 8.2   Loss Coefficients for Pipe Components $\left(h_L = K_L \frac{V^2}{2g} \right)$

Component	K_L	
a. Elbows		
Regular 90°, flanged	0.3	
Regular 90°, threaded	1.5	90° elbow
Long radius 90°, flanged	0.2	
Long radius 90°, threaded	0.7	
Long radius 45°, flanged	0.2	
Regular 45°, threaded	0.4	
b. 180° return bends		
180° return bend, flanged	0.2	
180° return bend, threaded	1.5	45° elbow
c. Tees		
Line flow, flanged	0.2	
Line flow, threaded	0.9	
Branch flow, flanged	1.0	180° return bend
Branch flow, threaded	2.0	
d. Union, threaded	0.08	
e. Valves		
Globe, fully open	10	
Angle, fully open	2	Tee
Gate, fully open	0.15	
Gate, $\frac{1}{4}$ closed	0.26	
Gate, $\frac{1}{2}$ closed	2.1	
Gate, $\frac{3}{4}$ closed	17	Tee
Swing check, forward flow	2	
Swing check, backward flow	∞	
Ball valve, fully open	0.05	
Ball valve, $\frac{1}{3}$ closed	5.5	Union
Ball valve, $\frac{2}{3}$ closed	210	

*See Fig. 8.31 for typical valve geometry.

(Data from Refs. 5, 10, 27; Images Source: Bruce Munson)

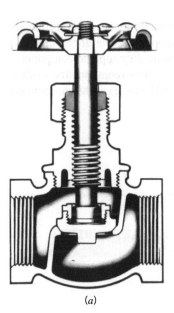

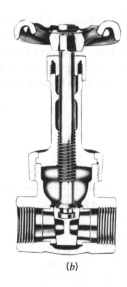

(a)

(b)

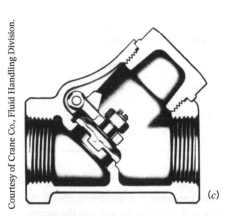

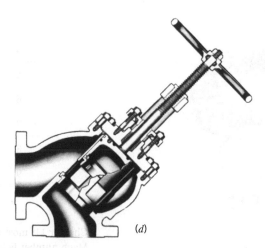

(c)

(d)

FIGURE 8.31 Internal structure of various valves: (a) globe valve, (b) gate valve, (c) swing check valve, (d) stop check valve.

flows. Opening of the valve reduces K_L, producing the desired flowrate. Typical cross sections of various types of valves are shown in **Fig. 8.31**. Some valves (such as the conventional globe valve) are designed for general use, providing convenient control between the extremes of fully closed and fully open. Others (such as a needle valve) are designed to provide very fine control of the flowrate. The check valve provides a diode type operation that allows fluid to flow in one direction only.

Loss coefficients for typical valves are given in Table 8.2. As with many system components, the head loss in valves is mainly a result of the dissipation of kinetic energy of a high-speed portion of the flow. This high speed, V_3, is illustrated in **Fig. 8.32**.

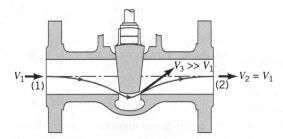

FIGURE 8.32 Head loss in a valve is due to dissipation of the kinetic energy of the large-velocity fluid near the valve seat.

With the exception of sudden expansions and exits, K_L values throughout this section are representative of typical values based on experimental measurements. The uncertainty can range from about 10 percent to more than 50 percent. In addition, they apply to components in isolation with either uniform or fully developed flow upstream. If two components are close together, the resulting losses usually are greater than the sum of the losses of the components in isolation.

EXAMPLE 8.6 | Minor Losses

Given The closed-circuit wind tunnel shown in **Fig. E8.6a** is a smaller version of that depicted in **Fig. E8.6b** in which air at standard conditions is to flow through the test section [between sections (5) and (6)] with a velocity of 200 m/s. The flow is driven by a fan that essentially increases the static pressure by the amount $p_1 - p_9$ that is needed to overcome the head losses experienced by the fluid as it flows around the circuit.

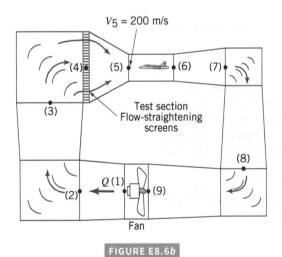

FIGURE E8.6a

FIGURE E8.6b

Find Estimate the value of $p_1 - p_9$ and the horsepower supplied to the fluid by the fan.

Solution The maximum velocity within the wind tunnel occurs in the test section (smallest area; see **Table E8.6**). Thus, the maximum Mach number of the flow is $Ma_5 = V_5/c_5$, where $V_5 = 200$ m/s and from Eq. 1.20 the speed of sound is $c_5 = (kRT_5)^{1/2} = \{1.4\,(287\,\text{J/kg}\cdot\text{K})[(273 + 59)\,\text{K}]\}^{1/2} = 365$ m/s. Thus, $Ma_5 = 200/365 = 0.548$. As was indicated in Chapter 3 and discussed fully in

TABLE E8.6		
Location	Area (m²)	Velocity (m/s)
1	22.0	36.4
2	28.0	28.6
3	35.0	22.9
4	35.0	22.9
5	4.0	200.0
6	4.0	200.0
7	10.0	80.0
8	18.0	44.4
9	22.0	36.4

Chapter 11, most flows can be considered as incompressible if the Mach number is less than about 0.3. Hence, we can use the incompressible formulas for this problem.

The purpose of the fan in the wind tunnel is to provide the necessary energy to overcome the net head loss experienced by the air as it flows around the circuit. This can be found from the energy equation between points (1) and (9) as

$$\frac{p_1}{\gamma} + \frac{V_1^2}{2g} + z_1 = \frac{p_9}{\gamma} + \frac{V_9^2}{2g} + z_9 + h_{L_{1-9}}$$

where $h_{L_{1-9}}$ is the total head loss from (1) to (9). With $z_1 = z_9$ and $V_1 = V_9$ this gives

$$\frac{p_1}{\gamma} - \frac{p_9}{\gamma} = h_{L_{1-9}} \tag{1}$$

Similarly, by writing the energy equation (Eq. 5.84) across the fan, from (9) to (1), we obtain

$$\frac{p_9}{\gamma} + \frac{V_9^2}{2g} + z_9 + h_p = \frac{p_1}{\gamma} + \frac{V_1^2}{2g} + z_1$$

where h_p is the actual head rise supplied by the pump (fan) to the air. Again since $z_9 = z_1$ and $V_9 = V_1$ this, when combined with Eq. 1, becomes

$$h_p = \frac{(p_1 - p_9)}{\gamma} = h_{L_{1-9}}$$

The actual power supplied to the air (horsepower, $\mathscr{P}_a$) is obtained from the fan head by

$$\mathscr{P}_a = \gamma Q h_p = \gamma A_5 V_5 h_p = \gamma A_5 V_5 h_{L_{1-9}} \tag{2}$$

Thus, the power that the fan must supply to the air depends on the head loss associated with the flow through the wind tunnel. To obtain a reasonable, approximate answer we make the following assumptions. We treat each of the four turning corners as a mitered bend with guide vanes so that from Fig. 8.31 $K_{L_{corner}} = 0.2$. Thus, for each corner

$$h_{L_{corner}} = K_L \frac{V^2}{2g} = 0.2 \frac{V^2}{2g}$$

where, because the flow is assumed incompressible, $V = V_5 A_5 / A$. The values of A and the corresponding velocities throughout the tunnel are given in Table E8.6.

We also treat the enlarging sections from the end of the test section (6) to the beginning of the nozzle (4) as a conical diffuser with a loss coefficient of $K_{L_{dif}} = 0.6$. This value is larger than that of a well-designed diffuser (see Fig. 8.28, for example). Since the wind tunnel diffuser is interrupted by the four turning corners and the fan, it may not be possible to obtain a smaller value of $K_{L_{dif}}$ for this situation. Thus,

$$h_{L_{dif}} = K_{L_{dif}} \frac{V_6^2}{2g} = 0.6 \frac{V_6^2}{2g}$$

The loss coefficients for the conical nozzle between sections (4) and (5) and the flow-straightening screens are assumed to be $K_{L_{noz}} = 0.2$ and $K_{L_{scr}} = 4.0$ (Ref. 13), respectively. We neglect the head loss in the test section, although this may not be practical with a test model present. Finally, we neglect major losses (friction) in the relatively short straight tunnel sections.

Thus, the total head loss is

$$h_{L_{1-9}} = h_{L_{corner7}} + h_{L_{corner8}} + h_{L_{corner2}} + h_{L_{corner3}} + h_{L_{dif}} + h_{L_{noz}} + h_{L_{scr}}$$

or

$$h_{L_{1-9}} = [0.2(V_7^2 + V_8^2 + V_2^2 + V_3^2)$$

$$+ 0.6\, V_6^2 + 0.2\, V_5^2 + 4.0\, V_4^2]/2g$$

$$= [0.2(80.0^2 + 44.2^2 + 28.6^2 + 22.9^2) + 0.6(200)^2$$

$$+ 0.2(200)^2 + 4.0(22.9)^2]\ m^2/s^2/[2(9.81\ m/s^2)]$$

or

$$h_{L_{1-9}} = 1838\ m$$

Hence, from Eq. 1 we obtain the pressure rise across the fan as

$$p_1 - p_9 = \gamma h_{L_{1-9}} = (1.22\ kg/m^3)(1838\ m)$$

$$= 21997\ N/m^2 = 21.997\ kPa \qquad (Ans)$$

From Eq. 2 we obtain the power added to the fluid as

$$\mathcal{P}_a = (1.22\ kg/m^3)(4.0\ m^2)(200\ m/s)(1838\ m)$$

$$= 17.598\ MW$$

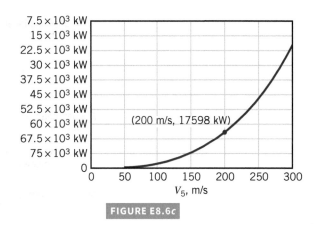

FIGURE E8.6c

or

$$\mathcal{P}_a = 17598\ kW \qquad (Ans)$$

Comments By repeating the calculations with various test section velocities, V_5, the results shown in **Fig. E8.6c** are obtained. Since the head loss varies as V_5^2 and the power varies as head loss times V_5, it follows that the power varies as the cube of the velocity. Thus, doubling the wind tunnel speed requires an eightfold increase in power.

With a closed-return wind tunnel of this type, all of the power required to maintain the flow is dissipated through viscous effects, with the energy remaining within the closed tunnel. If heat transfer across the tunnel walls is negligible, the air temperature within the tunnel will increase in time. For steady-state operations of such tunnels, it is often necessary to provide some means of cooling to maintain the temperature at acceptable levels.

It should be noted that the actual size of the motor that powers the fan must be greater than the calculated 17598 kW because the fan is not 100% efficient. The power calculated above is that needed by the fluid to overcome losses in the tunnel, excluding those in the fan. If the fan were 60% efficient, it would require a shaft power of $\mathcal{P} = 17598\ kW/(0.60) = 29330\ kW$ to run the fan. Determination of fan (or pump) efficiencies can be a complex problem that depends on the specific geometry of the fan. Introductory material about fan performance is presented in Chapter 12; additional material can be found in various references (Refs. 14, 15, 16, 37, for example).

It should also be noted that the above results are only approximate. Clever, careful design of the various components (corners, diffuser, etc.) may lead to improved (i.e., lower) values of the various loss coefficients, and hence lower power requirements. Since h_L is proportional to V^2, the components with the larger V tend to have the larger head loss. Thus, even though $K_L = 0.2$ for each of the four corners, the head loss for corner (7) is $(V_7/V_3)^2 = (80/22.9)^2 = 12.2$ times greater than it is for corner (3).

8.4.3 Noncircular Conduits

Many of the conduits that are used for conveying fluids are not circular in cross section. Although the details of the flows in such conduits depend on the exact cross-sectional shape, many round pipe results can be carried over, with slight modification, to flow in conduits of other shapes.

Theoretical results can be obtained for fully developed laminar flow in noncircular ducts, although the detailed mathematics often becomes rather cumbersome. For an arbitrary cross section, as is shown in **Fig. 8.33**, the velocity profile is a function of both y and $z\ [\mathbf{V} = u(y, z)\hat{\mathbf{i}}]$.

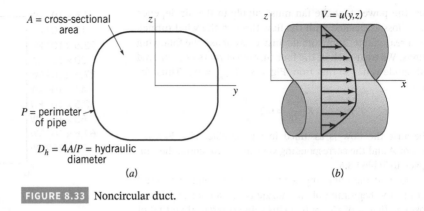

FIGURE 8.33 Noncircular duct.

This means that the governing equation from which the velocity profile is obtained (either the Navier–Stokes equations of motion or a force balance equation similar to that used for circular pipes, Eq. 8.6) is a partial differential equation rather than an ordinary differential equation. Although the equation is linear (for fully developed flow the convective acceleration is zero), its solution is not as straightforward as for round pipes. Typically the velocity profile is given in terms of an infinite series representation (Ref. 17).

Practical, easy-to-use results can be obtained as follows. Regardless of the cross-sectional shape, there are no inertia effects in fully developed laminar pipe flow. Thus, the friction factor can be written as $f = C/\text{Re}_h$, where the constant C depends on the particular shape of the duct, and Re_h is the Reynolds number, $\text{Re}_h = \rho V D_h / \mu$, based on the hydraulic diameter. The **hydraulic diameter** defined as $D_h = 4A/P$ is four times the ratio of the cross-sectional flow area divided by the wetted perimeter, P, of the pipe as is illustrated in Fig. 8.33. It represents a characteristic length that defines the size of a cross section of a specified shape. The factor of 4 is included in the definition of D_h so that for round pipes the diameter and hydraulic diameter are equal $[D_h = 4A/P = 4(\pi D^2/4)/(\pi D) = D]$. The hydraulic diameter is also used in the definition of the friction factor, $h_L = f(\ell/D_h)V^2/2g$, and the relative roughness, ε/D_h.

The values of $C = f\,\text{Re}_h$ for laminar flow have been obtained from theory and/or experiment for various shapes. Typical values are given in **Table 8.3** along with the hydraulic diameter.

> The hydraulic diameter is used for noncircular duct calculations.

TABLE 8.3 Friction Factors for Laminar Flow in Noncircular Ducts

Shape	Parameter	$C = f\,\text{Re}_h$
I. Concentric Annulus $D_h = D_2 - D_1$	D_1/D_2	71.8
	0.0001	80.1
	0.01	89.4
	0.1	95.6
	0.6	96.0
	1.00	
II. Rectangle $D_h = \dfrac{2ab}{a+b}$	a/b	
	0	96.0
	0.05	89.9
	0.10	84.7
	0.25	72.9
	0.50	62.2
	0.75	57.9
	1.00	56.9

(Data from Ref. 18)

Note that the value of C is relatively insensitive to the shape of the conduit. Unless the cross section is very "thin" in some sense, the value of C is not too different from its circular pipe value, $C = 64$. Once the friction factor is obtained, the calculations for noncircular conduits are identical to those for round pipes.

Calculations for fully developed turbulent flow in ducts of noncircular cross section are usually carried out by using the Moody chart data (or the Colebrook formula) for round pipes with the diameter replaced by the hydraulic diameter for both the Reynolds number ($\rho V D_h/\mu$) and the relative roughness (ε/D_h). Such calculations are usually accurate to within about 15%. If greater accuracy is needed, a more detailed analysis based on the specific geometry of interest is needed. It is important to note that actual cross-sectional area is used and *not* $(\pi/4)(D_h^2)$ when relating average velocity and flowrate (i.e., $Q = VA$).

> The Moody chart, developed for round pipes, can also be used for noncircular ducts.

EXAMPLE 8.7 | Noncircular Conduit

Given Air at a temperature of 49 °C and standard pressure flows from a furnace through an 0.2 m-diameter pipe with an average velocity of 3 m/s. It then passes through a transition section similar to the one shown in **Fig. E8.7** and into a square duct whose side is of length a. The pipe and duct surfaces are smooth ($\varepsilon = 0$). The head loss per foot is to be the same for the pipe and the duct.

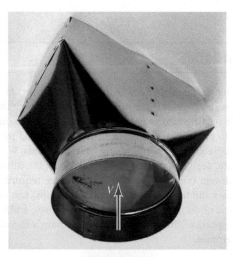

FIGURE E8.7

Find Determine the duct size, a.

Solution We first determine the head loss per foot for the pipe, $h_L/\ell = (f/D)V^2/2g$, and then size the square duct to give the same value. For the given pressure and temperature we obtain (from Table 1.7) $v = 1.76 \times 10^{-5}$ m²/s so that

$$\text{Re} = \frac{VD}{v} = \frac{(3\text{ m/s})(0.2\text{ m})}{1.76 \times 10^{-5}\text{ m}^2/\text{s}} = 34,090$$

With this Reynolds number and with $\varepsilon/D = 0$ we obtain the friction factor from Fig. 8.20 as $f = 0.022$ so that

$$\frac{h_L}{\ell} = \frac{0.022}{(0.2\text{ m})} \frac{(3\text{ m/s})^2}{2(9.81\text{ m/s}^2)} = 0.0504$$

Thus, for the square duct we must have

$$\frac{h_L}{\ell} = \frac{f}{D_h} \frac{V_s^2}{2g} = 0.0504 \qquad (1)$$

where

$$D_h = 4A/P = 4a^2/4a = a \qquad \text{and}$$

$$V_s = \frac{Q}{A} = \frac{\frac{\pi}{4}(0.2\text{ m})^2(3\text{ m})}{a^2} = \frac{0.0942}{a^2} \qquad (2)$$

is the velocity in the duct.

By combining Eqs. 1 and 2 we obtain

$$0.0504 = \frac{f}{a}\frac{(0.0942/a^2)^2}{2(9.81)}$$

or

$$a = 0.389\,f^{1/5} \qquad (3)$$

where a is in meter. Similarly, the Reynolds number based on the hydraulic diameter is

$$\text{Re}_h = \frac{V_s D_h}{v} = \frac{(0.0942/a^2)a}{1.76 \times 10^{-5}} = \frac{5352}{a} \qquad (4)$$

We have three unknowns (a, f, and Re_h) and three equations— Eqs. 3, 4, and either in graphical form the Moody chart (Fig. 8.20) or the Colebrook equation (Eq. 8.35a).

If we use the Moody chart, we can use a trial-and-error (iterative) solution as follows. As an initial attempt, assume the friction factor for the duct is the same as for the pipe. That is, assume $f = 0.022$. From Eq. 3 we obtain $a = 0.185$ m, while from Eq. 4 we have $\text{Re}_h = 3.05 \times 10^4$. From Fig. 8.20, with this Reynolds number and the given smooth duct we obtain $f = 0.023$, which does not quite agree with the assumed value of f. Hence, we do not have the solution. We try again, using the latest calculated value of $f = 0.023$ as our guess. The calculations are repeated until the guessed value of f agrees with the value obtained from Fig. 8.20. The final result (after only two iterations) is $f = 0.023$, $\text{Re}_h = 3.03 \times 10^4$, and

$$a = 0.186\text{ m} \qquad (Ans)$$

Comments Alternatively, we can use the Colebrook equation or the Haaland equation (rather than the Moody chart) to obtain the solution as follows. For a smooth pipe ($\varepsilon/D_h = 0$) the Colebrook equation, Eq. 8.35a, becomes

$$\frac{1}{\sqrt{f}} = -2.0 \log\left(\frac{\varepsilon/D_h}{3.7} + \frac{2.51}{\text{Re}_h\sqrt{f}}\right)$$

$$= -2.0 \log\left(\frac{2.51}{\text{Re}_h\sqrt{f}}\right) \qquad (5)$$

where from Eq. 3,

$$f = 0.269a^5 \tag{6}$$

If we combine Eqs. 4, 5, and 6 and simplify, Eq. 7 is obtained for a.

$$1.928a^{-5/2} = -2 \log (2.62 \times 10^{-4} a^{-3/2}) \tag{7}$$

By using a root-finding technique on a computer or calculator, the solution to Eq. 7 is determined to be $a = 0.187$ m, in agreement (given the accuracy of reading the Moody chart) with that obtained by the trial-and-error method given earlier.

Note that the length of the side of the equivalent square duct is $a/D = 0.187/0.2 = 0.935$, or approximately 93.5% of the diameter of the equivalent duct. It can be shown that this value, 93.5%, is a very good approximation for any pipe flow—laminar or turbulent. The cross-sectional area of the duct ($A = a^2 = 0.035$ m^2) is greater than that of the round pipe ($A = \pi D^2/4 = 0.0314$ m^2). Also, it takes less material to form the round pipe (perimeter $= \pi D = 0.64$ m) than the square duct (perimeter $= 4a = 0.75$ m). Circles are very efficient shapes.

8.5 PIPE FLOW EXAMPLES

Pipe systems may contain a single pipe with components or multiple interconnected pipes.

In the previous sections of this chapter, we discussed concepts concerning flow in pipes and ducts. The purpose of this section is to apply these ideas to the solutions of various practical problems. The application of the pertinent equations is straightforward, with rather simple calculations that give answers to problems of engineering importance. The main idea involved is to apply the energy equation between appropriate locations within the flow system, with the head loss written in terms of the friction factor and the minor loss coefficients. We will consider two classes of pipe systems: those containing a single pipe (whose length may be interrupted by various components), and those containing multiple pipes in parallel, series, or network configurations.

The Wide World of Fluids

New hi-tech fountains

Ancient Egyptians used fountains in their palaces for decorative and cooling purposes. Current use of fountains continues but with a hi-tech flair. Although the basic fountain still consists of a typical *pipe system* (i.e., pump, pipe, regulating valve, nozzle, filter, and basin), recent use of computer-controlled devices has led to the design of innovative fountains with special effects. For example, by using several rows of multiple nozzles, it is possible to program and activate control valves to produce water jets that resemble symbols, letters, or the time of day. Other fountains use specially designed nozzles to produce coherent, laminar streams of water that look like glass rods flying through the air. By using fast-acting control valves in a synchronized manner it is possible to produce mesmerizing three-dimensional patterns of water droplets. The possibilities are nearly limitless. With the initial artistic design of the fountain established, the initial engineering design (i.e., the capacity and pressure requirements of the nozzles and the size of the pipes and pumps) can be carried out. It is often necessary to modify the artistic and/or engineering aspects of the design in order to obtain a functional, pleasing fountain.

8.5.1 Single Pipes

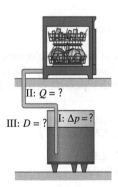

The nature of the solution process for pipe flow problems can depend strongly on which of the various parameters are independent parameters (the "given") and which is the dependent parameter (the "determine"). The three most common types of problems are shown in **Table 8.4** in terms of the parameters involved. We assume the pipe system is defined in terms of the length of pipe sections used and the number of elbows, bends, and valves needed to convey the fluid between the desired locations. In all instances we assume the fluid properties are given.

In a Type I problem we specify the desired flowrate or average velocity and determine the necessary pressure difference or head loss. For example, if a flowrate of 1.26×10^{-4} m^3/s is required for a dishwasher that is connected to the water heater by a given pipe system as shown in the adjacent figure, what pressure is needed in the water heater?

In a Type II problem we specify the applied driving pressure (or, alternatively, the head loss) and determine the flowrate. For example, how many m^3/s of hot water are

TABLE 8.4 Pipe Flow Types

Variable	Type I	Type II	Type III
a. Fluid			
Density	Given	Given	Given
Viscosity	Given	Given	Given
b. Pipe			
Diameter	Given	Given	Determine
Length	Given	Given	Given
Roughness	Given	Given	Given
c. Flow			
Flowrate or	Given	Determine	Given
Average Velocity			
d. Pressure			
Pressure Drop or	Determine	Given	Given
Head Loss			

Pipe flow problems can be categorized by what parameters are given and what is to be calculated.

supplied to the dishwasher if the pressure within the water heater is 413.7 kPa and the pipe system details (length, diameter, roughness of the pipe; number of elbows; etc.) are specified?

In a Type III problem we specify the pressure drop and the flowrate and determine the diameter of the pipe needed. For example, what diameter of pipe is needed between the water heater and dishwasher if the pressure in the water heater is 413.7 kPa (determined by the city water system) and the flowrate is to be not less than 1.26×10^{-4} m³/s (determined by the manufacturer)?

Several examples of these types of problems follow.

EXAMPLE 8.8 | Type I, Determine Pressure Drop

Given Water at 15.5 °C flows from the basement to the second floor through the 19 mm-diameter copper pipe (a drawn tubing) at a rate of $Q = 7.6 \times 10^{-4}$ m³/s and exits through a faucet of diameter 12.7 mm as shown in **Fig. E8.8a**.

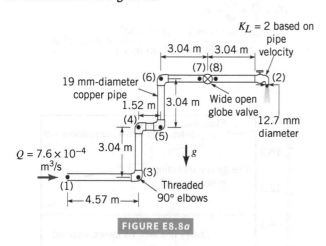

FIGURE E8.8a

Find Determine the pressure at point (1) if

a. all losses are neglected,

b. the only losses included are major losses, or

c. all losses are included.

Solution Since the fluid velocity in the pipe is given by $V_1 = Q/A_1 = Q/(\pi D^2/4) = (7.6 \times 10^{-4}$ m³/s$)/[\pi(0.019$ m$)^2/4] = 2.65$ m/s, and the fluid properties are $\rho = 999$ kg/m³ and $\mu = 0.0011$ Pa · s (see Table B.1), it follows that Re $= \rho V D/\mu = (999$ kg/m³$)(2.65$ m/s$)(0.019$ m$)/(0.0011$ Pa · s$) = 45727$. Thus, the flow is turbulent. The governing equation for either case (a), (b), or (c) is the energy equation given by Eq. 8.21,

$$\frac{p_1}{\gamma} + \alpha_1 \frac{V_1^2}{2g} + z_1 = \frac{p_2}{\gamma} + \alpha_2 \frac{V_2^2}{2g} + z_2 + h_L$$

where $z_1 = 0$, $z_2 = 6.08$ m, $p_2 = 0$ (free jet), $\gamma = \rho g = 9.8$ kN/m³, and the outlet velocity is $V_2 = Q/A_2 = (7.6 \times 10^{-4}$ m³/s$)/[\pi(0.0127$ m²$)/4] = 6$ m/s. We assume that the kinetic energy coefficients α_1 and α_2 are unity. This is reasonable because turbulent velocity profiles are nearly uniform across the pipe. Thus,

$$p_1 = \gamma z_2 + \frac{1}{2}\rho(V_2^2 - V_1^2) + \gamma h_L \tag{1}$$

where the head loss is different for each of the three cases.

a. If all losses are neglected ($h_L = 0$), Eq. 1 gives

$$p_1 = (9.81 \text{ kN/m}^3)(6.08 \text{ m})$$
$$+ \frac{999 \text{ kg/m}^3}{2}[(6 \text{ m/s})^2 - (2.65 \text{ m/s})^2]$$
$$= (59584 + 14474) \text{ N/m}^2 = 74058 \text{ N/m}^2$$

Comment Note that for this pressure drop, the amount due to elevation change (the hydrostatic effect) is $\gamma(z_2 - z_1) = 59.78$ kN/m^2 and the amount due to the increase in kinetic energy is $\rho(V_2^2 - V_1^2)/2 = 4.272$ kN/m^2.

b. If the only losses included are the major losses, the head loss is

$$h_L = f\frac{\ell}{D}\frac{V_1^2}{2g}$$

From Table 8.1 the roughness for a 19 mm-diameter copper pipe (drawn tubing) is $\varepsilon = 1.5 \times 10^{-6}$ m so that $\varepsilon/D = 8 \times 10^{-5}$. With this ε/D and the calculated Reynolds number (Re = 45,000), the value of f is obtained from the Moody chart as $f = 0.0215$. Note that the Colebrook equation (Eq. 8.35a) would give the same value of f. Hence, with the total length of the pipe as $\ell = (4.57 + 1.52 + 3.04 + 3.04 + 6.08)$ m = 18.25 m and the elevation and kinetic energy portions the same as for part (a), Eq. 1 gives

$$p_1 = \gamma z_2 + \frac{1}{2}\rho(V_2^2 - V_1^2) + \rho f\frac{\ell}{D}\frac{V_1^2}{2}$$

$$= (59584 + 14474) \text{ N/m}^2$$
$$+ (999 \text{ kg/m}^3)(0.0215)(18.25/0.019 \text{ m})\frac{(2.65 \text{ m/s})^2}{2}$$
$$= (59584 + 14474 + 72437) \text{ N/m}^2 = 146496 \text{ N/m}^2 \qquad (Ans)$$

Comment Of this pressure drop, the amount due to pipe friction is approximately $(147 \text{ kN/m}^2 - 74 \text{ kN/m}^2) = 73$ kN/m^2.

c. If major and minor losses are included, Eq. 1 becomes

$$p_1 = \gamma z_2 + \frac{1}{2}\rho(V_2^2 - V_1^2) + \rho f\frac{\ell}{D}\frac{V_1^2}{2} + \sum \rho K_L \frac{V^2}{2}$$

or

$$p_1 = 147 \text{ kN/m}^2 + \sum \rho K_L \frac{V^2}{2} \qquad (2)$$

where the 147 kN/m^2 contribution is due to elevation change, kinetic energy change, and major losses [part (b)], and the last term represents the sum of all of the minor losses. The loss coefficients of the components ($K_L = 1.5$ for each elbow and $K_L = 10$ for the wide-open globe valve) are given in Table 8.2 (except for the loss coefficient of the faucet, which is given in Fig. E8.8a as $K_L = 2$). Thus,

$$\sum \rho K_L \frac{V^2}{2} = (999 \text{ kg/m}^3)\frac{(2.65 \text{ m/s})^2}{2}[10 + 4(1.5) + 2]$$
$$= 63139.2 \text{ N/m}^2 \qquad (3)$$

Note that we did not include an entrance or exit loss because points (1) and (2) are located within the fluid streams, not within an attaching reservoir where the kinetic energy is zero. Thus, by combining Eqs. 2 and 3 we obtain the entire pressure drop as

$$p_1 = (147 + 63.14) \text{ kN/m}^2 = 210.14 \text{ kN/m}^2 \qquad (Ans)$$

This pressure drop calculated by including all losses should be the most realistic answer of the three cases considered.

Comments More detailed calculations will show that the pressure distribution along the pipe is as illustrated in **Fig. E8.8b** for cases (a) and (c)—neglecting all losses or including all losses. Note that not all of the pressure drop, $p_1 - p_2$, is a "pressure loss." The pressure change due to the elevation and velocity changes is completely reversible. The portion due to the major and minor losses is irreversible.

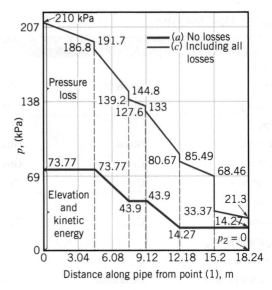

FIGURE E8.8b

This flow can be illustrated in terms of the energy line and hydraulic grade line concepts introduced in Section 3.7. As is shown in **Fig. E8.8c**, for case (a) there are no losses and the energy line (EL) is horizontal, one velocity head ($V^2/2g$) above the hydraulic grade line (HGL), which is one pressure head (γz) above the pipe itself. For cases (b) or (c) the energy line is not horizontal. Each bit of friction in the pipe or loss in a component reduces the available energy, thereby lowering the energy line. Thus, for case (a) the total head remains constant throughout the flow with a value of

$$H = \frac{p_1}{\gamma} + \frac{V_1^2}{2g} + z_1 = \frac{(74 \text{ kN/m}^2)}{(9.8 \text{ kN/m}^3)} + \frac{(2.65 \text{ m/s})^2}{2(9.81 \text{ m/s}^2)} + 0$$

$$= 7.93 \text{ m}$$

$$= \frac{p_2}{\gamma} + \frac{V_2^2}{2g} + z_2 = \frac{p_3}{\gamma} + \frac{V_3^3}{2g} + z_3 = \cdots$$

For case (c) the energy line starts at

$$H_1 = \frac{p_1}{\gamma} + \frac{V_1^2}{2g} + z_1$$

$$= \frac{(210.14 \text{ kN/m}^2)}{(9.8 \text{ kN/m}^3)} + \frac{(2.65 \text{ m/s})^2}{2(9.81 \text{ m/s}^2)} + 0 = 21.8 \text{ m}$$

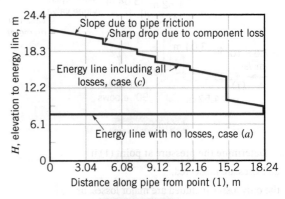

FIGURE E8.8c

and falls to a final value of

$$H_2 = \frac{p_2}{\gamma} + \frac{V_2^2}{2g} + z_2 = 0 + \frac{(6 \text{ m/s})^2}{2(9.81 \text{ m/s}^2)} + 6.08 \text{ m}$$

$$= 7.91 \text{ m}$$

The elevation of the energy line can be calculated at any point along the pipe. For example, at point (7), 15.2 m from point (1),

$$H_7 = \frac{p_7}{\gamma} + \frac{V_7^2}{2g} + z_7$$

$$= \frac{(68.47 \text{ kN/m}^2)}{(9.8 \text{ kN/m}^3)} + \frac{(2.65 \text{ m/s})^2}{2(9.81 \text{ m/s}^2)} + 6.08$$

$$= 13.42 \text{ m}$$

The head loss per foot of pipe is the same all along the pipe. That is,

$$\frac{h_L}{\ell} = f\frac{V^2}{2gD} = \frac{0.0215(2.65 \text{ m/s})^2}{2(9.81 \text{ m/s}^2)(0.019 \text{ m})} = 0.404 \text{ m/m}$$

Thus, the energy line is a set of straight-line segments of the same slope separated by steps whose height equals the head loss of the minor component at that location. As is seen from Fig. E8.8c, the globe valve produces the largest of all the minor losses.

Although the governing pipe flow equations are quite simple, they can provide very reasonable results for a variety of applications, as is shown in the next example.

EXAMPLE 8.9 | Type I, Determine Head Loss

Given A utility worker is fixing power lines in a manhole space at a busy intersection as shown in **Fig. E8.9a**. A major portion of car exhaust is carbon dioxide (CO_2), which has a density of 1.83 kg/m³. Since the density of air at standard conditions is 1.23 kg/m³, the carbon dioxide will tend to settle into the bottom of the manhole space and displace the oxygen, posing a very relevant danger of suffocation to utility workers. To avoid suffocation, a fan is used to supply fresh air from the surface into the manhole space as shown in **Fig. E8.9b**. The air is routed through an 0.203 m-diameter plastic hose that is 9.14 m in length.

Find Determine the horsepower needed by the fan to overcome the head loss with a required flowrate of 0.283 m³/s, when **(a)** neglecting minor losses and **(b)** including minor losses for one re-entrant entrance, one exit, and one 90° miter bend (no guide vanes).

Solution The purpose of the fan is to deliver air at a certain flowrate, which also requires the fan to overcome the net head loss within the pipe. This can be found from the energy equation between points (1) and (2) as

$$\frac{p_1}{\gamma} + \frac{V_1^2}{2g} + z_1 + h_p = \frac{p_2}{\gamma} + \frac{V_2^2}{2g} + z_2 + h_L$$

FIGURE E8.9a

MCCAIG/E+/Getty Images

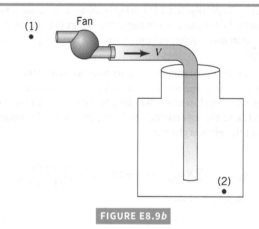

FIGURE E8.9b

where h_L is the total head loss from (1) to (2) and h_p is the head rise supplied by the fan to the air. Points (1) and (2) are assumed to be sufficiently away from the entrance and exit of the plumbing so that $p_1 = p_2 = V_1 = V_2 = 0$. Also, since the fluid is air, the change in elevation is negligible.

Thus, the energy equation reduces to $h_p = h_L$.

a. Neglecting minor losses it follows that

$$h_P = f\frac{\ell}{D}\frac{V^2}{2g} \qquad (1)$$

where V is the velocity within the hose. Therefore, we must use the Moody chart to find the friction factor, f. Since

$$V = \frac{Q}{A} = \frac{0.283 \text{ m}^3/\text{s}}{\frac{\pi}{4}(0.203 \text{ m})^2} = 8.75 \text{ m/s}$$

it follows that

$$\text{Re} = \frac{\rho VD}{\mu} = \frac{(1.23 \text{ kg/m}^3)(8.75 \text{ m/s})(0.203 \text{ m})}{1.8 \times 10^{-5} \text{ kg/m} \cdot \text{s}}$$

$$= 1.21 \times 10^5$$

From Table 8.1 the equivalent roughness, ε, for plastic is 0.0. Therefore, $\varepsilon/D = 0$. Using this information with the Moody chart, we find the friction factor, f, to be 0.017. We now have enough information to solve Eq. 1.

$$h_P = (0.017) \left[\frac{9.14 \text{ m}}{0.204 \text{ m}} \right] \left[\frac{(8.75 \text{ m/s})^2}{2(9.81 \text{ m/s}^2)} \right] = 2.97 \text{ m}$$

The actual power the fan must supply to the air (W, $\mathcal{P}_a$) can be calculated by

$$\mathcal{P}_a = \gamma Q h_P \qquad (2)$$

$$\mathcal{P}_a = (12.07 \text{ N/m}^3)(0.283 \text{ m}^3/\text{s})(2.97 \text{ m})$$

$$= 10.14 \text{ W}$$

Comment Typically, flexible hosing is used to route air into confined spaces. Minor losses were neglected in part **(a)**, but we need to be cautious in making this assumption with a relatively short overall pipe length.

b. For the additional minor losses, we need to include the entrance, exit, and bend. The loss coefficient for a reentrant entrance is 0.8 (Fig. 8.22), the loss coefficient for a 90° bend with no guide vanes is 1.1, and the loss coefficient for the exit is 1.0. Therefore, the total minor loss is given as

$$h_{L \text{ minor}} = \sum K_L \frac{V^2}{2g} = (0.8 + 1.1 + 1.0) \frac{(8.75 \text{ m/s})^2}{2(9.81 \text{ m/s}^2)}$$

$$h_{L \text{ minor}} = 11.21 \text{ m}$$

The power required to account for both the major and the minor losses is

$$\mathcal{P}_a = 10.14 \text{ W} + (12.07 \text{ N/m}^3)(0.283 \text{ m}^3/\text{s})(11.21)$$

$$= 48.43 \text{ W}$$

Comment Note that with the inclusion of minor losses, the horsepower required is more than four times what is needed for a straight pipe. It is easy to see that in this situation neglecting minor losses would provide an underestimation of the power required to pump 0.283 m³/s of air. Obviously, "minor" losses are not always minor! **Table E8.9** contains different power requirements for a few different pipe situations using the preceding flow properties. Keep in mind, these power requirements are what the fan must deliver to the air; when considering fan efficiency, even more power is required.

TABLE E8.9

Pipe Setup	$h_{L \text{ minor}}$ (m)	$h_{L \text{ major}}$ (m)	h_P (m)	$\mathcal{P}_a$ (kW)
Straight pipe	0	2.96	2.96	0.01
Entrance and exit	6.98	2.96	9.94	0.034
Entrance, exit, 1 miter bend	11.21	2.96	14.17	0.048
Entrance, exit, 2 miter bends	15.48	2.96	18.44	0.063
Entrance, exit, 4 miter bends	23.99	2.96	26.94	0.092

Caution also needs to be taken when placing the fan outside the confined space so that clean air is supplied and hazardous gases emitted from the manhole are not circulated back into the confined space. There is also the question of whether to have air forced into the confined space or have the contaminated air exhausted from the space and naturally draw in fresh air.

Some pipe flow problems require a trial-and-error solution technique.

Pipe flow problems in which it is desired to determine the flowrate for a given set of conditions (Type II problems) often require trial-and-error (i.e., iterative) or numerical root-finding techniques. This is because it is necessary to know the value of the friction factor to carry out the calculations, but the friction factor is a function of the unknown velocity (flowrate) in terms of the Reynolds number. The solution procedure is indicated in Example 8.10.

EXAMPLE 8.10 | Type II, Determine Flowrate

Given The fan shown in **Fig. E8.10a** is to provide airflow through the spray booth and fume hood so that workers are protected from harmful vapors and aerosols while mixing chemicals within the hood. For proper operation of the hood, the flowrate is to be between 0.00283 m³/s and 0.0057 m³/s. With the initial setup the flowrate is 0.00424 m³/s, the loss coefficient for the system is 5, and the duct is short enough so that major losses are negligible. It is proposed that when the factory is remodeled the 0.203 m-diameter galvanized iron duct will be 30.5 m long and the total loss coefficient will be 10.

Find Determine if the flowrate will be within the required 0.00283 m³/s to 0.0057 m³/s after the proposed remodeling. Assume that the head added to the air by the fan remains constant.

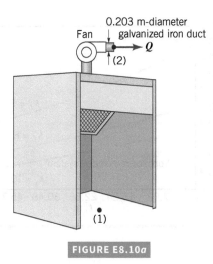

0.203 m-diameter
Fan galvanized iron duct

(2)

Q

(1)

FIGURE E8.10a

Solution We can determine the head that the fan adds to the air by considering the initial situation (i.e., before remodeling). To do this we write the energy equation between section (1) in the room and section (2) at the exit of the duct as shown in Fig. E8.10a.

$$\frac{p_1}{\gamma} + \frac{V_1^2}{2g} + z_1 + h_p = \frac{p_2}{\gamma} + \frac{V_2^2}{2g} + z_2 + h_L \qquad (1)$$

Since we are dealing with air, we can assume any change in elevation is negligible. We can also assume the pressure inside the room and at the exit of the duct is equal to atmospheric pressure and the air in the booth/room has zero velocity. Therefore, Eq. 1 reduces to

$$h_p = \frac{V_2^2}{2g} + h_L \qquad (2)$$

The diameter of the duct is given as 0.203 m, so the velocity at the exit can be calculated from the flowrate, where $V = Q/A = (0.00424 \text{ m}^3/\text{s})/[\pi(0.203 \text{ m})^2/4] = 7.86$ m/s. For the original configuration the duct is short enough to neglect major losses and only minor losses contribute to the total head loss. This head loss can be found from $h_{L, \text{minor}} = \sum K_L V^2/(2g) = 5(7.86 \text{ m/s})^2/[2(9.81 \text{ m/s}^2)] = 15.73$ m. With this information the simplified energy equation, Eq. 2, can now be solved to find the head added to the air by the fan.

$$h_p = \frac{(7.86 \text{ m/s})^2}{2(9.81 \text{ m/s}^2)} + 15.73 = 18.87 \text{ m}$$

The energy equation now must be solved with the new configuration after remodeling. Using the same assumptions as before gives the same reduced energy equation as shown in Eq. 2. With the increase in duct length to 30.5 m the duct is no longer short enough to neglect major losses. Thus,

$$h_p = \frac{V_2^2}{2g} + f\frac{\ell}{D}\frac{V^2}{2g} + \sum K_L\frac{V^2}{2g}$$

where $V_2 = V$ and $\sum K_L = 10$. We can now rearrange and solve for the velocity in m/s.

$$V = \sqrt{\frac{2gh_p}{1 + f\frac{\ell}{D} + \sum K_L}} = \sqrt{\frac{2(9.81 \text{ m/s}^2)(18.87 \text{ m})}{1 + f\left(\frac{30.5}{0.203}\right) + 10}}$$

$$= \sqrt{\frac{370}{11 + 150 f}} \qquad (3)$$

The value of f is dependent on Re, which is dependent on V, an unknown.

$$\text{Re} = \frac{\rho V D}{\mu} = \frac{(1.22 \text{ kg/m}^3)(V)(0.203 \text{ m})}{1.8 \times 10^{-5} \text{ kg/m} \cdot \text{s}}$$

or

$$\text{Re} = 13759 \, V \qquad (4)$$

where again V is in meter per second.

Also, since $\varepsilon/D = (0.00015 \text{ m})/(0.203 \text{ m}) = 0.00075$ (see Table 8.1 for the value of ε), we know which particular curve of the Moody chart is pertinent to this flow. Thus, we have three relationships (Eqs. 3, 4, and $\varepsilon/D = 0.00075$ curve of Fig. 8.20) from which we can solve for the three unknowns, f, Re, and V. This is done easily by an iterative scheme as follows.

It is usually simplest to assume a value of f, calculate V from Eq. 3, calculate Re from Eq. 4, and look up the new value of f in the Moody chart for this value of Re. If the assumed f and the new f do not agree, the assumed answer is not correct—we do not have the solution to the three equations. Although values of f, V, or Re could be assumed as starting values, it is usually simplest to assume a value of f because the correct value often lies on the relatively flat portion of the Moody chart for which f is quite insensitive to Re.

Thus, we assume $f = 0.019$, approximately the large Re limit for the given relative roughness. From Eq. 3 we obtain

$$V = \sqrt{\frac{370}{11 + 150(0.019)}} = 5.168 \text{ m/s}$$

and from Eq. 4

$$\text{Re} = 13759 \, (5.168) = 71106.5$$

With this Re and ε/D, Fig. 8.20 gives $f = 0.022$, which is not equal to the assumed solution of $f = 0.019$ (although it is close!). We try again, this time with the newly obtained value of $f = 0.022$, which gives $V = 5.09$ m/s and Re = 70,800. With these values, Fig. 8.20 gives $f = 0.022$, which agrees with the assumed value. Thus, the solution is $V = 5.09$ m/s, or

$$Q = VA = (5.09 \text{ m/s})\left(\frac{\pi}{4}\right)(0.203 \text{ m})^2 = 0.00275 \text{ m}^3/\text{s} \qquad (Ans)$$

Comment It is seen that operation of the system after the proposed remodeling will not provide enough airflow to protect workers from inhalation hazards while mixing chemicals within the hood. By repeating the calculations for different duct lengths and different total minor loss coefficients, the results shown in **Fig. E8.10b** are obtained, which give flowrate as a function of duct length for various values of the minor loss coefficient. It will be necessary to redesign the remodeled system (e.g., larger fan, shorter ducting, larger-diameter duct) so that the flowrate will be within the acceptable range. In many companies, teams of occupational safety and health experts and engineers work together during the design phase of remodeling (or when a new operation is being planned) to consider and prevent potential negative impacts on workers' safety and health in an effort called "Prevention through Design." They also may be required to ensure that exhaust from such a system exits the building away from human activity and into an area where it will not be drawn back inside the facility.

FIGURE E8.10*b*

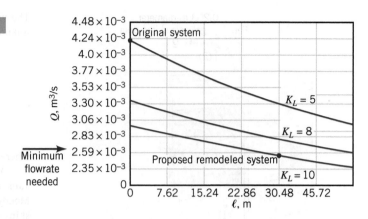

EXAMPLE 8.11 | Type II, Determine Flowrate

Given The turbine shown in **Fig. E8.11** extracts 37.25 kW from the water flowing through it. The 0.304 m-diameter, 91.4 m-long pipe is assumed to have a friction factor of 0.02. Minor losses are negligible.

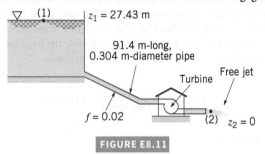

FIGURE E8.11

Find Determine the flowrate through the pipe and turbine.

Solution The energy equation (Eq. 8.21) can be applied between the surface of the lake [point (1)] and the outlet of the pipe as

$$\frac{p_1}{\gamma} + \frac{V_1^2}{2g} + z_1 = \frac{p_2}{\gamma} + \frac{V_2^2}{2g} + z_2 + h_L + h_T \qquad (1)$$

where h_T is the turbine head, $p_1 = V_1 = p_2 = z_2 = 0$, $z_1 = 27.43$ m, and $V_2 = V$, the fluid velocity in the pipe. The head loss is given by

$$h_L = f \frac{\ell}{D} \frac{V^2}{2g} = 0.02 \ \frac{91.4 \text{ m}}{(0.304 \text{ m})} \ \frac{V^2}{2(9.81 \text{ m/s}^2)} = 0.306 \ V^2 \text{ m}$$

where V is in m/s. Also, the turbine head is

$$h_T = \frac{\mathcal{P}_a}{\gamma Q} = \frac{\mathcal{P}_a}{\gamma(\pi/4)D^2 V}$$

$$= \frac{37250 \text{ W}}{(9.81 \text{ kN/m}^3)[(\pi/4)(0.304 \text{ m})^2 \ V]} = \frac{52.340}{V} \text{ m}$$

Thus, Eq. 1 can be written as

$$27.43 = \frac{V^2}{2(9.81)} + 0.306V^2 + \frac{52.340}{V}$$

or

$$7V^3 - 538V + 1027 = 0 \qquad (2)$$

where V is in m/s. The velocity of the water in the pipe is found as the solution of Eq. 2. Surprisingly, there are two real, positive roots: $V = 2.01$ m/s or $V = 7.59$ m/s. The third root is negative ($V = -9.6$ m/s) and has no physical meaning for this flow. Thus, the two acceptable flowrates are

$$Q = \frac{\pi}{4}D^2V = \frac{\pi}{4}(0.304 \text{ m})^2(2.01 \text{ m/s}) = 0.15 \text{ m}^3/\text{s} \qquad (Ans)$$

or

$$Q = \frac{\pi}{4}(0.304 \text{ m})^2(7.59 \text{ m/s}) = 0.56 \text{ m}^3/\text{s} \qquad (Ans)$$

Comments Either of these two flowrates gives the same power, $\mathcal{P}_a = \gamma Q h_T$. The reason for two possible solutions can be seen from the following. With the low flowrate ($Q = 0.15$ m³/s), we obtain the head loss and turbine head as $h_L = 1.23$ m and $h_T = 26$ m. Because of the relatively low velocity, there is a relatively small head loss and, therefore, a large head available for the turbine. With the large flowrate ($Q = 0.56$ m³/s), we find $h_L = 11.6$ m and $h_T = 6.86$ m. The high-speed flow in the pipe produces a relatively large loss due to friction, leaving a relatively small head for the turbine. However, in either case the product of the turbine head times the flowrate is the same. That is, the power extracted ($\mathcal{P}_a = \gamma Q h_T$) is identical for each case. Although either flowrate will allow the extraction of 37.25 kW from the water, the details of the design of the turbine itself will depend strongly on which flowrate is to be used. Such information can be found in Chapter 12 and various references about turbomachines (Refs. 14, 19, 20).

If the friction factor were not given, the solution to the problem would be lengthier. A trial-and-error solution similar to that in Example 8.10 would be required, along with the solution of a cubic equation.

In pipe flow problems for which the diameter is the unknown (Type III), an iterative or numerical root-finding technique is required. This is, again, because the friction factor is a function of the diameter—through both the Reynolds number and the relative roughness. Thus, neither Re $= \rho VD/\mu = 4\rho Q/\pi\mu D$ nor ε/D are known unless D is known.

Commercial pipe and tubing are only available in specific sizes. Pipe diameter calculation seldom corresponds exactly to a standard size. Specifying a standard pipe size is usually more economical than ordering a custom fabrication just to match your carefully computed diameter. Specifying the standard pipe size that has the next largest diameter is customary. This practice of rounding upward implies that there is little need to be extremely accurate in pipe sizing calculations. Be aware that the loss and/or flowrate does not equal the originally specified values when the larger pipe is used, so a recalculation of loss and flowrate may be necessary.

Examples 8.12 and 8.13 illustrate Type III problems.

EXAMPLE 8.12 | Type III Without Minor Losses, Determine Diameter

Given Air at standard temperature and pressure flows through a horizontal, galvanized iron pipe ($\varepsilon = 0.00015$ m) at a rate of 0.057 m^3/s. The pressure drop is to be no more than 3.4 kPa per 30.5 m of pipe.

Find Determine the minimum pipe diameter.

Solution We assume the flow to be incompressible with $\rho = 1.22$ kg/m^3 and $\mu = 1.85 \times 10^{-5}$ kg/m · s. Note that if the pipe were too long, the pressure drop from one end to the other, $p_1 - p_2$, would not be small relative to the pressure at the beginning, and compressible flow considerations would be required. For example, a pipe length of 61 m gives $(p_1 - p_2)/p_1 = [(3447\ \text{Pa})/(30.5\ \text{m})]$ (61 m)/101352 Pa = 0.068 = 6.8%, which is probably small enough to justify the incompressible assumption.

With $z_1 = z_2$ and $V_1 = V_2$, the energy equation (Eq. 8.21) becomes

$$p_1 = p_2 + f\frac{\ell}{D}\frac{\rho V^2}{2} \tag{1}$$

where $V = Q/A = 4Q/(\pi D^2) = 4(0.057\ \text{m}^3/\text{s})/\pi D^2$, or

$$V = \frac{0.073}{D^2}$$

where D is in meter. Thus, with $p_1 - p_2 = (3447\ \text{Pa})$ and $\ell = 30.5$ m Eq. 1 becomes

$$p_1 - p_2 = 3447\ \text{Pa}$$
$$= f\frac{(30.5\,\text{m})}{D}(1.22\,\text{kg/m}^3)\frac{1}{2}\left(\frac{0.073}{D^2}\frac{\text{m}}{\text{s}}\right)^2$$

or

$$D = 0.124\, f^{1/5} \tag{2}$$

where D is in m. Also Re $= \rho V D/\mu = (1.22\ \text{kg/m}^3)[(0.073/D^2)\ \text{m/s}]$ $D/(1.8 \times 10^{-5}\ \text{kg/m · s})$, or

$$\text{Re} = \frac{4947}{D} \tag{3}$$

and

$$\frac{\varepsilon}{D} = \frac{0.00015}{D} \tag{4}$$

Thus, we have four equations [Eqs. 2, 3, 4, and either the Moody chart, the Colebrook equation (8.35a) or the Haaland equation (8.35b)] and four unknowns (f, D, ε/D, and Re) from which the solution can be obtained by trial-and-error methods.

If we use the Moody chart, it is probably easiest to assume a value of f, use Eqs. 2, 3, and 4 to calculate D, Re, and ε/D, and

then compare the assumed f with that from the Moody chart. If they do not agree, try again. Thus, we assume $f = 0.02$, a typical value, and obtain $D = 0.124(0.02)^{1/5} = 0.056$ m, which gives $\varepsilon/D = 0.00015/0.056 = 0.0027$ and Re $= 4947/0.056 = 8.83 \times 10^4$. From the Moody chart we obtain $f = 0.027$ for these values of ε/D and Re. Since this is not the same as our assumed value of f, we try again. With $f = 0.027$, we obtain $D = 0.060$ m, $\varepsilon/D = 0.0026$, and Re $= 8.27 \times 10^4$, which in turn give $f = 0.027$, in agreement with the assumed value. Thus, the diameter of the pipe should be

$$D = 0.060\ \text{m} \tag{Ans}$$

Comment If we use the Colebrook equation (Eq. 8.35a) with $\varepsilon/D = 0.00015/0.124\, f^{1/5} = 0.00124/f^{1/5}$ and Re $= 4947/0.124\, f^{1/5} = 4.01 \times 10^4/f^{1/5}$, we obtain

$$\frac{1}{\sqrt{f}} = -2.0\log\left(\frac{\varepsilon/D}{3.7} + \frac{2.51}{\text{Re}\sqrt{f}}\right)$$

or

$$\frac{1}{\sqrt{f}} = -2.0\log\left(\frac{3.35 \times 10^{-4}}{f^{1/5}} + \frac{6.26 \times 10^{-5}}{f^{3/10}}\right)$$

By using a root-finding technique on a computer or calculator, the solution to this equation is determined to be $f = 0.027$, and hence $D = 0.060$ m, in agreement with the Moody chart method.

By repeating the calculations for various values of the flowrate, Q, the results shown in **Fig. E8.12** are obtained. Although an increase in flowrate requires a larger diameter pipe (for the given pressure drop), the increase in diameter is minimal. For example, if the flowrate is doubled from 0.028 m^3/s to 0.056 m^3/s, the diameter increases from 0.046 m to 0.058 m. Note that these exact diameters are probably not available, so size upward to the closest diameter.

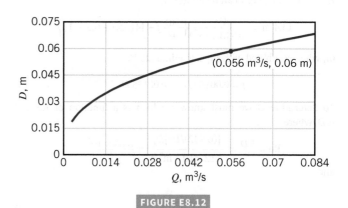

FIGURE E8.12

In the previous example we only had to consider major losses. In some instances the inclusion of major and minor losses can cause a slightly lengthier solution procedure, even though the governing equations are essentially the same. This is illustrated in Example 8.13.

EXAMPLE 8.13 | Type III with Minor Losses, Determine Diameter

Given Water at 15.6 °C ($v = 1.12 \times 10^{-6}$ m²/s, see Table 1.5) is to flow from reservoir A to reservoir B through a pipe of length 518 m and roughness 0.00015 m at a rate of $Q = 0.74$ m³/s as shown in **Fig. E8.13a**. The system contains a sharp-edged entrance and four flanged 45° elbows.

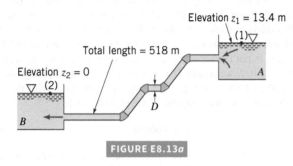

FIGURE E8.13a

Find Determine the pipe diameter needed.

Solution The energy equation (Eq. 8.21) can be applied between two points on the surfaces of the reservoirs ($p_1 = p_2 = V_1 = V_2 = z_2 = 0$) as follows:

$$\frac{p_1}{\gamma} + \frac{V_1^2}{2g} + z_1 = \frac{p_2}{\gamma} + \frac{V_2^2}{2g} + z_2 + h_L$$

or

$$z_1 = \frac{V^2}{2g}\left(f\frac{\ell}{D} + \Sigma K_L\right) \tag{1}$$

where $V = Q/A = 4Q/\pi D^2 = 4(0.74 \text{ m}^3/\text{s})/\pi D^2$, or

$$V = \frac{0.942}{D^2} \tag{2}$$

is the velocity within the pipe. (Note that the units on V and D are m/s and m, respectively.) The loss coefficients are obtained from Table 8.2 and Figs. 8.22 and 8.24 as $K_{L_{\text{ent}}} = 0.5$, $K_{L_{\text{elbow}}} = 0.2$, and $K_{L_{\text{exit}}} = 1$. Thus, Eq. 1 can be written as

$$13.4 \text{ m} = \frac{V^2}{2(9.81 \text{ m/s}^2)}\left\{\frac{581}{D}f + [4(0.2) + 0.5 + 1]\right\}$$

or, when combined with Eq. 2 to eliminate V,

$$f = 0.00152 D^5 - 0.00135 D \tag{3}$$

To determine D we must know f, which is a function of Re and ε/D, where

$$\text{Re} = \frac{VD}{v} = \frac{[(0.942)/D^2]D]}{1.12 \times 10^{-6} \text{ m}^2/\text{s}} = \frac{0.84 \times 10^6}{D} \tag{4}$$

and

$$\frac{\varepsilon}{D} = \frac{0.00015}{D} \tag{5}$$

where D is in meter. Again, we have four equations (Eqs. 3, 4, 5, and the Moody chart or the Colebrook equation) for the four unknowns D, f, Re, and ε/D.

Consider the solution by using the Moody chart. Although it is often easiest to assume a value of f and make calculations to determine if the assumed value is the correct one, with the inclusion of minor losses this may not be the simplest method. For example, if we assume $f = 0.02$ and calculate D from Eq. 3, we would have to solve a fifth-order equation. With only major losses (see Example 8.12), the term proportional to D in Eq. 3 is absent, and it is easy to solve for D if f is given. With both major and minor losses included, this solution for D (given f) would require more complication.

Thus, for this type of problem it is perhaps easier to assume a value of D, calculate the corresponding f from Eq. 3, and with the values of Re and ε/D determined from Eqs. 4 and 5, look up the value of f in the Moody chart (or the Colebrook equation). The solution is obtained when the two values of f are in agreement. A few rounds of calculation will reveal that the solution is given by

$$D \approx 0.497 \text{ m} \tag{Ans}$$

Comments Alternatively, we can use the Colebrook equation rather than the Moody chart to solve for D. This is easily done by using the Colebrook equation (Eq. 8.35a) with f as a function of D obtained from Eq. 3 and Re and ε/D as functions of D from Eqs. 4 and 5. The resulting single equation for D can be solved by using a root-finding technique on a computer or calculator to obtain $D = 0.497$ m. This agrees with the solution obtained using the Moody chart.

By repeating the calculations for various pipe lengths, ℓ, the results shown in **Fig. E8.13b** are obtained. As the pipe length increases it is necessary, because of the increased friction, to increase the pipe diameter to maintain the same flowrate.

It is interesting to attempt to solve this example if all losses are neglected so that Eq. 1 becomes $z_1 = 0$. Clearly from Fig. E8.13a, $z_1 = 13.4$ m. Obviously something is wrong. A fluid

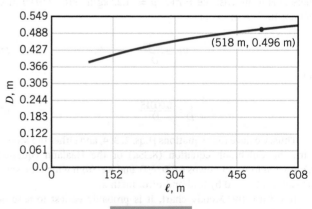

FIGURE E8.13b

cannot flow from one elevation, beginning with zero pressure and velocity, and end up at a lower elevation with zero pressure and velocity unless energy is removed (i.e., a head loss or a turbine) somewhere between the two locations. If the pipe is short (negligible friction) and the minor losses are negligible, there is still the kinetic energy of the fluid as it leaves the pipe and enters the reservoir. After the fluid meanders around in the reservoir for some time, this kinetic energy is lost and the fluid is stationary. No matter how small the viscosity is, the exit loss cannot be neglected. The same result can be seen if the energy equation is written from the free surface of the upstream tank to the exit plane of the pipe, at which point the kinetic energy is still available to the fluid. In either case the energy equation becomes $z_1 = V^2/2g$ in agreement with the inviscid results of Chapter 3 (the Bernoulli equation).

8.5.2 Multiple Pipe Systems

In many pipe systems there is more than one pipe involved. The complex system of tubes in our lungs (beginning as shown in the adjacent figure, with the relatively large-diameter trachea and ending in tens of thousands of minute bronchioles after numerous branchings) and the maze of pipes in a city's water distribution system are typical of such systems. The governing mechanisms for the flow in **multiple pipe systems** are the same as for the single pipe systems discussed in this chapter. However, because of the numerous unknowns involved, additional complexities may arise in solving for the flow in multiple pipe systems. Some of these complexities are discussed in this section.

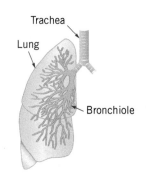

Trachea

Lung

Bronchiole

The Wide World of Fluids

Deepwater pipeline

Pipelines used to transport oil and gas are commonplace. But south of New Orleans, in deep waters of the Gulf of Mexico, a not-so-common *multiple pipe system* was recently installed. The new so-called Mardi Gras system of pipes is situated in water depths of 1310.6 to 2225 km. It transports oil and gas from five deepwater fields with the interesting names of Holstein, Mad Dog, Thunder Horse, Atlantis, and Na Kika. The deepwater pipelines connect with lines at intermediate water depths to transport the oil and gas to shallow-water fixed platforms and shore. The steel pipe used is 0.71 m in diameter with a wall thickness of 0.035 m. The thick-walled pipe is needed to withstand the large external pressure, which is about 22408 kPa at a depth of 2225 m. The pipe was installed in 73 m sections from a vessel the size of a large football stadium. The deepwater pipeline system has a total length of more than 724 km and the capability of transporting more than 158987.3 m^3 of oil per day and 4.25×10^7 m^3 of gas per day.

The simplest multiple pipe systems can be classified into series or parallel flows, as are shown in **Fig. 8.34**. The nomenclature is similar to that used in electrical circuits. Indeed, an analogy between fluid and electrical circuits is often made as follows. In a simple electrical circuit, there is a balance between the voltage (e), current (i), and resistance (R) as given by Ohm's law: $e = iR$. In a fluid circuit there is a balance between the pressure drop (Δp), the flowrate or velocity (Q or V), and the flow resistance as given in terms of the friction factor and minor loss coefficients (f and K_L). For a simple flow $[\Delta p = f(\ell/D)(\rho V^2/2)]$, it follows that $\Delta p = Q^2\widetilde{R}$, where $\widetilde{R}$, a measure of the resistance to the flow, is proportional to f.

The main differences between the solution methods used to solve electrical circuit problems and those for fluid circuit problems lie in the fact that Ohm's law is a linear equation (doubling the voltage doubles the current), while the fluid equations are generally nonlinear (doubling the pressure drop does not double the flowrate unless the flow is laminar). Thus, although some of the standard electrical engineering methods can be carried over to help solve fluid mechanics problems, others cannot.

One of the simplest multiple pipe systems is that containing pipes in *series*, as is shown in Fig. 8.34a. Every fluid particle that passes through the system passes through each of the pipes. Thus, the flowrate (but not the velocity) is the same in each pipe, and the head loss from

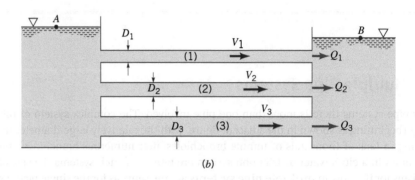

FIGURE 8.34   (a) Series and (b) parallel pipe systems.

point A to point B is the sum of the head losses in each of the pipes. The governing equations can be written as follows:

$$Q_1 = Q_2 = Q_3$$

and

$$h_{L_{A-B}} = h_{L_1} + h_{L_2} + h_{L_3}$$

(used in conjunction with the energy equation) where the subscripts refer to each of the pipes. In general, the friction factors will be different for each pipe because the Reynolds numbers ($\text{Re}_i = \rho V_i D_i / \mu$) and the relative roughnesses (ε_i / D_i) will be different. Thus

$$h_L = \sum h_{L,\text{major}} + \sum_{L,\text{minor}} = \sum f\left(\frac{L}{D}\right)\left(\frac{V^2}{2g}\right) + \sum K_L\left(\frac{V^2}{2g}\right)$$

or in terms of flowrate

$$h_L = \left(\sum \frac{f}{A^2}\frac{L}{D} + \sum \frac{K_L}{A^2}\right)\frac{Q^2}{2g}$$

If the flowrate is given, it is a straightforward calculation to determine the head loss or pressure drop (Type I problem). If the pressure drop is given and the flowrate is to be calculated (Type II problem), an iteration scheme is needed. In this situation none of the friction factors, f_i, are known, so the calculations may involve more trial-and-error attempts than for corresponding single pipe systems. The same is true for problems in which the pipe diameter (or diameters) is to be determined (Type III problems).

Another common multiple pipe system contains pipes in *parallel*, as is shown in Fig. 8.34b. In this system a fluid particle traveling from A to B may take any of the paths available, with the total flowrate equal to the sum of the flowrates in each pipe. However, by writing the energy equation between points A and B it is found that the head loss experienced by any fluid particle traveling between these locations is the same, independent of the path taken. Thus, the governing equations for parallel pipes are

$$Q = Q_1 + Q_2 + Q_3$$

and

$$h_{L_1} = h_{L_2} = h_{L_3}$$

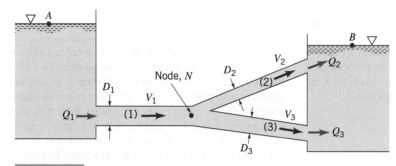

FIGURE 8.35 Multiple-pipe loop system.

Again, the method of solution of these equations depends on what information is given and what is to be calculated.

Another type of multiple pipe system called a *loop* is shown in **Fig. 8.35**. In this case the flowrate through pipe (1) equals the sum of the flowrates through pipes (2) and (3), or $Q_1 = Q_2 + Q_3$. As can be seen by writing the energy equation between the surfaces of each reservoir, the head loss for pipe (2) must equal that for pipe (3), even though the pipe sizes and flowrates may be different for each. That is,

$$\frac{p_A}{\gamma} + \frac{V_A^2}{2g} + z_A = \frac{p_B}{\gamma} + \frac{V_B^2}{2g} + z_B + h_{L_1} + h_{L_2}$$

for a fluid particle traveling through pipes (1) and (2), while

$$\frac{p_A}{\gamma} + \frac{V_A^2}{2g} + z_A = \frac{p_B}{\gamma} + \frac{V_B^2}{2g} + z_B + h_{L_1} + h_{L_3}$$

for fluid that travels through pipes (1) and (3). These can be combined to give $h_{L_2} = h_{L_3}$. This is a statement of the fact that fluid particles that travel through pipe (2) and particles that travel through pipe (3) all originate from common conditions at the junction (or node, N) of the pipes and all end up at the same final conditions.

The flow in a relatively simple looking multiple pipe system may be more complex than it appears initially. The branching system termed the *three-reservoir problem* shown in **Fig. 8.36** is such a system. Three reservoirs at known elevations are connected together with three pipes of known properties (length, diameter, and roughness). The problem is to determine the flowrates into or out of the reservoirs. If valve (1) were closed, the fluid would flow from reservoir B to C, and the flowrate could be easily calculated. Similar calculations could be carried out if valves (2) or (3) were closed with the others open.

With all valves open, however, it is not necessarily obvious which direction the fluid flows. For the conditions indicated in Fig. 8.36, it is clear that fluid flows from reservoir A because the other two reservoir levels are lower. Whether the fluid flows into or out of reservoir B depends on the elevation of reservoirs B and C and the properties (length, diameter, roughness) of the three pipes. In general, the flow direction is not obvious, and the solution process must include the determination of this direction. This is illustrated in Example 8.14.

For some pipe systems, the direction of flow is not known a priori.

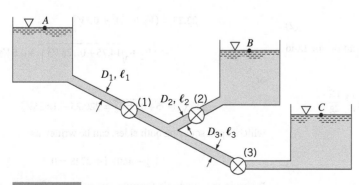

FIGURE 8.36 A three-reservoir system.

EXAMPLE 8.14 | Three-Reservoir, Multiple Pipe System

Given   Three reservoirs are connected by three pipes as are shown in **Fig. E8.14**. For simplicity we assume that the diameter of each pipe is 0.3 m, the friction factor for each is 0.02, and because of the large length-to-diameter ratio, minor losses are negligible.

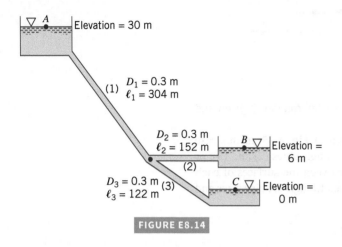

$D_1 = 0.3$ m
$\ell_1 = 304$ m
(1)

$D_2 = 0.3$ m
$\ell_2 = 152$ m
(2)

$D_3 = 0.3$ m (3)
$\ell_3 = 122$ m

Elevation = 30 m

Elevation = 6 m

Elevation = 0 m

FIGURE E8.14

Find   Determine the flowrate into or out of each reservoir.

Solution   It is not obvious which direction the fluid flows in pipe (2). However, we assume that it flows out of reservoir B, write the governing equations for this case, and check our assumption. The continuity equation requires that $Q_1 + Q_2 = Q_3$, which, since the diameters are the same for each pipe, becomes simply

$$V_1 + V_2 = V_3 \tag{1}$$

The energy equation for the fluid that flows from A to C in pipes (1) and (3) can be written as

$$\frac{p_A}{\gamma} + \frac{V_A^2}{2g} + z_A = \frac{p_C}{\gamma} + \frac{V_C^2}{2g} + z_C + f_1\frac{\ell_1}{D_1}\frac{V_1^2}{2g} + f_3\frac{\ell_3}{D_3}\frac{V_3^2}{2g}$$

By using the fact that $p_A = p_C = V_A = V_C = z_C = 0$, this becomes

$$z_A = f_1\frac{\ell_1}{D_1}\frac{V_1^2}{2g} + f_3\frac{\ell_3}{D_3}\frac{V_3^2}{2g}$$

For the given conditions of this problem we obtain

$$30 \text{ m} = \frac{0.02}{2(9.81 \text{ m/s}^2)}\frac{1}{(0.3 \text{ m})}\left[(304 \text{ m})V_1^2 + (122 \text{ m})V_3^2\right]$$

or

$$29 = V_1^2 + 0.4V_3^2 \tag{2}$$

where V_1 and V_3 are in m/s. Similarly the energy equation for fluid flowing from B and C is

$$\frac{p_B}{\gamma} + \frac{V_B^2}{2g} + z_B = \frac{p_C}{\gamma} + \frac{V_C^2}{2g} + z_C + f_2\frac{\ell_2}{D_2}\frac{V_2^2}{2g} + f_3\frac{\ell_3}{D_3}\frac{V_3^2}{2g}$$

or

$$z_B = f_2\frac{\ell_2}{D_2}\frac{V_2^2}{2g} + f_3\frac{\ell_3}{D_3}\frac{V_3^2}{2g}$$

For the given conditions this can be written as

$$11.6 = V_2^2 + 0.8V_3^2 \tag{3}$$

Equations 1, 2, and 3 (in terms of the three unknowns V_1, V_2, and V_3) are the governing equations for this flow, provided the fluid flows from reservoir B. It turns out, however, that there is no solution for these equations with positive, real values of the velocities. Although these equations do not appear to be complicated, there is no simple way to solve them directly. Thus, a trial-and-error solution is suggested. This can be accomplished as follows. Assume a value of $V_1 > 0$, calculate V_3 from Eq. 2, and then V_2 from Eq. 3. It is found that the resulting V_1, V_2, V_3 trio does not satisfy Eq. 1 for any value of V_1 assumed. There is no solution to Eqs. 1, 2, and 3 with real, positive values of V_1, V_2, and V_3. Thus, our original assumption of flow out of reservoir B must be incorrect.

To obtain the solution, assume the fluid flows into reservoirs B and C and out of A. For this case the continuity equation becomes

$$Q_1 = Q_2 + Q_3$$

or

$$V_1 = V_2 + V_3 \tag{4}$$

Application of the energy equation between points A and B and A and C gives

$$z_A = z_B + f_1\frac{\ell_1}{D_1}\frac{V_1^2}{2g} + f_2\frac{\ell_2}{D_2}\frac{V_2^2}{2g}$$

and

$$z_A = z_C + f_1\frac{\ell_1}{D_1}\frac{V_1^2}{2g} + f_3\frac{\ell_3}{D_3}\frac{V_3^2}{2g}$$

which, with the given data, become

$$23.23 = V_1^2 + 0.5\,V_2^2 \tag{5}$$

and

$$29.04 = V_1^2 + 0.4\,V_3^2 \tag{6}$$

Equations 4, 5, and 6 can be solved as follows. By subtracting Eq. 5 from 6 we obtain

$$V_3 = \sqrt{14.5 + 0.125\,V_2^2}$$

Thus, Eq. 5 can be written as

$$23.23 = (V_2 + V_3)^2 + 0.5V_2^2$$

$$= \left(V_2 + \sqrt{14.75 + 0.125\,V_2^2}\right)^2 + 0.5\,V_2^2$$

or

$$2V_2\sqrt{14.75 + 0.125V_2^2} = 220.25 - 1.625V_2^2 \tag{7}$$

which, upon squaring both sides, can be written as

$$V_2^4 - 460V_2^2 + 3748 = 0$$

By using the quadratic formula, we can solve for V_2^2 to obtain $V_2^2 = 0.745$. Thus, $V_2 = 0.863$ m/s and from Eq. 5, $V_1 = 4.74$ m/s. The corresponding flowrates are

$$Q_1 = A_1V_1 = \frac{\pi}{4}D_1^2V_1 = \frac{\pi}{4}(0.3\,\text{m})^2(4.74\,\text{m/s})$$

$$= 0.335\,\text{m}^3/\text{s from } A \qquad (Ans)$$

$$Q_2 = A_2V_2 = \frac{\pi}{4}D_2^2V_2 = \frac{\pi}{4}(0.3\,\text{m})^2(0.863\,\text{m/s})$$

$$= 0.06\,\text{m}^3/\text{s into } B \qquad (Ans)$$

and

$$Q_3 = Q_1 - Q_2 = (0.335 - 0.06)\,\text{m}^3/\text{s}$$

$$= 0.275\,\text{m}^3/\text{s into } C \qquad (Ans)$$

Note the slight differences in the governing equations depending on the direction of the flow in pipe (2)—compare Eqs. 1, 2, and 3 with Eqs. 4, 5, and 6.

Comment If the friction factors were not given, a trial-and-error procedure similar to that needed for Type II problems (see Section 8.5.1) would be required.

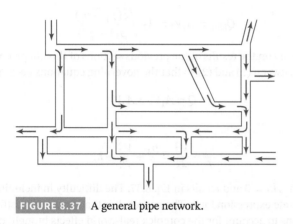

FIGURE 8.37 A general pipe network.

The ultimate in multiple pipe systems is a *network* of pipes such as that shown in **Fig. 8.37**. Networks like these often occur in city water distribution systems and other systems that may have multiple "inlets" and "outlets." The direction of flow in the various pipes is by no means obvious—in fact, it may vary in time, depending on how the system is used from time to time.

The solution for pipe network problems is often carried out by use of node and loop equations similar in many ways to that done in electrical circuits. For example, the continuity equation requires that for each *node* (the junction of two or more pipes) the net flowrate is zero. What flows into a node must flow out at the same rate. In addition, the net pressure difference completely around a *loop* (starting at one location in a pipe and returning to that location) must be zero. By combining these ideas with the usual head loss and pipe flow equations, the flow throughout the entire network can be obtained. Of course, trial-and-error solutions are usually required because the direction of flow and the friction factors may not be known. The addition of minor losses such as valves adds further complexity. Hardy Cross developed a clever calculation technique to perform the iterations (Ref. 35). Such a solution procedure using matrix techniques is ideally suited for computer use (Refs. 21, 22).

Pipe network problems can be solved using node and loop concepts.

It is often necessary to determine experimentally the flowrate in a pipe. In Chapter 3 we introduced various types of flow-measuring devices (Venturi meter, nozzle meter, orifice meter, etc.) and discussed their operation under the assumption that viscous effects were not important. In this section we will indicate how to account for the ever-present viscous effects in these flowmeters. We will also indicate other types of commonly used flowmeters.

8.6.1 Pipe Flowrate Meters

Three of the most common devices used to measure the instantaneous flowrate in pipes are the orifice meter, the nozzle meter, and the Venturi meter. As was discussed in Section 3.6.3, each of these meters operates on the principle that a decrease in flow area in a pipe causes an increase in velocity that is accompanied by a decrease in pressure. Correlation of the pressure

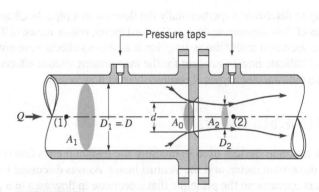

FIGURE 8.38 Typical pipe flowmeter geometry.

difference with the velocity provides a means of measuring the flowrate. In the absence of viscous effects and under the assumption of a horizontal pipe, application of the Bernoulli equation (Eq. 3.7) between points (1) and (2) shown in **Fig. 8.38** gave

$$Q_{ideal} = A_2V_2 = A_2\sqrt{\frac{2(p_1 - p_2)}{\rho(1 - \beta^4)}} \tag{8.37}$$

where $\beta = D_2/D_1$. Based on the results of the previous sections of this chapter, we anticipate that there is a head loss between (1) and (2) so that the governing equations become

$$Q = A_1V_1 = A_2V_2$$

and

$$\frac{p_1}{\gamma} + \frac{V_1^2}{2g} = \frac{p_2}{\gamma} + \frac{V_2^2}{2g} + h_L$$

The ideal situation has $h_L = 0$ and results in Eq. 8.37. The difficulty in including the head loss is that there is no accurate expression for it. The net result is that empirical coefficients are used in the flowrate equations to account for the complex real-world effects brought on by the nonzero viscosity. The coefficients are discussed in this section.

A typical **orifice meter** is constructed by inserting between two flanges of a pipe a flat plate with a hole, as shown in **Fig. 8.39**. The pressure at point (2) within the vena contracta is less than that at point (1). Nonideal effects occur for two reasons. First, the vena contracta area, A_2, is less than the area of the hole, A_o, by an unknown amount. Thus, $A_2 = C_c A_o$, where C_c is the contraction coefficient ($C_c < 1$). Second, the swirling flow and turbulent motion near the orifice plate introduce a head loss that cannot be calculated theoretically. Thus, an *orifice discharge coefficient*, C_o, is used to take these effects into account. That is,

$$Q = C_o Q_{ideal} = C_o A_o \sqrt{\frac{2(p_1 - p_2)}{\rho(1 - \beta^4)}} \tag{8.38}$$

where $A_o = \pi d^2/4$ is the area of the hole in the orifice plate. The value of C_o is a function of $\beta = d/D$ and the Reynolds number, $Re = \rho VD/\mu$, where $V = Q/A_1$. Typical values of C_o are given in **Fig. 8.40**. As shown by Eq. 8.38 and the adjacent figure, for a given value of C_o, the flowrate is proportional to the square root of the pressure difference. Note that the value of C_o depends on the specific construction of the orifice meter (e.g., the placement of the pressure taps, whether the orifice plate edge is square or beveled, etc.). Very precise conditions governing the construction of standard orifice meters have been established to provide the greatest accuracy possible (Refs. 23, 24).

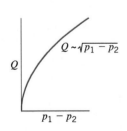

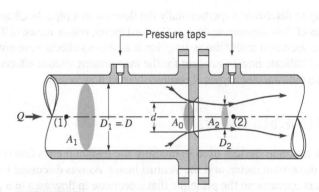

FIGURE 8.39 Typical orifice meter construction.

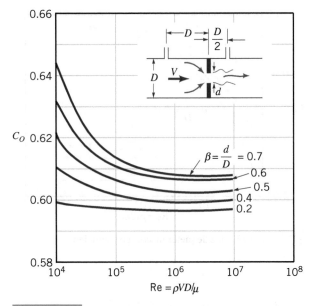

FIGURE 8.40 Orifice meter discharge coefficient.

Another type of pipe flowmeter that is based on the same principles used in the orifice meter is the **nozzle meter,** three variations of which are shown in **Fig. 8.41**. This device uses a contoured nozzle (typically placed between flanges of pipe sections) rather than a simple (and less expensive) plate with a hole as in an orifice meter. The resulting flow pattern for the nozzle meter is closer to ideal than the orifice meter flow. There is only a slight vena contracta and the secondary flow separation is less severe, but there still are viscous effects. These are accounted for by use of the *nozzle discharge coefficient, C_n,* where

The nozzle meter has less energy dissipated than the orifice meter.

$$Q = C_n Q_{\text{ideal}} = C_n A_n \sqrt{\frac{2(p_1 - p_2)}{\rho(1 - \beta^4)}} \qquad (8.39)$$

with $A_n = \pi d^2/4$. As with the orifice meter, the value of C_n is a function of the diameter ratio, $\beta = d/D$, and the Reynolds number, $\text{Re} = \rho VD/\mu$. Typical values obtained from experiments are shown in **Fig. 8.42**. Again, precise values of C_n depend on the specific details of the nozzle design. Accepted standards have been adopted (Ref. 24). Note that $C_n > C_o$; the nozzle meter has less energy dissipated than the orifice meter.

The most precise and most expensive of the three obstruction-type flowmeters is the **Venturi meter** shown in **Fig. 8.43** [G. B. Venturi (1746–1822)]. Although the operating principle for this device is the same as for the orifice or nozzle meters, the geometry of the Venturi meter is designed to reduce head losses to a minimum. This is accomplished

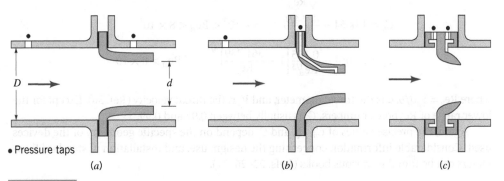

FIGURE 8.41 Typical nozzle meter construction.

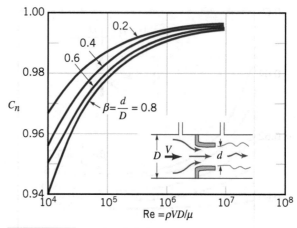

FIGURE 8.42 Nozzle meter discharge coefficient.

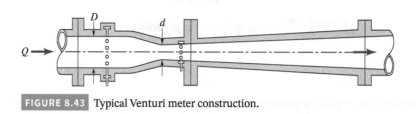

FIGURE 8.43 Typical Venturi meter construction.

by providing a relatively streamlined contraction (which eliminates separation ahead of the throat) and a very gradual expansion downstream of the throat (which eliminates separation in this decelerating portion of the device). Most of the head loss that occurs in a well-designed Venturi meter is due to friction losses along the walls rather than losses associated with separated flows and the inefficient mixing motion that accompanies such flow.

Thus, the flowrate through a Venturi meter is given by

$$Q = C_v Q_{ideal} = C_v A_T \sqrt{\frac{2(p_1 - p_2)}{\rho(1 - \beta^4)}} \tag{8.40}$$

The Venturi discharge coefficient is a function of the specific geometry of the meter.

where $A_T = \pi d^2/4$ is the throat area. The throat-to-pipe diameter ratio ($\beta = d/D$), the Reynolds number, and the shape of the converging and diverging sections of the meter are among the parameters that affect the value of C_v, the *Venturi discharge coefficient*. These values can typically be calculated from relations available for the particular type of Venturi meter used. For example, the standard American Society of Mechanical Engineers Venturi discharge coefficients can be determined with

$$C_v = 1.0054 - \frac{6.88}{\sqrt{Re_d}} \qquad 10^3 < Re_d < 5 \times 10^5$$

$$C_v = 1.0054 - \frac{4864.9}{Re_d} \qquad 5 \times 10^5 < Re_d < 8 \times 10^5$$

$$C_v = 1.0054 - \frac{0.185}{Re_d^{1/5}}\left[1 - \frac{361{,}239}{Re_d}\right]^{4/5} \qquad Re_d > 8 \times 10^5$$

where $Re_d = V_d d/v$, d is the throat diameter, and V_d is the throat velocity (Ref. 36). Except for the lower range of Reynolds numbers, C_v is usually between 0.97 and 0.99.

Again, the precise values of C_n, C_o, and C_v depend on the specific geometry of the devices used. Considerable information concerning the design, use, and installation of standard flowmeters can be found in various books (Refs. 23–26, 31).

EXAMPLE 8.15 | Nozzle Flowmeter

Given Ethyl alcohol flows through a pipe of diameter $D = 60$ mm in a refinery. The pressure drop across the nozzle meter used to measure the flowrate is to be $\Delta p = 4.0$ kPa when the flowrate is $Q = 0.003$ m³/s.

Find Determine the diameter, d, of the nozzle.

Solution From Table 1.5 the properties of ethyl alcohol are $\rho = 789$ kg/m³ and $\mu = 1.19 \times 10^{-3}$ N · s/m². Thus,

$$\text{Re} = \frac{\rho V D}{\mu} = \frac{4 \rho Q}{\pi D \mu}$$

$$= \frac{4(789 \text{ kg/m}^3)(0.003 \text{ m}^3/\text{s})}{\pi(0.06 \text{ m})(1.19 \times 10^{-3} \text{ N} \cdot \text{s/m}^2)} = 42,000$$

From Eq. 8.39 the flowrate through the nozzle is

$$Q = 0.003 \text{ m}^3/\text{s} = C_n \frac{\pi}{4} d^2 \sqrt{\frac{2(4 \times 10^3 \text{ N/m}^2)}{789 \text{ kg/m}^3(1 - \beta^4)}}$$

or

$$1.20 \times 10^{-3} = \frac{C_n d^2}{\sqrt{1 - \beta^4}} \tag{1}$$

where d is in meters. Note that $\beta = d/D = d/0.06$. Equation 1 and Fig. 8.42 represent two equations for the two unknowns d and C_n that must be solved by trial and error.

As a first approximation we assume that the flow is ideal, or $C_n = 1.0$, so that Eq. 1 becomes

$$d = \left(1.20 \times 10^{-3}\sqrt{1 - \beta^4}\right)^{1/2} \tag{2}$$

In addition, for many cases $1 - \beta^4 \approx 1$, so that an approximate value of d can be obtained from Eq. 2 as

$$d = (1.20 \times 10^{-3})^{1/2} = 0.0346 \text{ m}$$

Hence, with an initial guess of $d = 0.0346$ m or $\beta = d/D = 0.0346/0.06 = 0.577$, we obtain from Fig. 8.42 (using Re = 42,000) a value of $C_n = 0.972$. Clearly this does not agree with our initial assumption of $C_n = 1.0$. Thus, we do not have the solution to Eq. 1 and Fig. 8.42. Next we assume $\beta = 0.577$ and $C_n = 0.972$ and solve for d from Eq. 1 to obtain

$$d = \left(\frac{1.20 \times 10^{-3}}{0.972}\sqrt{1 - 0.577^4}\right)^{1/2}$$

or $d = 0.0341$ m. With the new value of $\beta = 0.0341/0.060 = 0.568$ and Re = 42,200, we obtain (from Fig. 8.42) $C_n \approx 0.972$ in agreement with the assumed value. Thus,

$$d = 34.1 \text{ mm} \tag{Ans}$$

Comment If numerous cases are to be investigated, it may be much easier to replace the discharge coefficient data of Fig. 8.42 by the equivalent equation, $C_n = \phi(\beta, \text{Re})$, and use a computer to iterate for the answer. Such equations are available in the literature (Ref. 24). This would be similar to using the Colebrook equation rather than the Moody chart for pipe friction problems.

By repeating the calculations, the nozzle diameters, d, needed for the same flowrate and pressure drop but with different fluids are shown in **Fig. E8.15**. The diameter is a function of the fluid viscosity because the nozzle coefficient, C_n, is a function of the Reynolds number (see Fig. 8.42). In addition, the diameter is a function of the density because of this Reynolds number effect and, perhaps more importantly, because the density is involved directly in the flowrate equation, Eq. 8.39. These factors all combine to produce the results shown in the figure.

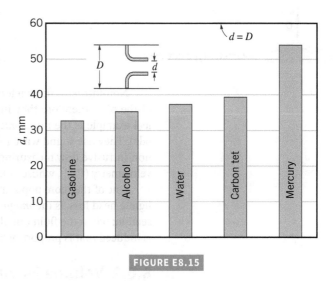

FIGURE E8.15

Numerous other devices are used to measure the flowrate in pipes. Many of these devices use principles other than the high-speed/low-pressure concept of the orifice, nozzle, and Venturi meters.

A quite common, accurate, and relatively inexpensive flowmeter is the *rotameter*, or variable area meter as is shown in **Fig. 8.44**. In this device a float is contained within a tapered, transparent metering tube that is attached vertically to the pipeline. As fluid flows through the meter (entering at the bottom), the float will rise within the tapered tube and reach an equilibrium height that is a function of the flowrate. This height corresponds to an equilibrium condition for which the net force on the float (buoyancy, float weight, fluid drag) is zero. A calibration scale in the tube provides the relationship between the float position and the flowrate.

Another useful pipe flowrate meter is a *turbine meter* as is shown in **Fig. 8.45**. A small, freely rotating propeller or turbine within the turbine meter rotates with an angular velocity that is a function of (nearly proportional to) the average fluid velocity in the pipe. This angular velocity is picked up magnetically and calibrated to provide a very accurate measure of the flowrate through the meter.

There are many types of flowmeters.

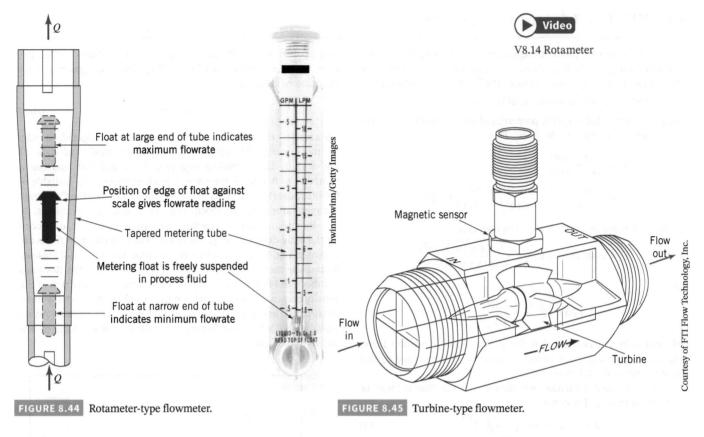

Float at large end of tube indicates maximum flowrate

Position of edge of float against scale gives flowrate reading

Tapered metering tube

Metering float is freely suspended in process fluid

Float at narrow end of tube indicates minimum flowrate

hwinnhwinn/Getty Images

Magnetic sensor

Flow out

Flow in

FLOW

Turbine

Courtesy of FTI Flow Technology, Inc.

FIGURE 8.44　Rotameter-type flowmeter.

FIGURE 8.45　Turbine-type flowmeter.

Non-intrusive flowmeters have become very common. These flowmeters operate outside of the pipe, therefore they induce no head loss and are relatively simple to install. One typical example is the *ultrasonic flowmeter* (available with either transit time or Doppler methods). They use sound wave reflection to measure the flow moving past them. Other common non-intrusive flow measuring devices are laser doppler velocimetry (LDV) and particle image velocimetry (PIV), whose details are beyond the scope of the current book.

One of the more popular flowmeters, which is not completely non-intrusive but has negligible head loss, is the *magnetic flowmeter*. These meters apply a magnetic field across a pipe section; when the fluid (an electrical conductor) passes through the magnetic field, a voltage is induced that is proportional to the flowrate.

8.6.2 Volume Flowmeters

Volume flowmeters measure volume rather than volume flowrate.

In many instances it is necessary to know the amount (volume or mass) of fluid that has passed through a pipe during a given time period, rather than the instantaneous flowrate. For example, we are interested in how many gallons of gasoline are pumped into the tank in our car rather than the rate at which it flows into the tank. Numerous quantity-measuring devices provide such information.

The *nutating disk meter* shown in **Fig. 8.46** is widely used to measure the net amount of water used in domestic and commercial water systems as well as the amount of gasoline delivered to your gas tank. This meter contains only one essential moving part and is relatively inexpensive and accurate. Its operating principle is very simple, but it may be difficult to understand its operation without actually inspecting the device firsthand. The device consists of a metering chamber with spherical sides and conical top and bottom. A disk passes through a central sphere and divides the chamber into two portions. The disk is constrained to be at an angle not normal to the axis of symmetry of the chamber. A radial plate (diaphragm) divides the chamber so that the entering fluid causes the disk to wobble (nutate), with fluid flowing alternately above or below the disk. The fluid exits the chamber after the disk has completed one wobble, which corresponds to a specific volume of fluid passing through the chamber. During each wobble of the disk, the pin attached to the tip of the center sphere, normal to the disk, completes one circle. The volume of fluid that has passed through the meter can be obtained by counting the number of revolutions completed.

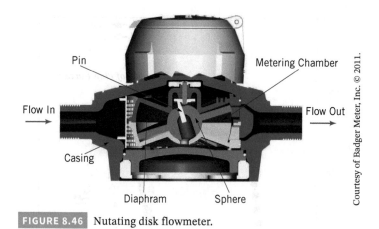

FIGURE 8.46 Nutating disk flowmeter.

Courtesy of Badger Meter, Inc. © 2011.

Another quantity-measuring device that is used for gas flow measurements is the *bellows meter*. It contains a set of bellows that alternately fill and empty as a result of the pressure of the gas and the motion of a set of inlet and outlet valves somewhat like the chambers and valves of a heart. For each cycle a known volume of gas passes through the meter. The common household natural gas meter (see adjacent photograph) is of this type.

The nutating disk meter (water meter) is an example of extreme simplicity—one cleverly designed moving part. The bellows meter (gas meter), on the other hand, is relatively complex—it contains many moving, interconnected parts. This difference is dictated by the application involved. One measures a common, safe-to-handle, relatively high-pressure liquid, whereas the other measures a relatively dangerous, low-pressure gas. Each device does its intended job very well.

There are numerous devices used to measure fluid flow, only a few of which have been discussed here. The reader is encouraged to review trustworthy Internet resources and the literature to gain familiarity with other useful, clever devices (Refs. 25, 26).

PiLens/Fotosearch LBRF/AGE Fotostock

The nutating disk meter has only one moving part; the bellows meter has a complex set of moving parts.

8.6.3 Multiphase Flow Measurement in Pipes

Multiphase slurry and pneumatic conveying systems are two other major applications of pipeline flows which transport wet and dry particles, respectively. In such systems measurement of discreet flow is done by two methods namely high-speed camera and pressure transducers.

In high-speed camera measurement, the particle flow is recorded using a high speed camera for known frame rates per second and these videos are then analyzed for flow velocity of particles using software like Tracker, etc. Similarly using methods of cross correlation for given set of pressure transducers fixed in a pipe network, the velocity of particles flows can be calculated by feeding analogue signal of the pressure transducers in DAQ softwares like LabVIEW as shown in Fig. 8.47. (Refs. 38, 39). In Fig. 8.47 the time lag of arrival of particles at positions 2 and 3, which are the known locations of pressure transducers, are recorded and based on the distance between the two pressure transducers (distance/time lag), the velocity of flow of particle (in plug) can be calculated.

8.6.4 Water Hammer and Their Measurements in Pipes

Water hammer phenomenon occurs in pipes (in particular penstock pipes in hydraulic power plants) when sudden closure of valves is done. This causes a sudden reduction in flow velocity through pipes resulting in sudden pressure rise known as water hammer. The water hammers are sustained by the pipes by using surge tanks which work as reservoir. The pressure determination in pipes under the action of water hammer is thus essential to keep a check on pressure rise to avoid its bursting.

The Joukowski equation is the equation which can be used to calculate this pressure rise in terms of the velocity change of fluid due to sudden valve closure and can be given as below (Refs. 40, 41, 42)

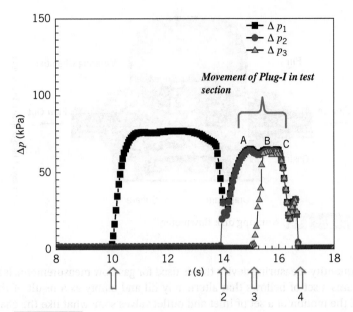

Pressure time signal graph for velocity measurement of particle through a circular pipe in solid-air multiphase system (Ref. 39).

$$\Delta p = \rho c \, \Delta V \qquad (8.41)$$

where Δp is the maximum pipe pressure rise in water hammer, ΔV is change in velocity due to water hammer, ρ is fluid density, and c is pressure wave travel speed.

Further the value of c can be calculated as (Refs. 40, 41, 42),

$$c = \sqrt{\frac{E}{\rho}}$$

$$\frac{1}{E} = \frac{1}{E_f} + \frac{D}{wE_p}$$

where E is composite elastic modulus, E_f is elastic modulus of fluid, and E_p is the elastic modulus of pipe material. The pressure wave travel time can be given as (Refs. 40, 41, 42)

$$tw = \frac{2L}{c}$$

where L is pipe length.

EXAMPLE 8.16 | Water Hammer

A pipe of 50 mm diameter carries water at 2 m/s. Find the rise in pressure due to sudden closure of the valve provided at the end of the pipe. The pipe can be treated as rigid and $E_{water} = 2.08 \times 10^9$ N/m².

Solution

Assuming that the sudden closure of the valve makes the water stationary from 2 m/s, we have,

$\Delta V = 2$ m/s and $\rho = 998$ kg/m³.

Thus,

$$C = \sqrt{\frac{E}{\rho}} = \sqrt{\frac{2.09 \times 10^9}{998}} = 1443.66 \text{ m/s}$$

Then, the rise in pressure due to water hammer can be calculated as

$$\Delta p = \rho c \Delta V = 998 \text{ kg/m}^3 \times 1443.66 \text{ m/s} \times 2 \text{ m/s} = 2881545 \text{ N/m}^2$$

Comment The value of Δp as obtained in Eq. (2) represents the intensity of water hammer created for a low range water flow velocity of 2 m/s, but the flow velocities through pipes can be as high as 15 m/s (or even higher) and the hammer effect shall then intensify accordingly. To have a better idea of range of water hammer pressure rise with respect to velocity of water flow through the above pipe of 50 mm diameter, a graph for Δp versus ΔV has been drawn in **Fig. E8.16**. The graph clearly indicates the linear relationship and for instance, establishes that 10 times rise in velocity of flow will result into a 10 times higher water hammer. Thus, the closing of valves should be avoided at high velocities of flow.

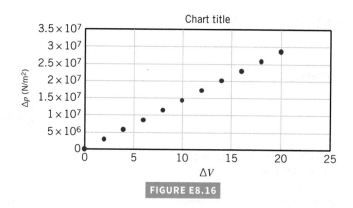

FIGURE E8.16

Surge Tanks Usually, the sudden closure of the valves situated at the end of the pipes sets water hammer upstream of the pipe valve for whole pipe length. Therefore, the whole pipe length upstream to the valve requires to be designed to sustain the high-pressure waves which increase the overall cost of the pipe network. To save this cost, a tank, called the surge tank is provided upstream to smaller pipe length as compared to full pipe length. The tank (2) as shown in Fig. 8.48 ensures that when the valve at 1 gets closed suddenly, the water is accumulated in the tank and thus the tank takes up the water hammer . This keeps the water hammer effect limited to the pipe length between 1 and 2 rather than affecting the whole pipe length.

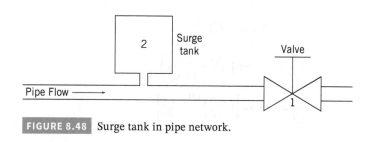

FIGURE 8.48 Surge tank in pipe network.

CHAPTER SUMMARY

This chapter discussed the flow of a viscous fluid in a pipe. General characteristics of laminar, turbulent, fully developed, and entrance flows are considered. Poiseuille's equation is obtained to describe the relationship among the various parameters for fully developed laminar flow.

Various characteristics of turbulent pipe flow are introduced and contrasted to laminar flow. It is shown that the head loss for laminar or turbulent pipe flow can be written in terms of the friction factor (for major losses) and the loss coefficients (for minor losses). In general, the friction factor is obtained from the Moody chart or the Colebrook formula and is a function of the Reynolds number and the relative roughness. The minor loss coefficients are a function of the flow geometry for each system component.

Analysis of noncircular conduits is carried out by use of the hydraulic diameter concept. Various examples involving flow in single pipe systems and flow in multiple pipe systems are presented. The inclusion of viscous effects and losses in the analysis of orifice, nozzle, and Venturi flowmeters is discussed.

The following checklist provides a study guide for this chapter. When your study of the entire chapter has been completed you should be able to

- write out meanings of the terms listed and understand each of the related concepts. These terms are particularly important and are set in **bold black** type in the text.

- determine which of the following types of flow will occur: entrance flow or fully developed flow; laminar flow or turbulent flow.

- calculate flow properties with the Poiseuille equation in appropriate situations and adhere to its limitations.

- explain the main properties of turbulent pipe flow and how they are different from or similar to laminar pipe flow.

- determine major losses in pipe systems with the Moody chart and the Colebrook equation.

- determine minor losses in pipe systems with minor loss coefficients.

- determine the head loss in noncircular conduits.

- incorporate major and minor losses into the energy equation to solve a variety of pipe flow problems, including Type I problems (determine the pressure drop or head loss), Type II problems (determine the flowrate), and Type III problems (determine the pipe diameter).

- solve problems involving multiple pipe systems.

- determine the flowrate through orifice, nozzle, and Venturi flowmeters as a function of the pressure drop across the meter.

KEY EQUATIONS

Entrance length	$\dfrac{\ell_e}{D} = 0.06\,\mathrm{Re}$ for laminar flow	(8.1)
	$\dfrac{\ell_e}{D} = 4.4\,(\mathrm{Re})^{1/6}$ for turbulent flow	(8.2)
Pressure drop for fully developed laminar pipe flow	$\Delta p = \dfrac{4\ell\tau_w}{D}$	(8.5)
Velocity profile for fully developed laminar pipe flow	$u(r) = \left(\dfrac{\Delta p D^2}{16\mu\ell}\right)\left[1 - \left(\dfrac{2r}{D}\right)^2\right] = V_c\left[1 - \left(\dfrac{2r}{D}\right)^2\right]$	(8.7)
Volume flowrate for fully developed laminar horizontal pipe flow	$Q = \dfrac{\pi D^4 \Delta p}{128\mu\ell}$	(8.9)
Friction factor for fully developed laminar pipe flow	$f = \dfrac{64}{\mathrm{Re}}$	(8.19)
Energy equation for incompressible, steady flow	$\dfrac{p_1}{\gamma} + \alpha_1\dfrac{V_1^2}{2g} + z_1 = \dfrac{p_2}{\gamma} + \alpha_2\dfrac{V_2^2}{2g} + z_2 + h_L$	(8.21)
Pressure drop for a horizontal pipe	$\Delta p = f\dfrac{\ell}{D}\dfrac{\rho V^2}{2}$	(8.33)
Head loss due to major losses	$h_{L\,major} = f\dfrac{\ell}{D}\dfrac{V^2}{2g}$	(8.34)
Colebrook formula	$\dfrac{1}{\sqrt{f}} = -2.0 \log\left(\dfrac{\varepsilon/D}{3.7} + \dfrac{2.51}{\mathrm{Re}\sqrt{f}}\right)$	(8.35a)
Haaland formula	$\dfrac{1}{\sqrt{f}} = -1.8 \log\left[\left(\dfrac{\varepsilon/D}{3.7}\right)^{1.11} + \dfrac{6.9}{\mathrm{Re}}\right]$	(8.35b)
Head loss due to minor losses	$h_{L\,minor} = K_L\dfrac{V^2}{2g}$	(8.36)
Volume flowrate for orifice, nozzle, or Venturi meter	$Q = C_i A_i\sqrt{\dfrac{2(p_1 - p_2)}{\rho(1 - \beta^4)}}$	(8.38, 8.39, 8.40)
Change in pressure due to water hammer in pipes	$\Delta p = \rho C \Delta V$	(8.41)

REFERENCES

1. Bernard, P. S., and Wallace, J. M., *Turbulent Flow—Analysis, Measurement, and Prediction*, Wiley, New York, 2002.
2. Panton, R. L., *Incompressible Flow*, 4th Ed., Wiley, New York, 2013.
3. Schlichting, H., and Gersten, K., *Boundary Layer Theory*, 9th Ed., Springer-Verlag, Berlin, 2017.
4. Gleick, J., *Chaos: Making a New Science*, Viking Penguin, New York, 1987.
5. White, F. M., *Fluid Mechanics*, 8th Ed., McGraw-Hill, New York, 2016.
6. Nikuradse, J., "Stomungsgesetz in Rauhen Rohren," *VDI-Forschungsch*, No. 361, 1933; or see NACA Tech Memo 1922.
7. Moody, L. F., "Friction Factors for Pipe Flow," *Transactions of the ASME*, Vol. 66, 1944.
8. Colebrook, C. F., "Turbulent Flow in Pipes with Particular Reference to the Transition Between the Smooth and Rough Pipe Laws," *Journal of the Institute of Civil Engineers London*, Vol. 11, 1939.
9. *ASHRAE Handbook of Fundamentals*, ASHRAE, Atlanta, 1981.
10. Streeter, V. L., ed., *Handbook of Fluid Dynamics*, McGraw-Hill, New York, 1961.
11. Sovran, G., and Klomp, E. D., "Experimentally Determined Optimum Geometries for Rectilinear Diffusers with Rectangular, Conical, or Annular Cross Sections," in *Fluid Mechanics of Internal Flow*, Sovran, G., ed., Elsevier, Amsterdam, 1967.
12. Runstadler, P. W., "Diffuser Data Book," Technical Note 186, Creare, Inc., Hanover, NH, 1975.
13. Laws, E. M., and Livesey, J. L., "Flow Through Screens," *Annual Review of Fluid Mechanics*, Vol. 10, Annual Reviews, Inc., Palo Alto, CA, 1978.

14 Balje, O. E., *Turbomachines: A Guide to Design, Selection and Theory*, Wiley, New York, 1981.

15 Wallis, R. A., *Axial Flow Fans and Ducts*, Wiley, New York, 1983.

16 Karassick, I. J. et al., *Pump Handbook*, 2nd Ed., McGraw-Hill, New York, 1985.

17 White, F. M., *Viscous Fluid Flow*, 3rd Ed., McGraw-Hill, New York, 2006.

18 Olson, R. M., *Essentials of Engineering Fluid Mechanics*, 4th Ed., Harper & Row, New York, 1980.

19 Dixon, S. L., *Fluid Mechanics of Turbomachinery*, 3rd Ed., Pergamon, Oxford, UK, 1978.

20 Finnemore, E. J., and Franzini, J. R., *Fluid Mechanics*, 10th Ed., McGraw-Hill, New York, 2002.

21 Streeter, V. L., and Wylie, E. B., *Fluid Mechanics*, 8th Ed., McGraw-Hill, New York, 1985.

22 Jeppson, R. W., *Analysis of Flow in Pipe Networks*, Ann Arbor Science Publishers, Ann Arbor, MI, 1976.

23 Bean, H. S., ed., *Fluid Meters: Their Theory and Application*, 6th Ed., American Society of Mechanical Engineers, New York, 1971.

24 "Measurement of Fluid Flow by Means of Orifice Plates, Nozzles, and Venturi Tubes Inserted in Circular Cross Section Conduits Running Full," Int. Organ. Stand. Rep. DIS-5167, Geneva, 1976.

25 Goldstein, R. J., ed., *Flow Mechanics Measurements*, 2nd Ed., Taylor and Francis, Philadelphia, 1996.

26 Benedict, R. P., *Measurement of Temperature, Pressure, and Flow*, 2nd Ed., Wiley, New York, 1977.

27 Hydraulic Institute, *Engineering Data Book*, 1st Ed., Cleveland Hydraulic Institute, 1979.

28 Harris, C. W., *University of Washington Engineering Experimental Station Bulletin*, 48, 1928.

29 Hamilton, J. B., *University of Washington Engineering Experimental Station Bulletin*, 51, 1929.

30 Miller, D. S., *Internal Flow Systems*, 2nd Ed., BHRA, Cranfield, UK, 1990.

31 Spitzer, D. W., ed., *Flow Measurement: Practical Guides for Measurement and Control*, Instrument Society of America, Research Triangle Park, NC, 1991.

32 Wilcox, D. C., *Turbulence Modeling for CFD*, DCW Industries, Inc., La Canada, California, 1994.

33 Mullin, T., ed., *The Nature of Chaos*, Oxford University Press, Oxford, UK, 1993.

34 Haaland, S. E., "Simple and Explicit Formulas for the Friction-Factor in Turbulent Pipe Flow," *Transactions of the ASME, Journal of Fluids Engineering*, Vol. 105, 1983.

35 Cross, H., "Analysis of Flow in Networks of Conduits or Conductors," *University of Illinois Bulletin No. 286*, November 1936.

36 *Performance Test Code 19.5—Flow Measurement*, American Society of Mechanical Engineering, New York, 2013.

37 Jorgensen, R., ed., *Fan Engineering*, Buffalo Forge Company, Buffalo, NY, 1983.

38 Rawat, A., and Haim K., "Detachment Velocity: A Borderline Between Different Types of Particulate Plugs," *Powder Technology*, Vol. 321, 2017: 293–300.

39 Rawat, A., and Haim K., "Modification and Validation of Particulate Plug-I Pressure Drop Model," *Powder Technology*, Vol. 347, 2019: 243–254.

40 Chaudhry, M. H., *Applied Hydraulic Transient*, 2nd Ed., Van Nostrand Reinhold, New York, 1987.

41 Houghtalen, R. J., Osman Akan, A. and Hwang, Ned H .C., *Fundamentals of Hydraulic Engineering System*, 4th Ed, Prentice Hall, Inc., Boston , 2010.

42 Wylie, E. B.and Victor L. Streeter. *Fluid Transients in Systems*, Prentice-Hall, Inc., Englewood Cliffs, NJ, 1993.

QUESTIONS AND PROBLEMS

Ⓒ Problem to be solved with aid of programmable calculator or computer.

Ⓞ Open-ended problem that requires critical thinking. These problems require various assumptions to provide the necessary input data. There are not unique answers to these problems.

Note: Unless specific values of required fluid properties are specified in the problem statement, use the values found in Tables 1.4–1.8 and in the tables in the Appendices.

Section 8.1 General Characteristics of Pipe Flow

8.1.1 Ⓞ Under normal circumstances is the airflow though your trachea (your windpipe) laminar or turbulent? List all assumptions and show all calculations.

8.1.2 (Video) Rainwater runoff from a parking lot flows through a 0.9 m-diameter pipe, completely filling it. Whether flow in a pipe is laminar or turbulent depends on the value of the Reynolds number (see Video V8.2). Would you expect the flow to be laminar or turbulent? Support your answer with appropriate calculations.

8.1.3 Blue and yellow streams of paint at 15 °C (each with a density of 824.6 kg/m³ and a viscosity 1000 times greater than water) enter a pipe with an average velocity of 1.2 m/s as shown in **Fig. P8.1.3**. Would you expect the paint to exit the pipe as green paint or separate streams of blue and yellow paint? Explain. Repeat the problem if the paint were "thinned" so that it is only 10 times more viscous than water. Assume the density remains the same.

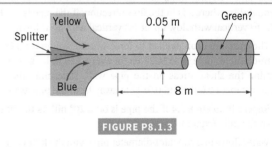

FIGURE P8.1.3

8.1.4 Air at 93 °C flows at standard atmospheric pressure in a pipe at a rate of 0.355 N/s. Determine the minimum diameter allowed if the flow is to be laminar.

8.1.5 To cool a given room it is necessary to supply 0.113 m³/s of air through an 0.2 m-diameter pipe. Approximately how long is the entrance length in this pipe?

8.1.6 Ⓒ The flow of water in a 3-mm-diameter pipe is to remain laminar. Plot a graph of the maximum flowrate allowed as a function of temperature for 0 < T < 100 °C.

8.1.7 The pressure distribution measured along a straight, horizontal portion of a 50-mm-diameter pipe attached to a tank is shown in the table below. Approximately how long is the entrance length? In the fully developed portion of the flow, what is the value of the wall shear stress?

x (m) (±0.01 m)	p (mm H$_2$O) (±5 mm)
0 (tank exit)	520
0.5	427
1.0	351
1.5	288
2.0	236
2.5	188
3.0	145
3.5	109
4.0	73
4.5	36
5.0 (pipe exit)	0

8.1.8 (See The Wide World of Fluids article titled "Nanoscale Flows," Section 8.1.1.) (a) Water flows in a tube that has a diameter of $D = 0.1$ m. Determine the Reynolds number if the average velocity is 10 diameters per second. (b) Repeat the calculations if the tube is a nanoscale tube with a diameter of $D = 100$ nm.

Section 8.2 Fully Developed Laminar Flow

8.2.1 For fully developed laminar pipe flow in a circular pipe, the velocity profile is given by $u(r) = 2(1 - r^2/R^2)$ in m/s, where R is the inner radius of the pipe. Assuming that the pipe diameter is 4 cm, find the maximum and average velocities in the pipe as well as the volume flowrate.

8.2.2 A viscous fluid flows in a 0.10-m-diameter pipe such that its velocity measured 0.012 m away from the pipe wall is 0.8 m/s. If the flow is laminar, determine the centerline velocity and the flowrate.

8.2.3 The wall shear stress in a fully developed flow portion of a 0.3 m-diameter pipe carrying water is 88.58 N/m^2. Determine the pressure gradient, $\partial p/\partial x$, where x is in the flow direction, if the pipe is (a) horizontal, (b) vertical with flow up, or (c) vertical with flow down.

8.2.4 The pressure drop needed to force water through a horizontal 0.025 m-diameter pipe is 4136.85 Pa for every 3.66 m length of pipe. Determine the shear stress on the pipe wall. Determine the shear stress at distances 0.0076 and 0.0127 m away from the pipe wall.

8.2.5 Repeat Problem 8.2.4 if the pipe is on a 20° hill. Is the flow up or down the hill? Explain.

8.2.6 Water flows in a constant-diameter pipe with the following conditions measured: At section (a) $p_a = 223.4$ kPa and $z_a = 17.3$ m; at section (b) $p_b = 204.78$ kPa and $z_b = 20.78$ m. Is the flow from (a) to (b) or from (b) to (a)? Explain.

8.2.7 For laminar flow in a round pipe of diameter D, at what distance from the centerline is the actual velocity equal to the average velocity?

8.2.8 Glycerin at 20 °C flows upward in a vertical 75-mm-diameter pipe with a centerline velocity of 1.0 m/s. Determine the head loss and pressure drop in a 10-m length of the pipe.

8.2.9 Water at 15.5 °C flows at a rate of 0.015 m^3/min through a 0.15 m I.D. plastic pipe. The pipe is 152.4 m long and rises a vertical height of 12.2 m over the 152.4 m. Find the pressure drop.

8.2.10 At time $t = 0$ the level of water in tank A shown in **Fig. P8.2.10** is 0.6 m above that in tank B. Plot the elevation of the water in tank A as a function of time until the free surfaces in both tanks are at the same elevation. Assume quasisteady conditions—that is, the steady pipe flow equations are assumed valid at any time, even though the flowrate does change (slowly) in time. Neglect minor losses. *Note:* Verify and use the fact that the flow is laminar.

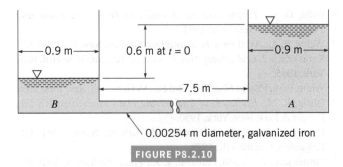

FIGURE P8.2.10

8.2.11 A fluid flows through a horizontal 0.0025 m-diameter pipe. When the Reynolds number is 1500, the head loss over a 6 m length of the pipe is 1.92 m. Determine the fluid velocity.

8.2.12 Asphalt at 50 °C, considered to be a Newtonian fluid with a viscosity 80,000 times that of water and a specific gravity of 1.09, flows through a pipe of diameter 0.05 m. If the pressure gradient is 361.93 kPa/m, determine the flowrate assuming the pipe is (a) horizontal; (b) vertical with flow up.

8.2.13 Oil of $SG = 0.87$ and a kinematic viscosity $\upsilon = 2.2 \times 10^{-4}$ m^2/s flows through the vertical pipe shown in **Fig. P8.2.13** at a rate of 4×10^{-4} m^3/s. Determine the manometer reading, h.

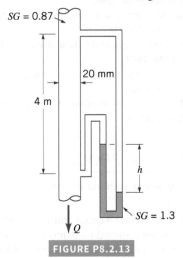

FIGURE P8.2.13

8.2.14 A liquid with $SG = 0.96$, $\mu = 9.2 \times 10^{-4}$ N · s/m^2, and vapor pressure $p_v = 1.2 \times 10^4$ N/m^2 (abs) is drawn into the syringe as is indicated in **Fig P8.2.14**. What is the maximum flowrate if cavitation is not to occur in the syringe?

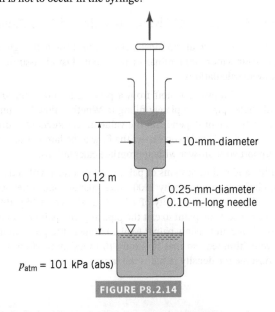

FIGURE P8.2.14

8.2.15 A 76 mm schedule 40 commercial steel pipe (with an actual inside diameter of 77.9 mm) carries 100 °C SAE 40 crankcase oil at the rate of 22.7 liter/min. The oil specific gravity is 0.89, and the absolute viscosity is 3.16×10^{-5} N · s/m². For the same pressure drop, what must the new pipe size be to carry the oil at 40.5 liter/min?

8.2.16 A 0.076 m schedule 40 commercial steel pipe (with an actual inside diameter of 0.0779 m) carries 100 °C SAE 40 crankcase oil at the rate of 3.78×10^{-4} m³/s. The oil specific gravity is 0.89, and the absolute viscosity is 3.16×10^{-5} N · s/m². Calculate the pipe size required to carry the same flowrate at approximately one-half the pressure drop of the 0.076 m pipe. Both pipes are horizontal.

8.2.17 Water at 20 °C flows down a vertical pipe with no pressure drop. Find the range of pipe diameters D (if any) for which the flow is definitely laminar.

8.2.18 A person is donating blood. The pint (~525 ml) bag in which the blood is collected is initially flat and is at atmospheric pressure. Neglect the initial mass of air in the 3.175 mm I.D., 1.22 m-long plastic tube carrying blood to the bag. The average blood pressure in the vein is 40 mm Hg above atmospheric pressure. Estimate the time required for the person to donate one pint of blood. Assume that blood has a specific gravity of 1.06 and a viscosity of 4.78×10^{-3} N · s/m². The needle's I.D. is 1.587 mm and the needle length is 0.05 m. The bag is 0.3 m below the needle inlet and the vein's I.D. is 3.175 mm. *Optional:* Donate a pint of blood and check your answer.

Section 8.3 Fully Developed Turbulent Flow

8.3.1 For oil ($SG = 0.86$, $\mu = 0.025$ Ns/m²) flow of 0.2 m³/s through a round pipe with diameter of 500 mm, determine the Reynolds number. Is the flow laminar or turbulent?

8.3.2 (Video) As shown in Video V8.9 and **Fig. P8.3.2**, the velocity profile for laminar flow in a pipe is quite different from that for turbulent flow. With laminar flow the velocity profile is parabolic; with turbulent flow at Re = 10,000 the velocity profile can be approximated by the power-law profile shown in the figure. (a) For laminar flow, determine at what radial location you would place a Pitot tube if it is to measure the average velocity in the pipe. (b) Repeat part (a) for turbulent flow with Re = 10,000.

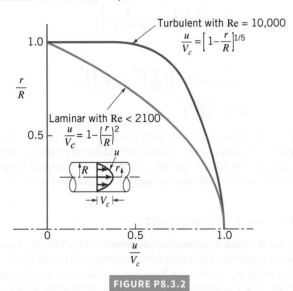

FIGURE P8.3.2

8.3.3 Water at 10 °C flows through a smooth 60-mm-diameter pipe with an average velocity of 8 m/s. Would a layer of rust of height 0.005 mm on the pipe wall protrude through the viscous sublayer? Justify your answer with appropriate calculations.

8.3.4 (Video) When soup is stirred in a bowl, there is considerable turbulence in the resulting motion (see Video V8.6). From a very simplistic standpoint, this turbulence consists of numerous intertwined swirls, each involving a characteristic diameter and velocity. As time goes by, the smaller swirls (the fine scale structure) die out relatively quickly, leaving the large swirls that continue for quite some time. Explain why this is to be expected.

8.3.5 Water at 15 °C flows through a 0.15 m-diameter pipe with an average velocity of 4.5 m/s. Approximately what is the height of the largest roughness element allowed if this pipe is to be classified as smooth?

Section 8.4.1 Major Losses

8.4.1 Water is pumped between two tanks as shown in **Fig. P8.4.1**. The energy line is as indicated. Is the fluid being pumped from A to B or B to A? Explain. Which pipe has the larger diameter: A to the pump or B to the pump? Explain.

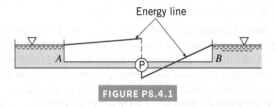

Energy line

FIGURE P8.4.1

8.4.2 A person with no experience in fluid mechanics wants to estimate the friction factor for 2.54 cm-diameter galvanized iron pipe at a Reynolds number of 8000. The person stumbles across the simple equation of $f = 64/Re$ and uses this to calculate the friction factor. Explain the problem with this approach and estimate the error.

8.4.3 During a heavy rainstorm, water from a parking lot completely fills an 0.45 m-diameter, smooth, concrete storm sewer. If the flowrate is 0.283 m³/s, determine the pressure drop in a 30 m horizontal section of the pipe. Repeat the problem if there is a 0.6 m change in elevation of the pipe per 30 m of its length.

8.4.4 Water flows through a horizontal plastic pipe with a diameter of 0.2 m at a velocity of 10 cm/s. Determine the pressure drop per meter of pipe and the power lost to the friction per meter of pipe.

8.4.5 Water flows downward through a vertical 10-mm-diameter galvanized iron pipe with an average velocity of 5.0 m/s and exits as a free jet. There is a small hole in the pipe 4 m above the outlet. Will water leak out of the pipe through this hole, or will air enter into the pipe through the hole? Repeat the problem if the average velocity is 0.5 m/s.

8.4.6 Air at standard conditions flows through an 0.2 m-diameter, 4.45 m-long, straight duct with the velocity versus head loss data indicated in the following table. Determine the average friction factor over this range of data.

V (m/min)	h (mm water)
1204	8.9
1137	8.13
1100	7.62
1045	6.86
1000	6.1
914	5.08
823	4.06

8.4.7 Water flows through a horizontal 60-mm-diameter galvanized iron pipe at a rate of 0.02 m³/s. If the pressure drop is 135 kPa per 10 m of pipe, do you think this pipe is (a) a new pipe, (b) an old pipe with a somewhat increased roughness due to aging, or (c) a very old pipe that is partially clogged by deposits? Justify your answer.

8.4.8 Water flows at a rate of 38 liters per minute in a new horizontal 0.02 m-diameter galvanized iron pipe. Determine the pressure gradient, $\Delta p/\ell$, along the pipe.

8.4.9 Carbon dioxide at a temperature of 0 °C and a pressure of 600 kPa (abs) flows through a horizontal 40-mm-diameter pipe with an average velocity of 2 m/s. Determine the friction factor if the pressure drop is 235 N/m² per 10-m length of pipe.

8.4.10 Blood (assume $\mu = 2.15 \times 10^{-3}$ N · s/m², $SG = 1.0$) flows through an artery in the neck of a giraffe from its heart to its head at a rate of 7.08×10^{-6} m³/s. Assume the length is 3 m and the diameter is 0.005 m. If the pressure at the beginning of the artery (outlet of the heart) is equivalent to 0.21 m Hg, determine the pressure at the end of the artery when the head is (a) 2.4 m above the heart or (b) 1.8 m below the heart. Assume steady flow. How much of this pressure difference is due to elevation effects, and how much is due to frictional effects?

8.4.11 A 40-m-long, 12-mm-diameter pipe with a friction factor of 0.020 is used to siphon 30 °C water from a tank as shown in **Fig. P8.4.11**. Determine the maximum value of h allowed if there is to be no cavitation within the hose. Neglect minor losses.

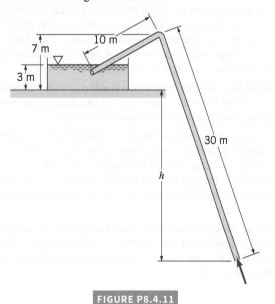

FIGURE P8.4.11

8.4.12 Gasoline flows in a smooth pipe of 40-mm diameter at a rate of 0.001 m³/s. If it were possible to prevent turbulence from occurring, what would be the ratio of the head loss for the actual turbulent flow compared to that if it were laminar flow?

8.4.13 A 0.9 m-diameter duct is used to carry ventilating air into a vehicular tunnel at a rate of 255 m³/min. Tests show that the pressure drop is 0.038 m of water per 457 m of duct. What is the value of the friction factor for this duct and the approximate size of the equivalent roughness of the surface of the duct?

8.4.14 H. Blasius correlated data on turbulent friction factor in smooth pipes. His equation $f_{smooth} \approx 0.3164$ Re$^{-1/4}$ is reasonably accurate for Reynolds numbers between 4000 and 10^5. Use this information for the following scenario. Water at 20 °C is to flow through a 3-cm I.D. plastic pipe at the rate of 0.001 m³/s. Find the incline angle of the pipe needed to make the static pressure constant along the pipe.

8.4.15 Von Karman suggested that the wholly turbulent friction factor be expressed by the equation

$$f = \frac{1}{4[0.57 - \log(\varepsilon/D)]^2}$$

where ε is the absolute roughness of the pipe. Compare the values predicted by this equation and those indicated on the Moody chart.

8.4.16 The Swamee and Jain formula for the friction factor is

$$f = \frac{0.25}{[\log(\varepsilon/3.7D + 5.74/\text{Re}^{0.9})]^2}$$

Compare this equation for $\varepsilon/D = 0.00001$, 0.0001, 0.001, and 0.01 and Reynolds numbers of 10^4, 10^5, 10^6, and 10^7 with the Moody chart and decide whether it is an acceptable replacement for the Colebrook formula.

8.4.17 The Haaland formula for the friction factor is

$$f = \frac{0.3086}{\{\log[6.9/\text{Re} + (\varepsilon/3.7D)^{1.11}]\}^2}$$

Compare this equation for f for $\varepsilon/D = 0.00001$, 0.0001, 0.001, and 0.01 and Reynolds numbers of 10^4, 10^5, 10^6, and 10^7 with the Moody chart and decide whether it is an acceptable replacement for the Colebrook formula.

8.4.18 The Churchill formula for the friction factor is

$$f = 8\left[\left(\frac{8}{\text{Re}}\right)^{12} + \frac{1}{(A+B)^{1.5}}\right]^{1/12}$$

where

$$A = \left\{-2.457 \ln\left[\left(\frac{7}{\text{Re}}\right)^{0.9} + \frac{\varepsilon}{3.7D}\right]\right\}^{16};$$

$$B = \left(\frac{37,530}{\text{Re}}\right)^{16}$$

Compare this equation for f for both the laminar and turbulent regimes for $\varepsilon/D = 0.00001$, 0.0001, 0.001, and 0.01 and Reynolds numbers of 10, 10^2, 10^3, 10^4, 10^5, 10^6, and 10^7 with the Moody chart and decide whether it is an acceptable replacement for the Colebrook formula.

Section 8.4.2 Minor Losses

8.4.19 Air at standard temperature and pressure flows through a 0.025 m-diameter galvanized iron pipe with an average velocity of 2.4 m/s. What length of pipe produces a head loss equivalent to (a) a flanged 90° elbow, (b) a wide-open angle valve, or (c) a sharp-edged entrance?

8.4.20 Given 90° threaded elbows used in conjunction with copper pipe (drawn tubing) of 0.02 m diameter, convert the loss for a single elbow to equivalent length of copper pipe for wholly turbulent flow.

8.4.21 To conserve water and energy, a "flow reducer" is installed in the shower head as shown in **Fig. P8.4.21**. If the pressure at point (1) remains constant and all losses except for that in the flow reducer are neglected, determine the value of the loss coefficient (based on the velocity in the pipe) of the flow reducer if its presence is to reduce the flowrate by a factor of 2. Neglect gravity.

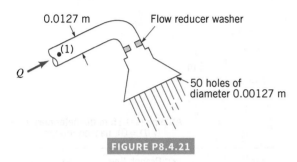

FIGURE P8.4.21

8.4.22 Water flows at a rate of 0.040 m³/s in a 0.12-m-diameter pipe that contains a sudden contraction to a 0.06-m-diameter pipe. Determine the pressure drop across the contraction section. How much of this pressure difference is due to losses and how much is due to kinetic energy changes?

8.4.23 Water flows from the container shown in **Fig. P8.4.23**. Determine the loss coefficient needed in the valve if the water is to "bubble up" 0.076 m above the outlet pipe. The entrance is slightly rounded.

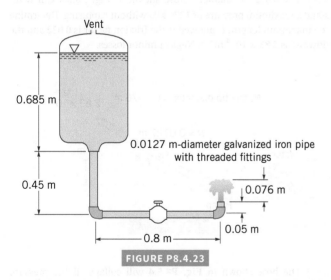

FIGURE P8.4.23

8.4.24 (See The Wide World of Fluids article titled "New Hi-tech Fountains," Section 8.5.) The fountain shown in **Fig. P8.4.24** is designed to provide a stream of water that rises $h = 3$ m to $h = 6$ m above the nozzle exit in a periodic fashion. To do this the water from the pool enters a pump, passes through a pressure regulator that maintains a constant pressure ahead of the flow control valve. The valve is electronically adjusted to provide the desired water height. With $h = 3$ m the loss coefficient for the valve is $K_L = 50$. Determine the valve loss coefficient needed for $h = 6$ m. All losses except for the flow control valve are negligible. The area of the pipe is 5 times the area of the exit nozzle.

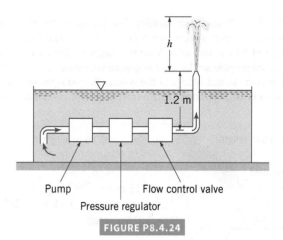

FIGURE P8.4.24

8.4.25 Water flows through the screen in the pipe shown in **Fig. P8.4.25** as indicated. Determine the loss coefficient for the screen.

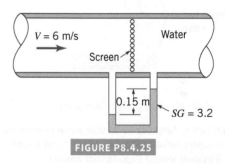

FIGURE P8.4.25

8.4.26 Air flows though the mitered bend shown in **Fig. P8.4.26** at a rate of 0.14 m³/s. To help straighten the flow after the bend, a set of 0.00635 m-diameter drinking straws is placed in the pipe as shown. Estimate the extra pressure drop between points (1) and (2) caused by these straws.

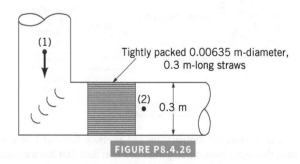

FIGURE P8.4.26

8.4.27 (Video) As shown in **Fig. P8.4.27**, water flows from one tank to another through a short pipe whose length is n times the pipe diameter. Head losses occur in the pipe and at the entrance and exit (see Video V8.10). Determine the maximum value of n if the major loss is to be no more than 10% of the minor loss and the friction factor is 0.02.

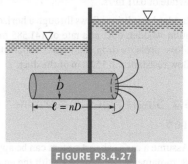

FIGURE P8.4.27

8.4.28 [Video] Water flows steadily through the 0.019 m-diameter galvanized iron pipe system shown in Video V8.15 and **Fig. P8.4.28** at a rate of 5.67×10^{-4} m³/s. Your boss suggests that friction losses in the straight pipe sections are negligible compared to losses in the threaded elbows and fittings of the system. Do you agree or disagree with your boss? Support your answer with appropriate calculations.

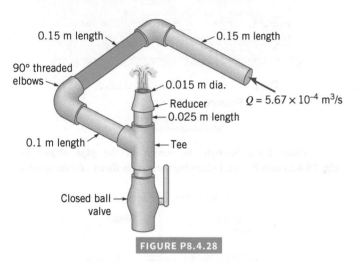

0.15 m length

0.15 m length

90° threaded elbows

0.015 m dia.

Reducer
0.025 m length

$Q = 5.67 \times 10^{-4}$ m³/s

0.1 m length

Tee

Closed ball valve

FIGURE P8.4.28

Section 8.4.3 Noncircular Conduits

8.4.29 Given two rectangular ducts with equal cross-sectional areas, but different aspect ratios (width/height) of 2 and 4, which will have the greater frictional losses? Explain your answer.

8.4.30 A viscous oil with a specific gravity $SG = 0.85$ and a viscosity of 0.10 Pa·s flows from tank A to tank B through the six rectangular slots indicated in **Fig. P8.4.30**. If the total flowrate is 30 mm³/s and minor losses are negligible, determine the pressure in tank A.

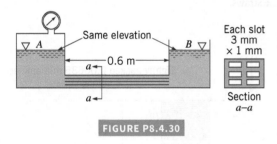

Same elevation

Each slot
3 mm
× 1 mm

A —0.6 m— B

a

a

Section
a–a

FIGURE P8.4.30

8.4.31 Air at standard temperature and pressure flows at a rate of 0.2 m³/s through a horizontal, galvanized iron duct that has a rectangular cross-sectional shape of 0.3 m by 0.15 m. Estimate the pressure drop per 61 m of duct.

8.4.32 Water at 20 °C flows through a concentric annulus of inner diameter $D_1 = 2.0$ cm and outer diameter $D_2 = 4.0$ cm. The surface roughness is 0.002 cm. Calculate the frictional pressure loss per unit length for a flowrate of 0.01 m³/s.

8.4.33 Air at standard conditions flows through a horizontal 0.3 m by 0.45 m rectangular wooden duct at a rate of 141.585 m³/min. Determine the head loss, pressure drop, and power supplied by the fan to overcome the flow resistance in 152.4 m of the duct.

Section 8.5.1 Single Pipes—Determine
Pressure Drop

8.5.1 [Video] Assume a car's exhaust system can be approximated as 4.26 m of 0.038 m-diameter cast-iron pipe with the equivalent of six

90° flanged elbows and a muffler (see Video V8.13). The muffler acts as a resistor with a loss coefficient of $K_L = 8.5$. Determine the pressure at the beginning of the exhaust system if the flowrate is 0.0028 m³/s, the temperature is 121 °C, and the exhaust has the same properties as air.

8.5.2 The pressure at section (2) shown in **Fig. P8.5.2** is not to fall below 413.68 kPa when the flowrate from the tank varies from 0 to 0.0283 m³/s and the branch line is shut off. Determine the minimum height, h, of the water tank under the assumption that (a) minor losses are negligible and (b) minor losses are not negligible.

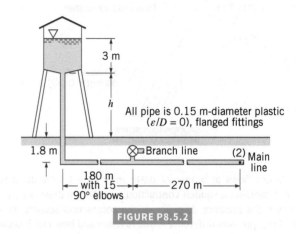

3 m

h

All pipe is 0.15 m-diameter plastic
($\varepsilon/D = 0$), flanged fittings

1.8 m

Branch line

(2) Main
line

180 m
with 15
90° elbows

270 m

FIGURE P8.5.2

8.5.3 Repeat Problem 8.5.2 with the assumption that the branch line is open so that half of the flow from the tank goes into the branch, and half continues in the main line.

8.5.4 The 0.0127 m-diameter hose shown in **Fig. P8.5.4** can withstand a maximum pressure of 1379 kPa without rupturing. Determine the maximum length, ℓ, allowed if the friction factor is 0.022 and the flowrate is 2.83×10^{-4} m³/s. Neglect minor losses.

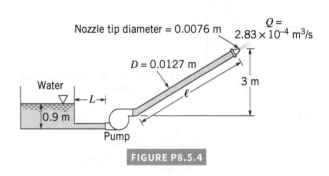

Nozzle tip diameter = 0.0076 m

$Q = 2.83 \times 10^{-4}$ m³/s

$D = 0.0127$ m

3 m

Water

L

ℓ

0.9 m

Pump

FIGURE P8.5.4

8.5.5 The hose shown in **Fig. P8.5.4** will collapse if the pressure within it is lower than 69 kPa below atmospheric pressure. Determine the maximum length, ℓ, allowed if the friction factor is 0.015 and the flowrate is 2.83×10^{-4} m³/s. Neglect minor losses.

8.5.6 According to fire regulations in a town, the pressure drop in a commercial steel horizontal pipe must not exceed 6.9 kPa per 45 m of pipe for flowrates up to 1.89 m³/min. If the water temperature is above 10 °C, can a 0.15 m-diameter pipe be used?

8.5.7 [Video] As shown in Video V8.14 and **Fig. P8.5.7**, water "bubbles up" 0.075 m above the exit of the vertical pipe attached to three horizontal pipe segments. The total length of the 0.019 m-diameter galvanized iron pipe between point (1) and the exit is 0.53 m. Determine the pressure needed at point (1) to produce this flow.

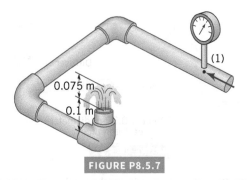

FIGURE P8.5.7

8.5.8 Water at 10 °C is pumped from a lake as shown in **Fig. P8.5.8**. If the flowrate is 0.011 m³/s, what is the maximum length inlet pipe, ℓ, that can be used without cavitation occurring?

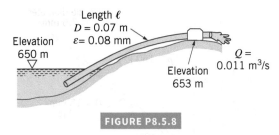

FIGURE P8.5.8

8.5.9 At a ski resort, water at 4.5 °C is pumped through a 0.0762 m-diameter, 610 m-long steel pipe from a pond at an elevation of 1306 m to a snow-making machine at an elevation of 1409 m at a rate of 0.00736 m³/s. If it is necessary to maintain a pressure of 1241 kPa at the snow-making machine, determine the horsepower added to the water by the pump. Neglect minor losses.

8.5.10 Crude oil having a specific gravity of 0.80 and a viscosity of 5.574×10^{-6} m²/s flows through a pumping station at a rate of 441.64 liter/m³. The oil then flows through 36573 m of 0.6 m I.D., horizontal, commercial steel pipe, enters another pumping station at 138 kPa, and leaves at a discharge pressure equal to the discharge pressure of the previous pumping station. Calculate the discharge pressures and the power supplied to the crude oil at the second pumping station. See **Fig. P8.5.10**.

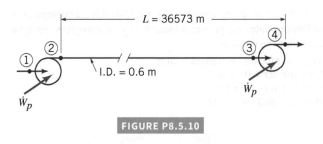

FIGURE P8.5.10

8.5.11 A motor-driven centrifugal pump delivers 15 °C water at the rate of 10 m³/min from a reservoir, through a 2500-m-long, 30-cm I.D. plastic pipe, to a second reservoir. The water level in the second reservoir is 40 m above the water level in the first reservoir. The pump efficiency is 75%. Find the motor output power. The pipe entrance is square edged.

8.5.12 An emergency flooding system for a nuclear reactor core is shown in **Fig. P8.5.12**. Find the power input required to flood the core at the rate of 19 m³/min. Assume a square-edged entrance, 15 °C water, and threaded connections.

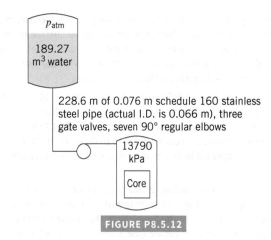

FIGURE P8.5.12

8.5.13 A hydraulic turbine takes water from a lake with the piping system shown in **Fig. P8.5.13**. Find the head h_t available to the turbine for flowrates of 141.58, 283.16, 424.75, and 566.34 m³/min.

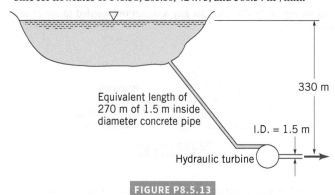

FIGURE P8.5.13

8.5.14 Water flows through a 0.05 m-diameter pipe with a velocity of 4.5 m/s as shown in **Fig. P8.5.14**. The relative roughness of the pipe is 0.004, and the loss coefficient for the exit is 1.0. Determine the height, h, to which the water rises in the piezometer tube.

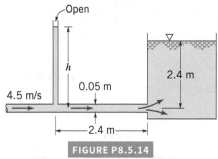

FIGURE P8.5.14

8.5.15 **Figure P8.5.15** shows the 15 °C water flowrates from the branches of a main supply line. Find the total pressure drop $(p_A - p_E)$ for soldered copper pipe. Assume that the loss coefficient for each tee is based on the threaded type.

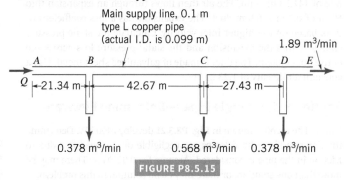

FIGURE P8.5.15

8.5.16 Water is pumped through a 60-m-long, 0.3-m-diameter pipe from a lower reservoir to a higher reservoir whose surface is 10 m above the lower one. The sum of the minor loss coefficient for the system is $K_L = 14.5$. When the pump adds 40 kW to the water, the flowrate is 0.20 m³/s. Determine the pipe roughness.

8.5.17 Natural gas ($\rho = 2.267$ kg/m³ and $v = 4.831 \times 10^{-6}$ m²/s) is pumped through a horizontal 0.15 m-diameter cast-iron pipe at a rate of 362.88 kg/hr. If the pressure at section (1) is 344.74 kPa (abs), determine the pressure at section (2) 14816 m downstream if the flow is assumed incompressible. Is the incompressible assumption reasonable? Explain.

8.5.18 (Video) As shown in **Fig. P8.5.18**, a standard household water meter is incorporated into a lawn irrigation system to measure the volume of water applied to the lawn. Note that these meters measure volume, not volume flowrate (see Video V8.15). With an upstream pressure of $p_1 = 344.74$ kPa, the meter registered that 3.4 m³ of water was delivered to the lawn during an "on" cycle. Estimate the upstream pressure, p_1, needed if it is desired to have 4.25 m³ delivered during an "on" cycle. List any assumptions needed to arrive at your answer.

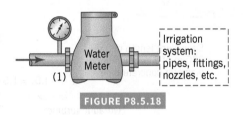

FIGURE P8.5.18

8.5.19 A fan is to produce a constant air speed of 40 m/s throughout the pipe loop shown in **Fig. P8.5.19**. The 3-m-diameter pipes are smooth, and each of the four 90° elbows has a loss coefficient of 0.30. Determine the power that the fan adds to the air.

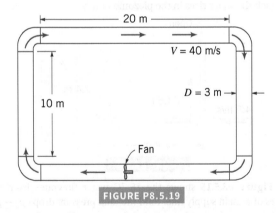

FIGURE P8.5.19

8.5.20 Air flows in a horizontal 30 m-long, 0.6 m × 0.6 m duct at the rate of 141.58 m³/min. The air then flows through an expansion into 60 m of 0.9 m × 0.9 m duct. The expansion has a loss coefficient of 0.80, based on the higher inlet velocity. Calculate the static pressure change across the expansion and the static pressure loss across the entire duct system. The duct is made of galvanized sheet metal. Use a constant air density of 1.23 kg/m³.

Section 8.5.1 Single Pipes—Determine Flowrate

8.5.21 The turbine shown in **Fig. P8.5.21** develops 400 kW. Determine the flowrate if (a) head losses are negligible or (b) head loss due to friction in the pipe is considered. Assume $f = 0.02$. *Note:* There may be more than one solution, or there may be no solution to this problem.

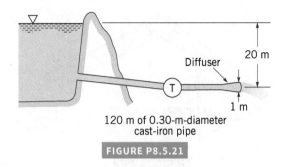

120 m of 0.30-m-diameter cast-iron pipe

FIGURE P8.5.21

8.5.22 Water flows from the nozzle attached to the spray tank shown in **Fig. P8.5.22**. Determine the flowrate if the loss coefficient for the nozzle (based on upstream conditions) is 0.75 and the friction factor for the rough hose is 0.11.

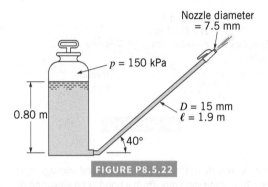

FIGURE P8.5.22

8.5.23 Water flows through the pipe shown in **Fig. P8.5.23**. Determine the net tension in the bolts if minor losses are neglected and the wheels on which the pipe rests are frictionless.

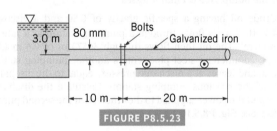

FIGURE P8.5.23

8.5.24 When the pump shown in **Fig. P8.5.24** adds 0.2 horsepower to the flowing water, the pressures indicated by the two gages are equal. Determine the flowrate.

Length of pipe between gages = 20 m
Pipe diameter = 0.03 m
Pipe friction factor = 0.03
Filter loss coefficient = 12

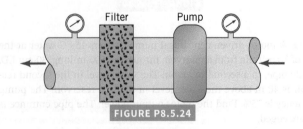

FIGURE P8.5.24

8.5.25 The pump shown in **Fig. P8.5.25** adds 25 kW to the water and causes a flowrate of 0.04 m³/s. Determine the flowrate expected if the pump is removed from the system. Assume $f = 0.016$ for either case and neglect minor losses.

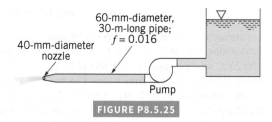

FIGURE P8.5.25

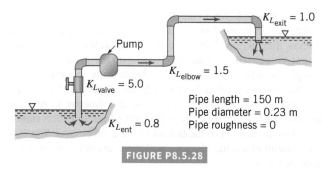

FIGURE P8.5.28

8.5.26 The vented storage tank shown in **Fig. P8.5.26** is used to refuel race cars at a race track. A total of 12 m of steel pipe (I.D. = 0.025 m), two 90° regular elbows, and a globe valve make up the system. Calculate the time needed to put 0.0757 m³ of fuel in a car tank. The pressures, p_2 and p_1, are equal, the connections are threaded, and the fuel has the properties of 20 °C normal octane ($v = 7.72 \times 10^{-7}$ m²/s).

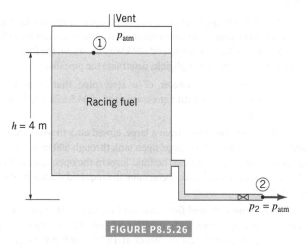

FIGURE P8.5.26

8.5.27 Calculate the water flowrate in the system shown in **Fig. P8.5.27**. The piping system includes four gate valves, two half-open globe valves, fourteen 90° regular elbows, and 76 m of 0.05 m schedule 40 commercial steel pipe (with an actual inside diameter of 0.0525 m). Assume threaded connections and a square-edged pipe entrance.

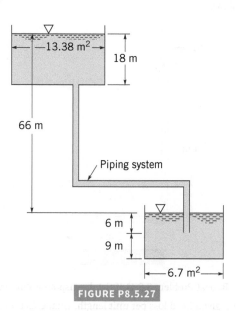

FIGURE P8.5.27

8.5.28 The pump shown in **Fig. P8.5.28** delivers a head of 75 m to the water. Determine the power that the pump adds to the water. The difference in elevation of the two ponds is 60 m.

8.5.29 For the standpipe system shown in **Fig. P8.5.29**, calculate the flowrate for $H = 1.2$ m, $D = 0.172$ m, $d = 0.03$ m, and $L = 1.2$ m. The fluid is 21.1 °C water. Assume steady flow and neglect the energy loss in the entrance nozzle. The pipe is commercial steel.

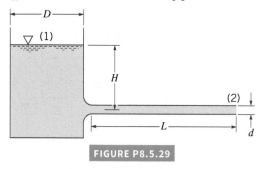

FIGURE P8.5.29

8.5.30 Water flows through two sections of the vertical pipe shown in **Fig. P8.5.30**. The bellows connection cannot support any force in the vertical direction. The 0.12 m-diameter pipe weights 3 N/m, and the friction factor is assumed to be 0.02. At what velocity will the force, F, required to hold the pipe be zero?

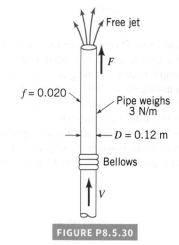

FIGURE P8.5.30

8.5.31 Water is circulated from a large tank, through a filter, and back to the tank as shown in **Fig. P8.5.31**. The power added to the water by the pump is 271 W. Determine the flowrate through the filter.

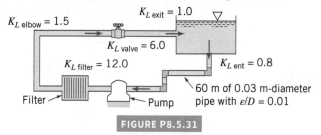

FIGURE P8.5.31

8.5.32 A thief siphoned 0.0567 m³ of gasoline from a gas tank in the middle of the night. The gas tank is 0.3 m wide, 0.6 m long, and 0.45 m high and was full when the thief started. The siphoning plastic tube has an inside diameter of 0.0125 m and a length of 1.2 m. Assume that at any instant of time, the steady-state mechanical energy equation is adequate to predict the gasoline flowrate through the tube. The gasoline level in the tank will drop 0.3 m. You may use the gasoline level after it has dropped 0.15 m to estimate the average gasoline flowrate. Use this flowrate to estimate the time needed to siphon the 0.0567 m^3 of gasoline. The siphon discharges at the level of the bottom of the gasoline tank. You may find it useful to use the Blasius equation for smooth pipes found in Problem 8.4.14.

8.5.33 Estimate the time required for the water depth in the reservoir shown in **Fig. P8.5.33** to drop from a height of 25 m to 5 m. The connections are threaded.

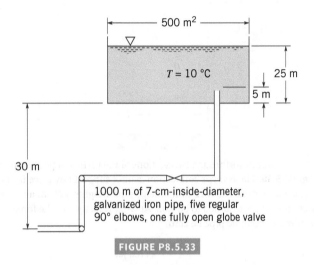

FIGURE P8.5.33

8.5.34 Sheldon and Leonard come home from a long day of studying to discover that their basement is flooded with water. They have a submersible pump for such an emergency and connect the pump discharge to a 3.6 m-long garden hose, as shown in **Fig. P8.5.34**. The garden hose may be considered a smooth plastic pipe with a 0.025 m inside diameter. If the initial water depth is 0.15 m and the pump has an input power of 186.5 W and an overall efficiency of 0.60, how long will it take to pump out the basement? The pump inlet piping length is negligible, and the garden hose outlet is 2.1 m above the basement floor.

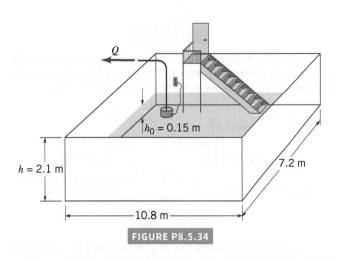

FIGURE P8.5.34

Section 8.5.1 Single Pipes—Determine Diameter

8.5.35 A company markets ethylene glycol antifreeze in half-gallon bottles. A machine fills and caps the bottles at a rate of 60 per minute. The 20 °C ethylene glycol ($\rho = 10886.16$ N/m^3, $v = 1.8 \times 10^{-5}$ m^2/s) is pumped to the machine from a tank 10.8 m away. The pressure at the discharge of the pump is 241.3 kPa, and the pressure at the inlet to the filling machine must be at least 103.4 kPa. Determine the diameter necessary for drawn copper pipe. The pipe has no elevation change.

8.5.36 A certain process requires 0.065 m^3/s of water to be delivered at a pressure of 206.84 kPa. This water comes from a large-diameter supply main in which the pressure remains at 413.7 kPa. If the galvanized iron pipe connecting the two locations is 60 m long and contains six threaded 90° elbows, determine the pipe diameter. Elevation differences are negligible.

8.5.37 Water is pumped between two large open reservoirs through 1.5 km of smooth pipe. The water surfaces in the two reservoirs are at the same elevation. When the pump adds 20 kW to the water, the flowrate is 1 m^3/s. If minor losses are negligible, determine the pipe diameter.

8.5.38 Determine the diameter of a steel pipe that is to carry 7.57 m^3/min of gasoline with a pressure drop of 34.5 kPa per 30 m of horizontal pipe.

8.5.39 Water is to be moved from a large, closed tank in which the air pressure is 137.9 kPa into a large, open tank through 600 m of smooth pipe at the rate of 0.085 m^3/s. The fluid level in the open tank is 45 m below that in the closed tank. Determine the required diameter of the pipe. Neglect minor losses.

8.5.40 A commercial steel flow channel in a heat exchanger has an equilateral triangle cross section with each side measuring 0.127 m and a length measuring 2.4 m. Water at 15 °C flowing through the channel has a pressure loss of 689.5 Pa. Find the water flowrate.

8.5.41 Ⓒ Rainwater flows through the galvanized iron downspout shown in **Fig. P8.5.41** at a rate of 0.006 m^3/s. Determine the size of the downspout cross section if it is a rectangle with an aspect ratio of 1.7 to 1 and it is completely filled with water. Neglect the velocity of the water in the gutter at the free surface and the head loss associated with the elbow.

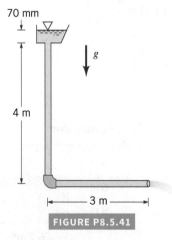

FIGURE P8.5.41

8.5.42 Ⓒ Repeat Problem 8.5.41 if the downspout is circular.

8.5.43 For a given head loss per unit length, what effect on the flowrate does doubling the pipe diameter have if the flow is (a) laminar, or (b) completely turbulent?

8.5.44 It is necessary to deliver 0.1275 m³/s of water from reservoir *A* to reservoir *B*, as shown in **Fig. P8.5.44**. The connecting piping consists of four fully open gate valves, twelve regular 90° elbows, one swing check valve, two fully open globe valves, and 900 m of commercial steel pipe. Find the minimum diameter for steel pipe needed to obtain this flowrate. All fittings and valves have flanged connections. Assume a square-edged entrance.

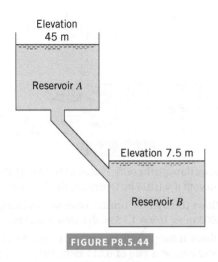

FIGURE P8.5.44

Section 8.5.2 Multiple Pipe Systems

8.5.45 A 10-m-long, 5.042-cm I.D. copper pipe has two fully open gate valves, a swing check valve, and a sudden enlargement to a 9.919-cm I.D. copper pipe. The 9.919-cm copper pipe is 5.0 m long and then has a sudden contraction to another 5.042-cm copper pipe. Find the head loss for a 20 °C water flowrate of 0.05 m³/s.

8.5.46 Air, assumed incompressible, flows through the two pipes shown in **Fig. P8.5.46**. Determine the flowrate if minor losses are neglected and the friction factor in each pipe is 0.015. Determine the flowrate if the 0.0127 m-diameter pipe were replaced by a 0.0254 m-diameter pipe. Comment on the assumption of incompressibility.

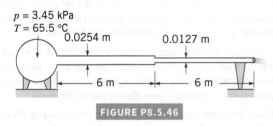

FIGURE P8.5.46

8.5.47 [C] Repeat Problem 8.5.46 if the pipes are galvanized iron and the friction factors are not known a priori.

8.5.48 [O] Estimate the power that the human heart must impart to the blood to pump it through the two carotid arteries from the heart to the brain. List all assumptions and show all calculations.

8.5.49 Normal octane at 20 °C ($\nu = 7.72 \times 10^{-7}$ m²/s) is to be delivered at a flowrate of 0.0113 m³/min through a 0.05 m schedule 40 commercial steel pipe (with an actual inside diameter of 0.0525 m). Energy losses are important, so consideration is being given to expanding the pipe to a 0.076 m schedule 40 commercial steel pipe (with an actual inside diameter of 0.078 m), using a 10° conical expansion. After a length ℓ of the 0.076 m pipe, it is decreased to a 0.05 m schedule 40 commercial steel pipe with a gradual contraction ($K_L = 0.10$, based on the smaller diameter pipe flow). Find the minimum length ℓ so that the energy

loss will be the same as in a 0.05 m pipe of length $\ell + 2$ (0.145 m). See **Fig. P8.5.49**.

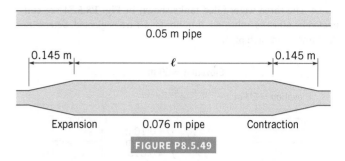

FIGURE P8.5.49

8.5.50 The flowrate between tank *A* and tank *B* shown in **Fig. P8.5.50** is to be increased by 30% (i.e., from *Q* to 1.30*Q*) by the addition of a second pipe (indicated by the dotted lines) running from node *C* to tank *B*. If the elevation of the free surface in tank *A* is 7.5 m above that in tank *B*, determine the diameter, *D*, of this new pipe. Neglect minor losses and assume that the friction factor for each pipe is 0.02.

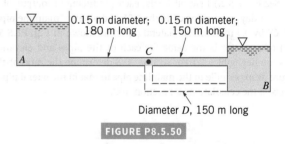

FIGURE P8.5.50

8.5.51 A 75 m-high building has a 0.154 m-diameter steel standpipe and a 30 m-long 0.065 m-diameter fire hose on each floor. The nearest fireplug is 30 m from the standpipe's ground-level connection. Assume that fire fighters connect a 0.15 m-diameter, 15 m-long fire hose from the fireplug to the fire truck and a 0.12 m-diameter, 15 m-long fire hose from the fire truck to the standpipe's ground-level connection. The National Fire Protection Association (NFPA) requires that a minimum pressure of 448.16 Pa be maintained at the connection of the 0.065 m-diameter hose and the standpipe while maintaining a flowrate of 1.89 m³/min through the fire hose. What pressure rise must the pump on the fire engine supply to satisfy the NFPA requirement for this building? The fire hydrant water pressure is 551.58 kPa and the water temperature is 15 °C. The connections are threaded.

8.5.52 With the valve closed, water flows from tank *A* to tank *B* as shown in **Fig. P8.5.52**. What is the flowrate into tank *B* when the valve is opened to allow water to flow into tank *C* also? Neglect all minor losses and assume that the friction factor is 0.02 for all pipes.

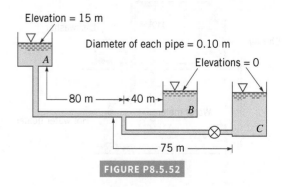

FIGURE P8.5.52

8.5.53 [C] Repeat Problem 8.5.52 if the friction factors are not known, but the pipes are steel pipes.

8.5.54 The three water-filled tanks shown in **Fig. P8.5.54** are connected by pipes as indicated. If minor losses are neglected, determine the flowrate in each pipe.

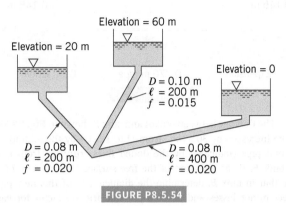

Elevation = 60 m
Elevation = 20 m
Elevation = 0

$D = 0.10$ m
$\ell = 200$ m
$f = 0.015$

$D = 0.08$ m
$\ell = 200$ m
$f = 0.020$

$D = 0.08$ m
$\ell = 400$ m
$f = 0.020$

FIGURE P8.5.54

8.5.55 (See The Wide World of Fluids article titled "Deepwater Pipeline," Section 8.5.2.) Five oil fields, each producing an output of Q barrels per day, connected to the 0.71 m-diameter "mainline pipe" (A–B–C) by 0.4 m-diameter "lateral pipes" as shown in **Fig. P8.5.55**. The friction factor is the same for each of the pipes and elevation effects are negligible. (a) For section A–B determine the ratio of the pressure drop per mile in the mainline pipe to that in the lateral pipes. (b) Repeat the calculations for section B–C.

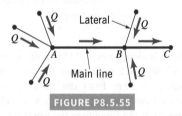

Lateral

Q

Q

Q

A B C

Main line

Q

FIGURE P8.5.55

8.5.56 As shown in **Fig. P8.5.56**, cold water ($T = 10$ °C) flows from the water meter to either the shower or the hot water heater. In the hot water heater it is heated to a temperature of 65.5 °C. Thus, with equal amounts of hot and cold water, the shower is at a comfortable 37.7 °C. However, when the dishwasher is turned on, the shower water becomes too cold. Indicate how you would predict this new shower temperature (assume the shower faucet is not adjusted). State any assumptions needed in your analysis.

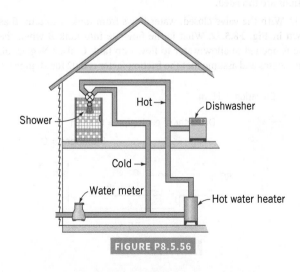

Shower

Hot

Dishwasher

Cold

Water meter

Hot water heater

FIGURE P8.5.56

Section 8.6 Pipe Flowrate Measurement

8.6.1 Water flows through the orifice meter shown in **Fig. P8.6.1** at a rate of 0.003 m³/s. If $d = 0.03$ m, determine the value of h and the pressure difference associated with this h.

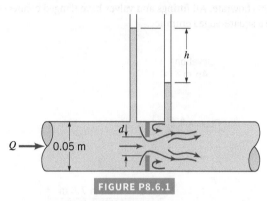

h

Q 0.05 m d

FIGURE P8.6.1

8.6.2 Water flows through the orifice meter shown in **Fig P8.6.1** such that $h = 0.48$ m with $d = 0.038$ m. Determine the flowrate.

8.6.3 Water flows through the orifice meter shown in **Fig. P8.6.1** at a rate of 0.003 m³/s. If $h = 1.15$ m, determine the value of d.

8.6.4 Water flows through a 40-mm-diameter nozzle meter in a 75-mm-diameter pipe at a rate of 0.015 m³/s. Determine the pressure difference across the nozzle if the temperature is (a) 10 °C or (b) 80 °C.

8.6.5 Gasoline flows through a 35-mm-diameter pipe at a rate of 0.0032 m³/s. Determine the pressure drop across a flow nozzle placed in the line if the nozzle diameter is 20 mm.

8.6.6 Air at 93.3 °C (366.4 K) and 413.7 kPa flows in a 0.1 m-diameter pipe at a rate of 2.313 N/s. Determine the pressure at the 0.05 m-diameter throat of a Venturi meter placed in the pipe.

8.6.7 A 0.06 m-diameter flow nozzle meter is installed in a 0.096 m-diameter pipe that carries water at 71.1 °C. If the air–water manometer used to measure the pressure difference across the meter indicates a reading of 0.93 m, determine the flowrate.

8.6.8 A 0.064-m-diameter nozzle meter is installed in a 0.097-m-diameter pipe that carries water at 60 °C. If the inverted air–water U-tube manometer used to measure the pressure difference across the meter indicates a reading of 1 m, determine the flowrate.

8.6.9 A 50-mm-diameter nozzle meter is installed at the end of a 80-mm-diameter pipe through which air flows. A manometer attached to the static pressure tap just upstream from the nozzle indicates a pressure of 7.3 mm of water. Determine the flowrate.

8.6.10 Water flows through the Venturi meter shown in **Fig. P8.6.10**. The specific gravity of the manometer fluid is 1.52. Determine the flowrate.

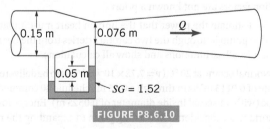

0.15 m   0.076 m Q

0.05 m

$SG = 1.52$

FIGURE P8.6.10

8.6.11 If the fluid flowing in Problem 8.6.10 were air, what would the flowrate be? Would compressibility effects be important? Explain.

8.6.12 [Video] The scale reading on the rotameter shown in Fig. P8.6.12 and Video V8.15 (also see **Fig. 8.6.11**) is directly proportional to the volumetric flowrate. With a scale reading of 2.6 the water bubbles up approximately 0.076 m. How far will it bubble up if the scale reading is 5.0?

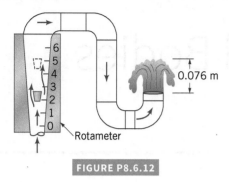

FIGURE P8.6.12

Lifelong Learning Problems

8.1 LL The field of biomedical engineering has undergone significant growth in recent years. Some universities have undergraduate and graduate programs in this field. Biomedical engineering applies engineering principles to help solve problems in the medical field for human health. Obtain information about biomedical engineering applications in blood flow. Summarize your findings in a brief report.

8.2 LL Data used in the Moody diagram were first published in 1944. Since then, there have been many innovations in pipe material, pipe design, and measurement techniques. Investigate whether there have been any improvements or enhancements to the Moody chart and friction factor relations. Summarize your findings in a brief report.

8.3 LL As discussed in Section 8.4.2, flow separation in pipes can lead to losses (we will also see in Chapter 9 that external flow separation is a significant issue). For external flows, many mechanisms have been devised to help mitigate and control flow separation from the surface, for example, from the wing of an airplane. Investigate either passive or active flow control mechanisms that can reduce or eliminate internal flow separation (e.g., flow separation in a diffuser). Summarize your findings in a brief report.

CHAPTER 9

Flow over Immersed Bodies

LEARNING OBJECTIVES

After completing this chapter, you should be able to:

- identify and discuss the features of external flow.
- explain the fundamental characteristics of a boundary layer, including laminar, transitional, and turbulent regimes.
- calculate boundary layer parameters for flow past a flat plate.
- explain the physical process of boundary layer separation.
- calculate the lift and drag forces for various objects.

In this chapter we consider various aspects of the flow over bodies that are completely immersed in a fluid. Examples include the flow of air around airplanes, automobiles, and falling snowflakes or the flow of water around submarines and fish. In these situations the object is completely surrounded by the fluid and the flows are termed *external flows*.

External flows involving air are often termed aerodynamics. Examples include an airplane flying through the atmosphere or a surface vehicle (e.g., car, truck, bicycle) moving through air. The fluid forces (primarily lift and drag) on these vehicles have become a very important topic. By correctly designing airplanes, cars, and trucks, it has become possible to greatly decrease the fuel consumption and improve the handling characteristics of the vehicle. Similar efforts have resulted in improved ships, whether they are surface vessels (surrounded by two fluids, air and water) or submersible vessels (surrounded completely by water).

Many practical situations involve flow past objects.

Other applications of external flows involve objects that are not completely surrounded by fluid, although they are placed in some external-type flow. For example, the proper design of a building (whether it is your house or a tall skyscraper) must include consideration of the various wind effects involved.

As with other areas of fluid mechanics, various approaches (theoretical, numerical, and experimental) are used to obtain information on the fluid forces developed by external flows. Theoretical (i.e., analytical) techniques can provide some of the needed information about such flows. However, because of the complexities of the governing equations, the flows, and the geometry of the objects involved, the amount of information obtained from purely theoretical methods is limited.

Much of the information about external flows comes from experiments carried out, for the most part, on scale models of the actual objects. Such testing includes the obvious wind tunnel testing of model vehicles, buildings, and even entire cities. In some instances the actual device, not a model, is tested in wind tunnels. **Figure 9.1a** shows a test of a vehicle in a wind tunnel. Better performance of cars, bikes, skiers, and numerous other objects has resulted from testing in wind tunnels. The use of water tunnels and towing tanks also provides useful information about the flow around ships and other objects. The information gathered from these experiments typically requires dimensional analysis for correlation and/or extrapolation. With advancement in computational fluid dynamics, or CFD, numerical methods are also capable of predicting external flows past objects. **Figure 9.1b** shows streamlines around a Formula 1 car as predicted by CFD. Appendix A provides an introduction to CFD.

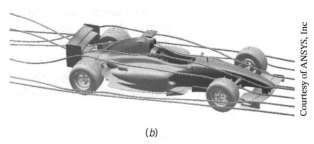

(a)   (b)

FIGURE 9.1 (a) Flow past a full-sized streamlined vehicle in a wind tunnel. (b) Predicted streamlines for flow past a Formula 1 race car as obtained by using computational fluid dynamics techniques.

In this chapter we consider characteristics of external flow past a variety of objects. We investigate the qualitative aspects of such flows and learn how to determine the various forces on objects surrounded by a moving liquid.

9.1 GENERAL EXTERNAL FLOW CHARACTERISTICS

A body immersed in a moving fluid experiences a resultant force due to the interaction between the body and the fluid surrounding it. In some instances (such as an airplane flying through still air) the fluid far from the body is stationary and the body moves through the fluid with velocity U. In other instances (such as the wind blowing past a building) the body is stationary and the fluid flows past the body with velocity U. In any case, we can fix the coordinate system in the body and treat the situation as fluid flowing past a stationary body with velocity U, the *upstream velocity*. For the purposes of this book, we will assume that the upstream velocity is constant in both time and location. That is, there is a uniform, constant velocity fluid flowing past the object. In actual situations this is often not true. For example, the wind blowing past a smokestack is nearly always turbulent and gusty (unsteady) and probably not of uniform velocity from the top to the bottom of the stack. Usually, from an engineering analysis viewpoint, the unsteadiness and nonuniformity are of minor importance.

Even with a steady, uniform upstream flow, the flow in the vicinity of an object may be unsteady. Examples of this type of behavior include the flutter that is sometimes found in the flow past airfoils (wings), the regular oscillation of telephone wires that "sing" in a wind, and the irregular turbulent fluctuations in the wake regions behind bodies.

The structure of an external flow and the ease with which the flow can be described and analyzed often depend on the nature of the body in the flow (including its orientation with the flow). Three general categories of bodies are shown in **Fig. 9.2**. They include (a) two-dimensional objects (infinitely long perpendicular to the page and of constant cross-sectional size and shape), (b) axisymmetric bodies (formed by rotating their cross-sectional shape about the axis of symmetry), and (c) three-dimensional bodies that may or may not possess a line or plane of

> For external flows it is usually easiest to use a coordinate system fixed to the object.

V9.1 Space Shuttle landing

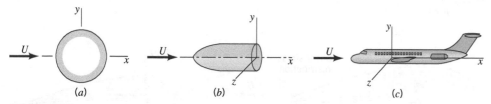

FIGURE 9.2 Flow classification: (a) two-dimensional, (b) axisymmetric, (c) three-dimensional.

symmetry. In practice there can be no truly two-dimensional bodies—nothing extends to infinity. However, many objects are sufficiently long so that the end effects are negligibly small.

Another classification of body shape can be made depending on whether the body is streamlined or blunt. The flow characteristics depend strongly on the amount of streamlining present. In general, *streamlined bodies* (e.g., airfoils, racing cars, etc.) have little effect on the surrounding fluid, compared with the effect that *blunt bodies* (e.g., parachutes, buildings, etc.) have on the fluid. Usually, but not always, it is easier to force a streamlined body through a fluid than it is to force a similar-sized blunt body at the same velocity. There are important exceptions to this basic rule.

9.1.1 Lift and Drag Concepts

A body interacts with the surrounding fluid through pressure and shear stresses.

When any body moves through a fluid, an interaction between the body and the fluid occurs; this effect can be given in terms of the forces at the fluid–body interface. These forces can be described in terms of the stresses—wall (surface) shear stresses on the body, τ_w, due to viscous effects and normal stresses due to the pressure, p. Typical shear stress and pressure distributions are shown in **Figs. 9.3a** and **9.3b**. Both τ_w and p vary in magnitude and direction along the surface.

It is often useful to know the detailed distribution of shear stress and pressure over the surface of the body, although such information is difficult to obtain. Many times, however, only the integrated or resultant effects of these distributions are needed. The resultant force in the direction of the upstream velocity is termed the **drag**, $\mathscr{D}$, and the resultant force normal to the upstream velocity is termed the **lift**, $\mathscr{L}$, as is indicated in **Fig. 9.3c**. Note that lift may be downward relative to the upstream velocity; in this case lift is termed downforce. For some three-dimensional bodies there may also be a side force that is perpendicular to the plane containing $\mathscr{D}$ and $\mathscr{L}$.

Drag is parallel to upstream velocity and lift is perpendicular.

The resultant of the shear stress and pressure distributions can be obtained by integrating the effect of these two quantities on the body surface as is indicated in **Fig. 9.4**. The x and y components of the fluid force on the small area element dA are

$$dF_x = (p\, dA) \cos \theta + (\tau_w\, dA) \sin \theta$$

and

$$dF_y = -(p\, dA) \sin \theta + (\tau_w\, dA) \cos \theta$$

Thus, the net x and y components of the force on the object are

$$\mathscr{D} = \int dF_x = \int p \cos \theta\, dA + \int \tau_w \sin \theta\, dA \tag{9.1}$$

and

$$\mathscr{L} = \int dF_y = -\int p \sin \theta\, dA + \int \tau_w \cos \theta\, dA \tag{9.2}$$

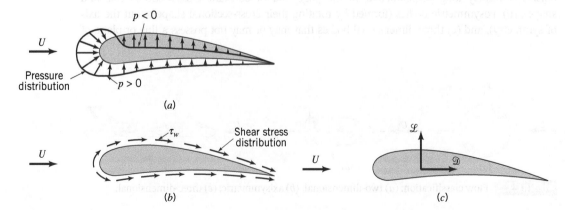

FIGURE 9.3 Forces from the surrounding fluid on a two-dimensional object: (*a*) pressure force, (*b*) viscous force, and (*c*) resultant force (lift and drag).

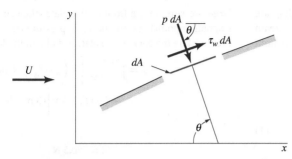

FIGURE 9.4 Pressure and shear forces on a small element of the surface of a body.

Of course, to carry out the integrations and determine the lift and drag, we must know the body shape (i.e., θ as a function of location along the body) and the distribution of τ_w and p along the surface. These distributions are often extremely difficult to obtain, either experimentally or theoretically. The pressure distribution can be obtained experimentally by use of a series of static pressure taps along the body surface. On the other hand, it is usually quite difficult to measure the wall shear stress distribution.

> **Lift and drag on a section of a body depend on the orientation of the surface.**

The Wide World of Fluids

Pressure-sensitive paint

For many years, the conventional method for measuring *surface pressure* has been to use static pressure taps consisting of small holes on the surface connected by hoses from the holes to a pressure measuring device. Pressure-sensitive paint (PSP) is now gaining acceptance as an alternative to the static surface pressure ports. The PSP material is typically a luminescent compound that is sensitive to the pressure on it and can be excited (visualized) by an appropriate light that is captured by special video imaging

equipment. Thus, it provides a quantitative measure of the surface pressure. One of the biggest advantages of PSP is that it is a global measurement technique, measuring pressure over the entire surface, as opposed to discrete points at the taps. PSP also has the advantage of being nonintrusive to the flow field. Although static pressure port holes are small, they do alter the surface and can slightly alter the flow, thus affecting downstream ports. In addition, the use of PSP eliminates the need for a large number of pressure taps and connecting tubes. This lowers measurement time and cost.

It is seen that both the shear stress and pressure force contribute to the lift and drag, since for an arbitrary body θ is neither zero nor 90° along the entire body. The exception is a flat plate aligned either parallel to the upstream flow ($\theta = 90°$) or normal to the upstream flow ($\theta = 0$) as is discussed in Example 9.1.

EXAMPLE 9.1 | Drag from Pressure and Shear Stress Distributions

Given Air at standard conditions flows past a flat plate as is indicated in **Fig. E9.1**. In case (a) the plate is parallel to the upstream flow, and in case (b) it is perpendicular to the upstream flow. The pressure and shear stress distributions on the surface are as indicated (obtained either by experiment or theory).

Find Determine the lift and drag on the plate.

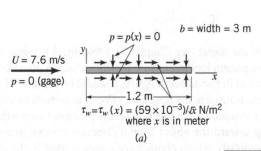

(a)

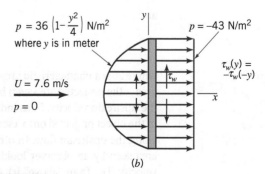

(b)

FIGURE E9.1a-b

Solution   For either orientation of the plate, the lift and drag are obtained from Eqs. 9.1 and 9.2. With the plate parallel to the upstream flow we have $\theta = 90°$ on the top surface and $\theta = 270°$ on the bottom surface so that the lift and drag are given by

$$\mathcal{L} = -\int_{\text{top}} p\, dA + \int_{\text{bottom}} p\, dA = 0$$

and

$$\mathcal{D} = \int_{\text{top}} \tau_w\, dA + \int_{\text{bottom}} \tau_w\, dA = 2\int_{\text{top}} \tau_w\, dA \qquad (1)$$

where we have used the fact that because of symmetry the shear stress distribution is the same on the top and the bottom surfaces, as is the pressure also [whether we use gage ($p = 0$) or absolute ($p = p_{\text{atm}}$) pressure]. There is no lift generated—the plate does not know up from down. With the given shear stress distribution, Eq. 1 gives

$$\mathcal{D} = 2\int_{x=0}^{1.2\text{m}} \left(\frac{59 \times 10^{-3}}{x^{1/2}}\, \text{N/m}^2\right)(3\text{ m})\, dx$$

or

$$\mathcal{D} = 0.77\text{ N} \qquad (Ans)$$

With the plate perpendicular to the upstream flow, we have $\theta = 0°$ on the front and $\theta = 180°$ on the back. Thus, from Eqs. 9.1 and 9.2

$$\mathcal{L} = \int_{\text{front}} \tau_w\, dA - \int_{\text{back}} \tau_w\, dA = 0$$

and

$$\mathcal{D} = \int_{\text{front}} p\, dA - \int_{\text{back}} p\, dA$$

Again there is no lift because the pressure forces act parallel to the upstream flow (in the direction of $\mathcal{D}$ not $\mathcal{L}$) and the shear stress is symmetrical about the center of the plate. With the given relatively large pressure on the front of the plate (the center of the plate is a stagnation point) and the negative pressure (less than the upstream pressure) on the back of the plate, we obtain the following drag

$$\mathcal{D} = \int_{y=-0.6\text{m}}^{0.6\text{ m}} \left[36\left(1 - \frac{y^2}{4}\right)\text{ N/m}^2\right.$$

$$\left. - (-43)\text{ N/m}^2\right](3\text{ m})\, dy$$

or

$$\mathcal{D} = 280.5\text{ N} \qquad (Ans)$$

Comments   Clearly there are two mechanisms responsible for the drag. On the ultimately streamlined body (a zero-thickness flat plate parallel to the flow) the drag is entirely due to the shear stress at the surface and, in this example, is relatively small. For the ultimately blunted body (a flat plate normal to the upstream flow) the drag is entirely due to the pressure difference between the front and back portions of the object and, in this example, is relatively large.

If the flat plate were oriented at an arbitrary angle relative to the upstream flow as indicated in **Fig. E9.1c**, there would be both a lift and a drag, each of which would be dependent on both the shear stress and the pressure. Both the pressure and shear stress distributions would be different for the top and bottom surfaces.

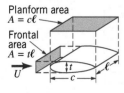

FIGURE E9.1c

Although Eqs. 9.1 and 9.2 are valid for any body, the difficulty in their use lies in obtaining the appropriate shear stress and pressure distributions on the body surface. Considerable effort has gone into determining these quantities, but because of the various complexities involved, such information is available only for certain simple situations. Note that even in Example 9.1, the shear stress and pressure distributions were given.

Without detailed information concerning the shear stress and pressure distributions on a body, Eqs. 9.1 and 9.2 cannot be used. The widely used alternative is to define dimensionless lift and drag coefficients and determine their approximate values by means of either a simplified analysis, some numerical technique, or an appropriate experiment. The **lift coefficient,** C_L, and **drag coefficient,** C_D, are defined as

| Lift coefficients and drag coefficients are dimensionless forms of lift and drag. |

$$C_L = \frac{\mathcal{L}}{\frac{1}{2}\rho U^2 A}$$

and

$$C_D = \frac{\mathcal{D}}{\frac{1}{2}\rho U^2 A}$$

where A is a characteristic area of the object (see Chapter 7). Typically, A is taken to be *frontal area*—the projected area seen by a person looking toward the object from a direction parallel to the upstream velocity, U, as indicated in the adjacent figure. It would be the area of the shadow of the object projected onto a screen normal to the upstream velocity as formed by a light shining along the upstream flow. In other situations A is taken to be the *planform area*—the projected area seen by an observer looking toward the object from a direction normal to the upstream velocity (i.e., from "above" it). Obviously, which characteristic area is used in the definition of the lift and drag coefficients must be clearly stated.

Planform area
$A = c\ell$
Frontal area
$A = t\ell$

9.1.2 Characteristics of Flow Past an Object

External flows past objects encompass an extremely wide variety of fluid mechanics phenomena. Clearly the character of the flow field is a function of the shape of the body and its orientation relative to the flow. Flows past relatively simple geometric shapes (e.g., a sphere or circular cylinder) are expected to have less complex flow fields than flows past a complex shape such as an airplane or a tree. However, even the simplest-shaped objects produce rather complex flows.

For a given-shaped object, the characteristics of the flow depend very strongly on various parameters such as size, orientation, speed, and fluid properties. As is discussed in Chapter 7, according to dimensional analysis arguments, the character of the flow should depend on the various dimensionless parameters involved. For typical external flows the most important of these parameters are the Reynolds number, $Re = \rho U \ell / \mu = U \ell / \upsilon$, the Mach number, $Ma = U/c$, and for flows with a free surface (i.e., flows with an interface between two fluids, such as the flow past a surface ship), the Froude number, $Fr = U/\sqrt{g\ell}$. (Recall that ℓ is some characteristic length of the object and c is the speed of sound.)

For the present, we consider how the external flow and its associated lift and drag vary as a function of the Reynolds number. Recall that the Reynolds number represents the ratio of inertial effects to viscous effects. In the absence of all viscous effects ($\mu = 0$), the Reynolds number is infinite. On the other hand, in the absence of all inertial effects (negligible mass or $\rho = 0$), the Reynolds number is zero. Clearly, any actual flow will have a Reynolds number between (but not including) these two extremes. The nature of the flow past a body depends strongly on whether $Re \gg 1$ or $Re \ll 1$.

Most external flows with which we are familiar are associated with moderately sized objects with a characteristic length on the order of $0.01\ m < \ell < 10\ m$. In addition, typical upstream velocities are on the order of $0.01\ m/s < U < 100\ m/s$, and the fluids involved are typically water or air. The resulting Reynolds number range for such flows is approximately $10 < Re < 10^9$. This is shown in the adjacent figure for air. As a rule of thumb, flows with $Re > 100$ are dominated by inertial effects, whereas flows with $Re < 1$ are dominated by viscous effects. Hence, most familiar external flows are dominated by inertia.

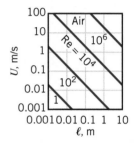

On the other hand, there are many external flows in which the Reynolds number is considerably less than 1, indicating in some sense that viscous forces are more important than inertial forces. The gradual settling of small particles of dirt in a lake or stream is governed by low Reynolds number flow principles because of the small diameter of the particles and their small settling speed. Similarly, the Reynolds number for objects moving through large-viscosity oils is small because μ is large. The general differences between small and large Reynolds number flow past streamlined and blunt objects can be illustrated by considering flows past two objects—one a flat plate parallel to the upstream velocity and the other a circular cylinder.

Flows past three flat plates of length ℓ with $Re = \rho U \ell / \mu = 0.1$, 10, and 10^7 are shown in **Fig. 9.5**. If the Reynolds number is small, the viscous effects are relatively strong and the plate affects the uniform upstream flow far ahead, above, below, and behind the plate. To reach that portion of the flow field where the velocity has been altered by less than 1% of its undisturbed value (i.e., $U - u < 0.01U$), we must travel relatively far from the plate. In low Reynolds number flows, the viscous effects are felt far from the object in all directions.

As the Reynolds number is increased (by increasing U, for example), the region in which viscous effects are important becomes smaller in all directions except downstream, as is shown in Fig. 9.5b. One does not need to travel very far ahead, above, or below the plate to reach areas in which the viscous effects of the plate are not felt. The streamlines are displaced from their original uniform upstream conditions, but the displacement is not as great as for the $Re = 0.1$ situation shown in Fig. 9.5a.

If the Reynolds number is large (but not infinite), the flow is dominated by inertial effects and the viscous effects are negligible everywhere except in a region very close to the plate and in the relatively thin **wake region** behind the plate, as shown in Fig. 9.5c. Since the fluid viscosity is not zero ($Re < \infty$), it follows that the fluid must stick to the solid surface (the no-slip boundary condition). There is a thin **boundary layer** region of thickness $\delta = \delta(x) \ll \ell$ (i.e., thin relative to the length of the plate) next to the plate in which the fluid velocity changes from the upstream value of $u = U$ to zero velocity on the plate. The thickness of this layer increases in the direction of flow, starting from zero at the forward or leading edge of the plate. The flow within the boundary layer may be laminar or turbulent, depending on various parameters involved.

For low Reynolds number flows, viscous effects are felt far from the object.

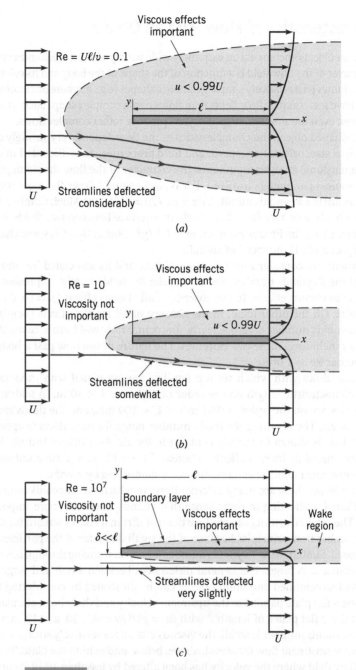

FIGURE 9.5 Character of the steady, viscous flow past a flat plate parallel to the upstream velocity: (a) low Reynolds number flow, (b) moderate Reynolds number flow, (c) large Reynolds number flow.

The streamlines of the flow outside of the boundary layer are nearly parallel to the plate. As we will see in the next section, the slight displacement of the external streamlines that are outside of the boundary layer is due to the thickening of the boundary layer in the direction of flow. The existence of the plate has very little effect on the streamlines outside of the boundary layer—either ahead, above, or below the plate. On the other hand, the wake region is due entirely to the viscous interaction between the fluid and the plate.

One of the great advancements in fluid mechanics occurred in 1904 as a result of the insight of Ludwig Prandtl (1875–1953), a German physicist and aerodynamicist. He conceived of the idea of the boundary layer—a thin region on the surface of a body in which viscous effects are very important and outside of which the fluid behaves essentially as if it were inviscid. Clearly the actual fluid viscosity is the same throughout; only the relative importance of the viscous effects (due to the velocity gradients) is different within or outside of the boundary layer. As is discussed in the next section, by using such a hypothesis it is possible to simplify the analysis of large Reynolds number flows, thereby allowing solution to external flow problems that are otherwise still unsolvable.

Thin boundary layers may develop in large Reynolds number flows.

As with the flow past the flat plate described above, the flow past a blunt object (such as a circular cylinder) also varies with Reynolds number. In general, the larger the Reynolds number, the smaller the region of the flow field in which viscous effects are important. For objects that are not sufficiently streamlined, however, an additional characteristic of the flow is observed. This is termed *flow separation* and is illustrated by the adjacent photograph and in **Fig. 9.6**.

Low Reynolds number flow (Re = $UD/v < 1$) past a circular cylinder is characterized by the fact that the presence of the cylinder and the accompanying viscous effects are felt throughout a relatively large portion of the flow field. As is indicated in Fig. 9.6a, for Re = $UD/v = 0.1$, the viscous effects are important several diameters in any direction from the cylinder. A somewhat surprising characteristic of this flow is that the streamlines are essentially symmetric about the center of the cylinder—the streamline pattern is the same in front of the cylinder as it is behind the cylinder.

As the Reynolds number is increased, the region ahead of the cylinder in which viscous effects are important becomes smaller, with the viscous region extending only a short distance ahead of the cylinder. The viscous effects are convected downstream, and the flow loses its upstream to downstream symmetry. Another characteristic of external flows becomes important—the flow separates from the body at the *separation location* as indicated in Fig. 9.6b. With the increase in Reynolds number, the fluid inertia becomes more important and at some location on the body, denoted the separation location, the fluid's inertia is such that it cannot follow the curved path around to the rear of the body. (This concept will be further explored in Section 9.2.6.) The result is a separation bubble behind the cylinder in which some of the fluid is actually flowing upstream, against the direction of the upstream flow (see the adjacent photograph showing laminar flow past an array of cylinders).

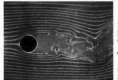

© Stanford University, with permission

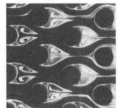

Photograph courtesy of ONERA, The French Aerospace Lab.

Flow past an array of cylinders.

Video

V9.2 Streamlined and blunt bodies

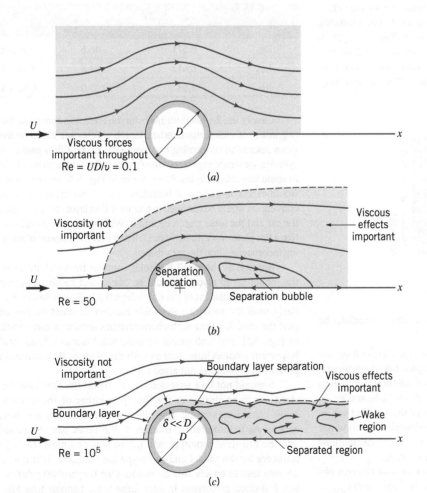

(a)

(b)

(c)

FIGURE 9.6 Character of the steady, viscous flow past a circular cylinder: (a) low Reynolds number flow, (b) moderate Reynolds number flow, (c) large Reynolds number flow.

V9.3 Human aero-
dynamic wake

At still larger Reynolds numbers, the area affected by the viscous forces is forced closer to the cylinder until it involves only a thin ($\delta \ll D$) boundary layer on the front portion of the cylinder and an irregular, unsteady (perhaps turbulent) wake region that extends far downstream of the cylinder. The fluid in the region outside of the boundary layer and wake region flows as if it were inviscid. Of course, the fluid viscosity is the same throughout the entire flow field. Whether or not viscous effects are important depends on which region of the flow field we consider. The velocity gradients within the boundary layer and wake regions are much larger than those in the remainder of the flow field. Since the shear stress (i.e., viscous effect) is the product of the fluid viscosity and the velocity gradient, it follows that viscous effects are confined to the boundary layer and wake regions.

Most familiar flows involve large Reynolds numbers.

The characteristics described in Figs. 9.5 and 9.6 for flow past a flat plate and a circular cylinder are typical of flows past streamlined and blunt bodies, respectively. The nature of the flow depends strongly on the Reynolds number, and in practice, many bodies exhibit both streamlined and blunt characteristics as illustrated in Example 9.2 (see Ref. 25 for many examples illustrating this behavior). Most familiar flows are similar to the large Reynolds number flows depicted in Figs. 9.5c and 9.6c, rather than the low Reynolds number flow situations. In the remainder of this chapter we will investigate more thoroughly these ideas and determine how to calculate the forces on immersed bodies.

EXAMPLE 9.2 | Characteristics of Flow Past Objects

Given It is desired to experimentally determine the various characteristics of flow past a car as shown in **Fig E9.2**. The following tests could be carried out: (a) $U = 20$ mm/s flow of glycerin past a scale model that is 34-mm tall, 100-mm long, and 40-mm wide, (b) $U = 20$ mm/s air flow past the same scale model, or (c) $U = 25$ m/s airflow past the actual car, which is 1.7-m tall, 5-m long, and 2-m wide.

Reynolds Number	(a) Model in Glycerin	(b) Model in Air	(c) Car in Air
Re_h	0.571	46.6	2.91×10^6
Re_b	0.672	54.8	3.42×10^6
Re_ℓ	1.68	137.0	8.56×10^6

Clearly, the Reynolds numbers for the three flows are quite different (regardless of which characteristic length we choose). Based on the previous discussion concerning flow past a flat plate or flow past a circular cylinder, we would expect that the flow past the actual car would behave in some way similar to the flows shown in Figs. 9.5c or 9.6c. That is, we would expect some type of boundary layer characteristic in which viscous effects would be confined to relatively thin layers near the surface of the car and the wake region behind it. Whether the car would act more like a flat plate or a cylinder would depend on the amount of streamlining incorporated into the car's design.

Because of the small Reynolds number involved, the flow past the model car in glycerin would be dominated by viscous effects, in some way reminiscent of the flows depicted in Figs. 9.5a or 9.6a. Similarly, with the moderate Reynolds number involved for the airflow past the model, a flow with characteristics similar to those indicated in Figs. 9.5b and 9.6b would be expected. Viscous effects would be important—not as important as with the glycerin flow, but more important than with the full-sized car.

It would not be a wise decision to expect the flow past the full-sized car to be similar to the flow past either of the models. The same conclusions result regardless of whether we use Re_h, Re_b, or Re_ℓ. As is indicated in Chapter 7, the flows past the model car and the full-sized prototype will not be similar unless the Reynolds numbers for the model and prototype are the same. It is not always an easy task to ensure this condition. One (expensive) solution is to test full-sized prototypes in very large wind tunnels (see Fig. 9.1), or small-scale models in smaller wind tunnels at higher velocity, as long as the Mach number is kept small.

© Stanford University, with permission.

FIGURE E9.2

Find Would the flow characteristics for these three situations be similar? Explain.

Solution The characteristics of flow past an object depend on the Reynolds number. For this instance we could pick the characteristic length to be the height, h, width, b, or length, ℓ, of the car to obtain three possible Reynolds numbers, $Re_h = Uh/v$, $Re_b = Ub/v$, and $Re_\ell = U\ell/v$. These numbers will be different because of the different values of h, b, and ℓ. Once we arbitrarily decide on the length we wish to use as the characteristic length, we must stick with it for all calculations when using comparisons between model and prototype.

With the values of kinematic viscosity for air and glycerin obtained from Tables 1.8 and 1.6 as $v_{air} = 1.46 \times 10^{-5}$ m²/s and $v_{glycerin} = 1.19 \times 10^{-3}$ m²/s, we obtain the following Reynolds numbers for the flows described.

9.2 BOUNDARY LAYER CHARACTERISTICS

As was discussed in the previous section, it is often possible to treat flow past an object as a combination of viscous flow in the boundary layer and inviscid flow elsewhere. If the Reynolds number is large enough, viscous effects are important only in the boundary layer regions near the object (and in the wake region behind the object). The boundary layer is the result of the no-slip boundary condition that requires the fluid to cling to any solid surface that it flows past. Outside of the boundary layer the velocity gradients normal to the flow are relatively small, and the fluid acts as if it were inviscid, even though the viscosity is not zero. A necessary condition for this structure of the flow is that the Reynolds number be large.

Large Reynolds number flow fields may be divided into viscous and inviscid regions.

9.2.1 Boundary Layer Structure and Thickness on a Flat Plate

There can be a wide variety in the size of a boundary layer and the structure of the flow within it. Part of this variation is due to the shape of the object on which the boundary layer forms. In this section we consider the simplest situation, one in which the boundary layer is formed on an infinitely long flat plate along which flows a viscous, incompressible fluid as is shown in **Fig. 9.7**. If the surface were curved or experiences a pressure gradient (e.g., a circular cylinder or an airfoil), the boundary layer structure would be more complex. Such flows are discussed in Section 9.2.6.

If the Reynolds number is sufficiently large, only the fluid in a relatively thin boundary layer on the plate will feel the effect of the plate. That is, except in the region next to the plate the flow velocity will be essentially $\mathbf{V} = U\hat{\mathbf{i}}$, the upstream velocity also known as the free-stream velocity. For the infinitely long flat plate extending from $x = 0$ to $x = \infty$, it is not obvious how to define the Reynolds number because there is no characteristic length. The plate has no thickness and is not of finite length!

For a finite length plate, it is clear that the plate length, ℓ, can be used as the characteristic length. For an infinitely long plate we use x, the coordinate distance along the plate from the leading edge, as the characteristic length and define the Reynolds number as $\text{Re}_x = Ux/v$. Thus, for any fluid or upstream velocity the Reynolds number will be sufficiently large for boundary layer type flow (i.e., Fig. 9.5c) if the plate is long enough. Physically, this means that the flow situations illustrated in Fig. 9.5 could be thought of as occurring on the same plate but should be viewed by looking at longer portions of the plate as we step away from the plate to see the flows in Fig. 9.5a, 9.5b, and 9.5c, respectively.

If the plate is sufficiently long, the Reynolds number $\text{Re} = U\ell/v$ is sufficiently large so that the flow takes on its boundary layer character (except very near the leading edge). The details of the flow field near the leading edge are lost to our eyes because we are standing so far from the plate that we cannot make out these details. On this scale (Fig. 9.5c) the plate has negligible effect on the fluid ahead of the plate. The presence of the plate is felt only in the relatively thin boundary layer and wake regions. As previously noted, Prandtl in 1904 was the first to hypothesize such a concept. It is one of the major turning points in fluid mechanics.

V9.4 Laminar boundary layer

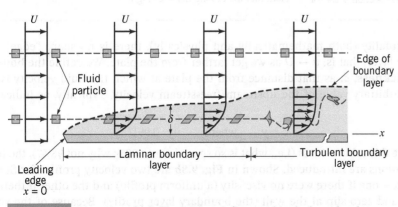

FIGURE 9.7 Distortion of a fluid particle as it flows within the boundary layer.

Fluid particles within the boundary layer experience viscous effects.

V9.5 Laminar/ turbulent transition

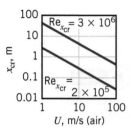

A better appreciation of the structure of the boundary layer flow can be obtained by considering what happens to a fluid particle that flows into the boundary layer. As is indicated in Fig. 9.7, a small rectangular particle retains its original shape as it flows in the uniform flow outside of the boundary layer. Once it enters the boundary layer, the particle begins to distort because of the velocity gradient within the boundary layer—the top of the particle has a larger speed than its bottom. The fluid particles do not rotate as they flow along outside the boundary layer, but they begin to rotate once they pass through the fictitious boundary layer surface and enter the world of viscous flow. The flow is said to be irrotational outside the boundary layer and rotational within the boundary layer. (In terms of the kinematics of fluid particles as is discussed in Section 6.1, the flow outside the boundary layer has zero vorticity, and the flow within the boundary layer has nonzero vorticity.)

At some distance downstream from the leading edge, the boundary layer flow becomes turbulent and the fluid particles become greatly distorted because of the random, irregular nature of the turbulence as discussed in Sections 8.3.1 and 8.3.2. One of the distinguishing features of turbulent flow is the occurrence of irregular mixing of fluid particles that range in size from the smallest fluid particles up to those comparable in size with the object of interest. For laminar flow, mixing occurs only on the molecular scale. This molecular scale is orders of magnitude smaller in size than typical size scales for turbulent flow mixing. The transition from a **laminar boundary layer** to a **turbulent boundary layer** occurs at a critical value of the Reynolds number, $\text{Re}_{x_{cr}}$, on the order of 2×10^5 to 3×10^6, depending on the roughness of the surface and the amount of turbulence in the upstream flow, as is discussed in Section 9.2.4. As shown in the adjacent figure, the location along the plate where the flow becomes turbulent, x_{cr}, moves toward the leading edge as the free-stream velocity increases.

The purpose of the boundary layer is to allow the fluid to change its velocity from the upstream (or free-stream) value of U to zero on the surface. Thus, $\mathbf{V} = 0$ at $y = 0$ and $\mathbf{V} \approx U\hat{\mathbf{i}}$ at the edge of the boundary layer, with the velocity profile, $u = u(x, y)$, bridging the boundary layer thickness, δ. This boundary layer characteristic occurs in a variety of flow situations, not just on flat plates. For example, boundary layers form on the surfaces of cars, on airplane wings, in the water running down the gutter of the street, and in the atmosphere as the wind blows across the surface of the Earth (land or water).

The Wide World of Fluids

The albatross: Nature's aerodynamic solution for long flights

The albatross is a phenomenal seabird that soars just above ocean waves, taking advantage of the local *boundary layer* to travel incredible distances with little to no wing flapping. This limited physical exertion results in minimal energy consumption and, combined with aerodynamic optimization, allows the albatross to easily travel 1000 km per day, with some tracking data showing almost double that amount. The albatross has high aspect ratio wings (up to 3.4 m in wingspan) and a lift-to-drag ratio ($\mathscr{L}/\mathscr{D}$) of approximately 27, both similar to high-performance sailplanes (see Section 9.2.4 for a discussion of high aspect ratio wings and Section 9.4.1 for a discussion of lift-to-drag ratio). With this aerodynamic configuration, the albatross then makes use of a technique called "dynamic soaring" to take advantage of the wind profile over the ocean surface. Based on the boundary layer profile, the albatross uses the rule of dynamic soaring, which is to climb when pointed upwind and dive when pointed downwind, thus constantly exchanging kinetic and potential energy. Though the albatross loses energy to drag, it can periodically regain energy due to vertical and directional motions within the boundary layer by changing local airspeed and direction. This is not a direct line of travel, but it does provide the most efficient method of long-distance flight.

V9.6 Boundary layer thickness

In actuality (both mathematically and physically), there is no sharp "edge" to the boundary layer; that is, $u \to U$ as we get farther from the plate. We define the **boundary layer thickness, δ,** as that distance from the plate at which the fluid velocity is within some arbitrary value of the upstream/free-stream velocity. Typically, as indicated in **Fig. 9.8a,**

$$\delta = y \qquad \text{where} \qquad u = 0.99U$$

To remove this arbitrariness (i.e., what is so special about 99%; why not 98%?), the following definitions are introduced. Shown in **Fig. 9.8b** are two velocity profiles for flow past a flat plate—one if there were no viscosity (a uniform profile) and the other if there were viscosity and zero slip at the wall (the boundary layer profile). Because of the velocity deficit, $U - u$, within the boundary layer, the flowrate across section b–b is less than that

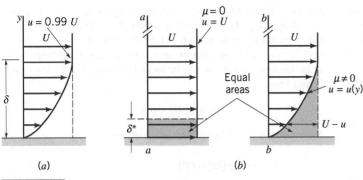

FIGURE 9.8 Boundary layer thickness: (a) standard boundary layer thickness, (b) boundary layer displacement thickness.

across section a–a. However, if we displace the plate at section a–a by an appropriate amount δ^*, the *boundary layer displacement thickness*, the flowrates across each section will be identical. This is true if

$$\delta^* bU = \int_0^\infty (U - u)b \, dy$$

where b is the plate width. Thus,

$$\delta^* = \int_0^\infty \left(1 - \frac{u}{U}\right) dy \qquad (9.3)$$

The displacement thickness represents the amount that the thickness of the body must be increased so that the fictitious uniform inviscid flow has the same mass flowrate properties as the actual viscous flow. It represents the outward displacement of the streamlines caused by the viscous effects on the plate. This idea allows us to simulate the presence that the boundary layer has on the flow outside of the boundary layer by adding the displacement thickness to the actual wall and treating the flow over the thickened body as an inviscid flow. The displacement thickness concept is illustrated in Example 9.3.

> The boundary layer displacement thickness is defined in terms of volumetric flowrate.

EXAMPLE 9.3 | Boundary Layer Displacement Thickness

Given Air flowing into a 0.6 m-square duct with a uniform velocity of 3 m/s forms a boundary layer on the walls as shown in **Fig. E9.3a**. The fluid within the core region (outside the boundary layers) flows as if it were inviscid. From advanced calculations it is determined that for this flow the boundary layer displacement thickness is given by

$$\delta^* = 0.0070 (x)^{1/2} \qquad (1)$$

where δ^* and x are in feet.

Find Determine the velocity $U = U(x)$ of the air within the duct but outside of the boundary layer.

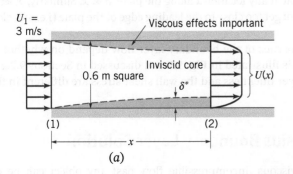

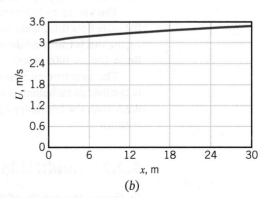

FIGURE E9.3

Solution If we assume incompressible flow (a reasonable assumption because of the low velocities involved), it follows that the volume flow-rate across any section of the duct is equal to that at the entrance (i.e., $Q_1 = Q_2$). That is,

$$U_1 A_1 = 3 \text{ m/s } (0.6 \text{ m})^2 = 1.1 \text{ m}^3/\text{s} = \int_{(2)} u \, dA$$

According to the definition of the displacement thickness, δ^*, the flow-rate across section (2) is the same as that for a uniform flow with velocity U through a duct whose walls have been moved inward by δ^*. That is,

$$1.1 \text{ m}^3/\text{s} = \int_{(2)} u \, dA = U(0.6 \text{ m} - 2\delta^*)^2 \tag{2}$$

By combining Eqs. 1 and 2 we obtain

$$1.1 \text{ m}^3/\text{s} = U(0.6 - 0.008x^{1/2})^2$$

or

$$U = \frac{1.1}{(0.6\text{m} - 0.008x^{1/2})^2} \text{m/s} \tag{Ans}$$

Comments Note that U increases in the downstream direction. For example, as shown in **Fig. E9.3b**, $U = 3.56$ m/s at $x = 30$ m. The viscous effects that cause the fluid to stick to the walls of the duct reduce the effective size of the duct, thereby (from conservation of mass principles) causing the fluid to accelerate (see also

Section 8.1.2). The pressure drop necessary to do this can be obtained by using the Bernoulli equation (Eq. 4.7) along the invis-cid streamlines from section (1) to (2). (Recall that this equation is not valid for viscous flows within the boundary layer. It is, however, valid for the inviscid flow outside the boundary layer.) Thus,

$$p_1 + \frac{1}{2}\rho U_1^2 = p + \frac{1}{2}\rho U^2$$

Hence, with $\rho = 1.20$ kg/m^3 and $p_1 = 0$ we obtain

$$
\begin{aligned}
p &= \frac{1}{2}\rho \left(U_1^2 - U^2 \right) \\
&= \frac{1}{2}(1.2 \text{ kg/m}^3) \\
&\quad \times \left[(3\text{m/s})^2 - \frac{(1.1)^2}{(0.6\text{m} - 0.008x^{1/2})^4 \text{ m}^4 \text{s}^2} \right]
\end{aligned}
$$

or

$$p = 0.6 \left[9 - \frac{1.2}{(0.6\text{m} - 0.008x^{1/2})^4} \right] \text{N/m}^2$$

For example, $p = -2.12$ N/m^2 at $x = 30$ m.

If it were desired to maintain a constant velocity along the centerline of this entrance region of the duct, the walls could be displaced outward by an amount equal to the boundary layer displacement thickness, δ^*.

Another boundary layer thickness definition, the *boundary layer momentum thickness*, Θ, is often used when determining the drag on an object. Again because of the velocity deficit, $U - u$, in the boundary layer, the momentum flux across section b–b in Fig. 9.8 is less than that across section a–a. This deficit in momentum flux for the actual boundary layer flow on a plate of width b is given by

$$\int \rho u (U - u) \, dA = \rho b \int_0^\infty u(U - u) \, dy$$

The boundary layer momentum thickness is defined in terms of momentum flux.

which by definition is the momentum flux in a layer of uniform speed U and thickness Θ. That is,

$$\rho b U^2 \Theta = \rho b \int_0^\infty u(U - u) \, dy$$

or

$$\Theta = \int_0^\infty \frac{u}{U}\left(1 - \frac{u}{U}\right) dy \tag{9.4}$$

All three boundary layer thickness definitions, δ, δ^*, and Θ, are of use in boundary layer analyses.

The boundary layer concept is based on the fact that the boundary layer is thin. For the flat plate flow this means that at any location x along the plate, $\delta \ll x$. Similarly, $\delta^* \ll x$ and Θ. Again, this is true if we do not get too close to the leading edge of the plate (i.e., not closer than $Re_x = Ux/v = 1000$ or so).

The structure and properties of the boundary layer flow depend on whether the flow is laminar or turbulent. As is illustrated in **Fig. 9.9** and discussed in Sections 9.2.2 through 9.2.5, both the boundary layer thickness and the wall shear stress are different in these two regimes.

9.2.2 Prandtl/Blasius Boundary Layer Solution

In theory, the details of viscous, incompressible flow past any object can be obtained by solving the governing Navier–Stokes equations discussed in Section 6.8.2. For steady,

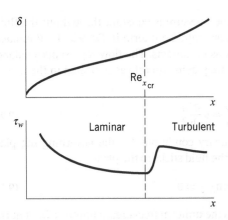

FIGURE 9.9 Typical characteristics of boundary layer thickness and wall shear stress for laminar and turbulent boundary layers.

two-dimensional laminar flows with negligible gravitational effects, these equations (Eqs. 6.127a, b, and c) reduce to the following:

$$u\frac{\partial u}{\partial x} + v\frac{\partial u}{\partial y} = -\frac{1}{\rho}\frac{\partial p}{\partial x} + v\left(\frac{\partial^2 u}{\partial x^2} + \frac{\partial^2 u}{\partial y^2}\right) \tag{9.5}$$

$$u\frac{\partial v}{\partial x} + v\frac{\partial v}{\partial y} = -\frac{1}{\rho}\frac{\partial p}{\partial y} + v\left(\frac{\partial^2 v}{\partial x^2} + \frac{\partial^2 v}{\partial y^2}\right) \tag{9.6}$$

which express Newton's second law. In addition, the conservation of mass equation, Eq. 6.31, for steady, two-dimensional incompressible flow is

$$\frac{\partial u}{\partial x} + \frac{\partial v}{\partial y} = 0 \tag{9.7}$$

The appropriate boundary conditions are that the fluid velocity (both u and v) far from the body is the upstream velocity and zero at solid body surfaces (i.e., the fluid sticks to the surface). Although the mathematical problem is well-posed, no one has obtained a general analytical solution to these equations applying to all flows past arbitrary body shapes! Although some analytical solutions exist for specific flows which allow simplifications to the equations, currently much work is being done to obtain numerical solutions to these governing equations for many flow geometries.

By using boundary layer concepts introduced in the previous sections, Prandtl was able to impose certain approximations (valid for large Reynolds number flows) and thereby to simplify the governing equations. In 1908, H. Blasius (1883–1970), one of Prandtl's students, was able to solve these simplified equations for the laminar boundary layer flow past a flat plate parallel to the flow. A brief outline of this technique and the results are presented in the following paragraphs. Additional details may be found in the literature (Refs. 1–3).

Since the boundary layer is thin, it is expected that the component of velocity normal to the plate is much smaller than that parallel to the plate and that the rate of change of any parameter (except pressure) across the boundary layer should be much greater than that along the flow direction. That is,

$$v \ll u \quad \text{and} \quad \frac{\partial}{\partial x} \ll \frac{\partial}{\partial y}$$

Physically, the flow is primarily parallel to the plate surface and any fluid property is convected downstream much more quickly than it is diffused across the streamlines. Pressure gradient normal to the surface is negligibly small.

With these assumptions it can be shown that the governing equations (Eqs. 9.5, 9.6, and 9.7) reduce to the following *boundary layer equations:*

$$\frac{\partial u}{\partial x} + \frac{\partial v}{\partial y} = 0 \tag{9.8}$$

$$u\frac{\partial u}{\partial x} + v\frac{\partial u}{\partial y} = -\frac{1}{\rho}\left(\frac{dp}{dx}\right) + v\frac{\partial^2 u}{\partial y^2} \tag{9.9a}$$

> The Navier–Stokes equations can be simplified for boundary layer flow analysis.

Although both these boundary layer equations and the original Navier–Stokes equations are nonlinear partial differential equations, there is a considerable difference between them. The y momentum equation has been eliminated, leaving only the original, unaltered continuity equation and a modified x momentum equation.

For boundary layer flow over a flat plate the pressure is constant throughout the fluid. Thus for flat plate, boundary layer flow, the pressure gradient term in Eq. 9.9a is eliminated, leaving only the x and y components of velocity as unknowns. The flow represents a balance between viscous and inertial effects, with pressure playing no role as indicated in the following flat plate boundary layer equation:

$$u\frac{\partial u}{\partial x} + v\frac{\partial u}{\partial y} = v\frac{\partial^2 u}{\partial y^2} \tag{9.9b}$$

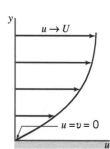

As shown within the adjacent figure, the boundary conditions for the governing flat plate boundary layer equations (9.8 and 9.9b) are that the fluid sticks to the plate

$$u = v = 0 \quad \text{on} \quad y = 0 \tag{9.10}$$

and that outside of the boundary layer the flow is the uniform free-stream flow $u = U$. That is,

$$u \to U \quad \text{as} \quad y \to \infty \tag{9.11}$$

Mathematically, the free-stream velocity is approached asymptotically as one moves away from the plate. Physically, the flow velocity is within 1% of the free-stream velocity at a distance of δ from the plate (as defined in the previous section).

In mathematical terms, the Navier–Stokes equations (Eqs. 9.5 and 9.6) and the continuity equation (Eq. 9.7) are elliptic equations, whereas the equations for boundary layer flow (Eqs. 9.8 and 9.9a or 9.9b) are parabolic equations. The nature of the solutions to these two sets of equations, therefore, is different. Physically, this fact translates to the idea that what happens downstream of a given location in a boundary layer cannot affect what happens upstream of that point. That is, whether the plate shown in Fig. 9.5c ends with length ℓ or is extended to length 2ℓ, the flow within the first segment of length ℓ will be the same. In addition, the presence of the plate has no effect on the flow ahead of the plate. On the other hand, ellipticity allows flow information to propagate in all directions, including upstream.

In general, the solutions of nonlinear partial differential equations (such as the boundary layer equations, Eqs. 9.8 and 9.9) are extremely difficult to obtain. However, by applying a clever coordinate transformation and change of variables, Blasius reduced the flat plate partial differential equations to an ordinary differential equation that he was able to solve. A brief description of this process is given in the following paragraphs. Additional details can be found in standard books dealing with boundary layer flow (Refs. 1, 2).

It can be argued that in dimensionless form the boundary layer velocity profiles on a flat plate should be similar regardless of the location along the plate. That is,

$$\frac{u}{U} = g\left(\frac{y}{\delta}\right)$$

where $g(y/\delta)$ is an unknown function to be determined. In addition, by applying an order of magnitude analysis of the forces acting on fluid within the boundary layer, it can be shown that the boundary layer thickness grows as the square root of x and inversely proportional to the square root of U. That is,

$$\delta \sim \left(\frac{vx}{U}\right)^{1/2}$$

Such a conclusion results from a balance between viscous and inertial forces within the boundary layer and from the fact that the velocity varies much more rapidly in the direction across the boundary layer than along it. (Details can be found in Ref. 4.)

Thus, we introduce the dimensionless *similarity variable* $\eta = (U/vx)^{1/2}y$ and the stream function $\psi = (vxU)^{1/2}f(\eta)$, where $f = f(\eta)$ is an unknown function. Recall from Section 6.2.3 that the velocity components for two-dimensional flow are given in terms of the stream function as $u = \partial\psi/\partial y$ and $v = -\partial\psi/\partial x$, which for this flow become

$$u = Uf'(\eta) \tag{9.12}$$

and

$$v = \left(\frac{vU}{4x}\right)^{1/2}(\eta f' - f) \tag{9.13}$$

The boundary layer equations can be written in terms of a similarity variable.

with the notation $(\ )' = d/d\eta$. We substitute Eqs. 9.12 and 9.13 into the governing equations, Eqs. 9.8 and 9.9b, to obtain (after considerable manipulation) the following nonlinear, third-order ordinary differential equation:

$$2f''' + ff'' = 0 \tag{9.14a}$$

As shown in the adjacent figure, the boundary conditions given in Eqs. 9.10 and 9.11 can be written as

$$f = f' = 0 \text{ at } \eta = 0 \quad \text{and} \quad f' \to 1 \text{ as } \eta \to \infty \tag{9.14b}$$

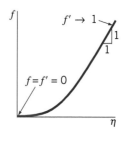

The original partial differential equations and boundary conditions have been reduced to an ordinary differential equation by use of the similarity variable η. The two independent variables, x and y, were combined into the similarity variable in a fashion that reduced the partial differential equations (and boundary conditions) to an ordinary differential equation. This type of reduction is not generally possible. For example, this method does not work on the full Navier–Stokes equations, although it does on the flat plate boundary layer equations (Eqs. 9.8 and 9.9b).

Although there is no known analytical solution to Eq. 9.14, it is relatively easy to integrate this equation on a computer with the results considered an exact solution. The dimensionless boundary layer profile, $u/U = f'(\eta)$, obtained by numerical solution of Eq. 9.14 (termed the Blasius solution), is sketched in **Fig. 9.10a** and is tabulated in **Table 9.1**. The velocity profiles at different x locations are similar in that there is only one curve necessary to describe the velocity at any point in the boundary layer. Because the similarity variable η contains both x and y, **Fig. 9.10b** shows that the actual velocity profiles are a function of both x and y. The profile at location x_1 is the same as that at x_2 except that the y coordinate is stretched by a factor of $(x_2/x_1)^{1/2}$. You can achieve similar changes in the boundary layer velocity profile by altering the kinematic viscosity, v, as shown in the adjacent figure.

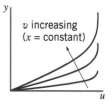

From the solution it is found that $u/U \approx 0.99$ when $\eta = 5.0$. Thus,

$$\delta = 5\sqrt{\frac{vx}{U}} \tag{9.15}$$

or

$$\frac{\delta}{x} = \frac{5}{\sqrt{Re_x}}$$

where $Re_x = Ux/v$. It can also be shown that the displacement and momentum thicknesses are given by

$$\frac{\delta^*}{x} = \frac{1.721}{\sqrt{Re_x}} \tag{9.16}$$

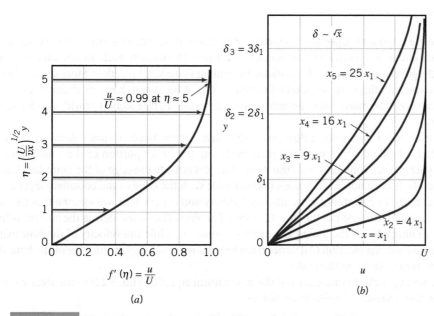

FIGURE 9.10 Blasius boundary layer profile: (a) boundary layer profile in dimensionless form using the similarity variable η, (b) similar boundary layer profiles at different locations along the flat plate.

TABLE 9.1 Laminar Flow along a Flat Plate (the Blasius Solution)

$\eta = y(U/\upsilon x)^{1/2}$	$f'(\eta) = u/U$	η	$f'(\eta)$
0	0	3.6	0.9233
0.4	0.1328	4.0	0.9555
0.8	0.2647	4.4	0.9759
1.2	0.3938	4.8	0.9878
1.6	0.5168	5.0	0.9916
2.0	0.6298	5.2	0.9943
2.4	0.7290	5.6	0.9975
2.8	0.8115	6.0	0.9990
3.2	0.8761	∞	1.0000

For large Reynolds numbers the boundary layer is relatively thin.

and

$$\frac{\Theta}{x} = \frac{0.664}{\sqrt{Re_x}} \tag{9.17}$$

As postulated, the boundary layer is thin provided that Re_x is large (i.e., $\delta/x \to 0$ as $Re_x \to \infty$).

With the velocity profile known, it is an easy matter to determine the wall shear stress, $\tau_w = \mu(\partial u/\partial y)_{y=0}$, where the velocity gradient is evaluated at the plate. The value of $\partial u/\partial y$ at $y = 0$ can be obtained from the Blasius solution to give

$$\tau_w = 0.332 U^{3/2} \sqrt{\frac{\rho\mu}{x}} \tag{9.18}$$

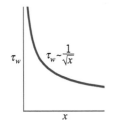

As indicated by Eq. 9.18 and illustrated in the adjacent figure, the shear stress decreases with increasing x because of the increasing thickness of the boundary layer—the velocity gradient at the wall decreases with increasing x. Also, τ_w varies as $U^{3/2}$, not as U as it does for fully developed laminar pipe flow. These variations are discussed in Section 9.2.3.

9.2.3 Momentum Integral Boundary Layer Equation for a Flat Plate

One of the important aspects of boundary layer theory is the determination of the drag caused by shear forces on a body. As was discussed in the previous section, such results can be obtained from the governing differential equations for laminar boundary layer flow. Since exact solutions are extremely difficult (or impossible) to obtain, it is of interest to have an alternative approximate method. The momentum integral method described in this section provides such an alternative.

We consider the uniform flow past a flat plate and the fixed control volume as shown in **Fig. 9.11**. The fluid adjacent to the plate makes up the lower portion of the control surface. The upper surface coincides with the streamline just outside the edge of the boundary layer at section (2). It need not (in fact, does not) coincide with the edge of the boundary layer except at section (2). In agreement with advanced theory and experiment, we assume that the pressure is constant throughout the flow field for a flat plate. The flow entering the control volume at the leading edge of the plate [section (1)] is uniform, while the velocity of the flow exiting the control volume [section (2)] varies from the upstream velocity at the edge of the boundary layer to zero velocity on the plate.

If we apply the x component of the momentum equation (Eq. 5.22) to the steady flow of fluid within the control volume we obtain

$$\sum F_x = \rho \int_{(1)} u\mathbf{V} \cdot \hat{\mathbf{n}}\, dA + \rho \int_{(2)} u\mathbf{V} \cdot \hat{\mathbf{n}}\, dA$$

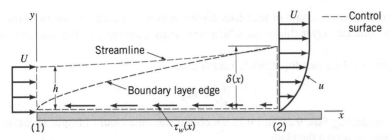

FIGURE 9.11 Control volume used in the derivation of the momentum integral equation for boundary layer flow.

where for a plate of width b

$$\sum F_x = -\mathcal{D} = -\int_{\text{plate}} \tau_w \, dA = -b \int_{\text{plate}} \tau_w \, dx \qquad (9.19)$$

and $\mathcal{D}$ is the drag that the plate exerts on the fluid. Note that the net force caused by the uniform pressure distribution does not contribute to this flow. Since the plate is solid and the upper surface of the control volume is a streamline, there is no flow through these areas. Thus,

$$-\mathcal{D} = \rho \int_{(1)} U(-U) \, dA + \rho \int_{(2)} u^2 \, dA$$

or

$$\mathcal{D} = \rho U^2 bh - \rho b \int_0^\delta u^2 \, dy \qquad (9.20)$$

> The drag on a flat plate depends on the velocity profile within the boundary layer.

Although the height h is not known, it is known that for conservation of mass the flowrate through section (1) must equal that through section (2), or

$$Uh = \int_0^\delta u \, dy$$

which can be written as

$$\rho U^2 bh = \rho b \int_0^\delta Uu \, dy \qquad (9.21)$$

Thus, by combining Eqs. 9.20 and 9.21 we obtain the drag in terms of the deficit of momentum flux across the outlet of the control volume as

$$\mathcal{D} = \rho b \int_0^\delta u(U - u) \, dy \qquad (9.22)$$

> Drag on a flat plate is related to momentum deficit within the boundary layer.

The idea of a momentum deficit is illustrated in the adjacent figures. If the flow were inviscid, the drag would be zero, since we would have $u \equiv U$ and the right-hand side of Eq. 9.22 would be zero. (This is consistent with the fact that $\tau_w = 0$ if $\mu = 0$.) Equation 9.22 points out the important fact that boundary layer flow on a flat plate is governed by a balance between shear drag (the left-hand side of Eq. 9.22) and a decrease in the momentum of the fluid (the right-hand side of Eq. 9.22). As x increases, δ increases and the drag increases. The thickening of the boundary layer is necessary to overcome the drag of the viscous shear stress on the plate. This is contrary to horizontal fully developed pipe flow in which the momentum of the fluid remains constant and the shear force is overcome by the pressure gradient along the pipe.

The development of Eq. 9.22 and its use was first put forth in 1921 by T. von Kármán (1881–1963), a Hungarian/German-American aerodynamicist. By comparing Eqs. 9.22 and 9.4 we see that the drag can be written in terms of the momentum thickness, Θ, as

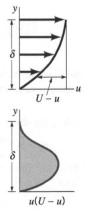

$$\mathcal{D} = \rho b U^2 \, \Theta \qquad (9.23)$$

Note that this equation is valid for laminar or turbulent flows.

The shear stress distribution can be obtained from Eq. 9.23 by differentiating both sides with respect to x to obtain

$$\frac{d\mathcal{D}}{dx} = \rho b U^2 \frac{d\Theta}{dx} \qquad (9.24)$$

The increase in drag per length of the plate, $d\mathscr{D}/dx$, occurs at the expense of an increase of the momentum boundary layer thickness, which represents a decrease in the momentum of the fluid.

Since $d\mathscr{D} = \tau_w b \, dx$ (see Eq. 9.19), it follows that

$$\frac{d\mathscr{D}}{dx} = b\tau_w \qquad (9.25)$$

Hence, by combining Eqs. 9.24 and 9.25 we obtain the *momentum integral equation* for the boundary layer flow on a flat plate

$$\tau_w = \rho U^2 \frac{d\Theta}{dx} \qquad (9.26)$$

Although no integral is visible in this form of the equation, recall that there is one "hiding" in the Θ term (see Eq. 9.4).

The usefulness of this relationship lies in the ability to obtain approximate results easily by using rather crude assumptions. For example, if we knew the detailed velocity profile in the boundary layer (e.g., the Blasius solution discussed in the previous section), we could evaluate either the right-hand side of Eq. 9.23 to obtain the drag or the right-hand side of Eq. 9.26 to obtain the shear stress. Fortunately, even a rather crude guess at the velocity profile will allow us to obtain reasonable drag and shear stress results from Eq. 9.26. This method is illustrated in Example 9.4.

EXAMPLE 9.4 | Momentum Integral Boundary Layer Equation

Given   Consider the laminar flow of an incompressible fluid past a flat plate at $y = 0$. The boundary layer velocity profile is approximated as $u = Uy/\delta$ for $0 \leq y \leq \delta$ and $u = U$ for $y > \delta$, as is shown in **Fig. E9.4**.

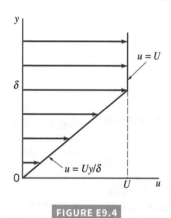

FIGURE E9.4

Find   Determine the shear stress by using the momentum integral equation. Compare these results with the Blasius results given by Eq. 9.18.

Solution   From Eq. 9.26 the shear stress is given by

$$\tau_w = \rho U^2 \frac{d\Theta}{dx} \qquad (1)$$

while for laminar flow we know that $\tau_w = \mu(\partial u/\partial y)_{y=0}$. For the assumed profile we have

$$\tau_w = \mu \frac{U}{\delta} \qquad (2)$$

and from Eq. 9.4

$$\Theta = \int_0^\infty \frac{u}{U}\left(1 - \frac{u}{U}\right) dy = \int_0^\delta \frac{u}{U}\left(1 - \frac{u}{U}\right) dy$$

$$= \int_0^\delta \left(\frac{y}{\delta}\right)\left(1 - \frac{y}{\delta}\right) dy$$

or

$$\Theta = \frac{\delta}{6} \qquad (3)$$

Note that as yet we do not know the value of δ (but suspect that it should be a function of x).

By combining Eqs. 1, 2, and 3 we obtain the following differential equation for δ:

$$\frac{\mu U}{\delta} = \frac{\rho U^2}{6} \frac{d\delta}{dx}$$

or

$$\delta \, d\delta = \frac{6\mu}{\rho U} dx$$

This can be integrated from the leading edge of the plate, $x = 0$ (where $\delta = 0$) to an arbitrary location x where the boundary layer thickness is δ. The result is

$$\frac{\delta^2}{2} = \frac{6\mu}{\rho U} x$$

or

$$\delta = 3.46 \sqrt{\frac{\upsilon x}{U}} \qquad (4)$$

Although the velocity profile is not actually the simple straight line we assumed, the approximate result given by Eq. 4 compares favorably with the (much more laborious to obtain) Blasius result given by Eq. 9.15.

The wall shear stress can also be obtained by combining Eqs. 1, 3, and 4 to give

$$\tau_w = 0.289 U^{3/2} \sqrt{\frac{\rho\mu}{x}} \qquad (Ans)$$

Again this approximate result is fairly close (within 13%) to the Blasius value of τ_w given by Eq. 9.18.

As is illustrated in Example 9.4, the momentum integral equation, Eq. 9.26, can be used along with an assumed velocity profile to obtain reasonable, approximate boundary layer results. The accuracy of these results depends on how closely the shape of the assumed velocity profile approximates the actual profile.

Thus, we consider a general velocity profile

$$\frac{u}{U} = g(Y) \quad \text{for} \quad 0 \le Y \le 1$$

and

$$\frac{u}{U} = 1 \quad \text{for} \quad Y > 1$$

Approximate velocity profiles are used in the momentum integral equation.

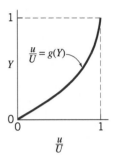

where the dimensionless coordinate $Y = y/\delta$ varies from 0 to 1 across the boundary layer. The dimensionless function $g(Y)$ can be any shape we choose, although it should be a reasonable approximation to the boundary layer profile, as shown in the adjacent figure. In particular, it should certainly satisfy the boundary conditions $u = 0$ at $y = 0$ and $u = U$ at $y = \delta$. That is,

$$g(0) = 0 \quad \text{and} \quad g(1) = 1$$

The linear function $g(Y) = Y$ used in Example 9.4 is one such possible profile. Other conditions, such as $dg/dY = 0$ at $Y = 1$ (i.e., $\partial u/\partial y = 0$ at $y = \delta$), and $d^2g/dY^2 = 0$ at $Y = 0$ (i.e., $\partial^2 u/\partial y^2 = 0$ at $y = 0$) could also be incorporated into the function $g(Y)$ to more closely approximate the actual profile.

For a given $g(Y)$, the drag can be determined from Eq. 9.22 as

$$\mathcal{D} = \rho b \int_0^\delta u(U - u)\, dy = \rho b U^2 \delta \int_0^1 g(Y)[1 - g(Y)]\, dY$$

or

$$\mathcal{D} = \rho b U^2 \delta C_1 \tag{9.27}$$

where the dimensionless constant C_1 has the value

$$C_1 = \int_0^1 g(Y)[1 - g(Y)]\, dY$$

Also, the wall shear stress can be written as

$$\tau_w = \mu \left.\frac{\partial u}{\partial y}\right|_{y=0} = \frac{\mu U}{\delta} \left.\frac{dg}{dY}\right|_{Y=0} = \frac{\mu U}{\delta} C_2 \tag{9.28}$$

where the dimensionless constant C_2 has the value

$$C_2 = \left.\frac{dg}{dY}\right|_{Y=0}$$

By combining Eqs. 9.25, 9.27, and 9.28 we obtain

$$\delta\, d\delta = \frac{\mu C_2}{\rho U C_1}\, dx$$

which can be integrated from $\delta = 0$ at $x = 0$ to give

$$\delta = \sqrt{\frac{2\upsilon C_2 x}{U C_1}}$$

or

$$\frac{\delta}{x} = \frac{\sqrt{2C_2/C_1}}{\sqrt{\text{Re}_x}} \tag{9.29}$$

By substituting this expression back into Eq. 9.28, we obtain

$$\tau_w = \sqrt{\frac{C_1 C_2}{2}}\, U^{3/2} \sqrt{\frac{\rho\mu}{x}} \tag{9.30}$$

To use Eqs. 9.29 and 9.30 we must determine the values of C_1 and C_2. Several assumed velocity profiles and the resulting values of δ are given in **Fig. 9.12** and **Table 9.2**. The more closely the assumed shape approximates the actual (i.e., Blasius) profile, the more accurate

Approximate boundary layer results are obtained from the momentum integral equation.

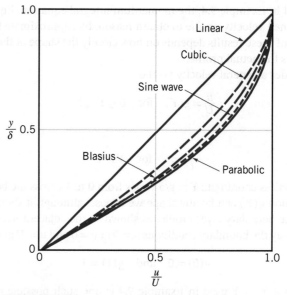

FIGURE 9.12 Typical approximate boundary layer profiles used in the momentum integral equation.

TABLE 9.2 Flat Plate Momentum Integral Results for Various Assumed Laminar Flow Velocity Profiles

Profile Character	$\delta \mathrm{Re}_x^{1/2}/x$	$c_f \mathrm{Re}_x^{1/2}$	$C_{Df} \mathrm{Re}_\ell^{1/2}$
a. Blasius solution	5.00	0.664	1.328
b. Linear $u/U = y/\delta$	3.46	0.578	1.156
c. Parabolic $u/U = 2y/\delta - (y/\delta)^2$	5.48	0.730	1.460
d. Cubic $u/U = 3(y/\delta)/2 - (y/\delta)^3/2$	4.64	0.646	1.292
e. Sine wave $u/U = \sin[\pi(y/\delta)/2]$	4.79	0.655	1.310

the final results. For any assumed profile shape, the functional dependence of δ and τ_w on the physical parameters ρ, μ, U, and x is the same. Only the constants are different. That is, $\delta \sim (\mu x/\rho U)^{1/2}$ or $\delta \mathrm{Re}_x^{1/2}/x = $ constant, and $\tau_w \sim (\rho \mu U^3/x)^{1/2}$, where $\mathrm{Re}_x = \rho U x/\mu$.

It is often convenient to use the dimensionless *local friction coefficient*, c_f, defined as

$$c_f = \frac{\tau_w}{\frac{1}{2}\rho U^2} \tag{9.31}$$

to express the wall shear stress. From Eq. 9.30 we obtain the approximate value

$$c_f = \sqrt{2C_1 C_2}\sqrt{\frac{\mu}{\rho U x}} = \frac{\sqrt{2C_1 C_2}}{\sqrt{\mathrm{Re}_x}}$$

while the Blasius solution result is given by

$$c_f = \frac{0.664}{\sqrt{\mathrm{Re}_x}} \tag{9.32}$$

These results are also indicated in Table 9.2.

For a flat plate of length ℓ and width b, the net friction drag, $\mathscr{D}_f$, can be expressed in terms of the dimensionless *friction drag* coefficient, C_{Df}, as

The friction drag coefficient is an integral of the local friction coefficient.

$$C_{Df} = \frac{\mathscr{D}_f}{\frac{1}{2}\rho U^2 b\ell} = \frac{b\int_0^\ell \tau_w \, dx}{\frac{1}{2}\rho U^2 b\ell}$$

or

$$C_{Df} = \frac{1}{\ell} \int_0^\ell c_f \, dx \tag{9.33}$$

We use the above approximate value of $c_f = (2C_1C_2\mu/\rho Ux)^{1/2}$ to obtain

$$C_{Df} = \frac{\sqrt{8C_1C_2}}{\sqrt{\mathrm{Re}_\ell}}$$

where $\mathrm{Re}_\ell = U\ell/\upsilon$ is the Reynolds number based on the plate length. The corresponding value obtained from the Blasius solution (Eq. 9.32) and shown in the adjacent figure gives

$$C_{Df} = \frac{1.328}{\sqrt{\mathrm{Re}_\ell}}$$

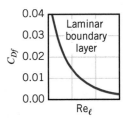

These results are also indicated in Table 9.2.

The momentum integral boundary layer method provides a relatively simple technique to obtain useful approximate boundary layer results. You may wonder why these approximate results are useful if we have the exact laminar flow flat plate boundary layer Blasius solution. As is discussed in Sections 9.2.5 and 9.2.6, this technique can be extended to boundary layer flows on curved surfaces (where the pressure and fluid velocity at the edge of the boundary layer are not constant) and to turbulent flows.

9.2.4 Transition from Laminar to Turbulent Flow

The analytical results given in Table 9.2 are restricted to laminar boundary layer flows along a flat plate with zero pressure gradient. They agree quite well with experimental results up to the point where the boundary layer flow becomes turbulent, which will occur for any free-stream velocity and any fluid provided the plate is long enough. This is true because the parameter that predicts the **transition** to turbulent flow is the Reynolds number—in this case the Reynolds number based on the distance from the leading edge of the plate, $\mathrm{Re}_x = Ux/\upsilon$.

The value of the Reynolds number at the transition location is a rather complex function of various parameters involved, including the roughness of the surface, the curvature of the surface (for example, a flat plate or a sphere), and some measure of the disturbances in the flow outside the boundary layer. On a flat plate with a sharp leading edge in a typical air-stream, the transition takes place at a distance x from the leading edge given by $\mathrm{Re}_{xcr} = 2 \times 10^5$ to 3×10^6. Unless otherwise stated, we will use $\mathrm{Re}_{xcr} = 5 \times 10^5$ in our calculations.

The actual transition from laminar to turbulent boundary layer flow may occur over a region of the plate, not at a specific single location. This occurs, in part, because of the spottiness of the transition. Typically, the transition begins at random locations on the plate in the vicinity of $\mathrm{Re}_x = \mathrm{Re}_{xcr}$. These spots grow rapidly as they are convected downstream until the entire width of the plate is covered with turbulent flow. The photo shown in **Fig. 9.13** illustrates this transition process.

The boundary layer on a flat plate will become turbulent if the plate is long enough.

V9.7 Transition on flat plate

Photograph courtesy of Brian Cantwell, Stanford University.

FIGURE 9.13 Turbulent spots and the transition from laminar to turbulent boundary layer flow on a flat plate. Flow from left to right.

The complex process of transition from laminar to turbulent flow involves the instability of the flow field. Small disturbances imposed on the boundary layer flow (e.g., from a vibration of the plate, a roughness of the surface, or a "wiggle" in the flow past the plate) will either grow (instability) or decay (stability), depending on where the disturbance is introduced into the flow. If these disturbances occur at a location with $\text{Re}_x < \text{Re}_{xcr}$ they will die out, and the boundary layer will return to laminar flow at that location. Disturbances imposed at a location with $\text{Re}_x > \text{Re}_{xcr}$ will grow and transform the boundary layer flow downstream of this location into turbulence. The study of the initiation, growth, and structure of these turbulent bursts or spots is an active area of fluid mechanics research.

Transition from laminar to turbulent flow also involves a noticeable change in the shape of the boundary layer velocity profile. Typical profiles obtained in the neighborhood of the transition location are indicated in **Fig. 9.14**. The turbulent profiles are flatter (see Sections 8.3.2 and 9.2.5 for explanations), have a larger velocity gradient at the wall, and produce a larger boundary layer thickness than do the laminar profiles.

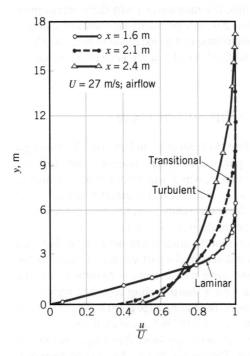

FIGURE 9.14 Typical boundary layer profiles on a flat plate for laminar, transitional, and turbulent flow (Ref. 1).

EXAMPLE 9.5 | Boundary Layer Transition

Given A fluid flows steadily past a flat plate with a velocity of $U = 3$ m/s.

Find At approximately what location will the boundary layer become turbulent, and how thick is the boundary layer at that point if the fluid is (a) water at 16 °C, (b) standard air, or (c) glycerin at 20 °C?

Solution For any fluid, the laminar boundary layer thickness on a flat plate is found from Eq. 9.15 as

$$\delta = 5 \sqrt{\frac{\nu x}{U}} \tag{1}$$

The boundary layer remains laminar up to

$$x_{cr} = \frac{\nu \, \text{Re}_{xcr}}{U}$$

Thus, if we assume $\text{Re}_{xcr} = 5 \times 10^5$ we obtain

$$x_{cr} = \frac{5 \times 10^5}{3 \text{ m/s}} \nu = 1.7 \times 10^5 \nu$$

and from Eq. (1)

$$\delta_{cr} \equiv \delta|_{x=x_{cr}} = 5 \left[\frac{\nu}{3} (1.7 \times 10^5 \nu) \right]^{1/2} = 1190 \, \nu$$

TABLE E9.5

Fluid	ν (m²/s)	x_{cr} (m)	δ_{cr} (m)
a. Water	1.12×10^{-6}	0.19	1.3×10^{-3}
b. Air	1.46×10^{-5}	2.482	1.7×10^{-2}
c. Glycerin	1.19×10^{-3}	202.3	1.4

(Ans)

where ν is in m²/s and x_{cr} and δ_{cr} are in meter. The values of the kinematic viscosity obtained from Tables 1.5 and 1.7 are listed in **Table E9.5** along with the corresponding x_{cr} and δ_{cr}.

Comment Laminar flow can be maintained on a longer portion of the plate if the kinematic viscosity (μ/ρ) is increased. However, the boundary layer flow eventually becomes turbulent, provided the plate is long enough. Similarly, the boundary layer thickness is greater if the kinematic viscosity is increased.

9.2.5 Turbulent Boundary Layer Flow

The structure of turbulent boundary layer flow is very complex, seemingly random, and irregular. It shares many of the characteristics described for turbulent pipe flow in Section 8.3. In particular, the velocity at any given location in the flow is unsteady in a random fashion. The flow can be thought of as a jumbled mix of intertwined eddies (or swirls) of different sizes (diameters and angular velocities). The adjacent photograph shows a laser-induced fluorescence visualization of a turbulent boundary layer on a flat plate (side view). The various fluid quantities involved (i.e., mass, momentum, energy) are convected downstream in the free-stream direction as in a laminar boundary layer. For turbulent flow they are also convected across the boundary layer (in the direction perpendicular to the plate) by the random transport of relatively small, finite-sized fluid masses associated with the turbulent eddies. There is considerable mixing involved with these finite-sized eddies—considerably more than is associated with the mixing found in laminar flow where it is confined to the molecular scale. Although there is considerable random motion of fluid particles perpendicular to the plate, there is very little net transfer of mass across the boundary layer—the largest flowrate by far is parallel to the plate.

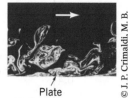

Plate

© J. P. Crimaldi, M. B. Wiley, & J. R. Koseff

Random transport of finite-sized fluid masses occurs within turbulent boundary layers.

There is, however, a considerable net transfer of x component of momentum perpendicular to the plate because of the random motion. Fluid masses moving toward the plate (in the negative y direction) have some of their excess momentum (they come from areas of higher velocity) removed by the plate. Conversely, masses moving away from the plate (in the positive y direction) gain momentum from the fluid (they come from areas of lower velocity). The net result is that the plate acts as a momentum sink, continually extracting momentum from the fluid. For laminar flows, such cross-stream transfer of these properties takes place solely on the molecular scale. For turbulent flow the randomness is associated with fluid mixing. Consequently, the shear force for turbulent boundary layer flow is considerably greater than it is for laminar boundary layer flow (see Section 8.3.2).

V9.8 Random flow but not turbulent

There are no "exact" solutions for turbulent boundary layer flow. As is discussed in Section 9.2.2, it is possible to solve the Prandtl boundary layer equations for laminar flow past a flat plate to obtain the Blasius solution (which is "exact" within the framework of the assumptions involved in the boundary layer equations). Since there is no precise expression for the shear stress in turbulent flow (see Section 8.3), solutions are not available for turbulent flow. However, as introduced in Appendix A considerable headway has been made in obtaining numerical (computer) solutions for turbulent flow by using approximate shear stress models. Also, progress is being made in the area of direct, full numerical integration of the basic governing equations, the Navier–Stokes equations.

Approximate turbulent boundary layer results can also be obtained by use of the momentum integral equation, Eq. 9.26, which is valid for either laminar or turbulent flat plate flow. What is needed for the use of this equation are reasonable approximations to the velocity profile $u = U g(Y)$, where $Y = y/\delta$ and u is the time-averaged velocity (the overbar notation, $\bar{u}$, of Section 8.3.2 has been dropped for convenience), and a functional relationship describing the wall shear stress. For laminar flow the wall shear stress was used as $\tau_w = \mu(\partial u/\partial y)_{y=0}$. In theory, such a technique should work for turbulent boundary layers also. However, as is discussed in Section 8.3, velocity fluctuations and hence turbulent shear stress contribute to the total shear stress; with no known exact model for turbulent stress, it is necessary to use some empirical relationship for the wall shear stress. This is illustrated in Example 9.6.

EXAMPLE 9.6 | Turbulent Boundary Layer Properties

Given Consider turbulent flow of an incompressible fluid past a flat plate. The boundary layer velocity profile is assumed to be $u/U = (y/\delta)^{1/7} = Y^{1/7}$ for $Y = y/\delta \le 1$ and $u = U$ for $Y > 1$ as shown in **Fig. E9.6**. This is a reasonable approximation of experimentally observed profiles, except very near the plate where this formula gives $\partial u/\partial y = \infty$ at $y = 0$. Note the differences between the assumed turbulent profile and the laminar profile. Also assume

that the shear stress agrees with the experimentally determined formula:

$$\tau_w = 0.0225\rho U^2 \left(\frac{v}{U\delta}\right)^{1/4} \tag{1}$$

Find Determine the boundary layer thicknesses δ, δ^*, and Θ and the wall shear stress, τ_w, as a function of x. Determine the friction drag coefficient, C_{Df}.

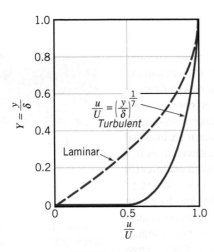

FIGURE E9.6

Solution Whether the flow is laminar or turbulent, it is true that the drag force is accounted for by a reduction in the momentum of the fluid flowing past the plate. The shear is obtained from Eq. 9.26 in terms of the rate at which the momentum boundary layer thickness, Θ, increases with distance along the plate as

$$\tau_w = \rho U^2 \frac{d\Theta}{dx} \qquad (1)$$

The boundary layer momentum thickness is obtained from Eq. 9.4 as

$$\Theta = \int_0^\infty \frac{u}{U}\left(1 - \frac{u}{U}\right) dy = \delta \int_0^1 \frac{u}{U}\left(1 - \frac{u}{U}\right) dY$$

By integration of the assumed velocity profile,

$$\Theta = \delta \int_0^1 Y^{1/7}(1 - Y^{1/7}) \, dY = \frac{7}{72}\delta \qquad (2)$$

where δ is an unknown function of x. By combining the assumed shear force dependence (Eq. 1) with Eq. 2, we obtain the following differential equation for δ:

$$0.0225\rho U^2 \left(\frac{v}{U\delta}\right)^{1/4} = \frac{7}{72}\rho U^2 \frac{d\delta}{dx}$$

or

$$\delta^{1/4}\, d\delta = 0.231\left(\frac{v}{U}\right)^{1/4} dx$$

This can be integrated from $\delta = 0$ at $x = 0$ to obtain

$$\delta = 0.370\left(\frac{v}{U}\right)^{1/5} x^{4/5} \qquad (3) \quad (Ans)$$

or in dimensionless form

$$\frac{\delta}{x} = \frac{0.370}{\mathrm{Re}_x^{1/5}}$$

Strictly speaking, the boundary layer near the leading edge of the plate is laminar, not turbulent, and the precise boundary condition should be the matching of the initial turbulent boundary layer thickness (at the transition location) with the thickness of the laminar boundary layer at that point. In practice, however, the laminar boundary layer often exists over a relatively short portion of the plate, and the error associated with starting the turbulent boundary layer with $\delta = 0$ at $x = 0$ can be negligible.

The displacement thickness, δ^*, and the momentum thickness, Θ, can be obtained from Eqs. 9.3 and 9.4 by integrating as follows:

$$\delta^* = \int_0^\infty \left(1 - \frac{u}{U}\right) dy = \delta \int_0^1 \left(1 - \frac{u}{U}\right) dY$$

$$= \delta \int_0^1 (1 - Y^{1/7})\, dY = \frac{\delta}{8}$$

Thus, by combining this with Eq. 3 we obtain

$$\delta^* = 0.0463\left(\frac{v}{U}\right)^{1/5} x^{4/5} \qquad (Ans)$$

You may have noticed that the assumed velocity profile is the one-seventh power-law introduced in Section 8.3.3. This assumed profile not only fails at the plate surface, but also fails at the outer edge of the boundary layer because the velocity gradient is not zero.

The given shear stress formula is actually obtained by transforming an experimentally determined formula for the friction factor in pipe flow (due to Blasius) into a form suitable for boundary layers (for example, by replacing pipe radius with boundary layer thickness).

Similarly, from Eqs. 2 and 3, the momentum thickness can be obtained as follows:

$$\Theta = \frac{7}{72}\delta = 0.0360\left(\frac{v}{U}\right)^{1/5} x^{4/5} \qquad (4) \quad (Ans)$$

The functional dependence for δ, δ^*, and Θ is the same; only the constants of proportionality are different. Typically, $\Theta < \delta^* < \delta$.

By combining Eqs. 1 and 3, we obtain the following result for the wall shear stress

$$\tau_w = 0.0225\rho U^2 \left[\frac{v}{U(0.370)(v/U)^{1/5} x^{4/5}}\right]^{1/4}$$

$$= \frac{0.0288\rho U^2}{\mathrm{Re}_x^{1/5}} \qquad (Ans)$$

This can be integrated over the length of the plate to obtain the friction drag on one side of the plate, $\mathcal{D}_f$, as

$$\mathcal{D}_f = \int_0^\ell b\tau_w \, dx = b(0.0288\rho U^2) \int_0^\ell \left(\frac{v}{Ux}\right)^{1/5} dx$$

or

$$\mathcal{D}_f = 0.0360\rho U^2 \frac{A}{\mathrm{Re}_\ell^{1/5}}$$

where $A = b\ell$ is the area of the plate. (This result can also be obtained by combining Eq. 9.23 and the expression for the momentum thickness given in Eq. 4.) The corresponding friction drag coefficient, C_{Df}, is

$$C_{Df} = \frac{\mathcal{D}_f}{\frac{1}{2}\rho U^2 A} = \frac{0.0720}{\mathrm{Re}_\ell^{1/5}} \qquad (Ans)$$

Comment Note that for the turbulent boundary layer flow the boundary layer thickness increases with x as $\delta \sim x^{4/5}$ and the shear stress decreases as $\tau_w \sim x^{-1/5}$. For laminar flow these dependencies are $x^{1/2}$ and $x^{-1/2}$, respectively. The random character of the turbulent flow causes a different structure of the flow.

Obviously the results presented in this example are valid only in the range of validity of the original data—the assumed velocity profile and shear stress. This range covers smooth flat plates with $5 \times 10^5 < \mathrm{Re}_\ell < 10^7$.

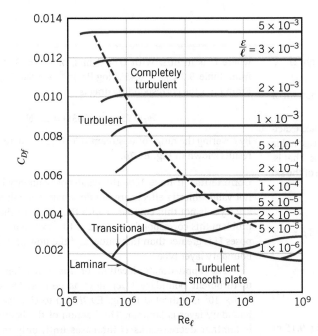

FIGURE 9.15 Friction drag coefficient for a flat plate parallel to the upstream flow (Ref. 15).

In general, the drag coefficient for a flat plate of length ℓ is a function of the Reynolds number, Re_ℓ, and the relative roughness, ε/ℓ. The results of numerous experiments covering a wide range of the parameters of interest are shown in **Fig. 9.15**. For laminar boundary layer flow the drag coefficient is a function of only the Reynolds number—surface roughness is not important. This is similar to laminar flow in a pipe. However, for turbulent flow, the surface roughness does affect the shear stress and, hence, the drag coefficient. This is similar to turbulent pipe flow in which the surface roughness may protrude into or through the viscous sublayer next to the wall and alter the flow in this thin, but very important, layer (see Section 8.4.1). Values of the roughness, ε, for different materials can be obtained from Table 8.1.

> The flat plate drag coefficient is a function of relative roughness and Reynolds number.

The drag coefficient diagram of Fig. 9.15 (boundary layer flow) shares many characteristics in common with the familiar Moody diagram (pipe flow) of Fig. 8.20, even though the mechanisms governing the mean flow are quite different. Fully developed horizontal pipe flow is governed by a balance between pressure forces and viscous forces. The fluid inertia remains constant throughout the flow. Boundary layer flow on a horizontal flat plate is governed by a balance between inertia effects and viscous forces. The pressure remains constant throughout the flow. (As is discussed in Section 9.2.6, for boundary layer flow on curved surfaces, the pressure is not constant.) On the other hand, the structure of the turbulence in both flows can be quite similar.

It is often convenient to have an equation for the drag coefficient as a function of the Reynolds number and relative roughness rather than the graphical representation given in Fig. 9.15. Although there is not one equation valid for the entire $Re_\ell - \varepsilon/\ell$ range, the equations presented in **Table 9.3** do work well for the conditions indicated.

TABLE 9.3 Equations for the Flat Plate Drag Coefficient (Ref. 1)

Equation	Flow Conditions
$C_{Df} = 1.328/(Re_\ell)^{0.5}$	Laminar flow
$C_{Df} = 0.455/(\log Re_\ell)^{2.58} - 1700/Re_\ell$	Transitional with $Re_{xcr} = 5 \times 10^5$
$C_{Df} = 0.455/(\log Re_\ell)^{2.58}$	Turbulent, smooth plate
$C_{Df} = [1.89 - 1.62 \log(\varepsilon/\ell)]^{-2.5}$	Completely turbulent

EXAMPLE 9.7 | Drag on a Flat Plate

Given The water ski shown in **Fig. E9.7a** moves through 21 °C water with a velocity U.

Find Estimate the drag caused by the shear stress on the bottom of the ski for $0 < U < 9$ m/s.

Solution Clearly the ski is not a flat plate, and it is not aligned exactly parallel to the upstream flow. However, we can obtain a reasonable approximation to the shear force by using the flat plate results. That is, the friction drag, $\mathcal{D}_f$, caused by the shear stress on the bottom of the ski (the wall shear stress) can be determined as

$$\mathcal{D}_f = \tfrac{1}{2}\rho U^2 \ell b C_{Df}$$

With $A = \ell b = 1\text{ m} \times 0.15\text{ m} = 0.15\text{ m}^2$, $\rho = 1000$ kg/m³, and $\mu = 1 \times 10^{-3}$ N · s/m² (see Table B.1) we obtain

$$\mathcal{D}_f = \tfrac{1}{2}(1000\text{ kg/m}^3)(0.15\text{ m}^2)U^2 C_{Df}$$
$$= 75\, U^2 C_{Df} \qquad (1)$$

where $\mathcal{D}_f$ and U are in Newton and m/s, respectively.

The friction coefficient, C_{Df}, can be obtained from Fig. 9.15 or from the appropriate equations given in Table 9.3. As we will see, for this problem, much of the flow lies within the transition regime where both the laminar and turbulent portions of the boundary layer flow occupy comparable lengths of the plate. We choose to use the values of C_{Df} from the table.

For the given conditions we obtain

$$\text{Re}_\ell = \frac{\rho U \ell}{\mu} = \frac{(1000\text{ kg/m}^3)(1\text{ m})U}{1 \times 10^{-3}\text{ N} \cdot \text{s/m}^2} = 1 \times 10^6\, U$$

where U is in m/s. With $U = 3$ m/s, or $\text{Re}_\ell = 3 \times 10^6$, we obtain from Table 9.3 $C_{Df} = 0.455/(\log \text{Re}_\ell)^{2.58} - 1700/\text{Re}_\ell = 3.1 \times 10^{-3}$. From Eq. 1 the corresponding drag is

$$\mathcal{D}_f = 75(3)^2(0.0031) = 2.1\text{ N}$$

By covering the range of upstream velocities of interest we obtain the results shown in **Fig. E9.7b**. *(Ans)*

Comments If Re $\lesssim 1000$ the results of boundary layer theory are not valid—inertia effects are not dominant enough and the boundary layer is not thin compared with the length of the plate. For our problem this corresponds to $U = 1 \times 10^{-3}$ m/s. For all practical purposes U is greater than this value, and the flow past the ski is of the boundary layer type.

The approximate location of the transition from laminar to turbulent boundary layer flow as defined by $\text{Re}_{cr} = \rho U x_{cr}/\mu = 5 \times 10^5$ is indicated in Fig. E9.7b. Up to $U = 0.4$ m/s the entire boundary layer is laminar. The fraction of the boundary layer that is laminar decreases as U increases until only the front 0.05 m is laminar when $U = 9$ m/s.

For anyone who has water skied, it is clear that it can require considerably more force to be pulled along at 9 m/s than the $2 \times 22\text{ N} = 44\text{ N}$ (two skis) indicated in Fig. E9.7b. As is discussed in Section 9.3, the total drag on an object such as a water ski consists of more than just the friction drag. Other components, including pressure drag and wave-making drag, add considerably to the total resistance.

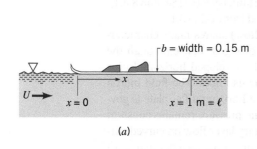

(a)

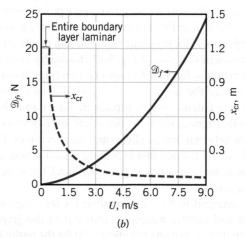

(b)

FIGURE E9.7

9.2.6 Effects of Pressure Gradient

The boundary layer discussions in the previous parts of Section 9.2 have dealt with flow along a flat plate in which the pressure is constant throughout the fluid. In general, when a fluid flows past an object other than a flat plate, the pressure field is not uniform. As shown in Fig. 9.6, if the Reynolds number is large, relatively thin boundary layers will develop along the surfaces. Within these layers the component of the pressure gradient in the streamwise direction (i.e., along the body surface) is not zero, although the pressure gradient normal to the surface is negligibly small. That is, if we were to measure the pressure while moving across the boundary layer from the body to the boundary layer edge, we would find that the pressure is essentially constant. However, the pressure does vary in the direction along the body surface if the body is curved, as shown in the adjacent figure. The variation in the **free-stream velocity**, U_{fs}, the fluid velocity at the edge of the boundary layer, is the cause of the pressure gradient in this direction. The characteristics of the entire flow (both within and outside of the boundary layer) are often highly dependent on the pressure gradient effects on the fluid within the boundary layer.

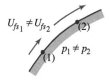

For a flat plate parallel to the upstream flow, the upstream velocity (that far ahead of the plate) and the free-stream velocity (that at the edge of the boundary layer) are equal—$U = U_{fs}$. This is a consequence of the negligible thickness of the plate. For bodies of nonzero thickness, these two velocities are different. This can be seen in the flow past a circular cylinder of diameter D. The upstream velocity and pressure are U and p_0, respectively. If the fluid were completely inviscid ($\mu = 0$), the Reynolds number would be infinite ($Re = \rho U D/\mu = \infty$) and the streamlines would be symmetrical, as are shown in **Fig. 9.16a**. The fluid velocity along the surface would vary from $U_{fs} = 0$ at the very front and rear of the cylinder (points A and F are stagnation points) to a maximum of $U_{fs} = 2U$ at the top and bottom of the cylinder (point C). This is also indicated in the adjacent figure. The pressure on the surface of the cylinder would be symmetrical about the vertical midplane of the cylinder, reaching a maximum value of $p_0 + \rho U^2/2$ (the stagnation pressure) at both the front and back of the cylinder, and a minimum of $p_0 - 3\rho U^2/2$ at the top and bottom of the cylinder. The pressure and free-stream velocity distributions are shown in **Figs. 9.16b** and **9.16c**. These characteristics can be obtained from potential flow analysis of Section 6.6.3.

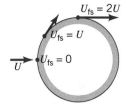

Because of the absence of viscosity (therefore, $\tau_w = 0$) and the symmetry of the pressure distribution for inviscid flow past a circular cylinder, it is clear that the drag on the cylinder is zero. The pressure distribution is symmetrical, so the "push" experienced in the flow direction on the front of the cylinder is balanced by the equal push opposite to the flow direction on the back. The predication of zero drag for an inviscid fluid even extends to a plate normal to the flow. Based on common experience, however, we know that there must be a net drag. Clearly, since there is no purely inviscid fluid, the reason for the observed drag must lie on the shoulders of the viscous effects.

To test this hypothesis, we could conduct an experiment by measuring the drag on an object (such as a circular cylinder) in a series of fluids with decreasing values of viscosity. To

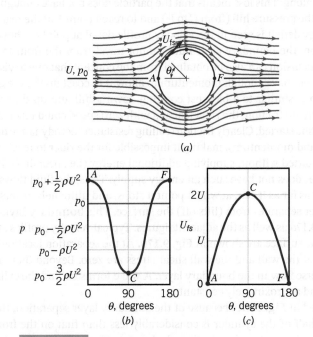

FIGURE 9.16 Inviscid flow past a circular cylinder:
(a) streamlines for the flow if there were no viscous effects,
(b) pressure distribution on the cylinder's surface,
(c) free-stream velocity on the cylinder's surface.

our initial surprise we would find that no matter how small we make the viscosity (provided it is not precisely zero) we would measure a finite drag, essentially independent of the value of μ. As was noted in Section 6.6.3, this leads to what has been termed *d'Alembert's paradox*—the drag on an object in an inviscid fluid is zero, but the drag on an object in a fluid with vanishingly small (but nonzero) viscosity is not zero.

The reason for the above paradox can be described in terms of the effect of the pressure gradient on boundary layer flow. Consider large Reynolds number flow of a real (viscous) fluid past a circular cylinder. As was discussed in Section 9.1.2, we expect the viscous effects to be confined to thin boundary layers near the surface at the base of which the fluid sticks to the surface ($\mathbf{V} = 0$)—a necessary condition for any real fluid ($\mu \neq 0$). The basic idea of boundary layer theory is that the boundary layer is thin enough so that it does not greatly disturb the flow outside the boundary layer. Based on this reasoning, for large Reynolds numbers the flow throughout most of the flow field would be expected to be as is indicated in Fig. 9.16*a*, the inviscid flow field.

The pressure distribution indicated in Fig. 9.16*b* is imposed on the boundary layer flow along the surface of the cylinder. In fact, there is negligible pressure variation across the thin boundary layer so that the pressure within the boundary layer is that given by the inviscid flow field. This pressure distribution along the cylinder is such that the stationary fluid at the nose of the cylinder ($U_{fs} = 0$ at $\theta = 0$) is accelerated to its maximum velocity ($U_{fs} = 2U$ at $\theta = 90°$) and then is decelerated back to zero velocity at the rear of the cylinder ($U_{fs} = 0$ at $\theta = 180°$). This is accomplished by a balance between pressure and inertia effects; viscous effects are absent for the inviscid flow outside the boundary layer.

Physically, in the absence of viscous effects, a fluid particle traveling from the front to the back of the cylinder coasts down the "pressure hill" from $\theta = 0$ to $\theta = 90°$ (from point A to C in Fig. 9.16*b*) and then back up the hill to $\theta = 180°$ (from point C to F) without any loss of energy. There is an exchange between kinetic and pressure energy, but there are no energy losses. The same pressure distribution is imposed on the viscous fluid within the boundary layer. The decrease in pressure in the direction of flow along the front half of the cylinder is termed a **favorable pressure gradient**. The increase in pressure in the direction of flow along the rear half of the cylinder is termed an **adverse pressure gradient**.

Consider a fluid particle within the boundary layer indicated in **Fig. 9.17a**. In its attempt to flow from A to F, it experiences the same pressure distribution as the particles in the free stream immediately outside the boundary layer—the inviscid flow field pressure. However, because of the viscous effects involved, the particle in the boundary layer experiences a loss of energy as it flows along. This loss means that the particle does not have enough energy to coast all of the way up the pressure hill (from C to F) and to reach point F at the rear of the cylinder. This kinetic energy deficit is seen in the velocity profile detail at point C, shown in Fig. 9.17*a*. Because of friction, the boundary layer fluid cannot travel from the front to the rear of the cylinder. (This conclusion can also be obtained from the concept that due to viscous effects the particle at C does not have enough momentum to allow it to coast up the pressure hill to F.)

The situation is similar to a bicyclist coasting down a hill and up the other side of the valley. If there were no friction, the rider starting with zero speed could reach the same height from which he or she started. Clearly friction (rolling resistance, aerodynamic drag, etc.) causes a loss of energy (and momentum), making it impossible for the rider to reach the height from which he or she started without supplying additional energy (i.e., pedaling). The fluid within the boundary layer does not have such an energy supply. Thus, the fluid flows against the increasing pressure as far as it can, at which point it stops, and ultimately, reverses its direction; the boundary layer separates from (lifts off) the surface. This **boundary layer separation** is indicated in Fig. 9.17*a* as well as the adjacent figures. Typical velocity profiles at representative locations along the surface are shown in **Fig. 9.17b**. At the separation location (profile D), the velocity gradient at the wall and the wall shear stress are zero. Beyond that location (from D to E) there is reverse flow in the boundary layer. A wake forms behind the cylinder where the pressure is low and approximately constant.

As is indicated in **Fig. 9.17c**, because of the boundary layer separation, the average pressure on the rear half of the cylinder is considerably less than that on the front half. Thus, a large pressure drag is developed, even though (because of small viscosity) the viscous shear drag may be quite small. D'Alembert's paradox is explained. No matter how small the viscosity, provided it is not zero, there will be a boundary layer that separates from the surface, giving a drag that is, for the most part, independent of the value of μ.

The pressure gradient in the external flow is imposed throughout the boundary layer fluid.

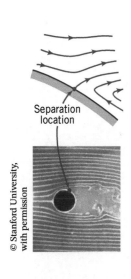

Separation
location

V9.9 Snow drifts

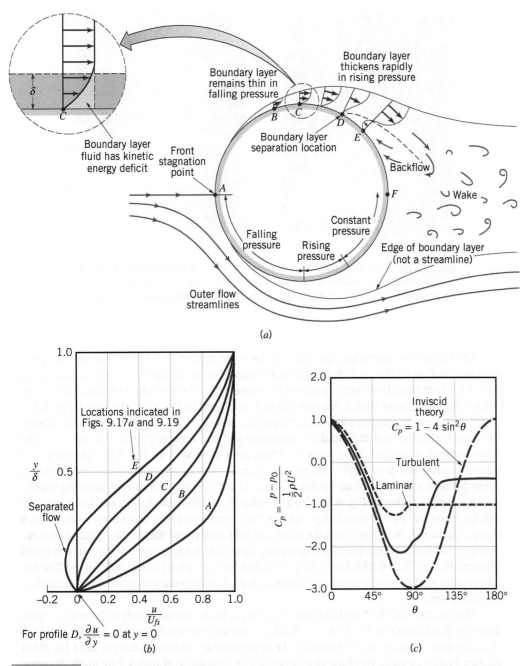

FIGURE 9.17 Boundary layer characteristics on a circular cylinder: (a) boundary layer separation location, (b) typical boundary layer velocity profiles at various locations on the cylinder, (c) surface pressure distributions for inviscid flow and boundary layer flow.

The Wide World of Fluids

Increasing truck km/liter

A large portion of the aerodynamic drag on semis (tractor-trailer rigs) is a result of the low pressure on the flat back end of the trailer. Researchers have recently developed a drag-reducing attachment that could reduce fuel costs on these big rigs by 10%. The device consists of a set of flat plates (attached to the rear of the trailer) that fold out into a tapered box shape, thereby making the originally flat rear of the trailer a somewhat more "aerodynamic" shape. Based on thorough wind tunnel testing and actual tests conducted with a prototype design used in a series of cross-country runs, it is estimated that trucks using the device could save approximately $6000 a year in fuel costs.

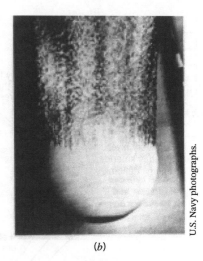

U.S. Navy photographs.

(a)   (b)

FIGURE 9.18 Effect of turbulence in boundary layer on delaying separation. An 0.21 m bowling ball is dropped into water at a speed of 7.62 m/s. (a) The ball is smooth; (b) the ball has a patch of sandpaper on the nose that causes transition.

Viscous effects within the boundary layer cause boundary layer separation.

The location of separation, the width of the wake region behind the object, and the pressure distribution on the surface depend on the nature of the boundary layer flow. Compared with a laminar boundary layer, a turbulent boundary layer flow has more kinetic energy and momentum associated with it because: (1) as is indicated in Fig. E9.6, the velocity profile is fuller, more nearly like the ideal uniform profile, and (2) there can be considerable energy associated with the swirling, random components of the velocity that do not appear in the time-averaged x component of velocity. Thus, as is indicated in Fig. 9.17c, the turbulent boundary layer can flow farther around the cylinder (farther up the pressure hill) before it separates than can the laminar boundary layer. These effects of turbulence delaying separation are clearly visible in **Fig. 9.18**. The two spheres are identical, except one has a patch of sandpaper on its nose. This sandpaper causes the boundary layer to become turbulent, which delays separation. Note that the wake is turbulent in both cases; there is no visible difference between the laminar and turbulent boundary layers on the spheres, but the wake is much different in size. The effect is lower drag. Note that because a flat plate parallel to the flow has no pressure gradient and thus no separation, a turbulent boundary layer would *increase* drag on a flat plate.

The structure of the flow field past a circular cylinder is completely different for a zero viscosity fluid than it is for a viscous fluid, no matter how small the viscosity is, provided it is not zero. This is due to boundary layer separation. Similar concepts hold for other shaped bodies as well. The flow past an airfoil at zero *angle of attack* (the angle between the upstream flow and the axis of the object) is shown in **Fig. 9.19a**; flow past the same airfoil at a 5° angle of attack is shown in **Fig. 9.19b**. Over the front portion of the airfoil the pressure decreases in the direction of flow—a favorable pressure gradient. Over the rear portion the pressure increases in the direction of flow—an adverse pressure gradient. The boundary layer velocity profiles at representative locations are similar to those indicated in Fig. 9.17b for flow past a circular cylinder. If the adverse pressure gradient is not too great (because the body is not too "thick" in some sense), the boundary layer fluid can flow into the slightly increasing pressure region (i.e., from C to the trailing edge in Fig. 9.19a) without separating from the surface. However, if the pressure gradient is too adverse (because the angle of attack is too large), the boundary layer will separate from the surface as indicated in Fig. 9.19b. Such situations can lead to the catastrophic loss of lift called *stall*, which is discussed in Section 9.4.

V9.10 Effect of body shape on pressure gradient

Streamlined bodies generally have no separated flow.

Streamlined bodies are generally those designed to eliminate (or at least to reduce) the effects of separation, whereas nonstreamlined (blunt) bodies generally have relatively large drag due to the low pressure in the separated regions (the wake). Although the boundary layer may be quite thin, it can appreciably alter the entire flow field because of boundary layer separation. These ideas are discussed in Section 9.3.

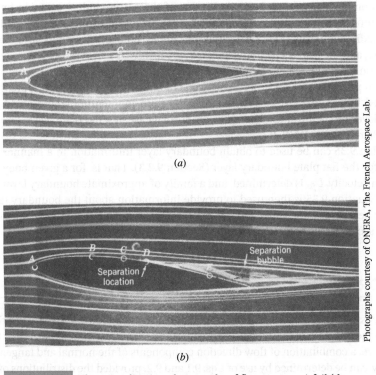

(a)

(b)

Photographs courtesy of ONERA, The French Aerospace Lab.

FIGURE 9.19 Flow visualization photographs of flow past an airfoil (the boundary layer velocity profiles for the points indicated are similar to those indicated in Fig. 9.17b): (a) zero angle of attack, no separation, (b) 5° angle of attack, flow separation. Dye in water.

9.2.7 Momentum Integral Boundary Layer Equation with Nonzero Pressure Gradient

The boundary layer results discussed in Sections 9.2.2 and 9.2.3 are valid only for boundary layers with zero pressure gradients. They correspond to the velocity profile labeled C in Fig. 9.17b. Boundary layer characteristics for flows with nonzero pressure gradients can be obtained from the nonlinear, partial differential boundary layer equations, Eqs. 9.8 and 9.9a, provided the pressure gradient is appropriately accounted for. Such an approach is beyond the scope of this book (Refs. 1, 2).

An alternative approach is to extend the momentum integral boundary layer equation technique (Section 9.2.3) so that it is applicable for flows with nonzero pressure gradients. The momentum integral equation for boundary layer flows with zero pressure gradient, Eq. 9.26, is a statement of the balance between the shear force on the plate (represented by τ_w) and rate of change of momentum of the fluid within the boundary layer [represented by $\rho U^2(d\Theta/dx)$]. For such flows the free-stream velocity is constant ($U_{fs} = U$). If the free-stream velocity is not constant [$U_{fs} = U_{fs}(x)$, where x is the distance measured along the curved body], the pressure will not be constant. This follows from the Bernoulli equation with negligible gravitational effects, since $p + \rho U_{fs}^2/2$ is constant along the streamlines outside the boundary layer. Thus,

$$\frac{dp}{dx} = -\rho U_{fs}\frac{dU_{fs}}{dx} \tag{9.34}$$

For a given body, the free-stream velocity and the corresponding pressure gradient on the surface can be approximated from inviscid flow techniques (potential flow) discussed in Section 6.7. (This is how the circular cylinder results of Fig. 9.16 were obtained.)

Flow in a boundary layer with nonzero pressure gradient is very similar to that shown in Fig. 9.11, except that the upstream velocity, U, is replaced by the free-stream velocity, $U_{fs}(x)$, and the pressures at sections (1) and (2) are not necessarily equal. By using the x component of the momentum equation (Eq. 5.22) with the appropriate shear forces and pressure forces acting on the control surface indicated in Fig. 9.11, the following integral momentum equation for boundary layer flows is obtained:

$$\tau_w = \rho\frac{d}{dx}\left(U_{fs}^2\Theta\right) + \rho\delta^* U_{fs}\frac{dU_{fs}}{dx} \tag{9.35}$$

Pressure gradient effects can be included in the momentum integral equation.

The derivation of this equation is similar to that of the corresponding equation for constant-pressure boundary layer flow, Eq. 9.26, although the inclusion of the pressure gradient effect brings in additional terms (Refs. 1, 2, 3). For example, both the boundary layer momentum thickness, Θ, and the displacement thickness, δ^*, are involved.

Equation 9.35, the general momentum integral equation for two-dimensional boundary layer flow, represents a balance between viscous forces (represented by τ_w), pressure forces (represented by $\rho U_{fs}\, dU_{fs}/dx = -dp/dx$), and the fluid momentum (represented by Θ, the boundary layer momentum thickness). In the special case of a flat plate, $U_{fs} = U = $ constant, and Eq. 9.35 reduces to Eq. 9.26.

Equation 9.35 can be used to obtain boundary layer information in a manner similar to that done for the flat plate boundary layer (Section 9.2.3). That is, for a given body shape the free-stream velocity, U_{fs}, is determined, and a family of approximate boundary layer profiles is assumed. Equation 9.35 is then used to provide information about the boundary layer thickness, wall shear stress, and other properties of interest. The details of this technique are not within the scope of this book (Refs. 1 and 3).

9.3　DRAG

As was discussed in Section 9.1, any object moving through a fluid will experience a drag, $\mathcal{D}$—a net force in the direction of flow due to the pressure and shear forces on the surface of the object. This net force, a combination of flow direction components of the normal and tangential forces on the body, can be determined by use of Eqs. 9.1 and 9.2, provided the distributions of pressure, p, and wall shear stress, τ_w, are known. Only in very rare instances can these distributions be determined analytically. The boundary layer flow past a flat plate parallel to the upstream flow as is discussed in Section 9.2 is one such case. Current advances in computational fluid dynamics, CFD (i.e., the use of computers to solve the governing equations of the flow field), have provided encouraging results for more complex shapes. However, much work in this area remains.

Most of the information pertaining to drag on objects is a result of numerous experiments with wind tunnels, water tunnels, towing tanks, and other ingenious devices that are used to measure the drag on scale models. As was discussed in Chapter 7, these data can be put into dimensionless form, and the results can be appropriately ratioed for prototype calculations. Typically, the result for a given-shaped object is a drag coefficient, C_D, where

$$C_D = \frac{\mathcal{D}}{\frac{1}{2}\rho U^2 A} \tag{9.36}$$

and C_D is a function of other dimensionless parameters such as Reynolds number, Re, Mach number, Ma, Froude number, Fr, and relative roughness of the surface, ε/ℓ. That is,

$$C_D = \phi(\text{shape, Re, Ma, Fr}, \varepsilon/\ell)$$

The character of C_D as a function of these parameters is discussed in this section.

9.3.1　Friction Drag

Friction drag, $\mathcal{D}_f$, is that part of the drag that is due directly to the shear stress, τ_w, on the object. It is a function of not only the magnitude of the wall shear stress but also of the orientation of the surface on which it acts. This is indicated by the factor $\tau_w \sin\theta$ in Eq. 9.1. If the surface is parallel to the upstream velocity, the entire shear force contributes directly to the drag. (In fact, the drag is due solely to the shear force.) This is true for the flat plate parallel to the flow as was discussed in Section 9.2. If the surface is perpendicular to the upstream velocity, the shear stress contributes nothing to the drag. Such is the case for a flat plate normal to the upstream velocity as was discussed in Section 9.1.

In general, the surface of a body will contain portions parallel to and normal to the upstream flow, as well as any direction in between. A circular cylinder is such a body. Because the viscosity of most common fluids is small, the contribution of the shear force to the overall drag on a body is often quite small. Such a statement should be worded in dimensionless terms. That is, because the Reynolds number of most familiar flows is quite large, the percent of the drag caused directly by the shear stress is often quite small. For highly streamlined bodies or for low Reynolds number flow, however, most of the drag may be due to friction drag.

The friction drag on a flat plate of width b and length ℓ oriented parallel to the upstream flow can be calculated from

$$\mathcal{D}_f = \tfrac{1}{2}\rho U^2 b\ell C_{Df}$$

where C_{Df} is the friction drag coefficient. The value of C_{Df}, given as a function of Reynolds number, $\mathrm{Re}_\ell = \rho U\ell/\mu$, and relative surface roughness, ε/ℓ, in Fig. 9.15 and Table 9.3, is a result of boundary layer analysis and experiments (see Section 9.2). Typical values of roughness, ε, for various surfaces are given in Table 8.1. As with the pipe flow discussed in Chapter 8, the flow is divided into two distinct categories—laminar or turbulent, with a transitional regime connecting them. The drag coefficient (and, hence, the drag) is not a function of the plate roughness if the flow is laminar. However, for turbulent flow the roughness does considerably affect the value of C_{Df}. As with pipe flow, this dependence is a result of the surface roughness elements protruding into or through the laminar sublayer (see Section 8.3).

> Friction (viscous) drag is the drag produced by viscous shear stresses.

Most objects are not flat plates parallel to the flow; instead, they are curved or variably angled surfaces along which the pressure varies. As was discussed in Section 9.2.6, this means that the boundary layer character, including the velocity gradient at the wall, is different for most objects from that for a flat plate. This can be seen in the change of shape of the boundary layer profile along the cylinder in Fig. 9.17b.

The precise determination of the shear stress along the surface of a curved body is quite difficult to obtain. Although approximate results can be obtained by a variety of techniques (Refs. 1, 2), these are outside the scope of this text. As is shown by the following example, if the shear stress is known, its contribution to the drag can be determined.

EXAMPLE 9.8 | Drag Coefficient Based on Friction Drag

Given A viscous, incompressible fluid flows past the circular cylinder shown in **Fig. E9.8a**. According to a more advanced theory of boundary layer flow, the boundary layer remains attached to the cylinder up to the separation location at $\theta \approx 108.8°$, with the dimensionless wall shear stress as is indicated in **Fig. E9.8b** (Ref. 1). The shear stress on the cylinder in the wake region, $108.8 < \theta < 180°$, is negligible.

Find Determine C_{Df}, the drag coefficient for the cylinder based on the friction drag only.

Solution The friction drag, $\mathcal{D}_f$, can be determined from Eq. 9.1 as

$$\mathcal{D}_f = \int \tau_w \sin\theta\, dA = 2\left(\frac{D}{2}\right)b\int_0^\pi \tau_w \sin\theta\, d\theta$$

where b is the length of the cylinder. Note that θ is in radians (not degrees) to ensure the proper dimensions of $dA = 2(D/2)b\,d\theta$. Thus,

$$C_{Df} = \frac{\mathcal{D}_f}{\tfrac{1}{2}\rho U^2 bD} = \frac{2}{\rho U^2}\int_0^\pi \tau_w \sin\theta\, d\theta$$

This can be put into dimensionless form by using the dimensionless shear stress parameter, $F(\theta) = \tau_w \sqrt{\mathrm{Re}}/(\rho U^2/2)$, given in Fig. E9.8b as follows:

$$C_{Df} = \int_0^\pi \frac{\tau_w}{\tfrac{1}{2}\rho U^2}\sin\theta\, d\theta = \frac{1}{\sqrt{\mathrm{Re}}}\int_0^\pi \frac{\tau_w \sqrt{\mathrm{Re}}}{\tfrac{1}{2}\rho U^2}\sin\theta\, d\theta$$

where $\mathrm{Re} = \rho UD/\mu$. Thus,

$$C_{Df} = \frac{1}{\sqrt{\mathrm{Re}}}\int_0^\pi F(\theta)\sin\theta\, d\theta \qquad (1)$$

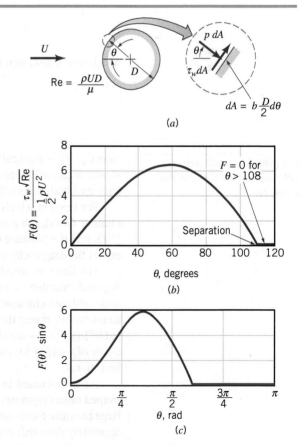

$$\mathrm{Re} = \frac{\rho UD}{\mu}$$

$$dA = b\frac{D}{2}d\theta$$

(a)

(b)

(c)

FIGURE E9.8

The function $F(\theta) \sin \theta$, obtained from Fig. E9.8b, is plotted in **Fig. E9.8c.** The necessary integration to obtain C_{Df} from Eq. 1 can be done by an appropriate numerical technique or by an approximate graphical method to determine the area under the given curve.

The result is $\displaystyle\int_0^\pi F(\theta) \sin \theta \, d\theta = 5.93$, or

$$C_{Df} = \frac{5.93}{\sqrt{\text{Re}}} \qquad\qquad (Ans)$$

Comments   Note that the total drag must include both the shear stress (friction) drag and the pressure drag. As we will see in Example 9.9, for the circular cylinder most of the drag is due to the pressure force.

The above friction drag result is valid only if the boundary layer flow on the cylinder is laminar. As is discussed in Section 9.3.3, for a smooth cylinder this means that $\text{Re} = \rho UD/\mu < 3 \times 10^5$. It is also valid only for flows that have a Reynolds number sufficiently large to ensure the boundary layer structure to the flow. For the cylinder, this means $\text{Re} > 100$.

9.3.2 Pressure Drag

Pressure (form) drag is the drag produced by normal stresses.

Pressure drag, $\mathscr{D}_p$, is that part of the drag that is due directly to the pressure, p, on an object. It is often referred to as *form drag* because of its strong dependency on the shape or form of the object. Pressure drag is a function of the magnitude of the pressure and the orientation of the surface element on which the pressure force acts. For example, the pressure force on either side of a flat plate parallel to the flow may be very large, but it does not contribute to the drag because it acts in the direction normal to the upstream velocity. On the other hand, the pressure force on a flat plate normal to the flow provides the entire drag.

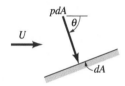

As previously noted, for most bodies, there are portions of the surface that are parallel to the upstream velocity, others normal to the upstream velocity, and the majority of which are at some angle in between, as shown in the adjacent figure. The pressure drag can be obtained from Eq. 9.1 provided a detailed description of the pressure distribution and the body shape is given. That is,

$$\mathscr{D}_p = \int p \cos \theta \, dA$$

which can be rewritten in terms of the *pressure drag coefficient*, C_{Dp}, as

$$C_{Dp} = \frac{\mathscr{D}_p}{\frac{1}{2}\rho U^2 A} = \frac{\int p \cos \theta \, dA}{\frac{1}{2}\rho U^2 A} = \frac{\int C_p \cos \theta \, dA}{A} \qquad (9.37)$$

The pressure coefficient is a dimensionless form of the pressure.

Here $C_p = (p - p_0)/(\rho U^2/2)$ is the *pressure coefficient*, where p_0 is a reference pressure. The magnitude of the reference pressure does not influence the drag directly because the net pressure force on a body is zero if the pressure is constant (i.e., p_0) on the entire surface.

For flows in which inertial effects are large relative to viscous effects (i.e., large Reynolds number flows), the pressure difference, $p - p_0$, scales directly with the dynamic pressure, $\rho U^2/2$, and the pressure coefficient is independent of Reynolds number. In such situations we expect the drag coefficient to be relatively independent of Reynolds number.

For flows in which viscous effects are large relative to inertial effects (i.e., very small Reynolds number flows), it is found that both the pressure difference and wall shear stress scale with the characteristic viscous stress, $\mu U/\ell$, where ℓ is a characteristic length. In such situations we expect the drag coefficient to be proportional to $1/\text{Re}$. That is, $C_D \sim \mathscr{D}/(\rho U^2/2) \sim (\mu U/\ell)/(\rho U^2/2) \sim \mu/\rho U\ell = 1/\text{Re}$. These characteristics are similar to the friction factor dependence of $f \sim 1/\text{Re}$ for laminar pipe flow and $f \sim$ constant for large Reynolds number flow (see Section 8.4).

As is discussed in Section 9.2.6 if the viscosity were zero, the pressure drag on any shaped object (symmetrical or not) in a steady flow would be zero. There perhaps would be large pressure forces on the front portion of the object, but there would be equally large (and oppositely directed) pressure forces on the rear portion. If the viscosity is not zero, the net pressure drag may be nonzero because of boundary layer separation. Example 9.9 illustrates this concept.

EXAMPLE 9.9 | Drag Coefficient Based on Pressure Drag

Given A viscous, incompressible fluid flows past the circular cylinder shown in Fig. E9.8a. The pressure coefficient on the surface of the cylinder (as determined from experimental measurements) is as indicated in **Fig. E9.9a**.

Find Determine the pressure drag coefficient for this flow. Combine the results of Examples 9.8 and 9.9 to determine the drag coefficient for a circular cylinder. Compare your results with those given in Fig. 9.23.

Solution The pressure (form) drag coefficient, C_{Dp}, can be determined from Eq. 9.37 as

$$C_{Dp} = \frac{1}{A}\int C_p \cos\theta \, dA = \frac{1}{bD}\int_0^{2\pi} C_p \cos\theta \, b\left(\frac{D}{2}\right)d\theta$$

or because of symmetry

$$C_{Dp} = \int_0^{\pi} C_p \cos\theta \, d\theta$$

where b and D are the length and diameter of the cylinder. To obtain C_{Dp}, we must integrate the $C_p \cos\theta$ function from $\theta = 0$ to $\theta = \pi$ radians. Again, this can be done by some numerical integration scheme or by determining the area under the curve shown in **Fig. E9.9b**. The result is

$$C_{Dp} = 1.17 \qquad\qquad (1) \quad (Ans)$$

Note that the positive pressure on the front portion of the cylinder ($0 \le \theta \le 30°$) and the negative pressure (less than the upstream value) on the rear portion ($90 \le \theta \le 180°$) produce positive contributions to the drag. The negative pressure on the front portion of the cylinder ($30 < \theta < 90°$) reduces the drag by pulling on the cylinder in the upstream direction. The positive area under the $C_p\cos\theta$ curve is greater than the negative area—there is a net pressure drag. In the absence of viscosity, these two contributions would be equal—there would be no pressure (or friction) drag.

The net drag on the cylinder is the sum of friction and pressure drag. Thus, from Eq. 1 of Example 9.8 and Eq. 1 of this example, we obtain the drag coefficient

$$C_D = C_{Df} + C_{Dp} = \frac{5.93}{\sqrt{Re}} + 1.17 \qquad (2) \quad (Ans)$$

This result is compared with the standard experimental value (obtained from Fig. 9.23) in **Fig. E9.9c**. The agreement is very good over a wide range of Reynolds numbers. For Re < 10 the curves diverge because the flow is not a boundary layer type flow—the shear stress and pressure distributions used to obtain Eq. 2 are not valid in this range. The drastic divergence in the curves for Re > 3×10^5 is due to the change from a laminar to turbulent boundary layer, with the corresponding change in the pressure distribution. This phenomenon was discussed in Section 9.2.6 and will be elaborated upon in Section 9.3.3.

Comment It is of interest to compare the friction drag to the total drag on the cylinder. That is,

$$\frac{\mathscr{D}_f}{\mathscr{D}} = \frac{C_{Df}}{C_D} = \frac{5.93/\sqrt{Re}}{(5.93/\sqrt{Re}) + 1.17} = \frac{1}{1 + 0.197\sqrt{Re}}$$

For Re = $10^3, 10^4$, and 10^5 this ratio is 0.138, 0.0483, and 0.0158, respectively. In other words, friction drag accounts for less than 15% of the total drag and can even be as low as 1%. Thus, most of the drag on the blunt cylinder is pressure drag—a result of the boundary layer separation.

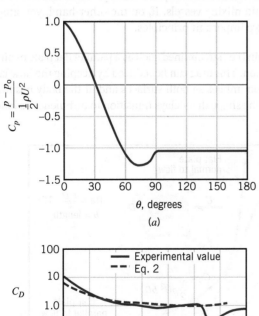

(a)

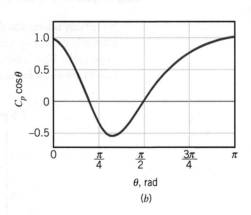

(b)

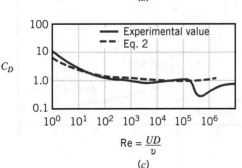

(c)

FIGURE E9.9

9.3.3 Drag Coefficient Data and Examples

As was discussed in previous sections, the net drag is produced by both pressure and shear stress effects. In most instances these two effects are considered together, and an overall drag coefficient, C_D, as defined in Eq. 9.36 is used. There is an abundance of such drag coefficient data available in the literature. This information covers incompressible and compressible viscous flows past objects of almost any shape of interest—both man-made and natural objects. In this section we consider a small portion of this information for representative situations. Additional data can be obtained from various sources (Refs. 5, 6).

Shape Dependence   Clearly the drag coefficient for an object depends on the shape of the object, with shapes ranging from those that are streamlined to those that are blunt. The drag on an ellipse with aspect ratio ℓ/D, where D and ℓ are the thickness and length parallel to the flow, illustrates this dependence. The drag coefficient $C_D = \mathcal{D}/(\rho U^2 bD/2)$, based on the frontal area, $A = bD$, where b is the length normal to the flow, is as shown in **Fig. 9.20**. The blunter the body, the larger the drag coefficient. With $\ell/D = 0$ (i.e., a flat plate normal to the flow) we obtain the flat plate value of $C_D = 1.9$. With $\ell/D = 1$ the corresponding value for a circular cylinder is obtained. As ℓ/D becomes larger, the value of C_D decreases.

For very large aspect ratios ($\ell/D \rightarrow \infty$) the ellipse behaves as a flat plate parallel to the flow. For such cases, the friction drag is greater than the pressure drag, and the value of C_D based on the frontal area, $A = bD$, would increase with increasing ℓ/D. (This occurs for larger ℓ/D values than those shown in the figure.) For such extremely thin bodies (i.e., an ellipse with $\ell/D \rightarrow \infty$, a flat plate, or very thin airfoils) it is customary to use the planform area, $A = b\ell$, in defining the drag coefficient. After all, it is the planform area on which the shear stress acts, rather than the much smaller (for thin bodies) frontal area. The ellipse drag coefficient based on the planform area, $C_D = \mathcal{D}/(\rho U^2 b\ell/2)$, is also shown in Fig. 9.20. Clearly the drag obtained by using either of these drag coefficients would be the same. They merely represent two different ways to package the same information.

From a design perspective, we often focus on minimizing drag, but this is not always the goal. Boundary layer separation and its subsequent wake promote mixing. This is desirable in heat exchange devices or fluid mixing vessels. If, on the other hand, you are designing low-drag shapes, keep in mind two important principles.

- If the body is long and thin (i.e., streamlined such as a parallel flat plate or airfoil), the drag is primarily caused by friction. This drag can be reduced by keeping the flow laminar as much as possible. This condition implies smooth surfaces and, if the body has thickness (like an airfoil), means selecting the shape that delays transition to turbulence as long as possible.

> The drag coefficient may be based on the frontal area or the planform area.

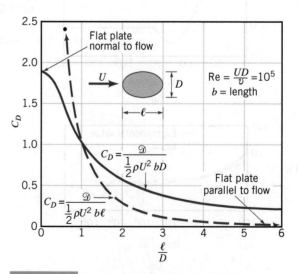

FIGURE 9.20 Drag coefficient for an ellipse with the characteristic area either the frontal area, $A = bD$, or the planform area, $A = b\ell$ (Ref. 5).

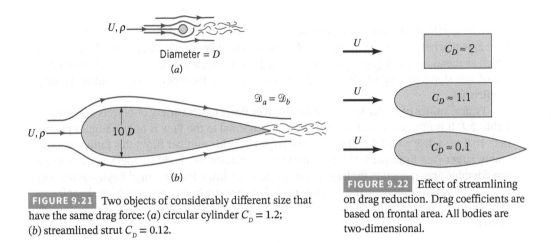

FIGURE 9.21 Two objects of considerably different size that have the same drag force: (a) circular cylinder $C_D = 1.2$; (b) streamlined strut $C_D = 0.12$.

FIGURE 9.22 Effect of streamlining on drag reduction. Drag coefficients are based on frontal area. All bodies are two-dimensional.

- If the body is blunt, the (high Reynolds number) drag is primarily pressure drag. This drag can be reduced by delaying separation as long as possible.

One way to delay boundary layer transition is to promote transition to a turbulent boundary layer (Fig. 9.17 and Fig. 9.18). This will be discussed more in the subsection titled "Surface Roughness." A better method is streamlining, that is, elongating the rear portion of the body. Incredibly the drag on the two two-dimensional objects drawn to scale in **Fig. 9.21** is the same. The width of the wake for the streamlined strut is very thin, on the order of that for the much smaller diameter circular cylinder. Consider the additional structural support offered with option (b) over option (a) with no added drag!

If the nose of the body is flat, nose rounding can also reduce the drag. **Figure 9.22** illustrates the benefits of nose rounding and rearward taper. The round nose-rear tapered body would have higher drag if it were reversed, with the tapered section pointing forward. The forward portion of a body can be quite blunt with a large (negative) pressure gradient, because separation does not occur in falling pressure. It is rearward (rising) pressure change that must be accomplished gradually. Be cautious not to add too much length due to taper; more surface area adds more friction drag, which could negate the benefits of reducing pressure drag.

Reynolds Number Dependence   Another parameter on which the drag coefficient can be very dependent is the Reynolds number. The main categories of Reynolds number dependence are (1) very low Reynolds number flow, (2) moderate Reynolds number flow (laminar boundary layer), and (3) very large Reynolds number flow (turbulent boundary layer). Examples of these three situations are discussed below.

Low Reynolds number flows (Re < 1), also known as creeping flows, result from small body sizes (e.g., dust or pulverized coal particles), large fluid viscosities (e.g., oil or grease), or very low density (e.g., high altitude flight). The mechanisms of drag are fundamentally different in low and high Reynolds number flow. The boundary layer concept, so valuable in high Reynolds number flow, is invalid at low Reynolds number. At low Reynolds number, viscous forces are not confined to a thin region near the body but are significant everywhere. That is, the boundary layer is so thick that it distorts the stream line pattern everywhere (see Fig. 9.6a). Low Reynolds number flows are governed by a balance of viscous and pressure forces, and inertia effects are negligibly small. In such instances the drag on a three-dimensional body is expected to be a function of the upstream velocity, U, the body size, ℓ, and the viscosity, μ. Thus, for a small grain of sand settling in a lake (see adjacent figure)

$$\mathcal{D} = f(U, \ell, \mu)$$

From dimensional considerations (see Section 7.7.1)

$$\mathcal{D} = C\mu\ell U \tag{9.38}$$

where the value of the constant C depends on the shape of the body. If we put Eq. 9.38 into dimensionless form using the standard definition of the drag coefficient, we obtain

$$C_D = \frac{\mathcal{D}}{\frac{1}{2}\rho U^2 \ell^2} = \frac{2C\mu\ell U}{\rho U^2 \ell^2} = \frac{2C}{\text{Re}}$$

Re < 1

$\mathcal{D} = f(U, \ell, \mu)$

where Re $= \rho U\ell/\mu$. The use of the dynamic pressure, $\rho U^2/2$, in the definition of the drag coefficient is somewhat misleading in the case of creeping flows (Re < 1) because it introduces inertia, which as mentioned is negligibly small compared to pressure and viscous effects. Nonetheless, use of this standard drag coefficient definition gives the 1/Re dependence for small Re drag coefficients.

Typical values of C_D for low Reynolds number flows past a variety of objects are given in **Table 9.4**. It is of interest that the drag on a disk normal to the flow is only 1.5 times greater than that on a disk parallel to the flow. For large Reynolds number flows this ratio is considerably larger (see Example 9.1). Streamlining (i.e., making the body slender) can produce a considerable drag reduction for large Reynolds number flows; for very small Reynolds number flows it can actually increase the drag because of an increase in the area on which shear forces act. For most objects, the low Reynolds number flow results are valid up to a Reynolds number of about 1.

> For very small Reynolds number flows, the drag coefficient varies inversely with the Reynolds number.

TABLE 9.4 Low Reynolds Number Drag Coefficients (Ref. 7) (Re $= \rho UD/\mu$, $A = \pi D^2/4$)

Object	$C_D = \mathscr{D}/(\rho U^2 A/2)$ (for Re $\leq$ 1)	Object	C_D
a. Circular disk normal to flow	20.4/Re	c. Sphere	24.0/Re
d. Circular disk parallel to flow	13.6/Re	d. Hemisphere	22.2/Re

EXAMPLE 9.10 | Low Reynolds Number Flow Drag

Given While workers spray paint on to the ceiling of a room, numerous small paint aerosols are dispersed into the air. Eventually these particles will settle out and fall to the floor or other surfaces. Consider a small spherical paint particle of diameter $D = 1 \times 10^{-5}$ m (or 10 μm) and specific gravity $SG = 1.2$.

Find Determine the time it would take this particle to fall 2.44 m from near the ceiling to the floor. Assume that the air within the room is motionless.

Solution A free-body diagram of the particle (relative to the moving particle) is shown in **Fig. E9.10a**. The particle moves downward with a constant velocity U that is governed by a balance between the weight of the particle, $\mathscr{W}$, the buoyancy force of the surrounding air, F_B, and the drag, $\mathscr{D}$, of the air on the particle.

From the free-body diagram, we obtain

$$\mathscr{W} = \mathscr{D} + F_B$$

where, if $\mathscr{V}$ is the particle volume

$$\mathscr{W} = \gamma_{\text{point}} \mathscr{V} = SG \gamma_{\text{H}_2\text{O}} \frac{\pi}{6} D^3 \qquad (1)$$

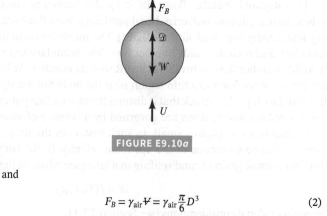

FIGURE E9.10a

and

$$F_B = \gamma_{\text{air}} \mathscr{V} = \gamma_{\text{air}} \frac{\pi}{6} D^3 \qquad (2)$$

We assume (because of the smallness of the object) that the flow will be creeping flow (Re < 1) with $C_D = 24/\text{Re}$ from Table 9.4 so that

$$\mathscr{D} = \frac{1}{2}\rho_{\text{air}} U^2 \frac{\pi}{4} D^2 C_D = \frac{1}{2}\rho_{\text{air}} U^2 \frac{\pi}{4} D^2 \left(\frac{24}{\rho_{\text{air}} UD/\mu_{\text{air}}} \right)$$

or

$$\mathscr{D} = 3\pi\mu_{air}UD \qquad (3)$$

We must eventually check to determine if the assumption that Re < 1 is valid. Equation 3 is called Stokes's law in honor of G. G. Stokes, a British mathematician and physicist. By combining Eqs. 1, 2, and 3, we obtain

$$SG\gamma_{H_2O}\frac{\pi}{6}D^3 = 3\pi\mu_{air}UD + \gamma_{air}\frac{\pi}{6}D^3$$

Then, solving for U,

$$U = \frac{D^2(SG\gamma_{H_2O} - \gamma_{air})}{18\mu_{air}} \qquad (4)$$

From Tables 1.5 and 1.7, we obtain $\gamma_{H_2O} = 9800$ N/m^3, $\gamma_{air} = 12.0$ N/m^3, and $\mu_{air} = 1.79 \times 10^{-5}$ N · s/m^2. Thus, from Eq. 4, we obtain

$$U = \frac{(10^{-5}\text{ m})^2[(1.2)(9800\text{ N/m}^3) - (12.0\text{ N/m}^3)]}{18(1.79 \times 10^{-5}\text{ N} \cdot \text{s/m}^2)}$$

or

$$U = 0.00365 \text{ m/s}$$

Let t_{fall} denote the time it takes for the particle to fall 2.44 m,

$$t_{fall} = \frac{2.44 \text{ m}}{0.00365 \text{ m/s}} = 668 \text{ s}$$

Therefore, it will take over 11 minutes for the paint particle to fall to the floor.

Since

$$\text{Re} = \frac{\rho DU}{\mu} = \frac{(1.23 \text{ kg/m}^3)(1 \times 10^{-5} \text{ m})(0.00365 \text{ m/s})}{1.79 \times 10^{-5} \text{ N} \cdot \text{s/m}^2}$$

$$= 0.00251$$

we see that Re < 1, and the form of the drag coefficient used is valid.

Comments By repeating the calculations for various particle diameters, the results shown in **Fig. E9.10b** are obtained. Note that very small particles fall extremely slowly. In fact, particles in the general size range that can be inhaled deeply into the lungs (on the order of 5 μm or less) will remain airborne most of the working day, thus continually exposing workers to potential hazards. It is often the smaller particles that cause the greatest health concerns, particularly to individuals exposed day after day for many years. In work environments that generate particle-laden air, proper engineering controls are needed to reduce inhalation exposures (e.g., appropriate ventilation) or, in extreme cases, personal protective equipment (PPE, e.g., respirator) may be required. PPE is considered a last resort because in practice it is the least effective means to protect workers. It is more effective and usually less expensive in the long run to design out or control a problem with engineering ingenuity—prevention through design.

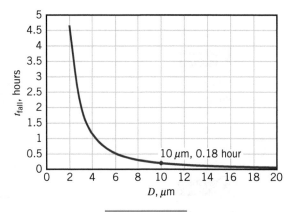

FIGURE E9.10b

Moderate Reynolds number flows tend to take on a boundary layer flow structure. For such flows past streamlined bodies, the drag coefficient tends to decrease slightly with Reynolds number. The $C_D \sim \text{Re}^{-1/2}$ dependence for a laminar boundary layer on a flat plate (see Table 9.3) is such an example. Moderate Reynolds number flows past blunt bodies generally produce drag coefficients that are relatively constant. The C_D values for the spheres and circular cylinders shown in **Fig. 9.23a** indicate this character in the range $10^3 < \text{Re} < 10^5$.

The structure of the flow field at selected Reynolds numbers indicated in Fig. 9.23a is shown in **Fig. 9.23b**. For a given object there is a wide variety of flow situations, depending on the Reynolds number involved. The curious reader is strongly encouraged to study the many beautiful photographs and videos of these (and other) flow situations found in Refs. 8 and 25.

For many shapes there is a sudden change in the character of the drag coefficient when the boundary layer becomes turbulent. This is illustrated in Fig. 9.15 for the flat plate and in Fig. 9.23 for the sphere and the circular cylinder. The Reynolds number at which this transition takes place is a function of the shape of the body.

For streamlined bodies, the drag coefficient increases when the boundary layer becomes turbulent because most of the drag is due to the shear force, which is greater for turbulent flow than for laminar flow. On the other hand, the drag coefficient for a relatively blunt object, such as a cylinder or sphere, actually decreases when the boundary layer becomes turbulent. As is discussed in Section 9.2.6, a turbulent boundary layer can travel further along the surface into the adverse pressure gradient on the rear portion of the cylinder before separation occurs. The result is a thinner wake and smaller pressure drag for turbulent boundary layer flow. This is indicated in Fig. 9.23 by the sudden decrease in C_D for $10^5 < \text{Re} < 10^6$. In a portion of this range

Flow past a cylinder can take on a variety of different structures.

 Video

V9.12 Karman vortex street

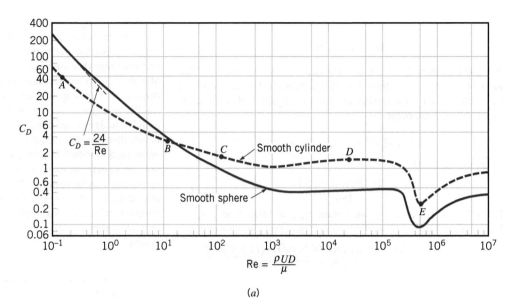

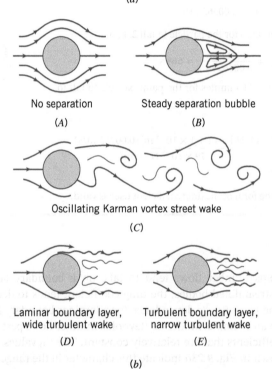

(a)

No separation   Steady separation bubble

(A)   *(B)*

Oscillating Karman vortex street wake

(C)

Laminar boundary layer,   Turbulent boundary layer,
wide turbulent wake   narrow turbulent wake

(D)   *(E)*

(b)

FIGURE 9.23   (a) Drag coefficient as a function of Reynolds number for a smooth circular cylinder and a smooth sphere. (b) Typical flow patterns for flow past a circular cylinder at various Reynolds numbers as indicated in (a).

V9.13 Flow past cylinder

V9.14 Oscillating sign

Video

V9.15 Flow past a flat plate

the actual drag (not just the drag coefficient) decreases with increasing speed. It would be very difficult to control the steady flight of such an object in this range—an increase in velocity requires a decrease in thrust (drag). In all other Reynolds number ranges the drag increases with an increase in the upstream velocity; even though C_D may decrease with Re, remember that drag calculation requires drag coefficient times velocity squared.

For extremely blunt bodies, like a flat plate perpendicular to the flow, the flow separates at the edge of the plate regardless of the nature of the boundary layer flow. Thus, the drag coefficient shows very little dependence on the Reynolds number. The drag coefficient for a flat plate perpendicular to the flow is about 1.9 for any Reynolds number.

The drag coefficients for a series of two-dimensional bodies of varying bluntness are given as a function of Reynolds number in **Fig. 9.24**. The characteristics described above are evident.

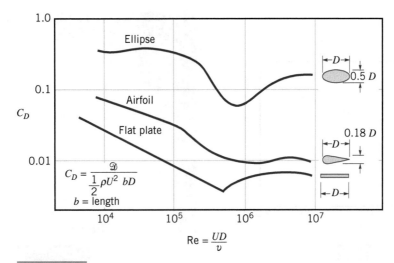

FIGURE 9.24 Character of the drag coefficient as a function of Reynolds number for objects with various degrees of streamlining, (two-dimensional flow) (Ref. 5).

EXAMPLE 9.11 | Terminal Velocity of a Falling Object

Given Hail is produced by the repeated rising and falling of ice particles in the updraft of a thunderstorm, as is indicated in **Fig. E9.11a**. When the hail becomes large enough, the aerodynamic drag from the updraft can no longer support the weight of the hail, and it falls from the storm cloud.

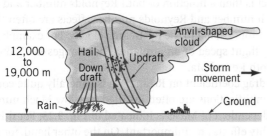

FIGURE E9.11a

Find Estimate the velocity, U, of the updraft needed to make $D = 4$-cm-diameter (i.e., "golf ball-sized") hail.

Solution As is discussed in Example 9.10, for steady-state conditions a force balance on an object falling through a fluid at its terminal velocity, U, gives

$$\mathcal{W} = \mathcal{D} + F_B$$

where $F_B = \gamma_{air}\Psi$ is the buoyant force of the air on the particle, $\mathcal{W} = \gamma_{ice}\Psi$ is the particle weight, and $\mathcal{D}$ is the aerodynamic drag. This equation can be rewritten as

$$\frac{1}{2}\rho_{air}U^2\frac{\pi}{4}D^2C_D = \mathcal{W} - F_B \qquad (1)$$

With $\Psi = \pi D^3/6$ and since $\gamma_{ice} \gg \gamma_{air}$ (i.e., $\mathcal{W} \gg F_B$), Eq. 1 can be simplified to

$$U = \left(\frac{4}{3}\frac{\rho_{ice}}{\rho_{air}}\frac{gD}{C_D}\right)^{1/2} \qquad (2)$$

By using $\rho_{ice} = 948$ kg/m^3, $\rho_{air} = 1.23$ kg/m^3, and $D = 4$ cm $= 0.04$ m, Eq. 2 becomes

$$U = \frac{4(948\,\text{kg/m}^3)(9.81\,\text{m/s}^2)(0.04\,\text{m})}{3(1.23\,\text{kg/m}^3)}$$

or

$$U = \frac{20.08}{\sqrt{C_D}} \qquad (3)$$

where U is in m/s. To determine U, we must know C_D. Unfortunately, C_D is a function of the Reynolds number (see Fig. 9.23), which is not known unless U is known. Thus, we must use an iterative technique similar to that done with the Moody chart for certain types of pipe flow problems (see Section 8.5).

From Fig. 9.23 we expect that C_D is on the order of 0.5. Thus, we assume $C_D = 0.5$ and from Eq. 3 obtain

$$U = \frac{20.08}{\sqrt{0.5}} = 28.4 \text{ m/s}$$

The corresponding Reynolds number (assuming $v = 1.46 \times 10^{-5}$ m^2/s) is

$$\text{Re} = \frac{UD}{v} = \frac{(28.4 \text{ m/s})(0.04 \text{ m})}{1.46\times10^{-5} \text{ m}^2/\text{s}} = 7.78\times10^4$$

For this value of Re we obtain from Fig. 9.23, $C_D = 0.5$. Thus, our assumed value of $C_D = 0.5$ was correct. The corresponding value of U is

$$U = 28.4 \text{ m/s} = 102.2 \text{ km/hr} \qquad (Ans)$$

Comments By repeating the calculations for various altitudes, z, above sea level (using the data given in Appendix C), the results shown in **Fig. E9.11b** are obtained. Because of the decrease in density with altitude, the hail falls even faster through the upper portions of the storm than when it hits the ground.

Clearly, an airplane flying through such an updraft would feel its effects (even if it were able to dodge the hail). As seen from Eq. 2, the larger the hail, the stronger the necessary updraft. Hailstones greater

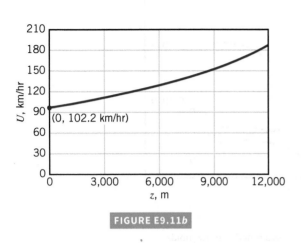

FIGURE E9.11*b*

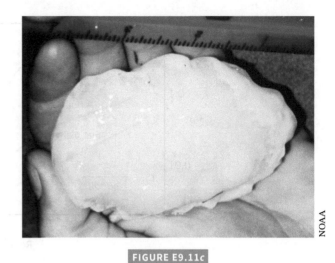

FIGURE E9.11*c*

than 15 cm in diameter have been reported. In reality, a hailstone is seldom spherical and often not smooth. The hailstone shown in **Fig. E9.11c** is larger than golf-ball sized and is clearly ellipsoidal in shape with additional smaller-scale surface roughness features. However, the calculated updraft velocities are in agreement with measured values.

Compressibility Effects

The above discussion is restricted to incompressible flows. If the velocity of the object is sufficiently large, compressibility effects become important and the drag coefficient becomes a function of the Mach number, $Ma = U/c$, where c is the speed of sound in the fluid. The introduction of Mach number effects complicates matters because the drag coefficient for a given object is then a function of both Reynolds number and Mach number—$C_D = \phi(Re, Ma)$. The Mach number and Reynolds number effects are often closely connected because both are directly proportional to the upstream velocity. For example, both Re and Ma increase with increasing flight speed of an airplane. The changes in C_D due to a change in U are due to changes in both Re and Ma.

> The drag coefficient is usually independent of Mach number for Mach numbers up to approximately 0.5.

The precise dependence of the drag coefficient on Re and Ma is generally quite complex (Ref. 12). However, the following simplifications are often justified. For low Mach numbers, the drag coefficient is essentially independent of Ma as is indicated in **Fig. 9.25**. For this situation, if Ma < 0.5 or so, compressibility effects are unimportant. On the other hand, for larger Mach number flows, the drag coefficient can be strongly dependent on Ma, with only secondary Reynolds number effects.

For most objects, values of C_D increase dramatically in the vicinity of Ma = 1 (i.e., transonic flow). This change in character, indicated by **Fig. 9.26**, is due to the existence of shock waves as indicated by the adjacent figure. Shock waves are extremely narrow regions in the flow field across which the flow parameters change in a nearly discontinuous manner, which are discussed in Chapter 11. Shock waves, which cannot exist in subsonic flows, provide a mechanism for the generation of drag that is not present in the relatively low-speed subsonic flows (see Fig. E11.3b in Example 11.3).

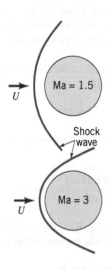

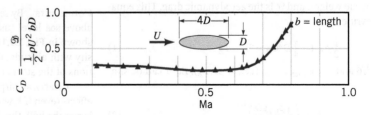

FIGURE 9.25 Drag coefficient as a function of Mach number for a two-dimensional object in subsonic flow (Ref. 5).

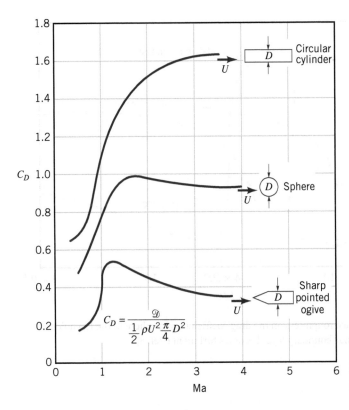

$$C_D = \dfrac{\mathscr{D}}{\frac{1}{2}\rho U^2 \frac{\pi}{4} D^2}$$

FIGURE 9.26 Drag coefficient as a function of Mach number for supersonic flow (Ref. 16).

The character of the drag coefficient as a function of Mach number is different for blunt bodies than for sharp bodies. As is shown in Fig. 9.26, sharp-pointed bodies develop their maximum drag coefficient in the vicinity of Ma = 1 (sonic flow), whereas the drag coefficient for blunt bodies increases with Ma far above Ma = 1. This behavior is due to the nature of the shock wave structure and the accompanying flow separation. The leading edges of wings for subsonic aircraft are usually quite rounded and blunt, while those of supersonic aircraft tend to be quite pointed and sharp. More information on these important topics can be found in standard texts about compressible flow and aerodynamics (Refs. 9, 10, and 23).

Surface Roughness

As is indicated in Fig. 9.15, the drag on a flat plate parallel to the flow is quite dependent on the surface roughness, provided the boundary layer flow is turbulent. In such cases the surface roughness protrudes through the laminar sublayer adjacent to the surface (see Section 8.4) and alters the wall shear stress. In addition to the increased turbulent shear stress, surface roughness can alter the Reynolds number at which the boundary layer flow becomes turbulent. Thus, a rough flat plate may have a larger portion of its length covered by a turbulent boundary layer than does the corresponding smooth plate. This also acts to increase the net drag on the plate.

In general, for streamlined bodies, the drag increases with increasing surface roughness. Great care is taken to design the surfaces of airplane wings to be as smooth as possible, since protruding rivets or screw heads can cause a considerable increase in the drag. On the other hand, for an extremely blunt body, such as a flat plate normal to the flow, the drag is independent of the surface roughness, since the shear stress is not in the upstream flow direction and contributes nothing to the drag.

> Depending on the body shape, an increase in surface roughness may increase or decrease drag.

For blunt bodies like a circular cylinder or sphere, an increase in surface roughness can actually cause a decrease in the drag. This is illustrated for a sphere in **Fig. 9.27**. As is discussed in Section 9.2.6, when the Reynolds number reaches the critical value (Re = 3×10^5 for a smooth sphere), the boundary layer becomes turbulent and the wake region behind the sphere becomes considerably narrower than if it were laminar (see Figs. 9.18 and 9.23b). The result is a considerable drop in pressure drag with a slight increase in friction drag, combining to give a smaller overall drag (and C_D).

The boundary layer can be tripped into turbulence at a smaller Reynolds number by using a rough-surfaced sphere. For example, the critical Reynolds number for a golf ball is

> Surface roughness can cause the boundary layer to become turbulent.

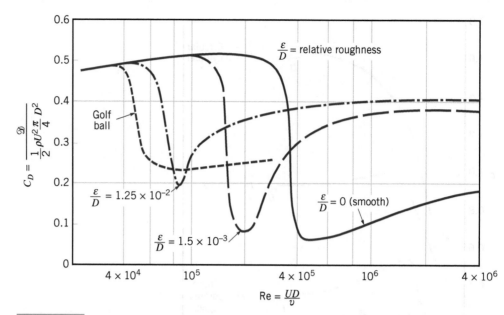

FIGURE 9.27　The effect of surface roughness on the drag coefficient of a sphere in the Reynolds number range for which the laminar boundary layer becomes turbulent (Ref. 5).

approximately $Re = 4 \times 10^4$. In the range $4 \times 10^4 < Re < 4 \times 10^5$, the drag on the standard rough (i.e., dimpled) golf ball is considerably less ($C_{Drough}/C_{Dsmooth} \approx 0.25/0.5 = 0.5$) than for the smooth ball. As is shown in Example 9.12, this is precisely the Reynolds number range for well-hit golf balls—hence, a reason for dimples on golf balls. The Reynolds number range for well-hit table tennis balls is less than $Re = 4 \times 10^4$. Thus, table tennis balls are smooth.

EXAMPLE 9.12 │ Effect of Surface Roughness

Given　A well-hit golf ball (diameter D = 4.3 cm, weight $\mathcal{W}$ = 0.44 N) can travel at U = 60 m/s as it leaves the tee. A well-hit table tennis ball (diameter D = 3.8 cm, weight $\mathcal{W}$ = 0.025 N) can travel at U = 20 m/s as it leaves the paddle.

Find　Determine the drag on a standard golf ball, a smooth golf ball, and a table tennis ball for the conditions given. Also determine the deceleration of each ball for these conditions.

Solution　For either ball, the drag can be obtained from

$$\mathcal{D} = \frac{1}{2}\rho U^2 \frac{\pi}{4} D^2 C_D \tag{1}$$

where the drag coefficient, C_D, is given in Fig. 9.27 as a function of the Reynolds number and surface roughness. For the golf ball in standard air

$$Re = \frac{UD}{\upsilon} = \frac{(60 \text{ m/s})(4.3/100 \text{ m})}{0.14 \times 10^{-4} \text{ m}^2/\text{s}} = 1.8 \times 10^5$$

while for the table tennis ball

$$Re = \frac{UD}{\upsilon} = \frac{(20 \text{ m/s})(3.8/100 \text{ m})}{(1.46 \times 10^{-5} \text{ m}^2/\text{s})} = 5.2 \times 10^4$$

The corresponding drag coefficients are C_D = 0.25 for the standard golf ball, C_D = 0.51 for the smooth golf ball, and

C_D = 0.50 for the table tennis ball. Hence, from Eq. 1 for the standard golf ball

$$\mathcal{D} = \frac{1}{2}(1.23 \text{ kg/m}^3)(60 \text{ m/s})^2 \frac{\pi}{4}(4.3 \times 10^{-2} \text{ m})^2 (0.25)$$

$$= 0.8 \text{ N} \tag{Ans}$$

for the smooth golf ball

$$\mathcal{D} = \frac{1}{2}(1.23 \text{ kg/m}^3)(60 \text{ m/s})^2 \frac{\pi}{4}(4.3 \times 10^{-2} \text{ m})^2 (0.51)$$

$$= 1.64 \text{ N} \tag{Ans}$$

and for the table tennis ball

$$\mathcal{D} = \frac{1}{2}(1.23 \text{ kg/m}^3)(20 \text{ m/s})^2 \frac{\pi}{4}(3.8 \times 10^{-2} \text{ m})^2 (0.50)$$

$$= 0.14 \text{ N} \tag{Ans}$$

The corresponding decelerations are $a = \mathcal{D}/m = g\mathcal{D}/\mathcal{W}$, where m is the mass of the ball. Thus, the deceleration relative to the acceleration of gravity, a/g (i.e., the number of g's deceleration) is $a/g = \mathcal{D}/\mathcal{W}$ or

$$\frac{a}{g} = \frac{0.8 \text{ N}}{0.44 \text{ N}} = 1.82 \text{ for the standard golf ball} \tag{Ans}$$

$$\frac{a}{g} = \frac{0.164\,\text{N}}{0.44\,\text{N}} = 3.73 \text{ for the smooth golf ball} \qquad (Ans)$$

$$\frac{a}{g} = \frac{0.14}{0.025} = 5.6 \text{ for the table tennis ball} \qquad (Ans)$$

Comments When considering drag force, don't be fooled by the magnitude of C_D alone. Notice that the drag force on the table tennis ball is considerably lower than on the standard golf ball, even though the drag coefficient is larger. This is a result of diameter difference and especially squared velocity difference. Note that there is a considerably smaller deceleration for the rough golf ball than for the smooth one. Because of its much larger drag-to-mass ratio, the table tennis ball slows down relatively quickly and does not travel as far as the golf ball. Note that with $U = 20$ m/s the standard golf ball has a drag of $\mathcal{D} = 0.09$ N and a deceleration of $a/g = 0.202$, considerably less than the $a/g = 4.77$ of the table tennis ball. Conversely, a table tennis ball hit from a tee at 60 m/s would decelerate at a rate of $a = 530$ m/s^2, or $a/g = 54.1$. It would not travel nearly as far as the golf ball.

By repeating the above calculations, the drag as a function of speed for both a standard golf ball and a smooth golf ball is shown in **Fig. E9.12**.

The Reynolds number range for which a rough golf ball has smaller drag than a smooth one (i.e., 4×10^4 to 3.6×10^5) corresponds to a flight velocity range of $14 < U < 122$ m/s. This is comfortably within the range of most golfers. (The fastest tee shot by top professional golfers is approximately 85 m/s.) As discussed in Section 9.4.2, the dimples (roughness) on a golf ball also help produce a lift (due to the spin of the ball) that allows the ball to travel farther than a smooth ball.

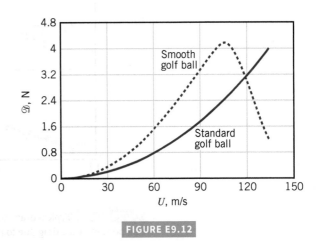

FIGURE E9.12

The Wide Word of Fluids

Dimpled baseball bats

For many years it has been known that dimples on golf balls can create a *turbulent boundary layer* and reduce the aerodynamic drag, allowing longer drives than with smooth balls. Thus, why not put dimples on baseball bats so that tomorrow's baseball sluggers can swing the bat faster and, therefore, hit the ball farther? MIT instructor Jeffery De Tullio pondered that question, performed experiments with dimpled bats to determine the answer, and received a patent for his dimpled bat invention. The result is that a batter can swing a dimpled bat approximately 3 to 5% faster than a smooth bat. Theoretically, this extra speed will translate to an extra 3 to 4.5 m distance on a long hit.

Froude Number Effects Another parameter on which the drag coefficient may be strongly dependent is the Froude number, $\text{Fr} = U/\sqrt{g\ell}$. As is discussed in Chapter 10, the Froude number is a ratio of the free-stream speed to a typical wave speed on the interface of two fluids, such as the surface of the ocean. An object moving on the surface, such as a ship, often produces waves that require a source of energy to generate. This energy comes from the ship and is manifest as a drag. [Recall that the rate of energy production (power) equals speed times force.] The nature of the waves produced often depends on the Froude number of the flow and the shape of the object—the waves generated by a water skier "plowing" through the water at a low speed (low Fr) are different than those generated by the skier "planing" along the surface at high speed (large Fr).

Thus, the drag coefficient for surface ships is a function of Reynolds number (viscous effects) and Froude number (wave-making effects); $C_D = \phi(\text{Re}, \text{Fr})$. As was discussed in Chapter 7, it is often quite difficult to run model tests under conditions similar to those of the prototype (i.e., same Re and Fr for surface ships). Fortunately, the viscous and wave effects can often be separated, with the total drag being the sum of the drag of these individual effects. A detailed account of this important topic can be found in standard texts (Ref. 11).

As is indicated in **Fig. 9.28**, the wave-making drag, $\mathcal{D}_w$, can be a complex function of the Froude number and the body shape. The rather "wiggly" dependence of the wave drag

V9.16 Jet ski

The drag coefficient for surface ships is a function of the Froude number.

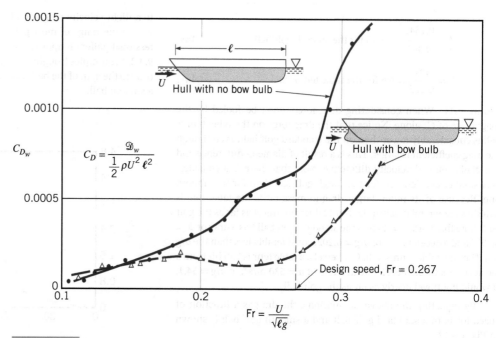

FIGURE 9.28 Typical drag coefficient data as a function of Froude number and hull characteristics for that portion of the drag due to the generation of waves (Ref. 21).

coefficient, $C_{Dw} = \mathcal{D}_w/(\rho U^2 \ell^2/2)$, on the Froude number shown is typical. It results from the fact that the structure of the waves produced by the hull is a strong function of the ship speed or, in dimensionless form, the Froude number. This wave structure is also a function of the body shape. For example, the bow wave, which is often the major contributor to the wave drag, can be reduced by use of an appropriately designed bulb on the bow, as is indicated in Fig. 9.28. In this instance the streamlined body (hull without a bulb) has more drag than the less streamlined one.

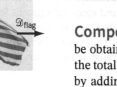

The drag on a complex body can be approximated as the sum of the drag on its parts.

Composite Body Drag Approximate drag calculations for a complex body can often be obtained by treating the body as a composite collection of its various parts. For example, the total force on a flag pole because of the wind (see the adjacent figure) can be approximated by adding the aerodynamic drag produced by the various components involved—the drag on the flag and the drag on the pole. In some cases considerable care must be taken in such an approach because of the interactions between the various parts. It may not be correct to merely add the drag of the components to obtain the drag of the entire object, although such approximations are often reasonable.

EXAMPLE 9.13 | Drag on a Composite Body

Given A 96 km/hr wind blows past the water tower shown in **Fig. E9.13a**.

Find Estimate the moment (torque), M, needed at the base to keep the tower from tipping over.

Solution We treat the water tower as a sphere resting on a circular cylinder and assume that the total drag is the sum of the drag from these parts. The free-body diagram of the tower is shown in **Fig. E9.13b**. By summing moments about the base of the tower, we obtain

$$M = \mathcal{D}_s\left(b + \frac{D_s}{2}\right) + \mathcal{D}_c\left(\frac{b}{2}\right) \qquad (1)$$

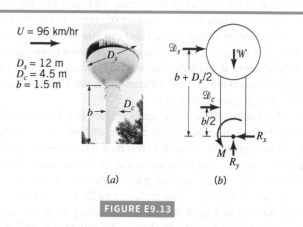

(a)   (b)

FIGURE E9.13

where

$$\mathcal{D}_s = \frac{1}{2}\rho U^2 \frac{\pi}{4} D_s^2 C_{Ds} \qquad (2)$$

and

$$\mathcal{D}_c = \frac{1}{2}\rho U^2 b D_c C_{Dc} \qquad (3)$$

are the drag on the sphere and cylinder, respectively. For standard atmospheric conditions, the Reynolds numbers are

$$\text{Re}_s = \frac{U D_s}{\upsilon} = \frac{(27 \text{ m/s})(12 \text{ m})}{0.14 \times 10^{-4} \text{ m}^2/\text{s}} = 2.31 \times 10^7$$

and

$$\text{Re}_c = \frac{U D_c}{\upsilon} = \frac{(27 \text{ m/s})(4.5 \text{ m})}{0.14 \times 10^{-4} \text{ m}^2/\text{s}} = 8.67 \times 10^6$$

The corresponding drag coefficients, C_{Ds} and C_{Dc}, can be approximated from Fig. 9.23 as

$$C_{Ds} \approx 0.3 \quad \text{and} \quad C_{Dc} \approx 0.7$$

Note that the value of C_{Ds} was obtained by an extrapolation of the given data to Reynolds numbers beyond those given

(a potentially dangerous practice!). From Eqs. 2 and 3 we obtain

$$\mathcal{D}_s = 0.5(1.23 \text{ kg/m}^3)(27 \text{ m/s})^2 \frac{\pi}{4}(12 \text{ m})^2(0.3)$$
$$= 15.2 \times 10^3 \text{ N}$$

and

$$\mathcal{D}_c = 0.5(1.23 \text{ kg/m}^3)(27 \text{ m/s})^2(15 \text{ m} \times 4.5 \text{ m})(0.7)$$
$$= 212 \times 10^3 \text{ N}$$

From Eq. 1 the corresponding moment needed to prevent the tower from tipping is

$$M = (15.2 \times 10^3 \text{ N})\left(15 \text{ m} + \frac{12}{2} \text{ m}\right) + (21.2 \times 10^3 \text{ N})\left(\frac{15}{2} \text{ m}\right)$$
$$= 4.78 \times 10^5 \text{ N} \cdot \text{m}$$

Comment The above result is only an estimate because (a) the wind is probably not uniform from the top of the tower to the ground, (b) the tower is not exactly a combination of a smooth sphere and a circular cylinder, (c) the cylinder is not of infinite length, (d) there will be some interaction between the flow past the cylinder and that past the sphere so that the net drag is not exactly the sum of the two, and (e) a drag coefficient value was obtained by extrapolation of the given data. However, such approximate results are often quite accurate.

The aerodynamic drag on automobiles provides an example of the use of adding component drag forces. The power required to move a car along a level street is used to overcome the rolling resistance and the aerodynamic drag. For speeds above approximately 48.3 km/hr, the aerodynamic drag becomes a significant contribution to the net propulsive force needed. The contributions of the drag due to various portions of cars (e.g., front end, windshield, roof, rear end, windshield peak, rear roof/trunk, and cowl) has been determined by numerous model and full-sized tests as well as by numerical calculations. As a result, it is possible to predict the aerodynamic drag on cars of a wide variety of body styles.

As is indicated in **Fig. 9.29**, the drag coefficient for cars has decreased rather continuously over the years. This reduction is a result of careful design of the shape and the details (such as window molding, rearview mirrors, etc.). An additional reduction in drag has been accomplished by a reduction of the projected area. The net result is a considerable increase in the gas mileage, especially at highway speeds. Considerable additional information about the aerodynamics of road vehicles can be found in the literature (Ref. 24).

V9.17 Drag on a truck

The Wide World of Fluids

At 5333 km/liter it doesn't cost much to "fill 'er up"

Typical gas consumption for a Formula 1 racer, a sports car, and a sedan is approximately 0.85 km/liter, 6.4 km/liter, and 12.8 km/liter, respectively. Thus, just how did the winning entry in the 2005 Shell Eco-Marathon achieve an incredible 5356.8 km/liter? To be sure, this vehicle is not as fast as a Formula 1 racer (although the rules require it to average at least 24 km/hr), and it can't carry as large a load as your family sedan can (the vehicle has barely

enough room for the driver). However, by using a number of clever engineering design considerations, this amazing fuel efficiency was obtained. The type (and number) of tires, the appropriate engine power and weight, the specific chassis design, and the design of the body shell are all important and interrelated considerations. To reduce *drag*, the aerodynamic shape of the high-efficiency vehicle was given special attention through theoretical considerations and wind tunnel model testing. The result is an amazing vehicle that can travel a long distance without hearing the usual "fill 'er up."

1920s, $C_D \approx 0.8$ 1970s, $C_D \approx 0.4$ 2000s, $C_D \approx 0.3$

FIGURE 9.29 The historical trend of streamlining automobiles to reduce their aerodynamic drag and increase their miles per gallon.

The effect of several important parameters (shape, Re, Ma, Fr, and roughness) on the drag coefficient for various objects has been discussed in this section. As stated previously, drag coefficient information for a very wide range of objects is available in the literature. Some of this information is given in **Figs. 9.30, 9.31,** and **9.32** for a variety of two- and three-dimensional,

Shape	Reference area A (b = length)	Drag coefficient $C_D = \dfrac{\mathcal{D}}{\frac{1}{2}\rho U^2 A}$	Reynolds number $Re = \rho UD/\mu$
Square rod with rounded corners	$A = bD$	$\begin{array}{c\|c} R/D & C_D \\ \hline 0 & 2.2 \\ 0.02 & 2.0 \\ 0.17 & 1.2 \\ 0.33 & 1.0 \end{array}$	$Re = 10^5$
Rounded equilateral triangle	$A = bD$	$\begin{array}{c\|c\|c} R/D & \rightarrow C_D & \leftarrow C_D \\ \hline 0 & 1.4 & 2.1 \\ 0.02 & 1.2 & 2.0 \\ 0.08 & 1.3 & 1.9 \\ 0.25 & 1.1 & 1.3 \end{array}$	$Re = 10^5$
Semicircular shell	$A = bD$	$\rightarrow 2.3$ $\leftarrow 1.1$	$Re = 2 \times 10^4$
Semicircular cylinder	$A = bD$	$\rightarrow 2.15$ $\leftarrow 1.15$	$Re > 10^4$
T-beam	$A = bD$	$\rightarrow 1.80$ $\leftarrow 1.65$	$Re > 10^4$
I-beam	$A = bD$	2.05	$Re > 10^4$
Angle	$A = bD$	$\rightarrow 1.98$ $\leftarrow 1.82$	$Re > 10^4$
Hexagon	$A = bD$	1.0	$Re > 10^4$
Rectangle	$A = bD$	$\begin{array}{c\|c} \ell/D & C_D \\ \hline \leq 0.1 & 1.9 \\ 0.5 & 2.5 \\ 0.65 & 2.9 \\ 1.0 & 2.2 \\ 2.0 & 1.6 \\ 3.0 & 1.3 \end{array}$	$Re = 10^5$

FIGURE 9.30 Typical drag coefficients for regular two-dimensional objects (Refs. 5, 6).

Shape	Reference area A	Drag coefficient C_D	Reynolds number $Re = \rho U D/\mu$
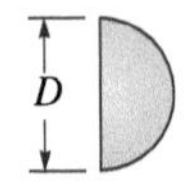 Solid hemisphere	$A = \dfrac{\pi}{4}D^2$	→ 1.17 ← 0.42	$Re > 10^4$
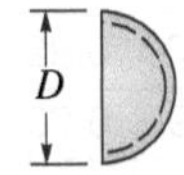 Hollow hemisphere	$A = \dfrac{\pi}{4}D^2$	→ 1.42 ← 0.38	$Re > 10^4$
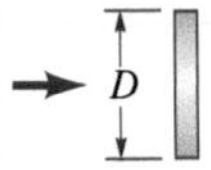 Thin disk	$A = \dfrac{\pi}{4}D^2$	1.1	$Re > 10^3$
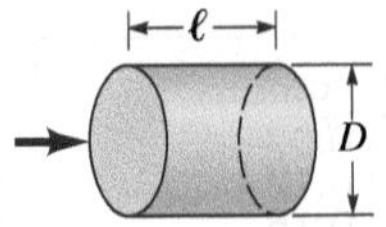 Circular rod parallel to flow	$A = \dfrac{\pi}{4}D^2$	ℓ/D C_D 0.5 1.1 1.0 0.93 2.0 0.83 4.0 0.85	$Re > 10^5$
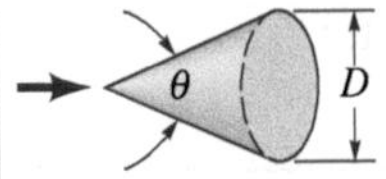 Cone	$A = \dfrac{\pi}{4}D^2$	θ, degrees C_D 10 0.30 30 0.55 60 0.80 90 1.15	$Re > 10^4$
Cube	$A = D^2$	1.05	$Re > 10^4$
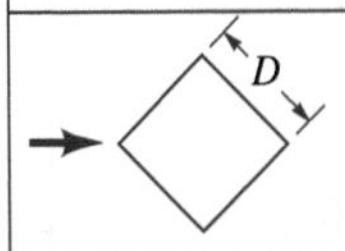 Cube	$A = D^2$	0.80	$Re > 10^4$
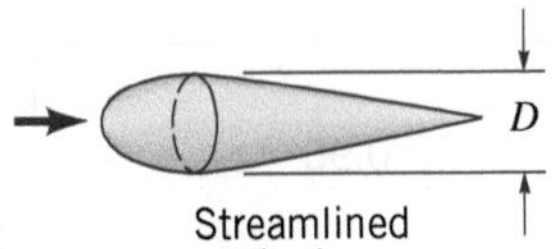 Streamlined body	$A = \dfrac{\pi}{4}D^2$	0.04	$Re > 10^5$

FIGURE 9.31 Typical drag coefficients for regular three-dimensional objects (Ref. 5).

natural and man-made objects. Recall that a drag coefficient of unity is equivalent to the drag produced by the dynamic pressure acting on an area of size A. That is, $\mathcal{D} = \tfrac{1}{2}\rho U^2 A C_D = \tfrac{1}{2}\rho U^2 A$ if $C_D = 1$. Typical nonstreamlined objects have drag coefficients on this order.

9.4 LIFT

As is indicated in Section 9.1, any object moving through a fluid will experience a net force of the fluid on the object. All objects experience a drag force, $\mathcal{D}$, which is parallel to the upstream flow. A force normal to the flow, lift $\mathcal{L}$, may be present only if the object is not symmetrical or if it does not produce a symmetrical flow field (such as flow around a rotating sphere). Considerable effort has been put forth to understand the various properties

Shape	Reference area	Drag coefficient C_D
 Parachute	Frontal area $A = \frac{\pi}{4}D^2$	1.4
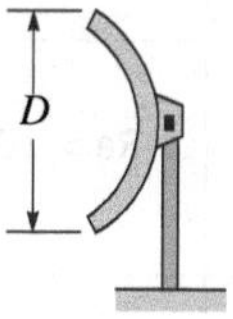 Porous parabolic dish	Frontal area $A = \frac{\pi}{4}D^2$	Porosity: 0 / 0.2 / 0.5 — → 1.42 / 1.20 / 0.82 — ← 0.95 / 0.90 / 0.80 — Porosity = open area/total area
 Average person	Standing / Sitting / Crouching	$C_DA = 0.84$ m^2 / $C_DA = 0.56$ m^2 / $C_DA = 0.23$ m^2
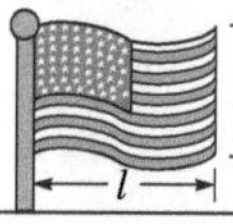 Fluttering flag	$A = \ell D$	ℓ/D : C_D — 1 : 0.07 — 2 : 0.12 — 3 : 0.15
 Empire State Building	Frontal area	1.4
 Six-car passenger train	Frontal area	1.8
Bikes — Upright commuter	$A = 0.5$ m^2	1.1
 Racing	$A = 3.36$ m^2	0.88
 Drafting	$A = 3.36$ m^2	0.50
 Streamlined	$A = 0.46$ m^2	0.12
Tractor-trailer trucks — Standard	Frontal area	0.96
With fairing	Frontal area	0.76
With fairing and gap seal	Frontal area	0.70
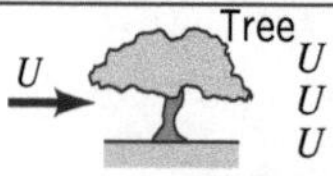 Tree — $U = 10$ m/s — $U = 20$ m/s — $U = 30$ m/s	Frontal area	0.43 — 0.26 — 0.20
 Dolphin	Wetted area	0.0036 at Re = 6 × 10^6 (flat plate has $C_{Df} = 0.0031$)
 Large birds	Frontal area	0.40

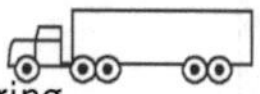

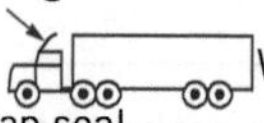

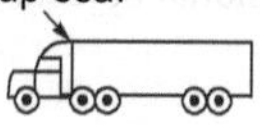

FIGURE 9.32 Typical drag coefficients for objects of interest (Refs. 5, 6, 14, 17).

of the generation of lift. Some objects, such as an airfoil on an airplane wing, a propeller, or turbine and compressor blades are designed to generate lift. Other objects are designed to reduce the lift generated. For example, the lift on a car tends to reduce the contact force between the wheels and the ground, causing reduction in traction and cornering ability. It is desirable to reduce this lift on a passenger car, or for the case of a racecar, reverse the lift to the downward direction (known as downforce).

9.4.1 Surface Pressure Distribution

The lift can be determined from Eq. 9.2 if the distributions of pressure and wall shear stress around the entire body are known. As is indicated in Section 9.1, such data are usually not known. Typically, the lift is given in terms of the lift coefficient,

$$C_L = \frac{\mathscr{L}}{\frac{1}{2}\rho U^2 A} \qquad (9.39)$$

which is obtained from experiments, advanced analysis, or numerical considerations. The lift coefficient is a function of the appropriate dimensionless parameters and, like the drag coefficient, can be written as

$$C_L = \phi(\text{shape, Re, Ma, Fr, } \varepsilon/\ell)$$

The Froude number, Fr, is important only if there is a free surface present, as with an underwater "wing" used to support a high-speed hydrofoil surface ship. Often the surface roughness, ε, is relatively unimportant in terms of lift—it has more of an effect on the drag. The Mach number, Ma, is of importance for relatively high-speed subsonic and supersonic flows (i.e., Ma > 0.8), and the Reynolds number effect is often not great. The most important parameter that affects the lift coefficient is the shape of the object. Considerable effort has gone into designing optimally shaped lift-producing devices. We will emphasize the effect of the shape on lift; the effects of the other dimensionless parameters can be found in the literature (Refs. 12, 13, and 23).

Most common lift-generating devices (e.g., airfoils, fans, spoilers on cars, etc.) operate in the large Reynolds number range in which the flow has a boundary layer character, with viscous effects confined to the boundary layers and wake regions. For such cases the wall shear stress, τ_w, contributes little to the lift. Most of the lift comes from the surface pressure distribution. A typical pressure distribution on a moving car is shown in **Fig. 9.33**. The distribution, for the most part, is consistent with simple Bernoulli equation analysis. Locations with high-speed flow (e.g., over the roof and hood) have low pressure, while locations with low-speed flow (e.g., on the grill and windshield) have high pressure. It is easy to believe that the integrated effect of this pressure distribution would provide a net upward force.

For objects operating in very low Reynolds number regimes (i.e., Re < 1), viscous effects are important, and the contribution of the shear stress to the lift may be as important as that of the pressure. Such situations include the flight of minute insects and the swimming of microscopic organisms. The relative importance of τ_w and p in the generation of lift in a typical large Reynolds number flow is shown in Example 9.14.

> The lift coefficient is a dimensionless form of the lift.

> The lift coefficient is a function of other dimensionless parameters.

> Usually most lift comes from pressure forces, not viscous forces.

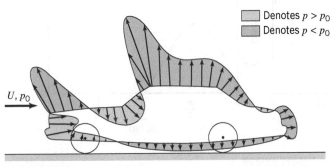

FIGURE 9.33 Pressure distribution on the surface of an automobile.

EXAMPLE 9.14 | Lift from Pressure and Shear Stress Distributions

Given When a uniform wind of velocity U blows past the semicircular building shown in **Fig. E9.14a,b,** the wall shear stress and pressure distributions on the outside of the building are as given previously in Figs. E9.8b and E9.9a, respectively.

Find If the pressure in the building is atmospheric (i.e., the value, p_0, far from the building), determine the lift coefficient and the lift on the roof.

Solution From Eq. 9.2 we obtain the lift as

$$\mathscr{L} = -\int p \sin \theta \, dA + \int \tau_w \cos \theta \, dA \qquad (1)$$

As is indicated in Fig. E9.14b, we assume that on the inside of the building the pressure is uniform, $p = p_0$, and that there is no shear stress. Thus, Eq. 1 can be written as

$$\mathscr{L} = -\int_0^\pi (p - p_0) \sin \theta \, b\left(\frac{D}{2}\right) d\theta$$
$$+ \int_0^\pi \tau_w \cos \theta \, b\left(\frac{D}{2}\right) d\theta$$

or

$$\mathscr{L} = \frac{bD}{2}\left[-\int_0^\pi (p - p_0) \sin \theta \, d\theta + \int_0^\pi \tau_w \cos \theta \, d\theta\right] \qquad (2)$$

where b and D are the length and diameter of the building, respectively, and $dA = b(D/2) \, d\theta$. Equation 2 can be put into

dimensionless form by using the dynamic pressure, $\rho U^2/2$, planform area, $A = bD$, and dimensionless shear stress

$$F(\theta) = \tau_w (\text{Re})^{1/2}/(\rho U^2/2)$$

to give

$$\mathscr{L} = \frac{1}{2}\rho U^2 A\left[-\frac{1}{2}\int_0^\pi \frac{(p - p_0)}{\frac{1}{2}\rho U^2} \sin \theta \, d\theta\right.$$
$$\left. + \frac{1}{2\sqrt{\text{Re}}}\int_0^\pi F(\theta) \cos \theta \, d\theta\right] \qquad (3)$$

From the data in Figs. E9.8b and E9.9a, the values of the two integrals in Eq. 3 can be obtained by determining the area under the curves of $[(p - p_0)/(\rho U^2/2)] \sin \theta$ versus θ and $F(\theta) \cos \theta$ versus θ plotted in **Figs. E9.14c** and **E9.14d**. The results are

$$\int_0^\pi \frac{(p - p_0)}{\frac{1}{2}\rho U^2} \sin \theta \, d\theta = -1.76$$

and

$$\int_0^\pi F(\theta) \cos \theta \, d\theta = 3.92$$

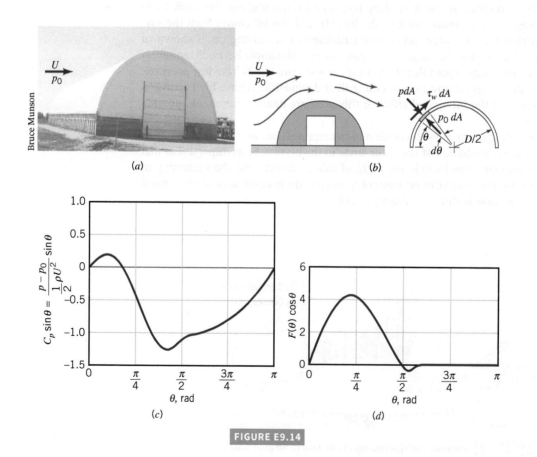

(a)   (b)

(c)   (d)

FIGURE E9.14

Thus, the lift is

$$\mathscr{L} = \frac{1}{2}\rho U^2 A\left[\left(-\frac{1}{2}\right)(-1.76) + \frac{1}{2\sqrt{\text{Re}}}(3.92)\right]$$

or

$$\mathscr{L} = \left(0.88 + \frac{1.96}{\sqrt{\text{Re}}}\right)\left(\frac{1}{2}\rho U^2 A\right) \qquad (Ans)$$

and

$$C_L = \frac{\mathscr{L}}{\frac{1}{2}\rho U^2 A} = 0.88 + \frac{1.96}{\sqrt{\text{Re}}} \qquad (4) \quad (Ans)$$

Comments Consider a typical situation with $D = 6$ m, $U = 10$ m/s, $b = 15$ m, and standard atmospheric conditions ($\rho = 1.23$ kg/m^3 and $v = 1.46 \times 10^{-5}$ m^2/s), which gives a Reynolds number of

$$\text{Re} = \frac{UD}{v} = \frac{(10 \text{ m/s})(6 \text{ m})}{1.46 \times 10^{-5} \text{ m}^2/\text{s}} = 4.11 \times 10^6$$

Hence, the lift coefficient is

$$C_L = 0.88 + \frac{1.96}{(4.11 \times 10^6)^{1/2}} = 0.88 + 0.001 = 0.881$$

Note that the pressure contribution to the lift coefficient is 0.88, whereas that due to the wall shear stress is only $1.96/(\text{Re}^{1/2}) = 0.001$. The Reynolds number dependency of C_L is quite minor. The lift is pressure dominated. Recall from Example 9.9 that this is also true for the drag on a similar shape.

From Eq. 4 with $A = 6$ m $\times$ 15 m $= 90$ m^2, we obtain the lift for the assumed conditions as

$$\mathscr{L} = \frac{1}{2}\rho U^2 A C_L$$
$$= \frac{1}{2}(1.23 \text{ kg/m}^3)(10 \text{ m/s})^2(90 \text{ m}^2)(0.881)$$

or

$$\mathscr{L} = 4876 \text{ N}$$

There is a considerable tendency for the building to lift off the ground. Clearly this is due to the object being nonsymmetrical. The lift force on a complete circular cylinder is zero, although the opposing fluid forces do tend to pull the upper and lower halves apart.

A typical device designed to produce lift does so by generating a pressure distribution that is different on the top and bottom surfaces. For large Reynolds number flows these pressure distributions are usually directly proportional to the dynamic pressure, $\rho U^2/2$, with viscous effects being of secondary importance. Hence, as indicated in the adjacent figure, for a given airfoil the lift is proportional to the square of the airspeed. Two airfoils used to produce lift are indicated in **Fig. 9.34**. Clearly the symmetrical one cannot produce lift unless the angle of attack, α, is nonzero. Because of the asymmetry of the nonsymmetric airfoil, the pressure distributions on the upper and lower surfaces are different, and a lift is produced even with $\alpha = 0$. Of course, there will be a certain value of α (less than zero for this case) for which the lift is zero. For this situation, the pressure distributions on the upper and lower surfaces are different, but their resultant (integrated) pressure forces will be equal and opposite.

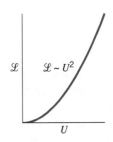

Since most airfoils are thin, it is customary to use the planform area, $A = bc$, in the definition of the lift coefficient. Here b is the length of the airfoil, or span, and c is the *chord length*—the length from the leading edge to the trailing edge as indicated in Fig. 9.34. Typical lift coefficients so defined are on the order of unity. That is, the lift force is on the order of the dynamic pressure times the planform area of the wing, $\mathscr{L} \approx (\rho U^2/2)A$. The *wing loading*, defined as the average lift per unit area of the wing, $\mathscr{L}/A$, therefore, increases with speed. For example, the wing loading of the 1903 Wright Flyer aircraft was 72 N/m^2, while for the present-day Boeing 747 aircraft it is 7200 N/m^2. The wing loading for a bumble bee is approximately 48 N/m^2 (Ref. 14).

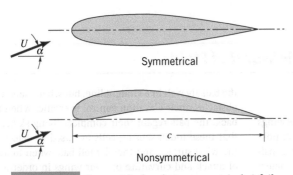

FIGURE 9.34 Symmetrical and nonsymmetrical airfoils.

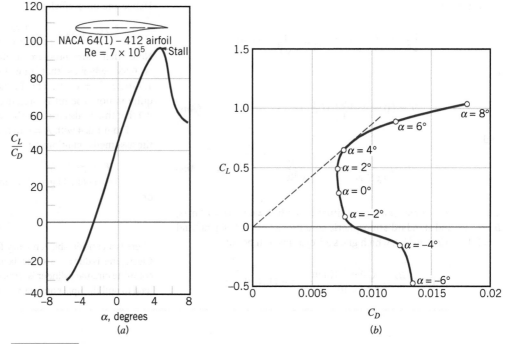

Video

V9.18 Stalled
airfoil

FIGURE 9.35 Two representations of the same lift and drag data for a typical airfoil: (*a*) lift-to-drag ratio as a function of angle of attack, with the onset of boundary layer separation on the upper surface indicated by the occurrence of stall, (*b*) the lift and drag polar diagram with the angle of attack indicated (Ref. 22).

At large angles of attack the boundary layer separates and the wing stalls.

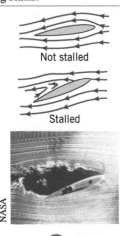

Video

V9.19 Bat flying

Although viscous effects and the wall shear stress contribute little to the direct generation of lift, they play an extremely important role in the design and use of lifting devices. This is because of the viscosity-induced boundary layer separation that can occur on nonstreamlined bodies such as airfoils that have too large an angle of attack (see Fig. 9.19). As is indicated in **Fig. 9.35, 9.36,** and **9.39,** up to a certain point, the lift coefficient increases rather steadily with the angle of attack, α. If α is too large, the boundary layer on the upper surface separates, the flow over the wing develops a wide, turbulent wake region, the lift decreases, and the drag increases. This condition, as indicated in the adjacent figures, is termed **stall**. Such conditions are extremely dangerous if they occur while the airplane is flying at a low altitude where there is not sufficient time and altitude to recover from the stall.

In many lift-generating devices the important quantity is the ratio of the lift to drag developed, $\mathcal{L}/\mathcal{D} = C_L/C_D$. Such information is often presented in terms of C_L/C_D versus α, as is shown in Fig. 9.35*a*, or in a *lift-drag polar* of C_L versus C_D with α as a parameter, as is shown in Fig. 9.35*b*. The most efficient angle of attack (i.e., largest C_L/C_D) can be found by drawing a line tangent to the $C_L - C_D$ curve from the origin, as is shown in Fig. 9.35*b*. High-performance airfoils generate lift that is perhaps 100 or more times greater than their drag. This translates into the fact that in still air they can glide a horizontal distance of 100 m for each 1 m drop in altitude.

The Wide World of Fluids

Bats feel turbulence

Researchers have discovered that at certain locations on the wings of bats, there are special touch-sensing cells with a tiny hair poking out of the center of the cell. These cells, which are very sensitive to air flowing across the wing surface, can apparently detect turbulence in the flow over the wing. If these hairs are removed, the bats fly well in a straight line, but when maneuvering to avoid obstacles, their elevation control is erratic. When the hairs grow back, the bats regain their complete flying skills. It is proposed that these touch-sensing cells are used to detect turbulence on the wing surface and thereby tell bats when to adjust the angle of attack and curvature of their wings in order to avoid stalling out in midair.

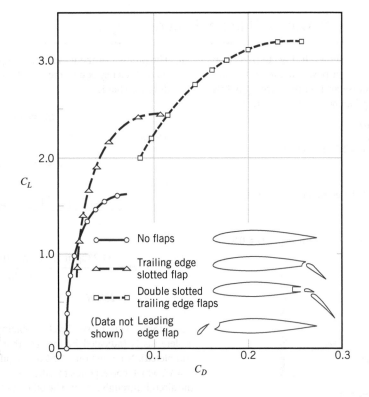

FIGURE 9.36 Typical lift
and drag alterations possible
with the use of various types
of flap designs (Ref. 18).

As is indicated in the preceding paragraph, the lift and drag on an airfoil can be altered by changing the angle of attack. This actually represents a change in the shape of the object. Other shape changes can be used to alter the lift and drag when desirable. In modern airplanes it is common to utilize leading edge and trailing edge flaps as is shown in Fig. 9.36. To generate the necessary lift during the relatively low-speed landing and takeoff procedures, the airfoil shape is altered by extending special flaps on the front and/or rear portions of the wing. Use of the flaps considerably enhances the lift, although it is at the expense of an increase in the drag (i.e., the airfoil is in a "dirty" configuration). This increase in drag is not of much concern during landing and takeoff operations—the decrease in landing or takeoff speed is more important than is a temporary increase in drag. During normal flight with the flaps retracted (i.e., the "clean" configuration), the drag is relatively small, and the needed lift force is achieved with the smaller lift coefficient and the larger dynamic pressure (higher speed).

The Wide World of Fluids

Learning from nature

For hundreds of years humans looked toward nature, particularly birds, for insight about flying. However, all early airplanes that closely mimicked birds proved to be unsuccessful. Only after much experimenting with rigid (or at least nonflapping) wings did human flight become possible. Recently, however, engineers have been turning to living systems—birds, insects, and other biological models—in an attempt to produce breakthroughs in aircraft design. Perhaps it is possible that nature's basic design concepts can be applied to airplane systems. For example, by morphing and

rotating their wings in three dimensions, birds have remarkable maneuverability that to date has no technological parallel. Birds can control the airflow over their wings by moving the feathers on their wingtips and the leading edges of their wings, providing designs that are more efficient than the flaps and rigid, pivoting tail surfaces of current aircraft (Ref. 14). On a smaller scale, understanding the mechanism by which insects dynamically manage unstable flow to generate lift may provide insight into the development of microscale air vehicles. With new hi-tech materials, computers, and automatic controls, aircraft of the future may mimic nature more than was once thought possible.

A wide variety of lift and drag information for airfoils can be found in standard aerodynamics books (Refs. 12, 13, 23).

EXAMPLE 9.15 | Lift and Power for Human-Powered Flight

Given In 1977 the *Gossamer Condor*, shown in **Fig. E9.15a**, won the Kremer prize by being the first human-powered aircraft to complete a prescribed figure-of-eight course around two turning points 0.5 mi apart (Ref. 19). The following data pertain to this aircraft:

$$\text{flight speed} = U = 4.6 \text{ m/s}$$

$$\text{wing size} = b = 29 \text{ m}, c = 2.3 \text{ m (average)}$$

$$\text{weight (including pilot)} = W = 934 \text{ N}$$

$$\text{drag coefficient} = C_D = 0.046 \text{ (based on planform area)}$$

$$\text{power train efficiency} = \eta$$

$$= \text{power to overcome drag/pilot power} = 0.8$$

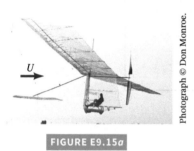

FIGURE E9.15a

Photograph © Don Monroe.

Find Determine

a. the lift coefficient, C_L, and

b. the power, $\mathcal{P}$, required by the pilot.

Solution

a. For steady flight conditions the lift must be exactly balanced by the weight, or

$$W = \mathcal{L} = \tfrac{1}{2}\rho U^2 A C_L$$

Thus,

$$C_L = \frac{2W}{\rho U^2 A}$$

where $A = bc = 29 \text{ m} \times 2.3 \text{ m} = 66.7 \text{ m}^2$, $W = 934$ N, and $\rho = 1.23 \text{ kg/m}^3$ for standard air. This gives

$$C_L = \frac{2(934\,\text{N})}{(1.23\,\text{kg/m}^3)(4.6\,\text{m/s})^2\,(66.7\,\text{m}^2)}$$

$$= 1.08 \qquad (Ans)$$

a reasonable number. The overall lift-to-drag ratio for the aircraft is $C_L/C_D = 1.08/0.046 = 23.5$.

b. The product of the power that the pilot supplies and the power train efficiency equals the useful power needed to overcome the drag, $\mathcal{D}$. That is,

$$\eta\mathcal{P} = \mathcal{D}U$$

where

$$\mathcal{D} = \tfrac{1}{2}\rho U^2 A C_D$$

Thus,

$$\mathcal{P} = \frac{\mathcal{D}U}{\eta} = \frac{\tfrac{1}{2}\rho U^2 A C_D U}{\eta} = \frac{\rho A C_D U^3}{2\eta} \qquad (1)$$

or

$$\mathcal{P} = \frac{(1.23\,\text{kg/m}^3)(66.7\,\text{m}^2)(0.046)(9.8\,\text{m/s})^3}{2(0.8)}$$

$$= 230 \text{ W} \qquad (Ans)$$

Comment This power level is obtainable by a well-conditioned athlete (as is indicated by the fact that the flight was successfully completed). Note that only 80% of the pilot's power (i.e., $0.8 \times 230 = 184$ W, which corresponds to a drag of $\mathcal{D} = 39.9$ N) is needed to force the aircraft through the air. The other 20% is lost because of the power train inefficiency.

By repeating the calculations for various flight speeds, the results shown in **Fig. E9.15b** are obtained. Note from Eq. 1 that for a constant drag coefficient, the power required increases as U^3—a doubling of the speed to 9 m/s would require an eightfold increase in power (i.e., 1.8 kW, well beyond the range of any human).

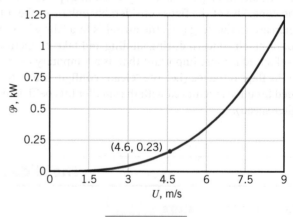

FIGURE E9.15b

9.4.2 Circulation

Inviscid flow analysis can be used to obtain ideal flow past airfoils.

Since viscous effects are of minor importance in the generation of lift, it should be possible to calculate the lift force on an airfoil by integrating the pressure distribution obtained from the equations governing inviscid flow past the airfoil. That is, the potential flow theory discussed in Chapter 6 should provide a method to determine the lift. Although the details are beyond the scope of this book, the following is found from such calculations (Ref. 4).

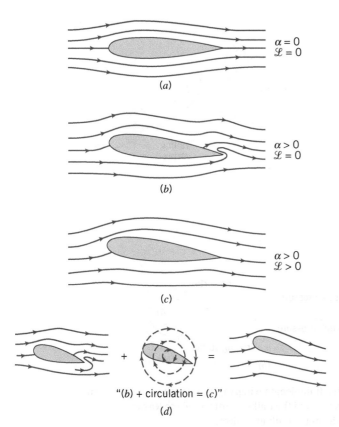

$\alpha = 0$
$\mathscr{L} = 0$

(a)

$\alpha > 0$
$\mathscr{L} = 0$

(b)

$\alpha > 0$
$\mathscr{L} > 0$

(c)

+ =

"(b) + circulation = (c)"

(d)

FIGURE 9.37 Inviscid flow past an airfoil: (a) symmetrical flow past the symmetrical airfoil at a zero angle of attack; (b) same airfoil at a nonzero angle of attack—no lift, flow near trailing edge not realistic; (c) same conditions as for (b) except circulation has been added to the flow—nonzero lift, realistic flow; (d) superposition of flows to produce the final flow past the airfoil.

The calculation of the inviscid flow past a two-dimensional airfoil gives a flow field as indicated in **Fig. 9.37**. The predicted flow field past an airfoil with no lift (i.e., a symmetrical airfoil at zero angle of attack, Fig. 9.37a) appears to be quite accurate (except for the absence of thin boundary layer regions). However, as is indicated in Fig. 9.37b, the calculated flow past the same airfoil at a nonzero angle of attack (but one small enough so that boundary layer separation would not occur) is not proper near the trailing edge. In addition, the calculated lift for a nonzero angle of attack is zero—in conflict with the known fact that such airfoils produce lift.

In reality, the flow should pass smoothly over the top surface as is indicated in Fig. 9.37c, without the strange behavior indicated near the trailing edge in Fig. 9.37b. As is shown in Fig. 9.37d, the unrealistic flow situation can be corrected by adding an appropriate clockwise swirling flow around the airfoil. The results are twofold: (1) The unrealistic behavior near the trailing edge is eliminated (i.e., the flow pattern of Fig. 9.37b is changed to that of Fig. 9.37c), and (2) the average velocity on the upper surface of the airfoil is increased while that on the lower surface is decreased. From the Bernoulli equation concepts (i.e., $p/\gamma + V^2/2g + z =$ constant), the average pressure on the upper surface is decreased and that on the lower surface is increased. The net effect is to change the original zero lift condition to that of a lift-producing airfoil.

The addition of the clockwise swirl is termed the addition of **circulation**. The amount of swirl (circulation) needed to have the flow leave the trailing edge smoothly is a function of the airfoil size and shape and can be calculated from potential flow (inviscid) theory (see Section 6.6.3 and Ref. 23). Although the addition of circulation to make the flow field physically realistic may seem artificial, it has well-founded mathematical and physical grounds. For example, consider the flow past a finite-length airfoil, as is indicated in **Fig. 9.38**. For lift-generating conditions the average pressure on the lower surface is greater than that on the upper surface. Near the tips of the wing this pressure difference will cause some of the fluid to attempt to migrate from the lower to the upper surface, as is indicated in Fig. 9.38b. At the same time, this fluid is swept downstream, forming a *trailing vortex* (swirl) from each

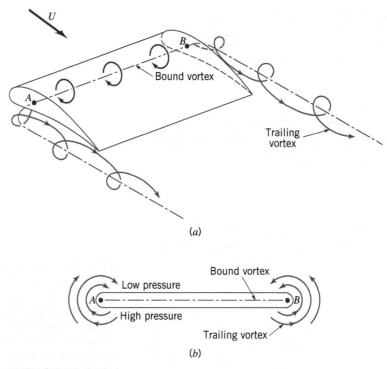

(a)

(b)

FIGURE 9.38 Flow past a finite-length wing: (a) the horseshoe vortex system produced by the bound vortex and the trailing vortices; (b) the leakage of air around the wing tips produces the trailing vortices.

V9.22 Wing tip vortices

NASA

wing tip (see Fig. 4.3). It is speculated that the reason some birds migrate in vee-formation is to take advantage of the updraft produced by the trailing vortex of the preceding bird. [It is calculated that for a given expenditure of energy, a flock of 25 birds flying in vee-formation could travel 70% farther than if each bird were to fly separately (Ref. 14).]

The trailing vortices from the right and left wing tips are connected by the *bound vortex* along the length of the wing. It is this vortex that generates the circulation that produces the lift. The combined vortex system (the bound vortex and the trailing vortices) is termed a horseshoe vortex. The strength of the trailing vortices (which is equal to the strength of the bound vortex) is proportional to the lift generated. Large aircraft (for example, a Boeing 747) can generate very strong trailing vortices that persist for a long time before viscous effects and instability mechanisms finally cause them to die out. Such vortices are strong enough to flip smaller aircraft out of control if they follow too closely behind the large aircraft. The adjacent photograph clearly shows a trailing vortex produced during a wake vortex study in which an airplane flew through a column of smoke.

The *wing tip vortices* are detrimental to wing performance, causing a decrease in lift and increase in drag, termed induced drag (Ref. 12). Increasing the aspect ratio, $\mathscr{A}$, of the wing reduces these effects. The aspect ratio is defined as the ratio of the square of the wing length to the planform area, $\mathscr{A} = b^2/A$. If the chord length, c, is constant along the length of the wing (a rectangular planform wing), this reduces to $\mathscr{A} = b/c$. **Figure 9.39** shows the effect of aspect ratio on the lift and drag coefficient for a typical finite wing. In general, the lift coefficient increases and the drag coefficient decreases with an increase in aspect ratio. Thus, long wings with a short chord are more efficient because the wing tip losses are lower than those for short wings with a long chord. High-performance soaring airplanes and highly efficient soaring birds (e.g., the albatross and sea gull) have long, narrow wings. Such wings, however, have considerable inertia that inhibits rapid maneuvers. Thus, highly maneuverable fighter or acrobatic airplanes and birds (e.g., the falcon) have small-aspect-ratio wings. Other methods of reducing the effects of induced drag include tapering a wing from the root to the tip or the addition of winglets (see The Wide World of Fluids).

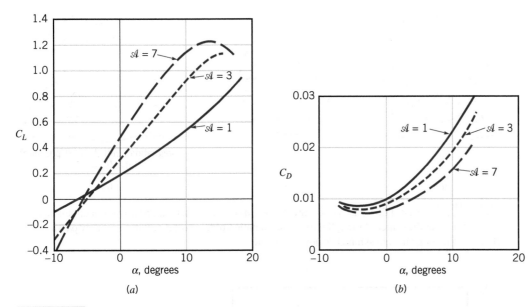

FIGURE 9.39 Typical lift and drag coefficient data as a function of angle of attack and the aspect ratio of a typical finite wing with the same airfoil shape: (a) lift coefficient, (b) drag coefficient.

The Wide World of Fluids

Why winglets?

Winglets, those upward turning ends of airplane wings, boost the performance by reducing induced drag. This is accomplished by reducing the strength of the wing tip vortices formed by the difference between the high pressure on the lower surface of the wing and the low pressure on the upper surface of the wing. (Spoilers on race cars have "end fences" to control this effect.) In essence, the winglet provides an effective increase in the aspect ratio of the wing without extending the wingspan. Winglets come in a variety of styles—the Airbus A320 has a very small upper and lower winglet; the Boeing 747-400 has a conventional, vertical upper winglet; and the Boeing Business Jet (a derivative of the Boeing 737) has an 2.4 m winglet with a curving transition from wing to winglet. Since the airflow around the winglets is quite complicated, the winglets must be carefully designed and tested for each aircraft; a poorly designed winglet can actually *increase* drag. In the past, winglets were more likely to be retrofitted to existing wings, but new airplanes are being designed with winglets from the start. Unlike tailfins on cars, winglets really do work.

As is indicated above, the generation of lift is directly related to the production of a swirl or vortex flow around the object. A nonsymmetric airfoil, by design, generates its own prescribed amount of swirl and lift. A symmetric object like a circular cylinder or sphere, which normally provides no lift, can generate swirl and lift if it rotates.

As is discussed in Section 6.6.3, the inviscid flow past a circular cylinder has the symmetrical flow pattern indicated in **Fig. 9.40a**. By symmetry the lift and drag are zero. However, if the cylinder is rotated about its axis in a stationary real ($\mu \neq 0$) fluid, the rotation will drag some of the fluid around, producing circulation about the cylinder as in **Fig. 9.40b**. When this circulation is combined with an ideal, uniform upstream flow, the flow pattern indicated in **Fig. 9.40c** is obtained. The flow is no longer symmetrical about the horizontal plane through the center of the cylinder; the average pressure is greater on the lower half of the cylinder than on the upper half, and a lift is generated. This effect is called the **Magnus effect**, after Heinrich Magnus (1802–1870), a German chemist and physicist who first investigated this phenomenon. (Note that the streamline pattern in Fig. 9.40c is idealized. In reality, a wake will be present causing less lift.) A similar lift is generated on a rotating sphere. It accounts for the various types of pitches in baseball (e.g., curve ball, "rising" fastball, sinker, etc.), the ability of a soccer player to hook the ball, and the hook or slice of a golf ball.

A spinning sphere or cylinder can generate lift.

Typical lift and drag coefficients for a smooth, spinning sphere are shown in **Fig. 9.41**. Although the drag coefficient is fairly independent of the rate of rotation, the lift coefficient

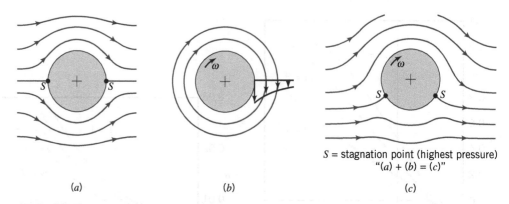

S = stagnation point (highest pressure)
"$(a) + (b) = (c)$"

(a)   (b)   (c)

FIGURE 9.40 Inviscid flow past a circular cylinder: (*a*) uniform upstream flow without circulation, (*b*) free vortex at the center of the cylinder, (*c*) combination of free vortex and uniform flow past a circular cylinder giving nonsymmetric flow and a lift.

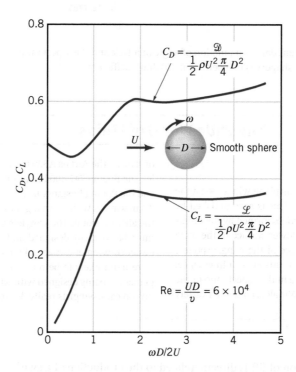

$$C_D = \frac{\mathcal{D}}{\frac{1}{2}\rho U^2 \frac{\pi}{4} D^2}$$

Smooth sphere

$$C_L = \frac{\mathcal{L}}{\frac{1}{2}\rho U^2 \frac{\pi}{4} D^2}$$

$$\text{Re} = \frac{UD}{v} = 6 \times 10^4$$

FIGURE 9.41 Lift and drag coefficients for a spinning smooth sphere (Ref. 20).

is strongly dependent on it. In addition (although not indicated in the figure), both C_L and C_D are dependent on the roughness of the surface. As was discussed in Section 9.3, in a certain Reynolds number range an increase in surface roughness actually decreases the drag coefficient. Similarly, an increase in surface roughness can increase the lift coefficient because the roughness helps drag more fluid around the sphere, increasing the circulation for a given angular velocity. Thus, a rotating, rough golf ball travels farther than a smooth one because the drag is less and the lift is greater. However, do not expect a severely roughed up (cut) ball to work better—extensive testing has gone into obtaining the optimum surface roughness for golf balls.

A dimpled golf ball has less drag and more lift than a smooth one.

EXAMPLE 9.16 | Lift on a Rotating Sphere

Given A table tennis ball weighing 2.45×10^{-2} N with diameter $D = 3.8 \times 10^{-2}$ m is hit at a velocity of $U = 12$ m/s with a back spin of angular velocity ω as is shown in **Fig. E9.16**.

Find What is the value of ω if the ball is to travel on a horizontal path, not dropping due to the acceleration of gravity?

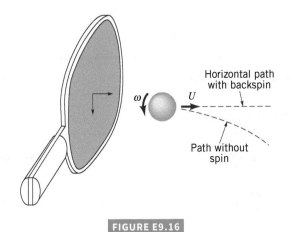

Horizontal path
with backspin

ω U

Path without
spin

Solution For horizontal flight, the lift generated by the spinning of the ball must exactly balance the weight, $\mathcal{W}$, of the ball so that

$$\mathcal{W} = \mathcal{L} = \tfrac{1}{2}\rho U^2 A C_L$$

or

$$C_L = \frac{2\mathcal{W}}{\rho U^2 (\pi/4) D^2}$$

where the lift coefficient, C_L, can be obtained from Fig. 9.39. For standard atmospheric conditions with $\rho = 1.23 \text{ kg/m}^3$ we obtain

$$C_L = \frac{2(2.45 \times 10^{-2} \text{ N})}{(1.23 \text{ kg/m}^3)(12 \text{ m/s})^2(\pi/4)(3.8 \times 10^{-2} \text{ m})^2}$$

$$= 0.244$$

which, according to Fig. 9.39, can be achieved if

$$\frac{\omega D}{2U} = 0.9$$

or

$$\omega = \frac{2U(0.9)}{D} = \frac{2(12 \text{ m/s})(0.9)}{3.8 \times 10^{-2} \text{ m}} = 568 \text{ rad/s}$$

Thus,

$$\omega = (568 \text{ rad/s})(60 \text{ s/min})(1 \text{ rev}/2\pi \text{ rad})$$
$$= 5420 \text{ rpm} \qquad\qquad (Ans)$$

Comment Is it possible to impart this angular velocity to the ball? With larger angular velocities the ball will rise and follow an upward curved path. Similar trajectories can be produced by a well-hit golf ball—rather than falling like a rock, the golf ball trajectory is actually curved up and the spinning ball travels a greater distance than one without spin. However, if topspin is imparted to the ball (as in an improper tee shot) the ball will curve downward more quickly than under the action of gravity alone—the ball is "topped" and a negative lift is generated. Similarly, rotation about a vertical axis will cause the ball to hook or slice to one side or the other.

Recall the deceleration of a table tennis ball in Example 9.12. As the velocity drops, the spin on the ball will no longer impart a horizontal path. The fact that C_D is higher than C_L as shown in Fig. 9.41 makes it difficult to impart lift. It is, in fact, impossible for a baseball pitcher to impart enough angular and translational velocity on a thrown baseball to cause it to rise. A "rising" fastball is an illusion to the hitter because it drops less than every other pitch witnessed.

CHAPTER SUMMARY

In this chapter the flow past objects is discussed. It is shown how the pressure and shear stress distributions on the surface of an object produce the net lift and drag forces on the object.

The character of flow past an object is a function of the Reynolds number. For large Reynolds number flows, a thin boundary layer forms on the surface. Properties of this boundary layer flow are discussed. These include the boundary layer thickness, whether the flow is laminar or turbulent, and the wall shear stress exerted on the object. In addition, boundary layer separation and its relationship to the pressure gradient are considered.

The drag, which contains portions due to friction (viscous) effects and pressure effects, is written in terms of the dimensionless drag coefficient. It is shown how the drag coefficient is a function of shape, with objects ranging from very blunt to very streamlined. Other parameters affecting the drag coefficient include the Reynolds number, Froude number, Mach number, and surface roughness.

The lift is written in terms of the dimensionless lift coefficient, which is strongly dependent on the shape of the object. Variation of the lift coefficient with shape is illustrated by the variation of an airfoil's lift coefficient with angle of attack.

The following checklist provides a study guide for this chapter. When your study of the entire chapter has been completed, you should be able to

- write out meanings of the terms listed and understand each of the related concepts. These terms are particularly important and are set in **bold black** type in the text.

- determine the lift and drag on an object from the given pressure and shear stress distributions on the object.

- for flow past a flat plate, calculate the boundary layer thickness, the wall shear stress, and the friction drag, and determine whether the flow is laminar or turbulent.

- explain the concept of the pressure gradient and its relationship to boundary layer separation.

- explain methods to reduce drag for both blunt and streamlined objects.

- for a given object, obtain the drag coefficient from appropriate tables, figures, or equations and calculate the drag on the object.

- explain why golf balls have dimples.

- for a given object, obtain the lift coefficient from appropriate figures and calculate the lift on the object.

KEY EQUATIONS

Drag force	$\mathcal{D} = \int dF_x = \int p\cos\theta\, dA + \int \tau_w \sin\theta\, dA$	(9.1)
Lift force	$\mathcal{L} = \int dF_y = -\int p\sin\theta\, dA + \int \tau_w \cos\theta\, dA$	(9.2)
Lift coefficient and drag coefficient	$C_L = \dfrac{\mathcal{L}}{\frac{1}{2}\rho U^2 A}, \quad C_D = \dfrac{\mathcal{D}}{\frac{1}{2}\rho U^2 A}$	(9.39), (9.36)
Boundary layer displacement thickness	$\delta^* = \displaystyle\int_0^\infty \left(1 - \dfrac{u}{U}\right) dy$	(9.3)
Boundary layer momentum thickness	$\Theta = \displaystyle\int_0^\infty \dfrac{u}{U}\left(1 - \dfrac{u}{U}\right) dy$	(9.4)
Blasius boundary layer thickness, displacement thickness, and momentum thickness for flat plate	$\dfrac{\delta}{x} = \dfrac{5}{\sqrt{\text{Re}_x}}, \quad \dfrac{\delta^*}{x} = \dfrac{1.721}{\sqrt{\text{Re}_x}}, \quad \dfrac{\Theta}{x} = \dfrac{0.664}{\sqrt{\text{Re}_x}}$	(9.15), (9.16), (9.17)
Blasius wall shear stress for flat plate	$\tau_w = 0.332 U^{3/2}\sqrt{\dfrac{\rho\mu}{x}}$	(9.18)
Drag on flat plate	$\mathcal{D} = \rho b U^2\, \Theta$	(9.23)
Blasius wall friction coefficient and friction drag coefficient for flat plate	$c_f = \dfrac{0.664}{\sqrt{\text{Re}_x}}, \quad C_{Df} = \dfrac{1.328}{\sqrt{\text{Re}_\ell}}$	(9.32)

REFERENCES

[1] Schlichting, H., and Gersten, K., *Boundary Layer Theory*, 9th Ed., Springer-Verlag, Berlin, 2017.

[2] Rosenhead, L., *Laminar Boundary Layers*, Oxford University Press, London, 1963.

[3] White, F. M., *Viscous Fluid Flow*, 3rd Ed., McGraw-Hill, New York, 2005.

[4] Currie, I. G., *Fundamental Mechanics of Fluids*, McGraw-Hill, New York, 1974.

[5] Blevins, R. D., *Applied Fluid Dynamics Handbook*, Van Nostrand Reinhold, New York, 1984.

[6] Hoerner, S. F., *Fluid-Dynamic Drag*, published by the author, Library of Congress No. 64,19666, 1965.

[7] Happel, J., *Low Reynolds Number Hydrodynamics*, Prentice-Hall, Englewood Cliffs, NJ, 1965.

[8] Van Dyke, M., *An Album of Fluid Motion*, Parabolic Press, Stanford, CA, 1982.

[9] Thompson, P. A., *Compressible-Fluid Dynamics*, McGraw-Hill, New York, 1972.

[10] Zucrow, M. J., and Hoffman, J. D., *Gas Dynamics*, *Vol. I*, Wiley, New York, 1976.

[11] Clayton, B. R., and Bishop, R. E. D., *Mechanics of Marine Vehicles*, Gulf Publishing Co., Houston, TX, 1982.

[12] Shevell, R. S., *Fundamentals of Flight*, 2nd Ed., Prentice-Hall, Englewood Cliffs, NJ, 1989.

[13] Kuethe, A. M., and Chow, C. Y., *Foundations of Aerodynamics, Bases of Aerodynamics Design*, 4th Ed., Wiley, New York, 1986.

[14] Vogel, J., *Life in Moving Fluids*, 2nd Ed., Willard Grant Press, Boston, 1994.

[15] White, F. M., *Fluid Mechanics*, 8th Ed., McGraw-Hill, New York, 2016.

[16] Vennard, J. K., and Street, R. L., *Elementary Fluid Mechanics*, 7th Ed., Wiley, New York, 1995.

[17] Gross, A. C., Kyle, C. R., and Malewicki, D. J., "The Aerodynamics of Human Powered Land Vehicles," *Scientific American*, Vol. 249, No. 6, 1983.

[18] Abbott, I. H., and Von Doenhoff, A. E., *Theory of Wing Sections*, Dover Publications, New York, 1959.

[19] MacReady, P. B., "Flight on 0.33 Horsepower: The Gossamer Condor," *Proc. AIAA 14th Annual Meeting* (Paper No. 78-308), Washington, D.C., 1978.

[20] Goldstein, S., *Modern Developments in Fluid Dynamics*, Oxford Press, London, 1938.

[21] Inui, T., "Wave-Making Resistance of Ships," *Transactions of the Society of Naval Architects and Marine Engineers*, Vol. 70, 1962.

[22] Abbott, I. H., von Doenhoff, A. E., and Stivers, L. S., Summary of Airfoil Data, NACA Report No. 824, Langley Field, VA, 1945.

[23] Anderson, J. D., *Fundamentals of Aerodynamics*, 6th Ed., McGraw-Hill, New York, 2016.

[24] Hucho, W. H., *Aerodynamics of Road Vehicles*, Butterworth–Heinemann, 1987.

[25] Homsy, G. M., et al., *Multimedia Fluid Mechanics*, 2nd Ed., CD-ROM, Cambridge University Press, New York, 2008.

QUESTIONS AND PROBLEMS

[C] Problem to be solved with aid of programmable calculator or computer.

[O] Open-ended problem that requires critical thinking. These problems require various assumptions to provide the necessary input data. There are not unique answers to these problems.

Note: Unless specific values of required fluid properties are specified in the problem statement, use the values found in Tables 1.4–1.8 and in the tables in the Appendices.

Section 9.1 General External Flow Characteristics

9.1.1 Assume that water flowing past the equilateral triangular bar shown in **Fig. P9.1.1** produces the pressure distributions indicated. Determine the lift and drag on the bar and the corresponding lift and drag coefficients (based on frontal area). Neglect shear forces.

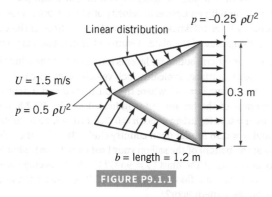

FIGURE P9.1.1

9.1.2 Fluid flows past the two-dimensional bar shown in **Fig. P9.1.2**. The pressures on the ends of the bar are as shown, and the average shear stress on the top and bottom of the bar is τ_{avg}. Assume that the drag due to pressure is equal to the drag due to viscous effects. (a) Determine τ_{avg} in terms of the dynamic pressure, $\rho U^2/2$. (b) Determine the drag coefficient for this object.

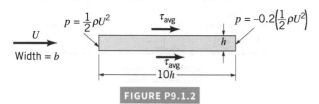

FIGURE P9.1.2

9.1.3 The average pressure and shear stress acting on the surface of the 1-m-square flat plate are as indicated in **Fig. P9.1.3**. Determine the lift and drag generated. Determine the lift and drag if the shear stress is neglected. Compare these two sets of results.

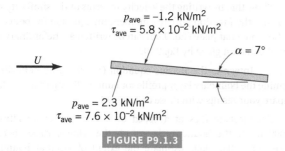

FIGURE P9.1.3

9.1.4 [C] The pressure distribution on the 1-m-diameter circular disk in **Fig. P9.1.4** is given in the table. Determine the drag on the disk.

r (m)	p (kN/m²)
0	4.34
0.05	4.28
0.10	4.06
0.15	3.72
0.20	3.10
0.25	2.78
0.30	2.37
0.35	1.89
0.40	1.41
0.45	0.74
0.50	0.0

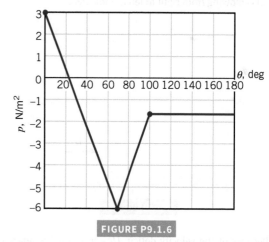

FIGURE P9.1.4

9.1.5 A nonspinning ball having a mass of 85 g is thrown vertically upward with a velocity of 0.45 m/s and has zero velocity at a height 76 m above the release point. Assume that the air drag on the ball is constant and find this constant "average" air drag. Neglect the buoyant force of air on the ball.

9.1.6 A 0.10-m-diameter circular cylinder moves through air with a speed U. The pressure distribution on the cylinder's surface is approximated by the three straight-line segments shown in **Fig. P9.1.6**. Determine the drag coefficient on the cylinder. Neglect shear forces.

FIGURE P9.1.6

9.1.7 Typical values of the Reynolds number for various animals moving through air or water are listed below. For which cases is inertia of the fluid important? For which cases do viscous effects dominate? For which cases would the flow be laminar or turbulent? Explain.

Animal	Speed	Re
(a) Large whale	10 m/s	300,000,000
(b) Flying duck	20 m/s	300,000
(c) Large dragonfly	7 m/s	30,000
(d) Invertebrate larva	1 mm/s	0.3
(e) Bacterium	0.01 mm/s	0.00003

9.1.8 Ⓞ Estimate the Reynolds numbers associated with the following objects moving through water. (a) a kayak, (b) a minnow, (c) a submarine, (d) a grain of sand settling to the bottom, (e) you swimming. Are viscous or inertial effects more dominant for each case?

9.1.9 Approximately how fast can the wind blow past a 6×10^{-3} m-diameter twig if viscous effects are to be of importance throughout the entire flow field (i.e., Re < 1)? Explain. Repeat for a 1×10^{-4} m-diameter hair and a 1.8 m-diameter smokestack.

9.1.10 Consider the following cases. (a) A small 0.15 m-long fish swims with a speed of 0.02 m/s. Would a boundary layer type flow be developed along the sides of the fish? Explain. (b) A 3.66 m-long kayak moves with a speed of 1.5 m/s. Would a boundary layer type flow be developed along the sides of the boat? Explain.

Section 9.2 Boundary Layer Characteristics

9.2.1 Water flows past a flat plate that is oriented parallel to the flow with an upstream velocity of 0.5 m/s. Determine the approximate location downstream from the leading edge where the boundary layer becomes turbulent. What is the boundary layer thickness at this location?

9.2.2 A viscous fluid flows past a flat plate such that the boundary layer thickness at a distance of 1.3 m from the leading edge is 12 mm. Determine the boundary layer thickness at distances of 0.20, 2.0, and 20 m from the leading edge. Assume laminar flow.

9.2.3 Water flows past a flat plate with an upstream velocity of $U = 0.02$ m/s. Determine the water velocity at a distance of 10 mm from the plate at distances of $x = 1.5$ m and $x = 15$ m from the leading edge.

9.2.4 The typical shape of small cumulus clouds is as indicated in **Fig. P9.2.4**. Based on boundary layer ideas, explain why it is clear that the wind is blowing from right to left as indicated.

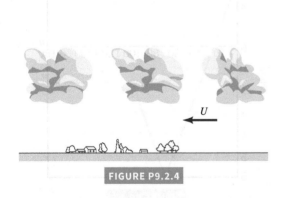

FIGURE P9.2.4

9.2.5 Because of the velocity deficit, $U - u$, in the boundary layer, the streamlines for flow past a flat plate are not exactly parallel to the plate. This deviation can be determined by use of the displacement thickness, δ^*. For air blowing past the flat plate shown in **Fig. P9.2.5**, plot the streamline $A-B$ that passes through the edge of the boundary layer ($y = \delta_B$ at $x = \ell$) at point B. That is, plot $y = y(x)$ for streamline $A-B$. Assume laminar boundary layer flow.

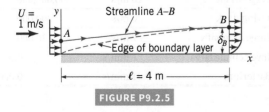

FIGURE P9.2.5

9.2.6 Air enters a square duct through a 3 m opening as is shown in **Fig. P9.2.6**. Because the boundary layer displacement thickness increases in the direction of flow, it is necessary to increase the cross-sectional size of the duct if a constant $U = 0.6$ m/s velocity is to be maintained outside the boundary layer. Plot a graph of the duct size, d, as a function of x for $0 \leq x \leq 3$ m if U is to remain constant. Assume laminar flow.

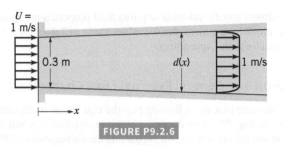

FIGURE P9.2.6

9.2.7 A smooth, flat plate of length $\ell = 6$ m and width $b = 4$ m is placed in water with an upstream velocity of $U = 0.5$ m/s. Determine the boundary layer thickness and the wall shear stress at the center and the trailing edge of the plate. Assume a laminar boundary layer.

9.2.8 An atmospheric boundary layer is formed when the wind blows over the Earth's surface. Typically, such velocity profiles can be written as a power law: $u = ay^n$, where the constants a and n depend on the roughness of the terrain. As is indicated in **Fig. P9.2.8**, typical values are $n = 0.40$ for urban areas, $n = 0.28$ for woodland or suburban areas, and $n = 0.16$ for flat open country (Ref. 23). (a) If the velocity is 6 m/s at the bottom of the sail on your boat ($y = 1.2$ m), what is the velocity at the top of the mast ($y = 9$ m)? (b) If the average velocity is 4.5 m/s on the tenth floor of an urban building, what is the average velocity on the sixtieth floor?

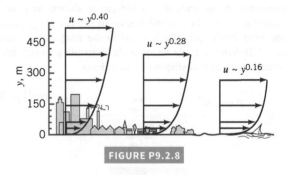

FIGURE P9.2.8

9.2.9 Ⓒ A 30-story office building (each story is 3.6 m tall) is built in a suburban industrial park. Plot the dynamic pressure, $\rho u^2/2$, as a function of elevation if the wind blows at hurricane strength (120 km/hr) at the top of the building. Use the atmospheric boundary layer information of Problem 9.2.9.

9.2.10 Show that by writing the velocity in terms of the similarity variable η and the function $f(\eta)$, the momentum equation for boundary layer flow on a flat plate (Eq. 9.9b) can be written as the ordinary differential equation given by Eq. 9.14.

9.2.11 Ⓒ Integrate the Blasius equation (Eq. 9.14) numerically to determine the boundary layer profile for laminar flow past a flat plate. Compare your results with those of Table 9.1.

9.2.12 An airplane flies at a speed of 640 km/hr at an altitude of 3000 m. If the boundary layers on the wing surfaces behave as those on a flat plate, estimate the extent of laminar boundary layer flow along the wing. Assume a transitional Reynolds number of

$Re_{xcr} = 5 \times 10^5$. If the airplane maintains its 640 km/hr speed but descends to sea-level elevation, will the portion of the wing covered by a laminar boundary layer increase or decrease compared with its value at 3000 m? Explain.

9.2.13 Ⓞ If the boundary layer on the hood of your car behaves as one on a flat plate, estimate how far from the front edge of the hood the boundary layer becomes turbulent. How thick is the boundary layer at this location?

9.2.14 A laminar boundary layer velocity profile is approximated by $u/U = [2 - (y/\delta)](y/\delta)$ for $y \leq \delta$, and $u = U$ for $y > \delta$. (a) Show that this parabolic profile satisfies the appropriate boundary conditions. (b) Use the momentum integral equation to determine the boundary layer thickness, $\delta = \delta(x)$. Compare the result with the exact Blasius solution.

9.2.15 Ⓒ Choose two of the velocity profiles and corresponding boundary layer parameter formulas presented in Table 9.2. Plot the thickness of the boundary layer and the wall shear stress versus x for values of x from zero to the laminar–turbulent transition point. The fluid is 20 °C water and the free-stream velocity is 10 m/s. Ask your instructor to specify which velocity profile models you should use.

9.2.16 Ⓒ Air at 138 kPa and 32 °C flows over a flat plate at 30 m/s. Find the value of x at which the Reynolds number is 10^4. Choose three velocity profile models from Table 9.2 and plot the velocity profiles (u versus y) on the same axes. Ask your instructor to specify which velocity profiles to choose.

9.2.17 A laminar boundary layer velocity profile is approximated by the two straight-line segments indicated in **Fig. P9.2.17**. Use the momentum integral equation to determine the boundary layer thickness, $\delta = \delta(x)$, and wall shear stress, $\tau_w = \tau_w(x)$. Compare these results with those in Table 9.2.

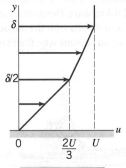

FIGURE P9.2.17

9.2.18 A laminar boundary layer velocity profile is approximated by $u/U = 2(y/\delta) - 2(y/\delta)^3 + (y/\delta)^4$ for $y \leq \delta$, and $u = U$ for $y > \delta$. (a) Show that this profile satisfies the appropriate boundary conditions. (b) Use the momentum integral equation to determine the boundary layer thickness, $\delta = \delta(x)$. Compare the result with the exact Blasius solution.

9.2.19 Consider 20 °C water flowing over a thin, wide, smooth flat plate aligned with the flow. The approach velocity is 60 km/hr and the Reynolds number based on the plate length is 2×10^5. Calculate the drag per unit width of the plate for each of the velocity profiles listed in Table 9.2. Next determine the drag coefficient C_D as used in Fig. 9.15. Which velocity profile(s) provide the best estimation of C_D? Why?

9.2.20 Ⓒ For a fluid of specific gravity $SG = 0.86$ flowing past a flat plate with an upstream velocity of $U = 5$ m/s, the wall shear stress on the flat plate was determined to be as indicated in the table given here.

Use the momentum integral equation to determine the boundary layer momentum thickness, $\Theta = \Theta(x)$. Assume $\Theta = 0$ at the leading edge, $x = 0$.

x (m)	τ_w (N/m²)
0	—
0.2	13.4
0.4	9.25
0.6	7.68
0.8	6.51
1.0	5.89
1.2	6.57
1.4	6.75
1.6	6.23
1.8	5.92
2.0	5.26

Section 9.3 Drag

9.3.1 Should a canoe paddle be made rough to get a "better grip on the water" for paddling purposes? Explain.

9.3.2 Two different fluids flow over two identical flat plates with the same laminar free-stream velocity. Both fluids have the same viscosity, but one is twice as dense as the other. What is the relationship between the drag forces for these two plates?

9.3.3 Fluid flows perpendicular to a flat plate with a drag force $\mathscr{D}_1$. If the free-stream velocity is doubled, will the new drag force, $\mathscr{D}_2$, be larger or smaller than $\mathscr{D}_1$ and by what amount?

9.3.4 A model is placed in an airflow at standard conditions with a given velocity and then placed in water flow at standard conditions with the same velocity. If the drag coefficients are the same between these two cases, how do the drag forces compare between the two fluids?

9.3.5 The drag coefficient for a newly designed hybrid car is predicted to be 0.21. The cross-sectional area of the car is 2.8 m². Determine the aerodynamic drag on the car when it is driven through still air at 25 m/s.

9.3.6 A 5-m-diameter parachute of a new design is to be used to transport a load from flight altitude to the ground with an average vertical speed of 3 m/s. The total weight of the load and parachute is 200 N. Determine the approximate drag coefficient for the parachute.

9.3.7 A 82 kg man parachutes from a plane using a hemispherical parachute in air at −7 °C and 101 kPa. Calculate the parachute diameter required for the man's terminal velocity not to exceed 6 m/s. Neglect the weight of the parachute.

9.3.8 A 70-kg soldier on a secret mission has to parachute from an airplane over the desert. It takes five seconds for him to release his parachute after he jumps. Taking the mass of the parachute as 5 kg, estimate the force acting on the soldier as he releases the parachute. The diameter of the parachute is 2.0 m. Neglect drag on the soldier's body prior to the opening of the parachute. The air is at 20 °C.

9.3.9 The aerodynamic drag on a car depends on the "shape" of the car. For example, the car shown in **Fig. P9.3.9** has a drag coefficient of 0.35 with the windows and roof closed. With the windows and roof open, the drag coefficient increases to 0.45. With the windows and roof open, at what speed is the amount of power needed to overcome aerodynamic drag the same as it is at 29 m/s with the windows and roof closed? Assume the frontal area remains the same. Recall that power is force times velocity.

Windows and roof closed: $C_D = 0.35$

Windows open; roof open: $C_D = 0.45$

9.3.10 An automobile engine has a maximum power output of 52 kW, which occurs at an engine speed of 2200 rpm. A 10% power loss occurs through the transmission and differential. The rear wheels have a radius of 40 m and the automobile is to have a maximum speed of 120 km/hr along a level road. At this speed, the power absorbed by the tires because of their continuous deformation is 20 kW. Find the maximum permissible drag coefficient for the automobile at this speed. The car's frontal area is 2 m².

9.3.11 A woman is riding a bicycle down an 18% slope. Her velocity is 25 km/hr into an oncoming 25-km/hr wind. The air is at 15 °C and 101 kPa. Assuming that the only forces affecting the speed are the weight and drag, calculate the drag coefficient if the frontal area is 0.6 m² and the combined mass is 54 kg. Speculate whether the rider is in the upright or racing position.

9.3.12 A baseball is thrown by a pitcher at 153 km/hr through standard air. The diameter of the baseball is 7 cm. Estimate the drag force on the baseball. Explain how the actual drag force may be different than your estimate.

9.3.13 A logging boat is towing a log at 1.5 m/s against a river current that is 1 m/s. The log is 0.5 m in diameter and 2 m long. (a) Estimate the power required if the axis of the log is parallel to the flow. (b) Estimate the power required if the axis of the log is perpendicular to the flow. (c) Explain any reason(s) why the actual power required may be different than your estimates.

9.3.14 How fast do small water droplets of 0.06-μm (6×10^{-8} m) diameter fall through the air under standard sea-level conditions? Assume the drops do not evaporate. Repeat the problem for standard conditions at 5000-m altitude.

9.3.15 Determine the drag on a small circular disk of 0.3 cm diameter moving 3×10^{-3} m/s through oil with a specific gravity of 0.87 and a viscosity 10,000 times that of water. The disk is oriented normal to the upstream velocity. By what percent is the drag reduced if the disk is oriented parallel to the flow?

9.3.16 The square, flat plate shown in **Fig. P9.3.16a** is cut into four equal-sized pieces and arranged as shown in **Fig. P9.3.16b**. Determine the ratio of the drag on the original plate [case (a)] to the drag on the plates in the configuration shown in (b). Assume laminar boundary flow. Explain your answer physically.

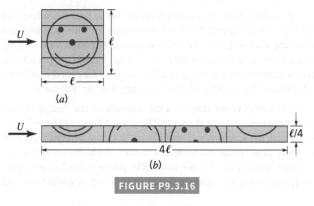

(a)

(b)

9.3.17 Water flows past a triangular flat plate oriented parallel to the free stream as shown in **Fig. P9.3.17**. Integrate the wall shear stress over the plate to determine the friction drag on one side of the plate. Assume laminar boundary layer flow.

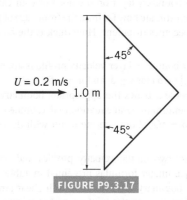

$U = 0.2$ m/s

45°

1.0 m

45°

9.3.18 For small Reynolds number flows, the drag coefficient of an object is given by a constant divided by the Reynolds number (see Table 9.4). Thus, as the Reynolds number tends to zero, the drag coefficient becomes infinitely large. Does this mean that for small velocities (hence, small Reynolds numbers) the drag is very large? Explain.

9.3.19 A rectangular cartop carrier of 0.5 m height, 1.2 m length (front to back), and 1.3 m width is attached to the top of a car. Estimate the additional power required to drive the car with the carrier at 105 km/hr through still air compared with the power required to drive only the car at 105 km/hr.

9.3.20 [Video] As shown in Video V9.2 and **Fig. P9.3.20a**, a kayak is a relatively streamlined object. As a first approximation in calculating the drag on a kayak, assume that the kayak acts as if it were a smooth, flat plate 5 m long and 0.6 m wide. Determine the drag as a function of speed and compare your results with the measured values given in **Fig. P9.3.20b**. Comment on reasons why the two sets of values may differ.

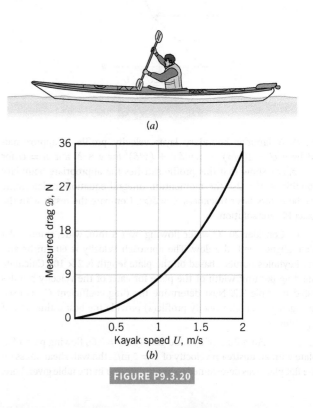

(a)

(b)

9.3.21 A coal barge 305 m long and 30.5 m wide is submerged at a depth of 4 m in 16 °C water. It is being towed at a speed of 19 km/hr. Estimate the friction drag on the barge.

9.3.22 A three-bladed helicopter blade rotates at 200 rpm. If each blade is 4 m long and 0.5 m wide, estimate the torque needed to overcome the friction on the blades if they act as flat plates.

9.3.23 A thin, smooth sign is attached to the side of a truck as is indicated in **Fig. P9.3.23**. Estimate the friction drag on the sign when the truck is driven at 89 km/hr.

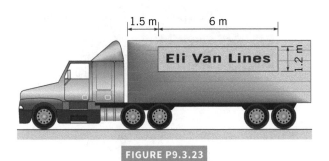

FIGURE P9.3.23

9.3.24 A 38.1-mm-diameter, 0.0245-N table tennis ball is released from the bottom of a 4-m-deep swimming pool. Assuming that the ball has reached its terminal velocity within 1 m, estimate the time required to rise the remaining distance to the surface.

9.3.25 A hot-air balloon roughly spherical in shape has a volume of 2000 m³ and a weight of 2.2 kN (including passengers, basket, balloon fabric, etc.). If the outside air temperature is 27 °C and the temperature within the balloon is 74 °C, estimate the rate at which it will rise under steady-state conditions if the atmospheric pressure is 101 kPa.

9.3.26 It is often assumed that "sharp objects can cut through the air better than blunt ones." Based on this assumption, the drag on the object shown in **Fig. P9.3.26** should be less when the wind blows from right to left than when it blows from left to right. Experiments show that the opposite is true. Explain.

FIGURE P9.3.26

9.3.27 An object falls at a rate of 30 m/s immediately prior to the time that the parachute attached to it opens. The final descent rate with the chute open is 3 m/s. Calculate and plot the speed of falling as a function of time from when the chute opens. Assume that the chute opens instantly, that the drag coefficient and air density remain constant, and that the flow is quasisteady.

9.3.28 Ⓞ Estimate the velocity with which you would contact the ground if you jumped from an airplane at an altitude of 1500 m and (a) air resistance is negligible, (b) air resistance is important, but you forgot your parachute, or (c) you use a 7.6 m diameter parachute.

9.3.29 As is discussed in Sections 9.2.6 and 9.3.3, the drag on a rough golf ball may be less than that on an equal-sized smooth ball. Does it follow that a 10-m-diameter spherical water tank resting on a 20-m-tall support should have a rough surface so as to reduce the moment needed at the base of the support when a wind blows? Explain.

9.3.30 A 12-mm-diameter cable is strung between a series of poles that are 50 m apart. Determine the horizontal force this cable puts on each pole if the wind velocity is 30 m/s.

9.3.31 A strong wind can blow a golf ball off the tee by pivoting it about point 1 as shown in **Fig. P9.3.31**. Determine the wind speed necessary to do this.

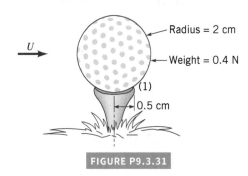

FIGURE P9.3.31

9.3.32 [Video] A 56 cm by 8.6 cm speed limit sign is supported on a 7.6 cm-wide, 1.5 m-long pole. Estimate the bending moment in the pole at ground level when a 48 km/hr wind blows against the sign (see Video V9.14). List any assumptions used in your calculations.

9.3.33 A 20-m/s wind blows against a 20-m-tall, 0.12-m-diameter flag pole. (a) Determine the anchoring moment at the base of the pole. (b) Determine the anchoring moment if a 2-m by 2.5-m flag is attached to the top of the pole. See Fig. 9.32 for drag coefficient data for flags.

9.3.34 Ⓞ During a flash flood, water rushes over a road as shown in **Fig. P9.3.34** with a speed of 19 km/hr. Estimate the maximum water depth, *h*, that would allow a car to pass without being swept away. List all assumptions and show all calculations.

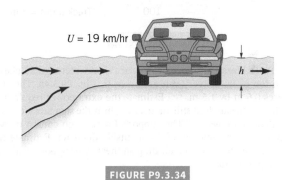

FIGURE P9.3.34

9.3.35 With the rider in the racing position, how much more power is required to pedal a bicycle at 24 km/hr into a 32 km/hr headwind than at 24 km/hr through still air? Pedaling at 24 km/hr in still air, how much more power is required for an upright position instead of a racing position?

9.3.36 Ⓞ Estimate the wind velocity necessary to knock over a 88 N garbage can that is 0.9 m tall and 0.6 m in diameter. List your assumptions.

9.3.37 On a day without any wind, your car consumes *x* gallons of gasoline when you drive at a constant speed, *U*, from point *A* to point *B* and back to point *A*. Assume that you repeat the journey, driving at the same speed, on another day when there is a steady wind blowing from *B* to *A*. Would you expect your fuel consumption to be less than, equal to, or greater than *x* gallons for this windy round-trip? Support your answer with appropriate analysis.

9.3.38 The structure shown in **Fig. P9.3.38** consists of three cylindrical support posts to which an elliptical flat plate sign is attached. Estimate the drag on the structure when a 22 m/s wind blows against it.

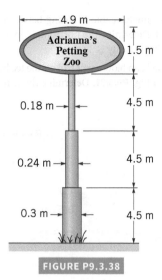

FIGURE P9.3.38

9.3.39 A 25000 kg truck coasts down a steep 7% mountain grade without brakes, as shown in **Fig. P9.3.39**. The truck's ultimate steady-state speed, V, is determined by a balance between weight, rolling resistance, and aerodynamic drag. Determine V if the rolling resistance for a truck on concrete is 1.2% of the weight and the drag coefficient based on frontal area is 0.76.

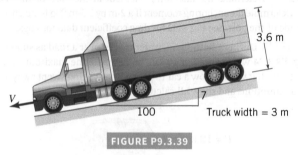

FIGURE P9.3.39

9.3.40 Phil's Pizza Parlor decides to place a thin, rectangular, plastic sign on top of its delivery van as shown in **Fig. P9.3.40**. The sign measures 0.6 m by 1.5 m. (a) Estimate the extra power required to drive the van in standard still air at 48 km/hr if the sign faces forward rather than sideways. (b) The supports for the sign consist of two steel pipes 25 cm long and of 3 cm outside diameter. Estimate the power required to overcome air drag on these two supports with the sign facing forward.

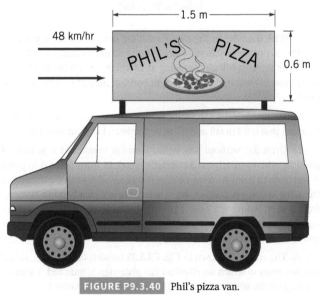

FIGURE P9.3.40 Phil's pizza van.

9.3.41 (Video) As shown in Video V9.17 and **Fig. P9.3.41**, the aerodynamic drag on a truck can be reduced by the use of appropriate air deflectors. A reduction in drag coefficient from $C_D = 0.96$ to $C_D = 0.70$ corresponds to a reduction of how many horsepower at a highway speed of 105 km/hr?

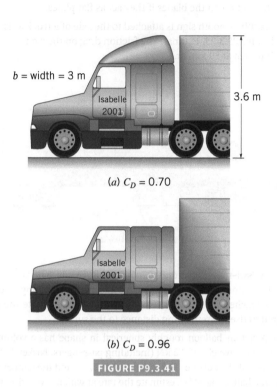

(a) $C_D = 0.70$

(b) $C_D = 0.96$

FIGURE P9.3.41

9.3.42 (C) A full-sized automobile has a frontal area of 2.2 m², and a compact car has a frontal area of 1.2 m². Both have a drag coefficient of 0.5 based on the frontal area. Find the horsepower required to move each automobile along a level road in still air at 88 km/hr. Assume that the power required to deform the tires continuously at this speed (called rolling resistance) is equal to the power to overcome the air resistance. Estimate the gas mileage of both automobiles if the energy supplied to the drive wheels is $\frac{1}{4}$ of that available in the fuel. A gallon of fuel has 136 kJ available energy.

9.3.43 Estimate the energy required for an average person (see Fig. 9.32) to run a 1.5 km in 4 minutes in still standard air. Compare your estimate if you instead modeled the person as a cylinder 1.8 m tall and 0.6 m in diameter.

9.3.44 (Video) As shown in Video V9.11 and **Fig. P9.3.44**, a vertical wind tunnel can be used for skydiving practice. Estimate the vertical wind speed needed if a 670 N person is to be able to "float" motionless when the person (a) curls up as in a crouching position or (b) lies flat. See Fig. 9.32 for appropriate drag coefficient data.

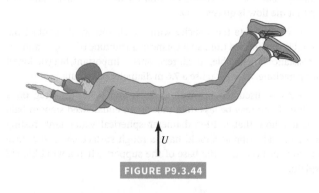

FIGURE P9.3.44

9.3.45 Compare the rise velocity of an 0.3 cm-diameter air bubble in water to the fall velocity of an 0.3 cm-diameter water drop in air. Assume each to behave as a solid sphere.

9.3.46 A 222 N box shaped like a 0.3 m cube falls from the cargo hold of an airplane at an altitude of 9000 m. If the drag coefficient of the falling box is 1.2, determine the time it takes for the box to hit the ocean. Assume that it falls at the terminal velocity corresponding to its current altitude, and use a standard atmosphere (Table C.1).

9.3.47 A 500-N cube of specific gravity $SG = 1.8$ falls through water at a constant speed U. Determine U if the cube falls (a) as oriented in **Fig. P9.3.47a**, (b) as oriented in **Fig. P9.3.47b**.

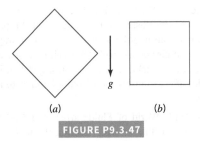

(a) (b)

FIGURE P9.3.47

9.3.48 O C The helium-filled balloon shown in **Fig. P9.3.48** is to be used as a wind-speed indicator. The specific weight of the helium is $y = 1.7$ N/m³, the weight of the balloon material is 0.9 N, and the weight of the anchoring cable is negligible. Plot a graph of θ as a function of U for $1 \leq U \leq 80$ km/hr. Would this be an effective device over the range of U indicated? Explain.

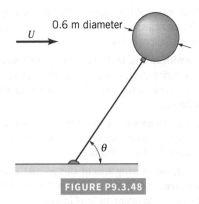

FIGURE P9.3.48

9.3.49 A 0.30-m-diameter cork ball ($SG = 0.21$) is tied to an object on the bottom of a river as is shown in **Fig. P9.3.49**. Estimate the speed of the river current. Neglect the weight of the cable and the drag on it.

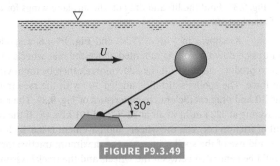

FIGURE P9.3.49

9.3.50 A shortwave radio antenna is constructed from circular tubing, as is illustrated in **Fig. P9.3.50**. Estimate the wind force on the antenna in a 100-km/hr wind.

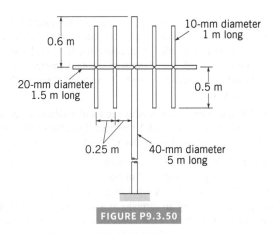

FIGURE P9.3.50

9.3.51 O Estimate the wind force on your hand when you hold it out of your car window while driving 88 km/hr. Repeat your calculations if you were to hold your hand out of the window of an airplane flying 880 km/hr.

9.3.52 O Estimate the energy that a runner expends to overcome aerodynamic drag while running a complete marathon race. This expenditure of energy is equivalent to climbing a hill of what height? List all assumptions and show all calculations.

9.3.53 A 2-mm-diameter meteor of specific gravity 2.9 has a speed of 6 km/s at an altitude of 50,000 m where the air density is 1.03×10^{-3} kg/m³. If the drag coefficient at this large Mach number condition is 1.5, determine the deceleration of the meteor.

9.3.54 Air flows past two equal sized spheres (one rough, one smooth) that are attached to the arm of a balance as is indicated in **Fig. P9.3.54**. With $U = 0$ the beam is balanced. What is the minimum air velocity for which the balance arm will rotate clockwise?

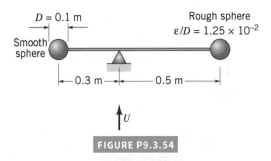

FIGURE P9.3.54

9.3.55 Video A 5 cm-diameter sphere weighing 0.6 N is suspended by the jet of air shown in **Fig. P9.3.55** and Video V3.2. The drag coefficient for the sphere is 0.5. Determine the reading on the pressure gage if friction and gravity effects can be neglected for the flow between the pressure gage and the nozzle exit.

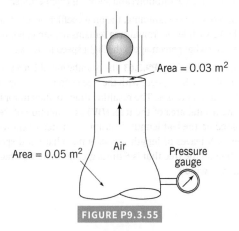

FIGURE P9.3.55

9.3.56 (Video) A smooth orange ball weighs 0.07 N (at sea level) and has a diameter of 4 cm. The discharge of a vacuum cleaner is directed upward and supports the ball 1.2 cm above the hose outlet similar to that shown in Video V3.2. If the hose inside diameter is 10 cm, estimate the volume flowrate through the vacuum cleaner. The air temperature is 38 °C.

9.3.57 A 96 km/hr wind blows against a football stadium scoreboard that is 11 m tall, 24 m wide, and 2.4 m thick (parallel to the wind). Estimate the wind force on the scoreboard. See Fig. 9.30 for drag data.

9.3.58 A marine location marker is a smoke-producing device usually dropped from an airplane and used to mark a reference point in the ocean. One is being tested in a wind tunnel to determine the drag force when it is carried by an airplane at a velocity of 90 m/s. The marker is a cylinder with flat ends, has a diameter of 0.13 m, and is 1.2 m long. Calculate the drag force on the marker in −12 °C air if the air flow is parallel to the cylinder's axis.

9.3.59 An airplane flies at 150 km/hr. (a) The airplane is towing a banner that is $b = 0.8$ m tall and $\ell = 25$ m long. If the drag coefficient based on area $b\ell$ is $C_D = 0.06$, estimate the power required to tow the banner. (b) For comparison, determine the power required if the airplane was instead able to tow a rigid flat plate of the same size. (c) Explain why one had a larger power requirement (and larger drag) than the other. (d) Finally, determine the power required if the airplane was towing a smooth spherical balloon with a diameter of 2 m.

9.3.60 The paint stirrer shown in **Fig. P9.3.60** consists of two circular disks attached to the end of a thin rod that rotates at 80 rpm. The specific gravity of the paint is $SG = 1.1$, and its viscosity is $\mu = 96 \times 10^{-2}\,\text{N} \cdot \text{s/m}^2$. Estimate the power required to drive the mixer if the induced motion of the liquid is neglected.

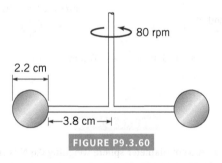

FIGURE P9.3.60

9.3.61 (O) If the wind becomes strong enough, it is "impossible" to paddle a canoe into the wind. Estimate the wind speed at which this will happen. List all assumptions and show all calculations.

9.3.62 By appropriate streamlining, the drag coefficient for an airplane is reduced by 12% while the frontal area remains the same. For the same power output, by what percentage is the flight speed increased?

9.3.63 As indicated in **Fig. P9.3.63**, the orientation of leaves on a tree is a function of the wind speed, with the tree becoming "more streamlined" as the wind increases. The resulting drag coefficient for the tree (based on the frontal area of the tree, HW) as a function of Reynolds number (based on the leaf length, L) is approximated as shown. Consider a tree with leaves of length $L = 0.09$ m. What wind speed will produce a drag on the tree that is 6 times greater than the drag on the tree in a 4.5 m/s wind?

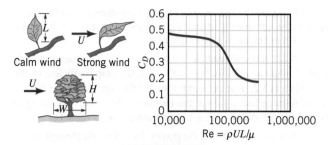

FIGURE P9.3.63

9.3.64 (See The Wide World of Fluids article "Dimpled Baseball Bats," Section 9.3.3.) How fast must a 9 cm-diameter, dimpled baseball bat move through the air in order to take advantage of drag reduction produced by the dimples on the bat? Although there are differences, assume the bat (a cylinder) acts the same as a golf ball in terms of how the dimples affect the transition from a laminar to a turbulent boundary layer.

9.3.65 (See The Wide World of Fluids article "At 5333 km/liter It Doesn't Cost Much to 'Fill 'er Up,'" Section 9.3.3.) (a) Determine the power it takes to overcome aerodynamic drag on a small (0.5 m² cross section), streamlined ($C_D = 0.12$) vehicle traveling 24 km/hr. (b) Compare the power calculated in part (a) with that for a large (3.4 m² cross-sectional area), nonstreamlined ($C_D = 0.48$) SUV traveling 27.63 km/liter on the interstate.

Section 9.4 Lift

9.4.1 A rectangular wing with an aspect ratio of 6 is to generate 4.4 kN of lift when it flies at a speed of 183 m/s. Determine the length of the wing if its lift coefficient is 1.0.

9.4.2 A 5.3 N kite with an area of 0.6 m² flies in a 18.3 m/s wind such that the weightless string makes an angle of 55° relative to the horizontal. If the pull on the string is 6.7 N, determine the lift and drag coefficients based on the kite area.

9.4.3 A Piper Cub airplane has a gross weight of 7784 N, a cruising speed of 185 km/hr, and a wing area of 17 m². Determine the lift coefficient of this airplane for these conditions.

9.4.4 A light aircraft with a wing area of 18.5 m² and a weight of 8.5 kN has a lift coefficient of 0.40 and a drag coefficient of 0.05. Determine the power required to maintain level flight.

9.4.5 An airplane weighs 1423 kN, has a wing area of 260 m², and has a wing length of 43 N. The atmospheric pressure and temperature are 101 kPa and 16 °C, respectively. The airplane is moving along a runway at 322 km/hr, and the wing has the lift and drag coefficients given in Fig. 9.39. Find the lift and drag on the airplane wings for an angle of attack of 10°.

9.4.6 (Video) As shown in Video V9.22 and **Fig. P9.4.6**, a spoiler (i.e., an upside-down airfoil) is mounted above the rear wheels of a race car to produce negative lift (i.e., downforce), thereby improving tractive force. The spoiler's airfoil is angled 10° with the race track and has lift and drag coefficient characteristics of Fig. 9.39. The race car is traveling at 322 km/hr in air at 32 °C and 101 kPa. (a) If the coefficient of friction between the wheels and pavement is 0.6, by how much would use of the spoiler increase the maximum tractive force that would be generated between the wheels and the track? Assume the airspeed past the spoiler equals the car speed and that the airfoil acts directly over the drive wheels. **(b)** How much drag is added to the car with the use of the spoiler's airfoil?

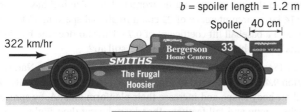

FIGURE P9.4.6

9.4.7 The wings of old airplanes are often strengthened by the use of wires that provided cross-bracing as shown in **Fig. P9.4.7**. If the drag coefficient for the wings was 0.020 (based on the planform area), determine the ratio of the drag from the wire bracing to that from the wings.

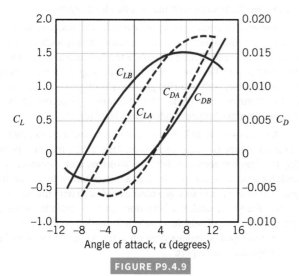

Speed: 113 km/hr
Wing area: 14 m²
Wire: length = 49 m
      diameter = 0.13 cm

FIGURE P9.4.7

9.4.8 A wing generates a lift $\mathcal{L}$ when moving through sea-level air with a velocity U. How fast must the wing move through the air at an altitude of 10,000 m with the same lift coefficient if it is to generate the same lift?

9.4.9 A design group has two possible wing designs (A and B) for an airplane wing. The planform area of either wing is 130 m² and each must provide a lift of 1,550,000 N.

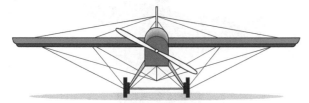

FIGURE P9.4.9

The airplane is to fly at 700 km/hr at an altitude of 10,000 m in the Standard Atmosphere. The lift and drag coefficients are shown in **Fig. P9.4.9**. Both sets of lift and drag coefficients give the total lift $\mathcal{L}$ and the total drag $\mathcal{D}$ on the airplane per unit area of the wing so that

$$\mathcal{L}_{\text{airplane}} = \frac{1}{2} C_L \rho A_{\text{wing}} U^2$$

and

$$\mathcal{D}_{\text{airplane}} = \frac{1}{2} C_D \rho A_{\text{wing}} U^2$$

where U is the airplane velocity. Which wing design would you recommend? Support your recommendation.

9.4.10 $\boxed{C}$ Air blows over the flat-bottomed, two-dimensional object shown in **Fig. P9.4.10**. The shape of the object, $y = y(x)$, and the fluid speed along the surface, $u = u(x)$, are given in the table. Determine the lift coefficient for this object.

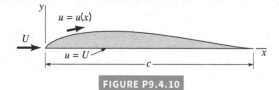

FIGURE P9.4.10

$x(\% \, c)$	$y(\% \, c)$	u/U
0	0	0
2.5	3.72	0.971
5.0	5.30	1.232
7.5	6.48	1.273
10	7.43	1.271
20	9.92	1.276
30	11.14	1.295
40	11.49	1.307
50	10.45	1.308
60	9.11	1.195
70	6.46	1.065
80	3.62	0.945
90	1.26	0.856
100	0	0.807

9.4.11 $\boxed{C}$ When air flows past the airfoil shown in **Fig. P9.4.11**, the velocity just outside the boundary layer, u, is as indicated. Estimate the lift coefficient for these conditions.

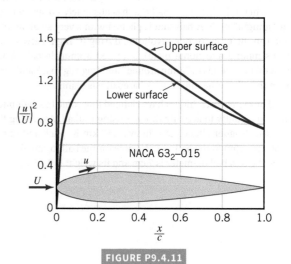

FIGURE P9.4.11

9.4.12 A Boeing 747 aircraft weighing 2580 kN when loaded with fuel and 100 passengers takes off with an airspeed of 225 km/hr. With the same configuration (i.e., angle of attack, flap settings, etc.), what is its takeoff speed if it is loaded with 372 passengers? Assume each passenger with luggage weighs 890 N.

9.4.13 Show that for unpowered flight (for which the lift, drag, and weight forces are in equilibrium) the glide slope angle, θ, is given by $\tan \theta = C_D/C_L$.

9.4.14 A sail plane with a lift-to-drag ratio of 25 flies with a speed of 80 km/hr. It maintains or increases its altitude by flying in thermals, columns of vertically rising air produced by buoyancy effects of nonuniformly heated air. What vertical airspeed is needed if the sail plane is to maintain a constant altitude?

9.4.15 If the lift coefficient for a Boeing 777 aircraft is 15 times greater than its drag coefficient, can it glide from an altitude of 9 km to an airport 129 km away if it loses power from its engines? Explain (see Problem 9.4.13).

9.4.16 Over the years there has been a dramatic increase in the flight speed (U), altitude (h), weight (W), and wing loading (W/A = weight divided by wing area) of aircraft. Use the data given in the table to determine the lift coefficient for each of the aircraft listed.

Aircraft	Year	W, N	U, km/hr	W/A, kPa²	h, km
Wright Flyer	1903	3.3	56	0.07	0
Douglas DC-3	1935	111	290	1.2	3
Douglas DC-6	1947	467	507	3.6	4.5
Boeing 747	1970	3559	917	7.2	9

9.4.17 If the required takeoff speed of a particular airplane is 54 m/s at sea level, what will be required at Denver (elevation 1524 m)? Use data given in Appendix C.

9.4.18 (Video) The landing speed of a winged aircraft such as the Space Shuttle is dependent on the air density (see Video V9.1). By what percent must the landing speed be increased on a day when the temperature is 43 °C compared to a day when it is 10 °C? Assume that the atmospheric pressure remains constant.

9.4.19 Commercial airliners normally cruise at relatively high altitudes (9 km to 11 km). Discuss how flying at this high altitude (rather than 3 km, for example) can save fuel costs.

9.4.20 A pitcher can pitch a "curve ball" by putting sufficient spin on the ball when it is thrown. A ball that has absolutely no spin will follow a "straight" path. A ball that is pitched with a very small amount of spin (on the order of one revolution during its flight between the pitcher's mound and home plate) is termed a knuckle ball. A ball pitched this way tends to "jump around" and "zig-zag" back and forth. Explain this phenomenon. *Note:* A baseball has seams.

9.4.21 For many years, hitters have claimed that some baseball pitchers have the ability to actually throw a rising fastball. Assuming that a top major leaguer pitcher can throw a 153 km/hr pitch and impart an 1800-rpm spin to the ball, is it possible for the ball to actually rise? Assume the baseball diameter is 7 cm and its weight is 0.15 kg.

9.4.22 Ⓒ A baseball leaves the pitcher's hand with horizontal velocity of 145 km/hr and travels a distance of 14 m. Neglect air drag and gravity, so the ball moves in a horizontal plane. The ball has a mass of 0.14 kg, a circumference of 23 cm, a rotational speed of 1600 rev/min, and a baseball lift coefficient of 0.75. How far does the baseball "break" in the horizontal plane in 16 °C, still air?

9.4.23 (See The Wide World of Fluids article "Learning from Nature," Section 9.4.1.) As indicated in **Fig. P9.4.23**, birds can significantly alter their body shape and increase their planform area, A, by spreading their wing and tail feathers, thereby reducing their flight speed. If during landing the planform area is increased by 50% and the lift coefficient increased by 30% while all other parameters are held constant, by what percent is the flight speed reduced?

FIGURE P9.4.23

9.4.24 (See The Wide World of Fluids article "Why Winglets?," Section 9.4.2.) It is estimated that by installing appropriately designed winglets on a certain airplane the drag coefficient will be reduced by 5%. For the same engine thrust, by what percent will the aircraft speed be increased by use of the winglets?

Lifelong Learning Problems

9.1 LL One of the "The Wide World of Fluids" articles in this chapter discusses pressure-sensitive paint—a new technique of measuring surface pressure. There have been other advances in fluid measurement techniques, particularly in velocity measurements. One such technique is particle image velocimetry, or PIV. Obtain information about PIV and its advantages. Summarize your findings in a brief report.

9.2 LL For typical aircraft flying at cruise conditions, it is advantageous to have as much laminar flow over the wing as possible because there is an increase in friction drag once the flow becomes turbulent. Various techniques have been developed to help promote laminar flow over the wing, both in airfoil geometry configurations as well as active flow control mechanisms. Obtain information on one of these techniques. Summarize your findings in a brief report.

9.3 LL We have seen in this chapter that streamlining an automobile can help to reduce the drag coefficient. One of the methods of reducing the drag has been to reduce the projected area. However, it is difficult for some road vehicles, such as a tractor-trailer, to reduce this projected area due to the storage volume needed to haul the required load. Over the years, work has been done to help minimize some of the drag on this type of vehicle. Obtain information on a method that has been developed to reduce drag on a tractor-trailer. Summarize your findings in a brief report.

CHAPTER 10

Open-Channel Flow

LEARNING OBJECTIVES

After completing this chapter, you should be able to:

- discuss the general characteristics of open-channel flow.
- use a specific energy diagram.
- apply appropriate equations to analyze uniform flow in an open channel.
- calculate key properties of a hydraulic jump.
- determine flowrates based on open-channel flow-measuring devices.

Open-channel flow involves the flow of a liquid in a channel or conduit that is not completely filled. A free surface exists between the flowing fluid (usually water) and fluid above it (usually the atmosphere). The main driving force for such flows is the fluid weight—gravity forces the fluid to flow downhill. Most open-channel flow results are based on correlations obtained from model and full-scale experiments. Additional information can be gained from analytical and numerical efforts.

Open-channel flows are essential to the world as we know it. The natural drainage of water through the numerous creek and river systems is an example of open-channel flow. Although the flow geometry for these systems is extremely complex, the resulting flow properties are of considerable economic, ecological, and recreational importance. Other examples of open-channel flows include the flow of rainwater in the gutters of our houses; the flow in canals, drainage ditches, sewers, and gutters along roads; the flow of small rivulets and sheets of water on a road during a rain; and the flow in the chutes of water rides in amusement parks.

Open-channel flow involves a free surface that can deform. Therefore, a brief introduction to the properties and characteristics of surface waves is included.

The purpose of this chapter is to investigate the basic features of open-channel flow. Because of the amount and variety of material available, only a brief introduction to the topic can be presented. Further information can be obtained from the references indicated.

V10.1 Offshore oil
drilling platform

10.1 GENERAL CHARACTERISTICS OF OPEN-CHANNEL FLOW

In our study of pipe flow (Chapter 8), we found that there are many ways to classify a flow—developing, fully developed, laminar, turbulent, and so on. For open-channel flow, the existence of a free surface allows additional types of flow. The extra freedom that allows the fluid to select its free-surface location and configuration (because it does not completely fill a pipe or conduit) allows important phenomena in open-channel flow that cannot occur in pipe flow. Some of the classifications of the flows are described below.

The manner in which the fluid depth, y, varies with time, t, and distance along the channel, x, is used to partially classify a flow. For example, the flow is *unsteady* or *steady* depending on

Open-channel flow can have a variety of characteristics.

whether the depth at a given location does or does not change with time. Some unsteady flows can be viewed as steady flows if the reference frame of the observer is changed. For example, a tidal bore (difference in water level) moving up a river is unsteady to an observer standing on the bank, but steady to an observer moving along the bank with the speed of the wave front of the bore. Other flows are unsteady regardless of the reference frame used. The complex, time-dependent, wind-generated waves on a lake are in this category. In this book we will consider only steady open-channel flows.

An open-channel flow is classified as *uniform flow* (UF) if the depth of flow does not vary along the channel ($dy/dx = 0$). Conversely, it is *nonuniform flow* or *varied flow* if the depth varies with distance ($dy/dx \neq 0$). Nonuniform flows are further classified as *rapidly varying flow* (RVF) if the flow depth changes considerably over a relatively short distance; $dy/dx \sim 1$. *Gradually varying flows* (GVF) are those in which the flow depth changes slowly with distance along the channel; $dy/dx \ll 1$. Examples of these types of flow are illustrated in **Fig. 10.1** and the adjacent photographs. The relative importance of the various forces involved (pressure, weight, shear, inertia) is different for the different types of flows.

As for any flow geometry, open-channel flow may be *laminar*, *transitional*, or *turbulent*, depending on various conditions involved. We can determine which type of flow exists by examining the Reynolds number, $Re = \rho V R_h / \mu$, where V is the average velocity of the fluid and R_h is the hydraulic radius of the channel (see Section 10.4). A general rule is that open-channel flow is laminar if $Re < 500$, turbulent if $Re > 12,500$, and transitional otherwise. The values of these dividing Reynolds numbers are only approximate—a precise knowledge of the channel geometry is necessary to obtain specific values. Since most open-channel flows involve water (which has a fairly small viscosity) and have relatively large characteristic lengths, it is rare to have laminar open-channel flows. For example, flow of 10 °C water ($v = 1.31 \times 10^{-6}$ m^2/s) with an average velocity of $V = 0.3$ m/s in a stream with a hydraulic radius of $R_h = 3.05$ m has $Re = V R_h / v = 7.1 \times 10^5$. The flow is turbulent. However, flow of a thin sheet of water down a driveway with an average velocity of $V = 0.076$ m/s such that $R_h = 0.0061$ (in such cases the hydraulic radius is approximately equal to the fluid depth; see Section 10.4) has $Re = 355$. The flow is laminar.

In some cases *stratified flows* are important. In such situations layers of two or more fluids of different densities flow in a channel. A layer of oil on water is one example of this type of flow. All of the open-channel flows considered in this book are *homogeneous flows*. That is, the fluid has uniform properties throughout.

Open-channel flows involve a free surface that can deform from its undisturbed, relatively flat configuration to form waves. Such waves move across the surface at speeds that depend on their size (height, length) and properties of the channel (depth, fluid velocity, etc.). The character of an open-channel flow may depend on the relationship between the velocity of the fluid and the speed of a wave with respect to the moving fluid. The dimensionless parameter that describes this behavior is termed the **Froude number**, $Fr = V/(g\ell)^{1/2}$, where ℓ is an appropriate characteristic length of the flow. This dimensionless parameter was introduced in Chapter 7 and is discussed more fully in Section 10.2. As shown in the adjacent figure, the special case of a flow with a Froude number of unity, $Fr = 1$, is termed a **critical flow**. If the Froude number is less than 1, the flow is **subcritical** (or *tranquil*). A flow with the Froude number greater than 1 is termed **supercritical** (or *rapid*).

Bruce Munson

Uniform flow

pierluigi palazzi/ 123RF

Rapidly varying flow

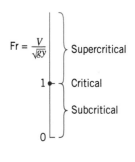
$Fr = \dfrac{V}{\sqrt{gy}}$

Supercritical

1 — Critical

Subcritical

0

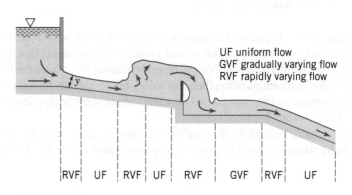

UF uniform flow
GVF gradually varying flow
RVF rapidly varying flow

RVF | UF | RVF | UF | RVF | GVF | RVF | UF

FIGURE 10.1 Classification of open-channel flow.

10.2 SURFACE WAVES

The distinguishing feature of flows involving a free surface (as in open-channel flows) is the opportunity for the free surface to deform into various shapes. The surface of a lake or the ocean is seldom "smooth as a mirror." It is usually distorted into ever-changing patterns associated with surface waves as shown in the adjacent photos. Some of these waves are very high, some barely ripple the surface; some waves are very long (the distance between wave crests), some are short; some are breaking waves that form whitecaps, others are quite smooth. Although a general study of this wave motion is beyond the scope of this book, an understanding of certain fundamental properties of simple waves is necessary for open-channel flow considerations. The interested reader is encouraged to use some of the excellent references available for further study about wave motion (Refs. 1, 2, 3, 14).

Bruce Munson

The Wide World of Fluids

Rogue waves

There is a long history of stories concerning giant rogue ocean waves that come out of nowhere and capsize ships. The movie *The Poseidon Adventure* (1972, 2006) is based on such an event. Although these giant, freakish waves were long considered fictional, recent satellite observations and computer simulations prove that, although rare, they are real. Such waves are single, sharply peaked mounds of water that travel rapidly across an otherwise relatively calm ocean. Although most ships are designed to withstand waves up to 15 m high, satellite measurements and data from offshore oil platforms indicate that such rogue waves can reach a height of 30 m. Although researchers still do not understand the formation of these large rogue waves, there are several suggestions as to how ordinary smaller waves can be focused into one spot to produce a giant wave. Additional theoretical calculations and wave tank experiments are needed to adequately grasp the nature of such waves. Perhaps it will eventually be possible to predict the occurrence of these destructive waves, thereby reducing the loss of ships and life because of them.

10.2.1 Wave Speed

Consider the situation illustrated in **Fig. 10.2a** in which a single elementary wave of small height, δy, is produced on the surface of a channel by suddenly moving the initially stationary end wall with speed δV. The water in the channel was stationary at the initial time, $t = 0$. A stationary observer will observe a single wave move down the channel with a **wave speed** c, with no fluid motion ahead of the wave and a fluid velocity of δV behind the wave. The motion is unsteady for such an observer. For an observer moving along the channel with speed c, the flow will appear steady as shown in **Fig. 10.2b**. To this observer, the fluid velocity will be $\mathbf{V} = -c\,\hat{\mathbf{i}}$ on the observer's right and $\mathbf{V} = (-c + \delta V)\hat{\mathbf{i}}$ to the left of the observer.

Video

V10.2 Filling your car's gas tank

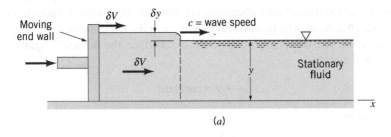

(a)

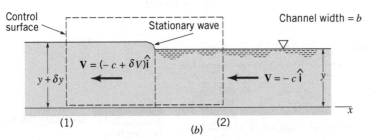

(b)

FIGURE 10.2 (a) Production of a single elementary wave in a channel as seen by a stationary observer. (b) Wave as seen by an observer moving with a speed equal to the wave speed.

The relationship between the various parameters involved for this flow can be obtained by application of the continuity and momentum equations to the control volume shown in Fig. 10.2b as follows. With the assumption of uniform one-dimensional flow, the continuity equation (Eq. 5.12) becomes

$$-cyb = (-c + \delta V)(y + \delta y)b$$

where b is the channel width. This simplifies to

$$c = \frac{(y + \delta y)\,\delta V}{\delta y}$$

or in the limit of small-amplitude waves with $\delta y \ll y$

$$c = y\frac{\delta V}{\delta y} \tag{10.1}$$

Similarly, the momentum equation (Eq. 5.22) is

$$\tfrac{1}{2}\gamma y^2 b - \tfrac{1}{2}\gamma(y + \delta y)^2 b = \rho bcy\,[(c - \delta V) - c]$$

where we have written the mass flowrate as $\dot{m} = \rho bcy$ and have assumed that the pressure variation is hydrostatic within the fluid. That is, the pressure forces on the channel cross sections (1) and (2) are $F_1 = \gamma y_{c1}A_1 = \gamma(y + \delta y)^2 b/2$ and $F_2 = \gamma y_{c2}A_2 = \gamma y^2 b/2$, respectively. If we again impose the assumption of small-amplitude waves [i.e., $(\delta y)^2 \ll y\,\delta y$], the momentum equation reduces to

$$\frac{\delta V}{\delta y} = \frac{g}{c} \tag{10.2}$$

A combination of Eqs. 10.1 and 10.2 gives the wave speed

$$\boxed{c = \sqrt{gy}} \tag{10.3}$$

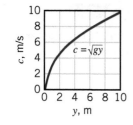

as indicated in the adjacent figure.

Equation 10.3 makes it clear that the speed of a small-amplitude wave is proportional to the square root of the depth of the fluid, y. The wave speed depends on the acceleration due to gravity, g, but not on the wave height, δy. The wave speed does not depend on the fluid's density because such wave motion is a balance between inertial effects (proportional to ρ) and weight or hydrostatic pressure effects (proportional to $\gamma = \rho g$). A ratio of these forces eliminates the common factor ρ but retains g. For extremely small waves (like those produced by insects on water as shown in **Video V10.3**), Eq. 10.3 is not valid because the effects of surface tension are significant.

The wave speed can be obtained from the continuity and momentum equations or the continuity and energy equations.

The wave speed can also be modeled by using the energy and continuity equations rather than the momentum and continuity equations as is done above. A simple wave on the surface is shown in **Fig. 10.3**. As seen by an observer moving with the wave speed, c, the flow is steady. Since the pressure is constant at any point on the free surface, the Bernoulli equation for this frictionless flow is simply

$$\frac{V^2}{2g} + y = \text{constant}$$

or by differentiating

$$\frac{V\,\delta V}{g} + \delta y = 0$$

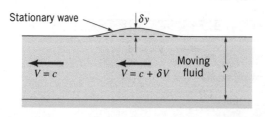

FIGURE 10.3 Stationary simple wave in a flowing fluid.

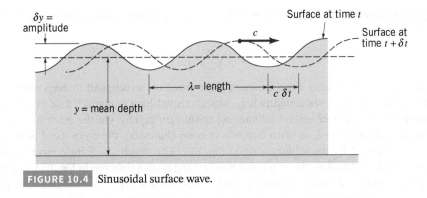

FIGURE 10.4 Sinusoidal surface wave.

Also, by differentiating the continuity equation, $Vy = \text{constant}$, we obtain

$$y\,\delta V + V\,\delta y = 0$$

We combine these two equations to eliminate δV and δy and use the fact that $V = c$ for this situation (the observer moves with speed c) to obtain the wave speed given by Eq. 10.3.

The above results are restricted to waves of small amplitude because we have assumed one-dimensional flow. That is, $\delta y/y \ll 1$. More advanced analysis and experiments show that the wave speed for finite-sized solitary waves exceeds that given by Eq. 10.3. To a first approximation, one obtains (Ref. 4)

$$c \approx \sqrt{gy}\left(1 + \frac{\delta y}{y}\right)^{1/2}$$

As indicated in the adjacent figure, the larger the amplitude, the faster the wave travels.

A more general description of wave motion can be obtained by considering continuous (not solitary) waves of sinusoidal shape as is shown in **Fig. 10.4**. By combining waves of various wavelengths, λ, and amplitudes, δy, it is possible to describe very complex surface patterns found in nature, such as the wind-driven waves on a lake. Mathematically, such a process consists of using a Fourier series (each term of the series represented by a wave of different wavelength and amplitude) to represent an arbitrary function (the free-surface shape).

A more advanced analysis of such sinusoidal surface waves of small amplitude shows that the wave speed varies with both the wavelength and fluid depth as (Ref. 1)

$$c = \left[\frac{g\lambda}{2\pi}\tanh\left(\frac{2\pi y}{\lambda}\right)\right]^{1/2} \tag{10.4}$$

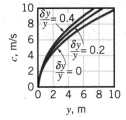

where $\tanh(2\pi y/\lambda)$ is the hyperbolic tangent of the argument $2\pi y/\lambda$. This dependence of wave speed on wavelength and fluid depth is depicted in **Fig. 10.5**. Where the water depth is

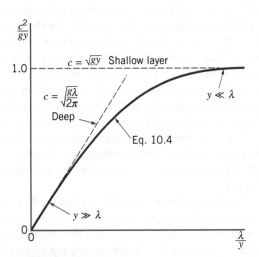

FIGURE 10.5 Wave speed as a function of wavelength.

much greater than the wavelength ($y \gg \lambda$, as in the ocean), the wave speed is independent of y and given by

$$c = \sqrt{\frac{g\lambda}{2\pi}}$$

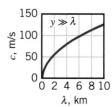

The adjacent figure shows the dependence of wave speed on wavelength in deep water. Note that waves with very long wavelengths [e.g., waves created by a tsunami ("tidal wave") with wavelengths on the order of several kilometers] travel very rapidly. On the other hand, if the fluid layer is shallow ($y \gg \lambda$, as often happens in open channels), the wave speed is given by $c = (gy)^{1/2}$, as derived for the solitary wave in Fig. 10.2. This result also follows from Eq. 10.4, since $\tanh(2\pi y/\lambda) \rightarrow 2\pi y/\lambda$ as $y/\lambda \rightarrow 0$. These two limiting cases are shown in Fig. 10.5. For moderate depth layers ($y \sim \lambda$), the results are given by the complete Eq. 10.4. Note that for a given fluid depth, the long wave travels the fastest. Hence, for our purposes we will consider the wave speed to be this limiting speed, $c = (gy)^{1/2}$.

The Wide World of Fluids

Tsunami, the nonstorm wave

A tsunami, often miscalled a "tidal wave," is a wave produced by a disturbance (for example, an earthquake, volcanic eruption, or meteorite impact) that vertically displaces the water column. Tsunamis are characterized as shallow-water waves, with long periods, very long wavelengths, and extremely large wave speeds. For example, the waves of the great December 2005, Indian Ocean tsunami traveled with speeds of 500–1000 m/s. Typically, these waves were of small amplitude in deep water far from land. Satellite radar measured the wave height as less than 1 m in these areas. However, as the waves approached shore and moved into shallower water, they slowed down considerably and reached heights up to 30 m. Because the rate at which a wave loses its energy is inversely related to its wavelength, tsunamis, with their wavelengths on the order of 100 km, not only travel at high speeds, but they also travel great distances with minimal energy loss. The furthest reported death from the Indian Ocean tsunami occurred approximately 8000 km from the epicenter of the earthquake that produced it. A more recent example is the tsunami that struck Japan in 2011.

10.2.2 Froude Number Effects

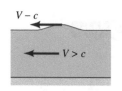

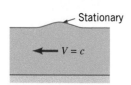

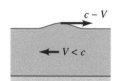

Consider an elementary wave traveling on the surface of a fluid, as in the adjacent figure and in Fig. 10.2a. If the fluid layer is stationary, the wave moves to the right with speed c relative to the fluid and the stationary observer. If the fluid is flowing to the left with velocity $V < c$, the wave (which travels with speed c relative to the fluid) will travel to the right with a speed of $c - V$ relative to a fixed observer. If the fluid flows to the left with $V = c$, the wave will remain stationary, but if $V > c$ the wave will be washed to the left with speed $V - c$.

The above ideas can be expressed in dimensionless form by use of the Froude number, $\text{Fr} = V/(gy)^{1/2}$, where we take the characteristic length to be the fluid depth, y. Thus, the Froude number, $\text{Fr} = V/(gy)^{1/2} = V/c$, is the ratio of the fluid velocity to the wave speed.

The following characteristics are observed when a wave is produced on the surface of a moving stream, as happens when a rock is thrown into a river. If the stream is not flowing, the wave spreads equally in all directions. If the stream is nearly stationary or moving in a tranquil manner (i.e., $V < c$), the wave can move upstream. Upstream locations are said to be in hydraulic communication with the downstream locations. That is, an observer upstream of a disturbance can tell that there has been a disturbance on the surface because that disturbance can propagate upstream to the observer. Viscous effects, which have been neglected in this discussion, will eventually damp out such waves far upstream. Such flow conditions, $V < c$, or $\text{Fr} < 1$, are termed *subcritical*.

On the other hand, if the stream is moving rapidly so that the flow velocity is greater than the wave speed (i.e., $V > c$), no upstream communication with downstream locations is possible. Any disturbance on the surface downstream from the observer will be washed farther downstream. Such conditions, $V > c$ or $\text{Fr} > 1$, are termed *supercritical*. For the special case of $V = c$ or $\text{Fr} = 1$, the upstream propagating wave remains stationary and the flow is termed *critical*.

EXAMPLE 10.1 | Surface Waves

Given At a certain location along the Rock River shown in **Fig. E10.1a**, the velocity, V, of the flow is a function of the depth, y, of the river as indicated in **Fig. E10.1b**. A reasonable approximation to these experimental results is

$$V = 5y^{2/3} \qquad (1)$$

where V is in m/s and y is in m.

FIGURE E10.1a

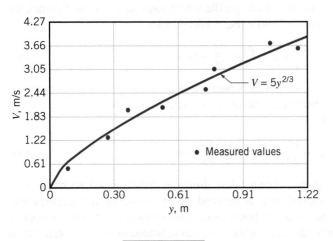

FIGURE E10.1b

Find For what range of water depth will a surface wave on the river be able to travel upstream?

Solution While the river travels to the left with speed V, the surface wave travels upstream (to the right) with speed

$c = (gy)^{1/2}$ relative to the water (not relative to the ground). Hence relative to the stationary ground, the wave travels to the right with speed

$$
\begin{aligned}
c - V &= (gy)^{1/2} - 5y^{2/3} \\
&= (9.81 \text{ m/s}^2 y)^{1/2} - 5y^{2/3} \qquad (2)
\end{aligned}
$$

For the wave to travel upstream, $c - V > 0$ so that from Eq. 2,

$$(9.81y)^{1/2} > 5y^{2/3}$$

or

$$y < 0.63 \text{ m} \qquad (Ans)$$

Comment As shown above, if the river depth is less than 0.63 m, its velocity is less than the wave speed and the wave can travel upstream against the current. This is consistent with the fact that if a wave is to travel upstream, the flow must be subcritical (i.e., Fr = $V/c < 1$). For this flow

$$
\begin{aligned}
\text{Fr} = V/c &= (5y^{2/3})/(gy)^{1/2} \\
&= 5y^{1/6}/(9.81 \text{ m/s}^2)^{1/2} \\
&= 1.59y^{1/6}
\end{aligned}
$$

This result is plotted in **Fig. E10.1c**. Note that in agreement with the above answer, for $y < 0.63$ m the flow is subcritical; the wave can travel upstream.

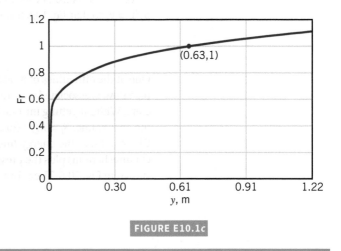

FIGURE E10.1c

The character of an open-channel flow may depend strongly on whether the flow is subcritical or supercritical. The characteristics of the flow may be completely opposite for subcritical flow than for supercritical flow. For example, as is discussed in Section 10.3, a "bump" on the bottom of a river (such as a submerged log) may cause the surface of the river to dip below the level it would have had if the log were not there, or it may cause the surface level to rise above its undisturbed level. Which situation will happen depends on the value of Fr. Similarly, for supercritical flows it is possible to produce steplike discontinuities in the fluid depth (called a hydraulic jump; see Section 10.7.1). For subcritical flows, however, changes in depth must be smooth and continuous. Certain open-channel flows, such as the broad-crested weir (Section 10.7.3), depend on the existence of critical flow conditions for their operation.

 Video

V10.5 Bicycle through a puddle

As strange as it may seem, there exist many similarities between the open-channel flow of a liquid and the compressible flow of a gas. The governing dimensionless parameter in each case is the fluid velocity, V, divided by a wave speed, the surface wave speed for open-channel flow or sound wave speed for compressible flow. Many of the differences between subcritical (Fr < 1) and supercritical (Fr > 1) open-channel flows have analogs in subsonic (Ma < 1) and supersonic (Ma > 1) compressible gas flow, where Ma is the Mach number. Some of these similarities are discussed in this chapter and in Chapter 11.

10.3 ENERGY CONSIDERATIONS

10.3.1 Energy Balance

A typical segment of an open-channel flow is shown in **Fig. 10.6**. The slope of the channel bottom (or *bottom slope*), $S_0 = (z_1 - z_2)/\ell$, is assumed constant over the segment shown. The fluid depths and velocities are $y_1, y_2, V_1,$ and V_2 as indicated. Note that the fluid depth is measured in the vertical direction and the distance x is horizontal. For most open-channel flows the value of S_0 is very small (the bottom is nearly horizontal). For example, the Mississippi River drops a distance of 448 m in its 3782 km length to give an average value of $S_0 = 0.000118$. In such circumstances the values of x and y are often taken as the distance along the channel bottom and the depth normal to the bottom, with negligibly small differences introduced by the two coordinate schemes.

With the assumption of a uniform velocity profile across any section of the channel, the one-dimensional energy equation for this flow (Eq. 5.84) becomes

$$\frac{p_1}{\gamma} + \frac{V_1^2}{2g} + z_1 = \frac{p_2}{\gamma} + \frac{V_2^2}{2g} + z_2 + h_L \tag{10.5}$$

where h_L is the head loss due to viscous effects between sections (1) and (2) and $z_1 - z_2 = S_0\ell$. Since the pressure is essentially hydrostatic at any cross section, we find that $p_1/\gamma = y_1$ and $p_2/\gamma = y_2$ so that Eq. 10.5 becomes

$$y_1 + \frac{V_1^2}{2g} + S_0\ell = y_2 + \frac{V_2^2}{2g} + h_L \tag{10.6}$$

One of the difficulties of analyzing open-channel flow, similar to that discussed in Chapter 8 for pipe flow, is associated with the determination of the head loss in terms of other physical parameters. Without getting into such details at present, we write the head loss in terms of the slope of the energy line, $S_f = h_L/\ell$ (often termed the *friction slope*), as indicated in Fig. 10.6. Recall from Chapter 4 that the energy line is located at a distance z (the elevation from some datum to the channel bottom) plus the pressure head (p/γ) plus the velocity head $(V^2/2g)$ above the datum as shown in Fig. 10.6. Therefore, Eq. 10.6 can be written as

$$y_1 - y_2 = \frac{(V_2^2 - V_1^2)}{2g} + (S_f - S_0)\ell \tag{10.7}$$

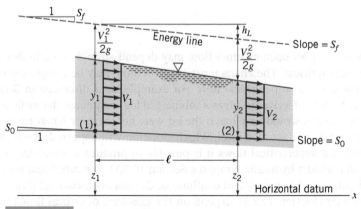

FIGURE 10.6 Typical open-channel geometry.

If there is no head loss, the energy line is horizontal ($S_f = 0$), and the total energy of the flow is free to shift between kinetic energy and potential energy in a conservative fashion. In the specific instance of a horizontal channel bottom ($S_0 = 0$) and negligible head loss ($S_f = 0$), Eq. 10.7 simply becomes

$$y_1 - y_2 = \frac{(V_2^2 - V_1^2)}{2g}$$

10.3.2 Specific Energy

The **specific energy** or specific head, E, defined as

$$E = y + \frac{V^2}{2g}$$

(10.8)

> The specific energy is the sum of potential energy and kinetic energy (per unit weight).

is often useful in open-channel flow considerations. The energy equation, Eq. 10.7, can be written in terms of E as

$$E_1 = E_2 + (S_f - S_0)\ell$$

(10.9)

If head losses are negligible, then $S_f = 0$ so that $(S_f - S_0)\ell = -S_0\ell = z_2 - z_1$ and the sum of the specific energy and the elevation of the channel bottom remains constant (i.e., $E_1 + z_1 = E_2 + z_2$, a statement of the Bernoulli equation).

If we consider a simple channel whose cross-sectional shape is a rectangle of width b, the specific energy can be written in terms of the flowrate per unit width, $q = Q/b = Vyb/b = Vy$, as

$$E = y + \frac{q^2}{2gy^2}$$

(10.10)

For a given channel of constant width, the value of q remains constant along the channel, although the depth, y, may vary. To gain insight into the flow processes involved, we consider the **specific energy diagram**, a graph of y and E, with q fixed, as shown in **Fig. 10.7**. The relationship between the flow depth, y, and the velocity head, $V^2/2g$, as given by Eq. 10.8, is indicated in the figure.

For given q and E, Eq. 10.10 can be rearranged into a cubic equation $[y^3 - Ey^2 + (q^2/2g) = 0]$ with three solutions, y_{sup}, y_{sub}, and y_{neg}. If the specific energy is large enough (i.e., $E > E_{min}$, where E_{min} is a function of q), two of the solutions are positive and the other, y_{neg}, is negative. The negative root, represented by the curved dashed line in Fig. 10.7, has no physical meaning and can be ignored. Thus, for a given flowrate and specific energy there are two possible depths, unless the vertical line from the E axis does not intersect the specific energy curve corresponding to the value of q given (i.e., $E < E_{min}$). These two depths are termed *alternate depths*.

> For a given value of specific energy, a flow may have two alternate depths.

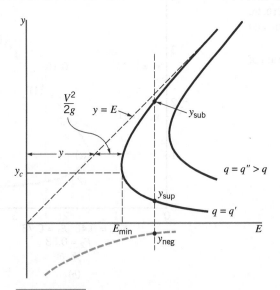

FIGURE 10.7 Specific energy diagram.

For large values of E the upper and lower branches of the specific energy diagram (y_{sub} and y_{sup}) approach $y = E$ and $y = 0$, respectively. These limits correspond to a very deep channel flowing very slowly ($E = y + V^2/2g \rightarrow y$ as $y \rightarrow \infty$ with $q = Vy$ fixed), or a very high-speed flow in a shallow channel ($E = y + V^2/2g \rightarrow V^2/2g$ as $y \rightarrow 0$).

As is indicated in Fig. 10.7, $y_{sup} < y_{sub}$. Thus, since $q = Vy$ is constant along the curve, it follows that $V_{sup} > V_{sub}$, where the subscripts "sub" and "sup" on the velocities correspond to the depths so labeled. The specific energy diagram consists of two portions divided by the E_{min} "nose" of the curve. We will show that the flow conditions at this location correspond to critical conditions (Fr = 1), those on the upper portion of the curve correspond to subcritical conditions (hence, the "sub" subscript), and those on the lower portion of the curve correspond to supercritical conditions (hence, the "sup" subscript).

To determine the value of E_{min}, we use Eq. 10.10 and set $dE/dy = 0$ to obtain

$$y_c = \left(\frac{q^2}{g}\right)^{1/3} \tag{10.11}$$

where the subscript "c" denotes conditions at E_{min}. By substituting this back into Eq. 10.10, we obtain

$$E_{min} = \frac{3y_c}{2}$$

By combining Eq. 10.11 and $V_c = q/y_c$, we obtain

$$V_c = \frac{q}{y_c} = \frac{(y_c^{3/2} g^{1/2})}{y_c} = \sqrt{gy_c}$$

or $Fr_c \equiv V_c/(gy_c)^{1/2} = 1$. Thus, critical conditions (Fr = 1) occur at the location of E_{min}. Since the fluid is deeper and the velocity smaller for the upper part of the specific energy diagram (compared with the conditions at E_{min}), such flows are subcritical (Fr < 1). Conversely, flows for the lower part of the diagram are supercritical. This relationship is depicted in the adjacent figure. Thus, for a given flowrate, q, if $E > E_{min}$ there are two possible depths of flow, one subcritical and the other supercritical.

It is often possible to determine various characteristics of a flow by considering the specific energy diagram. Example 10.2 illustrates this for a situation in which the channel bottom elevation is not constant.

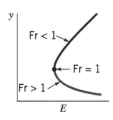

EXAMPLE 10.2 | Specific Energy Diagram—Quantitative

Given Water flows up a 0.15 m-tall ramp in a constant width, rectangular channel at a rate $q = 0.53$ m²/s as is shown in **Fig. E10.2a**. (For now disregard the "bump.") The upstream depth is 0.70 m, and viscous effects are negligible.

Find Determine the elevation of the water surface downstream of the ramp, $y_2 + z_2$.

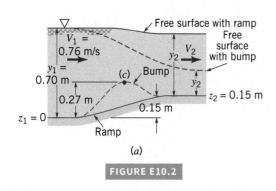

(a)

FIGURE E10.2

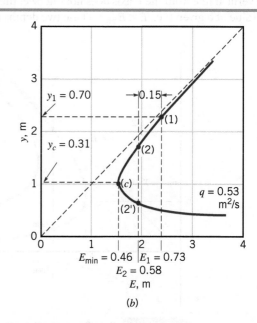

(b)

Solution With $S_0\ell = z_1 - z_2$ and $h_L = 0$, conservation of energy (Eq. 10.6 which, under these conditions, is actually the Bernoulli equation) requires that

$$y_1 + \frac{V_1^2}{2g} + z_1 = y_2 + \frac{V_2^2}{2g} + z_2$$

For the conditions given ($z_1 = 0, z_2 = 0.15$ m, $y_1 = 0.70$ m, and $V_1 = q/y_1 = 0.76$ m/s), this becomes

$$0.58 = y_2 + \frac{V_2^2}{19.6 \text{ m/s}^2} \tag{1}$$

where V_2 and y_2 are in m/s and m, respectively. The continuity equation provides the second equation

$$y_2 V_2 = y_1 V_1$$

or

$$y_2 V_2 = 0.53 \text{ m}^2/\text{s} \tag{2}$$

Equations 1 and 2 can be combined to give

$$y_2^3 - 0.58 y_2^2 + 0.156 = 0$$

which has solutions

$$y_2 = 0.52 \text{ m}, \quad y_2 = 0.19 \text{ m}, \quad \text{or} \quad y_2 = -0.142 \text{ m}$$

Note that two of these solutions are physically realistic, but the negative solution is meaningless. This is consistent with the previous discussions concerning the specific energy (recall the three roots indicated in Fig. 10.7). The corresponding elevations of the free surface are either

$$y_2 + z_2 = 0.52 \text{ m} + 0.15 \text{ m} = 0.67 \text{ m}$$

or

$$y_2 + z_2 = 0.19 \text{ m} + 0.15 \text{ m} = 0.34 \text{ m}$$

Which of these two flows is to be expected? This question can be answered by use of the specific energy diagram obtained from Eq. 10.10, which for this problem is

$$E = y + \frac{0.014}{y^2}$$

where E and y are in meter. The diagram is shown in **Fig. E10.2b**. The upstream condition corresponds to subcritical flow; the downstream condition is either subcritical or supercritical, corresponding to points 2 or 2′. Note that since $E_1 = E_2 + (z_2 - z_1) = E_2 + 0.15$ m, it follows that the downstream conditions are located 0.15 m to the left of the upstream conditions on the diagram.

With a constant-width channel, the value of q remains the same for any location along the channel. That is, all points for the flow from (1) to (2) or (2′) must lie along the $q = 0.53$ m²/s curve shown. Any deviation from this curve would imply either a change in q or a relaxation of the one-dimensional flow assumption. To stay on the curve and go from (1) around the critical point (point c) to point (2′) would require a reduction in specific energy to E_{min}. As is seen from Fig. E10.2a, this would require a specified elevation (bump) in the channel bottom so that critical conditions would occur

above this bump. The height of this bump can be obtained from the energy equation (Eq. 10.9) written between points (1) and (c) with $S_f = 0$ (no viscous effects) and $S_0\ell = z_1 - z_c$. That is, $E_1 = E_{min} - z_1 + z_c$. In particular, since $E_1 = y_1 + 0.513/y_1^2 = 0.73$ m and $E_{min} = 3y_c/2 = 3(q^2/g)^{1/3}/2 = 0.46$ m, the top of this bump would need to be $z_c - z_1 = E_1 - E_{min} = 0.73$ m $- 0.58$ m $= 0.27$ m above the channel bottom at section (1). The flow could then accelerate to supercritical conditions (Fr$_{2'} > 1$) as is shown by the free surface represented by the dashed line in Fig. E10.2a.

Since the actual elevation change (a ramp) shown in Fig. E10.2a does not contain a bump, the downstream conditions will correspond to the subcritical flow denoted by (2), not the supercritical condition (2′). Without a bump on the channel bottom, the state (2′) is inaccessible from the upstream condition state (1). Such considerations are often termed the *accessibility of flow regimes*. Thus, the surface elevation is

$$y_2 + z_2 = 0.67 \text{ m} \tag{Ans}$$

Note that since $y_1 + z_1 = 0.70$ m and $y_2 + z_2 = 0.67$ m, the elevation of the free surface decreases as it goes across the ramp.

Comment If the flow conditions upstream of the ramp were supercritical, the free-surface elevation and fluid depth would increase as the fluid flows up the ramp. This is indicated in **Fig. E10.2c** along with the corresponding specific energy diagram, as is shown in **Fig. E10.2d**. For this case the flow starts at (1) on the lower (supercritical) branch of the specific energy curve and ends at (2) on the same branch with $y_2 > y_1$. Since both y and z increase from (1) to (2), the surface elevation, $y + z$, also increases. Thus, flow up a ramp is different for subcritical than it is for supercritical conditions.

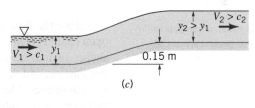

(c)

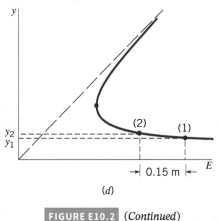

(d)

FIGURE E10.2 (Continued)

10.4 UNIFORM FLOW

Many channels are designed to carry fluid at a uniform depth and velocity all along their length. Irrigation canals are frequently of uniform depth and cross section for considerable lengths. Natural channels such as rivers and creeks are seldom of uniform shape, although a reasonable approximation to the flowrate in such channels can often be obtained by assuming uniform flow. Uniform flow is the open channel analog to fully developed pipe flow. In this section we will discuss various aspects of such flows.

Uniform flow ($dy/dx = 0$) can be accomplished by adjusting the bottom slope, S_0, so that it precisely equals the slope of the energy line, S_f. That is, $S_0 = S_f$. From an energy point of view, uniform flow is achieved by a balance between the potential energy lost by the fluid as it coasts downhill and the energy that is dissipated by viscous effects (head loss) associated with shear stresses throughout the fluid. Similar conclusions can be reached from a force balance analysis as discussed in Section 10.4.1.

10.4.1 Uniform Flow Approximations

We consider fluid flowing in an open channel of constant cross-sectional size and shape such that the depth remains constant as is indicated in **Fig. 10.8**. The area of the section is A, and the **wetted perimeter** (i.e., the length of the perimeter of the cross section in contact with the fluid) is P. The interaction between the fluid and the atmosphere at the free surface is assumed negligible so that this portion of the perimeter is not included in the definition of the wetted perimeter.

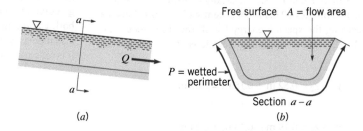

(a)   (b)

FIGURE 10.8 Uniform flow in an open channel.

The wall shear stress acts on the wetted perimeter of the channel.

Since the fluid must adhere to the solid surfaces, the actual velocity distribution in an open channel is not uniform. Some typical velocity profiles measured in channels of various shapes are indicated in **Fig. 10.9a**. The maximum velocity is often found somewhat below the free surface, and the fluid velocity is zero on the wetted perimeter, where a wall shear stress, τ_w, is developed. This shear stress is seldom uniform along the wetted perimeter, with typical variations as are indicated in **Fig. 10.9b**.

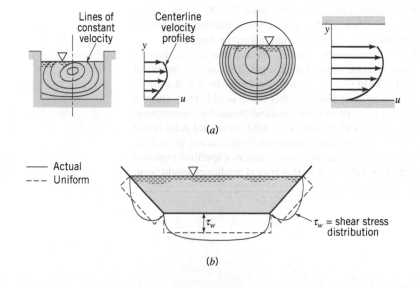

FIGURE 10.9 Typical velocity and shear stress distributions in an open channel: (a) velocity distribution throughout the cross section, (b) shear stress distribution on the wetted perimeter.

Fortunately, reasonable analytical results can be obtained by assuming a uniform velocity profile, V, and a constant wall shear stress, τ_w. Similar assumptions were made for pipe flow situations (Chapter 8), with the friction factor being used to determine the head loss.

10.4.2 The Chezy and Manning Equations

The basic equations used to determine the uniform flowrate in open channels were derived many years ago. Continual refinements have taken place to obtain better values of the empirical coefficients involved. The result is a semiempirical equation that provides useful engineering results. A more refined analysis is perhaps not warranted because of the complexity and uncertainty of the flow geometry (i.e., channel shape and the irregular makeup of the wetted perimeter, particularly for natural channels).

Under the assumptions of steady uniform flow, the x component of the momentum equation (Eq. 5.22) applied to the control volume indicated in **Fig. 10.10** simply reduces to

$$\sum F_{x_n} = \rho Q(V_2 - V_1) = 0$$

since $V_1 = V_2$. There is no acceleration of the fluid, and the momentum flux across section (1) is the same as that across section (2). The flow is governed by a simple balance between the forces in the direction of the flow. Thus, $\Sigma F_x = 0$, or

$$F_1 - F_2 - \tau_w P\ell + \mathcal{W} \sin \theta = 0 \tag{10.12}$$

where F_1 and F_2 are the hydrostatic pressure forces across either end of the control volume, as shown in the adjacent figure. Because the flow is at a uniform normal depth $(y_{n1} = y_{n2})$, it follows that $F_1 = F_2$ so that these two forces do not contribute to the force balance. The term $\mathcal{W} \sin \theta$ is the component of the fluid weight that acts down the slope, and $\tau_w P\ell$ is the shear force on the fluid, acting up the slope as a result of the interaction of the water and the channel's wetted perimeter. Thus, Eq. 10.12 becomes

$$\tau_w = \frac{\mathcal{W} \sin \theta}{P\ell} = \frac{\mathcal{W} S_0}{P\ell}$$

where we have used the approximation that $\sin \theta \approx \tan \theta = S_0$, since the bottom slope is typically very small (i.e., $S_0 \ll 1$). Likewise, for a very small channel slope the distinction between x_n, y_n-coordinates aligned with the channel bottom as shown in Figure 10.10, and the x, y-coordinates shown in Figure 10.6, is a distinction without a practical difference. Since $\mathcal{W} = \gamma A\ell$, the force balance equation becomes

$$\tau_w = \frac{\gamma A\ell S_0}{P\ell} = \gamma R_h S_0 \tag{10.13}$$

where the **hydraulic radius** is defined as $R_h = A/P$. Note that contrary to the relationship between the radius and diameter of a circle in which the radius is one-half of the diameter, the hydraulic radius is one-quarter of the hydraulic diameter $(R_h = D_h/4)$.

For steady, uniform depth flow in an open channel, there is no fluid acceleration.

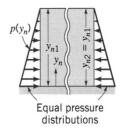

Equal pressure distributions

For uniform depth, channel flow is governed by a balance between friction and weight.

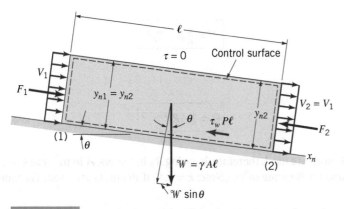

FIGURE 10.10 Control volume for uniform flow in an open channel.

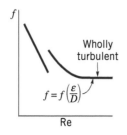

Most open-channel flows are turbulent. In fact, typical Reynolds numbers are quite large, well above the transitional value and into the wholly turbulent regime. As was discussed in Chapter 8, and shown in the adjacent figure, for very large Reynolds number pipe flows (wholly turbulent flows), the friction factor, f, is found to be independent of Reynolds number, and dependent only on the relative roughness, ε/D, of the pipe surface. For such cases, the wall shear stress is proportional to the dynamic pressure, $\rho V^2/2$, and independent of the Reynolds number. That is,

$$\tau_w = K\rho\frac{V^2}{2}$$

where K is a constant dependent upon only the roughness of the pipe.

It is not unreasonable that similar shear stress dependencies occur for the large Reynolds number open-channel flows. In such situations, Eq. 10.13 becomes

$$K\rho\frac{V^2}{2} = \gamma R_h S_0$$

or

$$V = C\sqrt{R_h S_0} \tag{10.14}$$

where the constant C is termed the *Chezy coefficient* and Eq. 10.14 is termed the **Chezy equation**. This equation, one of the oldest in fluid mechanics, was developed in 1768 by A. Chezy (1718–1798), a French engineer who designed a canal for the Paris water supply. The value of the Chezy coefficient, which must be determined by experiments, is not dimensionless but has the dimensions of (length)$^{1/2}$ per time (i.e., the square root of the units of acceleration).

From a series of experiments it was found that the slope dependence of Eq. 10.14 $\left(V \sim S_0^{1/2}\right)$ is reasonable but that the dependence on the hydraulic radius is more nearly $V \sim R_h^{2/3}$ rather than $V \sim R_h^{1/2}$. In 1889, R. Manning (1816–1897), an Irish engineer, developed the following somewhat modified equation for open-channel flow to more accurately describe the R_h dependence:

$$V = \frac{R_h^{2/3} S_0^{1/2}}{n} \tag{10.15}$$

Equation 10.15 is termed the **Manning equation**, and the parameter n is the **Manning resistance coefficient** (usually shortened to "Manning's n"). Its value is dependent on the surface material of the channel's wetted perimeter and is obtained from experiments. It is not dimensionless, having the units of s/m$^{1/3}$.

As is discussed in Chapter 7, any correlation should be expressed in dimensionless form, with the coefficients that appear being dimensionless coefficients, such as the friction factor for pipe flow or the drag coefficient for flow past objects. Thus, Eq. 10.15 should be expressed in dimensionless form. Unfortunately, the Manning equation is so widely used and has been used for so long that it will continue to be used in its dimensional form with a coefficient, n, that is not dimensionless. The values of n found in the literature (such as **Table 10.1**) were developed for SI units.

Thus, the velocity in uniform flow in an open channel is obtained from the Manning equation written as

The Manning equation is used to obtain the velocity or flow-rate in an open channel.

$$\boxed{V = \frac{\kappa}{n} R_h^{2/3} S_0^{1/2}} \tag{10.16}$$

and

$$\boxed{Q = \frac{\kappa}{n} A R_h^{2/3} S_0^{1/2}} \tag{10.17}$$

where $\kappa = 1$ if SI units are used, therefore, by using R_h in meters, A in m^2, and $\kappa = 1$, the average velocity is m/s and the flowrate m^3/s. (Note: $\kappa = 1.49$ if BG units are used. The value 1.49 is the

TABLE 10.1 Values of the Manning Coefficient, n (Ref. 6)

Wetted Perimeter	n	Wetted Perimeter	n
A. Natural channels		**D. Artificially lined channels**	
Clean and straight	0.030	Glass	0.010
Sluggish with deep pools	0.040	Brass	0.011
Major rivers	0.035	Steel, smooth	0.012
B. Floodplains		Steel, painted	0.014
		Steel, riveted	0.015
Pasture, farmland	0.035	Cast iron	0.013
Light brush	0.050	Concrete, finished	0.012
Heavy brush	0.075	Concrete, unfinished	0.014
Trees	0.15	Planed wood	0.012
C. Excavated earth channels		Clay tile	0.014
		Brickwork	0.015
Clean	0.022	Asphalt	0.016
Gravelly	0.025		
Weedy	0.030	Corrugated metal	0.022
Stony, cobbles	0.035	Rubble masonry	0.025

cube root of the number of feet per meter: $(3.281 \text{ ft/m})^{1/3} = 1.49$. By using R_h in feet, A in ft^2, and $\kappa = 1.49$, the average velocity is ft/s and the flowrate ft^3/s.)

Typical values of the Manning coefficient are presented in Table 10.1. As expected, the rougher the wetted perimeter, the larger the value of n. For example, the roughness of floodplain surfaces increases from pasture to brush to tree conditions. So does the corresponding value of the Manning coefficient. Thus, for a given depth of flooding, the flowrate varies with floodplain roughness as depicted in the adjacent figure.

Precise values of n are often difficult to obtain. Except for artificially lined channel surfaces like those found in new canals or flumes, the channel surface structure may be quite complex and variable. There are various methods used to obtain a reasonable estimate of the value of n for a given situation (Ref. 5). For the purpose of this book, the values from Table 10.1 are sufficient. Note that the error in Q is directly proportional to the error in n. A 10% error in the value of n produces a 10% error in the flowrate. Considerable effort has been put forth to obtain the best estimate of n, with extensive tables of values covering a wide variety of surfaces (Ref. 7). It should be noted that the values of n given in Table 10.1 are valid only for water as the flowing fluid.

Both the friction factor for pipe flow and the Manning coefficient for channel flow are parameters that relate the wall shear stress to the makeup of the bounding surface. Thus, various results are available that describe n in terms of the equivalent pipe friction factor, f, and the surface roughness, ε (Ref. 8). For our purposes we will use the values of n from Table 10.1.

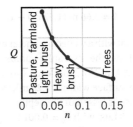

V10.7 Uniform channel flow

10.4.3 Uniform Flow Examples

A variety of interesting and useful results can be obtained from the Manning equation. The examples in this section illustrate some of the typical considerations.

The main parameters involved in uniform flow are the size and shape of the channel cross section (A, R_h), the slope of the channel bottom (S_0), the character of the material lining the channel bottom and walls (n), and the average velocity or flowrate (V or Q). Although the Manning equation is a rather simple equation, the ease of using it depends in part on which variables are known and which are to be determined.

Determination of the flowrate of a given channel with flow at a given depth (often termed the *normal flowrate* for *normal depth*, sometimes denoted y_n) is obtained from a straightforward calculation as is shown in Example 10.3.

EXAMPLE 10.3 | Uniform Flow, Determine Flowrate

Given   Water flows in the canal of trapezoidal cross section shown in **Fig. E10.3a**. The bottom drops 0.43 m per 304.8 m of length. The canal is lined with new finished concrete.

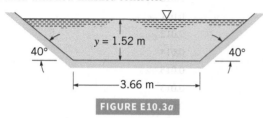

FIGURE E10.3a

Find   Determine

a. the flowrate and

b. the Froude number for this flow.

Assume uniform flow

Solution

a. From Eq. 10.17,

$$Q = \frac{\kappa}{n} A R_h^{2/3} S_0^{1/2} \qquad (1)$$

where $\kappa = 1$, since the dimensions are given in SI units. For a depth of $y = 1.52$ m, the flow area is

$$A = 3.66 \text{ m} (1.52 \text{ m}) + 1.52 \text{ m} \left(\frac{1.52}{\tan 40°} \text{ m}\right) = 8.34 \text{ m}^2$$

so that with a wetted perimeter of $P = 3.66 \text{ m} + 2(1.52/\sin 40° \text{ m}) = 8.41$ m, the hydraulic radius is determined to be $R_h = A/P = 0.99$ m. Note that even though the channel is quite wide (the free-surface width is 7.28 m), the hydraulic radius is only 0.99 m, which is less than the depth.

Thus, with $S_0 = 0.0014$, Eq. 1 becomes

$$Q = \frac{1}{n} (8.34 \text{ m}^2)(0.99 \text{ m})^{2/3}(0.0014)^{1/2} = \frac{0.31}{n}$$

where Q is in m^3/s.

From Table 10.1, we obtain $n = 0.012$ for the finished concrete. Thus,

$$Q = \frac{0.31}{0.012} = 25.91 \text{ m}^3/\text{s} \qquad (Ans)$$

b. The Froude number based on the maximum depth for the flow can be determined from $Fr = V/(gy)^{1/2}$. For the finished concrete case,

$$Fr = 0.804 \qquad (Ans)$$

The flow is subcritical.

Comments   The corresponding average velocity, $V = Q/A$, is 3.1 m/s. It does not take a very steep slope [$S_0 = 0.0014$ or $\theta = \tan^{-1}(0.0014) = 0.080°$] for this velocity.

By repeating the calculations for various surface types (i.e., various Manning coefficient values), the results shown in **Fig. E10.3b** are obtained. Note that the increased roughness causes a decrease in the flowrate. This is an indication that for the turbulent flows involved, the wall shear stress increases with surface roughness.

The same results would be obtained for the channel if its size were given in meters. We would use the same value of n but set $\kappa = 1$ for this SI units situation.

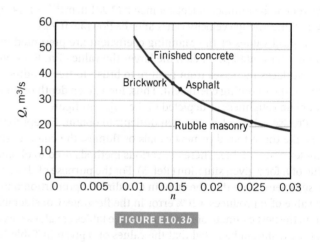

FIGURE E10.3b

In some instances a trial-and-error or iteration method must be used to solve for the dependent variable. This is often encountered when the flowrate, channel slope, and channel material are known, and the flow depth is to be determined as illustrated in Examples 10.4 to 10.7.

EXAMPLE 10.4 | Uniform Flow, Determine Normal Depth

Given   Water flows in the channel shown in Fig. E10.3a at a rate of $Q = 10.0$ m^3/s. The canal lining is weedy.

Find   Determine the depth of the flow.

Solution   In this instance neither the flow area nor the hydraulic radius is known, although they can be written in terms of the depth, y. The flowrate is given in m^3/s, the bottom width is 3.66 m and the area is

$$A = y\left(\frac{y}{\tan 40°}\right) + 3.66y = 1.19y^2 + 3.66y$$

where A and y are in square meters and meters, respectively. Also, the wetted perimeter is

$$P = 3.66 + 2\left(\frac{y}{\sin 40°}\right) = 3.11y + 3.66 \text{ m}$$

so that

$$R_h = \frac{A}{P} = \frac{1.19y^2 + 3.66y}{3.11y + 3.66} \text{ m}$$

where R_h and y are in meters. Thus, with $n = 0.030$ (from Table 10.1), Eq. 10.17 can be written as

$$Q = 10 = \frac{\kappa}{n} A R_h^{2/3} S_0^{1/2} \, m^3/s$$

$$= \frac{1.0}{0.030} (1.19y^2 + 3.66y) \left(\frac{1.19y^2 + 3.66y}{3.11y + 3.66} \right)^{2/3}$$

$$\times (0.0014)^{1/2}$$

which can be rearranged into the form

$$(1.19y^2 + 3.66y)^5 - 515(3.11y + 3.66)^2 = 0 \qquad (1)$$

where y is in meters. The solution of Eq. 1 can be easily obtained by use of a simple rootfinding numerical technique or by trial-and-error methods. The only physically meaningful root of Eq. 1 (i.e., a positive, real number) gives the solution for the normal flow depth at this flowrate as

$$y = 1.50 \, m \qquad (Ans)$$

Comment By repeating the calculations for various flowrates, the results shown in **Fig. E10.4** are obtained. Note that the water depth is not linearly related to the flowrate. That is, if the flowrate is doubled, the depth is not doubled.

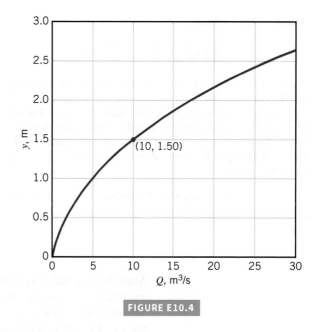

FIGURE E10.4

In Example 10.4 we found the uniform depth for a given flowrate. Since the equation for this depth is a nonlinear equation (usually a polynomial), it may be that there is more than one solution to the problem. For a given channel there may be two or more depths that carry the same flowrate. Although this is not normally so, it can and does happen, as is illustrated by Example 10.5.

EXAMPLE 10.5 | Uniform Flow, Maximum Flowrate

Given Water flows in a round pipe of diameter D at a depth of $0 \le y \le D$, as is shown in **Fig. E10.5a**. The pipe is laid on a constant slope of S_0, and the Manning coefficient is n.

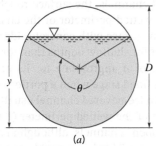

(a)

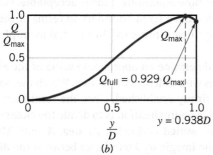

(b)

FIGURE E10.5

Find

a. At what depth does the maximum flowrate occur?

b. Show that for certain flowrates there are two depths possible with the same flowrate. Explain this behavior.

Solution

a. According to the Manning equation (Eq. 10.17), the flowrate is

$$Q = \frac{\kappa}{n} A R_h^{2/3} S_0^{1/2} \qquad (1)$$

where S_0, n, and κ are constants for this problem. From geometry it can be shown that

$$A = \frac{D^2}{8} (\theta - \sin \theta)$$

where θ, the angle indicated in Fig. E10.5a, is in radians. Similarly, the wetted perimeter is

$$P = \frac{D\theta}{2}$$

so that the hydraulic radius is

$$R_h = \frac{A}{P} = \frac{D(\theta - \sin \theta)}{4\theta}$$

Therefore, Eq. 1 becomes

$$Q = \frac{\kappa}{n} S_0^{1/2} \frac{D^{8/3}}{8(4)^{2/3}} \left[\frac{(\theta - \sin \theta)^{5/3}}{\theta^{2/3}} \right]$$

This can be written in terms of the flow depth by using $y = (D/2)[1 - \cos(\theta/2)]$.

A graph of flowrate versus flow depth, $Q = Q(y)$, has the characteristic indicated in **Fig. E10.5b**. In particular, the maximum flowrate, Q_{max}, does not occur when the pipe is full; $Q_{full} = 0.929 Q_{max}$. It occurs when $y = 0.938D$, or $\theta = 5.28$ rad $= 303°$. Thus,

$$Q = Q_{max} \text{ when } y = 0.938D \qquad (Ans)$$

b. For any $0.929 < Q/Q_{max} < 1$ there are two possible depths that give the same Q. The reason for this behavior can be seen by considering the gain in flow area, A, compared to the increase in wetted perimeter, P, for $y \approx D$. The flow area increase for an increase in y is very slight in this region, whereas the increase in wetted perimeter, and hence the increase in shear force holding back the fluid, is relatively large. The net result is a decrease in flowrate as the depth increases.

Comment For most practical problems, the slight difference between the maximum flowrate and full pipe flowrates is negligible, particularly in light of the usual uncertainty of the value of n.

Analysis of a circular conduit (pipe) running partially full is a somewhat unique problem. First, one must be careful to use the correct geometric formulas for A and P depending on whether the conduit is more than or less than half full. Second, it has been shown that Manning's n is not constant but varies with depth (Ref. 9).

The Wide World of Fluids

Done without GPS or lasers

Two thousand years before the invention of such tools as the Global Positioning System (GPS) or laser surveying equipment, Roman engineers were able to design and construct structures that made a lasting contribution to Western civilization. For example, one of the best surviving examples of Roman aqueduct construction is the Pont du Gard, an aqueduct that spans the Gardon River near Nîmes, France. This aqueduct is part of a circuitous, 50-km-long open channel that transported water to the Roman Colony at Nîmes from a spring located 20 km away. The spring is only 14.6 m above the point of delivery, giving an average *bottom slope* of only 3×10^{-4}. It is obvious that to carry out such a project, the Roman understanding of hydraulics, surveying, and construction was well advanced.

For many open channels, the surface roughness varies across the channel.

In many constructed channels and in most natural channels, the surface roughness (and, hence, the Manning coefficient) varies along the wetted perimeter of the channel. A drainage ditch, for example, may have a rocky bottom surface with concrete side walls to prevent erosion. Thus, the effective n will be different for shallow depths than for deep depths. Similarly, a river channel may have one value of n appropriate for its normal channel and another very different value of n during its flood stage when a portion of the flow occurs across fields or through floodplain woods. An ice-covered channel usually has a different value of n for the ice than for the remainder of the wetted perimeter (Ref. 7). (Strictly speaking, such ice-covered channels are not "open" channels, although analysis of their flow is often based on open-channel flow equations. This is acceptable, since the ice cover is often thin enough so that it represents a fixed boundary in terms of the shear stress resistance, but it cannot support a significant pressure differential as in pipe flow situations.)

A variety of methods have been used to determine an appropriate value of the effective roughness of channels that contain subsections with different values of n. Which method gives the most accurate, easy-to-use results is not firmly established, since the results are nearly the same for each method (Ref. 5). A reasonable approximation is to divide the channel cross section into N subsections, each with its own wetted perimeter, P_i, area, A_i, and Manning coefficient, n_i. The P_i values do not include the imaginary boundaries between the different subsections. The total flowrate is assumed to be the sum of the flowrates through each section. This technique is illustrated by Example 10.6.

EXAMPLE 10.6 | Uniform Flow, Variable Roughness

Given Water flows along the drainage canal having the properties shown in **Fig. E10.6a**. The bottom slope is $S_0 = 1 \text{ m}/500 \text{ m} = 0.002$.

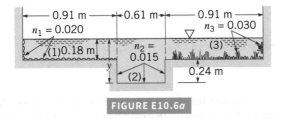

FIGURE E10.6a

Find Estimate the flowrate when the depth in the center channel is $y = 0.18 \text{ m} + 0.24 \text{ m} = 0.42 \text{ m}$.

Solution We divide the cross section into three subsections as is indicated in Fig. E10.6a and write the flowrate as $Q = Q_1 + Q_2 + Q_3$, where for each section

$$Q_i = \frac{1}{n_i} A_i R_{h_i}^{2/3} S_0^{1/2}$$

The appropriate values of A_i, P_i, R_{hi}, and n_i are listed in **Table E10.6**. Note that the imaginary portions of the perimeters between sections (denoted by the vertical dashed lines in Fig. E10.6a) are not included in the P_i. That is, for section (2)

$$A_2 = 0.61 \text{ m} \ (0.24 \text{ m} + 0.18 \text{ m}) = 0.26 \text{ m}^2$$

and

$$P_2 = 0.61 \text{ m} + 2(0.24 \text{ m}) = 1.1 \text{ m}$$

so that

$$R_{h_2} = \frac{A_2}{P_2} = \frac{0.26 \,\text{m}^2}{1.1 \,\text{m}} = 0.24 \text{ m}$$

Thus, the total flowrate is

$$Q = Q_1 + Q_2 + Q_3 = (0.002)^{1/2}$$
$$\times \left[\frac{(0.26 \,\text{m}^2)(0.24 \,\text{m})^{2/3}}{0.015} + \frac{(0.17 \,\text{m}^2)(0.15 \,\text{m})^{2/3}}{0.020} \right.$$
$$\left. + \frac{(0.17 \,\text{m}^2)(0.15 \,\text{m})^{2/3}}{0.030} \right]$$

TABLE E10.6

i	$A_i \,(\text{m}^2)$	$P_i \,(\text{m})$	$R_{hi} \,(\text{m})$	n_i
1	0.17	1.1	0.15	0.020
2	0.26	1.1	0.24	0.015
3	0.17	1.1	0.15	0.030

or

$$Q = 0.48 \text{ m}^3/\text{s} \qquad (Ans)$$

Comments If the entire channel cross section were considered as one flow area, then $A = A_1 + A_2 + A_3 = 0.59 \text{ m}^2$ and $P = P_1 + P_2 + P_3 = 3.3 \text{ m}$, or $R_h = A/P = 0.59 \text{ m}^2/3.3 \text{ m} = 0.18 \text{ m}$. The flowrate is given by Eq. 10.17, which can be written as

$$Q = \frac{1}{n_{\text{eff}}} A R_h^{2/3} S_0^{1/2}$$

where n_{eff} is the effective value of n for this channel. With $Q = 0.48 \text{ m}^3/\text{s}$ as determined above, the value of n_{eff} is found to be

$$n_{\text{eff}} = \frac{1 A R_h^{2/3} S_0^{1/2}}{Q}$$
$$= \frac{1(0.59)(0.59)^{2/3}(0.002)^{1/2}}{0.48} = 0.039$$

As expected, the effective roughness (Manning's n) is between the minimum ($n_2 = 0.015$) and maximum ($n_3 = 0.030$) values for the individual subsections.

By repeating the calculations for various depths, y, the results shown in **Fig. E10.6b** are obtained. Note that there are two distinct portions of the graph—one when the water is contained entirely within the main, center channel ($y < 0.24 \text{ m}$) and the other when the water overflows into the side portions of the channel ($y > 0.24 \text{ m}$).

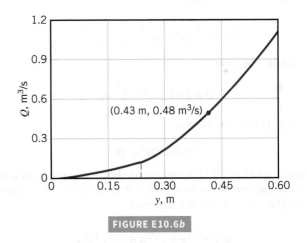

FIGURE E10.6b

One type of problem often encountered in open-channel flows is that of determining the *best hydraulic cross section*, defined as the section of the minimum area for a given flowrate, Q, slope, S_0, and roughness coefficient, n. By using $R_h = A/P$ we can write Eq. 10.17 as

$$Q = \frac{\kappa}{n} A \left(\frac{A}{P} \right)^{2/3} S_0^{1/2} = \frac{\kappa}{n} \frac{A^{5/3} S_0^{1/2}}{P^{2/3}}$$

which can be rearranged as

$$A = \left(\frac{nQ}{\kappa S_0^{1/2}} \right)^{3/5} P^{2/5}$$

> For a given flowrate, the channel of minimum area is denoted as the best hydraulic cross section.

where the quantity in the parentheses is a constant for a given application. Thus, a channel with minimum A is one with a minimum P, so that both the amount of excavation needed and the amount of material to line the surface are minimized by the best hydraulic cross section.

The best hydraulic cross section possible is that of a semicircular channel. No other shape has as small a wetted perimeter for a given area. It is often desired to determine the best shape

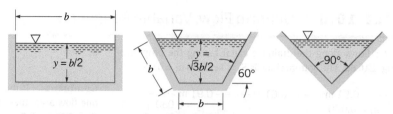

FIGURE 10.11 Best hydraulic cross sections for a rectangle, a 60° trapezoid, and a triangle.

for a class of cross sections. The results for rectangular, trapezoidal (with 60° sides), and triangular shapes are shown in **Fig. 10.11**. For example, the best hydraulic cross section for a rectangle is one whose depth is half its width; for a triangle it is a 90° triangle.

EXAMPLE 10.7 | Uniform Flow—Best Hydraulic Cross Section

Given Water flows uniformly in a rectangular channel of width b and depth y.

Find Determine the aspect ratio, b/y, for the best hydraulic cross section.

Solution The uniform flow is given by Eq. 10.17 as

$$Q = \frac{\kappa}{n} A R_h^{2/3} S_0^{1/2} \tag{1}$$

where $A = by$ and $P = b + 2y$, so that $R_h = A/P = by/(b + 2y)$. We rewrite the hydraulic radius in terms of A as

$$R_h = \frac{A}{(2y + b)} = \frac{A}{(2y + A/y)} = \frac{Ay}{(2y^2 + A)}$$

so that Eq. 1 becomes

$$Q = \frac{\kappa}{n} A \left(\frac{Ay}{2y^2 + A} \right)^{2/3} S_0^{1/2}$$

This can be rearranged to give

$$A^{5/2} y = K(2y^2 + A) \tag{2}$$

where $K = (nQ/k S_0^{1/2})^{3/2}$ is a constant. The best hydraulic section is the one that gives the minimum A for all y. That is, $dA/dy = 0$. By differentiating Eq. 2 with respect to y, we obtain

$$\frac{5}{2} A^{3/2} \frac{dA}{dy} y + A^{5/2} = K \left(4y + \frac{dA}{dy} \right)$$

which, with $dA/dy = 0$, reduces to

$$A^{5/2} = 4Ky \tag{3}$$

With $K = A^{5/2}y/(2y^2 + A)$ from Eq. 2, Eq. 3 can be written in the form

$$A^{5/2} = \frac{4A^{5/2}y^2}{(2y^2 + A)}$$

which simplifies to $y = (A/2)^{1/2}$. Thus, because $A = by$, the best hydraulic cross section for a rectangular shape has a width b and a depth

$$y = \left(\frac{A}{2} \right)^{1/2} = \left(\frac{by}{2} \right)^{1/2}$$

or

$$2y^2 = by$$

That is, the rectangle with the best hydraulic cross section is twice as wide as it is deep, or

$$b/y = 2 \tag{Ans}$$

Comments A rectangular channel with $b/y = 2$ will give the smallest area (and smallest wetted perimeter) for a given flowrate. Conversely, for a given area, the largest flowrate in a rectangular channel will occur when $b/y = 2$. For $A = by = $ constant, if $y \to 0$ then $b \to \infty$, and the flowrate is small because of the large wetted perimeter $P = b + 2y \to \infty$. The maximum Q occurs when $y = b/2$. However, as seen in **Fig. E10.7**, the maximum represented by this optimal configuration is a rather weak one. For example, for aspect ratios between 1 and 4, the flowrate is within 96% of the maximum flowrate obtained with the same area and $b/y = 2$.

An alternate but equivalent method to obtain the aforementioned answer is to use the fact that $dR_h/dy = 0$, which follows from Eq. 1 using $dQ/dy = 0$ (constant flowrate) and $dA/dy = 0$ (best hydraulic cross section has minimum area). Differentiation of $R_h = Ay/(2y^2 + A)$ with constant A gives $b/y = 2$ when $dR_h/dy = 0$.

The best hydraulic cross section can be calculated for other shapes in a similar fashion. The results (given here without proof) for rectangular, trapezoidal (with 60° sides), and triangular shapes are shown in Fig. 10.11.

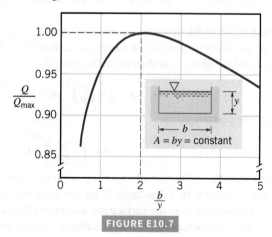

FIGURE E10.7

10.5 MOST EFFICIENT CHANNEL SECTION

An inspection of the Manning equation reveals that, everything else remaining constant, the discharge carried under the normal flow condition will increase with decreasing wetted perimeter. Thus, for a given cross-sectional area, the channel section having the shortest wetted perimeter will have the maximum conveyance capacity. Such a channel section is called the **most efficient channel section**. Therefore, a key factor in designing a channel is to maximize its conveyance capabilities while minimizing its perimeter. The **hydraulically most efficient section** is defined as a channel with the least wetted perimeter for a given conveyance capability. Therefore, the costs of a channel will generally be near minimum if it is designed as the hydraulically most efficient section. In this section we discuss for designing least cost trapezoidal and triangular channels accounting for lining and excavation costs.

The Manning's equation may be combined with continuity equation to give

$$Q = \frac{1}{n}AR_h^{2/3}S_0^{1/2} \tag{10.18}$$

This can be rearranged to give

$$Q = \frac{A^{5/3}S_0^{1/2}}{nP^{2/3}} \tag{10.19}$$

where Q is flow rate, n is Manning's roughness coefficient, S_0 is bed slope, A is cross sectional area and P is wetted perimeter. Thus, for given values of roughness coefficient and bed slope, the discharge is maximum for a given area of cross-sectional when wetted perimeter is minimum. Such a cross-section is known as the most efficient channel section.

10.5.1 Trapezoidal Channel Section

Consider a trapezoidal section of side slope m:1 as shown in **Fig. 10.12a**.

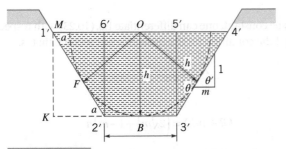

FIGURE 10.12a Most efficient trapezoidal channel section.

For this section, the wetted perimeter P can be given as

$$P = 1'-2'+2'-3'+3'-4'$$

$$= 1'-2'+B+3'-4$$

$$1'-2' = 3'-4'$$

$$P = B+2\times(3'-4') = B+2\sqrt{(3'-5')^2 + (5'-4')^2}$$

$$= B+2\sqrt{(h^2)+(h\tan\theta)^2}$$

$$= B+2\sqrt{(h)^2+\left(\frac{h}{\tan\theta'}\right)^2}$$

$$= B+2\sqrt{(h)^2+\left(\frac{h}{(1/m)}\right)^2}$$

$$= B + 2\sqrt{(h)^2 + (mh)^2}$$

$$= B + 2h\sqrt{(m)^2 + 1} \tag{10.20}$$

$$A = \frac{1}{2}\Big[(1'-4') + (2'-3')\Big](3'-5')$$

$$= \frac{1}{2}\Big[(1'-6') + (2'-3') + (5'-4') + (2'-3')\Big](3'-5')$$

As
$$1'-6' = 5'-4'$$

$$= \frac{1}{2}[(5'-4') + (2'-3') + (5'-4') + (2'-3')](3'-5')$$

$$= \frac{1}{2}\Big[2(2'-3') + 2(5'-4')\Big](3'-5')$$

$$= [(2'-3') + (5'-4')](3'-5')$$

Thus
$$A = (B + mh)h \tag{10.21}$$

and
$$B = \frac{A}{h} - mh$$

From Eq, 10.20

$$P = \frac{A}{h} - mh + 2h\sqrt{m^2 + 1} \tag{10.22}$$

$$P = \frac{A}{h} + h\left(2\sqrt{m^2 + 1} - m\right) \tag{10.23}$$

Considering A and n to be constant, one can differentiate Eq. (10.23) with respect to h and set the differential to zero to get the condition for the most efficient section, that is,

$$-\frac{A}{h^2} + 2\sqrt{m^2 + 1} - m = 0 \tag{10.24}$$

$$(B + mh) = \left(2\sqrt{m^2 + 1} - m\right)h \tag{10.25}$$

$$B = 2\left(\sqrt{m^2 + 1} - m\right)h \tag{10.26}$$

Thus, the top width $T = B + 2mh = 2\sqrt{m^2 + 1} \cdot h$, that is, the top width is equal to twice the length of the sloping side.

Referring to triangle OMF in Fig. (10.12)

$$OF = OM \sin \theta$$

$$OF = \frac{T}{2} \cdot \frac{1}{\sqrt{m^2 + 1}} = \frac{2\sqrt{m^2 + 1} \cdot h}{2} \frac{1}{\sqrt{m^2 + 1}}$$

$$OF = h$$

The hydraulic radius R of the section is given by

$$R_h = \frac{(B + mh)h}{\left(B + 2\sqrt{m^2 + 1} \cdot h\right)}$$

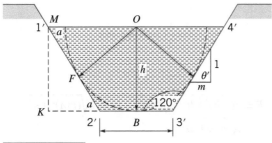

FIGURE 10.12*b* Most efficient trapezoidal channel section.

By substituting for *B* from Eq. (10.26), we get

$$R_h = \frac{h}{2}$$

Note: If we consider expression for T ($OM = O{-}1' = T/2$), OF and channel flow height h together, it clearly manifests that the most efficient channel section will have its proportions in such a manner so that it may inscribe a semicircle in it as shown in Fig. 10.12*a*.

If the side slope is assumed to be a variable and h and A are constant, the condition for the most efficient section becomes $dP/dm = 0$. Differentiating Eq. (10.23), with respect to m?

$$h \frac{2}{2\sqrt{m^2+1}} - 2m - h = 0$$

$$2m = \sqrt{m^2+1}$$

$$n = 1/\sqrt{3}$$

$$\tan\theta' = \frac{1}{m} = \sqrt{3}$$

which means

$$\theta' = 60°$$

Thus, from Fig. 10.12*b*, we can deduce that, if B = slant length (3′–4′), then the most efficient trapezoidal channel shall be half of the regular hexagon.

10.5.2 Triangular Channel Section

Consider a triangular section of side slope m:1 as shown in Fig. 10.13.

$$\tan\theta' = \frac{1}{m} = \frac{1}{\tan\theta}$$

which implies $\tan\theta = m$

Thus, from triangle 1′–O–3′, we have

$$\tan\theta = m = \frac{O{-}3'}{h}$$

$$O{-}3' = mh$$

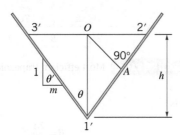

FIGURE 10.13 Most efficient triangular channel section.

Thus, area

$$A = \frac{1}{2}(2mh)h = mh^2$$

and wetted perimeter

$$P = 1'-2'+1'-3'$$

$$P = \sqrt{(mh)^2 + h^2} + \sqrt{(mh)^2 + h^2} = 2\sqrt{(mh)^2 + h^2}$$
$$= 2\sqrt{m^2 + 1} \cdot h$$

$$P = 2\sqrt{m^2 + 1} \cdot \left(A/m\right)^{1/2}$$

$$P^2 = \frac{4\sqrt{m^2 + 1}}{m}A = 4\left(m + \frac{1}{m}\right)A \qquad (10.27)$$

For the section to be most efficient, wetted perimeter should be a minimum for given cross sectional area or $(dp/dm) = 0$. Differentiating Eq. (10.27) with respect to m, we get

$$2P\frac{dP}{dm} = 0 = \left(4 - \frac{4}{m^2}\right)A$$

$$m = 1.0 \text{ (as negative value of } m \text{ will not be acceptable)}$$

that is, $\tan\theta = 45°$

Therefore, the overall central angle (angle $3'-1'-2'$, i.e. 2θ) shall be 90°.

Thus, a triangular section with a central angle of 90° is the most efficient triangular section.

Further, for $m = 1$,

$$A = h^2 \quad \text{and} \quad P = 2\sqrt{2}h$$

Thus, the hydraulic radius of triangular section flow

$$R_h = \frac{A}{P} = \frac{h^2}{2\sqrt{2}h} = \frac{h}{2\sqrt{2}}$$

Also, the top width of the triangular channel

$$T = 2'-3' = 2mh$$

That is, $T = 2h$.

EXAMPLE 10.8 | Most Efficient Trapezoidal Section

Given A most efficient trapezoidal channel section that carries a discharge of 30 m³/s with its longitudinal slope as 0.0009 and Manning's coefficient n as 0.018.

Find Normal depth, bed width and side slope.

Solution The most efficient trapezoidal channel section is as shown in **Fig. E10.8**, where $m = \frac{1}{\sqrt{3}}$ and $B = 3'-4'$

From triangle $3'-4'-5'$

$$\sin 60° = \frac{h}{3'-4'}$$

that is

$$3'-4' = \frac{h}{\sin 60°}$$

which means

$$3'-4' = B = \frac{h}{\sqrt{3}/2}$$

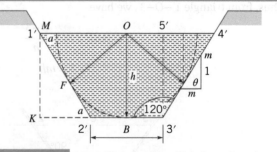

FIGURE E10.8 Most efficient trapezoidal channel section.

that is

$$B = \frac{2h}{\sqrt{3}} \qquad (1)$$

Now from Eqs. (10.21) and (10.23) we have following relationships for area (A) and wetted perimeter (P)

$$A = (B + mh)h \qquad (2)$$

and

$$P = \frac{A}{h} + h\left(2\sqrt{(m^2)+1} - m\right) \qquad (3)$$

Thus, for most efficient trapezoidal channel Eqs. (2) and (3) can be reduced to

$$A = \sqrt{3}h^2 \text{ and } P = 2\sqrt{3}h$$

From Eq.(10.19) the flow rate Q can be given as

$$Q = \frac{A^{5/3} S_0^{1/2}}{n P^{2/3}} \qquad (4)$$

On putting the values of A and P for most efficient trapezoidal channel flow from above, we have

$$Q = \frac{\left(\sqrt{3}h^2\right)^{5/3} S_0^{1/2}}{n\left(2\sqrt{3}h\right)^{2/3}}$$

which implies

$$\frac{Qn}{h^{8/3} S_0^{1/2}} = 1.091 \qquad (5)$$

Substituting $Q = 30$ m³/s, $n = 0.018$ and $S_0 = 0.0009$ in Eq. (5),

$$\frac{30 \times (0.018)}{h^{8/3} (0.0009)^{1/2}} = 1.091$$

which implies that the normal depth $h = 2.86$ m $\qquad$ (Ans)

and therefore, the channel width

$$B = \frac{2h}{\sqrt{3}} = 3.302 \text{ m} \qquad (Ans)$$

and side slope of most efficient channel

$$m = \frac{1}{\sqrt{3}} = 0.57735 \qquad (Ans)$$

Comments Similar to the way ratio $\dfrac{Qn}{h^{8/3} S_0^{1/2}}$ has been calculated for most efficient trapezoidal channel flow in the current problem, the ratio can also be calculated for other most efficient channel sections (viz. triangular, rectangular, etc.). The values of this ratio for triangular and rectangular channel sections are 0.5 and 1.260, respectively.

EXAMPLE 10.9 | Most Efficient Triangular Section

Given A triangular irrigation canal caries a discharge of 20 m³/s at bed slope of 1/5000. The side slope of the canal is 1:1 and Manning's coefficient n is 0.018.

Find The central depth of the most efficient channel flow.

Solution For most efficient triangular channel flow, the side slope m should be 1. Incidentally in the given problem, the slope 1:m is given as 1:1. Therefore, for the given conditions, the corresponding dimensions and flow rate will also be associated to the most efficient channel flow.

The other given parameters are:

$$Q = 20 \text{ m}^3/\text{s}$$

$$S_0 = 1/5000, n = 0.018.$$

From rearranged Manning's equation (10.19), we have

$$Q = \frac{(A^{5/3} S_0^{1/2})}{n P^{2/3}} \qquad (1)$$

For most efficient triangular channel section flow ($m = 1$), we have following relationships for area (A) and wetted perimeter (P)

$$A = mh^2 = h^2$$

$$P = 2(\sqrt{m^2 + 1})h = 2\sqrt{2}h$$

From Eq. (1), we can write

$$20 = \frac{(h^2)^{5/3}\left(\dfrac{1}{5000}\right)^{1/2}}{0.018(2\sqrt{2h})^{2/3}} \qquad (2)$$

which implies that the central depth

$$h = 4.37 \text{ m} \qquad (Ans)$$

Comment A minor rearrangement in Eqs. (2) & (3), will help get a relationship as

$$Q = 0.393 \, h^{(8/3)} \qquad (3)$$

A graph can be plotted between flowrate (Q) and central depth (h) of the most efficient triangular channel flow for given value of n and S_0 as shown in **Fig. E10.9** as represented by Eq. (3).

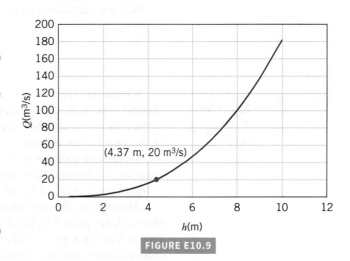

FIGURE E10.9

10.6 GRADUALLY VARIED FLOW

In many situations the flow in an open channel is not uniform. This can occur for several reasons: The bottom slope is not constant, the cross-sectional shape and area vary in the flow direction, or there is some obstruction across a portion of the channel. Physically, the difference between the component of weight and the shear forces in the direction of flow produces a change in the fluid momentum that requires a change in velocity and, from continuity considerations, a change in depth.

In this section we consider gradually varying flows. For such flows, $dy/dx \ll 1$ and it is reasonable to impose the one-dimensional velocity assumption. At any section the total head is $H = V^2/2g + y + z$ and the energy equation (Eq. 10.5) becomes

$$H_1 = H_2 + h_L$$

where h_L is the head loss between sections (1) and (2).

As shown in Figure 10.6, the slope of the energy line is $dH/dx = -(dh_L/dx) = -S_f$ and the slope of the channel bottom is $dz/dx = -S_0$ (downward slope is considered positive). Thus since

$$\frac{dH}{dx} = \frac{d}{dx}\left(\frac{V^2}{2g} + y + z\right) = \frac{V}{g}\frac{dV}{dx} + \frac{dy}{dx} + \frac{dz}{dx}$$

we obtain

$$-\frac{dh_L}{dx} = \frac{V}{g}\frac{dV}{dx} + \frac{dy}{dx} - S_0$$

or

$$\frac{V}{g}\frac{dV}{dx} + \frac{dy}{dx} = S_0 - S_f \qquad (10.28)$$

For a given flowrate per unit width, q, in a rectangular channel of constant width b, we have $V = q/y$ or by differentiation

$$\frac{dV}{dx} = -\frac{q}{y^2}\frac{dy}{dx} = -\frac{V}{y}\frac{dy}{dx}$$

so that the kinetic energy term in Eq. 10.28 becomes

$$\frac{V}{g}\frac{dV}{dx} = -\frac{V^2}{gy}\frac{dy}{dx} = -\mathrm{Fr}^2\frac{dy}{dx} \qquad (10.29)$$

where $\mathrm{Fr} = V/(gy)^{1/2}$ is the local Froude number of the flow. Substituting Eq. 10.29 into Eq. 10.28 and simplifying gives

$$\boxed{\frac{dy}{dx} = \frac{S_0 - S_f}{1 - \mathrm{Fr}^2}} \qquad (10.30)$$

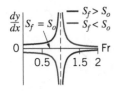

We see that the rate of change of fluid depth, dy/dx, depends on the local slope of the channel bottom, S_0, the slope of the energy line, S_f, and the Froude number, Fr. As depicted in the adjacent figure, the value of dy/dx can be either negative, zero, or positive, depending on the values of these three parameters. That is, the channel flow depth may be constant or it may increase or decrease in the flow direction, depending on the values of S_0, S_f, and Fr. The behavior of subcritical flow may be the opposite of that for supercritical flow, as shown by the denominator, $1 - \mathrm{Fr}^2$, of Eq. 10.30.

Although in the derivation of Eq. 10.30 we assumed q is constant (i.e., a rectangular channel), Eq. 10.30 is valid for channels of any constant cross-sectional shape, provided the Froude number is interpreted properly (Ref. 3). In this book we will consider only rectangular cross-sectional channels when using this equation.

In order to proceed further with Eq. 10.30, we need to relate $S_f(= h_L/\ell)$ to other problem variables such as V, R_h, and roughness (n). An exact relationship would be very complicated because the flow is most likely turbulent and the shear stress is not uniformly distributed. In the absence of more accurate information, the following model is used: we assume that the energy loss at any location along the channel is the same as the energy loss that occurs in a uniform flow with the same values of the flow variables (i.e., V, R_h, n). The fluid velocity would then be given by the Manning equation (Eq. 10.15). The friction slope is found by noting that for uniform flow, the friction slope and the channel slope are equal, $S_0 = S_f$, so we can write the Manning equation as

$$V = \frac{\kappa}{n} R_h^{2/3} S_f^{1/2}$$

Solving for the friction slope gives

$$S_f = \frac{n^2}{\kappa^2} \left(\frac{V^2}{R_h^{4/3}} \right) \tag{10.31}$$

If we substitute the equations for S_f, V, R_h, and Fr into Eq. 10.30, we get

$$\frac{dy}{dx} = \frac{S_0 - \left(\frac{n^2}{\kappa^2} \right) \left(\frac{Q^2}{b^2 y^2} \right) \left(\frac{by}{b + 2y} \right)^{-4/3}}{1 - \frac{Q^2}{g b^2 y^3}} \tag{10.32}$$

where Q (flowrate), g (acceleration of gravity), and κ (1 for SI, 1.49 for BG) are constants and b (channel width), n (Manning resistance coefficient), and S_0 (channel slope) are known functions of x (distance along the channel).

A solution of this differential equation gives the fluid depth (or surface profile) as a function of position along the channel. It is usually not possible to integrate the differential equation analytically; however, a numerical solution is quite feasible. Using a computer, the channel can be divided into many small steps for accurate computation. A pencil-and-paper calculation using only a few reaches, called the *unit step method*, can be used to calculate the profile with acceptable accuracy. An interesting feature of either method is that for supercritical flow, the calculations begin at a known point and proceed downstream, while for subcritical flow, calculations begin at a point and proceed upstream. See Refs. 3, 5, and 6 for details.

10.7 RAPIDLY VARIED FLOW

In many open channels, flow depth changes occur over a relatively short distance so that $dy/dx \sim 1$. Such **rapidly varied flow** conditions are often quite complex and difficult to analyze precisely. Fortunately, many useful approximate results can be obtained by using a simple one-dimensional model along with appropriate experimentally determined coefficients when necessary. In this section we discuss several of these flows.

Some rapidly varied flows occur in constant area channels for reasons that are not immediately obvious. The hydraulic jump is one such case. As can be seen in the photograph in **Fig. 10.14**, the flow may change from a relatively shallow, high-speed condition into a relatively deep, low-speed condition within a horizontal distance of just a few flow depths. Other rapidly varied flows may be due to a sudden change in the channel geometry such as the flow in an expansion or contraction section of a channel as sketched in **Fig. 10.15**.

In many cases the flow depth may change significantly in a short distance.

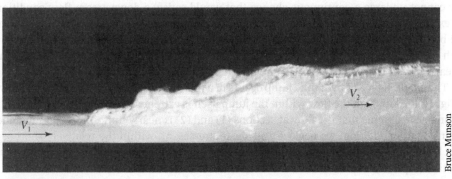

FIGURE 10.14 Hydraulic jump.

Bruce Munson

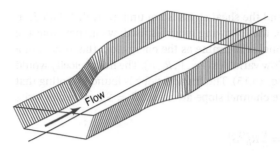

FIGURE 10.15 Rapidly varied flow may occur in a channel transition section.

V10.9 Bridge pier scouring

V10.10 Big Sioux River Bridge collapse

V10.8 Erosion in a channel

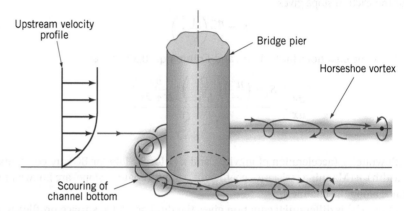

FIGURE 10.16 The complex three-dimensional flow structure around a bridge pier.

In such situations the flow field is often two- or three-dimensional. There may be regions of flow separation, flow reversal, or unsteady oscillations of the free surface. For the purpose of some analyses, these complexities can be neglected and a simplified analysis can be useful. In other cases, however, it is the complex details of the flow that are the most important property of the flow; any analysis must include their effects. The scouring of a river bottom in the neighborhood of a bridge pier, as depicted in **Fig. 10.16**, is such an example. A one- or two-dimensional model of this flow would not be sufficient to describe the complex structure of the flow that is responsible for the erosion near the foot of the bridge pier.

Many open-channel flow-measuring devices are based on principles associated with rapidly varied flows. Among these devices are broad-crested weirs, sharp-crested weirs, critical flow flumes, and sluice gates. The operation of such devices is discussed in sections 10.7.2 and 10.7.3.

10.7.1 The Hydraulic Jump

Under certain conditions it is possible that the fluid depth will change very rapidly over a short length of the channel without any change in the channel configuration. Such changes in depth can be approximated as a discontinuity in the free-surface elevation ($dy/dx \to \infty$). For reasons discussed below, this step change in depth is always from a shallow to a deeper depth—always a step up, never a step down.

Physically, this near discontinuity, called a **hydraulic jump**, occurs because the upstream flow cannot undergo a continuous change that would satisfy a downstream flow condition. For example, a sluice gate may require that the conditions immediately downstream of the gate be supercritical flow, while obstructions in the channel on the downstream end of the reach may require that the flow be subcritical. The hydraulic jump provides the mechanism (a nearly discontinuous one) to make the transition between the two types of flow.

The simplest type of hydraulic jump occurs in a horizontal, rectangular channel as shown in **Fig. 10.17**. Although the flow within the jump itself is extremely complex and agitated, it is reasonable to assume that the flow at sections (1) and (2) is nearly uniform, steady, and one-dimensional. In addition, we neglect wall shear stresses, τ_w, within the relatively short segment between these two sections. Under these conditions the x component of the momentum equation (Eq. 5.22) for the control volume indicated can be written as

$$F_1 - F_2 = \rho Q(V_2 - V_1) = \rho V_1 y_1 b(V_2 - V_1)$$

A hydraulic jump is a steplike increase in fluid depth in an open channel.

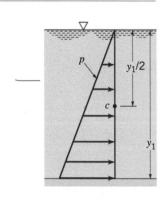

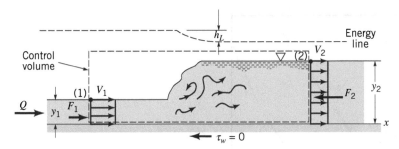

FIGURE 10.17 Hydraulic jump geometry.

where, as depicted in the figure in the margin, the pressure force at either section is hydrostatic. That is, $F_1 = p_{c1}A_1 = \gamma y_1^2 b/2$ and $F_2 = p_{c2}A_2 = \gamma y_2^2 b/2$, where $p_{c1} = \gamma y_1/2$ and $p_{c2} = \gamma y_2/2$ are the pressures at the centroids of the channel cross sections and b is the channel width (see Section 2.8). Thus, the momentum equation becomes

$$\frac{y_1^2}{2} - \frac{y_2^2}{2} = \frac{V_1 y_1}{g}(V_2 - V_1) \tag{10.33}$$

In addition to the momentum equation, we have the conservation of mass equation (Eq. 5.12)

$$y_1 b V_1 = y_2 b V_2 = Q \tag{10.34}$$

and the energy equation (Eq. 5.84)

$$y_1 + \frac{V_1^2}{2g} = y_2 + \frac{V_2^2}{2g} + h_L \tag{10.35}$$

The head loss, h_L, in Eq. 10.35 is due to the dissipation associated with the violent turbulent mixing that occurs within the jump itself.

Clearly Eqs. 10.33, 10.34, and 10.35 have a solution $y_1 = y_2$, $V_1 = V_2$, and $h_L = 0$. This represents the case of no jump. Since these are nonlinear equations, it may be possible that more than one solution exists. The other solutions can be obtained as follows. By combining Eqs. 10.33 and 10.34 to eliminate V_2 we obtain

$$\frac{y_1^2}{2} - \frac{y_2^2}{2} = \frac{V_1 y_1}{g}\left(\frac{V_1 y_1}{y_2} - V_1\right) = \frac{V_1^2 y_1}{g y_2}(y_1 - y_2)$$

which can be simplified by cancelling a common nonzero factor $y_1 - y_2$ from each side to give

$$\left(\frac{y_2}{y_1}\right)^2 + \left(\frac{y_2}{y_1}\right) - 2\text{Fr}_1^2 = 0$$

where $\text{Fr}_1 = V_1/\sqrt{g y_1}$ is the upstream Froude number. Using the quadratic formula we obtain

$$\frac{y_2}{y_1} = \frac{1}{2}\left(-1 \pm \sqrt{1 + 8\,\text{Fr}_1^2}\right)$$

Clearly the solution with the minus sign is not possible (it would give a negative y_2/y_1). Thus,

$$\boxed{\frac{y_2}{y_1} = \frac{1}{2}\left(-1 + \sqrt{1 + 8\,\text{Fr}_1^2}\right)} \tag{10.36}$$

This depth ratio, y_2/y_1, across the hydraulic jump is shown as a function of the upstream Froude number in **Fig. 10.18**. The portion of the curve for $\text{Fr}_1 < 1$ is dashed in recognition of the fact that to have a hydraulic jump the flow must be supercritical. That is, the solution as given by Eq. 10.36 must be restricted to $\text{Fr}_1 \geq 1$, for which $y_2/y_1 \geq 1$. This can be shown by consideration of the energy equation, Eq. 10.35, as follows. The dimensionless head loss, h_L/y_1, can be obtained from Eq. 10.35 as

$$\boxed{\frac{h_L}{y_1} = 1 - \frac{y_2}{y_1} + \frac{\text{Fr}_1^2}{2}\left[1 - \left(\frac{y_1}{y_2}\right)^2\right]} \tag{10.37}$$

where, for given values of Fr_1, the values of y_2/y_1 are obtained from Eq. 10.36. The dashed line in Fig. 10.18 shows the dimensionless head loss predicted by Eq. 10.37 for $\text{Fr}_1 < 1$. Because a negative head loss would violate the second law of thermodynamics (viscous effects always dissipate mechanical energy), it is not possible to produce a hydraulic jump with $\text{Fr}_1 < 1$. The head loss across the jump corresponds to the lowering of the energy line shown in Fig. 10.17.

A flow must be supercritical (Froude number > 1) to produce a hydraulic jump. This is analogous to the compressible flow discussed in Chapter 11 in which it is shown that the flow

V10.11 Hydraulic jump in a river

The depth ratio across a hydraulic jump depends on the Froude number only.

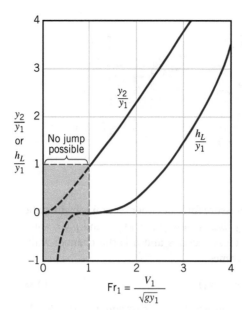

FIGURE 10.18 Depth ratio and dimensionless head loss across a hydraulic jump as a function of upstream Froude number.

of a gas must be supersonic (Mach number > 1) to produce the discontinuity called a shock wave. However, the fact that a flow is supercritical (or supersonic) does not guarantee the production of a hydraulic jump (or shock wave). The mathematically trivial solution $y_1 = y_2$ and $V_1 = V_2$ is also possible.

The fact that there is an energy loss across a hydraulic jump is useful in many situations. For example, the relatively large amount of energy contained in the fluid flowing down the spillway of a dam like that shown in the adjacent photograph could cause damage to the channel below the dam. By placing suitable flow control objects in the channel downstream of the spillway, it is possible (if the flow is supercritical) to produce a hydraulic jump on the apron of the spillway and thereby dissipate a considerable portion of the energy of the flow (Ref. 13). That is, the dam spillway produces supercritical flow, and the channel downstream of the dam requires subcritical flow. The resulting hydraulic jump provides the means to change the character of the flow.

Hydraulic jumps dissipate energy.

The Wide World of Fluids

Grand Canyon rapids building

Virtually all of the rapids in the Grand Canyon were formed by rock debris carried into the Colorado River from side canyons. Severe storms wash large amounts of sediment into the river, building debris fans that narrow the river. This debris forms crude dams that back up the river to form quiet pools above the rapids. Water exiting the pool through the narrowed channel can reach supercritical conditions and produce *hydraulic jumps* downstream. Since the configuration of the jumps is a function of the flowrate, the

difficulty in running the rapids can change from day to day. Also, rapids change over the years as debris is added to or removed from the rapids. For example, Crystal Rapid, one of the notorious rafting stretches of the river, changed very little between the first photos of 1890 and those of 1966. However, a debris flow from a severe winter storm in 1966 greatly constricted the river. Within a few minutes the configuration of Crystal Rapid was completely changed. The new, immature rapid was again drastically changed by a flood in 1983. While Crystal Rapid is now considered full grown, it will undoubtedly change again, perhaps in 100 or 1000 years.

EXAMPLE 10.10 | Hydraulic Jump

Given Water on the horizontal apron of the 30.5 m-wide spillway shown in **Fig. E10.10a** has a depth of 0.18 m and a velocity of 5.5 m/s.

Find Determine the depth, y_2, after the jump, the Froude numbers before and after the jump, Fr_1 and Fr_2, and the power dissipated, $\mathscr{P}_d$, within the jump.

Solution Conditions across the jump are determined by the upstream Froude number

$$Fr_1 = \frac{V_1}{\sqrt{gy_1}} = \frac{5.5 \text{ m/s}}{[(9.81 \text{ m/s}^2)(0.18 \text{ m})]^{1/2}} = 4.14 \qquad (Ans)$$

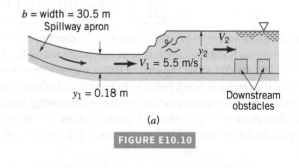

FIGURE E10.10

Thus, the upstream flow is supercritical, and it is possible to generate a hydraulic jump as sketched.

From Eq. 10.36 we obtain the depth ratio across the jump as

$$\frac{y_2}{y_1} = \frac{1}{2}\left(-1 + \sqrt{1 + 8\,\text{Fr}_1^2}\,\right)$$
$$= \frac{1}{2}\left[-1 + \sqrt{1 + 8\,(4.1)^2}\,\right] = 5.32$$

or

$$y_2 = 5.32\,(0.18\text{ m}) = 0.97\text{ m} \qquad (Ans)$$

Since $Q_1 = Q_2$, or $V_2 = (y_1 V_1)/y_2 = 0.18$ m $(5.5$ m/s$)/0.97$ m $= 1.02$ m/s, it follows that

$$\text{Fr}_2 = \frac{V_2}{\sqrt{gy_2}} = \frac{1.03\text{ m/s}}{[(9.81\text{ m/s}^2)(0.97\text{ m})]^{1/2}} = 0.33 \qquad (Ans)$$

As is true for any hydraulic jump, the flow changes from supercritical to subcritical flow across the jump.

The power (energy per unit time) dissipated, $\mathcal{P}_d$, by viscous effects within the jump can be determined from the head loss as (see Eq. 5.85)

$$\mathcal{P}_d = \gamma Q h_L = \gamma b y_1 V_1 h_L \qquad (1)$$

where h_L is obtained from Eq. 10.35 or 10.37 as

$$h_L = \left(y_1 + \frac{V_1^2}{2g}\right) - \left(y_2 + \frac{V_2^2}{2g}\right) = \left[0.18\text{ m} + \frac{(5.5\text{ m/s})^2}{2(9.81\text{ m/s}^2)}\right]$$
$$- \left[0.97\text{ m} + \frac{(1.03\text{ m/s})^2}{2(9.81\text{ m/s}^2)}\right]$$

or

$$h_L = 0.69\text{ m}$$

Thus, from Eq. 1,

$$\mathcal{P}_d = 371.5\text{ kW} \qquad (Ans)$$

Comments This power, which is dissipated within the highly turbulent motion of the jump, is converted into an increase in water temperature, T. That is, $T_2 > T_1$. Although the power dissipated is considerable, the difference in temperature is not great because the flowrate is quite large.

By repeating the calculations for the given flowrate $Q_1 = A_1 V_1 = b_1 y_1 V_1 = 30.58$ m³/s but with various upstream depths, y_1, the results shown in **Fig. E10.10b** are obtained. Note that a slight change in water depth can produce a considerable change in energy dissipated. Also, if $y_1 > 0.47$ m the flow is subcritical ($\text{Fr}_1 < 1$) and no hydraulic jump can occur.

The hydraulic jump flow process can be illustrated by use of the specific energy concept introduced in Section 10.3. Equation 10.33 can be written in terms of the specific energy, $E = y + V^2/2g$, as $E_1 = E_2 + h_L$, where $E_1 = y_1 + V_1^2/2g = 1.72$ m and $E_2 = y_2 + V_2^2/2g = 1.15$ m. As is discussed in Section 10.3, the specific energy diagram for this flow can be obtained by using $V = q/y$, where

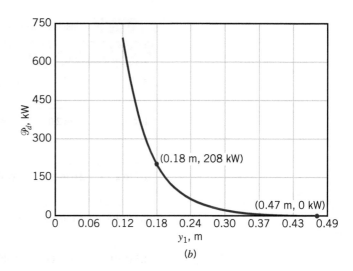

(b)

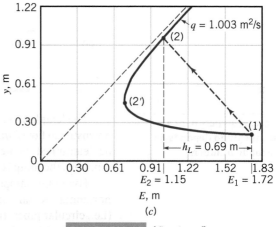

FIGURE E10.10 (Continued)

$$q = q_1 = q_2 = \frac{Q}{b} = y_1 V_1 = 0.183\text{ m }(5.5\text{ m/s})$$
$$= 1.003\text{ m}^2/\text{s}$$

Thus,

$$E = y + \frac{q^2}{2gy^2} = y + \frac{(1.003\text{ m}^2/\text{s})^2}{2(9.81\text{ m/s}^2)y^2} = y + \frac{0.051}{y^2}$$

where y and E are in feet. The resulting specific energy diagram is shown in **Fig. E10.10c**. Because of the head loss across the jump, the upstream and downstream values of E are different. In going from state (1) to state (2), the fluid does not proceed along the specific energy curve and pass through the critical condition at state 2′. Rather, it jumps from (1) to (2), as is represented by the dashed line in the figure. From a one-dimensional consideration, the jump is a discontinuity. In actuality, the jump is a complex three-dimensional flow incapable of being represented as a discontinuous change on the one-dimensional specific energy diagram.

The actual structure of a hydraulic jump is a complex function of Fr_1, even though the depth ratio and head loss are given quite accurately by a simple one-dimensional flow analysis (Eqs. 10.36 and 10.37). A detailed investigation of the flow indicates that there are essentially five types of surface and jump conditions. The classification of these jumps is presented in **Table 10.2**, along with sketches of the structure of the jump. For flows that are barely supercritical, the jump is more like a standing wave, without a nearly step change in depth. In some Froude number ranges the jump is unsteady, with regular periodic oscillations traveling downstream. (Recall that the wave cannot travel upstream against the supercritical flow.)

TABLE 10.2 Classification of Hydraulic Jumps (Ref. 11)

Fr_1	y_2/y_1	Classification	Sketch
<1	1	Jump impossible	
1 to 1.7	1 to 2.0	Standing wave or undulant jump	
1.7 to 2.5	2.0 to 3.1	Weak jump	
2.5 to 4.5	3.1 to 5.9	Oscillating jump	
4.5 to 9.0	5.9 to 12	Stable, well-balanced steady jump; insensitive to downstream conditions	
>9.0	>12	Rough, somewhat intermittent strong jump	

The actual structure of a hydraulic jump depends on the Froude number.

V10.12 Hydraulic jump in a sink

The length of a hydraulic jump (the distance between the nearly uniform upstream and downstream flows) may be of importance in the design of channels. Although its value cannot be determined theoretically, experimental results indicate that over a wide range of Froude numbers, the jump is approximately seven downstream depths long (Ref. 5).

Hydraulic jumps can occur in a variety of channel flow configurations, not just in horizontal, rectangular channels as discussed above. Jumps in nonrectangular channels (i.e., circular pipes, trapezoidal canals) behave in a manner quite like those in rectangular channels, although the details of the depth ratio and head loss are somewhat different from jumps in rectangular channels.

Other common types of hydraulic jumps include those that occur in sloping channels, as is indicated in **Fig. 10.19a** and the submerged hydraulic jumps that can occur just downstream of a sluice gate, as is indicated in **Fig. 10.19b**. Details of these and other jumps can be found in standard open-channel flow references (Refs. 3 and 5).

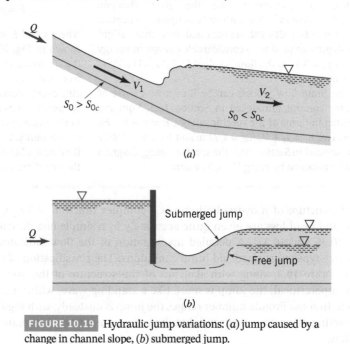

FIGURE 10.19 Hydraulic jump variations: (a) jump caused by a change in channel slope, (b) submerged jump.

10.7.2 Sharp-Crested Weirs

A weir is an obstruction on a channel bottom over which the fluid must flow. It provides a convenient method of determining the flowrate in an open channel by making of a single depth measurement. A **sharp-crested weir** is essentially a vertical sharp-edged flat plate placed across the channel in a way such that the fluid must flow across the sharp edge drop into the pool downstream of the weir plate, as is shown in **Fig. 10.20**. The specific shape of the flow area in the plane of the weir plate is used to designate the type of weir. Typical shapes include the rectangular weir, the triangular weir, and the trapezoidal weir, as depicted in **Fig. 10.21**.

The complex nature of the flow over a weir makes it impossible to obtain precise analytical expressions for the flow as a function of other parameters, such as the weir height, P_w, **weir head**, H, the fluid depth upstream, and the geometry of the weir plate (angle θ for triangular weirs or aspect ratio, b/H, for rectangular weirs). The flow structure is far from one-dimensional, exhibiting a variety of interesting flow phenomena.

Gravitational force and inertia are the dominant influences on flow over a weir. From a highly simplified point of view, gravity accelerates the fluid from its free-surface elevation upstream of the weir to a larger velocity as it flows down the hill formed by the nappe. Although viscous and surface tension effects are usually of secondary importance, such effects cannot be entirely neglected (Ref. 11). Generally, experimentally determined coefficients are used to account for these effects.

As a first approximation, we assume that the velocity profile upstream of the weir plate is uniform and that the pressure within the nappe is atmospheric. In addition, we assume that the fluid flows horizontally over the weir plate with a nonuniform velocity profile, as shown in **Fig. 10.22**. With $p_B = 0$ the Bernoulli equation for flow along the arbitrary streamline A–B can be written as

$$\frac{p_A}{\gamma} + \frac{V_1^2}{2g} + z_A = (H + P_w - h) + \frac{u_2^2}{2g} \tag{10.38}$$

where h is the distance that point B is below the free surface. We do not know the location of point A from which came the fluid that passes over the weir at point B. However, since the total head for any particle along the vertical section (1) is the same, $z_A + p_A/\gamma + V_1^2/2g = H + P_w + V_1^2/2g$,

A sharp-crested weir can be used to determine the flowrate.

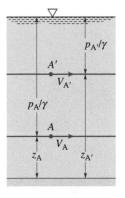

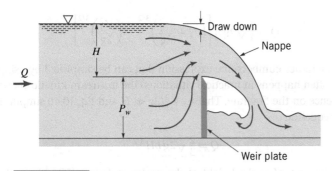

FIGURE 10.20 Sharp-crested weir geometry.

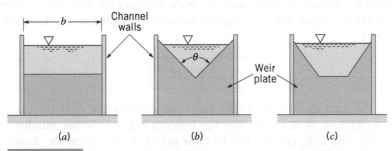

FIGURE 10.21 Sharp-crested weir plate geometry: (a) rectangular, (b) triangular, (c) trapezoidal.

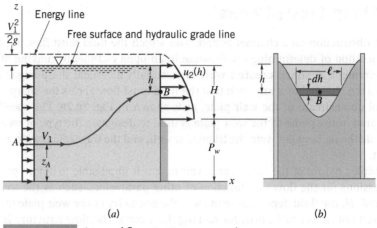

FIGURE 10.22   Assumed flow structure over a weir.

the specific location of A (i.e., A or A' shown in the adjacent figure) is not needed, and the velocity of the fluid over the weir plate is obtained from Eq. 10.38 as

$$u_2 = \sqrt{2g\left(h + \frac{V_1^2}{2g}\right)}$$

The flowrate can be calculated from

$$Q = \int_{(2)} u_2\, dA = \int_{h=0}^{h=H} u_2\ell\, dh \tag{10.39}$$

where $\ell(h)$ is the cross-channel width of a strip of the weir area, as is depicted in **Fig. 10.22b**. For a rectangular weir, ℓ is constant. For other weirs, such as triangular or circular weirs, the value of ℓ is known as a function of h.

For a rectangular weir, $\ell = b$, and the flowrate becomes

$$Q = \sqrt{2g}\, b \int_0^H \left(h + \frac{V_1^2}{2g}\right)^{1/2} dh$$

or

$$Q = \frac{2}{3}\sqrt{2g}\, b \left[\left(H + \frac{V_1^2}{2g}\right)^{3/2} - \left(\frac{V_1^2}{2g}\right)^{3/2}\right] \tag{10.40}$$

Equation 10.40 is a rather cumbersome expression that can be simplified by using the fact that with $P_w \gg H$ (as often happens in practical situations) the upstream kinetic energy has a negligibly small influence on the flowrate. That is, $V_1^2/2g \ll H$ and Eq. 10.40 simplifies to the basic rectangular weir equation

$$Q = \frac{2}{3}\sqrt{2g}\, bH^{3/2} \tag{10.41}$$

A weir coefficient is used to account for nonideal conditions excluded in the simplified analysis.

Note that the weir head, H, is the height of the upstream free surface above the crest of the weir. As is shown in Fig. 10.20, because of the drawdown effect, H is not the distance of the free surface above the weir crest as measured directly above the weir plate.

Because of the numerous approximations made to obtain Eq. 10.41, an experimentally determined correction factor is used to improve the accuracy in predictions of flowrate. Thus, the final form is

$$\boxed{Q = C_{wr} \frac{2}{3}\sqrt{2g}\, bH^{3/2}} \tag{10.42}$$

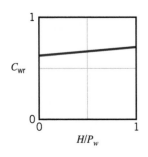

where C_{wr} is the rectangular weir coefficient. From dimensional analysis it is expected that C_{wr} is a function of Reynolds number (viscous effects), Weber number (surface tension effects), and H/P_w (geometry). In most practical situations, the Reynolds and Weber number effects are negligible, and the following correlation, shown in the adjacent figure, can be used (Refs. 4, 7):

$$C_{wr} = 0.611 + 0.075\left(\frac{H}{P_w}\right) \tag{10.43}$$

More precise values of C_{wr} can be found in the literature, if needed (Refs. 3 and 14).

The triangular sharp-crested weir is often used for flow measurements, particularly for measuring flowrates over a wide range of values. For small flowrates, the head, H, for a rectangular weir would be very small and the flowrate could not be measured accurately. However, with the triangular weir, the flow width decreases as H decreases so that even for small flowrates, reasonable heads are developed. Accurate results can be obtained over a wide range of Q.

The triangular weir equation can be obtained from Eq. 10.39 by using

$$\ell = 2(H - h)\tan\left(\frac{\theta}{2}\right)$$

where θ is the angle of the V-notch (see Figs. 10.29 and 10.30). After carrying out the integration and again neglecting the upstream kinetic energy ($V_1^2/2g \ll H$), we obtain

$$Q = \frac{8}{15}\tan\left(\frac{\theta}{2}\right)\sqrt{2g}\,H^{5/2}$$

An experimentally determined triangular weir coefficient, C_{wt}, is used to account for the effects neglected in the analysis so that

$$Q = C_{wt}\frac{8}{15}\tan\left(\frac{\theta}{2}\right)\sqrt{2g}\,H^{5/2} \tag{10.44}$$

Typical values of C_{wt} for triangular weirs are in the range of 0.58 to 0.62, as is shown in **Fig. 10.23**. Note that although C_{wt} and θ are dimensionless, the value of C_{wt} is a function of the weir head, H, which is a dimensional quantity. Although using dimensional parameters is not recommended (see the dimensional analysis discussion in Chapter 7), tradition dictates that such parameters are often used for open-channel flow.

The above results for sharp-crested weirs are valid provided the area under the nappe is ventilated to atmospheric pressure. Although this is not a problem for triangular weirs, for rectangular weirs it is sometimes necessary to provide ventilation tubes to ensure atmospheric pressure in this region. In addition, depending on downstream conditions, it is possible to obtain the submerged weir operation, as shown in **Fig. 10.24**. Clearly the flowrate will be different for these situations from that given by Eqs. 10.40 and 10.42.

Video

V10.13 Triangular weir

Video

V10.14 Low-head dam

Flowrate over a weir depends on whether the nappe is free or submerged.

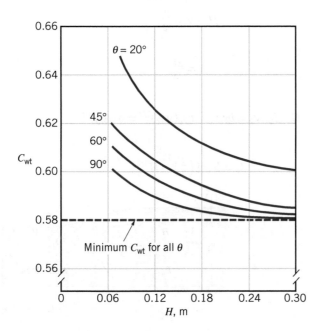

FIGURE 10.23 Weir coefficient for triangular sharp-crested weirs (Ref. 9).

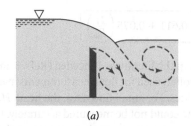
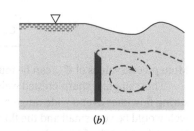

(a)　　　(b)

FIGURE 10.24 Flow conditions over a weir without a free nappe: (*a*) plunging nappe, (*b*) submerged nappe.

10.7.3 Broad-Crested Weirs

A **broad-crested weir** is a structure in an open channel that has a horizontal crest above which the pressure in the fluid approximates the hydrostatic pressure distribution. A typical configuration is shown in **Fig. 10.25**. Generally, to ensure proper operation, these weirs are restricted to the range $0.08 < H/L_w < 0.50$. These conditions are drawn to scale in the adjacent figure. For long weir blocks (H/L_w less than 0.08), head losses across the weir cannot be neglected. On the other hand, for short weir blocks (H/L_w greater than 0.50), the streamlines of the flow over the weir block are not horizontal. Although broad-crested weirs can be used in channels of any cross-sectional shape, we restrict our attention to rectangular channels.

$H/L_w = 0.08$

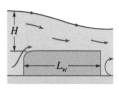

$H/L_w = 0.50$

The operation of a broad-crested weir is based on the fact that nearly uniform critical flow is achieved in the short reach above the weir block. (If $H/L_w < 0.08$, viscous effects are important, and the flow is subcritical over the weir.) If the kinetic energy of the upstream flow is negligible, then $V_1^2/2g \ll y_1$ and the upstream specific energy is $E_1 = V_1^2/2g + y_1 \approx y_1$. Observations show that as the flow passes over the weir block, it accelerates and reaches critical conditions, $y_2 = y_c$ and $\mathrm{Fr}_2 = 1$ (i.e., $V_2 = c_2$), corresponding to the nose of the specific energy curve (see Fig. 10.7). The flow does not accelerate to supercritical conditions. To do so would require the ability of the downstream fluid to communicate with the upstream fluid to let it know that there is an end of the weir block. Since waves cannot propagate upstream against a critical flow, this information cannot be transmitted. The flow remains critical, not supercritical, across the weir block.

The Bernoulli equation can be applied between point (1) upstream of the weir and point (2) over the weir where the flow is critical to obtain

$$H + P_w + \frac{V_1^2}{2g} = y_c + P_w + \frac{V_c^2}{2g}$$

or, if the upstream velocity head is negligible,

$$H - y_c = \frac{(V_c^2 - V_1^2)}{2g} = \frac{V_c^2}{2g}$$

However, since $V_2 = V_c = (gy_c)^{1/2}$, we find that $V_c^2 = gy_c$ so that we obtain

$$H - y_c = \frac{y_c}{2}$$

The broad-crested weir is governed by critical flow across the weir block.

---

or

$$y_c = \frac{2H}{3}$$

Thus, the flowrate is

$$Q = by_2V_2 = by_cV_c = by_c(gy_c)^{1/2} = b\sqrt{g}\,y_c^{3/2}$$

or

$$Q = b\sqrt{g}\left(\frac{2}{3}\right)^{3/2} H^{3/2}$$

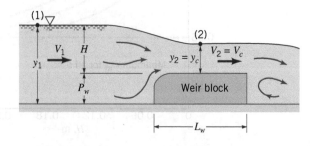

FIGURE 10.25 Broad-crested weir geometry.

Again an empirical weir coefficient is used to account for the various real-world effects not included in the above simplified analysis. That is,

$$Q = C_{wb} b \sqrt{g} \left(\frac{2}{3}\right)^{3/2} H^{3/2} \tag{10.45}$$

where approximate values of C_{wb}, the broad-crested weir coefficient shown in the adjacent figure, can be obtained from the equation (Ref. 6)

$$C_{wb} = 1.125 \left(\frac{1 + H/P_w}{2 + H/P_w}\right)^{1/2} \tag{10.46}$$

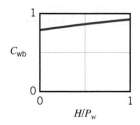

EXAMPLE 10.11 | Sharp-Crested and Broad-Crested Weirs

Given Water flows in a rectangular channel of width $b = 2$ m with flowrates between $Q_{min} = 0.02$ m³/s and $Q_{max} = 0.60$ m³/s. This flowrate is to be measured by using either (a) a rectangular sharp-crested weir, (b) a triangular sharp-crested weir with $\theta = 90°$, or (c) a broad-crested weir. In all cases the bottom of the flow area over the weir is a distance $P_w = 1$ m above the channel bottom.

Find Plot a graph of $Q = Q(H)$ for each weir and comment on which weir would be best for this application.

Solution

a. For the rectangular weir with $P_w = 1$ m, Eqs. 10.42 and 10.43 give

$$Q = C_{wr}\frac{2}{3}\sqrt{2g}bH^{3/2}$$

$$= \left(0.611 + 0.075\frac{H}{P_w}\right)\frac{2}{3}\sqrt{2g}\,bH^{3/2}$$

Thus,

$$Q = (0.611 + 0.075H)\frac{2}{3}\sqrt{2(9.81 \text{ m/s}^2)}(2 \text{ m})H^{3/2}$$

or

$$Q = 5.91(0.611 + 0.075H)H^{3/2} \tag{1}$$

where H and Q are in meters and m³/s, respectively. The results from Eq. 1 are plotted in **Fig. E10.11**.

b. Similarly, for the triangular weir, Eq. 10.44 gives

$$Q = C_{wt}\frac{8}{15}\tan\left(\frac{\theta}{2}\right)\sqrt{2g}\,H^{5/2}$$

$$= C_{wt}\frac{8}{15}\tan(45°)\sqrt{2(9.81 \text{ m/s}^2)}\,H^{5/2}$$

or

$$Q = 2.36\,C_{wt}H^{5/2} \tag{2}$$

where H and Q are in meters and m³/s and C_{wt} is obtained from Fig. 10.23. For example, with $H = 0.20$ m $= 0.656$ ft, we find $C_{wt} = 0.584$, or $Q = 2.36\,(0.584)(0.20)^{5/2} = 0.0247$ m³/s. The triangular weir results are also plotted in Fig. E10.9.

c. For the broad-crested weir, Eqs. 10.40 and 10.41 give

$$Q = C_{wb}b\sqrt{g}\left(\frac{2}{3}\right)^{3/2}H^{3/2}$$

$$= 1.125\left(\frac{1 + H/P_w}{2 + H/P_w}\right)^{1/2}b\sqrt{g}\left(\frac{2}{3}\right)^{3/2}H^{3/2}$$

Thus, with $P_w = 1$ m

$$Q = 1.125\left(\frac{1 + H}{2 + H}\right)^{1/2}(2 \text{ m})\sqrt{9.81 \text{ m/s}^2}\left(\frac{2}{3}\right)^{3/2}H^{3/2}$$

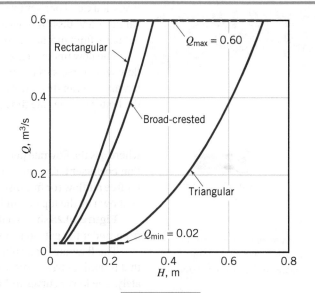

FIGURE E10.11

or

$$Q = 3.84\left(\frac{1 + H}{2 + H}\right)^{1/2}H^{3/2} \tag{3}$$

where, again, H and Q are in meters and m³/s. This result is also plotted in Fig. E10.9.

Comments Although it appears as though any of the three weirs would work well for the upper portion of the flowrate range, neither the rectangular nor the broad-crested weir would be very accurate for small flowrates near $Q = Q_{min}$ because of the small head, H, at these conditions. The triangular weir, however, would allow reasonably large values of H at the lowest flowrates. The corresponding heads with $Q = Q_{min} = 0.02$ m³/s for rectangular, triangular, and broad-crested weirs are 0.0312, 0.182, and 0.0375 m, respectively.

In addition, as discussed in this section, for proper operation the broad-crested weir geometry is restricted to $0.08 < H/L_w < 0.50$, where L_w is the weir block length. From Eq. 3 with $Q_{max} = 0.60$ m³/s, we obtain $H_{max} = 0.349$. Thus, we must have $L_w > H_{max}/0.5 = 0.698$ m to maintain proper critical flow conditions at the largest flowrate in the channel. However, with $Q = Q_{min} = 0.02$ m³/s, we obtain $H_{min} = 0.0375$ m. Thus, we must have $L_w < H_{min}/0.08 = 0.469$ m to ensure that frictional effects are not important. Clearly, these two constraints on the geometry of the weir block, L_w, are incompatible.

A broad-crested weir will not function properly under the wide range of flowrates considered in this example. The sharp-crested triangular weir would be the best of the three types considered, provided the channel can handle the $H_{max} = 0.719$-m head.

FIGURE 10.26 Three variations of underflow gates: (*a*) vertical gate, (*b*) radial gate, (*c*) drum gate.

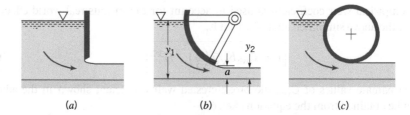

(*a*)　　　　(*b*)　　　　(*c*)

10.7.4 Underflow (Sluice) Gates

A variety of **underflow gate** (or **sluice gate**) structures is available for flowrate control at the crest of an overflow spillway (as shown in the adjacent photograph), or at the entrance of an irrigation canal or river from a lake. Three types are illustrated in **Fig. 10.26**. Each has certain advantages and disadvantages in terms of costs of construction, ease of use, and the like, although the basic fluid mechanics involved is the same in all instances.

The flow under a gate is said to be free outflow when the fluid issues as a jet of supercritical flow with a free surface open to the atmosphere as shown in Fig. 10.26. In such cases it is customary to write this flowrate as the product of the distance, a, between the channel bottom and the bottom of the gate times the convenient reference velocity $(2gy_1)^{1/2}$. That is,

$$q = C_d a \sqrt{2gy_1} \tag{10.47}$$

where q is the flowrate per unit width. The discharge coefficient, C_d, is a function of the contraction coefficient, $C_c = y_2/a$, and the depth ratio y_1/a. Typical values of the discharge coefficient for free outflow (or free discharge) from a vertical sluice gate are typically between 0.55 and 0.60 as shown by the top line in **Fig. 10.27** (Ref. 3).

Figure 10.28 shows a flow in which downstream conditions cause the jet of water issuing from under the gate to be overlaid by a mass of water that is quite turbulent.

The flowrate for a submerged (or drowned) gate can be obtained from the same equation that is used for free outflow (Eq. 10.47), provided the discharge coefficient is modified appropriately. The lower curves in Fig. 10.27 show typical values of C_d for a drowned outflow. Consider flow for a given gate and upstream conditions (i.e., given y_1/a) corresponding to a vertical line in the figure. With $y_3/a = y_1/a$ (i.e., $y_3 = y_1$) there is no head to drive the flow so that $C_d = 0$ and the fluid is stationary. For a given upstream depth (y_1/a fixed), the value of C_d increases with decreasing y_3/a until the maximum value of C_d is reached. This maximum corresponds to the free discharge conditions and is represented by the free outflow line so labeled in Fig. 10.27. For values of y_3/a that give C_d values between zero and its maximum, the jet from the gate is overlaid (drowned) by the downstream water and the flowrate is therefore reduced when compared with a free discharge situation. Similar results are obtained for the radial gate and drum gate.

Video

V10.15 Spillway gate

The flowrate from an underflow gate depends on whether the outlet is free or drowned.

Video

V10.16 Unsteady under and over

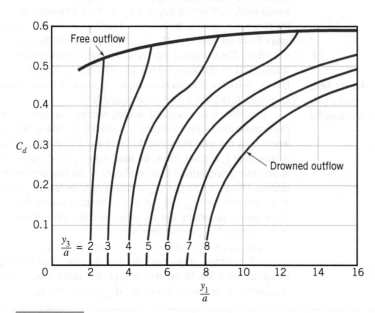

FIGURE 10.27 Typical discharge coefficients for underflow gates (Ref. 3).

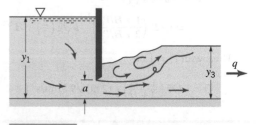

FIGURE 10.28 Drowned outflow from a sluice gate.

EXAMPLE 10.12 | Sluice Gate

Given Water flows under the sluice gate shown in **Fig. E10.12**. The channel width is $b = 6.1$ m, the upstream depth is $y_1 = 1.82$ m, and the gate is $a = 0.30$ m off the channel bottom.

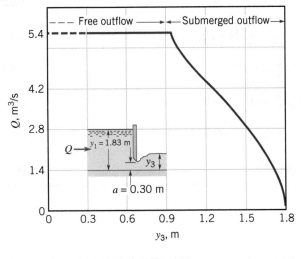

FIGURE E10.12

Find   Plot a graph of flowrate, Q, as a function of y_3.

Solution   From Eq. 10.47 we have

$$Q = bq = baC_d \sqrt{2gy_1}$$
$$= 6.1 \text{ m} (0.3 \text{ m})C_d \sqrt{2(9.81 \text{ m/s}^2)(1.82)}$$

or

$$Q = 11.12C_d \text{ m}^3/\text{s} \qquad (1)$$

The value of C_d is obtained from Fig. 10.27 along the vertical line $y_1/a = 6$. For $y_3 = 1.82$ m (i.e., $y_3/a = 6 = y_1/a$) we obtain $C_d = 0$, indicating that there is no flow when there is no head difference across the gate. The value of C_d increases as y_3/a decreases, reaching a maximum of $C_d = 0.56$ when $y_3/a = 3.2$. Thus, with $y_3 = 3.2a = 0.98$ m

$$Q = 11.12 \times 0.56 \text{ m}^3/\text{s} = 6.22 \text{ m}^3/\text{s}$$

The flowrate for $0.98 \text{ m} \le y_3 \le 1.82 \text{ m}$ is obtained from Eq. 1 and the C_d values of Fig. 10.26 with the results as indicated in Fig. E10.10.

Comment   For $y_3 < 0.98$ m the flowrate is independent of y_3, and the outflow is a free (not submerged) outflow. For such cases the inertia of the water flowing under the gate is sufficient to produce free outflow even with $y_3 > a$.

CHAPTER SUMMARY

This chapter discussed various aspects of flows in an open channel. A typical open-channel flow is driven by the component of gravity in the direction of flow. The character of such flows can be a strong function of the Froude number, which is the ratio of the fluid speed to the free-surface wave speed. The specific energy diagram is used to provide insight into the flow processes involved in open-channel flow.

Uniform flow is achieved by a balance between the potential energy lost by the fluid as it coasts downhill and the energy dissipated by viscous effects. Alternately, it represents a balance between weight and friction forces. The Manning equation models the dependence of the flowrate on the channel slope, the channel cross-sectional geometry, and the roughness of the channel's surfaces. Values of the empirical Manning coefficient used in the Manning equation are dependent on the surface material roughness. The Manning equation is not dimensionally homogeneous.

For **gradually varied flow**, the fluid depth changes slowly along the length of the channel. The fluid depth is described by a first order ordinary differential equation, which can be solved numerically.

The hydraulic jump is an example of nonuniform depth open-channel flow. If the Froude number of a flow is greater than 1, the flow is supercritical, and a hydraulic jump may occur. The momentum and continuity equations are used to obtain the relationship between the upstream Froude number and the depth ratio across the jump. The energy dissipated in the jump and the head loss can then be determined from the energy equation.

The use of weirs to measure the flowrate in an open channel is discussed. The relationships between the flowrate and the weir head are developed for both sharp-crested and broad-crested weirs.

The following checklist provides a study guide for this chapter. When your study of the entire chapter has been completed, you should be able to

- write out meanings of the terms listed here and understand each of the related concepts. These terms are particularly important and are set in **bold black** type in the text.

- determine the Froude number for a given flow and explain the concepts of subcritical, critical, and supercritical flows.

- plot and interpret the specific energy diagram for a given flow.

- use the Manning equation to analyze uniform flow in an open channel.

- calculate properties such as the depth ratio and the head loss for a hydraulic jump.

- determine the flowrates over sharp-crested weirs, broad-crested weirs, and under underflow gates.

KEY EQUATIONS

Froude number	$\text{Fr} = V/(gy)^{1/2}$	
Wave speed	$c = \sqrt{gy}$	(10.3)

Specific energy	$E = y + \dfrac{V^2}{2g}$	(10.8)
Manning equation	$V = \dfrac{\kappa}{n} R_h^{2/3} S_0^{1/2}$	(10.16)
Hydraulic jump depth ratio	$\dfrac{y_2}{y_1} = \dfrac{1}{2}\left(-1 + \sqrt{1 + 8\,\mathrm{Fr}_1^2}\right)$	(10.36)
Hydraulic jump head loss	$\dfrac{h_L}{y_1} = 1 - \dfrac{y_2}{y_1} + \dfrac{\mathrm{Fr}_1^2}{2}\left[1 - \left(\dfrac{y_1}{y_2}\right)^2\right]$	(10.37)
Rectangular sharp-crested weir	$Q = C_{wr}\dfrac{2}{3}\sqrt{2g}\,bH^{3/2}$	(10.42)
Triangular sharp-crested weir	$Q = C_{wt}\dfrac{8}{15}\tan\left(\dfrac{\theta}{2}\right)\sqrt{2g}\,H^{5/2}$	(10.44)
Broad-crested weir	$Q = C_{wb}\,b\sqrt{g}\left(\dfrac{2}{3}\right)^{3/2}H^{3/2}$	(10.45)
Underflow (sluice) gate	$q = C_d\,a\sqrt{2gy_1}$	(10.47)

REFERENCES

[1] Currie, C. G., and Currie, I. G., *Fundamental Mechanics of Fluids*, 3rd Ed., Marcel Dekker, New York, 2003.

[2] Stoker, J. J., *Water Waves*, Interscience, New York, 1957.

[3] Henderson, F. M., *Open Channel Flow*, Macmillan, New York, 1966.

[4] Rouse, H., *Elementary Fluid Mechanics*, Wiley, New York, 1946.

[5] French, R. H., *Open Channel Hydraulics*, McGraw-Hill, New York, 1992.

[6] Chow, V. T., *Open Channel Hydraulics*, McGraw-Hill, New York, 1959.

[7] Blevins, R. D., *Applied Fluid Dynamics Handbook*, Van Nostrand Reinhold, New York, 1984.

[8] Daugherty, R. L., and Franzini, J. B., *Fluid Mechanics with Engineering Applications*, McGraw-Hill, New York, 1977.

[9] Camp, T. R., "Design of Sewers to Facilitate Flow," *Sewage Works Journal*, 18(3), 1946.

[10] Lenz, A. T., "Viscosity and Surface Tension Effects on V-Notch Weir Coefficients," *Transactions of the American Society of Civil Engineers*, Vol. 108, 759–802, 1943.

[11] White, F. M., *Fluid Mechanics*, 5th Ed., McGraw-Hill, New York, 2003.

[12] U.S. Bureau of Reclamation, Research Studies on Stilling Basins, Energy Dissipators, and Associated Appurtenances, Hydraulic Lab Report Hyd.-399, June 1, 1955.

[13] Wallet, A., and Ruellan, F., *Houille Blanche*, Vol. 5, 1950.

[14] Spitzer, D. W., ed., *Flow Measurement: Practical Guides for Measurement and Control*, Instrument Society of America, Research Triangle Park, NC, 1991.

QUESTIONS AND PROBLEMS

Ⓒ Problem to be solved with aid of programmable calculator or computer.

Ⓞ Open-ended problem that requires critical thinking. These problems require various assumptions to provide the necessary input data. There are not unique answers to these problems.

Note: Unless specific values of required fluid properties are specified in the problem statement, use the values found in Tables 1.4–1.8 and in the tables in the Appendices.

Section 10.2 Surface Waves

10.2.1 On a distant planet small-amplitude waves travel across a 1-m-deep pond with a speed of 5 m/s. Determine the acceleration of gravity on the surface of that planet.

10.2.2 Determine the critical depth for a flow of 200 m³/s through a rectangular channel of 10-m width. If the water flows 3.8-m deep, is the flow supercritical? Explain.

10.2.3 Determine the minimum depth in a 3-m-wide rectangular channel if the flow is to be subcritical with a flowrate of $Q = 60$ m³/s.

10.2.4 Do shallow waves propagate at the same speed in all fluids? Explain why or why not.

10.2.5 Waves on the surface of a tank are observed to travel at a speed of 2 m/s. How fast would these waves travel if (a) the tank were in an elevator accelerating downward at a rate of 4 m/s², (b) the tank accelerates horizontally at a rate of 9.81 m/s², (c) the tank were aboard the orbiting Space Shuttle? Explain.

10.2.6 In flowing from section (1) to section (2) along an open channel, the water depth decreases by a factor of 2 and the Froude number changes from a subcritical value of 0.5 to a supercritical value of 3.0. Determine the channel width at (2) if it is 3.65 m wide at (1).

10.2.7 The flowrate in a 15 m-wide, 0.61 m-deep river is $Q = 5.3$ m³/s. Is the flow subcritical or supercritical?

10.2.8 A trout jumps, producing waves on the surface of a 0.8-m-deep mountain stream. If it is observed that the waves do not travel upstream, what is the minimum velocity of the current?

10.2.9 Observations at a shallow sandy beach show that even though the waves several hundred yards out from the shore are not parallel to the beach, the waves often "break" on the beach nearly parallel to the shore as indicated in **Fig. P10.2.9**. Explain this behavior based on the wave speed $c = (gy)^{1/2}$.

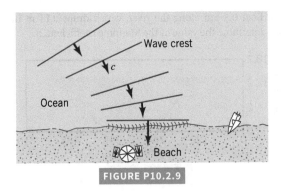

FIGURE P10.2.9

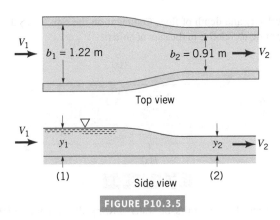

FIGURE P10.3.5

10.2.10 (See Fluids in the News article titled "Tsunami, the Nonstorm Wave," Section 10.2.1.) An earthquake causes a shift in the ocean floor that produces a tsunami with a wavelength of 100 km. How fast will this wave travel across the ocean surface if the ocean depth is 3000 m?

Section 10.3 Energy Considerations

10.3.1 Water flows in a 10-m-wide open channel with a flowrate of 5 m³/s. Determine the two possible depths if the specific energy of the flow is $E = 0.6$ m.

10.3.2 Water flows in a rectangular channel with a flowrate per unit width of $q = 2.5$ m²/s. Plot the specific energy diagram for this flow. Determine the two possible depths of flow if $E = 2.5$ m.

10.3.3 Water flows in a 1.52 m-wide rectangular channel with a flowrate of $Q = 0.85$ m³/s and an upstream depth of $y_1 = 0.76$ m as is shown in **Fig. P10.3.3**. Determine the flow depth and the surface elevation at section (2).

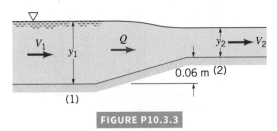

FIGURE P10.3.3

10.3.4 Water flows over the bump in the bottom of the rectangular channel shown in **Fig. P10.3.4** with a flowrate per unit width of $q = 4$ m²/s. The channel bottom contour is given by = 0.2, where and x are in meters. The water depth far upstream of the bump is = 0.4 m. Plot a graph of the water depth, $y = y(x)$, and the surface elevation, $z = z(x)$, for −4 m × 4 m. Assume one-dimensional flow.

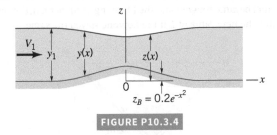

FIGURE P10.3.4

10.3.5 Water in a rectangular channel flows into a gradual contraction section as is indicated in **Fig. P10.3.5**. If the flowrate is $Q = 0.71$ m³/s and the upstream depth is $y_1 = 0.61$ m, determine the downstream depth, y_2.

10.3.6 A channel has a rectangular cross section, a width of 40 m, and a flowrate of 4000 m³/s. The normal water depth is 20 m. The flow then encounters a 4.0-m-high dam. Find the water depth directly above the dam if the flow is critical. Assume frictionless flow.

10.3.7 A rectangular channel has a gradual contraction in width from 18 m to 9 m and a bed level drop of 0.15 m below the upstream channel bed, which the increased velocity caused by the contraction scoured. If the upstream and downstream flow depths are both 3.1 m, compute the discharge in the channel.

10.3.8 Water flows in a rectangular channel with a flowrate per unit width of $Q = 1.5$ m²/s and a depth of 0.5 m at section (1). The head loss between sections (1) and (2) is 0.03 m. Plot the specific energy diagram for this flow and locate states (1) and (2) on this diagram. Is it possible to have a head loss of 0.06 m? Explain.

10.3.9 Water flows in a horizontal, rectangular channel with an initial depth of 1 m and an initial velocity of 4 m/s. Determine the depth downstream if losses are negligible. Note that there may be more than one solution.

10.3.10 A smooth transition section connects two rectangular channels as shown in **Fig. P10.3.10**. The channel width increases from 1.83 to 2.13 m, and the water surface elevation is the same in each channel. If the upstream depth of flow is 0.91 m, determine h, the amount the channel bed needs to be raised across the transition section to maintain the same surface elevation.

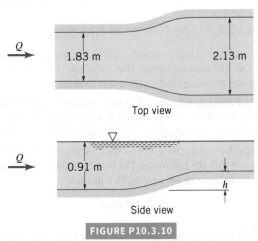

FIGURE P10.3.10

10.3.11 Water flows over a bump of height $h = h(x)$ on the bottom of a wide rectangular channel as is indicated in **Fig. P10.3.11**. If energy losses are negligible, show that the slope of the water surface is given by $dy/dx = -(dh/dx)/[1 - (V^2/gy)]$, where $V = V(x)$ and $y = y(x)$ are the local velocity and depth of flow. Comment on the sign (i.e., <0, =0, or >0) of dy/dx relative to the sign of dh/dx.

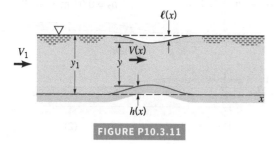

FIGURE P10.3.11

10.3.12 Consider 2.83 m³/s of water flowing down a rectangular channel measuring 3 m wide. The normal depth is 0.91 m. A 1.22-m-diameter pier is located in the channel. Find the water depth as it flows past the pier. Assume frictionless flow.

10.3.13 Water flows in the river shown in **Fig. P10.3.13** with a uniform bottom slope. The total head at each section is measured by using Pitot tubes as indicated. Determine the value of z_4 if the flow is uniform depth.

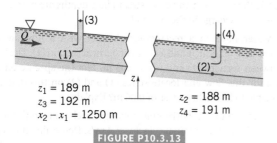

$z_1 = 189$ m
$z_3 = 192$ m
$x_2 - x_1 = 1250$ m

$z_2 = 188$ m
$z_4 = 191$ m

FIGURE P10.3.13

10.3.14 Supercritical, uniform flow of water occurs in a 5.0-m-wide, rectangular, horizontal channel. The flow has a depth of 1.5 m and a flowrate of 45.0 m³/s. The water flow encounters a 0.25-m rise in the channel bottom. Find the normal depth after the rise in the channel bottom. Is the flow after the rise subcritical, critical, or supercritical? Assume frictionless flow.

Section 10.4.2 The Chezy and Manning Equations

10.4.1 Water flows in a 5-m-wide channel with a speed of 2 m/s and a depth of 1 m. The channel bottom slopes at a rate of 1 m per 1000 m. Determine the Manning coefficient for this channel.

10.4.2 The following data are taken from measurements on Indian Fork Creek: $A = 26$ m², $P = 16$ m, and $S_0 = 0.02$ m/62 m. Determine the average shear stress on the wetted perimeter of this channel.

10.4.3 Consider laminar flow down a wide rectangular channel making an angle θ with the horizontal. The fluid has kinematic viscosity υ and the volume flowrate per unit width is given by

$$q = \frac{gy^3 \sin \theta}{3\upsilon}$$

where y is the fluid depth perpendicular to the channel bottom. Find Manning's n for the flow.

10.4.4 Given a trapezoidal channel, 3 m wide with side slope parameter, $m = 2$. The channel has a bottom slope $S_0 = 0.004$ m/m, and Manning's roughness coefficient, $n = 0.030$. Determine the normal depth, y_0, that produces a discharge of $Q_0 = 20$ m³/s.

10.4.5 At a particular location, the cross section of the Columbia River is as indicated in **Fig. P10.4.5**. If on a day without wind it takes

5 min to float 0.8 km along the river, which drops 0.14 m in that distance, determine the value of the Manning coefficient, n.

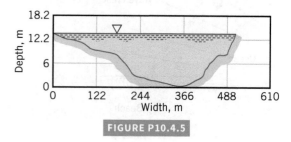

FIGURE P10.4.5

10.4.6 A uniform flow of 3115 m³/s is measured in a natural channel that is approximately rectangular in shape with a 808 m width and 5.34 m depth. The water-surface elevation drops 0.11 m per mile. Based on the computed Manning coefficient, n, characterize the type of natural channel observed. Also compute the Froude number and determine whether the flow is subcritical or supercritical.

Section 10.4.3 Uniform Flow—Determine Flowrate

10.4.7 A 2-m-diameter pipe made of finished concrete lies on a slope of 1-m elevation change per 1000-m horizontal distance. Determine the flowrate when the pipe is half full.

10.4.8 By what percent is the flowrate reduced in the rectangular channel shown in **Fig. P10.4.8** because of the addition of the thin center board? All surfaces are of the same material.

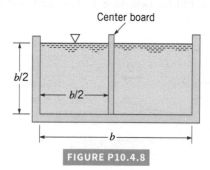

FIGURE P10.4.8

10.4.9 A large trapezoidal channel cut through stone has side slopes of 1:1 and a bed width of 89 m. Find the uniform flow in the channel when the flow depth is 9.2 m and the bed slopes 0.30 m.

10.4.10 When the channel of triangular cross section shown in **Fig. P10.4.10** was new, a flowrate of Q caused the water to reach $L = 2$ m up the side as indicated. After considerable use, the walls of the channel became rougher and the Manning coefficient, n, doubled. Determine the new value of L if the flowrate stayed the same.

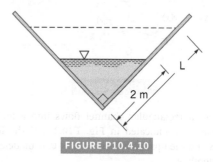

FIGURE P10.4.10

10.4.11 An unfinished concrete rectangular channel is 5 m wide and has a slope of 0.50°. The water is 0.5 m deep. Find the discharge rate for uniform flow.

10.4.12 A trapezoidal channel with a bottom width of 3.0 m and sides with a slope of 2:1 (horizontal:vertical) is lined with fine gravel ($n = 0.020$) and is to carry 10 m³/s. Can this channel be built with a slope of $S_0 = 0.00010$ if it is necessary to keep the velocity below 0.75 m/s to prevent scouring of the bottom? Explain.

10.4.13 Water flows in a 2-m-diameter finished concrete pipe so that it is completely full and the pressure is constant all along the pipe. If the slope is $S_0 = 0.005$, determine the flowrate by using open-channel flow methods. Compare this result with that obtained by using pipe flow methods of Chapter 8.

10.4.14 A round concrete storm sewer pipe used to carry rainfall run-off from a parking lot is designed to be half full when the rainfall rate is a steady 0.025 m/hr. Will this pipe be able to handle the flow from a 0.05 m/hr rainfall without water backing up into the parking lot? Support your answer with appropriate calculations.

10.4.15 Find the discharge per unit width for a wide channel having a bottom slope of 0.00015. The normal depth is 0.003 m. Assume laminar flow and justify the assumption. The fluid is 20 °C water.

10.4.16 Water flows down a wide rectangular channel having Manning's $n = 0.015$ and bottom slope = 0.0015. Find the rate of discharge and normal depth for critical flow conditions.

10.4.17 Determine the flow depth for the channel shown in **Fig. P10.4.17** if the flowrate is 15 m³/s.

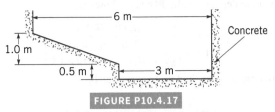

FIGURE P10.4.17

10.4.18 At a given location, under normal conditions a river flows with a Manning coefficient of 0.030, and a cross section as indicated in **Fig. P10.4.18a**. During flood conditions at this location, the river has a Manning coefficient of 0.040 (because of trees and brush in the floodplain) and a cross section as shown in **Fig. P10.4.18b**. Determine the ratio of the flowrate during flood conditions to that during normal conditions.

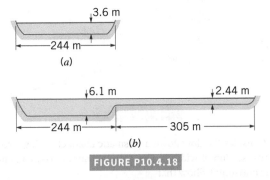

FIGURE P10.4.18

10.4.19 The channel in **Fig. P10.4.19** has two floodplains as shown. Find the discharge if the center channel is lined with brick and the two floodplains are lined with cobblestones. The slope S_0 is 0.00025.

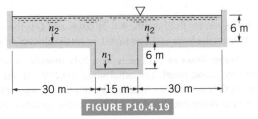

FIGURE P10.4.19

Section 10.4.3 Uniform Flow—Determine Depth or Size

10.4.20 An old, rough-surfaced, 2-m-diameter concrete pipe with a Manning coefficient of 0.025 carries water at a rate of 5.0 m³/s when it is half full. It is to be replaced by a new pipe with a Manning coefficient of 0.012 that is also to flow half full at the same flowrate. Determine the diameter of the new pipe.

10.4.21 Four sewer pipes of 0.5-m diameter join to form one pipe of diameter D. If the Manning coefficient, n, and the slope are the same for all of the pipes, and if each pipe flows half full, determine D.

10.4.22 The spillway of a dam is 6.1 m wide and has a flowrate of 142 m³/s. The spillway makes an angle of 30° with the horizontal. Find the vertical water depth down the spillway. See Problem 10.6.2.

10.4.23 The flowrate in the clay-lined channel ($n = 0.025$) shown in **Fig. P10.4.23** is to be 8.5 m³/s. To prevent erosion of the sides, the velocity must not exceed 1.52 m/s. For this maximum velocity, determine the width of the bottom, b, and the slope, S_0.

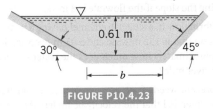

FIGURE P10.4.23

10.4.24 The rate of discharge through the canal in **Fig. P10.4.24** is to be 10 m³/s. Find the width b and the bottom slope. The velocity will not exceed 5.0 m/s.

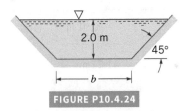

FIGURE P10.4.24

10.4.25 A rectangular, unfinished concrete channel of 8.5 m width is laid on a slope of 1.5 m/km. Determine the flow depth and Froude number of the flow if the flowrate is 11.3 m³/s.

10.4.26 An engineer is to design a channel lined with planed wood to carry water at a flowrate of 2 m³/s on a slope of 10 m/800 m. The channel cross section can be either a 90° triangle or a rectangle with a cross section twice as wide as its depth. Which would require less wood and by what percent?

10.4.27 Find the diameter required for reinforced concrete pipe laid at a slope of 0.001 and required to carry a uniform flow of 0.55 m³/s when the depth is 75% of the diameter.

10.4.28 A major river is divided into three parts or courses—the upper course, the middle course, and the lower course. The slope is 13 m/km in the upper course, 2 m/km in the middle course, and 0.2 m/km in the lower course. All three courses have rectangular cross sections. The upper course has a normal depth of 12 m. The river width is 37 m. Find the normal depths of the middle and lower courses.

10.4.29 An 2.44 m-diameter concrete drainage pipe that flows half full is to be replaced by a concrete-lined V-shaped open channel having an interior angle of 90°. Determine the depth of fluid that will exist in the V-shaped channel if it is laid on the same slope and carries the same discharge as the drainage pipe.

10.4.30 A rectangular brick-lined channel has a bottom slope of 0.0025 and is designed to carry a uniform water flowrate of 8.5 m³/s.

Would the channel need fewer bricks if the channel were 0.61 m wide, 1.83 m wide, or 3.05 m wide? Explain.

10.4.31 Water flows uniformly at a depth of 1 m in a channel that is 5 m wide as shown in **Fig. P10.4.31**. Further downstream, the channel cross section changes to that of a square of width and height b. Determine the value of b if the two portions of this channel are made of the same material and are constructed with the same bottom slope.

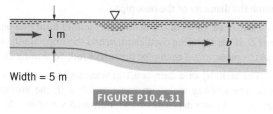

FIGURE P10.4.31

Section 10.4.3 Uniform Flow—Determine Slope

10.4.32 Water flows 1 m deep in a 2-m-wide finished concrete channel. Determine the slope if the flowrate is 3 m³/s.

10.4.33 Uniform flow in a sluggish channel having a nearly rectangular cross section that is 152 m wide and 5.03 m deep carries a flow of 234 m³/s. Approximately how much does the water surface elevation drop along a river mile?

10.4.34 To prevent weeds from growing in a clean earthen-lined canal, it is recommended that the velocity be no less than 0.76 m/s. For the symmetrical canal shown in **Fig. P10.4.34**, determine the minimum slope needed.

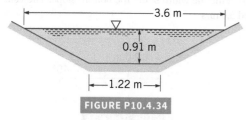

FIGURE P10.4.34

10.4.35 The symmetrical channel shown in **Fig. P10.4.34** is dug in sandy loam soil with $n = 0.020$. For such surface material it is recommended that to prevent scouring of the surface the average velocity be no more than 0.53 m/s. Determine the maximum slope allowed.

10.4.36 **Figure P10.4.36** shows a cross section of an aqueduct that carries water at 50 m³/s. The value of Manning's n is 0.015. Find the bottom slope.

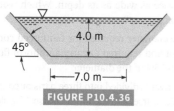

FIGURE P10.4.36

10.4.37 The depth downstream of a sluice gate in a rectangular wooden channel of width 5 m is 0.60 m. If the flowrate is 18 m³/s, determine the channel slope needed to maintain this depth. Will the depth increase or decrease in the flow direction if the slope is (a) 0.02; (b) 0.01?

10.4.38 A 15.2 m-long aluminum gutter (Manning coefficient $n = 0.011$) on a section of a roof is to handle a flowrate of 0.0042 m³/s during a heavy rainstorm. The cross section of the gutter is shown in **Fig. P10.4.38**. Determine the vertical distance that this gutter must be pitched (i.e., the difference in elevation between the two ends of the gutter) so that the water does not overflow the gutter. Assume uniform depth channel flow.

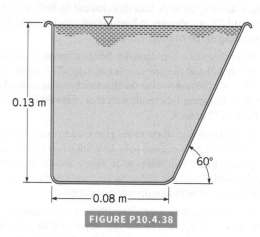

FIGURE P10.4.38

Section 10.6 Gradually Varied Flow

10.6.1 Consider the flow down a prismatic channel having a trapezoidal cross section of base width b and top width $b + 2y \cos \theta \cot \phi$. The channel bottom makes an angle θ with the horizontal, and y is the vertical fluid depth (see **Fig. P10.6.1**). Show that

$$\frac{dy}{dx} = \frac{\tan \theta - (n^2 Q^2 / A^2 R_h^{4/3} \kappa^2)}{1 - \left[\dfrac{Q^2}{A^2 g y \cos \theta} \right] \left[\dfrac{b + 2y \cos \theta \cot \phi}{b + y \cos \theta \cot \phi} \right]}$$

Note that the average fluid velocity is given by

$$V = \frac{Q}{A} = \frac{Q}{y \cos \theta (b + y \cos \theta \cot \phi)}$$

Discuss the form of the equation for small values of θ ($\theta \sim 1°$).

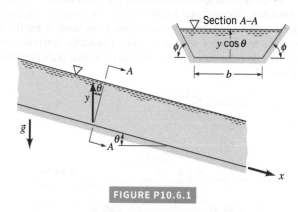

FIGURE P10.6.1

10.6.2 Consider the flow down a prismatic channel having a rectangular cross section of width b. The channel bottom makes an angle θ with the horizontal. Show that

$$\frac{dy}{dx} = \frac{\tan \theta - (n^2 Q^2) / (A^2 R_h^{4/3} \kappa^2)}{1 - Q^2 / (A^2 g y \cos^2 \theta)}$$

where y is the vertical fluid depth, $A = by \cos \theta$, and Q is the volume flowrate. Discuss the form of the equation for small values of θ ($\theta \sim 1°$).

10.6.3 Ⓒ Water flows at 0.42 m³/s in a 0.91 m-wide rectangular clean-earth irrigation canal. The canal slope is 0.275°. At one point, the water depth is 0.91 m. (a) Accurately compute the water depth at a location 61 m downstream. (b) Is the flow at the upstream location subcritical or supercritical? At the downstream location? (c) Sketch the canal and the water surface profile.

Section 10.7.1 The Hydraulic Jump

10.7.1 Water flows upstream of a hydraulic jump with a depth of 0.5 m and a velocity of 6 m/s. Determine the depth of the water downstream of the jump.

10.7.2 A 5.0-m-wide channel has a slope of 0.004, a 8.0-m^3/s water flowrate, and a water depth 1.5 m after a hydraulic jump. Find the water depth before the jump.

10.7.3 The water depths upstream and downstream of a hydraulic jump are 0.3 and 1.2 m, respectively. Determine the upstream velocity and the power dissipated if the channel is 50 m wide.

10.7.4 A hydraulic jump at the base of a spillway of a dam is such that the depths upstream and downstream of the jump are 0.90 and 3.6 m, respectively. If the spillway is 10 m wide, what is the flowrate over the spillway?

10.7.5 A rectangular channel 3.0 m wide has a flowrate of 5.0 m^3/s with a normal depth of 0.50 m. The flow then encounters a dam that rises 0.25 m above the channel bottom. Will a hydraulic jump occur? Justify your answer.

10.7.6 A rectangular flume is 0.5 m wide and water flows at a rate of 1.0 m^3/s and a depth of 0.50 m. Find the depth after a hydraulic jump and the power loss in the jump.

10.7.7 Water flows in a rectangular channel with velocity $V = 6$ m/s. A gate at the end of the channel is suddenly closed so that a wave (a moving hydraulic jump) travels upstream with velocity $V_w = 2$ m/s. Determine the depths ahead of and behind the wave. Note that this is an unsteady problem for a stationary observer. However, for an observer moving to the left with velocity V_w, the flow appears as a steady hydraulic jump.

10.7.8 A hydraulic jump occurs in a 4 m-wide rectangular channel at a point where the slope changes from 3 m per 100 m upstream of the jump to h m per 100 m downstream of the jump. The depth and velocity of the uniform flow upstream of the jump are 0.5 m/s and 8 m/s, respectively. Determine the value of h if the flow downstream of the jump is to be uniform flow.

10.7.9 (See The Wide World of Fluids article titled "Grand Canyon Rapids Building." Section 10.7.1.) During the flood of 1983, a large hydraulic jump formed at "Crystal Rapid" on the Colorado River. People rafting the river at that time report "entering the rapid at almost 14 m/s, hitting a 6.1 m-tall wall of water, and exiting at about 4.5 m/s." Is this information (i.e., upstream and downstream velocities and change in depth) consistent with the principles of a hydraulic jump? Show calculations to support your answer.

Section 10.7.2 Sharp-Crested Weirs

10.7.10 A rectangular sharp-crested weir is used to measure the flowrate in a 3.05 m-wide channel. It is desired to have the upstream channel flow depth be 1.83 m when the flowrate is 1.42 m^3/s. Determine the height, P_w, of the weir plate.

10.7.11 Water flows over a 2-m-wide rectangular sharp-crested weir. Determine the flowrate if the weir head is 0.1 m and the upstream channel depth is 1 m.

10.7.12 Water flows over the sharp-crested weir shown in **Fig. P10.7.12**. The weir plate cross section consists of a semicircle and

a rectangle. Plot a graph of the estimated flowrate, Q, as a function of head, H. List all assumptions and show all calculations.

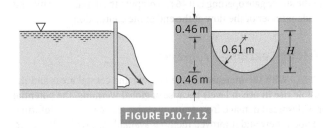

FIGURE P10.7.12

10.7.13 Water flows over a broad-crested weir that has a width of 4 m and a height of $P_w = 1.5$ m. The free-surface well upstream of the weir is at a height of 0.5 m above the surface of the weir. Determine the flowrate in the channel and the minimum depth of the water above the weir block.

10.7.14 An engineering laboratory experiment uses a triangular weir in an open channel to measure flowrate. The nominal weir angle is 90°. In a certain test, the head of water above the weir was 0.1 m. The uncertainty in the weir angle is ±2° and the uncertainty in the depth measurement is ±0.0013 m. Compute the flowrate and the uncertainty in the flowrate.

Section 10.7.4 Underflow (Sluice) Gates

10.7.15 Water flows under a sluice gate in a 18 m-wide finished concrete channel as is shown in **Fig. P10.7.15**. Determine the flowrate. If the slope of the channel is 0.762 m/61 m, will the water depth increase or decrease downstream of the gate? Assume $C_c = y_2/a = 0.65$. Explain.

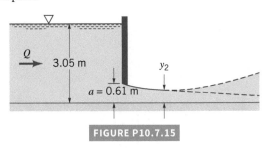

FIGURE P10.7.15

10.7.16 A water-level regulator (not shown) maintains a depth of 2.0 m downstream from a 10-m-wide drum gate as shown in **Fig. P10.7.16**. Plot a graph of flowrate, Q, as a function of water depth upstream of the gate, y_1, for $2.0 \le y_1 \le 5.0$ m.

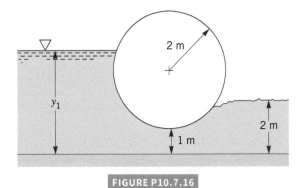

FIGURE P10.7.16

10.7.17 A flow of 24.7 m^3/s passes under a sluice gate in a rectangular channel having a gradual contraction in width from 24.3 m to 15.8 m. The channel bed has scoured to a level that is 0.152 m, below the upstream channel bed, owing to the increased velocity caused by the sluice gate and the contraction. If the upstream flow depth is 1.22 m and the sluice gate opening is 0.46 m, compute the depth, velocity, and Froude number of the flow at the end of the contraction.

Lifelong Learning Problems

10.1 LL With the increased usage of low-lying coastal areas and the possible rise in ocean levels because of global warming, the potential for widespread damage from tsunamis is increasing. Obtain information about new and improved methods available to predict the occurrence of these damaging waves and how to better use coastal areas so that massive loss of life and property does not occur. Summarize your findings in a brief report.

10.2 LL Recent photographs from NASA's Mars Orbiter Camera on the Mars Global Surveyor provide new evidence that water may still flow on the surface of Mars. Obtain information about the possibility of current or past open-channel flows on Mars and other planets or their satellites. Summarize your findings in a brief report.

10.3 LL Hydraulic jumps are normally associated with water flowing in rivers, gullies, and other such relatively high-speed open channels. However, recently, hydraulic jumps have been used in various manufacturing processes involving fluids other than water (such as liquid metal solder) in relatively small-scale flows. Obtain information about new manufacturing processes that involve hydraulic jumps as an integral part of the process. Summarize your findings in a brief report.

CHAPTER 11

Compressible Flow

LEARNING OBJECTIVES

After completing this chapter, you should be able to:

- explain speed of sound and Mach number and their practical significance.

- distinguish between incompressible and compressible flows and determine when the approximations associated with assuming fluid incompressibility are acceptable.

- describe important features of different categories of compressible flow.

- discuss various responses of compressible flow to area change in a duct.

- discuss and recognize choking in compressible flow.

- solve problems involving isentropic and nonisentropic flows including flows with area change, flows with normal shocks, flows with friction, and flows with heating or cooling.

- appreciate the similarities between compressible flow of gases and open-channel flow of liquids.

- move on to advanced study of compressible flow.

Most first courses in fluid mechanics concentrate on constant density (incompressible) flows. In earlier chapters of this book, we mainly considered incompressible flow behavior. The model of an incompressible fluid is convenient because when constant density and constant (including zero) viscosity are assumed, problem solutions are greatly simplified. Also, fluid incompressibility allows us to build on the Bernoulli equation as was done, for example, in Chapter 5. Examples in Chapters 5–10 should have convinced you that nearly incompressible flows are common in everyday experiences.

Any study of fluid mechanics would, however, be incomplete without a brief introduction to compressible flow. Fluid compressibility is very important in numerous engineering applications of fluid mechanics. For example, the measurement of high-speed flow velocities requires compressible flow theory. The flows in gas turbine engine components are generally compressible. Many aircraft fly fast enough to encounter compressible flow.

The variation of fluid density for compressible flows requires a relationship between density and other fluid properties, especially pressure and temperature. This relationship, the fluid equation of state, often unimportant for incompressible flows, is vital in the analysis of compressible flows. Also, temperature variations for compressible flows are usually significant, and thus the energy equation is necessary. Curious phenomena can occur with compressible flows. For example, we can have fluid acceleration because of friction, fluid deceleration in a converging duct, fluid temperature decrease with heating, and the formation of abrupt discontinuities in flows across which fluid properties change appreciably.

For simplicity, in this introductory study of compressibility effects we mainly consider the steady, one-dimensional, constant (including zero) viscosity, compressible flow of an ideal gas. We limit our study to compressibility due to high-speed flow. In this chapter, one-dimensional flow refers to flow involving uniform distributions of fluid properties over any flow cross-sectional area. Both frictionless ($\mu = 0$) and frictional ($\mu \neq 0$) compressible flows are considered. If the change in density associated with a change of pressure is considered a measure of compressibility, our experience suggests that gases and vapors are much more compressible than liquids. We focus our attention on the compressible flow of a gas because such flows occur often. We limit our discussion to ideal gases, since the equation of state for an ideal gas is uncomplicated, yet representative of actual gases at pressures and temperatures of engineering interest, and because the flow trends associated with an ideal gas are generally applicable to other compressible fluids.

Another advantage of emphasizing ideal gases is that the equations used for solving many problems are closed-form algebraic equations. A disadvantage is that there are many equations and they are almost all "messy" (nonlinear, and complicated in appearance). In practice, calculations are often made with the aid of tables of numerical values or by reading values from graphs.

Because of the large number of equations and because compressible flow analysis has its own special "jargon," it is easy for students to get lost. The most important objective is to understand the physics of compressible flow rather than how to manipulate the myriad equations, charts, and tables. Toward this end, we will explore the unique phenomena of each type of compressible flow qualitatively, using the minimum number of basic equations before developing and using the computational equations, charts, and tables.

An excellent film about compressible flow is available (see Ref. 1). You really should watch this film, ideally both *before* and *after* studying this chapter.

> **We consider mostly ideal gas flows.**

11.1   IDEAL GAS THERMODYNAMICS

V11.1 Lighter flame

Concepts from thermodynamics and concepts from fluid mechanics are equally important in compressible flow analysis. Before we begin the study of compressible flow, we should consider a few basic concepts from thermodynamics. If you have already studied thermodynamics, this section will serve as a review. In this section, we present all the concepts and equations needed to understand the compressible flow of an ideal gas but the developments and explanations are brief. Consult a thermodynamics textbook such as Ref. 2 for more detail.

Before we can develop compressible flow equations, we must learn how to evaluate ideal gas property changes. The equation of state for an ideal gas is

$$\rho = \frac{P}{RT} \tag{11.1}$$

We have already discussed fluid pressure, p, density, ρ, and temperature, T, in earlier chapters. The gas constant, R, is a constant for each distinct ideal gas or mixture of ideal gases, where

$$R = \frac{\lambda}{M_{\text{gas}}} \tag{11.2}$$

With this notation, λ is the universal gas constant and M_{gas} is the molecular weight of the ideal gas or gas mixture. Listed in Tables 1.7 and 1.8 are values of the gas constants of some common gases. Knowing any two of pressure, temperature, or density, we can calculate the third.

For an ideal gas, **internal energy**, $\check{u}$, is part of the stored energy of the gas as explained in Section 5.3 and is a function of temperature only (Ref. 2). Thus, the ideal gas specific heat at constant volume, c_v, can be expressed as

$$c_v = \left(\frac{\partial \check{u}}{\partial T} \right)_v = \frac{d\check{u}}{dT} \tag{11.3}$$

where the subscript v on the partial derivative refers to differentiation at constant specific volume, $v = 1/\rho$. For a particular ideal gas, c_v is a function of temperature only. Equation 11.3 can be rearranged to yield

$$d\check{u} = c_v\, dT$$

Thus,

$$\check{u}_2 - \check{u}_1 = \int_{T_1}^{T_2} c_v\, dT \tag{11.4}$$

Equation 11.4 allows us to evaluate the change in internal energy, $\check{u}_2 - \check{u}_1$, associated with ideal gas flow from section (1) to section (2). For simplicity, we can assume that c_v is constant for a particular ideal gas and obtain from Eq. 11.4

$$\boxed{\check{u}_2 - \check{u}_1 = c_v(T_2 - T_1)} \tag{11.5}$$

Actually, c_v for a particular gas varies with temperature (see Ref. 2). However, for moderate changes in temperature, the constant c_v assumption is reasonable.

The fluid property **enthalpy**, $\check{h}$, is defined as

$$\check{h} = \check{u} + \frac{p}{\rho} \tag{11.6}$$

It combines internal energy, $\check{u}$, and pressure energy (flow work), p/ρ, and is useful when working with the energy equation (Eq. 5.69). For an ideal gas, we have already stated that

$$\check{u} = \check{u}(T)$$

From the equation of state (Eq. 11.1)

$$\frac{p}{\rho} = RT$$

Thus, it follows that

$$\check{h} = \check{h}(T)$$

Since for an ideal gas, enthalpy is a function of temperature only, the ideal gas specific heat at constant pressure, c_p, can be expressed as

$$c_p = \left(\frac{\partial \check{h}}{\partial T}\right)_p = \frac{d\check{h}}{dT} \tag{11.7}$$

where the subscript p on the partial derivative refers to differentiation at constant pressure, and c_p is a function of temperature only. The rearrangement of Eq. 11.7 leads to

$$d\check{h} = c_p \, dT$$

and

$$\check{h}_2 - \check{h}_1 = \int_{T_1}^{T_2} c_p \, dT \tag{11.8}$$

Equation 11.8 allows us to evaluate the change in enthalpy, $\check{h}_2 - \check{h}_1$, associated with ideal gas flow from section (1) to section (2). For simplicity, we can assume that c_p is constant for a specific ideal gas and obtain from Eq. 11.8

$$\boxed{\check{h}_2 - \check{h}_1 = c_p(T_2 - T_1)} \tag{11.9}$$

For moderate temperature changes, specific heat values can be considered constant.

As is true for c_v, the value of c_p for a given gas varies with temperature. Nevertheless, for moderate changes in temperature, the constant c_p assumption is reasonable.

From Eqs. 11.5 and 11.9 we see that changes in internal energy and enthalpy are related to changes in temperature. We turn our attention now to developing relationships for determining c_v and c_p. Combining Eqs. 11.6 and 11.1 we get

$$\check{h} = \check{u} + RT \tag{11.10}$$

Differentiating Eq. 11.10 with respect to temperature leads to

$$\frac{d\check{h}}{dT} = \frac{d\check{u}}{dT} + R \tag{11.11}$$

From Eqs. 11.3, 11.7, and 11.11 we conclude that

$$c_p - c_v = R \tag{11.12}$$

Equation 11.12 indicates that the difference between c_p and c_v is constant for each ideal gas regardless of temperature. Also $c_p > c_v$. The **specific heat ratio**, k, is defined as

$$k = \frac{c_p}{c_v} \tag{11.13}$$

so combining Eqs. 11.12 and 11.13 leads to

$$c_p = \frac{kR}{k-1} \tag{11.14}$$

The four gas constants are interrelated.

and

$$c_v = \frac{R}{k-1} \tag{11.15}$$

Actually, c_p, c_v, and k are all somewhat temperature dependent for any ideal gas. We will assume constant values for these variables in this book. Values of k and R for some commonly used gases at nominal temperatures are listed in Tables 1.7 and 1.8. These tabulated values can be used with Eqs. 11.14 and 11.15 to determine the values of c_p and c_v. Example 11.1 demonstrates how internal energy and enthalpy changes can be calculated for a flowing ideal gas having constant c_p and c_v.

Density, pressure, temperature, internal energy, and enthalpy are all important properties that describe the state of an ideal gas but they do not form a complete set. One more property is necessary. This property is the **entropy** and is given the symbol s for entropy per unit mass. The units of s are J/kgK in SI, ft-lb/slug °R in BG and Btu/lbm °R in EE. One useful verbal definition of entropy is as follows: *Entropy is the property that does not change in any process which has no friction and no heat transfer.* Such a process is said to be *reversible* (no friction) and *adiabatic* (no heat transfer). A shorter label for such a process is **isentropic** (*iso* means "same"). It follows that entropy does change if a process has friction or heat transfer, or both.

It can be shown (Ref. 2) that a proper mathematical definition of entropy is

$$s_2 - s_1 = \int_1^2 \frac{\delta q_{\text{reversible}}}{T} \tag{11.16}$$

Changes in entropy are important because they are related to loss of available energy.

where "1" and "2" are the endpoints of a process, $q_{\text{reversible}}$ is the heat transfer per unit mass that would occur in a reversible (frictionless) process between "1" and "2", and T is the absolute temperature during the process. Note that entropy is defined only in terms of its *change* in a process. The friction loss of available energy causes the entropy change in a flow process to exceed the actual heat transfer (δq) divided by the temperature

$$ds \geq \frac{\delta q}{T} \tag{11.17}$$

Because entropy is a property, it can be related to other fluid properties. For any pure substance, including ideal gases, the "first $T\,ds$ equation" is (see Ref. 2)

$$T\,ds = d\breve{u} + p\,d\left(\frac{1}{\rho}\right) \tag{11.18}$$

where T is absolute temperature, s is entropy, $\breve{u}$ is internal energy, p is absolute pressure, and ρ is density. Differentiating Eq. 11.6 leads to

$$d\breve{h} = d\breve{u} + p\,d\left(\frac{1}{\rho}\right) + \left(\frac{1}{\rho}\right)dp \tag{11.19}$$

By combining Eqs. 11.18 and 11.19, we obtain

$$T \, ds = d\check{h} - \left(\frac{1}{\rho}\right) dp \tag{11.20}$$

Equation 11.20 is often referred to as the "second $T \, ds$ equation." For an ideal gas, Eqs. 11.1, 11.3, and 11.18 can be combined to yield

$$ds = c_v \frac{dT}{T} + \frac{R}{1/\rho} d\left(\frac{1}{\rho}\right) \tag{11.21}$$

and Eqs. 11.1, 11.7, and 11.20 can be combined to yield

$$ds = c_p \frac{dT}{T} - R \frac{dp}{p} \tag{11.22}$$

> **Changes in entropy can be related to changes in temperature, pressure, and density.**

If c_p and c_v are assumed to be constant for a given gas, Eqs. 11.21 and 11.22 can be integrated to get

$$\boxed{s_2 - s_1 = c_v \ln \frac{T_2}{T_1} + R \ln \frac{\rho_1}{\rho_2}} \tag{11.23}$$

and

$$\boxed{s_2 - s_1 = c_p \ln \frac{T_2}{T_1} - R \ln \frac{p_2}{p_1}} \tag{11.24}$$

Equations 11.23 and 11.24 allow us to calculate the change of entropy of an ideal gas flowing from one section to another with constant specific heat values (c_p and c_v).

EXAMPLE 11.1 | Internal Energy, Enthalpy, Density, and Entropy for an Ideal Gas

Given Air flows steadily between two sections in a long, straight portion of 0.10 m-diameter pipe as is indicated in **Fig. E11.1**. The uniformly distributed temperature and pressure at each section are $T_1 = 300$ K, $p_1 = 689500$ N/m², and $T_2 = 252$ K, $p_2 = 126868$ N/m².

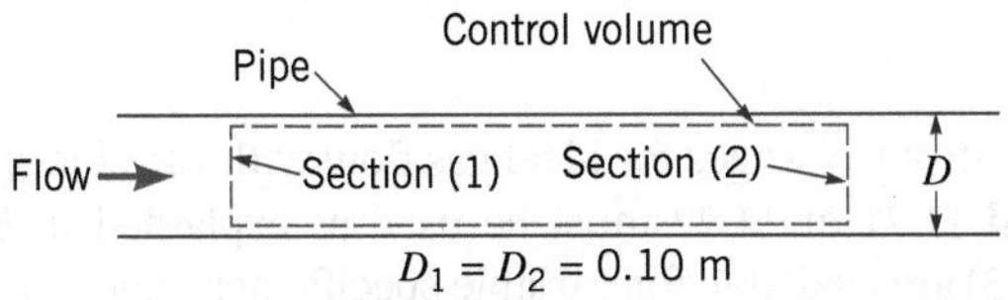

FIGURE E11.1

Find Calculate the **(a)** change in internal energy between sections (1) and (2), **(b)** change in enthalpy between sections (1) and (2), **(c)** change in density between sections (1) and (2), and **(d)** change in entropy between sections (1) and (2).

Solution

a. Assuming air behaves as an ideal gas, we can use Eq. 11.5 to evaluate the change in internal energy between sections (1) and (2). Thus

$$\check{u}_2 - \check{u}_1 = c_v (T_2 - T_1) \tag{1}$$

From Eq. 11.15 we have

$$c_v = \frac{R}{k - 1} \tag{2}$$

and from Table 1.7, $R = 287$ J/kg · K and $k = 1.4$. Throughout this book, we use the nominal values of k for common gases listed in Tables 1.7 and consider these values as being representative.

From Eq. 2 we obtain

$$c_v = \frac{287}{(1.4 - 1)} \text{ J/kg} \cdot \text{K}$$
$$= 718 \text{ J/kg} \cdot \text{K} \tag{3}$$

Combining Eqs. 1 and 3 yields

$$\check{u}_2 - \check{u}_1 = c_v (T_2 - T_1) = 718 \text{ J/kg} \cdot \text{K}$$
$$\times (252 \text{ K} - 300 \text{ K})$$
$$= -34464 \text{ J/kg} \cdot \text{K} \tag{Ans}$$

b. For enthalpy change we use Eq. 11.9. Thus

$$\check{h}_2 - \check{h}_1 = c_p (T_2 - T_1) \tag{4}$$

where since $k = c_p/c_v$ we obtain

$$c_p = kc_v = (1.4)[718 \text{ J/kg} \cdot \text{K}]$$
$$= 1005.2 \text{ J/kg} \cdot \text{K} \tag{5}$$

From Eqs. 4 and 5 we obtain

$$\check{h}_2 - \check{h}_1 = c_p (T_2 - T_1)$$
$$= 1005.2 \text{ J/kg} \cdot \text{K} \times (252 \text{ K} - 300 \text{ K})$$
$$= -48249.6 \text{ J/kg} \cdot \text{K} \tag{Ans}$$

c. For density change we use the ideal gas equation of state (Eq. 11.1) to get

$$\rho_2 - \rho_1 = \frac{p_2}{RT_2} - \frac{p_1}{RT_1} = \frac{1}{R}\left(\frac{p_2}{T_2} - \frac{p_1}{T_1}\right) \tag{6}$$

Using the pressures and temperatures given in the problem statement we calculate from Eq. 6

$$\rho_2 - \rho_1 = \frac{1}{287\,\text{J/kg}\cdot\text{K}}$$

$$\times \left[\frac{126868\ \text{N/m}^2}{252\ \text{K}} - \frac{689500\ \text{N/m}^2}{300\ \text{K}}\right]$$

or

$$\rho_2 - \rho_1 = -6.25\ \text{N/m}^3 \qquad (Ans)$$

d. We can calculate the entropy change between sections by using either Eq. 11.23 or Eq. 11.24. We use both to demonstrate that the same result is obtained either way.
From Eq. 11.23,

$$s_2 - s_1 = c_v \ln \frac{T_2}{T_1} + R \ln \frac{\rho_1}{\rho_2} \tag{7}$$

To evaluate $s_2 - s_1$ from Eq. 7 we need the density ratio, ρ_1/ρ_2, which can be obtained from the ideal gas equation of state (Eq. 11.1) as

$$\frac{\rho_1}{\rho_2} = \left(\frac{p_1}{T_1}\right)\left(\frac{T_2}{p_2}\right) \tag{8}$$

and thus from Eqs. 7 and 8,

$$s_2 - s_1 = c_v \ln \frac{T_2}{T_1} + R \ln\left[\left(\frac{p_1}{T_1}\right)\left(\frac{T_2}{p_2}\right)\right] \tag{9}$$

By substituting values already identified in the problem statement into Eq. 9 with

$$\left(\frac{p_1}{T_1}\right)\left(\frac{T_2}{p_2}\right) = \left(\frac{689500}{300\ \text{K}}\right)\left(\frac{252\ \text{K}}{126868\ \text{N/m}^2}\right) = 4.56$$

we get

$$s_2 - s_1 = [718\ \text{J/kg}\cdot\text{K}] \ln\left(\frac{252\ \text{K}}{300\ \text{K}}\right)$$

$$+ [287\ \text{J/kg}\cdot\text{K}] \ln 4.56$$

or

$$s_2 - s_1 = 310.2\ \text{N}\cdot\text{m/kg}\cdot\text{K} \qquad (Ans)$$

From Eq. 11.24,

$$s_2 - s_1 = c_p \ln \frac{T_2}{T_1} - R \ln \frac{p_2}{p_1} \tag{10}$$

By substituting known values into Eq. 10 we obtain

$$s_2 - s_1 = [1005.2\ \text{J/kg}\cdot\text{K}] \ln\left(\frac{252\ \text{K}}{300\ \text{K}}\right)$$

$$- [287\ \text{J/kg}\cdot\text{K}] \ln\left(\frac{126868}{689500}\right)$$

or

$$s_2 - s_1 = 310.2\ \text{N}\cdot\text{m/kg}\cdot\text{K} \qquad (Ans)$$

Comment   There is a significant change in density when compared with the upstream density

$$\rho_1 = \frac{p_1}{RT_1} = \frac{126868\ \text{N/m}^2}{(287\ \text{J/kg}\cdot\text{K})(300\ \text{K})}$$

$$= 1.47\ \text{kg/m}^3$$

As anticipated, both Eqs. 11.23 and 11.24 yield the same result for the entropy change, $s_2 - s_1$ The entropy change is positive although the change of all other properties is negative.
Note that, both the pressures and temperatures used must be absolute.

If internal energy, enthalpy, and entropy changes for ideal gas flow with variable specific heats are desired, Eqs. 11.4, 11.8, and 11.21 or 11.22 must be used as explained in Ref. 2. Detailed tables (see, for example, Ref. 3) are available for variable specific heat calculations.

For the isentropic flow of an ideal gas with constant c_p and c_v, we get from Eqs. 11.23 and 11.24

$$c_v \ln \frac{T_2}{T_1} + R \ln \frac{\rho_1}{\rho_2} = c_p \ln \frac{T_2}{T_1} - R \ln \frac{p_2}{p_1} = 0 \tag{11.25}$$

By combining Eq. 11.25 with Eqs. 11.14 and 11.15 we obtain

$$\left(\frac{T_2}{T_1}\right)^{k/(k-1)} = \left(\frac{\rho_2}{\rho_1}\right)^k = \left(\frac{p_2}{p_1}\right) \tag{11.26}$$

a useful relationship between temperature, density, and pressure for an ideal gas undergoing an isentropic process. From Eq. 11.26 we can conclude that

$$\boxed{\frac{p}{\rho^k} = \text{constant}} \tag{11.27}$$

for an ideal gas with constant c_p and c_v flowing isentropically, a result already used without proof earlier in Chapters 1, 4, and 5.

The Wide World of Fluids

Hilsch tube (Ranque vortex tube)

Years ago (around 1930) a French physics student (George Ranque) discovered that appreciably warmer and colder portions of *rapidly swirling airflow* could be separated in a simple apparatus consisting of a tube open at both ends into which was introduced, somewhere in between the two openings, swirling air at *high pressure*. Warmer air near the outer portion of the swirling air flowed out one open end of the tube through a simple valve and colder air near the inner portion of the swirling air flowed out the opposite end of the tube. Rudolph Hilsch, a German physicist, improved on this discovery (ca. 1947). Hot air temperatures of 127 °C and cold air temperatures of −46 °C have been claimed in an optimized version of this apparatus. Thus far the inefficiency of the process has prevented it from being widely adopted for practical use.

11.2 STAGNATION PROPERTIES

The thermodynamic state of a fluid particle is defined by its properties ($p, \rho, T, \breve{u}, \breve{h}, s$); but we must also know the velocity of the particle and, possibly, its position in a gravitational field.

The thermodynamic properties are called *static properties*; they are the values that would be measured by instruments that are static *with respect to the fluid*. The static properties represent the molecular structure of the fluid and obey all equations of state and other property-related laws and thermodynamic equations. The particle's velocity and elevation are specified separately.

A useful approach is to combine the (static) thermodynamic properties with the velocity and elevation to obtain equivalent thermodynamic properties that represent the total (thermodynamic *and* mechanical) state. This is done by using **stagnation properties** which are *the properties that the fluid would obtain if it were brought to a condition of zero velocity and zero elevation in a frictionless process with no heat transfer and no work.*

We can obtain equations for stagnation properties with the aid of the adjacent figure which shows a (possibly imaginary) process that brings a fluid particle to rest and zero elevation. Applying the energy equation to the streamline between 1 and 0 gives

$$q + w_s = \breve{h}_0 - \breve{h}_1 + \frac{V_0^2}{2} - \frac{V_1^2}{2} + gz_0 - gz_1$$

where

$$q = \frac{\dot{Q}_{\text{net in}}}{\dot{m}} \quad \text{and} \quad w_s = \frac{\dot{W}_{\text{shaft net in}}}{\dot{m}}$$

By definition, q, w_s, V_0, and z_0 are all zero, so we get

$$\breve{h}_0 = \breve{h}_1 + \frac{V_1^2}{2} + gz_1$$

where $\breve{h}_0$ is called the *stagnation enthalpy*. In compressible flow, potential energy and gravity force are almost always negligible, and the definition of stagnation enthalpy is usually simplified to

$$\breve{h}_0 \equiv \breve{h} + \frac{V^2}{2} \tag{11.28}$$

The subscript 1 has been dropped to obtain a general definition of stagnation enthalpy. Note that we did not need to use the specification of frictionless flow to obtain the stagnation enthalpy.

If the fluid is an ideal gas with constant specific heat, Eq. 11.28 becomes

$$c_p T_0 = c_p T + \frac{V^2}{2}$$

and the *stagnation temperature* is given by

$$T_0 = T + \frac{V^2}{2c_p} \tag{11.29}$$

We can obtain equations for the *stagnation pressure* and *stagnation density* by noting that, by definition, the stagnation process is isentropic (frictionless with no heat transfer). For an ideal gas,

$$\frac{p_0}{p} = \left(\frac{T_0}{T}\right)^{k/(k-1)}$$

(11.30)

and

$$\frac{\rho_0}{\rho} = \left(\frac{T_0}{T}\right)^{1/(k-1)}$$

(11.31)

Dividing Eq. 11.29 by T, we obtain

$$\frac{T_0}{T} = 1 + \frac{V^2}{2c_p T}$$

(11.32)

Substituting into Eqs. 11.30 and 11.31, we have

$$p_0 = p\left(1 + \frac{V^2}{2c_p T}\right)^{k/(k-1)}$$

(11.33)

and

$$\rho_0 = \rho\left(1 + \frac{V^2}{2c_p T}\right)^{1/(k-1)}$$

(11.34)

Equations 11.28–11.34 show that the stagnation properties are determined by the static properties *and* the fluid velocity. Because any fluid particle always has a pressure, a temperature, and a velocity (including $V = 0$), any fluid particle can be described by both its static properties and its stagnation properties. We may say that "the fluid has a pressure of 100 kPa and a stagnation pressure of 125 kPa." The static properties actually are the properties of the fluid, and the stagnation properties are the properties that the fluid would have if it were brought to zero velocity in an isentropic process with no work. If (and only if) the fluid particle has zero velocity, its static and stagnation properties are equal. As we will discover in subsequent sections, stagnation properties are useful as reference even if no stagnation location exists in the flow.

Stagnation properties represent the maximum possible values of the fluid properties unless energy is added to the fluid by heat transfer or work. The postulated stagnation process represents the ideal conversion of the fluid's kinetic and potential energies into pressure and internal energy.

Stagnation properties are sometimes called *total properties*. Because the process of stagnation (that is, bringing the fluid to rest) can be accomplished by a process with friction as well as by a frictionless process, there is some possible ambiguity in using the term *stagnation* in cases where viscous flow processes are considered. This ambiguity can also be remedied by use of the term "isentropic stagnation properties."

11.3   MACH NUMBER AND SPEED OF SOUND

The **Mach number**, Ma, was introduced in Chapters 1 and 7 as a dimensionless measure of compressibility in a fluid flow. In this and subsequent sections, we develop some useful relationships involving the Mach number. The Mach number is defined as the ratio of the local flow velocity, V, to the local **speed of sound**, c.

Mach number is the ratio of local flow and sound speeds.

$$\text{Ma} = \frac{V}{c}$$

What we perceive as sound generally consists of weak pressure pulses that move through air with a Mach number of one. When our eardrums respond to a succession of these pulses, we hear sounds.

To derive an equation for the speed of sound, we analyze the one-dimensional fluid mechanics of an infinitesimally thin, weak pressure pulse moving through a fluid at rest (see **Fig. 11.1a**). Ahead of the pressure pulse, the fluid velocity is zero, and the fluid pressure and

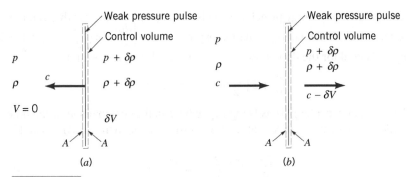

FIGURE 11.1 (a) Weak pressure pulse moving through a fluid at rest. (b) The flow relative to a control volume containing a weak pressure pulse.

density are p and ρ. Behind the pressure pulse, the fluid velocity has changed by an amount δV, and the pressure and density of the fluid have also changed by amounts δp and $\delta \rho$. We select an infinitesimally thin control volume that moves with the pressure pulse as is sketched in Fig. 11.1a. The speed of the weak pressure pulse is considered constant and in one direction only; thus, our control volume is inertial.

For an observer moving with this control volume (**Fig. 11.1b**), it appears as if fluid is entering the control volume through surface area A with speed c at pressure p and density ρ and leaving the control volume through surface area A with speed $c - \delta V$, pressure $p + \delta p$, and density $\rho + \delta \rho$. When the continuity equation (Eq. 5.16) is applied to the flow through this control volume, the result is

$$\rho A c = (\rho + \delta \rho) A (c - \delta V) \tag{11.35}$$

or

$$\rho c = \rho c - \rho\, \delta V + c\, \delta \rho - (\delta \rho)(\delta V) \tag{11.36}$$

> The changes in fluid properties across a sound wave are extremely small compared to their local values.

Since $(\delta \rho)(\delta V)$ is much smaller than the other terms in Eq. 11.36, we drop it from further consideration and keep

$$\rho\, \delta V = c\, \delta \rho \tag{11.37}$$

The linear momentum equation (Eq. 5.29) can also be applied to the flow through the control volume of Fig. 11.1b. The result is

$$-c\rho c A + (c - \delta V)(\rho + \delta \rho)(c - \delta V) A = pA - (p + \delta p) A \tag{11.38}$$

Frictional forces are negligibly small because the edges of the control volume have no area. We again neglect higher order terms such as $(\delta V)^2$ compared to $c\, \delta V$ and combine Eqs. 11.35 and 11.38 to get

$$-c\rho c A + (c - \delta V)\rho A c = -\delta p A$$

or

$$\rho\, \delta V = \frac{\delta p}{c} \tag{11.39}$$

From Eqs. 11.37 (continuity) and 11.39 (linear momentum) we obtain

$$c^2 = \frac{\delta p}{\delta \rho}$$

or

$$c = \sqrt{\frac{\delta p}{\delta \rho}} \tag{11.40}$$

This expression for the speed of sound results from application of the conservation of mass and linear momentum principles to the flow through the control volume of Fig. 11.1b. These principles were similarly used in Section 10.2.1 to obtain an expression for the speed of waves traveling on the surface of a liquid in a channel.

The conservation of energy principle can also be applied to the flow through the control volume of Fig. 11.1b. If the energy equation expressed as $-\left[\dfrac{dp}{\rho} + d\left(\dfrac{V^2}{2}\right) + g\,dz\right] = \delta(\text{loss})$ $=\left(T\,ds - \delta q_{\text{netin}}\right)$ is used for the flow through this control volume, the result is

$$\frac{\delta p}{\rho} + \delta\left(\frac{V^2}{2}\right) + g\,\delta z = \delta(\text{loss}) \tag{11.41}$$

For gas flow we can consider $g\,\delta z$ as being negligibly small in comparison to the other terms in the equation. Also, if we assume that the flow is frictionless, then $\delta(\text{loss}) = 0$ and Eq. 11.41 becomes

$$\frac{\delta p}{\rho} + \frac{(c - \delta V)^2}{2} - \frac{c^2}{2} = 0$$

or, neglecting $(\delta V)^2$ compared to $c\,\delta V$, we obtain

$$\rho\,\delta V = \frac{\delta p}{c} \tag{11.42}$$

By combining Eqs. 11.37 (continuity) and 11.42 (energy) we again find that

$$c = \sqrt{\frac{\delta p}{\delta \rho}}$$

which is identical to Eq. 11.40. Thus, the linear momentum and conservation of energy principles lead to the same result. If we further assume that the frictionless flow through the control volume of Fig. 11.1b is adiabatic (no heat transfer), then the flow is isentropic. In the limit, as δp and $\delta \rho$ become vanishingly small ($\delta p \to \delta \rho \to 0$)

$$c = \sqrt{\left(\frac{\partial p}{\partial \rho}\right)_s} \tag{11.43}$$

where the subscript s is used to designate that the partial differentiation occurs at constant entropy.

The bulk modulus of elasticity, E_v, of any fluid including liquids is defined as (see Section 1.7.1)

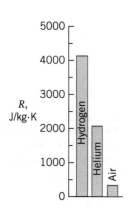

R, J/kg·K

$$E_v = \frac{dp}{d\rho/\rho} = \rho\left(\frac{\partial p}{\partial \rho}\right)_s \tag{11.44}$$

Thus, in general, from Eqs. 11.43 and 11.44,

$$c = \sqrt{\frac{E_v}{\rho}} \tag{11.45}$$

Speed of sound is larger in fluids that are more difficult to compress.

Equation 11.43 informs us that we can calculate the speed of sound by determining the partial derivative of pressure with respect to density at constant entropy. For the isentropic flow of an ideal gas (with constant c_p and c_v), we learned earlier (Eq. 11.27) that

$$p = (\text{constant})\rho^k$$

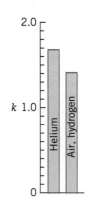

k

and thus

$$\left(\frac{\partial p}{\partial \rho}\right)_s = (\text{constant})\,k\rho^{k-1} = \frac{p}{\rho^k}k\rho^{k-1} = k\frac{p}{\rho} = kRT \tag{11.46}$$

Thus, for an ideal gas

$$c = \sqrt{kRT} \tag{11.47}$$

From Eq. 11.47 and the adjacent charts for R and k, we conclude that for a given temperature, the speed of sound, c, in hydrogen and in helium, is higher than in air.

Values of the speed of sound are tabulated in Tables B.1 and B.2 for water and in Tables B.3 and B.4 for air. From experience we know that air is more easily compressed than water. Note from the values of c in Tables B.1 through B.4 and the adjacent graph that the speed of sound in air is much less than it is in water. From Eq. 11.44, we can conclude that if a fluid is truly incompressible, its bulk modulus would be infinitely large, as would be the speed of sound in that fluid. Thus, an incompressible flow is always an approximation of reality.

If the Mach number is large, the effects of fluid compressibility are significant, but if it is small, compressibility effects are negligible. If we square the Mach number and assume an ideal gas, we get

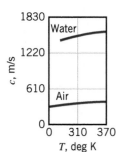

c, m/s

$$\text{Ma}^2 = \frac{V^2}{kRT} = \frac{\rho V^2}{kp} \tag{11.48}$$

The Wide World of Fluids

Sonification

The normal human ear is capable of detecting even very subtle sound patterns produced by *sound waves*. Most of us can distinguish the bark of a dog from the meow of a cat or the roar of a lion, or identify a person's voice on the telephone before they identify who is calling. The number of "things" we can identify from subtle sound patterns is enormous. Combine this ability with the power of computers to transform the information from sensor transducers into variations in pitch, rhythm, and volume and you have *sonification*, the representation of data in the form of sound. With this emerging technology, pathologists may soon learn to "hear" abnormalities in tissue samples, engineers may "hear" flaws in gas turbine engine blades being inspected, and scientists may "hear" a desired attribute in a newly invented material. Perhaps the concept of hearing the trends in data sets may become as commonplace as seeing them. Analysts may listen to the stock market and make decisions. Of course, none of this can happen in a vacuum.

The criteria for neglecting fluid compressibility suggested in Sections 1.7.2 and 4.8.1 are equivalent to specifying that the Mach number is small.

The importance of Mach number as a parameter in compressible flow can be illustrated by considering the ideal gas equations for stagnation properties. The stagnation-to-static-temperature ratio, Eq. 11.29, gives

$$\frac{T_0}{T} = 1 + \frac{V^2}{2c_p T}$$

Multiplying and dividing the second term on the right by kR and using Eqs. 11.12, 11.47, and 11.48, we obtain

$$\frac{V^2}{2c_p T} = \frac{k-1}{2}\left(\frac{V^2}{c^2}\right) = \left(\frac{k-1}{2}\right)\text{Ma}^2 \tag{11.49}$$

so

$$\boxed{\frac{T_0}{T} = 1 + \frac{k-1}{2}\text{Ma}^2} \tag{11.50}$$

Using Eqs. 11.33 and 11.34, we get

$$\boxed{\frac{p_0}{p} = \left(1 + \frac{k-1}{2}\text{Ma}^2\right)^{k/(k-1)}} \tag{11.51}$$

and

$$\boxed{\frac{\rho_0}{\rho} = \left(1 + \frac{k-1}{2}\text{Ma}^2\right)^{1/(k-1)}} \tag{11.52}$$

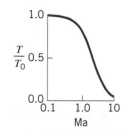

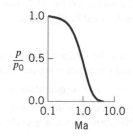

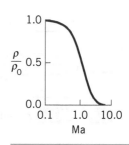

For isentropic flows the temperature, pressure, and density ratios are functions of the Mach number.

The adjacent figures show the variation of these ratios with Mach number for a gas with $k = 1.4$ (the value for air). Note that the ratios are graphed in the form (static property)/(stagnation property). The graphs show that pressure is most strongly affected by compressibility and least by temperature.

Because of the exponential form of Eqs. 11.51 and 11.52, calculations involving these equations are tedious. Traditionally, numerical values of the property ratios as functions of Mach number are evaluated using graphs or tables. Appendix D presents both graphs and brief tables for a gas with $k = 1.4$.

When using graphs or tables, an issue with notation arises. The expression $\frac{p}{p_0}$ might refer to the ratio of two distinct quantities, p and p_0, or to the function $\left(1 + \frac{k-1}{2}\text{Ma}^2\right)^{-\frac{k}{k-1}}$ or to a single numerical value from a graph, table (Appendix D), or software. In this book we shall resolve the dilemma as follows: When solving problems,

- $\frac{p}{p_0}$ will refer to either the concept "static-to-stagnation pressure ratio" or to the ratio of two distinct quantities $\left(\text{if } p = 6 \text{ and } p_0 = 10, \text{ then } \frac{p}{p_0} = \frac{6}{10}\right)$;

- $\frac{p}{p_0}$[Ma] will refer to the Mach number function $\left(\frac{p}{p_0}[\text{Ma}] = \left(1 + \frac{k-1}{2}\text{Ma}^2\right)^{-\frac{k}{k-1}}\right)$;

- $\left.\frac{p}{p_0}\right)_z$ will refer to a single number found from a graph, table, or software for argument "z", such as $\left.\frac{p}{p_0}\right)_{\text{Ma}=1} = 0.528$.

This notation will apply to other ratios such as $\frac{T}{T_0}$ and $\frac{A}{A^*}$. Note that T/T_0 and p/p_0 (numbers less than 1.0) are tabulated and graphed with Ma as a independent parameter in Appendix D. There is no column or graph for density ratio because

$$\frac{\rho}{\rho_0} = \left(\frac{p}{p_0}\right)\bigg/\left(\frac{T}{T_0}\right)$$

The remaining variable (A/A^*) will be explained later.

Even use of the tables can be cumbersome because a different set of tables must be prepared for each different value of k. Fortunately, Eqs. 11.50–11.52 can be programmed into a computer or calculator. Alternately, any number of compressible flow function calculators can be found on the Internet or as a smartphone app.

EXAMPLE 11.2 | Mach Number, Stagnation Properties, and Mach Number Tables

Given An airliner cruises at 250 m/s at an altitude of 10 km above sea level as depicted in **Fig. E11.2**.

Find

a. Calculate the airliner's Mach number and the stagnation pressure and temperature relative to the airliner.

b. Compute the same quantities for an airliner speed of 500 m/s.

c. Compute the same quantities for a speed of 250 m/s but change the altitude to 1 km above sea level.

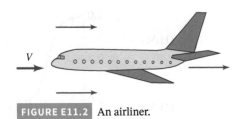

FIGURE E11.2 An airliner.

Solution

a. We assume that the atmospheric conditions are those of Standard Atmosphere. The atmospheric properties can be calculated from Eqs. 2.7–2.9 or read from Fig. 2.6 or Table C.1:

$$p_{10\text{ km}} = 26.50 \text{ kPa} \quad \text{and} \quad T_{10\text{ km}} = 223.3 \text{ K}$$
$$p_{1\text{ km}} = 89.88 \text{ kPa} \quad \text{and} \quad T_{1\text{ km}} = 281.7 \text{ K}$$

The Mach number is

$$\text{Ma} = \frac{V}{c}$$

From Eq. 11.47,

$$c = \sqrt{kRT} = 20.05\sqrt{T(\text{K})} \text{ m/s}$$

At 10 km,

$$c = 20.05\sqrt{223.3} \text{ m/s} = 299.6 \text{ m/s}$$

Thus when $V = 250$ m/s,

$$\text{Ma} = \frac{250 \text{ m/s}}{299.6 \text{ m/s}}$$
$$\text{Ma} = 0.834 \text{ at 10 km and 250 m/s} \qquad (Ans)$$

The airliner is cruising at subsonic speed.

From Table D.I, the pressure and temperature ratios at this Mach number are

$$\left.\frac{p}{p_0}\right)_{\text{Ma}=0.834} = 0.633 \quad \text{and} \quad \left.\frac{T}{T_0}\right)_{\text{Ma}=0.834} = 0.878$$

The static properties are those of the atmosphere. The stagnation properties relative to the airliner are

$$p_0 = \frac{p_{10\text{ km}}}{p/p_0)} \quad \text{and} \quad T_0 = \frac{T_{10\text{ km}}}{T/T_0)}$$

Substituting, we have

$$p_0 = \frac{26.50 \text{ kPa}}{0.633}$$
$$p_0 = 41.86 \text{ kPa at 10 km and 250 m/s} \qquad (Ans)$$

Then

$$T_0 = \frac{223.3 \text{ K}}{0.878}$$
$$T_0 = 254.3 \text{ K at 10 km and 250 m/s} \qquad (Ans)$$

b. Next, consider 500 m/s at 10 km. The speed of sound and the static pressure and temperature are still the values appropriate to the 10-km altitude. The Mach number is

$$\text{Ma} = \frac{V}{c} = \frac{500 \text{ m/s}}{299.6 \text{ m/s}}$$
$$\text{Ma} = 1.669 \text{ at 10 km and 500 m/s} \qquad (Ans)$$

This is a supersonic cruise condition (Ma > 1). At this Mach number, the pressure and temperature ratios are

$$\left.\frac{p}{p_0}\right)_{\text{Ma}=1.669} = 0.212 \quad \text{and} \quad \left.\frac{T}{T_0}\right)_{\text{Ma}=1.669} = 0.642$$

Calculating p_0 and T_0, we have

$$p_0 = \frac{26.50 \text{ kPa}}{0.212}$$

or

$$p_0 = 125.0 \text{ kPa at 10 km and 500 m/s} \qquad (Ans)$$

and

$$T_0 = \frac{223.3 \text{ K}}{0.642 \text{ K}}$$

or

$$T_0 = 347.8 \text{ K at 10 km and 500 m/s} \qquad (Ans)$$

c. Finally, consider 250 m/s at 1 km. The speed of sound is now

$$c = \sqrt{kRT} = 20.05\sqrt{281.7 \text{ K}} \text{ m/s} = 336.5 \text{ m/s}$$

The Mach number is

$$\text{Ma} = \frac{V}{c} = \frac{250 \text{ m/s}}{336.5 \text{ m/s}}$$

$$\text{Ma} = 0.743 \text{ at 1 km and 250 m/s} \qquad (Ans)$$

Again, this is a subsonic cruise condition. To calculate the stagnation pressure and temperature, we must use the atmospheric properties at 1 km. For some variety we use Eqs. 11.50 and 11.51. They give

$$p_0 = p\left(1 + \frac{k-1}{2}\text{Ma}^2\right)^{k/(k-1)} = 89.88 \text{ kPa}[1 + 0.2(0.743)^2]^{3.5}$$

so

$$p_0 = 129.7 \text{ kPa at 1 km and 250 m/s} \qquad (Ans)$$

and

$$T_0 = T\left(1 + \frac{k-1}{2}\text{Ma}^2\right) = 281.7 \text{ K}[1 + 0.2(0.743)^2]$$

so

$$T_0 = 312.8 \text{ K at 1 km and 250 m/s} \qquad (Ans)$$

Comments The Mach number changed with altitude, even for the same velocity. The reason is that the temperature and, hence, the speed of sound changed. The stagnation properties changed with speed and altitude. A pocket calculator or computer is often easier to use than to look up and interpolate values in a table. Calculating the temperature ratio first is usually best because this ratio is then raised to various powers to obtain pressure and density ratios. For rapid, though imprecise, calculations, use Fig. D.I.

11.4   COMPRESSIBLE FLOW REGIMES

Experience has demonstrated that compressibility can have a large influence on important flow variables. For example, in **Fig. 11.2** the variation of drag coefficient with Reynolds number and Mach number is shown for airflow over a sphere. It is clear from this figure that any effects of compressibility become more pronounced as the Mach number increases. The discussion in

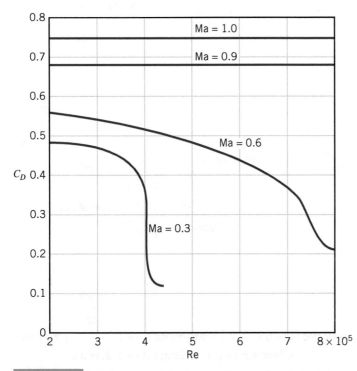

FIGURE 11.2 The variation of the drag coefficient of a sphere with Reynolds number and Mach number.

Section 4.8.1 indicates that compressibility effects increase as the fluid velocity (and hence Mach number) increases. Clearly compressibility effects can be of considerable importance.

To further illustrate some curious features of compressible flow, a simplified example is considered. Imagine the emission of weak pressure pulses from a point source. These pressure waves are spherical and expand radially outward from the point source at the speed of sound, c. If a pressure wave is emitted at different times, t_{wave}, we can determine where several waves will be at a common instant of time, t, by using the relationship

$$r = (t - t_{\text{wave}})c$$

where r is the radius of the sphere-shaped wave emitted at time $= t_{\text{wave}}$. For a stationary point source, the symmetrical wave pattern shown in **Fig. 11.3a** results.

The wave pattern from a moving source is not symmetrical.

When the point source moves to the left with a constant velocity, V, the wave pattern is no longer symmetrical. **Figures 11.3b, 11.3c, and 11.3d** illustrate the wave patterns at $t = 3$ s for different values of V. Also shown with a "+" are the positions of the moving point source at values of time, t, equal to 0 s, 1 s, 2 s, and 3 s. Knowing where the point source has been at different instants is important because it indicates to us where the different waves originated.

From the pressure wave patterns of Fig. 11.3, we can draw some useful conclusions. Before doing this we should recognize that if instead of moving the point source to the left, we held the point source stationary and moved the fluid to the right with velocity V, the resulting pressure wave patterns would be identical to those indicated in Fig. 11.3.

When the point source and the fluid are stationary, the pressure wave pattern is symmetrical (Fig. 11.3a) and an observer anywhere in the pressure field would hear the same sound

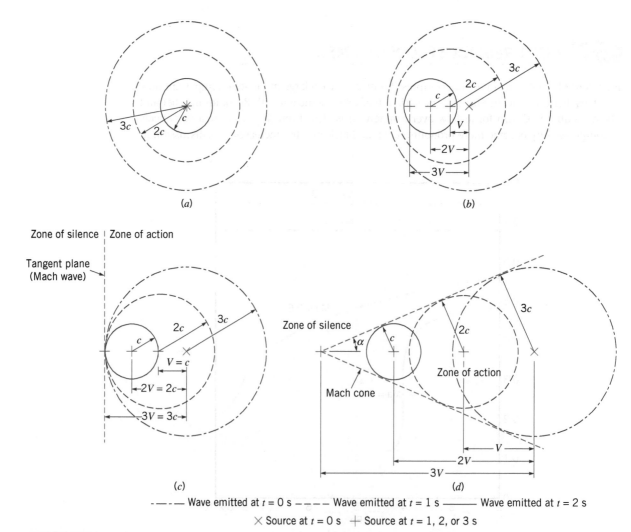

FIGURE 11.3 (a) Pressure waves at $t = 3$ s, $V = 0$; (b) pressure waves at $t = 3$ s, $V < c$; (c) pressure waves at $t = 3$ s, $V = c$; (d) pressure waves at $t = 3$ s, $V > c$.

frequency from the point source. When the velocity of the point source (or the fluid) is very small in comparison with the speed of sound, the pressure wave pattern will still be nearly symmetrical. The speed of sound in an incompressible fluid is infinitely large. Thus, the stationary point source and stationary fluid situation is representative of incompressible flows. For truly incompressible flows, the communication of pressure information throughout the flow field is unrestricted and instantaneous ($c = \infty$).

When the point source moves in fluid at rest (or when fluid moves past a stationary point source), the pressure wave patterns vary in asymmetry, with the extent of asymmetry depending on the ratio of the point source (or fluid) velocity and the speed of sound. When $V/c < 1$, the wave pattern is similar to the one shown in Fig. 11.3b. This flow is **subsonic** and compressible. A stationary observer will hear a higher frequency sound ahead of the source and a lower frequency behind it because the wave pattern is asymmetrical. We call this phenomenon the *Doppler effect*. Pressure information can still travel unrestricted throughout the flow field, but not symmetrically or instantaneously.

When $V/c = 1$, pressure waves are not present ahead of the moving point source. The flow is **sonic**. If you were positioned to the left of the moving point source, you would not hear the point source until it was coincident with your location. For flow moving past a stationary point source at the speed of sound ($V/c = 1$), the pressure waves are all tangent to a plane that is perpendicular to the flow and that passes through the point source. The concentration of pressure waves in this tangent plane suggests the formation of an abrupt pressure variation across the plane. This plane is often called a **Mach wave**. Note that communication of pressure information is restricted to the region of flow downstream of the Mach wave. The region of flow upstream of the Mach wave is called the *zone of silence*, and the region of flow downstream of the tangent plane is called the *zone of action*.

When $V > c$, the flow is **supersonic**, and the pressure wave pattern resembles the one depicted in Fig. 11.3d. A cone (**Mach cone**) that is tangent to the pressure waves can be constructed to represent the Mach wave that separates the zone of silence from the zone of action. The communication of pressure information is restricted to the zone of action. From the sketch of Fig. 11.3d, we can see that the angle of this cone, α, is given by

$$\sin \alpha = \frac{c}{V} = \frac{1}{\mathrm{Ma}} \tag{11.53}$$

This relationship between Mach number, Ma, and Mach cone angle, α, shown in the adjacent figure, is valid for $V/c > 1$ only. The concentration of pressure waves at the surface of the Mach cone suggests an abrupt (but still infinitesimal) pressure and density variation across the cone surface. An abrupt density change can be visualized in a flow field by using special optical techniques. Examples of flow visualization methods include the schlieren, shadowgraph, and interferometer techniques (see Ref. 4). A schlieren photo of a flow for which $V > c$ is shown in **Fig. 11.4**. The airflow through the row of compressor blade airfoils is as shown with the arrow. The flow enters supersonically ($\mathrm{Ma}_1 = 1.14$) and leaves subsonically ($\mathrm{Ma}_2 = 0.86$). The center two airfoils have pressure tap hoses connected to them. Regions of significant changes in fluid density appear in the supersonic portion of the flow. Also, the region of separated flow on each

V11.3 Jet noise

V11.4 Speed boat

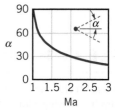

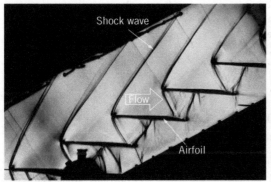

Photograph courtesy of TAGT-1.2 Cascade, R. Fuchs, H. A. Schreiber, W. Steinert, German Aerospace Center (DLR), Institute of Propulsion Technology, 1996.

FIGURE 11.4 The schlieren visualization of flow (supersonic to subsonic) through a row of compressor airfoils.

V11.5 Compressible flow visualization

airfoil is visible. It should be noted that waves seen in the photograph are not exactly infinitesimal Mach waves but weak finite strength shock waves generated by the finite-size airfoils.

This discussion about pressure wave patterns suggests the following categories of fluid flow:

1. **Incompressible flow:** Ma $\leq$ 0.3. Unrestricted, nearly symmetrical and instantaneous pressure communication.

2. **Compressible subsonic flow:** 0.3 < Ma < 1.0. Unrestricted but noticeably asymmetrical pressure communication.

3. **Compressible supersonic flow:** Ma $\geq$ 1.0. Formation of Mach wave; pressure communication restricted to zone of action.

In addition to the above-mentioned categories of flows, two other regimes are commonly referred to: namely, **transonic flows** (0.9 $\leq$ Ma $\leq$ 1.2) and **hypersonic flows** (Ma > 5). Modern aircraft are mainly powered by gas turbine engines with transonic internal flows. When a spacecraft reenters the Earth's atmosphere, the flow is hypersonic. Future aircraft may be expected to operate from subsonic to hypersonic speeds.

The Wide World of Fluids

Supersonic and compressible flows in gas turbines

Modern gas turbine engines commonly involve compressor and turbine blades that are moving so fast that the fluid flows over the blades are locally *supersonic* (see Fig. 11.4). Density varies considerably in these flows so they are also considered to be *compressible*. *Shock waves* can form when these supersonic flows are sufficiently decelerated. Shocks formed at blade leading edges or on blade surfaces can interact with other blades and shocks and seriously affect blade aerodynamic and structural performance. It is possible to have supersonic flows past blades near the outer diameter of a rotor with *subsonic flows* near the inner diameter of the same rotor. These rotors are considered to be *transonic* in their operation. Very large aero gas turbines can involve thrust levels exceeding 45359 kg. Two of these engines are sufficient to carry over 350 passengers halfway around the world at high subsonic speed.

EXAMPLE 11.3 | Mach Cone

Given An aircraft cruising at 1000-m elevation, z, above you moves past in a flyby. It is moving with a Mach number equal to 1.5 and the ambient temperature is 20 °C.

Find How many seconds after the plane passes overhead do you expect to wait before you hear the aircraft?

Solution Since the aircraft is moving supersonically (Ma > 1), we can imagine a Mach cone originating from the forward tip of the craft as is illustrated in **Fig. E11.3a**. A photograph of this phenomenon is shown in **Fig. E11.3b**. When the surface of the cone reaches the observer, the "sound" of the aircraft is perceived. The angle α in Fig. E11.4 is related to the elevation of the plane, z, and the ground distance, x, by

$$\alpha = \tan^{-1}\frac{z}{x} = \tan^{-1}\frac{1000}{Vt} \tag{1}$$

Also, assuming negligible change of Mach number with elevation, we can use Eq. 11.53 to relate Mach number to the angle α. Thus,

$$\text{Ma} = \frac{1}{\sin\alpha} \tag{2}$$

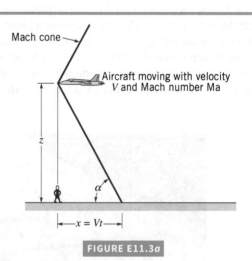

Mach cone

Aircraft moving with velocity V and Mach number Ma

z

α

$x = Vt$

FIGURE E11.3a

Combining Eqs. 1 and 2 we obtain

$$\text{Ma} = \frac{1}{\sin[\tan^{-1}(1000/Vt)]} \tag{3}$$

The speed of the aircraft can be related to the Mach number with

$$V = (\text{Ma})c \tag{4}$$

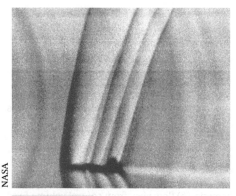

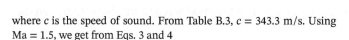

FIGURE E11.3b NASA Schlieren photograph of shock waves from a T-38 aircraft at Mach 1.1, 3962 m.

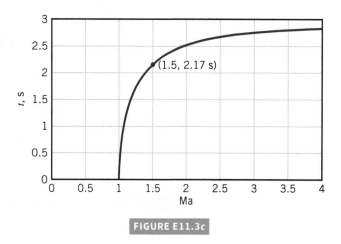

FIGURE E11.3c

where c is the speed of sound. From Table B.3, $c = 343.3$ m/s. Using Ma = 1.5, we get from Eqs. 3 and 4

$$1.5 = \frac{1}{\sin\left\{\tan^{-1}\left[\dfrac{1000 \text{ m}}{(1.5)(343.3 \text{ m/s})t}\right]\right\}}$$

or

$$t = 2.17 \text{ s} \qquad\qquad (Ans)$$

Comment By repeating the calculations for various values of Mach number, Ma, the results shown in **Fig. E11.3c** are obtained. Note that for subsonic flight (Ma < 1) there is no delay since the sound travels faster than the aircraft. You can hear a subsonic aircraft approaching.

The answer above is only an estimate. In reality, the wave emanating from the aircraft will be a finite strength shock wave and will have a shallower angle. You would hear the sound of the aircraft as a "sonic boom."

11.5 SHOCK WAVES

In the preceding section, we saw that introducing an infinitesimal disturbance into a supersonic flow generates a stationary (with respect to the disturbance) Mach wave. Placing a finite disturbance (say, a sharp wedge or cone) in a steady supersonic flow generates a finite-strength wave as shown in the adjacent photographs.

Finite-strength disturbance waves are called *shock waves*. Shock waves are extremely thin, and fluid properties and velocity change by large amounts (tens or even hundreds of percent) across a shock wave. Shock waves are actually rather common occurrences; examples of shock waves include explosion waves from an extra-powerful Fourth of July firecracker, the crack of a whip, sonic "booms," which are actually the shock waves generated by supersonic aircraft, and water hammer, which can occur when a water faucet or valve is opened or closed rapidly.

Shock waves may be moving, like the wave that results from rapidly closing a valve in a pipe, or stationary, as in flow over a stationary wedge or blunt body. In either case, a shock wave adjusts the flow to some *downstream* condition. A *normal shock* is perpendicular to the upstream flow. An *oblique shock* is inclined at a constant angle to the upstream flow. A *curved shock* has a variable angle between the wave and the upstream flow. Flows that contain oblique or curved shocks are at least two-dimensional because the shock changes the direction of the flow. One-dimensional flow can contain only normal shocks. For the sake of simplicity, our consideration is limited to normal shocks. For the same reason, we consider shocks only in an ideal gas.

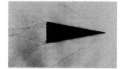

Oblique shock in supersonic flow over a right circular cone.

Courtesy of Launch and Flight Division. Ballistic Research Laboratory/ARRADCOM, Aberdeen Proving Ground, MD.

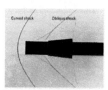

Curved shock standing ahead of a blunt body in supersonic flow.

Courtesy of Launch and Flight Division, Ballistic Research Laboratory/ARRADCOM, Aberdeen Proving Ground, MD.

11.5.1 Normal Shock

Figure 11.5a shows a normal shock in a duct. The shock may be moving or stationary with respect to the duct. We develop analysis that applies to either case by using a control volume fixed to the shock. Velocities and stagnation properties are defined relative to the shock. **Figure 11.5b** shows the flow into and out of the control volume. Subscripts x and y indicate upstream and

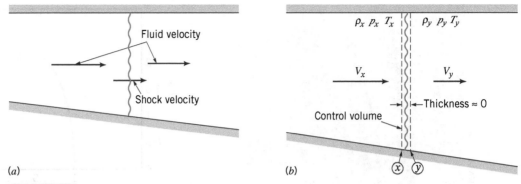

(a)   (b)

FIGURE 11.5 Normal shock in a duct: (a) flow schematic; (b) flow view from control volume fixed to shock.

downstream positions, respectively. We assume that the velocity is uniform. We assume that the shock has zero thickness, so

$$A_x = A_y = A$$

and the volume of the control volume is zero.

> **Normal shock waves are assumed to be infinitesimally thin discontinuities.**

The continuity equation is

$$\rho_x A V_x = \rho_y A V_y \tag{11.54}$$

From the ideal gas law,

$$\rho = \frac{p}{RT} \tag{11.55}$$

We can write the velocity as

$$V = \text{Ma}(c) = \text{Ma}\sqrt{kRT} \tag{11.56}$$

Substituting into Eq. 11.54 and cancelling A, k, and R, we obtain

$$\frac{p_x \text{Ma}_x}{\sqrt{T_x}} = \frac{p_y \text{Ma}_y}{\sqrt{T_y}}$$

or

$$\frac{\text{Ma}_y}{\text{Ma}_x} = \frac{p_x}{p_y}\sqrt{\frac{T_y}{T_x}} \tag{11.57}$$

The linear momentum equation for the flow direction x is

$$\sum F = \rho_y V_y^2 A - \rho_x V_x^2 A$$

Because the sides of the control volume in contact with the duct wall have zero thickness, there is no force on them. The only forces are from pressure on the inlet and outlet planes, so the equation becomes

$$p_x A - p_y A = \rho_y V_y^2 A - \rho_x V_x^2 A$$

From Eq. 11.48,

$$\rho V^2 = kp\,\text{Ma}^2$$

and the momentum equation reduces to

$$p_x(1 + k\text{Ma}_x^2) = p_y(1 + k\text{Ma}_y^2)$$

V11.8 Blast waves

or

$$\frac{p_y}{p_x} = \frac{1 + k\text{Ma}_x^2}{1 + k\text{Ma}_y^2} \tag{11.58}$$

Next, we apply the general energy equation,

$$\dot{Q} + \dot{W}_s = \dot{m}\left(\check{h}_y - \check{h}_x + \frac{V_y^2}{2} - \frac{V_x^2}{2}\right)$$

Because the control volume has zero thickness, there can be neither heat transfer nor work, so the energy equation reduces to

$$\check{h}_y + \frac{V_y^2}{2} = \check{h}_x + \frac{V_x^2}{2}$$

or, alternatively, in terms of stagnation enthalpy,

$$\check{h}_{0_y} = \check{h}_{0_x} \tag{11.59}$$

For an ideal gas, with constant c_p the energy equation becomes

$$T_{0_y} = T_{0_x} \tag{11.60}$$

Substituting Eq. 11.50 into Eq. 11.60, we have

$$T_y\left(1 + \frac{k-1}{2}\mathrm{Ma}_y^2\right) = T_x\left(1 + \frac{k-1}{2}\mathrm{Ma}_x^2\right)$$

or

$$\frac{T_y}{T_x} = \frac{1 + [(k-1)/2]\mathrm{Ma}_x^2}{1 + [(k-1)/2]\mathrm{Ma}_y^2} \tag{11.61}$$

Equations 11.57, 11.58, and 11.61 relate the ratio of temperature and pressure across a shock to the Mach numbers upstream and downstream from the shock. If we select a single parameter, say, upstream Mach number Ma_x, we can express the other three parameters in terms of the selected parameter. To do this, we square Eq. 11.57:

$$\frac{\mathrm{Ma}_y^2}{\mathrm{Ma}_x^2} = \left(\frac{p_x}{p_y}\right)^2\left(\frac{T_y}{T_x}\right)$$

Then we substitute Eqs. 11.58 and 11.61 into the right side. The result is a quadratic equation relating Ma_y^2 to Ma_x^2; k appears in the equation as a parameter. Because the equation is quadratic, there are two solutions for Ma_y^2:

$$\boxed{\mathrm{Ma}_y^2 = \mathrm{Ma}_x^2} \tag{11.62}$$

and

$$\boxed{\mathrm{Ma}_y^2 = \frac{\mathrm{Ma}_x^2 + 2/(k-1)}{[2k/(k-1)]\,\mathrm{Ma}_x^2 - 1}} \tag{11.63}$$

The first solution seems trivial, but it shows that the Mach number need not change discontinuously at a plane. This solution is valid if there is no shock. If there is a shock, the second solution, Eq. 11.63, gives the downstream Mach number in terms of the upstream Mach number. For any realistic value of k (that is, $k > 1$), calculations with Eq. 11.63 show that Ma_x and Ma_y always lie on opposite sides of $\mathrm{Ma} = 1.0$; that is, if $\mathrm{Ma}_x > 1$, then $\mathrm{Ma}_y < 1$ and vice versa.

We can now substitute Eqs. 11.62 and 11.63 into the pressure ratio and temperature ratio equations, Eqs. 11.58 and 11.61. The constant Mach number solution gives pressure and temperature ratios equal to unity; that is, there can be no discontinuous change of pressure and temperature at a plane if there is no shock at the plane. If there is a shock, the pressure and temperature ratios are

Ratios of thermodynamic properties across a normal shock are functions of the Mach numbers.

$$\boxed{\frac{p_y}{p_x} = \frac{2k}{k+1}\mathrm{Ma}_x^2 - \frac{k-1}{k+1}} \tag{11.64}$$

and

$$\boxed{\frac{T_y}{T_x} = \left(1 + \frac{k-1}{2}\mathrm{Ma}_x^2\right)\left(\frac{2k}{k-1}\mathrm{Ma}_x^2 - 1\right)\left[\frac{2(k-1)}{(k+1)^2\,\mathrm{Ma}_x^2}\right]}$$

(11.65)

A shock may change stagnation properties as well as static properties. According to Eq. 11.60, the stagnation temperature ratio across a shock is

$$\boxed{\frac{T_{0_y}}{T_{0_x}} = 1}$$

(11.66)

We obtain the stagnation pressure ratio across a shock by first expressing stagnation pressure in terms of static pressure and Mach number. Using Eq. 11.51, we have

$$\frac{p_{0_y}}{p_{0_x}} = \frac{p_y\left\{1 + [(k-1)/2]\,\mathrm{Ma}_y^2\right\}^{k/(k-1)}}{p_x\left\{1 + [(k-1)/2]\,\mathrm{Ma}_x^2\right\}^{k/(k-1)}}$$

We reduce this expression to an equation involving only only Ma_x^2 and k by substituting Eq. 11.64 for the static pressure ratio and Eq. 11.63 for Ma_y^2 to get

$$\boxed{\frac{p_{0_y}}{p_{0_x}} = \left(\frac{k+1}{2}\mathrm{Ma}_x^2\right)^{\frac{k}{k-1}}\left(1 + \frac{k-1}{2}\mathrm{Ma}_x^2\right)^{\frac{-k}{(k-1)}}\left(\frac{2k}{k+1}\mathrm{Ma}_x^2 - \frac{k-1}{k+1}\right)^{\frac{-1}{(k-1)}}}$$

(11.67)

Across a normal shock the values of some parameters increase, some remain constant, and some decrease.

Equations 11.63–11.67 enable us to calculate properties on the downstream side of a shock if we know the upstream properties and Mach number Ma_x. **Table 11.1** indicates typical results of calculations made with the equations. Note the last column, which tabulates results for a (hypothetical) shock that occurs in a subsonic flow. Our discussion of the different nature of subsonic and supersonic flow indicates that discontinuous changes of fluid properties do not occur in subsonic flow. To prove this, consider the second law of thermodynamics, in mathematical form:

$$s_y - s_x \geq \int_x^y \frac{\delta q}{T}$$

The flow changes from supersonic to subsonic across a normal shock.

With no heat transfer to or from the gas as it passes through a shock wave,

$$s_y - s_x \geq 0$$

TABLE 11.1 Relative Changes of Properties as Predicted by the Normal Shock Equations

	$\mathrm{Ma}_x > 1$ (Supersonic Upstream)	$\mathrm{Ma}_x < 1$ (Subsonic Upstream)
Ma_y	< 1 (subsonic)	> 1 (Not physically possible) (supersonic)
$\dfrac{p_y}{p_x}$	> 1	< 1 (Not physically possible)
$\dfrac{T_y}{T_x}$	> 1	< 1 (Not physically possible)
$\dfrac{T_{0_y}}{T_{0_x}}$	= 1	= 1
$\dfrac{p_{0_y}}{p_{0_x}}$	< 1	> 1 (Not physically possible)
$s_y - s_x$	> 0	< 0 (Not physically possible)

Working with a dimensionless entropy change is convenient. Dividing Eq. 11.24 by R gives

$$\frac{s_y - s_x}{R} = \frac{k}{k-1} \ln\left(\frac{T_y}{T_x}\right) - \ln\left(\frac{p_y}{p_x}\right)$$

Substituting Eqs. 11.64 and 11.65 for the pressure and temperature ratios yields an expression for entropy change in terms of Mach number. By selecting various values of Ma_x, we can determine $(s_y - s_x)$. The entropy change is negative for all subsonic inlet Mach numbers. Shock waves can exist only if the upstream flow (relative to the shock) is supersonic. Shock waves always increase the static pressure, density, and temperature and decrease the stagnation pressure (relative to the shock).

If you need to calculate property changes across a shock wave, you can use Eqs. 11.63–11.67; however, these equations are quite tedious. This is especially true if you know, say, T_y/T_x and are trying to calculate Mach number. Appendix D has tables and a graph of the normal shock functions for $k = 1.4$. A quick and accurate calculation can be made with a compressible flow function calculator from the Internet or as a smartphone app (or perhaps programmed yourself on a computer or programmable calculator).

EXAMPLE 11.4 | Stagnation Pressure Drop Across a Normal Shock

Given Designers involved with fluid mechanics work hard at minimizing loss of available energy in their designs. Shock waves produce significant losses when they occur. In addition to an increase in entropy, shocks result in a reduction in stagnation pressure.

Find For normal shocks, show that the stagnation pressure drop (and thus loss) is larger for higher Mach numbers.

Solution We assume that air ($k = 1.4$) behaves as a typical gas and use Table D.S to respond to the above-stated requirements. Since

$$\frac{p_{0_x} - p_{0_y}}{p_{0_x}} = 1 - \frac{p_{0_y}}{p_{0_x}}$$

we can construct **Table E11.4.1** with values of p_{0_y}/p_{0_x} from Table D.S.

Comment When the Mach number of the flow entering the shock is low, say $Ma_x = 1.2$, the flow across the shock is nearly isentropic and the loss in stagnation pressure is small. However, as shown in **Fig. E11.4**, at larger Mach numbers, the entropy change across the normal shock rises dramatically and the stagnation pressure drop across the shock is appreciable. If a shock occurs at $Ma_x = 2.5$, only about 50% of the upstream stagnation pressure is recovered.

In devices where supersonic flows occur, for example, high-performance aircraft engine inlet ducts and high-speed wind tunnels, designers attempt to prevent shock formation, or if shocks must occur, they design the flow path so that shocks are positioned where they are weak (small Mach number).

TABLE E11.4.1

Ma_x	p_{0_y}/p_{0_x}	$\dfrac{p_{0_x} - p_{0_y}}{p_{0_x}}$
1.0	1.0	0
1.2	0.99	0.01
1.5	0.93	0.07
2.0	0.72	0.28
2.5	0.50	0.50
3.0	0.33	0.67
3.5	0.21	0.79
4.0	0.14	0.86
5.0	0.06	0.94

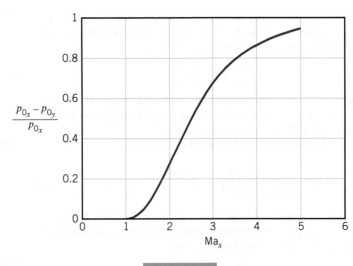

FIGURE E11.4

Of interest also is the static pressure rise that occurs across a normal shock. These static pressure ratios, p_y/p_x, obtained from Table D.S are shown in **Table E11.4.2** for a few Mach numbers. For a developing boundary layer, any pressure rise in the flow direction is considered as an adverse pressure gradient that can possibly cause flow separation (see Section 9.2.6). Thus, shock–boundary layer interactions are of great concern to designers of high-speed flow devices.

TABLE E11.4.2

Ma_x	p_y/p_x
1.0	1.0
1.2	1.5
1.5	2.5
2.0	4.5
3.0	10
4.0	18
5.0	29

EXAMPLE 11.5 | Supersonic Flow Pitot Tube

Given A total pressure probe is inserted into a supersonic air flow. A shock wave forms just upstream of the impact hole and head as illustrated in **Fig. E11.5**. The probe measures a total pressure of 413700 Pa. The stagnation temperature at the probe head is measured with a thermocouple and found to be 556 K. The static pressure upstream of the shock is measured with a wall tap to be 82740 Pa.

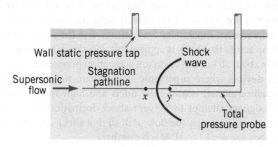

FIGURE E11.5

Find Determine the Mach number and velocity of the flow.

Solution We assume that the flow along the stagnation pathline is isentropic except across the shock. Also, the shock is treated as a normal shock. Thus, in terms of the data we have

$$\frac{p_{0_y}}{p_x} = \left(\frac{p_{0_y}}{p_{0_x}}\right)\left(\frac{p_{0_x}}{p_x}\right) \tag{1}$$

where p_{0_y} is the stagnation pressure measured by the probe, and p_x is the static pressure measured by the wall tap. The stagnation pressure upstream of the shock, p_{0_x}, is not measured.

Combining Eqs. 1, 11.51, and 11.67 we obtain

$$\frac{p_{0_y}}{p_x} = \frac{\{[(k+1)/2]\,Ma_x^2\}^{k/(k-1)}}{\{[2k/(k+1)]\,Ma_x^2 - [(k-1)/(k+1)]\}^{1/(k-1)}} \tag{2}$$

which is called the *Rayleigh Pitot-tube formula*. Values of p_{0_y}/p_x from Eq. 2 are considered important enough to be included in Fig. D.S for $k = 1.4$. Thus, for $k = 1.4$ and

$$\frac{p_{0_y}}{p_x} = \frac{413700\ \text{Pa}}{82740\ \text{Pa}} = 5$$

we use Fig. D.S (or Eq. 2) to ascertain that

$$Ma_x = 1.9 \tag{Ans}$$

To determine the flow velocity we need to know the static temperature upstream of the shock, since

$$V_x = Ma_x c_x = Ma_x\sqrt{kRT_x} \tag{3}$$

The stagnation temperature downstream of the shock was measured and found to be

$$T_{0_y} = 556\ \text{K}$$

Since the stagnation temperature remains constant across a normal shock,

$$T_{0_x} = T_{0_y} = 556\ \text{K}$$

For the isentropic flow upstream of the shock, Eq. 11.50 or Fig. D.I can be used. For $Ma_x = 1.9$,

$$\frac{T_x}{T_{0_x}} = 0.59$$

or

$$T_x = (0.59)(556\ \text{K}) = 328\ \text{K}$$

With Eq. 3 we obtain

$$V_x = 1.87\sqrt{(287\ \text{J/kg}\cdot\text{K})(328\ \text{K})(1.4)}$$
$$= 679\ \text{m/s} \tag{Ans}$$

Comment Application of the incompressible flow Pitot tube results (see Section 4.5) would give highly inaccurate results because of the presence of a shock with subsequent large pressure and density changes involved.

11.6 ISENTROPIC FLOW

In this section, we consider in further detail steady, one-dimensional, isentropic flow. As explained earlier, one-dimensionality implies velocity and fluid property change in the streamwise direction only. We consider flows through ducts with uniformly distributed velocities and fluid properties at each cross-section. Much of what we develop can also apply to the flow of a fluid particle along its pathline or flow through an infinitesimal streamtube.

Isentropic flow has constant entropy and was discussed briefly in Section 11.1, where we learned that adiabatic and frictionless (reversible) flow is the most common form of isentropic flow. Some ideal gas relationships for isentropic flows were developed in Section 11.1. An isentropic flow is not achievable with actual fluids because of friction. Nonetheless, the study of isentropic flow is useful because it helps us to gain an understanding of actual compressible flow phenomena. In many situations, the isentropic flow model is sufficiently accurate to use for engineering analysis and design.

Flow with no friction and no heat transfer is isentropic.

11.6.1 Steady Isentropic Flow of an Ideal Gas

For a one-dimensional flow of a compressible fluid we make the following assumptions:

- The flow is steady.
- There are no body forces (gravity, electromagnetic, or others)
- There is no friction.
- There is no heat transfer.
- The flow does not pass through a shock wave.

The last three assumptions taken together imply that the flow is adiabatic and reversible, therefore, the entropy of the fluid is constant:

$$s = \text{Constant}$$

or, in terms of two locations in the flow

$$s_1 = s_2$$

For steady flow, the energy equation is

$$q + w_{\text{shaft}} = \left(\check{h}_2 + \frac{V_2^2}{2} \right) - \left(\check{h}_1 + \frac{V_1^2}{2} \right)$$

According to our assumptions, q is zero, but w_{shaft} is not necessarily zero, and the energy equation becomes

$$w_{\text{shaft}} = \left(\check{h}_2 + \frac{V_2^2}{2} \right) - \left(\check{h}_1 + \frac{V_1^2}{2} \right)$$

When we use the definition of stagnation enthalpy,

$$w_{\text{shaft}} = \check{h}_{0_2} - \check{h}_{0_1}$$

If the fluid is an ideal gas,

$$w_{\text{shaft}} = c_p(T_{0_2} - T_{0_1}) \tag{11.68}$$

Doing work on the fluid increases its stagnation temperature, even if there is no heat transfer.

Now we consider changes of stagnation pressure in isentropic flow. By definition of the stagnation state,

$$s_{0_1} = s_1 \quad \text{and} \quad s_{0_2} = s_2$$

For isentropic flow,

$$s_{0_2} - s_{0_1} = s_2 - s_1 = 0$$

thus

$$\frac{p_{0_2}}{p_{0_1}} = \left(\frac{T_{0_2}}{T_{0_1}}\right)^{c_p/R} = \left(\frac{T_{0_2}}{T_{0_1}}\right)^{k/(k-1)} \tag{11.69}$$

Using Eq. 11.68, we can also write

$$\frac{p_{0_2}}{p_{0_1}} = \left(1 - \frac{w_{\text{shaft}}}{c_p T_{0_1}}\right)^{k/(k-1)} \tag{11.70}$$

A situation of considerable interest in compressible flow is the case with no shaft work. The following statement applies to this case: *In isentropic flow with no shaft work, all stagnation properties are constant.*

This special case ($w_{\text{shaft}} = 0$) is so important in compressible flow that most engineers and authors use the phrase "isentropic flow" to mean "isentropic flow with no work." We will do the same, thus we add the assumption that there is no work to the assumptions of no heat transfer and no friction.

Equations relating property and velocity changes in isentropic flow of an ideal gas can easily be developed from the stagnation property definitions and the constancy of stagnation properties. Consider, for example, the calculation of the change in temperature between two points in an isentropic flow. Because T_0 is constant, we can write

$$T_{0_1} = T_1 + \frac{V_1^2}{2c_p} = T_2 + \frac{V_2^2}{2c_p} = T_{0_2} \tag{11.71}$$

Using Mach number instead of velocity is sometimes convenient. From Eq. 11.50.

$$\frac{T_2}{T_1} = \frac{T_{0_2}\{1 + [(k-1)/2]\,\text{Ma}_1^2\}}{T_{0_1}\{1 + [(k-1)/2]\,\text{Ma}_2^2\}}$$

but, as T_0 is constant,

$$\frac{T_2}{T_1} = \frac{1 + [(k-1)/2]\,\text{Ma}_1^2}{1 + [(k-1)/2]\,\text{Ma}_2^2} \tag{11.72}$$

A third approach involves use of compressible flow functions (from tables, graphs, or software) relating static-stagnation property ratios and Mach number. We calculate the temperature ratio by

$$\frac{T_2}{T_1} = \frac{T}{T_0}\bigg)_{\text{Ma}_2} \bigg/ \frac{T}{T_0}\bigg)_{\text{Ma}_1} \tag{11.73}$$

Recall that

$$\frac{T}{T_0}\bigg)_{\text{Ma}}$$

is a numerical value read from a table or graph (such as those in Appendix D) or computed by software at Mach number Ma. In a fourth method, we make the calculation in two steps:

$$T_{0_2} = T_{0_1} = \frac{T_1}{\dfrac{T}{T_0}\bigg)_{\text{Ma}_1}}$$

and

$$T_2 = T_{0_2}\left[\frac{T}{T_0}\bigg)_{\text{Ma}_2}\right] \tag{11.74}$$

Any of Eqs. 11.71–11.74 could be used to relate temperature and velocities (or Mach numbers) at point 1 and 2. Which equation is most convenient depends on the exact information available. Do not fall into the trap of always using Mach number functions; direct calculations are sometimes easier. Similarly, pressure and velocity (or Mach number) in isentropic flow can be related by any of the following equations:

$$p_1\left(1 + \frac{V_1^2}{2c_pT_1}\right)^{k/(k-1)} = p_2\left(1 + \frac{V_2^2}{2c_pT_2}\right)^{k/(k-1)} \tag{11.75}$$

$$\frac{p_2}{p_1} = \frac{\{1 + [(k-1)/2]\,\text{Ma}_1^2\}^{k/(k-1)}}{\{1 + [(k-1)/2]\,\text{Ma}_2^2\}^{k/(k-1)}} \tag{11.76}$$

$$\frac{p_2}{p_1} = \frac{p}{p_0}\bigg)_{\text{Ma}_2} \bigg/ \frac{p}{p_0}\bigg)_{\text{Ma}_1} \tag{11.77}$$

$$p_2 = p_{0_2}\left[\frac{p}{p_0}\bigg)_{\text{Ma}_2}\right] \tag{11.78}$$

where

$$p_{0_2} = p_{0_1} = \frac{p_1}{\dfrac{p}{p_0}\bigg)_{\text{Ma}_1}}$$

Note that stagnation conditions do not need to be present in the flow to use stagnation properties. They can simply serve as reference properties to determine static properties and other properties of interest. This is illustrated in Example 11.6.

EXAMPLE 11.6 | Use of Isentropic Flow Functions, Graphs, and Tables

Given Air flows in a duct as shown in **Fig. E11.6**. At one point in the flow, the pressure, temperature, and velocity are 344750 Pa, 854 K, and 152 m/s. At a downstream point, the Mach number is 1.80. The flow is steady and isentropic.

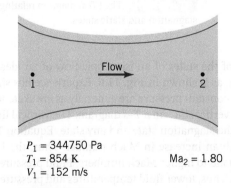

$P_1 = 344750$ Pa
$T_1 = 854$ K
$V_1 = 152$ m/s
$\text{Ma}_2 = 1.80$

FIGURE E11.6 Flow between two points in a duct.

Find Air pressure, temperature, and velocity at the downstream point. Do the calculation twice, once using isentropic flow equations and once using Table D.I or Fig. D.I.

Solution We first do the calculations with isentropic flow equations. The energy equation for adiabatic flow with no work between 1 and 2 is

$$c_pT_1 + \frac{V_1^2}{2} = c_pT_2 + \frac{V_2^2}{2}$$

The velocity at 2 is

$$V_2 = \text{Ma}_2c_2 = \text{Ma}_2\sqrt{kRT_2}, \quad \text{so} \quad c_pT_1 + \frac{V_1^2}{2} = \left(c_p + \frac{kR\,\text{Ma}_2^2}{2}\right)T_2$$

or

$$T_2 = \frac{c_pT_1 + V_1^2/2}{c_p + kR\,\text{Ma}_2^2/2}$$

Substituting numbers, we have

$$T_2 = \frac{(1005\ \text{J/kg·K})(854\ \text{K}) + \dfrac{(152)^2\ \text{m}^2/\text{s}^2}{2(9.81\ \text{N·m/s}^2)}}{1005\ \text{J/kg·K} + \dfrac{1.4(287\ \text{J/kg·K})(1.80)^2}{2}}$$

$$T_2 = 519\ \text{K} \tag{Ans}$$

The velocity at 2 is

$$V_2 = \text{Ma}_2c_2 = \text{Ma}_2\sqrt{kRT_2} = (1.80)\left[\sqrt{(1.4)\times(2.87\ \text{J/kg·K})(519\ \text{K})}\right]$$

$$V_2 = 822\ \text{m/s} \tag{Ans}$$

We compute the pressure at 2 from the isentropic process equation, Eq. 11.26:

$$p_2 = p_1\left(\frac{T_2}{T_1}\right)^{k/(k-1)} = 344750\ \text{Pa}\left(\frac{519}{854}\right)^{3.5}$$

$$p_2 = 60324\ \text{Pa} \tag{Ans}$$

Now we repeat the calculation using Table D.I and Fig. D.I. The Mach number at 1 is

$$Ma_1 = \frac{V_1}{c_1} = \frac{V_1}{\sqrt{kRT_1}} = \frac{152 \text{ m/s}}{\sqrt{(1.4)(287 \text{ J/kg} \cdot \text{K})(854 \text{ K})}} = 0.26$$

Next, we use Table D.I to obtain the pressure and temperature ratio functions at 1 and at 2. At Ma = Ma_1 = 0.43

$$\left.\frac{p}{p_0}\right)_{Ma=0.43} = 0.8806 \quad \text{and} \quad \left.\frac{T}{T_0}\right)_{Ma=0.43} = 0.9643$$

(Graph D.I can also be used to find values for p/p_0 and T/T_0 but the values would be less precise, 0.88 and 0.96 respectively)
   At Ma = Ma_2 = 1.80,

$$\left.\frac{p}{p_0}\right)_{Ma=1.80} = 0.1740 \quad \text{and} \quad \left.\frac{T}{T_0}\right)_{Ma=1.80} = 0.6068$$

Because the flow is isentropic,

$$p_2 = p_1 \left.\frac{p}{p_0}\right)_{Ma_2} \Big/ \left.\frac{p}{p_0}\right)_{Ma_1} = 344750 \text{ Pa}\left(\frac{0.1740}{0.8806}\right)$$

$$p_2 = 68120 \text{ Pa} \qquad (Ans)$$

Now

$$T_2 = T_1 \left.\frac{T}{T_0}\right)_{Ma_2} \Big/ \left.\frac{T}{T_0}\right)_{Ma_1} = 854 \text{ K}\left(\frac{0.6068}{0.9643}\right)$$

$$T_2 = 537.4 \text{ K} \qquad (Ans)$$

The velocity at 2 is

$$V_2 = Ma_2\sqrt{kRT_2} = (1.80)\left[\sqrt{(1.4)(2.87 \text{ J/kg} \cdot \text{K})(519 \text{ K})}\right]$$

$$V_2 = 822 \text{ m/s} \qquad (Ans)$$

Comments   Of course the answers determined by using Table D.I are the same as those determined by using equations. We leave the decision to you as to which method is easier.

Note that the stagnation conditions do not exist explicitly in the flow, yet stagnation properties served as reference properties to calculate other properties of interest.

Always convert pressures and temperatures to absolute units as the first step in solving a compressible flow problem.

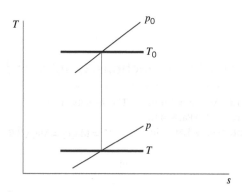

FIGURE 11.6   The (T–s) diagram relating stagnation and static states.

A very useful means of keeping track of the states of an isentropic flow of an ideal gas is a **temperature–entropy (T–s) diagram**, as is shown in **Fig. 11.6**. Experience has shown (see, for example, Refs. 2 and 3) that lines of constant pressure are generally as are sketched in Fig. 11.6. An isentropic flow is confined to a vertical line on a T–s diagram. The vertical line in Fig. 11.6 is representative of flow between the stagnation state and any state. Equation 11.72 shows that fluid temperature decreases with an increase in Mach number. Thus, the lower temperature levels on a T–s diagram correspond to higher Mach numbers. Fluid pressure also decreases with an increase in Mach number. Thus, lower fluid temperatures and pressures are associated with higher Mach numbers.

11.6.2 Incompressible Flow and the Bernoulli Equation

One of the most widely used and abused equations in fluid mechanics is the Bernoulli equation:

$$p + \frac{1}{2}\rho V^2 + \gamma z = \text{Constant}$$

The assumptions used in deriving the Bernoulli equation are similar to those we just used in deriving the isentropic flow equations, except that whereas the Bernoulli equation assumes incompressible flow, the isentropic equations assume adiabatic flow of an ideal gas with no shocks and no work. In this section, we show that, for a very small Mach number, the isentropic equation for pressure becomes identical to the Bernoulli equation. In addition, we develop a criterion for deciding whether a gas flow can be treated as incompressible.

Consider a steady flow with no shear stress, shaft work, or heat transfer. Under these conditions, the stagnation pressure is constant. We also assume that elevation changes are negligible. If the fluid is incompressible, we can calculate the pressure at any location from the Bernoulli equation:

$$p = p_0 - \frac{1}{2}\rho V^2 \tag{11.79}$$

If the fluid is compressible and an ideal gas, the static pressure and stagnation pressure are related by

$$p_0 = p\left(1 + \frac{k-1}{2}\mathrm{Ma}^2\right)^{k/(k-1)} \tag{11.80}$$

If we restrict consideration to Mach numbers less than 1, we can expand the Mach number term into an infinite series using the binomial theorem:

$$(1 + x)^n = 1 + nx + \frac{n(n-1)}{2!}x^2 + \frac{n(n-1)(n-2)}{3!}x^3 + \cdots \tag{11.81}$$

Putting

$$x = \frac{k-1}{2}\mathrm{Ma}^2 \quad \text{and} \quad n = \frac{k}{k-1}$$

we get

$$\left(1 + \frac{k-1}{2}\mathrm{Ma}^2\right)^{k/(k-1)} = 1 + \frac{k}{2}\mathrm{Ma}^2 + \frac{k}{8}\mathrm{Ma}^4 + \text{H.O.T.}$$

Compressibility effects are more important at higher Mach numbers.

where higher-order terms (H.O.T.) involves higher *even* powers of Ma. Substituting into Eq. 11.81 gives

$$p_0 = p\left(1 + \frac{k\mathrm{Ma}^2}{2} + \frac{k\mathrm{Ma}^4}{8} + \text{H.O.T.}\right)$$

Because we are interested in small values of Ma, the higher-order terms are negligible, and we can write the equation

$$p_0 \approx p + \frac{kp\mathrm{Ma}^2}{2}\left(1 + \frac{\mathrm{Ma}^2}{4}\right) \tag{11.82}$$

From Eq. 11.48, we have

$$kp\mathrm{Ma}^2 = \rho V^2$$

Substituting and rearranging, we get

$$p \approx p_0 - \frac{1}{2}\rho V^2\left(1 + \frac{\mathrm{Ma}^2}{4}\right) \tag{11.83}$$

If the Mach number is small, then $\mathrm{Ma}^2/4$ is small compared to 1, and we can write

$$p \approx p_0 - \frac{1}{2}\rho V^2 \tag{11.84}$$

therefore the Bernoulli equation is an approximation to the isentropic flow pressure relation for small Mach number. The accuracy of this approximation depends on the "smallness" of the Mach number. Equation 11.83 shows that the error is proportional to $\mathrm{Ma}^2/4$ at low Mach numbers. If we want to limit the error in using the Bernoulli equation for calculation of pressure to no more than, say, 2 percent, then

$$\mathrm{Ma} \leq \sqrt{4(0.02)} = 0.283$$

There is nothing unique about 2 percent error. For rough estimates, 5 percent error may be acceptable, in which case the Mach number must be less than about 0.45. The most widely used criterion for the boundary between compressible and incompressible flow is: *A flow with Ma < 0.3 everywhere can be assumed to be incompressible.*

Recall that we arrived at essentially this same criterion in Section 4.8.1.

11.6.3 The Critical State

We have shown that the nature of a compressible flow changes drastically when the fluid velocity exceeds the speed of sound. The flow state that corresponds to exactly sonic flow is called the *critical state*. The fluid properties at the critical state are called the *critical properties*. Consider acceleration of a gas from zero velocity to the local sonic speed in an isentropic process. As the fluid's velocity increases, its temperature and speed of sound decrease. The stagnation temperature is constant for this process, so we can write

$$T_0 = T\left(1 + \frac{k-1}{2}\text{Ma}^2\right)$$

At the critical state, the Mach number is unity. Using the superscript * to denote conditions at the critical state, we have

$$T_0 = T^*\left(1 + \frac{k-1}{2}(1)^2\right)$$

or

$$\frac{T^*}{T_0} = \frac{2}{k+1} \tag{11.85}$$

For the specified isentropic process, the pressure and density at the critical state are given by

$$\frac{p^*}{p_0} = \left(\frac{T^*}{T_0}\right)^{k/(k-1)} = \left(\frac{2}{k+1}\right)^{k/(k-1)} \tag{11.86}$$

and

$$\frac{\rho^*}{\rho_0} = \left(\frac{T^*}{T_0}\right)^{1/(k-1)} = \left(\frac{2}{k+1}\right)^{1/(k-1)} \tag{11.87}$$

For an isentropic flow, properties at the critical state are uniquely determined by the stagnation properties and k. The critical properties T^*, p^*, and ρ^* are constant if the stagnation properties are constant.

The stagnation and critical pressures and temperatures are shown on the T–s diagram of **Fig. 11.7.**

The stagnation and critical states are at the same entropy level.

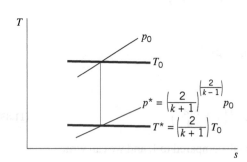

FIGURE 11.7 The relationship between the stagnation and critical states.

11.7 ONE-DIMENSIONAL FLOW IN A VARIABLE AREA DUCT

A compressible flow in a duct may respond to four different types of inputs: cross-sectional area variation, viscous friction between the fluid and the duct wall, heat addition or removal, and shaft work. Work exchange is the subject of Chapter 12 so we will remove it from further consideration here. For simplicity and clarity, we will consider each of the remaining inputs in the absence of the other two. Here we will examine the effects of variation of cross-sectional area in the absence of friction or heat; the flow will be *frictionless* and *adiabatic* (meaning no heat addition or removal). With neither friction nor heat transfer, the flow will be isentropic except for any shocks that are present.

When fluid flows steadily through a duct that has a flow cross-sectional area that varies with axial distance, the conservation of mass (continuity) equation

$$\dot{m} = \rho A V = \text{constant} \tag{11.88}$$

can be used to relate the velocity and density to the area. For incompressible flow, the fluid density remains constant and the flow velocity from section to section varies inversely with cross-sectional area. However, when the flow is compressible, density, cross-sectional area, and flow velocity can all vary from section to section. We proceed to determine how fluid properties and flow velocity change with axial location in a variable area duct when the flow through the duct is steady and isentropic. Our preliminary analysis will be done *without* assuming that the fluid is an ideal gas.

11.7.1 General Considerations

In Chapter 4, Newton's second law was applied to the inviscid (frictionless) and steady flow of a fluid particle. For the streamwise direction, the result (Eq. 4.5) for either compressible or incompressible flows is

$$dp + \frac{1}{2}\rho \, d(V^2) + \gamma \, dz = 0 \tag{11.89}$$

The frictionless flow from section to section through a control volume is also described by Eq. 11.89, if the flow is one-dimensional, because every particle of fluid will have the same experience. For gas flow, the potential energy term, $\gamma \, dz$, can be dropped because of its small size in comparison to the other terms. Thus, an appropriate equation of motion in the streamwise direction for the steady, one-dimensional, and isentropic (adiabatic and frictionless) flow of gas is obtained from Eq. 11.89 as

$$\frac{dp}{\rho V^2} = -\frac{dV}{V} \tag{11.90}$$

If we form the logarithm of both sides of the continuity equation (Eq. 11.88), the result is

$$\ln \rho + \ln A + \ln V = \text{constant}$$

Differentiating we get

$$\frac{d\rho}{\rho} + \frac{dA}{A} + \frac{dV}{V} = 0$$

or

$$-\frac{dV}{V} = \frac{d\rho}{\rho} + \frac{dA}{A}$$

Now we combine this result and Eq. 11.90 together with $d\rho = dp/(dp/d\rho)$ to obtain

$$\frac{dp}{\rho V^2}\left(1 - \frac{V^2}{dp/d\rho}\right) = \frac{dA}{A} \tag{11.91}$$

Since the flow being considered is isentropic, the speed of sound is related to variations of pressure with density by

$$c = \sqrt{\left(\frac{\partial p}{\partial \rho}\right)_s}$$

This equation, combined with the definition of Mach number

$$\text{Ma} = \frac{V}{c}$$

and Eq. 11.91 yields

$$\frac{dp}{\rho V^2}(1 - \text{Ma}^2) = \frac{dA}{A}$$

> Density, cross-sectional area, and velocity may all vary for compressible flow in a duct.

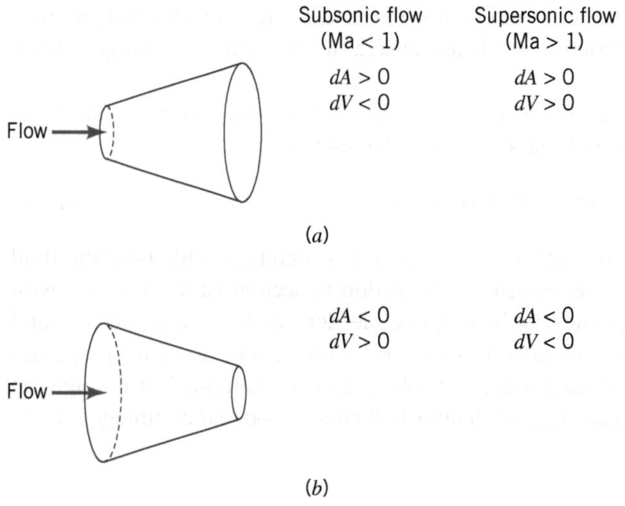

Subsonic flow | Supersonic flow
(Ma < 1) | (Ma > 1)

$dA > 0$ $dA > 0$
$dV < 0$ $dV > 0$

Flow →

(a)

$dA < 0$ $dA < 0$
$dV > 0$ $dV < 0$

Flow →

(b)

FIGURE 11.8 (a) A diverging duct. (b) A converging duct.

Combining this result with Eq. 11.90 gives

$$\frac{dV}{V} = -\frac{dA}{A}\frac{1}{(1 - \text{Ma}^2)} \tag{11.92}$$

We can use Eq. 11.92 to conclude that when the flow is subsonic (Ma < 1), velocity and section area changes are in opposite directions. In other words, the area increase associated with subsonic flow through a diverging duct like the one shown in **Fig. 11.8a** is accompanied by a velocity decrease. Subsonic flow through a converging duct (see **Fig. 11.8b**) involves an increase of velocity. These trends are consistent with incompressible flow behavior, which we described earlier in this book, for instance, in Chapters 4 and 8.

Equation 11.92 also serves to show us that when the flow is supersonic (Ma > 1), velocity and area changes are in the same direction. A diverging duct (Fig. 11.8a) will accelerate a supersonic flow. A converging duct (Fig. 11.8b) will decelerate a supersonic flow. These trends are the opposite of what happens for incompressible and subsonic compressible flows. These observations are summarized in **Table 11.2**.

In isentropic flow, an increase in velocity always corresponds to a Mach number increase and vice versa; therefore a *converging* passage always drives the Mach number *toward* one and a *diverging* passage always drives the Mach number *away* from one. These conditions imply that the Mach number cannot pass through one (either accelerating or decelerating) in a passage that is only convergent or only divergent. If we consider the process of accelerating a gas from rest or a low subsonic Mach number in a simply converging passage, the *maximum*

> A converging duct will decelerate a supersonic flow and accelerate a subsonic flow.

TABLE 11.2 Effects of area change in isentropic compressible flow

		Ma < 1	Ma > 1
Converging passage	Velocity (V)	Increases	Decreases
	Temperature (T)	Decreases	Increases
	Pressure (p)	Decreases	Increases
	Density (ρ)	Decreases	Increases
$dA < 0$	Passage acts as a	Nozzle	Diffuser
Diverging passage	Velocity (V)	Decreases	Increases
	Temperature (T)	Increases	Decreases
	Pressure (p)	Increases	Decreases
	Density (ρ)	Increases	Decreases
$dA > 0$	Passage acts as a	Diffuser	Nozzle

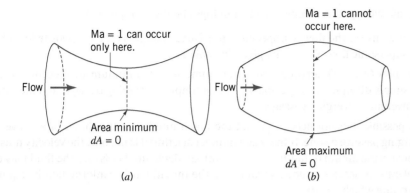

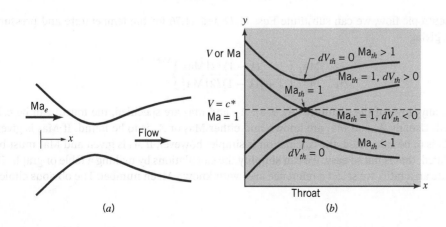

FIGURE 11.9 (*a*) A converging–diverging duct. (*b*) A diverging–converging duct.

possible Mach number that can be achieved is one. This maximum value can occur *only* at the end of the converging passage.

Now suppose that the Mach number is one at some point in a flow. Since infinite acceleration ($dV \to \pm\infty$) is not possible unless there is a shock wave, which is specifically excluded by the assumptions, Eq. 11.92 implies that

$$dA = 0 \quad \text{when Ma} = 1$$

In isentropic flow, sonic flow (Ma = 1) can occur only at an area minimum (**Fig. 11.9**). The occurrence of sonic flow at an area maximum is excluded by the considerations identified in the preceding paragraph. To accelerate a fluid from rest (or a subsonic Mach number) to a supersonic Mach number, a *converging–diverging nozzle* is required. A rocket engine such as the one shown in the adjacent figure uses a converging–diverging nozzle to accelerate hot gases to a very large Mach number with corresponding high velocity in order to generate sizeable thrust.

A converging–diverging duct is required to accelerate a flow from subsonic to supersonic flow conditions.

Finally, suppose that a compressible fluid is flowing in a passage that actually has an area minimum (called a *throat*). Is it always possible to conclude that the Mach number is one at the throat? To investigate this question, we rearrange Eq. 11.92 as follows:

$$\frac{dA}{A} = (\text{Ma}^2 - 1)\frac{dV}{V}$$

If dA is zero, then two options are possible:

$$\text{Ma} = 1 \quad \text{or} \quad dV = 0$$

Simply knowing that $dA = 0$ does not allow us to determine which condition occurs. **Figure 11.10** illustrates possible velocity (or Mach number) distributions in a passage with a minimum area. If the throat Mach number is not one, the velocity must have a local maximum or minimum at the throat ($dV_{th} = 0$). This behavior is observed in a Venturi tube in incompressible flow as discussed in Chapters 4 and 8. If the throat Mach number is one, dV_{th}, can be either positive or negative, and the velocity can either increase or decrease downstream from the throat, depending on downstream conditions.

FIGURE 11.10 Possible velocity (or Mach number) distributions in passage with an area minimum.

The following points summarize our findings about isentropic flow:

- The response of the flow to a specific type of area change is exactly opposite for subsonic and supersonic flow. Details are given in Table 11.2.

- Sonic flow (Ma = 1) can occur only at a minimum area. Minimum areas occur at the inlet of a simply diverging passage, the outlet of a simply converging passage, and the throat of a converging–diverging passage.

- It is possible, but not necessary, that the Mach number at the throat of a converging–diverging passage be one. If the Mach number at a throat is not one, the velocity must pass through a maximum or minimum. If the throat Mach number is one, the fluid may either accelerate or decelerate downstream from the throat. Further information is required to determine which occurs.

We used a completely general property relation (the second Tds equation), so these conclusions are valid for one-dimensional isentropic flow of any compressible fluid, ideal gas or not.

11.7.2 Isentropic Flow of an Ideal Gas with Area Change

By assuming that the fluid is an ideal gas, we can obtain working equations for flow in variable-area ducts. The adjacent figure illustrates the general flow geometry. The typical problem involves relating changes of fluid properties and velocity to the passage area. Calculation of mass flowrate or force on the duct walls is also important. For isentropic flow, we can use Eqs. 11.71–11.78 and Table D.I or Fig. D.I to relate pressure, temperature, velocity, and Mach number for flow between any two planes in the duct. All we need for a complete analysis are equations relating Mach number to passage area and mass flowrate to Mach number, flow properties, and passage area.

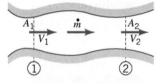

Isentropic flow between planes 1 and 2 in a variable-area duct.

Area–Mach Number Relation

To obtain an equation relating area and Mach number, we use the continuity equation for a control volume with inflow at plane 1 and outflow at plane 2 (see the adjacent figure):

$$\rho_1 V_1 A_1 = \rho_2 V_2 A_2$$

For an ideal gas,

$$\rho = \frac{p}{RT} \quad \text{and} \quad V = \text{Ma}(c) = \text{Ma}\sqrt{kRT}$$

Substituting these equations into the continuity equation gives

$$\sqrt{\frac{k}{R}} \left(\frac{p_1 A_1 \text{Ma}_1}{\sqrt{T_1}} \right) = \sqrt{\frac{k}{R}} \left(\frac{p_2 A_2 \text{Ma}_2}{\sqrt{T_2}} \right)$$

Canceling k and R and solving for the area ratio, we have

$$\frac{A_2}{A_1} = \frac{\text{Ma}_1}{\text{Ma}_2} \left(\frac{p_2}{p_1} \right)^{-1} \left(\frac{T_2}{T_1} \right)^{1/2}$$

For isentropic flow, we can substitute Eqs. 11.72 and 11.76 for the temperature and pressure ratios, giving

$$\frac{A_2}{A_1} = \frac{\text{Ma}_1}{\text{Ma}_2} \left\{ \frac{1 + [(k-1)/2]\text{Ma}_2^2}{1 + [(k-1)/2]\text{Ma}_1^2} \right\}^{(k+1)/2(k-1)}$$

If any three of the quantities A_1, Ma_1, A_2, and Ma_2 are specified, the fourth can be calculated. Usually A_1 and Ma_1 are known and either Ma_2 or A_2 is to be found. If Ma_2 is given and A_2 is to be calculated, the calculation is simple; however, if A_2 is given and Ma_2 must be calculated, that is not so easy. We can simplify the calculations by making a table or graph. To tabulate area ratio, we select a reference area with known Mach number. The obvious choice is the area corresponding to a Mach number of one. We set $Ma_1 = 1$, $A_1 = A^*$ and drop the subscript 2 to get

$$\frac{A}{A^*} = \frac{1}{Ma}\left[\frac{2}{k+1}\left(1 + \frac{k-1}{2}Ma^2\right)\right]^{(k+1)/2(k-1)} \qquad (11.93)$$

In Eq. 11.93, A is the area at the location where the Mach number is Ma, and A^* (the critical area) is the area that *would correspond to* a Mach number of one. Note that A^* is a reference area; the actual area need not be A^* anywhere in the flow. Values of A/A^* for $k = 1.4$ are included in the isentropic flow table and the isentropic flow function graphs in Appendix D. The relationship between A/A^* and Ma for any value of k can usually be obtained from "calculators" on the Internet or a smartphone. The area ratio function can be used to relate areas and Mach numbers at two planes in a duct by

The ratio of flow area to the critical area is a useful concept for isentropic duct flow.

$$\frac{A_2}{A_1} = \frac{A}{A^*}\bigg)_{Ma_2}\bigg/\frac{A}{A^*}\bigg)_{Ma_1} \qquad (11.94)$$

A plot of the area ratio function versus Mach number (the adjacent figure) reveals some interesting information and confirms some of the observations made in the preceding section. First, we note that A/A^* is always greater than one except at Ma $= 1$, where A/A^* is 1. This confirms that in isentropic flow, the Mach number can be one only at the minimum area. Next, we note that if Mach number is given, there is a single corresponding area ratio, so the problem of calculating the area required to obtain a given Mach number seems easy. We must be careful, however. If the initial Mach number (Ma_1) is less than one and the desired final Mach number (Ma_2) is greater than one, a minimum area exactly equal to A^* must occur between planes 1 and 2; otherwise the flow cannot pass through Ma $= 1$.

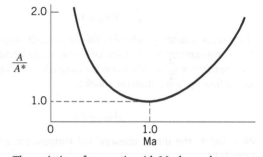

The variation of area ratio with Mach number for isentropic flow of an ideal gas ($k = 1.4$, linear coordinate scales).

Finally, we note that if A/A^* is known, there are *two* corresponding Mach numbers. Choice of the proper Mach number for a known area ratio requires more information than just the area ratio itself. Generally, we must know the pressure or some other item of information in addition to the area ratio to obtain the Mach number. We can sometimes logically eliminate one of the possibilities. If we know that the flow is subsonic at plane 1 and that a minimum area exactly equal to A^* does not occur between planes 1 and 2, the flow at 2 cannot be supersonic.

Before concluding this section on area ratio, we offer a few comments and warnings. The critical area A^* is the area necessary to accelerate or decelerate a flow to sonic conditions. We do not necessarily identify A^* with any *real* area in a flow passage; it is a reference condition, just like the stagnation state is a reference condition. If we know that the flow is sonic somewhere in a passage, the passage area equals the critical area at the sonic location. We also know that it must be the minimum area if the flow is isentropic. However, the converse is not always true. If there is a minimum area in a flow passage, we *cannot* assume that the minimum area is equal to A^* without further evidence that the flow actually is sonic at the minimum area. Stated simply, in a passage with a throat, the throat area (A_{th}) does not necessarily equal the critical area (A^*) unless other evidence indicates it is.

EXAMPLE 11.7 | Area Ratio Function and Flow in a Converging–Diverging Passage

Given The small laboratory wind tunnel (similar to one used in Ref. 1) shown in **Fig. E11.7** draws air from the atmosphere. The pressure and temperature of the laboratory air are 100 kPa and 20 °C (293.3 K). Schlieren photographs reveal a shock wave in the nozzle, as shown.

Find The air temperature and pressure at the throat and at planes 1, 2, and 3. Assume isentropic flow except for the shock.

Solution The passage is converging–diverging, so we must be alert to the possibilities of sonic flow at the throat and supersonic flow in the diverging section. There is a shock in the diverging section of the

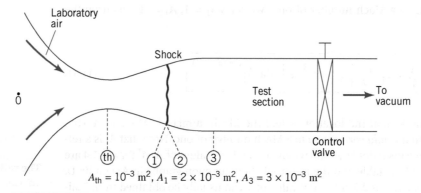

$A_{th} = 10^{-3} \text{ m}^2$, $A_1 = 2 \times 10^{-3} \text{ m}^2$, $A_3 = 3 \times 10^{-3} \text{ m}^2$

FIGURE E11.7 Small laboratory wind tunnel.

nozzle so the flow is supersonic at 1. Because the flow is subsonic (Ma $\approx$ 0) in the laboratory, the air must pass through Ma = 1 somewhere upstream from plane 1. The *only* place this can occur is the throat, so

$$\text{Ma}_{th} = 1.0$$

The process is isentropic ahead of the shock, so the stagnation properties are constant up to the shock and equal to the laboratory pressure and temperature (since Ma $\approx$ 0 in the laboratory air). The critical area for the flow upstream from the shock is

$$A^*_{0 \rightarrow 1} = A_{th}$$

We calculate the throat pressure and temperature with the aid of Table D.I.

$$p_{th} = p_0 \left(\frac{p}{p_0}\right)_{\text{Ma}=1} = 100 \text{ kPa } (0.5283)$$

or

$$p_{th} = 52.8 \text{ kPa} \qquad (Ans)$$

and

$$T_{th} = T_0 \left(\frac{T}{T_0}\right)_{\text{Ma}=1} = 293.2 \text{ K } (0.8333)$$

or

$$T_{th} = 244.3 \text{ K} \qquad (Ans)$$

The flow at plane 1, just ahead of the shock, has

$$A^*_1 = A_{th} = 1.0 \times 10^{-3} \text{ m}^2$$

The area ratio for plane 1 is

$$\frac{A_1}{A^*} = \frac{2 \times 10^{-3} \text{ m}^2}{1 \times 10^{-3} \text{ m}^2} = 2.000$$

From Table D.I, we find the *supersonic* Mach number corresponding to this area ratio:

$$\text{Ma}_1 = 2.20$$

The pressure and temperature ratios (from Table D.I) are

$$\left.\frac{p}{p_0}\right)_{\text{Ma}=2.20} = 0.0935 \quad \text{and} \quad \left.\frac{T}{T_0}\right)_{\text{Ma}=2.20} = 0.5081$$

We now calculate

$$p_1 = 0.0935 \, p_{0_1} = 0.0935 \, (100 \text{ kPa})$$

or

$$p_1 = 9.35 \text{ kPa} \qquad (Ans)$$

and

$$T_1 = 0.5081 T_{0_1} = 0.5081 \, (293.2 \text{ K})$$

or

$$T_1 = 149.0 \text{ K} \qquad (Ans)$$

Plane 2 lies downstream from the shock. We must use the shock functions in Table D.S to relate properties at plane 2 to those at plane 1. For Ma$_1$ = 2.20,

$$\left.\frac{p_y}{p_x}\right)_{\text{Ma}_x=2.20} = 5.480,$$

$$\left.\frac{T_y}{T_x}\right)_{\text{Ma}_x=2.20} = 1.857, \quad \text{and} \quad \text{Ma}_y = 0.547$$

Then

$$p_2 = \frac{p_y}{p_x}\right) p_1 = (5.480)(9.35 \text{ kPa})$$

or

$$p_2 = 51.2 \text{ kPa} \qquad (Ans)$$

and

$$T_2 = \frac{T_y}{T_x}\right) T_1 = (1.857)(149.0 \text{ K})$$

or

$$T_2 = 276.7 \text{ K} \qquad (Ans)$$

To calculate the properties at plane 3, we note that the flow from plane 2 to plane 3 is isentropic, although with different values of p_0 and A^* than the flow upstream from the shock. In particular, A^*_3, which is equal to A^*_2, *is not* equal to the throat area. We calculate the area ratio function at plane 3 from

$$\frac{A}{A^*}\right)_3 = \left(\frac{A_3}{A_2}\right)\left(\frac{A}{A^*}\right)_{\text{Ma}_2}$$

From Table D.I

$$\left.\frac{A}{A^*}\right)_{\text{Ma}=0.547} = 1.259$$

so

$$\frac{A}{A^*}\right)_3 = \left(\frac{3 \times 10^{-3} \text{ m}^2}{2 \times 10^{-3} \text{ m}^2}\right)(1.259) = 1.888$$

This value of A/A^* corresponds to a supersonic Mach number and a subsonic Mach number. As there is no throat between planes 2 and 3, the flow at plane 3 must be subsonic. From Table D.I,

$$Ma_2 = 0.547, \quad \frac{p}{p_0}\bigg)_{Ma=0.547} = 0.8159$$

$$Ma_3 = 0.326, \quad \frac{p}{p_0}\bigg)_{Ma=0.326} = 0.9290, \quad \text{and} \quad \frac{T}{T_0}\bigg)_{Ma=0.326} = 0.9792$$

We calculate p_3:

$$p_3 = p_2\frac{p}{p_0}\bigg)_3 \bigg/ \frac{p}{p_0}\bigg)_2 = 51.2 \text{ kPa} \left(\frac{0.9290}{0.8159}\right)$$

or

$$p_3 = 58.3 \text{ kPa} \qquad (Ans)$$

The expression for T_3 is

$$T_3 = T_0 \left(\frac{T}{T_0}\right)_3$$

but as the shock does not affect T_0,

$$T_3 = (293.2 \text{ K})(0.9792)$$

or

$$T_3 = 287.1 \text{ K} \qquad (Ans)$$

Comments Note the logic used to decide that A^* upstream of the shock is equal to A_{th}, the geometric area of the throat. The value of A^* on the down-stream side of the shock can be calculated as

$$A_2^* = \frac{A_2}{(A/A^*)_{Ma_2}} = 1.59 \times 10^{-3} \text{ m}^2$$

Note that $A_2^* > A_1^*$.

There are alternative ways to manipulate the functions and tables to obtain the answers. For example, we could use

$$T_3 = T_2\frac{T}{T_0}\bigg)_3 \bigg/ \frac{T}{T_0}\bigg)_2 \quad \text{and} \quad p_3 = p_{0_1}\frac{p_{0_y}}{p_{0_x}}\bigg)_{Ma_1} \left(\frac{p}{p_0}\right)_3$$

instead of the equations we used.

This example demonstrates how to use the shock functions and tables to patch together different isentropic flows at a shock.

Mass Flow Relations and Choking Several alternative equations can be used for calculation of mass flowrate in a duct. The "primitive variable" equation for mass flowrate is

$$\boxed{\dot{m} = \rho V A} \qquad (11.95)$$

If the fluid is an ideal gas,

$$\rho = \frac{p}{RT} \quad \text{and} \quad V = Ma(c) = Ma\sqrt{kRT}$$

and an alternative equation for mass flowrate is

$$\boxed{\dot{m} = \sqrt{\frac{k}{R}}\left(\frac{pAMa}{\sqrt{T}}\right)} \qquad (11.96)$$

Both Eqs. 11.95 and 11.96 evaluate the mass flowrate in terms of static properties. By substituting

$$p = p_0\left[1 + \left(\frac{k-1}{2}\right)Ma^2\right]^{-k/(k-1)} \quad \text{and} \quad T = T_0\left[1 + \left(\frac{k-1}{2}\right)Ma^2\right]^{-1}$$

we obtain the mass flowrate in terms of stagnation properties:

$$\boxed{\dot{m} = \sqrt{k}\left(\frac{p_0 A}{\sqrt{RT_0}}\right)Ma\left[1 + \left(\frac{k-1}{2}\right)Ma^2\right]^{-(k+1)/2(k-1)}} \qquad (11.97)$$

This equation has certain advantages for isentropic flow because p_0 and T_0 are constants. Any of Eqs. 11.95–11.97 can be used to calculate mass flowrate. The choice between them depends on what data are available.

By setting $Ma = 1$ and $A = A^*$ in Eq. 11.97, we get

$$\boxed{\dot{m} = \frac{p_0 A^*}{\sqrt{RT_0}}\sqrt{k}\left(\frac{2}{k+1}\right)^{(k+1)/2(k-1)}} \qquad (11.98)$$

Equation 11.98 shows that A^* is constant in steady isentropic flow with constant values of $\dot{m}$, p_0, T_0, R, and k.

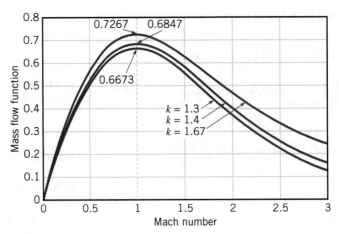

FIGURE 11.11 Dimensionless mass flow function $[(\dot{m}\sqrt{RT_0})/(p_0 A)]$.

We can obtain a dimensionless mass flow function from Eq. 11.97:

$$\frac{\dot{m}\sqrt{RT_0}}{p_0 A} = \sqrt{k}\,\text{Ma}\left(1 + \frac{k-1}{2}\text{Ma}^2\right)^{-(k+1)/2(k-1)} \tag{11.99}$$

A plot of this function is shown in **Fig. 11.11**. The key fact about Fig. 11.11 for this discussion is the occurrence of the maximum at Ma $= 1$:

$$\left.\frac{\dot{m}\sqrt{RT_0}}{p_0 A}\right)_{\text{max}} = f(k\ \text{only})$$

We can draw several conclusions from this fact.

- If p_0 and T_0 are fixed (isentropic flow), a specified mass flowrate can be forced through a limiting minimum area. No smaller area can pass that flow.

- If p_0 and T_0 are fixed (isentropic flow), any particular fixed-geometry duct, with fixed minimum area, can pass only a certain maximum mass flowrate. This maximum mass flowrate occurs when the Mach number at the minimum area is one. Once the Mach number at the minimum area becomes one, it is not possible to increase the mass flowrate.

- It is possible to increase the limiting mass flowrate in a fixed geometry duct or to force a specified mass flow through a smaller area by *increasing* $p_0/\sqrt{T_0}$. Doing work on a gas increases both p_0 and T_0; however, the ratio $p_0/\sqrt{T_0}$ is usually increased by compression.

The preceding statements describe the phenomenon called *choking.*

> *The choking phenomenon occurs in compressible duct flow when the local Mach number reaches 1 at the minimum area in the duct. When this occurs, the mass flowrate through the duct cannot be increased unless the ratio of stagnation pressure to square root of stagnation temperature is increased.*

> **Choked flow occurs when the Mach number is 1.0 at the minimum cross-sectional area.**

EXAMPLE 11.8 | Design of a Compressible Flow Passage

Given The duct between the turbine and exhaust nozzle of a jet engine is 0.60 m in diameter as shown in **Fig. E11.8a**. The gas flowing in the duct has stagnation pressure and temperature of 275 kPa and 700 K, respectively. The gas flowrate is 82 kg/s.

Find The relevant areas of a nozzle for expanding the gas to standard atmospheric pressure. Calculate the force required to hold the nozzle to the duct. The gas has $k = 1.4$ and $R = 265$ N $\cdot$ m/kg $\cdot$ K.

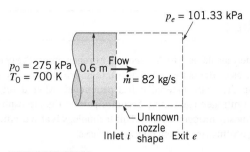

FIGURE E11.8a Schematic diagram of nozzle design problem.

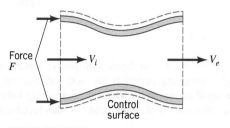

FIGURE E11.8b Control volume for nozzle force analysis.

Solution First establish what the nozzle geometry should be. The desired static-stagnation pressure ratio at the exit plane is

$$\left.\frac{p}{p_0}\right)_e = \frac{p_e}{p_{0_e}} = \frac{101.33 \text{ kPa}}{275.0 \text{ kPa}} = 0.3684$$

From Table D.I, the desired exit Mach number is

$$\text{Ma}_e = 1.285$$

We assume that the inlet Mach number will be less than 1, so we need a minimum area (throat) between the entrance and the exit. We calculate the throat area from Eq. 11.98, with $k = 1.4$:

$$\dot{m} = 0.6847 \frac{p_0 A^*}{\sqrt{RT_0}}$$

For $A_{th} = A^*$,

$$A_{th} = \frac{\dot{m}\sqrt{RT_0}}{0.6847 p_0} = \frac{82 \text{ kg/s}\sqrt{(265 \text{ N} \cdot \text{m/kg} \cdot \text{K})(700 \text{ K})}}{0.6847(275,000 \text{ N/m}^2)}$$

$$A_{th} = 0.1876 \text{ m}^2 \qquad\qquad (Ans)$$

The inlet area of the nozzle is

$$A_i = \frac{\pi}{4} D_i^2 = \frac{\pi}{4}(0.6 \text{ m})^2$$

$$A_i = 0.2874 \text{ m}^2 \qquad\qquad (Ans)$$

The nozzle exit area is that required to produce a Mach number of 1.285. Thus,

$$A_e = A_{th}\frac{A}{A^*}\Big)_{\text{Ma}_e}$$

From Table D.I,

$$\frac{A}{A^*}\Big)_{\text{Ma}=1.285} = 1.0601 \quad \text{so} \quad A_e = 0.1876 \text{ m}^2 (1.0601)$$

$$A_e = 0.1989 \text{ m}^2 \qquad\qquad (Ans)$$

To calculate the force, we apply the linear momentum equation to a control volume enclosing the nozzle, as shown in **Fig. E11.8b**. We assume that the anchoring force, F, acts in the flow direction. The momentum equation gives

$$F = (p_e + \rho_e V_e^2)A_e - (p_i + \rho_i V_i^2)A_i + p_{\text{atm}}(A_i - A_e)$$

From Eq. 11.48,

$$\rho V^2 = kp\text{Ma}^2$$

so

$$F = p_e A_e(1 + k\text{Ma}_e^2) - p_i A_i(1 + k\text{Ma}_i^2) + p_{\text{atm}}(A_i - A_e)$$

We find Ma_i and p_i from

$$\frac{A}{A^*}\Big)_{\text{Ma}_i} = \frac{A_i}{A_{th}} = \frac{0.2874 \text{ m}^2}{0.1876 \text{ m}^2} = 1.5319$$

From Table D.I (using subsonic Mach number),

$$\text{Ma}_i = 0.419 \quad \text{and} \quad \frac{p}{p_0}\Big)_i = 0.8862$$

Then

$$p_i = (0.8862)(275 \text{ kPa}) = 243.7 \text{ kPa}$$

The force F is

$$F = (101,303 \text{ N/m}^2)(0.1989 \text{ m}^2)[1 + 1.4(1.285)^2]$$

$$- (243,700 \text{ N/m}^2)(0.2874 \text{ m}^2)[1 + 1.4(0.419)^2]$$

$$+ 101,330 \text{ N/m}^2(0.2874 \text{ m}^2 - 0.1989 \text{ m}^2)$$

$$= -11,500 \text{ N}$$

$$F = 11,500 \text{ N} \qquad \text{Opposite the flow direction.} \qquad (Ans)$$

Comments A "force function,"

$$\frac{F}{F^*} \equiv \frac{pA(1 + k\text{Ma}^2)}{p^* A^*(1 + k)} = \frac{1 + k\text{Ma}^2}{\text{Ma}\sqrt{2(k + 1)\left(1 + \frac{k-1}{2}\text{Ma}^2\right)}}$$

is listed in some compressible flow tables (Ref. 4).

We treated p_{atm} and p_e as separate entities in the momentum equation even though they are equal for this problem. In some cases, the pressure at the exit of a supersonic nozzle is not equal to the pressure of the surrounding atmosphere (Section 11.7.4).

The nozzle flow was calculated as isentropic because the exit Mach number was to be supersonic; therefore no shocks should occur in the nozzle.

The flow in the nozzle is choked. The given flowrate (82 kg/s) with the given stagnation properties (275 kPa, 700 K) cannot pass through any area smaller than 0.1876 m^2.

The Wide World of Fluids

Liquid knife

A supersonic stream of liquid nitrogen is capable of cutting through engineering materials like steel and concrete. Originally developed at the Idaho National Engineering Laboratory for cutting open barrels of waste products, this technology is now more widely available. The fast-moving nitrogen enters the cracks and crevices of the material being cut, then expands rapidly and breaks up the solid material it has penetrated. After doing its work, the nitrogen gas simply becomes part of the atmosphere, which is mostly nitrogen already. This technology is also useful for stripping coatings even from delicate surfaces.

11.7.3 Operation of a Converging Nozzle

We can show how the information about compressible flow in variable-area ducts fits together by examining the flow in a converging nozzle. **Figure 11.12** shows a converging nozzle that exhausts gas from a large reservoir to a variable-pressure receiver. We assume that the pressure and temperature in the reservoir are constant. We can vary the pressure in the exhaust region (called the *back pressure*) by means of the control valve that connects the receiver to a downstream vacuum pump. By opening or closing the control valve, we decrease or increase the back pressure and therefore expect to increase or decrease the nozzle flowrate. We assume that there is no wall friction and no heat transfer or work.

We can make several preliminary observations about the flow in the nozzle.

- The nozzle is only convergent, so the flow cannot pass *through* $\text{Ma} = 1$.
- The flow at the nozzle inlet is subsonic ($\text{Ma} \approx 0$), so the flow in the entire nozzle is subsonic, with the possible exception of the nozzle exit.
- The flow cannot be supersonic in the nozzle, so there can be no shocks and the flow is isentropic everywhere in the nozzle. The stagnation properties are constant and equal to the gas properties in the reservoir.
- The maximum possible Mach number in the nozzle is one. This value can occur only at the nozzle exit (minimum area).
- There is a maximum possible mass flowrate that can occur. This maximum is determined by the gas constants (k and R), the values of the reservoir (stagnation) properties, and the exit area. The maximum occurs only when the Mach number is one at the nozzle exit.

The flow through the nozzle is described in terms of the pressure and Mach number distribution in the nozzle and the (dimensionless) mass flowrate ($\dot{m}\sqrt{RT_0}/p_0A_e$), all as functions of the ratio of back pressure to supply pressure and k. We also plot the ratio of pressure in the nozzle exit plane (p_e) to the back pressure (p_e/p_B). The various curves are shown in **Fig. 11.13**.

The curves and points labeled *a* correspond to a closed control valve. There is no flow. The pressure equals the reservoir pressure everywhere, and the Mach number is zero everywhere.

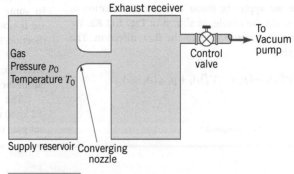

Exhaust receiver

Gas Pressure p_0 Temperature T_0

To Vacuum pump

Control valve

Supply reservoir Converging nozzle

FIGURE 11.12 Apparatus for studying flow in a converging nozzle.

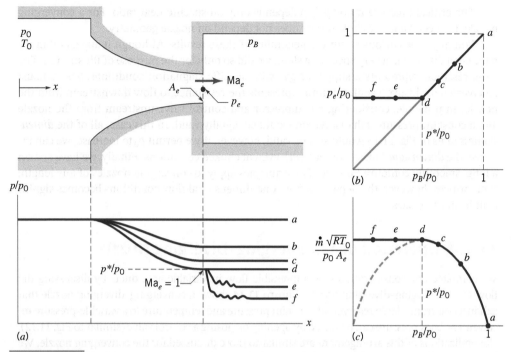

FIGURE 11.13 Performance of a converging nozzle at various pressure ratios.

Case *b* corresponds to a slight opening of the control valve. The back pressure is less than the supply pressure and there is flow. The gas accelerates from the reservoir to the nozzle exit, and the pressure drops. Minimum pressure and maximum Mach number are at the nozzle exit. The pressure at the nozzle exit is equal to the back pressure. If we know the ratio p_B/p_0, the equations

$$\frac{p_e}{p_0} = \frac{p_B}{p_0} \quad \text{and} \quad \text{Ma}_e = \text{Ma})_{p_e/p_0}$$

enable us to determine the Mach number at the nozzle exit. We can then calculate the mass flow at the exit (and hence in the entire nozzle) from Eq. 11.97 or 11.99.

Case *c* is similar to case *b*, except that a larger control valve opening permits a lower back pressure with correspondingly higher mass flow and exit Mach number and a lower exit pressure. Note that the maximum Mach number still occurs at the nozzle exit and is less than 1.

In case *d*, the control valve has been opened a sufficient amount to bring the exit Mach number to a value of 1.0. The exit pressure now equals the critical pressure (p^*) and also equals the back pressure. Knowing $\text{Ma}_e = 1.0$ allows calculation of the mass flowrate.

Case *e* corresponds to opening the control valve farther than in case *d*. When we try this, we find that no changes occur in the nozzle. In case *d*, we reached the limit of the nozzle's capability. The exit Mach number cannot exceed 1, the exit pressure cannot drop below the critical pressure, which is determined solely by the supply pressure and the specific heat ratio (Eq. 11.86), and the mass flow cannot exceed the choked value. The only difference between case *d* and case *e* is that the back pressure and exit pressure are no longer equal. The flow must adjust to the lower back pressure *after* leaving the nozzle. The downstream flow is multidimensional, so the pressure curve is shown as a wavy line downstream from the nozzle exit. Opening the control valve with further lowering of the back pressure does not change the nozzle flow.

At first, these results may seem strange; however, they have a simple physical explanation. Once the exit flow becomes sonic, pressure waves from the exhaust region cannot propagate upstream into the nozzle and supply reservoir to cause any flow adjustment.

There are two *flow regimes* in a simple converging nozzle: *unchoked flow* and *choked flow*. Which flow regime occurs in a given case depends on the relative values of the back pressure and critical pressure.

- If $p_B/p_0 > p^*/p_0$, the flow is unchoked.
- If $p_B/p_0 \leq p^*/p_0$, the flow is choked.

The critical pressure ratio (p^*/p_0) depends only on specific heat ratio. For a converging nozzle, the occurrence of choked flow does not depend on nozzle geometry.

You might be curious about the generality of these results. Although it appears that we have been discussing a very specific system, we did so only for the purpose of illustration. The large reservoir represents a supply of gas with specific stagnation conditions. The exhaust receiver/control valve/vacuum pump represents the resistance to flow downstream from the nozzle. In principle, considering a compressor and control valve upstream from the nozzle and a constant-pressure exhaust region would be equally valid. In this case, all of the *dimensionless* plots of Fig. 11.13 would still be valid; however, if we permit p_0 to increase, we can increase the *dimensional* mass flowrate ($\dot{m}$), even at choked conditions. Finally, we have shown a long, smoothly contoured nozzle. The principles apply to converging nozzles of any length, even orifices; however, the departure from one-dimensional flow conditions becomes significant in short nozzles.

11.7.4 Operation of a Converging–Diverging Nozzle

We complete our consideration of compressible flow in variable-area ducts by discussing the flow in a converging–diverging nozzle. **Figure 11.14** shows a converging–diverging nozzle that exhausts gas from a large reservoir at constant pressure and temperature to a variable-pressure receiver. The back pressure can be varied by opening or closing a control valve (similar to Fig. 11.12). Generalizations of this arrangement are similar to those discussed for the converging nozzle. We assume that there is no wall friction, heat transfer, or work.

Preliminary observations about the flow in the converging–diverging nozzle are as follows:

- The nozzle is convergent–divergent, so the flow can pass through Ma = 1. The flow also can be subsonic everywhere.
- If Ma = 1 anywhere, it must be at the throat.
- There may be supersonic flow in the divergent portion of the nozzle, so there may be shocks in the flow. If there are shocks, the flow is not completely isentropic.
- If there are no shocks, the flow is isentropic. If there are shocks, the flow from the reservoir up to the first shock is isentropic. Flow downstream from a shock also is isentropic but with different values of entropy, stagnation pressure, and critical area.
- The maximum possible Mach number that can occur anywhere in the passage corresponds to acceleration of the fluid in an isentropic process from the reservoir to the nozzle exit. The maximum possible Mach number can occur only at the exit and is determined by the ratio of exit area to throat area.
- The maximum possible mass flowrate in the nozzle is determined by the gas constants, the reservoir properties, and the minimum area, which occurs at the throat.

The flow in the converging–diverging nozzle can be described by plots of pressure and Mach number distributions in the nozzle, the dimensionless mass flowrate ($\dot{m}\sqrt{RT_0}/p_{0_1}A_{th}$), and

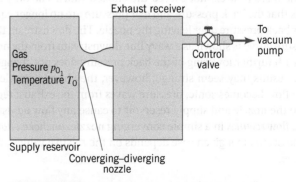

FIGURE 11.14 Apparatus for studying flow in a converging–diverging nozzle.

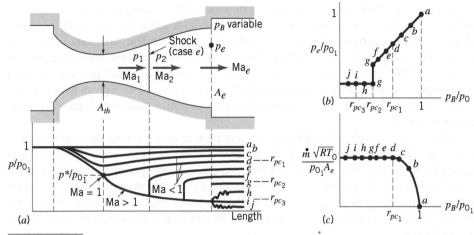

FIGURE 11.15 Performance of converging–diverging nozzle at various pressure ratios.

the ratio of nozzle exit pressure to supply pressure (p_e/p_{0_1}), all as functions of the ratio of back pressure to supply pressure (p_B/p_{0_1}).

The nozzle performance curves are shown in **Fig. 11.15**. The case labeled a corresponds to complete closure of the control valve. There is no flow, the pressure is uniform, and the Mach number is zero everywhere.

Curves and points labeled b represent a slightly open control valve. The gas accelerates in the convergent portion of the nozzle, and the pressure drops. The throat Mach number is considerably less than 1 (subsonic flow). The flow decelerates in the divergent portion of the nozzle, and pressure rises. The pressure and Mach number distributions are roughly symmetrical about the throat. There are no shocks because the flow is subsonic everywhere. The flow at the nozzle exit is subsonic, and the exit and back pressures are equal. The exhaust Mach number can be found from

$$\left.\frac{p}{p_0}\right)_{\mathrm{Ma}_e} = \frac{p_B}{p_{0_1}}$$

and the mass flow can be calculated from Eq. 11.97. A value of A^* can be calculated from Eq. 11.93 or Eq. 11.98 applied at the exhaust plane. A^* will be less than A_{th}.

Curve c represents a slightly larger control valve opening. The situation is qualitatively similar to case b, but the mass flow is larger. The pressure and Mach number distributions are still roughly symmetrical about the throat. The maximum Mach number occurs at the throat.

In case d, the control valve has been opened just enough to bring the throat Mach number exactly to 1.0. The flow is still subsonic everywhere except exactly at the throat, where Ma = 1.0. The flow is isentropic, but now

$$p_{th} = p^* \quad \text{and} \quad A_{th} = A^*$$

The mass flow has reached its maximum possible value and the flow has just become choked. The pressure rises downstream from the throat, so the back pressure at which the converging–diverging nozzle chokes is greater than p^*. The pressure ratio that causes choking to first occur in a converging–diverging nozzle is called the *first critical pressure ratio* (r_{pc_1}) and can be calculated by

$$r_{pc_1} = \left.\frac{p}{p_0}\right)_{A/A^*=A_e/A_{th},\,\mathrm{Ma}<1} \tag{11.100}$$

that is, by finding the pressure ratio corresponding to the subsonic value of A/A^*.

What happens when the valve is opened beyond the case d point? Because the throat is choked, the mass flow cannot be increased, and conditions upstream of the throat cannot be affected. There is a response in the flow downstream from the throat because, at case d, the downstream flow is subsonic. When the valve is opened and the back pressure is lowered, the fluid begins to accelerate as it enters the divergent portion of the nozzle; that is, the flow becomes supersonic. If the control valve is opened only slightly past the case d point, the downstream resistance is too large for complete acceleration in the divergent portion of the nozzle, so a shock occurs in the divergent portion.

A variety of flow situations can occur for flow in a converging–diverging duct.

Curves and points labeled *e* represent the flow for a slightly more open control valve. The flow accelerates in the convergent portion of the nozzle, reaches sonic speed at the throat, and accelerates to supersonic speed downstream from the throat. The supersonic acceleration terminates in a shock wave. Downstream from the shock, the flow experiences subsonic deceleration and exits the passage with $Ma_e < 1$. The exit pressure and back pressure are equal. The exact position of the shock depends on the back pressure, and for a given back pressure, is fixed. The mass flowrate for case *e*, as well as for all lower values of back pressure, is the same as for case *d*.

Opening the control valve and lowering the back pressure causes the shock to move downstream (case *f*). Note that once shock passes a plane, the flow *up to* that plane is no longer affected by lowering the back pressure. As the valve is opened farther, the shock is eventually drawn to the nozzle exit. The flow accelerates isentropically all the way from the reservoir to the exit shock. Exactly at the exit, the pressure jumps to the back pressure as the fluid exits through the shock. This situation is shown as case *g*. The exit pressure is double-valued at case *g*.

Lowering the back pressure further causes the shock to move out of the nozzle and become multidimensional (case *h*). The flow accelerates isentropically from the reservoir to the exit. The gas exits the nozzle supersonically with Ma_e corresponding to the nozzle's exit-to-throat area ratio. The exit pressure is determined only by the stagnation pressure (p_{0_1}) and the exit Mach number and is not equal to the back pressure. The gas adjusts to the back pressure externally. If we continue to lower the back pressure, it eventually reaches equality with the exit pressure (case *i*). At this condition, there is no pressure adjustment in the exiting gas and the flow is said to be *perfectly expanded*. Lowering the pressure further (case *j*) requires external expansion pressure adjustments but does not affect the nozzle flow.

We can summarize this information about converging–diverging nozzle flow as follows. There are four regimes of flow in a converging–diverging nozzle.

- *Venturi regime* (cases *a–d*). The flow is subsonic and isentropic everywhere. The flow accelerates in the convergent portion and decelerates in the divergent portion. Maximum Mach number and minimum pressure occur at the throat.

- *Shock regime* (cases *d–g*). The flow is subsonic in the convergent portion, sonic at the throat, and partly supersonic in the divergent portion. The acceleration terminates in a shock that stands in the divergent portion at a location determined by the exact value of the back pressure. The flow experiences subsonic deceleration from the shock to the exit. The flow is choked.

- *Overexpanded regime* (cases *g–i*). The flow accelerates throughout the nozzle. The throat flow is sonic, and the exit flow is supersonic. The pressure of the gas increases to the back pressure downstream from the nozzle exit.

- *Underexpanded regime* (cases *i–j*). This case is similar to the overexpanded regime, except that the external pressure adjustments are expansive rather than compressive.

The boundaries between the four flow regimes are indicated by curves *d*, *g*, and *i*. The ratios of back pressure to supply pressure that correspond to these cases are called the *first critical pressure ratio* (r_{pc_1}), the *second critical pressure ratio* (r_{pc_2}), and the *third critical pressure ratio* (r_{pc_3}). These ratios are calculated in the following manner.

- First critical pressure ratio: See Eq. 11.100.
- Second critical pressure ratio: Assume that a shock stands at the nozzle exit and calculate the ratio of *downstream static* to *upstream stagnation* pressure:

$$r_{pc_2} = \left.\frac{p_y}{p_x}\right)_{Ma_e} \times r_{pc_3} \tag{11.101}$$

where $\left.\dfrac{p_y}{p_x}\right)_{Ma_e}$ is the shock static pressure ratio function and Ma_e is given by

$$Ma_e = Ma)_{A/A^*=A_e/A_{th}, Ma>1}$$

Video

V11.6 Rocket engine start-up

Video

V11.7 Supersonic nozzle flow

- Third critical pressure ratio:

$$r_{pc_3} = \frac{p}{p_0}\bigg)_{Ma_e} \qquad (11.102)$$

Calculation of these ratios is easier than it looks. Note that the three critical pressure ratios for a converging–diverging nozzle depend only on the nozzle geometry and the specific heat ratio of the flowing gas. The first step in calculating flow through a converging–diverging nozzle is to classify the flow. You do so by calculating the three critical pressure ratios and comparing them to the actual ratio of back pressure to supply pressure. After you have identified the proper flow regime, the rest of the problem falls into place easily.

The Wide World of Fluids

Rocket nozzles

To develop the massive thrust needed for Space Shuttle liftoff, the gas leaving the rocket nozzles had to be moving supersonically. For this to happen, the nozzle flow path must first converge, then diverge as shown in the adjacent figure. Entering the nozzle at very high pressure and temperature, the gas accelerates in the converging portion of the nozzle until the flow chokes at the nozzle throat. Downstream of the throat, the gas further accelerates in the diverging portion of the nozzle (area ratio of 77.5 to 1), finally exiting into the atmosphere supersonically. At launch, the static pressure of the gas flowing from the nozzle exit is less than atmospheric and so the flow is overexpanded. At higher elevations where the atmospheric pressure is much less than at launch level, the static pressure of the gas flowing from the nozzle exit is greater than atmospheric and so now the flow is underexpanded. The result is expansion or divergence of the exhaust gas plume as it exits into the atmosphere.

EXAMPLE 11.9 | Converging and Converging–Diverging Nozzles

Given A thick-walled pressure vessel has a hole in its side. **Figure E11.9** shows two possible configurations for the hole. The pressure and temperature of the air inside the vessel are 137900 Pa and 311 K, respectively. The pressure outside the vessel is 100667 Pa.

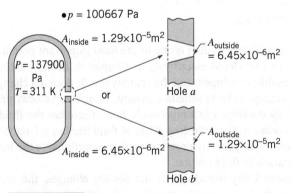

FIGURE E11.9 Thick-walled pressure vessel with two possible hole configurations.

Find The leak rate of air for each hole configuration.

Solution First recognize that hole *a* acts like a simply converging nozzle and hole *b* acts like a converging–diverging nozzle. The "converging portion" of hole *b* is formed by streamlines within the tank.

Both holes have the same minimum area and the supply temperature and pressure are the same, so the maximum possible flowrate is the same for both holes. If both holes are choked, they have the same flowrate. We first check both holes for choking.

Hole a

Hole *a* will be choked if

$$\frac{p_B}{p_{0_1}} < \frac{p^*}{p_0} = 0.5283$$

For hole *a*,

$$\frac{p_B}{p_{0_1}} = \frac{100667}{137900} = 0.730$$

Hole *a* will not be choked.

Hole b

Hole *b* will be choked if

$$\frac{p_B}{p_{0_1}} < r_{pc_1}$$

where

$$r_{pc_1} = \frac{p}{p_0}\bigg)_{A/A^*} \quad \text{where } A/A^* = \frac{A_e}{A_{th}} = 1.29 \times 10^{-5}/6.45 \times 10^{-6} = 2.0, Ma < 1$$

From Table D.I

$$r_{pc_1} = 0.9371$$

The back pressure ratio is still 0.730, so hole b is choked. Next, calculate the mass flow through hole b using Eq. 11.98, with $k = 1.4$ and Ma = 1:

$$\dot{m}_b = 0.6847 \left(\frac{p_0 A_{\min}}{\sqrt{RT_0}} \right)$$

$$= 0.6847 \left[\frac{(137900 \text{ Pa})(6.45 \times 10^{-6} \text{ m}^2)}{\sqrt{287 \text{ J/kg} \cdot \text{K}(311 \text{ K})}} \right]$$

$$\dot{m}_b = 0.00203 \text{ kg/s} \qquad (Ans)$$

To calculate the leak rate in hole a, we need the Mach number at the exit. For

$$\frac{p_e}{p_0} = \frac{p_b}{p_0} = 0.730$$

Table D.I gives

$$\text{Ma}_e = 0.686$$

Using Eq. 11.99, we have

$$\dot{m}_a = \frac{(137900 \text{ Pa})(6.45 \times 10^{-6} \text{ m}^2)}{\sqrt{(287 \text{ J/kg} \cdot \text{K})(311 \text{ K})}}$$
$$\times \sqrt{1.4}(0.686)[1 + (0.2)(0.686)^2]^{-3}$$

$$\dot{m}_a = 0.00240 \text{ kg/s} \qquad (Ans)$$

Comments These answers are only approximations because the flow through real holes in the wall of a pressure vessel would probably not be one-dimensional. The third and second critical pressure ratios for hole b are

$$r_{pc_3} = \left. \frac{p}{p_0} \right)_{A/A^*} \text{ where } A/A^* = \frac{A_e}{A_{th}} = 2, \text{Ma} > 1 = r_{pc_3} = 0.0935$$

$$(\text{note: Ma}_e = 2.20)$$

and

$$r_{pc_2} = \left. \frac{p_y}{p_x} \right)_{\text{Ma}=2.20} \times r_{pc_3}$$

Using Table D.S, we have

$$r_{pc_2} = (5.480)(0.0935) = 0.512$$

Hole b is operating in the shock regime.

11.8 CONSTANT-AREA DUCT FLOW WITH FRICTION

We now examine the effects of wall friction in one-dimensional compressible flow in a duct. To simplify our discussion, we limit consideration to flow in ducts of constant cross-sectional area. We assume that the flow is one-dimensional, with the velocity varying only with duct length but *not* with distance across the duct. This situation is somewhat artificial because a transverse velocity gradient is always associated with shear stress. The situation that we have assumed is most closely approximated by turbulent flow. The single value of fluid velocity associated with any duct cross section should be interpreted as an average velocity.

11.8.1 Preliminary Consideration: Comparison with Incompressible Duct Flow

Flow of an incompressible fluid in a constant-area duct is one of the most important practical problems in fluid mechanics and was considered extensively in Chapter 8. Incompressible flow in long ducts eventually becomes fully developed, and the velocity profile does not change shape. Continuity requires that the average velocity remain constant, so neither velocity profile shape nor magnitude vary in fully developed incompressible flow. Therefore the fluid's kinetic energy and momentum are constant. The retarding effect of fluid friction is balanced by a pressure drop in the flow direction. This pressure drop is directly proportional to the irreversible loss of mechanical energy caused by fluid friction.

What changes does fluid compressibility introduce? If the density changes, the average velocity must also change to satisfy the continuity equation. Changing the average velocity changes the magnitude of the entire velocity profile. More important, if the velocity changes, the fluid's kinetic energy and momentum also change. Even if the flow is fully developed, the retarding effect of fluid friction is balanced by *both* pressure *and* momentum changes. Although fluid friction causes an irreversible loss of mechanical energy, this loss cannot be interpreted in terms of a (static) pressure drop. Pressure changes in *compressible* flow, *cannot* be calculated from the Darcy–Weisbach equation. In compressible flow, friction appears explicitly only in the momentum equation. The effects of friction on flow can be determined only by simultaneous solution of the equations of continuity, momentum, energy, and state.

We must use the general energy equation, so we make an explicit assumption about work and heat transfer. In keeping with the approach of considering inputs one at a time, in addition to constant area, we assume that there is no shaft work and that there is no heat transfer. Adiabatic flow with friction in a constant area duct is called *Fanno flow* after Gino Girolamo Fanno, an Italian mechanical engineer who developed the analysis.

If the duct is reasonably short, the adiabatic flow assumption is reasonable. Because compressibility effects usually occur only if the fluid velocity is large and because, on dimensional grounds, mechanical energy loss increases as the square of the velocity, ducts in which friction and compressibility are both significant are usually kept short and heat transfer can be neglected.

> Fanno flow involves wall friction with no heat transfer and constant cross-sectional area.

11.8.2 The Fanno Line

As we did with variable-area flow, we will first investigate what we can learn about Fanno flow without specifying any particular fluid. Consider the steady, one-dimensional, and adiabatic flow of a compressible fluid through the constant area duct shown in **Fig. 11.16**. This is Fanno flow. For the control volume indicated, the energy equation (Eq. 5.69) leads to

$$\dot{m}\left[\check{h}_2 - \check{h}_1 + \frac{V_2^2 - V_1^2}{2} + g\left(z_2 - z_1\right)\right] = \dot{Q}_{\substack{net \\ in.}} + \dot{W}_{\substack{shaft \\ net\ in}}$$

0 (negligibly small for gas flow)

0 (flow is adiabatic)

0 (no compressors or turbines)

or

$$\check{h} + \frac{V^2}{2} = \check{h}_0 = \text{constant} \tag{11.103}$$

where $\check{h}_0$ is the stagnation enthalpy, which is constant for flow with no work and no heat transfer.

The continuity equation for the control volume shown in Fig. 11.16 is

$$\rho_1 V_1 A_1 = \rho_2 V_2 A_2$$

Since the area is constant, this can be reduced to

$$\rho V = G = \text{constant} \tag{11.104}$$

where G, the density-velocity product, is called the *mass flow density*. Combining Eqs. 11.103 and 11.104 we get

$$\check{h} + \frac{(\rho V)^2}{2\rho^2} = \check{h}_0$$

or

$$\check{h} + \frac{G^2}{2\rho^2} = \check{h}_0 \tag{11.105}$$

With G and $\check{h}_0$ constant, Eq. 11.105 is a relationship between $\check{h}$ and ρ (or $\check{h}$ and v, since $v = 1/\rho$), which, for specific values of G and $\check{h}_0$, can be graphed as a line, as shown in the adjacent figure. This line is called a *Fanno line*. It represents all possible states of the fluid for specific values of G and $\check{h}_0$.

For any simple compressible fluid, if two properties are known, the other properties can be determined. If we pick several points on the $\check{h} - \rho$ curve, we can determine corresponding values of temperature and entropy and graph the Fanno line on a T–s diagram, as shown in **Fig. 11.17**.

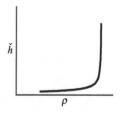

Enthalpy-Density Curve for Fanno Flow

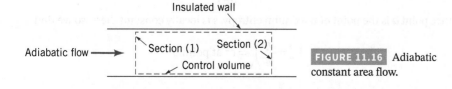

FIGURE 11.16 Adiabatic constant area flow.

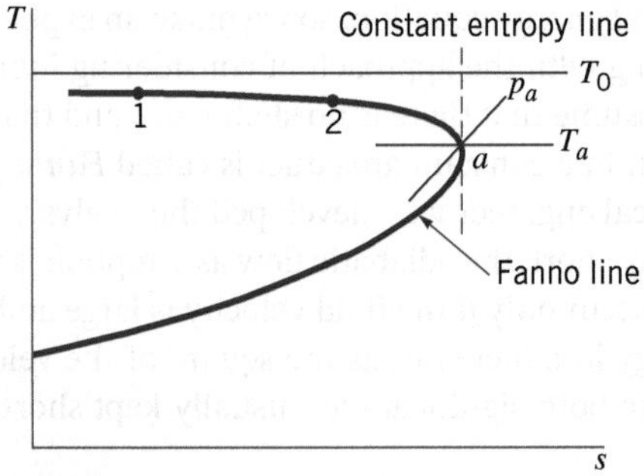

FIGURE 11.17 The T–s diagram for Fanno flow.

We can observe many important features of Fanno flow by examining Fig. 11.17. Any flow state, such as that at Section 1 in Fig. 11.16, can be shown as a single point on the corresponding Fanno line. The action of friction between the duct wall and the fluid between Section 1 and Section 2 causes the fluid properties to change and the state point on the Fanno line to move from point 1 to point 2. The second law of thermodynamics tells us that entropy cannot decrease in an adiabatic process so the state point must always move from left to right, toward maximum entropy, point a.

Entropy increases in Fanno flows because of wall friction.

The fluid kinetic energy (actually, $V^2/2c_p$) is represented by the vertical distance between the "T_0" line and the Fanno line. We can draw two conclusions from this. First, the "upper branch" of the Fanno line represents slower speeds and the "lower branch" represents faster speeds. Second, along the lower branch, friction causes the fluid to slow down, while along the upper branch, friction actually causes the fluid to speed up! Both of these actions are different from incompressible flow where the fluid speed remains constant in fully developed flow.

Point a, the point of maximum entropy, is apparently quite significant. We proceed to find the fluid speed at that point. Fanno flow is governed by the continuity equation, Eq. 11.104, and the energy equation, Eq. 11.103. Differentiating these equations,

$$\rho \, dV + V \, d\rho = 0 \tag{11.106}$$

and

$$d\check{h} + V \, dV = 0 \tag{11.107}$$

From Eq. 11.106

$$dV = -V \frac{d\rho}{\rho} \tag{11.108}$$

Substituting Eq. 11.108 into Eq. 11.107 gives

$$d\check{h} = V^2 \frac{d\rho}{\rho} \tag{11.109}$$

From the second Tds equation, Eq. 11.20

$$d\check{h} = T \, ds + \frac{dp}{\rho} \tag{11.110}$$

The equations above, Eqs. 11.106–11.110, apply anywhere along the Fanno line. At the maximum entropy point, point a, $ds = 0$ and Eq. 11.110 becomes

$$d\check{h} = \frac{dp}{\rho} \tag{11.111}$$

Substituting Eq. 11.111 into 11.109 and rearranging yields

$$V^2 = \frac{dp}{d\rho}\bigg)_a$$

But since point a is the point of maximum entropy, s is locally constant there, so we find

$$V_a^2 = \frac{\partial p}{\partial \rho}\bigg)_s = c^2 \text{ at point } a$$

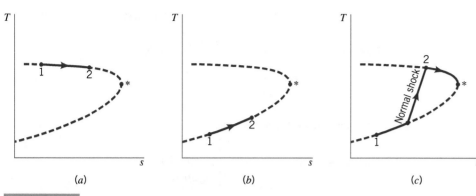

FIGURE 11.18 (a) Subsonic Fanno flow. (b) Supersonic Fanno flow. (c) Normal shock occurrence in Fanno flow.

That is, *the fluid speed equals the sonic speed and* Ma = 1 *at the point of maximum entropy*. It follows that the flow described by the upper (lower speed) branch is subsonic and the flow described by the lower (higher speed) branch is supersonic. We can replace the "a" designation of the maximum entropy point with the more familiar designation of sonic flow, *, thus the temperature is T^*, the pressure is p^*, and so on.

Because the entropy can only increase in an adiabatic process, Fanno flow is always driven toward Ma = 1 by friction. It also follows that the flow cannot pass through Ma = 1. This is another manifestation of *choking*; when the flow reaches sonic speed in a duct, the flow cannot respond to further frictional forces. This in turn means that sonic speed can only occur at the *end* of a frictional duct. One might ask "What happens when a Fanno flow is choked and more frictional duct is added?" The answer is that the mass flowrate $\dot{m}$ and the mass flow density G will be reduced and the flow will be described by another Fanno line, one shifted to the right.

From the Fanno line itself, the effect of friction on the fluid temperature and the fluid entropy can be noted. The effect of friction on static pressure is not surprising; when velocity increases, pressure decreases and vice versa. Lastly, although the flow cannot pass through the point of maximum entropy, it is possible to "jump" from the supersonic branch to the subsonic branch by passing through a normal shock wave.

Some examples of Fanno flow behavior are shown in **Fig. 11.18**. A case involving subsonic Fanno flow that is accelerated by friction to a higher Mach number without choking is illustrated in Fig. 11.18a. A supersonic flow that is decelerated by friction to a lower Mach number without choking is illustrated in Fig. 11.18b. In Fig. 11.18c, an abrupt change from supersonic to subsonic flow in the Fanno duct is represented. This sudden deceleration occurs across a standing normal shock.

The qualitative aspects of Fanno flow that we have already discussed are summarized in **Table 11.3** and **Fig. 11.19**.

Friction accelerates a subsonic Fanno flow.

TABLE 11.3 Summary of Fanno Flow Behavior

Parameter	Flow	
	Subsonic Flow	**Supersonic Flow**
Stagnation temperature	Constant	Constant
Ma	Increases (maximum is 1)	Decreases (minimum is 1)
Friction	Accelerates flow	Decelerates flow
Pressure	Decreases	Increases
Temperature	Decreases	Increases
Stagnation pressure	Decreases	Decreases

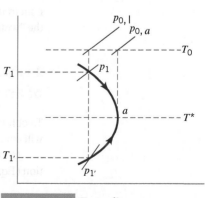

FIGURE 11.19 Fanno line.

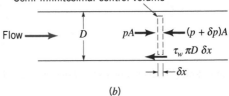

Section (1)   Section (2)

FIGURE 11.20 (a) Finite control volume. (b) Semi-infinitesimal control volume.

Movement of the state point along the Fanno line is caused by friction and governed by the linear momentum equation. Applying this equation (Eq. 5.22) to the flow through the control volume sketched in **Fig. 11.20a** gives

$$p_1 A_1 - p_2 A_2 - R_x = \dot{m}(V_2 - V_1)$$

where R_x is the frictional force exerted by the inner pipe wall on the fluid. Since $A_1 = A_2 = A$ and $\dot{m} = \rho A V = $ constant (i.e., $\rho V = \rho_1 V_1 = \rho_2 V_2$), we obtain

$$p_1 - p_2 - \frac{R_x}{A} = \rho V(V_2 - V_1) \tag{11.112}$$

The differential form of Eq. 11.112, which is valid for Fanno flow through the semi-infinitesimal control volume shown in **Fig. 11.20b**, is

$$-dp - \frac{\tau_w \pi D \, dx}{A} = \rho V \, dV \tag{11.113}$$

The wall shear stress, τ_w, is related to the wall friction factor, f, by Eq. 8.20 as

Friction forces in Fanno flow are given in terms of the friction factor.

$$f = \frac{8\tau_w}{\rho V^2} \tag{11.114}$$

By substituting Eq. 11.114 and $A = \pi D^2/4$ into Eq. 11.113, we obtain

$$-dp - f\rho \frac{V^2}{2} \frac{dx}{D} = \rho V \, dV \tag{11.115}$$

Dividing by p gives the differential momentum equation in dimensionless form

$$\frac{dp}{p} + \frac{\rho V^2}{2} \frac{f}{p} \frac{dx}{D} + \frac{\rho}{p} \frac{d(V^2)}{2} = 0 \tag{11.116}$$

which relates fluid properties (p, ρ) and velocity to the duct friction parameter, $f \, dx/D$. For a noncircular duct, the diameter D is replaced by the hydraulic diameter, $D_h = 4A/P$, where P is the "wetted" perimeter.

11.8.3 Adiabatic Frictional Flow (Fanno Flow) of an Ideal Gas

To obtain quantitative results, it is necessary to specify a particular working fluid. As usual, we will consider an ideal gas.

Combining the ideal gas equation of state (Eq. 11.1), the ideal gas speed-of-sound equation (Eq. 11.46), and the Mach number definition with Eq. 11.116 leads to

$$\frac{dp}{p} + \frac{f}{2} \text{Ma}^2 k \frac{dx}{D} + k \frac{\text{Ma}^2}{2} \frac{d(V^2)}{V^2} = 0 \tag{11.117}$$

Since $V = \text{Ma}\, c = \text{Ma}\sqrt{kRT}$, then

$$V^2 = \text{Ma}^2 kRT$$

or

$$\frac{d(V^2)}{V^2} = \frac{d(\text{Ma}^2)}{\text{Ma}^2} + \frac{dT}{T} \tag{11.118}$$

The application of the energy equation (Eq. 5.69) to Fanno flow gives $T + V^2/2c_p = T_0 = constant$. Differentiating and dividing by T gives

$$\frac{dT}{T} + \frac{d(V^2)}{2c_p T} = 0 \tag{11.119}$$

Substituting Eqs. 11.14, 11.47, into Eq. 11.119 yields

$$\frac{dT}{T} + \frac{k-1}{2}\text{Ma}^2 \frac{d(V^2)}{V^2} = 0 \tag{11.120}$$

which can be combined with Eq. 11.118 to form

$$\frac{d(V^2)}{V^2} = \frac{d(\text{Ma}^2)/\text{Ma}^2}{1 + [(k-1)/2]\,\text{Ma}^2} \tag{11.121}$$

We can combine the continuity and state equations with Eq. 11.118 to get

$$\frac{dp}{p} = \frac{1}{2}\frac{d(V^2)}{V^2} - \frac{d(\text{Ma}^2)}{\text{Ma}^2} \tag{11.122}$$

Consolidating Eqs. 11.122 and 11.117 leads to

$$\frac{1}{2}(1 + k\,\text{Ma}^2)\frac{d(V^2)}{V^2} - \frac{d(\text{Ma}^2)}{\text{Ma}^2} + \frac{fk}{2}\text{Ma}^2\frac{dx}{D} = 0 \tag{11.123}$$

Finally, incorporating Eq. 11.121 into Eq. 11.123 yields

$$\frac{(1 - \text{Ma}^2)\,d(\text{Ma}^2)}{\{1 + [(k-1)/2]\text{Ma}^2\}k\text{Ma}^4} = f\frac{dx}{D} \tag{11.124}$$

Equation 11.124 can be integrated from one section to another in a Fanno flow duct. We elect to use the critical (*) state as a reference and to integrate Eq. 11.124 from an upstream state to the critical state. Thus

$$\int_{\text{Ma}}^{\text{Ma}^*=1} \frac{(1 - \text{Ma}^2)\,d(\text{Ma}^2)}{\{1 + [(k-1)/2]\text{Ma}^2\}k\text{Ma}^4} = \int_{\ell}^{\ell^*} f\frac{dx}{D} \tag{11.125}$$

where ℓ is length measured from an arbitrary but fixed upstream reference location to a section in the Fanno flow. For an approximate solution, we can assume that the friction factor is constant at an average value over the integration length, $\ell^* - \ell$. We also consider a constant value of k. Thus, we obtain from Eq. 11.125

> For Fanno flow, the Mach number is a function of the distance to the critical state.

$$\boxed{\frac{1}{k}\frac{(1 - \text{Ma}^2)}{\text{Ma}^2} + \frac{k+1}{2k}\ln\left\{\frac{[(k+1)/2]\,\text{Ma}^2}{1 + [(k-1)/2]\,\text{Ma}^2}\right\} = \frac{f(\ell^* - \ell)}{D}} \tag{11.126}$$

For a given value of k, values of $f(\ell^* - \ell)/D$ can be tabulated as a function of Mach number for Fanno flow. For example, values of $f(\ell^* - \ell)/D$ for air ($k = 1.4$) Fanno flow are graphed as a function of Mach number in Fig. D.F in Appendix D and in the adjacent figure and tabulated in Table D.F. Note that the critical state does not have to exist in the actual Fanno flow being considered, since for any two sections in a given Fanno flow

$$\boxed{\frac{f(\ell^* - \ell_2)}{D} - \frac{f(\ell^* - \ell_1)}{D} = \frac{f}{D}(\ell_1 - \ell_2)} \tag{11.127}$$

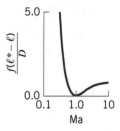

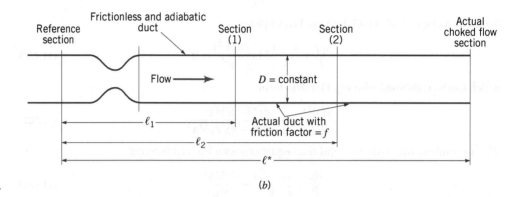

Frictionless and adiabatic duct

Reference section — Section (1) — Section (2) — Imagined choked flow section

Flow ⟶ D = constant

Actual duct with friction factor = f Imagined duct friction factor = f

ℓ_1

ℓ_2

ℓ^*

(a)

Frictionless and adiabatic duct

Reference section — Section (1) — Section (2) — Actual choked flow section

Flow ⟶ D = constant

Actual duct with friction factor = f

ℓ_1

ℓ_2

ℓ^*

(b)

FIGURE 11.21 (a) Unchoked Fanno flow. (b) Choked Fanno flow.

For Fanno flow, the length of duct needed to produce a given change in Mach number can be determined.

The sketch in **Fig. 11.21** illustrates the physical meaning of Eq. 11.127. Recall that for a noncircular duct, D is replaced by D_h.

For a given Fanno flow (constant specific heat ratio, duct diameter, and friction factor) the length of duct required to change the Mach number from Ma_1 to Ma_2 can be determined from Eqs. 11.126 and 11.127, a graph such as Fig. D.F, a table such as Table D.F, or even software. To get the values of other fluid properties in Fanno flow we need to develop more equations.

By consolidating Eqs. 11.118 and 11.120 we obtain

$$\frac{dT}{T} = -\frac{(k-1)}{2\{1 + [(k-1)/2]\,\mathrm{Ma}^2\}}\, d(\mathrm{Ma}^2) \tag{11.128}$$

Integrating Eq. 11.128 from any state upstream in a Fanno flow to the critical (*) state leads to

$$\boxed{\frac{T}{T^*} = \frac{(k+1)/2}{1 + [(k-1)/2]\,\mathrm{Ma}^2}} \tag{11.129}$$

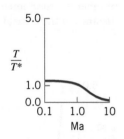

$\dfrac{T}{T^*}$

The velocity ratio is

$$\frac{V}{V^*} = \frac{\mathrm{Ma}\sqrt{kRT}}{\sqrt{kRT^*}} = \mathrm{Ma}\sqrt{\frac{T}{T^*}} \tag{11.130}$$

Substituting Eq. 11.129 into 11.130 yields

$$\boxed{\frac{V}{V^*} = \left\{ \frac{[(k+1)/2]\,\mathrm{Ma}^2}{1 + [(k-1)/2]\,\mathrm{Ma}^2} \right\}^{1/2}} \tag{11.131}$$

$\dfrac{V}{V^*}$

Equations 11.129 and 11.131 are graphed for air in the adjacent figures.

From the continuity equation we get for Fanno flow

$$\frac{\rho}{\rho^*} = \frac{V^*}{V} \tag{11.132}$$

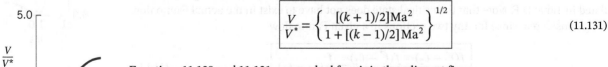

Combining 11.129 and 11.131 results in

$$\frac{\rho}{\rho^*} = \left\{ \frac{1 + [(k-1)/2]\,\text{Ma}^2}{[(k+1)/2]\text{Ma}^2} \right\}^{1/2} \tag{11.133}$$

The ideal gas equation of state (Eq. 11.1) leads to

$$\frac{p}{p^*} = \frac{\rho}{\rho^*}\frac{T}{T^*} \tag{11.134}$$

and merging Eqs. 11.134, 11.133, and 11.129 gives

$$\boxed{\frac{p}{p^*} = \frac{1}{\text{Ma}} \left\{ \frac{(k+1)/2}{1 + [(k-1)/2]\,\text{Ma}^2} \right\}^{1/2}} \tag{11.135}$$

This relationship is graphed for air in the adjacent figure.
Finally, the stagnation pressure ratio can be written as

$$\frac{p_0}{p_0^*} = \left(\frac{p_0}{p}\right)\left(\frac{p}{p^*}\right)\left(\frac{p^*}{p_0^*}\right) \tag{11.136}$$

which by use of Eqs. 11.51 and 11.135 yields

$$\boxed{\frac{p_0}{p_0^*} = \frac{1}{\text{Ma}}\left[\left(\frac{2}{k+1}\right)\left(1 - \frac{k-1}{2}\text{Ma}^2\right)\right]^{[(k+1)/2(k-1)]}} \tag{11.137}$$

For Fanno flow, thermodynamic and flow properties can be calculated as a function of Mach number.

Values of $f(\ell^* - \ell)/D$, T/T^*, V/V^*, p/p^*, and p_0/p_0^* for Fanno flow of air ($k = 1.4$) are graphed as a function of Mach number in Fig. D.F of Appendix D. The usefulness of Fig. D.F and Table D.F is illustrated in Examples 11.10 and 11.11.
See Ref. 7 for additional information about compressible internal flow.

EXAMPLE 11.10 | The Fanno Line

Given Air $\left(k = 1.4, R = 287 \text{ J/kg} \cdot \text{K}\right)$ enters an insulated constant cross-section duct with the following properties: $T_0 = 288.38$ K, $T_1 = 286.08$ K, $p_1 = 98599$ Pa.

Find Plot a Fanno line on a T–s diagram

Solution We will plot the Fanno line from a table of values of T and $s - s_1$ where, from Eq. 11.22, the entropy change is given by

$$s - s_1 = c_p \ln\left(\frac{T}{T_1}\right) - R\ln\left(\frac{p}{p_1}\right) \text{ with } c_p = \frac{kR}{k-1} = \frac{(1.4)(287\text{J}/\text{kg}\cdot\text{K})}{1.4 - 1}$$

$= 1005$ J/kg · K. Perhaps the simplest way to generate a table is to use the Mach number as a parameter. For any value of the Mach number,

$$\frac{T}{T_1} = \frac{\dfrac{T}{T^*}\Big)_{\text{Ma}}}{\dfrac{T}{T^*}\Big)_{\text{Ma}_1}} \text{ and } \frac{p}{p_1} = \frac{\dfrac{p}{p^*}\Big)_{\text{Ma}}}{\dfrac{p}{p^*}\Big)_{\text{Ma}_1}} \text{ where } \frac{T}{T^*}\Big)_{\text{Ma}} \text{ and } \frac{p}{p^*}\Big)_{\text{Ma}} \text{ are Fanno}$$

flow functions given by Table D.F or Fig. D.F or Eqs. 11.129 and 11.135.

The initial value Ma_1 is found from

$$\frac{T_0}{T}\Big)_{\text{Ma}_1} = \frac{T_{0_1}}{T_1} = \frac{288.38}{286.08}$$

$$= 1.0080 \text{ so that } \text{Ma}_1 = \left[\frac{2}{k-1}\left(\frac{T_{0_1}}{T_1} - 1\right)\right]^{\frac{1}{2}} = \left[\frac{2}{1.4-1}(1.0080 - 1)\right]^{\frac{1}{2}}$$

$= 0.20$. From Table D.F $\dfrac{p}{p^*}\Big)_{\text{Ma}_1} = 5.4555$ and $\dfrac{T}{T^*}\Big)_{\text{Ma}_1} = 1.1905$.

Now choosing Ma = 0.4, from Table D.F $\dfrac{p}{p^*}\Big)_{\text{Ma}=0.4} = 2.6958$ and

$\dfrac{T}{T^*}\Big)_{\text{Ma}=0.4} = 1.1628$ so that $T = 286.08$ K $\left(\dfrac{1.1628}{1.1905}\right) = 279.42$ K and

$$s - s_1 = (1005 \text{ J/kg} \cdot \text{K})\ln\left(\frac{1.1628}{1.1905}\right) - (287 \text{ J/kg} \cdot \text{K})\ln\left(\frac{2.6958}{5.4555}\right)$$

$$= 178.65 \text{ N} \cdot \text{m/kg} \cdot \text{K}$$

Repeating these calculations for other Mach numbers (using a spreadsheet) gives **Table E11.10**.

TABLE E11.10

Ma	T/T^*	p/p^*	T/T_1	p/p_1	T [K]	$s - s_1$ [N · m/kg · K]
0.1	1.1976	10.9435	1.0060	2.0060	287.80	193.26
0.4	1.1628	2.6958	0.9767	0.4942	279.43	178.12
0.6	1.1194	1.7634	0.9403	0.3232	269.00	261.46
0.8	1.0638	1.2893	0.8936	0.2363	255.65	300
1	1.0000	1.0000	0.8400	0.1833	240.31	310.54
1.2	0.9317	0.8044	0.7826	0.1474	223.89	301.79
1.4	0.8621	0.6632	0.7241	0.1216	207.17	279.02
1.6	0.7937	0.5568	0.6667	0.1021	190.72	246.05
2	0.6667	0.4082	0.5600	0.0748	160.21	159.75
2.5	0.5333	0.2921	0.4480	0.0535	128.17	31.52
3.5	0.3478	0.1685	0.2922	0.0309	83.58	−240.25

The Fanno line is plotted in **Figure E11.10**.

Comments As can be seen from the figure and the table, the Mach number at the "nose" of the Fanno line is 1. The upper branch represents subsonic flow and the lower branch represents supersonic flow. According to the second law of thermodynamics, since Fanno flow is an adiabatic process, the entropy can only increase; therefore, the flow state must always move toward the "nose." Movement along the Fanno line is caused by friction in the flow, so friction always drives the Mach number toward one.

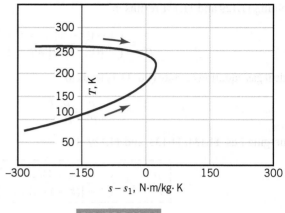

FIGURE E.11.10

EXAMPLE 11.11 | Choked and Nonchoked Fanno Flow

Given Atmospheric air ($T_0 = 288$ K, $p_0 = 101$ kPa) is drawn through a converging nozzle into a 2-m-long, 0.1-m-diameter circular duct as shown in **Fig. E11.11**. The friction factor for the duct is 0.02. Assume that the flow is adiabatic (without heat transfer) everywhere and that friction is negligible in the nozzle flow.

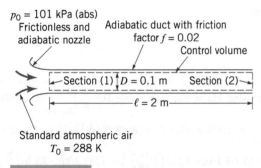

FIGURE E11.11 Adiabatic flow in a long duct

Find

a. What is the maximum mass flowrate in this system? For this flowrate, determine the static and stagnation temperatures and pressures at the duct inlet and at the exit.

b. What will be the mass flowrate if the duct is shortened to 1 m and the same duct exit pressure is maintained?

c. What back pressure would be required to obtain the flowrate from part (a) through the shorter duct of part (b)?

Solution

a. The flow in the nozzle will be isentropic flow with area change and the flow in the duct will be Fanno flow. When the pressure at the duct exit (back pressure) is lowered, flow will begin in the system. As the pressure decreases, the flowrate will increase. For sufficiently high back pressures, the flow will be subsonic throughout. For any flowrate, the velocity will increase in the nozzle because of the area reduction and it

will increase further in the duct because of friction. The highest velocity will always occur at the end of the duct. Eventually, as the back pressure is lowered, the velocity at the end of the duct will reach the speed of sound. The flow will then become choked and further decreases in back pressure will have no effect on the nozzle-duct system. The Mach number at the end of the duct will be 1; the Mach number at any other location in the duct (including the duct entrance/nozzle exit) will be subsonic.

The key to solving this problem is to determine the Mach number at section 1, the nozzle exit/duct inlet. At section 1

$$\left. f\frac{\ell^* - \ell_1}{D} \right)_1 = f\frac{\ell}{D} = 0.02 \frac{2m}{0.1m} = 0.4 \qquad (1)$$

because the flow is choked (Ma = 1) at section 2. We can now determine Ma_1 three different ways: we could solve Eq. 11.126 (which would be quite unpleasant); we could use Fig. D.F (which would be quick but with limited precision); or we could use Table D.F (which would be precise but somewhat cumbersome). Here we will use the figure or the table, either of which gives

$$Ma_1 = 0.63$$

We now calculate p_1, p_{0_1}, T_1, and T_{0_1} by considering the nozzle and using isentropic flow functions. The stagnation properties are those in the atmosphere upstream from the nozzle

$$p_{0_1} = 101 \text{ kPa and } T_{0_1} = 288 \text{ K}$$

From Graph D.I or Table D.I

$$\left. \frac{p}{p_0} \right)_1 = 0.76 \text{ and } \left. \frac{T}{T_0} \right)_1 = 0.93$$

Then

$$p_1 = (0.76)(101 \text{ kPa})$$
$$= 77 \text{ kPa(abs) and } T_1 = (0.93)(288 \text{ K}) = 268 \text{ K} \qquad (Ans)$$

The mass flowrate can be calculated using any one of Eqs. 11.95–11.97, with 11.96 being easiest for the data available

$$\dot{m} = \sqrt{\frac{k}{R}} \frac{pA\text{Ma}}{\sqrt{T}}$$

$$= \sqrt{\frac{1.4}{286.9 \text{ Nm/kgK}}} \frac{(77,000 \text{ N/m}^2) \frac{\pi}{4}(0.1 \text{ m})^2(0.63)}{\sqrt{268 \text{ K}}}$$

$$\left(\frac{\text{kg} \cdot \text{m}^2}{\text{N} \cdot \text{m} \cdot \text{s}^2}\right)^{\frac{1}{2}} = 1.65 \text{ kg/s} \qquad (Ans)$$

To calculate the air properties at section 2, we use Fanno flow functions and note that $\text{Ma}_2 = 1$ so all properties at 2 are critical properties. From Fig. D.F or Table D.F at $\text{Ma}_1 = 0.63$

$$\left.\frac{p}{p^*}\right)_1 = 1.7 \quad \left.\frac{T}{T^*}\right)_1 = 1.1 \quad \left.\frac{p_0}{p_0^*}\right)_1 = 1.16$$

Then

$$p_2 = p^* = \frac{p_1}{\frac{p}{p^*}\big)_1} = \frac{77 \text{ kPa}}{1.7} = 45 \text{ kPa} \quad \text{and}$$

$$T_2 = T^* = \frac{T_1}{\frac{T}{T^*}\big)_1} = \frac{268 \text{ K}}{1.1} = 240 \text{ K} \qquad (Ans)$$

and

$$p_{0_2} = p_0^* = \frac{p_{0_1}}{\frac{p_0}{p_0^*}\big)_1} = \frac{101 \text{ kPa}}{1.16} = 87.1 \text{ kPa} \quad \text{and}$$

$$T_{0_2} = T_0^* = T_{0_1} = 288 \text{ K} \; (T_0 \text{ is constant in Fanno flow}) \qquad (Ans)$$

b. Now, $\ell = 1$ m and $f\frac{\ell}{D} = 0.2$ Also $p_{\text{Back}} = 45$ kPa,

$$p_{0_1} = 101 \text{ kPa}, \quad \text{and} \quad T_0 = 288 \text{ K}$$

The solution to this problem hinges on whether the flow is still choked at the duct exit. If it is, we can proceed just like we did for part a. If not, we would probably have to iterate to try to match the exit pressure to the back pressure. We begin by assuming the flow is choked. We can check this assumption by comparing the duct exit pressure to the back pressure.

Taking $f\frac{\ell^* - \ell_1}{D} = 0.2$ gives $\text{Ma}_1 = 0.70, \frac{p}{p^*}\big)_1 = 1.5$ and $\frac{p}{p_0}\big)_1 = 0.72$. Then we can calculate

$$p_2 = p^* = p_0 \frac{p}{p_0}\big)_1 \bigg/ \frac{p}{p^*}\big)_1 = (101 \text{ kPa})\left(\frac{0.72}{1.5}\right) = 48.5 \text{ kPa (abs)}$$

We see that the back pressure (45 kPa) is less than the exit plane pressure (48.5 kPa), so the flow will be choked at the exit plane and the flow will adjust to the backpressure in a multidimensional process downstream from the duct exit. To calculate the flowrate, we will use Eq. 11.96 again but this time at the exit. The exit temperature is

$$T_2 = T_0 \frac{T_2}{T_{0_2}} = T_0 \frac{T^*}{T_0} = T_0\left(\frac{2}{k+1}\right) = (288 \text{ K})(0.833) = 240 \text{ K}$$

and the mass flowrate is

$$\dot{m} = \sqrt{\frac{k}{R}} \frac{pA\text{Ma}}{\sqrt{T}}$$

$$= \sqrt{\frac{1.4}{286.9 \text{ Nm/kgK}}} \frac{(48,500 \text{ N/m}^2) \frac{\pi}{4}(0.1 \text{ m})^2(1.0)}{\sqrt{240 \text{ K}}}$$

$$\times \left(\frac{\text{kg} \cdot \text{m}^2}{\text{N} \cdot \text{m} \cdot \text{s}^2}\right)^{\frac{1}{2}} = 1.73 \text{ kg/s} \qquad (Ans)$$

The flowrate goes up when the duct length (and hence flow resistance) goes down. This should be no surprise.

c. Now we know $\dot{m} = 1.65 \frac{\text{kg}}{\text{s}}$ and $f\frac{\ell}{D} = 0.2$ and we wish to find the exit plane pressure. Because the desired flowrate (1.65 kg/s) is less than the maximum (choking) flowrate (1.73 kg/s), we know that the flow is not choked and the exit plane pressure and the back pressure are equal. Because the inlet stagnation properties and the flowrate are the same as in part (a), we know that conditions at the nozzle exit/duct entrance are

$$\text{Ma}_1 = 0.63 \quad f\frac{\ell^* - \ell_1}{D} = 0.4 \quad p_1 = 77 \text{ kPa} \quad \frac{p}{p^*}\big)_1 = 1.7$$

Then

$$f\frac{\ell^* - \ell_2}{D} = f\frac{\ell^* - \ell_1}{D} - f\frac{\ell}{D} = 0.4 - 0.2 = 0.2$$

From Fig. D.F or Table D.F

$$\text{Ma}_2 = 0.70 \quad \frac{p}{p^*}\big)_2 = 1.5$$

so

$$p_2 = p_{\text{Back}} = p_1 \frac{p}{p^*}\big)_2 \bigg/ \frac{p}{p^*}\big)_1 = 77 \text{ kPa} \frac{1.5}{1.7} = 68 \text{ kPa(abs)} \qquad (Ans)$$

Comments For a given duct cross-section and inlet stagnation pressure, the back pressure and duct friction $\left(\text{in the form } f\frac{\ell}{D}\right)$ determine the flowrate. Both provide resistance to the flow. The back pressure controls the flow only if it is above the choking value. This example verifies that decreasing the duct friction leads to a larger flowrate even if the duct is choked and that raising the backpressure (above the choking value) decreases the flowrate. The reverse is also true.

In part (c), we calculated the back pressure that would be necessary to produce a specified flowrate in a particular duct. How that pressure is achieved was of no concern. The problem of determining the flowrate for a specified back pressure (greater than the choking value) is more complicated, usually requiring iteration.

An important question for many practical applications is how to find f. From dimensional analysis, we expect that

$$f = f\left(\text{Re}, \text{Ma}, \frac{\varepsilon}{D}\right)$$

Based on a rather small amount of experimental evidence, dependence of f on Ma appears to be weak, at least for subsonic flow; accordingly, values of f for Ma = 0 (incompressible flow) are usually sufficient. For fully developed flow, we obtain values for f from the Moody chart (Fig. 8.20), the Colebrook equation, Eq. 8.35a, or the laminar flow friction factor equation, Eq. 8.19.

Because f depends on Reynolds number, our assumption of a constant value along the duct seems questionable. The Reynolds number can be written

$$\text{Re} = \frac{\rho V D}{\mu} = \frac{\dot{m}}{A}\frac{D}{\mu} = \frac{\dot{m}}{\mu}\frac{D}{A} \tag{11.138}$$

The mass flowrate, diameter, and cross-sectional area are all constant, so the variation of Reynolds number along the duct results only from viscosity. For a gas, viscosity varies rather slowly with temperature ($\mu \sim T^{0.7}$). In addition, the dependence of f on Re is weak in turbulent flow, which is the usual condition at velocities high enough for compressibility to be significant so the assumption of constant f is surprisingly good.

The following example illustrates determination of the friction factor during a Fanno flow calculation.

EXAMPLE 11.12 | Duct Sizing in Fanno Flow

Given An astronaut breathes oxygen supplied to a spacesuit through an umbilical cord (**Fig. E11.12a**). The umbilical is 7 m long. The oxygen is supplied from a storage tank at 100 kPa and 283 K. The suit pressure is 20 kPa.

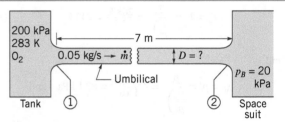

FIGURE E11.12b Schematic diagram of space suit umbilical.

For a circular duct, from Eq. 11.138,

$$\text{Re} = \frac{\dot{m}}{\mu}\frac{D}{\pi D^2/4} = \frac{4\dot{m}}{\pi \mu D}$$

Substituting numbers where possible, we have

$$\text{Re} = \frac{(4)(0.05 \text{ kg/s})}{(\pi)(1.97 \times 10^{-5} \text{ N} \cdot \text{s/m}^2)D} = \frac{3230}{D} \tag{1}$$

where D is in meters.

To calculate the flowrate at plane 1 we use Eq. 11.97. As we know $\dot{m}$ but do not know the area A, we rearrange the equation and substitute $A = \pi D^2/4$ to get the following:

$$D^2 = \frac{4}{\pi}\left(\frac{\dot{m}\sqrt{RT_0}}{p_{0_1}\sqrt{k}}\right)\left[\frac{1}{\text{Ma}_1}\left(1 + \frac{k-1}{2}\text{Ma}_1^2\right)^{(k+1)/2(k-1)}\right]$$

We can speed up the calculations considerably if we recognize that the quantity in brackets is

$$\left(\frac{k+1}{2}\right)^{(k+1)/2(k-1)} A/A^*)_{\text{Ma}_1} = 1.728\, A/A^*)_{\text{Ma}_1}$$

for $k = 1.4$. Thus

$$D^2 = \frac{4}{\pi}\left[\frac{(0.05 \text{ kg/s})\sqrt{(260 \text{ N} \cdot \text{m/kg} \cdot \text{K})(283 \text{ K})}}{(200,000 \text{ N/m}^2)\sqrt{1.4}}\right](1.728)A/A^*)_{\text{Ma}}$$

and

$$D = \sqrt{D^2} = 0.0112\left(\sqrt{A/A^*)_{\text{Ma}_1}}\right) \tag{2}$$

FIGURE E11.12a Astronaut being supplied with oxygen through an umbilical.

Find The required inside diameter of the umbilical to supply 0.05 kg/s of oxygen. The inside of the umbilical is lined with a material with roughness of 0.01 mm.

Solution **Figure E11.12b** is a schematic diagram of the duct flow problem.

The flow in the umbilical obviously is subsonic (no converging–diverging nozzle at the inlet). The properties of oxygen are $R = 260 \text{ N} \cdot \text{m/kg} \cdot \text{K}$, $k = 1.4$, and $\mu = 1.97 \times 10^{-5} \text{ N} \cdot \text{s/m}^2$.

The ratio of back pressure to supply pressure is

$$\frac{p_{Back}}{p_{0_1}} = \frac{20 \text{ kPa}}{100 \text{ kPa}} = 0.20$$

To determine the diameter, we must use Eqs. 1 and 2, together with the isentropic flow table (for A/A^*), the Fanno flow table (to relate duct friction and Mach number), and the Moody Chart or Colebrook formula (for f). Clearly this will require iteration. It is most convenient to use Ma_1 as the iteration parameter. We begin with

$$Ma_1^{(1)} = 0.5$$

The superscript (1) indicates that this is our first guess. From Table D.I

$$A/A^*)_{Ma_1} = 1.340$$

From Eq. 2,

$$D = 0.0112\sqrt{1.340} = 0.013 \text{ m}$$

From Eq. 1,

$$Re = \frac{3230}{0.013} = 2.48 \times 10^5$$

Also,

$$\frac{\varepsilon}{D} = \frac{1 \times 10^{-5} \text{ m}}{1.3 \times 10^{-2} \text{ m}} = 0.00077$$

From Fig. 8.20 (the Moody chart),

$$f = 0.020$$

The duct length parameter is

$$f\frac{\ell}{D} = (0.020)\left(\frac{7 \text{ m}}{0.013 \text{ m}}\right) = 10.77$$

Next, we calculate the choking pressure ratio for this duct length parameter. We let Ma_1^* be the inlet Mach number corresponding to choking. From Table D.F for $f(\ell^* - \ell_1)/D = 10.77$,

$$Ma_1^* = 0.227 \quad \text{and} \quad \frac{p}{p^*}\bigg)_{Ma_1^*} = 4.801$$

For Table D.I

$$\frac{p}{p_0}\bigg)_{Ma_1^*} = 0.9648$$

The choking pressure ratio is

$$\left(\frac{p_B^*}{p_{0_1}}\right) = \frac{\dfrac{p_1}{p_{0_1}}}{\dfrac{p_1}{p^*}} = \frac{0.9648}{4.801} = 0.201$$

Because

$$\frac{p_{Back}}{p_{0_1}} = 0.10 < \frac{p_B^*}{p_{0_1}} = 0.201$$

the duct seems to be choked. Setting $f(\ell^* - \ell_1)/D = f(\ell)/D = 10.77$, we find that the "new" Mach number is

$$Ma_{1,new} = Ma_1^*$$

As $Ma_1^* = 0.227 \neq 0.5 = Ma_1^{(1)}$, we now try

$$Ma_1^{(2)} = 0.227$$

From Table D.I,

$$A/A^*)_{Ma_1} = 2.629$$

From Eqs. 1 and 2, $D = 0.0182$ m and $Re = 1.79 \times 10^5$. Also,

$$\frac{\varepsilon}{D} = \frac{10^{-5} \text{ m}}{1.82 \times 10^{-2} \text{ m}} = 5.49 \times 10^{-4}$$

and, from the Moody chart

$$f = 0.0193$$

This gives

$$f(\ell)/D = 0.0193\left(\frac{7 \text{ m}}{0.0182 \text{ m}}\right) = 7.38$$

From Tables D.F and D.I,

$$Ma_1^* = 0.264, \quad p/p^*)_{Ma_1^*} = 4.117, \quad \text{and} \quad p/p_0)_{Ma_1^*} = 0.9526$$

so

$$\frac{p_B^*}{p_{0_1}} = \frac{0.9526}{4.117} = 0.231 \quad (> 0.2)$$

The duct seems to be choked. We now repeat the procedure with

$$Ma_1^{(3)} = 0.264$$

The results of this calculation are

$$D = 0.0169 \text{ m}, \quad Re = 1.93 \times 10^5, \quad \frac{\varepsilon}{D} = 5.92 \times 10^{-4}$$

$$f = 0.019, \quad f(\ell)/D = 7.994, \quad Ma_1^* = 0.256, \quad \text{and}$$

$$\frac{p_B^*}{p_{0_1}} = 0.225 \quad \text{(Choked)}$$

For the next iteration

$$Ma_1^{(4)} = 0.256$$

The results for this trial are

$$D = 0.0172 \text{ m}, \quad Re = 1.89 \times 10^5, \quad \frac{\varepsilon}{D} = 5.81 \times 10^{-4}$$

$$f = 0.019, \quad f(\ell)/D = 7.855, \quad Ma_1^* = 0.258, \quad \text{and}$$

$$\frac{p_B^*}{p_{0_1}} = 0.226 \quad \text{(Choked)}$$

It is clear that the Mach number, to two decimal places, converges to

$$Ma_1 = 0.26$$

For this Mach number, the calculated parameters are

$$D = 0.017 \text{ m}, \quad R = 1.90 \times 10^5$$

$$\frac{\varepsilon}{D} = 5.88 \times 10^{-4}, \quad f = 0.019, \quad f(\ell)/D = 7.82$$

$$\text{and} \quad \frac{p_B^*}{p_{0_1}} = 0.226 \quad \text{(The flow is choked at the duct exit.)}$$

We specify that the duct size is

$$D = 0.017 \text{ m} = 1.7 \text{ cm} \qquad (Ans)$$

Comments Just as with rigid pipes, the actual duct would probably not be exactly 1.7 cm in diameter. Considering the conservative design approach used in a space program, the actual duct may be twice the size, say 3.5–4.0 cm or even larger. The system would be fitted with some sort of control valve or pressure regulator.

Don't be fooled into thinking this flow could be modeled as incompressible because the inlet Mach number is low (0.26). The Mach number at the spacesuit is 1, and the flow is choked. Compressibility is very significant.

A Note about Supersonic Flow Our discussion has concentrated on subsonic frictional flow. A discussion of supersonic frictional flow would be interesting but of limited practical value compared to the effort expended for three reasons. First, Eq. 11.137 shows that loss of stagnation pressure is very large if the Mach number is greater than 1, so supersonic flow is usually avoided if friction is significant. Second, interaction between shocks and boundary layers is a complex multidimensional process. Third, the maximum duct length in supersonic flow is so short ($f(\ell^* - \ell_1)/D < 0.822$) that the flow is never fully developed. The one-dimensional Fanno flow model is not sufficiently accurate to be of value. If you are interested in discussions of supersonic Fanno flow, see Reference 4.

11.9 FRICTIONLESS FLOW IN A CONSTANT-AREA DUCT WITH HEATING OR COOLING

Compressibility is often significant in flows with heating or cooling. Examples include flow in steam generators, gas turbine combustion chambers, and moisture evaporation or condensation in a high-speed air stream. In order to focus on the effects of heating or cooling alone, we assume that the flow is steady, without friction, and the duct area is constant. As we have done implicitly in previous sections, we also assume that the fluid has fixed composition and constant specific heat. This is called *Rayleigh Flow* after the British physicist J. W. Strutt, Lord Rayleigh.

Because of the nature of real flows with heating or cooling, the Rayleigh flow model is further from reality than isentropic flow or Fanno flow. Heat exchange with a surface in contact with a fluid is intimately involved with fluid friction. Combustion of fuel in air results in a change of fluid composition, as does evaporation or condensation. Combustion typically results in high temperatures, which lead to changes in c_p, c_v, and k. Nevertheless, the Rayleigh flow model is reasonably accurate when the temperature difference between a surface and the fluid is large so that significant heat transfer occurs over a short length of duct, when the fuel/air ratio in combustion is small, and when the ratio of liquid mass evaporated or condensed to fluid mass is small. Even if these conditions are not met, Rayleigh flow yields useful insight into the effect of heating or cooling on a compressible flow.

11.9.1 The Rayleigh Line

Consider steady, one-dimensional, frictionless flow through the finite control volume shown in **Fig. 11.22**. Writing the linear momentum equation gives

$$p_1 A - p_2 A + \overset{0}{\cancel{F_f}} = \dot{m}(V_2 - V_1) = \rho V A (V_2 - V_1)$$

Dividing by the area and regrouping

$$p_1 + (\rho V)V_1 = p_2 + (\rho V)V_2 \tag{11.139}$$

The continuity equation for the control volume is

$$\rho_1 V_1 A = \rho_2 V_2 A$$

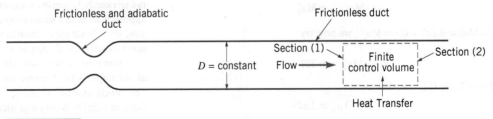

Frictionless and adiabatic duct   Frictionless duct

D = constant   Flow   Section (1)   Finite control volume   Section (2)

Heat Transfer

FIGURE 11.22 Rayleigh flow.

Dividing by the area

$$\rho V = G = \text{constant} \tag{11.140}$$

where G, the density-velocity product, is called the *mass flow density*. Combining Eqs. 11.139 and 11.140, we get

$$p + \frac{G^2}{\rho} = \text{constant} = I \tag{11.141a}$$

where I is called the *Impulse function*. Using the specific volume $v = \frac{1}{\rho}$, we can write

$$p + G^2 v = I \tag{11.141b}$$

With G and I constant, Eq. 11.141b is a relationship between p and v, which, for specific values of G and I, can be graphed as a straight line, as shown in the adjacent figure. This line is called a *Rayleigh line*. It represents all possible states of the fluid for specific values of G and I.

For any simple compressible fluid, if two properties are known, the other properties can be determined. If we pick several points on the p-v curve, we can determine corresponding values of temperature and entropy and graph the Rayleigh line on a T–s diagram, as shown in **Fig. 11.23**.

We can observe many important features of Rayleigh flow by examining Fig. 11.23. Any flow state, such as that at Section 1 in Fig. 11.22, can be shown as a single point on the corresponding Rayleigh line. The addition or removal of heat between Section 1 and Section 2 causes the fluid properties to change and the state point on the Rayleigh line to move along the line. Since there is no friction, the second law of thermodynamics, Eq. 5.101, in the form

$$ds = \frac{\delta q}{T}$$

tells us that the state point will move to the right ($ds > 0$) if the fluid is heated ($\delta q > 0$) and the state point will move to the left if the fluid is cooled ($\delta q < 0$). Thus heating always drives the fluid toward maximum entropy (point a) and cooling drives the fluid away from the maximum entropy point.

What is the fluid speed at point a, the point of maximum entropy? Rayleigh flow is governed by the continuity equation, Eq. 11.140, and the linear momentum equation, Eq. 11.141a.

Differentiating the momentum equation

$$dp - \frac{G^2}{\rho^2} d\rho = 0$$

Using Eq. 11.140

$$V^2 = \frac{dp}{d\rho} \tag{11.141c}$$

Equation 11.141c gives the flow speed at any point on the Rayleigh line. At point a, $ds = 0$ and the entropy is locally constant, therefore

$$V_a^2 = \left(\frac{\partial p}{\partial \rho}\right)_s = c^2 \text{ at point } a$$

That is, *the fluid speed equals the sonic speed* and Ma = 1 *at the point of maximum entropy.* It follows that the flow described by the upper branch of the Rayleigh line is subsonic and the

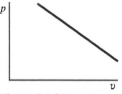

p

v

The Rayleigh Line in p-v Coordinates

The maximum entropy state on the Rayleigh line corresponds to sonic conditions.

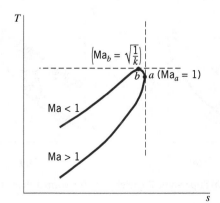

T

$\left(\text{Ma}_b = \sqrt{\frac{1}{k}}\right)$

$b \bullet a \ (\text{Ma}_a = 1)$

Ma < 1

Ma > 1

s

FIGURE 11.23 Rayleigh line.

TABLE 11.4 Summary of Rayleigh Flow Characteristics

	Heating		Cooling	
	Subsonic	Supersonic	Subsonic	Supersonic
V	Increase	Decrease	Decrease	Increase
Ma	Increase	Decrease	Decrease	Increase
T	Increase for $0 \le \text{Ma} \le \sqrt{1/k}$	Increase	Decrease for $0 \le \text{Ma} \le \sqrt{1/k}$	Decrease
	Decrease for $\sqrt{1/k} \le \text{Ma} \le 1$		Increase for $\sqrt{1/k} \le \text{Ma} \le 1$	
T_0	Increase	Increase	Decrease	Decrease
P	Decrease	Increase	Increase	Decrease
p_0	Decrease	Decrease	Increase	Increase

flow described by the lower branch of the Rayleigh line is supersonic. We can replace the "a" designation of the maximum entropy point with the more familiar designation of sonic flow, *, thus the temperature is T^*, the pressure is p^*, and so on. (It must be strongly stated that the * properties for Rayleigh flow are different from the * properties for isentropic flow and Fanno flow.)

Rayleigh flow is always driven *toward* Ma = 1 by heating and *away from* Ma = 1 by cooling. The flow can pass through Ma = 1 but only by heating to precisely Ma = 1 and then switching to cooling; however, this would be nearly impossible in practice. Furthermore, if we restrict the process to heat addition, it is not possible to pass beyond point a; from any particular starting location, there is a maximum amount of heating possible. This is yet another manifestation of choking, in this case *thermal choking*; when the flow is heated to sonic speed in a duct, the flow cannot respond to further heating. There is no limit to cooling, other than absolute zero. One might ask "What happens when a Rayleigh flow is choked and more heat is added?" The answer is that the mass flowrate $\dot{m}$ and the mass flow density G will be reduced and the flow will be described by a different Rayleigh line, one shifted to the right.

There are two final points to note about the Rayleigh line. First, the subsonic branch has a point of maximum temperature (point b in Fig. 11.23). In an ideal gas, $\text{Ma}_b = \sqrt{\frac{1}{k}}$. Between this point and the maximum-entropy point, the temperature actually drops when heat is added! Lastly, although it is impossible for the flow to pass through the point of maximum entropy, it is possible to "jump" from the supersonic branch to the subsonic branch by passing through a normal shock wave. A summary of the qualitative aspects of Rayleigh flow is given in **Table 11.4** and **Fig. 11.24**.

Fluid temperature reduction can accompany heating in a subsonic Rayleigh flow.

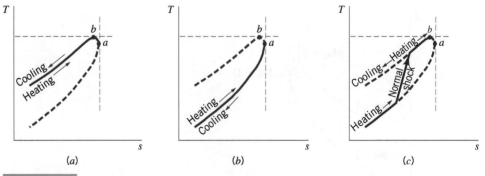

FIGURE 11.24 (*a*) Subsonic Rayleigh flow. (*b*) Supersonic Rayleigh flow. (*c*) Normal shock in a Rayleigh flow.

11.9.2 Frictionless Flow of an Ideal Gas with Heating or Cooling (Rayleigh Flow)

To obtain quantitative results, it is necessary to specify a particular working fluid. As usual, we will consider an ideal gas.

We elect to use the state of the fluid at point a of Fig. 11.23 as the reference state. As shown earlier, the Mach number at point a is 1 so, as done for isentropic flow and Fanno flow, we will designate this state by the superscript (*). Even though the Rayleigh flow being considered may not choke and state a is not achieved by the flow, this reference state is useful.

If we apply the linear momentum equation to Rayleigh flow between any upstream section and the section, actual or imagined, where state * is attained, we get

$$p + \rho V^2 = p^* + \rho^* V^{2*}$$

or

$$\frac{p}{p^*} + \frac{\rho V^2}{p^*} = 1 + \frac{\rho^*}{p^*} V^{2*} \tag{11.142}$$

By substituting the ideal gas equation of state (Eq. 11.1) into Eq. 11.142 and making use of the ideal gas speed-of-sound and the definition of Mach number, we obtain

$$\boxed{\frac{p}{p^*} = \frac{1+k}{1+k\mathrm{Ma}^2}} \tag{11.143}$$

This relationship is graphed for air ($k = 1.4$) in the adjacent figure.

From the ideal gas equation of state (Eq. 11.1), we conclude that

$$\frac{T}{T^*} = \frac{p}{p^*}\frac{\rho^*}{\rho} \tag{11.144}$$

Conservation of mass (Eq. 11.140) with constant A gives

$$\frac{\rho^*}{\rho} = \frac{V}{V^*} \tag{11.145}$$

which when combined with Eqs. 11.47 (ideal gas speed of sound) and the Mach number definition gives

$$\frac{\rho^*}{\rho} = \mathrm{Ma}\sqrt{\frac{T}{T^*}} \tag{11.146}$$

Combining Eqs. 11.144 and 11.146 leads to

$$\frac{T}{T^*} = \left(\frac{p}{p^*}\mathrm{Ma}\right)^2 \tag{11.147}$$

which when combined with Eq. 11.143 gives (also see adjacent figure)

$$\boxed{\frac{T}{T^*} = \left[\frac{(1+k)\mathrm{Ma}}{1+k\mathrm{Ma}^2}\right]^2} \tag{11.148}$$

From Eqs. 11.145, 11.146, and 11.148 we see that

$$\boxed{\frac{\rho^*}{\rho} = \frac{V}{V^*} = \mathrm{Ma}\left[\frac{(1+k)\mathrm{Ma}}{1+k\mathrm{Ma}^2}\right]} \tag{11.149}$$

This relationship is graphed for air ($k = 1.4$) in the adjacent figure.

The energy equation (Eq. 5.69) tells us that because of the heat transfer involved in Rayleigh flows, the stagnation temperature varies. The heat and the stagnation temperature are related by

$$q_{1-2} = c_p(T_{02} - T_{01}) \tag{11.150}$$

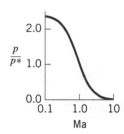

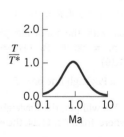

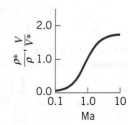

We note that

$$\frac{T_0}{T_0^*} = \left(\frac{T_0}{T}\right)\left(\frac{T}{T^*}\right)\left(\frac{T^*}{T_0^*}\right) \tag{11.151}$$

We can use Eq. 11.50 (developed earlier for steady, isentropic, ideal gas flow) to evaluate T_0/T and T^*/T_0^* because these two temperature ratios, by definition of the stagnation state, involve isentropic processes. Equation 11.148 can be used for T/T^*. Thus, consolidating Eqs. 11.151, 11.50, and 11.148, we obtain

$$\boxed{\frac{T_0}{T_0^*} = \frac{2(k+1)\,\text{Ma}^2\left(1 + \dfrac{k-1}{2}\text{Ma}^2\right)}{(1 + k\text{Ma}^2)^2}} \tag{11.152}$$

This relationship is graphed for air ($k = 1.4$) in the adjacent figure.

Finally, we observe that

$$\frac{p_0}{p_0^*} = \left(\frac{p_0}{p}\right)\left(\frac{p}{p^*}\right)\left(\frac{p^*}{p_0^*}\right) \tag{11.153}$$

We can use Eq. 11.51 developed earlier for steady, isentropic, ideal gas flow to evaluate p_0/p and p^*/p_0^*, because these two pressure ratios, by definition, involve isentropic processes. Equation 11.143 can be used for p/p^*. Together, Eqs. 11.51, 11.143, and 11.153 give

$$\boxed{\frac{p_0}{p_0^*} = \frac{(1+k)}{(1 + k\text{Ma}^2)}\left[\left(\frac{2}{k+1}\right)\left(1 + \frac{k-1}{2}\text{Ma}^2\right)\right]^{k/(k-1)}} \tag{11.154}$$

This relationship is graphed for air ($k = 1.4$) in the adjacent figure.

Values of p/p^*, T/T^*, ρ/ρ^* or V/V^*, T_0/T_0^*, and p/p_0^* are graphed in Fig. D.R of Appendix D and tabulated in Table D.R as a function of Mach number for Rayleigh flow with $k = 1.4$.

See Ref. 7 for a more advanced treatment of internal flows with heat transfer.

EXAMPLE 11.13 | The Rayleigh Line

Given Air $\left(k = 1.4, R = 287 \text{ J/kg} \cdot \text{K}\right)$ enters a constant cross-section duct with the following properties: $T_0 = 288.38$ K, $T_1 = 286.08$ K, $p_1 = 98599$ Pa (These are the same conditions as Example 11.10)

Find Plot a Rayleigh line on a T–s diagram

Solution We will plot the Rayleigh line from a table of values of T and $s - s_1$ where, from Eq. 11.24, the entropy change is given by

$$s - s_1 = c_p \ln\left(\frac{T}{T_1}\right) - R \ln\left(\frac{p}{p_1}\right) \text{ with } c_p$$

$$= \frac{kR}{k-1} = \frac{(1.4)(287 \text{ J/kg·K})}{1.4-1} = 1005 \text{ J/kg} \cdot \text{K}$$

Perhaps the simplest way to generate a table is to use the Mach number as a parameter. For any value of the Mach number,

$$\frac{T}{T_1} = \frac{\dfrac{T}{T^*}\bigg)_{\text{Ma}}}{\dfrac{T}{T^*}\bigg)_{\text{Ma}_1}} \text{ and } \frac{p}{p_1} = \frac{\dfrac{p}{p^*}\bigg)_{\text{Ma}}}{\dfrac{p}{p^*}\bigg)_{\text{Ma}_1}} \text{ where } \frac{T}{T^*}\bigg)_{\text{Ma}} \text{ and } \frac{p}{p^*}\bigg)_{\text{Ma}}$$

are Rayleigh flow functions given by Eqs. 11.148 and 11.143 or Table D.R or Fig. D.R.

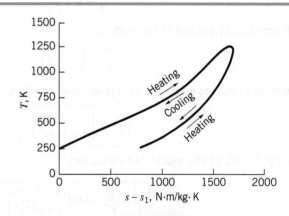

FIGURE E11.13

The initial value Ma_1 is found from $\dfrac{T_0}{T}\bigg)_{\text{Ma}_1} = \dfrac{T_{0_1}}{T_1} = \dfrac{288.38}{286.08} = 1.0080$ so that $\text{Ma}_1 = \left[\dfrac{2}{k-1}\left(\dfrac{T_{0_1}}{T_1} - 1\right)\right]^{\frac{1}{2}} = \left[\dfrac{2}{1.4-1}(1.0080 - 1)\right]^{\frac{1}{2}}$

$= 0.20$. From Table D.R $\dfrac{p}{p^*}\bigg)_{\text{Ma}_1} = 2.2727$ and $\dfrac{T}{T^*}\bigg)_{\text{Ma}_1} = 0.2066$.

Now choosing Ma = 0.4, from Table D.R $\left.\dfrac{p}{p^*}\right)_{\text{Ma}=0.4} = 1.9608$

and $\left.\dfrac{T}{T^*}\right)_{\text{Ma}=0.4} = 0.6152$ so that $T = 286.08\,\text{K}\,\dfrac{0.6152}{0.2066} = 852\,\text{K}$ and $s - s_1$

$= 1005\,\text{J/kg}\cdot\text{K}\ln\dfrac{0.6152}{0.2066} - 287\,\text{J/kg}\cdot\text{K}\ln\dfrac{2.2727}{1.9608} = 1054.2\,\text{N}\cdot\text{m/kg}\cdot\text{K}.$

Repeating these calculations for other Mach numbers (using a spreadsheet) gives the table below:

The Rayleigh line is plotted in **Figure E11.13**.

Comments As can be seen from the figure and the table, the Mach number at the "nose" of the Rayleigh line is 1. The upper branch represents subsonic flow and the lower branch represents supersonic flow. Movement along the Rayleigh line is caused by heating or cooling. Heating always drives the Mach number toward one.

Ma	T_2/T^*	T_1/T^*	p_2/p^*	p_1/p^*	T_2/T_1	p_2/p_1	T [K]	$s - s_1$ [N·m/kg·K]
0.2	0.207	0.207	2.273	2.273	1.000	1.000	286.3	0
0.4	0.615	0.207	1.961	2.273	2.977	0.863	851.8	1138.4
0.6	0.917	0.207	1.596	2.273	4.437	0.702	1269.3	1594.8
0.8	1.025	0.207	1.266	2.273	4.963	0.557	1420	1777.4
1	1.000	0.207	1.000	2.273	4.840	0.440	1384.4	1820.4
1.2	0.912	0.207	0.796	2.273	4.413	0.350	1262.7	1793.6
1.4	0.805	0.207	0.641	2.273	3.898	0.282	1115.3	1729.1
1.6	0.702	0.207	0.524	2.273	3.396	0.230	971.8	1648.6
1.8	0.609	0.207	0.434	2.273	2.947	0.191	843.5	1557.3
2	0.529	0.207	0.364	2.273	2.560	0.160	732.3	1466.0
2.5	0.379	0.207	0.246	2.273	1.833	0.108	524.3	1245.8
3	0.280	0.207	0.176	2.273	1.357	0.078	388.0	1036.4
3.5	0.214	0.207	0.132	2.273	1.037	0.058	296.3	848.5

EXAMPLE 11.14 | Rayleigh Flow in a Combustion Chamber

Given Air enters a constant-area combustion chamber at 222 K and 82740 Pa. The air inlet velocity is 122 m/s. Fuel is burned in the chamber, providing 116 kJ/kg of heat input.

Find

a. Determine the pressure, temperature, Mach number, and fluid velocity at the exit.

b. What is the maximum possible heat that could be added?

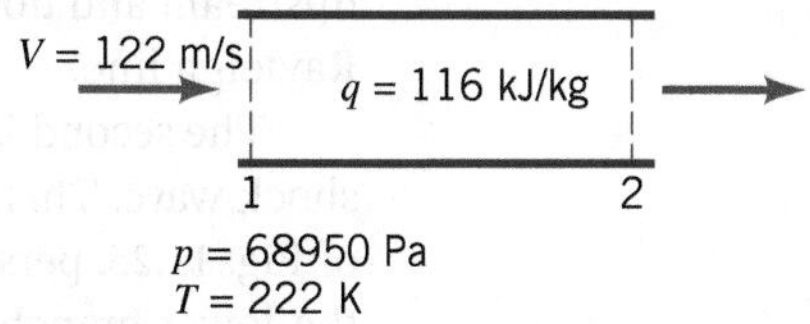

Solution

a. A schematic of the combustion chamber is shown in **Figure E11.14**.

At section 1

$c = \sqrt{kRT}$

$= \sqrt{1.4(287\,\text{J}/\text{kg}\cdot\text{K})(222\,\text{K})}$

$= 299\,\text{m/s}$

$\text{Ma}_1 = \dfrac{222}{299} = 0.74$

$T_{0_1} = \dfrac{T_1}{\left.\dfrac{T}{T_0}\right)_{\text{Ma}_1}} = \dfrac{222}{0.9675} = 229\,\text{K}$

$\left.\dfrac{p}{p^*}\right)_{\text{Ma}_1} = 1.943 \quad \left.\dfrac{T}{T^*}\right)_{\text{Ma}_1} = 0.6345 \quad \left.\dfrac{T_0}{T_0^*}\right)_{\text{Ma}_1} = 0.5465$

FIGURE E11.14 A combustion chamber.

The energy equation can be used to find the downstream stagnation temperature

$T_{0_2} = T_{0_1} + \dfrac{q}{c_p} = 229\,\text{K} + \dfrac{116\times10^3\,\text{J/kg}}{1005\,\text{J/kg}\cdot\text{K}} = 344\,\text{K}$

Now

$\dfrac{T_{0_2}}{T_0^*} = \left.\dfrac{T_{0_2}}{T_{0_1}}\dfrac{T_0}{T_0^*}\right)_1 = \dfrac{344}{229}(0.5465) = 0.8219$

From Table D.R at

$\left.\dfrac{T_0}{T_0^*}\right)_2 = 0.8217; \quad \left.\dfrac{p}{p^*}\right)_2 = 1.5904,$

$\left.\dfrac{T}{T^*}\right)_2 = 0.9196 \quad \text{and} \quad \text{Ma}_2 = 0.602 \quad \quad (Ans)$

Therefore

$$p_2 = p_1 \frac{\left.\dfrac{p}{p^*}\right)_2}{\left.\dfrac{p}{p^*}\right)_1} = 68950 \text{ Pa} \frac{1.5902}{1.9428} = 56436 \text{ Pa}$$

$$T_2 = T_1 \frac{\left.\dfrac{T}{T^*}\right)_2}{\left.\dfrac{T}{T^*}\right)_1} = 222 \text{ K} \frac{0.9196}{0.6345} = 322 \text{ K} \qquad (Ans)$$

$$V_2 = \text{Ma}_2 \sqrt{kRT_2}$$
$$= 0.602 \sqrt{1.4(287 \text{ J/kg·K})(322 \text{ K})} = 217 \text{ m/s}$$

(Ans)

b. The maximum possible heat addition would cause thermal choking, with the Mach number at section 2 equal to 1.0. This would correspond to $T_{0_2} = T_0^*$. From part (a)

$$T_{0_1} = 229 \text{ K} \quad \text{and} \quad \left.\frac{T_0}{T_0^*}\right)_1 = 0.5465$$

Then

$$T_{0_2,\text{max}} = T_0^* = T_{0_1} \frac{1}{\left.\dfrac{T_0}{T_0^*}\right)_1} = \frac{229 \text{ K}}{0.5465} = 419 \text{ K}$$

Finally

$$q_{\text{max}} = c_p\left(T_0^* - T_{0_1}\right) = 1005 \text{ J/kg·K}(419 \text{ K} - 229 \text{ K})$$
$$= 191 \text{ kJ/kg} \qquad (Ans)$$

Comments   Note that in part (a), heat addition leads to a substantial velocity increase. In Rayleigh flow, there is no way to determine the length of the duct.

11.9.3 Rayleigh Lines, Fanno Lines, and Normal Shocks

The energy equation for Fanno flow and the momentum equation for Rayleigh flow are valid for flow across normal shocks.

From the analyses of the Rayleigh line in Section 11.9.1, the Fanno line in Section 11.8.1, and the normal shock in Section 11.5, it is apparent that the steady flow of an ideal gas across a normal shock is governed by some of the same equations used for describing both Fanno and Rayleigh flows (energy equation for Fanno flows, momentum equation for Rayleigh flow, and continuity equation for both). Thus, for a given density–velocity product (ρV), gas (R, k), and conditions at the inlet of the normal shock $(T_x, p_x, \text{ and } s_x)$, the conditions downstream of the shock (state y) will be on both a Fanno line and a Rayleigh line that pass through the inlet state (state x), as is illustrated in **Fig. 11.25**. In other words, the upstream and downstream states are represented by the intersection of a Fanno line and a Rayleigh line.

The second law of thermodynamics requires that entropy must increase across a normal shock wave. This law and sketches of the Fanno line and Rayleigh line intersection, like that of Fig. 11.25, persuade us to conclude that flow across a normal shock can only proceed from the lower branch intersection to the upper branch intersection; that is, from supersonic to subsonic flow. Also, we observe that a shock can occur in either Fanno flow or Rayleigh flow, but only from supersonic flow to subsonic flow.

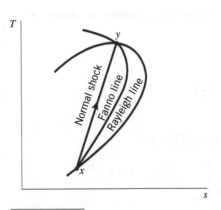

FIGURE 11.25 The relationship between a normal shock and Fanno and Rayleigh lines.

11.10 ANALOGY BETWEEN COMPRESSIBLE AND OPEN-CHANNEL FLOWS

During a first course in fluid mechanics, students rarely study both open-channel flows (Chapter 10) and compressible flows. This is unfortunate because these two kinds of flows are strikingly similar in several ways. Furthermore, the analogy between open-channel and compressible flows is useful because important two-dimensional compressible flow phenomena can be simply and inexpensively demonstrated with a shallow, open-channel flow field in a *ripple tank* or *water table*.

The propagation of weak pressure pulses (sound waves) in a compressible flow can be considered to be comparable to the movement of small-amplitude gravity waves on the surface of an open-channel flow. In each case—one-dimensional compressible flow and open-channel flow—the influence of flow velocity on wave pattern is similar. When the flow velocity is less than the wave speed, wave fronts can move upstream of the wave source and the flow is subsonic (compressible flow) or subcritical (open-channel flow). When the flow velocity is equal to the wave speed, wave fronts cannot move upstream of the wave source and the flow is sonic (compressible flow) or critical (open-channel flow). When the flow velocity is greater than the wave speed, the flow is supersonic (compressible flow) or supercritical (open-channel flow). Normal shocks can occur in supersonic compressible flows. Hydraulic jumps can occur in supercritical open-channel flows. Comparison of the characteristics of normal shocks (Section 11.5) and hydraulic jumps (Section 10.7.1) suggests a strong resemblance and thus analogy between the two phenomena.

For compressible flows a meaningful dimensionless variable is the Mach number, where

> Compressible gas flows and open-channel liquid flows are strikingly similar in several ways.

$$Ma = \frac{V}{c}$$

In open-channel flows, an important dimensionless variable is the Froude number, where

$$Fr = \frac{V_{oc}}{\sqrt{gy}} \tag{11.155}$$

The velocity of the channel flow is V_{oc}, the acceleration of gravity is g, and the depth of the flow is y. Since the speed of a small-amplitude wave on the surface of an open-channel flow, c_{oc}, is (see Section 10.2.1)

$$c_{oc} = \sqrt{gy} \tag{11.156}$$

we conclude that

$$Fr = \frac{V_{oc}}{c_{oc}} \tag{11.157}$$

From these equations we see the similarity between Mach number (compressible flow) and Froude number (open-channel flow).

For compressible flow, the continuity equation is

$$\rho AV = \text{constant} \tag{11.158}$$

where V is the flow velocity, ρ is the fluid density, and A is the flow cross-sectional area. For an open-channel flow, conservation of mass leads to

$$ybV_{oc} = \text{constant} \tag{11.159}$$

where V_{oc} is the flow velocity, and y and b are the depth and width of the open-channel flow. Comparing Eqs.11.158 and 11.159 we note that if flow velocities are considered similar and flow area, A, and channel width, b, are considered similar, then compressible flow density, ρ, is analogous to open-channel flow depth, y. This analogy between depth and density can be used to interpret water table results in terms of compressible flow results.

It should be pointed out that the similarity between Mach number and Froude number is generally not exact. If compressible flow and open-channel flow velocities are considered to be similar, then it follows that for Mach number and Froude number similarity the wave speeds c and c_{oc} must also be similar.

From the development of the equation for the speed of sound in an ideal gas (see Eqs. 11.43 and 11.46) we have for the compressible flow

$$c = \sqrt{(\text{constant})k\rho^{k-1}} \qquad (11.160)$$

From Eqs. 11.160 and 11.156, we see that if y is to be completely similar to ρ as suggested by comparing Eqs. 11.158 and 11.159, then k should be equal to 2. Typically $k = 1.3$ or 1.4 or 1.66, not 2 (see Tables 1.7 and 1.8). This limitation to exactness is, however, usually not serious enough to compromise the benefits of the analogy between compressible and open-channel flows.

11.11 TWO-DIMENSIONAL SUPERSONIC FLOW

A very brief introduction to two-dimensional supersonic flow is included here so that you are aware of the basic principles that are associated with this type of flow. We begin with a consideration of supersonic flow over a wall with a small change of direction as sketched in **Fig. 11.26**.

We apply the component of the linear momentum equation (Eq. 5.22) parallel to the Mach wave to the flow across the Mach wave (see Eq. 11.53 for the definition of a Mach wave). The result is that the component of velocity parallel to the Mach wave is constant across the Mach wave. That is, $V_{t1} = V_{t2}$. Thus, from the simple velocity triangle construction indicated in Fig. 11.26, we conclude that the flow accelerates because of the change in direction of the flow. If several changes in wall direction are involved as shown in **Fig. 11.27**, then the supersonic flow accelerates (expands) because of the changes in flow direction across the Mach waves (also called *expansion* waves). Each Mach wave makes an appropriately smaller angle α with the upstream wall because of the increase in Mach number that occurs with each direction change (see Section 11.4). A rounded expansion corner may be considered as a series of infinitesimal changes in direction. Conversely, even sharp corners are actually rounded when viewed on a small enough scale. Thus, expansion fans as illustrated in **Fig. 11.28** are commonly used for supersonic flow around a "sharp" corner. If the flow across the Mach waves is considered to be isentropic, then the increase in flow speed is accompanied by a decrease in static pressure. Therefore, for the flow of an ideal gas, the stagnation properties remain unchanged as the flow passes through an expansion fan.

When the change in supersonic flow direction involves the change in wall orientation that is sketched in **Fig. 11.29**, compression rather than expansion occurs. The flow decelerates and the static pressure increases across the Mach wave. For several changes in wall direction, as indicated in **Fig. 11.30**, several Mach waves occur, each at an appropriately larger angle α with the upstream wall. A rounded compression corner may be considered as a series of infinitesimal changes in direction and even apparently sharp corners are actually rounded. Mach waves or compression waves can coalesce to form an oblique shock wave as shown in **Fig. 11.31**. Unlike the expansion fan in which each Mach wave is of infinitesimal strength, the coalescence of Mach waves produces an oblique shock across which there is a change of static and stagnation properties (and hence entropy) and Mach number.

Supersonic flows accelerate across expansion Mach waves.

 Video

V11.9 Wedge oblique shocks and expansions

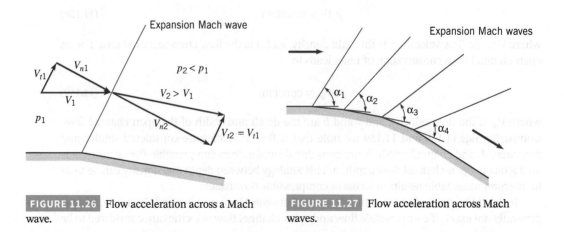

FIGURE 11.26 Flow acceleration across a Mach wave.

FIGURE 11.27 Flow acceleration across Mach waves.

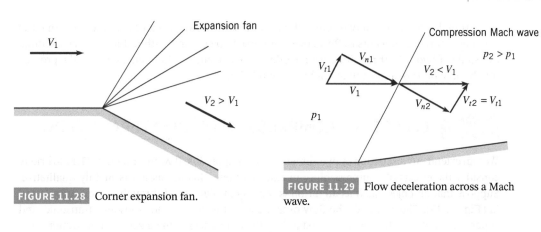

FIGURE 11.28 Corner expansion fan.

FIGURE 11.29 Flow deceleration across a Mach wave.

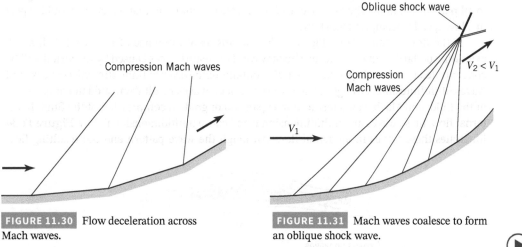

FIGURE 11.30 Flow deceleration across Mach waves.

FIGURE 11.31 Mach waves coalesce to form an oblique shock wave.

 Video

V11.10 Bullet oblique shocks

 Video

V11.11 Space Shuttle oblique shocks

The above discussion of compression waves can be usefully extended to supersonic flow impinging on an object. For example, for supersonic flow incident on a sharp wedge-shaped leading edge (see **Fig. 11.32**), an attached oblique shock can form as suggested in Fig. 11.32*a*. For the same incident Mach number but with a larger wedge angle, a detached curved shock as sketched in Fig. 11.32*b* can result. In Example 11.5, we considered flow along a stagnation pathline across a detached curved shock to be identical to flow across a normal shock wave.

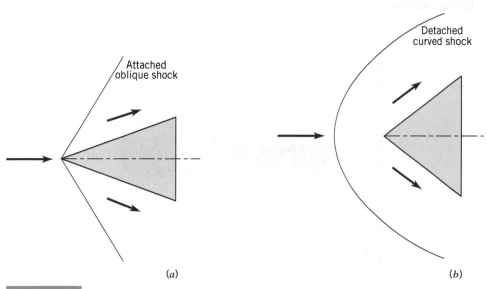

(a)   (b)

FIGURE 11.32 Supersonic flow over a wedge: (*a*) Smaller wedge angle results in attached oblique shock. (*b*) Large wedge angle results in detached curved shock.

From this brief look at two-dimensional supersonic flow, one can easily conclude that the extension of these concepts to flows over immersed objects and within ducts can be exciting, especially if two- and three-dimensional effects are considered. References 4, 5, and 6 provide much more on this subject than could be included here.

11.12 EFFECTS OF COMPRESSIBILITY IN EXTERNAL FLOW

We can extend our discussion of two-dimensional compressible flow from Section 11.11 to briefly consider the effects of compressibility in external flow. This discussion is mainly qualitative. Suppose that an object such as an airfoil is moving at high speed through a compressible fluid (**Fig. 11.33**). The nature of the flow over the airfoil is different for subsonic, transonic, and supersonic motion. The key to understanding the differences is the discussion of Section 11.4. If the flow is subsonic everywhere, disturbances introduced by the airfoil motion are propagated to all parts of the fluid. The flow pattern (Fig. 11.33a) is similar to that which we would expect in a low-speed (incompressible) flow.

If the flow is supersonic (Fig. 11.33b), the upstream influence of the airfoil is limited, extending no farther than a leading shock wave. The disturbances caused by the airfoil will be concentrated in waves. In general, the fluid contains both waves of finite strength (shocks) and waves of infinitesimal strength (Mach waves). Shock waves result from sudden compression of the fluid, and Mach waves result from expansion or gradual compression of the fluid. In external flow, these waves are multidimensional and cause multidimensional flow. **Figure 11.34** illustrates these waves in an actual flow. Although the wave pattern and the resulting flow

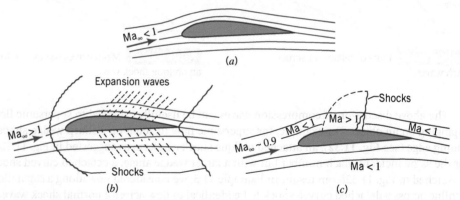

FIGURE 11.33 Airfoil in compressible flow: (a) subsonic flow; (b) supersonic flow; (c) transonic flow.

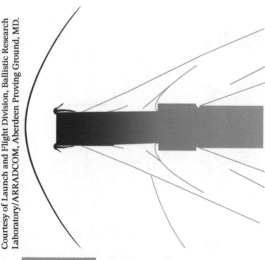

Courtesy of Launch and Flight Division, Ballistic Research Laboratory/ARRADCOM, Aberdeen Proving Ground, MD.

FIGURE 11.34 Supersonic flow over an object is influenced by a complex wave pattern.

appear to be complex, two- and three-dimensional supersonic flow can be analyzed by first analyzing the basic waves (oblique shocks and Mach waves) and then "patching" the waves together to obtain a complete flow field (Refs. 4–6).

Figure 11.33c illustrates transonic flow, in which some regions of the field are subsonic and some regions are supersonic. Shock waves exist in the supersonic portion of the flow. Analysis of transonic flow is very difficult because it is neither entirely like subsonic flow nor entirely like supersonic flow.

Dimensional analysis of compressible external flow indicates that drag and lift coefficients depend on Mach number as well as Reynolds number:

$$C_D = C_D(\alpha, \text{Re}, \text{Ma}_\infty) \quad \text{and} \quad C_L = C_L(\alpha, \text{Re}, \text{Ma}_\infty)$$

Obtaining test data in which both Reynolds number and Mach number are controlled is very difficult (see Section 7.9.2). Most studies concentrate on the effects of Reynolds number alone (usually at low Mach number; see Chapter 9) or on the effects of Mach number alone. For high Reynolds number,

$$C_D \approx C_D(\alpha, \text{Ma}_\infty \text{ only}) \quad \text{and} \quad C_L \approx C_L(\alpha, \text{Ma}_\infty \text{ only})$$

Figure 9.26 illustrates the effect of Mach number on drag coefficient for three axisymmetric bodies. Note the following important effects:

- The drag coefficient is roughly constant for very low and very high Mach numbers.
- The drag coefficient at supersonic speeds is higher than the drag coefficient at low subsonic speeds.
- The drag coefficient rises sharply near $\text{Ma} = 1$.
- The value of the Mach number at which the "drag rise" occurs is pushed closer to 1 by making the body more "streamlined."

Some of these effects can be explained as follows. In supersonic flow, energy is required to maintain the wave pattern, so the drag is higher than in subsonic flow with no waves. The drag rise near $\text{Ma} = 1$ is associated with the sudden appearance of (nearly) normal shock waves in locally supersonic regions of the flow. The (slight) drag drop after $\text{Ma} = 1$ corresponds to a relative weakening of these shock waves as they become inclined to the flow. The drag rise delay for a more streamlined body is due to the fact that the first appearance of locally supersonic flow and shock waves is delayed by streamlining.

The value of Ma_∞ at which the highest local Mach number in the field first becomes 1 is called the *critical Mach number*. Note that, as the Mach number near the maximum body thickness is always greater than the free-stream Mach number, the critical Mach number is always less than 1. The "drag rise" occurs at a Mach number slightly higher than the critical Mach number. The drag rise that occurs near "Mach One" is the "sound barrier" that had to be overcome before supersonic flight was possible.

The effects of Mach number on lift and drag at subsonic and supersonic (but *not* transonic) speed are illustrated by the following equations (Refs. 4–5).

- Prandtl–Glauert rules for subsonic flow:

$$C_D \approx \frac{C_{D\,\text{incomp}}}{\sqrt{1 - \text{Ma}_\infty^2}} \tag{11.161}$$

$$C_L \approx \frac{C_{L\,\text{incomp}}}{\sqrt{1 - \text{Ma}_\infty^2}} \tag{11.162}$$

where $C_{D\,\text{incomp}}$ and $C_{L\,\text{incomp}}$ are the values for incompressible flow.

- Ackeret rules for supersonic flow:

$$C_D \approx \frac{\text{Constant}}{\sqrt{\text{Ma}_\infty^2 - 1}} \tag{11.163}$$

$$C_L \approx \frac{\text{Constant}}{\sqrt{\text{Ma}_\infty^2 - 1}} \tag{11.164}$$

These rules apply only to *thin, two-dimensional* objects. The equations obviously cannot be valid when Ma$_\infty$ is 1 and become less accurate as Ma$_\infty$ approaches 1. The angle of attack must be constant and, strictly speaking, so must the Reynolds number. The "constant" in the Ackeret rules for supersonic flow is not the incompressible flow value of the corresponding coefficient because supersonic flow is not qualitatively or quantitatively similar to incompressible flow. The "incompressible" flow coefficients and the "constants" in the Ackeret rules are functions of body shape and Reynolds number. You should consider Eqs. 11.161–11.164 as illustrative, rather than directly useful for calculations.

EXAMPLE 11.15 | Aerodynamic Force Calculations in Compressible Flow

Given The astronaut in Example 11.12 enters a spherical reentry capsule and returns to Earth. The reentry capsule is 2 m in diameter and has a mass of 1250 kg. At one point during the reentry, the capsule's altitude is 30 km, and it is traveling with a speed of 2300 m/s at an angle of 20° below the horizontal as shown in **Fig. E11.15a**.

Find Calculate the acceleration of the capsule in g's. Also estimate the temperature on the windward surface of the capsule.

Solution The capsule acceleration is computed from Newton's second law:

$$\vec{a} = \frac{\sum \vec{F}}{m} = \frac{\vec{F}_R}{m}$$

Figure E11.15b shows a free-body diagram of the capsule. The x axis is aligned with the direction of travel. The mass m includes the mass of the capsule and the mass of the astronaut inside. The astronaut's mass is not given, so we have to assume a reasonable value, say, 80 kg. Then

$$m = 1250 \text{ kg} + 80 \text{ kg} = 1330 \text{ kg}$$

The forces on the capsule are the gravity force $\vec{W}$ and the aerodynamic drag, $\vec{\mathcal{D}}$. The magnitude of $\vec{W}$ is

$$W = mg$$

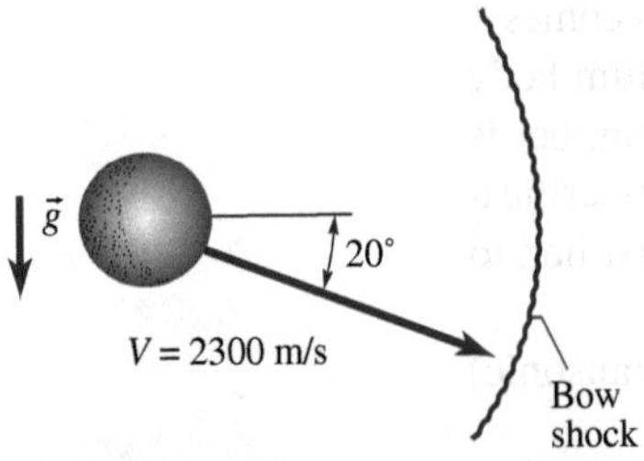

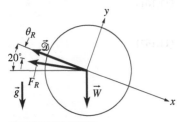

FIGURE E11.15a Spherical reentry capsule in flight.

FIGURE E11.15b Free-body diagram of reentry capsule. $\vec{F}_R$ is the resultant force.

At an altitude of 30 km, the value of g is found in Table C.1, or

$$g = 9.72 \text{ m/s}^2$$

which gives

$$W = (1330 \text{ kg})(9.72 \text{ m/s}^2) = 1.29 \times 10^4 \text{ N}$$

The magnitude of the drag force is

$$\mathcal{D} = C_D \left(\frac{1}{2}\rho V^2\right)\left(\frac{\pi}{4}D^2\right)$$

where D is the capsule diameter. From Eq. 11.48, we have

$$\rho V^2 = kp\text{Ma}^2, \quad \text{so} \quad \mathcal{D} = \frac{\pi}{8} C_D kp\text{Ma}^2 D^2$$

Figure 9.26 gives drag coefficients as a function of Mach number for spheres. To compute the Mach number, we need the temperature. From Table C.1, $T = 226.5$ K and $p = 1.20$ kPa at a 30-km altitude. Then

$$c = \sqrt{kRT} = (20.05 \sqrt{226.5}) \text{ m/s} = 302 \text{ m/s}$$

and

$$\text{Ma} = \frac{V}{c} = \frac{2300 \text{ m/s}}{302 \text{ m/s}} = 7.62$$

From Fig. 9.26,

$$C_D \approx 0.9$$

so

$$\mathcal{D} = \frac{\pi}{8}(0.9)(1.4)(1.20 \times 10^3 \text{ N/m}^2)(7.62)^2(2 \text{ m})^2 = 1.38 \times 10^5 \text{ N}$$

Summing forces in the y direction (see Fig. E11.15b), we have

$$\sum F_y = W \cos 20° = -1.21 \times 10^4 \text{ N}$$

Summing forces in the x direction, we have

$$\sum F_x = W \sin 20° - \mathcal{D}$$

$$= (1.29 \times 10^4 \text{ N}) \sin 20° - 13.8 \times 10^4 \text{ N} = -13.4 \times 10^4 \text{ N}$$

The resultant force has magnitude

$$F_R = \sqrt{(1.21)^2 + (13.4)^2} \times 10^4 \text{ N} = 13.5 \times 10^4 \text{ N}$$

The angle of the resultant force is

$$\theta_R = \tan^{-1}\left(\frac{F_y}{F_x}\right) = \tan^{-1}\left(\frac{-1.21}{-13.4}\right) = 5.2°$$

The acceleration has the same direction as the resultant force; that is, it is directed away from the direction of travel and $(20° − 5.2°) = 14.8°$ *above* the horizontal. The magnitude of the acceleration is

$$a = \frac{F_R}{m} = \frac{135{,}000 \text{ N}}{1330 \text{ kg}} = 101.5 \text{ m/s}^2$$

Expressing the answer in g's $(1 \; g = 9.807 \text{ m/s}^2)$, we have

$$a = 10.35 \; g\text{'s, opposite direction of travel}$$
$$\text{and } 14.8° \text{ above the horizontal} \qquad (Ans)$$

The temperature on the windward surface of the reentry capsule is approximately equal to the stagnation temperature relative to the capsule. Note that the bow shock does not affect the stagnation temperature. From Eq. 11.50,

$$T_0 = T\left(1 + \frac{k-1}{2}\text{Ma}^2\right) = 226.5 \text{ K}\,[1 + 0.2(7.62)^2]$$

$$\underset{\text{surface}}{T_{\text{windward}}} \approx T_0 = 2860 \text{ K} \qquad (Ans)$$

Comments Note that the strong bow shock ahead of the reentry capsule had no effect on our calculations. All the objects in Fig. 9.26 generate shocks when traveling at supersonic speed. The Mach number must be interpreted as the value far ahead of the object (i.e., upstream from the shock).

The deceleration of 10.35 g's is a little severe. A peak value of about 7 g's is generally taken as the safe upper limit for sustained exposure of human beings. Note that the standard g, 9.807 m/s², is used.

Our surface temperature estimate is too high for two reasons. First, nonideal gas effects are important at such high temperatures, so the true value of T_0 is about 10 percent lower than the calculated value. Second, the surface temperature is somewhat lower than the stagnation temperature of the gas. The true temperature of the surface is called the *recovery temperature* and is calculated by (Ref. 4).

$$T_{\text{surf}} = T\left[1 + r\left(\frac{k-1}{2}\right)\text{Ma}^2\right]$$

where r, the recovery factor, is about 0.9. Combining recovery factor and nonideal gas effects yields a surface temperature estimate of

$$\underset{\text{surface}}{T_{\text{windward}}} \approx 2350 \text{ K}$$

Obviously, some sort of heat shield is necessary to prevent an astronaut roast!

CHAPTER SUMMARY

This chapter considers the flow of fluid with substantial changes in density associated with high speeds. While the flow of liquids may most often be considered incompressible over a wide range of speeds, the flow of gases and vapors may involve substantial fluid density changes at higher speeds. At lower speeds, with Mach number less than about 0.3, gas and vapor flows may be treated as incompressible.

Since fluid density and other fluid property changes are significant in compressible flows, property relationships are important. Ideal gases have well-defined fluid property relationships and allow useful conclusions to be made about compressible flows.

The Mach number is a key variable in compressible flow. Most easily understood as the ratio of the local speed of flow and the speed of sound in the flowing fluid, it is a measure of the extent to which the flow is compressible. It is used to define regimes of compressible flows, which range from subsonic (Mach number less than 1) to supersonic (Mach number greater than 1) to hypersonic (Mach number greater than 5).

Liquids are nearly incompressible so the speed of sound is large and Mach number is very small in liquid flows.

There are profound differences between subsonic and supersonic flows. In subsonic flow, disturbances at a point are propagated throughout the entire flow field, but in supersonic flow, disturbances affect only a limited region downstream from the point itself. In supersonic flow, the effect of a specific point is concentrated in a *Mach wave*. Finite disturbances in supersonic flow often result in a *shock wave*—a discontinuous jump in fluid properties and velocity.

The notion of an isentropic (constant entropy) flow is introduced. By definition, isentropic flow is adiabatic (no heat transfer) and frictionless. Isentropic *stagnation properties* are defined in order to account for the effects of the fluid's velocity on the thermodynamic properties. The temperature–entropy (T–s) diagram is introduced as an aid in visualizing property changes in compressible flow.

The important practical problem of compressible flow in a duct is considered next. Three different inputs are examined: area variation, friction, and heat addition or removal. In all cases, the phenomenon of *choking* is observed. Choking occurs when the Mach number reaches 1 at some location in a duct. Choking limits the amount of area decrease, duct friction, or heat addition that can occur.

Three major nonisentropic compressible flows considered in this chapter are Fanno flows, Rayleigh flows, and flows across normal shock waves. Unusual outcomes include the conclusions that friction can accelerate a subsonic Fanno flow, heating can result in fluid temperature reduction in a subsonic Rayleigh flow, and a flow can decelerate from supersonic flow to subsonic flow across a shock wave. The value of temperature–entropy (T–s) diagramming of flows to better understand them is demonstrated.

Numerous formulas describing a variety of ideal gas compressible flows are derived. These formulas can be easily solved with computers. However, to provide the learner with a better grasp of the details of a compressible flow process, a graphical or tabular approach is often used.

The striking analogy between compressible and open-channel flows leads to a brief discussion of the usefulness of a ripple tank or water table to simulate compressible flows.

Expansion and compression Mach waves associated with two-dimensional compressible flows are introduced, as is the formation of oblique shock waves from compression Mach waves.

The following checklist provides a study guide for this chapter. When your study of the entire chapter is completed, you should be able to

- write out the meanings of the terms listed here and understand each of the related concepts. These terms are particularly important and are set in **bold black** type in the text.

- estimate the change in ideal gas properties in a compressible flow.

- calculate Mach number value for a specific compressible flow.

- estimate when a flow may be considered incompressible and when it must be considered compressible to preserve accuracy.

- estimate details of isentropic flows of an ideal gas though converging, diverging, and converging–diverging passages.

- explain the phenomenon of choked flow.

- estimate details of Fanno and Rayleigh flows and flows across normal shock waves.

- explain the analogy between compressible and open-channel flows.

KEY EQUATIONS

Ideal gas equation of state	$\rho = \dfrac{p}{RT}$	(11.1)
Internal energy change	$\breve{u}_2 - \breve{u}_1 = c_v(T_2 - T_1)$	(11.5)
Enthalpy	$\breve{h} = \breve{u} + \dfrac{p}{\rho}$	(11.6)
Enthalpy change	$\breve{h}_2 - \breve{h}_1 = c_p(T_2 - T_1)$	(11.9)
Specific heat difference	$c_p - c_v = R$	(11.12)
Specific heat ratio	$k = \dfrac{c_p}{c_v}$	(11.13)
Specific heat at constant pressure	$c_p = \dfrac{kR}{k-1}$	(11.14)
Specific heat at constant volume	$c_v = \dfrac{R}{k-1}$	(11.15)
First Tds equation	$T\,ds = d\breve{u} + p\,d\left(\dfrac{1}{\rho}\right)$	(11.18)
Second Tds equation	$T\,ds = d\breve{h} - \left(\dfrac{1}{\rho}\right)dp$	(11.20)
Entropy change in ideal gas	$s_2 - s_1 = c_v \ln \dfrac{T_2}{T_1} + R \ln \dfrac{\rho_1}{\rho_2}$	(11.23)
Entropy change in ideal gas	$s_2 - s_1 = c_p \ln \dfrac{T_2}{T_1} - R \ln \dfrac{p_2}{p_1}$	(11.24)
Isentropic process in ideal gas	$\dfrac{p}{\rho^k} = \text{constant}$	(11.27)
Speed of sound	$c = \sqrt{\left(\dfrac{\partial p}{\partial \rho}\right)_s}$	(11.43)
Speed of sound in liquid	$c = \sqrt{\dfrac{E_v}{\rho}}$	(11.45)
Speed of sound in gas	$c = \sqrt{kRT}$	(11.47)
Mach cone angle	$\sin \alpha = \dfrac{c}{V} = \dfrac{1}{\text{Ma}}$	(11.53)
Mach number	$\text{Ma} = \dfrac{V}{c}$	
Isentropic flow	$\dfrac{T}{T_0} = \dfrac{1}{1 + [(k-1)/2]\text{Ma}^2}$	(11.50)
Isentropic flow	$\dfrac{p}{p_0} = \left\{\dfrac{1}{1 + [(k-1)/2]\text{Ma}^2}\right\}^{k/(k-1)}$	(11.51)
Isentropic flow	$\dfrac{\rho}{\rho_0} = \left\{\dfrac{1}{1 + [(k-1)/2]\text{Ma}^2}\right\}^{1/(k-1)}$	(11.52)

Normal shock	$\mathrm{Ma}_y^2 = \dfrac{\mathrm{Ma}_x^2 + [2/(k-1)]}{[2k/(k-1)]\mathrm{Ma}_x^2 - 1}$	(11.63)
Normal shock	$\dfrac{p_y}{p_x} = \dfrac{2k}{k+1}\,\mathrm{Ma}_x^2 - \dfrac{k-1}{k+1}$	(11.64)
Normal shock	$\dfrac{T_y}{T_x} = \left(1 + \dfrac{k-1}{2}\mathrm{Ma}_x^2\right)\left(\dfrac{2k}{k-1}\mathrm{Ma}_x^2 - 1\right)\left[\dfrac{2(k-1)}{(k+1)^2\,\mathrm{Ma}_x^2}\right]$	(11.65)
Normal shock	$\dfrac{p_{0_y}}{p_{0_x}} = \left(\dfrac{k+1}{2}\mathrm{Ma}_x^2\right)^{\frac{k}{k-1}}\left(1 + \dfrac{k-1}{2}\mathrm{Ma}_x^2\right)^{\frac{-k}{(k-1)}}\left(\dfrac{2k}{k+1}\mathrm{Ma}_x^2 - \dfrac{k-1}{k+1}\right)^{\frac{-1}{(k-1)}}$	(11.67)
Isentropic flow-critical temperature ratio	$\dfrac{T^*}{T_0} = \dfrac{2}{k+1}$	(11.85)
Isentropic flow-critical pressure ratio	$\dfrac{p^*}{p_0} = \left(\dfrac{2}{k+1}\right)^{k/(k-1)}$	(11.87)
Isentropic flow with area change	$\dfrac{dV}{V} = -\dfrac{dA}{A}\dfrac{1}{(1-\mathrm{Ma}^2)}$	(11.92)
Isentropic flow with area change	$\dfrac{A}{A^*} = \dfrac{1}{\mathrm{Ma}}\left\{\dfrac{1 + [(k-1)/2]\mathrm{Ma}^2}{1 + [(k-1)/2]}\right\}^{(k+1)/[2(k-1)]}$	(11.93)
Fanno flow	$\dfrac{1}{k}\dfrac{(1-\mathrm{Ma}^2)}{\mathrm{Ma}^2} + \dfrac{k+1}{2k}\ln\left\{\dfrac{[(k+1)/2]\mathrm{Ma}^2}{1 + [(k-1)/2]\mathrm{Ma}^2}\right\} = \dfrac{f(\ell^* - \ell)}{D}$	(11.126)
Fanno flow	$\dfrac{T}{T^*} = \dfrac{(k+1)/2}{1 + [(k-1)/2]\mathrm{Ma}^2}$	(11.129)
Fanno flow	$\dfrac{V}{V^*} = \left\{\dfrac{[(k+1)/2]\mathrm{Ma}^2}{1 + [(k-1)/2]\mathrm{Ma}^2}\right\}^{1/2}$	(11.131)
Fanno flow	$\dfrac{p}{p^*} = \dfrac{1}{\mathrm{Ma}}\left\{\dfrac{(k+1)/2}{1 + [(k-1)/2]\mathrm{Ma}^2}\right\}^{1/2}$	(11.135)
Fanno flow	$\dfrac{p_0}{p_0^*} = \dfrac{1}{\mathrm{Ma}}\left[\left(\dfrac{2}{k+1}\right)\left(1 + \dfrac{k-1}{2}\mathrm{Ma}^2\right)\right]^{[(k+1)/2(k-1)]}$	(11.137)
Rayleigh flow	$\dfrac{p}{p^*} = \dfrac{1+k}{1 + k\mathrm{Ma}^2}$	(11.143)
Rayleigh flow	$\dfrac{T}{T^*} = \left[\dfrac{(1+k)\,\mathrm{Ma}}{1 + k\mathrm{Ma}^2}\right]^2$	(11.148)
Rayleigh flow	$\dfrac{\rho^*}{\rho} = \dfrac{V}{V^*} = \mathrm{Ma}\left[\dfrac{(1+k)\mathrm{Ma}}{1 + k\mathrm{Ma}^2}\right]$	(11.149)
Rayleigh flow	$\dfrac{T_0}{T_0^*} = \dfrac{2(k+1)\mathrm{Ma}^2\left(1 + \dfrac{k-1}{2}\mathrm{Ma}^2\right)}{(1 + k\mathrm{Ma}^2)^2}$	(11.152)
Rayleigh flow	$\dfrac{p_0}{p_0^*} = \dfrac{(1+k)}{(1 + k\mathrm{Ma}^2)}\left[\left(\dfrac{2}{k+1}\right)\left(1 + \dfrac{k-1}{2}\mathrm{Ma}^2\right)\right]^{k/(k-1)}$	(11.154)

REFERENCES

[1] Coles, D., "Channel Flow of a Compressible Fluid," Available on the Internet at *http://web.mit.edu/hml/ncfmf.html* and also on You Tube®. Summary description of film in *Illustrated Experiments in Fluid Mechanics, The NCFMF Book of Film Notes,* MIT Press, Cambridge, MA, 1972.

[2] Moran, M. J., Shapiro, H. N., Boettner, D. D., and Bailey, M. B., *Fundamentals of Engineering Thermodynamics,* 9th Ed., Wiley, New York, 2016.

[3] Keenan, J. H., Chao, J., and Kaye, J., *Gas Tables,* 2nd Ed., Wiley, New York, 1980.

[4] Shapiro, A. H., *The Dynamics and Thermodynamics of Compressible Fluid Flow,* Vol. 1, Wiley, New York, 1953.

[5] Liepmann, H. W., and Roshko, A., *Elements of Gasdynamics,* Dover Publications, 2002.

[6] Anderson, J. D., Jr., *Modern Compressible Flow with Historical Perspective,* 3rd Ed., McGraw-Hill, New York, 2003.

[7] Greitzer, E. M., Tan, C. S., and Graf, M. B., *Internal Flow Concepts and Applications,* Cambridge University Press, UK, 2007.

QUESTIONS AND PROBLEMS

Ⓒ Problem to be solved with aid of programmable calculator or computer.

Ⓞ Open-ended problem that requires critical thinking. These problems require various assumptions to provide the necessary input data. There are not unique answers to these problems.

Note: Unless specific values of required fluid properties are specified in the problem statement, use the values found in Tables 1.4–1.8 and in the tables in the Appendices.

Section 11.1 Ideal Gas Thermodynamics

11.1.1 Distinguish between flow of an ideal gas and inviscid flow of a fluid.

11.1.2 Air flows steadily between two sections in a duct. At section (1), the temperature and pressure are $T_1 = 80$ °C, $p_1 = 301$ kPa(abs), and at section (2), the temperature and pressure are $T_2 = 180$ °C, $p_2 = 181$ kPa(abs). Calculate the (a) change in internal energy between sections (1) and (2), (b) change in enthalpy between sections (1) and (2), (c) change in density between sections (1) and (2), (d) change in entropy between sections (1) and (2). How would you estimate the loss of available energy between the two sections of this flow?

11.1.3 Consider the flow process in **Fig. P11.1.3**. Does the fluid flow from left to right or from right to left? Justify your answer. Assume as ideal gas with constant specific heats.

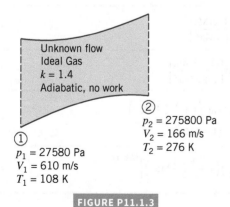

11.1.4 Ⓞ Video As demonstrated in Video V11.1 fluid density differences in a flow may be seen with the help of a schlieren optical system. Discuss what variables affect fluid density and the different ways in which a variable density flow can be achieved.

11.1.5 Video Describe briefly how a schlieren optical visualization system (Videos V11.1 and V11.5, also Fig. 11.4) works. How else might density changes in a fluid flow be made visible to the eye?

11.1.6 Methane is compressed adiabatically from 200 kPa(abs) and 50 °C to 400 kPa(abs). What is the minimum compressor exit temperature possible? Explain.

11.1.7 Air expands adiabatically through a turbine from a pressure and temperature of 1241100 Pa, 890 K to a pressure of 101357 Pa. If the actual temperature change is 85% of the ideal temperature change, determine the actual temperature of the expanded air and the actual enthalpy and entropy differences across the turbine.

11.1.8 An expression for the value of c_p for carbon dioxide as a function of temperature is

$$c_p = 286 - \frac{1.15 \times 10^5}{T} + \frac{2.49 \times 10^6}{T^2}$$

where c_p is in (N · m)/(kg · K) and T is in K. Compare the change in enthalpy of carbon dioxide using the constant value of c_p (see Table 1.7) with the change in enthalpy of carbon dioxide using the expression above, for $T_2 - T_1$ equal to (a) 5.6 K, (b) 556 K, (c) 1668 K. Set $T_1 = 300$ K.

11.1.9 Air flows in a 15-cm-diameter horizontal pipe. At section 1, $p = 600$ kPa, $T = 70$ °C, and $V = 35$ m/s. At section 2: $T = 42$ °C and $V = 115$ m/s. Determine (a) The pressure at section 2 (b) The total friction force on the pipe wall (c) The heat transfer

11.1.10 An air heater in a large coal-fired steam generator heats fresh air entering the steam generator by cooling flue gas leaving the steam generator. 1.26×10^{-4} kg/s of air at 311 K and 1.39×10^{-4} kg/s of flue gas at 655 K enters the air heater. The flue gas leaves at 427.5 K. Flue gas has $c_p = 1.09$ kJ/kg · K and $k = 1.39$. Pressure changes are small and may be neglected. Calculate the temperature of the air leaving the air heater and the total entropy change for the process.

Section 11.2 Stagnation Properties

11.2.1 The air velocity in the duct in **Fig. P11.2.1** is 229 m/s. The air static temperature is 311 K. Use the mercury manometer measurement to calculate the static and stagnation pressures of the flowing air.

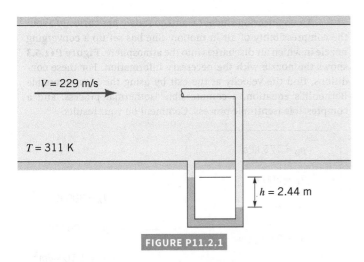

$V = 229$ m/s

$T = 311$ K

$h = 2.44$ m

FIGURE P11.2.1

11.2.2 Air is flowing in a duct as shown in **Fig. P11.2.2**. A Pitot-static tube and a thermocouple are inserted into the flow stream as shown. Calculate the air velocity and the air-mass flowrate.

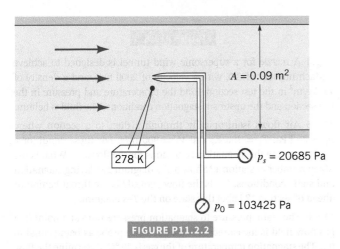

$A = 0.09$ m^2

278 K

$p_s = 20685$ Pa

$p_0 = 103425$ Pa

FIGURE P11.2.2

11.2.3 Steam (H$_2$O vapor) flows in a pipeline in a power station. The steam pressure is 1034250 Pa, its temperature is 533 K, and it flows with velocity 229 m/s. Calculate the stagnation pressure and stagnation temperature. If you are familiar with Steam Tables or steam property software, use these tools to make an "exact" calculation. If you are not familiar with these tools, model the steam as an ideal gas with molecular weight of 18 and $k = 1.3$.

Section 11.3 Mach Number and Speed of Sound

11.3.1 If the observed speed of sound in steel is 5300 m/s, determine the bulk modulus of elasticity of steel in N/m^3. The density of steel is nominally 7790 kg/m^3. How does your value of E_v for steel compare with E_v for water at 15.6 °C? Compare the speeds of sound in steel, water, and air at standard atmospheric pressure and 15 °C and comment on what you observe.

11.3.2 An airplane is flying at a flight (or local) Mach number of 0.72 at 10,000 m in the Standard Atmosphere. Find the ground speed (a) if the air is not moving relative to the ground and (b) if the air is moving at 20 km/hr in the opposite direction from the airplane.

11.3.3 The stagnation pressure in a Mach 2 wind tunnel operating with air is 800 kPa. A 1.5 cm-diameter sphere positioned in the wind tunnel has a drag coefficient of 0.95. Calculate the drag force on the sphere.

11.3.4 Calculate the speed of sound in air, helium, and hydrogen. The temperature is 294 K.

11.3.5 Determine the Mach number of a car moving in standard air at a speed of (a) 11 m/s, (b) 25 m/s, and (c) 45 m/s.

11.3.6 A total pressure probe is inserted into a supersonic airflow. A shock wave forms just upstream of the impact hole. The probe measures a total pressure of 500 kPa(abs). The stagnation temperature at the probe head is 500 K. The static pressure upstream of the shock is measured with a wall tap to be 100 kPa(abs). From these data, estimate the Mach number and velocity of the flow.

11.3.7 The stagnation pressure indicated by a Pitot tube mounted on an airplane in flight is 45 kPa(abs). If the aircraft is cruising in standard atmosphere at an altitude of 10,000 m, determine the speed and Mach number involved.

Section 11.4 Compressible Flow Regimes

11.4.1 (Video) A schlieren photo of a bullet moving through air (see Video V11.10) at 101357 Pa and 293 K indicates a Mach cone angle of 28°. How fast was the bullet moving in m/s?

11.4.2 At a given instant of time, two pressure waves, each moving at the speed of sound, emitted by a point source moving with constant velocity in a fluid at rest, are shown in **Fig. P11.4.2**. Determine the Mach number involved and indicate with a sketch the instantaneous location of the point source.

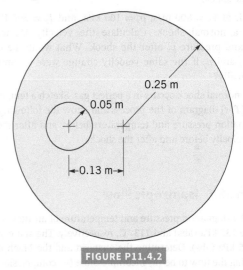

0.25 m

0.05 m

0.13 m

FIGURE P11.4.2

11.4.3 If a new Boeing 787 Dreamliner cruises at a Mach number of 0.87 at an altitude of 9144 m, how fast is this in m/s?

11.4.4 Explain how you could vary the Mach number but not the Reynolds number in airflow past a sphere. For a constant Reynolds number of 300,000, estimate how much the drag coefficient will increase as the Mach number is increased from 0.3 to 1.0.

11.4.5 Air flows in a constant-area, insulated duct. The air enters the duct at 288 K, 344950 Pa, and Ma = 0.45. At a downstream location, the Mach number is one. Find: (a) The pressure and temperature at the downstream location (b) The change in specific entropy (c) The frictional force if the duct is circular and 0.30 m in diameter.

Section 11.5 Shock Waves

11.5.1 A normal shock occurs in a stream of oxygen. The oxygen flows at Ma = 1.8 and the upstream pressure and temperature are 103425 Pa and 303 K. (a) Calculate the following on the downstream side of the shock: static pressure, stagnation pressure, static temperature, stagnation temperature, static density, and velocity. (b) If the Mach number is doubled to 3.6, what will be the resulting values of the parameters listed in part (b)?

11.5.2 [Video] The Pitot tube on a supersonic aircraft cruising at an altitude of 9144 m senses a stagnation pressure of 82740 Pa. If the atmosphere is considered standard, determine the airspeed and Mach number of the aircraft. A shock wave is present just upstream of the probe impact hole.

11.5.3 An aircraft cruises at a Mach number of 2.0 at an altitude of 15 km. Inlet air is decelerated to a Mach number of 0.4 at the engine compressor inlet. A normal shock occurs in the inlet diffuser upstream of the compressor inlet at a section where the Mach number is 1.2. For isentropic diffusion, except across the shock, and for standard atmosphere, determine the stagnation temperature and pressure and the static pressure of the air entering the engine compressor.

11.5.4 At some point for air flow in a duct, $p = 137900$ Pa, $T = 277$ K, and $V = 152$ m/s. Can a normal shock occur at this point?

11.5.5 A normal shock propagates at 610 m/s into the still air in a tube. The temperature and pressure of the air are 300 K and 101357 Pa before "hit" by the shock. Calculate the air temperature, pressure, and velocity after the shock, the stagnation temperature and pressure *relative to the shock* ahead of and behind the shock, and the stagnation temperature and pressure *relative to the tube* ahead of and behind the shock.

11.5.6 Air at $V_x = 800$ m/s, $p_x = 100$ kPa, and $T_x = 300$ K passes through a normal shock. Calculate the velocity V_y, temperature T_y, and pressure p_y after the shock. What would be the values of T_y and p_y if the same velocity change were accomplished isentropically?

11.5.7 A normal shock occurs in a perfect gas. Sketch a temperature–entropy (T–s) diagram of the process and show the following: Static and stagnation pressure and temperature before and after the shock and gas velocity before and after the shock.

Section 11.6 Isentropic Flow

11.6.1 The stagnation pressure and temperature of air flowing past a probe are 135 kPa (abs) and 113 °C, respectively. The static air pressure is 85 kPa (abs). Determine the airspeed and the Mach number considering the flow to be (a) incompressible (b) compressible.

11.6.2 Air flows isentropically through a duct as shown in **Fig. P11.6.2**. For the conditions shown, find the Mach number at both stations 1 and 2 and the flowrate. The diameter is not necessarily constant.

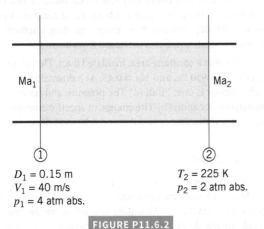

$D_1 = 0.15$ m
$V_1 = 40$ m/s
$p_1 = 4$ atm abs.

$T_2 = 225$ K
$p_2 = 2$ atm abs.

FIGURE P11.6.2

11.6.3 An engineering student wants to satisfy her curiosity about the compressibility of air in motion. She has set up a converging nozzle in which air discharges into the atmosphere. **Figure P11.6.3** shows the nozzle with the necessary information. For these conditions, find the velocity at the exit by using the incompressible Bernoulli's equation, a compressible isothermal process, and a compressible isentropic process. Comment on your results.

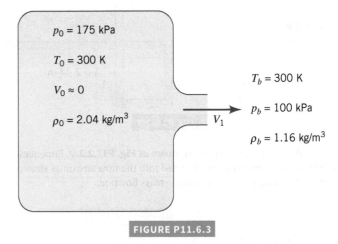

$p_0 = 175$ kPa
$T_0 = 300$ K
$V_0 \approx 0$
$\rho_0 = 2.04$ kg/m^3

$T_b = 300$ K
$p_b = 100$ kPa
$\rho_b = 1.16$ kg/m^3
V_1

FIGURE P11.6.3

11.6.4 A nozzle for a supersonic wind tunnel is designed to achieve a Mach number of 3.0, with a velocity of 2000 m/s, and a density of 1.0 kg/m^3 in the test section. Find the temperature and pressure in the test section and the upstream stagnation conditions. The fluid is helium.

11.6.5 Air flows isentropically through a duct to a section where $p_1 = 25$ kPa, $T_1 = 300$ K, and $V_1 = 900$ m/s. For these conditions: (a) Determine the stagnation conditions for the flow. (b) What is the Mach number at station 1? Show a T–s diagram displaying stagnation and static conditions. (c) Is the flow choked? Is the throat behind or ahead of section 1? Label this state on the T–s diagram.

11.6.6 The static pressure to stagnation pressure ratio at a point in a gas flow field is measured with a Pitot-static probe as being equal to 0.6. The stagnation temperature of the gas is 20 °C. Determine the flow speed in m/s and the Mach number if the gas is air. What error would be associated with assuming that the flow is incompressible?

Section 11.7 One-Dimensional Flow in a Variable Area Duct

11.7.1 Air flows steadily and isentropically from standard atmospheric conditions to a receiver pipe through a converging duct. The cross-sectional area of the throat of the converging duct is 4.65×10^{-3} m^2. Determine the mass flowrate through the duct if the receiver pressure is (a) 68950 Pa, (b) 34475 Pa. Sketch temperature–entropy diagrams for situations (a) and (b). Verify results obtained with values from the appropriate graph in Appendix D with calculations involving ideal gas equations. Is condensation of water vapor a concern? Explain.

11.7.2 Helium at 292.5 K and 101357 Pa in a large tank flows steadily and isentropically through a converging nozzle to a receiver pipe. The cross-sectional area of the throat of the converging passage is 4.65×10^{-3} m^2. Determine the mass flowrate through the duct if the receiver pressure is (a) 68950 Pa, (b) 34475 Pa. Sketch temperature–entropy diagrams for situations (a) and (b).

11.7.3 An ideal gas is to flow isentropically from a large tank where the air is maintained at a temperature and pressure of 288 K and 551800 Pa to standard atmospheric discharge conditions. Describe in general terms the kind of duct involved and determine the duct exit Mach number and velocity in m/s if the gas is air.

11.7.4 The flow blockage associated with the use of an intrusive probe can be important. Determine the percentage increase in section velocity corresponding to a 0.5% reduction in flow area due to probe blockage for airflow if the section area is 1.0 m^2, $T_0 = 20\ °C$, and the unblocked flow Mach numbers are (a) Ma = 0.2, (b) Ma = 0.8, (c) Ma = 1.5, (d) Ma = 3.0.

11.7.5 At a certain point in a pipe, air flows steadily with a velocity of 150 m/s and has a static pressure of 70 kPa and a static temperature of 4 °C. The flow is adiabatic and frictionless. (a) Calculate the maximum possible reduction in area and the following quantities for that minimum area: stagnation pressure, stagnation temperature, static pressure, static temperature, velocity, and Mach number. (b) Calculate the quantities listed in (a) at a point where the area is 15% smaller than the *initial* area.

11.7.6 A tank of oxygen has a hole of area 0.5 cm^2 in its wall. The temperature of the oxygen in the tank is 25 °C. Calculate the rate (kg/s) at which oxygen leaks to the atmosphere for tank pressures of 135 kPa and 375 kPa. Assume frictionless flow.

11.7.7 The gas entering a rocket nozzle has a stagnation pressure of 1500 kPa and a stagnation temperature of 3000 °C. The rocket is traveling in the still Standard Atmosphere at 30,000 m. Find the throat and exit area for a flowrate of 10 kg/s. Assume $k = 1.35$, $R = 287.0$ N · m/kg · K. The gas is perfectly, expanded to the ambient pressure.

11.7.8 Air flows in the channel in **Fig. P11.7.8**. Determine the Mach number, static pressure, and stagnation pressures at station 3. Assume isentropic flow except for the normal shock wave.

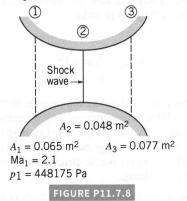

$A_2 = 0.048\ m^2$

$A_1 = 0.065\ m^2$ $A_3 = 0.077\ m^2$
$Ma_1 = 2.1$
$p1 = 448175\ Pa$

FIGURE P11.7.8

11.7.9 A jet engine is to be designed for an altitude of 12,000 m, where the atmospheric pressure is 19.3 kPa. The jet nozzle has a supersonic exit Mach number and is perfectly expanded. The stagnation pressure and temperature of the gas are 100 kPa and 600 °C. The flowrate of gas is 45 kg/s. Calculate the throat area, exit area, and exit velocity. Use $k = 1.4$ and $R = 260$ J/kg · K for the gas.

11.7.10 Air is flowing in the converging–diverging nozzle shown in **Fig. P11.7.10**. Determine the three critical pressure ratios and the Mach numbers immediately upstream and immediately downstream from the shock.

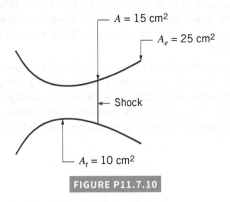

$A = 15\ cm^2$

$A_e = 25\ cm^2$

← Shock

$A_t = 10\ cm^2$

FIGURE P11.7.10

11.7.11 A convergent–divergent nozzle has an exit throat area ratio of 3.0. It is to be supplied with air. Find: (a) The first, second, and third critical pressure ratios; (b) The exit plane Mach number in each case; (c) The throat Mach number in each case.

11.7.12 An ideal gas flows isentropically through a converging-diverging nozzle. At a section in the converging portion of the nozzle, $A_1 = 0.1\ m^2$, $p_1 = 600$ kPa(abs), $T_1 = 20\ °C$, and $Ma_1 = 0.6$. For section (2) in the diverging part of the nozzle, determine A_2, p_2, and T_2 if $Ma_2 = 3.0$ and the gas is air.

11.7.13 Air is supplied to a convergent–divergent nozzle from a reservoir where the pressure is 100 kPa. The air is then discharged through a short pipe into another reservoir where the pressure can be varied. The cross-sectional area of the pipe is twice the area of the throat of the nozzle. Friction and heat transfer may be neglected throughout the flow. If the discharge pipe has constant cross-sectional area, determine the range of static pressure in the pipe for which a normal shock will stand in the divergent section of the nozzle. If the discharge pipe tapers so that its cross-sectional area is reduced by 25%, show that a normal shock cannot be drawn to the end of the divergent section of the nozzle. Find the maximum strength of shock (as expressed by the upstream Mach number) that can be formed.

Section 11.8 Constant-Area Duct Flow with Friction

11.8.1 Air flows adiabatically between two sections in a constant area pipe. At upstream section (1), $p_{0,1} = 689500$ Pa, $T_{0,1} = 332$ K, and $Ma_1 = 0.5$. At downstream section (2), the flow is choked. Estimate the magnitude of the force per unit cross-sectional area exerted by the inside wall of the pipe on the fluid between sections (1) and (2).

11.8.2 Supersonic airflow enters an adiabatic, constant area pipe (inside diameter = 0.1 m) with $Ma_1 = 2.0$. The pipe friction factor is 0.02. If a standing normal shock is located right at the pipe exit, and the Mach number just upstream of the shock is 1.2, determine the length of the pipe.

11.8.3 Consider the flow of air through the piping system shown in **Fig. P11.8.3**. If the system is choked, determine the location where the Mach number is 1. Calculate the pressure ratio p_B/p_{0_1} that causes choking in this system.

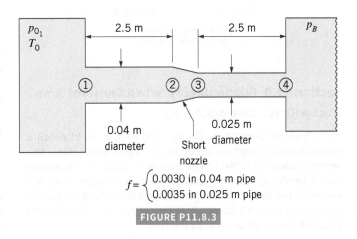

$$f = \begin{cases} 0.0030 \text{ in } 0.04 \text{ m pipe} \\ 0.0035 \text{ in } 0.025 \text{ m pipe} \end{cases}$$

FIGURE P11.8.3

11.8.4 Estimate the maximum mass flowrate of air that can be passed by the duct shown in **Fig. P11.8.4** for $\mu = 2.69 \times 10^{-7}$ N · s/m^2.

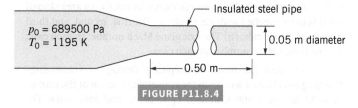

FIGURE P11.8.4

11.8.5 Waste gas (CO_2) is vented to outer space from a spacecraft through a circular pipe 0.2 m long. The pressure and temperature in the spacecraft are 35 kPa and 25 °C. The gas must be vented at the rate of 0.01 kg/s. The friction factor for the flow in the pipe is given by

$$f = 64/Re, \quad Re < 5000,$$

$$f = 0.013, \quad Re > 5000, \quad Re = \frac{4\dot{m}}{\pi\mu D}$$

The viscosity (μ) is 4×10^{-4} N · s/m^2. Determine the required pipe diameter.

11.8.6 Air enters a 4-cm-square galvanized steel duct with $p_0 = 150$ kPa, $T_0 = 400$ K, and $V_1 = 120$ m/s. (*Note:* $\mu = 2.2 \times 10^{-5}$ N · s/m^2). (a) Compute the maximum possible duct length for these conditions. (b) If the actual duct length is 0.75 times the maximum value, calculate the mass flowrate, the exit pressure, and the stagnation pressure. (c) If the actual duct length is 1.3 times the maximum value for the stated conditions, compute the new mass flowrate and inlet velocity. Assume a low back pressure and use the same value of f as used in the maximum possible duct length case.

11.8.7 **Figure P11.8.7** shows an insulated pipe attached to a tank of air. Estimate the maximum mass flowrate that the pipe could exhaust from the tank.

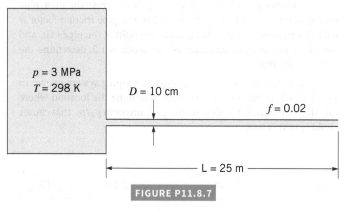

$p = 3$ MPa
$T = 298$ K
$D = 10$ cm
$f = 0.02$
$L = 25$ m

FIGURE P11.8.7

Section 11.9 Frictionless Flow in a Constant-Area Duct with Heating or Cooling

11.9.1 Standard atmospheric air [$T_0 = 288$ K, $p_0 = 101$ kPa(abs)] is drawn steadily through an isentropic converging nozzle into a frictionless diabatic ($q = 500$ kJ/kg) constant area duct. For maximum flow, determine the values of static temperature, static pressure, stagnation temperature, stagnation pressure, and flow velocity at the inlet [section (1)] and exit [section (2)] of the constant area duct. Sketch a temperature-entropy diagram for this flow.

11.9.2 An ideal gas enters [section (1)] a frictionless, constant area duct with the following properties: $T_0 = 293$ K; $p_0 = 101$ kPa (abs) and $Ma_1 = 0.2$.

For Rayleigh flow, determine corresponding values of fluid temperature and entropy change for various levels of pressure and plot the Rayleigh line if the gas is helium.

11.9.3 Air enters a length of constant area pipe with $p_1 = 200$ kPa(abs), $T_1 = 500$ K, and $V_1 = 400$ m/s. If 500 kJ/kg of energy is removed from the air by frictionless heat transfer between sections (1) and (2), determine p_2, T_2, and V_2. Sketch a temperature–entropy diagram for the flow between sections (1) and (2).

11.9.4 Air enters a frictionless, constant area duct with Ma = 2.5, $T_0 = 20$ °C, and $p_0 = 101$ kPa(abs). The gas is decelerated by heating until a normal shock occurs where the local Mach number is 1.3. Downstream of the shock, the subsonic flow is accelerated with heating until it exits with a Mach number of 0.9. Determine the static temperature and pressure, the stagnation temperature and pressure, and the fluid velocity at the duct entrance, just upstream and downstream of the normal shock, and at the duct exit. Sketch the temperature–entropy diagram for this flow.

11.9.5 Air enters a 15-cm pipe with velocity 120 m/s, 1 atmosphere pressure, and $T = 100$ °C. How much heat must be added to bring the air to the maximum static temperature? What are the values of temperature and pressure at this point?

11.9.6 Show that for Rayleigh flow, the maximum amount of heat that may be added to the gas is given by:

$$\frac{q_{max}}{c_p T_1} = \frac{(Ma_1^2 - 1)^2}{2(k+1)Ma_1^2}$$

11.9.7 Air is stored in a tank where the pressure is 275800 Pa and the temperature is 277 K. A converging-diverging nozzle with an exit-to-throat area ratio of 2.5 attaches the tank to a duct where heat is exchanged with the air. The exit pressure is 103425 Pa and a normal shock stands at the exit of the nozzle. Determine the magnitude and direction of the heat exchange.

11.9.8 Prove that, in Rayleigh flow, the Mach number at the point of maximum temperature is $1/\sqrt{k}$.

Lifelong Learning Problems

11.1 LL Is there a limit to how fast an object can move through the atmosphere? Explain.

11.2 LL Discuss the similarities between hydraulic jumps in open-channel flow and shock waves in compressible flow. Explain how this knowledge can be useful.

11.3 LL (See The Wide World of Fluids article titled "Hilsch Tube [Ranque Vortex Tube]," Section 11.1.) Explain why a Hilsch tube works and cite some high and low gas temperatures actually achieved. What is the most important limitation of a Hilsch tube, and how can it be overcome?

11.4 LL (See The Wide World of Fluids article titled "Supersonic and Compressible Flows in Gas Turbines," Section 11.3.) Using typical physical dimensions and rotation speeds of manufactured gas turbine rotors, consider the possibility that supersonic fluid velocities relative to blade surfaces are possible. How do designers use this knowledge?

CHAPTER 12

Turbomachines

LEARNING OBJECTIVES

After completing this chapter, you should be able to:

- explain how a turbomachine works.
- describe the basic differences between a turbine and a pump.
- discuss the importance of minimizing losses in a turbomachine.
- select an appropriate turbomachine for a particular application.
- determine the operating point for a turbomachine installed in a system.

- sketch typical turbomachine blades and velocity diagrams.
- apply similarity laws to predict the effects of size and/or speed on a family of turbomachines.
- perform engineering work or undertake advanced study involving the fluid mechanics of turbomachinery (e.g., design, development, research).

In previous chapters we often used "black boxes" to represent fluid machines such as pumps or turbines. The purpose of this chapter is to study the fluid mechanics of these devices when they are turbomachines.

Pumps and turbines (often turbomachines) are used in a wide variety of configurations. In general, pumps add energy to the fluid—they do work on the fluid to move and/or increase the pressure of the fluid; turbines extract energy from the fluid—the fluid does work on them. The term "pump" will be used to refer to all pumping machines, including *pumps*, *fans*, *blowers*, and *compressors*.

The major components of a typical turbomachine are a rotor attached to an input/output shaft and blades (sometimes called "buckets" or "vanes") attached to that rotor to form flow passages. (The rotor is often called an *impeller* [in a pump], a *runner* [in a hydraulic turbine], or a *wheel* [in a fan].) A fluid that is moving can force rotation and produce shaft power. In this case we have a turbine. On the other hand, we can exert a shaft torque, typically with a motor, and by using blades on a rotor, force the fluid to move. In this case we have a pump. **Figure 12.1** shows the turbine and compressor (pump) rotors of an automobile turbocharger. Examples of turbomachine-type pumps include simple window fans, propellers on ships or airplanes, squirrel-cage fans on home furnaces, axial-flow water pumps used in deep wells, and compressors in automobile turbochargers. Examples of turbines include the turbine portion of gas turbine engines on aircraft, steam turbines used to drive generators at electrical generation stations, and the small, high-speed air turbines that power dentist drills.

Turbomachines serve in an enormous array of applications in our daily lives and thus play an important role in modern society. These machines can have a high-power density (power per unit volume), relatively few moving parts, and reasonable efficiency.

> Turbomachines are dynamic fluid machines that add (for pumps) or extract (for turbines) flow energy.

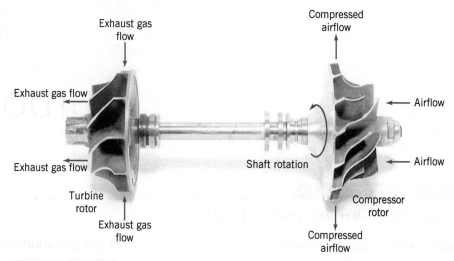

FIGURE 12.1 Automotive turbocharger turbine and compressor rotors. (Source: Tonglint Turbo Technologies.)

The following sections provide an introduction to the fluid mechanics of these important machines. References 1–6 are a few examples of the many books that offer much more information about turbomachines.

12.1 INTRODUCTION

Turbomachines exchange fluid energy and mechanical energy using blades mounted on a rotor.

Turbomachines are mechanical devices consisting of several *blades* attached to a *rotor* that either extract energy from a fluid (turbine) or add energy to a fluid (pump) as a result of dynamic interactions between the device and the fluid. Although the actual design and construction of these devices often require considerable insight and effort, their basic operating principles are not very complicated.

The working fluid can be either a gas (as with a window fan or a gas turbine engine), a vapor (steam turbine), or a liquid (as with the water pump on a car or a turbine at a hydroelectric power plant). While the basic operating principles are the same whether the fluid is a liquid or a gas, important differences in the fluid dynamics involved can occur. For example, cavitation may be an important design consideration when liquids are involved if the pressure at any point within the flow is reduced to the vapor pressure. Compressibility effects may be important when gases are involved if the Mach number becomes large enough.

Many turbomachines contain some type of housing or casing that surrounds the rotor, thus forming an internal flow passageway through which the fluid flows (see **Fig. 12.2**). Others, such as a windmill or a window fan, are unducted. Some turbomachines include stationary blades or vanes in addition to rotor blades. These stationary vanes can be arranged to accelerate the flow with a drop in pressure and thus serve as nozzles. Or these vanes can be set to slow the flow with a pressure rise and thus act as diffusers.

Turbomachines are classified as axial-flow, mixed-flow, or radial-flow machines.

Turbomachines are classified as **axial-flow, mixed-flow,** or **radial-flow** machines depending on the predominant direction of the fluid motion relative to the rotor axis as the fluid passes the blades (see Fig. 12.2). For an axial-flow machine the fluid maintains a significant axial-flow component from the inlet to outlet of the rotor. For a radial-flow machine the flow across the blades involves a substantial radial-flow component at the rotor inlet, exit, or both. In other machines, designated as mixed-flow machines, there may be significant radial- and axial-flow velocity components for the flow through the rotor row. Each type of machine has advantages and disadvantages for different applications and in terms of fluid-dynamic performance.

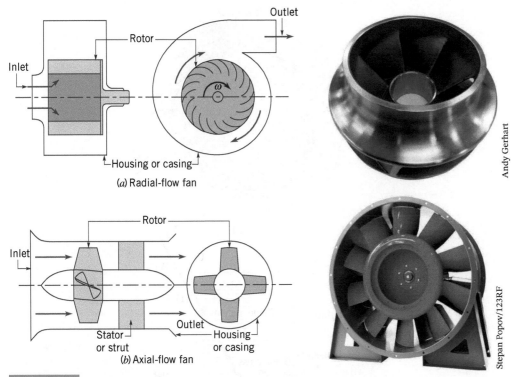

(a) Radial-flow fan

(b) Axial-flow fan

FIGURE 12.2   (a) A radial-flow turbomachine, (b) an axial-flow turbomachine.

12.2   BASIC ENERGY CONSIDERATIONS

An understanding of the work transfer in turbomachines can be obtained by considering the basic operation of a household fan (pump) and a windmill (turbine). Although the actual flows in such devices are very complex (i.e., three-dimensional and unsteady), the essential phenomena can be illustrated by use of simplified flow considerations and velocity diagrams.

Consider a fan blade driven at constant angular velocity, ω, by a motor as is shown in **Fig. 12.3a**. We denote the blade speed as $U = \omega r$, where r is the radial distance from the axis of the fan. The absolute fluid velocity (that seen by a person sitting stationary at the table on which the fan rests) is denoted **V**, and the relative velocity (that seen by a person riding on the fan blade) is denoted **W**. As shown by the adjacent figure, the actual (absolute) fluid velocity is the vector sum of the relative velocity and the blade velocity

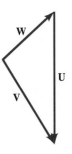

$$\boxed{\mathbf{V} = \mathbf{W} + \mathbf{U}} \tag{12.1}$$

A simplified sketch of the fluid velocity as it "enters" and "exits" the fan at radius r is shown in **Fig. 12.3b**. The shaded surface labeled a–b–c–d is a portion of the cylindrical surface (including a "slice" through the blade) shown in Fig. 12.3a. In this section, and throughout this chapter, we assume for simplicity that the flow moves smoothly along the blade so that relative to the moving blade the velocity is parallel to the leading and trailing edges (points 1 and 2) of the blade. In other words, we assume that the fluid is *perfectly guided* by the blade. Obviously, this assumption becomes more accurate as the ratio of blade length to blade-to-blade spacing increases. For now we assume that the fluid enters and leaves the fan at the same distance from the axis of rotation; thus, $U_1 = U_2 = \omega r$. In actual turbomachines, the entering and leaving flows are not necessarily tangent to the blades, and the fluid pathlines can change radius. These considerations are important at design and off-design operating conditions. Interested readers are referred to Refs. 1, 2, and 3 for more information about these aspects of turbomachine flows.

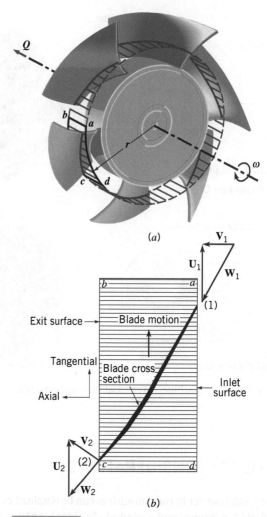

FIGURE 12.3 Idealized flow through a fan: (*a*) fan
blade geometry; (*b*) absolute velocity, **V**; relative
velocity, **W**; and blade velocity, **U** at the inlet and exit
of the fan blade section.

With this information we can construct the **velocity diagrams** shown in Fig. 12.3*b*.
Note that this view is from the top of the fan, looking radially down toward the axis of rotation. The motion of the blade is toward the top of the page; the motion of the incoming air
is assumed to be directed along the axis of rotation. The important concept to grasp from
this sketch is that the fan blade (because of its shape and motion) "pushes" the fluid, causing it to change direction. The absolute velocity vector, **V**, is turned during its flow across
the blade from section (1) to section (2). Initially the fluid had no component of absolute
velocity in the direction of the motion of the blade, the θ (or tangential) direction. When the
fluid leaves the blade, this tangential component of absolute velocity is nonzero. For this to
occur, the blade must push on the fluid in the tangential direction. That is, the blade exerts
a tangential force on the fluid in the direction of the motion of the blade. This tangential
force and the blade motion are in the same direction—the blade does work on the fluid. This
device is a pump.

On the other hand, consider the windmill shown in **Fig. 12.4*a*** (also see **Video V12.1**).
Rather than the rotor being driven by a motor, the blades move in the direction of the force exerted on each blade by the wind blowing through the rotor. We again note that because of the
blade shape and motion, the absolute velocity vectors at sections (1) and (2), V_1 and V_2, have
different directions as depicted in **Figures 12.4*b*** and *c*. For this to happen, the blades must
have pushed up on the fluid—opposite to the direction of blade motion. Alternatively, because

When blades move because of
the fluid force, we have a turbine; when blades move fluid,
we have a pump.

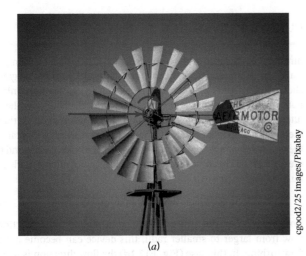

(a)

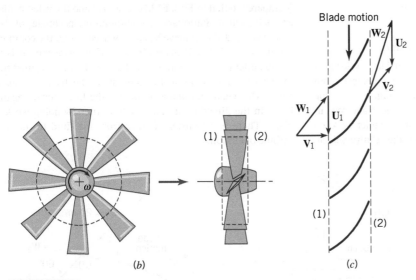

FIGURE 12.4 Idealized flow through a windmill: (a) windmill; (b) windmill blade geometry; (c) absolute velocity, **V**; relative velocity, **W**; and blade velocity, **U** at the inlet and exit of the windmill blade section.

of equal and opposite forces (action/reaction) the fluid must have pushed on the blades in the direction of their motion—the fluid does work on the blades. This extraction of energy from the fluid is the purpose of a turbine.

These examples involve work transfer to or from a flowing fluid in two axial-flow turbomachines. Similar concepts hold for other turbomachines including mixed-flow and radial-flow configurations.

The Wide World of Fluids

Current from currents

The use of large, efficient *wind turbines* to generate electrical power is becoming more commonplace throughout the world. "Wind farms" containing numerous turbines located at sites that have proper wind conditions can produce a significant amount of electrical power. Recently, researchers have been investigating the possibility of harvesting the power of ocean currents and tides by using current turbines that function much like wind turbines. Rather than being driven by wind, they derive energy from naturally occurring flows such as rivers and ocean currents that occur at many locations in the 70% of the Earth's surface that is water. Clearly, a 2.5 m/s tidal current is not as fast as a 70 km/hr wind driving a wind turbine. However, turbine power output is proportional to the fluid density, and seawater is more than 800 times more dense than air. Therefore, significant power can be extracted from slow, but massive, ocean currents. One promising configuration involves blades twisted in a helical pattern. This technology may provide electrical power that is both ecologically friendly and economically viable.

EXAMPLE 12.1 | Basic Difference Between a Pump and a Turbine

Given The rotor shown in **Fig. E12.1a** rotates at a constant angular velocity $\omega = 100$ rad/s. Although the fluid initially approaches the rotor in an axial direction, the flow across the blades is primarily outward (see Fig. 12.2a). Measurements indicate that the absolute velocity at the inlet and outlet are $V_1 = 12$ m/s and $V_2 = 15$ m/s, respectively.

Find Is this device a pump or a turbine?

Solution To answer this question, we need to know if the tangential component of the force of the blade on the fluid is in the direction of the blade motion (a pump) or opposite to it (a turbine). We assume that the blades are tangent to the incoming relative velocity and that the relative flow leaving the rotor is tangent to the blades as shown in **Fig. E12.1b**. We calculate the inlet and outlet blade speeds as

$$U_1 = \omega r_1 = (100 \text{ rad/s})(0.1 \text{ m}) = 10 \text{ m/s}$$

and

$$U_2 = \omega r_2 = (100 \text{ rad/s})(0.2 \text{ m}) = 20 \text{ m/s}$$

With the known, absolute fluid velocity and blade velocity at the inlet, we can draw the velocity diagram (the graphical representation of Eq. 12.1) at that location as shown in **Fig. E12.1c**. Note that we have assumed that the absolute flow at the blade row inlet is radial (i.e., the direction of V_1 is radial). At the outlet we know the blade velocity, U_2; the outlet speed, V_2; and the relative velocity

direction, β_2 (because of the blade geometry). Therefore, we can graphically (or trigonometrically) construct the outlet velocity diagram as shown in the figure. By comparing the velocity diagrams at the inlet and outlet, it can be seen that as the fluid flows across the blade row, the absolute velocity vector turns in the direction of the blade motion. At the inlet there is no component of absolute velocity in the direction of rotation; at the outlet this component is not zero. That is, the blade pushes and turns the fluid in the direction of the blade motion, thereby doing work on the fluid, adding energy to it.

This device is a pump. (Ans)

Comment On the other hand, by reversing the direction of flow from larger to smaller radii, this device can become a radial-flow turbine. In this case (**Fig. E12.1d**) the flow direction is reversed (compared to that in Figs. E12.1a, b, and c) and the velocity diagrams are as indicated. Stationary vanes around the perimeter of the rotor would be needed to achieve V_1 as shown. Note that the component of the absolute velocity, V, in the direction of the blade motion is smaller at the outlet than at the inlet. The blade must push against the fluid in the direction opposite the motion of the blade to cause this. Hence (by equal and opposite forces), the fluid pushes against the blade in the direction of blade motion, thereby doing work on the blade. There is a transfer of work from the fluid to the blade—a turbine operation.

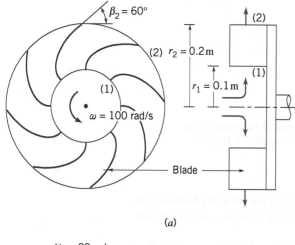

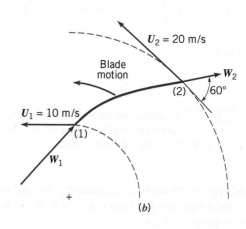

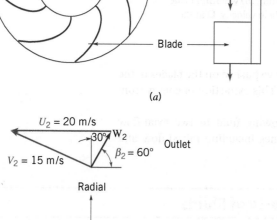

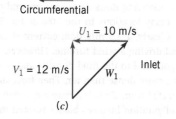

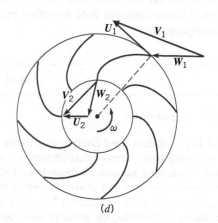

12.3 ANGULAR MOMENTUM CONSIDERATIONS

In the previous section we indicated how work transfer to or from a fluid flowing through a pump or a turbine occurs by interaction between moving rotor blades and the fluid. Because all of these turbomachines involve the rotation of a rotor about a central axis, it is appropriate to discuss their performance in terms of torque and angular momentum.

Recall that work can be written as force times distance or as torque times angular displacement. Hence, if the shaft torque (the torque that the shaft applies to the rotor) and the rotation of the rotor are in the same direction, energy is transferred from the shaft to the rotor and from the rotor to the fluid—the machine is a pump. Conversely, if the torque exerted by the shaft on the rotor is opposite to the direction of rotation, the energy transfer is from the fluid to the rotor—a turbine. The amount of shaft torque (and, hence, rate of shaft work) can be obtained from the moment-of-momentum equation derived formally in Section 5.2.3.

> When shaft torque and rotation are in the same direction, we have a pump; otherwise we have a turbine.

Consider a fluid particle traveling outward through the rotor in the radial-flow machine shown in Figs. E12.1a, b, and c. For now, assume that the particle enters the rotor with a radial velocity only (i.e., no "swirl"). After being acted upon by the rotor blades during its passage from the inlet [section (1)] to the outlet [section (2)], this particle exits with radial (r) and circumferential (θ) components of velocity. Thus, the particle enters with no angular momentum about the axis of rotation but leaves with nonzero angular momentum about that axis. (Recall that the axial component of angular momentum for a particle is its mass times the distance from the axis times the θ component of absolute velocity.)

In a turbomachine a series of particles (a continuum) passes through the rotor. Thus, the moment-of-momentum equation applied to a control volume as derived in Section 5.2.3 is valid. For steady flow (or for turbomachine rotors with steady-in-the-mean or steady-on-average cyclical flow), Eq. 5.42 gives

> V12.2 Self-propelled lawn sprinkler

$$\sum (\mathbf{r} \times \mathbf{F}) = \int_{cs} (\mathbf{r} \times \mathbf{V}) \rho \mathbf{V} \cdot \hat{\mathbf{n}}\, dA$$

Recall that the left side of this equation represents the sum of the external torques (moments) acting on the contents of the control volume, and the right side is the net rate of flow of moment-of-momentum (**angular momentum**) through the control surface.

The axial component of this equation applied to the one-dimensional simplification of flow through a turbomachine rotor with section (1) as the inlet and section (2) as the outlet results in

$$T_{shaft} = -\dot{m}_1(r_1 V_{\theta 1}) + \dot{m}_2(r_2 V_{\theta 2}) \tag{12.2}$$

where T_{shaft} is the **shaft torque** applied to the contents of the control volume. The "−" is associated with mass flowrate into the control volume, and the "+" is used with the outflow. The sign of the V_θ component depends on the direction of V_θ and the blade motion, U. If V_θ and U are in the same direction, then V_θ is positive. The sign of the torque exerted by the shaft on the rotor, T_{shaft}, is positive if T_{shaft} is in the same direction as rotation, and negative otherwise.

As seen from Eq. 12.2, the shaft torque is directly proportional to the mass flowrate, $\dot{m} = \rho Q$. (It takes considerably more torque and power to pump water than to pump air with the same volume flowrate.) The torque also depends on the tangential component of the absolute velocity, V_θ. Equation 12.2 is often called the **Euler turbomachine equation**, a name also frequently assigned to Eqs. 12.4, 12.5, and 12.8.

> The Euler turbomachine equation is derived from the axial component of the moment-of-momentum equation.

Also recall that the **shaft power**, $\dot{W}_{shaft}$, is related to the shaft torque and angular velocity by

$$\dot{W}_{shaft} = T_{shaft}\, \omega \tag{12.3}$$

By combining Eqs. 12.2 and 12.3 and using the fact that $U = \omega r$, we obtain

$$\dot{W}_{shaft} = -\dot{m}_1(U_1 V_{\theta 1}) + \dot{m}_2(U_2 V_{\theta 2}) \tag{12.4}$$

Again, the value of V_θ is positive when V_θ and U are in the same direction and negative otherwise. Also, $\dot{W}_{shaft}$ is positive when the shaft torque and ω are in the same direction and negative otherwise. Thus, $\dot{W}_{shaft}$ is positive when power is supplied to the contents of the control volume

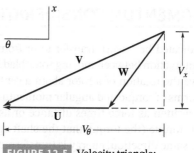

FIGURE 12.5 Velocity triangle: **V** = absolute velocity, **W** = relative velocity, **U** = blade velocity.

(pumps) and negative otherwise (turbines). This outcome is consistent with the sign convention involving the work term in the energy equation considered in Chapter 5 (see Eq. 5.67).

Finally, in terms of work per unit mass, $w_{shaft} = \dot{W}_{shaft}/\dot{m}$, we obtain

$$w_{shaft} = -U_1 V_{\theta 1} + U_2 V_{\theta 2} \qquad (12.5)$$

where we have used the fact that by conservation of mass, $\dot{m}_1 = \dot{m}_2$. Equations 12.3, 12.4, and 12.5 are the basic equations for pumps or turbines whether the machines are radial-, mixed-, or axial-flow devices and for compressible and incompressible flows. Note that neither the axial nor the radial component of velocity enter into the specific work (work per unit mass) equation.

Another useful but more laborious form of Eq. 12.5 can be obtained by writing the right-hand side in a slightly different form based on the velocity diagrams at the entrance or exit as shown generically in **Fig. 12.5**. The velocity component V_x is the generic through-flow component of velocity, and it can be axial, radial, or in-between depending on the rotor configuration. From the large right triangle we note that

$$V^2 = V_\theta^2 + V_x^2$$

or

$$V_x^2 = V^2 - V_\theta^2 \qquad (12.6)$$

From the small right triangle we note that

$$V_x^2 + (V_\theta - U)^2 = W^2 \qquad (12.7)$$

By combining Eqs. 12.6 and 12.7 we obtain

$$V_\theta U = \frac{V^2 + U^2 - W^2}{2}$$

which when written for the inlet and exit and combined with Eq. 12.5 gives

$$w_{shaft} = \frac{V_2^2 - V_1^2 + U_2^2 - U_1^2 - (W_2^2 - W_1^2)}{2} \qquad (12.8)$$

Turbomachine work is related to changes in absolute, relative, and blade velocities.

Thus, the power and the shaft work per unit mass can be calculated from the speed of the blade, U, the absolute fluid speed, V, and the fluid speed relative to the blade, W. This is an alternative to using fewer components of the velocity as suggested by Eq. 12.5. Equation 12.8 contains more terms than Eq. 12.5; however, it is an important concept equation because it shows that work is associated with a change in kinetic energy ($V^2/2$), work done by "centrifugal force" ($U^2/2$), and reaction to acceleration or diffusion of the flow in the blade passage ($W^2/2$). Thus, the most effective pump will have a kinetic energy increase, flow radially outward, and flow decelerated within blade passages. The most effective turbine would have kinetic energy decrease, flow radially inward, and flow accelerated within blade passages. Because of the general nature of the velocity diagram in Fig. 12.5, Eq. 12.8 is applicable for axial-, radial-, and mixed-flow rotors.

12.4 THE CENTRIFUGAL PUMP

One of the most common radial-flow turbomachines is the **centrifugal pump.** This type of pump has two main components: an *impeller* attached to a rotating shaft, and a stationary *casing*, *housing*, or *volute* enclosing the impeller. The impeller consists of a number of blades (usually curved), also sometimes called *vanes*, arranged in a regular pattern around the shaft. **Figure 12.6** shows the essential features of a centrifugal pump. As the impeller rotates, fluid is sucked in through the *eye* of the casing and flows radially outward. Energy is added to the fluid by the rotating blades, and both pressure and absolute velocity are increased as the fluid flows from the eye to the periphery of the blades. For the simplest type of centrifugal pump, the fluid discharges directly into a volute-shaped casing. The casing shape reduces the velocity as the fluid leaves the impeller, and this decrease in kinetic energy is converted into an increase in pressure. The volute-shaped casing, with its increasing area in the direction of flow, is used to produce an essentially uniform velocity distribution as the fluid moves around the casing into the discharge opening. For large centrifugal pumps, a different design is often used in which diffuser guide vanes surround the impeller. The diffuser vanes decelerate the flow as the fluid is directed into the pump casing. This type of centrifugal pump is referred to as a *diffuser pump*.

Impellers are generally of two types. For one configuration the blades are arranged on a hub or backing plate and are open on the other (casing or shroud) side. A typical *open impeller* is shown in **Fig. 12.7a.** For the second type of impeller, called an *enclosed* or *shrouded* impeller, the blades are covered on both hub and shroud ends as shown in **Fig. 12.7b.**

V12.3 Windshield washer pump

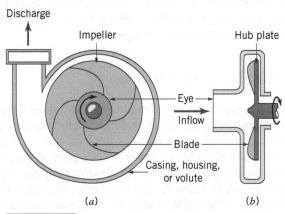

(a) (b)

FIGURE 12.6 Schematic diagram of basic elements of a centrifugal pump.

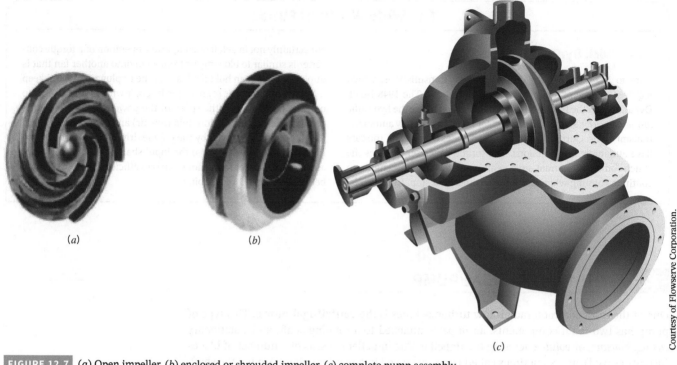

FIGURE 12.7 (*a*) Open impeller, (*b*) enclosed or shrouded impeller, (*c*) complete pump assembly.

Pump impellers can also be *single* or *double suction*. For the single-suction impeller the fluid enters through the eye on only one side of the impeller, whereas for the double-suction impeller the fluid enters the impeller along its axis from both sides. The double-suction arrangement reduces end thrust on the shaft, and also, since the net inlet flow area is larger, inlet velocities are reduced. **Figure 12.7c** shows a double suction pump assembly.

Pumps can be *single* or *multistage*. For a single-stage pump, only one impeller is mounted on the shaft, whereas for multistage pumps, several impellers are mounted on the same shaft. The stages operate in series; that is, the discharge from the first stage flows into the eye of the second stage, the discharge from the second stage flows into the eye of the third stage, and so on. The flowrate is the same through all stages, but each stage develops an additional pressure rise. Thus, a large discharge pressure, or head, can be developed by a multistage pump.

> **Centrifugal pumps involve radially outward flows.**

Centrifugal pumps come in a variety of arrangements (open or shrouded impellers, volute or diffuser casings, single- or double-suction, single- or multistage), but the basic operating principle remains the same. Work is done on the fluid by the rotating blades (centrifugal action and tangential blade force acting on the fluid over a distance), creating a large increase in kinetic energy as well as an increase in pressure of the fluid flowing through the impeller. The kinetic energy is converted into a further increase in pressure as the fluid flows from the impeller into the casing enclosing the impeller. A simplified theory describing the behavior of the centrifugal pump is developed in the following section.

12.4.1 Theoretical Considerations

Although flow through a pump is very complex (unsteady and three-dimensional), the basic theory of operation of a centrifugal pump can be developed by considering the average one-dimensional flow of the fluid as it passes between the inlet and the outlet sections of the rotating impeller. As shown in **Fig. 12.8**, for a typical blade passage, the absolute velocity, V_1, of the fluid entering the passage is the vector sum of the velocity of the blade, U_1, rotating in a circular path with angular velocity ω, and the relative velocity, W_1, within the blade passage so that $V_1 = W_1 + U_1$. Similarly, at the exit $V_2 = W_2 + U_2$. Note that $U_1 = r_1\omega$ and $U_2 = r_2\omega$. Fluid velocities are taken to be average velocities over the inlet and exit sections of the blade passage. The relationship between the various velocities is shown graphically in Fig. 12.8. (In Fig. 12.8,

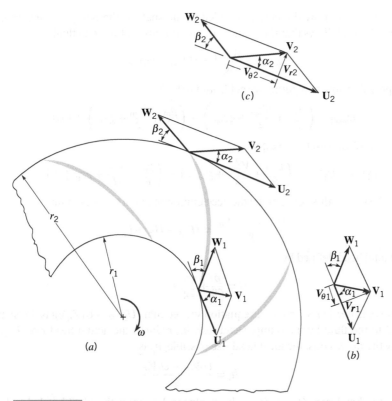

FIGURE 12.8 Velocity diagrams at the inlet and exit of a centrifugal pump impeller.

the velocity vectors are shown in their "true" position, with each vector originating at a single point [fluid particle] at the impeller inlet and outlet. These velocity diagrams must be equivalent to the triangular velocity diagrams used in the preceding section. Both sets of diagrams follow the parallelogram rule for vector addition to show the decomposition of the velocity vectors entering and leaving the blade passages into a component in the direction of blade row rotation and a component tangential to the trailing edge of the blade.

As discussed in Section 12.3, Newton's second law in the form of the moment-of-momentum equation, applied to a control volume enclosing the flow passage between two adjacent blades, provides a relationship between the torque required to drive the pump's shaft, T_{shaft}, and the flow between the blades.

$$T_{\text{shaft}} = \dot{m}(r_2 V_{\theta 2} - r_1 V_{\theta 1}) \tag{12.9}$$

or

$$\boxed{T_{\text{shaft}} = \rho Q(r_2 V_{\theta 2} - r_1 V_{\theta 1})} \tag{12.10}$$

where $V_{\theta 1}$ and $V_{\theta 2}$ are the tangential components of the absolute velocities, $\mathbf{V}_1$ and $\mathbf{V}_2$ (see Figs. 12.8b,c).

For a rotating shaft, the power transferred, $\dot{W}_{\text{shaft}}$, is given by

$$\dot{W}_{\text{shaft}} = T_{\text{shaft}} \omega$$

and, therefore, from Eq. 12.10

$$\dot{W}_{\text{shaft}} = \rho Q \omega (r_2 V_{\theta 2} - r_1 V_{\theta 1})$$

Expressing the component of velocity in the direction of blade rotation as the product of the radius and the angular velocity ($U_1 = r_1 \omega$, $U_1 = r_1 \omega$), the power transferred is given by

$$\dot{W}_{\text{shaft}} = \rho Q(U_2 V_{\theta 2} - U_1 V_{\theta 1}) \tag{12.11}$$

Centrifugal pump impellers have an increase in blade velocity along the flow path ($U_2 > U_1$).

Equation 12.11 shows how the power supplied to the shaft of the pump is transferred to the flowing fluid. It also follows that the shaft work per unit mass of flowing fluid is

$$w_{shaft} = \frac{\dot{W}_{shaft}}{\rho Q} = U_2 V_{\theta 2} - U_1 V_{\theta 1} \tag{12.12}$$

For incompressible flow in a pump, we get from Eq. 5.82

$$w_{shaft} = \left(\frac{p_{out}}{\rho} + \frac{V_{out}^2}{2} + g z_{out}\right) - \left(\frac{p_{in}}{\rho} + \frac{V_{in}^2}{2} + g z_{in}\right) + loss$$

Combining Eq. 12.12 with this, we get

$$U_2 V_{\theta 2} - U_1 V_{\theta 1} = \left(\frac{p_{out}}{\rho} + \frac{V_{out}^2}{2} + g z_{out}\right) - \left(\frac{p_{in}}{\rho} + \frac{V_{in}^2}{2} + g z_{in}\right) + loss$$

Dividing both sides of this equation by the acceleration of gravity, g, we obtain

$$\frac{U_2 V_{\theta 2} - U_1 V_{\theta 1}}{g} = H_{out} - H_{in} + h_L$$

where H is total head defined by

$$H = \frac{p}{\rho g} + \frac{V^2}{2g} + z$$

and $h_L = loss/g$ is head loss. From this equation we see that $(U_2 V_{\theta 2} - U_1 V_{\theta 1})/g$ is the shaft work head added to the fluid by the pump. Head loss, h_L, reduces the actual head rise, $H_{out} - H_{in}$, achieved by the fluid. Thus, the ideal head rise possible, h_i, is

$$h_i = \frac{U_2 V_{\theta 2} - U_1 V_{\theta 1}}{g} \tag{12.13}$$

The pump actual head rise is less than the pump ideal head rise by an amount equal to the head loss in the pump.

The actual head rise, $H_{out} - H_{in} = h_a$, is always less than the ideal head rise, h_i, by an amount equal to the head loss, h_L, in the pump. Some additional insight into the meaning of Eq. 12.13 can be obtained by using the following alternate version (see Eq. 12.8 and the associated discussion):

$$h_i = \frac{1}{2g}\left[(V_2^2 - V_1^2) + (U_2^2 - U_1^2) + (W_1^2 - W_2^2)\right] \tag{12.14}$$

A useful relationship between the flowrate and the pump ideal head rise can be obtained as follows. Often the flow entering the impeller has no component of velocity in the direction of blade rotation, no *swirl* ($V_{\theta 1} = 0$, $\alpha_1 = 90°$ in Fig. 12.8b). In this case, Eq. 12.13 reduces to

$$h_i = \frac{U_2 V_{\theta 2}}{g} \tag{12.15}$$

From Fig. 12.8c

$$\cot \beta_2 = \frac{U_2 - V_{\theta 2}}{V_{r2}}$$

so that Eq. 12.15 can be expressed as

$$h_i = \frac{U_2^2}{g} - \frac{U_2 V_{r2} \cot \beta_2}{g} \tag{12.16}$$

The flowrate, Q, is related to the radial component of the absolute velocity through the equation

$$Q = 2\pi r_2 b_2 V_{r2} \tag{12.17}$$

where b_2 is the impeller blade height at the radius r_2. Thus, combining Eqs. 12.16 and 12.17 yields

$$h_i = \frac{U_2^2}{g} - \frac{U_2 \cot \beta_2}{2\pi r_2 b_2 g} Q \tag{12.18}$$

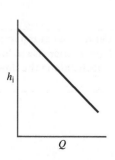

The adjacent figure clearly shows that, for a specific impeller and rotational speed, the ideal head rise h_i as modeled by Eq. 12.18 is a maximum for no flow through the pump and decreases linearly with increasing flowrate Q. For actual pumps, the blade angle β_2 falls in the range of 15°–35°, with a normal range of 20° < β_2 < 25°, and with 15° < β_1 < 50° (Ref. 11). Blades with β_2 < 90° are called *backward curved*, whereas blades with β_2 > 90° are called *forward curved* and blades with $\beta_2 = 90°$ are called *radial*. Pumps are not usually designed with forward-curved vanes because such pumps tend to experience unstable flow. Only small and inexpensive pumps are designed with radial vanes.

EXAMPLE 12.2 | Centrifugal Pump Performance Based on Inlet/Outlet Velocities

Given Water is pumped at the rate of 88 liter/s through a centrifugal pump operating at a speed of 1750 rpm. The impeller has a uniform blade height, b, of 5 cm with $r_1 = 4.0$ cm and $r_2 = 18$ cm, and the exit blade angle β_2 is 23° (see Fig. 12.8). Assume ideal flow conditions and that the tangential velocity component, $V_{\theta 1}$, of the water entering the blade is zero ($\alpha_1 = 90°$).

Find Determine **(a)** the tangential velocity component, $V_{\theta 2}$, at the exit, **(b)** the ideal head rise, h_i, and **(c)** the power, $\dot{W}_{shaft}$, transferred to the fluid. Discuss the difference between ideal and actual head rise. Is the power, $\dot{W}_{shaft}$, ideal or actual? Explain.

Solution

a. At the exit the velocity diagram is as shown in Fig. 12.8c, where $\mathbf{V}_2$ is the absolute velocity of the fluid, $\mathbf{W}_2$ is the relative velocity, and $\mathbf{U}_2$ is the tip velocity of the impeller with

$$U_2 = r_2\omega = (18 \times 10^{-2}\,\text{m})(2\pi\,\text{rad/rev})\frac{(1750\,\text{rpm})}{(60\,\text{s/min})}$$

$$= 33\,\text{m/s}$$

Since the flowrate is given, it follows that

$$Q = 2\pi r_2 b_2 V_{r2}$$

or

$$V_{r2} = \frac{Q}{2\pi r_2 b_2}$$

$$= \frac{88\,\text{liter/s}}{(1000\,\text{kg/m}^3)(2\pi)(0.18\,\text{m})(0.05\,\text{m})}$$

$$= 1.6\,\text{m/s}$$

From Fig. 12.8c we see that

$$\cot \beta_2 = \frac{U_2 - V_{\theta 2}}{V_{r2}}$$

so that

$$V_{\theta 2} = U_2 - V_{r2}\cot\beta_2$$

$$= (33 - 1.6\cot 23°)\,\text{m/s}$$

$$= 29\,\text{m/s} \tag{Ans}$$

b. From Eq. 12.15 the ideal head rise is given by

$$h_i = \frac{U_2 V_{\theta 2}}{g} = \frac{(33\,\text{m/s})(29\,\text{m/s})}{9.81\,\text{m/s}^2}$$

$$= 98\,\text{m} \tag{Ans}$$

Alternatively, from Eq. 12.16, the ideal head rise is

$$h_i = \frac{U_2^2}{g} - \frac{U_2 V_{r2}\cot\beta_2}{g}$$

$$= \frac{(33\,\text{m/s})^2}{9.81\,\text{m/s}^2} - \frac{(33\,\text{m/s})(1.6\,\text{m/s})\cot 23°}{9.81\,\text{m/s}^2}$$

$$= 98\,\text{m} \tag{Ans}$$

c. From Eq. 12.11, with $V_{\theta 1} = 0$, the power transferred to the fluid is given by the equation

$$\dot{W}_{shaft} = \rho Q U_2 V_{\theta 2}$$

$$= \frac{(1000)(88\,\text{liter/s})(33\,\text{m/s})(29\,\text{m/s})}{[(\text{kg}\cdot\text{m/s}^2)/\text{N}](1000)(60\,\text{s/min})}$$

$$= 84.5\,\text{kW} \tag{Ans}$$

Note that the ideal head rise and the power transferred to the fluid are related through the relationship

$$\dot{W}_{shaft} = \rho g Q h_i$$

Comment It should be emphasized that results given in the previous equation involve the ideal head rise. The actual head-rise performance characteristics of a pump are usually determined by experimental measurements obtained in a testing laboratory. The actual head rise is always less than the ideal head rise for a specific flowrate because of the loss of available energy associated with actual flows. Also, it is important to note that even if actual values of U_2 and V_{r2} are used in Eq. 12.16, the ideal head rise is calculated. The only idealization used in this example problem is that the exit flow angle is identical to the blade angle at the exit. If the actual exit flow angle was made available in this example, it could have been used in Eq. 12.16 to calculate the ideal head rise.

The pump power, $\dot{W}_{shaft}$, is the actual power required to achieve a blade speed of 33 m/s, a flowrate of 88 liter/s, and the tangential velocity, $V_{\theta 2}$, associated with this example. If pump losses could somehow be reduced to zero (every pump designer's dream), the actual and ideal head rise would have been identical at 98 m. As is, the ideal head rise is 98 m and the actual head rise something less, probably about 84 m.

Figure 12.9 shows the ideal head versus flowrate curve (Eq. 12.18) for a centrifugal pump with backward-curved vanes ($\beta_2 < 90°$). Since there are simplifying assumptions (i.e., zero losses) associated with the equation for h_i, we would expect that the actual rise in head of fluid, h_a, would be less than the ideal head rise, and this is indeed the case. As shown in Fig. 12.9, the h_a versus Q curve lies below the ideal head-rise curve and shows a nonlinear variation with Q. The differences between the two curves (as represented by the shaded areas between the curves) arise from several sources. These differences include losses due to fluid skin friction in the blade passages, which vary as Q^2, and other losses due to such factors as flow separation, impeller blade-casing clearance flows, and other three-dimensional flow effects. Near the design flowrate, some of these other losses are minimized.

Centrifugal pump design is a highly developed field, with much known about pump theory and design procedures (see, for example, Refs. 4–6). However, due to the general complexity of flow through a centrifugal pump, the actual performance of the pump cannot be

> Ideal and actual head-rise levels differ by the head losses in the pump.

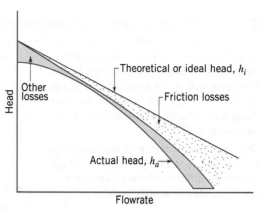

FIGURE 12.9 Effect of losses on the pump head–flowrate curve.

accurately predicted on a completely theoretical basis as indicated by the data of Fig. 12.9. Actual pump performance is determined experimentally through tests on the pump. From these tests, pump characteristics are determined and presented as **pump performance curves**. It is this information that is most helpful to the engineer responsible for incorporating pumps into a given flow system.

12.4.2 Pump Performance Characteristics

The actual head rise, h_a, and the pump flowrate, Q, are determined experimentally using an arrangement similar to the one shown in **Fig. 12.10**. The flowrate is measured by a flowmeter of the type discussed in Sec. 8.6. The pump head rise is

$$h_a = \frac{p_2 - p_1}{\gamma} + z_2 - z_1 + \frac{V_2^2 - V_1^2}{2g} \tag{12.19}$$

with sections (1) and (2) at the pump inlet and exit, respectively. Typically, the differences in elevation and velocity heads are small so that

$$h_a \approx \frac{p_2 - p_1}{\gamma} \tag{12.20}$$

The power, $\mathcal{P}_f$, transferred to the fluid is given by the equation

$$\mathcal{P}_f = \gamma Q h_a \tag{12.21}$$

and this quantity, expressed in terms of horsepower, is traditionally called the *water horsepower*. Thus,

$$\mathcal{P}_f = \text{water horsepower} = \frac{\gamma Q h_a}{550} \tag{12.22}$$

with γ expressed in N/m^3, Q in m^3/s, and h_a in m. Note that the γ appearing in Eq. 12.22 must be the specific weight of the fluid moving through the pump.

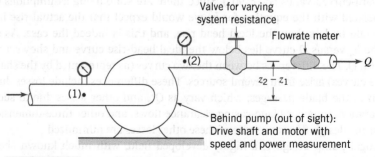

FIGURE 12.10 Typical experimental arrangement for determining the performance of a pump.

In addition to the head or power transferred to the fluid, the **overall efficiency**, η, is of interest, where

$$\eta = \frac{\text{power gained by the fluid}}{\text{shaft power driving the pump}} = \frac{\mathcal{P}_f}{\dot{W}_{\text{shaft}}} \tag{12.23}$$

The denominator of this relationship represents the total power applied to the shaft of the pump and is expressed in kilowatt (kW). Thus,

$$\eta = \frac{\gamma Q h_a}{\text{kW}}$$

The overall pump efficiency is diminished by the *hydraulic losses* in the pump, as previously discussed, and further reduced by the *mechanical losses* in the bearings and seals. There may also be some power loss due to leakage of the fluid between the back surface of the impeller hub plate and the casing, or through other pump components. This leakage contribution to the overall efficiency is called the *volumetric loss*. Therefore, the overall efficiency is the product of the *hydraulic efficiency*, η_h, the *mechanical efficiency*, η_m, and the *volumetric efficiency*, η_v, so that $\eta = \eta_h \eta_m \eta_v$.

> Overall pump efficiency is the ratio of power actually gained by the fluid to the shaft power supplied.

Performance characteristics for a given pump geometry and operating speed are usually given in the form of plots of h_a, η, and bhp versus Q (commonly referred to as *capacity*) as illustrated in **Fig. 12.11**. Actually, only two curves are needed since h_a, η, and bhp are related through Eq. 12.23. For convenience, all three curves are usually displayed in a single graph. Note that for the pump characterized by the data of Fig. 12.11, the head curve continuously rises as the flowrate decreases, and in this case the pump is said to have a *rising head* curve. As shown in the adjacent figure, pumps may also have $h_a - Q$ curves that initially rise as Q is decreased from the design value and then fall with a continued decrease in Q. These pumps have a *falling* (or *drooping*) *head* curve. The head developed by the pump at zero discharge is called the *shutoff head*, and it represents the rise in pressure head across the pump with the discharge valve closed. Since there is no flow with the valve closed, the related efficiency is zero, and the power supplied by the pump is simply dissipated as heat. Although centrifugal pumps can be operated for short periods of time with the discharge valve closed, damage will occur due to overheating and large mechanical stresses if operated for an extended period of time with the valve closed.

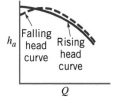

As can be seen from Fig. 12.11, as the discharge is increased from zero the brake horsepower increases, with a subsequent fall as the maximum discharge is approached. As previously noted, with h_a and bhp known, the efficiency can be calculated. As shown in Fig. 12.11, the efficiency is a function of the flowrate and reaches a maximum value at some particular value of the flowrate, commonly referred to as the *normal* or *design* flowrate or capacity for the

FIGURE 12.11 Typical performance characteristics for a centrifugal pump of a given size operating at a constant impeller speed.

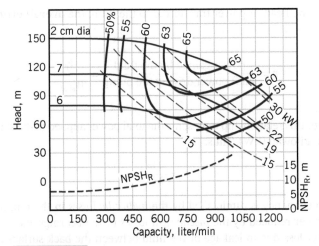

FIGURE 12.12 Performance curves for a two-stage centrifugal pump operating at 3500 rpm. Data given for three different impeller diameters.

pump. The points on the various curves corresponding to the maximum efficiency are denoted as the *best efficiency points* (BEP). When selecting a pump for a particular application, it is usually desirable to have the pump operate near its maximum efficiency. Thus, performance curves of the type shown in Fig. 12.11 are very important to the engineer responsible for the selection of pumps for a particular flow system. Matching the pump to a particular flow system is discussed in Section 12.4.4.

Pump performance characteristics from pump manufacturers are also presented in charts of the type shown in **Fig. 12.12**. Impellers with different diameters may be used in a given casing, so performance characteristics for several impeller diameters can be provided with corresponding lines of constant efficiency and brake horsepower as illustrated in Fig. 12.12. Thus, the same information can be obtained from this type of graph as from the curves shown in Fig. 12.11.

It is to be noted that an additional curve is given in Fig. 12.12, labeled $NPSH_R$, which stands for *net positive suction head required*. As discussed in the following section, the significance of this curve is related to conditions on the suction side of the pump, which must also be carefully considered when selecting and positioning a pump.

12.4.3 Net Positive Suction Head (NPSH)

On the suction side of a pump, low pressures are commonly encountered, with the concomitant possibility of cavitation occurring within the pump. As discussed in Section 1.8, cavitation occurs when the liquid pressure at a given location is reduced to the vapor pressure of the liquid. When this occurs, vapor bubbles form (the liquid starts to "boil"); this phenomenon can cause a loss in efficiency as well as structural damage to the pump. To characterize the potential for cavitation, the difference between the total head on the suction side, near the pump impeller inlet, $p_s/\gamma + V_s^2/2g$, and the liquid vapor pressure head, p_v/γ, is used. This difference is called the net positive suction head (NPSH)

> Cavitation, which may occur in liquid pumps and hydraulic turbines, should be avoided.

$$\text{NPSH} = \frac{p_s}{\gamma} + \frac{V_s^2}{2g} + z_s - \frac{p_v}{\gamma} \tag{12.24}$$

The position reference for the elevation head usually passes through the centerline of the pump impeller inlet so that $z_s = 0$.

There are actually two values of NPSH of interest. The first is the *required* NPSH, denoted $NPSH_R$, that must be maintained, or exceeded, so that cavitation will not occur. The lowest pressure in the pump inlet occurs in the impeller eye, not at the suction flange, so it is necessary to determine the $NPSH_R$ by experiment. This is the curve shown in Fig. 12.12. Pumps are tested to determine the value for $NPSH_R$, as defined by Eq. 12.24, by either directly detecting

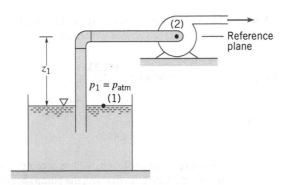

FIGURE 12.13 Schematic of a pump installation in which the pump must lift fluid from one level to another.

cavitation or observing a change in the head–flowrate curve or the efficiency (Ref. 7). The second value for NPSH of concern is the *available* NPSH, denoted NPSH_A, which represents the suction head that actually occurs for the particular flow system. This value can be determined experimentally, or calculated if the system parameters are known. A typical flow system is shown in **Fig. 12.13**. The energy equation applied between the free liquid surface, where the pressure is atmospheric, p_{atm}, and a point on the suction side of the pump near the impeller inlet yields

$$\frac{p_{atm}}{\gamma} + z_1 = \frac{p_s}{\gamma} + \frac{V_s^2}{2g} + \sum h_L$$

where $\sum h_L$ represents head losses between the free surface and the pump inlet. The elevation, z_1, is measured from the reference plane, usually passing through the pump shaft. For the situation shown in Fig. 12.13, z_1 is a negative number. Thus, the head available at the pump inlet is

$$\frac{p_s}{\gamma} + \frac{V_s^2}{2g} = \frac{p_{atm}}{\gamma} + z_1 - \sum h_L$$

so that

$$\boxed{\text{NPSH}_A = \frac{p_{atm}}{\gamma} + z_1 - \sum h_L - \frac{p_v}{\gamma}} \tag{12.25}$$

For this calculation, absolute pressures are normally used because the vapor pressure is usually specified as an absolute pressure. For proper pump operation it is necessary that

$$\text{NPSH}_A \geq \text{NPSH}_R$$

It is noted from Eq. 12.25 that as the height of the pump above the fluid surface is increased (i.e., z_1 becomes more negative), the NPSH_A is decreased. Therefore, there is some maximum height that the liquid can be lifted above which the pump cannot operate without cavitation. The specific value depends on the head losses and the value of the vapor pressure. It is further noted that if the supply tank or reservoir is *above* the pump, z_1 will be positive in Eq. 12.25, and the NPSH_A will increase as this height is increased.

> For proper pump operation, the available net positive suction head must be greater than the required net positive suction head.

EXAMPLE 12.3 | Net Positive Suction Head

Given A centrifugal pump is to be placed above a large, open water tank, as shown in Fig. 12.13, and is to pump water at a rate of $1.4 \times 10^{-2} \text{ m}^3/\text{s}$. At this flowrate the required net positive suction head, NPSH_R, is 4.5 m, as specified by the pump manufacturer. The water temperature is 30 °C and atmospheric pressure is 101.3 kPa. Assume that the major head loss between the tank and the pump inlet is due to a filter at the pipe inlet having a minor loss coefficient $K_L = 20$. Other losses can be neglected. The pipe on the suction side of the pump has a diameter of 10 cm.

Find Determine the maximum height that the pump can be located above the water surface without cavitation. If you were required to place a valve in the flow path, would you place it upstream or downstream of the pump? Why?

Solution From Eq. 12.25 the available net positive suction head, $NPSH_A$, is given by the equation

$$NPSH_A = \frac{p_{atm}}{\gamma} + z_1 - \sum h_L - \frac{p_v}{\gamma}$$

and the limiting value for z_1 will occur when $NPSH_A = NPSH_R$. Thus,

$$(-z_1)_{max} = \frac{p_{atm}}{\gamma} - \sum h_L - \frac{p_v}{\gamma} - NPSH_R \qquad (1)$$

Since the only head loss to be considered is the loss

$$\sum h_L = K_L \frac{V^2}{2g}$$

with

$$V = \frac{Q}{A} = \frac{1.4 \times 10^{-2} \text{ m}^3/\text{s}}{(\pi/4)(10 \times 10^{-2} \text{ m})^2} = 1.8 \text{ m/s}$$

it follows that

$$\sum h_L = \frac{(20)(1.8 \text{ m/s})^2}{2(9.8 \text{ m/s}^2)} = 3.1 \text{ m}$$

The water vapor pressure at 30 °C is 4.2 kPa and $\gamma = 9.77$ kN/m³. Equation (1) can now be written as

$$(-z_1)_{max} = \frac{101.3 \text{ kPa}}{9.77 \text{ kN/m}^3}$$
$$- 3 \text{ m} - \frac{4.2 \text{ kPa}}{9.77 \text{ kN/m}^3} - 4.5 \text{ m}$$
$$= 2.43 \text{ m} \qquad (Ans)$$

Thus, to prevent cavitation, with its accompanying poor pump performance and long-term damage, the pump should not be located higher than 2.43 m above the water surface.

Comment If the valve is placed upstream of the pump, it would now operate with a lower inlet pressure because of the additional upstream loss and could now suffer cavitation with its usually negative consequences. So, placing the valve on the downstream side of the pump is normally the better choice.

12.4.4 System Characteristics, Pump-System Matching, and Pump Selection

The system equation relates the head required by the system to the flowrate through the system.

A typical flow system in which a pump is used is shown in **Fig. 12.14**. The mechanical energy equation applied between points (1) and (2) indicates that

$$h_a = (p_2 - p_1)/\gamma z_2 - z_1 + \sum h_L \qquad (12.26)$$

where h_a is the actual head that must be supplied by the pump, and $\sum h_L$ represents all friction losses in the pipe and minor losses for pipe fittings and valves. From our study of pipe flow, we know that h_L varies approximately as the flowrate squared; that is, $h_L \propto Q^2$ (see Section 8.4). For the system shown, the net pressure change is zero, but this is not always true. Thus, Eq. 12.26 can be written in the form

$$h_a = z_2 - z_1 + KQ^2 \qquad (12.27)$$

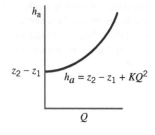

where K depends on the pipe sizes and lengths, friction factors, and minor loss coefficients. Equation 12.27 is the **system equation** and shows how the actual head gained by the fluid from the pump is related to the system parameters. The adjacent figure displays the **system curve**, showing the relationship between required head and the system expressed in the system equation. In this case the system parameters include the change in elevation head, $z_2 - z_1$, and the losses due to friction as expressed by KQ^2. Each flow system has its own specific system equation. If the flow is laminar, the pipe frictional losses will be proportional to Q rather than Q^2 (see Section 8.2).

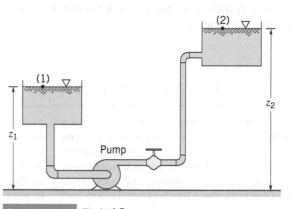

FIGURE 12.14 Typical flow system.

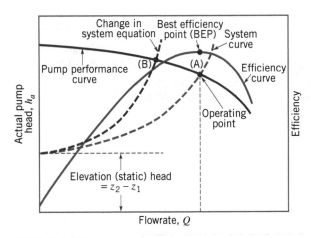

FIGURE 12.15 Utilization of the system curve and the pump performance curve to obtain the operating point for the system.

There is also a unique relationship between the actual pump head gained by the fluid and the flowrate, which is governed by the pump design (as indicated by the pump performance curve). To determine the *operating point* when a specific pump is installed in a specific system, it is necessary to utilize both the system curve, as determined by the system equation, and the pump performance curve. If both curves are plotted on the same graph, as illustrated in **Fig. 12.15**, their intersection (point A) represents the operating point for the system. That is, this point gives the head and flowrate that satisfy both the system equation and the pump equation. On the same graph the pump efficiency is shown. Ideally, we want the operating point to be near the best efficiency point (BEP) for the pump. For a given pump, it is clear that as the system equation changes, the operating point will shift. For example, if the pipe friction increases due to pipe wall fouling, the system curve changes, resulting in the operating point A shifting to point B in Fig. 12.15 with a reduction in flowrate and efficiency. The following example shows how the system and pump characteristics can be used to determine an operating point and decide if a particular pump is suitable for a given application.

> The intersection of the pump performance curve and the system curve is the operating point.

The Wide World of Fluids

Space Shuttle fuel pumps

The fuel *pump* of your car engine is vital to its operation. Similarly, the fuels (liquid hydrogen and oxygen) of each Space Shuttle main engine (there are three per shuttle) relied on multistage *turbopumps* to get from storage tanks to main combustors. High pressures were utilized throughout the pumps to avoid cavitation. The pumps, some *centrifugal* and

some *axial*, were driven by axial-flow, *multistage turbines*. Pump speeds were as high as 35,360 rpm. The liquid oxygen was pumped from 0.7 MPa to 51 MPa, the liquid hydrogen from 0.21 MPa to 50 MPa. Liquid hydrogen and oxygen flowrates of about 65000 liter/min and 23000 liter/min, respectively, were achieved. These pumps could empty your home swimming pool in seconds. The hydrogen went from −252.7 °C in storage to +3315.5 °C in the combustion chamber!

EXAMPLE 12.4 | Use of Pump Performance Curves

Given Water is to be pumped from one large, open tank to a second large, open tank as shown in **Fig. E12.4a**. The pipe diameter throughout is 15 cm, and the total length of the pipe between the pipe entrance and exit is 60 m. Minor loss coefficients for the entrance, exit, and the elbow are shown, and the friction factor for the pipe can be assumed constant and equal to 0.02.

A certain centrifugal pump having the performance characteristics shown in **Fig. E12.4b** is suggested as a good pump for this flow system.

Find With this pump, what would be the flowrate between the tanks? Do you think this pump would be a good choice?

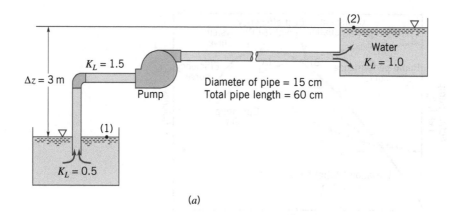

(a)

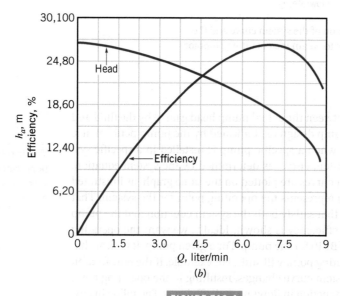

(b)

FIGURE E12.4

Solution   Application of the energy equation between the two free surfaces, points (1) and (2) as indicated, gives

$$\frac{p_1}{\gamma} + \frac{V_1^2}{2g} + z_1 + h_a = \frac{p_2}{\gamma} + \frac{V_2^2}{2g} + z_2$$
$$+ f\frac{\ell}{D}\frac{V^2}{2g} + \sum K_L \frac{V^2}{2g} \qquad (1)$$

Thus, with $p_1 = p_2 = 0$, $V_1 = V_2 = 0$, $\Delta z = z_2 - z_1 = 3$ m, $f = 0.02$, $D = 15/100$ m, and $\ell = 60$ m, Eq. 1 becomes

$$h_a = 3 + \left[0.02\,\frac{60\text{ m}}{15/100\text{ m}} + (0.5 + 1.5 + 1.0)\right]\frac{V^2}{2(9.81\text{ m/s}^2)} \qquad (2)$$

where the specified minor loss coefficients have been used. Since

$$V = \frac{Q}{A} = \frac{Q(\text{m}^3/\text{s})}{(\pi/4)(15/100\text{ m})^2}$$

Eq. 2 can be expressed as

$$h_a = 3 + 1795\,Q^2 \qquad (3)$$

where Q is in m³/s, or with Q in liters per minute

$$h_a = 3 + 5 \times 10^{-7} Q^2 \qquad (4)$$

Equation 3 or 4 represents the system equation for this particular flow system and reveals how much actual head the fluid will need from the pump to maintain a certain flowrate. Performance data shown in Fig. E12.4b shows the actual head provided by this particular pump when it operates at a certain flowrate. Thus, when Eq. 4 is plotted on the same graph with performance data, the intersection of the two curves represents the operating point for the pump and the system. This combination is shown in **Fig. E12.4c** with the intersection observed to occur at

$$Q = 6000 \text{ liter/min} \qquad (Ans)$$

with the corresponding actual head gained equal to 21 m.

Another concern is whether the pump is operating efficiently at the operating point. As can be seen from Fig. E12.4c, although this is not peak efficiency, which is about 86%, it is close (about 84%). Thus, this pump would be a satisfactory choice, assuming the 6000 liter/min flowrate is at or near the desired flowrate.

The power needed to drive the pump is

$$\dot{W}_{\text{shaft}} = \frac{\gamma Q h_a}{\eta}$$
$$= \frac{(9810 \text{ N/m}^3)(6000 \text{ liter/min}) \times 10^{-3} \times 21 \text{ m}}{0.84 \times 60 \text{ s/min}}$$
$$= 24500 \text{ m} \cdot \text{N/s} = 24.5 \text{ kW}$$

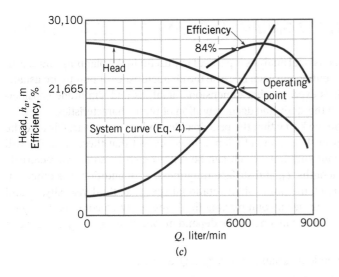

(c)

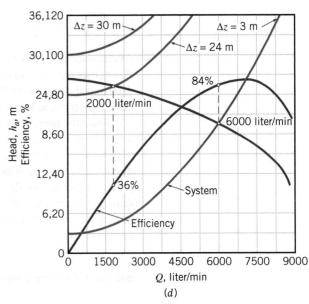

(d)

FIGURE E12.4 (*Continued*)

Comment By repeating the calculations for $\Delta z = z_2 - z_1 = 24$ m and 30 m (rather than the given 3 m), the results shown in **Fig. E12.4d** are obtained. Although the given pump could be used with $\Delta z = 24$ m (provided that the 2000 liter/min flowrate produced is acceptable), it would not be an acceptable pump for this application because its efficiency would be only 36%. Energy could be saved by using a different pump with a performance curve that more nearly matches the new system requirements (i.e., higher efficiency at the operating condition). On the other hand, the given pump would not work at all for $\Delta z = 30$ m because its maximum head ($h_a = 26.8$ m when $Q = 0$) is not enough to lift the water 30 m, let alone overcome head losses. This is shown in Fig. E12.4d by the fact that for $\Delta z = 30$ m the system curve and the pump performance curve do not intersect.

Note that head loss within the pump itself was accounted for with the pump efficiency, η. Thus, $h_s = h_a / \eta$, where h_s is the pump shaft work head and h_a is the actual head rise experienced by the flowing fluid.

If an engineer prefers an analytical procedure instead of a graphical procedure, all that needs to be done is to fit a least squares curve to the pump curve (a second order curve will usually suffice). Upon equating the expression for the pump curve to the expression for the system curve, the engineer has only to solve a polynomial equation. Of course a personal computer, scientific calculator, or even a smart phone makes these mathematical tasks almost trivial.

Pumps can be arranged in series or in parallel to provide for additional head or flow capacity. When two pumps are placed in *series*, the resulting pump performance curve is obtained by adding heads at the same flowrate. As illustrated in **Fig. 12.16a**, for two identical pumps in series, both the actual head gained by the fluid and the flowrate are increased, but neither will be doubled if the system curve remains the same. The operating point is at (A) for one pump and moves to (B) for two pumps in series. For two identical pumps in *parallel*, the combined performance curve is obtained by adding flowrates at the same head, as shown in **Fig. 12.16b**. As illustrated, the flowrate for the system will not be doubled with the addition of two pumps in parallel (if the same system curve applies). However, for a relatively flat system curve, as shown in Fig. 12.16b, a significant increase in flowrate can be obtained as the operating point moves from point (A) to point (B).

> For two pumps in series, add heads; for two in parallel, add flowrates.

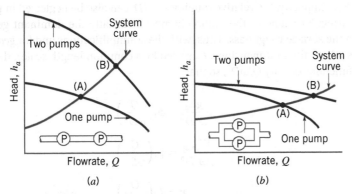

FIGURE 12.16 Effect of operating pumps in (a) series and (b) in parallel.

12.5 DIMENSIONLESS PARAMETERS AND SIMILARITY LAWS

As discussed in Chapter 7, dimensional analysis is particularly useful in the planning and execution of experiments and in presenting the data. Because the characteristics of pumps are usually determined experimentally, it is expected that dimensional analysis and similitude considerations will prove to be useful in the study and documentation of these characteristics.

From the previous section we know that the principal dependent pump variables are the actual head rise, h_a, shaft power, $\dot{W}_{shaft}$, and efficiency, η. We expect that these variables will depend on the geometrical configuration, which can be represented by some characteristic diameter, D, other pertinent lengths, ℓ_i, and surface roughness, ε. In addition, the other important variables are flowrate, Q, the pump shaft rotational speed, ω, fluid viscosity, μ, and fluid density, ρ. We will only consider incompressible fluids presently, so compressibility effects need not concern us yet. Thus, any one of the dependent variables h_a, $\dot{W}_{shaft}$, and η can be expressed as

$$\text{Dependent variable} = f(D, \ell_i, \varepsilon, Q, \omega, \mu, \rho)$$

and a straightforward application of dimensional analysis leads to

$$\text{Dependent pi term} = \phi\left(\frac{\ell_i}{D}, \frac{\varepsilon}{D}, \frac{Q}{\omega D^3}, \frac{\rho \omega D^2}{\mu}\right) \tag{12.28}$$

The dependent pi term involving the head is usually expressed as $C_H = gh_a/\omega^2 D^2$, where gh_a is the actual head rise in terms of energy per unit mass, rather than simply h_a, which is energy per unit weight. This dimensionless parameter is called the **head rise coefficient** or, more simply, the **head coefficient**. The dependent pi term involving the shaft power is expressed as $C_{\mathscr{P}} = \dot{W}_{shaft}/\rho \omega^3 D^5$, and this standard dimensionless parameter is termed the **power coefficient**. The rotational speed, ω, which appears in these dimensionless groups is expressed in rad/s. The final dependent pi term is the efficiency, η, which is already dimensionless. Thus, in terms of dimensionless parameters the performance characteristics are expressed as

$$C_H = \frac{gh_a}{\omega^2 D^2} = \phi_1\left(\frac{\ell_i}{D}, \frac{\varepsilon}{D}, \frac{Q}{\omega D^3}, \frac{\rho \omega D^2}{\mu}\right)$$

$$C_{\mathscr{P}} = \frac{\dot{W}_{shaft}}{\rho \omega^3 D^5} = \phi_2\left(\frac{\ell_i}{D}, \frac{\varepsilon}{D}, \frac{Q}{\omega D^3}, \frac{\rho \omega D^2}{\mu}\right)$$

$$\eta = \frac{\rho g Q h_a}{\dot{W}_{shaft}} = \phi_3\left(\frac{\ell_i}{D}, \frac{\varepsilon}{D}, \frac{Q}{\omega D^3}, \frac{\rho \omega D^2}{\mu}\right)$$

The last pi term in each of the above equations is a form of Reynolds number that represents the relative influence of viscous effects. When the pump flow involves high Reynolds numbers, as is usually the case, experience has shown that the effect of the Reynolds number can be neglected. (This high Reynolds number effect can be observed in the Moody diagram, Fig. 8.20.) For simplicity, the relative roughness, ε/D, can also be neglected in pumps because the highly irregular shape of the pump chamber is usually the dominant geometric factor rather than the surface roughness. Thus, with these simplifications and for *geometrically similar* pumps (all pertinent dimensions, ℓ_i, scaled by a common length scale), the dependent pi terms are functions of only $Q/\omega D^3$, so that

$$\frac{gh_a}{\omega^2 D^2} = \phi_1\left(\frac{Q}{\omega D^3}\right) \tag{12.29}$$

$$\frac{\dot{W}_{shaft}}{\rho \omega^3 D^5} = \phi_1\left(\frac{Q}{\omega D^3}\right) \tag{12.30}$$

$$\eta = \phi_3\left(\frac{Q}{\omega D^3}\right) \tag{12.31}$$

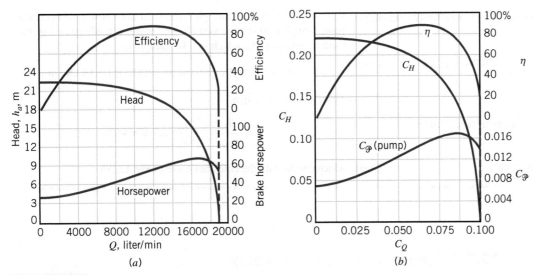

FIGURE 12.17 Typical performance data for a centrifugal pump: (a) characteristic curves for a 30.5 cm. centrifugal pump operating at 1000 rpm, (b) dimensionless characteristic curves. (Source: Data from Rouse, H., *Elementary Mechanics of Fluids*, Wiley, New York, 1946.)

The dimensionless parameter $C_Q = Q/\omega D^3$ is called the **flow coefficient**. These three equations provide the desired similarity relationships among a family of geometrically similar pumps. If two pumps from the family are operated at the same value of flow coefficient

$$\left(\frac{Q}{\omega D^3}\right)_1 = \left(\frac{Q}{\omega D^3}\right)_2 \tag{12.32}$$

it then follows that

$$\left(\frac{gh_a}{\omega^2 D^2}\right)_1 = \left(\frac{gh_a}{\omega^2 D^2}\right)_2 \tag{12.33}$$

$$\left(\frac{\dot{W}_{shaft}}{\rho\omega^3 D^5}\right)_1 = \left(\frac{\dot{W}_{shaft}}{\rho\omega^3 D^5}\right)_2 \tag{12.34}$$

$$\eta_1 = \eta_2 \tag{12.35}$$

where the subscripts 1 and 2 refer to any two pumps from the family of geometrically similar pumps.

With these so-called **pump scaling laws** it is possible to experimentally determine the performance characteristics of one pump in the laboratory and then use these data to predict the corresponding characteristics for other pumps within the family under different operating conditions. **Figure 12.17a** shows some typical curves obtained for a centrifugal pump. **Figure 12.17b** shows the results plotted in terms of the dimensionless coefficients, C_Q, C_H, $C_\mathcal{P}$, and η. From these curves the performance of different-sized, geometrically similar pumps can be predicted, as can the effect of changing speeds on the performance of the pump from which the curves were obtained. It is to be noted that the efficiency, η, is related to the other coefficients through the relationship $\eta = C_Q C_H C_\mathcal{P}^{-1}$. This follows directly from the definition of η.

> Pump scaling laws relate geometrically similar pumps.

EXAMPLE 12.5 | Use of Pump Scaling Laws

Given An 20 cm-diameter centrifugal pump operating at 1200 rpm is geometrically similar to the 30.5 cm-diameter pump having the performance characteristics of Figs. 12.17a and 12.17b while operating at 1000 rpm. The working fluid is water at 15 °C.

Find For peak efficiency, predict the discharge, actual head rise, and shaft power (kW) for this smaller pump.

Solution As is indicated by Eq. 12.31, for a given efficiency the flow coefficient has the same value for a given family of pumps. From Fig. 12.17b we see that at peak efficiency $C_Q = 0.0625$. Thus, for the 20 cm pump

$$Q = C_Q \omega D^3$$
$$= (0.0625)(1200/60 \text{ rev/s})(2\pi \text{ rad/rev})(20/100 \text{ m})^3$$
$$Q = 0.063 \text{ m}^3/\text{s} \tag{Ans}$$

or in terms of L/min

$$Q = (0.063 \text{ m}^3/\text{s})(1000 \text{ liter/m}^3)(60 \text{ s/min})$$
$$= 3780 \text{ liter/min} \tag{Ans}$$

The actual head rise and the shaft horsepower can be determined in a similar manner since at peak efficiency $C_H = 0.19$ and

$C_{\mathcal{P}} = 0.014$, so that with $\omega = 1200$ rev/min (1 min/60 s)(2π rad/rev) = 126 rad/s

$$h_a = \frac{C_H \omega^2 D^2}{g} = \frac{(0.19)(126 \text{ rad/s})^2 (20/100 \text{ m})^2}{9.81 \text{ m/s}^2} = 12.3 \text{ m} \qquad (Ans)$$

and

$$\dot{W}_{shaft} = C_{\mathcal{P}} \rho \omega^3 D^5$$
$$= (0.014)(1000 \text{ kg/m}^3)(126 \text{ rad/s})^3 (20/100 \text{ m})^5$$
$$= 8.96 \text{ kN} \cdot \text{m/s}$$

$$\dot{W}_{shaft} = 8.96 \text{ kN} \cdot \text{m/s} = 8.96 \text{ kW} \qquad (Ans)$$

Comment   This last result gives the shaft horsepower, which is the power supplied to the pump shaft. The power actually gained by the fluid is equal to $\gamma Q h_a$, which in this example is

$$\mathcal{P}_f = \gamma Q h_a = (9800 \text{ N/m}^3)(0.063 \text{ m}^3/\text{s})(12.3 \text{ m}) = 7.6 \text{ kN} \cdot \text{m/s}$$

Thus, the efficiency, η, is

$$\eta = \frac{\mathcal{P}_f}{\dot{W}_{shaft}} = \left(\frac{7.6}{8.96} \right) \times 100 = 84.8\%$$

which checks with the efficiency curve of Fig. 12.17b.

12.5.1 Special Pump Scaling Laws

Effects of changes in pump operating speed and impeller diameter are often of interest.

Two special cases related to pump similitude commonly arise. In the first case we are interested in how a change in the operating speed, ω, for a *given pump*, affects pump characteristics. It follows from Eq. 12.32 that for the same flow coefficient (and therefore the same efficiency) with $D_1 = D_2$ (the same pump)

$$\frac{Q_1}{Q_2} = \frac{\omega_1}{\omega_2} \qquad (12.36)$$

The subscripts 1 and 2 now refer to the same pump operating at two different speeds at the same flow coefficient. Also, from Eqs. 12.33 and 12.34 it follows that

$$\frac{h_{a1}}{h_{a2}} = \frac{\omega_1^2}{\omega_2^2} \qquad (12.37)$$

and

$$\frac{\dot{W}_{shaft1}}{\dot{W}_{shaft2}} = \frac{\omega_1^3}{\omega_2^3} \qquad (12.38)$$

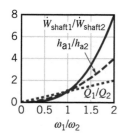

Thus, for a given pump operating at a given flow coefficient, the flowrate varies directly with speed, the head varies as the speed squared, and the power varies as the speed cubed. These effects of angular velocity variation are illustrated in the adjacent figure. These scaling laws are useful in estimating the effect of changing pump speed when some data are available from a pump test obtained by operating the pump at a particular speed.

A question that often arises in connection with application of Eqs. 12.37 and 12.38 is does the operating point when the pump rotates at speed ω_1 transform directly to the operating point when the pump rotates at speed ω_2? Since the operating point depends on the system characteristics as well as the pump characteristics, the answer is generally "no." There is one very important exception. If the system is unchanged and the system equation is of the form $h_s = KQ^2$ (i.e., no elevation or pressure head) then the operating point does transform directly.

In the second special case we are interested in how a change in the impeller diameter, D, of a geometrically similar family of pumps, operating at a *given speed*, affects pump characteristics. As before, it follows from Eq. 12.32 that for the same flow coefficient with $\omega_1 = \omega_2$

$$\frac{Q_1}{Q_2} = \frac{D_1^3}{D_2^3} \qquad (12.39)$$

Similarly, from Eqs. 12.33 and 12.34

$$\frac{h_{a1}}{h_{a2}} = \frac{D_1^2}{D_2^2} \qquad (12.40)$$

and

$$\frac{\dot{W}_{shaft1}}{\dot{W}_{shaft2}} = \frac{D_1^5}{D_2^5} \qquad (12.41)$$

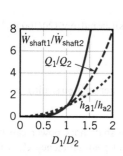

Thus, for a family of geometrically similar pumps operating at a given speed and the same flow coefficient, the flowrate varies as the diameter cubed, the head varies as the diameter squared, and the power varies as the diameter raised to the fifth power. These strong effects of diameter variation are illustrated in the adjacent figure. These scaling relationships are based on the condition that, as the impeller diameter is changed, all other important geometric variables are properly scaled to maintain geometric similarity. This type of geometric scaling is not always possible due to practical difficulties associated with manufacturing the pumps. It is common practice for manufacturers to put impellers of different diameters in the same size pump casing. In this case, complete geometric similarity is not maintained, and the scaling relationships expressed in Eqs. 12.39, 12.40, and 12.41 will not, in general, be valid. However, experience has shown that if the impeller diameter change is not too large, less than about 20%, these scaling relationships can still be used to estimate the effect of a change in the impeller diameter. The pump similarity laws expressed by Eqs. 12.36 through 12.41 are sometimes referred to as the *pump affinity laws*.

> Pump affinity laws relate the same pump at different speeds or geometrically similar pumps at the same speed.

The effects of viscosity and surface roughness have been neglected in the foregoing similarity relationships. However, it has been found that as the pump size decreases, these effects more significantly influence efficiency because of smaller clearances and blade size. An approximate, empirical relationship to estimate the influence of diminishing size on efficiency is

$$\frac{1 - \eta_2}{1 - \eta_1} \approx \left(\frac{D_1}{D_2}\right)^n \tag{12.42}$$

The value of n is usually given as 1/4 or 1/5; however, values as small as 1/10 and as large as 1/2 have been proposed (Ref. 9).

In general, it is to be expected that the similarity laws will not be very accurate if tests on a model pump with water are used to predict the performance of a prototype pump with a highly viscous fluid, such as oil, because at the much smaller Reynolds number associated with the oil flow, the fluid physics involved is different from the higher Reynolds number flow associated with water. Some empirical methods of correcting for Reynolds number effects are discussed in Ref. 10.

12.5.2 Specific Speed

A useful dimensionless parameter can be obtained by eliminating diameter D between the flow coefficient and the head rise coefficient. This is accomplished by raising the flow coefficient to an exponent (1/2) and dividing this result by the head coefficient raised to another exponent (3/4) so that

$$\frac{(Q/\omega D^3)^{1/2}}{(gh_a/\omega^2 D^2)^{3/4}} = \frac{\omega\sqrt{Q}}{(gh_a)^{3/4}} = N_s \tag{12.43}$$

The dimensionless parameter N_s is called the **specific speed**. Specific speed varies with the flow coefficient just as the other coefficients and efficiency discussed earlier do. However, it is customary to specify a value of specific speed at the flow coefficient corresponding to peak efficiency only, i.e., at the BEP. For pumps with low Q and high h_a, the specific speed is low compared to a pump with high Q and low h_a. Centrifugal pumps typically are low-capacity, high-head pumps, and therefore have low specific speeds.

Specific speed as defined by Eq. 12.43 is dimensionless, and therefore independent of the system of units used in its evaluation as long as a consistent unit system is used. However, in the United States a modified, dimensional form of specific speed, N_{sd}, is commonly used, where

$$\boxed{N_{sd} = \frac{\omega(\text{rpm})\sqrt{Q(\text{gpm})}}{[h_a(\text{ft})]^{3/4}}} \tag{12.44}$$

In this case N_{sd} is said to be expressed in *U.S. customary units*. Typical values of N_{sd} are in the range $500 < N_{sd} < 4000$ for centrifugal pumps. Both N_s and N_{sd} have the same physical meaning, but their magnitudes will differ by a constant conversion factor ($N_{sd} = 2733N_s$) when ω in Eq. 12.43 is expressed in rad/s.

Each family or class of pumps has a particular range of values of specific speed associated with it. Thus, pumps that have low-capacity, high-head characteristics will have specific speeds that are smaller than pumps that have high-capacity, low-head characteristics. The specific speed is very useful to engineers and designers. If the required head, flowrate, and speed are specified, it is possible to select an appropriate (most efficient) type of pump

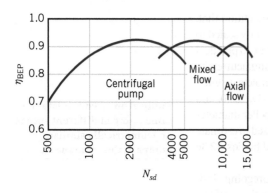

for a particular application. For example, as shown in the adjacent figure, as the specific speed, N_{sd}, increases beyond about 2000 the peak efficiency, η, of the purely radial-flow centrifugal pump starts to fall off, and other types of pump designs may be preferred. In addition to the centrifugal pump, the *axial-flow* pump is widely used. As discussed in Section 12.6, in an axial-flow pump the direction of flow is primarily parallel to the rotating shaft rather than radial as in the centrifugal pump. Axial-flow pumps are essentially high-capacity, low-head pumps, and therefore have large specific speeds ($N_{sd} > 9000$). *Mixed-flow* pumps combine features of both radial-flow and axial-flow pumps and have intermediate values of specific speed. **Figure 12.18** illustrates how the specific speed changes as the configuration of the pump changes from centrifugal or radial to axial.

12.5.3 Suction Specific Speed

With an analysis similar to that used to obtain the specific speed, the *suction specific speed*, S_s, can be expressed as

$$S_s = \frac{\omega\sqrt{Q}}{[g(\text{NPSH}_R)]^{3/4}} \tag{12.45}$$

> Specific speed may be used to approximate what general pump geometry (axial, mixed, or radial) to use for maximum efficiency.

where h_a in Eq. 12.43 has been replaced by the required net positive suction head (NPSH_R). This dimensionless parameter is useful in determining the required operating conditions on the suction side of the pump. As was true for the specific speed, N_s, the value for S_s commonly used is for peak efficiency. For a family of geometrically similar pumps, S_s should have a fixed value. If this value is known, then the NPSH_R can be estimated for other pumps within the same family operating at different values of ω and Q.

As noted for N_s, the suction specific speed as defined by Eq. 12.45 is also dimensionless, and the value for S_s is independent of the system of units used. However, as was the case for specific speed, in the United States a modified dimensional form for the suction specific speed, designated as S_{sd}, is commonly used, where

$$S_{sd} = \frac{\omega\,(\text{rpm})\,\sqrt{Q\,(\text{gpm})}}{[\text{NPSH}_R\,(\text{ft})]^{3/4}} \tag{12.46}$$

For double-suction pumps the discharge, Q, in Eq. 12.46 is one-half the total discharge.

Typical values for S_{sd} fall in the range of 7000 to 12,000 with a "nominal" value of approximately 7900 (Ref. 12). If S_{sd} is specified, Eq. 12.46 can be used to estimate the NPSH_R for a given set of operating conditions. However, this calculation would generally only provide an approximate value for the NPSH_R, and the actual determination of the NPSH_R for a particular pump should be made through a direct measurement whenever possible. Note that $S_{sd} = 2733\,S_s$, with ω expressed in rad/s in Eq. 12.45.

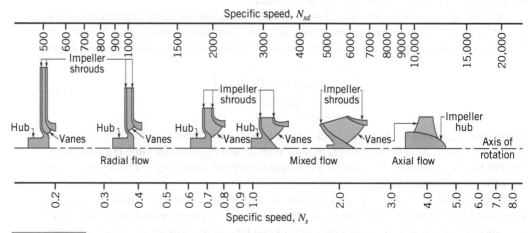

FIGURE 12.18 Variation in specific speed at maximum efficiency with type of pump. (Source: Adapted from Hydraulic Institute, *Hydraulic Institute Standards*, 14th Edition, Hydraulic Institute, Cleveland, Ohio, 1983, used with permission. Hydraulic Institute, Parsippany, New Jersey, 07045 www.pumps.org.)

12.6 AXIAL-FLOW AND MIXED-FLOW PUMPS

As noted previously, centrifugal pumps are radial-flow machines that operate most efficiently for applications requiring high heads at relatively low flowrates. This head–flowrate combination typically yields specific speeds (N_{sd}) that are less than approximately 4000. For many applications, such as those associated with drainage and irrigation, high flowrates at low heads are required and centrifugal pumps are not suitable. In this case, axial-flow pumps are commonly used. This type of pump consists essentially of a propeller confined within a cylindrical casing. Axial-flow pumps are often called *propeller pumps*. For this type of pump the flow is primarily in the axial direction (parallel to the axis of rotation of the shaft), as opposed to the radial flow found in the centrifugal pump. Whereas the head developed by a centrifugal pump includes a contribution due to centrifugal action, the head developed by an axial-flow pump is due primarily to the tangential force exerted by the rotor blades on the fluid. A schematic of an axial-flow pump arranged for vertical operation is shown in **Fig. 12.19**. The rotor is connected to a motor through a shaft, and as it rotates (usually at a relatively high speed) the fluid is sucked in through the inlet. Typically the fluid discharges through a row of fixed stator (guide) vanes used to straighten the flow leaving the rotor. Some axial-flow pumps also have inlet guide vanes upstream of the rotor row, and some are multistage in which pairs (*stages*) of rotating blades (*rotor blades*) and fixed vanes (*stator blades*) are arranged in series. Axial-flow pumps usually have specific speed N_s in excess of 3.3. Very high specific speed pumps do not have stator vanes.

> Axial-flow pumps often have alternating rows of stator blades and rotor blades.

The definitions and broad concepts that were developed for centrifugal pumps are also applicable to axial-flow pumps. The actual flow characteristics, however, are quite different. In **Fig. 12.20** typical head, power, and efficiency characteristics are compared for a centrifugal pump and an axial-flow pump. It is noted that at design capacity (maximum efficiency) the head and brake horsepower are the same for the two pumps selected. But as the flowrate decreases, the power input to the centrifugal pump falls to 134 kW at shutoff, whereas for the axial-flow pump the power input increases to 388 kW at shutoff. This characteristic of the axial-flow pump can cause overloading of the drive motor if the flowrate is reduced significantly from the design capacity. It is also noted that the head curve for the axial-flow pump is much steeper than that for the centrifugal pump. Thus, with axial-flow pumps there will be a large change in head with a small change in the flowrate, whereas for the centrifugal pump, with its relatively flat head curve, there will be only a small change in head with large changes in the flowrate. It is further observed from Fig. 12.20 that, except at design capacity, the efficiency of the axial-flow pump is lower than that of the centrifugal pump. To improve operating characteristics, some axial-flow pumps are constructed with adjustable blades.

The characteristics of centrifugal and axial-flow pumps have important ramifications for a strategy for starting the pumps. A centrifugal pump should be started with the system closed because the pump requires less power with no flow. An axial-flow pump, on the other hand, should be started with the system wide open to avoid the power spike at zero flow. For applications requiring specific speeds intermediate to those for centrifugal and axial-flow

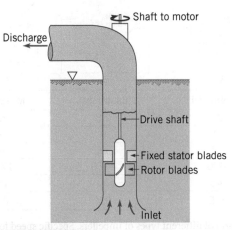

FIGURE 12.19 Schematic diagram of an axial-flow pump arranged for vertical operation.

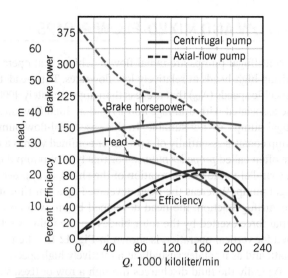

FIGURE 12.20 Comparison of performance characteristics for a centrifugal pump and an axial-flow pump, each rated 159 kiloliter/min at a 5 m head. (Source: Data from Kristal, F. A., and Annett, F. A., *Pumps: Types, Selection, Installation, Operation, and Maintenance*, McGraw-Hill, New York, 1953.)

pumps, mixed-flow pumps have been developed that operate efficiently in the specific speed range $1.5 < N_s < 3.3$. As the name implies, the flow in a mixed-flow pump has both a radial and an axial component. **Figure 12.21** shows some typical data for centrifugal, mixed-flow, and axial-flow pumps, each operating with the same flowrate. These data indicate that as we proceed from the centrifugal pump to the mixed-flow pump to the axial-flow pump, the specific speed increases, the head decreases, the speed increases, the impeller diameter decreases, and the eye diameter increases.

The dimensionless parameters and scaling relationships developed in the previous sections apply to all three types of pumps—centrifugal, mixed-flow, and axial-flow—because the dimensional analysis used is not restricted to a particular type of pump. Additional information about pumps can be found in Refs. 4, 7, 9, 13, and 14.

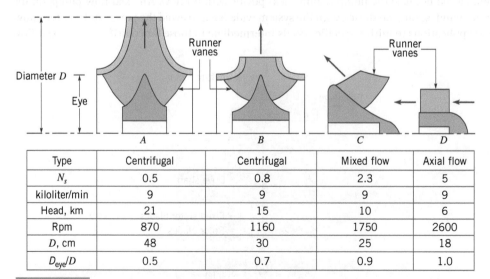

Type	Centrifugal	Centrifugal	Mixed flow	Axial flow
N_s	0.5	0.8	2.3	5
kiloliter/min	9	9	9	9
Head, km	21	15	10	6
Rpm	870	1160	1750	2600
D, cm	48	30	25	18
D_{eye}/D	0.5	0.7	0.9	1.0

FIGURE 12.21 Comparison of different types of impellers. Specific speed for centrifugal pumps based on single suction and identical flowrate. (Source: Adapted from Kristal, F. A., and Annett, F. A., *Pumps: Types, Selection, Installation, Operation, and Maintenance*, McGraw-Hill, New York, 1953.)

12.7 TURBINES

As discussed in Section 12.2, turbines are devices that extract energy from a flowing fluid. The geometry of turbines is such that the fluid exerts a torque on the rotor in the direction of its rotation. The shaft power generated is available to drive generators or other devices.

In the following sections we discuss mainly the operation of hydraulic turbines (those for which the working fluid is water) and to a lesser extent gas and steam turbines (those for which the density of the working fluid may be much different at the inlet than at the outlet). No matter the working fluid, turbines can be broadly classified into two types: **impulse turbines** and **reaction turbines**. These classifications are based on the mechanism of the fluid interaction with the turbine blades. In an impulse turbine, the force on the blades is produced solely by turning the fluid, without appreciable pressure drop in the blade passage, with all of the pressure drop occurring in a fixed nozzle. The force on the vane in Examples 5.10 and 5.17 is solely from impulse. In a reaction turbine, some of the fluid-vane force is from fluid turning and some of the force is a reaction to acceleration of the fluid relative to the vane. In reaction blading, a pressure drop occurs in both a fixed nozzle and the moving vane. Turbine blading is characterized by the *degree of reaction (R)*, which is the ratio of the drop in static pressure (or enthalpy) across the moving blade to the overall drop in static pressure (or enthalpy) across the fixed nozzle plus the moving blade. Impulse turbines have $R = 0$ while reaction turbines typically have $0.1 < R < 0.7$.

The **Pelton wheel** shown in **Fig. 12.22** is a classical example of an impulse turbine. In these machines the total head of the incoming fluid (the sum of the pressure head, velocity head, and elevation head) is converted into a large velocity head at the exit of the supply nozzle (or nozzles if a multiple nozzle configuration is used). Both the pressure drop across the bucket (blade) and the change in relative speed of the fluid across the bucket are negligible. The space surrounding the rotor is not completely filled with fluid. It is the impulse of the individual jets of fluid striking the buckets that generates the torque.

For reaction turbines, on the other hand, the rotor is surrounded by a casing (or volute), which is completely filled with the working fluid. There is both a pressure drop and a fluid relative speed change across the rotor. As shown for the radial-inflow turbine in **Fig 12.23**, guide vanes act as nozzles to accelerate the flow and turn it in the appropriate direction as the fluid enters the rotor. Thus, part of the pressure drop occurs across the guide vanes and part occurs across the rotor. In many respects the operation of a reaction turbine is similar to that of a pump "flowing backward," although such oversimplification can be quite misleading.

Both impulse and reaction turbines can be analyzed using the moment-of-momentum principles discussed in Section 12.3. In general, impulse turbines are high-head, low-flowrate devices, while reaction turbines are low-head, high-flowrate devices.

The two basic types of hydraulic turbines are impulse and reaction.

12.7.1 Impulse Turbines

Although there are various types of impulse turbine designs, perhaps the easiest to understand is the Pelton wheel (see **Fig. 12.24**). Lester Pelton (1829–1908), an American mining engineer during the California gold mining days, is responsible for many of the still-used features of

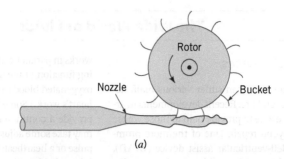

(a)

Courtesy of Voith Hydro, York, PA.

(b)

FIGURE 12.22 (a) Schematic diagram of a Pelton wheel turbine, (b) photograph of a Pelton wheel turbine.

this type of turbine. The Pelton wheel is essentially an improved water wheel. In one way, the Pelton wheel is not typical of other turbomachines; it is neither axial flow nor radial flow; it is "tangential flow." A Pelton wheel is most efficient when operated with a large head (for example, a water source from a lake located significantly above the turbine nozzle), which is converted into a relatively large velocity at the exit of the nozzle. Among the many design considerations for such a turbine are the head loss that occurs in the pipe (the penstock) transporting the water to the turbine, the design of the nozzle, and the design of the buckets on the rotor.

As shown in Fig. 12.24, a high-speed jet of water strikes the Pelton wheel buckets and is deflected. The water enters and leaves the control volume surrounding the wheel as free jets (at atmospheric pressure). In addition, a person riding on the bucket would note that the

Pelton wheel turbines operate most efficiently with a larger head and lower flowrates.

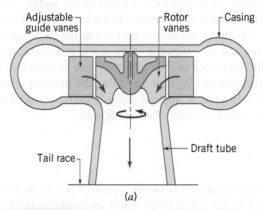

(a)

FIGURE 12.23 (a) Schematic diagram of a reaction turbine, (b) photograph of a reaction turbine.

Courtesy of Voith Hydro, York, PA.

(b)

FIGURE 12.23 *(Continued)*

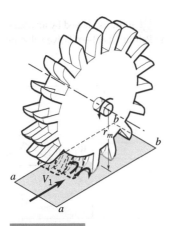

FIGURE 12.24 Details of Pelton wheel turbine.

speed of the water does not change as it slides across the buckets (assuming viscous effects are negligible). That is, the magnitude of the relative velocity does not change, but its direction does. The change in direction of the velocity of the fluid jet causes a torque on the rotor, resulting in a power output from the turbine.

Design of the optimum shape of the buckets to obtain maximum power is a very difficult matter. Ideally, the fluid enters and leaves the control volume shown in **Fig. 12.25** with no radial component of velocity. The buckets would ideally turn the relative velocity vector through 180°, but physical constraints dictate that β, the angle of the exit edge of the blade, is less than 180°. Thus, the fluid leaves with an axial component of velocity as shown in **Fig. 12.26**.

The inlet and exit velocity diagrams at the arithmetic mean radius, r_m, are assumed to be as shown in **Fig. 12.27**. To calculate the torque and power, we must know the tangential components of the absolute velocities at the inlet and exit. From Fig. 12.27 we see that

$$V_{\theta 1} = V_1 = W_1 + U \tag{12.47}$$

and

$$V_{\theta 2} = W_2 \cos \beta + U \tag{12.48}$$

 Video

V12.4 Pelton wheel lawn sprinkler

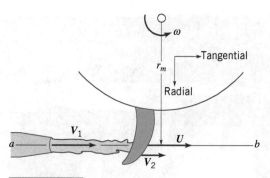

FIGURE 12.25 Ideal fluid velocities for a Pelton wheel turbine.

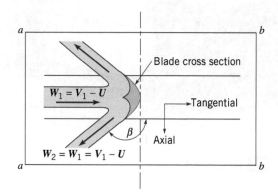

FIGURE 12.26 Flow as viewed by an observer riding on the Pelton wheel—relative velocities.

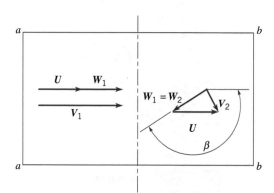

FIGURE 12.27 Inlet and exit velocity diagrams for a Pelton wheel turbine.

In Pelton wheel analyses, we assume the relative speed of the fluid is constant (no friction).

Thus, with the assumption that $W_1 = W_2$, we can combine Eqs. 12.47 and 12.48 to obtain

$$V_{\theta 2} - V_{\theta 1} = (U - V_1)(1 - \cos \beta) \tag{12.49}$$

This change in tangential component of velocity combined with Eqs. 12.2 and 12.4 gives

$$T_{\text{shaft}} = \dot{m} r_m (U - V_1)(1 - \cos \beta)$$

where $\dot{m} = \rho Q$ is the mass flowrate through the turbine. Since $U = \omega R_m$, it follows that

$$\dot{W}_{\text{shaft}} = T_{\text{shaft}}\, \omega = \dot{m} U(U - V_1)(1 - \cos \beta) \tag{12.50}$$

These results are plotted in **Fig. 12.28** along with typical experimental results. Note that $V_1 > U$ (i.e., the jet impacts the bucket), and $\dot{W}_{\text{shaft}} < 0$ (i.e., the turbine extracts power from the fluid).

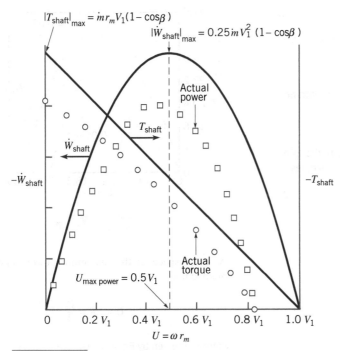

$|T_{shaft}|_{max} = \dot{m} r_m V_1(1-\cos\beta)$

$|\dot{W}_{shaft}|_{max} = 0.25 \dot{m} V_1^2 (1-\cos\beta)$

Actual power

T_{shaft}

$\dot{W}_{shaft}$

$-\dot{W}_{shaft}$

$-T_{shaft}$

$U_{max\ power} = 0.5 V_1$

Actual torque

| 0 | 0.2 V_1 | 0.4 V_1 | 0.6 V_1 | 0.8 V_1 | 1.0 V_1 |

$U = \omega r_m$

FIGURE 12.28 Typical theoretical and experimental power and torque for a Pelton wheel turbine as a function of bucket speed.

Several interesting points can be noted from these results. First, the power is a function of β. However, a typical value of $\beta = 165°$ (rather than the optimum 180°) results in a relatively small (less than 2%) reduction in power since $1 - \cos 165° = 1.966$, compared to $1 - \cos 180° = 2$. Second, although the torque is maximum when the wheel is stopped ($U = 0$), there is no power under this condition—to extract power one needs both force and motion. Third, the power output is a maximum when

$$U|_{max\ power} = \frac{V_1}{2} \tag{12.51}$$

This can be shown by using Eq. 12.50 and solving for U that gives $d\dot{W}_{shaft}/dU = 0$. A bucket speed one-half the speed of the fluid coming from the nozzle gives maximum power. Fourth, the maximum speed occurs when $T_{shaft} = 0$ (i.e., the load is completely removed from the turbine, as would happen if the shaft connecting the turbine to the generator were to break and frictional torques were negligible). For this case $U = \omega R = V_1$, the turbine is "freewheeling," and the water simply passes across the rotor without any force on the buckets.

EXAMPLE 12.6 | Pelton Wheel Turbine Characteristics

Given Water to drive a Pelton wheel is supplied through a pipe from a lake as indicated in **Fig. E12.6a**. The head loss due to friction in the pipe is important, but minor losses can be neglected.

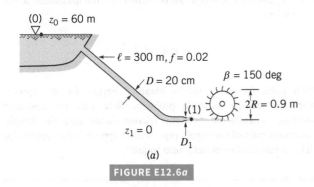

FIGURE E12.6a

Find

a. Determine the nozzle diameter, D_1, that will give the maximum power output.

b. Determine the maximum power and the angular velocity of the rotor at the conditions found in part (a).

Solution

a. As indicated by Eq. 12.50, the power output depends on the flow-rate, $Q = \dot{m}/\rho$, and the jet speed at the nozzle exit, V_1, both of which depend on the diameter of the nozzle, D_1, and the head loss associated with the supply pipe. That is

$$\dot{W}_{shaft} = \rho Q U(U - V_1)(1 - \cos\beta) \tag{1}$$

The nozzle exit speed, V_1, can be obtained by applying the energy equation (Eq. 5.84) between a point on the lake surface (where $V_0 = p_0 = 0$) and the nozzle outlet (where $z_1 = p_1 = 0$) to give

$$z_0 = \frac{V_1^2}{2g} + h_L \qquad (2)$$

where the head loss is given in terms of the friction factor, f, as (see Eq. 8.34)

$$h_L = f \frac{\ell}{D} \frac{V^2}{2g}$$

The speed, V, of the fluid in the pipe of diameter D is obtained from the continuity equation

$$V = \frac{A_1 V_1}{A} = \left(\frac{D_1}{D}\right)^2 V_1$$

We have neglected minor losses associated with the pipe entrance and the nozzle. With the given data, Eq. 2 becomes

$$z_0 = \left[1 + f \frac{\ell}{D} \left(\frac{D_1}{D}\right)^4\right] \frac{V_1^2}{2g} \qquad (3)$$

or

$$V_1 = \left[\frac{2gz_0}{1 + f \frac{\ell}{D}\left(\frac{D_1}{D}\right)^4}\right]^{1/2}$$

$$= \left[\frac{2(9.8 \text{ m/s}^2)(60 \text{ m})}{1 + 0.02\left(\dfrac{300 \text{ m}}{20/100 \text{ m}}\right)\left(\dfrac{D_1}{20/100}\right)^4}\right]^{1/2}$$

$$= \frac{34.3}{\sqrt{1 + 18750 D_1^4}} \qquad (4)$$

where D_1 is in meter.

By combining Eqs. 1 and 4 and using $Q = \pi D_1^2 V_1/4$, we obtain the power as a function of D_1 and U as

$$\dot{W}_{\text{shaft}} = \frac{5.02 \times 10^4 \, U D_1^2}{\sqrt{1 + 18750 D_1^4}} \left[U - \frac{113.5}{\sqrt{1 + 152 D_1^4}}\right] \qquad (5)$$

where U is in meter per second and $\dot{W}_{\text{shaft}}$ is in N · m/s. These results are plotted as a function of U for various values of D_1 in **Fig. E12.6b**.

As shown by Eq. 12.51, the maximum power (in terms of its variation with U) occurs when $U = V_1/2$, which, when used with Eqs. 4 and 5, gives

$$\dot{W}_{\text{shaft}} = -\frac{15 \times 10^6 D_1^2}{(1 + 18750 D_1^4)^{3/2}} \qquad (6)$$

The maximum power possible occurs when $d\dot{W}_{\text{shaft}}/dD_1 = 0$, which according to Eq. 6 can be found as

$$\frac{d\dot{W}_{\text{shaft}}}{dD_1} = -15 \times 10^6 \left[\frac{2 D_1}{(1 + 18750 D_1^4)^{3/2}}\right.$$

$$\left. - \left(\frac{3}{2}\right)\frac{4 \times 18750 D_1^5}{(1 + 18750 D_1^4)^{5/2}}\right] = 0$$

or

$$37250 \, D_1^4 = 1$$

Thus, the nozzle diameter for maximum power output is

$$D_1 = 0.07 \text{ m} \qquad (Ans)$$

b. The corresponding maximum power can be determined from Eq. 6 as

$$\dot{W}_{\text{shaft}} = \frac{15 \times 10^6 (0.07)^2}{[1 + 18750(0.07)^4]^{3/2}} = -4.2 \times 10^4 \text{ N} \cdot \text{m/s}$$

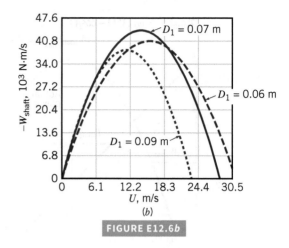

FIGURE E12.6b

The rotor speed at the maximum power condition can be obtained from

$$U = \omega R = \frac{V_1}{2}$$

where V_1 is given by Eq. 4. Thus,

$$\omega = \frac{V_1}{2R} = \frac{\dfrac{34.3}{\sqrt{1 + 18750(0.07)^4}} \text{ m/s}}{2 \times 0.9 \text{ m}}$$

$$= 31.6 \text{ rad/s} \times 1 \text{ rev}/2\pi \text{ rad} \times 60 \text{ s/min}$$

$$= 302 \text{ rpm} \qquad (Ans)$$

Comment The reason that an optimum diameter nozzle exists can be explained as follows. A larger-diameter nozzle will allow a larger flowrate but will produce a smaller jet velocity because of the head loss within the supply side. A smaller-diameter nozzle will reduce the flowrate but will produce a larger jet velocity. Since the power depends on a product combination of flowrate and jet velocity (see Eq. 1), there is an optimum-diameter nozzle that gives the maximum power.

These results can be generalized (i.e., without regard to the specific parameter values of this problem) by considering Eqs. 1 and 3 and the condition that $U = V_1/2$ to obtain

$$\dot{W}_{\text{shaft} \, | \, U=V_1/2} = -\frac{\pi}{16}\rho(1 - \cos\beta)$$

$$\times (2gz_0)^{3/2} D_1^2 \Big/ \left(1 + f\frac{\ell}{D^5}D_1^4\right)^{3/2}$$

By setting $d\dot{W}_{\text{shaft}}/dD_1 = 0$, it can be shown that the maximum power occurs when

$$D_1 = D \Big/ \left(2f\frac{\ell}{D}\right)^{1/4}$$

which gives the same results obtained earlier for the specific parameters of the example problem. Note that the optimum condition depends only on the friction factor and the length-to-diameter ratio of the supply pipe. What happens if the supply pipe is frictionless or of essentially zero length?

A second type of impulse turbine that is widely used (most often with air as the working fluid) is shown in **Fig. 12.29**. A circumferential series of fluid jets strikes the rotating blades which, as with the Pelton wheel, alter both the direction and magnitude of the absolute velocity. As with the Pelton wheel, the inlet and exit pressures (i.e., on either side of the rotor) are equal, and the magnitude of the relative velocity is unchanged as the fluid slides across the blades (if frictional effects are negligible).

Typical inlet and exit velocity diagrams are shown in **Fig. 12.30**. As discussed in Section 12.2, in order for the absolute velocity of the fluid to be changed as indicated during its passage across the blade, the blade must push on the fluid in the direction opposite of the blade motion. Hence, the fluid pushes on the blade in the direction of the blade's motion—the fluid does work on the blade (a turbine). This turbine is an axial-flow device.

Dentist drill turbines are usually of the impulse class.

 Video

V12.5 Dental drill

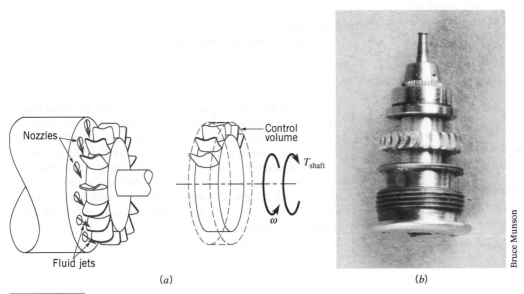

FIGURE 12.29 (*a*) A multinozzle, non-Pelton wheel impulse turbine commonly used with air as the working fluid, (*b*) dental drill.

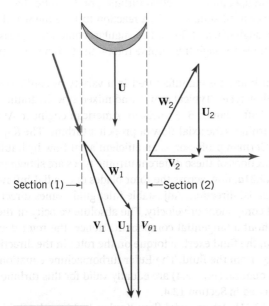

FIGURE 12.30 Inlet and exit velocity triangles for the impulse turbine shown in Fig. 12.29.

Bruce Munson

EXAMPLE 12.7 | Axial-Flow Impulse Turbine (Dental Drill)

Given An air turbine used to drive the high-speed drill used by your dentist is shown in Fig. 12.29 (also see **Video V12.5**). Air exiting from the upstream nozzle holes forces the turbine blades to move in the direction shown. The turbine rotor speed is 300,000 rpm, the tangential component of velocity out of the nozzle is twice the blade speed, and the tangential component of the absolute velocity out of the rotor is zero. The radius to the blade tip is 0.43 cm, and the radius to the blade base is 0.34 cm.

Find Estimate the shaft energy per unit mass of air flowing through the turbine.

Solution We use the fixed, nondeforming control volume that includes the turbine rotor and the fluid in the rotor blade passages at an instant of time (see Fig. 12.29a). The only torque acting on this control volume is the shaft torque. For simplicity we analyze this problem using an arithmetic mean radius, r_m, where

$$r_m = \frac{1}{2}(r_0 + r_i)$$

A sketch of the velocity diagrams at the rotor entrance and exit is shown in Fig. 12.30.

Application of Eq. 12.5 gives

$$w_{shaft} = -U_1 V_{\theta 1} + U_2 V_{\theta 2} \tag{1}$$

where w_{shaft} is shaft energy per unit of mass flowing through the turbine. From the problem statement, $V_{\theta 1} = 2U$ and $V_{\theta 2} = 0$, where

$$U = \omega r_m = (300{,}000 \text{ rev/min})(1 \text{ min/60 s})(2\pi \text{ rad/rev})$$
$$\times (0.43 \text{ cm} + 0.34 \text{ cm})/2(100 \text{ cm/m}) \tag{2}$$
$$= 121 \text{ m/s}$$

is the mean-radius blade velocity. Thus, Eq. (1) becomes

$$w_{shaft} = -U_1 V_{\theta 1} = -2U^2 = -2(121 \text{ m/s})^2$$
$$= -29300 \text{ m}^2/\text{s}^2$$
$$= -29 \text{ kN} \cdot \text{m/kg} \tag{Ans}$$

Comment For each kg of air passing through the turbine there is 29 kN·m/kg of energy available at the shaft to drive the drill. However, because of fluid friction, the actual amount of energy given up by each kg of air will be greater than the amount available at the shaft. How much greater depends on the efficiency of the fluid-mechanical energy transfer between the fluid and the turbine blades.

Recall that the shaft power, $\dot{W}_{shaft}$, is given by

$$\dot{W}_{shaft} = \dot{m} w_{shaft}$$

Hence, to determine the power, we need to know the mass flow-rate, $\dot{m}$, which depends on the size and number of the nozzles. Although the energy per unit mass is large (i.e., 29 kN·m/kg), the flowrate is small, so the power is not "large." Obviously, compressibility effects are likely at the speeds involved in this problem.

Turbine wheels of similar design are often used as the first stage of power-generating steam turbines. Of course, both the size of the wheel and the magnitude of the power extracted are very much larger than those in this dental drill.

12.7.2 Reaction Turbines

Reaction turbines are best suited for higher flowrate and lower head situations.

As indicated in the previous section, impulse turbines are best suited (i.e., most efficient) for lower-flowrate and higher-head operations. Reaction turbines, on the other hand, are best suited for higher-flowrate and lower-head situations. In a reaction turbine the working fluid completely fills the passageways through which it flows. The angular momentum, pressure, and velocity of the fluid decrease as it flows through the turbine rotor—the turbine rotor extracts energy from the fluid.

As with pumps, turbines are manufactured in a variety of configurations—radial-flow, mixed-flow, and axial-flow types. Typical radial- and mixed-flow hydraulic turbines are called *Francis turbines*, named after James B. Francis, an American engineer. At very low heads the most efficient type of turbine is the axial-flow or propeller turbine. The *Kaplan turbine*, named after Victor Kaplan, a German professor, is an efficient axial-flow hydraulic turbine with adjustable blades. Cross sections of these different turbine types are shown in **Fig. 12.31**.

As shown in Fig. 12.31a, flow across the rotor blades of a radial-inflow turbine has a major component in the radial direction. Adjustable inlet guide vanes direct the water into the rotor with a tangential component of velocity. The absolute velocity of the water leaving the rotor is essentially without a tangential component. Hence, the rotor decreases the angular momentum of the fluid, the fluid exerts a torque on the rotor in the direction of rotation, and the rotor extracts energy from the fluid. The Euler turbomachine equation (Eq. 12.2) and the corresponding power equation (Eq. 12.4) are equally valid for this turbine as they are for the centrifugal pump discussed in Section 12.4.

As shown in Fig. 12.31b, for an axial-flow Kaplan turbine, the fluid flows through the inlet guide vanes, called *wicket gates,* and achieves a tangential velocity in a vortex (swirl)

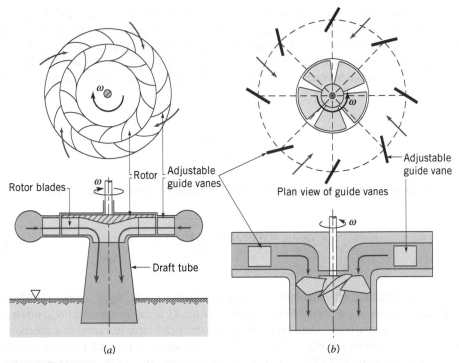

FIGURE 12.31 (a) Typical radial-flow Francis turbine, (b) typical axial-flow Kaplan turbine.

motion before it reaches the rotor. Flow across the rotor contains a major axial component. Both the inlet guide vanes and the turbine blades can be adjusted by changing their setting angles to produce the best match (optimum output) for the specific operating conditions. For example, the operating head available may change from season to season and/or the flowrate through the rotor may vary.

Pumps and turbines are often thought of as the "inverse" of each other. Pumps add energy to the fluid; turbines remove energy. The propeller on an outboard motor (a pump) and the propeller on a Kaplan turbine are in some ways geometrically similar, but they perform opposite tasks. Similar comparisons can be made for centrifugal pumps and Francis turbines. In fact, some large pump/turbines at "pumped storage" hydroelectric power plants are designed to be run as turbines during high-power demand periods (i.e., during the day) and as pumps to resupply the upstream reservoir from the downstream reservoir during low-demand times (i.e., at night). Thus, a pump type often has its corresponding turbine type. However, is it possible to have the "inverse" of a Pelton wheel turbine—an impulse pump?

As with pumps, incompressible flow turbine performance is often specified in terms of appropriate dimensionless parameters. The flow coefficient, $C_Q = Q/\omega D^3$, the head coefficient, $C_H = gh_a/\omega^2 D^2$, and the power coefficient, $C_\mathscr{P} = \dot{W}_{shaft}/\rho\omega^3 D^5$, are defined in the same way for pumps and turbines. On the other hand, turbine efficiency, η, is the inverse of pump efficiency. That is, the efficiency is the ratio of the shaft power output to the power available in the flowing fluid, or

> Actual head available for a turbine, h_a, is always greater than shaft work head, h_s, because of head loss, h_L, in the turbine.

$$\eta = \frac{\dot{W}_{shaft}}{\rho g Q h_a}$$

For geometrically similar turbines and for negligible Reynolds number and surface roughness effects, the relationships between the dimensionless parameters are given functionally by those shown in Eqs. 12.29, 12.30, and 12.31. That is,

$$C_H = \phi_1(C_Q), \qquad C_\mathscr{P} = \phi_2(C_Q), \quad \text{and} \quad \eta = \phi_3(C_Q)$$

where the functions ϕ_1, ϕ_2, and ϕ_3 are dependent on the type of turbine involved. Also, for turbines the efficiency, η, is related to the other coefficients according to $\eta = C_\mathscr{P}/C_H C_Q$.

As indicated above, the design engineer has a variety of turbine types available for any given application. It is necessary to determine which type of turbine would best fit the job (i.e., be most efficient) before detailed design work is attempted. As with pumps, the use of a specific speed parameter can help provide this information. For hydraulic turbines, the rotor diameter D is eliminated between the flow coefficient and the power coefficient to obtain the *power specific speed*

$$N_s' = \frac{\omega \sqrt{\dot{W}_{shaft}/\rho}}{(gh_a)^{5/4}}$$

We use the more common, but not dimensionless, definition of power specific speed

$$N_{sd}' = \frac{\omega\,(\mathrm{rpm}) \sqrt{\dot{W}_{shaft}\,(\mathrm{kW})}}{[h_a\,(\mathrm{m})]^{5/4}} \qquad (12.52)$$

Specific speed may be used to approximate what kind of turbine geometry (axial to radial) would operate most efficiently.

That is, N_{sd}' is calculated with angular velocity, ω, in rpm; shaft power, $\dot{W}_{shaft}$, in power (kW), and actual head available, h_a, in meter. Note that the flowrate, Q, does not appear in N_{sd}'; it is a dependent variable for turbines. Optimum turbine efficiency (for large turbines) as a function of specific speed is displayed in **Fig. 12.32**. Also shown are representative rotor and casing cross sections. Note that impulse turbines are best at low specific speeds; that is, when operating with large heads and small flowrate. The other extreme is axial-flow turbines, which are the most efficient type if the head is low and if the flowrate is large. For intermediate values of specific speeds, radial- and mixed-flow turbines offer the best performance.

The data shown in Fig. 12.32 are meant only to provide a guide for turbine-type selection. The actual turbine efficiency for a given turbine depends very strongly on the detailed design of the turbine. Considerable analysis, testing, and experience are needed to produce an efficient turbine. However, the data of Fig. 12.32 are representative. Much additional information can be found in the literature.

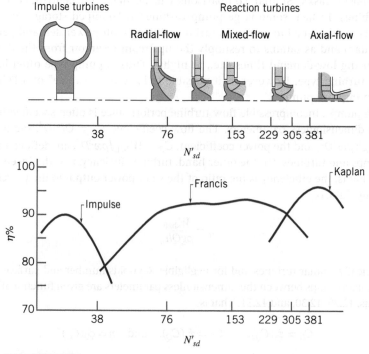

FIGURE 12.32 Typical turbine cross sections and maximum efficiencies as a function of specific speed.

The Wide World of Fluids

Cavitation damage in hydraulic turbines

The occurrence of cavitation in hydraulic pumps seems to be an obvious possibility since low-suction pressures are expected. Cavitation damage can also occur in hydraulic turbines, even though they do not seem obviously prone to this kind of problem. Local acceleration of liquid over blade surfaces can be sufficient to result in local pressures low enough to cause fluid vaporization or cavitation. Further along the flow path, the fluid can decelerate rapidly enough with accompanying increase in local pressure to make cavitation bubbles collapse with enough intensity to cause blade surface damage or damage to the draft tube in the form of material erosion. Over time, this erosion can be severe enough to require repair or replacement, which is very expensive.

EXAMPLE 12.8 | Use of Specific Speed to Select Turbine Type

Given A hydraulic turbine is to operate at an angular velocity of 6 rev/s, a flowrate of 0.28 m³/s, and a head of 6 m.

Find What type of turbine should be selected? Explain.

Solution The most efficient type of turbine to use can be obtained by calculating the specific speed, N'_{sd}, and using the information of Fig. 12.32. To use the dimensional form of the specific speed indicated in Fig. 12.32, we must convert the given data into the appropriate units. For the rotor speed we get

$$\omega = 6 \text{ rev/s} \times 60 \text{ s/min} = 360 \text{ rpm}$$

To estimate the shaft power, we assume all of the available head is converted into power and multiply this amount by an assumed efficiency (94%).

$$\dot{W}_{shaft} = \gamma Q z \eta = (9800 \text{ N/m}^3)(0.28 \text{ m}^3/\text{s}) \left[\frac{6 \text{ m} \,(0.94)}{1 \text{ N} \cdot \text{m/s} \cdot \text{W}} \right]$$

$$\dot{W}_{shaft} = 15.5 \text{ kW}$$

Thus for this turbine,

$$N'_{sd} = \frac{\omega \sqrt{\dot{W}_{shaft}}}{(h_a)^{5/4}} = \frac{(360 \text{ rpm}) \sqrt{15.5 \text{ kW}}}{(6 \text{ m})^{5/4}} = 151$$

According to the information of Fig. 12.32,

A mixed-flow Francis turbine would
probably give the highest efficiency and
an assumed efficiency of 0.94 is appropriate. *(Ans)*

Comment What would happen if we wished to use a Pelton wheel for this application? Note that with only a 6 m head, the maximum jet velocity, V_1, obtainable (neglecting viscous effects) would be

$$V_1 = \sqrt{2gz} = \sqrt{2 \times 9.8 \text{ m/s}^2 \times 6 \text{ m}} = 10.8 \text{ m/s}$$

As shown by Eq. 12.51, for maximum efficiency of a Pelton wheel the jet velocity is ideally two times the blade velocity. Thus, $V_1 = 2\omega R$, or the wheel diameter, $D = 2R$, is

$$D = \frac{V_1}{\omega} = \frac{10.8 \text{ m/s}}{(6 \text{ rev/s} \times 2\pi \text{ rad/rev})} = 0.29 \text{ m}$$

To obtain a flowrate of $Q = 0.28 \text{ m}^3/\text{s}$ at a velocity of $V_1 = 10.8 \text{ m/s}$, the jet diameter, d_1, must be given by

$$Q = \frac{\pi}{4} d_1^2 V_1$$

or

$$d_1 = \left[\frac{4Q}{\pi V_1} \right]^{1/2} = \left[\frac{4(0.28 \text{ m}^3/\text{s})}{\pi (10.8 \text{ m/s})} \right]^{1/2} = 0.18 \text{ m}$$

A Pelton wheel with a diameter of $D = 0.29$ m supplied with water through a nozzle of diameter $d_1 = 0.18$ m is not a practical design. Typically $d_1 \ll D$ (see Fig. 12.22). By using multiple jets it would be possible to reduce the jet diameter. However, even with 8 jets, the jet diameter would be 0.06 m, which is still too large (relative to the wheel diameter) to be practical. Hence, the above calculations reinforce the results presented in Fig. 12.32—a Pelton wheel would not be practical for this application. If the flowrate were considerably smaller, the specific speed could be reduced to the range where a Pelton wheel would be the type to use (rather than a mixed-flow reaction turbine).

It might seem that turbine speed is selected arbitrarily but it is not. Since most large hydraulic turbines drive electric generators, the rotational speed is determined by the number of magnetic poles in the generator and the frequency desired (60 Hz in North America and 50 Hz in Europe).

12.8 FANS

When the fluid to be moved is air, or some other gas or vapor, *fans* are commonly used. Types of fans vary from the small fan used for cooling desktop computers to large fans used in many industrial applications such as ventilating large buildings and providing mechanical draft for fossil fuel-fired steam generators in power stations. Fans typically operate at relatively low rotation speeds (a few hundred to a few thousand rpm) and are capable of moving large volumes of gas (more than a million m³/hr). Although the fluid of interest is a gas, the change in gas density through the fan does not usually exceed 7%, which for air represents a change in pressure of only about 7 kPa (Ref. 15). Thus, in dealing with fans, the gas density is assumed

Fans are used to move air and other gases and vapors.

to be constant, and the flow analysis is based on incompressible flow concepts. Because of the low-pressure rise involved, fans are often constructed of lightweight sheet metal. Fans are also called *blowers*, *boosters*, and *exhausters*, depending on the location within the system; that is, blowers are located at the system entrance, exhausters are at the system exit, and boosters are at some intermediate position within the system. Turbomachines used to produce larger changes in gas density and pressure than are possible with fans are called *compressors* (see Section 12.9.1).

As is the case for pumps, fan designs include centrifugal (radial-flow) fans, as well as mixed-flow and axial-flow (propeller) fans, and the analysis of fan performance closely follows that previously described for pumps. The shapes of typical performance curves for centrifugal and axial-flow fans are quite similar to those shown in Fig. 12.18 for centrifugal and axial-flow pumps. However, fan performance data are often given in terms of pressure rise, either static or total, rather than the more conventional head rise commonly used for pumps. Fan static pressure rise, the most commonly used measure of fan performance, is a hybrid quantity, being the difference between outlet static pressure and inlet total pressure.

Scaling relationships for fans are the same as those developed for pumps, that is, Eqs. 12.32 through 12.35 apply to fans as well as pumps. As noted above, for fans it is common to replace the head, h_a, in Eq. 12.33 with pressure head, $p_a/\rho g$, so that Eq. 12.33 becomes

$$\left(\frac{p_a}{\rho \omega^2 D^2}\right)_1 = \left(\frac{p_a}{\rho \omega^2 D^2}\right)_2 \tag{12.53a}$$

where p_a is either fan total pressure rise or fan static pressure rise and, as before, the subscripts 1 and 2 refer to any two fans from the family of geometrically similar fans. Specific speed for fans is defined by Eq. 12.43, with gh_a replaced by p_a/ρ. As for pumps, U.S. industry uses a dimensional specific speed N_{sdf}, where

$$N_{sdf} = \frac{\omega\,(\text{rpm})\sqrt{Q\,(\text{cfm})}}{[p_a\,(\text{in. H}_2\text{O})]^{3/4}} \tag{12.53b}$$

and p_a is relative to standard air density. Centrifugal fans have specific speed in the range $2000 < N_{sdf} < 20{,}000$ and axial fans lie in the range $50{,}000 < N_{sdf} < 200{,}000$.

Additional information about fans can be found in Refs. 15–18. Equations 12.53, 12.32, and 12.34 are called the *fan laws* and can be used to scale performance characteristics between members of a family of geometrically similar fans.

The Wide World of Fluids

Hi-tech ceiling fans

Energy savings of up to 25% can be realized if thermostats in air-conditioned homes are raised by a few degrees. This can be accomplished by using ceiling fans and taking advantage of the increased sensible cooling brought on by air moving over skin. If the energy used to run the fans can be reduced, additional energy savings can be realized. Most ceiling fans use flat, fixed pitch, nonaerodynamic *blades* with uniform chord length. Because the tip of a paddle moves through air faster than its root does, airflow over such fan blades is lowest near the hub and highest at the tip. By making the fan blade more propeller-like, it is possible to have a more uniform, efficient distribution. However, since ceiling fans are restricted by law to operate at less than 200 rpm, ordinary airplane propeller design is not appropriate. After considerable design effort, a highly efficient ceiling fan capable of delivering the same airflow as the conventional design with only half the power has been successfully developed and marketed. The fan blades are based on the slowly turning prop used in the *Gossamer Albatross*, the human-powered aircraft that flew across the English Channel in 1979.

12.9 COMPRESSIBLE FLOW TURBOMACHINES

Compressible flow turbomachines are in many ways similar to the incompressible flow pumps and turbines described in previous portions of this chapter. The main difference is that the density of the fluid (a gas or vapor) changes significantly from the inlet to the outlet of compressible flow machines. This added feature sometimes has interesting consequences, benefits, and complications.

Compressors are like pumps, adding energy to the fluid, causing a significant pressure rise and a corresponding significant increase in density and temperature. Compressible flow turbines, on the other hand, remove energy from the fluid, causing a lower pressure and a smaller density and temperature at the outlet than at the inlet. The information provided earlier about basic energy considerations (Section 12.2) and basic angular momentum considerations (Section 12.3) is directly applicable to these turbomachines.

As discussed in Chapter 11, compressible flow study requires an understanding of the principles of thermodynamics. Similarly, an in-depth analysis of compressible flow turbomachines requires use of various thermodynamic concepts. For example, the fluid temperature change is of little consequence in liquid pumps and turbines, but it is of critical importance in compressible flow machines. In this section we provide only a brief discussion of some of the general properties of compressors and compressible flow turbines. The interested reader is encouraged to read some of the excellent references available for further information (e.g., Refs. 1–3, 19–21).

12.9.1 Compressors

Turbocompressors operate with the continuous compression of gas flowing through the device. Since there is a significant pressure and density increase, there is also a considerable temperature increase.

Radial-flow (or centrifugal) compressors are essentially centrifugal pumps (see Section 12.4) that use a gas (rather than a liquid) as the working fluid. They are typically high pressure rise, low flowrate, and are axially compact turbomachines. A photograph of a centrifugal compressor rotor is shown in **Fig. 12.33**.

The amount of compression is typically given in terms of the *total pressure ratio*, $PR = p_{02}/p_{01}$, where the pressures are absolute. Thus, a radial-flow compressor with $PR = 3.0$ can compress standard atmospheric air from 101.3 kPa to 3.0×101.3 kPa $= 304$ kPa, with a corresponding temperature rise of about 104 °C.

Higher pressure ratios can be obtained by using *multiple-stage* devices in which flow from the outlet of the preceding stage proceeds to the inlet of the following stage. If each stage has the same pressure ratio, PR, the overall pressure ratio after n stages is PR^n. Thus, as shown in the adjacent figure, a four-stage compressor with individual stage $PR = 2.0$ can compress standard air from $p_{0\,in} = 101.3$ kPa to $p_{0\,out} = 2^4 \times 101.3$ kPa $= 1620$ kPa. Adiabatic (i.e., no heat transfer) compression of a gas causes an increase in temperature and requires more work than isothermal (constant temperature) compression of a gas. An interstage cooler (i.e., an intercooler heat exchanger) as shown in **Fig. 12.34** can be used to reduce the compressed gas temperature entering the second stage and thus the work required.

Relative to centrifugal water pumps, radial compressors of comparable size rotate at much higher speeds, typically on the order of tens of thousands of revolutions per minute. It is not uncommon for the rotor blade tip speed and the speed of the absolute flow leaving the impeller to be greater than the speed of sound. Such high speeds generate very large values of head, typically a few miles (of air) (see Eq. 12.14).

Multistaging is common in high-pressure ratio compressors.

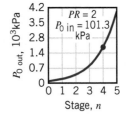

FIGURE 12.33 Centrifugal compressor rotor.

Photograph courtesy of Concepts NREC.

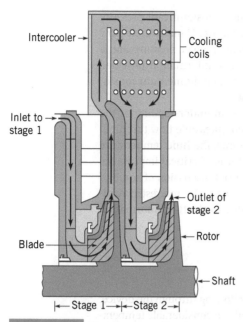

FIGURE 12.34 Two-stage centrifugal compressor with an intercooler.

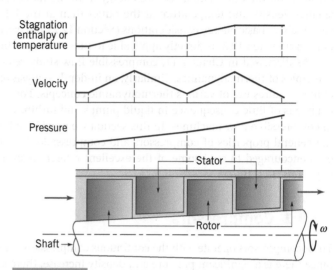

FIGURE 12.35 Enthalpy, velocity, and pressure distribution in an axial-flow compressor.

The axial-flow compressor is the other widely used configuration. This type of turbomachine has a lower pressure rise per stage and a higher flowrate, and is more radially compact than a centrifugal compressor. As shown in **Fig. 12.35**, axial-flow compressors usually consist of several stages, with each stage containing a rotor/stator row pair. For an 11-stage compressor, a pressure ratio of $PR = 1.2$ per stage gives an overall pressure ratio of $p_{02}/p_{01} = 1.2^{11} = 7.4$. As the gas is compressed and its density increases, a smaller annulus cross-sectional area is required and the flow channel size and blade height decrease from the inlet to the outlet of the compressor. The typical jet aircraft engine uses an axial-flow compressor as one of its main components (see **Fig. 12.36** and Ref. 22).

An axial-flow compressor can include a set of *inlet guide vanes* upstream of the first rotor row. These guide vanes optimize the relative velocity into the first rotor row by directing

Axial-flow compressor multistaging requires less frontal area but more length than centrifugal compressors.

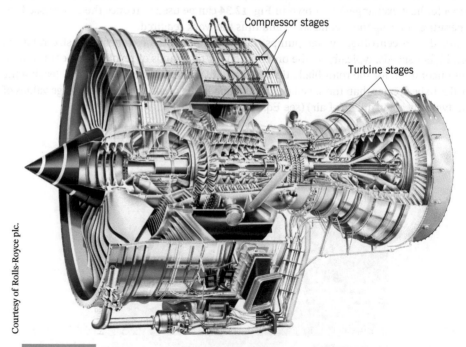

Courtesy of Rolls-Royce plc.

FIGURE 12.36 Rolls-Royce Trent 900 three-shaft propulsion system.

the flow away from the axial direction. *Rotor blades* push on the gas in the direction of blade motion and to the rear, adding energy (like in an axial pump) and moving the gas through the compressor. The *stator blade* rows act as diffusers, turning the fluid back toward the axial direction and increasing the static pressure. The stator blades cannot add energy to the fluid because they are stationary. Typical pressure, velocity, and enthalpy distributions along the axial direction are shown in Fig. 12.35. [If you are not familiar with the thermodynamic concept of enthalpy (see Section 11.1), you may replace "enthalpy" by temperature as a qualitative approximation.] The degree of reaction of the compressor stage is equal to the ratio of the rise in static enthalpy or temperature achieved across the rotor to the enthalpy or temperature rise across the entire stage. Most modern compressors involve 50% or higher reaction.

Video

V12.6 Flow in a compressor stage

The blades in an axial-flow compressor are airfoils carefully designed to produce appropriate lift and drag forces on the flowing gas. As occurs with airplane wings, compressor blades can stall (see Section 9.4). When the flowrate is decreased from the design amount, the velocity diagram at the entrance of the rotor row indicates that the relative flow meets the blade leading edge at larger angles of incidence than the design value. When the angle of incidence becomes too large, blade stall can occur and the result is *compressor surge* or *stall*—unstable flow conditions that can cause excessive vibration, noise, poor performance, and possible damage to the machine. The lower flowrate bound of compressor operation is related to the beginning of these instabilities (see **Fig. 12.37**).

> Compressor blades can stall, and unstable flow conditions can subsequently occur.

Other important compressible flow phenomena such as shock waves (see Section 11.5.3) and choked flow (see Section 11.4.2) occur commonly in compressible flow turbomachines. These phenomena are very sensitive to even very small changes or variations of geometry. Shock strength is kept low to minimize shock loss, and choked flows limit the upper flowrate boundary of machine operation. See Fig. 12.37, where choking is indicated by the vertical portion of the pressure ratio and efficiency lines.

The experimental performance data for compressors are systematically summarized with parameters prompted by dimensional analysis. As mentioned earlier, total pressure ratio, p_{02}/p_{01}, is used instead of the head-rise coefficient associated with pumps, blowers, and fans.

Two different types of efficiency, *isentropic efficiency* and *polytropic efficiency*, are used to characterize compressor performance. Each of these compressor efficiencies involves a ratio of ideal work to actual work required to accomplish the compression. The isentropic efficiency involves a ratio of the ideal work required with an adiabatic and frictionless (no loss) compression process to the actual work required to achieve the same total pressure rise. The polytropic efficiency involves a ratio of the ideal work required to achieve the actual end state

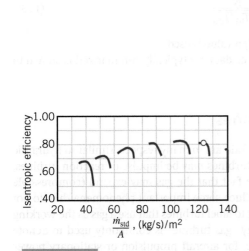

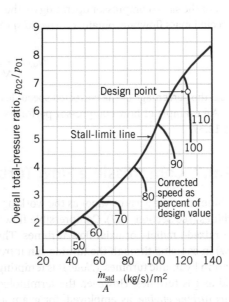

FIGURE 12.37 Performance characteristics of an axial-flow compressor (Source: Data from Johnson, I. A., and Bullock, R. D., eds., Aerodynamic Design of Axial-Flow Compressors, NASA SP-36, National Aeronautics and Space Administration, Washington, D.C., 1965.)

of the compression with a polytropic and frictionless process between the actual beginning and end stagnation states across the compressor and the actual work involved between these same states. A more detailed explanation of these efficiencies is beyond the scope of this text. Those interested in learning more about these parameters should study any of several available books on turbomachines (for example, Refs. 2 and 3).

The flow parameter commonly used for compressors is based on the following dimensionless grouping from dimensional analysis

$$\frac{\dot{m}\sqrt{kRT_{01}}}{D^2 p_{01}}$$

where R is the gas constant, $\dot{m}$ the mass flowrate, k the specific heat ratio, T_{01} the stagnation temperature at the compressor inlet, D a characteristic length, typically rotor diameter, and p_{01} the stagnation pressure at the compressor inlet.

To account for variations in test conditions, the following strategy is employed. We set

$$\left(\frac{\dot{m}\sqrt{kRT_{01}}}{D^2 p_{01}}\right)_{test} = \left(\frac{\dot{m}\sqrt{kRT_{01}}}{D^2 p_{01}}\right)_{std}$$

where the subscript "test" refers to a specific test condition and "std" refers to the standard atmosphere ($p_0 = 101.3$ kPa, $T_0 = 288$ K) condition. When we consider a given compressor operating on a given working fluid (so that R, k, and D are constant), the above equation reduces to

$$\dot{m}_{std} = \frac{\dot{m}_{test}\sqrt{T_{01\,test}/T_{0\,std}}}{p_{01\,test}/p_{0\,std}} \tag{12.54}$$

In essence, $\dot{m}_{std}$ is the compressor-test mass flowrate "corrected" to the standard atmosphere inlet condition. The *corrected compressor mass flowrate*, $\dot{m}_{std}$, is used instead of flow coefficient. Often, $\dot{m}_{std}$ is divided by A, the frontal area of the compressor flow path or by D^2.

While for pumps, blowers, and fans, rotor speed was accounted for in the flow coefficient, it does not appear in the corrected mass flowrate. Thus, for compressors, rotor speed needs to be accounted for with an additional group. This dimensionless group is

$$\frac{ND}{\sqrt{kRT_{01}}}$$

which is called the *machine Mach number.*

For the same compressor operating on the same gas, we eliminate D, k, and R and, as with corrected mass flowrate, obtain a *corrected speed*, N_{std}, where

$$N_{std} = \frac{N}{\sqrt{T_{01}/T_{std}}} \tag{12.55}$$

Often, the percentage of the corrected speed design value is used.

An example of how compressor performance data are typically summarized is shown in Fig. 12.37.

12.9.2 Compressible Flow Turbines

A gas turbine engine generally consists of a compressor, a combustor, and a turbine.

Turbines that use a gas or vapor as the working fluid are in many respects similar to hydraulic turbines (see Section 12.7). Compressible flow turbines may be impulse or reaction turbines, and mixed-, radial-, or axial-flow turbines. The fact that the gas may expand (compressible flow) in coursing through the turbine can introduce some important phenomena that do not occur in hydraulic turbines. (*Note:* It is tempting to label turbines that use a gas as the working fluid as gas turbines. However, the terminology "gas turbine" is commonly used to denote a *gas turbine engine*, as employed, for example, for aircraft propulsion or stationary power generation. As shown in Fig. 12.36, these engines typically contain a compressor, combustion chamber, and turbine.)

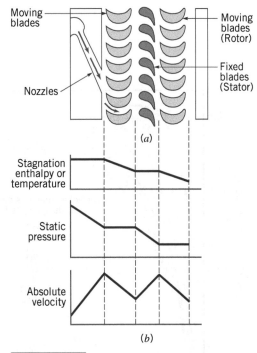

FIGURE 12.38 Enthalpy, velocity, and pressure distribution in a two-stage impulse turbine.

Although for compressible flow turbines the axial-flow type is common, the radial-inflow type is also used for various purposes. As shown in Fig. 12.1, the turbine that drives the typical automobile turbocharger compressor is a radial-inflow type. The main advantages of the radial-inflow turbine are: (1) It is robust and durable, (2) it is axially compact, and (3) it can be relatively inexpensive. A radial-flow turbine usually has a lower efficiency than an axial-flow turbine, but lower initial costs or configuration limitations may be a compelling incentive in choosing a radial-flow turbine over an axial-flow one.

Axial-flow turbines are widely used compressible flow turbines. Steam turbines used in electrical generating plants and marine propulsion and the turbines used in gas turbine engines are usually of the axial-flow type. Often they are multistage turbomachines, although single-stage compressible flow turbines are also produced. They may be either an impulse type or a reaction type. With compressible flow turbines, the ratio of static enthalpy (or temperature) drop across the rotor to this drop across the stage, rather than the ratio of static pressure differences, is used to determine reaction. Strict impulse (zero pressure drop) turbines have slightly negative reaction; the static enthalpy or temperature actually increases across the rotor. Zero-reaction turbines involve no change of static enthalpy or temperature across the rotor but do involve a slight pressure drop.

The early stages of steam and gas turbines are usually axial-flow impulse type, similar to Fig. 12.30 but considerably larger. A two-stage, axial-flow impulse turbine is shown in **Fig. 12.38a**. The gas accelerates through the supply nozzles, has some of its energy removed by the first-stage rotor blades, accelerates again through the second-stage nozzle row, and has additional energy removed by the second-stage rotor blades. As shown in **Fig. 12.38b**, the static pressure remains constant across the rotor rows. Across the second-stage nozzle row, the static pressure decreases, absolute velocity increases, and the stagnation enthalpy (temperature) is constant. Flow across the second rotor is similar to flow across the first rotor. Because the working fluid is a gas, the significant decrease in static pressure across the turbine results in a significant decrease in density—the flow is compressible. Hence, more detailed analysis of this flow must incorporate various compressible flow concepts developed in Chapter 11. Interesting phenomena such as shock waves and choking due to sonic conditions at the "throat" of the flow passage between blades can occur because of compressibility effects. The interested reader is encouraged to consult the various references available (e.g., Refs. 2, 3, and 21) for fascinating applications of compressible flow principles in turbines.

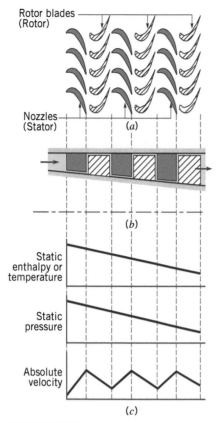

FIGURE 12.39 Enthalpy, pressure, and velocity distribution in a three-stage reaction turbine.

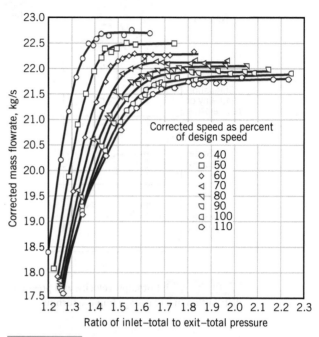

FIGURE 12.40 Typical compressible flow turbine performance "map." (Source: Data from Glassman, A. J., ed., Turbine Design and Application, Vol. 3, NASA SP-290, National Aeronautics and Space Administration, Washington, D.C., 1975.)

The rotor and nozzle blades in a three-stage, axial-flow reaction turbine, such as would appear in the later stages of steam and gas turbines, are shown in **Fig. 12.39a**. The axial variations of pressure and velocity are shown in **Fig. 12.39c**. Both the stationary and rotor blade (passages) act as flow-accelerating nozzles. That is, the static pressure and enthalpy (temperature) decrease in the direction of flow for both the fixed and the rotating blade rows. This distinguishes the reaction turbine from the impulse turbine (see Fig. 12.38b). Energy is removed from the fluid by the rotors only (the stagnation enthalpy or temperature is constant across the adiabatic flow stators).

Because of the reduction of static pressure in the downstream direction, the gas expands, and the flow passage area must increase from the inlet to the outlet of this turbine. This is seen in **Fig. 12.39b**.

Performance data for compressible flow turbines are summarized with the help of parameters derived from dimensional analysis. Isentropic and polytropic efficiencies (see Refs. 2, 3, and 21) are commonly used, as are inlet-to-outlet total pressure ratios (p_{01}/p_{02}), corrected rotor speed (see Eq. 12.55), and corrected mass flowrate (see Eq. 12.54). **Figure 12.40** shows a compressible flow turbine performance "map." Note the occurrence of choking (the curves become flat).

> Turbine performance maps are used to display complex turbine characteristics.

CHAPTER SUMMARY

Various aspects of turbomachine flow are considered in this chapter. The connection between fluid angular momentum change and shaft torque is key to understanding how turbopumps and turbines operate.

The shaft torque associated with change in the axial component of angular momentum of a fluid as it flows through a pump or turbine is described in terms of the inlet and outlet velocity diagrams. Velocity diagrams help us understand the relationships between the absolute, relative, and blade velocities.

Performance characteristics for centrifugal pumps are discussed. Standard dimensionless pump parameters, similarity laws, and the concept of specific speed are presented for use in pump analysis. How to use pump performance curves and the system curve for proper pump operation and selection is presented. A brief discussion of axial-flow and mixed-flow pumps is given.

An analysis of impulse turbines is provided, with emphasis on the Pelton wheel turbine. For impulse turbines there is negligible pressure difference across the blade; the torque is a consequence of the change in direction of the fluid jet striking the blade. Radial-flow and axial-flow reaction turbines are briefly discussed. Brief discussions of fans and compressible flow machines are presented.

The following checklist provides a study guide for this chapter. When your study of the entire chapter has been completed, you should be able to

- write out meanings of the terms listed here and explain each of the related concepts. These terms are particularly important and are set in **bold black** type in the text.
- draw appropriate velocity diagrams for flows entering and leaving specified pump or turbine configurations.
- estimate the actual shaft torque, actual shaft power, and ideal pump head rise for a given centrifugal pump configuration.

- use pump performance curves and the system curve to predict pump performance in a specified system.
- predict the performance characteristics for one pump or fan based on the performance of another pump or fan of the same family using scaling laws.
- use specific speed to determine whether a radial-flow, mixed-flow, or axial-flow machine would be most appropriate for a specified situation.
- estimate the actual shaft torque and actual shaft power for flow through an impulse turbine.
- estimate the actual shaft torque and actual shaft power for a specified reaction turbine.
- use specific speed to determine whether an impulse or a reaction turbine would be more appropriate for a given situation.
- read a performance map for a compressor or compressible flow turbine and point out operating limits.

KEY EQUATIONS

Vector addition of velocities	$\mathbf{V} = \mathbf{W} + \mathbf{U}$	(12.1)
Shaft torque	$T_{shaft} = -\dot{m}_1(r_1 V_{\theta 1}) + \dot{m}_2(r_2 V_{\theta 2})$	(12.2)
Shaft power	$\dot{W}_{shaft} = T_{shaft}\,\omega$	(12.3)
Shaft power	$\dot{W}_{shaft} = -\dot{m}_1(U_1 V_{\theta 1}) + \dot{m}_2(U_2 V_{\theta 2})$	(12.4)
Shaft work	$w_{shaft} = U_2 V_{\theta 2} - U_1 V_{\theta 1}$	(12.12)
Pump ideal head rise	$h_i = \dfrac{U_2 V_{\theta 2} - U_1 V_{\theta 1}}{g}$	(12.13)
Pump actual head rise	$h_a = \dfrac{p_2 - p_1}{\gamma} + z_2 - z_1 + \dfrac{V_2^2 - V_1^2}{2g}$	(12.19)
Pump similarity relationship	$\dfrac{gh_a}{\omega^2 D^2} = \phi_1\!\left(\dfrac{Q}{\omega D^3}\right)$	(12.29)
Pump similarity relationship	$\dfrac{\dot{W}_{shaft}}{\rho\omega^3 D^5} = \phi_2\!\left(\dfrac{Q}{\omega D^3}\right)$	(12.30)
Pump similarity relationship	$\eta = \phi_3\!\left(\dfrac{Q}{\omega D^3}\right)$	(12.31)
Pump scaling law	$\left(\dfrac{Q}{\omega D^3}\right)_1 = \left(\dfrac{Q}{\omega D^3}\right)_2$	(12.32)
Pump scaling law	$\left(\dfrac{gh_a}{\omega^2 D^2}\right)_1 = \left(\dfrac{gh_a}{\omega^2 D^2}\right)_2$	(12.33)
Pump scaling law	$\left(\dfrac{\dot{W}_{shaft}}{\rho\omega^3 D^5}\right)_1 = \left(\dfrac{\dot{W}_{shaft}}{\rho\omega^3 D^5}\right)_2$	(12.34)
Pump scaling law	$\eta_1 = \eta_2$	(12.35)
Specific speed (pumps)	$N_{sd} = \dfrac{\omega(\mathrm{rpm})\sqrt{Q(\mathrm{gpm})}}{[h_a\,(\mathrm{ft})]^{3/4}}$	(12.44)
Specific speed (turbines)	$N'_{sd} = \dfrac{\omega\,(\mathrm{rpm})\sqrt{\dot{W}_{shaft}\,(\mathrm{bhp})}}{[h_a\,(\mathrm{ft})]^{5/4}}$	(12.52)

REFERENCES

[1] Cumpsty, N. A., *Jet Propulsion*, 2nd Ed., Cambridge University Press, Cambridge, UK, 2003.

[2] Saravanamuttoo, H. I. H., Rogers, G. F. C., and Cohen, H., *Gas Turbine Theory*, 5th Ed., Prentice-Hall, Saddle River, NJ, 2001.

[3] Wilson, D. G., and Korakianitis, T., *The Design of High-Efficiency Turbomachinery and Gas Turbines*, 2nd Ed., Prentice-Hall, Saddle River, NJ, 1998.

[4] Stepanoff, H. J., *Centrifugal and Axial Flow Pumps*, 2nd Ed., Wiley, New York, 1957.

[5] Wislicenus, G. F., *Preliminary Design of Turbopumps and Related Machinery*, NASA Reference Publication 1170, 1986.

[6] Neumann, B., *The Interaction Between Geometry and Performance of a Centrifugal Pump*, Mechanical Engineering Publications Limited, London, 1991.

[7] Garay, P. N., *Pump Application Desk Book*, Fairmont Press, Lilburn, GA, 1990.

[8] Rouse, H., *Elementary Mechanics of Fluids*, Wiley, New York, 1946.

[9] Kittredge, C. P., "Estimating the Efficiency of Prototype Pumps From Model Tests," ASME Paper 67-WA-FE6, 1967.

[10] Wright, T., and Gerhart, P. M., *Fluid Machinery: Application, Selection, and Design,* CRC-Taylor and Francis, Boca Raton, FL, 2010.

[11] Hydraulic Institute, *Hydraulic Institute Standards*, 14th Ed., Hydraulic Institute, Cleveland, OH, 1983.

[12] Heald, C. C., ed., *Cameron Hydraulic Data*, 17th Ed., Ingersoll-Rand, Woodcliff Lake, NJ, 1988.

[13] Kristal, F. A., and Annett, F. A., *Pumps: Types, Selection, Installation, Operation, and Maintenance*, McGraw-Hill, New York, 1953.

[14] Karassick, I. J., et al., *Pump Handbook*, 3rd Ed., McGraw-Hill, New York, 2000.

[15] Stepanoff, A. J., *Turboblowers*, Wiley, New York, 1955.

[16] Jorgensen, R., ed., *Fan Engineering*, 9th Ed., Howden Buffalo, Buffalo, New York, 1999 (also available as a CD-ROM).

[17] Wallis, R. A., *Axial Flow Fans and Ducts*, Wiley, New York, 1983.

[18] Reason, J., "Fans," *Power*, Vol. 127, No. 9, 103–128, 1983.

[19] Cumpsty, N. A., *Compressor Aerodynamics*, Longman Scientific & Technical, Essex, UK, and Wiley, New York, 1989.

[20] Johnson, I. A., and Bullock, R. D., eds., *Aerodynamic Design of Axial-Flow Compressors*, NASA SP-36, National Aeronautics and Space Administration, Washington, D.C., 1965.

[21] Glassman, A. J., ed., *Turbine Design and Application*, Vol. 3, NASA SP-290, National Aeronautics and Space Administration, Washington, D.C., 1975.

[22] Saeed Farokhi, *Aircraft Propulsion*, Wiley, New York, 2009.

QUESTIONS AND PROBLEMS

© Problem to be solved with aid of programmable calculator or computer.

Ⓞ Open-ended problem that requires critical thinking. These problems require various assumptions to provide the necessary input data. There are not unique answers to these problems.

Note: Unless specific values of required fluid properties are specified in the problem statement, use the values found in Tables 1.4–1.8 and in the tables in the Appendices.

Section 12.1 Introduction and Section 12.2 Basic Energy Considerations

12.1.1 Air (assumed incompressible) flows across the rotor shown in **Fig. P12.1.1** such that the magnitude of the absolute velocity increases from 15 m/s to 25 m/s. Measurements indicate that the absolute velocity at the inlet is in the direction shown. Determine the direction of the absolute velocity at the outlet if the fluid puts no torque on the rotor. Is the rotation CW or CCW? Is this device a pump or a turbine?

12.1.2 The measured shaft torque on the turbomachine shown in **Fig. P12.1.2** is -60 N · m when the absolute velocities are as indicated. Determine the mass flowrate. What is the angular velocity if the magnitude of the shaft power is 1800 N · m/s? Is this machine a pump or a turbine? Explain.

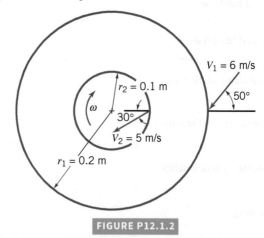

FIGURE P12.1.2

12.1.3 Uniform horizontal sheets of water of 3-mm thickness issue from the slits on the rotating manifold shown in **Fig. P12.1.3**. The velocity relative to the arm is a constant 3 m/s along each slit. Determine the torque needed to hold the manifold stationary. What would the angular velocity of the manifold be if the resisting torque is negligible?

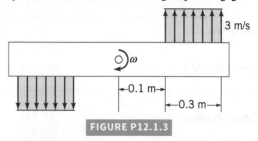

FIGURE P12.1.3

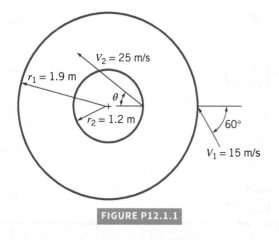

FIGURE P12.1.1

12.1.4 Sketched in **Fig. P12.1.4** are the upstream [section (1)] and downstream [section (2)] velocity triangles at the arithmetic mean radius for flow through an axial-flow turbomachine rotor. The axial component of velocity is 15 m/s at sections (1) and (2). **(a)** Label each velocity vector appropriately. Use **V** for absolute velocity, **W** for relative velocity, and **U** for blade velocity. **(b)** Are you dealing with a turbine or a fan? **(c)** Calculate the work per unit mass involved. **(d)** Sketch a reasonable blade section. Do you think that the actual blade exit angle will need to be less or greater than 15°? Why?

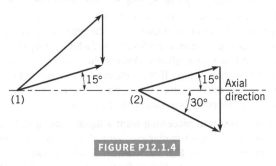

FIGURE P12.1.4

Section 12.4 The Centrifugal Pump
Section 12.4.1 Theoretical Considerations

12.4.1 The radial component of velocity of water leaving the centrifugal pump sketched in **Fig. P12.4.1** is 14 m/s. The magnitude of the absolute velocity at the pump exit is 28 m/s. The fluid enters the pump rotor radially. Calculate the shaft work required per unit mass flowing through the pump.

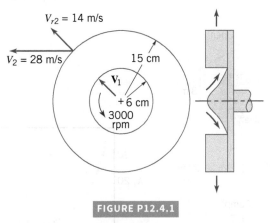

FIGURE P12.4.1

12.4.2 Water enters a centrifugal pump with an absolute velocity $V_1 = 10$ m/s in the radial direction and leaves with an absolute velocity V_2, which makes an angle of $\varphi_2 = 60°$ with the radial direction, as shown in **Fig. P12.4.2**. The impeller width (perpendicular to the paper) is $b = 0.125$ m, $R_1 = 0.125$ m, and $R_2 = 0.35$ m. Find the input torque T required to drive the pump if there are no friction losses.

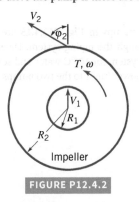

FIGURE P12.4.2

12.4.3 A centrifugal pump impeller is rotating at 1200 rpm in the direction shown in **Fig. P12.4.3**. The flow enters parallel to the axis of rotation and leaves at an angle of 30° to the radial direction. The absolute exit velocity, V_2, is 28 m/s. **(a)** Draw the velocity triangle for the impeller exit flow. **(b)** Estimate the torque necessary to turn the impeller if the fluid is water. What will the impeller rotation speed become if the shaft breaks?

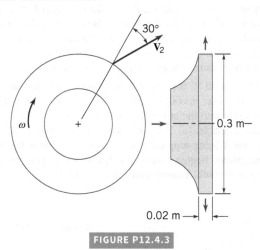

FIGURE P12.4.3

12.4.4 A centrifugal radial water pump has the dimensions shown in **Fig. P12.4.4**. The volume rate of flow is 0.007 m³/s, and the absolute inlet velocity is directed radially outward. The angular velocity of the impeller is 960 rpm. The exit velocity as seen from a coordinate system attached to the impeller can be assumed to be tangent to the vane at its trailing edge. The hydraulic efficiency is 82%, and the mechanical efficiency is 96%. Calculate the power required to drive the pump.

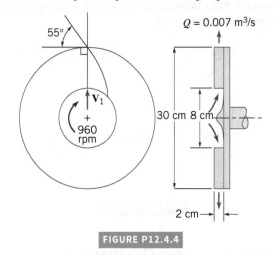

FIGURE P12.4.4

Section 12.4.2 Pump Performance Characteristics

12.4.5 Water is pumped with a centrifugal pump, and measurements made on the pump indicate that for a flowrate of 15 liter/s the required input power is 4.5 kW. For a pump efficiency of 62%, what is the actual head rise of the water being pumped?

12.4.6 The performance characteristics of a certain centrifugal pump are determined from an experimental setup similar to that shown in Fig. 12.10. When the flowrate of a liquid ($SG = 0.9$) through the pump is 7.5 liter/s, the pressure gage at (1) indicates a vacuum of 95 mm of mercury and the pressure gage at (2) indicates a pressure of 80 kPa. The diameter of the pipe at the inlet is 110 mm and at the exit it is 55 mm. If $z_2 - z_1 = 0.5$ m, what is the actual head rise across the pump? Explain how you would estimate the pump motor power requirement.

Section 12.4.3 Net Positive Suction Head (NPSH)

12.4.7 In Example 12.3, how will the maximum height, z_1, that the pump can be located above the water surface change if the water temperature is decreased to 4 °C?

12.4.8 A centrifugal pump with a 0.17 m-diameter impeller has the performance characteristics shown in Fig. 12.12. The pump is used to pump water at 38 °C, and the pump inlet is located 4 m above the open water surface. When the flowrate is 0.01 m³/s, the head loss between the water surface and the pump inlet is 2 m of water. Would you expect cavitation in the pump to be a problem? Assume standard atmospheric pressure. Explain how you arrived at your answer.

12.4.9 Water at 40 °C is pumped from an open tank through 200 m of 50-mm-diameter smooth horizontal pipe as shown in **Fig. P12.4.9** and discharges into the atmosphere with a velocity of 3 m/s. Minor losses are negligible. (a) If the efficiency of the pump is 70%, how much power is being supplied to the pump? (b) What is the NPSH$_A$ at the pump inlet? Neglect losses in the short section of pipe connecting the pump to the tank. Assume standard atmospheric pressure.

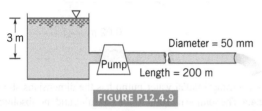

3 m

Diameter = 50 mm

Pump

Length = 200 m

FIGURE P12.4.9

12.4.10 The centrifugal pump is not self-priming. That is, if the water is drained from the pump and pipe as shown in **Fig. P12.4.10(a)**, the pump will not draw the water into the pump and start pumping when the pump is turned on. However, if the pump is primed (i.e., filled with water as in **Fig. P12.4.10(b)**), the pump does start pumping water when turned on. Explain this behavior.

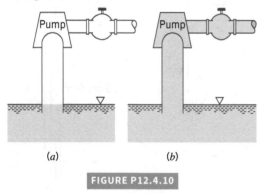

Pump   Pump

(a)   (b)

FIGURE P12.4.10

Section 12.4.4 System Characteristics, Pump-System Matching, and Pump Selection

12.4.11 A centrifugal pump having a head-capacity relationship given by the equation $h_a = 180 - 6.10 \times 10^{-4}Q^2$, with h_a in meter when Q is in m³/s, is to be used with a system similar to that shown in Fig. 12.14. For $z_2 - z_1 = 15$ m, what is the expected flowrate if the total length of constant diameter pipe is 183 m and the fluid is water?

Assume the pipe diameter to be 10 cm and the friction factor to be equal to 0.02. Neglect all minor losses.

12.4.12 A centrifugal pump having a 15 cm-diameter impeller and the characteristics shown in Fig. 12.7 is to be used to pump gasoline through 1220 m of commercial steel 8 cm-diameter pipe. The pipe connects two reservoirs having open surfaces at the same elevation. Determine the flowrate. Do you think this pump is a good choice? Explain.

12.4.13 A centrifugal pump having the characteristics shown in Example 12.4 is used to pump water between two large open tanks through 30 m of 20 cm-diameter pipe. The pipeline contains four regular flanged 90° elbows, a check valve, and a fully open globe valve. Other minor losses are negligible. Assume the friction factor $f = 0.02$ for the 30 m section of pipe. If the static head (difference in height of fluid surfaces in the two tanks) is 9 m, what is the expected flowrate? Do you think this pump is a good choice? Explain.

12.4.14 In a chemical processing plant a liquid is pumped from an open tank, through a 0.1-m-diameter vertical pipe, and into another open tank as shown in **Fig. P12.4.14(a)**. A valve is located in the pipe, and the minor loss coefficient for the valve as a function of the valve setting is shown in **Fig. P12.4.14(b)**. The pump head-capacity relationship is given by the equation $h_a = 52.0 - 1.01 \times 10^3Q^2$ with h_a in meters when Q is in m³/s. Assume the friction factor $f = 0.02$ for the pipe, and all minor losses, except for the valve, are negligible. The fluid levels in the two tanks can be assumed to remain constant. (a) Determine the flowrate with the valve wide open. (b) Determine the required valve setting (percent open) to reduce the flowrate by 50%.

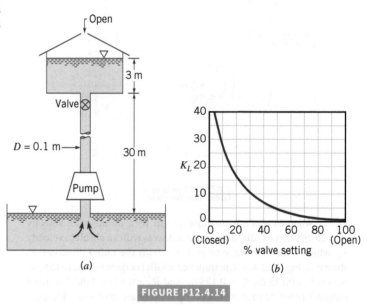

Open

3 m

Valve

D = 0.1 m

30 m

Pump

(a)

K_L

% valve setting

(Closed)   (Open)

(b)

FIGURE P12.4.14

12.4.15 Two of the pumps in **Fig. P12.4.15** are operated in series to supply water through the piping system. Determine the flowrate through the piping system for 10 °C water and screwed connections. Then find the total power input to the two pumps.

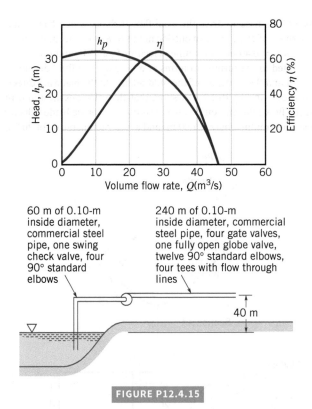

60 m of 0.10-m inside diameter, commercial steel pipe, one swing check valve, four 90° standard elbows

240 m of 0.10-m inside diameter, commercial steel pipe, four gate valves, one fully open globe valve, twelve 90° standard elbows, four tees with flow through lines

40 m

FIGURE P12.4.15

12.4.16 ○ Water is pumped between the two tanks described in Example 12.4 once a day, 365 days a year, with each pumping period lasting two hours. The water levels in the two tanks remain essentially constant. Estimate the annual cost of the electrical power needed to operate the pump if it were located in your city. You will have to make a reasonable estimate for the efficiency of the motor used to drive the pump. Due to aging, it can be expected that the overall resistance of the system will increase with time. If the operating point shown in Fig. E12.4c changes to a point where the flowrate has been reduced to 63 liter/s, what will be the new annual cost of operating the pump? Assume that the cost of electrical power remains the same.

Section 12.5 Dimensionless Parameters and Similarity Laws

12.5.1 A centrifugal pump having an impeller diameter of 1 m is to be constructed so that it will supply a head rise of 200 m at a flowrate of 4.1 m³/s of water when operating at a speed of 1200 rpm. To study the characteristics of this pump, a 1/5 scale, geometrically similar model operated at the same speed is to be tested in the laboratory. Determine the required model discharge and head rise. Assume that both model and prototype operate with the same efficiency (and therefore the same flow coefficient).

12.5.2 Do the head–flowrate data shown in Fig. 12.12 appear to follow the similarity laws as expressed by Eqs. 12.39 and 12.40? Explain.

12.5.3 A centrifugal fan operating in a duct has the dimensionless parameters

$$C_Q = \frac{Q}{\omega D^3} \quad \text{and} \quad C_H = \frac{\Delta p}{\rho \omega^2 D^2}$$

where C_Q is a flow coefficient, C_H is a head coefficient, Q is the volume flowrate, ω is the fan speed, D is the fan diameter, ρ is the fluid density,

and Δp is the fan pressure rise. **Figure P12.5.3** shows this fan's performance curve in dimensional form for a fan speed of $\omega_{15} = 1500$ rpm. Find the fan operating points (Q and Δp) for $\omega_{30} = 3000$ rpm and corresponding to points 1, 3, and 5 at $\omega_{15} = 1500$ rpm. Assume similarity between 1500 rpm and 3000 rpm.

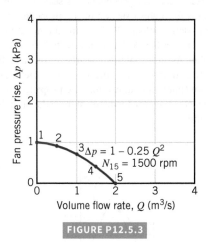

$3 \Delta p = 1 - 0.25 Q^2$
$N_{15} = 1500$ rpm

FIGURE P12.5.3

12.5.4 A centrifugal pump has the performance characteristics of the pump with the 15 cm-diameter impeller described in Fig. 12.7. Note that the pump in this figure is operating at 3500 rpm. What is the expected head gained if the speed of this pump is reduced to 2800 rpm while operating at peak efficiency?

12.5.5 A prototype fan has a 6 m diameter, an inlet pressure of 100 kPa, an inlet temperature of 21 °C, and a speed of 90 rpm. A $\frac{1}{10}$-scale model of the fan has the same inlet pressure and temperature, an inlet power of 1 kW, a flowrate of 0.1 m³/s, and a speed of 1800 rpm. Find the corresponding input power and flowrate of the prototype fan. Neglect Reynolds number effects.

12.5.6 In a certain application, a pump is required to deliver 0.3 m³/s against a 90 m head when operating at 1200 rpm. What type of pump would you recommend?

12.5.7 A centrifugal pump operates at 300 rpm to deliver 20 °C lubricating oil. A $\frac{1}{5}$-size, geometrically similar pump delivering 15 °C water is used to model the larger pump. How fast should the smaller pump run? Discuss the accuracy of the result.

Section 12.6 Axial-Flow and Mixed-Flow Pumps

12.6.1 A certain axial-flow pump has a specific speed of $N_s = 5.0$. If the pump is expected to deliver 190 liter/s when operating against a 4.5 m head, at what speed (rpm) should the pump be run? Draw a sketch of the pump impeller (front and side views).

12.6.2 A certain pump is known to have a capacity of 3 m³/s when operating at a speed of 60 rad/s against a head of 20 m. Based on the information in Fig. 12.18, would you recommend a radial-flow, mixed-flow, or axial-flow pump? Draw a sketch of the pump impeller (front and side views).

12.6.3 The system resistance for a pipeline is given by $\Delta p_{sys} = 2.0 Q^2$, where Δp_{sys} is the pressure rise required of a pump to deliver the flowrate Q through the piping system. A pump has the pressure–rise–flow characteristic given by $\Delta p_p = 30.0 - 3.0 Q^2$. In both curves, Δp is in kPa and Q is in m³/s. Find the pump input power if this pump is placed in this piping system and the pump overall efficiency is 90%.

12.6.4 The axial-flow pump shown in Fig. 12.19 is designed to move 0.3 m³/s of water over a head rise of 1.5 m of water. Estimate the motor power requirement and the $U_2 V_{\theta 2}$ needed to achieve this flowrate on a continuous basis. Comment on any cautions associated with where the pump is placed vertically in the pipe.

12.6.5 A propeller-driven airplane is traveling with a velocity $V_1 = 321$ km/h (89 m/s). The propeller diameter is 3 m and it rotates at 3000 rpm. **Figure P12.6.5** shows a propeller cross-sectional profile and the velocity diagrams for a short section of the propeller at a radius of 1 m. The outlet relative velocity W_2 is assumed tangent to the propeller at the outlet, so $\beta_1' = \beta_2 = 50°$. The air density is constant as it flows over the propeller. Assume that the flow area for the mass flowrate interacting with this short section of the propeller is the same upstream and downstream (inlet and outlet) from the propeller (i.e., $A_1 = A_2$). Produce the velocity diagram downstream from the propeller by finding U, W_2, V_2, and α_2.

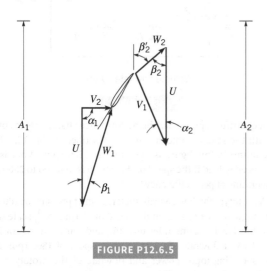

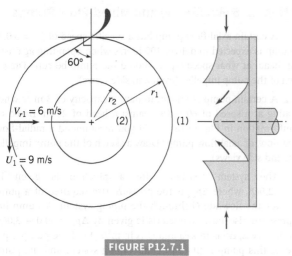

FIGURE P12.6.5

Section 12.7 Turbines (also see Sec. 12.3)

12.7.1 An inward flow radial turbine (see **Fig. P12.7.1**) involves a nozzle angle, α_1, of 60° and an inlet rotor tip speed, U_1, of 9 m/s. The ratio of rotor inlet to outlet diameters is 2.0. The radial component of velocity remains constant at 6 m/s through the rotor, and the flow leaving the rotor at section (2) is without angular momentum. If the flowing fluid is water and the stagnation pressure drop across the rotor is 110 kPa, determine the loss of available energy across the rotor and the efficiency involved.

FIGURE P12.7.1

12.7.2 The frictionless converging stationary nozzle of the hydraulic turbine shown in **Fig. P12.7.2** has an inlet pressure $p_0 = 480$ kPa, a negligible inlet velocity V_0, and an exit pressure $p_1 = 101$ kPa. The

velocity V_1 is used to drive the axial flow turbine, which has a rotational speed of 185.7 rpm, an outside radius of $R = 1.20$ m, and a blade height of $h = 0.40$ m. Determine velocities V_1 and V_2 and the power transmitted to the turbine in terms of α_1 and α_2. The fluid density is constant and the velocities of the fluid relative to the blade are directed as shown in Fig. P12.7.2. Assume that the velocities are uniform over the flow inlet and outlet areas and use an average blade velocity at the blade midheight ($r = 1.00$ m) and a water temperature of 20 °C.

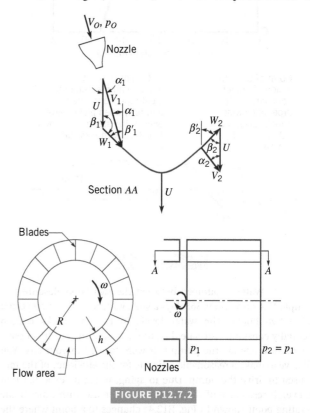

FIGURE P12.7.2

12.7.3 A water turbine wheel rotates at the rate of 100 rpm in the direction shown in **Fig. P12.7.3**. The inner radius, r_2, of the blade row is 30 cm, and the outer radius, r_1, is 60 cm. The absolute velocity vector at the turbine rotor entrance makes an angle of 20° with the tangential direction. The inlet blade angle is 60° relative to the tangential direction. The blade outlet angle is 120°. The flowrate is 0.3 m³/s. For the flow tangent to the rotor blade surface at inlet and outlet, determine an appropriate constant blade height, b, and the corresponding power available at the rotor shaft. Is the shaft power greater or less than the power lost by the fluid? Explain.

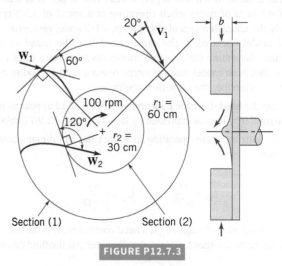

FIGURE P12.7.3

12.7.4 A sketch of the arithmetic mean radius blade sections of an axial-flow water turbine stage is shown in **Fig. P12.7.4**. The rotor speed is 1500 rpm. (a) Sketch and label velocity triangles for the flow entering and leaving the rotor row. Use **V** for absolute velocity, **W** for relative velocity, and **U** for blade velocity. Assume flow enters and leaves each blade row at the blade angles shown. (b) Calculate the work per unit mass delivered at the shaft.

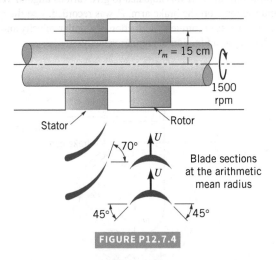

FIGURE P12.7.4

12.7.5 **Figure P12.7.5** shows a piping system with frictional losses of $h_{L1-2} = 4.0Q^2$, with h_{L1-2} in meter and Q in m³/s. The turbine performance characteristics are given by $h_t = 20 + 12.0Q$, where h_t is the turbine head in meter and Q is in m³/s. Find the flowrate Q.

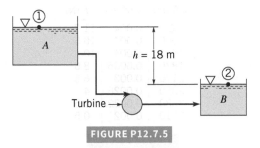

FIGURE P12.7.5

12.7.6 (Video) A small Pelton wheel is used to power an oscillating lawn sprinkler as shown in **Video V12.4** and **Fig. P12.7.6**. The arithmetic mean radius of the turbine is 2.5 cm, and the exit angle of the blade is 135° relative to the blade motion. Water is supplied through a single 5 mm-diameter nozzle at a speed of 15 m/s. Determine the flowrate, the maximum torque developed, and the maximum power developed by this turbine.

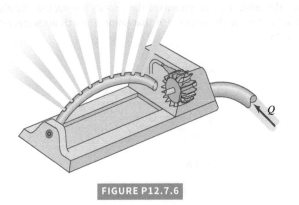

FIGURE P12.7.6

12.7.7 A windmill has an approach velocity $V_1 = 38$ km/h (10.5 m/s) and a blade diameter of 30 m. The blade rotates at 15 rpm. **Figure P12.7.7** shows a blade cross-sectional profile and the velocity diagrams for a short section of the blade at a radius of $r = 8$ m. The outlet relative velocity W_2 is tangent to the blade at the outlet so $\beta_2 = \beta'_2 = 30°$. The air density is constant as it flows over the blade. Assume that the flow area for the mass flowrate that interacts with this short section of the blade is the same upstream and downstream of the blade (i.e., $A_1 = A_2$). Find the velocity diagram downstream of the blade by finding U, W_2, V_2, and α_2.

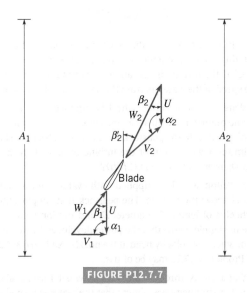

FIGURE P12.7.7

12.7.8 The single-stage, axial-flow turbomachine shown in **Fig. P12.7.8** involves water flow at a volumetric flowrate of 9 m³/s. The rotor revolves at 600 rpm. The inner and outer radii of the annular flow path through the stage are 0.46 and 0.61 m, and $\beta_2 = 60°$. The flow entering the rotor row and leaving the stator row is axial when viewed from the stationary casing. Is this device a turbine or a pump? Estimate the amount of power transferred to or from the fluid.

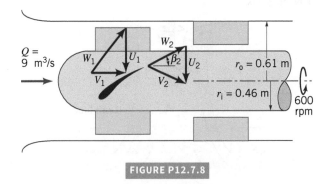

FIGURE P12.7.8

12.7.9 (Video) For an air turbine of a dentist's drill like the one discussed in Example 12.7 and shown in **Video V12.5**, calculate the average blade speed associated with a rotational speed of 350,000 rpm. Estimate the air pressure needed to run this turbine.

12.7.10 Water for a Pelton wheel turbine flows from the headwater and through the penstock as shown in **Fig. P12.7.10**. The effective friction factor for the penstock, control valves, and the like is 0.032, and the diameter of the jet is 0.20 m. Determine the maximum power output.

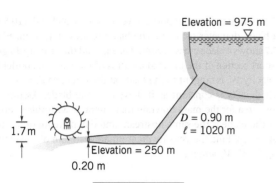

12.7.11 A Pelton wheel has a diameter of 2 m and develops 500 kW when rotating 180 rpm. What is the average force of the water against the blades? If the turbine is operating at maximum efficiency, determine the speed of the water jet from the nozzle and the mass flowrate.

12.7.12 Water to run a Pelton wheel is supplied by a penstock of length ℓ and diameter D with a friction factor f. If the only losses associated with the flow in the penstock are due to pipe friction, show that the maximum power output of the turbine occurs when the nozzle diameter, D_1, is given by $D_1 = D/(2f\ell/D)^{1/4}$.

12.7.13 A Pelton wheel is supplied with water from a lake at an elevation H above the turbine. The penstock that supplies the water to the wheel is of length ℓ, diameter D, and friction factor f. Minor losses are negligible. Show that the power developed by the turbine is maximum when the velocity head at the nozzle exit is $2H/3$. *Note:* The result of Problem 12.7.12 may be of use.

12.7.14 Water flows through the Pelton wheel turbine shown in Fig. 12.25. For simplicity we assume that the water is turned 180° by the blade. Show, based on the energy equation (Eq. 5.84), that the maximum power output occurs when the absolute velocity of the fluid exiting the turbine is zero.

12.7.15 A 1-m-diameter Pelton wheel rotates at 300 rpm. Which of the following heads (in meters) would be best suited for this turbine: (a) 2, (b) 5, (c) 40, (d) 70, or (e) 140? Explain.

12.7.16 Draft tubes as shown in **Fig. P12.7.16** are often installed at the exit of Kaplan and Francis turbines. Explain why such draft tubes are advantageous.

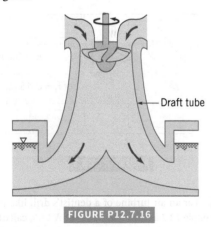

12.7.17 Turbines are to be designed to develop 22 MW while operating under a head of 21 m and an angular velocity of 60 rpm. What type of turbine is best suited for this purpose? Estimate the flowrate needed.

12.7.18 Water at 28 bar is available to operate a turbine at 1750 rpm. What type of turbine would you suggest to use if the turbine should have an output of approximately 150 kW?

12.7.19 It is desired to produce 37 MW with a head of 15 m and an angular velocity of 100 rpm. How many turbines would be needed if the specific speed is to be (a) 50, (b) 100?

12.7.20 [C] Test data for the small Francis turbine shown in **Fig. P12.7.20** is given in the following table. The test was run at a constant 10 m head just upstream of the turbine. The Prony brake on the turbine output shaft was adjusted to give various angular velocities, and the force on the brake arm, F, was recorded. Use the given data to plot curves of torque as a function of angular velocity and turbine efficiency as a function of angular velocity.

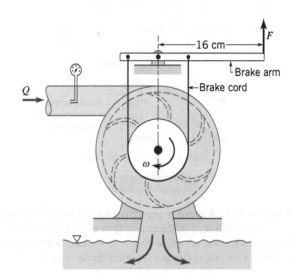

ω (rpm)	Q (m³/s)	F (N)
0	0.004	11.7
1000	0.004	10.5
1500	0.004	10
1870	0.0035	8.5
2170	0.003	6.5
2350	0.0026	4
2580	0.0022	1.5
2710	0.002	0.5

Section 12.8 Fans

12.8.1 For the fan of both Examples 5.19 and 5.28 discuss what fluid flow properties you would need to measure to estimate fan efficiency.

12.8.2 A lossless motor drives the fan shown in **Fig. P12.8.2** at 40 Hz. The power input to the motor is 40 amps at 440 volts. For the geometry shown, what is the discharge flowrate of air through the fan? Assume that the tangential component of the velocity leaving the impeller is equal to that of the impeller at that point. The exit air temperature $T_2 = 15\,°C$.

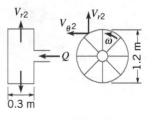

12.8.3 A centrifugal fan has a power input of 25 kW, an inner radius of $R_1 = 0.5$ m, an outer radius of $R_2 = 1.0$ m, and delivers 100 kg/s of air. There are no friction losses, the air inlet absolute velocity has no tangential component, and the outlet absolute velocity has a tangential component equal to the blade velocity at the outer radius R_2. The rotor depth (i.e., blade height) is 1.0 m. What is the required rotational speed of the rotor?

12.8.4 An axial fan operating at 1000 rpm has the characteristics shown in **Fig. P12.8.4**. It delivers 15 °C atmospheric air through a 50-cm I.D., galvanized, sheet-metal, horizontal duct having a length of 175 m and seven 90° long-radius elbows. For constant air density what is the flowrate if the duct discharges to the atmosphere?

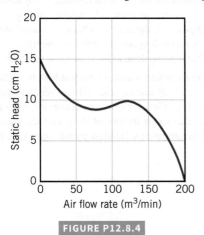

FIGURE P12.8.4

12.8.5 A model fan with wheel diameter 80 cm is tested at a speed of 1750 rpm. The test fluid is air with density 1.2 kg/m³. At its BEP, the fan produces 4 m³/s at total pressure rise of 20 cm H₂O. A geometrically similar fan is to handle 95 m³/s of flue gas with density 0.8 kg/m³ and 75 cm H₂O total pressure rise. Determine the required size and speed of the flue gas fan. Note any assumptions and/or limitations.

Section 12.9 Compressible Flow Turbomachines

12.9.1 Obtain photographs/images of a variety of turbo-compressor rotors and categorize them as axial-flow or radial-flow compressors. Explain briefly how they are used. Note any unusual features. Repeat for compressible flow turbines.

12.9.2 An axial flow compressor stage shown in **Fig. P12.9.2** has the inlet and outlet velocity diagrams shown. Calculate the work per unit mass. Quantities are $U_1 = U_2 = U = 230$ m/s, $V_1 = 135$ m/s, $W_1 = 270$ m/s, $\alpha_1 = 90°$, $V_2 = 165$ m/s, $W_2 = 190$ m/s, and $\alpha_2 = 53.8°$.

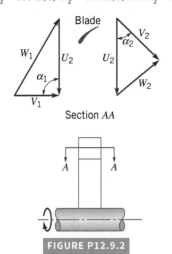

FIGURE P12.9.2

12.9.3 The axial flow gas turbine stage shown in **Fig. P12.9.3** has a mean blade radius of $R = 12$ cm, a rotational speed of 15,000 rpm, a mass flowrate of 4 kg/s, $W_1 = 295$ m/s, $V_1 = 475$ m/s, $\alpha_1 = 20°$, $W_2 = 265$ m/s, $V_2 = 147$ m/s, and $\alpha_2 = 99.4°$. Find the power transmitted from the gas to the turbine blade.

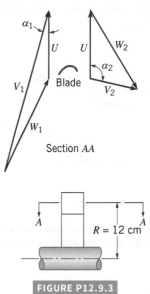

FIGURE P12.9.3

12.9.4 A centrifugal air compressor has a rotor inner diameter of $D_1 = 5$ cm, a rotor outer diameter of $D_2 = 16$ cm, a rotor depth of 25 cm, and a rotor rotational speed of 3600 rpm. The fluid relative velocities (W) are purely radial. The compressor delivers an air mass flowrate of 0.45 kg/s with $T_1 = 21$ °C, $p_1 = 101$ kPa, $T_2 = 115$ °C, and $p_2 = 228$ kPa. Find the power transferred by the rotor to the air and the inlet relative velocity W_1.

12.9.5 The axial flow steam turbine rotor shown in **Fig. P12.9.5** has a blade outer radius $R_o = 75$ cm, a blade inner radius $R_i = 60$ cm, a steam inlet pressure $p_1 = 1400$ kPa, a steam inlet density $\rho_1 = 4.7$ kg/m³, and an inlet absolute velocity $V_1 = 305$ m/s making an angle of 70° with the axial direction. The steam outlet pressure $p_2 = 350$ kPa and outlet density is 1.6 kg/m³. $\beta_2 = 40°$. The rotor rotates at 3600 rpm. Using a blade velocity at the blade midheight ($R = 70$ cm), find the power transferred from the steam to the rotor.

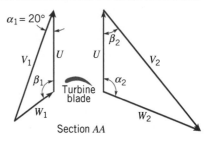

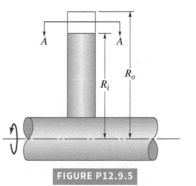

FIGURE P12.9.5

12.9.6 The **Fig. P12.9.6** shows a nozzle vane and a rotor blade for an axial flow gas turbine stage. The blade speed is 240 m/s. The absolute velocity leaving the stage is identical to the absolute velocity entering the stage and both are purely in the axial direction. **(a)** Draw and label the velocity diagrams. Show values for all three velocities and the absolute and relative angles on each diagram. **(b)** Calculate the work per unit mass for this stage.

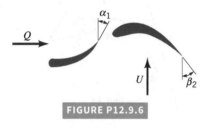

FIGURE P12.9.6

Lifelong Learning Problems

12.1 LL What do you think are the major unresolved fluid dynamics problems associated with gas turbine engine compressors? For gas turbine engine high-pressure and low-pressure turbines? For jet engine fans?

12.2 LL Pick a specific day (any day will do). During that day note and record at least ten different instances when the operation of one or more turbomachines makes your life better or easier or more comfortable. (If the machine operates more-or-less continuously, like a furnace fan, note only once.) In each case, record the type of machine (e.g., "axial-flow fan"). If you don't know, make your best guess. Then, for each machine, estimate its flowrate, pressure change, and power input or output.

12.3 LL What are current efficiencies achieved by the following categories of turbomachines? **(a)** Wind turbines; **(b)** hydraulic turbines; **(c)** power plant steam turbines; **(d)** aircraft gas turbine engines; **(e)** natural gas pipeline compressors; **(f)** home vacuum cleaner blowers; **(g)** laptop computer cooling fan; **(h)** irrigation pumps; **(i)** dentist drill air turbines. What is being done to improve these devices?

12.4 LL Perform an online search. Identify at least ten different pump manufacturers. For each, choose at least three different types of pumps in their product line. Find performance curves for these pumps and estimate the specific speed for five of them and the suction specific speed for five different pumps. Do these pumps agree with Fig. 12.18?

APPENDIX A

Computational Fluid Dynamics

A.1   INTRODUCTION

What Is Computational Fluid Dynamics (CFD)?

As with many questions related to fluid mechanics, this question is much easier to ask than answer. It is not simply the use of computers to model the behavior of fluid components and systems. Going all the way back to hydrostatics, using a computer to perform the arithmetic required to compute the force on a submerged surface from the hydrostatic pressure distribution is not CFD. Using a numerical method to approximate the ordinary differential equation that results from using the Bernoulli and continuity equations to model tank draining or filling is not CFD. Using a computer to compute the friction factor for pipe flow from the implicit Colebrook formula is not CFD. The common trait shared by these examples is that the general continuum-level model of fluid flow as represented by the Navier–Stokes equations has been greatly simplified using assumptions justified by the application before the computational power of the computer has been brought to bear. The generally applicable equations describing the behavior of fluids have been simplified to an explicit algebraic equation, a fairly simple ordinary differential equation, and an implicit algebraic equation, respectively. In all three instances, a reasonably useful answer could be obtained with a pencil, a piece of paper, a calculator, and a modest amount of effort. In all three instances the specialized physics of the problem of interest is already deeply imbedded in the model before a computational method is employed.

Computational fluid dynamics (CFD) is the simulation of fluid flow using the vast computational power of modern computers to obtain approximate solutions to generally applicable models of fluid behavior (e.g., the Navier–Stokes equations, Eqs. 6.127a, b, c and the continuity equation, Eq. 6.27) for a wide variety of boundary and initial conditions. CFD replaces the partial differential equations with discretized algebraic equations that approximate the partial differential equations. These algebraic equations are then numerically solved to obtain flow field values at discrete points in space and/or time. Because the Navier–Stokes equations enforce fundamental principles at every point in the flow field, an analytical solution to these equations would provide the solution for an infinite number of points in the flow. The CFD simulation solves for the relevant flow variables only at the discrete points, which make up the grid or mesh of the solution (discussed in more detail below). Interpolation schemes must be used to obtain values at non-grid point locations.

In general, most applications of CFD take the same basic approach. For each application, the specific tasks executed and tools employed likely depend on the complexity of the flow, the "answers" being sought, available computer resources, available expertise in CFD, and whether a commercially available CFD package is used, or a problem-specific CFD algorithm is developed. In today's market, there are many commercial CFD codes available to solve a wide variety of problems. However, if the intent is to conduct a thorough investigation of a specific fluid flow problem in a research environment, it is possible that taking the time to develop a problem-specific algorithm may be the best choice in the long run.

CFD can be thought of as a numerical experiment. In a typical fluids experiment, an experimental model is built, measurements of the flow interacting with that model are made, and the results are analyzed. In CFD, the building of the model is replaced with the formulation of the governing equations and the development of the numerical algorithm. The process of obtaining measurements is replaced with using a computer and software to implement the algorithm to simulate the flow. Of course, the analysis of results is common ground for both techniques.

In a very real sense, CFD represents a "marriage" of the disciplines of fluid mechanics, numerical analysis, and computer science. Therefore a thorough examination of CFD topics is well beyond the scope of this textbook. This appendix highlights some of the more important topics in CFD, but is only intended as a brief introduction. The topics include discretization of the governing equations, mesh generation, boundary conditions, turbulence models, solving the equations, application of CFD, and some representative examples.

A.2 A VERY SIMPLE EXAMPLE

We begin our study of CFD with a very simple example. Although modern CFD analyses are developed very formally, the development of this example will be more intuitive. After developing nearly all of the important concepts of CFD in this example, we will be prepared to discuss the features of the large and powerful CFD codes currently available.

EXAMPLE A.1 | Flow in a Converging Channel

Given Water flows in the converging channel shown in **Fig. EA.1a**.

Find Determine the pressure distribution on the upper (flat) wall.

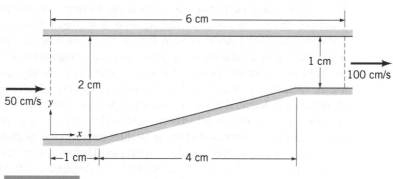

FIGURE EA.1a A converging channel.

Solution Determining the pressure distribution requires solution of the continuity and the Navier–Stokes equations. To keep this example simple, we make the following assumptions:

1. The flow is steady.
2. The flow is repeatable in planes parallel to the paper (i.e., it is two-dimensional).
3. Shear stress is negligible, so the fluid is assumed to have zero viscosity.
4. The fluid is incompressible.

With these assumptions the governing equations (Eqs. 6.27, 6.127a, b, c) reduce to

$$\frac{\partial u}{\partial x} + \frac{\partial v}{\partial y} = 0$$

$$u\frac{\partial u}{\partial x} + v\frac{\partial u}{\partial y} = -\frac{1}{\rho}\frac{\partial p}{\partial x}$$

$$u\frac{\partial v}{\partial x} + v\frac{\partial v}{\partial y} = -\frac{1}{\rho}\frac{\partial p}{\partial y}$$

In so-called *primitive variables* (u, v, p) there are three equations in three unknowns—and two of the equations are nonlinear partial differential equations. Fortunately, from Section 6.4.2, a solution to the two momentum equations is simply Bernoulli's equation

$$\frac{p}{\rho} + \frac{u^2 + v^2}{2} = \text{Constant}$$

Introducing a stream function ψ such that $u = \dfrac{\partial \psi}{\partial y}$ and $v = -\dfrac{\partial \psi}{\partial x}$ (see Eq. 6.37), the continuity equation is automatically satisfied and $\dfrac{\partial^2 \psi}{\partial x^2} + \dfrac{\partial^2 \psi}{\partial y^2} = -\zeta_z$. Finally, because the flow originates in a region of uniform velocity where $\zeta_z = 0$ and the fluid has zero viscosity, $\zeta_z = 0$ everywhere (see Section 6.4.3). We can solve the problem by first solving $\dfrac{\partial^2 \psi}{\partial x^2} + \dfrac{\partial^2 \psi}{\partial y^2} = 0$, then by finding u and v from ψ, and, finally, calculating the requested pressure distribution from Bernoulli's equation.

The task now is to solve the partial differential equation (PDE)

$$\frac{\partial^2 \psi}{\partial x^2} + \frac{\partial^2 \psi}{\partial y^2} = 0 \tag{1}$$

Of course we must provide a set of boundary conditions in order to specify the geometry of this particular channel. Boundary conditions for ψ are straightforward; we specify values for ψ at all boundaries. Recalling that for two-dimensional flow, the difference between the values of ψ between two streamlines is the volume flow between the streamlines, we can write the boundary conditions as:

Along the inlet: $\psi = 50y$ $x = 0; 0 \le y \le 2$

Along the upper wall: $\psi = 100$ $0 \le x \le 6; y = 2$

Along the outlet: $\psi = 100(y - 1)$ $x = 6; 1 \le y \le 2$

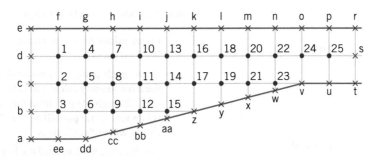

FIGURE EA.1b Channel with grid.

Along the bottom wall: $\psi = 0$

$0 \leq x \leq 1; y = 0, 1 \leq x \leq 5;$

$y = 0.25(x - 1), 5 \leq x \leq 6; y = 1$

It might be possible to solve the equation analytically using, say, a complex variable transformation. If we were able to solve it analytically, the benefit would be that we could determine the velocities at *any* point in the channel. The disadvantages would be that we would be in for some complicated mathematics and the operations necessary to calculate values at any point would likely be tedious.

Let's decide that we don't need the pressure and velocities at every point; values at several specific points will be satisfactory. We seek a numerical (computational) solution. To get started, we need to do two things immediately: establish the points in the field where we will determine values of ψ and put the PDE in a form that allows us to use algebraic equations to calculate values of ψ for every point that we define.

We start by defining a *grid* or *mesh* that is overlaid on the flow field. Such a grid is shown in **Fig. EA.1b**. For this case, we use a uniform ($\Delta x = \Delta y = 0.5$ cm) Cartesian (square) grid. The points at which we will compute ψ are the points at the intersection of the grid lines and are numbered from 1 to 25. The grid lines at the far left and the far right are the channel inlet and the channel outlet, respectively. These points and the points on the walls are assigned letters rather than numbers; ψ is known at these points and they will provide boundary conditions.

In order to calculate 25 unknown values of ψ, we will need 25 equations. Since Eq. (1) applies at every point in the field, we can obtain exactly 25 equations by applying it at each of the 25 points. To do so, we must approximate Eq. (1) using the value of ψ at a point together with values at several surrounding points.

Fig. EA.1c shows a typical point in the field, labeled P, and its neighboring points, labeled N, E, S, and W.* We approximate the first derivative of ψ by either

$$\left.\frac{\partial \psi}{\partial x}\right)_P \approx \frac{\psi_E - \psi_P}{\Delta x} \tag{2}$$

or

$$\left.\frac{\partial \psi}{\partial x}\right)_P \approx \frac{\psi_P - \psi_W}{\Delta x} \tag{3}$$

These two approximations, called a *forward difference* and a *backward difference*, become more accurate as Δx becomes smaller. Although intended as approximations of the derivative at P, they are better approximations for the derivative at point PE midway

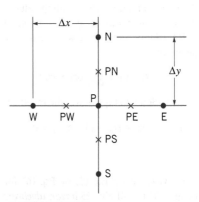

FIGURE EA.1c A typical point (P) with four surrounding points.

between P and E and point PW, midway between P and W. We obtain an approximation to the second derivative by

$$\frac{\partial^2 \psi}{\partial x^2} = \frac{\partial}{\partial x}\left(\frac{\partial \psi}{\partial x}\right) \approx \frac{\partial \psi/\partial x)_{PE} - \partial \psi/\partial x)_{PW}}{\Delta x}$$

Equations 2 and 3 are good approximations for the derivative at points PE and PW, so

$$\frac{\partial^2 \psi}{\partial x^2} \approx \frac{(\psi_E - \psi_P)/\Delta x - (\psi_P - \psi_W)/\Delta x}{\Delta x} = \frac{\psi_E + \psi_W - 2\psi_P}{(\Delta x)^2} \tag{4}$$

This approximation for the second derivative is most accurate at P, because it is "centered" there.

By similar reasoning, the y derivative is approximated by

$$\frac{\partial^2 \psi}{\partial y^2} \approx \frac{\psi_N + \psi_S - 2\psi_P}{(\Delta y)^2}$$

and Eq. (1) is approximated by

$$\left(\frac{1}{\Delta x}\right)^2 (\psi_E + \psi_W - 2\psi_P) + \left(\frac{1}{\Delta y}\right)^2 (\psi_N + \psi_S - 2\psi_P) = 0$$

If $\Delta x = \Delta y$ (a square grid), we get

$$\left(\frac{1}{\Delta x}\right)^2 (\psi_E + \psi_W + \psi_N + \psi_S - 4\psi_P) = 0$$

or, as Δx is finite,

$$\psi_N + \psi_E + \psi_S + \psi_W - 4\psi_P = 0 \tag{5}$$

Eq. (5) can be written for any interior point in the field (after we change the PNESW notation to a global point-numbering system). For points adjacent to the boundaries, at least one of the surrounding

---

*The notation comes from the points on a compass.

FIGURE EA.1d   Regular point (S) lies outside a boundary.

points is on a boundary and the corresponding value of ψ is known. Sometimes the point located a distance Δx away may lie *outside* the boundary (**Fig. EA.1d**), in which case we find a value for ψ at the point by extrapolation:

$$\psi_S = \psi_P + \frac{\Delta x}{\Delta x_B}(\psi_B - \psi_P)$$

Substituting this expression into Eq. (5) gives the modified equation

$$\psi_N + \psi_E + a\psi_B + \psi_W - (3 + a)\psi_P = 0 \qquad (6)$$

where $a = \dfrac{\Delta x}{\Delta x_B}$.

Now we can write either Eq. (5) or Eq. (6) for each of the 25 points in the grid. This will give 25 *linear, algebraic* equations in 25 unknowns (the 25 ψ's). A few of the equations are:

$$75 + 100 + \psi_4 + \psi_2 - 4\psi_1 = 0$$

$$\psi_2 + \psi_4 + \psi_6 + \psi_8 - 4\psi_5 = 0$$

$$\psi_9 + \psi_{11} + \psi_{15} + 2(0) - (3 + 2)\psi_{12} = 0$$

Solving the 25 equations will give the 25 values of ψ. The equations can be represented as

$$\mathbf{AX = B}$$

where **A** is a 25 × 25 matrix, **X** is the column vector of 25 ψ's, and **B** is the vector of known values assembled from known boundary values of ψ. The matrix **A** is sparse; no row contains more than 5 entries (typically, four "1's" and one "−4"). Even an introductory course in numerical methods will introduce the student to several well-established algorithms for solving such a system of equations. Today, a solution for the present problem can be quickly obtained using a personal computer, a programmable calculator, or even a smart phone.

Here we solve the equations by *Gauss–Sidel iteration*. The method begins by rewriting the equations in a form where the value of ψ at the central point is written in terms of values at the surrounding points, for example

$$\psi_1 = (75 + 100 + \psi_4 + \psi_2)/4$$

$$\psi_5 = (\psi_2 + \psi_4 + \psi_6 + \psi_8)/4$$

$$\psi_{12} = (\psi_9 + \psi_{11} + \psi_{15} + 0)/5$$

Now we proceed in a series of steps

- Guess values for all unknown ψ_s. (For our problem, $\psi = 50$ is a good guess.)
- "Visit" each point in turn calculating a new value of ψ_P using the most recently calculated values for ψ on the right side. If we move through the grid from top to bottom and left to right, the values for ψ_N and ψ_W will be "new" values, and the values for ψ_E and ψ_S will be "old" values.
- Continue the process until the largest change in ψ_P is as small as you want it to be.

This process can be easily programmed for a computer or programmable calculator.

This process works well on a spreadsheet. Simply type the equation for any ψ in terms of the surrounding ψ's in the appropriate cell. The software will object that a circular reference occurs; simply turn on "iterations" or switch to manual recalculation. **Fig. EA.1e** shows the converged solution as calculated in a spreadsheet.

These results are mildly interesting but this array of numbers is not particularly informative. One thing we could do is plot several streamlines. This would be done by picking several values for ψ (values of 25, 50, and 75 would be convenient), locating the point in each column where that value of ψ occurs, and connecting the points. Interpolation within a column would be required to determine the location of each point.

The stated objective of this problem is to determine the pressure distribution on the upper wall. To do this we must find the velocity along the upper wall and then find the pressure from Bernoulli's equation. The velocities at the upper wall are given by

$$u = \frac{\partial \psi}{\partial y}; v = 0$$

Returning for the moment to the PNESW notation (see Fig. EA.1c), an approximate value for u can be computed from (see Eqs. (2) and (3))

$$u_P \approx \frac{\psi_P - \psi_S}{\Delta y}$$

where P is a point on the upper wall (P = e, f, g, ... r) and S is the first point in the field (S = d, 1, 5, 7, ... 25, s). See Fig. EA.1b. The pressure is computed in terms of the pressure coefficient

$$C_p = \frac{p - p_{ref}}{\frac{1}{2}\rho V_{ref}^2} = 1 - \frac{u^2 + v^2}{V_{ref}^2}$$

where the reference pressure and velocity are those at the channel inlet. To illustrate, for point j

$$u_j = \frac{\psi_j - \psi_{13}}{\Delta y} = \frac{100 - 68.538}{0.5} = 62.92$$

$$C_{p,j} = 1 - \frac{u_j^2 + v_j^2}{V_{in}^2} = 1 - \frac{62.92^2 + 0^2}{50^2} = -0.584$$

The velocity and pressure coefficient on the upper wall are graphed in **Fig. EA.1f**. The computed points have been connected by a smooth curve although we have values only at the discrete points.

	A	B	C	D	E	F	G	H	I	J	K	L	M	N	O	P
1																
2							Upper Wall of Channel									
3			100.000	−100.000	−100.000	−100.000	−100.000	−100.000	−100.000	−100.000	−100.000	−100.000	−100.000	−100.000	−100.000	Channel
4			75.000	74.404	73.579	72.327	70.641	68.538	66.004	62.979	59.419	55.381	51.435	50.359	50.000	Outlet
5		Channel	50.000	49.038	47.583	45.090	41.697	37.507	32.501	26.492	19.316	10.671	0.000	0.000	0.000	
6		Inlet	25.000	24.166	22.626	18.754	13.549	7.294	0.000							
7			0.000	0.000	0.000											
8																

FIGURE EA.1e   Converged solution for values of ψ calculated by an electronic spreadsheet.

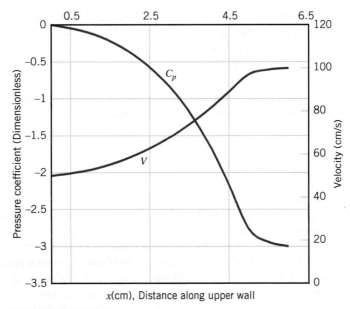

FIGURE EA.1f Velocity and pressure distribution along upper wall of a converging channel.

Comments Our results are approximate. There are two things we could do immediately to improve the accuracy. First, we could use a finer grid, say $\Delta x = \Delta y = 0.25$ cm. This would give about 100 points in the field, requiring us to solve 100 equations in 100 unknowns. The second accuracy improvement is a little less obvious. We used a simple backward difference to compute the velocity. The resulting value would be more representative of the velocity at a point halfway between the wall and the first grid point than it would the velocity at the wall. We could obtain a value that better represents the wall velocity by using a *three-point (backward) difference* at the wall. The formula uses the point at the wall, the first point away from the wall, and the point beyond that. In PNESW notation we have

$$u \approx \frac{3\psi_P - 4\psi_S + \psi_{S+1}{}^*}{2\Delta y}$$

Note that it was necessary to specify boundary conditions that completely surround the flow field. Knowing the inlet and wall boundary conditions is not especially difficult, but the outlet boundary is not so clear. From continuity, we know that the average velocity at the outlet must be 100 cm/s but it would not necessarily be uniform. It may be better to "extend" the channel further downstream to assure that the outlet velocity is truly uniform.

Computing the flow field in this example was straightforward and successful for one very important reason: our assumptions about the flow led to a *linear* PDE for the stream function. This led to a set of linear algebraic equations for calculating ψ. Although we chose an iterative method for solving the equations, iteration was not really necessary; a direct solution via Gaussian elimination or matrix inversion was an option. Because the equations are linear and the boundary conditions are well-posed, a unique solution to the problem was guaranteed. Things become far more difficult when the nonlinear Navier–Stokes equations are to be solved.

---

*This formula can be derived by fitting a parabola to the points P, S, and S+1, and evaluating the derivative at point P (i.e., $y = 0$).

Example A.1 demonstrates the steps that are required to perform a CFD simulation. A review of the example will identify the following steps:

- Identify the flow system to be simulated and determine the features to be included in the simulation. Stated another way, the capabilities of the CFD code and the flow system model must be matched. Will the simulation be fully three-dimensional? Steady or unsteady? Incompressible? Laminar, turbulent, or inviscid?

- Discretize the partial differential equations, converting them into a set of algebraic equations.

- Discretize the flow field using an appropriate grid (or *mesh*).

- Specify the boundary conditions.

- Select a *turbulence model* (if the flow is or is expected to be turbulent).

- Solve the system of algebraic equations (usually by iteration).

- Process the solution and display the results in useful form. Called *post-processing*, this step generates graphs, vector plots, and, often, color graphics visualizations of the flow field, pressure field and/or temperature field.
- Investigate and verify the accuracy of the solution.

A few of these steps are self-explanatory. The following sections discuss some options for the others and, in some cases, their consequences.

A.3   DISCRETIZATION

The process of *discretization* involves developing a set of algebraic equations (based on discrete points in the flow domain) to be used in place of the partial differential equations. Of the various discretization techniques available for the numerical solution of the governing differential equations, the following types are most common: (1) the finite element method, (2) the boundary element method, (3) the finite difference method, and (4) the finite volume method. In each of these methods, the continuous flow field (i.e., velocity and pressure as a function of space and time) is described in terms of discrete (rather than continuous) values at prescribed locations and times. Through these techniques the differential equations are replaced by a set of (usually nonlinear) algebraic equations that can be solved with the aid of a computer.

For the *finite element* method, the flow field is broken into a set of small fluid elements (frequently triangular areas if the flow is two-dimensional, or small volume elements if the flow is three-dimensional). A number of *nodes* are defined for each element, always one at each vertex (corner) and sometimes elsewhere such as the center of each side. The field variables (u, v, p, etc.) are approximated within each element by an algebraic function containing the nodal values. The approximating functions are then substituted into the appropriate equations (mass, momentum, energy). Because the functions are approximations, the conservation equations are not exactly satisfied; a set of *residuals* is produced. The residuals are then multiplied by a weighting function and integrated over the element. The computational equations are developed by minimizing the weighted residual for each element and then connecting together all of the elements in the field (a process called *assembly*). This process produces a set of N algebraic equations in N unknowns where N is the number of unknown nodal values. These equations are solved with the aid of a computer.

The number, size, and shape of elements are dictated in part by the particular flow geometry and flow conditions for the problem at hand. As the number of elements increases (as is necessary for flows with complex boundaries), the number of simultaneous algebraic equations that must be solved increases rapidly. Problems involving one million to ten million (or more) elements are not uncommon in today's CFD community, particularly for complex three-dimensional geometries. The finite element method is extremely popular in computational solid mechanics but is not as widely used in CFD. Further information about this method can be found in Refs. 1 and 2.

For the *boundary element method*, the boundary of the flow field (not the entire flow field as in the finite element method) is broken into discrete segments (Ref. 3) and appropriate singularities such as sources, sinks, doublets, and vortices are distributed along these boundary elements. The strengths and type of the singularities are chosen so that the appropriate boundary conditions are obtained on the boundary elements. For points in the flow field not on the boundary, values for the quantities of interest are computed by adding the contributions from the singularities on the boundaries. Although the details of this method are mathematically sophisticated, it may (depending on the particular problem) require less computational time and space than the finite element method. Typical boundary elements and their associated singularities (vortices) for two-dimensional flow past an airfoil are shown in **Fig. A.1**. Such use of the boundary element method in aerodynamics is often termed the *panel method* in recognition of the fact that each element plays the role of a panel on the airfoil surface (Ref. 4). Because the boundary element method simulates the flow field using sources, sinks, and vortices, it can only be used to simulate inviscid (potential) flows.

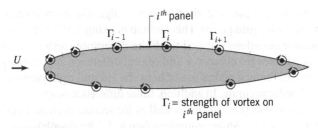

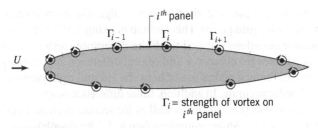

Γ_i = strength of vortex on
i^{th} panel

FIGURE A.1 Panel method for flow past an airfoil.

The *finite difference method* for computational fluid dynamics is perhaps the most easily understood of the four methods listed above. The finite difference method was used somewhat informally in Example AE.1. For this method the flow field is dissected into a set of grid points and the continuous functions (velocity, pressure, etc.) are approximated by discrete values of these functions calculated at the grid points. Derivatives of the functions are approximated using the differences between the function values at local grid points divided by the grid spacing. Using a Taylor series expansion to relate the value of a variable at one point to its value at a neighboring point provides a strong mathematical foundation for developing the required approximations and for understanding the limitations of the approximation (see Ref. 5).

For example, consider a rectangular grid applied to a flow domain as shown in **Fig. A.2**.

The grid stencil shows five grid points in *x*–*y* space with the center point being labeled as *i, j*. Note that the PNESW notation of Example AE.1 has been replaced by double subscript notation. The index notation is used as subscripts on variables to signify location. For example, $u_{i+1,j}$ is the *u* component of velocity at the first point to the right of the center point *i, j*. The grid spacing in the *i* and *j* directions is given as Δx and Δy, respectively.

To find an algebraic approximation to a first derivative term such as $\partial u/\partial x$ at the *i, j* grid point, consider a Taylor series expansion written for *u* at *i* + 1 as

$$u_{i+1,j} = u_{i,j} + \left(\frac{\partial u}{\partial x}\right)_{i,j}\frac{\Delta x}{1!} + \left(\frac{\partial^2 u}{\partial x^2}\right)_{i,j}\frac{(\Delta x)^2}{2!} + \left(\frac{\partial^3 u}{\partial x^3}\right)_{i,j}\frac{(\Delta x)^3}{3!} + \cdots \qquad (A.1)$$

Solving for the underlined term in the above equation results in the following:

$$\left(\frac{\partial u}{\partial x}\right)_{i,j} = \frac{u_{i+1,j} - u_{i,j}}{\Delta x} + O(\Delta x) \qquad (A.2)$$

where $O(\Delta x)$ contains higher order terms proportional to Δx, $(\Delta x)^2$, and so forth. Equation A.2 represents a forward difference equation to approximate the first derivative using values at *i* + 1, *j* and *i, j* along with the grid spacing in the *x* direction. Obviously in solving for the $\partial u/\partial x$ term we have ignored higher order terms such as the second and third derivatives present in Eq. A.1. This process is termed *truncation* of the Taylor series expansion. The lowest order term that was truncated included $(\Delta x)^2$. Notice that the first derivative term contains Δx. When solving for the first derivative, all terms on the right-hand side were divided by Δx. Therefore, the term $O(\Delta x)$ signifies that this equation has error of "order (Δx)," which is due to the neglected terms in the Taylor series and is called truncation error. Hence, the forward difference is termed first-order accurate.

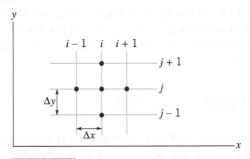

FIGURE A.2 Standard rectangular grid.

Thus, we can transform a partial derivative into an algebraic expression involving values of the variable at neighboring grid points. This method of using the Taylor series expansions to obtain discrete algebraic equations is called the finite difference method. Similar procedures can be used to develop backward difference and central difference approximations of the first derivative. The central difference makes use of both the left and right points (i.e., $i - 1, j$ and $i + 1, j$) and is second-order accurate. In addition, finite difference equations can be developed for the other spatial directions (i.e., $\partial u/\partial y$) as well as for second derivatives ($\partial^2 u/\partial x^2$), which are also contained in the Navier–Stokes equations (see Ref. 5 for details).

Applying this method to all terms in the governing equations approximates the differential equations as a set of algebraic equations involving the physical variables at the grid points (i.e., $u_{i,j}$, $p_{i,j}$ for $i = 1, 2, 3, \ldots$ and $j = 1, 2, 3, \ldots$, etc.). This set of equations is then solved by appropriate numerical techniques. The larger the number of grid points used, the larger the number of equations that must be solved. For further information on the finite difference method, see Refs. 5 and 6.

The concept behind the *finite volume method* might be obvious to students of fluid mechanics. In this method, the flow field is divided into a large number of tiny control volumes. In the formal development of the method, the continuity and Navier–Stokes equations are integrated over a volume and a mathematical expression called the Gauss divergence theorem is used to convert the volume integrals of momentum and shear stress into surface integrals. The resulting algebraic equation(s) are identical to those we would obtain by applying the continuity and momentum equations from Chapter 5 using the tiny control volume. Early finite volume codes computed the velocity at the faces of the control volume and the pressure at the center and were more-or-less restricted to rectangular (in 2-D) and rectangular parallelepiped (in 3-D) volumes. More recent codes compute both velocity and pressure at the center of the volume and can use triangular (2-D) and tetrahedral (3-D) volumes to better fit irregular geometries. As in the finite element and finite difference methods, applying the finite volume formulation results in a large number of algebraic equations for the field variables. See Refs. 7 and 8 for further details of the finite volume method.

If the flow is unsteady, it is necessary to discretize time as well as space. Time appears in the governing equations only as a first derivative and a simple first-order forward difference is most often used

$$\partial u/\partial t \approx \left(u_{i,j}^{n+1} - u_{i,j}^{n} \right)/\Delta t \tag{A.3}$$

where superscripts identify the time level and Δt is the time step. The most significant issue in unsteady problems is not discretizing the time derivative; it is whether to evaluate the remainder of the terms at the old time level (n) (an *explicit method*) or the new time level ($n+1$) (an *implicit method*.) There are advantages and disadvantages in making one choice or the other and there are very significant consequences with respect to the numerical methods to be employed.

A student of CFD should realize that the discretization of the governing equations yields algebraic equations that are an approximation to the original partial differential equation. Along with this approximation comes some amount of error. This type of error is termed truncation error because the Taylor series expansion used to represent a derivative is "truncated" at some reasonable point and the higher order terms are ignored. The truncation errors tend to zero as the grid is refined by making Δx and Δy smaller, so grid refinement is one method of reducing this type of error. Another type of unavoidable numerical error is the so-called round-off error. This type of error is due the limitation that only a finite number of digits can be used to represent the value of a quantity. Engineering students can run into round-off errors from their calculators if they plug values into the equations at an early stage of the solution process. Fortunately, for most CFD cases, if the algorithm is set up properly, round-off errors are usually negligible.

A.4 THE COMPUTATIONAL GRID

The result of a CFD simulation is a set of values for variables of interest (e.g., velocity, pressure) at a finite number of points in the flow domain. The process by which the location of these points is defined varies with the discretization being employed. For example, a finite-difference model directly defines the locations of the points in space. It is frequently useful to connect these points with lines to form a "mesh" or "grid" that covers the flow domain. We often refer to the process of defining the points as "generating" a mesh or grid. A finite-volume discretization requires that the flow field be divided into small volumes. In this case, the boundaries of the volumes define the mesh. Again, the process of defining the volumes is usually called grid or mesh generation. As will be explored in more detail in a following section, points at which the values of variables are computed may, or may not, be at the intersection of grid lines. Discretization of the flow field for a finite element model is again referred to as mesh generation, again the mesh is formed by element boundaries, and again the points at which values of variables are computed may or may not be the intersections of grid lines. The type of grid developed for a given problem can have a significant impact on the numerical simulation, including the accuracy of the solution. The grid must represent the geometry correctly and accurately, since an error in this representation can have a significant effect on the solution.

The grid must also have sufficient resolution to capture the relevant flow physics, otherwise they will be lost. This particular requirement is problem dependent. For example, if a flow field has small-scale structures, the grid resolution must be sufficient to capture these structures. It is usually necessary to increase the number of grid points or volumes (i.e., use a finer mesh) where large gradients are to be expected, such as in the boundary layer near a solid surface. The same can also be said for the temporal resolution. The time step, Δt, used for unsteady flows must be smaller than the smallest time scale of the flow features being investigated.

Generally, the types of grids fall into two categories: structured and unstructured, depending on whether or not there exists a systematic pattern of connectivity of the grid points with their neighbors. As the name implies, a *structured grid* has some type of regular, coherent structure to the mesh layout that can be defined mathematically. The simplest structured grid is a uniform rectangular grid, as shown in **Fig. A.3a** and used in Example A.1. However, structured grids are not restricted to rectangular geometries. **Fig. A.3b** shows a structured grid wrapped around a parabolic surface. Notice that grid points are clustered near the surface (i.e., grid spacing in normal direction decreases as one moves toward the surface) to help capture the steep flow gradients found in the boundary layer region. This type of variable grid spacing is used wherever there is a need to increase grid resolution and is termed grid stretching.

For the *unstructured grid*, the grid cell arrangement is irregular and has no systematic pattern. The grid cell geometry usually consists of various-sized triangles for two-dimensional problems and tetrahedrals for three-dimensional grids. An example of an unstructured grid is shown in **Fig. A.4**. Unlike structured grids, for an unstructured grid each grid cell and the connection information to neighboring cells is defined separately. This increases the computer code's complexity as well as the memory space required. The advantage to an unstructured grid is that it can be applied to complex geometries, where structured grids would have difficulty. The finite difference method is usually restricted to structured grids whereas the finite volume and finite element methods can use either structured or unstructured grids.

Other grids include hybrid, moving, and adaptive grids. A grid that uses a combination of grid elements (rectangles, triangles, etc.) is termed a *hybrid grid*. As the name implies, the *moving grid* is helpful for flows involving a time-dependent geometry. If, for example, the

VA.1 Dynamic grid

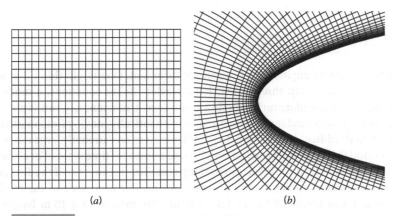

FIGURE A.3 Structured grids. (*a*) Rectangular grid. (*b*) Grid around a parabolic surface.

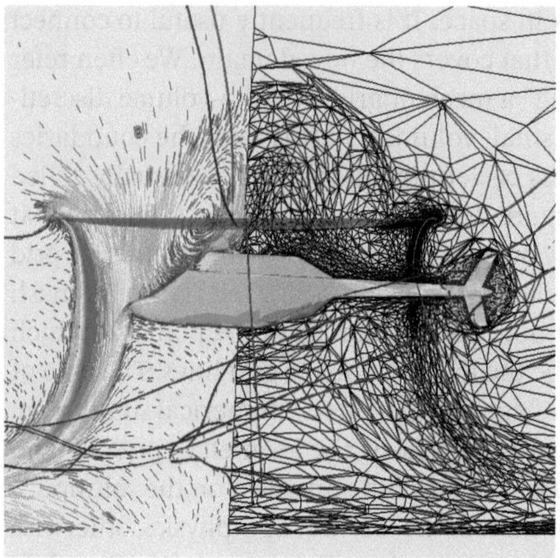

FIGURE A.4 Anisotropic adaptive mesh for flow induced by the rotor of a helicopter in hover, with ground effect. Left: flow; Right: grid.

problem involves simulating the flow within a pumping heart or the flow around a flapping wing, a mesh that moves with the geometry is desired. The nature of the *adaptive grid* lies in its ability to literally adapt itself during the simulation. For this type of grid, while the CFD code is working toward a converged solution, the grid will adapt itself to place additional grid resources in regions of steep gradients. Such a grid is particularly useful when a new problem arises and the user is not sure where to refine the grid to capture steep gradients.

A.5 BOUNDARY CONDITIONS

The Navier–Stokes equations (see Eqs. 6.127a, b, c), are valid for all Newtonian fluid flow problems. Thus, if the same equations are solved for all types of problems, how is it possible to achieve different solutions for different types of flows involving different flow geometries? The answer lies in the *boundary conditions* of the problem. The boundary conditions are what distinguishes one flow field from another (for example, flow past an automobile and flow past a person running) and result in a solution unique to the specified flow geometry.

It is critical to specify the correct boundary conditions so that the CFD simulation is a well-posed problem and an accurate representation of the physical problem. Poorly defined boundary conditions can affect the accuracy of the solution. One of the most common boundary conditions used for simulation of viscous flow is the no-slip condition, as discussed in Section 1.6. Thus, for example, for two-dimensional external or internal flows, the x and y components of velocity (u and v) are set to zero at the stationary wall to satisfy the no-slip condition. Other boundary conditions that must be appropriately specified involve inlets, outlets, far-field, wall gradients, etc. It is important to select the correct physical boundary condition for the problem and to correctly implement this boundary condition in the numerical simulation.

A.6 TURBULENCE MODELS

The majority of flows of engineering interest are turbulent. One might think that with the ability to solve the unsteady, three-dimensional Navier–Stokes equations, a modern CFD code would be able to calculate turbulent flows with ease. Unfortunately, this is not the case. Turbulence consists of an indeterminate number of chaotically moving fluid masses, which are loosely termed "eddies" (see Section 8.3). Eddies exist in an extremely large range of sizes (called *scales*) and their motion has an extremely large number of time scales. For air flowing at 10 m/s in a 0.1-m-diameter duct, the size of the largest eddies are the order of the duct diameter (0.1 m) and the smallest eddies are the order of 10^{-5} m. The time scale of the motion varies from about 1 s to about 10^{-4} s. To fully capture the motion in a 10-m length of this duct would require a computation with 10^{11} grid points (control volumes) and at least 10^{5} time steps.

Fortunately, in most circumstances engineers have no need for that level of detail; information on the average velocities, pressure, and shear stress is sufficient. As explained in Section 8.3, we tackle the problem by breaking the velocity and other flow variables into the sum of an average value and a fluctuating component

$$u = \bar{u} + u' \quad \text{where} \quad \bar{u} = \frac{1}{T}\int_t^{t+T} u\, dt \quad \text{and} \quad \overline{u'} = 0$$

To develop a set of equations for computing the average quantities, we substitute this formulation for the quantities of interest into the continuity and Navier–Stokes equations and average the equations (steady in the mean, incompressible flow is assumed here). The result is

$$\frac{\partial \bar{u}}{\partial x} + \frac{\partial \bar{v}}{\partial y} + \frac{\partial \bar{w}}{\partial z} = 0 \tag{A.4}$$

$$\rho\left(\bar{u}\frac{\partial \bar{u}}{\partial x} + \bar{v}\frac{\partial \bar{u}}{\partial y} + \bar{w}\frac{\partial \bar{u}}{\partial z}\right) = -\frac{\partial p}{\partial x} + \frac{\partial}{\partial x}\left(\mu\frac{\partial \bar{u}}{\partial x} - \rho\overline{u'u'}\right) + \frac{\partial}{\partial y}\left(\mu\frac{\partial \bar{u}}{\partial y} - \rho\overline{u'v'}\right)$$
$$+ \frac{\partial}{\partial z}\left(\mu\frac{\partial \bar{u}}{\partial z} - \rho\overline{u'w'}\right) \tag{A.5a}$$

$$\rho\left(\bar{u}\frac{\partial \bar{v}}{\partial x} + \bar{v}\frac{\partial \bar{v}}{\partial y} + \bar{w}\frac{\partial \bar{u}}{\partial z}\right) = -\frac{\partial p}{\partial y} + \frac{\partial}{\partial x}\left(\mu\frac{\partial \bar{v}}{\partial x} - \rho\overline{u'v'}\right) + \frac{\partial}{\partial y}\left(\mu\frac{\partial \bar{v}}{\partial y} - \rho\overline{v'v'}\right)$$
$$+ \frac{\partial}{\partial z}\left(\mu\frac{\partial \bar{v}}{\partial z} - \rho\overline{v'w'}\right) \tag{A.5b}$$

$$\rho\left(\bar{u}\frac{\partial \bar{w}}{\partial x} + \bar{v}\frac{\partial \bar{w}}{\partial y} + \bar{w}\frac{\partial \bar{w}}{\partial z}\right) = -\frac{\partial p}{\partial z} + \frac{\partial}{\partial x}\left(\mu\frac{\partial \bar{w}}{\partial x} - \rho\overline{u'w'}\right) + \frac{\partial}{\partial y}\left(\mu\frac{\partial \bar{w}}{\partial y} - \rho\overline{v'w'}\right)$$
$$+ \frac{\partial}{\partial z}\left(\mu\frac{\partial \bar{w}}{\partial z} - \rho\overline{w'w'}\right) \tag{A.5c}$$

The last three equations are called the *Reynolds Averaged Navier–Stokes (RANS) equations*. The nine extra terms on the right side ($-\rho\overline{u'v'}$, $-\rho\overline{v'^2}$, etc.) are called *Reynolds Stresses* (see Section 8.3).

At first glance it would seem that the problem of computing average quantities in turbulent flow is straightforward; all we need to do is solve the RANS equations. A second look reveals an enormous problem—we have introduced six new unknowns into the problem (the Reynolds stresses) but we have no new equations! This is the *closure problem* of turbulent flow. Just like it is beyond current capability to solve the full unsteady Navier–Stokes equations for a turbulent flow, it is beyond current capability to calculate the Reynolds stresses. (A little reflection will reveal both are really the same task.) The only path forward is to construct and use a *turbulence model*; a mathematical relationship for the estimation of turbulence properties (in this case, the Reynolds stresses) in terms of available characteristics of the flow field. These models are typically based on plausible assumptions about the physics of turbulence and always require input from experimental measurements.

Many different turbulence models have been proposed and most CFD codes permit the user to choose among several (typically about 10 choices are available). Many models are based on the idea of an *eddy viscosity*, η, defined by a relationship of the form (see Section 8.3.2 and Eq. 8.27):

$$\tau_{\text{turb},xy} = -\rho\overline{u'v'} = \eta\left(\frac{\partial \bar{u}}{\partial y} + \frac{\partial \bar{v}}{\partial x}\right)$$

Note that this is not a physical law and the eddy viscosity is not a fluid property; it varies strongly with flow conditions. The core assumption in the model is that, at any point, the stress is proportional to the mean strain rate.

The eddy viscosity is typically constructed from a relationship based on dimensional reasoning

$$\eta = \rho \times \textit{length scale} \times \textit{velocity scale}$$

Eddy viscosity (and other) turbulence models are characterized by the number of differential equations that must be solved to determine the eddy viscosity or the Reynolds stresses. The most common example of a *zero equation model* is the mixing length model (see Eq. 8.28) where

$$\textit{length scale} = \ell_m \quad \text{and} \quad \textit{velocity scale} = \ell_m\left|\frac{\partial \bar{u}}{\partial y}\right|$$

In this model, the mixing length must be prescribed in terms of the dimensions of the problem and/or the characteristics of the mean flow.

The most widely used turbulence model in engineering CFD is the two equation $k - \varepsilon$ model, where k is the *turbulence kinetic energy* ($k = \frac{1}{2}(\overline{u'^2} + \overline{v'^2} + \overline{w'^2})$) and ε is the dissipation of mechanical energy caused by the turbulence, $\varepsilon = \rho \dfrac{d(\text{loss})}{dt}$. The eddy viscosity is computed using

$$\eta = \rho C_\eta \frac{k^2}{\varepsilon}$$

where C_η is a constant determined from experimental data. Values for k and ε are computed by solving partial differential equations of the form

$$\overline{u}\frac{\partial \phi}{\partial x} + \overline{v}\frac{\partial \phi}{\partial y} + \overline{w}\frac{\partial \phi}{\partial z} = Diffusion\ of\ \phi + Production\ of\ \phi - Destruction\ of\ \phi$$

where ϕ stands for k or ε. The three terms on the right side involve averages of second- and third-order products of fluctuating quantities, which themselves must be modeled.

There are a number of other turbulence models available. Another popular two-equation model uses the "specific dissipation," ω ($\omega = \varepsilon/k$), in place of ε. In addition to eddy viscosity models, there are Reynolds stress models that solve equations, both differential and algebraic, for the Reynolds stresses themselves. There are also *large eddy simulation* models, which actually compute the details of the larger eddies and model the smaller scales.

Most CFD codes use a special method for applying boundary conditions for turbulent flow near a wall. In order to apply the no-slip condition at the wall, an extremely fine grid would be required to properly resolve the viscous sublayer (see Section 8.3.3). To avoid this requirement, so-called *wall functions* are used. The first grid point away from the wall is assumed to lie in the overlap region and Eq. 8.30 is used to determine the wall shear stress

$$\frac{u_p}{\left(\frac{\tau_w}{\rho}\right)^{\frac{1}{2}}} = 2.44 \ln\left(\frac{y_p}{\upsilon}\left[\frac{\tau_w}{\rho}\right]^{\frac{1}{2}}\right) + 5.0$$

For each of these models of turbulent flow, the model's form is somewhat ad hoc, several constants like C_η require input from experimental data, and the model works well for some cases but not for others. It should be apparent that turbulence modeling keeps CFD from being an exact science and demands both intelligent choices and a certain degree of skepticism from the user of CFD. See (Ref. 9) for further information about turbulence models and wall functions.

A.7   SOLVING THE EQUATIONS

After the grid is defined, the partial differential equations discretized (including any PDEs used to model turbulence), and the boundary conditions specified, a set of algebraic equations remains to be solved. If the flow is modeled as inviscid, with either velocity potential or stream function as the unknown, the algebraic equations will be linear. For viscous flows, modeled by either the Navier–Stokes equations or by the boundary layer approximation, the equations will be nonlinear and, in order to solve them, the equations must be linearized in some way. The nonlinearity is typically handled by iteration, requiring that the linearized equations be solved at each iteration step.

In most instances, the number of equations to be solved will be quite large, between three and ten equations for each grid point or volume, with thousands of grid points.[1] No matter how large, the system of equations can be written in matrix notation as

$$\mathbf{Ax} = \mathbf{b}$$

---

[1]An exception is the boundary element method, where the number of elements, hence equations, is typically the order of tens or hundreds.

where $\mathbf{A}$ is the $n \times n$ matrix of coefficients, n is the number of equations, $\mathbf{x}$ is a column vector of unknowns such as u, v, k, and $\mathbf{b}$ is a column vector of known values derived from problem geometry, boundary conditions, fluid properties, etc.

In general, there are two methods for solving systems of linear algebraic equations: direct methods and iterative methods. A direct method solves the equations in a definite number of steps, which can be determined from the number of equations. Probably the most well-known direct method is Gaussian elimination. In this method, the matrix is converted to an upper triangular matrix (all zeros below the diagonal) by successively adding or subtracting multiples of rows to eliminate the coefficients below the diagonal. After the upper triangular matrix is constructed, the equations can be solved "bottom-to-top" by back substitution.

Another well-known direct method is to calculate the inverse of matrix $\mathbf{A}$, $\mathbf{A}^{-1}$. Determining the inverse of a matrix requires several steps. Many "canned" software programs and even spreadsheets can be used to determine the inverse of matrices of moderate size. The equations are then solved by

$$\mathbf{x} = \mathbf{A}^{-1}\mathbf{b}$$

There are many other methods for direct solution of systems of equations; consult a linear algebra textbook (Ref. 10) for more information.

Direct solution methods are seldom used in CFD codes. Although the deterministic character of a direct method may be appealing, these methods require a very large number of arithmetic operations and are subject to problems such as the propagation of round-off error. The matrices that arise in CFD are generally very large and very sparse, meaning that any row or column contains a large number of zeros.[2] A direct method would waste a considerable amount of computer time and storage by manipulating zeros.

Iterative methods are generally more efficient in CFD codes. An advantage of iterative methods is that the iterations required because the discretized equations are nonlinear are done simultaneously with the iterations for solving the system of equations.

In general, the steps in an iterative method for solving $\mathbf{Ax} = \mathbf{b}$ are:

- Rearrange the equations so that for each, one element from $\mathbf{x}$ appears by itself on the left side of the equation and the right side is populated with other elements of $\mathbf{x}$, elements of $\mathbf{A}$, and elements of $\mathbf{b}$. Different iterative methods will produce different right-side expressions.
- Compute a new value (a new iterate) for each element of $\mathbf{x}$.
- If the change between iterates is sufficiently small by some measure, an approximate solution has been computed. If the change is not sufficiently small, repeat the two preceding steps until it is sufficiently small (known as convergence) or the solution process fails.

In Example A.1, we used the Gauss–Sidel (iterative) method where the most recently available values of the unknowns are used on the right side. The *Jacobi method* substitutes new values on the right side only after a complete pass through the entire system of equations.

An obvious question is "How do we know when to stop iterating?" With a small system of equations, such as might be solved in a spreadsheet, the iterations are usually stopped when the largest change in any unknown is less than some value, say 0.01%. With a spreadsheet that visually displays the results of each iteration, the iterations may be continued until no change in any variable is observed, which means the change is smaller than the number of digits being displayed. In a large system, such as occurs in a CFD calculation, a more complex method is used. At any step in the iteration, the current set of variables, say u_i, v_i, and k_i can be substituted into the discretized equations, in this case, the x-momentum, y-momentum, and turbulence model equations. Because the solution is not exact, each equation at each point will not be satisfied; there will be a *residual* r_i (for, say, the x-momentum equation) $r_{x,i} = (right\ side\ of\ x\text{-}momentum\ equation) - (left\ side\ of\ x\text{-}momentum\ equation)$. A total residual is calculated from $R_x = \sum_{i=1}^{n} |r_{x,i}|$.

---

[2]Again, the boundary element method is an exception.

The total residual for each equation is tracked and when the largest of them drops to an acceptable level, the iteration is considered to have converged. It is not unusual for this process to require hundreds or even thousands of iterations.

Sometimes, it is desirable to speed up an iteration to achieve convergence more rapidly, at other times an iteration must be slowed to avoid instability. This is done using a process called *relaxation*. At any iteration step, we can write

$$u_i = \lambda u_i^{\text{new}} + (1 - \lambda) u_i^{\text{old}}$$

where u_i is any variable and λ is a *relaxation parameter*. If $0 \leq \lambda \leq 1$ the calculation is said to be *underrelaxed* and if $1 \leq \lambda \leq 2$ the calculation is said to be *overrelaxed*. The choice of λ for a given simulation is usually made by trial and error.

Finally, some CFD codes use a *multigrid method* to speed convergence. In this method, iterations are done alternately on finer and coarser grids. For details, see Refs. (7 and 8).

A.8   SOME UNEXPECTED COMPLICATIONS

The process of developing a CFD code requires many choices. The developer must choose a discretization method, a type of grid, turbulence model(s), and more. For example, the seemingly simple task of approximating the first derivative, $\frac{\partial \varphi}{\partial x}$, requires choosing between a forward difference, a backward difference, a central difference, a three-point forward difference (see Example A.1), a three-point backward difference—the list goes on. None of these approximations is intrinsically correct; if one takes the limit as Δx goes to zero, all of the approximations become the (exact) first derivative. Clearly the best choices are those that produce the most accurate flow predictions. Different choices yield different sets of equations to solve and one should expect to get different answers. Usually this would be acceptable because the answers are approximations anyway.

Sometimes different choices have more serious consequences; some seemingly correct approximations may not work at all. "Not working" can take on several forms: failure to converge; instability; producing nonphysical results. CFD has provento be especially sensitive to the selection of proper formulations, and several "obvious" approaches have proven to be totally unsatisfactory. This section discusses some of the special formulations that were discovered only by trial and error.

Conservative Form   The (Reynolds averaged) Navier–Stokes equations (Eqs. 6.127a, b, c; A.5a, b, c) are said to be in *nonconservative form*. Recall that these equations were derived by applying Newton's Second Law to a fluid element (see Fig. 6.11). If we add the continuity equation in the form of Eq. 6.31 to each of Eqs. 127a, b, c and gather terms, the equations can be written (shown here for the x-component only) as

$$\frac{\partial(\rho u)}{\partial t} + \frac{\partial}{\partial x}(\rho u^2) + \frac{\partial}{\partial y}(\rho u v) + \frac{\partial}{\partial z}(\rho u w) = -\frac{\partial p}{\partial x} + \mu \left(\frac{\partial^2 u}{\partial x^2} + \frac{\partial^2 u}{\partial y^2} + \frac{\partial^2 u}{\partial z^2} \right) \qquad \text{(A.6)}$$

Equation A.6 is the x-momentum equation in *conservative form* and is the form of the equation we would get by applying the momentum equation to an infinitesimal control volume instead of an infinitesimal fluid element. Experience has shown that in some discretization schemes the conservative form of the equations must be used or the calculations will not converge due to failure to conserve mass, momentum, and energy throughout the flow field.

Staggered Grid   In CFD, the purpose of the grid is to define a number of discrete locations where the flow variables (u, p, and so on) are to be computed. In the finite difference method, these locations are typically at the intersection of the grid lines, and in the finite volume method, the obvious location is the center of the finite volume. In Example A.1, the grid points were defined by the intersection of the grid lines, and the stream function was calculated at those points. When solving the more complicated Navier–Stokes equations, one's natural inclination would be to define and calculate all variables (u, v, w, p, k, ε) at the same location. Like many other "obvious" choices in CFD, collocating all variables at a single

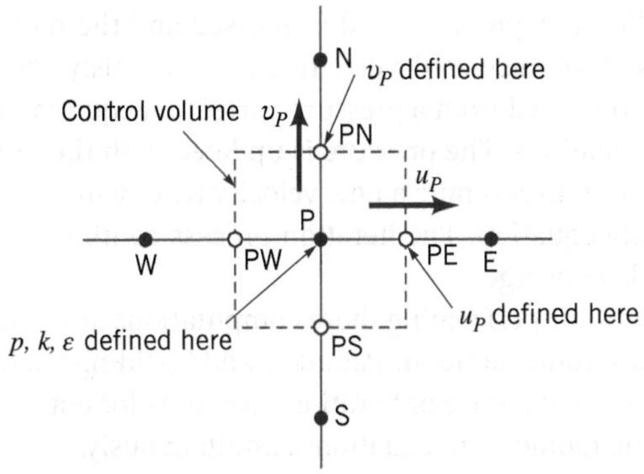

FIGURE A.5 Staggered grid arrangement.

point can lead to serious difficulties; in this case calculating an unrealistic pressure field. One solution to this difficulty is to use a *staggered grid*. Consider the two-dimensional grid layout shown in **Fig. A.5**. The finite difference method approximates the PDEs at point P; the finite volume method considers mass and momentum balances for the control volume surrounding point P. In a staggered grid, scalar quantities such as pressure and turbulence kinetic energy are defined and computed at point P, the "center" of the grid element, but velocities are defined and computed at the "halfway" locations PE and PN (i.e., at the faces of the surrounding control volume). Note that the fluid velocities at points PW and PS are associated with points (volumes) W and S. Many CFD codes use the staggered grid but others choose to use higher-order difference approximations on a collocated grid to avoid challenges that arise for the staggered-grid such as accurate modeling of complex geometries.

Upwind Differencing Consider the transport of some property φ out of the control volume shown in Fig. A.5 (φ could be temperature, momentum, turbulence kinetic energy, or any other property that is carried by the flow). Also, suppose the flow is from left to right. The obvious discrete expression for the flow of φ (per unit area) out from the control volume is $\rho u_P \left(\frac{\varphi_P + \varphi_E}{2} \right)$, where the term in parentheses approximates the value of φ at the control volume face. This formulation is called a central difference. At times this "obvious" formulation leads to difficulties ("wiggles" in the solution). Imagine that the flow is very strong. It stands to reason that the value of φ upstream (upwind) of the control volume face would have a stronger influence on the value of φ there than the value downstream. This effect is simulated by using only φ_P in place of $\frac{\varphi_P + \varphi_E}{2}$. On the other hand, if the flow is from right to left (u_P negative), then only φ_E would be used. This strategy is called *upwind differencing* or *donor cell differencing*.[3] It effectively eliminates "wiggles" in the flow solution.

Upwind differencing does have a negative side. It tends to smooth out or "smear" sharp changes in the flow, an effect called *false diffusion*. This can be mostly eliminated by blending an upwind difference and a central difference, based on the strength of the flow transport.

Pressure-Velocity Coupling The (Reynolds averaged) Navier–Stokes equations together with the continuity equation (Eqs. 6.27; 6.127a, b, c; A4; A.5a, b, c) are coupled partial differential equations; in particular, all components of velocity appear in all of the equations. After the equations are discretized, each variable is computed from a different equation. The obvious approach is to compute u using the x-momentum equation, v using the y-momentum equation, and w using the z-momentum equation. This leaves the continuity equation available to compute p—but pressure does not appear in the continuity equation![4] This knotty problem is usually solved by a *pressure correction algorithm*. This is an iterative method that

---

[3] One of the early developers of upwind differencing has called it the "pigpen method," noting that the odor is much stronger downwind from a pig pen than it is upwind!

[4] If the flow is compressible, density ρ does appear in the continuity equation and p is related to ρ and temperature. The energy equation must be included to calculate temperature, and continuity can be used to compute density and pressure then determined. Here we are considering incompressible flow.

generally proceeds as follows. A pressure field is guessed and the momentum equations are solved for the velocities. The calculated velocities do not satisfy the continuity equation. A pressure correction is computed from a pressure correction equation derived from the continuity and momentum equations. The pressure is updated with the pressure correction, and the new pressure field is used to compute a new velocity field, which, hopefully, comes closer to satisfying the continuity equation. The iteration process continues until both the velocity field and the pressure field converge.

Many of the algorithms for performing these computations are based on or are similar to the SIMPLE algorithm (an ironic name) of Patankar and Spalding (Ref. 11). One helpful feature of this method is that the sequence of iterations accounts for both the pressure correction and the nonlinearity of the momentum equations simultaneously.

A.9 VERIFICATION AND VALIDATION

Verification and validation of the simulation are critical steps in the CFD process. This is a necessary requirement for CFD, particularly because it is possible to have a converged solution that is nonphysical. Verification seeks to ensure that the software is correct. That means that it correctly implements the algorithms, models, and other features as intended by the author(s). It does not ensure that the simulation produced by the software is a good representation of a real flow. Validation seeks to ensure that the simulation of the flow of interest is indeed a good representation of that flow. This is frequently accomplished by comparison of the software's simulations to data from other sources, such as experiment measurements, for a flow that it similar to the flow of interest.

Fig. A.6 shows the streamlines for viscous flow past a circular cylinder at a specific instant after it was impulsively started from rest. The lower half of the figure represents the results of a finite difference calculation; the upper half of the figure represents the photograph from an experiment of the same flow situation. It is clear that the numerical and experimental results agree quite well. For any CFD simulation, several levels of testing need to be accomplished before one can have confidence in the solution. The most important verification to be performed is a grid convergence study. In its simplest form, it consists of proving that further refinement of the grid (i.e., increasing the number of grid points) does not alter the final solution. When this has been achieved, you have a grid-independent solution. Other verification factors that need to be investigated include the suitability of the convergence criterion, whether the time step is adequate for the time scale of the problem, and comparison of CFD solutions to existing data, at least for baseline cases. Even when using a commercial CFD code that has been validated on many problems in the past, the CFD practitioner still needs to verify the results through such measures as grid-dependence testing.

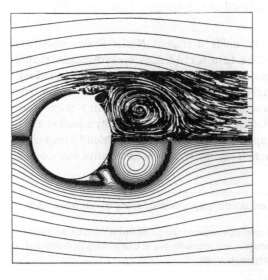

FIGURE A.6 Streamlines for flow past a circular cylinder at a short time after the flow was impulsively started. The upper half is a photograph from a flow visualization experiment. The lower half is from a finite difference calculation. (Source: From Hall, E. J., and Pletcher, R. H., Simulation of Time Dependent, Compressible Viscous Flow Using Central and Upwind-Biased Finite-Difference Techniques, Technical Report HTL-52, CFD-22, College of Engineering, Iowa State University, 1990.)

A.10 APPLICATION OF CFD

In the early stages of CFD, research and development were primarily driven by the aerospace and nuclear industries. Today, CFD is still used as a research tool, but it also has found a place in industry as a design tool. There is now a wide variety of industries that make at least some use of CFD, including automotive, power generation, HVAC, naval, civil, chemical, biological, and others. Industries are using CFD as an engineering tool that complements experimental and theoretical work in fluid dynamics.

A.10.1 Advantages of CFD

There are many advantages to using CFD for simulation of fluid flow. One of the most important advantages is the realizable savings in time and cost for engineering design. In the past, coming up with a new engineering design often meant somewhat of a trial-and-error method of building and testing multiple prototypes prior to finalizing the design. Using CFD can help engineers significantly improve a design before testing models and before building prototypes. This translates to a significant savings in time and cost. It should be noted that CFD is not meant to replace experimental testing, but rather to work in conjunction with it. Experimental testing is still a necessary component of engineering design. Other advantages include the ability of CFD to: (1) provide flow information in regions that would be difficult to test experimentally, (2) simulate real flow conditions, (3) conduct large parametric tests on new designs in a shorter time, and (4) enhance visualization of complex flow phenomena.

A good example of the advantages of CFD is shown in **Fig. A.7**. Researchers used a type of CFD approach called "large eddy simulation" to simulate the fluid dynamics of a tornado as it encounters a debris field and begins to pick up sand-sized particles. A full animation of this tornado simulation can be accessed by visiting WileyPLUS. The motivation for this work is to investigate whether there are significant differences in the fluid mechanics

VA.2 tornado
simulation

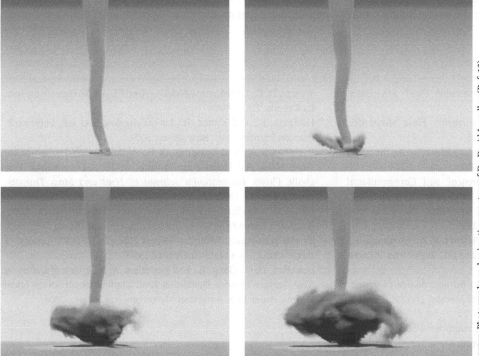

(Source: Photographs and animation courtesy of Dr. David Lewellen (Ref. 13) and Paul Lewellen, West Virginia University.)

FIGURE A.7 Results from a large eddy simulation showing the visual appearance of the debris and funnel cloud from a simulated medium swirl F3-F4 tornado. The funnel cloud is translating at 15 m/s and is ingesting 1-mm-diameter "sand" from the surface as it encounters a debris field. Please visit the book web site to access a full animation of this tornado simulation.

when debris particles are present. Historically it has been difficult to obtain comprehensive experimental data throughout a tornado, so CFD is helping us understand the complex fluid dynamics involved in such a flow.

A.10.2 A Warning

An important point that a beginning CFD user should understand is that one cannot treat the computer as a "magic black box" when performing flow simulations. It is quite possible to obtain a fully converged solution for the CFD simulation, but achieving convergence is no guarantee that the results are physically correct. This is why it is important to have a good understanding of the flow physics and how they are modeled. Any numerical technique (including those discussed above), no matter how simple in concept, contains many hidden subtleties and potential problems. For example, it may seem reasonable that a finer grid would ensure a more accurate numerical solution. While this may be true, it is not always so straightforward; a variety of stability or convergence problems may occur. In such cases the numerical "solution" obtained may exhibit unreasonable oscillations or the numerical result may "diverge" to an unreasonable (and incorrect) result. Other problems that may arise include (but are not limited to): (1) difficulties in dealing with the nonlinear terms of the Navier–Stokes equations, (2) difficulties in modeling turbulent flows, (3) convergence issues, (4) difficulties in obtaining a quality grid for complex geometries, and (5) managing resources, both time and computational, for complex problems such as unsteady three-dimensional flows.

APPENDIX SUMMARY

In CFD, there are many different numerical schemes, grid techniques, etc. Each has advantages and disadvantages. A great deal of care must be used in obtaining approximate numerical solutions to the equations of fluid motion. The process is not as simple as the often-heard "just let the computer do it." Remember that CFD is a tool and as such needs to be used appropriately to produce meaningful results and, in the wrong hands, it may do more harm than good. The general field of computational fluid dynamics, in which computers and numerical methods are combined to solve fluid flow problems, represents an extremely important subject area in advanced fluid mechanics. Considerable progress has been made in the past few years, but much remains to be done. The reader is encouraged to consult the available literature.

REFERENCES

[1] Baker, A. J., *Finite Element Computational Fluid Mechanics*, McGraw-Hill, New York, 1983.

[2] Carey, G. F., and Oden, J. T., *Finite Elements: Fluid Mechanics*, Prentice-Hall, Englewood Cliffs, NJ, 1986.

[3] Brebbia, C. A., and Dominguez, J., *Boundary Elements: An Introductory Course*, McGraw-Hill, New York, 1989.

[4] Moran, J., *An Introduction to Theoretical and Computational Aerodynamics*, Wiley, New York, 1984.

[5] Anderson, J. D., *Computational Fluid Dynamics: The Basics with Applications*, McGraw-Hill, New York, 1995.

[6] Tannehill, J. C., Anderson, D. A., and Pletcher, R. H., *Computational Fluid Mechanics and Heat Transfer*, 2nd Ed., Taylor and Francis, Washington, D.C., 1997.

[7] Versteeg, H., and Malalasekera, W., *An Introduction to Computational Fluid Dynamics: The Finite Volume Method*, 2nd Ed., Pearson Prentice Hall, New Jersey, 2007.

[8] Tu, J., Yeoh, G. H., and Liu, C., *Computational Fluid Dynamics: A Practical Approach*, 2nd Ed., Butterworth-Heinmann, Paperback, United Kingdom, 2012.

[9] Wilcox, D. C., *Turbulence Modeling for CFD*, DCW Industries, Inc., La Canada, CA, 1994.

[10] Hoffman, K., and Kunze, R., *Linear Algebra*, 2nd Ed., Paperback, Pearson Prentice Hall, New Jersey, 2009.

[11] Patankar, S. V., and Spalding, D. B., A Calculation Procedure for Heat, Mass, and Momentum Transfer in Three-Dimensional Parabolic Flows, *International Journal of Heat and Mass Transfer*, Vol. 15, p. 1787, 1972.

[12] Hall, E. J., and Pletcher, R. H., *Simulation of Time Dependent, Compressible Viscous Flow Using Central and Upwind-Biased Finite-Difference Techniques*, Technical Report HTL-52, CFD-22, College of Engineering, Iowa State University, 1990.

[13] Lewellen, D. C., Gong, B., and Lewellen, W. S., *Effects of Debris on Near-Surface Tornado Dynamics*, 22nd Conference on Severe Local Storms, Paper 15.5, American Meteorological Society, 2004.

Physical Properties of Fluids

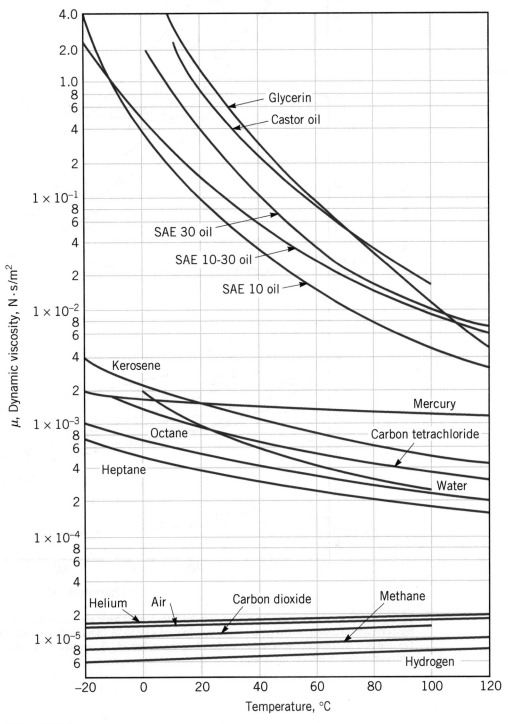

FIGURE B.1 Dynamic (absolute) viscosity of common fluids as a function of temperature. To convert to BG units of lb · s/ft² multiply N · s/m² by 2.089×10^{-2}. (Source: From Curves from R. W. Fox and A. T. McDonald, *Introduction to Fluid Mechanics*, 3rd Ed., Wiley, New York, 1985. Reproduced with permission of John Wiley & Sons.)

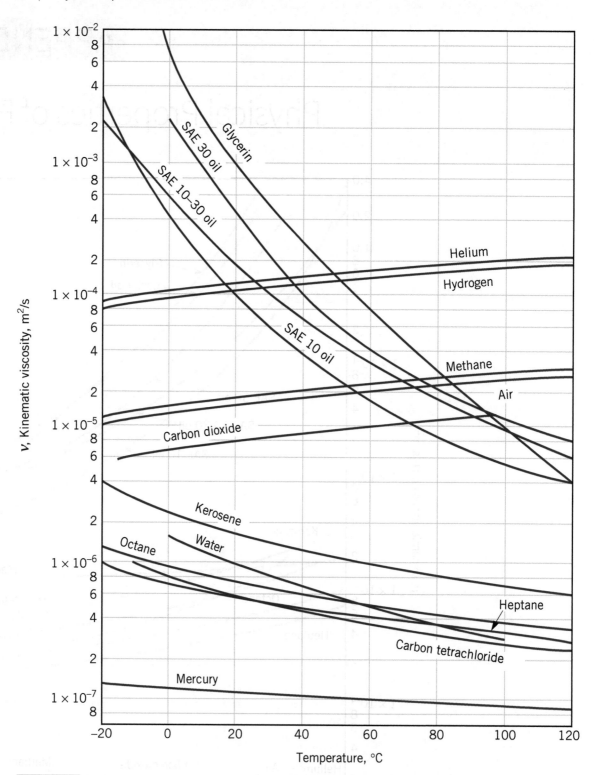

FIGURE B.2 Kinematic viscosity of common fluids (at atmospheric pressure) as a function of temperature. To convert to BG units of ft²/s multiply m²/s by 10.76. (Source: From Curves from R. W. Fox and A. T. McDonald, *Introduction to Fluid Mechanics*, 3rd Ed., Wiley, New York, 1985. Reproduced with permission of John Wiley & Sons.)

TABLE B.1 Physical Properties of Water (SI Units)

Temperature (°C)	Density, ρ (kg/m³)	Specific Weight[a], γ (kN/m³)	Dynamic Viscosity, μ (N · s/m²)	Kinematic Viscosity, ν (m²/s)	Surface Tension, σ (N/m)	Vapor Pressure, p_v [N/m²(abs)]	Speed of Sound, c (m/s)
0	999.9	9.806	1.787 E − 3	1.787 E − 6	7.56 E − 2	6.105 E + 2	1403
5	1000.0	9.807	1.519 E − 3	1.519 E − 6	7.49 E − 2	8.722 E + 2	1427
10	999.7	9.804	1.307 E − 3	1.307 E − 6	7.42 E − 2	1.228 E + 3	1447
20	998.2	9.789	1.002 E − 3	1.004 E − 6	7.28 E − 2	2.338 E + 3	1481
30	995.7	9.765	7.975 E − 4	8.009 E − 7	7.12 E − 2	4.243 E + 3	1507
40	992.2	9.731	6.529 E − 4	6.580 E − 7	6.96 E − 2	7.376 E + 3	1526
50	988.1	9.690	5.468 E − 4	5.534 E − 7	6.79 E − 2	1.233 E + 4	1541
60	983.2	9.642	4.665 E − 4	4.745 E − 7	6.62 E − 2	1.992 E + 4	1552
70	977.8	9.589	4.042 E − 4	4.134 E − 7	6.44 E − 2	3.116 E + 4	1555
80	971.8	9.530	3.547 E − 4	3.650 E − 7	6.26 E − 2	4.734 E + 4	1555
90	965.3	9.467	3.147 E − 4	3.260 E − 7	6.08 E − 2	7.010 E + 4	1550
100	958.4	9.399	2.818 E − 4	2.940 E − 7	5.89 E − 2	1.013 E + 5	1543

Sources: Data from *Handbook of Chemistry and Physics*, 69th Ed., CRC Press, 1988; R. D. Blevins, Applied Fluid Dynamics Handbook, Van Nostrand Reinhold Co., Inc., New York, 1984.
[a]Density and specific weight are related through the equation $\gamma = \rho g$. For this table, $g = 9.807$ m/s².

TABLE B.2 Physical Properties of Water (BG/EE Units)

Temperature (°F)	Density, ρ (slugs/ft³)[a]	Specific Weight[b], γ (lb/ft³)	Dynamic Viscosity, μ (lb · s/ft²)	Kinematic Viscosity, ν (ft²/s)	Surface Tension[c], σ (lb/ft)	Vapor Pressure, p_v [lb/in.²(abs)]	Speed of Sound, c (ft/s)
32	1.940	62.42	3.732 E − 5	1.924 E − 5	5.18 E − 3	8.854 E − 2	4603
40	1.940	62.43	3.228 E − 5	1.664 E − 5	5.13 E − 3	1.217 E − 1	4672
50	1.940	62.41	2.730 E − 5	1.407 E − 5	5.09 E − 3	1.781 E − 1	4748
60	1.938	62.37	2.344 E − 5	1.210 E − 5	5.03 E − 3	2.563 E − 1	4814
70	1.936	62.30	2.037 E − 5	1.052 E − 5	4.97 E − 3	3.631 E − 1	4871
80	1.934	62.22	1.791 E − 5	9.262 E − 6	4.91 E − 3	5.069 E − 1	4819
90	1.931	62.11	1.500 E − 5	8.233 E − 6	4.86 E − 3	6.979 E − 1	4960
100	1.927	62.00	1.423 E − 5	7.383 E − 6	4.79 E − 3	9.493 E − 1	4995
120	1.918	61.71	1.164 E − 5	6.067 E − 6	4.67 E − 3	1.692 E + 0	5049
140	1.908	61.38	9.743 E − 6	5.106 E − 6	4.53 E − 3	2.888 E + 0	5091
160	1.896	61.00	8.315 E − 6	4.385 E − 6	4.40 E − 3	4.736 E + 0	5101
180	1.883	60.58	7.207 E − 6	3.827 E − 6	4.26 E − 3	7.507 E + 0	5195
200	1.869	60.12	6.342 E − 6	3.393 E − 6	4.12 E − 3	1.152 E + 1	5089
212	1.860	59.83	5.886 E − 6	3.165 E − 6	4.04 E − 3	1.469 E + 1	5062

Sources: Data from *Handbook of Chemistry and Physics*, 69th Ed., CRC Press, 1988; R. D. Blevins, Applied Fluid Dynamics Handbook, Van Nostrand Reinhold Co., Inc., New York, 1984.
[a]To obtain EE units (lbm/ft³) multiply by 32.174.
[b]Density and specific weight are related through the equation $\gamma = \rho g$. For this table, $g = 32.174$ ft/s².
[c]In contact with air.

TABLE B.3   Physical Properties of Air at Standard Atmospheric Pressure (SI Units)

Temperature (°C)	Density, ρ (kg/m³)	Specific Weight[a], γ (N/m³)	Dynamic Viscosity, μ (N · s/m²)	Kinematic Viscosity, ν (m²/s)	Specific Heat Ratio, k (—)	Speed of Sound, c (m/s)
−40	1.514	14.85	1.57 E − 5	1.04 E − 5	1.401	306.2
−20	1.395	13.68	1.63 E − 5	1.17 E − 5	1.401	319.1
0	1.292	12.67	1.71 E − 5	1.32 E − 5	1.401	331.4
5	1.269	12.45	1.73 E − 5	1.36 E − 5	1.401	334.4
10	1.247	12.23	1.76 E − 5	1.41 E − 5	1.401	337.4
15	1.225	12.01	1.80 E − 5	1.47 E − 5	1.401	340.4
20	1.204	11.81	1.82 E − 5	1.51 E − 5	1.401	343.3
25	1.184	11.61	1.85 E − 5	1.56 E − 5	1.401	346.3
30	1.165	11.43	1.86 E − 5	1.60 E − 5	1.400	349.1
40	1.127	11.05	1.87 E − 5	1.66 E − 5	1.400	354.7
50	1.109	10.88	1.95 E − 5	1.76 E − 5	1.400	360.3
60	1.060	10.40	1.97 E − 5	1.86 E − 5	1.399	365.7
70	1.029	10.09	2.03 E − 5	1.97 E − 5	1.399	371.2
80	0.9996	9.803	2.07 E − 5	2.07 E − 5	1.399	376.6
90	0.9721	9.533	2.14 E − 5	2.20 E − 5	1.398	381.7
100	0.9461	9.278	2.17 E − 5	2.29 E − 5	1.397	386.9
200	0.7461	7.317	2.53 E − 5	3.39 E − 5	1.390	434.5
300	0.6159	6.040	2.98 E − 5	4.84 E − 5	1.379	476.3
400	0.5243	5.142	3.32 E − 5	6.34 E − 5	1.368	514.1
500	0.4565	4.477	3.64 E − 5	7.97 E − 5	1.357	548.8
1000	0.2772	2.719	5.04 E − 5	1.82 E − 4	1.321	694.8

Source: Data from R. D. Blevins, *Applied Fluid Dynamics Handbook*, Van Nostrand Reinhold Co., Inc., New York, 1984.

[a]Density and specific weight are related through the equation $\gamma = \rho g$. For this table $g = 9.807$ m/s².

TABLE B.4 Physical Properties of Air at Standard Atmospheric Pressure (BG/EE Units)

Temperature (°F)	Density, ρ (slugs/ft^3)[a]		Specific Weight[b], γ (lb/ft^3)		Dynamic Viscosity, μ (lb · s/ft^2)		Kinematic Viscosity, ν (ft^2/s)		Specific Heat Ratio, k (—)	Speed of Sound, c (ft/s)
−40	2.939	E − 3	9.456	E − 2	3.29	E − 7	1.12	E − 4	1.401	1004
−20	2.805	E − 3	9.026	E − 2	3.34	E − 7	1.19	E − 4	1.401	1028
0	2.683	E − 3	8.633	E − 2	3.38	E − 7	1.26	E − 4	1.401	1051
10	2.626	E − 3	8.449	E − 2	3.44	E − 7	1.31	E − 4	1.401	1062
20	2.571	E − 3	8.273	E − 2	3.50	E − 7	1.36	E − 4	1.401	1074
30	2.519	E − 3	8.104	E − 2	3.58	E − 7	1.42	E − 4	1.401	1085
40	2.469	E − 3	7.942	E − 2	3.60	E − 7	1.46	E − 4	1.401	1096
50	2.420	E − 3	7.786	E − 2	3.68	E − 7	1.52	E − 4	1.401	1106
60	2.373	E − 3	7.636	E − 2	3.75	E − 7	1.58	E − 4	1.401	1117
70	2.329	E − 3	7.492	E − 2	3.82	E − 7	1.64	E − 4	1.401	1128
80	2.286	E − 3	7.353	E − 2	3.86	E − 7	1.69	E − 4	1.400	1138
90	2.244	E − 3	7.219	E − 2	3.90	E − 7	1.74	E − 4	1.400	1149
100	2.204	E − 3	7.090	E − 2	3.94	E − 7	1.79	E − 4	1.400	1159
120	2.128	E − 3	6.846	E − 2	4.02	E − 7	1.89	E − 4	1.400	1180
140	2.057	E − 3	6.617	E − 2	4.13	E − 7	2.01	E − 4	1.399	1200
160	1.990	E − 3	6.404	E − 2	4.22	E − 7	2.12	E − 4	1.399	1220
180	1.928	E − 3	6.204	E − 2	4.34	E − 7	2.25	E − 4	1.399	1239
200	1.870	E − 3	6.016	E − 2	4.49	E − 7	2.40	E − 4	1.398	1258
300	1.624	E − 3	5.224	E − 2	4.97	E − 7	3.06	E − 4	1.394	1348
400	1.435	E − 3	4.616	E − 2	5.24	E − 7	3.65	E − 4	1.389	1431
500	1.285	E − 3	4.135	E − 2	5.80	E − 7	4.51	E − 4	1.383	1509
750	1.020	E − 3	3.280	E − 2	6.81	E − 7	6.68	E − 4	1.367	1685
1000	8.445	E − 4	2.717	E − 2	7.85	E − 7	9.30	E − 4	1.351	1839
1500	6.291	E − 4	2.024	E − 2	9.50	E − 7	1.51	E − 3	1.329	2114

Source: Data from R. D. Blevins, *Applied Fluid Dynamics Handbook*, Van Nostrand Reinhold Co., Inc., New York, 1984.
[a]To obtain EE units (lbm/ft^3) multiply by 32.174.
[b]Density and specific weight are related through the equation $\gamma = \rho g$. For this table $g = 32.174$ ft/s^2.

APPENDIX C

Properties of the U.S. Standard Atmosphere

TABLE C.1 Properties of the U.S. Standard Atmosphere (SI Units)

Altitude (m)	Temperature (°C)	Acceleration of Gravity, g (m/s²)	Pressure, p [N/m²(abs)]		Density, ρ (kg/m³)		Dynamic Viscosity, μ (N · s/m²)	
−1,000	21.50	9.810	1.139	E + 5	1.347	E + 0	1.821	E − 5
0	15.00	9.807	1.013	E + 5	1.225	E + 0	1.789	E − 5
1,000	8.50	9.804	8.988	E + 4	1.112	E + 0	1.758	E − 5
2,000	2.00	9.801	7.950	E + 4	1.007	E + 0	1.726	E − 5
3,000	−4.49	9.797	7.012	E + 4	9.093	E − 1	1.694	E − 5
4,000	−10.98	9.794	6.166	E + 4	8.194	E − 1	1.661	E − 5
5,000	−17.47	9.791	5.405	E + 4	7.364	E − 1	1.628	E − 5
6,000	−23.96	9.788	4.722	E + 4	6.601	E − 1	1.595	E − 5
7,000	−30.45	9.785	4.111	E + 4	5.900	E − 1	1.561	E − 5
8,000	−36.94	9.782	3.565	E + 4	5.258	E − 1	1.527	E − 5
9,000	−43.42	9.779	3.080	E + 4	4.671	E − 1	1.493	E − 5
10,000	−49.90	9.776	2.650	E + 4	4.135	E − 1	1.458	E − 5
15,000	−56.50	9.761	1.211	E + 4	1.948	E − 1	1.422	E − 5
20,000	−56.50	9.745	5.529	E + 3	8.891	E − 2	1.422	E − 5
25,000	−51.60	9.730	2.549	E + 3	4.008	E − 2	1.448	E − 5
30,000	−46.64	9.715	1.197	E + 3	1.841	E − 2	1.475	E − 5
40,000	−22.80	9.684	2.871	E + 2	3.996	E − 3	1.601	E − 5
50,000	−2.50	9.654	7.978	E + 1	1.027	E − 3	1.704	E − 5
60,000	−26.13	9.624	2.196	E + 1	3.097	E − 4	1.584	E − 5
70,000	−53.57	9.594	5.221	E + 0	8.283	E − 5	1.438	E − 5
80,000	−74.51	9.564	1.052	E + 0	1.846	E − 5	1.321	E − 5

Source: Data abridged from *U.S. Standard Atmosphere, 1976*, U.S. Government Printing Office, Washington, D.C.

TABLE C.2 Properties of the U.S. Standard Atmosphere (BG/EE Units)

Altitude (ft)	Temperature (°F)	Acceleration of Gravity, g (ft/s^2)	Pressure, p [lb/in.^{2}(abs)]	Density, ρ (slugs/ft^3)[a]		Dynamic Viscosity, μ (lb · s/ft^2)	
−5,000	76.84	32.189	17.554	2.745	E − 3	3.836	E − 7
0	59.00	32.174	14.696	2.377	E − 3	3.737	E − 7
5,000	41.17	32.159	12.228	2.048	E − 3	3.637	E − 7
10,000	23.36	32.143	10.108	1.756	E − 3	3.534	E − 7
15,000	5.55	32.128	8.297	1.496	E − 3	3.430	E − 7
20,000	−12.26	32.112	6.759	1.267	E − 3	3.324	E − 7
25,000	−30.05	32.097	5.461	1.066	E − 3	3.217	E − 7
30,000	−47.83	32.082	4.373	8.907	E − 4	3.107	E − 7
35,000	−65.61	32.066	3.468	7.382	E − 4	2.995	E − 7
40,000	−69.70	32.051	2.730	5.873	E − 4	2.969	E − 7
45,000	−69.70	32.036	2.149	4.623	E − 4	2.969	E − 7
50,000	−69.70	32.020	1.692	3.639	E − 4	2.969	E − 7
60,000	−69.70	31.990	1.049	2.256	E − 4	2.969	E − 7
70,000	−67.42	31.959	0.651	1.392	E − 4	2.984	E − 7
80,000	−61.98	31.929	0.406	8.571	E − 5	3.018	E − 7
90,000	−56.54	31.897	0.255	5.610	E − 5	3.052	E − 7
100,000	−51.10	31.868	0.162	3.318	E − 5	3.087	E − 7
150,000	19.40	31.717	0.020	3.658	E − 6	3.511	E − 7
200,000	−19.78	31.566	0.003	5.328	E − 7	3.279	E − 7
250,000	−88.77	31.415	0.000	6.458	E − 8	2.846	E − 7

Source: Data abridged from *U.S. Standard Atmosphere, 1976*, U.S. Government Printing Office, Washington, D.C.

[a]To obtain EE units (lbm/ft^3), multiply by 32.174.

Compressible Flow Functions for an Ideal Gas with $k = 1.4$

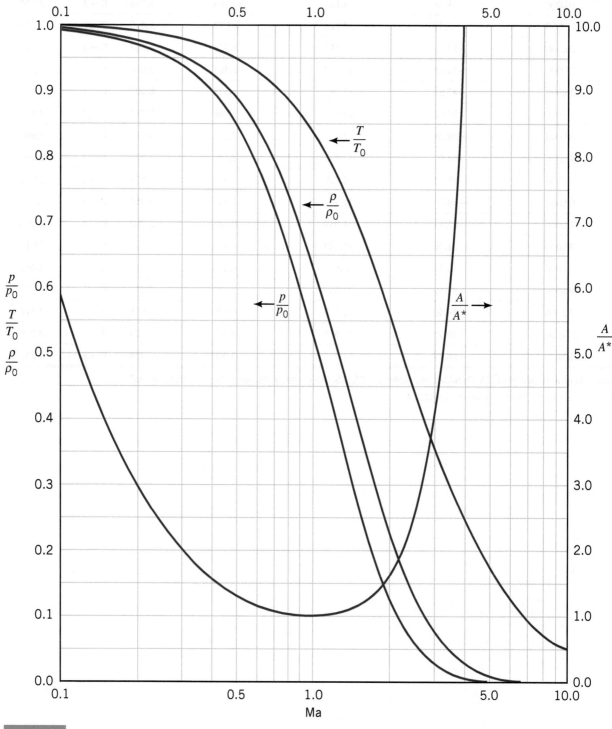

FIGURE D.1 Isentropic flow of an ideal gas with $k = 1.4$. (Source: Courtesy of Dr. Bruce A. Reichert.)

TABLE D.1 Isentropic Flow Functions for an Ideal Gas with $k = 1.4$

Ma	T/T_0	p/p_0	ρ/ρ_0	A/A^*	Ma	T/T_0	p/p_0	ρ/ρ_0	A/A^*
0.0000	1.0000	1.0000	1.0000	∞	2.0500	0.5433	0.1182	0.2176	1.7600
0.0500	0.9995	0.9983	0.9988	11.5914	2.1000	0.5313	0.1094	0.2058	1.8369
0.1000	0.9980	0.9930	0.9950	5.8218	2.1500	0.5196	0.1011	0.1946	1.9185
0.1500	0.9955	0.9844	0.9888	3.9103	2.2000	0.5081	0.0935	0.1841	2.0050
0.2000	0.9921	0.9725	0.9803	2.9635	2.2500	0.4969	0.0865	0.1740	2.0964
0.2500	0.9877	0.9575	0.9694	2.4027	2.3000	0.4859	0.0800	0.1646	2.1931
0.3000	0.9823	0.9395	0.9564	2.0351	2.3500	0.4752	0.0740	0.1556	2.2953
0.3500	0.9761	0.9188	0.9413	1.7780	2.4000	0.4647	0.0684	0.1472	2.4031
0.4000	0.9690	0.8956	0.9243	1.5901	2.4500	0.4544	0.0633	0.1392	2.5168
0.4500	0.9611	0.8703	0.9055	1.4487	2.5000	0.4444	0.0585	0.1317	2.6367
0.5000	0.9524	0.8430	0.8852	1.3398	2.5500	0.4347	0.0542	0.1246	2.7630
0.5500	0.9430	0.8142	0.8634	1.2549	2.6000	0.4252	0.0501	0.1179	2.8960
0.6000	0.9328	0.7840	0.8405	1.1882	2.6500	0.4159	0.0464	0.1115	3.0359
0.6500	0.9221	0.7528	0.8164	1.1356	2.7000	0.4068	0.0430	0.1056	3.1830
0.7000	0.9107	0.7209	0.7916	1.0944	2.7500	0.3980	0.0398	0.0999	3.3377
0.7500	0.8989	0.6886	0.7660	1.0624	2.8000	0.3894	0.0368	0.0946	3.5001
0.8000	0.8865	0.6560	0.7400	1.0382	2.8500	0.3810	0.0341	0.0896	3.6707
0.8500	0.8737	0.6235	0.7136	1.0207	2.9000	0.3729	0.0317	0.0849	3.8498
0.9000	0.8606	0.5913	0.6870	1.0089	2.9500	0.3649	0.0293	0.0804	4.0376
0.9500	0.8471	0.5595	0.6604	1.0021	3.0000	0.3571	0.0272	0.0762	4.2346
1.0000	0.8333	0.5283	0.6339	1.0000	3.1000	0.3422	0.0234	0.0685	4.6573
1.0500	0.8193	0.4979	0.6077	1.0020	3.2000	0.3281	0.0202	0.0617	5.1210
1.1000	0.8052	0.4684	0.5817	1.0079	3.3000	0.3147	0.0175	0.0555	5.6286
1.1500	0.7908	0.4398	0.5562	1.0175	3.4000	0.3019	0.0151	0.0501	6.1837
1.2000	0.7764	0.4124	0.5311	1.0304	3.5000	0.2899	0.0131	0.0452	6.7896
1.2500	0.7619	0.3861	0.5067	1.0468	3.6000	0.2784	0.0114	0.0409	7.4501
1.3000	0.7474	0.3609	0.4829	1.0663	3.7000	0.2675	0.0099	0.0370	8.1691
1.3500	0.7329	0.3370	0.4598	1.0890	3.8000	0.2572	0.0086	0.0335	8.9506
1.4000	0.7184	0.3142	0.4374	1.1149	3.9000	0.2474	0.0075	0.0304	9.7990
1.4500	0.7040	0.2927	0.4158	1.1440	4.0000	0.2381	0.0066	0.0277	10.7188
1.5000	0.6897	0.2724	0.3950	1.1762	4.5000	0.1980	0.0035	0.0174	16.5622
1.5500	0.6754	0.2533	0.3750	1.2116	5.0000	0.1667	0.0019	0.0113	25.0000
1.6000	0.6614	0.2353	0.3557	1.2502	5.5000	0.1418	0.0011	0.0076	36.8690
1.6500	0.6475	0.2184	0.3373	1.2922	6.0000	0.1220	0.0006	0.0052	53.1798
1.7000	0.6337	0.2026	0.3197	1.3376	6.5000	0.1058	0.0004	0.0036	75.1343
1.7500	0.6202	0.1878	0.3029	1.3865	7.0000	0.0926	0.0002	0.0026	104.1429
1.8000	0.6068	0.1740	0.2868	1.4390	7.5000	0.0816	0.0002	0.0019	141.8415
1.8500	0.5936	0.1612	0.2715	1.4952	8.0000	0.0725	0.0001	0.0014	190.1094
1.9000	0.5807	0.1492	0.2570	1.5553	8.5000	0.0647	0.0001	0.0011	251.0862
1.9500	0.5680	0.1381	0.2432	1.6193	9.0000	0.0581	0.0000	0.0008	327.1893
2.0000	0.5556	0.1278	0.2300	1.6875	10.0000	0.0476	0.0000	0.0005	535.9375

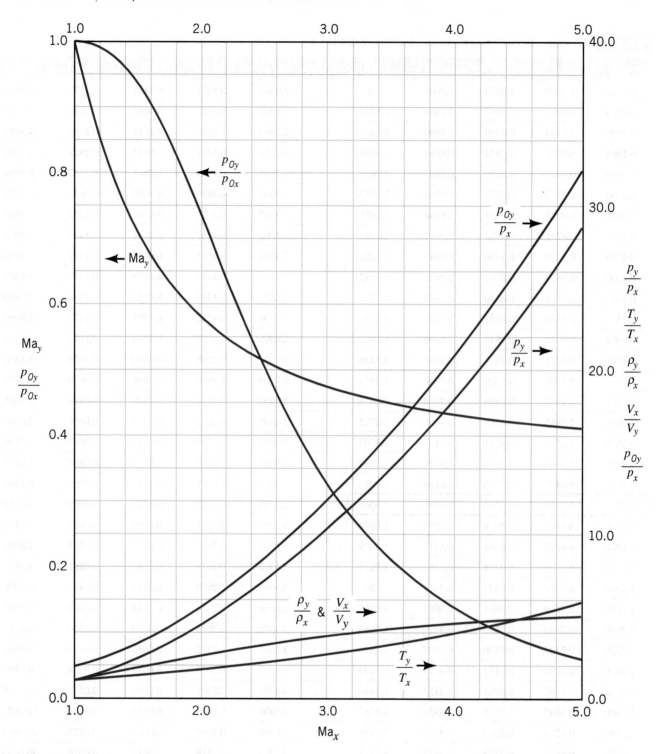

FIGURE D.S Normal shock flow of an ideal gas with $k = 1.4$. (Source: Courtesy of Dr. Bruce A. Reichert.)

TABLE D.S   Normal Shock Functions for an Ideal Gas with $k = 1.4$

Ma_x	Ma_y	p_y/p_x	T_y/T_x	$\rho_y/\rho_x = V_x/V_y$	p_{0y}/p_{0x}	Ma_x	Ma_y	p_y/p_x	T_y/T_x	$\rho_y/\rho_x = V_x/V_y$	p_{0y}/p_{0x}
1.00	1.0000	1.0000	1.0000	1.0000	1.0000	3.05	0.4723	10.6863	2.7383	3.9025	0.3145
1.05	0.9531	1.1196	1.0328	1.0840	0.9999	3.10	0.4695	11.0450	2.7986	3.9466	0.3012
1.10	0.9118	1.2450	1.0649	1.1691	0.9989	3.15	0.4669	11.4096	2.8598	3.9896	0.2885
1.15	0.8750	1.3763	1.0966	1.2550	0.9967	3.20	0.4643	11.7800	2.9220	4.0315	0.2762
1.20	0.8422	1.5133	1.1280	1.3416	0.9928	3.25	0.4619	12.1563	2.9851	4.0723	0.2645
1.25	0.8126	1.6563	1.1594	1.4286	0.9871	3.30	0.4596	12.5383	3.0492	4.1120	0.2533
1.30	0.7860	1.8050	1.1909	1.5157	0.9794	3.35	0.4573	12.9263	3.1142	4.1507	0.2425
1.35	0.7618	1.9596	1.2226	1.6028	0.9697	3.40	0.4552	13.3200	3.1802	4.1884	0.2322
1.40	0.7397	2.1200	1.2547	1.6897	0.9582	3.45	0.4531	13.7196	3.2472	4.2251	0.2224
1.45	0.7196	2.2863	1.2872	1.7761	0.9448	3.50	0.4512	14.1250	3.3151	4.2609	0.2129
1.50	0.7011	2.4583	1.3202	1.8621	0.9298	3.55	0.4492	14.5363	3.3839	4.2957	0.2039
1.55	0.6841	2.6363	1.3538	1.9473	0.9132	3.60	0.4474	14.9533	3.4537	4.3296	0.1953
1.60	0.6684	2.8200	1.3880	2.0317	0.8952	3.65	0.4456	15.3763	3.5245	4.3627	0.1871
1.65	0.6540	3.0096	1.4228	2.1152	0.8760	3.70	0.4439	15.8050	3.5962	4.3949	0.1792
1.70	0.6405	3.2050	1.4583	2.1977	0.8557	3.75	0.4423	16.2396	3.6689	4.4262	0.1717
1.75	0.6281	3.4063	1.4946	2.2791	0.8346	3.80	0.4407	16.6800	3.7426	4.4568	0.1645
1.80	0.6165	3.6133	1.5316	2.3592	0.8127	3.85	0.4392	17.1263	3.8172	4.4866	0.1576
1.85	0.6057	3.8263	1.5693	2.4381	0.7902	3.90	0.4377	17.5783	3.8928	4.5156	0.1510
1.90	0.5956	4.0450	1.6079	2.5157	0.7674	3.95	0.4363	18.0363	3.9694	4.5439	0.1448
1.95	0.5862	4.2696	1.6473	2.5919	0.7442	4.00	0.4350	18.5000	4.0469	4.5714	0.1388
2.00	0.5774	4.5000	1.6875	2.6667	0.7209	4.10	0.4324	19.4450	4.2048	4.6245	0.1276
2.05	0.5691	4.7363	1.7285	2.7400	0.6975	4.20	0.4299	20.4133	4.3666	4.6749	0.1173
2.10	0.5613	4.9783	1.7705	2.8119	0.6742	4.30	0.4277	21.4050	4.5322	4.7229	0.1080
2.15	0.5540	5.2263	1.8132	2.8823	0.6511	4.40	0.4255	22.4200	4.7017	4.7685	0.0995
2.20	0.5471	5.4800	1.8569	2.9512	0.6281	4.50	0.4236	23.4583	4.8751	4.8119	0.0917
2.25	0.5406	5.7396	1.9014	3.0186	0.6055	4.60	0.4217	24.5200	5.0523	4.8532	0.0846
2.30	0.5344	6.0050	1.9468	3.0845	0.5833	4.70	0.4199	25.6050	5.2334	4.8926	0.0781
2.35	0.5286	6.2763	1.9931	3.1490	0.5615	4.80	0.4183	26.7133	5.4184	4.9301	0.0721
2.40	0.5231	6.5533	2.0403	3.2119	0.5401	4.90	0.4167	27.8450	5.6073	4.9659	0.0667
2.45	0.5179	6.8363	2.0885	3.2733	0.5193	5.00	0.4152	29.0000	5.8000	5.0000	0.0617
2.50	0.5130	7.1250	2.1375	3.3333	0.4990	5.50	0.4090	35.1250	6.8218	5.1489	0.0424
2.55	0.5083	7.4196	2.1875	3.3919	0.4793	6.00	0.4042	41.8333	7.9406	5.2683	0.0297
2.60	0.5039	7.7200	2.2383	3.4490	0.4601	6.50	0.4004	49.1250	9.1564	5.3651	0.0211
2.65	0.4996	8.0262	2.2902	3.5047	0.4416	7.00	0.3974	57.0000	10.4694	5.4444	0.0154
2.70	0.4956	8.3383	2.3429	3.5590	0.4236	7.50	0.3949	65.4583	11.8795	5.5102	0.0113
2.75	0.4918	8.6562	2.3966	3.6119	0.4062	8.00	0.3929	74.5000	13.3867	5.5652	0.0085
2.80	0.4882	8.9800	2.4512	3.6636	0.3895	8.50	0.3912	84.1250	14.9911	5.6117	0.0064
2.85	0.4847	9.3096	2.5067	3.7139	0.3733	9.00	0.3898	94.3333	16.6927	5.6512	0.0050
2.90	0.4814	9.6450	2.5632	3.7629	0.3577	9.50	0.3886	105.1250	18.4915	5.6850	0.0039
2.95	0.4782	9.9862	2.6206	3.8106	0.3428	10.00	0.3876	116.5000	20.3875	5.7143	0.0030
3.00	0.4752	10.3333	2.6790	3.8571	0.3283	∞	0.3780	∞	∞	6.0000	0.0000

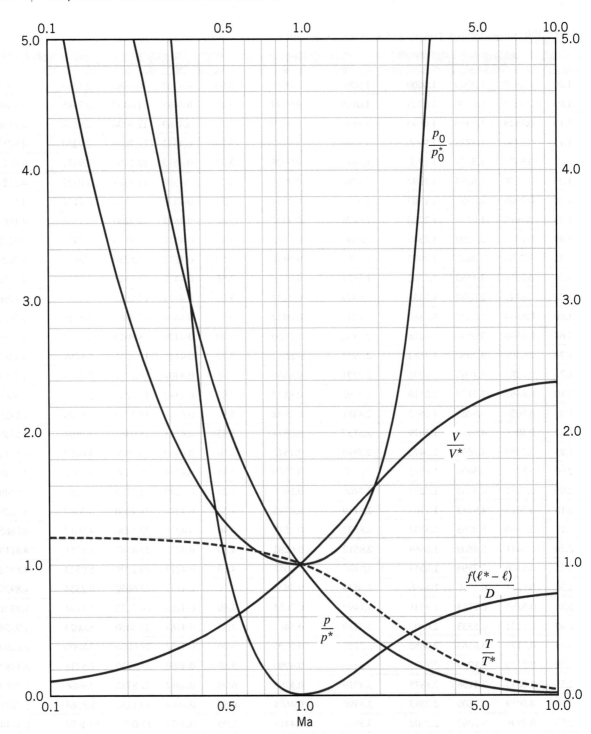

FIGURE D.F Fanno flow of an ideal gas with $k = 1.4$. (Source: Courtesy of Dr. Bruce A. Reichert.)

TABLE D.F Fanno Flow Functions for an Ideal Gas with $k = 1.4$

Ma	$f(\ell^*-\ell)/D$	p/p^*	T/T^*	p_0/p_0^*	$\rho/\rho^* = V^*/V$	Ma	$f(\ell^*-\ell)/D$	p/p^*	T/T^*	p_0/p_0^*	$\rho/\rho^* = V^*/V$
0.00	∞	∞	1.2000	∞	∞	1.60	0.1724	0.5568	0.7937	1.2502	0.7016
0.05	280.0203	21.9034	1.1994	11.5914	18.2620	1.65	0.1902	0.5342	0.7770	1.2922	0.6876
0.10	66.9216	10.9435	1.1976	5.8218	9.1378	1.70	0.2078	0.5130	0.7605	1.3376	0.6745
0.15	27.9320	7.2866	1.1946	3.9103	6.0995	1.75	0.2250	0.4929	0.7442	1.3865	0.6624
0.20	14.5333	5.4554	1.1905	2.9635	4.5826	1.80	0.2419	0.4741	0.7282	1.4390	0.6511
0.25	8.4834	4.3546	1.1852	2.4027	3.6742	1.85	0.2583	0.4562	0.7124	1.4952	0.6404
0.30	5.2993	3.6191	1.1788	2.0351	3.0702	1.90	0.2743	0.4394	0.6969	1.5553	0.6305
0.35	3.4525	3.0922	1.1713	1.7780	2.6400	1.95	0.2899	0.4234	0.6816	1.6193	0.6211
0.40	2.3085	2.6958	1.1628	1.5901	2.3184	2.00	0.3050	0.4082	0.6667	1.6875	0.6124
0.45	1.5664	2.3865	1.1533	1.4487	2.0693	2.10	0.3339	0.3802	0.6376	1.8369	0.5963
0.50	1.0691	2.1381	1.1429	1.3398	1.8708	2.20	0.3609	0.3549	0.6098	2.0050	0.5821
0.55	0.7281	1.9341	1.1315	1.2549	1.7092	2.30	0.3862	0.3320	0.5831	2.1931	0.5694
0.60	0.4908	1.7634	1.1194	1.1882	1.5753	2.40	0.4099	0.3111	0.5576	2.4031	0.5580
0.65	0.3246	1.6183	1.1065	1.1356	1.4626	2.50	0.4320	0.2921	0.5333	2.6367	0.5477
0.70	0.2081	1.4935	1.0929	1.0944	1.3665	2.60	0.4526	0.2747	0.5102	2.8960	0.5385
0.75	0.1273	1.3848	1.0787	1.0624	1.2838	2.70	0.4718	0.2588	0.4882	3.1830	0.5301
0.80	0.0723	1.2893	1.0638	1.0382	1.2119	2.80	0.4898	0.2441	0.4673	3.5001	0.5225
0.85	0.0363	1.2047	1.0485	1.0207	1.1489	2.90	0.5065	0.2307	0.4474	3.8498	0.5155
0.90	0.0145	1.1291	1.0327	1.0089	1.0934	3.00	0.5222	0.2182	0.4286	4.2346	0.5092
0.95	0.0033	1.0613	1.0165	1.0021	1.0440	3.50	0.5864	0.1685	0.3478	6.7896	0.4845
1.00	0.0000	1.0000	1.0000	1.0000	1.0000	4.00	0.6331	0.1336	0.2857	10.7188	0.4677
1.05	0.0027	0.9443	0.9832	1.0020	0.9605	4.50	0.6676	0.1083	0.2376	16.5622	0.4559
1.10	0.0099	0.8936	0.9662	1.0079	0.9249	5.00	0.6938	0.0894	0.2000	25.0000	0.4472
1.15	0.0205	0.8471	0.9490	1.0175	0.8926	5.50	0.7140	0.0750	0.1702	36.8690	0.4407
1.20	0.0336	0.8044	0.9317	1.0304	0.8633	6.00	0.7299	0.0638	0.1463	53.1798	0.4357
1.25	0.0486	0.7649	0.9143	1.0468	0.8367	6.50	0.7425	0.0548	0.1270	75.1343	0.4317
1.30	0.0648	0.7285	0.8969	1.0663	0.8123	7.00	0.7528	0.0476	0.1111	104.1429	0.4286
1.35	0.0820	0.6947	0.8794	1.0890	0.7899	7.50	0.7612	0.0417	0.0980	141.8415	0.4260
1.40	0.0997	0.6632	0.8621	1.1149	0.7693	8.00	0.7682	0.0369	0.0870	190.1094	0.4239
1.45	0.1178	0.6339	0.8448	1.1440	0.7503	8.50	0.7740	0.0328	0.0777	251.0862	0.4221
1.50	0.1361	0.6065	0.8276	1.1762	0.7328	9.00	0.7790	0.0293	0.0698	327.1893	0.4207
1.55	0.1543	0.5808	0.8105	1.2116	0.7166	∞	0.8215	0.0000	0.0000	∞	0.4082

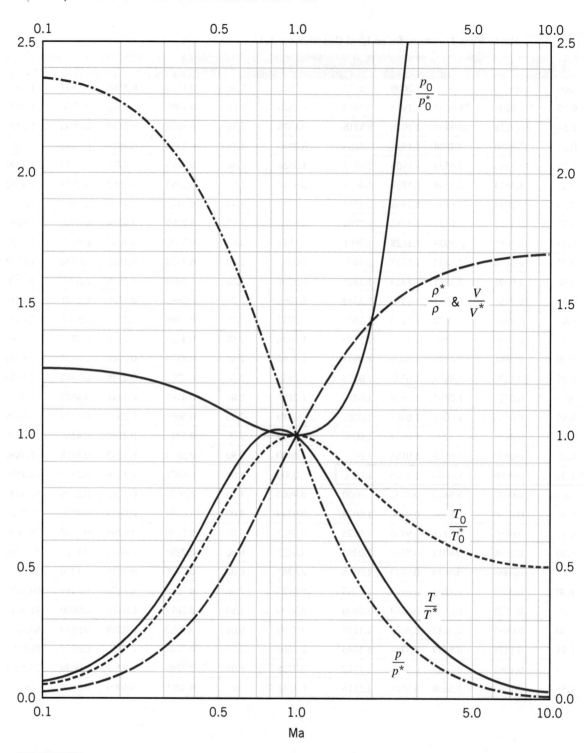

FIGURE D.R Rayleigh flow of an ideal gas with $k = 1.4$. (Source: Courtesy of Dr. Bruce A. Reichert.)

TABLE D.R Rayleigh Flow Functions for an Ideal Gas with $k = 1.4$

Ma	T_0/T_0^*	p/p^*	T/T^*	p_0/p_0^*	$\rho/\rho^* = V^*/V$	Ma	T_0/T_0^*	p/p^*	T/T^*	p_0/p_0^*	$\rho/\rho^* = V^*/V$
0.00	0.0000	2.4000	0.0000	1.2679	∞	1.60	0.8842	0.5236	0.7017	1.1756	0.7461
0.05	0.0119	2.3916	0.0143	1.2657	167.2500	1.65	0.8718	0.4988	0.6774	1.2066	0.7364
0.10	0.0468	2.3669	0.0560	1.2591	42.2500	1.70	0.8597	0.4756	0.6538	1.2402	0.7275
0.15	0.1020	2.3267	0.1218	1.2486	19.1019	1.75	0.8478	0.4539	0.6310	1.2767	0.7194
0.20	0.1736	2.2727	0.2066	1.2346	11.0000	1.80	0.8363	0.4335	0.6089	1.3159	0.7119
0.25	0.2568	2.2069	0.3044	1.2177	7.2500	1.85	0.8250	0.4144	0.5877	1.3581	0.7051
0.30	0.3469	2.1314	0.4089	1.1985	5.2130	1.90	0.8141	0.3964	0.5673	1.4033	0.6988
0.35	0.4389	2.0487	0.5141	1.1779	3.9847	1.95	0.8036	0.3795	0.5477	1.4516	0.6929
0.40	0.5290	1.9608	0.6151	1.1566	3.1875	2.00	0.7934	0.3636	0.5289	1.5031	0.6875
0.45	0.6139	1.8699	0.7080	1.1351	2.6409	2.10	0.7741	0.3345	0.4936	1.6162	0.6778
0.50	0.6914	1.7778	0.7901	1.1141	2.2500	2.20	0.7561	0.3086	0.4611	1.7434	0.6694
0.55	0.7599	1.6860	0.8599	1.0940	1.9607	2.30	0.7395	0.2855	0.4312	1.8860	0.6621
0.60	0.8189	1.5957	0.9167	1.0753	1.7407	2.40	0.7242	0.2648	0.4038	2.0451	0.6557
0.65	0.8683	1.5080	0.9608	1.0582	1.5695	2.50	0.7101	0.2462	0.3787	2.2218	0.6500
0.70	0.9085	1.4235	0.9929	1.0431	1.4337	2.60	0.6970	0.2294	0.3556	2.4177	0.6450
0.75	0.9401	1.3427	1.0140	1.0301	1.3241	2.70	0.6849	0.2142	0.3344	2.6343	0.6405
0.80	0.9639	1.2658	1.0255	1.0193	1.2344	2.80	0.6738	0.2004	0.3149	2.8731	0.6365
0.84515	0.9796	1.2000	1.0286	1.0116	1.1667	2.90	0.6635	0.1879	0.2969	3.1359	0.6329
0.85	0.9810	1.1931	1.0285	1.0109	1.1600	3.00	0.6540	0.1765	0.2803	3.4245	0.6296
0.90	0.9921	1.1246	1.0245	1.0049	1.0977	3.50	0.6158	0.1322	0.2142	5.3280	0.6173
0.95	0.9981	1.0603	1.0146	1.0012	1.0450	4.00	0.5891	0.1026	0.1683	8.2268	0.6094
1.00	1.0000	1.0000	1.0000	1.0000	1.0000	4.50	0.5698	0.0818	0.1354	12.5023	0.6039
1.05	0.9984	0.9436	0.9816	1.0012	0.9613	5.00	0.5556	0.0667	0.1111	18.6339	0.6000
1.10	0.9939	0.8909	0.9603	1.0049	0.9277	5.50	0.5447	0.0554	0.0927	27.2113	0.5971
1.15	0.9872	0.8417	0.9369	1.0109	0.8984	6.00	0.5363	0.0467	0.0785	38.9459	0.5949
1.20	0.9787	0.7958	0.9118	1.0194	0.8727	6.50	0.5297	0.0399	0.0673	54.6830	0.5932
1.25	0.9689	0.7529	0.8858	1.0303	0.8500	7.00	0.5244	0.0345	0.0583	75.4138	0.5918
1.30	0.9580	0.7130	0.8592	1.0437	0.8299	7.50	0.5200	0.0301	0.0509	102.2875	0.5907
1.35	0.9464	0.6758	0.8323	1.0594	0.8120	8.00	0.5165	0.0265	0.0449	136.6235	0.5898
1.40	0.9343	0.6410	0.8054	1.0777	0.7959	8.50	0.5135	0.0235	0.0399	179.9236	0.5891
1.45	0.9218	0.6086	0.7787	1.0983	0.7815	9.00	0.5110	0.0210	0.0356	233.8840	0.5885
1.50	0.9093	0.5783	0.7525	1.1215	0.7685	10	0.5070	0.0170	0.0290	381.6149	0.5875
1.55	0.8967	0.5500	0.7268	1.1473	0.7568	∞	∞	0.0000	0.0000	∞	0.5833

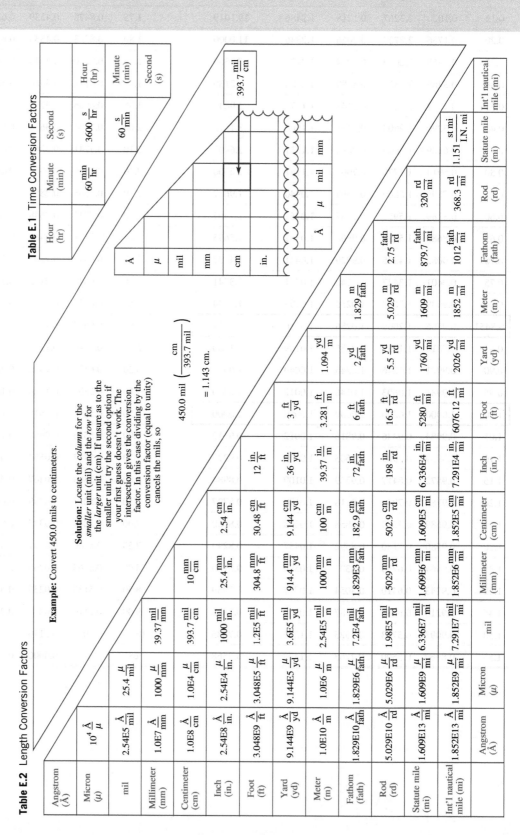

Table E.1 Time Conversion Factors

	Hour (hr)	Minute (min)	Second (s)	
		Hour (hr)
		$60\frac{min}{hr}$	$3600\frac{s}{hr}$	Minute (min)
			$60\frac{s}{min}$	Second (s)

Table E.2 Length Conversion Factors

Example: Convert 450.0 mils to centimeters.

Solution: Locate the *column* for the *smaller* unit (mil) and the *row* for the *larger* unit (cm). If unsure as to the smaller unit, try the second option if your first guess doesn't work. The intersection gives the conversion factor. In this case dividing by the conversion factor (equal to unity) cancels the mils, so

$$450.0\ mil\left(\frac{cm}{393.7\ mil}\right)$$
$$= 1.143\ cm.$$

$$393.7\frac{mil}{cm}$$

Table E.3 Area Conversion Factors

	Square statute mile (mi²)	Acre	Square meter (m²)	Square yard (yd²)	Square foot (ft²)	Square inch (in²)	Square centimeter (cm²)
Square centimeter (cm²)	$2.589\text{E}10\ \frac{cm^2}{mi^2}$	$4.047\text{E}7\ \frac{cm^2}{acre}$	$1.0\text{E}4\ \frac{cm^2}{m^2}$	$8361\ \frac{cm^2}{yd^2}$	$929.0\ \frac{cm^2}{ft^2}$	$6.452\ \frac{cm^2}{in^2}$	
Square inch (in²)	$4.014\text{E}9\ \frac{in^2}{mi^2}$	$6.272\text{E}6\ \frac{in^2}{acre}$	$1550\ \frac{in^2}{m^2}$	$1296\ \frac{in^2}{yd^2}$	$144\ \frac{in^2}{ft^2}$		
Square foot (ft²)	$2.788\text{E}7\ \frac{ft^2}{mi^2}$	$4.356\text{E}4\ \frac{ft^2}{acre}$	$10.76\ \frac{ft^2}{m^2}$	$9\ \frac{ft^2}{yd^2}$			
Square yard (yd²)	$3.098\text{E}6\ \frac{yd^2}{mi^2}$	$4840\ \frac{yd^2}{acre}$	$1.196\ \frac{yd^2}{m^2}$				
Square meter (m²)	$2.589\text{E}6\ \frac{m^2}{mi^2}$	$4047\ \frac{m^2}{acre}$					
Acre	$640\ \frac{acre}{mi^2}$						
Square statute mile (mi²)							

Table E.4 Volume Conversion Factors

	Milliliter (ml) (cubic centimeter)	Cubic inch (in³)	U.S. fluid ounce (fl oz)	U.S. liquid quart (qt)	Liter (L)	U.S. liquid gallon (gal)	Cubic foot (ft³)	Cubic yard (yd³)	Cubic meter (m³)	Cubic statute mile (mi³)
Milliliter (ml) (cubic centimeter)		$16.39\ \frac{ml}{in^3}$	$29.574\ \frac{ml}{fl\ oz}$	$946.4\ \frac{ml}{qt}$	$1000\ \frac{ml}{L}$	$3785.4\ \frac{ml}{gal}$	$2.83\text{E}4\ \frac{ml}{ft^3}$	$7.646\text{E}5\ \frac{ml}{yd^3}$	$1.0\text{E}6\ \frac{ml}{m^3}$	$4.166\text{E}15\ \frac{ml}{mi^3}$
Cubic inch (in³)			$1.805\ \frac{in^3}{fl\ oz}$	$57.75\ \frac{in^3}{qt}$	$61.02\ \frac{in^3}{L}$	$231\ \frac{in^3}{gal}$	$1728\ \frac{in^3}{ft^3}$	$4.666\text{E}4\ \frac{in^3}{yd^3}$	$6.102\text{E}4\ \frac{in^3}{m^3}$	$2.544\text{E}14\ \frac{in^3}{mi^3}$
U.S. fluid ounce (fl oz)				$32\ \frac{fl\ oz}{qt}$	$33.81\ \frac{fl\ oz}{L}$	$128\ \frac{fl\ oz}{gal}$	$957.5\ \frac{fl\ oz}{ft^3}$	$2.585\text{E}4\ \frac{fl\ oz}{yd^3}$	$3.381\text{E}4\ \frac{fl\ oz}{m^3}$	$1.409\text{E}14\ \frac{fl\ oz}{mi^3}$
U.S. liquid quart (qt)					$1.057\ \frac{qt}{L}$	$4\ \frac{qt}{gal}$	$29.92\ \frac{qt}{ft^3}$	$807.9\ \frac{qt}{yd^3}$	$1.057\text{E}3\ \frac{qt}{m^3}$	$4.404\text{E}11\ \frac{qt}{mi^3}$
Liter (L)						$3.7854\ \frac{L}{gal}$	$28.3\ \frac{L}{ft^3}$	$764.6\ \frac{L}{yd^3}$	$1000\ \frac{L}{m^3}$	$4.166\text{E}12\ \frac{L}{mi^3}$
U.S. liquid gallon (gal)							$7.48\ \frac{gal}{ft^3}$	$202.0\ \frac{gal}{yd^3}$	$264.2\ \frac{gal}{m^3}$	$1.101\text{E}12\ \frac{gal}{mi^3}$
Cubic foot (ft³)								$27\ \frac{ft^3}{yd^3}$	$35.31\ \frac{ft^3}{m^3}$	$1.472\text{E}11\ \frac{ft^3}{mi^3}$
Cubic yard (yd³)									$1.308\ \frac{yd^3}{m^3}$	$5.452\text{E}9\ \frac{yd^3}{mi^3}$
Cubic meter (m³)										$4.166\text{E}9\ \frac{m^3}{mi^3}$
Cubic statute mile (mi³)										

Table E.5 Mass Conversion Factors

	Kilogram (kg)	Pound (lbm) (avdp.)	Ounce (oz) (avdp.)	Gram (g)	Grain (gr)
Slug	$14.594\ \dfrac{kg}{slug}$	$32.174\ \dfrac{lbm}{slug}$	$514.8\ \dfrac{oz}{slug}$	$1.459E4\ \dfrac{g}{slug}$	$2.252E5\ \dfrac{gr}{slug}$
Kilogram (kg)		$2.2046\ \dfrac{lbm}{kg}$	$35.27\ \dfrac{oz}{kg}$	$1000\ \dfrac{g}{kg}$	$1.543E4\ \dfrac{gr}{kg}$
Pound (lbm) (avdp.)			$16\ \dfrac{oz}{lbm}$	$453.6\ \dfrac{g}{lbm}$	$7000\ \dfrac{gr}{lbm}$
Ounce (oz) (avdp.)				$28.35\ \dfrac{g}{oz}$	$437.5\ \dfrac{gr}{oz}$
Gram (g)					$15.43\ \dfrac{gr}{g}$

Table E.6 Force Conversion Factors

	Dyne	Poundal (pdl)	Newton (N)	Pound (lb) (avdp.)	Short ton	Metric ton	Long ton
Poundal (pdl)	$1.383E4\ \dfrac{dyne}{pdl}$						
Newton (N)	$E5\ \dfrac{dyne}{N}$	$7.233\ \dfrac{pdl}{N}$					
Pound (lb) (avdp.)	$4.448E5\ \dfrac{dyne}{lb}$	$32.174\ \dfrac{pdl}{lb}$	$4.448\ \dfrac{N}{lb}$				
Short ton	$8.896E8\ \dfrac{dyne}{s\ ton}$	$6.435E4\ \dfrac{pdl}{s\ ton}$	$8896\ \dfrac{N}{s\ ton}$	$2000\ \dfrac{lb}{s\ ton}$			
Metric ton	$9.807E8\ \dfrac{dyne}{m\ ton}$	$7.094E4\ \dfrac{pdl}{m\ ton}$	$9808\ \dfrac{N}{m\ ton}$	$2205\ \dfrac{lb}{m\ ton}$	$1.102\ \dfrac{s\ ton}{m\ ton}$		
Long ton	$9.964E8\ \dfrac{dyne}{l\ ton}$	$7.207E4\ \dfrac{pdl}{l\ ton}$	$9964\ \dfrac{N}{l\ ton}$	$2240\ \dfrac{lb}{l\ ton}$	$1.12\ \dfrac{s\ ton}{l\ ton}$	$1.016\ \dfrac{m\ ton}{l\ ton}$	

Table E.7 Pressure Conversion Factors

	atm	Bar	psi	in. Hg (32 °F)	ft H_2O (39.2 °F)	in. H_2O (39.2 °F)	mm Hg (0 °C)	Torr	Pascal $\left(\frac{N}{m^2}\right)$	$\frac{dyne}{cm^2}$
atm		1.0132 $\frac{bar}{atm}$	14.696 $\frac{psi}{atm}$	29.92 $\frac{in.\ Hg}{atm}$	33.90 $\frac{ft\ H_2O}{atm}$	406.8 $\frac{in.\ H_2O}{atm}$	760 $\frac{mm\ Hg}{atm}$	760 $\frac{Torr}{atm}$	1.013E5 $\frac{Pa}{atm}$	1.032E6 $\frac{dyne/cm^2}{atm}$
Bar			14.504 $\frac{psi}{bar}$	29.53 $\frac{in.\ Hg}{bar}$	33.46 $\frac{ft\ H_2O}{bar}$	401.5 $\frac{in.\ H_2O}{bar}$	750.1 $\frac{mm\ Hg}{bar}$	750.1 $\frac{Torr}{bar}$	1.0E5 $\frac{Pa}{bar}$	1.0E6 $\frac{dyne/cm^2}{bar}$
psi				2.036 $\frac{in.\ Hg}{psi}$	2.307 $\frac{ft\ H_2O}{psi}$	27.68 $\frac{in.\ H_2O}{psi}$	51.71 $\frac{mm\ Hg}{psi}$	51.71 $\frac{Torr}{psi}$	6.895E3 $\frac{Pa}{psi}$	6.895E4 $\frac{dyne/cm^2}{psi}$
in. Hg (32 °F)					1.133 $\frac{ft\ H_2O}{in.\ Hg}$	13.60 $\frac{in.\ H_2O}{in.\ Hg}$	25.4 $\frac{mm\ Hg}{in.\ Hg}$	25.4 $\frac{Torr}{in.\ Hg}$	3386 $\frac{Pa}{in.\ Hg}$	3.386E4 $\frac{dyne/cm^2}{in.\ Hg}$
ft H_2O (39.2 °F)						12 $\frac{in.\ H_2O}{ft\ H_2O}$	22.42 $\frac{mm\ Hg}{ft\ H_2O}$	304.8 $\frac{Torr}{ft\ H_2O}$	2989 $\frac{Pa}{ft\ H_2O}$	2.989E4 $\frac{dyne/cm^2}{ft\ H_2O}$
in. H_2O (39.2 °F)							1.868 $\frac{mm\ Hg}{in.\ H_2O}$	1868 $\frac{Torr}{in.\ H_2O}$	249.1 $\frac{Pa}{in.\ H_2O}$	2491 $\frac{dyne/cm^2}{in.\ H_2O}$
mm Hg (0 °C)								1.00 $\frac{Torr}{mm\ Hg}$	133.3 $\frac{Pa}{mm\ Hg}$	1333 $\frac{dyne/cm^2}{mm\ Hg}$
Torr									1.00 $\frac{Pa}{Torr}$	1000 $\frac{dyne/cm^2}{Torr}$
Pascal $\left(\frac{N}{m^2}\right)$										10 $\frac{dyne/cm^2}{Pa}$
$\frac{dyne}{cm^2}$										

Table E.8 Velocity Conversion Factors

	$\frac{ft}{min}$ (fpm)	$\frac{cm}{s}$	$\frac{m}{min}$	$\frac{in.}{s}$	$\frac{km}{hr}$	$\frac{ft}{s}$ (fps)	$\frac{statute\ mile}{hr}$ (mph)	Knots	$\frac{m}{s}$
$\frac{ft}{min}$ (fpm)		1.969 $\frac{fpm}{cm/s}$	3.281 $\frac{fpm}{m/min}$	5 $\frac{fpm}{in./s}$	54.68 $\frac{fpm}{km/hr}$	60 $\frac{fpm}{fps}$	88 $\frac{fpm}{mph}$	101.3 $\frac{fpm}{knot}$	196.9 $\frac{fpm}{m/s}$
$\frac{cm}{s}$			1.667 $\frac{cm/s}{m/min}$	2.54 $\frac{cm/s}{in./s}$	27.78 $\frac{cm/s}{km/hr}$	30.48 $\frac{cm/s}{fps}$	44.7 $\frac{cm/s}{mph}$	51.44 $\frac{cm/s}{knot}$	100 $\frac{cm/s}{m/s}$
$\frac{m}{min}$				1.524 $\frac{m/min}{in./s}$	16.67 $\frac{m/min}{km/hr}$	18.29 $\frac{m/min}{fps}$	26.82 $\frac{m/min}{mph}$	30.87 $\frac{m/min}{knot}$	60 $\frac{m/min}{m/s}$
$\frac{in.}{s}$					10.94 $\frac{in./s}{km/hr}$	12 $\frac{in./s}{fps}$	17.6 $\frac{in./s}{mph}$	20.26 $\frac{in./s}{knot}$	39.37 $\frac{in./s}{m/s}$
$\frac{km}{hr}$						1.097 $\frac{km/hr}{fps}$	1.609 $\frac{km/hr}{mph}$	1.852 $\frac{km/hr}{knot}$	3.6 $\frac{km/hr}{m/s}$
$\frac{ft}{s}$ (fps)							1.47 $\frac{fps}{mph}$	1.688 $\frac{fps}{knot}$	3.281 $\frac{fps}{m/s}$
$\frac{statute\ mile}{hr}$ (mph)								1.151 $\frac{mph}{knot}$	2.237 $\frac{mph}{m/s}$
Knots									1.944 $\frac{knot}{m/s}$
$\frac{m}{s}$									

Table E.9 Absolute Viscosity Conversion Factors

$\dfrac{\text{lb}\cdot\text{s}}{\text{in.}^2}$	$\dfrac{\text{lb}\cdot\text{s}}{\text{ft}^2}$	$\dfrac{\text{lbm}}{\text{in.}\cdot\text{s}}$	$\dfrac{\text{lbm}}{\text{ft}\cdot\text{s}}$	$\dfrac{\text{N}\cdot\text{s}}{\text{m}^2}$	$\dfrac{\text{kg}}{\text{m}\cdot\text{s}}$	Poise $\left(\dfrac{\text{g}}{\text{cm}\cdot\text{s}}\right)$	Centipoise	
	$144\,\dfrac{\text{lb}\cdot\text{s/ft}^2}{\text{lb}\cdot\text{s/in}^2}$	$386.1\,\dfrac{\text{lbm/in.}\cdot\text{s}}{\text{lb}\cdot\text{s/in}^2}$	$4633\,\dfrac{\text{lbm/ft}\cdot\text{s}}{\text{lb}\cdot\text{s/in}^2}$	$6895\,\dfrac{\text{N}\cdot\text{s/m}^2}{\text{lb}\cdot\text{s/in}^2}$	$6895\,\dfrac{\text{kg/m}\cdot\text{s}}{\text{lb}\cdot\text{s/in}^2}$	$6.895\text{E}4\,\dfrac{\text{poise}}{\text{lb}\cdot\text{s/in}^2}$	$6.895\text{E}6\,\dfrac{\text{centipoise}}{\text{lb}\cdot\text{s/in}^2}$	$\dfrac{\text{lb}\cdot\text{s}}{\text{in.}^2}$
		$2.681\,\dfrac{\text{lbm/in.}\cdot\text{s}}{\text{lb}\cdot\text{s/ft}^2}$	$32.174\,\dfrac{\text{lbm/ft}\cdot\text{s}}{\text{lb}\cdot\text{s/ft}^2}$	$47.88\,\dfrac{\text{N}\cdot\text{s/m}^2}{\text{lb}\cdot\text{s/ft}^2}$	$47.88\,\dfrac{\text{kg/m}\cdot\text{s}}{\text{lb}\cdot\text{s/ft}^2}$	$478.8\,\dfrac{\text{poise}}{\text{lb}\cdot\text{s/ft}^2}$	$4.788\text{E}4\,\dfrac{\text{centipoise}}{\text{lb}\cdot\text{s/ft}^2}$	$\dfrac{\text{lb}\cdot\text{s}}{\text{ft}^2}$
			$12\,\dfrac{\text{lbm/ft}\cdot\text{s}}{\text{lbm/in.}\cdot\text{s}}$	$17.86\,\dfrac{\text{N}\cdot\text{s/m}^2}{\text{lbm/in.}\cdot\text{s}}$	$17.86\,\dfrac{\text{kg/m}\cdot\text{s}}{\text{lbm/in.}\cdot\text{s}}$	$178.6\,\dfrac{\text{poise}}{\text{lbm/in.}\cdot\text{s}}$	$1.786\text{E}4\,\dfrac{\text{centipoise}}{\text{lbm/in.}\cdot\text{s}}$	$\dfrac{\text{lbm}}{\text{in.}\cdot\text{s}}$
				$1.488\,\dfrac{\text{N}\cdot\text{s/m}^2}{\text{lbm/ft}\cdot\text{s}}$	$1.488\,\dfrac{\text{kg/m}\cdot\text{s}}{\text{lbm/ft}\cdot\text{s}}$	$14.88\,\dfrac{\text{poise}}{\text{lbm/ft}\cdot\text{s}}$	$1488\,\dfrac{\text{centipoise}}{\text{lbm/ft}\cdot\text{s}}$	$\dfrac{\text{lbm}}{\text{ft.}\cdot\text{s}}$
					$1\,\dfrac{\text{kg/m}\cdot\text{s}}{\text{N}\cdot\text{s/m}^2}$	$10\,\dfrac{\text{poise}}{\text{N}\cdot\text{s/m}^2}$	$1000\,\dfrac{\text{centipoise}}{\text{N}\cdot\text{s/m}^2}$	$\dfrac{\text{N}\cdot\text{s}}{\text{m}^2}$
						$10\,\dfrac{\text{poise}}{\text{kg/m}\cdot\text{s}}$	$1000\,\dfrac{\text{centipoise}}{\text{kg/m}\cdot\text{s}}$	$\dfrac{\text{kg}}{\text{m}\cdot\text{s}}$
							$100\,\dfrac{\text{centipoise}}{\text{poise}}$	Poise $\left(\dfrac{\text{g}}{\text{cm}\cdot\text{s}}\right)$
								Centipoise

Table E.10 Kinematic Viscosity Conversion Factors

Centistoke					
Stoke (cm²/s)	$100\,\dfrac{\text{centistoke}}{\text{stoke}}$				
$\dfrac{\text{in.}^2}{\text{s}}$	$645.2\,\dfrac{\text{centistoke}}{\text{in.}^2\text{/s}}$	$6.452\,\dfrac{\text{stoke}}{\text{in.}^2\text{/s}}$			
$\dfrac{\text{ft}^2}{\text{s}}$	$9.29\text{E}4\,\dfrac{\text{centistoke}}{\text{ft}^2\text{/s}}$	$929\,\dfrac{\text{stoke}}{\text{ft}^2\text{/s}}$	$144\,\dfrac{\text{in}^2\text{/s}}{\text{ft}^2\text{/s}}$		
$\dfrac{\text{m}^2}{\text{s}}$	$1.0\text{E}6\,\dfrac{\text{centistoke}}{\text{m}^2\text{/s}}$	$10{,}000\,\dfrac{\text{stoke}}{\text{m}^2\text{/s}}$	$929\,\dfrac{\text{in.}^2\text{/s}}{\text{m}^2\text{/s}}$	$10.76\,\dfrac{\text{ft}^2\text{/s}}{\text{m}^2\text{/s}}$	
	Centistoke	Stoke (cm²/s)	$\dfrac{\text{in.}^2}{\text{s}}$	$\dfrac{\text{ft}^2}{\text{s}}$	$\dfrac{\text{m}^2}{\text{s}}$

TABLE E.11 Temperature Conversions

Convert from	To	
Celsius (°C)	Fahrenheit (°F)	$T_F = 1.8\,T_C + 32°$
Fahrenheit (°F)	Celsius (°C)	$T_C = (5/9)\,T_F - 32°$
Rankine (°R)	Kelvin (K)	$T_K = (0.5556)\,T_R$
Kelvin (K)	Rankine (°R)	$T_R = (1.8)\,T_K$
Celsius (°C)	Kelvin (K)	$T_K = T_C + 273.15$
Fahrenheit (°F)	Rankine (°R)	$T_R = T_F + 459.67$

Table E.12 Volume Flowrate Conversion Factors

	m³/s	ft³/s	Liter/s	ft³/min	gal/min (gpm)	cm³/s
m³/s		$35.61\ \dfrac{\text{ft}^3/\text{s}}{\text{m}^3/\text{s}}$	$1000\ \dfrac{\text{L/s}}{\text{m}^3/\text{s}}$	$2119\ \dfrac{\text{ft}^3/\text{min}}{\text{m}^3/\text{s}}$	$15{,}850\ \dfrac{\text{gpm}}{\text{m}^3/\text{s}}$	$60\text{E}6\ \dfrac{\text{cm}^3/\text{s}}{\text{m}^3/\text{s}}$
ft³/s			$28.32\ \dfrac{\text{L/s}}{\text{ft}^3/\text{s}}$	$60\ \dfrac{\text{ft}^3/\text{min}}{\text{ft}^3/\text{s}}$	$449\ \dfrac{\text{gpm}}{\text{ft}^3/\text{s}}$	$2.832\text{E}4\ \dfrac{\text{cm}^3/\text{s}}{\text{ft}^3/\text{s}}$
Liter/s				$2.119\ \dfrac{\text{ft}^3/\text{min}}{\text{L/s}}$	$15.85\ \dfrac{\text{gpm}}{\text{L/s}}$	$1000\ \dfrac{\text{cm}^3/\text{s}}{\text{L/s}}$
ft³/min					$7.481\ \dfrac{\text{gpm}}{\text{ft}^3/\text{min}}$	$471.9\ \dfrac{\text{cm}^3/\text{s}}{\text{ft}^3/\text{min}}$
gal/min (gpm)						$63.09\ \dfrac{\text{cm}^3/\text{s}}{\text{gpm}}$
cm³/s						

Table E.13 Power Conversion Factors

	ft·lb/min	g·cal/min	Btu/hr	Watt (W)	ft·lb/s	g·cal/s	Btu/min	Horsepower (hp)	Kilowatt (kW)	Btu/s
ft·lb/min										
g·cal/min	$3.086\ \dfrac{\text{ft·lb/min}}{\text{g·cal/min}}$									
Btu/hr	$12.96\ \dfrac{\text{ft·lb/min}}{\text{Btu/hr}}$	$4.20\ \dfrac{\text{g·cal/min}}{\text{Btu/hr}}$								
Watt (W)	$44.25\ \dfrac{\text{ft·lb/min}}{\text{W}}$	$14.34\ \dfrac{\text{g·cal/min}}{\text{W}}$	$3.4134\ \dfrac{\text{Btu/hr}}{\text{W}}$							
ft·lb/s	$60\ \dfrac{\text{ft·lb/min}}{\text{ft·lb/s}}$	$19.44\ \dfrac{\text{g·cal/min}}{\text{ft·lb/s}}$	$4.629\ \dfrac{\text{Btu/hr}}{\text{ft·lb/s}}$	$1.356\ \dfrac{\text{W}}{\text{ft·lb/s}}$						
g·cal/s	$185.2\ \dfrac{\text{ft·lb/min}}{\text{g·cal/s}}$	$60\ \dfrac{\text{g·cal/min}}{\text{gm·cal/s}}$	$14.29\ \dfrac{\text{Btu/hr}}{\text{g·cal/s}}$	$4.184\ \dfrac{\text{W}}{\text{g·cal/s}}$	$3.086\ \dfrac{\text{ft·lb/s}}{\text{g·cal/s}}$					
Btu/min	$778\ \dfrac{\text{ft·lb/min}}{\text{Btu/min}}$	$252.0\ \dfrac{\text{g·cal/min}}{\text{Btu/min}}$	$60\ \dfrac{\text{Btu/hr}}{\text{Btu/min}}$	$17.57\ \dfrac{\text{W}}{\text{Btu/min}}$	$12.96\ \dfrac{\text{ft·lb/s}}{\text{Btu/min}}$	$4.200\ \dfrac{\text{g·cal/s}}{\text{Btu/min}}$				
Horsepower (hp)	$3.3\text{E}4\ \dfrac{\text{ft·lb/min}}{\text{hp}}$	$1.069\text{E}4\ \dfrac{\text{g·cal/min}}{\text{hp}}$	$2542\ \dfrac{\text{Btu/hr}}{\text{hp}}$	$745.7\ \dfrac{\text{W}}{\text{hp}}$	$550\ \dfrac{\text{ft·lb/s}}{\text{hp}}$	$6.416\text{E}5\ \dfrac{\text{g·cal/s}}{\text{hp}}$	$42.44\ \dfrac{\text{Btu/min}}{\text{hp}}$			
Kilowatt (kW)	$4.425\text{E}4\ \dfrac{\text{ft·lb/min}}{\text{kW}}$	$1.432\text{E}4\ \dfrac{\text{g·cal/min}}{\text{kW}}$	$3413.4\ \dfrac{\text{Btu/hr}}{\text{kW}}$	$1000\ \dfrac{\text{W}}{\text{kW}}$	$737.6\ \dfrac{\text{ft·lb/s}}{\text{kW}}$	$238.7\ \dfrac{\text{g·cal/s}}{\text{kW}}$	$56.83\ \dfrac{\text{Btu/min}}{\text{kW}}$	$1.341\ \dfrac{\text{hp}}{\text{kW}}$		
Btu/s	$4.666\text{E}4\ \dfrac{\text{ft·lb/min}}{\text{Btu/s}}$	$1.512\text{E}4\ \dfrac{\text{g·cal/min}}{\text{Btu/s}}$	$3600\ \dfrac{\text{Btu/hr}}{\text{Btu/s}}$	$1054\ \dfrac{\text{W}}{\text{Btu/s}}$	$778\ \dfrac{\text{ft·lb/s}}{\text{Btu/s}}$	$252.0\ \dfrac{\text{g·cal/s}}{\text{Btu/s}}$	$60\ \dfrac{\text{Btu/min}}{\text{Btu/s}}$	$1.414\ \dfrac{\text{hp}}{\text{Btu/s}}$	$1.054\ \dfrac{\text{kW}}{\text{Btu/s}}$	

Table E.14 Energy Conversion Factors

From \ To	erg (dyne·cm)	ft·poundal	in.·lb	Joule (J) (N·m)	ft·lb	g·cal	Liter·atm	Btu	ft³·atm	kcal	hp·hr
kWh	3.6E13	8.543E7	3.186E7	3.6E6	2.655E6	8.606E5	3.553E4	3414	1255	860.6	1.341
hp·hr	2.684E13	6.370E7	2.376E7	2.685E6	1.98E6	6.416E5	2.649E4	2546	935.9	641.6	
kcal	4.184E10	9.929E4	3.703E4	4184	3086	1000	41.29	3.966	1.459		
ft³·atm	1.0133E9	6.809E4	2.539E4	2869	2116	6855	28.32	2.721			
Btu	1.054E10	25020	9332	1054	778	252.0	10.41				
Liter·atm	3.578E7	2405	896.8	101.3	74.74	24.2					
g·cal	4.184E7	99.29	37.03	4.184	3.086						
ft·lb	1.356E7	32.174	12	1.356							
Joule (J) (N·m)	1.0E7	23.73	8.850								
in.·lb	1.130	2.681									
ft·poundal	4.214E5										

Table E.15 Specific Energy Conversion Factors

From \ To	erg/g	ft·lb/slug	Joule/kg	ft·lb/lbm	Btu/slug	Btu/lbm	g·cal/g	kcal/kg
ft·lb/slug	929.4							
Joule/kg	1.0E4	10.76						
ft·lb/lbm	2.989E4	32.174	2.990					
Btu/slug	7.224E5	778	72.23	24.17				
Btu/lbm	2.324E7	2.5020E4	2324	778	32.174			
g·cal/g	4.184E7	4.504E4	4184	1400	57.90	1.800		
kcal/kg	4.184E7	4.504E4	4184	1400	57.90	1.800	1.0	